Fieldbook of Natural History

second edition

Fieldbook of Natural History

second edition

By E. LAURENCE PALMER
Late Professor Emeritus of Nature and Science Education
Cornell University

Revised by H. SEYMOUR FOWLER
Chairman, Science Education Faculty
The Pennsylvania State University
University Park, Pennsylvania

McGRAW-HILL BOOK COMPANY
New York / St. Louis / San Francisco / Düsseldorf
Johannesburg / Kuala Lumpur / London / Mexico
Montreal / New Delhi / Panama / Paris / São Paulo
Singapore / Sydney / Tokyo / Toronto

Library of Congress Cataloging in Publicati

Palmer, Ephraim Laurence, date
    Fieldbook of natural history.

    1. Natural history.   I. Fowler, Ho
Seymour, date   II. Title.
QH45.2.P34 1974   500.9   73-18290
ISBN 0-07-048425-2
ISBN 0-07-048196-2 (text ed.)

FIELDBOOK OF NATURAL HISTORY

1234567890KPKP7987654

This book was set in Times Roman by Rocappi, Inc.
The editors were William J. Willey and Richard S. Laufer;
the designer was J. Paul Kirouac;
the production supervisor was Sam Ratkewitch.
The drawings were done by Vantage Art, Inc.
Kingsport Press, Inc., was printer and binder.

Preparation of a revision of the first edition of E. Laurence Palmer's *Fieldbook of Natural History* was, indeed, a monumental task. Dr. Palmer collected material for the fieldbook during a lifetime of dedicated service to the study of field natural history. The success of his initial effort has been reflected in the continuous demand for copies of the first edition for more than 25 years. The general pattern, style, and format of the first edition required little change.

After 25 years, Palmer's comments in the Preface to the first edition still hold true in large part. The social sciences have emerged as leaders in the reform movements. The natural and physical sciences plus their applied fields, technology and engineering, are now held suspect by a broad spectrum of the population. The fruits of their efforts are respected and yet feared. The practitioners in these fields too often are accepted not for their contributions to mankind but because of man's fear of the products of their efforts. Perhaps in the two decades since the first edition was published, the superficial conceit of the social sciences has been strengthened. Certainly the academic arrogance of the natural and physical sciences has not diminished. Technology, with its continued successes, has made man its slave and certainly made him ask why he has become an infinitesimal cog in technology's giant machine. Also, thinking man questions the materialistic contributions to that life-style which has been foisted upon him by the fruits of those who practice in the field of natural and physical sciences, as well as in engineering and technology. The gulf of misunderstanding between the scientists and technologists and the man who uses their wares becomes wider and wider. There is a need to help man understand himself and the environment in which he finds himself without making him master the expertise of the natural scientist, the engineer, and the technologist. To help accomplish this, there is a need for a book which places man in his environment and recognizes him as a part of that environment but does not force him to learn a whole new vocabulary useful only to the expert who may employ it to include and exclude certain groups.

The new interest in the environment and the recreational use of the out-of-doors demands a readable and understandable guide, without the specificity of the expert's jargon but including generalized information at a level of sophistication of understanding which is usable by those not expert in the field. To do this and satisfy all users is, indeed, a difficult task.

What has been discussed, to this point, is preamble to an explanation of why a revision of the fieldbook was in order at this time. To use what may be considered a trite phrase, there is nothing so inevitable as change. There have been changes over the past two decades. The second edition addresses itself to those changes.

First, as has been mentioned earlier, there is a renewed interest in ecology, the environment, the out-of-doors as a recreational resource, and man's place on his planet. To be more exact in describing specific content changes, the second edition incorporates the following modifications: data gathered from space exploration has been incorporated; mineral resource information has been revised on the basis of current data; a more contemporary classification of plants (after Bold) and animals has been included; many more species of plants and animals have been described; information included in the section on the earth, its geology and weather, has been expanded and revised.

An attempt has been made to include what may be termed practical information wherever possible. Common forms in commonplace settings have been incorporated. The practice of using simpler language when it is appropriate has been continued. In other words, technical terms have been avoided whenever possible.

In conclusion, the author of the revision has attempted to maintain everything which was considered acceptable from the first edition. Content reviewers have helped me locate portions of the original manuscript which needed attention. It is hoped that the second edition will be judged a work useful to the beginner on the natural history road and acceptable to the specialist in each of the divisions treated. The test of the second edition is its acceptability in the year 2000, when, it is hoped, the second edition will have survived as long as the first did.

H. Seymour Fowler

It is probably trite to begin by stating that it is impossible to acknowledge the contributions of all persons to the second edition of the fieldbook, but this acknowledgment is given with all sincerity.

The late Dr. E. Laurence Palmer made it all possible. I was introduced to the fieldbook, in sections, more than 25 years ago when I was a graduate student and secondary school teacher. Some 8 years ago, the early planning for the revision was initiated. At that time, Dr. Palmer was most helpful in describing the many changes that he proposed to incorporate in the second edition. Most of those changes have been included.

Dr. Katherine V. Palmer, widow of Dr. E. L. Palmer, has been most helpful and cooperative throughout the project. Through her thoughtfulness much of the artwork from the first edition was made available for the second edition. Also, Mrs. Palmer took complete charge of the mollusks section and suggested all changes in that section that were incorporated in the second edition. Photographs of the rocks and minerals were provided by Wards' National Science Establishment and Professor Charles Thornton, Pennsylvania State University.

A revision of a work like the *Fieldbook of Natural History* is impossible without careful reviews of the substantive material done by experts in the particular specialties represented by the various sections of the fieldbook. Many helpful comments, suggestions, criticisms, and corrections of errors were provided by numerous authorities in the fields described in the list which follows.

Astronomy   Dr. Paul W. Hodge, University of Washington

Rocks and minerals   Dr. Richard V. Dietrich, Central Michigan University

Weather   Professor Elliott Abrams, The Pennsylvania State University

Lower plants through conifers   Dr. Melvin Fuller, University of Georgia

Monocotyledons   Professor James E. Perley, College of Wooster

Dicotyledons   Dr. Warren H. Wagner, University of Michigan

The lower animal phyla through the arthropods   Professor Lion F. Gardner, Rutgers University; Dr. Ke Chung Kim of The Pennsylvania State University prepared the final and more complete critique.

Fish   Dr. Robert Butler, The Pennsylvania State University

Amphibians and reptiles   Dr. Edward Bellis, The Pennsylvania State University

Birds   Dr. Douglas Lancaster, Laboratory of Ornithology, Cornell University

Mammals   Dr. William Burt, University of Colorado Museum

There were many instances when changes suggested by the reviewers were not incorporated in the second edition. In each such case I requested assistance from my colleagues at The Pennsylvania State University. If evidence for modifications of the reviewer's comments was compelling, the original reviewer's suggestions were modified. Many scientists of The Pennsylvania State University gave suggestions and evidence for alternative choices. So many persons were helpful that I hesitate to list any for fear of missing many others. To all I am deeply grateful, and I have expressed my sincere appreciation to each of them personally.

The final decision to include or exclude suggestions was mine. Therefore, I accept the complete responsibility for all alterations of the original manuscript. When the reader agrees with the changes, I accept credit; where he cannot accept the change, I respect his decision to do just that.

Many new and substitute line drawings are included in the second edition. These are the work of Rae D. Chambers, who is a competent artist and a qualified biologist. Her interest in the revision and enthusiasm for her contribution to the second edition are most appreciated.

No major revision of a respected work can be accomplished without the complete cooperation of an author's family. It is impossible to express the magnitude of credit due an author's wife. She is the so-called "textbook widow" during the progress of a manuscript. Therefore, I am deeply grateful to my wife for accepting, graciously, all of the inconveniences imposed upon her by the completion of the second edition of Palmer's *Fieldbook of Natural History*.

Many students in my teacher-education classes in New York, Oregon, Iowa, Virginia, Texas, Korea, Panama Canal Zone and, in particular, Pennsylvania served as recipients of information which is included in the second edition of the fieldbook. Their responses to that input were carefully weighed, and their reactions dictated, in part, decisions to include or exclude some of the information for the second edition.

It is always difficult to be able to trace the origin of facts, ideas, concepts, or principles one uses in his writing. I am sure that my association with a large number of persons in college and university teaching, in the professional societies and with government agencies has provided information for the second edition. The publications of many state and federal agencies have proven valuable and important sources and resources. The classification at the division level in the plants section of the fieldbook is that which appears in Harold C. Bold's *Morphology of Plants* and permission to use the names of the Divisions was given by Harper and Row, Publishers. Permission was granted by University of Michigan Press to paraphrase some statements found in Alexander H. Smith's *Mushroom Hunter's Field Guide*. A. J. McClane's *McClane's Standard Fishing Encyclopedia* served as a useful resource and permission was granted by Holt, Rinehart and Winston, Inc., to reprint selected paraphrased material from that volume. In the rocks and minerals section of the second edition, I have used paraphrased selected statements from Frederick P. Hough's *A Field Guide to Rocks and Minerals,* and Houghton Mifflin Company, publishers of the book, have provided permission to use the material. Dover Publications, Inc., have generously provided permission to include selected comments concerning food habits of animals first reported in their publication *American Wildlife and Plants, A Guide to Wildlife and Food Plants,* by Alexander C. Martin and Herbert S. Zim. Selected illustrations in the mammals section are from *Palmer's Fieldbook of the Mammals* by E. Lawrence Palmer, copyright © 1957 by E. Lawrence Palmer, reprinted by permission of the publishers, E. P. Dutton & Co., Inc. The second edition of the fieldbook does not directly compete with any of the publications listed.

H. Seymour Fowler

History, education, economics, and similar increasingly popular social sciences have on occasion shown us the error of our ways.
They have defined problems that we must face in our increasingly
complex civilization. Science has provided us with facts that may
be used in solving these problems, and engineering and management have suggested ways of using these facts effectively. There
has sometimes been a superficial conceit in the social sciences, an
academic bigotry in the natural and physical sciences, and an im-
personal, mechanistic routine in the technical fields that have failed
to help the layman and beginning student use his immediate material environment in meeting the practical problems of the day.
Much of the literature available to the average person has become
too technical, too highly specialized to serve a practical purpose to
him. It is to help meet this situation that this book has been written.
It is hoped that this combination of philosophy, facts, and techniques may help us all enjoy doing what must be done, when it must
be done wherever we may be. This should lead to a sound citizen-
ship, a rational conservation policy, and a happy life.

During the more than a decade that has been devoted to its actual
preparation, a number of changes have taken place that bear on the
nature of the book. At one time it was thought best to confine the
content to biological matters, but there seemed to be a need for a
book that would also give information on the stars, rocks, and minerals. The restrictions imposed by meeting this need have eliminated some two hundred plants and animals on which the material
had already been prepared as well as a section on historic geology
and on galls.

This book is not a textbook in botany, zoology, geology, or astronomy in the strictest sense. Nor is it a manual for the identification
of most of the objects considered in those sciences. Students with
special interests in those fields will no doubt find superior offerings
elsewhere of a more comprehensive nature. However, few books
on plants and animals seem to recognize that a cow is an animal or
corn is a plant, and most of them ignore the existence of familiar
domesticated species. The author has included some domesticated and economic forms, in spite of the advice of his publisher,
because it is his belief that people want to know about cows, corn,
cod, and chickens. In general, he has attempted to choose those
things found most commonly in an unmodified or still identifiable
form in the field, stream, or wood lot, in the grocery or fruit store, or
behind the kitchen sink.

There will be many who will disagree with some of the selections for
inclusion in the book. To these persons the author may only say
that were he to rewrite the book he might make inclusions and
rejections similar to those they have in mind.

The forms here included, or rejected in spite of their preparation, represent to the author the things that have most interested him, his students, and his friends in more than a third of a century of teaching field natural history from New England to Hawaii.

Throughout the book wherever appropriate a uniform treatment has been followed of presenting a picture and description for identification supplemented by data on range and relationships, life history, ecology, and economic importance.

Technical terms have been avoided wherever simpler language was adequate. Generally, line drawings have been used because they offer superior opportunity for emphasis of distinctive characters.

Occasionally, to increase the number of kinds of things considered, two or more species have been listed in the descriptive columns. In these cases the names have been given numbers, which have been repeated where significant in the printed material below.

In summary, the author has endeavored to write the kind of book he would have liked to have had available when he began the study of natural history as a youngster, as well as a book he can use now and in his old age to add the research and experience of others to what he may see for himself in his experiences with natural history. It is his earnest desire that the book may offer an equal appeal to you.

It should be remembered that no book of this type can give complete data on matters of distribution, life history and variation. The life history or reproduction material usually appearing in the third paragraph of descriptive matter selects what may be a typical reproductive sequence. The actual story may vary greatly because of climatic and other conditions. In the Southern Hemisphere for example the dates suggested would hardly be appropriate. Here a reasonably suitable modification may be made by adding six months to the suggested dates. Incidentally it seems that between 70 and 80% of the species considered may be found for study in New Zealand and it is probable that this may hold for some other Southern Hemisphere countries. This is not much different from the percentage that would prevail in any part of continental United States.

E. Laurence Palmer

In the preparation of encyclopedic material such as is found in this book, one of the major regrets is that space forbids a just documentation of all the sources of inspiration and information that have been called upon knowingly or otherwise. Since the author has lived a year in the tropics, a year on the Pacific Coast, some years in the Middle West, and many years in the East and South, it is difficult for him to know in some instances just what is the source of this or that bit of information. To all his named and unnamed collaborators, therefore, the author acknowledges a debt of gratitude. Except for the marine species in part, the author has had personal field or zoo experience with almost all the creatures here considered. Without apology he has naturally called upon these experiences in writing this book.

However, certain sources have been of such direct assistance that the author wishes to acknowledge his indebtedness. Roughly, these are those who have given actual information, those who have provided the illustrations, those who have reviewed the manuscript, those who have consented to its preparation in connection with regular professional duties, and those who have helped prepare the manuscript for the printed form.

The publications of the Federal and state governments have been drawn upon freely in providing material supplementary to that based on personal experience. Particularly helpful have been the bulletins of the United States Department of Agriculture and the various services interested in natural resources, the bulletins of the various state colleges of agriculture and of forestry, as well as various state surveys in the field of natural history. Professional journals have also provided many useful data. The major journals of each of the fields of interest here considered have been consulted freely in the preparation of this material.

Lectures in colleges and at professional meetings have added a surprising volume of material. The extent to which standard texts have been directly or indirectly involved is naturally difficult to determine.

At the beginning, it was hoped to limit the illustrations to the work of three artists, but this plan was of necessity altered. However, the major work in each of the original fields agreed upon represents the efforts of three young women who began the work in 1936. Practically all the drawings of plants are from the pen of Mrs. Elizabeth Burckmyer, who has stayed with the project from beginning to end. Most of the drawings of vertebrates were made by M. Hope Sawyer, daughter of Edmund Sawyer, the bird artist. The excellent drawings of invertebrates begun by Velma Knox were continued by Mae Geiger. Unfortunately, it was not possible for these two artists to complete their section. It became possible through purchase to use many of the drawings of birds and mammals made by the late

Louis Agassiz Fuertes and some of the drawings of plants by the late Mary Eaton. Other artists represented in these pages, in the order of the amount of work done, include Clara Garrett, who did some of the work on birds, E. M. Reilly, Jr., and Heinz Meng, who drew some of the illustrations of invertebrates, and Stephen Collins, who made seven of the drawings of mammals. A few drawings in each category have been made by the author. Credit is gladly given Dr. Leon Hausman for permission to use his drawings representing the character of hairs. Photographs of the marine shells are by the author's wife, Dr. Katherine Van Winkle Palmer, who also made approximately half of the photographs of rocks and minerals. The remainder of the rock and mineral photographs were produced with the cooperation of Ward's Natural Science Establishment of Rochester, New York, who have extensive files in this field. Without the cooperation of all these persons this book could not have been completed. Their help is gladly acknowledged.

Criticisms of prepared material before a book appears are of inestimable value to an author, and it is hoped that through careful survey errors in this book have been reduced to a minimum. The various sections have had the benefit of the criticism and comment of many of the author's associates, as follows:

Astronomy Prof. S. L. Boothroyd
Rocks and Minerals Prof. W. Storrs Cole
Fungi Prof. H. H. Whetzel
Flowering-plant Nomenclature Prof. R. T. Clausen
Mollusks Dr. Katherine V. Palmer
Other Invertebrates Profs. J. D. Hood, W. T. M. Forbes, and
  J. C. Bradley
Fish Prof. E. C. Raney
Amphibians and Reptiles Prof. A. H. Wright
Birds Prof. A. A. Allen
Wild Mammals Prof. W. J. Hamilton, Jr.
Domestic Mammals Profs. E. S. Savage and G. W. Salisbury

Acknowledgment is due the administration of the New York State College of Agriculture at Cornell University for permission to undertake the book while regularly employed by the institution and for use of material originally published in the Cornell Rural School Leaflet. Much of the material is rewritten from special inserts that have appeared in *Nature Magazine,* of which the author has been Nature Education Director since 1925. Permission to use this material and the helpful criticisms of readers of *Nature Magazine* have molded the nature of the book considerably. This cooperation has been thoroughly appreciated.

The author's wife, Dr. Katherine Van Winkle Palmer, has helped see the material through the pains of publication and has been most charitable in giving priority to the book when legitimate demands on her time existed.

For the kindly cooperation of all who have been identified in any way with the project thanks are gladly given.

E. Laurence Palmer

ERRATUM

Palmer and Fowler
FIELDBOOK OF NATURAL HISTORY, 2/e
ISBN 0-07-048196-2 (TEXT EDITION)
ISBN 0-07-048425-2

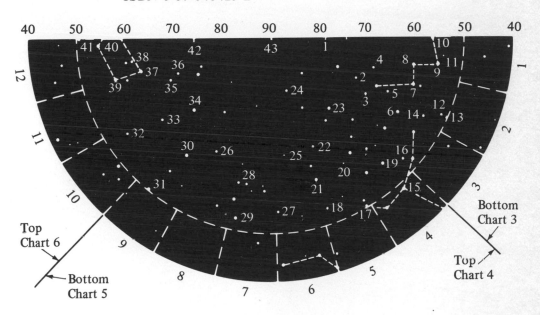

A preview of the contents of this book will show that it has some bearing on each of the major fields with which natural history deals: the universe itself, our solar system, the earth and the rocks, minerals, plants, and animals represented in practically any normal environment in which man lives.

As a preliminary guide, a synopsis of the major groups covered is presented here in the hope that it may indicate the scope of the book and be a convenience in finding the different groups quickly. Information needed to use the major sections is provided in the initial treatment of each section.

Fieldbook of Natural History

second edition

THE SKY AT NIGHT

On p. 5 there is a general map of the stars one might see either with the naked eye or with a low-power glass at any time of the year in the temperate parts of the Northern Hemisphere.

To use this map most effectively, trace or cut out the frame shown on p. 3. Then cut out the indicated portion of the frame. Now place the frame over the map in such a way that the proper date is adjacent to the hour when the observations are to be made. The stars and the map and the horizons should now be oriented.

After you have become acquainted with the major stars and constellations with the aid of this map, you will probably wish to know more about some of them. The maps on pp. 6–15 will help you. Charts III–X are continuations of the areas adjacent to the margins of Charts I and II as indicated. With these maps you may learn interesting facts, and perhaps some myths, about most of the stars you can see in the sky above you at night.

In reading about these stars in the text that accompanies each map, you may find some confusing terms. For example, you find that the North Star, Polaris, Star 43, Chart I, is a triple star, with certain distances, certain magnitudes, and periods. This means that what you see as a point of light as Polaris is really three stars. You learn that it is at a distance of 190 light-years, which means that it took light 190 years to reach the earth from this group. Since light travels at the rate of about 186,000 miles a second, or about 6 million million miles in a year, we find it easier to speak of the distance in terms of light-years than in actual miles. If you wish to get an approximation in miles, you may do so by multiplying the designated light-years by about 6 million million. You will soon give up this bit of mathematics as unnecessary because such numbers seem meaningless.

The brilliance of stars is spoken of as the magnitude. When we speak of stars' magnitudes, we do not mean size. For instance, the smaller the number given to indicate magnitude, the brighter is the star. First-magnitude stars are the brightest, and stars with a magnitude of 6 or more cannot be seen by the unaided eye. The most brilliant stars have a brightness greater than magnitude 1 and are therefore spoken of as having a minus magnitude. Sirius, for example, has a magnitude of −1.6.

When we say that a star is variable, we mean a number of things, but chiefly we refer to variation in brilliance. This variation in brilliance may have several causes, which we will not discuss here.

The term *period* is used to indicate the time required for one star to complete a revolution around another or the time for one complete cycle of variation of brilliance.

Constellations are rather arbitrarily accepted groups of stars. In general, they represent some figure as described in earlier times. Some of the major stars of the constellations are connected by dotted lines in the maps. Constellations help us locate stars.

Planets, of course, are not stars and cannot be found in these maps. The nature of the planets is given in the table on pp. 18–20, but their position cannot be indicated here. With the help of the constellations that you can locate with these maps, a good almanac will enable you to recognize the more conspicuous planets to be found at different times of the year.

Since it is probable that you may wish to locate a star whose name you have heard, a directory of the stars considered in these maps follows. See pp. 3–4. In the directory, IX, 15 means that the star is Star 15 on Chart IX. This book is not concerned with all of the myths about the stars or constellations. However, some source myths help us to understand and locate and appreciate the stars which make up a constellation. If we recognize the constellations, we can map the heavens for our own viewing.

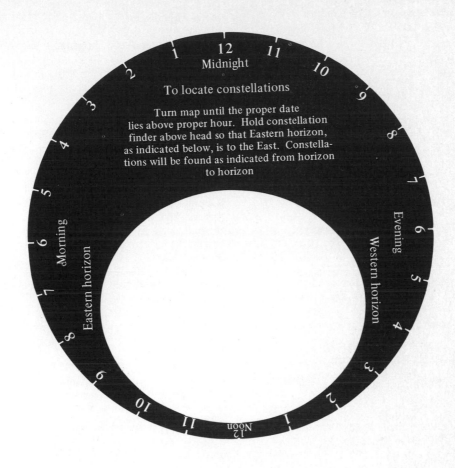

To locate constellations

Turn map until the proper date lies above proper hour. Hold constellation finder above head so that Eastern horizon, as indicated below, is to the East. Constellations will be found as indicated from horizon to horizon

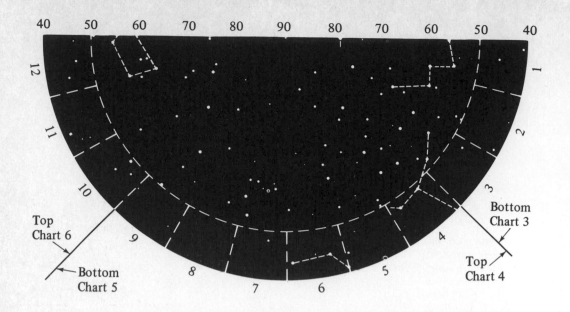

CHART 1 POLAR CONSTELLATION CHART

Cepheus. Represents a famous king in mythology, the husband of Cassiopeia. This constellation can be imagined in king form if one star represents the head, and two nearby the hands of the king. Cepheus is not easy to locate; it is just ahead of Cassiopeia as they move around Polaris. Star 1, period 166.2 years.

Cassiopeia. One of the easy-to-locate, evident constellations. Appears as letter M or W depending on location in the sky. Cassiopeia was a beautiful queen of Ethiopia, the wife of King Cepheus, according to some legends. Because constellation appears to move around the polestar, it is a good approximate timekeeper for experienced observers. Constellation is supposed to approximate a chair on which the beautiful queen sits. Star 2, double, magnitudes 4.7, 7.2; period 52.95 years; Star 3, triple, magnitudes 4.2, 7.1, 8.1; Star 4, triple, magnitudes 4.4, 8.9, 9.5; Star 5, magnitude 3.4; Star 6, gold-green, double, magnitudes 6.1, 6.6; Star 7, magnitude 2.8; Star 8, magnitude 2.2; Star 9, gold-red, double, magnitudes 3.6, 7.9; period 507.6 years; Star 10, magnitude 2.4; distance 5.7 light-years; Star 11, Shedir, variable, magnitudes 3.6–7.9; distance 47 light-years.

Perseus. Perseus, it was believed, was a handsome, brave young prince who killed the Gorgon Medusa. Perseus flies through the heavens with the head of Medusa behind him. Star 12, double spiral nebula joined with a ring; Star 13, double; periods 126.5 days, 63.2 days; Star 14, double, variable, magnitudes 7.5–11 and 8; period 320 days; Star 15, Mirfak, magnitude 1.9; period 4.09 days; Star 16, magnitude 3.1; Star 17, magnitude 4.5; period 1.5 days.

Camelopardus, the Giraffe. Star 18, double, magnitudes 4.4, 11.3; period 3.8 days; Star 19, double, magnitudes 4.8, 9.2; Star 20, nebula; Star 21, magnitude 4.7; Star 22, magnitude 4.3; period 7.9 days; Star 23, magnitude 4.7; Star 24, double, magnitudes 4.5, 8; Star 25, triple, magnitudes 6.7, 7.9, 10.9; Star 26, triple, magnitudes 6.2, 8, 9.2.

Auriga, the Charioteer, represents Neptune, perhaps, driving his chariot drawn by sea horses. Some mythologists think Auriga represents a deformed son of Vulcan and Minerva who invented a four-horse chariot. There is a brilliant star, Capella, a first-magnitude star in the constellation. Supposedly Capella represents the ewe which nursed the child, Jupiter. Some of the ancients believed they saw in Auriga the figure of a man carrying a goat over his back with the kids in his arms. Star 27, magnitude 2.1.

Lynx. Composed of faint stars. Some persons believe that its name originates from the fact that only a person with eyes like a lynx can locate and see the constellation. Is described as "in a barren region." Star 28, triple, magnitudes 5.2, 6.1, 7.4; Star 29, variable, magnitudes 7–13.8; variation period 379.2 days.

Ursa Major, the Great Bear. The most famous of the circumpolar constellations; sometimes referred to as the Big Dipper. The best skymark for starting constellation study. In the latitude of the U.S., the dipper never sets. See it highest in the sky in early evenings of the spring and lowest in the autumn. An American Indian legend alludes to the belief that the constellation represents an animal looking for a place to lie down. Two stars at the fore (lip) of the dipper are often referred to as the pointer stars because they point to the polestar (Polaris), the star at the end of the handle, or of the tail, of Ursa Minor.

The largest clock in the sky. Supposedly the large bear is being chased around the sky—as you will see if you observe the Bear (Dipper) for about 3-4 hours in a night. The constellation is like a great clock which moves backward (counterclockwise). Observers must remember that the pointers (you choose them) do not follow a regimented position night by night. This relates, of course, to earth motions. Some observers can actually tell time at night by observing the Big Bear, or Big Dipper. Observers should remember that the constellation assumes a given position about 4 minutes earlier than it did the night before. We must remember that the constellation, Big Dipper or Big Bear, only appears to move in the sky. Actually, the earth is rotating below the star group and the star groups (constellations) "appear" to move. This apparent motion is caused by the rotation of the earth on its axis. Of course, the careful observer must remember also the revolution of the earth around the sun. Polaris has another significance to observers. Actually, the latitude of any place *in the Northern Hemisphere* is equal to the number of degrees Polaris is above the horizon at that location. Star 30, magnitude 3.5; Star 31, variable, magnitudes 8.4 to over 13; period 247 days; Star 32, double, magnitudes 5, 5.6; period 99.7 years; Star 33, double, magnitudes 3.8, 9; Star 34, double, magnitudes 5, 8; Star 35, magnitude 4.9; period 11.58 days; Star 36, variable, magnitudes 7-13; variation period 302 days; Star 37, Dubhe, magnitude 2; distance 70 light-years; Star 38, double, magnitudes 5.8, 7.1; period 71.9 years; Star 39, magnitude 2.4; period 0.31 days; distance 60 light-years; Star 40, magnitude 2.5; Star 41, spiral nebula.

Draco, the Dragon. Look between the two bears (the Big Dipper and the Little Dipper). There, you find the form of this constellation. Numerous legends are available to explain Draco. In spite of the discrepancies in the legends, Draco appears nightly between the two Bears or two Dippers. Star 42, double, magnitudes 7, 8.3; period 42 years.

Ursa Minor, the Small Bear. The polestar is the end of the handle of the Little Dipper, or Ursa Minor. A difficult constellation to make out. The handle of the Little Dipper curves away from the polestar—up from it in early evening in summer and down from it in early evening in winter. Remember, the two dippers (Ursa Major and Ursa Minor; the Big Dipper and Little Dipper) appear to pour into one another. The polestar, Polaris, 43. This is really a triple star. The major group has a magnitude of 2.1, is yellow, is 190 light-years distant, and is 1,000 times as brilliant as our sun. Two of the units turn around each other in a period of 3.96 days, and these in turn move about the third in a period of 11.9 years. Another unit has a magnitude of 9.

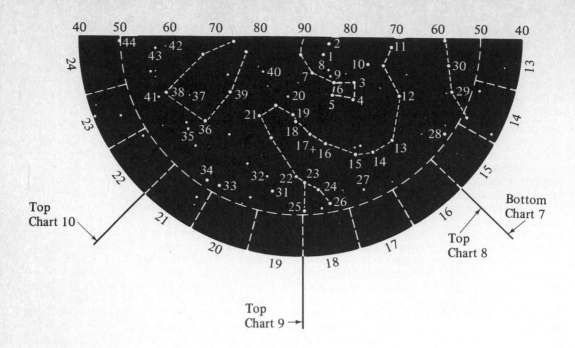

CHART 2

Camelopardus, the Giraffe.   Sometimes spelled Camelopardalis.   Situated in a region where stars are few, between Polaris (the polestar) and Auriga.   Star 1, variable, magnitudes 7.9–13.7; variation period 269.5 days; Star 2, double, magnitudes 5, 6.

Ursa Minor, the Small Bear.   Star 3, Kochab, the polestar in A.D. 140; magnitude 2.2; distance, 230 light-years.   Stars 4, 5, 6, parts of the Little Dipper; Star 7, magnitude 4.4; period 39.5 days; Star 8, double, magnitudes 7, 8; variation period 115 years; Stars 9 and 10 variable, magnitudes over 7.

Draco, the Dragon.   Star 11, magnitude 3.9; Star 12, Thuban, magnitude 3.64; period 51.38 days; the polestar when pyramids were built more than 5,000 years ago; Star 13, magnitude 3.5; Star 14, triple, magnitudes 7.5, 7.7, 9; Star 15 double, magnitudes 2.9, 8.1; Star 16, magnitude 3.2; Star 17, magnitude 4.87; period 5.28 days (the pole of the ecliptic is indicated by a cross almost south of this star); Star 18, double, magnitudes 4.8, 6.5; Star 19, magnitude 3.6; period 281.8 days; Star 20, period 4.12 days; Star 21, double, magnitudes 4, 7.6; Star 22, triple, magnitudes 4.7, 7.7, 7; Star 23, variable, magnitudes 7.5–12.2; variation period 426 days; Star 24, double, magnitudes 4.6, 4.6; Star 25, magnitude 2.4, distance 74 light-years; Star 26, double, magnitudes 5, 5.1; Star 27, double, magnitudes 5, 6.

Bootes, the Herdsman.   Frequently referred to as the Driver or the Herdsman; lies south and east of Ursa Major;   Star 28, double, magnitudes 4, 12.

Ursa Major, the Great Bear.   Star 29, Mizar, the first double discovered, magnitudes 2.1, 4.2; period 20.23 days; distance, 80 light-years; Star 30, magnitude 1.68; 56 light-years away; period 4.15 years.

Cygnus, the Swan.   Also known as the Northern Cross.   Appears, to some, as a swan in flight.   It is of interest—in the Northern Hemisphere—that this constellation, at Christmastime, appears on the northwestern horizon like a starry cross set up and above our earth.   A good pointer star in the constellation is Deneb, the bright first-magnitude star at the top of the cross.   Star 31, double, magnitudes 5, 9; Star 32, double, magnitudes 6.8, 7.4; Star 33, variable, magnitudes 1.5–1.9 in 10 days; Star 34, triple, magnitudes 7, 9, 10.1.

Cepheus.   Star 35, triple, magnitudes 6.3, 7.9, 8; Star 36, magnitude 2.6; distance 36 light-years; Star 37, double, magnitudes 4.7, 6.5; Star 38, variable, magnitudes 3.6–4.3; variation period 5 days 8 hours 48 minutes; Star 39, double, magnitudes 3, 7; period 0.19 days; Star 40, variable, magnitudes 7.9–13; variation period 486 days; Star 41, double, magnitudes 9.3, 10.8; period 54.9 years; distance, 12.5 light-years.

Cassiopeia.   Star 42, triple, magnitudes 7, 8.5, 9; Star 43, quadruple, magnitudes 4.8, 10, 7.4, 8.9; period 6 days; Star 44, variable, magnitudes 5.3–12.7; variation period 431.6 days.

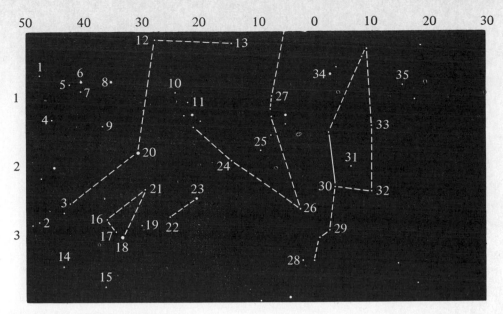

CHART 3

Constellation *Cassiopeia*, Star 1, magnitude 5.02, period 1.7 days.

Stars 2 to 12 are in the constellation *Andromeda*. Is outlined by a row of four very bright stars. Look for these four bright stars beginning in the northeast corner of the square of Pegasus. Andromeda, Cassiopeia, Cepheus, Perseus, and Pegasus form what was called the Royal Family in legends as old as 4,000 years. The Andromeda Nebula is in the constellation and can be seen as a fuzzy patch of light; faint but visible. Andromeda Nebula is a galaxy of stars over 2 million light-years away. Meteor showers in this region occur between November 23 and 27. Star 2, double, magnitudes 6.4, 7.3; period 204.7 years. Star 3, double, magnitudes 2.3, 5.4, is a cerulean blue and gold star with a period of 55 years. Star 4, double, magnitudes 4.9, 6.5, is made of two blazing suns. Star 5 is an oval pearl-white nebula. Star 6 is a great spiral nebula and the nearest of the spiral nebulae to the earth. It is 900,000 light-years distant and 50,000 light-years in diameter. On a reasonable assumption, it takes about 17 million years for it to complete a single revolution and it produces 1½ billion times the light produced by our sun. It has a mass of the order of 1 billion times the mass of our sun. Star 7 has a magnitude of 4.4 and a period of 4.28 days. Star 8, double, magnitudes 6, 9, has a period of 143.7 days; Star 9, magnitude 2.4; Star 10, magnitude 4.3; period 17.76 days; Star 11, double, magnitudes 6.1, 6.7; period 109.07 years. Star 12, Alpheratz, has a magnitude of 2.2 and a period of 96.67 days. Its temperature is about 11,000°C.

Star 13, Algenib, in the constellation *Pegasus*, an ancient winged horse of mythology with four bright stars which are easily seen. Stars of Pegasus and Andromeda together appear as giant dipper. Has a magnitude of 2.9. One side of the Great Square of Pegasus is limited by these last two stars, 12 and 13.

Perseus, Star 14, a beautiful cluster 20,000 light-years distant; Star 15, a double, magnitudes 5.6, 6.7; period 33.3 years. The constellation *Triangulum* includes a number of interesting stars. Triangle is isosceles in shape, lies on its side, just below Andromeda and above Aries. Stars 16, 17, and 18 are almost in line and have magnitudes of 3.1, 5, and 4, respectively. Star 19, double, magnitudes 5.6, 6.4, is sapphire-gold. Star 20 is a nebula 900,000 light-years away with a period of 160,000 years. Star 21 has a magnitude of 3.5 and a period of 1.7 days. In the constellation *Aries*, Star 22, Hamal, the Shepherd's Star, magnitude 2.2; Star 23, magnitude 2.7; period 107 days. The Ram of ancient mythology, a faithful sheep which was rewarded for its heroic deeds by being placed in the heavens by Jupiter. Stars 24, 25, 26, and 27 are in *Pisces*. Inconspicuous zodiacal constellation; supposedly represents two fishes with tails joined by ribbon.

The constellation *Cetus, the Whale*, includes at least eight interesting stars. Star 28 is a double, magnitudes 3, 6.8. Star 29 is Mira, whose magnitude varies from 1.7–9.6. It has a variation period of 332 days and is 165 light-years distant. It is a globe of gas 250 million miles in diameter and about 300 times the size of our sun. Star 30 is a dark nebula. Star 31, a nebula, is moving through space at the velocity of 1,240 miles a second. Star 32 is a dark nebula. Star 33 has a magnitude of 4. Star 34 is a double, magnitudes 5.2, 6.4; period 6.88 years. Star 35 has a magnitude of 2.2.

9

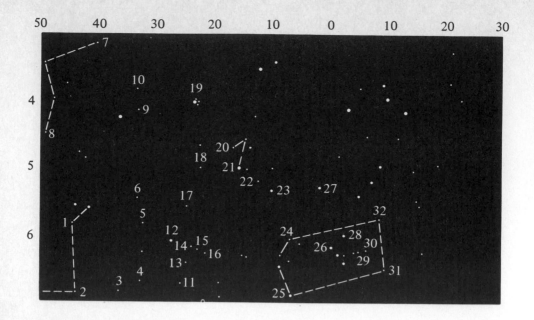

CHART 4

Capella, Star 1, is the most conspicuous one in *Auriga*. Its magnitude is 0.21; period 104 days. It is 150 times as bright as the sun, has a surface temperature of 10,400°F., and is 47 light-years distant. Star 2, magnitude 2.7; period 3.96 days; mass, 17 times that of the sun; Star 3, magnitude 5.3; period 28.2 days. Star 4 is a mass of 20,000 stars. Star 5, triple. Star 6 is a large sun, magnitude 2.9.

Perseus, Algol, Star 7, the Ghoul, or Demon Star; magnitude 2.2–3.2; period 2.87 days; mass, ⅔ that of the sun; 160 times as bright as the sun. Stars 8, 9, and 10, periods of 284, 6.9, and 4.4 days, respectively.

Taurus, the Bull, is always interesting. Has long curving horns, and fiery Aldebaran is its eye. Head made up of V-shaped cluster. Bull seems to retreat continuously from the mighty hunter *Orion.* With field glasses the Pleiades, a cluster of stars, is visible. In mythology, the Pleiades are the seven sisters. Famous in many legends. Actually Pleiades cluster has several hundred stars. Taurus is a zodiacal constellation. *The Pleiades* and the star Aldebaran are the most important. Star 11, magnitude 4.5; period 5.9 days; Star 12, magnitude 1.8; Star 13, magnitude 5; period 27.8 days; Star 14, double, magnitudes 5.8, 6.6; blue-white; Star 15, "crab nebula"; Star 16, magnitude 3; period 138 days; Star 17, a nebular cluster; Star 18, double, magnitudes 4.3, 7.2; period 15 days. Alcyone, Star 19, a star in the Pleiades, has a magnitude of 3 and is millions of miles in diameter. The cluster is made of over 500 giant suns and is 30 light-years in diameter. The volume of each of the five brightest stars of the cluster is at least 800 times that of the sun. The cluster is 325 light-years distant, or 2½ times more remote than *the Hyades.* Star 20, magnitude 5.6; period 8.4 days. Aldebaran, Star 21, the "eye" of the Bull, the Hyades, magnitude 1.1, has a diameter of 40 million miles.

Orion, the Hunter. Considered by many the finest constellation in the heavens. Easily recognized by three bright stars in Orion's belt. The mighty hunter. Stars of the constellation are referred to in the Bible frequently. Stars 22 and 23, periods 56 and 16.61 years; Bellatrix, Star 24, the Female Warrior, magnitude 1.7; Betelgeuse, Star 25, magnitude 0.9; period 6 years; surface temperature 4700°F; 1,200 times as bright as the sun. Its volume is 27 million times and its diameter 300 times that of the sun. It is 192 light-years distant and has an average density of about ¹⁄₁,₀₀₀ of an atmosphere. Star 26, greenish; double, magnitudes 2, 6.8; period 5.7 days; Star 27, magnitude 3.8; period 3.7 days; Star 28, double, magnitudes 3.4, 5; period 8 days; Star 29, a great nebula; 650 light-years distant; diameter 5 or 6 light-years, or 30 or 40 million million miles. It is so rare that if the air left in the supposed vacuum of an electric light globe were expanded to the size of our capitol at Washington, it would approximate the condition prevailing there. Star 30, double nebula, magnitudes 2.8, 7.3; period 29 days; Star 31, Saiph, magnitude 2.2; Star 32, Rigel, magnitude 0.3; distance 543 light-years; blue-white; 18,000 times as bright as our sun; temperature 28,800°F.

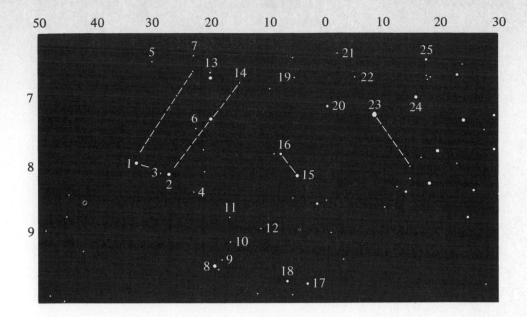

CHART 5

The constellation *Gemini* is responsible for the expression "By Jiminy." Known as the twins; favorites of ancient mariners. Castor, Star 1, is an important double star, magnitudes 2, 2.8, each part six times as brilliant as the sun would be at an equal distance; orange-blue; 42 light-years distant; period 306 years. Pollux, Star 2, the other of the "twins," magnitude 1.2; diameter 14 million miles; 33 light-years distant; Star 3, magnitude 4.6; period 19.6 days; Star 4, variable, magnitudes 8–13; variation period 288 days; Stars 5 and 6, barely visible; Star 7, variable, magnitudes 3.2–4; variation period 231 days. The June 21–July 20 group in the zodiacal series.

The cluster Praesepe, Star 8, 363 suns; often referred to as a "beehive." The units vary in magnitude from 6 to 12, there being 173 in the main cluster. The cluster is about 600 light-years distant. Star 9, variable, magnitudes 8.4 to over 14; variation period 305 days; Star 10, variable, magnitudes 7.5–13; variation period 256 days; Star 11, double, magnitudes 5.3, 6.3; period 60 years; Star 12, variable, magnitudes 6–11.3; variation period 362 days; Star 13, magnitude 4.4; period 131 days; Alhena, Star 14, the "Circlet," magnitude 1.9; period 2,175 days.

Procyon, Star 15, is the major star in the constellation of *Canis Minor,* or *the Little Dog.* In mythology thought to belong to the hunter Orion, which it follows. It is 10.6 light-years distant; 5½ times as bright as the sun; pale topaz color; surface temperature 14,400°F. Except for Sirius, it is the closest to the earth of the larger stars. It is a double, magnitudes 5, 13.5. Star 16, variable, magnitudes 7.5–13.

Star 17 is in the constellation *Hydra* and has a period of 256 days. Star 18, double, magnitudes 3.7, 5.2; period 15.3 years; distance 135 light-years; mass 3½ times the sun.

The constellation *Monoceros, the Unicorn,* includes stars 19 to 23. Comparatively a new constellation, so few legends exist. Star 19 is a cluster; Star 20 a very bright cluster, magnitude 6. Star 21, variable, magnitudes 7.2–12.9; variation period 332 days. Star 22, double, with masses 86 and 72 times (?) that of the sun. The magnitude is 6.6, the period 14.4 days, and the distance 10,000 light-years. Star 23 is a cluster with a ruby-colored star in the center.

The constellation *Canis Major,* or *Great Dog,* is conspicuous because of Star 24, Sirius, the Scorcher. This star is a double, magnitudes −1.6, 8.5; period 49.3 years; distance 8.9 light-years; 30 times as bright as the sun; mass 3.4 times that of the sun; surface temperature 19,800°F; weight, on the average 1 ton for every cubic inch. Star 25 has a magnitude of 1.9 and a period of 6 years.

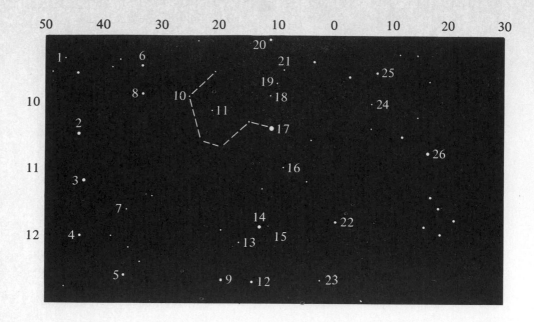

CHART 6

Ursa Major, the constellation of *the Great Bear,* extends into the area here represented and includes stars 1 to 5. Star 1, magnitude 5.7; period 15.98 days; Star 2, magnitude 3.5; Star 3, magnitude 4.8; period 15.8 days; Star 4, magnitude 6.6; a large oblong nebula. Star 5 is known as the Runaway Star. Its velocity through space is such that every 2 minutes it moves a distance equal to the path of the moon about our earth. No other known star has such a velocity. In spite of this terrific speed, it apparently moves but the breadth of the moon is 300 years. This is because of its great distance from us, 21.7 light-years.

Star 6, in the constellation *Lynx,* has a magnitude of 3.3.

The constellation *Leo Minor* contains Stars 7 and 8; Star 7, magnitude 7.6; distance 7.9 light-years; Star 8, variable, magnitudes 7–13; variation period 370.5 days. The constellation *Leo* has as its most interesting star Regulus. Star 9, double, magnitudes 4.5, 8.4; period 71.7 days; Star 10, magnitude 3.1; Star 11, variable, magnitudes 8.6–13; variation period 273.1 days. Star 12 is Denebola, a star of 2.2 magnitude, which is 25 light-years distant from the earth. Star 13 is a luminous nebula and Star 14 is a spiral nebula. Star 15, double, magnitudes 4, 7, shines with an amber-turquoise color. Star 16, magnitude 3.8; period 12.3 days. Regulus, Star 17, double, magnitudes 1.3, 8.4, the chief star in the constellation, is 72.4 light-years from the earth. It has a diameter of 3 million miles and is therefore about 3½ times as great in diameter as our sun. Regulus appears as a jewel in the handle of a sickle outlined by about six stars. It lies almost exactly on the ecliptic, or the path made by the sun across the sky. It forms one angle of a conspicuous triangle, the other angles being at the stars Procyon and Pollux. Regulus appears conspicuously white. Star 18, also in the constellation Leo, is a variable star, magnitudes 4.6–10.5; variation period 312.8 days. Star 19, magnitude 3.7; period 14.49 days. An ancient constellation in mythology, associated with the seasonal movement of the sun. Supposedly controlled or warned of the inundations of the Nile in Egypt.

Stars 20 and 21 are in the constellation *Cancer, the Crab.* A zodiacal constellation. Star 20 has a magnitude of 5.1 and a period of 6.39 days; Star 21, double.

Star 22 is an oval nebula with a star center in the constellation *Leo;* Star 23 in the constellation *Virgo* has a magnitude of 3.8. An ancient constellation, the virgin, a zodiacal constellation. This area has more nebulae than any other similar area in the heavens. Star 24, in the constellation *Sextans,* double, magnitudes 5.8, 6.1; period 72.76 years; Star 25, in the constellation *Hydra,* magnitude 2.2; distance 226 light-years. Star 26 is an elliptic, pale, steel-blue nebula.

12

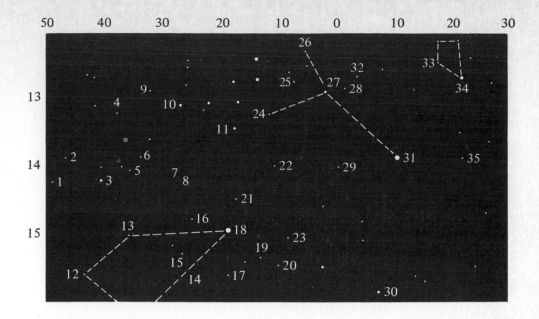

CHART 7

Canes Venatici. Mythology tells us these are the hunting dogs of Bootes, the Herdsman. They lead Bootes in the chase of the Great Bear about the Pole. Star 1, Alcaid, has a magnitude of 1.9 and is 108 light-years distant. Star 2, spiral nebula of enormous size, occupies a space equal to 20 million times the orbit of Neptune about the sun; though whirling incredibly fast, it takes over 45 million years to make one revolution. Star 3, variable, magnitudes 6.1 to 12.7; variation period 326 days; Star 4, Cor Caroli, double, magnitudes 2.9, 5.7; period 5.5 days; Star 5, double, magnitudes 5, 5.8; period 220 years; Star 6, double, magnitudes 7.2, 7.7; period 48 years; Star 7, a globular cluster visible with the unaided eye on a clear night; composed of several hundred thousand suns; 60,000 light-years distant from the earth. This distance is so great that were our sun magnified twice and placed in the center of this cluster, it would be too small to be visible on a photographic plate. It requires light 65 years to cross its angular diameter. Star 8, double, magnitudes 7.6, 8; period, 199 years.

Coma Berenices. Stars 9, 10, and 11 barely visible. Commonly called Berenice's hair because this group of stars plus others represent beautiful tresses of an ancient queen of Egypt.

Bootes, the Hunter. Star 12, double, magnitudes 7.5, 8; period, 97.93 years; Star 13, magnitude 4.83; period, 211.95 days; Star 14, double, magnitudes, 4.8, 6.4; period, 159.54 days; Star 15, variable, magnitudes, 6.6 to 12.9; variation period, 223.3 days; Star 16, double, magnitudes, 4, 8; period, 9.6 days; Star 17, double, magnitudes, 4.8, 6.4; variation period, 159.54 years; Star 18, Arcturus, magnitude, 0.2; diameter, 21 million miles, or 24½ times that of the sun; 40 light-years distant; 1/5,000 as dense as water. From Arcturus to the handle of the Big Dipper is 40°. Star 19, double, magnitudes 4.4, 4.8; period, 130 years; Star 20, double, magnitudes 7.8, 8.8; period, 238 years; Star 21, double, magnitudes 2, 8; period, 497.1 days; Star 22, double, magnitudes 6.3, 6.3; period, 23 years; Star 23, double, magnitudes 5.5, 6.8. Many legends relate to this constellation. Thought, by some, to be Atlas because of the figure and because of its nearness to the Pole; it seems to hold up the heavens. Some referred to it as Bear Driver because it appears to chase the Great Bear around the Pole.

Virgo, the Virgin. Star 24, magnitude 3; Star 25, variable, magnitudes 6.4-12.1; variation period 145.47 days; Star 26, variable, magnitudes 8.7-13.5; variation period 339 days; Star 27, variable, magnitudes 8-12; variation period 438 days; Star 28, double, magnitudes 3.6, 3.7; period 182.3 years; Star 29, magnitude 3.4; Star 30, variable, magnitudes 4.8-6.2; variation period 2.327 days. This star has an eclipse whose duration is 13 hours. Star 31, Spica, magnitude 1.2; white; period 4.01 days; 227 light-years distant from the earth; Star 32, double, magnitudes 8.6, 14.5; period, 218.8 days. An ancient constellation; the virgin, a zodiacal constellation.

Corvus, the Crow. Apparently almost all mythology describes this constellation in the form of a bird. Stars 33 and 34, in *Hydra.*

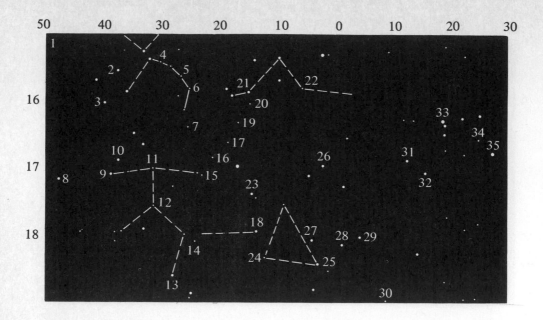

CHART 8

Bootes. Star 1, double, magnitudes 5.3, 6.2; period 204.74 years; Star 2, double double, magnitudes 4.4, 6.5, and 7.2, 7.8; period of last two, 244.37 years.

Northern Crown, Corona Borealis. Star 3, variable, magnitudes 7.2-12; variation period 358 days; Star 4, double, magnitudes 5.6, 6; period 41.5 years; Star 5, Gemma, magnitude 2.3; period 17.36 days; distance 62 light-years; Star 6, double, magnitudes 4, 7; period 87.8 years; Star 7, variable, magnitudes 7.3-14.2; period 486 days. Legend tells us this is the crown given to an ancient maiden who was helpful to her husband. Some American Indians believed it to represent a group of sisters.

Hercules. Star 8, nebula; Star 9, magnitude 3.6; Star 10, variable, magnitudes 7.8-13.5; variation period 280.2 days; Star 11, double, magnitudes 3, 6.5; period 34.46 years; Star 12, magnitude 3.9; period 4 days; Star 13, double, magnitudes 3.5, 9.5; period 43.23 years; Star 14, variable, magnitudes 8-12.5; variation period 221 days; Star 15, blue-green nebula; Star 16, magnitude 2.81; period 410 days; Star 17, double, magnitudes 3.8, 8.2; Star 18, double, magnitudes 3.1, 6.1; diameter 185,800,000 miles; 21.7 light-years distant; Star 19, variable, magnitudes 8.6-14.8; variation period 317.7 days. Constellation named for ancient Theban hero of mythology. Jupiter, it is said, placed Hercules in the heavens.

Serpens. Star 20, variable, magnitudes 5.6-13; variation period 357.3 days; Star 21, double, magnitudes 3.7, 9.1; Star 22, Serpentis, magnitude 2.8. Stars of this constellation mingle with those of previous constellation, Serpens.

Ophiuchus. Star 23, variable, magnitudes 7.3-12.6; variation period 308.3 days; Star 24, magnitude 2.1; distance 60 light-years; Star 25, magnitude 2.9; Star 26, double, magnitudes 4, 6.1; period 134 years; Star 27, magnitude 4.4; Star 28, double, magnitudes 4.9, 6.2; period 46 years; Star 29, magnitude 4.6; period 26.27 days; Star 30, double, magnitudes 5.3, 6; period 223.82 years; Star 31, variable, magnitudes 7-10.5; variation period 302.5 days; Star 32, variable, magnitudes 3.8-13; variation period 233.8 days; Star 33, magnitude 2.9; period 6.8 days; Star 34, double, magnitudes 3, 8; period 0.247 days; Star 35, Antares, magnitude 1.2; 360 light-years distant from the earth; diameter about 450 times that of the sun. If the sun and its system were to occupy the same space, the earth would be buried throughout its orbit, as would also Mars. 121 million million worlds could crowd into Antares. It is a mammoth globe of incandescent gas whose density is 0.00001 that of water. It is said that it would take 1⅓ million earths to make one sun, 27 million suns to make one Betelgeuse, and 91 million suns to make one Antares.

Scorpius. Stars 33, 34, 35 are in the constellation *Scorpius,* the *Scorpion.* Is a zodiacal constellation. Sun enters this constellation about October 10. To the west of Scorpius lies Libra (a zodiacal constellation) designated as the claws of the scorpion by the ancients.

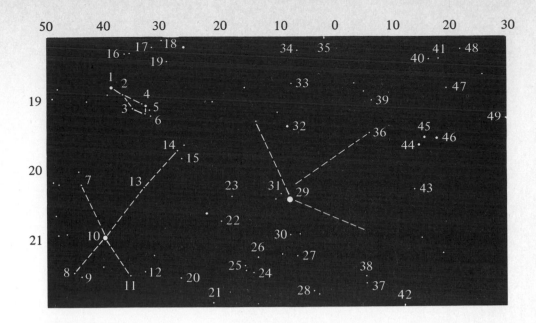

CHART 9

Lyra, the Lyre. Star 1, Vega, magnitude 0.1, brightest first-magnitude star in northern heavens, about 50 times as bright as our sun; distinctly blue-white. 1,500 Vegas would give more starlight than all the stars in our heavens; 26 light-years distant, or about 8 million times as far from the earth as the earth is from the sun. Our solar system is moving toward Vega at the rate of 400 million miles a year. In about A.D. 12,000, Vega will be our polestar just as Polaris now is. Star 2, magnitude 4.3; period 4.3 days; Star 3, double, magnitudes 4.5, 6.5; period 245 days; Star 4, double, magnitudes 3.6, 7; period 12.9 days; Star 5, a ring nebula; Star 6, double, magnitudes 5.2, 8.7; period 45.8 years; Star 7, double, magnitudes 3, 7.9. Constellation high in eastern sky throughout the summer. In mythology, constellation represents musical instrument of Apollo or Mercury, ancient gods. Vega in this constellation was the first star to be photographed, in 1850.

Cygnus, the Swan. Star 8, Deneb, double, magnitudes 1.3, 12; distance 650 light-years. Though at a great distance, it is very brilliant; even though it seems less brilliant than Sirius, it is 10,000 times as bright as our sun. Star 9, magnitude 4.7; period 2.8 days; Star 10, magnitude 2.3; distance 540 light-years; Star 11, double, magnitudes 5, 6.3; Star 12, the eastern arm of the Northern Cross, magnitude 2.6; Star 13, variable, magnitudes 4–13.5; probably larger than Mira; variation period 406 days; Star 14, double, magnitudes 8.2, 8.2; period 243.9 years; Star 15, Albireo, double, magnitudes 3, 5.3.

Hercules. Stars 16 to 19 not conspicuous. The kneeler, a famous Theban hero of mythology.

Vulpecula, the Fox. Stars 20 and 21 not conspicuous; Star 22, the Giant Dumbbell Nebula. Lies just south of Cygnus, the Swan.

Sagitta, the Arrow. Star 23, double, magnitudes 5.4, 6.4; period 25.2 years.

Delphinus, the Dolphin. Star 24, double, magnitudes 4, 5; Star 25, variable, magnitudes 8.4–12; variation period 277.5 days. Star 26, double, magnitudes 4, 5; period 26.8 years; Star 27, nebula; Star 28, double, magnitudes 5.8, 6.3; period 97.4 years. Also Job's Coffin, a group of five faint stars.

Aquila, the Eagle. Star 29, Altair, blue-white, magnitude 0.9, or nine times as bright as our sun; 15.4 light-years distant; Star 30, variable, magnitudes 7.6–13; variation period 284.4 days. Star 31, magnitude 2.8; distance 112 light-years; Star 32, variable, magnitudes 5.8–12; variation period 355 days; Star 33, variable, magnitudes 6.5–9; variation period 335 days; Star 34, nebula; Star 35, double, magnitudes 4.1, 6; period 87.8 years; Star 36, not visible to naked eye. In the Milky Way, look for three stars in a line; middle star brightest.

Aquarius: Stars 37 and 38. In mythology, the Water Carrier, a symbol of the rainy season. Look for four stars forming the water jar. A zodiacal group; appears between February 16 and March 12. *Scutum:* Stars 39, 40, and 41. *Capricornus:* Star 42. A zodiacal group appearing between January 19 and February 16. *Sagittarius, the Archer:* Stars 43 to 49. Zodiacal constellation appearing between December 18 and January 19.

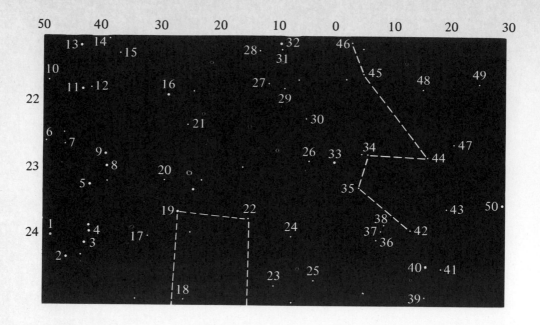

CHART 10

The area here covered is relatively poorly supplied with important stars.

In the constellation *Andromeda,* we have Star 1, a rose-colored star, magnitude 5; Star 2, magnitude 4; period 20.54 days; Star 3, a bright blue-green, elliptic nebula; Star 4, magnitude 5.9; period 3.22 days.

The constellation *Lacerta, the Lizard,* shows Star 5, variable, magnitudes 8.3–14.5; variation period 298.7 days; Star 6, a cluster; Star 7, magnitude 4.6; period 2.6 days; Star 8, a double double with magnitudes of 6.5, 6.5, and 8.5, 10; Star 9, variable, magnitudes 8–12.5. An inconspicuous group of stars between Cepheus on the north and Pegasus on the south.

The constellation *Cygnus* shows Star 10, a cluster; Star 11, a variable, magnitudes 8.4–12; variation period 103.6 days; Star 12, a triple, magnitudes 7, 7.9, 11; Star 13, a nebula; Star 14, double, magnitudes 5.3, 5.9; brightness 6/100 as bright as the sun; Star 15, double, magnitudes 3.8, 8; period 47 years; Star 16, double, red-gold to purple; magnitudes 4, 5.

Pegasus is represented by Star 17, a double, magnitudes 5.5, 7.5; Star 18, double, magnitudes 5.8, 11; period 26.4 years; Star 19, Scheat, whose diameter, 150 million miles, would enclose 5 million suns; distance 541 light-years; magnitude 2.2; Star 20, magnitude 3.1; period 818 days; Star 21, magnitude 3.6; period 10.21 days; Star 22, Markab, magnitude 2.6; distance 272 light-years; Star 23, double, magnitudes 6.9, 7.3; Star 24, variable, magnitudes 7.8–13; variation period 317.5 days; Star 25, a glowing ember, magnitude 6.2; Star 26, a double, magnitudes 5.8, 7.2; Star 27, a globular cluster; Star 28, a variable, magnitudes 8–12; variation period 262 days; Star 29, a triple, contrasting colors; magnitudes 2.5, 8.8, 11.5; Star 30, variable, magnitudes 8.2–14; variation period 303 days; Star 31, double, magnitudes 5.3, 5.4; period 5.7 years; Star 32, a triple, magnitudes 4.1, 5.7, 11; Star 33, double, magnitudes 4, 4.1; Star 34, double, magnitudes 5.6, 5.7; Star 35, triple, magnitudes 7, 7.5, 8; Star 36, double, magnitudes 7, 8; Star 37, double, magnitudes 7, 8; Star 38, double, magnitudes 4, 8.5; Stars 39 to 47 either invisible to the naked eye or nearly so, and relatively unimportant.

Capricornus, Star 48, magnitude 3; Star 49, magnitude 5. With the exception of Cancer, the most inconspicuous of the zodiacal constellations.

Fomalhaut, Star 50, magnitude 1.3 and distance 25.51 light-years, is seen in autumn in the constellation *Pisces Australis,* or *Southern Fish.*

APPROXIMATE ANATOMY OF ATMOSPHERE				
Major Units	Temp. °F / Distance Mi.	Temp. °C / Distance Km.	Major Units And Characteristics
Magnetosphere 400–40,000 Mi. 644–64,375 Km.	2,256 @ 450	1,245 @ 720	G Layer — 450 Mi. And Up Or 720 Km. And Up
Ionosphere 50–400 Mi. 80–644 Km.	1,150 @ 100	631 @ 160	F Layer — 150–450 Mi. Or 241–720 Km. Reflects Broadcasts @ 75 Mi. or 120 Km.
Mesosphere 25–50 Mi. 40–80 Km.	−130 @ 50; 28 @ 30	−90 @ 80; −2 @ 48	E Layer — 70–150 Mi. Or 112–240 Km. Dust Belt @ 60 Mi. or 96 Km.
Stratosphere 10–25 Mi. 16–40 Km.	−45 @ 20	−43 @ 32	Noctilucent Clouds At 50 Mi. or 80 Km. D Layer 40–70 Mi. Or 64–112 Km.
	−76 @ 12; −68 @ 10	−60 @ 20; −55 @ 16	Sound Reflected @ 20 Mi. or 32 Km. Ozone Layer 12–20 Mi. Or 19–32 Km. Ultra-Violet Barrier
Tropopause	10 Mi.	16 Km.	Sulphate Layer 10–12 Mi. Or 16–19 Km. May Affect Rainfall
Troposphere 0–10 Mi. 0–16 Km.	0 @ 4; 60 @ Surface	−18 @ 6.4; 16 @ Surface	Cosmic Rays Affected @ 7 Mi. Or 12 Km.

INTERPLANETARY SPACE

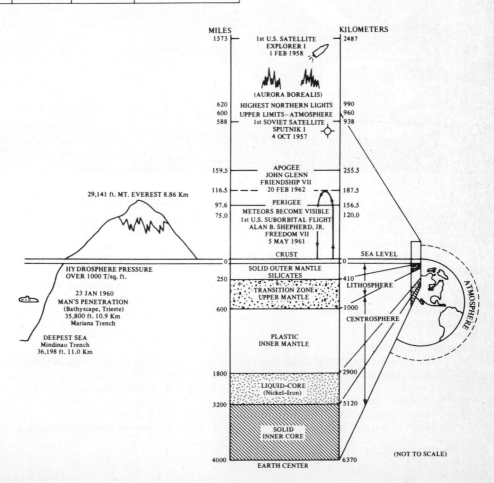

MILES / KILOMETERS

1573 / 2487 — 1st U.S. SATELLITE EXPLORER I 1 FEB 1958

(AURORA BOREALIS)

620 / 990 — HIGHEST NORTHERN LIGHTS
600 / 960 — UPPER LIMITS – ATMOSPHERE
588 / 938 — 1st SOVIET SATELLITE SPUTNIK I 4 OCT 1957

159.5 / 255.5 — APOGEE JOHN GLENN FRIENDSHIP VII
116.5 / 187.5 — 20 FEB 1962
97.6 / 156.5 — PERIGEE
75.0 / 120.0 — METEORS BECOME VISIBLE
1st U.S. SUBORBITAL FLIGHT ALAN B. SHEPHERD, JR. FREEDOM VII 5 MAY 1961

0 / 0 — CRUST — SEA LEVEL

29,141 ft. MT. EVEREST 8.86 Km.

HYDROSPHERE PRESSURE OVER 1000 T/sq. ft.

23 JAN 1960 MAN'S PENETRATION (Bathyscape, Trieste) 35,800 ft. 10.9 Km Mariana Trench

DEEPEST SEA Mindinao Trench 36,198 ft. 11.0 Km

250 / 410 — SOLID OUTER MANTLE SILICATES — LITHOSPHERE
600 / 1000 — TRANSITION ZONE UPPER MANTLE — CENTROSPHERE

PLASTIC INNER MANTLE

1800 / 2900 — LIQUID-CORE (Nickel–Iron)
3200 / 5120

SOLID INNER CORE

4000 / 6370 — EARTH CENTER

ATMOSPHERE

(NOT TO SCALE)

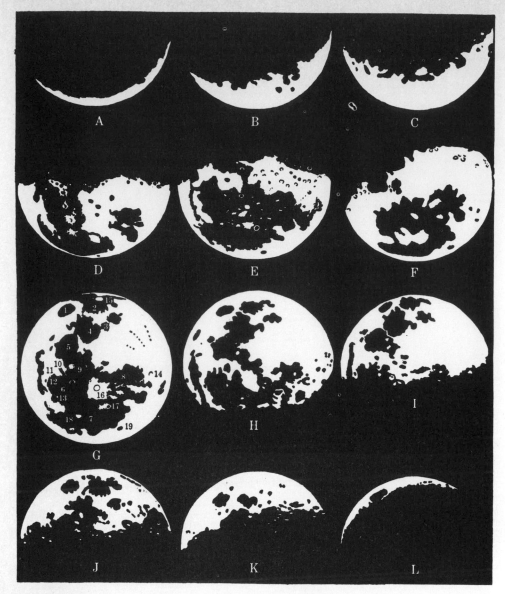

Areas on the Moon

1	Sea of Crises, 70,000 sq mi, 280 by 360 mi	11	Alps Mountains
2	Sea of Fecundity	12	Plain of Plato, 60 mi across
3	Sea of Nectar	13	Bay of Rainbows
4	Sea of Tranquillity	14	Crater of Tycho, 54 mi across, 17,000 ft deep
5	Sea of Serenity	15	Crater of Langrenus, 90 mi across, 10,000 ft high
6	Sea of Showers	16	Crater of Copernicus, 56 mi across, 11,000 ft high
7	Sea of Storms	17	Crater of Kepler, 22 mi across, 10,000 ft high
8	Sea of Clouds	18	Crater of Aristarchus
9	Apennine Mountains	19	Crater of Grimaldi, 148 by 129 mi
10	Caucasus Mountains		

Phases of the Moon

A	Waxing crescent, in west at sunset, 3¾ days	G	Full moon, in east at sunset, 14½ days
B	Waxing crescent, high in west at sunset, 5½ days	H	Gibbous moon, in east after sunset, 18 days
C	First quarter, in south at sunset, 6¼ days	I	Gibbous moon, in east later, 20 days
D	First quarter, toward east at sunset, 7¼ days	J	Last quarter, in east later yet, 20½ days
E	Gibbous moon, farther east at sunset, 9¾ days	K	Waning crescent, late at night, 24 days
F	Gibbous moon, still farther east at sunset, 11¾ days	L	Waning crescent, at dawn, 27 days

18

	Sun	Mercury	Venus
Location	On the average, 93 million miles from the earth, the distance varying because earth moves in an ellipse. An average star near one edge of ($\frac{1}{5}$ of distance across) the Milky Way, a lens-shaped galaxy containing about 200 billion stars.	From 43 million to 28 million miles from sun. 49½ million miles from earth when between earth and sun and 136 million miles when sun is between earth and Mercury. $\frac{2}{5}$ earth-sun distance; designated an inner planet.	Average 67 million miles from sun, 26 million miles from earth when between earth and sun, and 160 million when sun is between earth and Venus. $\frac{3}{4}$ earth-sun distance; designated an inner planet.
Size	Surface area, over 12,000 times that of earth; volume, over 1,300,000 times as much; weight, 333,000 times as much, about 2 octillion tons. About 1½ times as dense as water and $\frac{1}{4}$ as dense as the earth. Diameter, 864,000 miles, or over 109 times that of earth	Area, $\frac{1}{7}$ that of earth, or about equal to Asia and Africa combined. Volume, $\frac{1}{18}$; density, $\frac{4}{5}$; mass, $\frac{1}{20}$ that of earth. Diameter, less than $\frac{1}{2}$ that of earth, or 3,009 miles	Area, about $\frac{9}{10}$ that of earth; mass, about $\frac{4}{5}$; density, 0.87 that of earth. Diameter, 7,700 miles, or about equal that of earth
Light and other relations	While light comes to the earth from the sun in 8 minutes 19 seconds, sound would take 14 years to travel the distance if there were atmosphere to carry it. Loses 4,200,000 metric tons per second according to carbon cycle theory. Intensity 600,000 times that of moon. Earth receives small fraction, $\frac{1}{2}$ trillionth, perhaps, of energy transmitted by the sun. Energy produced by nuclear fusion process; hydrogen to helium through several intervening steps. Visible white light from sun is really a mixture of colors of the spectrum, violet through red.	Light from sun reaches planet in 4 minutes; because of dark rocks only $\frac{1}{16}$ is reflected. Brilliance is faint, at its best about equaling Vega. Appears orange	Sunlight reaches planet in 5 minutes 50 seconds; because of bright cloud surfaces, $\frac{3}{5}$ is reflected. Most brilliant of the planets; at best, 15 times brilliance of Sirius. Ordinarily bright enough to be distinguished from nearby stars. Locate by knowing constellation in which it appears. Appears yellow.
Movement	Completes a turn on its axis in about 25 days. With the earth at the sun's center, the moon would be about halfway between the center and outside of the sun. Sun also moves with the movement of other bodies in the Milky Way.	Moves 23 to 35 miles per second in orbit for a year of 87 days 23 hours 15 minutes, with axis 0° to perpendicular to orbit plane and 7° to right angle to ecliptic	Moves 22 miles per second in orbit 1.9 times that of Mercury for year of 224.7 days, with axis 0° perpendicular to orbit plane and 3°5′ to right angle to ecliptic
Sidereal day		59 days	243 days retrograde
Climate	Temperature at the surface, about 11,000°F or 6000°C and much hotter inside. Heat from the sun evenly distributed over the earth would melt each year an ice covering 114 ft thick.	One side bright, with temperature of 600°F; other, dark, with temperature of −449.4°F. Little or no atmosphere	One side hot and one side cold, as in Mercury. Receives twice as much heat and light as the earth does. Densely cloudy, with clouds many miles deep. Approximate mean temperature of 700°F
Weight variation compared to earth	100 lb on earth would weigh 2,760 lb on sun	100 lb on earth would equal 31 lb on Mercury	100 lb on earth would equal 85 lb on Venus
Moons	Has no moons	Has no moons	Has no moons

Earth	Earth's moon	Mars	Jupiter
Varies from 91½ million miles from the sun in January to 94½ million miles in July. About 93 million miles as an average	About 238,840 miles from the earth, or about 30 times diameter of earth at equator	Average 141½ million miles from sun. From 35 to 248 million miles from the earth, varying because of positions on their orbits. Distance from sun about 1½ earth-sun distances. Designated an inner planet with Mercury, Venus, and earth.	Averages about 483 million miles from sun. Between 390 and 576 million miles from the earth. About 5 earth-sun distances
Earth's surface would appear smooth to its nearest neighbor, though it appears rough to us. Volume, 259 billion cubic miles; weight, 6 sextillion tons; density, 5.53 times that of water. Diameter, 7,918 miles. Area, 196,950,000 sq mi. Very nearly spherital	Area of surface, 1/14 that of earth. Volume, 1/49; mass, 0.0122; density, just over half that of earth, about 3.5 times that of water. Diameter, 2,163 miles, or about 1/4 that of earth. The pull of the moon on the earth is the major cause of our tides	Area, 2/7 that of the earth. Shows conspicuous polar caps and dark markings and craters. Volume, 1/7; mass, 1/8–1/10; density, 0.73 that of earth, 3.92 times the density of water. Diameter, 4,200 miles	Surface 131.3 times; volume, 1,312 times; mass, 317 times; density, 0.24 times that of earth. Diameter through polar region, 82,880 miles and through equatorial region, 88,640 miles
Sunlight reaches earth in 8 minutes 19 seconds, nearly ½ being reflected because of clouds and dust in atmosphere. Earth intercepts one two-billionth of sun's radiant energy or power equal to that produced by burning 21 billion tons of bituminous coal every hour	When earth is directly between sun and moon, moon is eclipsed; moon eclipses sun when it is between sun and earth. A new moon appears when moon is roughly between earth and sun. About 1/14 the light received is reflected	Sunlight reaches Mars in 12 minutes 30 seconds, because of cloudless atmosphere only 1/7 of the light received is reflected. The third most brilliant planet. Appears ruddy. Ordinarily bright enough to see if one knows constellation in which it occurs; appears red or ruddy.	Sunlight reaches planet in 43 minutes 20 seconds; because of cloud masses, 9/20 of that received is reflected. Second most brilliant planet, appearing a little more brilliant than brightest star. Appears yellow
Moves 18½ miles per second in orbit 2.6 that of Mercury for a year of 365.24 days, with axis 23½° from right angle to orbit plane	Moves about the earth in 29 days 12 hours 44 minutes 2.86 seconds, turning on axis so that the same side faces the earth at all times. Earth reflects light to moon	Moves 15 miles per second in orbit 3.9 that of Mercury for year of 687 days, with axis at 23°59′ to right angle to orbit plane	Moves 8 miles per second in orbit 13.4 times that of Mercury, for a year of 11.86 earth years with axis at 3°6′ to perpendicular to orbit plane
23 hours 56 minutes		24 hours 37 minutes	9 hours 53 minutes
At least 100 miles of atmosphere, all but 4 percent within 15 miles of solid surface; atmosphere acts as a blanket against cold and diffuses light. Energy equal to 1½ horsepower per sq yd reaches the earth from the sun. Percentage composition of gases remains rather constant in spite of man's activities	Side toward sun always lighted and hot; side not lighted very cold—200°F. Little or no atmosphere. So-called seas are not water but are smoother and therefore appear as darker regions because of regularly reflected light. Would undoubtedly be uninhabitable for man	Average 20°F. Water would boil at 111°F. Relatively small amount of water present, as compared to earth. Thin and cloudless atmosphere generally, except for haze over poles during their winters	Receives about 1/27 as much heat and light per unit of area as does the earth. Average temperature −200°F. Covered with cloud mass at least 1,000 miles deep, composed of ammonia, methane, and hydrogen. Occasional red probably nitrous oxide
	100 lb on earth would weigh 16½ lb on the moon	100 lb on earth would weigh 38 lb on Mars	100 lb on earth would weigh 264 lb on Jupiter
Has one moon. See next column	Has no moons	Two moons. Phobos, about 10 miles in diameter and 5.850 miles from planet center, has period of 7 hours 39 minutes, rises in west and sets in east. Deimos, about 5 miles in diameter and 14,650 miles from center of Mars, rises in east and, 65⅔ hours later, sets in west	Twelve moons, all rising in east and setting in west. Io, about size of earth's moon, and Europa, slightly smaller; Ganymede, a little larger than, and Callisto about the size of Mercury. Four moons can be seen with good field glasses. Moons as large or larger than earth's moon

	Saturn	Uranus	Neptune	Pluto
Location	Nearly 10 times the distance of the earth from the sun, or 887 million miles. From 734 to 1,017 million miles from the earth, depending on place in orbit. Referred to as an outer planet	1,781,900,000 miles from the sun. Average 1,800,000,000 miles from the earth. Referred to as an outer planet. 20 times earth-sun distance	2,791,600,000 miles from the sun; about 2,800,000,000 miles from the earth. Referred to as an outer planet. 30 times earth-sun distance	Averages 4 billion miles from the sun; at its minimum distance, nearer the sun than Neptune is. Most eccentric orbit of any planet. Referred to as an outer planet. 40 times earth-sun distance.
Size	Surface area 81.4 times; volume, 734 times; mass, 95 times, that of earth. Mostly liquid, $\frac{1}{8}$ the density of earth; would float on water. Solid part relatively small. Diameter, 66,300 polar and 74,100 equatorial miles	Surface area, 16 times; volume, 60 times; mass, 14.66 times; density, 0.25 times that of the earth. Density, 1.27 that of water, about $\frac{1}{4}$ that of earth. Diameter, 30,900 miles	Surface area, 18 times; volume, 72 times; mass, 17.16 times; density, 0.3 that of the earth, 1.6 that of water. Diameter, 32,900 miles	Surface area, volume, density, and mass not known. Diameter, probably about that of Mars
Light and other relations	Sunlight reaches the planet in 1 hour 19 minutes 30 seconds; $\frac{2}{5}$ of that received is reflected. Brilliance of first-magnitude star, with dull, steady, orange light. Usually observer can locate if he knows constellation in which planet occurs. Appears yellow.	Sunlight reaches planet in 2 hours 31 minutes 35 seconds, about $\frac{9}{20}$ of that received being reflected. Appears sea green in color and at best is just visible to naked eye. Can be located with good binoculars if constellation in which it occurs is known. Appears green.	Sunlight reaches the planet in 4 hours 10 minutes 35 seconds, and about $\frac{1}{2}$ is reflected. Seen only with telescope and appears much like Uranus. Can be located with good binoculars if constellation in which it occurs is known. Appears yellow.	Sunlight reaches planet in about 5 hours 30 minutes; amount reflected is not known. Barely visible on clear night with 15-inch telescope
Movement	Moves at 6 miles a second on orbit 24.6 times that of Mercury, with axis tipped at 26°48′ to perpendicular to orbit plane for a year of 29.46 earth years	Moves 4 miles per second in orbit 49.5 times that of Mercury for a year of 84 earth years 7 days, with axis probably 98° from right angle to orbit plane	Moves at 3.4 miles per second in orbit 77.5 times that of Mercury, with axis inclined by unknown amount from perpendicular to orbit plane for a year of 164.8 earth years	Moves at 1.8 miles per second in orbit $\frac{4}{3}$ that of Neptune; orbit more inclined to plane of earth's orbit than that of other planets; year equals 248 earth years
Sidereal day	10 hours 15 minutes	10 hours 50 minutes	15 hours	6.25 days
Climate	Receives about $\frac{1}{90}$ as much heat and light per unit of area as does the earth; temperature about −300°F. Covered by deep cloudy atmosphere with ammonia snow much like that of Jupiter, and with brilliant white clouds of unknown composition appearing about once in 10 years	Temp. about −366°F. Receives about $\frac{1}{360}$ as much heat and light per unit of area as does the earth. Vast cloud-laden atmosphere containing gaseous methane and probably nitrogen and hydrogen. Clouds are ammonia snow, possibly with liquid droplets of ammonia and methane in suspension	Receives about $\frac{1}{900}$ as much heat and light per unit of area as does the earth. Atmosphere in general similar to that of Uranus	Nature of climate not known. Probably has no atmosphere
Weight variation	100 lb on earth would weigh about 117 lb on Saturn	100 lb on earth would weigh 90 lb on Uranus	100 lb on earth would weigh 89 lb on Neptune	Weight of 100 lb on Pluto not known
Moons	Ten moons. Titan about equal to earth's moon; Rhea and Iapetus, each about 1,100 miles in diameter; five, from 800 to 300 miles through; two, smaller. All rise in east and set in west. Three flat concentric rings less than 50 miles thick circle planet, appearing on edge about once in 15 years. Inner edges about 8,000 miles, outer edge 86,000 miles from planet center.	Five moons. Ariel, 120,000 miles from planet, encircles it in 2.5 days; Umbriel, 167,000 miles, in 4.1 days; Titania, 273,000 miles, in 8.7 days; and Oberon, 365,000 miles, in 13.5 days. The fifth, discovered in 1948, is 81,000 miles distant from Uranus	Two moons, largest about the size of earth's, rise in east and set in west, going around the planet once every 6 days	No moons known

THE MINERAL KINGDOM

The Plant Kingdom and the Animal Kingdom are dependent on the Mineral Kingdom for continued existence. It is important that the Mineral Kingdom be reasonably understood by the one member of the Animal Kingdom able to modify his environment to his advantage or disadvantage.

At least two things of importance stand out in this planning. Some parts of the Mineral Kingdom are essential to our way of life and are present on the earth in limited quantities. It is reasonable to assume that we cannot renew our mineral resources by drawing on other parts of the universe, and many of our dwindling mineral resources are definitely not renewable. Each generation of human beings will find itself facing a new set of problems brought into prominence by dwindling resources, new demands created by scientific discovery, and the discovery of new uses of resources now considered useless. World economics, politics, and happiness are dependent largely on the use or abuse made of the resources at hand in different parts of the world. It therefore behooves us all to be able to recognize at least a few of the important mineral resources at hand and to understand their interrelationships with each other and with plants and animals.

Much as soap and pie and bread are generic names for happy or unhappy combinations of the materials of which they are made, so are most of the rocks we recognize combinations of minerals in different proportions and subjected to highly variable treatment. No two pies or soaps are alike and similarly no two granites or sandstones are alike. Two granites may have identical chemical composition but one may have been subjected to a vastly different combination of heat and pressure. Those differences will be reflected in the rocks.

In the identification of minerals in the following tables at least two sets of standards will have to be recognized: differences in hardness and differences in fusibility. These are explained here.

In the table of hardness given below, known as the Mohs' scale, a material will scratch any material with a lower number, i.e., topaz will scratch quartz but not corundum.

Table of Hardness

 1 Talc, scratched by thumbnail
 2 Gypsum, scratched by thumbnail
 3 Calcite, scratched by a copper penny
 4 Fluorite
 5 Apatite, scratched by point of good steel knife and by glass
 6 Feldspar
 7 Quartz
 8 Topaz
 9 Corundum
10 Diamond

Six minerals are chosen as standard in measuring fusibility in a blowpipe flame.

Table of Fusibility

 1 Stibnite, very easily fusible
 2 Chalcopyrite, easily fusible
 3 Almandite garnet, of medium fusibility
 4 Actinolite, intense heating necessary
 5 Orthoclase, prolonged heating necessary
 6 Bronzite, very hard to obtain any evidence of fusibility

Streak characteristics may be demonstrated by observing the color made by rubbing the mineral to be tested on a piece of unglazed porcelain.

Other characteristics of minerals useful in identification are:

1 Color: First property noted by beginner; diagnostic for some minerals, e.g., blue of azurite and green of malachite. Both indicate copper minerals.
2 Luster: Appearance in reflected light; this characteristic permits a division into two main classes—the metallic lusters and the nonmetallic lusters; for example, nonmetallic ones are further described as: (1) vitreous—meaning that mineral looks like glass; (2) greasy—meaning that the luster has a brilliant, yet almost greasy paper look such as sulfur; (3) resinous—like rosin; (4) silky—indicative of a finely fibrous mineral in parallel bands of threadlike needles such as asbestos; and (5) pearly—resembling reflections from flat surfaces of an iridescent shell such as mica.
3 Specific gravity: Weight of the mineral compared to the weight of an equal volume of water; e.g., a specific gravity of 3.0 means the substance is three times as heavy as water. This

difference relates to our experience of "feeling the weight" of specimens when we say they are either light or heavy.

4 Cleavage: Described as the tendency of a mineral to break in smooth flat planes. This property is related to the atomic arrangement of the crystal.

5 Fracture: Refers to the character of a broken surface; e.g., minerals may have an uneven fracture, an irregular grainy fracture, or a curving shell-like fracture spoken of as conchoidal.

6 Diaphaneity, referred to in some guides as translucency: Minerals vary from transparent to opaque.

7 Other properties: A few minerals may be identified because they are magnetic, are easily dissolved in water or some common solution, have a taste, or fluoresce (show a distinctive color in ultraviolet light).

ATMOSPHERE

Atmospheric Water

Water vapor and moisture in atmosphere mean the same thing. The amount present varies from practically nothing to 4 percent. As vapor condenses, it becomes visible as fogs and clouds that may evaporate and vanish. Increased barometric pressure and a lowering of temperature may speed condensation.

Because it is lighter, moist air is displaced by heavier dry air and rises. These changes contribute considerably to the making of weather and of humid and arid areas on the earth and consequently to the nature and abundance of life in different places.

Snow, Sleet, Glaze, Hail

When water vapor condenses in air at a temperature below 32°F, *snow* is formed. When the condensed rain freezes in the air after it has formed into drops, *sleet* is formed; when rain comes in contact with sufficiently colder objects, it becomes *glaze*. *Hail* is formed by frozen rain being forced upward by strong rising columns of air into regions where ice is added to the existent pellets.

Snow crystals are 6-sided, of an infinite variety of types, the types usually indicative of the conditions under which the crystals are formed: they usually fall to the ground through the air by the force of gravity much as sediments of sand and silt are settled out of water or dust out of air.

Weather

Water vapor frequently condenses as clouds. Different types of clouds may appear in the sky by themselves or in combination with other clouds. The forecasts following each cloud description apply when that cloud type appears by itself. Generally, the forecasts based on middle and low clouds are most valid for only a few hours. With a little practice a viewer can classify clouds. Clouds are classified on the following basis:

1　Form and appearance
　　A　Stratus (stratiform): Sheetlike, layerlike
　　B　Cumulus: Piled up, rolled, cottony, with cauliflowerlike appearance when fully developed
2　Height
　　A　High clouds: Average height, 20,000 to 30,000 ft above the earth; 4 to 6 miles
　　B　Middle clouds: 6,500 to 20,000 ft above the earth; 1½ to 4 miles
　　C　Low clouds: From near the earth to 6,500 ft; less than 1½ miles
　　D　Clouds of vertical development: 1,500 to 65,000 ft; ¼ to 12 miles

HIGH CLOUDS
A　Cirrus: Look like wisps and feathers, mare's tails
　　　　　Composed of ice crystals
　　　　　Produce no shade
　　　　　Height about 6 miles
　　　　　Forecast: Fair weather
B　Cirrocumulus: Referred to as a mackerel sky
　　　　　Composed of ice crystals
　　　　　Height about 4 miles
　　　　　Forecast: Fair weather
C　Cirrostratus: High, milky, veillike
　　　　　Composed of ice crystals
　　　　　Little shade; halos common
　　　　　Height about 5 to 6 miles
　　　　　Forecast: Not fair, but no rain

MIDDLE LAYER CLOUDS
A　Altocumulus: High, cotton ball-like
　　　　　May be isolated globular patches
　　　　　Some shade; may or may not obscure the sun, depending on thickness
　　　　　Composed of water droplets or ice crystals, depending on height and the time of
　　　　　　year
　　　　　Average height 2 miles; may occur much higher to almost 4 miles
　　　　　Forecast: Probably fair but may produce occasional snow in winter and rain at
　　　　　　other times
B　Altostratus: Overcast, thick or thin, medium gray in color, may become iron gray
　　　　　Composed of water droplets or ice crystals
　　　　　Produce considerable shade but no halos
　　　　　Forecast: Steady light rain, or snow

LOW LAYER CLOUDS
A　Nimbostratus: Low, thick, dark gray, appear threatening
　　　　　Bases wet-looking and ragged; uniform gray when snow is falling
　　　　　Contain water droplets and ice crystals
　　　　　Height—near the earth's surface to 2 miles high
　　　　　Forecast: Rain, usually steady, or snow
B　Stratocumulus: Low, look like cotton balls
　　　　　Contain water droplets
　　　　　Widespread across the sky
　　　　　Height to 1½ miles
　　　　　Forecast: Chance of rain: sometimes break up with sunshine thereafter
C　Stratus: Uniform, thin clouds of indefinite shape
　　　　　Look like a fog lifted above the ground
　　　　　Sky appears gray
　　　　　Contain water droplets which dissipate quickly with the sun's heat
　　　　　Height to 1 mile
　　　　　Forecast: May rain (drizzle) or snow (no accumulation likely)
　　　　　On some days there may be stratus clouds near the ground and altostratus clouds
　　　　　　above. This combination of clouds may be associated with steady rain. The problem
　　　　　　with these forecasts is that lower clouds often hide higher clouds.

CLOUDS OF VERTICAL DEVELOPMENT

A   Cumulus clouds: Cottonlike, fluffy, woolly, flat base, domed top, rounded sides
                    Contain water droplets
                    May appear in all levels, from 1½ to 4 miles
                    Forecast: Fair weather

B   Cumulonimbus clouds: Thunderheads, massive, cauliflower-shaped, great vertical development,
                    flat dark bottom, anvil-shaped
                    Height from 1½ to 5 miles
                    Contain ice crystals and water droplets
                    Forecast: Heavy showers of rain or snow: thundershowers.   May have
                    hail

FOG

A cloud, the base of which rests on the earth
Usually water droplets but may be composed of ice crystals

CLOUD TYPES

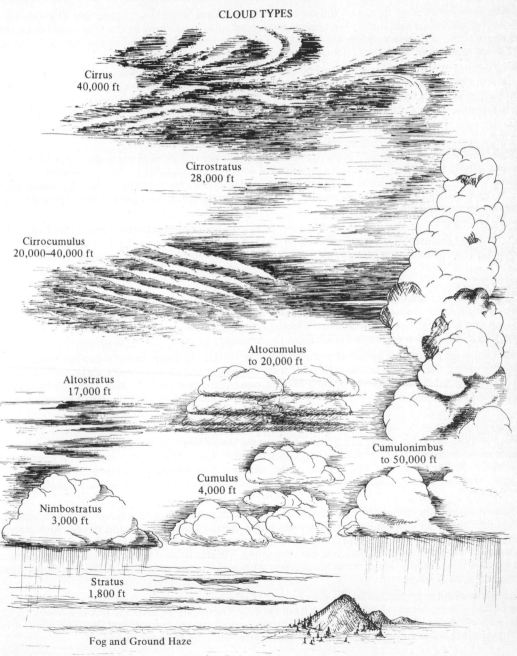

Cirrus
40,000 ft

Cirrostratus
28,000 ft

Cirrocumulus
20,000–40,000 ft

Altocumulus
to 20,000 ft

Altostratus
17,000 ft

Cumulonimbus
to 50,000 ft

Cumulus
4,000 ft

Nimbostratus
3,000 ft

Stratus
1,800 ft

Fog and Ground Haze

HYDROSPHERE

Surface Water

Nearly ¾ of the surface of the earth is covered with water. From this surface water evaporates, leaving mineral deposits in an ever-increasing concentration. Moving to the lower levels of the surface of the earth, the water carves this surface violently by *gully erosion* or inconspicuously by *sheet erosion*. When it moves, it carries with it the more buoyant materials at hand.

Erosion control practices beginning on a 3 percent slope are designed to hold desirable plant foods in position. Sudden and violent descent of surface water provides power that may support industry and may relieve man and other animals of labor. A sustained yield of power is attained by dams that hold back surpluses and incidentally also serve to prevent floods and to help sustain navigation and water supply.

Snowbanks

When snowflakes come to rest on the ground, they may form something comparable to *sedimentary rocks* like sandstone and shale. Successive deposits of snow may show as strata in snowbanks, just as strata show in ordinary sedimentary rocks. Changing temperatures and varied deposits produce readable changes in the banks. Dust and dirt also may appear in definite layers.

Snowbanks serve splendidly as a thermal blanket over the earth modifying otherwise sudden temperature changes that might be fatal to plants like winter wheat. They also provide a cover that protects mice from hawks and other predators. They interfere with or assist transportation and, to some degree, delay the runoff of waters that may have fallen on the earth as snow.

Groundwater

Water that falls on the surface of the earth evaporates, runs off, or penetrates the porous portions. Eventually it reaches an impervious region and accumulates variously. As *groundwater,* it may constitute a reserve available for the use of plants and animals able to reach it when the upper earth surfaces become dry. At different depths, the chemical nature and the temperature of groundwater may vary considerably.

The varied ability of plants to reach the *water table* may well determine the plants to be found in a given area. Deep-rooted plants may appear to thrive under conditions fatal to adjacent shallow-rooted plants. Burrowing animals may reach water not available to nonburrowers.

Ice

When snow has continued to accumulate, the weight of the upper layers may well change or metamorphose the snow in the lower layers. Pressure may well change snow to ice as it does in glaciers on sloping mountaintops or on sloping rooftops. The ice gradually makes its way downward, carrying with it earthy materials from the mountain or roof.

Since ice is 92 percent as heavy as water, it floats in water. In many rapidly flowing streams, ice may carve the stream beds considerably. It may also pluck and carve rocks of gorge walls or grind the shores of lakes and oceans. It may blanket ponds, shutting off light and upsetting biological balances that may have been established, and it may hold life captive under water or in snowbanks.

Water

Water is fluid between 32 and 212°F solidifying at the lower figure and vaporizing at the upper. In centigrade, these points are 0° for freezing and 100° for boiling. Water expands when it freezes and ice has a specific gravity of 0.92 instead of the standard 1 of water. Water is colorless, tasteless, transparent and in its natural impure state readily dissolves such substances as lime and salt (see pp. 38 and 44).

Because water is the great solvent of nature, it occurs with many kinds of solutes. Distilled water is of course free from impurities; rain may bear sulfur, dusts of many sorts, and oxygen and carbon dioxide absorbed from air in descending; flowing waters may dissolve lime carbonate, sulfur, lithia, and salt from rocks; seawater may become very saline. Water is hard when it contains ions such as calcium or magnesium, which form precipitates with many negative ions or radicals such as those found in natural soaps.

The water cycle is important in the distribution of important compounds in nature. Falling rain may remove some elements of impurities from the air. These may be added to through contact with the soil and rocks or may be concentrated through evaporation of water and the precipitation of compounds that have been in solution and through movements of water in soil.

Water of course is the basic material in which living protoplasm, whether in plants or in animals, functions. It is obvious that it also is useful as a source of power, as a means of transportation in fighting fire, in the removal and elimination of wastes, in encouraging or in discouraging different chemical actions, and as an unexcelled drink.

LITHOSPHERE AND/OR MANTLE

Fragmented Sediments

Gravels

Mechanical analysis: Diameter of fine gravel, 1-2 mm; granule gravel, 2-4 mm; pebble gravel, 4-64 mm; cobble gravel, 64-256 mm; boulder gravel, 256 to over 1,000 mm. Units may be either rounded and waterworn or angular and unworn. Technical soil standards allow 7-8 percent of loam in fine gravel classification.

Some geologists classify gravel units under "size of pea" as sand. Sands and gravels at point of origin are *residual*. Those carried by gravity are *colluvial;* those carried by streams, *alluvial;* those formed in lakes, *lacustrine;* those in the sea, *marine;* those by ice, *glacial*. Loose gravity-assembled gravels are commonly angular *talus*.

Consolidated gravels with rounded surfaces make *conglomerates* while consolidated angular talus or angular gravels make *breccias*. These may under suitable conditions be metamorphosed into *quartzite conglomerates, quartzite breccias,* and other rocks. Gravels because of their weight are naturally settled out of flowing streams before the finer sands, which in turn settle out before the silts.

Gravels are used commercially in road building and in the making of concrete (a strong mixture is 1 part of cement to 2 parts of sand to 4 parts of gravel, which yields concrete that resists 1,500 lb crushing weight per sq in. but has only 400 lb tensile strength and weighs about 150 lb per cu ft). Gravel provides for drainage about buildings and has other uses.

Sands

Mechanical analysis: Diameter of coarse sand, 1-0.5 mm; medium sand, 0.25-0.5 mm; fine sand, 0.1-0.25 mm; very fine sand, 0.05-0.1 mm. Allowed percent of loam and clay: coarse sand, 16.8 percent; medium sand, 13 percent; fine sand, 37 percent; very fine sand, 44 percent. Specific heat: 0.185-0.198. Specific gravity: around 2.65. In addition to above groups, gravelly sands and loamy sands are recognized.

Soil technologists use the above as standards. Some geologists consider units up to "size of pea" as sand. *Residual sands* may be angular while *waterworn sands* are rounded; *windborne sands* like *loess* are angular, or, if in desert formations, are more perfectly rounded. Chemically, sands may be of any rock material such as quartz or limestone, the nature varying accordingly.

Sands are almost invariably produced by the breaking down of larger units and a re-sorting of units according to size by gravity, water, or wind. The longer or more violent the sorting and wearing process, the smaller the resultant units. They may be consolidated into sandstones of different types. In spite of their actual weight, they make *light* soils as compared with the lighter-weight clays.

Quartz sand is used for making glass, polishing sand, molding sand, hourglass sand (from deserts), drinking-water filters, ballast for boats and balloons, furnaces, for confining or for extinguishing fires. Since sands do not hold water well, they are mixed with clays to lighten soils. They vary much in their durability. Some contribute to the acidity of soils.

Silts and Clays

Silts range in diameter from 0.002 to 0.05 mm while *clay* particles are below 0.002 mm in diameter. *Loess* is fine-grained, wind-blown dust that has settled. *Gumbo* is a fine, silty soil that when wet becomes sticky, almost mucilaginous and impervious and soapy. *Loams* are mixtures of sands, clays, and organic matter. Silt plus clay plus water makes *mud*.

Silts, clays, and muds may be moved by air, water, gravity, or ice as are the larger sands and gravels and are similarly classified on this basis. They tend to compact and prevent the entrance of air and water or, when dry, to contract and crack in large units unless *flocculated* by substances like lime that bring small units together around independent centers.

Silts, clays, and muds may consolidate to form *shales* of homogeneous or heterogeneous nature, of varying degrees of hardness, of highly variable chemical nature depending on the basic materials involved; these may metamorphose to form *slates* or similar metamorphic rocks. Evaporation, capillarity, gravity, and exhaustion by plant use may change the nature of silts and clays.

Properly managed, these fine inorganic particles may make the agriculture of an area. Improperly managed, they may break it by being carried away by water or wind, or by sealing over soils more suitable for plant growth. Many clays are basic in ceramics, in brickmaking, in the arts. Some are useful in polishing and some as sources of important component minerals.

Chemical Sediments

Siliceous Deposits

Quartz sand of course is siliceous. Disregarding this, we find much silica deposited variously and rather continuously from many sources. Anyone recognizes the readiness with which water carries lime in solution or otherwise. Few recognize that approximately $\frac{4}{7}$ as much silica as calcium bicarbonate is carried by water in one way or another.

Sources of siliceous deposits include fine silica particles deposited from suspension or from colloids in fresh water or more commonly probably in salt water. Again, silica may be deposited from the bodies of animals or of plants. Spicules in sponge skeletons, portions of shells, shells of microscopic animals, "glass" shells of diatoms are but a few of the organic sources of silica.

Diatoms reproduce rapidly, about once every 5 days; so an **uninhibited** one might yield over 60 tons in a year. In the Southern Hemisphere, over 10 million sq miles of sea bottom are covered with diatomaceous oozes. In California, there is a bed nearly ½ mile thick of deposits of these plants. In a short time, a ridge nearly 6 in. thick and 20 miles long was deposited on our Washington coast.

Siliceous earths may contribute to formation of flints and cherts. Deposition at present is probably greatest where fresh waters flow into salt water. Fine polishing powders, waterglass, tooth pastes, fireproof packings, soaps, scouring mixtures, and many explosives include as essentials some of the siliceous materials here briefly discussed.

Limestone Deposits

Limestone deposits are commonly but not always light-colored. They effervesce with hydrochloric acid. At about 900°C, they free carbon dioxide and leave calcium oxide which, exposed to air, takes up water and carbon dioxide and again becomes hard calcium carbonate. Types include coral, coquina, hydraulic limestone, chalk, onyx, travertine, bog lime, marl, and so on.

Bog lime commonly is precipitated from plants; *chalk,* formed by shells of Foraminifera; *coquina,* by cemented shell fragments; *coral,* by cementing of coral particles with plant action involved; *travertine* and *onyx,* by evaporation of water from saturated lime waters; *encrinal* limestone, from crinoids, and so on. Many are readily dissolved in water and then precipitated out.

Limestone deposits may originate from fragmentation of larger limestone units, by deposition from mixtures brought together by currents, by mixtures of volcanic gases with water, by mixtures arising from actions of springs, by breaking of shells or skeletons of animals, by action of marine algae like *Lithothamnius* or freshwater algae such as *Chara* (p. 65), by bacterial action and evaporation.

(Continued)

Limes are used in making mortar, hydraulic limestone that hardens under water like portland cement, in making portland cement, in building stones, in road making, in industry, in the flocculation of clay soils making them lighter, in correcting acidity of sour soils, in part as plant food, and in assisting plants to take up food from soil.

Plants and Plant Products in Soils and Rocks
Vigorous plant growth penetrates soils or some rocks letting in air, water, and organic matter and changing the chemical nature of the soil. Dead plants or plant material may become a part of the soil adding to the humus that is the basis of our agricultural wealth.

In *bogs,* acid conditions prevent complete decay and plant remains accumulate at the rate of possibly 1 ft in 10 years. As this *peat* is put under pressure of new material above, it may be compressed. Raw peat 20 ft deep might produce 1 ft of *bituminous coal.* Probably 300 years of plant growth is involved in development of basic material for 1 ft of coal.

Bituminous coal may be metamorphosed into *anthracite.* It is estimated that virgin North America had an average of 9 in. of topsoil and that without management 1 in. might be added in 500 years. With encouragement of plant growth, we can probably speed this up to form $1/2$ in. of topsoil from subsoil in 1 year. *Oil shales* and petroleum products are of marine origin. Oil may collect in shales formed as freshwater deposits but they originate in a saltwater situation.

Plant materials or other materials of organic origin in earth and rocks yield us peat, lignite, coals, oils, and petroleum that provide the heat, power, and light that make our civilization possible. Since some of these resources are not renewable as are topsoils, their wise use is a social responsibility that cannot be avoided.

Soil
Soils bring together a mixture of inorganic mineral matter and partly decomposed organic residues. Soils form, in part, from rocks which have disintegrated and decayed through processes of weathering. For man, soil is the most important product of the weathering process. Soils differ from area to area, in quantity and quality. When the thin upper layer of the mantle has decomposed and been altered sufficiently by the action of living organisms, and plants can be supported by this altered mantle, we can assume a soil is present. The decayed organic matter in the soil is called *humus.* Soils have texture, which refers to the size of the grains or particles (sand, silt, clay), and structure, which refers to the arrangement of the particles in aggregates. Fine-grained soils are heavy soils and water moves through them slowly whereas coarse soils are light soils and water moves through them readily. Soils develop in response to climate and living things acting on the parent material over time. The parent material may be any one of the classes of rocks; igneous, sedimentary, or metamorphic. These rocks must, of necessity, be altered. Some soils develop in place, are residual; while others, such as glacial soils, loessal soils, alluvial, or colluvial soils, develop from materials which have been transported by glaciers, or the wind, or water, or gravity.

Soils are classified broadly based on the prevailing climate and its associated vegetation. For example, *laterite* soils are developed in hot, humid climates under conditions of intense weathering. The soil material is high in oxides of iron and aluminum; red to yellow in color and leached. *Chernozem* soils are formed in temperate, subhumid climates, typically under grassland vegetation. These soils are black, granular in structure, fertile, and moderately leached. *Podzol* soils are developed in cool, moist climates under hardwood or mixed hardwood-conifer woods. The soil is gray, low in organic matter, typically acid, leached, and underlain by a clayey subsoil.

SEDIMENTARY ROCKS

Conglomerates
Conglomerates are cemented masses of waterworn pebbles, sands, and gravels. The component sands and gravels may or may not have been sorted and may or may not have a common origin. Consolidated angular fragments are distinguished as *breccia.* Some conglomerates are called puddingstones.

The nature of the component parts of conglomerate or of breccia of course determines the nature of the aggregate, whether this applies to the cement or to the cemented parts. Conglomerates are commonly named from their component parts, as *quartz conglomerate.*

Since the units in conglomerates are waterworn, it is safe to assume that conglomerates indicate their formation under water. The larger the units, the more probable it is that they were formed under swiftly flowing water. They are therefore useful in interpreting the geologic past of an area.

Some relatively fine-grained conglomerates with varying colors are cut and polished to make attractive stone objects. They may be used as building stones where the cements are well established. The commoner cements are iron oxide, lime carbonate, and silica. These often determine the nature and durability of the resultant conglomerate.

Sandstone
Where sand is cemented together but the individual grains are still identifiable, we have *sandstone.* Sandstone may be sufficiently porous to hold or contain a quarter of its volume of water.

Grit is composed of angular, coarse-grained quartz sand; *arkose,* of angular quartz and feldspar; and *graywacke* of quartz, feldspar, and some other mineral. Compacted silts make *siltstone.*

On the basis of the cements involved, sandstones are classified as *ferruginous* if the cement is iron oxide; *calcareous,* if lime carbonate; *siliceous,* if silica; *argillaceous,* if clay; and *micaceous,* if mica.

Sandstones that are porous make excellent water filters, and great reserves of water are stored underground in porous sandstones. Sandstones are used in building materials, the celebrated "brownstone fronts" of New York City being sandstone. They may metamorphose into quartzites.

Shale
Consolidated muds of various origins laid down in layers that usually split readily are *shales.* They vary from those that are uniform to those that are not, those that are hard to those that are soft, those that are well-consolidated to those that are not. They may or may not contain many extraneous materials such as sand, organic remains, and so on.

Calcareous shales have more or less lime carbonate in them; *arenaceous shales,* more or less sand; *ferruginous shales* (usually red), more or less iron oxide; *argillaceous shales,* little besides clay; *bituminous shales,* more or less plant and animal remains; *carbonaceous shales* (usually black); *fossiliferous shales,* many fossil remains; and *oil shales,* more or less oil.

Shales are not ordinarily durable when exposed to the elements. They readily disintegrate to form muds that may contribute to the making of new shales. They may on the other hand become more consolidated and metamorphosed to form *slates* or similar metamorphic rocks. They are for the most part no doubt formed originally of water-deposited silts and muds.

Shales have little use as building stones, but crushed they may be used in road building, though here they are inferior to harder materials. They are mixed with lime in the making of portland cement. Their presence underground may contribute to the development of springs, and when porous they may affect the water table.

Limestone
Deposits of lime carbonate in the form of marl, shell remains, bog lime, and so on may be consolidated to form limestones, the nature of the limestone depending considerably on the nature of the contributing limy material. Fine particles definitely make limestone with qualities different from those formed of large particles, and impurities make their own particular changes.

Limestone, of course, effervesces with hydrochloric acid as does the mineral constituent calcite. It may be more or less readily dissolved in water, affecting the quality of the water and the nature of the movement of water underground and of the underground structure. Pure limestone is white but gray, green, brown, black, or red forms are found. It metamorphoses into *marble.*

(Continued)

Shale

Conglomerate

Breccia

Sand

Silt–loam

Sandstone

Limestone

29

Limestone material has almost always been deposited under water, either fresh or salt. *Chalk,* a fine-grained limestone, may be caused by Foraminifera; *coquina,* by shell fragments; *lithographic limestone* is a fine, dense limestone with conchoidal fracture that is resonant when struck.

Limestone is the basis of most portland cement. Fired at 900°, it frees a gas carbon dioxide. The remaining calcium oxide, mixed with shale and ground, gives us cement that when mixed with water hardens into portland cement used in building houses, roads, and even ships. Limestones may be used directly as building stone.

Coal and Petroleum
Plants and animals of the past prevented in various ways from complete decomposition have left remains that may be consolidated into rocks of a sedimentary nature or may have freed oils and similar products. *Bituminous* or soft coal is more compacted than *lignite* or brown coal, which in turn is more compacted than *peat.* In all of these, plant remains may often be easily identified.

Bituminous coals are heated in closed ovens to drive off the gas, oil, and other materials derived from them. One ton of soft coal may yield 1,400 lb of *coke,* 19 lb of ammonium sulfate, 10,000 cu ft of illuminating gas, 2 gal of oil, and 7 gal of tar.

Cannel coal is a variety of coal derived probably from spore accumulations. Coal is usually associated with fresh water deposits; petroleum, with salt water. Bituminous coals may metamorphose into *anthracite* coal. Lignite crumbles; soft coal breaks easily; hard coal fractures conchoidally.

Coal of course is the basis of much industry. Coke is essential in the making of steel and the tar by-products are the basis of the coal-tar industry that provides us with medicines, dyes, plastics, and so on. Some by-products are valuable as agricultural fertilizers. Soft coal is the world's most abundant, most widely distributed.

Petroleum clearly is related to sedimentary rocks. Its accumulation appears to be part of normal processes of sedimentation. Most oil accumulated as an organic accumulation in sedimentary rocks deposited on the sea floor. Perhaps petroleum results from the accumulation of dead remains of many marine forms on the sea floor, rapid burial in sediments, and natural distillation. Other factors which appear important are a natural enclosed basin where circulation of water is slow and oxidation is slowed down. Reservoir rocks are most efficient in accumulation of petroleum; e.g., 60 percent of petroleum reserves are in sandstone and about 40 percent in some form of limestone and dolomite and their relatives. Probably oil formed in source rocks and then migrated to more permeable strata.

Some Other Sedimentary Rocks
Dolomite is formed when from 25 to 50 percent of the calcium in a limestone is replaced by magnesium. If something less than ¼ of the calcium is replaced, we name the rock a *dolomitic limestone.* Both dolomitic limestone and dolomite are less soluble than ordinary limestone and so weather more slowly. *Marl* is formed from the porous masses of shells and shell fragments that accumulate on the bottoms of many freshwater lakes. The lime-secreting alga *Chara* produces marl. Some marine sediments which presumably formed at the outer margin of shale mud and contain a mixture of clay and finely divided shell fragments are also called marl. *Oolite* is a rock consisting of small concentric built-up particles which look like fish roe. Oolites may be calcareous, siliceous, ferruginous, or phosphatic. *Tufas* are calcareous deposits formed about the mouths of springs; *travertines* are deposits of calcite in limestone caves. *Chert* is a compact, dense siliceous material occurring both as distinct layers and as roundish nodules distributed through beds of other rocks, chiefly limestone. *Flint,* a dark gray to black chert, is essentially silica. *Agate* is a banded variety of chert. *Taconite* is a iron-rich chert; an iron ore; found, for example, in the Lake Superior region. *Diatomaceous earth* was formed from minute plants, diatoms, that have siliceous skeletons. When they die, their siliceous cell coverings accumulate in beds of shale. *Phosphate rock* developed in the sea, was raised with uplifting of the earth's surface. The phosphate rocks are important sources of fertilizer.

METAMORPHIC ROCKS

Quartzite
Metamorphosed sandstone in which the grains have been so firmly metamorphosed or cemented that they lose their entity may be *quartzite.* Such rocks are very heavy and splinter when broken or show a conchoidal fracture. They lack almost wholly the porous nature of the basic sandstone; when broken, the sand grains split, rather than separating from their adjacent grains.

Quartzites are composed mostly of quartz, but this may be accompanied by mica, feldspar, and chlorite. In such cases, the quartzite is distinguished by the compounding of these names with quartzite.

The sand particles in quartzite may have been cemented together with silicon dioxide. Under heat and great pressure, the material becomes almost homogeneous. Quartzite is almost impossible to shape to a desired form.

Quartzites make good road ballast that may last almost indefinitely. Quartzite composed of pure quartz may be ground and used in the making of glass. Quartzites are relatively abundant on the rockbound coasts of New England.

Schists
Schist comes from the Greek word *schistos,* meaning cleft, or *schizein,* meaning to split. Probably the most widely occurring of the metamorphic rocks. For example, slates grade into schists with increasing grain size. All schists include tabular, flaky, or even fibrous minerals in their composition. Many schists, such as the familiar flagstone, split easily into blocks. In Europe, flagstones are used in courtyards and castle walls; in the U.S. they are used for fireplaces, patios, and barbecue grills. Schists may be made from a wide variety of rocks by recrystallization under high temperature and pressure. Mudstones, muddy sandstone, basalt, and other dark igneous rocks and clayey limestones may be metamorphosed into schist.

While schists break in a plane in one direction, they break irregularly at right angles to that plane. In their coarser forms, they may grade into *gneiss* and in the finer forms into *slate.*

Schist varieties are named after their component parts such as *garnet, mica, chlorite, talc, hornblende* schists and they vary greatly in color. Exposed to weathering, the color and texture change in a year may be startling.

Schist bases for dams may be dangerous because water leaching and action may undermine dam foundations and cause trouble. A few national disasters have been traced to structures on such rocks.

Gneiss (definitely banded or streaked, coarsely granular)
In these rocks, the elements must be coarse enough to be identified as such with the naked eye. Banding may be continuous, interrupted, straight, curved, or even twisted; the color may be red, brown, gray, green, or black although some are almost white.

The usual minerals in gneiss are quartz, feldspar, and mica or hornblende. Different gneisses are named from the materials from which they are derived as *biotite, gniess, gabbro, muscovite, syenite, hornblende, diorite, garnet,* and *granite gneiss.* Names descriptive of structure are *foliated, banded,* and *augen* (eye-shaped masses) *gneiss.*

In gneisses, the components may be in part represented by the streaks or bands, the metamorphosis not completely changing the evidence of the source of the material. Gneisses originate deep in the earth and become exposed by wearing away of the upper coverings.

Gneisses appear abundantly in the West in the Rockies, the Cascade and the Sierra Nevada ranges, over most of Canada, in New England, along the Appalachian areas, and in the Lake Superior region. They are heavy, make excellent roads and ballast, are durable, and figure generously in heavy construction work.

Marble
Marble is metamorphosed limestone in which the component units have lost their entity, just as sand grains lose their entity

(Continued)

Schist

Marble

Quartzite

Gneiss

Slate

Felsite

in changing from sandstone to quartzite. Broadly speaking, the term *marble* may include magnesium carbonate of *dolomitic limestone* with the usual calcium carbonate of the *typical limestone.*

When pure, marble is white. Marble does not split in a plane as do the schists. It polishes beautifully but acid in rain may cause it to weather badly in a few generations.

Marble is quarried in such widely separated areas as Italy, Washington, California, Georgia, and Vermont. In comparison with such metamorphic rocks as the quartzites, it is cut readily. Impurities in it are responsible for more varieties than can here be mentioned.

Marble is a popular building stone. It is one of the commonest stones used for monuments but an old cemetery will show how durable it may be over a span of a century. It is also used in the arts. In modern building, it is being replaced to some extent by commercial substitutes made of ground materials cemented into shape.

Products of Metamorphosed Organic Materials
Anthracite coal is bituminous coal that has been subjected to heat and pressure and so metamorphosed in the process that the individual plant elements have lost their entity and the fracture is conchoidal. Flame from such coal is lower and cleaner and possibly hotter, and anthracite fires require less frequent attention. Anthracite has a high luster and is hard and always black.

Organic material under certain conditions, usually associated with marine situations, undergoes a series of changes that yield petroleum, natural gas, tar, paraffin, asphalt, crude oil, and other common substances. It is believed they were formed from life on sea bottoms that accumulated and was held in the muds that became shales.

Asphalt is a solid residue such as is found in California and Texas, left when some of the gases and oils have been removed from crude petroleum. Pennsylvania oils usually have as a base paraffin rather than asphalt. Oil, being lighter than water, rises above it and is therefore often found held captive above water underground.

The importance of anthracite, petroleum, tar, asphalt, and similar material is too obvious to need review. Even graphite, used either as a lubricant or in some electrical equipment, jet used in making ornaments, or diamonds used for cutting hard materials may be a part of a cycle in which living things may have had a part.

Slate
Metamorphosed shales split more perfectly than do the original shales. Because of the great variety of minerals in muds and clays we can expect a great variety of metamorphosed shales. They differ from the sandstones, marbles, and the quartzites in that the particles are too small to be identified by the eye.

In addition to slates composed of clays, there are those with varying amounts of calcite, magnetite, chlorite, graphite, and so on; some have varying amounts of sandy materials mixed with the finer units. Other slates have silica, lime, or bituminous materials and are named accordingly.

Colors of slates are usually determined by the materials originally in them. Greens come from chlorides; reds and blues, from iron compounds; blacks, from carbonaceous material; and pale colors possibly from siliceous materials. A slate may become a *mica schist* if metamorphism continues, or in an intermediate state, with fine foliation and lustrous cleavage surfaces, *phyllite.*

Slates were used in schools in the early days and to some extent are still used as blackboards; however, it is now common to use slate wastes or even to grind slaty materials and cement them to make the products available commercially. Raw slate shingles have been largely replaced by shingles made of slate waste ground and cemented. These wastes are used in fire prevention.

IGNEOUS ROCKS
Intrusive Rocks

Granite
Composed primarily of quartz, orthoclase feldspar, and mica. There may or may not be hornblende. *Granite* is a general name applied to light-colored granular rocks of this general composition and nature.

The fine-grained, dense rock with alkali feldspar and quartz is *rhyolite,* a variety of *felsite;* if porphyritic, it is *rhyolite-porphyry.*

Granites are formed deep in the earth and become exposed gradually by erosion of mountaintops or otherwise. When they disintegrate with weathering, the mica is weathered first, then the feldspar. These contribute to the formation of muds and clays, leaving the quartz to form sand. Granites do not weather rapidly and are resistant to the acid action that affects marble.

Granites occur in great masses as bedrock or in fragments that have been broken loose and rolled into cobbles, boulders, or pebbles. They are cut with difficulty, take a good and permanent polish, make excellent building stones of a permanent nature, and are used in buildings, walls, roads, for fill, and so on.

Syenite
Light-colored rocks with a salt-and-pepper effect, like a granite without any quartz in it. They are composed essentially of orthoclase feldspar and mica with or without hornblende or augite. Syenite may have many other minerals than those mentioned as typical, including plagioclase feldspar, magnetite, and apatite. Acidic.

In syenite, the mineral units are usually large enough to be distinguished by the naked eye and the coarseness of the rock as a whole is essentially as outlined under the classification of granites. If the plagioclase feldspar approaches in amount the orthoclase present, the rock is a *monzonite;* if it exceeds the orthoclase, the syenite becomes a *diorite,* another deep-seated rock.

Syenite was formed in batholiths under mountain ranges; it becomes exposed only gradually through the erosion of softer overlying materials. It commonly remains as an erect uneroded core in the form of a batholith. Since, like granite, it is intrusive, it may have metamorphosed somewhat the rocks into which it made its way or which were adjacent to it. If it is dense, it tends to become *trachyte,* a type of felsite.

Syenites are not so common as granites. They may be more easily carved and worked than granites. Syenite finds general use as fill and in road making but since it is less abundant than granite it is naturally less well known.

Diorite
Medium-dark-colored rock, granular, with salt-and-pepper effect. Like syenite, it lacks quartz. In diorite, plagioclase feldspar replaces the orthoclase feldspar that is essential in granite and in syenite. Mica should usually be present, and there may or may not be hornblende. Of course, other minerals may be present to some degree; among these commonly are apatite and magnetite. If quartz should occur, the mixture is called *quartz diorite.* Basic.

As in syenite, there is variation in size of the mineral units, and the degree of coarseness is essentially that described under granite. A diorite that is dense, stony, not glassy or dark gray, green, or black is called *felsite* or *dacite* (lime-soda feldspar and quartz) and one that is porphyritic is *dacite porphyry.* Dacite is a shallow rock. *Hornblende diorite* is darker.

Like granite and syenite, diorites originate deep in the earth and reach the surface by gradual erosion of the covering rocks. Since they have intruded into the covering rock blanket, they may appear as batholiths or mountain cores in old mountain ranges. Augite may be present in considerable proportion; if this exceeds the hornblende the rock becomes *gabbro.* Weathering is slow.

The uses of diorite are much the same as those of syenite. Diorites are much less common and widely distributed than are the granites although they may be found in the eastern United States, Canada, and the Rocky Mountain areas. As is the case with all rocks, there are intergradations with other kinds due to the presence of different minerals.

Gabbro

Dark rocks with a coarse salt-and-pepper effect may be known as *gabbro*. Composed primarily of plagioclase feldspar and pyroxene. If olivine is also present, then the gabbro is classified as *olivine* gabbro. A dense equivalent with unusually small crystals is classified as *diabase*.

While gabbro is made essentially of plagioclase feldspar and pyroxene, it may be classified on the basis of the relative abundance of the two. Thus a gabbro with a great predominance of plagioclase feldspar is called *anorthosite;* one with a predominance of pyroxene is called *pyroxenite*. Other minerals may appear together with these in varying numbers and abundance.

Gabbro, like granite, is usually formed deep in the earth and has cooled slowly over great periods of time. However, in the case of gabbro, some of the upper material may on occasion extrude itself from the protective blanket, cool quickly, and appear as a denser rock. This is known as trap rock. It, of course, is much finer grained than the typical gabbro.

These dark rocks are, where abundant, used commonly as building stone, in road making, for heavy fill, and for similar important purposes.

Extrusive Rocks

Obsidian, Pumice, Pitchstone, Scorias, Trachylyte

Glassy rocks, either light or dark in color or mixed; generally if broken they may have a rounded or conchoidal fracture. Some are dull. Chemical composition varies with the different kinds but crystals are absent or too small to be detected as they are in other rocks here discussed.

Most common of these rocks is *obsidian,* which looks much like a hard, colored glass that shows shell-like ridges on the breaks. It has a hardness of 6 and may vary from black through green or red to a translucent condition. *Pitchstone* has a more resinous luster and *pearlstone* or *perlite* has definite rounded masses that persist.

Where gases formed cavities in cooling lavas, we find different kinds of extrusive rocks. If the cavities are small and uniform, we get glassy vesicular *pumice;* if they are large, *scoria,* the former often being light and the latter dark. If the cavities have been filled with other minerals, we get *amygdaloid*. All these are lavas that have cooled quickly either as a result of volcanic explosion or by pouring out.

The color of these extrusive rocks is determined largely by the chemical and physical nature of the rocks. Fine ash or *tuff* or *volcanic ash* represents the finer dusts that may be suspended in the air longer than most extrusive materials.

Diorite

Lava

Scoria

Syenite

Trachylyte

Granite

Gabbro

Obsidian

33

HEMATITE

Fe_2O_3. Metallic iron, 70 percent

Specific gravity: 4.9 to 6.5

Hardness: 5.5 to 6.5

Streak: Reddish-brown

Luster: Brilliant metallic to submetallic to dull

Cleavage: None

Fracture: May show a fine granular surface or be coarser

Color: Steel-gray to iron-black, to red, to red-brown, to black. An earthy kind, dull in luster, is the red ocher used in making paint. Red streak is the most important test in distinguishing dark compact varieties from limonite

Fusibility: 5 to 5½. Burned on coal becomes magnetic and with soda may become a magnetic powder. Gives little or no water in closed tube and becomes black and magnetic under blowpipe flame

Soluble in strong hydrochloric acid but action is slow. Varieties include *Clinton iron ore* (red); *red hematite* (soft ochraceous powder); *specular hematite* (brilliant crystals); and others

Crystals not common in ores but when found are rhombohedral

Found with other iron ores, more particularly in turgite and limonite, in sedimentary and metamorphic rocks

A chief source of iron. Metallic iron has a specific gravity of 7.8, a melting point of 1520°C, a boiling point of 2450°C. Commonly alloyed with chromium, manganese, molybdenum, tungsten, vanadium, silicon, nickel, and cobalt. Carbon is added in making steel. Carbon steel softens or loses its temper at 500°F; in steels containing tungsten and cobalt this is raised to 1000°F. Steel is probably the basic metal of our modern civilization with the automobile, railroad, and construction accounting for most of the demand.

Hematite

MAGNETITE

Magnetic Iron Oxide, Lodestone
Fe_3O_4. Metallic iron, 72.4 percent

Specific gravity: 4.9 to 5.18

Hardness: 5.5 to 6.5

Streak: Black

Luster: Metallic to submetallic to dull

Cleavage: Not distinct; parallel to octohedron, may be distinct

Fracture: Uneven, conchoidal, brittle

Color: Iron-black

Fusibility: 5 to 5½. Hard to fuse; in open flame is oxidized to nonmagnetic hematite, magnetic before heating

Soluble in hydrochloric acid

May be massive, with or without laminated structure, or as sand

Found mostly in metamorphic rocks or black sands associated with apatite, hornblende, chlorite, pyrite, zircon, ilmenite

See hematite for characteristics of metallic iron. Magnetite is used somewhat as a source of metallic iron. The variety known as *lodestone* is a natural magnet. Found as an accessory mineral in many igneous rocks. Emery is an intimate mixture of magnetite and corundum. In the U.S., magnetite has been mined in the Adirondacks of New York, in Pennsylvania, and in Utah.

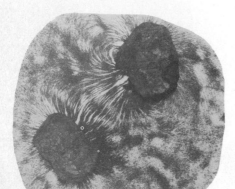

Magnitite

SIDERITE

$FeCO_3$. Iron, 48.3 percent

Specific gravity: 3.8 to 3.9

Hardness: 3.5 to 4

Streak: White to pale yellow

Luster: Vitreous to pearly

Cleavage: In 3 directions, rhombohedral, perfect at 73° and 107°

Fracture: Conchoidal and uneven

Color: Gray, yellow, brown, black, or white, often discolored

Fusibility: 4½ to 5. Becomes black and magnetic when heated, breaks down and yields carbon dioxide, but fuses only with difficulty

Dissolves with conspicuous effervescence in hot hydrochloric acid; solution has the reaction for ferrous iron

Crystallization: hexagonal, rhombohedral, or granular

Found in veins, beds, concretions in shale, limestone, coal; associated with minerals of silver, pyrite, cryolite, and dolomite

Very common in ore veins which were formed at low and medium temperatures. Often associated with calcite, barite, and sulfides. It forms large concretions (clay ironstone) in sedimentary rocks.

Siderite

PYRITE

Fool's Gold
FeS_2. Sulfur, 53.3 percent; iron, 46.7 percent

Specific gravity: 4.9 to 5.2

Hardness: 6 to 6.5

Streak: Greenish-black to brownish-black. Marcasite: yellow

Luster: Metallic

Cleavage: None, brittle. Marcasite: cleavage poor, prismatic

Fracture: Uneven

Color: Pale to brass-yellow, tarnishes brown, sometimes iridescent

Fusibility: 2½ to 3. Becomes magnetic in open flame. Burned on coal, it gives a blue flame from the sulfur. Burned in a closed tube, there is a sublimate of sulfur and a magnetic residue

Soluble in cold nitric acid but not in hydrochloric acid. In marcasite, sulfur separates in cold nitric acid

Crystallization: In form of isometric cubes, often with striations

Found in various rocks, often with gold and silver ores, in schists, concretions, clay, shale, and coal. Marcasite, and at times pyrite, is the source of acid-mine drainage from both deep and open-pit mining of anthracite and bituminous coal. In a complex series of chemical reactions the FeS_2 reacts with oxygen and water to form, among other products, sulfuric acid. The sulfuric acid pollutes streams and impoundments

Used primarily as a source of sulfur and in the manufacture of sulfuric acid, in the manufacture of paper. Iron pyrite is not a common source of metallic iron.

Pyrite

CHALCOCITE

Copper Glance
Cu_2S. Copper, 79.8 percent; sulfur, 20.2 percent

Specific gravity: 5.5 to 5.8

Hardness: 2.5 to 3, for crystals

Streak: Black to lead-gray

Luster: Metallic in crystal; dull in soft forms

Cleavage: Poor

Fracture: Conchoidal

Color: Black to lead-gray; may be blue- or green-coated. Gray color distinguishes it from related copper sulfides

Fusibility: 2½. Gives off sulfur fumes heated in open tube. Melts to sputtering globule on charcoal. After roasting and moistening with HCl, gives a copper flame reaction

Crystallization: Usually massive, orthorhombic

Found with pyrite, chalcopyrite, galena, and similar minerals

Copper metal has specific gravity of 8.4 to 8.9, melts at 1000 to 1200°C, boils from 1980 to 2310°C. Alloyed with zinc, makes *brass;* with tin and zinc, *bronze;* with zinc and nickel, *German silver;* with aluminum, *aluminum bronze.* Next to silver, copper is the best conductor of electricity. Used in electrical equipment, in making wire and wire cloth, in building construction, in automobiles, in ships, and in ammunition, in paints, printing, plating.

Chalcocite

CHROMITE

Chromium Trioxide
$FeCr_2O_4$. Chromium trioxide, 68 percent

Specific gravity: 4.3 to 4.6

Hardness: 5.5

Streak: Dark brown

Luster: Dull to submetallic

Cleavage: None or indistinct, parting in 4 directions, 70½° and 109½°

Fracture: Uneven, conchoidal

Color: Iron-black to brownish-black

Fusibility: Almost infusible. Edges rounded in the reducing flame and some magnetic properties may develop. Decomposed by fusion with potassium or sodium bisulfate

Insoluble in acids

Crystallization: Usually massive or in granules and compact

Found in peridotite, serpentine, in black sands, and in platinum placers associated with olivine, talc, chlorite, and magnetite. Pure chromite is rare because magnesium usually substitutes for some ferrous iron, and aluminum and ferric iron substitute for chromium

Chromium metal has specific gravity of 6.7 to 7.1, melts at 1650°C, boils at 2200°C, becomes magnetic at − 15°C, is harder than iron or cobalt.

(Continued)

Chromite

Used principally in alloy steels to make them harder even though chromium may be only 1 percent of the finished steel. Stainless steel is chromium steel in which the chromium is about 12 to 30 percent. Chromium is also used in furnaces, in plating plumbing fixtures, in paints, dyes, and printing. Because of its improvement of steel and the restricted existence of the ore, it is considered a strategic mineral. Rhodesia, Turkey, the Philippines, and Africa are leading producers.

GALENA

Lead Glance
PbS. Lead, 86.6 percent; sulfur, 13.4 percent

Specific gravity: 7.4 to 7.6

Hardness: 2.5 to 2.75

Streak: Lead-gray

Luster: Metallic

Cleavage: In 3 directions, at 90°, brittle

Fracture: Even

Color: Lead-gray

Fusibility: 2. Burned on charcoal, forms lead globules and a yellow sublimate bordered by white; gives off sulfur fumes in open tube

Soluble in strong nitric acid leaving a white powdery precipitate; all lead salts are poisonous

Crystallization: isometric

Found associated with sphalerite, pyrite, barite, chalcopyrite, fluorite; some lead ores rich in silver. Principle ore of lead. Widely distributed on all continents of the world, mined in many countries. In the U.S., Missouri, Idaho, and Utah have important sources

Metallic lead has specific gravity of 11.25, a melting point of 327°C, and a boiling point of 1525°C. Nearly ⅓ of the lead in commercial use is in storage batteries; other important uses including roofing, plumbing, paints, cable covering. Alloyed with tin and antimony it makes *type metal;* with tin, *solder;* with tin, antimony or bismuth and copper, *pewter.* Lead is used as a low-friction metal in heavy bearings. Crystals were used as the "detector" in old-time crystal radios.

Galena

SPHALERITE

Zinc Ore, Black Jack, Zinc Blende
ZnS. Zinc 67 percent and iron, manganese, cadmium, or others

Specfic gravity: 4

Hardness: 3.5 to 4

Streak: Brown to yellow to white

Luster: Resinous to adamantine

Cleavage: Perfect; dodecahedral, in 6 directions, 60°-90°-120°

Fracture: Conchoidal, brittle

Color: Yellow, brown, black, green, red, or even white

Fusibility: 5. Magnetic in oxidizing flame. Greenish to deep-green flame in reducing flame with soda on coal. Zinc can be rolled into sheets

Decomposed by hot hydrochloric acid. Zinc is soluble in soda and potash and in dilute nitric acid

Crystallization: Usually granular and compact, tetragonal

Found with galena, pyrite, chalcopyrite, fluorite, siderite, and with silver and gold ores. Occurs in veins and replacement deposits in limestones. Chief ore mineral of zinc; so mined on every continent. U.S. chief producer. Cleavage and luminescence make it of considerable mineralogical interest. Sometimes fluoresces orange in ultraviolet light

Zinc metal has specific gravity of 6.9 to 7.1, melts at 419°C, and boils at 918°C. Chemical element 30; atomic weight, 65.38. Ductile, malleable, gray metal. Freshly polished, it produces bluish-white luster. Alloyed with copper makes *brass;* with tin and copper, *bronze;* with nickel and copper, *German silver.* Is brittle at red heat but malleable and ductile at low liquid temperature. Sulfuric acid is a by-product of smeltering of sphalerite. Zinc is used in galvanizing iron, in solder, in storage batteries, and in dyes. Die casting alloys account for a large percentage of zinc used in U.S.

Sphalerite

CASSITERITE

Tin Stone, Tin Ore
SnO_2. Tin, about 78.6 percent

Specific gravity: 6.8 to 7.1

Hardness: 6 to 7

Streak: White to grayish-brown

Luster: Adamantine, greasy, dull

Cleavage: Indistinct, brittle, poor

Fracture: Usually uneven, but sometimes smooth, subconchoidal

Color: Variable from brown to black, or rarely yellow, red, gray, or white

Fusibility: With intense heat with soda on charcoal leaves the white sublimate SnO_2. When fused with soda, sulfur, and charcoal, gives metallic globule with yellow coating

Yields metallic tin with zinc and hydrochloric acid. When the tin is rubbed it yields a characteristic color. Acid acts slowly on tin. Ore is decomposed by fusion with potassium hydroxide in nickel crucible

Crystallization: Tetragonal with thick prismatic crystals

Found in granite, gneiss, or pegmatite, commonly massive or in kidney forms or in rounded granules (see cut) as stream tin in placers. Most important ore of tin. Now recovered from placer or residual deposits in Malay States, Indonesia, and the Congo; mined in Bolivia. U.S. is a have-not nation as far as this chemical element is concerned

Metallic tin is weak, silvery white, with specific gravity of 7.3, melts at 232°C, begins to volatilize at 1600°C and to boil at 2270°C. At 100°C, it is highly ductile and may be made into wire but at − 39°C for at least 14 hours it becomes brittle. Not affected by exposure to air; a poor conductor of electricity and of heat. Used mostly in plating iron to prevent rust for use in cans, roofing, and utensils. A common alloy in *white metal, solid silver, babbitt, type metal;* used in dyes, polishing mixtures, and as filler for silk. Tin chemicals and compounds find extensive use in electroplating, ceramic, and plastics industries.

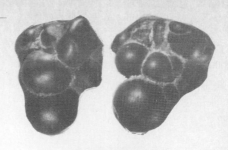
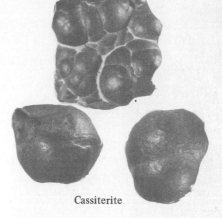

Cassiterite

CORUNDUM

Al_2O_3. Aluminum, 52.9 percent

Specific gravity: 3.9 to 4.1

Hardness: 9

Streak: Uncolored

Luster: Vitreous to pearly and adamantine

Cleavage: Basal and rhombohedral, with twinning common

Fracture: Conchoidal to uneven

Color: Gray, yellow, brown, red, white, blue

Fusibility: Decomposed by fusion with potassium hydroxide in a nickel crucible. Infusible and unaltered by soda. Powder becomes blue after being heated a long time with cobalt nitrate

Dissolves slowly in borax and salt of phosphorus to make a clear glass that is colorless when no iron is present. Emery contains magnetite, hematite, or spinel

Crystallization: Hexagonal, with prisms and rounded forms common

Found in crystalline rocks such as gneiss, granite, schist, slate, and limestone associated with chlorite, tourmaline, and Kyanite

Corundum and emery are among the best abrasives for whetstones, grinding wheels for polishing, stone cutting, and glass etching. Gems of the same material are the transparent *blue sapphire,* the green *Oriental emerald,* the yellow *Oriental topaz,* and the red *ruby.* Although emery has been replaced by synthetic abrasives in many instances, it is still used because of its abrasive and polishing properties by lapidarists and in the manufacture of such optical equipment as lenses and prisms. Marketed as familiar emery paper, emery cloth, and emery wheels.

Corundum

CINNABAR

Mercury
HgS

Specific gravity: 8.1 to 8.2

Hardness: 2 to 2.5

Streak: Scarlet

Luster: Adamantine. In pure form cinnabar is vermilion red; if impure, luster dull and is brownish-red in color

Cleavage: Poor, in 3 directions, prismatic, at 60° and 120°

Fracture: Uneven

Fusibility: 1½. Volatile, heated in a closed tube forms a black sublimate; 1 part dry cinnabar to 4 parts dry soda heated gently in a closed tube yields minute globules

Cinnabar

(Continued)

Clean copper dipped in a mercury solution yields a silvery metallic mirror coating

Found in veins in sandstone and limestone associated with pyrite, marcasite, barite, quartz, sulfur, and opal. Important mines in Spain

Mercury has specific gravity of 13.595. It solidifies at −38.37°C and boils at 360°C. Solid mercury is malleable, ductile, can be sliced, and has a specific gravity of 14.19. Mercury has a uniform expansion and contraction with temperature change and unites readily with many other metals. It is used in making mirrors, thermometers, vapor lamps, in electrical equipment, in medicine, in the extraction of heavy metals like gold from pulverized mixed ores. Mercury compounds are toxic; cause real problems to living things when concentrated in waterways. Mercury and almost all of its compounds are quite poisonous to man and animals. Mercury, the metal, and many of its compounds should be handled only with precautions.

GOLD

Gold
Au

Specific gravity: 15.6 to 19.3

Hardness: 2.5 to 3

Streak: Yellow-gold or lighter

Luster: Metallic

Cleavage: None

Fracture: Hackly

Color: Golden-yellow or paler

Fusibility: 2½ to 3. Fuses at 1100°C. Highly ductile and malleable at ordinary temperatures; 1 g can be drawn into 2 miles of wire and 1 oz beaten to cover 300 sq ft

Not affected by any single acids, but aqua regia and telluric acid will dissolve it

Crystallization: Rarely in perfect form. Isometric

Found scattered in plates, scales, and grains in ore rocks whether igneous, metamorphic, or sedimentary. Frequently in quartz veins and in stream deposits

Pure gold, chemical element 79, deep yellow, a very soft and very dense metal, has a specific gravity of 19.32. It melts at 1063°C and boils at 2500°C. It is greenish, may be welded, is a good conductor of heat and electricity but not so good as silver, and alloys freely with a number of other metals. It is used in coins and in jewelry. Gold coins are generally 9 parts gold and 1 part copper. *White gold* is an alloy with platinum, silver, or nickel and zinc; *purple,* with aluminum; *blue,* with iron or steel; *red,* with copper; and *green,* with silver and cadmium or with silver, copper, and cadmium. Gold used to be considered evidence of industry and of a sound economy. Concentrates in stream beds, either in small flakes or in larger nuggets, from which it can be recovered by panning. Nuggets become more rounded as they are carried by the stream.

Gold ore

HALITE

Salt, Sodium Chloride, Rock Salt
NaCl

Specific gravity: 2.1 to 2.6

Hardness: 2 to 2.5

Streak: White

Luster: Vitreous

Cleavage: In 3 directions, cubical, perfect

Fracture: Conchoidal

Color: Colorless, white, red, blue, or yellow

Fusibility: 1½. In flame, gives a persistent yellow color that is very brilliant

Is readily dissolved in water and of course has a salty taste

Crystallization: In isometric cubes

Found in sedimentary rocks, usually in strata, and at edges of bodies of evaporating salt water associated with gypsum. Salt is recovered by mining or by introducing water to dissolve the salt beds and pumping up the solution of salt to evaporating impoundments. Worldwide in occurrence; mined in New York State, Michigan, New Mexico, and Louisiana

Salt has a variety of uses as a preservative of food, in the making of glass and of explosives, in dyes, in metallurgy, in making baking soda, sal soda, caustic soda, washing soda, and in medicine. It is used to melt snow, in refrigeration processes, to absorb moisture from atmosphere, in softening water, and in many other ways. Important over man's long history as a part of his diet and in his commerce.

Halite

GRAPHITE

Black Lead, Plumbago
C. Often with iron and clay

Specific gravity: 1.9 to 2.3

Hardness: 1

Streak: Grayish-black and shiny

Luster: Metallic

Cleavage: Perfect, in 1 direction. Feels greasy. Mineral can be split into thin, flexible, but nonelastic sheets

Fracture: Thin, flexible plates

Color: Iron-black to dark steel-gray

Fusibility: Infusible before blowpipe but burns at high temperatures

Insoluble in acids

Crystallization: In 6-sided tabular crystals. Easily separated and broken

Found in masses or flakes in beds or seams in schists, limestones, granites, or sometimes in clay rocks

Used as lubricant, in making lead pencils and polishes, for prevention of oxidation and wearing of metal surfaces, and as positive electrodes in dry cells, in pipe cements, in much electrical equipment, in manufacture of bleaching powders and of alkalies, in the making of crucibles for electric furnaces, in foundry molds, in electrotyping, and in glazing. Used as nuclear moderator. Both graphite and diamond are nearly perfect forms of the element carbon. Graphite, because of spacing of its atoms, is opaque and one of the softest minerals; diamond has closer-spaced atoms, so has a higher specific gravity; yet is transparent and the hardest substance known. European name for graphite, *plumbago,* means black in reference to graphites used in making "lead" of lead pencils.

Graphite

SULFUR

Sulfur
S

Specific gravity: 2.05 to 2.09

Hardness: 1.5 to 2.5

Streak: White

Luster: Resinous

Cleavage: Poor

Fracture: Uneven to conchoidal. Most natural crystals are orthorhombic

Color: Sulfur-yellow to gray, brown, green, or red

Fusibility: 1. Burns freely with a bluish flame giving off fumes of sulfur dioxide

Not soluble in water or the usual test acids but is soluble in carbon bisulfide

Commonly in masses but may be in orthorhombic pyramids or tablets. Chiefly associated with volcanic rocks. In U.S. commercial deposits may be in sediments. May be contaminated with clay or bitumen. Volcanic sulfur may contain selenium, arsenic, or other elements

Found in combination with ores bearing iron, copper, lead, zinc, or in a pure form

Sulfur melts into a yellow sticky liquid at 160°C but begins melting at 108°C. It burns at 270°C with a bluish flame. Sulfur is used in many commercial ways including the manufacture of sulfuric acid, paper, explosives, fireworks, matches, insecticides, rubber, bleaching mixtures, in medicines, and as fertilizer. In time of war, it is necessary for the manufacture of much ammunition. Chemical element 16; atomic weight 32. Rhombic sulfur, also called brimstone and alpha sulfur. Sulfur is a fairly common deposit of late-stage volcanic activity. Commercially important deposits may have been formed from gypsum.

Sulfur

APATITE

$Ca_5(F, Cl)(PO_4)_3$

Specific gravity: 3.17 to 3.23

Hardness: 5

Streak: White

Luster: Vitreous to somewhat resinous

Cleavage: Poor, basal

Fracture: Conchoidal to uneven and brittle

Color: Green to blue-green, to brown, to flesh-colored, to yellow

Fusibility: About 5. Fuses in blowpipe flame at thin edges giving orange color to flame, but if moistened with sulfuric acid the flame is blue-green

Soluble in nitric acid and in hydrochloric acid; with sulfuric acid yields the precipitate calcium sulfate

Apatite

(Continued)

Crystallization: Hexagonal crystals.

A source of phosphorus that has many uses, particularly as a basis of fertilizer. Pure metallic phosphorus is white to transparent, waxy, with specific gravity of 1.8 at 0°C. It melts at 44.3°C and boils at 287°C. In moist air, it bursts into flame at 34°C. Related red phosphorus flames at 350°C. Constituent of rocks which may frequently be the source of phosphorus needed by plants. Vertebrate bones have essentially an apatite composition and structure. Widely distributed.

FLUORITE

Fluorspar, Calcium Fluoride
CaF_2. Fluorine, 48.9 percent

Specific gravity: 3.01 to 3.25

Hardness: 4

Streak: White

Luster: Vitreous and splendent. Colorless, black, white, brown, and various shades of the last three colors mentioned

Cleavage: Perfect, in 4 directions, octahedral

Fracture: Conchoidal to splintery, brittle

Fusibility: 3. Gives calcium flame of red and forms an alkaline enamel when fused in blowpipe flame; in closed tube, glows and flies to pieces. Fused with a salt of phosphorus, it etches glass

Mixed with sulfuric acid yields hydrofluoric acid gas or flames that may etch glass

Crystallization: Isometric, cubes or approximate cubes, massive. Often fluorescent (usually purple) under ultraviolet light

Found in seams or veins associated with ores of silver and gold; with quartz, calcite, and others

Fluorine is a greenish-yellow gas that boils at −187°C and has a specific gravity of 1.14. Hydrofluoric acid is used as a wood preservative, in the gas freon used in mechanical refrigerators, in the refining of gold, silver, copper, and lead, in the manufacture of aluminum from bauxite, in glass making, and in the electrical industry.

Fluorite

GARNET

Aluminum silicates with magnesium, iron, and manganese; or calcium silicates with chromium, aluminum, and iron

Specific gravity: 3.15 to 4.3

Hardness: 6.7 to 7.5

Streak: White

Luster: Vitreous in crystals; resinous when massive

Cleavage: No true cleavage

Fracture: Uneven to subconchoidal

Color: Red, yellow, brown, green, black, white

Fusibility: 3 to 4. Fuses variously depending on the kind; uvarovite almost infusible, while almandite yields a magnetic globule; decomposed by fusion with sodium and potassium carbonate

Crystallization: Isometric, dodecahedrons and trapezohedrons

Found mostly in schists and gneisses with crystals relatively common. A relatively common mineral. Found in metamorphic rocks, associated with chromium deposits and with metamorphosed limestones

Used commercially as an abrasive, but the finer stones are of gem quality including the *red carbuncle,* the rose and purple brilliant *rhodolite* and *uvarovite,* which take a high polish and are of various colors. Their powders are used in leather polishes and in wood polishing.

Garnet

CHALCEDONY

A Waxy Quartz
SiO_2

Specific gravity: 2.65

Hardness: 7

Streak: White

Luster: Vitreous, waxy, dull

Cleavage: None, brittle and tough

Fracture: Conchoidal

Color: White, gray, brown, to black

Fusibility: Fuses with equal amounts of soda on platinum wire to form a clear glass

Differs from ordinary quartz crystal in that it is not in the form of the typical hexagonal prisms but is usually with rounded surfaces

Crystallization: Usually dense

Chalcedony

Found in thin seams and bands or as crusts inside caves or cavities; often found by breaking cobbles. Usually occurs as crusts with rounded surface. Common amateur rock collector's item throughout the world

Many are almost of gem quality. *Agate* is variegated, banded, or clouded chalcedony; *amethyst,* purple or blue; *bloodstone,* red; *cat's eye,* opalescent and variable; *heliotrope,* green dotted with red; *jasper,* yellow, red, or brown; and *moss agate,* with mosslike discoloration. *Flint* is an impure chalcedony formed in concretions and *chert* impure chalcedony that has a splintery fracture. *Onyx* is a variegated chalcedony and *agatized wood* is formed when chalcedony has replaced wood to the finest detail. Chert, like flint, is formed in nodules

Chalcedony was the basis of the Stone Age civilization much as iron is of that today. Sometimes listed as a form of quartz; not a special entry in some mineral lists.

QUARTZ, SILICA

$Si O_2$

Specific gravity: 2.65

Hardness: 7

Streak: White

Luster: Vitreous, greasy. Ordinarily colorless and transparent

Cleavage: Indistinct and brittle

Fracture: Conchoidal

Fusibility: Fuses with difficulty. With equal amounts of soda on platinum wire fuses to form a clear glass

May be etched by hydrofluoric acid, is slightly soluble in water but more particularly in alkaline water

Crystallization: Usually prismatic, hexagonal

Constitutes about 12 percent of the mineral matter of the earth's crust and is possibly our most abundant mineral

Quartz glass permits passage of ultraviolet light. Rock crystals are essential in radio communication. The many varieties include *rose quartz, smoky quartz, topaz,* and *opal.* Source of silica for manufacture of glass. Chalcedony (above) is the name given to microcrystalline quartz. Used as prisms in some optical instruments.

Quartz

BIOTITE

Black Mica
$K(Mg, Fe)_3 AlSi_3 O_{10}(OH)_2$

Specific gravity: 2.7 to 3.1

Hardness: 2.5 to 3

Streak: Uncolored

Luster: Pearly or silky to vitreous or splendent

Cleavage: Perfect, into thin elastic plates

Fracture: Little shown

Color: Green to black

Fusibility: 5. Heated strongly in closed tube, gives off little water. Blowpipe flame whitens and fuses thin edges. May react as iron after fusion with borax, soda, or salt of phosphorus

Boiling sulfuric acid decomposes it completely, leaving thin silica scales as a residue

Crystallization: Monoclinic, tabular, or in short prisms

Found as a constituent of granites, gneisses, and pegmatites, usually associated with quartz, orthoclase, hornblende, plagioclase, and others

Used somewhat as is muscovite but is less resistant to weathering. Vermiculite, which resembles mica, has widespread application as lightweight constituent in making concrete and plaster; also used in insulation, as a plant-growing medium, and as a base for outdoor barbecue grills.

Biotite

MUSCOVITE

White Mica, Potash Mica
$KAl_3 Si_3 O_{10}(OH)_2$; often with more or less impurity of many other elements

Specific gravity: 2.75 to 3

Hardness: 2 to 2.5

Streak: Uncolored

Luster: Vitreous to pearly or silky

Cleavage: Perfect basal, plates. Thin, flexible, and elastic cleavage sheets

Fracture: Ragged, poor, tough

Color: Colorless, gray, brown, green-brown

Muscovite

(Continued)

Fusibility: 4½ to 5. Gives off water if heated in closed tube. Under blowpipe flame, whitens and fuses on the thin edges to a yellowish glass

Insoluble in acids, not decomposed even by boiling in concentrated sulfuric acid

Crystallization: Monoclinic, rhombic, tabular, foliar

Found commonly in granites, pegmatites, schists, and gneisses associated with quartz, hornblende. Best developed in pegmatite dikes

Formerly used for windowpanes and for panes in furnaces, also in electrical devices and as a lubricant. It is resistant to weathering and may persist under conditions that eliminate associated minerals in different rocks. Used in fireproof paints and shingles. Sometimes plates of muscovite are found that are a yard across. Also formerly marketed in some places as Christmas tree "snow"; now being replaced by substitutes.

ORTHOCLASE

Potassium Feldspar, Potash Feldspar
$KAlSi_3O_8$

Specific gravity: 2.57

Hardness: 6

Streak: Uncolored

Luster: Pearly

Cleavage: Two at right angles

Fracture: Uneven and conchoidal

Color: Flesh-colored, white, red, pale yellow, or gray

Fusibility: 5. Fuses quietly; yields a potassium (pale violet) flame when burned with gypsum; fuses with difficulty to a transparent glass

Insoluble in most common acids

Crystallization: Monoclinic; crystals often prismatic

Found in such igneous rocks as granites, and syenites associated with quartz, muscovite, biotite, hornblende, and plagioclase feldspars. The plagioclase feldspars include soda orthoclase. Orthoclase breaks at right angles while plagioclase breaks at oblique angles. Orthoclase is commonest in granites rich in quartz while plagioclase is common in quartz-poor granites. Plagioclase may show fine striae on flat faces. Used in ceramic and glass industry. Microcline, a "close relative," is the characteristic feldspar of coarse granite. Green variety of microcline used in jewelry under the name *amazonstone*. Another form of feldspar, *moonstone*, reflects a bluish-green sheen.

Orthoclase

PLAGIOCLASE FELDSPAR

Soda-Potash and Soda-Lime Feldspars
$NaAlSi_3O_8$ to $CaAl_2Si_2O_8$

Specific gravity: 2.65 to 2.67

Hardness: 6 to 6.5

Streak: Uncolored to white

Luster: Pearly

Cleavage: 2 directions

Fracture: Uneven, may show fine striations on cleavage faces

Color: White, colorless, bluish, green, gray, or reddish

Fusibility: 3½ to 4. Shows a strong sodium flame when burned with gypsum; fuses quietly to form a clear, glasslike enamel

Is acted on only slightly by acids commonly used for testing minerals. *Labradorite,* a plagioclase feldspar, is affected slightly by hydrochloric acid

Crystallization: Usually compact. Crystals triclinic

Found in such rocks as the granites, syenite, and diorites associated with quartz, orthoclase, biotite, and others

The plagioclase feldspars break at oblique angles in contrast with the right-angled breaks of orthoclase feldspars. In the plagioclases, there are potash-soda plagioclases such as *microcline* and soda-lime plagioclases such as *albite, oligoclase, andesine, labradorite, bytownite,* and *anorthite,* in which the specific gravity increases from 2.62 to 2.65 in albite to 2.74 to 2.76 in anorthite. In some references on minerals, feldspars are not subdivided into Orthoclase and Plagioclase as separate forms. Instead both forms appear under one heading, feldspar.

Plagioclase feldspar

PYROXENES

A Calcium-Magnesium-Iron Silicate
A mineral group. Parallels amphibole series. Silicates of Mg, Al, Fe, Mn, Ca, Na, Li

Specific gravity: 3.1 to 3.6

Hardness: 5 to 6

Streak: Greenish-brown to gray to white

Pyroxenes

Luster: Vitreous to dull

Cleavage: Imperfect, prismatic, in 2 directions lengthwise 87° and 93°

Fracture: Uneven

Color: Light to dark green

Fusibility: 4. Fuses quietly or with little tumescence into a shining black glass that may sometimes be green or brown. If much iron is present, the mass may be magnetic

Chemical differences of related rocks may be reflected in reactions and in behavior in flame test

Crystallization: Monoclinic. Orthorhombic, typically. Stout prisms almost rectangular

Found in grains rather commonly in igneous rocks and in crystalline limestone associated with garnet, chlorite, amphiboles, magnetite

A representative pyroxene group includes *enstatite, augite, acmite,* and *rhodonite,* the latter being important because of its manganese. In the *pyroxenes,* the cleavage faces meet at angles of 87° and 93° while in the *amphiboles* they meet at angles of 56° and 124°. One of most important groups of rock-forming minerals; a common constituent of rocks formed at high temperatures. May be the major constituents of basalts and gabbros. Jadeite, the more valued of two kinds of true jade, is a pyroxene. One form is used for carvings in China.

HORNBLENDE

[A species of Amphibole Family]
A Calcium-Iron-Magnesium Silicate
$CaNa(Mg, Fe)_4(Al, Fe, Ti)_3Si_6O_{22}(O, OH)_2$

Specific gravity: 2.0 to 3.4

Hardness: 5 to 6

Streak: Gray to dark gray; sometimes white mark or scratch on streak plate

Luster: Vitreous to pearly. Some silky

Cleavage: Prismatic in 2 directions, 56° & 124° to each other

Fracture: Uneven and splintery. Subconchoidal

Color: Dark to bright green, blue-green to black or brown

Fusibility: 3 to 4. Fuses to a magnetic mass that is sometimes globular and is usually shiny, usually somewhat intumescent, giving the characteristic sodium flame

Not acted on by acids. May lose luster and become earthy, changing to chlorite, epidote, calcite, and quartz

Crystallization: Monoclinic, orthorhombic, prismatic, may be granular

Found commonly in grains showing parallel sides and ragged ends associated with quartz, feldspars, and biotite

A typical amphibole rock found widely distributed in rocks particularly in New England, the Piedmont area to the south, and the Blue Ridge area. It is a primary rock mineral. Has a widespread occurrence in metamorphic and igneous rocks.

Hornblende

OLIVINE

Chrysolite, Peridot, a Ferromagnesia Silicate
$(Mg, Fe)_2SiO_4$. Actually a series of minerals with compositions ranging from pure magnesium silicate (forsterite) through (chrysolite), a magnesium-iron silicate, to fayalite; the iron silicate

Specific gravity: 3.2 to 3.6

Hardness: 6.5 to 7

Streak: White to yellowish-white

Luster: Vitreous. Nonmetallic luster. Named olivine because of its typical olive-green color

Cleavage: No good cleavage

Fracture: Uneven and conchoidal, brittle

Fusibility: Loses its color but rarely fuses in flame; those forms that are rich in iron, like hyalosiderite, may fuse to make magnetic globules. Gives off little or no water in closed tube

Is decomposed by sulfuric acid and by hydrochloric acid by the separation of the silica

Crystallization: Prismatic or tabular, orthorhombic, granular

Found in gabbro, basalt, and peridotite where it may sometimes represent 50 percent of the rock; also in schists and in igneous rocks. Abundant in many dark and heavy rocks, especially those formed deep in the earth's crust. Theoretically, never found with free quartz. Found in crystalline limestones. Chiefly a rock-forming mineral in igneous rocks. A typical mineral in nonmetallic meteorites. May be altered to serpentine. Usually identified by its color and occurrence. One of the first minerals to form upon crystallization of a silicate magma.

Olivine

Calcite

CALCITE

Calc Spar

$CaCO_3$

Specific gravity: 2.7

Hardness: 3

Streak: White

Luster: Vitreous to dull

Cleavage: Perfect, in 3 directions, rhombohedral

Fracture: Conchoidal

Color: White, colorless, pale gray, yellow, red, green, blue

Fusibility: Gives a calcium flame when burned with hydrochloric acid, the color being reddish-yellow. Infusible but becomes alkaline

Effervesces with hydrochloric acid, giving off carbon dioxide

Crystallization: Hexagonal

Found in limestones, marble, chalk, and marl, in veins, many of these being almost wholly calcite

Calcite is next to quartz in importance. Geologically, its chief function is in cementing particles of other minerals together. Man uses it for cement, for fertilizer, to flocculate clay bringing particles together permitting a more even distribution of air in soil. *Iceland spar* is a colorless, transparent calcite of value in optical instruments and in polarizing microscopes. *Chalk* is soft, white, and earthy. Other forms are *travertine, calc-tufa.* In the earth's crust, one of the most common and widespread minerals. Found in a variety of sedimentary and metamorphic rocks. Well-developed crystals occur in veins. Forms deposits in streams and in caves as stalactites and stalagmites. Calcite acted upon by water and CO_2 forms soluble $Ca(HCO_3)_2$. This may help produce hard water. Phenomenon of double refraction was first observed in crystals of calcite and the discovery of polarization of light resulted.

DOLOMITE

Pearl Spar

$CaMg(CO_3)_2$. Carbonate rock which contains more than 50 percent dolomite is commonly called dolomite, the rock, or dolostone

Specific gravity: 2.8 to 2.9

Hardness: 3.5 to 4

Streak: White or gray

Luster: Vitreous to pearly

Cleavage: In 3 directions, perfect 74° and 106°, rhombohedral blocks

Fracture: Conchoidal to uneven

Color: White and variously tinted

Fusibility: Gives a yellow to orange flame when burned with hydrochloric acid, is infusible in blowpipe flame

Effervesces in hot hydrochloric acid but not in cold

Crystallization: Hexagonal and rhombohedral with the faces often curved

Found as dolomitic limestone and marble associated with serpentine, gypsum, talc, and with ores of lead and zinc

Dolomitic limestone is harder and less soluble in acids than true limestones. Stone is used in building operations, in metallurgy as a flux and as a source of magnesium. Fossils in dolomite tend to exhibit poor preservation of structural details. Dolomite and dolomitic limestones are known in rocks from all ages. Many dolomites are replaced limestones.

Dolomite

GYPSUM

$CaSO_4 \cdot 2H_2O$

Specific gravity: 2.3

Hardness: 1.5 to 2. Easily scratched by fingernail

Streak: White

Luster: Vitreous, pearly, silky, or nonmetallic

Cleavage: In selenite and satin spar, easy, in 2 directions

Fracture: Splintery. Platy fragments are flexible but not elastic

Color: Colorless, white, gray, pearly, yellow, or red

Fusibility: 3. Gives reddish-yellow flame. Heated in closed tube gives off water and becomes opaque and dull. Crushed bead is alkaline

Crystallization: Flat, tabular, prismatic, or lenticular, monoclinic

Found in sedimentary and other rocks in seams

Used in making of portland cement to retard setting, used similarly in plaster, also used in insulation, stucco, wallboard, tile, lath, in the making of glass and pottery. Commonest sulfate mineral; deposited from seawater or brines from salt lakes. Gypsum is heated in kettles or kilns at temperatures about 200°C to remove some of the water of crystallization. Product is plaster of paris.

Gypsum

SERPENTINE

Antigorite, a Hydrous Magnesium Silicate
$Mg_3Si_2O_5(OH)_4$

Specific gravity: 2.5 to 2.65

Hardness: 2.5 to 5

Streak: White and glassy

Luster: Silky, waxy, greasy

Cleavage: None to fibrous

Fracture: Conchoidal to splintery

Fusibility: 5 to 6. Yields water when heated in a closed tube. Blowpipe flame fuses thin splinters at the edges. Normally is affected by average flame

Crystallization: Usually fibrous, foliated or massive or granular

Found in such rocks as peridotites, gabbros, and others commonly rich in magnesium often accompanied by dolomite, calcite, or magnesite

Used in rocks as building stones or monuments; ground and used in fireproof devices including clothing for fighting fires. *Asbestos* is a fibrous serpentine known as *chrysolite* used in making tiles, in electrical devices, to insulate heat pipes. Excellent nonconductor of heat, sound, and electricity. Closely related nickel-rich serpentines are important ores of that metal (garnierite). Commonly used, easily worked, in decorative carvings. May be present in rocks with a greasy feel.

Serpentine

TALC

$Mg_3Si_4O_{10}(OH)_2$

Specific gravity: 2.7

Hardness: 1. Softest mineral on Mohs' scale of hardness

Streak: Usually but not always white

Luster: Pearly, at least on the cleavage edge

Fracture: Similar to cleavage. Cleavage micaceous; can be easily cut

Fusibility: 5. Exfoliates with heat and fuses with difficulty; may yield water when heated intensely in a closed tube. If it fuses, it forms a white enamel. Heated with cobalt nitrate, it turns pale pink

Is not easily decomposed by acids. Has a greasy feeling

Crystallization: Orthorhombic or monoclinic or foliate or massive

Found filling spaces between other minerals in such rocks as serpentine, schists, and dolomite or in large deposits

Used in furnace lining, laundry tubs, sinks, switchboards, table tops, tank walls, in ground form in making of fireproofing, as filler for glazed papers, and for many other important things in modern civilization. A massive form of talc is soapstone. Commonly quarried from metamorphosed rocks. Talcum powder is a product from talc. Small pieces, called French chalk, are used by tailors to mark cloth.

Talc

KAOLINITE

Kaolin, China Clay
$Al_2Si_2O_5(OH)_4$

Specific gravity: 2.6 to 2.63

Hardness: 2 to 2.5

Streak: White

Luster: Pearly, dull

Cleavage: Crystals, perfect, flexible, inelastic plates

Fracture: Earthy

Fusibility: Heated in closed tube yields water. Moistened with cobalt solution and heated develops a blue color

Not soluble in acids

Crystallization: Monoclinic, usually compact noncrystalline mass

Found where rocks, particularly the feldspars, have decayed or in bed of some of the iron ores, particularly in Pennsylvania

Used in the manufacture of china and porcelain and in the making of tile and bricks. If it is impure, especially if it has iron in it, it becomes brownish or red in the baking process. It is the basis of much of our clay. Gives earthy odor when breathed upon. It is also used extensively as a filler in rubber products and for coating and filling paper products.

Kaolinite

CHALCOPYRITE

Copper Pyrites
$CuFeS_2$

Specific gravity: 4.1 to 4.3

Hardness: 3.5 to 4.0

Streak: Greenish-black

Luster: Metallic, may have brass-yellow color or iridescent tarnish

Cleavage: One poor, seldom observable, to none

Fracture: Uneven

Color: Usually massive with metallic luster and brass-yellow, often with a bronze or iridescent tarnish; greenish-blue crystal

Fusibility: On charcoal fuses to black, magnetic globule, touched with HCl tints flame with blue flash. Solution with strong nitric acid is green; ammonia causes precipitate of red iron hydroxide and blue solution remains

Crystallization: Sphenoidal crystals resemble tetrahedrons. Crystals common, often large; faces frequently uneven and tarnished in iridescent hues. Massive crystals in rocks and associated with sulfide veins

Most widely occurring of the primary copper ore minerals; many secondary copper minerals have been derived from it by alteration. An original constituent of igneous rocks; occurs in pegmatite dikes, as a product of contact metamorphism, and in metallic veins. Frequently associated with pyrite. Like iron pyrite is called fool's gold. Unlike sectile gold, chalcopyrite is brittle. Not as hard as iron pyrites; chalcopyrite, unlike iron pyrites, can be scratched with a knife. Sometimes mined for its associated impurities, gold or silver. Associated with nickel deposits in several locations. Structure closely related to mineral sphalerite. In the U.S. it is mined in Montana, Arizona and Tennessee.

Chalcopyrite

CARNOTITE

Uranium Ore, Radium Ore
$K_2(UO_2)_2(VO_4)_2 \cdot 3H_2O$

Specific gravity: 4 to 5

Hardness: 1 to 2

Streak: Leaves colored mark on streak plate. Yellow mineral color is characteristic

Luster: Earthy; bright yellow color to lemon-yellow to greenish-yellow; opaque. Nonmetallic

Cleavage: Crystal plates said to have basal cleavage, but not visible; powderlike, crumbles easily

Fracture: None; powdery, scaly. Loose granular form is usual; sometimes occurs as micalike crystals. Microscopic crystals are monoclinic

Fusibility: Infusible. Powder turns red-brown when added to hot HNO_3; when dissolved forms a green solution. Residue of evaporation of acid is fluorescent

Found associated with sandstone deposits in western Colorado and eastern Utah. Found in a conglomerate at Jim Thorpe, Pennsylvania. At one time it was the chief source of U.S. radium ores; now important source of uranium; also a source of vanadium. Occurs as a secondary mineral resulting from action of groundwater. Sometimes associated with dinosaur bones and petrified wood. The prospector armed with a Geiger counter hunts for carnotite.

Carnotite

CLASSIFICATION OF LIVING THINGS

Living things are grouped (classified) by man on the basis of structural similarities and differences into categories, somewhat like a hierarchy. The hierarchy of categories is reported as: kingdom, phylum or division, class, order, family, genus, species. There are subcategories of several of the major categories. Some taxonomists believe in fewer groups within the categories and some believe in dividing categories into ever-differing smaller groups.

There are at least five different classification systems in use (or in existence in printed form) today. These different classification systems are based, in part, on the manner in which they specify the makeup of kingdoms. All of the systems include or exclude individual forms by comparing structural similarities and/or differences. For the purposes of this fieldbook we shall recognize two kingdoms, Plantae and Animalia.

KINGDOM PLANTAE

Traditionally various divisions of plants (biologist's term which is equivalent to *phyla* when used for animals) have been separated into two groups: Thallophyta—either unicellular or multicellular plants that show little tissue differentiation and have no stems, roots, or leaves; and Embryophyta—multicellular plants which have multicellular reproductive structures and in which the early stages of embryonic development occur while the embryo is maintained within the female reproductive organ. This, in early classification systems, provided two subkingdoms. Now, botanists use the two names of these subkingdoms (Thallophyta and Embryophyta) only for convenience. Also, it was customary to group members of the thallophytes in two categories based on presence of chlorophyll (algae) and lack of chlorophyll (fungi). These, too, are terms of convenience and so are not afforded recognition by taxonomists as taxons. In this book we shall employ the classification system that divides the Plant Kingdom into three subkingdoms. Within each subkingdom there are many divisions.

The kingdom Plantae is divided into three subkingdoms:
1 Subkingdom *Prokaryota,* which includes the blue-green algae and the bacteria.
2 Subkingdom *Mycota,* which includes the slime molds, algalike fungi, sac fungi, club fungi, and imperfect fungi.
3 Subkingdom *Chlorota,* which includes the green algae, the euglenoids, the brown algae, the yellow-green algae, the golden-brown algae, diatoms, the dinoflagellates, the red algae, the stoneworts, the liverworts, the mosses, psilopsids, the club mosses, the horsetails, ferns, cycads, ginkgos, conifers, the flowering plants.

DIVISION SCHIZOMYCOTA CLASS SCHIZOMYCETES [Bacteria]

Most persons regard these minute forms as essentially harmful and certainly too small to see. They may wonder why bacteria are included in a book on field natural history where presumably a compound microscope will not be available. We look across a valley and see masses of trees making a forest without being able to determine from a distance just what kinds of trees are in the forest. Similarly, we may see masses of scum on vinegar, strings of rusty gelatin lining water pipes at their outlet, or swellings on the roots of clover. These are either masses of bacteria or of their products or evidence that certain bacteria are in a given place. If we can make a census of mammals of a forest by observing the tracks without ever seeing a mammal, why can we not consider the evidence of the existence of bacteria in the outdoors? The field naturalist should have some recognition of the evidence of the existence of bacteria, which are probably the smallest living things (unless one considers viruses as living forms) and which have lived on the earth probably longer than any other living things. More than 3000 species.

One philosophical bacteriologist has estimated that for every harmful bacterial organism to be found in the world there are approximately 50 others that are harmless or essentially helpful. To live happy, successful lives, we must learn how to use the helpful forms and to avoid those that are dangerous.

The term *microbe* cannot be used as a synonym for bacteria. Microbe includes not only the bacteria but the molds, the protozoa, and other microscopic plants or animals. Many of these microbes have a prodigious reproductive capacity. While a young pig may double its weight in a month, a young yeast (not bacteria) can double its weight in 2 hours. A yeast-fermenting tank can produce in a single day the protein equivalent of eight pigs a day. It is quite possible that in the future we may be eating bacteria and yeasts instead of pork and beef and do so with less strain on ourselves and on other living things. Let us survey some of the important roles played by bacteria.

Bacteria and the Soil
An acre of rich farm soil might well support three woodchucks weighing about 30 lb. The same soil might well support 240 lb of bacteria and an additional 240 lb of fungi and microscopic animals. By far the greater proportion of the bacteria would be usefully engaged in changing the compounds of the soil. Some of these would be found in the roots of legumes and many might be engaged in fixing free nitrogen or taking part in the changes that make various foods available to plants and to animals. The same soil might also contain tetanus bacilli that, once transferred to our bodies, might cause lockjaw and death. Other bacteria in the soil might play a part in fixing or in freeing iron, sulfur, or phosphorus, all essential to living things.

Bacteria and the Air
Even the free open air may be the carrier of some kinds of bacteria. Sometimes these are relatively dry, but more often they are in small drops of liquid suspended in the air. Liquids from a person with tuberculosis may be discharged as sputum or in other ways. When these liquids dry, they may free *Mycobacterium tuberculosis* into the air. If these reach a healthy person before they are themselves killed by high temperature or by sunlight, they may cause the disease in the healthy person. Similarly, *Streptococcus,* which causes streptococcic sore throat, or death, may be spread. Sunlight, high temperatures, and dryness are fatal to a majority of the dangerous bacteria found in the air and elsewhere.

Bacteria in General
Bacteria which cause disease in man, his animals, or his plants are called germs, although not all disease-producing microorganisms are bacteria. Some bacterial diseases of worldwide importance are bubonic plague, cholera, diphtheria, syphilis, gonorrhea, leprosy, scarlet fever, tetanus, tuberculosis, typhoid fever, meningitis, bacterial dysentery, bacterial pneumonia, whooping cough, strep throat, boils, and abscesses. In nations with good, modern, efficient medical services, these diseases are no longer serious threats to the public health. In their host, man, bacteria may occur in such great numbers as to put a drain on his system and interfere with normal functions. Bacteria may actually destroy man's cells and tissues or produce toxins or poisons. On the other hand, many bacteria are beneficial—as in the decay process which recycles chemical elements. Bacteria in man's intestine synthesize some vitamins. Several vitamins, acetic acid (vinegar), acetone, butanol, and lactic acid are produced in commercial form by bacteria. These organisms are utilized in the commercial preparation of certain skins such as leather goods and in the curing of tobacco. Certain dairy products, e.g., butter and some cheeses, are produced by bacterial action. It is hoped that bacterial forms will be found which function as substitutes for harmful chemical pesticides. A new field in the drug industry based on chemicals from the Actinomycetes group of bacteria has given us some of our most potent weapons against some bacterial diseases; e.g., Streptomycin, Aureomycin, Terramycin, and Neomycin.

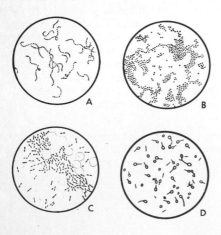

CLASS SCHIZOMYCETES [Bacteria]

Disease-causing bacteria
A *Cause of streptococcic throat*
B *Cause of pus*
C *Cause of pneumonia*
D *Cause of lockjaw*

DIVISION SCHIZOMYCOTA. CLASS SCHIZOMYCETES

Crown Gall of Peach
Agrobacterium tumifaciens
Found on peach, cherry, almond, apricot, plum, prune, chestnut, walnut, apple, tomato, potato, beet. Intercommunicable.

Fire Blight of Pear
Bacillus amylovorus
Worst on pear but also bad on apple, quince, hawthorn, plum, apricot, mountain ash, and other plants.

Fire Blight of Pear

Crown Gall of Peach

DIVISION CYANOPHYCOPHYTA [Blue-Green Algae]

Either unicellular or filamentous organisms; sometimes as a single cell or as a colony of cells. Occur in numerous and varied habitats. Many in fresh water, few marine species; some in hot springs, others as partner with fungus in lichens. Common in soils. Observed on damp rocks in woodlands, on flowerpots, and on tree bark. Blooms of blue-greens develop on ponds or lakes which contain high concentration of organic matter rich in nitrogenous compounds. Decomposition of algae causes odor and uses oxygen dissolved in water, causing conditions intolerable to animal forms. 7,500 species or more.

CLASS MYXOPHYCEAE

Order Hormongales./Family Oscillatoriaceae

Oscillatoria
Oscillatoria sp.
Long, slender, unbranched filaments of cells commonly with long axis across the filament, usually blue-green, but some relatives are red. Found on moist banks and cliffs or in clear or putrid water. Common on bottoms of temporary puddles; frequently found floating with attached mud on surface of the water of puddle. May grow in stagnant water; in pools, ditches and occurs intermingled with other forms as a chief component of plankton. Appears to bloom on the surface of some dry regions after several weeks of continuous rainfall. American genera, 11; species, 30. Related species give redness to Red Sea. Blue-green color due to chlorophyll, carotin, xanthophyll, and phycocyanin, not in definite bearers. Considered an index of contamination in water, either by self or by associated organisms. Has characteristic oscillating movement. Mass of plants in laboratory dish will creep up sides of container.

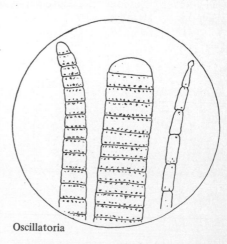

Oscillatoria

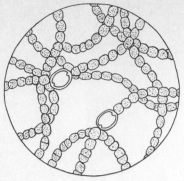

Nostoc

DIVISION CYANOPHYCPHYTA. CLASS MYXOPHYCEAE

Order Hormongales./Family Nostocaceae

Nostoc
Nostoc sp.
Under microscope appears like necklace of round cells embedded in gobs of jelly. Some species cause slipperiness of underwater rocks; others float to water surface or cling to rocks or wet earth. One species survives 3 ft underground; another, in a lake at 60-ft depth. American genera, 6; species, 22. Some colonies may be 2 ft or more across but most appear as floating jelly masses that appear to be muddy. May provide some index of pollution. One species forms colonies as large or larger than a hen's egg. Sometimes forms brown layers on the bottoms of pools or in swamps. May be associated with a fungus in a mutualistic relationship called a lichen. Some species prefer bottoms of fast moving clear mountain streams. Widespread genus.

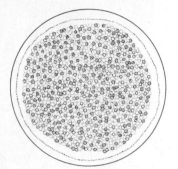

Microcystis

Order Chroococcales./Family Chroococcaceae

Microcystis
Microcystis sp.
Essentially planktonic. May cause "water blooms" in hard-water lakes. Cells marble-shaped. May occur in microscopic or macroscopic form; in macroscopic form, enclosed gas bubbles may be seen. Produce such dense surface scums on lakes and reservoirs that they indirectly cause death of fish by suffocation. Occur in abundance in warm months of the year. Considered an indicator genus of water pollution. Toxins produced purported to have killed cattle and some water birds.

SUBKINGDOM CHLOROTA
DIVISION CHLOROPHYCOPHYTA [The Green Algae]

This group is of special interest to biologists because it is believed that higher plants arose from them; thus they represent the only division of algae which do not represent an evolutionary dead end. Probably more than 7,000 species. Also called grass-green algae. Species of green algae in the U.S. outnumber the combined species of all other fresh-water algae. If one collects algae, he is almost assured to have in his collections some representatives of this division. Found in both still and fast-moving waters; common as fresh-water plankton. There are marine forms and species that inhabit brackish water. Generally most abundant in late spring and early autumn. May form feltlike masses on surface of damp soils; also grow within soils. May form as mats on tree trunks, exterior woodwork, and moist brickwork; found in deep fresh water, in brine lakes, and as the alga of a lichen. Some species cause red snow. May be unicellular, colonial, or filamentous; swimming, floating, or attached.

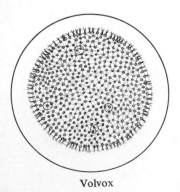

Volvox

CLASS CHLOROPHYCEAE

Order Volvocales./Family Volvocaceae

Volvox
Volvox sp.
A spherical green colony of cells, one cell in thickness; 500 or more cells per colony; motile. Daughter colonies, spherical in shape, form and break away from original colony. Few cells in colony are capable of forming daughter colonies. Sexual reproduction present. More frequently found in permanent rather than temporary ponds. Globular colony usually easily seen with unaided eye; may appear as a little green sphere. Occurs where nitrogenous compounds are plentiful; may be responsible for blooms for short periods in summer months. Frequently included in introductory biology text as an example of motile green alga shaped like a green ball.

DIVISION CHLOROPHYCOPHYTA. CLASS CHLOROPHYCEAE

Order Ulotrichales./Family Ulotrichaceae

Ulothrix
Ulothrix sp.
Commonly unbranched strings of green cells unlike most other genera that branch. Filaments break into new strings, may produce four swimming bodies per cell. United States genera, seven. Some species favor cold water, some warm; some running water, some quiet. Temperature has little effect on reproduction. Female cells swarm about a day; males, nearer a week when temperature is below 50°F. Most typical order of the Chlorophyceae. Most species are attached at some point. Some species appear in early spring, disappear, and reappear in fall. One species, on West Coast, among other habitats, grows along waterline on fishing boats. Important member of "producers" in aquatic food chains. May be a member of algae producing yellow snow. Microscopic.

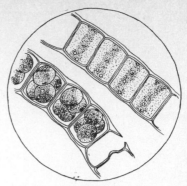

Ulothrix

Order Ulotrichales./Family Protococcaceae

Protococcus
Protococcus sp.
Simple green cells that divide into two and then into four. Probably commonest alga in the world and is worldwide in distribution. About five species. Found in air, on trees, or rocks, in soil, on wet woods, in lichens, and elsewhere. Affected by humid winds, making tree bark greenest on side toward damp winds, not necessarily toward the north. Usually found on side of tree protected from prevailing winds as this alga absorbs water from moist air. In some references, referred to as *Pleurococcus*. Frequently used as an illustration in introductory biology texts. Distributed by wind, water, and insects. Reported surviving winter temperatures of −40°C. May cause, with other algae, yellow snow. Rarely occurs in places where annual rainfall is less than 20 in. Individual cells, microscopic; forms large green macroscopic masses on various moist surfaces. Basis of food for lichens that are basis of food for reindeer and other animals that are basis of food for human beings.

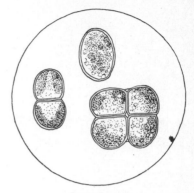

Protococcus

Order Ulotrichales./Family Cladophoraceae

Cladophora
Cladophora sp.
Branched strings of cells, with length of each cell rarely over eight times its breadth. Reproduces by division, by nonsexual and by sexual spores, rarely forming large balls. United States genera, branched alga, two; unbranched, two. Worldwide, found in fresh, salt, or brackish water, sometimes 150 ft below water surface. Common in freshwaterways, supplying there some aeration and much basic food for valuable aquatic animals. May, on occasion, contribute to water pollution when too abundant. Many species perennial. One of the largest genera of algae. May occur as alga partner in lichen. Feels harsh to the touch if concentrated.

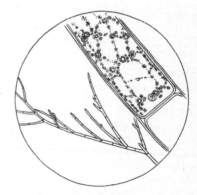

Cladophora

Order Ulotrichales./Family Ulvaceae

Sea Grass
Enteromorpha sp.
Axis of plant a hollow tube made up of one layer of cells, sometimes described as tubular, and in single branches. May be compressed. From 4 in. to 2 ft long and often referred to as grass or sea grass. Occurs as strings, ribbons; may be flattened or thickened or have inflated branches. Found on both the Atlantic Coast and the Pacific Coast of the U.S. Occurs in a wide diversity of places. Common green marine alga of piers and boat bottoms. In fresh water streams and in lakes; frequently associated with other algae as temporary floating mats. Mostly restricted to salt water, although certain species are able to live in brackish water; can adapt to fresh waters. Some survive attached to bottoms of boats which travel from fresh water to salt water. Used as food species in some commercial fish-raising projects. In salt marshes, may be conspicuous in late summer. May pose problems if abundant in irrigation ditches.

Sea Grass

Sea Moss, Bryopsis

DIVISION CHLOROPHYCOPHYTA. CLASS CHLOROPHYCEAE

Order Siphonales./Family Bryopsidaceae

Sea Moss, Bryopsis
Bryopsis corticulans

Species described is found on West Coast of U.S. Generic name is descriptive; bryon meaning moss and opsis meaning appearance; so moss-like, really more fern-like, in appearance. Plant a fern-like, repeatedly branched dark green or blackish green alga. Dioecious. Usually symmetrical in appearance. Several shoots may develop from a common base: Plants 2–8 in. long. Attractive.

Found attached to rocks in shallow water along shorelines. May be found on piers or attached to muddy bottoms near shore. Reproduces sexually.

Some species of this genus occur on the Pacific Coast from British Columbia to southern California. Four or more species described for the Atlantic Coast. Generally speaking, genus is more common in warmer waters.

Merman's Shaving Brush

Order Siphonales./Family Codiaceae

Merman's Shaving Brush, Neptune's Shaving Brush
Penicillus capitatus

A stalk with rootlike structures and a terminal tuft of free filaments collected together in one plant is probably responsible for the origin of this calcified alga's common name, Neptune's Shaving Brush. The plants' rootlike structures commonly anchor the plant to shoreline regions with sandy or muddy bottoms. Species listed here occurs in warm, quiet water off the Florida coast as well as around Bermuda and the West Indies. Stalks always erect when submerged. Reported as probably of Atlantic Coast origin in most distribution lists.

Plant bodies are readily preserved as fossils because of calcification of plant. Authors report that no part of the life cycle of this plant is of economic importance. Certainly of esthetic value as the plants look like beautiful, miniature, green shaving brushes.

Mermaid's Cup

Order Siphonales./Family Dasycladaceae

Mermaid's Cup, Mermaid's Wine Goblet
Acetabularia crenulata

Common; highly attractive green alga, sometimes in quiet water from southern Florida and west to Texas. Only 1 to 3 in. tall; appears like a greenish-white mushroom or a disjointed, extended umbrella. Occurs frequently as a carpet in shallow waters. A lime-encrusted genus confined to warm waters, e.g., the Mediterranean and the Indo-Pacific regions. Each erect "plant stem" bears one or more whorls of branched sterile side extensions. Only basal portion of plant survives throughout year. Reproduction by gametes. Two species *crenulata* and *mediterranea,* used in rather famous experiments to demonstrate function of nucleus in reproduction. These experiments made possible because portions of one species may be grafted on or within portions of the other species. Commonly referred to as illustrative material in beginning biology textbooks. Has been used in important studies on morphogenesis, nuclear function, nuclear-cytoplasmic relationship. This plant responds to direction of light, growing toward light source.

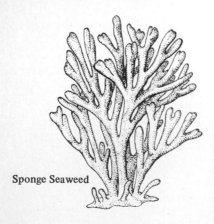

Sponge Seaweed

Order Siphonales./Family Codiaceae

Sponge Seaweed, Codium
Codium fragile

Made up of a spongy thallus, plant body; anchored by a basal disk of rootlike structures; variable in form; appears as wormlike threads which are both erect and branched. One species appears as a flat cushion, another as a round ball. Some species prostrate and some more or less erect with lobelike branches. Variable in shape. Dioecious; described as elongated, cylindrical, twice branched. Marine forms, attached to substratum by rootlike structures, rhizoids. Several species of *Codium* occur along the Pacific Coast and on the Atlantic Coast from Cape Cod, Mass., southward. Frequently on tops and sides of rocks at or near low-tide level.

Reproduce vegetatively; and sexually. Used as a food source by man along with many other genera and termed *limu* in Hawaii; generally eaten raw. May be boiled with fish or shrimp, or used to wrap an animal which is to be cooked underground.

DIVISION CHLOROPHYCOPHYTA [Green Algae].
CLASS CHLOROPHYCEAE [Green Algae]

Order Oedogoniales./Family Oedogoniaceae

Oedogonium
Oedogonium sp.
String of barrellike cells with small, green color bearers. Reproduce by
breaking, by asexual spores, by single cell acting as egg with many swimming males, or by males producing small plants that yield sperms.
United States genera, 3; species to over 250. This genus unbranched and
thus unlike others. Attached or free-floating in still or flowing water.
Fruits May to July in north, after which strings may break. Plants commonly covered with lime. May begin and end life as a single filament or
may multiply to form masses which float to the surface and are dirty
yellow green in color. Microscopic as individual cells but macroscopic
in a mass. Can be collected in almost any season in the temperate zone.
May cause "yellow snow." May appear in certain markets, e.g., India, in
dried packets.

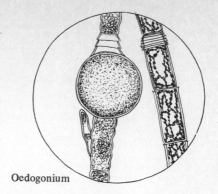

Oedogonium

Order Ulvales./Family Ulvaceae

Sea Lettuce
Ulva lactuca
Length to over 2 ft. Width to over 6 in. Broad, thin, with lobed, folded
edges thickest at base. Attached with inconspicuous holdfast. Bright
green. Highly variable in shape, with related *U. latissima* appearing as
floating pale green sheets to 10 ft long and nearly as broad.

Related *Monastroma*, of single cell layer; with *Ulva*, a double layer; and
Enteromorpha, with layers separated to make a hollow. *Ulra*, on exposed
rocks or wood, from Gulf of California to Alaska and from Gulf of Mexico to James Bay. Pacific Coast species, 13.

Some reproduction by fragmentation. Asexual reproduction through
freeing of four to eight swimming spores from a cell. Sexual reproduction by union of two similar swimming cells, eight of which come from a
single cell. United cells germinate into irregularly threadlike plants.
Sporophyte stage differs from smaller-celled gametophyte stage.

Ulva chiefly marine but may be found in brackish water. Related *Enteromorpha* grows well on bottoms of river boats that alternate between fresh
water and salt water. Some of the family grow in brine lakes where
salinity is much higher than in the ocean. *Ulva* survives considerable
temperature range.

Ulva may foul shores causing bad odor after storm. May be boiled and
seasoned and eaten as green laver or green slack though its use as food is
vastly inferior to other algae here discussed. Apparently the commercial
possibilities of sea lettuces have not been studied exhaustively. Occurs in
fresh water also; free-floating form. In Hawaii may be used as constituent of ceremonial feast.

Sea Lettuce

Order Chlorococcales./Family Hydrodictyaceae

Water Net
Hydrodictyon sp.
Small, fine, green nets easily recognized by naked eye. Reproduce by
breaking of nets or by small complete nets formed in individual cells by
spores swarming just after daybreak. United States genera, in family, 4;
species, two. Found in still, permanent ponds of fresh water over most of
United States. When undisturbed, nets over 1 ft long may develop.
Number of cells in a net once formed remains the same with new nets
formed by breaking of old. Each net of about 100 cells. Large masses of
Water Net may clog irrigation ditches and cause problems in small ponds
or lakes. Colonies macroscopic, made up of several thousand plants.
Many species of this genus may be indicative of comparatively pure water.

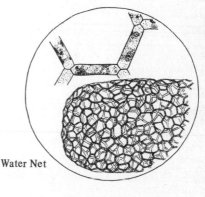

Water Net

Order Chlorococcales./Family Oocystaceae

Chlorella
Chlorella sp.
Small, spherical or ellipsoidal cells. Microscopic. Reproduces by forming several spores which burst out of the cell. Classification of species
difficult. Occurs free-living and within some invertebrates; makes fresh
water sponge green. Has been experimented with extensively as a potential alga protein source for man. Has been demonstrated that algae cells
have a high vitamin content and over 50 percent protein. Some say that
dried *Chlorella* tastes like raw pumpkin or lima beans. In experimental
tanks, *Chlorella* may yield 20 tons to the acre; some botanists predict 100
tons/acre yield. Also used in study of process of photosynthesis. Found
in many different habitats; terrestrial, media rich in organic matter as, for
example, sewage waters, on bark of trees, damp walls and flowerpots in
greenhouses, in aquaria, and soil; may be associated with a fungus in a
lichen.

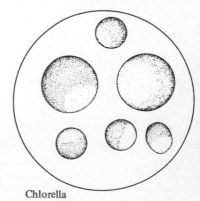

Chlorella

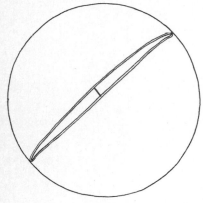

Pondscum

Desmids

DIVISION CHLOROPHYCOPHYTA [Green Algae].
CLASS CHLOROPHYCEAE

Order Zygmentales./Family Zygmentaceae

Pondscum
Spirogyra sp.

Unbranched strings, usually bright green and somewhat slimy or slippery, often enclosing gas bubbles in the floating or attached masses. Each cell with a spring-shaped coiled green body. One of the commonest of the green algae in fresh water. Reproduce by breaking of strings, by joining of adjacent cells or of cells in adjacent strings, induced by changing ratio of surface and volume of water, temperature, or light. A pollutant. Called "frog's spit" by youngsters. Forms "clouds" of green, cottony growth usually in quiet water. In warm pools, plants may decompose and produce unpleasant odors. Can be collected at almost any time of the year in temperate zone.

Order Zygmentales./Family Desmidaceae

Desmids
Closterium sp. *et al.*

Microscopic, of almost all imaginable shapes but all with cells with vertical pores through walls. *Closterium* (shown) is one of 65 species found in hard waters. Some 21 United States genera commonly found free-floating in lakes and streams where hydrogen-ion content is between 5 and 6. Of the genera, 13 are solitary cell types. Reproduces by dividing or sexually by union of adjacent cells, producing a cell that resists unfavorable conditions and divides to make new individuals. Genus is common throughout U.S. With magnification observer may see granules of gypsum in vacuoles in cell. Common in late spring and early summer.

DIVISION EUGLENOPHYCOPHYTA [The Euglenoids]

In some classification systems, these organisms are considered as members of the Protista; they include both plantlike and animallike primitive organisms. In the classification system accepted here, Euglenophycophyta are given division status along with 19 other divisions in the kingdom Plantae. About 400 species are known. They are plantlike in that many species which are included possess chlorophyll and so are photosynthetic; they are animallike in that they lack a cell wall and are highly motile. Species which lack chlorophyll are, like animals, heterotrophic. In older classifications, zoologists have accepted the forms as animals and as flagellated protozoa, while botanists have regarded them as algae. Their habitats are varied, ranging from fresh water through to soil; on various damp surfaces, and even within the intestinal tracts of some animals.

CLASS EUGLENOPHYCEAE

Order Euglenales./Family Euglenaceae

Euglena
Euglena sp.
Microscopic, free-swimming, unicellular; with numerous ovoid, star-shaped or platelike chloroplasts which are grass-green; eyespot or red pigment spot usually evident. Single, "tail," flagellum. Famous as an organism included as an example of a plant-animal in many introductory biology texts. Common in waters rich in organic matter. May color water a dark green. May be the organism producing green color in stagnant water in late summer or a velvety surface film or bloom on water. Species of Euglena may give red color or green color to snow in various parts of the world and some make seas appear yellow. Some species of Euglena have proven useful as assay organisms for such vitamins as B_{12}.

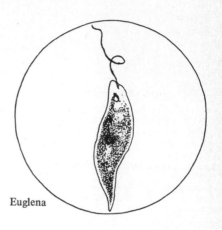

Euglena

DIVISION PHAEOPHYCOPHYTA

The Brown Algae. Almost exclusively marine forms. Most plants in this division are called seaweeds. Most common along rocky coasts of cooler parts of the oceans; seen in abundance attached to rocks at low tide. Some species form mats and are floating. All multicellular, most macroscopic. Possess chlorophyll. Some have a branching filamentous thallus (plant body). Large kelps of the sea belong in this division. Some 225 genera and 1,400 species have been described.

CLASS PHAEOPHYCEAE

Order Laminariales./Family Alariaceae

Henware
Alaria esculenta
Blade broad, leathery, to 10 ft long and 10 in. wide, with small, lateral, winglike leaflets below the broad, terminal blade, the leaflets being without midribs; commonly with wavy margins torn and broken to midrib. Stem compressed, to 1 ft long, ½ in. wide. Holdfast branched and rather elaborate. One Alaskan species is more than 75 feet long.

Differs from *Laminaria* in possession of midrib in blade. Essentially northern. Commonest on exposed rocky coasts below tide levels. From Long Island north to Hudson Bay and Newfoundland. Nine Pacific Coast species from California to Alaska.

Perennial, but each year blade is replaced by new growth. Ordinary plant seen is sporophyte or spore-producing, the sexual plants being microscopic and bearing either male or female sex organs with the female filaments being the simpler and usually bearing one terminal egg that develops into the sporophyte when fertilized.

Named *Alaria* from the Latin *ala* meaning wing and referring to the winglike leaflets below the blade. The sporophylls that bear the spores taking part in asexual reproduction are shed annually. Sporangia are borne on the margins of the sporophylls. Sexual plants do not resemble each other closely.

Yields the food known as "kombu" from the northernmost Japanese Islands where it is used with meats and sauces or poured over rice or used as tea. Is eaten in Iceland, in Scotland, and in Ireland. Known as "henware," "murlins," "badderlocks," and "daberlocks" in these areas. A commercial source of potash.

Henware

Devil's Shoelace

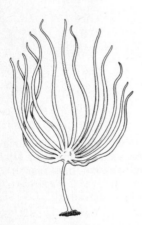

Fan Kelp

Sea Colanders

DIVISION PHAEOPHYCOPHYTA. CLASS PHAEOPHYCEAE

Order Laminariales./Family Chordaceae

Devil's Shoelace
Chorda filum
Fronds, with basal disk, ropelike, slender, less than 1 in. in diameter but frequently over 12 ft long and reported to 40 ft. When young, the surface is coated with delicate colorless hairs, but when mature the tips are commonly partly decayed. Whole mass of these plants looks like a writhing collection of snakes. Brown, with air cavity in middle. Species name, *filum*, once used for generic name and means string.

Common on stones and shells below tide level, held rather erect by internal air cavity. May reach water surface and there rot. Ranges from New Jersey to Baffin Bay, through the arctic regions and into Hudson Bay and Baffin Bay.

Asexual spores are borne in special structures somewhat conelike in shape on the surface of the blade. From these spores, produced in late summer and autumn, arise the conspicuous annual plants we see. Sexual reproduction is effected through microscopic structures not easily seen.

Apparently, from the distribution notes, this plant does not favor the warmer seas as do so many of the marine algae but in masses it does lend character to northern marine gardens not duplicated in southern areas. It is reported as growing in beds in the North Sea and British Channel 20 miles long and under 600 ft wide.

The plants adhere to mounting paper when they are made into herbarium specimens. Old fronds should be allowed to dry before mounting, but the young ones may be floated onto the paper, to which they will eventually fasten themselves if dried under pressure.

Order Laminariales./Family Laminariaceae

Fan Kelp, Devil's Apron
Laminaria digitata
Blade of to 30 or more fingerlike segments each to 3 to 4 ft long and growing from a common point or expanded area at the end of a stout stalk that is to 5 ft long, flattened above and in section showing concentric growth rings but no mucilage ducts. Holdfast fibrous.

At least 10 species on Pacific Coast of United States with *L. sinclairii* found from Vancouver Island to Obispo County, Calif. *L. digitata* ranges from Staten Island to Hudson Bay, Nova Scotia, and New Brunswick most commonly on exposed rocky bottom below low-tide levels.

Perennial. Asexual zoospores and sexual gametes are essentially alike. For some time, it was not believed that sexual reproduction took place in *Laminaria* but it is now well established. The male and female sex cells are essentially alike.

Named from Latin *lamina* meaning leaf or blade. Yields phycocolloid laminarin, a reserve carbohydrate; and fucoidin, a water-soluble carbohydrate of no recognized commercial value. Rich in iodine that constitutes 0.49 percent (dry weight computation) of whole plant; also considered high in certain vitamins and minerals.

In U.S. algin is produced from *Laminaria* and *Macrocystis*. Algin is used as a stabilizer in dairy products. Japanese who eat it are free of goiter. Dried known as "kombu"; shredded and dyed green, as "ao"; crisped over fire, as "hoira kimbu"; with sugar icing as "kwashi." Tons harvested in Japan and Russia for fertilizer.

Sea Colanders
Agarum cribrosum
Blade to over 9 ft long and 1 ft wide, abundantly perforated, heart-shaped at base when mature and with a median midrib to 1½ in. wide at the base. Stalk stout, cylindrical, to 2 in. long and ⅓ in. through. Blade unrolls from 2 conelike scrolls and looks tattered. Holdfast slender, spreading.

Appears in closely related forms on Atlantic and Pacific Coasts. This species from Bering Sea to San Juan Island, Wash., with the related *A. fimbriatum* from Puget Sound to San Pedro. *A. cribrosum,* from Massachusetts to Ellesmere Island.

Fruiting areas are irregular in shape, irregularly placed, and appear darker and thicker than the regular vegetative tissues, usually more abundant near the midrib. Spore cases usually elliptical in outline. Forms found growing in shallow water or tide pools are dwarfed in comparison with deep-water plants.

The common name colander probably refers to the fact that the blade resembles the kitchen colander in its sievelike appearance. The genus *Agarum* refers to the Malayan word *agar-agar* or edible seaweed.

While the plant no doubt has many of the uses recognized for its close relatives, the literature does not often list it as a commercially valuable plant. It should be assumed from its genetic name that it is edible and that it may be important as a source of agar-agar.

DIVISION PHAEOPHYCOPHYTA. CLASS PHAEOPHYCEAE

Order Laminariales./Family Lessoniaceae

Ribbon Kelp, Bladder Kelp
Nereocystis luetkeana

Length probably not over 150 ft. in spite of published reports of larger size. A single large globular float is developed at the end of the long flexible stem and at the base of the relatively few but long flat leaf blades. Stem shows outer layer of colored cells around a storage area that surrounds a central pith.

Found on our Pacific Coast and is definitely associated with rocky shoals and reefs, found in water from 10 to 100 ft deep but at its best where there are strong currents or heavy wave action. Found from Aleutians to California.

Gamete-bearing cells are found in dense masses on certain parts of the plant. The freed gametes were formerly taken for zoospores or asexual spores, but apparently they produce small plants that yield the regular sperms and eggs from which the huge plant develops after the egg is fertilized.

Nereocystis beds yield potash. Of the dry weight of the plant, about 25 percent is potassium chloride, which at the outbreak of World War I was worth $40 a ton but which has since increased in value. The plants have some value as food for cattle, as well as being used for raw fertilizers.

The common names include bladder kelp. The scientific name refers to Nereis, the daughter of the sea god, Nereus. So closely are the plants associated with rocky reefs that fishermen and sailors using shallow-draft boats recognize them as important navigation guides. Among largest and most conspicuous of all algae. Sailors and fishermen call accumulations of this form kelp beds! Large masses break off and float in Alaskan waters. Candied dried plant sold by trade name Seatron. Plant also called sea otter's cabbage.

Ribbon Kelp

Giant or Vine Kelp
Macrocystis pyrifera

Length reported to 2,000 ft. making it probably the longest plant in the world. Blades bear thin, wrinkled, olive leaflets, with egg-sized bladders at point of attachment. Stem rarely over ¼ in. in diameter. Leaves 2 to 4 ft long, and 3 to 4 in. wide. Holdfast branched; about 1 bushel in size.

Grows in water 20 to 70 ft deep, forming dense beds where conditions are right. Restricted to our Pacific Coast from Alaska to California and on to Asia and Kamchatka. Bladder-supported leaves may float at surface.

Named from the relatively large floats. Yields new plants in essentially the same manner as was described for *Laminaria*. Some areas may be well-populated by fragmentation of the huge plants with the floats supporting the fragments where they may continue growth.

Yields the phycocolloid algin, alginic acid and sodium alginate, ammonium alginate, calcium alginate and chromium alginate, also much potash and iodine. Of the whole plant, some 0.28 percent is iodine. Algin was first separated in 1883, using sodium carbonate for digestion of the plant material.

Giant Kelp

Harvested commercially in California. Harvested by cutting the floating parts of kelp a few feet below water's surface. Algin produced from this plant is used as an emulsifying, stabilizing, and suspending agent in ice cream, syrups, puddings, salad dressings, toothpaste, lotions, film emulsions, and paints. Used as supplement in cattle feed to combat certain deficiency diseases, e.g., those in which iron or iodine is required by the animal. Reported to contain medicinally important quantities of iodine, iron, copper, calcium, phosphorus, sodium, potassium, magnesium, sulfur, manganese, aluminum, zinc, and chlorine.

Order Fucales./Family Fucaceae

Rockweed
Ascophyllum nodosum

Length to nearly 10 ft but usually much smaller. Erect, from a disk-shaped holdfast. Main stem and main branches flattened, with egg-shaped, air-filled bladders to over 1 in. long and ⅔ as wide, with usually but one to a branch. With flattened, club-shaped branches to 1 in. long, solitary or grouped. Individual plants live more than one season.

Commonly found growing with *Fucus* on rocky coasts or on rocks emerging from muddy bottoms below the tide levels. Also on sheltered coasts at about mean sea level; associated with barnacles. Ranges from New Jersey to Newfoundland and Baffin Island, Labrador, and Cumberland Sound. Also in Bermuda.

In reproduction, the branchlets change into, or are replaced by, the sexual reproductive tissues that are either male or female. They are yellowish, drop from the main plant in spring following the winter's fruiting, and are replaced by new branchlets. The eggs and sperms are different in size and four eggs are in each egg sac.

Rockweed

(Continued)

In quiet water, these plants are more slender than in rough water. They occur in salt marshes and in brackish water entangled with other plants or poorly attached to shells and other such objects. The plants appear to be made of cartilage. The broken or shed fruiting braches often are found in great numbers.

Of little or no economic importance, though it is possible that they have some small value as a fertilizer. The plant turns black on drying, has larger air chambers than *Fucus,* and is in general tougher than *Fucus,* as might be indicated by its larger size. Well adapted to laboratory study of brown alga type specimen in inland institutions because it is frequently available from local fish markets. In parts of Europe, this alga is used as stock feed which is probably as good as hay, oats, and potato tops.

Bladder Wrack, Rockweed
Fucus vesiculosus
Length of sprays to 2 ft with distinct midrib through all the branches, rather regularly forked, flattened, olive-green, with swollen bladderlike structures filled with gelatin. Fronds tough, flexible, leathery. Holdfasts branching, rootlike, leathery, with suckerlike ends. Plant coarse, ribbon-like, covered with mucilage. Specific name means "full of bladders."

One of the world's most widely distributed groups, being found in the arctic and temperate oceans of the world, but this species not found abundantly south of New York on our Atlantic Coast though it appears from North Carolina to Newfoundland and Baffin Bay. Intertidal seaweed. Occurs frequently in great masses on stones and rocks; more frequently in quiet waters.

Apparently there are no asexual spores produced. Sexual reproduction results from union of sperms and eggs produced on different plants from autumn to spring in the southern part of the range. Male receptacles are reddish when opened while the females are olive-green, but both are near branching tips.

Yields the phycocolloid fucoidin, a polysaccharide first extracted by Kylin from *Fucus* and *Laminaria* in 1913. Like laminarin, it as yet has no well-established commercial value. It probably occurs in nature as a calcium salt and, on hydrolysis, gives rise to fucose. Because of mucilage, plants do not dry quickly.

Plants are collected by seashore farmers and spread on the ground as fertilizer. In Europe used to manufacture an algal meal used as stock feed. Have some value as a source of iodine and at one time were an important source of commercial potash. A medicine used in treating obesity has been taken from *Fucus,* and it has been used to some extent in making cosmetics.

Order Fucales./Family Sargassaceae

Gulfweed
Sargassum filipendula
Length to nearly 1 yd. Erect, with main stems smooth, sparingly forked, and bearing alternate stalked leaves that are to 3 to 4 in. long and ⅓ in. wide, thin and sometimes forked but usually with saw-toothed edges. Stems also bear spurs ending in bladders that are to ⅕ in. in diameter. Holdfast conical, spreading, lobed.

This species grows attached to shells and stones in relatively quiet water from low-tide level to 100 ft deep. There are some 150 species of *Sargassum* in the world, mostly about Australia and Japan, with 17 in Pacific United States. This species, Florida to Nantucket.

Probably perennial, bearing fruit at maturity that arise from union of different-sized sex cells with but one egg to an egg cell. Fruiting bodies are cylindrical and more or less forked. *S. fluitans* and *S. natans* of our Atlantic seas reproduce only by fragmentation, no fruiting bodies being known.

Some believe name comes from Spanish *sargozo* meaning seaweed, others believe from Portuguese *sargaco*. The Atlantic Sargasso Sea, long famous as a dead spot in days of sailing, which lies between lat. 25 to 31°N and long. 40 to 70°W, is not composed primarily of sargassum. Where Gulfweed accumulates, masses originate through movement of the Gulf Stream current. One authority does not agree that a Sargasso Sea exists. Yields the phycocolloid algin, alginic acid, and alginates of sodium, ammonia, calcium, and chromium.

Japanese fishermen eat the tender tips of the floating plants of their region. Because of relatively small size of this species, it is not likely to be of great commercial value compared with other algin producers such as *Laminaria* and *Macrocystis* and similar plants. An ingredient of raw algae, "limu," eaten on the islands of Hawaii. In Japan one species of *Sargassum* is used for New Year's decoration—like English use of holly.

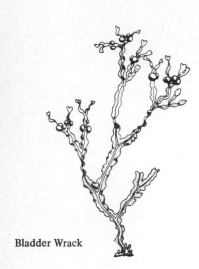

Bladder Wrack

Gulfweed

DIVISION CHRYSOPHYCOPHYTA [Yellow-Green Algae, Golden-Brown Algae, Diatoms]

A variety of forms are combined to make up this division. The members have yellow or brown pigments which predominate; outer shells of silicon. May have metallic luster because of accumulation of oils. Include motile or immotile unicellular species, and colonial forms; some are filamentous. Common in fresh water and marine situations as well as in soil. Some are part of the plankton of ponds and lakes, some coat stems and leaves of water plants and cover stones in streams and make them slippery. Diatoms have accumulated over geologic time as diatomaceous earth. About 9,200 species.

CLASS XYANTHOPHYCEAE

Order Heterosiphonales./Family Botrydiaceae
Botrydium sp.
A terrestrial alga that grows chiefly on drying muddy banks of streams and ponds; also on bare, damp soil especially along borders of damp paths. When found at all, usually abundant enough to cover substrate. Also occurs on damp soil under greenhouse benches. Appears as tiny, green, balloonlike or pear-shaped growths. Produces underground root-like (rhizoidal) branches. Visible with unaided eye at times; easily seen with 10X hand lens. Occurs in warm months generally. After a rain may cover the soil of a whole cultivated field in plains or prairies; disappears when soil dries.

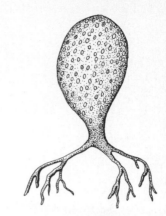

Botrydium

Order Heterosiphonales./Family Vaucheriaceae

Vaucheria, Moss Felt
Vaucheria sp.
Dense or weak mats of light yellow-green, slimy, branched filaments, without cross walls, attached or floating. Reproduce by breaking, by swimming spores, or by large eggs fertilized by sperms from bent male sex organ. Of seven known groups of the genus, four are found in United States with specific differences based largely on nature of male sex organ. Swimming spores may be easily induced to form by bringing land forms into water, aquatic forms into darkness, or running-water forms into quiet water. On damp soil; on rocks, in flowing water; occasionally in mats floating on ponds. Filaments can be seen with unaided eye. Common in greenhouses. May be collected at almost any time of the year in temperate zones. A collection of *Vaucheria* may contain thousands of associated small plants and animals.

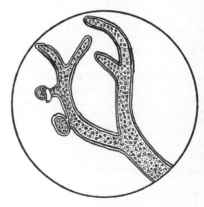

Vaucheria

CLASS BACILLARIOPHYCEAE [the Diatoms]

Order Pennales./Family Diatomaceae

Diatoms
Commonly appear as jellylike, slippery strings or masses clinging to underwater supports in fresh water. Essentially, masses of glass boxes enclosing plants reproduce by dividing and growing new valves to old. May remain dormant for years (48 years recorded). Forms with bilaterally symmetric valves, in order Pennales; with circular or irregular valves, in order Centrales. Excellent silt anchors, source of food for fish, and basis of diatomaceous earth. In some classifications considered a subclass of the Diatomaceae. Organisms of great biological and economic importance. Occur both as fresh-water and marine planktonic and benthic forms. May encrust aquatic vegetation. Single-celled, colonial, or in strands. Walls contain silicon dioxide. Marine forms essentially cold-water forms. Most prevalent in spring and fall months. If abundant, give water a fishy odor.

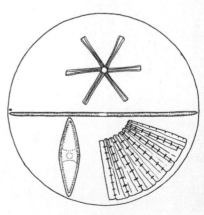

Diatoms

DIVISION PYRROPHYCOPHYTA [Dinoflagellates]

Almost all forms are solitary; most swim by movement of two flagella. In good microscope mounts, flagella can be seen; one trailing out behind cell in motion. Cell divided by furrow from which flagella appear. Cell wall, when present, made up of plates which are firm—appears armored. Contains pigment, chlorophyll, plus other pigments. Pigmented eyespot present. Some protozo-ologists claim the dinoflagellates as Protozoa. Abundant in fresh and salt water. Frequently an important part of plankton. One form produces "red tide" in the Gulf of Mexico; occurs in such great numbers at times as to cause destruction of fish. Some marine forms luminescent. About 100 genera and 900 species reported.

CLASS DINOPHYCEAE

Order Dinokontae./Family Ceratiaceae

Ceratium
Ceratium sp.
Has characteristics of other dinoflagellates plus what appear to be long horns as extensions of the cell both forward and backward. Microscopic. Cell wall heavy. Fresh-water species with plates. Conspicuous girdle (groove) around midsection of cell. Produces aplanospores which resist drying and remain viable for many years. Reproduces sexually after a conjugation tube is formed between two individuals. Largely marine forms. Only two species in fresh water. Occurs mixed with other algae or in open waters of lakes. In lakes may produce a bloom in great numbers which changes color of water to coffee-brown.

DIVISION RHODOPHYCOPHYTA [Red Algae]

Mostly marine seaweeds; few species in fresh water or on land. Often at greater depths than brown algae. Most multicellular, attached to substratum; few species unicellular. Important source of colloids used commercially; agar used in culturing bacteria, suspending agents used in chocolate milk and puddings; stabilizers in some ice creams, some cheeses, some salad dressings. Added to icings, creams, marshmallows as moisture retainer. Contain chlorophyll a; and some forms possess chlorophyll d, which is not found in any other group of plants. Also contain phycocyanins as well as phycoerythrins; the latter give many of the plants their red color. Not all red algae are red; for instance, some are black. About 400 genera and 4,000 species are known. Very abundant, sometimes at great depths in tropical seas; some precipitate calcium carbonate and so are important in reef formation.

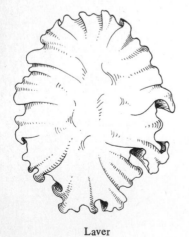

Laver

CLASS RHODOPHYCEAE

Order Bangiales./Family Bangiaceae

Laver
Porphyra umbilicalis
When young, narrow; but when mature, like a broad, somewhat elastic membrane, fragile, often only one cell in thickness, purple to olive to brownish-purplè but always with a smooth, satiny sheen. Attached by a small obscure holdfast at the center. Edges frilled or wavy in appearance. Whole plant appears slippery. Described as ruffled.

On rocks or piling from the low-tide to mid-tide lines. This species from New Jersey to Hudson Bay and Newfoundland; closely related forms are in both hemispheres from Polar seas to Mediterranean, Australia, Tasmania, Cape Horn, and Cape of Good Hope. Common on intertidal rocks along the whole Pacific Coast.

Reproduction: asexually, by means of spores freed to swim from rather large continuous areas of the blade; sexually, by spores freed from plants that may be either male or female or both; in the latter case, male and female portions are generally separated, with the males coming from a marginal band and the females from over the surface.

An annual plant that may be found at almost any time of the year. The name is derived from a word meaning purple dye. The pressed plant does not adhere to paper when dry. Used in England as food as "laver"; in Ireland as "sloke"; in Hawaii, as "limu"; and in Japan as "amanori" or "amori." Boiled and eaten with lemon juice, spice, or butter. In South Wales fishwives sell "laver bread." It has been extensively harvested in California.

DIVISION RHODOPHYCOPHYTA [Red Algae].
CLASS RHODOPHYCEAE

Order Gelidiales./Family Gelidiaceae

Tengusa
Gelidium corneum
Frond flat and horny, to 4 in. high, narrow, erect, branching, in one plane being 2 to 3 times pinnate and with sections composed of somewhat compressed cylinders. Very variable in shape, size, and color, being red to reddish-brown to purplish-red and when exposed becoming a dirty white. Tip of branchlets club-shaped.

The Atlantic Coast species is *G. crinale* that reaches a height of 3 in., while the West Coast *G. corneum* is 1 in. higher. Grows in tufts on mud-covered rocks and on other algae. Most abundant on West Coast from California to Lower California, Mexico. Inhabits rocky intertidal and infratidal areas of Pacific Coast.

Asexual spore cases are found in the surface of rather special areas or on special branches. Sexual plants are like those bearing asexual spores in general, with the male elements produced from the surface of considerable areas and the female borne on inner tissues and spread through the general nutritive tissue.

Yields agar (gelose), a phycocolloid, agarinic acid, and the salts sodium agarinate, potassium agarinate, calcium agarinate, and magnesium agarinate. In Japan the species of most importance has been *G. amansii.* Gelidium agar in 1 percent water solution sets at 35 to 50°C to form firm gel and melts at 80 to 100°C. Harvested by divers in southern California for making agar, especially during World War II, when demand for agar was great and Japanese supply was cut off.

Probably the most important of all the agarophytes as source of agar necessary for bacteriological work, as medicine to relieve constipation, as food in form of jellies, soups, sauces, and desserts. Gives rigidity to soft canned fish, improving marketability, after shipping. Known as "Oriental isinglass." Most important of the edible seaweeds in Japan is a Tengusa.

Tengusa

Order Cryptonemiales./Family Corallimaceae

Common Coralline
Corallina officinalis
Like slender, rigid, jointed, bushy, lime-encrusted, easily broken coral, to about 3 in. in height. The joints are cylindrical but flattened above and wedge-shaped. Color varies from reddish-purple to gray-green to almost chalky white. There is some flexibility at the joints. Branches are at least three times branched. Genus name comes from plant's resemblance to a coral animal; appears like a fountain in growth.

Common on Atlantic Coast, particularly from New York north to Newfoundland, growing mostly on rocks in fairly deep water up to area between tides. Also found in tidal pools. Likely to be more matted in rougher waters. A large, robust Pacific Coast form, var. *chilensis* occurs along the whole Pacific Coast.

In members of this family the reproductive organs are sunken in the crust of the plant or on enlarged terminal branches; in this species they are hornless, except in those bearing the male sex organs. There is a small pore at the end of the sex-bearing regions.

Found throughout the year, but so variable that many have suggested different species names for the forms that different environments have developed. These plants are not mucilage-covered and therefore do not adhere to mounting paper when made into herbarium specimens.

Probably of little direct economic importance to man. It is believed that the corallines growing frequently on coral reefs add to the limes of such reefs and so help them grow.

Common Coralline

Order Gigartinales./Family Gracilariaceae

Ceylon Moss
Gracilaria confervoides
Frond bushy, arising from a disklike holdfast that frequently breaks, freeing the whole plant. Branches to 0.1 in. in diameter, dividing repeatedly and variously but not so conspicuously evenly as in some close relatives. Firm, fleshy, and tapering to fine tips at the free ends. Dull purple-red to purple to green.

This species in shallow, warm, quiet bays attached on loose materials near bottom, from the tropics and Florida to Prince Edward Island. *G. multipartita* or *G. foliifera* is much branched and found on both coasts. Other relatives in Africa, Australia, Britain.

Asexual spores borne in structures scattered over the branchlets, these structures being oval in shape and developed definitely from surface cells. Male element also produced from surface cells. Female element borne on structures on special branches but after fertilization these develop into conspicuous numerous hemispheres.

(Continued)

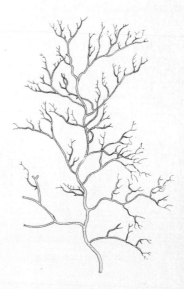

Ceylon Moss

DIVISION RHODOPHYCOPHYTA [Red Algae].
CLASS RHODOPHYCEAE

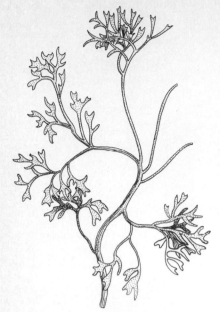

Phyllophora

In the New England area, the plant is crisply bushy. Yields gracilaria agar, agarinic acid, and the salts sodium agarinate, potassium agarinate, calcium agarinate, and magnesium agarinate. *Gracilaria* was the original source of commercial agar, whose uses have been discussed elsewhere in this description.

Gracilaria is the chief commercial source of agar produced in the vicinity of Beaufort, N.C., and no doubt figures in commercial activities elsewhere. It is used in bacteriological work, in dental impressions, in health foods, and in baking. Exploited as a commercial source of agar although the product is inferior to that obtained from species of *Gelidium*.

Order Gigartinales./ Family Phyllophoraceae

Phyllophora
Phyllophora membranifolia
Fronds to 20 in. long but usually shorter, with short, stubby, wedge-shaped, forked, or cleft branches each under 1 in. long. Several fronds, sometimes as many as 20, arise from a single disk-shaped holdfast. Like a firm membrane that is dull purple or clear red but with some brighter areas.

Found commonly growing on stones in shallow tidal pools or in water several feet deep but mostly in warmer waters; usually well below tidal range. This species ranges from New Jersey to New Brunswick, Prince Edward Island, and Cumberland Sound.

Asexual spores produced in rather large, slightly raised spots near the bases of the blades with surrounding cells obviously radiating. Male element produced on small, brightly colored blades and attached to the edges of these blades. Female element borne in stalked structures usually attached to stem. Exceedingly variable in form.

A perennial plant. Found throughout the year but commonly fruiting in autumn and winter. Yields an agaroid that is possibly a kind of agar though it apparently has not been studied exhaustively or put into much commercial use in the United States. At least three species represented on our Atlantic Coast. Three species occur on the California coast, and are scarce, deep-water forms.

From Odessa and the Black Sea, it is reported that a species close to ours yields a kind of agar, but as to any further commercial use we have not been informed.

Order Gigartinales./ Family Gigartinaceae

Irish Moss, Carrageen
Chondrus crispus
Fronds to about 6 in. high, arising in clumps from a disklike holdfast, with flattened stems that divide and subdivide to form rather broad fans, there being 5 to 6 subdivisions and the divisions taking place when the stems are about ⅛ to ½ in. wide. Stems thick, tough, leathery, olive to dark purple to jet black. Called Irish moss because it is a common red alga on the coast of Ireland.

Most abundant on our Atlantic Coast from Maine to Carolina but also common on to Newfoundland, growing throughout the year on wood, shells, and rocks from mid-tide line to considerable depth. Does well in highly varied environments. Gathered at low tide from rock ledges; brought to shore, washed, spread out to dry before marketing.

Asexual spores produced in obvious cases formed below the surface of the tissue. Male elements borne in conspicuous cases on the surface of young branches with the female elements borne on other branches from the inner surface, with the fruiting bodies to 0.1 in. across and swollen laterally.

Fruiting occurs during the summer with the asexual activity more common in late autumn. Color varies from pale in shallow pools to purple in deep pools to iridescent in the sun. Named from Greek word for cartilage. Extracted carrageen in 3 percent water solution gels at 27 to 30°C; and in 5 percent solution, at 40 to 41°C. Ash, 20 percent.

Of the water-free matter, 65 percent is gelatin. Used in making gloss soaps, jellies, salad dressings, toothpaste, lotions, and syrups, puddings, cosmetics, shaving soaps, shoe polish, blanc mange pudding; for curing leather and sizing cloth. Industry centers at Scituate, Mass. Yields phycocolloid, carrageenin. Irish moss has been collected for more than 125 years in some regions of the Massachusetts shore. Japanese use the gelatin of this seaweed for shampooing the hair.

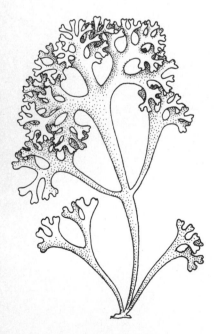

Irish Moss, Carrageen

DIVISION RHODOPHYCOPHYTA [Red Algae].
CLASS RHODOPHYCEAE

Batters
Gigartina stellata

Fronds to 6 in. high and 2 in. broad, in thin, flat, gelatinous, leathery, wavy, wedge-shaped divisions that fork at the base only to subdivide again in the same plane at a higher level. Unlike Irish moss, has tufted protuberances on the concave side of the fronds. Dark purple to black and rigid when dried. Species *harveyana* is common on Pacific Coast; occurs off Oregon and California. Genus common on Pacific Coast at low levels on almost any exposed, rocky shore. Reddish blades and characteristic wartlike growths on blade surfaces provide easy means of identification. Some species of the genus take on much different form.

Common in the tropics and from Massachusetts north to Nova Scotia and Newfoundland, growing with Irish moss in shallow tide pools, on exposed rocky shores but always near low-water mark. Found at any time of the year.

Perennial. Asexual spores borne from immersed, irregular cases of the inner surface. Male elements appear from cases in terminal, closely branched stems and from surface cells. Female elements borne in special crowded clumps or branches, and from areas under the surface. When fertilized, develop into fleshy units.

Fruiting takes place in winter. Dried plants do not adhere to mounting paper. Yields the phycocolloid carrageenin and carrageenic acid and the salts potassium carrageenate and calcium carrageenate; together with Irish moss, is the principal source of these materials. Known as the polysaccharide, carrageenic extract.

Carrageenin from *Gigartina* and *Chondrus* has a higher ash content (20 percent) than agar and requires 3 percent rather than 1 percent concentration in water to gel. It gels at a lower temperature than agar. In New Zealand, may be called Irish moss.

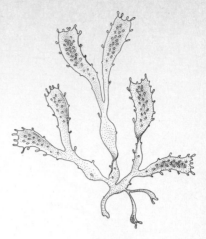

Batters

Order Rhodymeniales./Family Rhodymeniaceae

Dulse, Sea Kale
Rhodymenia palmata

Fronds to nearly 2 ft long, with an inconspicuous stalk arising from a small disklike holdfast and bearing blades that break into two or more divisions with the forks to 6 in. wide but usually narrower. Like leathery membranes that are purple or red in color. Frequently shows numerous smaller fronds near base.

Found from the mid-tide zone to deep water but in the southern part of the range restricted to the deeper water. This species ranges from New Jersey to Baffin Bay and Ellesmere Island and in Europe is found south to the Mediterranean. Worldwide genus; almost all red in color and found between low and high tide lines. Genus much more widely distributed along Pacific Coast of the United States. One species, *pacifica*, very common along coast of California and southern Oregon. Same species occurs in Washington in the Puget Sound region.

Asexual spores are produced in scattered cases that appear as darker spots on the blade and are also indicated by a thickening of the outer layer. These structures appear in the winter season. In this species, no true sexual reproduction seems to have been discovered. Related species have sexual reproduction that results in a swollen fruit with a loose case.

When dried, this plant has a faint odor of violets. It is eaten in widely separated parts of the world. Icelanders dry it, make a flour of it, eat it raw, or boil it in milk. Scots use it for chewing tobacco, as a relish, for medicine, or boiled in milk as food. Irish know it as "dillesk," boil it in milk, chew it, and eat it with fish. On Isle of Wight used to induce perspiration through fevers.

In time of famine in Ireland, dulse and potatoes formed a diet that saved this important part of the world's population. In Norway, it is known as sheep's-weed because sheep are extremely fond of it and seek it at low tide when it is exposed.

Dulse

Order Ceramiales./Family Ceramiaceae

Red Ceramium
Ceramium rubrum

Fronds like large, leafless, bushy structures rising to over 1 ft in height, forking and reforking with the ends finally clawlike and incurving, with a "bark" of small cells forming rings at joints and making obscure bands, obviously lime-encrusted and the whole bush arising from a basal disk. Deep red, with some variation. Caliperlike tips are used in identification of genus.

Common in almost all seas of the globe, growing on rocks, on plants, on wood and on shells from between the tidal levels to moderately deep water. Found throughout the year. This species is found from the tropics and Florida to Newfoundland. Species *pacificum* common from southern California to Vancouver Island.

(Continued)

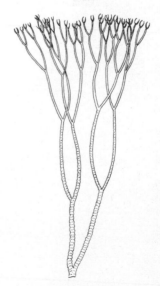

Red Ceramium

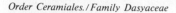

Chenille Weed

Asexual spores are produced from 1 to 3 rows of cells around the joints or scattered and indicated by a slight swelling. Male elements produced from crowded tufts on the surfaces of the younger, more slender branchlets. Female elements borne on lateral branchlets that become protected by 3 to 5 incurving branches.

Fruiting on this species takes place in the summer months. Many varieties or even species have been suggested but these are joined by such closely integrated series that it is doubtful if they are valid. Differences recognized include variation in branching, in region where fruits are borne, in stoutness, and even in color.

It is doubtful if this alga is of any great economic importance to man. It is believed that because of its incrustation of lime it may be of assistance to corals in the development of coral reefs.

Order Ceramiales. / Family Dasyaceae

Chenille Weed
Dasya pedicellata
Fronds to 10 ft long, with the cylindrical stem covered with fine hairs that give the whole plant a velvety appearance. Stems long and sinuous rather than bushy, with branches sparingly divided. Color brilliant red or pink but when taken out of water, appears like strings of purple jelly in masses.

Found just below the low-tide mark, usually in protected waters where the tide flows. This species ranges from Florida to Nantucket and extends on south into the tropics. It is rather abundant in the New York Bay area. One species, *pacifica*, occurs locally on California coast. In any location, usually found 3 to 12 ft below low-water level. Probably 70 or more species in the genus.

Fruits from midsummer to autumn. Individual plants produce either asexual spores or the male element or the female element. Asexual spores are borne in distinctive curved structures. Male cells are slender, pointed, and usually filament-tipped. Egg-bearing units are usually stalked and 0.1 in. through.

The plant sticks to paper well when mounted as an herbarium specimen, but because of its delicate nature should be subjected only to light pressure. It is considered a good late autumn alga and in the Woods Hole area is obvious as late as October. Dasya means hairy and the name is appropriate for this plant.

Apparently there is no obvious economic value to be attached to this plant unless one recognizes its beauty; it certainly adds to what might otherwise be an uninteresting bit of environment. It is probably good food for some marine animals and provides them with some shelter from their enemies.

Order Ceramiales. / Family Rhodomelaceae

Polysiphonia
Polysiphonia fibrillosa
Fronds to 10 in. or slightly longer, appearing like pinkish or reddish slippery "worms" hanging attached to exposed rocks, shells, wood, or, in fact, as an epiphyte on other algae, at low tide or erect or weakly floating in shallow water. Profusely and irregularly branched, coarse at the base but fine at the tips. Light or dark brownish-red to black.

Over 100 species found on both our Atlantic and our Pacific coasts, most commonly seen on stones and vegetation in shallow water but may be found in deep water as well. Not an abundant alga and in some New England areas·is rare. New Jersey to New Brunswick.

Asexual reproduction, through spores borne on somewhat spirally distorted branchlets. Male elements borne in spindle-shaped structures at the ends of the branchlets. Female elements borne in short-stalked egg-shaped or spherical structures and forming a relatively large fruiting structure. Fruits in summer.

This is a summer annual that favors or rather does best in warm bays. The genus is represented by many species, of which approximately a dozen are found along our Atlantic Coast, in which the rather obvious differences are based on type of branching, stoutness, and types of fruiting areas.

Apparently little economic importance can be attached to this rather beautiful red alga. It merely helps make a tidal area look as we think such a region should. When mounted on paper, the plant appears to be surrounded by a peculiar "mist," owing probably to fine hairlike coverings not noticeable otherwise. Frequently used as an object of study by beginning students of red algae.

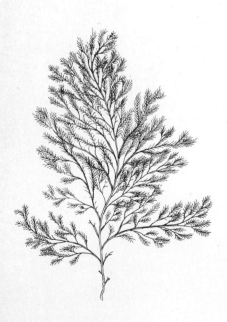

Polysiphona

DIVISION CHAROPHYTA

The stoneworts or brittleworts are sometimes also classified in the division Chlorophycophyta. Represented by two genera, *Nitella* and *Chara*. Only one class, Charophyceae; with one order, Charales; and one family, Characeae. Both genera grow in muddy or sandy bottoms of clear lakes and ponds. Also found in slow-flowing limestone streams. Some species have capacity to precipitate calcium carbonate from waters and they form encrustations over their surfaces. Found as fossils. Worldwide in distribution. Harsh to touch; so given name, stoneworts.

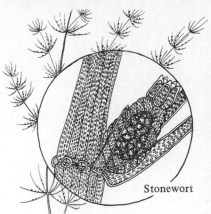

Stonewort

CLASS CHAROPHYCEAE

Order Charales./Family Characeae

Chara, Stonewort
Chara sp.
Branched, with branches of two kinds, one being somewhat needlelike and arranged in whorls and the other providing axes for the divisions. Brash, commonly lime-covered, with "fruits" borne in axils of shorter branches. Relatively common in springs and ponds with high lime content; widely distributed. *Nitella* is more slender and fragile than *Chara* and without cell-covered internodes. Presence in water apparently limits development of mosquito larvae. Classification controversial.

Ill-smelling (garlic or skunk ordor). Macroscopic. Worldwide in distribution. Marl and other kinds of calcareous deposits may be formed largely by *Chara* over long periods of time. May grow to 4 ft in length.

SUBKINGDOM MYCOTA

Plants with no stems, roots, or leaves, yet of great importance to man. Subkingdom contains the slime molds and the true fungi (the algal fungi, the sac fungi, and the club fungi). Fungi include the yeasts, rusts, smuts, mildews, molds, and mushrooms. In some classification systems the two groups, slime molds and true fungi, are separated so that the latter three are included in a single division, the Eumycota. The subkingdom Mycota is characterized by lack of chlorophyll and, therefore, heterotrophic nutrition. Members have a cellular organization like that of the subkingdom Chlorota. A major difference is, of course, that they lack plastids. Included in the subkingdom are five divisions: Myxomycota, Phycomycota, Ascomycota, Basidiomycota, and Deuteromycota.

DIVISION MYXOMYCOTA [Slime Molds]

Characterized especially by their vegetative phase, which is unique both as to morphology and as to nutrition. Vegetative phase, the plasmodium, occurs on plants or soil; frequently in woodlands; almost always in moist dark places. Sometimes considered "plantlike animals" or "animallike plants." Also have been considered protozoa by some classifiers and fungi by others. Probably some 400 species worldwide now known.

In plasmodium stage of life history, they may move like a simple animal, e.g., an amoeba. Eventually the plasmodium dries and may form fruiting bodies freeing spores like many other plants. Air currents disperse spores. Slime molds are common but because of their small size are not commonly recognized.

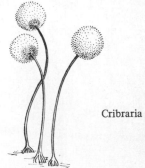

Cribraria

CLASS MYXOMYCETES

[Subclass Endosporeae]
Order Cribrariales./Family Cribrariaceae

Cribraria
Cribraria dictydioides
Fruiting stage appears to be much like a bunch of light-colored spheres on the ends of tapering, slender, dull brown stems that may be erect or bent. Abundant on rotten logs or planks, particularly of oak or pine, from Pennsylvania to Nebraska. *Cribraria* is a widely distributed genus throughout the U.S. Species *intricata*, variety *dictydioides*, of some authors; 1.5 to 3.0 mm total height. Plasmodium lead-colored or brownish-black. Spore-forming bodies gregarious, often form large colonies, ocher-brown to yellowish-brown.

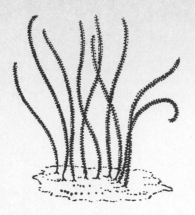

Stemonitis

Order Amaurochaetales./Family Stemonitaceae

Stemonitis
Stemonitis webberi
Fruiting stage appears to be like a cluster of rusty brown feathers to ⅖ in. long, naked at base and with midrib shining jet black and expanded at base; spores pale, reddish-purple. May be found on rotting wood, particularly abundantly in the Mississippi Valley and to the west. Species *splendens*, variety *webberi*, of some authors. Plasmodium white, rarely pale yellow. Genus widely distributed.

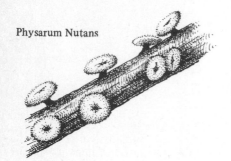

Physarum Nutans

Order Amaurochaetales./Family Physaraceae

Physarum
Physarum nutans
Plasmodium form bright yellow or yellowish-gray, sometimes greenish, to 1.5 mm in height. Spore-bearing bodies form in a mass, are erect, stalked, white, grayish-white, or iridescent. Spore stalk awl-shaped, yellow, olive, or black. Spores brownish-violet. Found on rotten wood in damp woods. Common throughout U.S. More than 60 species in this genus.

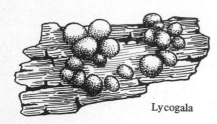

Lycogala

Order Cribrariales./Family Lycogalaceae

Lycogala
Lycogala epidendrum
Fruiting stage appears to be irregular olive to black, somewhat warted, depressed spheres to ⅖ in. through, breaking irregularly usually near the top leaving thin, tough case. Mass of spores pinkish-gray or pink; becoming ocher color with age. Commonest slime mold in world, found everywhere on decaying woods. Plasmodium coral-red, rarely white, cream-colored, or yellow. Genus widely distributed throughout U.S.

Hermitrichia Sp.

Order Cribrariales./Family Trichiaceae

Hemitrichia
Hemitrichia clavata
Plasmodium watery white or rose-red. Height to 3 mm. Spore-bearing body, erect, not solitary or clustered, develops close mass; to 1.5 mm high, shiny ocher or olive-yellow in color. Stalk of spore body cylindrical; olive or reddish-brown or nearly black. Found on dead wood in damp places; common and abundant throughout U.S. Spore-bearing bodies vary in shape from season to season.

DIVISION PHYCOMYCOTA [The Algal Fungi]

The name of the division is due to the fact that some mycologists used to think the members were derived from algal progenitors which had lost their chlorophyll. As a result of this loss of chlorophyll, they were forced to assume a parasitic or saprophytic mode of life, it was argued. Most of the forms have nonseptate (without cross walls) mycelium. This, too, was used as an argument that the Phycomycota are like some algal forms. Other mycologists believe Phycomycota and certain algae developed independently. The Division contains such groups as chytrids and Blastocladiales which are water- or soil-inhabiting forms; the water forms are frequently parasitic on algae and other water molds and some vertebrates and the soil forms are parasitic on vascular plants; also contains the white rusts and downy mildews; bread molds, fly fungi, and animal traps—fungi and commensal forms from the alimentary canal of certain insects. Some are marine forms. Many are important decay agents which help in recycling elements.

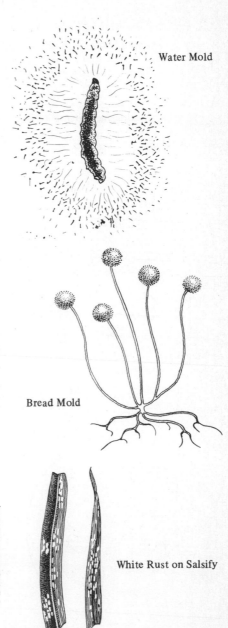

Water Mold

Bread Mold

White Rust on Salsify

CLASS OOMYCETES

Order Saprolegniales./Family Saprolegniaceae

Water Mold
Saprolegnia sp.
Appears most commonly as a gray fuzz on dead animals in water though frequently may appear on such fish as goldfish in a crowded aquarium. Causes diseases in fish and fish eggs and may do significant damage in fish hatcheries. Members of the order are almost wholly aquatic, living mostly on dead animal matter or sometimes on plant matter. There are some 100 species in 16 genera. Reproduce by division, by swimming spores, and by an apparent or real sexual reproduction according to different authorities. Controlled by salt bath. In hatcheries formalin is used to control fungus. Many species can be easily isolated and cultivated in the laboratory. Collect a jar of pond water; add a few dead flies, split boiled wheat or corn grains, and wait for gray-white growth to appear on the things added. *Saprolegnia* invades the skin of fish, consumes their scales, skin, and flesh, and finally kills the fish. The species *parasitica* parasitizes newts, frogs, and eels as well as many fish. Probably most effective parasite when host is injured or in a closed, crowded condition.

CLASS PHYCOMYCETES

Order Mucorales./Family Mucoraceae

Bread Mold, Black Bread Mold
Rhizopus sp.
Appears commonly as a gray, webby mold on bread, spotted with black dots. Most abundant on decaying organic matter in humid environment where matter is rich in starches and sugars, but some species are parasitic on other fungi. Some 150 species have been described in the order. Reproduces by division, by freeing of air-borne spores from black-dot sporangia, or by regular sexual reproduction resulting from joining of elements. Serves as a decay hastener. Some species of *Rhizopus* used in manufacture of fumaric acid and for some steps in the manufacture of cortisone. One form produces alcohol; others lactic acid. Some forms are parasitic on strawberries. Life cycle frequently used as illustrative of molds in beginning biology courses.

Order Peronosporales./Family Albuginaceae

White Rust on Salsify
Albugo sp.
Members of genus common as white blister patches on many kinds of plants, particularly the mustards. There are at least 300 species in the genus, all being parasitic on higher plants, nonaquatic, with air-borne reproductive spores and a complex sexual reproduction of different-sized reproductive cells. Important relatives include *Phytophthora infestans*, which causes late blight of potato, and *Pythium* sp., which causes common "damping off" disease of many garden plant seedlings. A chemical fungicide, Hydromercurichlorophenol, is suggested as a seed treatment to combat infections by fungi. It is a fungicide registered with USDA and should be used only while carrying out the precautions proposed by the USDA. Precautions suggested will aid in combating unnecessary environmental pollution.

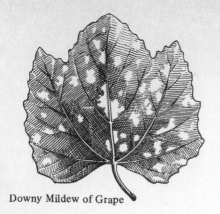

Downy Mildew of Grape

DIVISION PHYCOMYCOTA. CLASS PHYCOMYCETES

Order Peronosporales./Family Peronosporaceae

Downy Mildew of Grape
Plasmopora viticola
Appears as downy spot on any green part of grape, changing from pale green to yellow to brown with definite border, or as brown rot of fruits, in which case fruit wrinkles but does not become dry and hard as with black rot caused by *Guignardia*, which may cause 50 percent crop loss. Effective control is by early spraying with bordeaux mixture or ammoniacal copper carbonate or by similar spray after harvest to protect vines. Known also as "brown" or "gray" rot. Several chemical fungicides have been reported effective by their manufacturers and they are registered with the USDA: Captan or CuO applied as a dust; $CuSO_4$, basic as a spray; and Ferbam and Dithane applied in either dust or spray form. Recommendations of producers should be followed to avoid pollution of the environment with unwanted chemicals.

DIVISION ASCOMYCOTA

A very large assemblage of fungi including the yeasts, cup fungi, morels, and truffles. Some forms are skin parasites on man. Also included are numerous plant pathogens such as the organisms which cause Dutch elm disease, chestnut blight, powdery mildews, smuts, and leaf curl in several plants.

Known by one distinguishing feature, the ascus, or saclike structure containing ascospores. Some of the spore-producing fruiting bodies are large; e.g., cup fungi, morels, saddle fungi, and truffles.

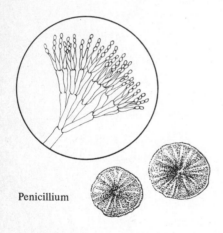

Penicillium

CLASS ASCOMYCETES

Order Aspergilliales./Family Aspergilliaceae

Penicillium
Penicillium sp.
Common and cosmopolitan; called green molds and blue molds, common on citrus fruits, jellies, preserves. Spores travel in the air. Some species attack and destroy fruits; e.g., apples in storage. Destroy leather and natural fibers in fabrics. Sometimes destroy silage in silos. Produce organic acids, e.g., citric, fumaric, gallic, gluconic, and oxalic. Species *roqueforti* and *camemberti* form cheeses carrying their names. Species *notatum* and *chrysogenum* sources of wonder drug, penicillin.

In life cycle, mycelium produces erect branches with broomlike extensions that form conidia which are black, blue or green and give the mass of the mold its characteristic color. Spores are produced and liberated. In some species sexual reproduction takes place also.

Order Saccharomycetales./Family Schizosaccharomycetaceae

Yeast
Saccharomyces cerevisiae
Microscopic, unicellular; cell wall with chitin; possesses a minute nucleus. In presence of sugars and relatively high temperatures, yeasts multiply rapidly. Yeasts divide by fission process and by budding. Chemical reaction forms carbon dioxide bubbles in bread dough and alcohol of alcoholic beverages such as beer and wines. In either case, process is one of fermentation of sugar. In U.S. source of sugar most frequently used for industrial alcohol is blackstrap molasses. Where malted grain drinks are preferred, corn, rye, wheat, or barley may be the source of sugars. May become a good source of protein as food supplement. Yeast proteins contain many essential amino acids. Source of B vitamins. Sold as powder and in rectangular solids together with starch as inert matter. Yeasts are also considered destructive in certain food industries such as the soft cheeses to which they impart a "yeasty" flavor.

Individual yeast cells appear almost colorless; yet colonies may be white, cream-colored, or tinged with brownish pigments.

Yeast

DIVISION ASCOMYCOTA. CLASS ASCOMYCETES

Order Helotiales./Family Phacidiaceae

Black Spot of Maple
Rhytisma acerinum
Appears as thick, irregular black spots that shine like tar, usually to ⅔ in. across, on leaves in late summer causing leaves to fall early and definitely weakening the tree. Ascocarps overwinter in fallen dead maple leaves. Ascospores form tubes which invade the maple leaf through stomata (openings in lower epidermis of the leaf). Red maple and silver maple most commonly attacked, though Norway maple and sycamore maples also suffer in Europe. In spring, black spots in fallen leaves develop spores that are shot into air; so control is essentially that of burning infected leaves, though bordeaux spray helps infected nursery stock. Several fungicides reported effective against black spot of maple have been registered with the USDA. They are Copper-Bordeaux mixture applied as a spray as the leaves unfold in the spring, Phenylmercuric Ammonium Acetate (PMA) applied as a spray, and $CuSO_4$, basic as a spray or dust. All fungicides should be employed only in the quantity suggested by their manufacturers to avoid environmental pollution.

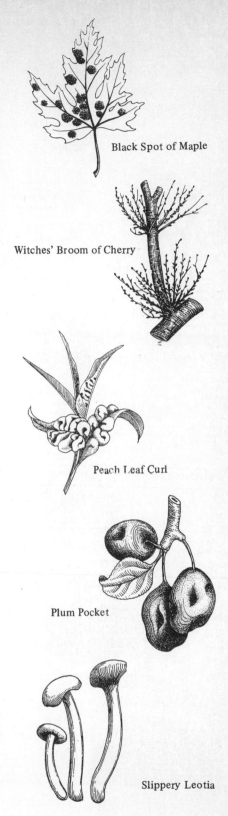

Black Spot of Maple

Order Taphrinales./Family Taphrinaceae

Witches' Broom of Cherry
Taphrina cerasi
Appears as broomlike growths on English sweet cherry, sour cherry, wild black cherry, wild red cherry, choke cherry, and several kinds of plums. Fungus lives in twigs from year to year, attacking both bark and wood, stimulating bud growth, and causing deformity. After infected leaves fall, a whitish surface may develop on underside that may help spread disease ruining branches for fruit bearing. Some witches' brooms are caused by fungi and mites.

Witches' Broom of Cherry

Peach Leaf Curl
Taphrina deformans
Leaves puff up and curl, with blade becoming thickened; leaves turn from green through yellow to red and finally to silver and may drop, stripping the tree. Found on peach, causing peach blister and appearing first in spring when leaves begin to puff. Annual infestation may cause death of tree and in some sections may seriously interfere with peach crop. Bordeaux spray or copper sulfate or lime sulfur is used, the last when insects are troublesome. Disease occurs wherever peaches are grown; still does considerable damage but now is better controlled than in the past. Many fungicides, registered with the USDA, have been reported as effective against peach leaf curl by their manufacturers. They are: sprays using lime sulfur or calcium polysulfides, copper carbonate, sodium polysulfide, and copper sulfate. Copper oxide applied in the dormant period, PMA, also Ferbam and Copper Tetra Copper Calcium Oxychloride applied as a spray or dust. Any fungicide should be applied as directed by its manufacturer and with a keen awareness of its potential for environmental pollution.

Peach Leaf Curl

Plum Pocket
Taphrina pruni
Appears after fruits fall; infected fruits become pale yellow, then covered with powder, then pockets turn black, and the fruit falls. Fruit is ruined, and pit becomes an empty thin shell. Infected trees may bear pockets year after year without nearby trees being affected even though every fruit on the diseased tree is affected badly. Treatment, which may not be effective, consists of cutting and burning infected parts or using lime-sulfur spray. Copper Bordeaux Mixture Spray used in the redbud or popcorn stage is recommended as treatment by some manufacturers. Winters in twigs.

Plum Pocket

Order Pezizales./Family Geoglossaceae

Slippery Leotia
Leotia lubrica
Found on ground on decaying wood in forest usually growing in clusters. Fruiting bodies are to over 2 in. tall, like slender tenpin, with expanded irregular sphere at top, hollow when old, slippery, and olive-yellow. Stem not grooved as in the helvellas, more or less uniform in diameter throughout length, and either straight or curved. While the plant is edible and of good food value, it is not likely ever to be popular or of economic importance. Some authors place in family Heliotaceae.

Slippery Leotia

Morel, Sponge Mushroom

Morel, Sponge Mushroom
Morchella sp.
Erect fruit body to 6 in. tall, hollow, pale flesh-colored, brittle, tender, moist, not deeply penetrating ground. Top portion, pileus, looks like a sponge. Found over most of United States in woodlands, mostly where there has been fire, from May to June but highly variable in different years. Spores shed best during damp and rainy weather. Excel all other fungi in food value but apparently not adaptable to controlled cultivation and may be infested with insects. When washed and fried in butter, they are delicious. No method has been found to produce morels commercially. None poisonous; easily identified, so good specimens for beginning "mushroom hunter." May be confused with some species of *Helvella*.

Helvella

Helvella
Helvella lacunosa
Helvellas are usually smaller than the morels, appearing like deformed or sick morels, with smooth irregular caps and with long pitted stems. Sometimes called saddle fungi or false morel. In this species the lower margin of the cap is close to the stem or grown to it. Fruiting tops appear in late summer and fall and plant is of wide geographic distribution. Related *Gyromytra* has cap with brainlike convolutions. Might be mistaken for a deformed morel and may be poisonous. Poisonous substance is helvellic acid. Genus *Helvella* common in the woods, growing in soil among old leaves; some grow in orchards and in open fields.

White Helvella

White Helvella
Helvella crispa
One of most common species of saddle fungi or false morels. Much like *H. lacunosa* except that lower margin of cap is free from deeply furrowed stem and cap is white or whitish. Occurs in woods in summer and autumn. Contains poison, helvellic acid, which can cause death or make partaker violently ill. Should not be confused with edible morel which it resembles. Morel has "spongelike" top; that of *Helvella* is folded. If in doubt about the identity, don't gather the plant for eating.

Shield–like Peziza

Shield-like Peziza
Peziza coutellata
Found on damp rotten logs from July to October, generally plentiful and beautiful when examined closely. Fruiting bodies become flat with age, are vermilion in center and paler at margin, hairy toward the margin, with straight black hairs. Fruiting bodies usually grouped in clusters of relatively small numbers and most evident in damp parts of season. Of no apparent direct economic importance. Found in February on West Coast.

Scarlet Cup

Scarlet Cup
Peziza coccinea
Found on old sticks in open or in damp woods in early spring or late winter on ground, scattered or in groups and spectacular. Irregularly cup-shaped. Stem, if any, short. Cup to 2 in. across, with entire margin, with brilliant red interior and with flesh-colored pinkish-white outside. Fruits edible but almost too beautiful to eat. If allowed to stand quietly in warm dry room, clouds of spores may be freed by simply tapping the fruiting portion.

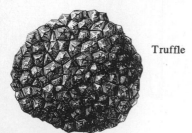

Truffle

Truffle
Tuber aestivum
Epicureans prize truffles highly as food in continental Europe. Occur in rather open forests. Mainly under species of *Quercus*, oak. Ascocarps are subterranean. Ascospores are liberated only when ascocarp (underground part) decays or is broken by animals. Truffles of commerce native to France and Italy. Trained dogs and pigs locate the truffles and dig them up. People then gather them and sell them in the markets. Order is represented by several species in the U.S., mostly from Pacific Coast states. Most U.S. species produce small ascocarps not used as food by man.

DIVISION ASCOMYCOTA. CLASS ASCOMYCETES

Order Helotiales./Family Sclerotiniaceae

Brown Rot of Plum
Sclerotinia fructicola
Common as brown rot on apples, plums, cherries, peaches, apricots, quinces, and pears, producing mummied fruits and wrinkled limbs. Parts of affected host become wrinkled and dry, showing first as brown rot. In spring affected fruits produce diseased fruit bodies that resemble stalked cups. Very destructive disease favored by moist, cloudy weather. Controlled somewhat by removal and burning of affected parts and by use of lime-sulfur spray. Tar-oil spraying, as for aphid infestation, and used late in dormant period, gives partial control by destroying spore cushions. Lime-sulfur spray is suggested by some fungicide manufacturers as effective if applied a few days before harvest of fruits.

Brown Rot of Plum

Order Erysiphales or Perisporiales./Family Erysiphaceae

Powdery Mildew of Lilac
Microsphaerea alni
Appears as powdery white spots something like flour on the leaves of lilac. In late season, black dots appear scattered among the gray powder. These are fruit bodies that rupture to free reproductive units. In the North this is an unimportant parasite of lilac, but in the South practically an entire crop of pecans has been known to have been destroyed by this one fungus. Burning of infected refuse and spraying with bordeaux in early stages is standard control. Also recommended as sprays to combat powdery mildews are dinocap, colloidal sulfur, and wettable sulfur. In addition, the following fungicides have been reported by the USDA as being effective: Lime Sulfur and Pipron applied at the time foliage appears; Folpet and Sulfur applied as a dust before or at the time symptoms of the disease appear, and Zineb applied as a dust spray. For pecans, a fungicide reported is: A mixture of 5.2 parts by weight (83.9 percent) of ammoniates [Ethylene Bis (dithiocarbomato)] Zinc with 1 part by weight (16.1 percent Ethylenebis [dithiocarbamic] Acid Bimolecular and Trimolecular Cyclic Anhydrosulfides and Disulfides [Polyram] solutions applied by hydraulic equipment or by aircraft or by mist blower equipment. Treatment is begun when buds are opening. It is well to follow manufacturer's recommendations to avoid undue pollution of the environment by fungicides.

Mildew on Lilac

Order Hypocreales./Family Clavicipitaceae

Ergot
Claviceps purpurea
Appears as black or purplish body several times as long as the fruit of rye, wheat, oats, barley, quack or other cereals or wild grass that are affected. Many forage grasses also unfortunately may be infected. New plants are entered at flowering time. Grain and forage are ruined for fodder, since infected plants may cause abortion and other ill effects in cattle and other mammals. Used medicinally in treatment of migraine headaches. Too dangerous for home use. Fungus contains a number of poisonous alkaloids. Sclerotia, hard resistant body formed as ergot; used to prepare a powerful abortifacient, also used to control hemorrhage during childbirth. Burning is control. Infected seeds float in 20 percent salt solution. Deep plowing, to 10 in. of old infected field, to bury infected heads is recommended as control measure. Careful inspection of seed for presence of infection also recommended. Epidemics of ergot poisoning in A.D. 944 and 1090 in France have been described as Saint Anthony's fire which, in the convulsive type, led to excruciating pain and a sensation of burning or fire. Limbs which became gangrenous were described as being eaten up by the holy fire and blackened like charcoal. Saint Anthony's powers were invoked for the aid of victims of ergot poisoning.

Ergot

Insect Fungus
Cordyceps sp.
Appears as an "ergotlike" or club-shaped growth from caterpillars or pupae of various insects. Some species are found growing on different species of insects, while other species of the fungus may live on one or more species of insects. The fungus is parasitic, thus killing living insects and possibly being useful because of this. Some species live as parasites on other fungi that live underground. Fungus grows throughout the body of the insect and, in one author's expression, converts the insect's body into a mycelial mummy. It rarely is of sufficient abundance or importance to be considered seriously. One species may cause minor epidemic among grubs of June beetles. Use of genus in biological warfare against insects not successful.

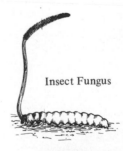

Insect Fungus

Order Hypocreales./Family Nectriaceae

Maple Canker
Nectria cinnabarina
Appears as serious cankers that appear pinkish on the roots and branches of many woody plants including currant, horse chestnut, maple, apple, and quince. Mycelium attacks and destroys sapwood of tree causing the bark to die of desiccation (canker formation). Young twigs are killed; this form of disease is called dieback. The organism gains entrance

Canker

(Continued)

through a wound and eventually kills its host. Flesh-colored or pinkish hemispheres appear on bark, changing to chocolate brown, girdling the tree, and killing it; causes dark-green discoloration of sapwood. Control is by removal and destruction of infected areas and by painting of wounds, although some mycologists argue this method is ineffective. In severe infections by *Nectria,* spray with Copper-Bordeaux mixture just before leaf fall, at about 50 percent leaf fall, and in spring when buds start to fall. Manufacturer of fungicide PMA recommends PMA in two applications; May and June.

Order Hypocreales. / Family Phyllachoraceae

Sooty Spot on Clover
Phyllachora trifolii
Appears as pale dots on upper leaf surfaces accompanied by black dots on lower surface on clovers such as red, crimson, alsike, and closely related species. The black spots closely resemble those made by the rust fungi and found in three stages widely distributed over the United States on at least four species of clover. While the sooty spot disease may cause much damage to clover forage, there seem to be no known effective control measures.

Order Dothidiales. / Family Dothidiaceae

Black Knot of Plum
Dibotryon morbosum
Appears as heavy, roughened, black knots on affected parts of both wild and cultivated forms of cherry and plum affecting the fine twigs or the trunk or limbs. In cherry the knot becomes finely honeycombed, and it may in any case give off gummy secretions. Affected areas may be to 1 ft long. Disease is very dangerous and destructive. Control is by pruning out all affected areas before January when spores are freed, burning diseased parts, and using bordeaux in March. Certain fungicides: Lime Sulfur solution applied in the green tip stage and Copper Sulfate basic applied at the same time are suggested by manufacturers as effective fungicides on plum.

Order Sphaeropsidales. / Family Microsphaerellaceae

Leaf Blight of Strawberry
Mycosphaerella fragariae
Appears first as red or purplish spots that enlarge and finally become white with red or purple borders, irregularly distributed. Found on leaves of strawberries most abundantly at about time of flowering. While it affects most varieties, Marshall and Brandywine seem to be somewhat resistant. Control measures consist of using resistant forms, planting healthy plants only, removing affected leaves, and using bordeaux just before flowering. In certain of the artificial classification systems, some members of the genus are placed in Deuteromycetes, the imperfect fungi. Several fungicides have been reported as effective by their manufacturers in control of leaf blight of strawberry: Copper Sulfate, Basic applied as a spray or dust in preblossom stage and after leaves form; Diodine as a spray or dust when new growth appears; and Copper Tetra Copper Oxychloride as a dust; PMA in solution in spring dormant stages or delayed dormant stage and treatment repeated in postharvest time, or in the fall; Phenylmercuric monoethanol ammonium acetate in aqueous form applied in spring when new growth is visible in the crown and repetition of the treatment in the fall dormant period before mulching. While applying any fungicide, all necessary precautions should be used to avoid environmental pollution.

Order Pseudosphaeriales. / Family Pleosporaceae

Apple Scab
Venturia inequalis
Appears on leaves and fruits of apple as velvet-green spots that may eventually produce cracks and generally deform the fruit. A closely allied and somewhat similar disease of pear may affect the twigs and flowers as well as the leaves and fruits. Where apples are grown for market, strong bordeaux should be used just before the flowers open, just after the petals fall, and again a week or two later, depending in part on the weather. Weather which favors opening of apple buds also favors development of ascospores, infecting agent, of this fungus. Spores spread by raindrops. May find one stage of the life cycle listed as genus *Spilocacea.* Sprays with captan, diodine acetate, or wettable sulfur recommended as treatment. Many other fungicides reported effective by their manufacturers: Polyram, Flypet, Glyodin, PMA, DNOC, Dithane, Lime Sulfur, Copper Sulfate in dormant stage; and Ferbam and Copper-Bordeaux in prebloom and early flowering stages. It is important to follow manufacturer's recommended use suggestions to avoid unnecessary environmental pollution.

Black Mold of Clover

Black Knot of Plum

Leaf Blight of Strawberry

Apple Scab

DIVISION ASCOMYCOTA. CLASS ASCOMYCETES

Order Sphaeriales./Family Xylariaceae

Dead-man's Finger
Xylaria sp.
Appears like an irregular, dark to black club to 3 in. long and with a maximum diameter of a little over 1 in., with a white corky interior, which shows when the brittle top is broken and exposes visible spore cases. Found on old woods in woodlands particularly around decaying stumps. Genus is common throughout both tropical and temperate lands. One species, *polymorpha,* decays the roots of living or recently dead hardwood trees and will attack some apple trees. It is highly variable in size, color, and form. While at least two species are associated with the rot of the wood of apple trees, the plants are not considered dangerous.

Dead-man's Finger

Order Phacidiales./Family Phacidiaceae

Shot-hole Disease in Cherry
Coccomyces sp.
Leaves at first appear to be shot full of holes because of the dropping out of the fungus-destroyed tissue, tissue that is transparent before the holes appear. Disease is found on the leaves of cherries, plums, and their kind except in the case of a few resistant varieties. Disease organism is frequently considered as *Cylindrosporium padi* of the imperfect fungi. May cause much loss. Control is early spring bordeaux spray. Many chemical fungicides have been reported as effective against the infection. All should be used with caution.

Shot-hole Disease of Cherry

The Lichens
Ascomycota, or Basidiomycota

These plant bodies are made up of a fungus, the mycobiont, and an alga, the phycobiont. For many years botanists considered them as "one plant." The lichen may represent a biological relationship called mutualism; that is, each partner benefits from the relationship. The alga produces food for itself and the fungus, and in return the fungus provides protection for itself and the alga from such adverse conditions as drought. This idea of a mutualistic relationship between the fungus and alga in the lichen is not held as fact by some botanists who maintain that the relationship is one of the host-parasite variety, in which the fungus is a weak parasite on its algal host. It has also been suggested that the alga is really a slave (helotism) of the fungus. Recent research indicates that the fungus can be separated from its partner (the alga) under laboratory conditions at least and that the fungus will "parasitize" its host if cells of the algal host are available, and, furthermore, that the two partners will subsequently grow together as in their natural condition.

Since the fungus partner appears to be the dominant partner in the lichen, we shall, for convenience sake, place the lichens discussed with either the Ascomycota or the Basidiomycota, depending upon the classification of the lichen's fungal partner.

CLASS ASCOLICHENES

[Subclass Gymnocarpeae]
Order Lecanorales./Family Usnaceae

Hair-like Usnea
Usnea longissima
Appears like hanging gray threads that may be to 5 ft long, round in cross section, soft and flexible. Often inhabited by insects. Found on tree branches and bark in the northern United States, Alaska, and British Columbia, especially in the humid mountainous regions. The interior is a tough strand surrounded by a spongy material. Fruit bodies are small disks with a few marginal threads. The name of the genus is derived from the Arabic term meaning Lichen. Algal host is *Protococcus.*

Hanging Moss

Old-man's Beard

Old-man's Beard
Usnea barbata
Apparently made of numerous branching, radiating fibers of the same color, often green to straw-colored, with units round in cross section. Found hanging on bark and branches of living or dead trees in all parts of North America and many parts of the world. Fruiting bodies shieldlike with a lighter disk. Fibers to over 4 in. long. Described by Theophrastus, who lived in 300 B.C., as being useful to stimulate growth of hair though no such merit is now recognized. Algal host is *Protococcus*.

Order Lecanorales./Family Parmeliaceae

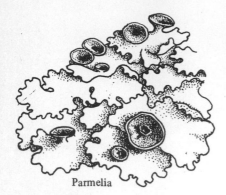

Parmelia
Parmelia sp.
Appears like sheets of crumpled leather attached by black "roots" to rocks or trees. If torn, will show white fibers in the center under a hand lens. Sometimes has a powdery appearance owing to ruptures associated with freeing of the reproductive organs. Fruiting bodies are in the form of disks or cups, usually with a raised center, that appear scattered over the lichen's surface. Plant is an encrusted lichen that grows slowly and is at best on a long-undisturbed base. Helps make soil. Algal host is *Protococcus*. Widespread genus, species of which are distributed throughout the U.S.

Parmelia

[Subclass Gymnocarpeae]
Order Lecanorales./Family Physciaceae

Physcia
Physcia leucomela
Appears like narrow, leaflike structures irregularly divided, smooth, with marginal hairlike structures, white beneath, and generally more or less tangled into irregular masses. Some botanists put this lichen in the family Physciaceae in an order Discolichenes, an order to which all the lichens here considered belong and in all of which the alga involved is a *Protococcus*. This lichen also listed as *Anaptychia leucomela* by some authors. Species of both genera widespread throughout U.S.

Physcia

Order Lecanorales./Family Cladoniaceae

Pixie Cup Cladonia
Cladonia pyxidata
Appears to be somewhat scalelike, bearing at intervals grayish, scaly fruit bodies that are hollow and seem like stalked, expanding, erect cups that rise to a height of 1 in. and look like plugged funnels with widespread margins. Plants are found on the earth or on such old dry wood as fence rails. The algal host is *Pleurococcus*. One species, *rangiferina,* reported as the famous reindeer moss of the north. Genus is widespread throughout the U.S. Common name refers to the trumpet-shaped (pixie-cup-shaped) fruiting bodies.

Pixie–cup Cladonia

Scarlet-crested Cladonia, British Soldiers
Cladonia cristatella
A scalelike broken base produces a number of erect, branched, hollow standards that terminate in bright red, sometimes glistening knobs that are the fruiting bodies. The plant is in most conspicuous fruiting condition in early spring, when it is greener, redder, and obviously more active. It occurs on the ground or on dead wood. It has a common name of British soldiers, referring possibly to the red "uniform." Plant is excellent for indoor terrariums. Algal host is *Pleurococcus*. Distributed throughout the U.S. east of the Rocky Mountains.

British Soldiers

Order Lecanorales./Family Gyrophoraceae

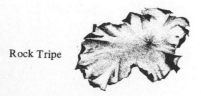

Rock Tripe

Rock Tripe
Umbilicaria mammulata
Occurs on overhanging cliffs, near water or on rocks or soil in the forest; common on large boulders in open woods. When wet, these leathery brown shields lie flat held by a stout cord *(Umbilicaria)* at their centers. If atmosphere dries, edges of thallus (plant body) curl, exposing black sooty colored underside which forms black tubes over the rocks. Margin of the thallus frequently torn. Plant might be considered a "living hygrometer" in the region where it grows.

Thallus foliose, that is to say, flat and more or less leaflike; 1 to 9 in. in diameter. Body attached to substrate by an umbilicus made up of plant tissue. Spore-bearing bodies within the plant body or shortly stalked; may be black, usually in an elongated or roundish fold. Plant body large to very large, thick, smooth, or rarely rough; usually rough, ashy, or ashy brown; may be whitish or bluish appearing to have a whitish bloom. Appears like a patch peeling away from a surface. Underside brown to blackish appearing curled up and above plant body as changes occur. Has strong dark brown or black rootlike structures, rhizines.

Some confusion as to classification in different references. *U. mammulata* commonest species in eastern North America: *U. proboscidea* on exposed rocks and talus in boreal, arctic, or alpine regions; *U. phaea* common on boulders and cliffs in the West from southern California to British Columbia.

Umbilicate lichens have been credited with saving the lives of Arctic explorers; also of hunters and trappers in far north Canada and Alaska. Called "tripe de roche" by Sir John Franklin, Arctic explorer, whose life it was supposed to have saved. In Japan, the foliose rock tripes called *iwatake,* are eaten in salads or fried in deep fat; are considered a delicacy. Rock tripes are boiled to extract a gelatinous soup thickener. When dry, these lichens are very brittle; so if collected in this state, they should be first moistened before being removed from substrate.

Associated alga is *Protococcus.*

DIVISION BASIDIOMYCOTA

The most advanced division of fungi, called club fungi, includes those with largest fruiting bodies; e.g., bracket fungi grow to 20 in. (50 cm) or more in diameter and the giant puffball may become 60 in. (150 cm) across; some mushrooms have caps 8 in. (20 cm) in diameter.   Huge numbers of spores are produced by each fruiting body.   Spores produced on basidia.   Basidiospores, upon germination, give rise to septate (cross-walled) mycelia with uninucleated cells.   When two compatible hyphae from mycelia meet, nuclei pass from one to the other and a binucleate mycelium is produced.   From the binucleate mycelium is produced the thallus (plant body).   Basidiomycota contain two groups: one contains such forms as rusts and smuts and the jelly fungi; the second group contains the mushrooms, shelf fungi, coral fungi, earthstars, stinkhorns, bird's nest fungi, and puffballs.

CLASS HETEROBASIDIOMYCETES

Order Ustilaginales./Family Ustilaginaceae

Corn Smut
Ustilago maydis
Appears first as large, glistening, white blisters that later become dark and burst to free masses of dark spores that are spread by the wind. Unfortunately no symptoms of the infection appear on the host (corn ear) until the smut balls are formed. Found on ears, stems, and leaves of corn particularly during or after a moist season, especially in a long-growing or rich soil or where a crop is too crowded. May destroy as much as $\frac{2}{3}$ of a crop. Control is effected by removal of infected parts from crop, avoiding use of infected manure, and by treatment of seed. Best controlled by a rotation that does not permit growing of corn more than once in 3 years on same land. Fungicidal seed dressings not effective.

Corn Smut

Oat Smut

Oat Smut
Ustilago avenae
Common example of a smut fungus which can be found almost anywhere. Causes heads of oats to become black and prevents grain formation. Enters plant at time of seed germination and when oat is in flower by means of windblown spores. Appears to mature at about time of flowering of oat. Smut develops very slowly while the seed is maturing and the mycelium remains dormant in the seed. When seed germinates, mycelium of smut grows in tissues of seedling. While it affects the whole plant, it is observable superficially only in the heads. Control is effected by dipping seed in 1 pt of formalin to 30 gal of water for 10 minutes to destroy spores, then drying. Can be controlled by organomercury seed dressings. These controls should be used with caution because of danger of accumulation of mercury as an environmental poison.

Onion Smut

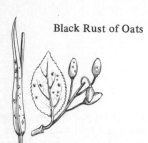

Black Rust of Oats

Cedar Apple Rust

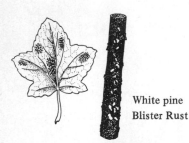

White pine
Blister Rust

Trembling
Fungus

DIVISION BASIDIOMYCOTA. CLASS HETEROBASIDIOMYCETES

Order Ustilaginales./Family Tilletiaceae

Onion Smut
Urocystis cepulae
Appears as dark streaks on bulb and leaves of onion. These break, freeing spores into the ground, and these infect new onion plants grown in that ground. Infection takes place independent of climatic conditions once the soil has become contaminated. Common control practices include care in transplanting healthy plants into clean soil, avoiding repeated use of same soil for similar crop, or drilling soil with 2 parts of sulfur to 1 part of air-slaked lime. In the Channel Islands off England a very dilute solution of formaldehyde is dribbled into the onion rows at planting time. This appears to be an effective pest control measure. Manufacturers of fungicides in the U.S. report Captan and Thiran applied as dust and Nabam and Formaldehyde applied as drench in rows as effective control measures. Special precautions should be exercised to avoid environmental pollution when using chemical fungicides.

Order Uredinales./Family Pucciniaceae

Black Rust of Grain
Puccinia graminis
On the leaves of European barberry may be found scars on upper surface and clusters of yellow cups on under surface. Spores from these blow to grain and develop a plant body that goes through the plant, weakening it and eventually breaking the surface of many important cereals and many forage grasses, appearing as yellow or reddish streaks, whose spores further spread disease. Later, black spots develop to start cycle again on barberry. Dispersal of spores by winds takes place over immense distances; e.g., from Mexico and the Mississippi basin to the prairie provinces of Canada. Same species causes wheat rust, about which much has been written. One eradication technique involved the elimination of barberry host in vicinity of wheat. Another essential step in combating *graminis* infection is the continuous development of new varieties resistant to the fungus. Sulfur dust applied at the first sign of the disease has been reported as an effective fungicide and is registered as a treatment with the USDA.

Cedar Apple, Apple Rust
Gymnosporangium sp.
Rusty spots may appear on the leaves of apple on underside as clusters of cups with corresponding yellow spots on upper surface. These injure health of apple and free spores that blow to red cedar where, after development, great orange jellylike structures appear in spring and early summer. Spores from these are carried to apple and start cycle again. Fruits of apples may be affected. Control may be helped by removal of infected red cedars. Has been reported that Thiram, or TMTD, a tetramethyl derivative of thiuramdisulfhide, is recommended as a spray treatment against the fungus. Also reported is the use of Copper—Bordeaux Mixture applied from June through August. No fungicidal treatment should be employed without precautions against unnecessary environmental pollution.

Order Uredinales./Family Melampsoraceae

White-pine Blister Rust
Cronartium ribicola
Pine infection appears in spring as yellow-orange blisters on young twigs, older branches, or sometimes on the trunks of white pine. From these the spores are carried to currant or a related plant, where, after development, structures are developed that spread spores back to pine again to start a new infection. Control by removal of infected areas and destruction of currants and gooseberry (*Ribes* sp.) bushes in vicinity of pines. Disease may completely destroy white pines in area. Disease introduced from Europe. Attacks both eastern and two western species of commercial white pines in the U.S. A chemical fungicide 3-[2-(3,5-Dimethyl-z^2-Oxocyclohexyl)-z^2-Hydroxethyl] Glutarimide [cycloheximide] [Acti-Di-one[R]] is reported effective by its manufacturer if sprayed on each canker site and over the surrounding bark area.

Order Tremellales./Family Tremellaceae

Trembling Fungus
Tremella frondosa
Appears like soft, clammy, yielding folds of jellylike material up to 4 in. high and to 5 in. wide, brown and glistening. Found on rotten wood at any time from fall to spring, with various species differing in color and in shape. Spores are given off from the surface since there are no pores or gills. This species is sometimes known as "witches' butter"; while it is edible, it is of no great economic importance as food or plant disease in the U.S. Some species of this genus, however, are used extensively as food by the Chinese.

DIVISION BASIDIOMYCOTA. CLASS HETEROBASIDIOMYCETES

Funnel-shaped Fungus
Phlogiotis hevelloides
Appear like erect stems with shoehorn tips, or like irregular funnels—usually undeveloped on one side, reddish-yellow, smooth, clammy, gelatinous, to 4 in. tall. Found growing on the ground in woods, most commonly from dead roots or other rotting wood, and becoming very hard when dried artificially. Most commonly found developed at best in autumn. Not considered as edible or of any economic importance. Probably serve as an eliminator of dead wood.

Funnel–shaped Fungus

CLASS HOMOBASIDIOMYCETES

[Subclass Hymenomycetes]
Order Hymeniales./Family Clavariaceae

Coral Fungus
Clavaria sp.
Upright, repeatedly branching, with tips always pointing upward in contrast with downward-pointing hydnums. Height to 8 in., brittle, pink, red, orange, or yellow. Widely distributed. Spores borne by whole upper surface and may be carried by insects. Some grow from ground, others from rotting wood. None are poisonous; many are edible, few delicious. Usually tough; some have disagreeable flavor. Should be cooked slowly, then fried in butter to taste like noodles or put in soups. One species, *botrytis,* collected for food in the Pacific Northwest and in our southeastern states, is a late summer and fall species. Appears as massive fruiting bodies in arcs or fairy rings, often under rhododendrons. Species *cinerea* in the Lake States and to the east is common on duff of low moist conifer woods and is described as edible and popular. Where found, species generally occur on hardwoods.

Coral Fungus

[Subclass Hymenomycetes]
Order Hymeniales./Family Hydnaceae

Coral Hydnum
Hydnum coralloides
From a common stem many long branches arise; from anywhere beneath the surface of these branches hang spines to $\frac{1}{4}$ in. long. Single fruiting body may weigh over 1 lb. Found only on decaying hardwoods, while related bearshead hydnum *H. caput-ursi* is found on living or on dead hardwoods and has long teeth grouped from beneath ends of branches. Remarkably free from insect pests; both species are edible. May be bitter, but not if boiled slowly.

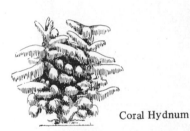

Coral Hydnum

Hedgehog Fungus
Hydnum repandum
Pore surface like preceding but under 4 in. tall, with white, buff, or yellow cap with downward-hanging, white, brittle, spore-bearing teeth. Found commonly in woods, on ground, variable, but usually with supporting stem in middle though some are from one side. Considered one of the very best table mushrooms and safe because no closely resembling species are poisonous. *H. putidum* has a coarse, spongy stem and when fresh, has an offensive odor. All species may be found repeatedly in one spot. Species *imbricatum* of sandy oak woods in Great Lakes region, in the Rocky Mountains, and along the Pacific Coast, and species *scabrosum* in conifer and hardwood forests such as scrub oak forests of the central states and in the southeastern states, are not edible and may be poisonous.

Hedgehog Fungus

Order Hymeniales./Family Cantharellaceae

Craterellus, Trumpet of Death, Horn of Plenty
Craterellus cornucopioides
Has appearance of a deep funnel with a broad hornlip. Undersurface of horn is the fruiting surface, which is uneven, dark, smoky in color. Found on the ground in the woods most commonly in late summer and autumn. Common woodland fungus along edges of old roads, trails, and in exposed places in hardwood forests from eastern North America through the Great Lakes States. Differs from *Cantharellus* in that the undersurface is smooth when the plants are young while in the canterelles it is covered with folds. This species is darker than *Cantharellus cantherellus* and has a more uneven undersurface than that species. Both are edible. Considered choice but doesn't change its dark color when cooked. Apparently it is about like *Cantharellus cibarius.*

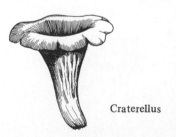

Craterellus

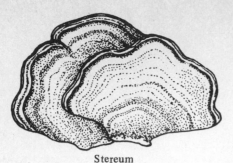

Stereum

Suspicious Boletus

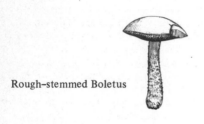

Rough–stemmed Boletus

Edible Boletus

Cone–like Boletus

DIVISION BASIDIOMYCOTA. CLASS HOMOBASIDIOMYCETES

[Subclass Hymenomycetes]
Order Polyporales. / Family Thellephoraceae

Stereum
Stereum versicolor
Thin, woody, leathery shelves that grow crowded together with upper surface distinctly marked with eccentric rows of different shades of gray to make easily recognized zones. While it somewhat resembles *Polyporus versicolor* (p. 79), the undersurface of this species is smooth, even, and not composed of fine, closely crowded pores. Fruiting bodies woody and therefore not edible. Plant serves as a wood rot. Silver-leaf disease of forest and shade trees is caused by related *S. purpureum.*

Order Agaricales. / Family Boletaceae

Suspicious Boletus
Boletus felleus
Cap to 8 in. across; nearly flat, smooth, yellow-brown, with white flesh that may turn pink where wounded. Tubes on underside are white when young but flesh-colored when mature. Found on or near decayed logs, usually near hemlock or spruce but sometimes in open fields. Most common in July, on through September when the season is wet and warm. This species is too bitter to be edible, but it is doubtful whether it is poisonous. Some forms are to 6 in. tall. A widespread genus from coast to coast. Three species, *mirabilis, edulis,* and *eastwoodiae* (poisonous), should be of interest to western collectors.

Rough-stemmed Boletus
Boletus scaber
Cap white, red, brown, or smoky, from 1 to 3 or more inches across, with flesh that is soft and white but turns dark when bruised. The free ends of the downward-pointing tubes form a white convex surface, and the tubes are free from each other or nearly so. Stem rough-appearing, giving the species its name. Found in woods or in grassy places, on the ground. Grows mostly under birch. This species is edible and lacks the bitter taste of *B. felleus.* By some, it is considered to be excellent food.

Order Agaricales. / Family Boletaceae

Edible Boletus
Boletus edulis
Fruiting body umbrella-form, with cap from 4 to 6 in. across, broad, smooth, grayish-red to brown, with flesh that is white or yellow and is red under the skin. Tubes white at first, but as they mature they become greenish or yellowish; free from each other or nearly so. Stem rather stout, swollen at base and to 7 to 8 in. long.

Members of this genus differ from *Fistulina* and *Polyporus* in that pore-bearing layer easily separates from flesh of cap. While these layers will usually separate from cap in *Boletinus* and in *Strobilomyces,* they do not do so easily. This species is found in woods or in open places particularly in warm weather from July through September. Found on the ground in each case.

Spores shed downward through pores in cap when mature. Thousands of spores may be expelled from a single pore, and there are probably thousands of pores in some caps. Spores are wind-borne. When they fall on a suitable food supply, they develop into the plant that bears the fruits.

Usually not found in fruit in great numbers at any one spot. Widely distributed both in America and in Europe. When eaten raw, it tastes like a nut or has a distinct sweetish flavor. Since tubes may become slimy with age, they may be discarded and the rest of the cap saved for eating.

One of the more desirable fungi for food and in Europe is considered a distinct delicacy. The caps are sometimes sliced and dried for later use in soups or as a regular mushroom dish. They may be eaten raw or fried in butter to give body to a soup much as noodles do. A "choice" species. Some imported dried from Poland. Of special interest to western collectors.

Cone-like Boletus
Strobilomyces strobilaceus
Fruiting body like a coarse umbrella, with cap to 4 in. across, soft and spongy, with surface covered with coarse, hairy, upright, blackish-brown scales, with margin scale-fringed and undersurface crowded, with pores that are white at first but changing to black when wounded and having angular openings. Stem to 5 in. long and to ¾ in. thick; brittle.

Strobilomyces differs from *Boletus* and other common fleshy pore-bearing stool-shaped mushrooms in that tubes are white, are at first covered by veil, and do not separate easily from cap as they do in *Boletus.* Flesh of pine-cone fungus, which is at first a pale tan when broken, becomes a dark blue in few minutes and exudes a reddish-brown juice. Pores to 1 in. long. Stem less woolly than cap.

Spores shed downward from pores when fruiting body is in its prime in fall and summer. While mature fruit body may become soft and decayed, old one may become dried, black, and resemble a piece of dead wood. Spores dark brown in mass, nearly globular, and roughened.

Found on ground, not usually in great numbers but over most parts of country and into Canada. Stem and cap are favored hosts for larvae of many insects, and so specimens should be examined carefully before they are used as food by man. Rather rapid color change may be used as some index of relative freshness of plant.

Not easy to confuse this plant with any poisonous species. Not likely to be sufficiently abundant anywhere at any time to make up an important part of the volume of any collection. This is the most common one in United States and Canada. Related species are found in Europe.

Sulfur-colored Polyporus

Order Polyporales./Family Polyporaceae

Sulphur-colored Polyporus
Polyporus sulphureus
Fruit body appears as bright red shelf or series of shelves, the lower of which are usually larger with undersurfaces crowded with round pores. Older shelves show radiating furrows and ridges that create a fanlike effect for whole fruiting body, and in the older shelf the rim is thin.

Family includes a variety of forms, some umbrellalike, as in *Strobilomyces* and *Boletus,* and many shelflike or even platelike but all marked by crowded pores on undersurface of the fruiting body. Related beefsteak mushroom, *Fistulina hepatica,* has soft, fleshy shelf that is juicy and blood-red when young but tough and hard when old. Other genera and species in family are always hard or woody.

Spores shed from downward-pointing pores, new layers of which may be formed in succeeding years in suitable weather conditions at 4½ million per sq in. per hour, or 375 million per week. Wood rotted by fungus resembles red-brown charcoal, with concentric and radial cracks filled with large sheets of fungus.

Lives on woody plants that are alive or dead, favoring cone bearers, oak, cherry, maple, chestnut, butternut, walnut, alder, locusts, apple, pear, but is also found on other trees. It fruits in fall when summer droughts have passed and appears year after year on same logs.

Fruiting shelves cannot be confused with any poisonous mushroom and if collected young, parboiled in salt water, and fried in butter have a delicious taste. Older shelves are tough. The plant is, of course, a pest to a tree and it may be desirable to burn infected logs to protect healthy trees.

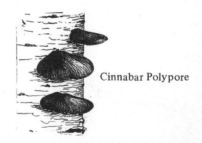

Cinnabar Polypore

Cinnabar Polypore
Polyporus cinnabarinus
Conspicuous, vermilion-colored shelves that are dry, more or less spongy, to nearly 4 in. across, with light or yellowish-red flesh. Pores are, of course, densely crowded on lower surface. Found most commonly on the dead brances of oak, cherry, or maple, particularly late in autumn. Shelves are merely fruiting bodies of a plant that is buried in wood acting as the host. Too woody to be considered edible. Nor considered an important plant pest.

Yellowish Wood Rot
Polyporus versicolor
Rather thin, leathery, somewhat flexible shelves, plain and sometimes rigid, velvety above with adjacent zones of different shades of coloration showing conspicuously. Pores very minute. Found commonly crowded on dead logs at any time of year. Destructive to certain trees even though it attacks dead wood only. Since it destroys wood quickly, it reduces strength of whole tree, which eventually falls through weakness. It resembles *Stereum*. The enzyme, ligninase, instrumental in decomposition of lignin has been reported found in this species.

Yellowish
Wood
Rot

White Butt Rot
Fomes applanatus
Shelves hard and woody and flexible only at growing edge temporarily. Top is marked with eccentric ridges marking year's growth. Pore surface beneath always points downward and is white unless rubbed. Found growing on living or dead trees, particularly beech, maple, birch, oak, and poplar, at any time of year. Fruiting bodies increase in size in succeeding years. Plant is highly destructive to valuable trees. One species, *pinicola,* causes extensive damage to pulpwood. No control of infection known. May release 350,000 spores each second; each spore potentially the beginning of a new site of infection.

White Butt Rot

Yellowish
Sapwood Rot

Favolus

Daedalea

Artist's Fungus

Beefsteak Mushroom

Lenzites

DIVISION BASIDIOMYCOTA. CLASS HOMOBASIDIOMYCETES

DIVISION BASIDIOMYCOTA. CLASS HOMOBASIDIOMYCETES

Yellowish Sapwood Rot
Fomes fomentarius
Exceptionally hard, thick shelves that look somewhat like the hoof of a horse and do not extend far from surface of host. Each succeeding year's growth is marked by an increase in size of the fruit body as a new layer is added at bottom; many fruiting bodies may appear relatively close together on a badly infected tree. Found on beech and birch most commonly and of course can be found at any time of year. Recognized as a serious beech and birch pest.

Favolus
Favolus canadensis
Usually fruiting body is supported by a stalk from one side. Cap tough, fleshy, thin with pores large, angularly elongated, and with their greatest width extending radially from the stalk. Plant common on fallen branches particularly of hickory. Fruiting bodies are most common from September until the frost. Interesting because pores begin to suggest gills. Name refers to honeycomb appearance of undersurface of cap.

Daedalea
Daedalea quercina
Pale, wood-colored shelves of considerable thickness where in contact with the host. Pores at first round but with maturity become apparently stretched to form definite labyrinths. Not easily confused with other species. Found on oak stumps or fallen oak logs at any time of year, often with many fruit bodies rather close together. Shelves may be to 4 in. or even more broad. Since plant does not attack living trees, it is not considered of economic importance.

Artist's Fungus
Ganoderma applanatum
Fan-shaped to irregular, woody (actually feels like soft wood) fruiting body; upper surface grayish to nearly white or rarely black; often appears with concentric lines which represent "annual rings" or year's growth. Commonest of our forest fungi, occurs principally on hardwood trees; in particular on fallen trees and old stumps. Fruiting body may become 75 cm, or 30 in., across. It is estimated that one such spore fruit produces 30 billion spores per day, few of which germinate. Does not attack healthy, uninjured trees. Almost any young person familiar with the woods has broken the fungus body from a tree and carved his initials on its "underside." It makes a good "natural canvas" on which drawings can be scratched with sharp or blunt instruments. Leaders of young woodland field trippers should look for this species as it makes for a good object lesson.

Beefsteak Mushroom
Fistulina hepatica
Easily recognized as great red shelves supported by stalks from one side. With pores on undersurface yellowish and distinct, relatively inflexible. Shelf has a tendency to turn up at edge showing yellow undersurface contrasting with red upper. Shelf is about 8 by 14 in., and stem may be wanting. Found on dead trunks of oak, chestnut, and other hardwoods from June through September whenever the season is a wet one. Edible but sour. Upper surface more or less liver color; resembles raw meat, hence its name. Common across the U.S. but most abundant in southeastern states.

Lenzites
Lenzites betulina
Corky, leathery brackets that are firm and do not show growth zones on upper surface, of uniform color, dirty grayish-white, with pores almost gill-like and certainly not typically tubelike. Found in a variety of forms on many kinds of trees but usually more abundant on birch; of widespread distribution. Some botanists incorrectly consider this a member of the Agaricaceae, because of the gill-like pores. It is apparently of no great economic importance.

DIVISION BASIDIOMYCOTA. CLASS HOMOBASIDIOMYCETES

Order Agaricales./Family Lepiotaceae

Parasol Mushroom
Lepiota procera
Height to 8 in. Cap to 5 in. across, brownish spotted, but usually darkest at tip, which is often well-raised. Gills free from stem, white, soft, and dry. Cap easily separates from stem, which has a swollen base, with white flesh and a free separate ring but no basal cup. Found in thin woods, by roadsides, in open pastures from July through September. In older plants cap may be cracked at margin. Edible.

Order Agaricales./Family Amanitaceae

Fly Amanita
Amanita muscaria
Beautiful, with 6-in. cap, changing from spherical to convex to plane, yellow to deep orange, with striated margin, with sticky surface covered with thick angular persistent scales. Stem white or yellow, with ring and conspicuous cup. Most variable *Amanita* especially in color from one geographic region to another. Found from June to November on ground, in open or in woods, rarely in great numbers. Poisonous. Some persons parboil and "remove poisons." This practice not recommended. Young buttons may be mistaken for young puffballs. Check by splitting lengthwise with knife. If *Amanita*, cap, gills, and beginning of stalk will be visible. In the East, found commonly under aspen and pine; in the Far West under spruce and fir.

Orange Amanita
Amanita caesarea
Cap red or yellow or orange, bell-shaped, with gills yellow and free from the stem, which is hollow, arises from a free-bordered cup, and bears a broad collarlike ring. Cap is generally naked or bears broad, thin patches, has a smooth surface except at margin, where it has definite striations; often more than a foot in diameter in a closely related species, *calyptroderma*. Color deepest at apex. Found in woods, on ground in spring and summer. Is edible but should ordinarily be avoided because of its close resemblance to deadly poisonous species. Common in south-eastern states. *A. calyptroderma* occurs in October and November on Pacific Coast, in particular in northern California under oak and pine cover.

Destroying Angel
Amanita verna
Cap white, egg-shaped at first but becoming flat, sticky. Stem erect, stuffed when young and hollow when mature, tall; arises from a cup and bears a broad collarlike ring, longer than width of cap. Found in woods from spring to fall; in hardwood and mixed stands east of the Great Plains. Probably our most poisonous fungus. Poison becomes evident about 40 hours after it is taken, causing pain, muscular paralysis and death in 2 to 3 days or earlier. First aid probably useless. Precaution: Never eat a white *Amanita*.

Order Agaricales./Family Tricholomataceae

Honey-colored Armillaria
Armillaria mellea
Fruiting bodies to 12 in. tall, with cap to 8 in. across and stem to nearly 1 in. in diameter. Usually honey-colored but sometimes white or dull reddish-brown, being darker in center of cap, which is usually covered with dark brown to blackish scales though these may be absent. Usually many fruits in clusters.

Differs from *Amanita* and *Lepiota* in that cap does not break cleanly and easily from stem, the tissues of the two being continuous. Stem usually bears a ring (annulus) though this is not always distinct. No cup (volva). Presence of annulus separates it from related *Tricholoma* and *Clitocybe*, in both of which stem and cap are grown together. Worldwide in distribution.

Spores shed in summer, attack roots of host trees, entering bark and spreading under bark until roots are killed, and spreading until eventually tree is killed. Growth may take place for many years before new fruiting bodies are produced, the hidden tissue often appearing as cords or strands. Fruits in August and September before the leaves have changed color.

Host plants include many cone bearers as well as forest trees and small fruits. In newly cleared lands, the diseased roots of forest trees may infect newly planted orchard trees or other cultivated woody plants. The dark subterranean cords may spread infection from tree to tree.

Economically injurious to important woody plants, destroying timber and reducing fruit production. The fruit bodies are considered edible and also choice by some persons. Often appears in large numbers after first killing frosts and may be canned for use as food. Should be of interest to collectors in western U.S.

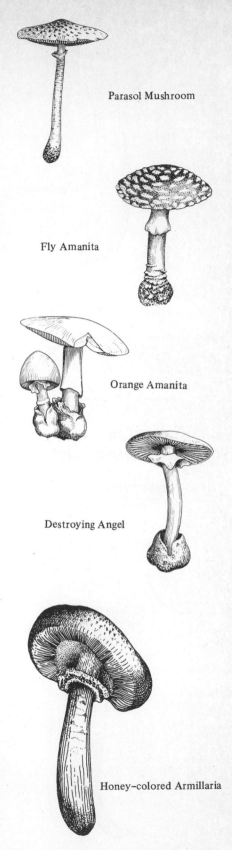

Parasol Mushroom

Fly Amanita

Orange Amanita

Destroying Angel

Honey–colored Armillaria

Jack-o-lantern

Clitocybe, Jack-o-lantern
Clitocybe illudens
Fruiting body borne in clusters arising often from a common point, with stems to 5 in. long and ½ in. thick. Cap to 5 in. across with raised portion in center when young, or depressed when old; margin often waxy. Whole plant a bright saffron yellow though flesh inside is soft and white or yellow.

Lacks both ring and cup; stem and cap are grown firmly together. Unlike otherwise rather similar *Tricholoma,* gills at their bases extend down onto stem. Many species, including edible, mealy-surfaced *C. laccata,* short-stemmed giant clitocybe, ivory clitocybe, the funnel-shaped, the lacquered, the cloudy, the fragrant, the clustered clitocybes.

Like *Armillaria,* some clitocybes may spread through cords from root to root underground. The spores are shed from the fruiting bodies and can attack wounded woody plants and there develop new plant bodies that may eventually bear other fruit bodies. Spores, white.

C. illudens is most conspicuous at base of decaying stumps and at night is often luminescent. Wood being rotted by this fungus may be collected and used to supply an eerie light in a tent while camping, though the luminescence is most conspicuous in wet weather.

A wood-destroying fungus. Supposed to be poisonous or unwholesome, though bad element is apparently destroyed or removed by boiling. Soldiers in North Africa have written of excitement caused at night by steady glow of infected roots cut in trenches in front lines.

Poisonous, causing nausea and vomiting; apparently not fatal. Oak seems to be its favorite substratum. Species *illudens* common east of Great Plains; *olearis* in California.

Velvet-stemmed Collybia
Collybia velutipes
Fruiting bodies appear in crowded masses on trunks of trees with stems bent upward, to 5 in. long, yellow when young but densely covered with brown velvet when old and bearing caps that are to 3 in. across, at first red-brown, then tan, then glistening with sticky gelatinous substance. Gills, white, narrow, crowded.

Collybia differs from *Mycena* in the cap margin in *Collybia* is at first bent inward rather than straight, gills of neither are decurrent, and although cap and stem are grown together, they are of obviously different texture. White-spored, with relatively thin gills. Related *C. radicata* grows on ground with flat cap and rootlike base to stem. Both common and well-distributed. Genus widely distributed; *C. butyracea* in pine plantations across the U.S.; *C. familia* on conifer logs in northern and western U.S.; *C. acervata* common in the Pacific Northwest; and *C. umbonata* under redwoods in northern California.

Spores from this plant may be shed from the fresh fruit bodies from earliest spring even into January, it being one of the fungi appearing most widely spaced through the year. Fruit appears either from dead trees or from wounds or dead spots in living trees.

Plant may injure living trees by destroying the old wood that provides support. A favored species is basswood though it is found frequently on other hardwoods. The fruiting is most vigorous in late fall and winter; late records are probably of persistent fall specimens.

Plant injurious to woody plants but fruiting bodies provide excellent food for man. They may be collected and eaten fresh, fried in fat, or canned and used later in season. The related *C. radicata* is also edible but not usually so abundant.

Velvet-stemmed Collybia

Rooting Collybia
Collybia radicata
With white to gray caps that are sticky when wet. Stem has a cartilaginous bark that is somewhat brittle and extends into a slender root nearly equal to the length of the stem above ground. Gills uncrowded and grown fast to stem. Cap to 3 in. across, stem to 8 in. tall and to ⅓ in. thick. Flesh white. Found on ground in woodlands or at their edges usually in tufts into late fall even after the frosts and usually around old stumps. Edible and not confused with bad forms. *C. acervata* of special interest to collectors in western U.S.

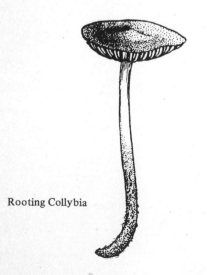

Rooting Collybia

DIVISION BASIDIOMYCOTA. CLASS HOMOBASIDIOMYCETES

Mycena
Mycena galericulata

Height to 5 in., but caps are rarely over 1 in. across and stems are slender and to ⅛ in. through. Cap has distinct elevation near center, is bell-shaped, brown or gray. Gills extend downward on stem by means of a tooth, are white or flesh-colored and not crowded. Stem hollow, with hairy base, cartilaginous. Abundant in woods generally in dense clusters from September until after frost. Edible, but too delicate to provide a substantial meal. Relative, *M. leaiana,* has an orange-yellow cap, yellow gills with orange border, a yellow to orange stalk; spring and early summer mushroom in hardwood stands east of the Great Plains.

Mycena

Bell Omphalia
Xeromphalia campanella

Height to 2 in. Cap to 1 in. across; with distinct depression in center, smooth, rusty yellow. Gills run down stem at point of attachment. Stem watery, hairy at the base, hollow, brownish, slender, and often curved upward rather than straight, lighter at top. Found in woods in damp locations, usually abundant from summer through fall. These mushrooms are delicious though they are difficult to collect in quantities to make a meal.

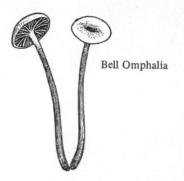

Bell Omphalia

Oyster Mushroom
Pleurotus ostreatus

Cap may be over 8 in. across, with top white or ivory and smooth, usually developed unevenly to one side of a short thick stem, or there may be no stem. Gills thin, white, soft, spongy and extend down onto stem; thicker than in *Hygrophorus* and not waxy. No ring or cup. Found usually on hardwoods. Caps are usually infected by active beetles that crawl in gills. Caps decay rapidly in wet weather. Edible, particularly when young; may be attached directly to wood. Fruits in spring and fall; during cool, wet summers also. In Rocky Mountains and the Pacific Coast region is abundant on alder, maple, and cottonwood. In West may be found late in the season on poplars along irrigation ditches.

Oyster Mushroom

Elm Mushroom
Pleurotus ulmarius

Much like oyster mushroom, but with stem often central. Gills do not run down stem at point of attachment, and stem is generally conspicuously thickened at base. Gills and cap usually not so heavily populated with insects and slugs as are those of oyster mushroom. This species lives on elms living or dead, and may often be collected from city street trees in quantity enough to make a good meal. When properly prepared, are delicious.

Elm Mushroom

Order Agaricales./Family Hygrophoraceae

Hygrophorus
Hygrophorus sp.

Cap sticky, waxy, either sooty, orange, yellow, or other colors. *H. russula* has sticky cap. Gills relatively few, waxy, and broadest nearest cap, in some species extending down stem and grown fast to it. No ring and no "death cup." While gills appear to be waxy or watery and have a surface that can be peeled off relatively easily, they do not bleed when ruptured. Fruiting bodies are found growing in woods and on lawns. Apparently there are no species that are dangerous to eat. *H. russula* considered a good edible form by some collectors; occurs in open grassy woods, in particular oak forests east of the Great Plains. May form fairy rings.

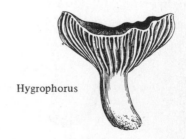

Hygrophorus

Delicious Milky Cap

Pungent Russula

Canterella

DIVISION BASIDIOMYCOTA. CLASS HOMOBASIDIOMYCETES

Order Agaricales./Family Russulaceae

Delicious Milky Cap
Lactarius deliciosus
Fruiting body a stoollike body that, when bruised, gives off a milky or colored juice. Cap sticky when moist, to 4½ in. across, convex when young but funnel-shaped when old; mottled orange to red or changing to green with juice similarly colored. Flesh soft, spongy, brittle, white to orange. Gills extend down stem. Stem to 3 in. long.

White-spored, resembling *russula* but differing primarily in possession of colored or milky juice, which exudes freely in plant or more particularly in broken gills. Gills brittle, rigid, may or may not extend down stem in different species. Important species include the scented, the flesh-colored, the watery, the peppery, the indigo, the coral-colored, and the fleecy milky cap.

Spores shed from July to November; landing on damp ground in coniferous woodlands, may germinate to develop a subterranean plant body that eventually will bear new fruiting bodies. Fruiting bodies not commonly clustered in great numbers as is common with those mushrooms that may be found on stumps or on woody plants.

Food evidently associated with evergreen woods. While this species and some others are edible, a number of species are not, and some are poisonous. Stem frosted or covered with bloom at first, spotted with orange and turning green where touched in this species.

Some consider this species a favored mushroom but others disagree. Since there are dangerous or inedible species in the genus, it is safer to avoid eating any of the milky-juiced mushrooms unless more exhaustive data than are here given are available to make sure of identification. Widespread; in pine forests on well-drained humus from Rocky Mountains through the Lake States and into eastern North America; abundant along the Pacific Coast in the fall.

Pungent Russula
Russula emetica
Fruiting body umbrellalike, with cap to 3 in. broad, sunken in center, sticky when moist, smooth, brittle, with skin peeling easily to show faint reddish flesh. Cap red, but paler with age. Gills not crowded, some may be forked near stem, which is to 3 in. long and to ¾ in. thick, solid or pithy but fragile. Much of flesh is white. While-spored.

Resembles *Lactarius* but lacking its milky or colored juice when broken and being, like it, ordinarily rather brittle. Few russulas "bleed" slightly and, if anything, caps of russulas are more brittle than those of *Lactarius* owing in part to the bunching of tissues, a character not visible without a microscope. Many russulas are edible, but a few are disagreeable and this one is reported by some to be poisonous.

Fruiting bodies appear from July to October, from the plant body hidden in rotten logs or in the forest floor. This species is not common in the open, and some russulas favor pines and spruces, while some grow in sand and some are at their best in open hardwood forests.

Species name *emetica* implies that this plant may serve as an emetic. Certainly, it has a strong and distinctive taste, but there are large numbers of persons who eat it regularly and enjoy it, although some books label it as poisonous. Some go so far as to say that there are no poisonous russulas.

Economic importance is small, and since fruits usually do not appear in great numbers at given place, they are not likely to appear in a mess of mushrooms unless mixed with others. For this reason, it is possible that they have been considered dangerous because of their strong taste and because they may be mixed with dangerous plants.

Order Canthrellales./Family Canthrellaceae

Chanterelle
Cantharellus cibarius
Fruiting body like wavy-margined cup, only slightly depressed in center and with rather coarse, uncrowded gills extending down stem from beneath cap. Cap to 3 in. across, fleshy, smooth, firm, like yellow of an egg. Stem to 2 in. long, solid, flexible. Whole plant fragrantly scented.

Blunt-edged gills extend down onto stem distinguishing this genus from other white-spored fungi. Three common species include orange-yellow *C. aurantiacus* of doubtful food value, the vermilion-red *C. cinnabarinus,* and this species, both of which are edible. Color usually sufficient identification for these, though gills of *C. cibarius* are fewer branched and more decurrent than in *C. aurantiacus.*

In July and August, fruit bodies appear singly or scattered in groups on ground, usually in coniferous woodlands or in hardwood forests. *C. aurantiacus* grows in abundance from rotten wood, probably saprophytic. While this is classed as a white-spored mushroom, spores are faintly yellow and elliptical.

Orange-yellow chanterelle is found from July to September; the yellow, from July through August; and the red from July through October.   Yellow is particularly well-known and favored in Europe as well as in the United States and Canada.

Fragrance and nutty taste as well as abundance of chanterelle has made it famous and popular world over but more particularly in days of ancient Rome.   To some, nutty flavor is somewhat bitter; and by some, orange-yellow species is considered as poisonous.   *C. cibarius* an edible and choice specimen in the gourmet trade.   Found throughout U.S. in forested areas.   In the Pacific Northwest it fruits in the fall; in California in the late fall or winter.

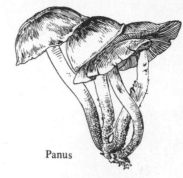

Fairy-ring Mushroom

Order Agaricales./Family Tricholomataceae

Fairy-ring Mushroom
Marasmius oreades
Cap to 3 in. across, tan or ivory, smooth, with rounded raised center, somewhat leathery in dry weather but reviving when wet, easily separating from the slender, tough to 4-in. stem.   Flesh firm and white.   *Collybia,* otherwise much like it, will not revive when wilted.   Found usually in rings in grassy fields or on lawns or orchards from spring to fall and from year to year.   "Pest" on some golf courses; sometimes found in pine plantations.   In the Pacific Northwest it may be a problem on lawns. Ring said by some to be caused by exhaustion of available food from center.   Name fairy ring comes from old superstition that mushrooms growing in a circle represent the path of dancing fairies.   Immediately within ring, grass may be greener because of availability of nitrogen from dying mycelium.   Dried, sliced, and then cooked in soups are delicious. Popular as an edible variety; however, can be confused with some poisonous varieties which are of similar size.

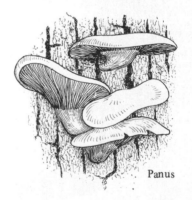

Panus

Panus
Panus stipticus
Leathery caps, with lateral stem or no stem, with gills entire, and with an acute edge.   In this species, plant is cinnamon in color, most unpleasant to taste, and poisonous.   Like *Marasmius,* this species will revive after it has been wilted by drying.   It appears abundantly on old logs and stumps from early fall to winter and plant body buried in wood is phosphorescent and glows in the dark.   Applied to a wound it contracts blood vessels and stops bleeding, styptic.

Panus
Panus rudis
Cap an extension of stem, covered with a fuzz that is thick and reddish but faded with age.   Gills pale tan, extend down onto stem almost to its base.   Cap usually not over 2 in. across but caps are usually crowded in considerable numbers and a mass may sometimes nearly cover exposed part of a tree trunk or log.   Plant leathery, attacked rather freely by beetles, not injured by drying, and in spite of toughness, may make good soup or gravy flavoring.

Panus

Order Agaricales./Class Coprinaceae

Inky-cap Mushroom
Coprinus atramentarius
Cap egg-shaped at first, smooth except for a few spotty scales in center, gray-brown or yellowish but black when mature.   Gills closely crowded when young and break down into an inky fluid when mature.   This is a form of self-digestion and mushrooms with this feature are placed in *Coprinus.*   Takes its common name, inky-cap, from the black, "ink-colored" fluid produced in self-digestion.   Stem slender, smooth, hollow, whitish, with a temporary ring, to 4 in. long.   A layer of dark threads covers the base of the stalk.   It occurs on ground, in clumps, in woods, or in open as on lawns, from late summer through fall.   If cooked immediately after harvesting, it is excellent or can be made into a catsup that is delicious. No poisonous *Coprinus* species.   It is reported that a few people experience a peculiar type of intoxication if they eat the mushrooms after drinking alcoholic beverages.   A widespread species found in spring or fall during cool wet weather.

Inky-cap
Mushroom

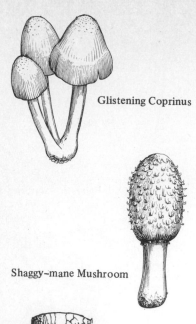

Glistening Coprinus

Shaggy–mane Mushroom

Schizophyllum

Cortinarius

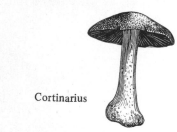

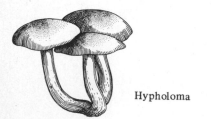

Hypholoma

DIVISION BASIDIOMYCETES. CLASS HOMOBASIDIOMYCETES

Order Agaricales./Class Coprinaceae

Glistening Coprinus
Coprinus micaceus
Caps much like those of inky cap; but are smaller and when young, are covered with glistening micalike scales, are bell-shaped and definitely striated: tawny yellow, brownish-yellow, or gray when young but become dirty in appearance and finally black when the crowded gills break down into a black ink and shed the spores. Stem smooth, slender, fragile, to 3 in. long and cap to 2 in. broad. Occurs in great numbers from stumps of old trees on lawns or in fields. Delicious when promptly and properly cooked. Found throughout North America. Fruits in spring and fall. Common in cities.

Shaggy-mane Mushroom
Coprinus cometus
Cap at first oblong to cylindrical, whitish, with scattered yellow scales but splits at margin, flattens, and becomes shaggy, to 3 in. long before it expands. Gills white, closely crowded, turn pink, red, purple, and finally black dissolving into a dripping ink. Stem smooth, hollow, long and when young with a ring; to 5 in. long and to ⅓ in. through. Found in meadows and on lawns but not crowded. Excellent as food if taken when young or used as flavoring when older. Best if collected in button stage before gills have darkened. Fruits in spring and fall.

Order Agaricales./Family Agaricaceae

Schizophyllum
Schizophyllum alneum or *commune*
Dry, corky, fanlike shelves, with gills that are split at the edges and appear to be double, to 1½ in. across, with a downy surface; shrivels when dry and recovers when moist; without a stem and with white spores; may appear gray, then a purplish-brown, and then even white. Found commonly growing on old wood such as twigs and branches, particularly on maple, but may be found at any time of year and in most parts of the world. Hardly a parasite, since it serves to destroy waste. *S. commune* is worldwide in distribution. Easily recognized and cultivated. Used extensively for the study of heredity and nuclear behavior.

Order Agaricales./Family Cortinariaceae

Cortinarius
Cortinarius violaceus
Cap nearly flat, after being convex in youth; dark violet, covered with numerous tufts of hairy scales, to 4 in. across, with violet flesh and with a cobwebby interior that is distinct from skin of cap. Gills powdery with spores but become rusty with age, attached to stem and notched at stem end. Stem swollen at base, to 5 in. long, solid. Found for most part in woods in late summer or early fall, scattered or solitary. Edible but not exceptionally popular; however, considered choice by some. Found in protected overage coniferous forests of northern regions and the western mountains. Of particular interest to collectors in western states of U.S.

Hypholoma
Hypholoma sublateritium
Cap continuous with stem, brick-red but often paler at margin, convex or flat, smooth, with white or yellow flesh, to 3 in. across, with a veil that remains attached to margin of cap but with no ring. Stem smooth, fibrous, to 3½ in. long, rusty, tapering to base. Found often in large clusters around old stumps with many stems arising from a common point, from September through to winter. Considered as edible fungus but is usually insect fouled and bitter.

Pholiota
Pholiota praecox
Cap is unbrella-shaped, to 2 in. across, convex or plane, slightly sticky
when moist, white or pale tan, with incurved margin. Gills are crowded,
unequal, widest at center, from dirty white or gray to rusty. Stem is to 3
in. long and to ¼ in. through, spongy inside or hollow. Found on lawns,
in meadows, and along roadsides from May to June, while related fatty
pholiota *P. adiposa* is found on tree trunks from September through frost.
A mass of early pholiota may weigh a pound and make a family's meal.
Species of this genus widespread; *squarrosoides* in hardwood forests from
eastern North America through the Lake States to the Pacific Northwest;
a similar distribution but around conifers as well as hardwoods for *squarrosa.*

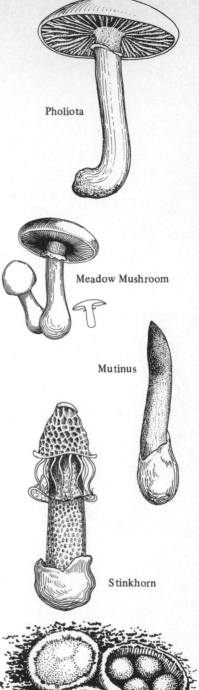

Pholiota

Order Agaricales./Family Agaricaceae

Meadow Mushroom
Agaricus campestris
Cap to 4 in. across, hemispherical, then convex, then expanded, soft,
satiny white, with much incurved margin at first. Gills free from stem,
thin, crowded, white, then pink, then blackish-brown. Stem to 3 in. long
and to ⅔ in. through, with ring usually persistent. Flesh thick, firm,
white, with good flavor and scent. Cap and stem easily separate but cap
leaves a ring. May resemble some poisonous lepiotas. Among species
of this genus are the horse mushroom, the flat-top, the sylvan, the reddish-brown, and others. Commercial plantations are established in cellars and tunnels where blocks of plant body "spawn" are planted in manure and rich soil and maintained at 50 to 60°F. Commercial
mushrooms should develop in less than 10 weeks after beds are established if properly managed and soil, temperature, and humidity are satisfactory. Yield in summer is usually small but in fall large. Minor spore
differences are recognized between commercial and wild forms. This
species is probably eaten more than any other fungus though not so important as yeasts, rusts, smuts, and the like. Fruits in pastures, meadows,
and appears from late summer into fall; found throughout the U.S.

Meadow Mushroom

[Subclass Gasteromycetes]
Order Phallales./Family Phallaceae

Mutinus or Mutinus Stinkhorn
Mutinus caninus
Like a slender pointed cylinder, with a red-tipped or a flesh-colored
tipped end. Apparently comes from a small egg the skin of which remains attached around the base. The erect cylinder is definitely spongy
and looks something like soft flesh. Saprophytic, living on dead plant
matter and is therefore of no serious importance. Found in damp thickets, woodlands, old gardens, usually in July and August. Fruiting body
does not last long.

Mutinus

Stink-horn Fungus
Dictyophora duplicata
A beautiful but horribly foul-smelling plant appearing like a large egg
that produces a pitted cylindrical stem that is to 9 in. long and to 2 in. in
diameter, crowned by a slightly larger, more pitted cap from which hangs
an open lacework structure that supports a free ring. Found more or less
singly in open meadows and grasslands but more particularly around old
sawdust piles in summer and fall. Apparently of no economic importance and too offensive in odor to be popular.

Stinkhorn

Order Nidulariales./Family Nidulariaceae

Bird's-nest Fungus
Crucibulum vulgare
Like minute birds' nests that are about ⅓ in. across, at first covered by a
thin membrane that breaks, revealing the "eggs" nestling in the shallow,
yellowish nest. So-called "eggs" rupture to free reproductive cells.
"Eggs" are actually spore packages. Found sometimes in great profusion
on old sticks, wooden bridges, fence rails, and other unused wood. Undoubtedly saprophytic and of no serious destructive properties. The related family Sphaerobolaceae produces but one egg per fruit body instead
of many. One species, *levis,* is reported throughout the U.S.

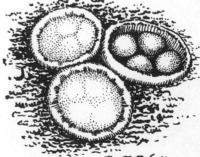

Bird's nest Fungus

DIVISION BASIDIOMYCOTA. CLASS HOMOBASIDIOMYCETES

[Subclass Gasteromycetes]
Order Lycoperdales./Family Geastraceae

Earthstar
Geastrum sp.

Earth Star

Looks like a sphere with an opening at top from which spores are freed and supported by a star-shaped base that may vary its position considerably in accordance with weather. Actually a puffball in which outer layer splits in radial fissures and then breaks away and spreads out to form a star-shaped pattern. It opens in dry weather, raising plant from ground, and closes in wet weather, bringing spore-bearing sphere again close to ground. Found on the ground in woodlands or in open grassy territory. It apparently is of no great economic importance though interesting.

Order Lycoperdales./Family Lycoperdaceae

Lilac Puffball
Lycoperdon cyathiforme

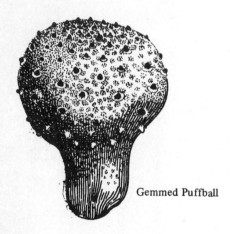

Lilac Puffball

Like slightly flattened globes that are to 6 in. in diameter, whitish, pinkish, then brown, and in the upper areas cracking and freeing the dark brown to purple spores. When old, the fruiting bodies appear to be scooped-out cups. It is sometimes called the beaker-shaped puffball because of the stout stem that supports the expanded beaker-shaped upper part. When young, the fruiting bodies are white both on the outside and inside. They are edible when young and excellent. Do not eat if interior of puffball is beginning to turn yellow. One species, *pyriforme,* is called the pear-shaped puffball. Common on rotting wood throughout the U.S.

Gemmed Puffball
Lycoperdon gemmatum

Top-shaped spheres that are to 3 in. high and to 2 in. wide, with a relatively stout stem that penetrates the ground and may be longer than the ball above the ground is through. When young, balls are warty on outside and surface changes from fine-scaled to fuzzy, to warty, to smooth.

Most of interior of all members of this genus at maturity breaks down into masses of light spores. Some species differences based on the way outer walls break to free these spores. In some, whole wall seems to crumble into bits while in others breaks are along more or less definite lines.

In addition to details given for *Lycoperdon,* it should be pointed out that in this plant there is a tremendous loss in reproductive process and an infinitely small proportion of spores ever develop into plants that themselves bear spores. This is true of most of fungi, of course, but more obvious with dark-spored forms.

These puffballs have been recommended for use as quenchers for flow of blood. It is doubtful if they have any merit in this connection though some fungi do have styptic properties, as has already been pointed out. Probably they would do little damage if used on a wound.

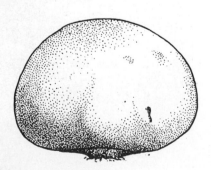

Gemmed Puffball

While this species is edible when young, it lacks the agreeable flavor of the giant puffball. Like other puffballs, it should be eaten only when flesh is firm and white. The cooking procedure is the same as that outlined for the giant puffball.

Giant Puffball
Calvatia giganteum

White or whitish balls, when young, varying in size from one to a few inches through in *Lycoperdon* to a maximum of 5 ft high and 4 ft wide in *Calvatia,* and in shape from spheres, to somewhat elongated vases, to depressed spheres with surfaces smooth or otherwise. Fruiting bodies above ground.

Related *Scleroderma* has underground fruit bodies thick-walled and with black or purple interior. Lycoperdales include all true puffballs and earthstars. *Calvatia,* the giant puffball, breaks the fruit body wall irregularly. Some, with papery walls, are classified under genus *Bovista.* Related earthstar *Geaster* breaks fruit body to form a many pointed star that opens and closes in dry and wet weather.

Lycoperdon species usually rupture more or less papery fruit body to free air borne spores through definite area at top. Spores wind-borne. Fruit bodies white throughout at first, but later black. A 5 lb 3 oz puffball yields about 7 trillion spores; if each matured, second generation would be 800 times the size of the earth.

Plants are saprophytes living on decaying plant material either in open grassy fields, in wooded areas, or in cultivated lands. One species, *L. gemmatum,* is often found in considerable numbers on well-rotted logs, though pear-shaped *L. pyriforme* is even more partial to old wood.

All puffballs with white interiors and fruit bodies above ground are edible and some are delicious. Larger calvatias may make many meals for a family. Fruits may be sliced, parboiled in salt water, and fried in butter or in hot fat to make a delicious dish. Should be eaten only when collected white.

Giant Puffball

DIVISION BASIDIOMYCOTA. CLASS HOMOBASIDIOMYCETES

Order Sclerodermatales./Family Sclerodermataceae

Common Scleroderma
Scleroderma sp.

Look like hard-skinned, scabby spheres or at least somewhat spherical, with a firm greenish-yellow interior that shows when broken. In some species the exterior is orange but in most it is a greenish- or a yellowish-brown or dark brown. It is found commonly in pastures and meadows particularly where the soil is poor, generally from August on through November. Usually, these fungi do not exceed 3 in. in diameter. Not poisonous but not good as food. Commonest of hard-skinned puffballs. One species, *aurantium,* occurs occurs during summer and fall on humus and rotten logs in both conifer and hardwood forests throughout northern, eastern, and central U.S.

Common Scleroderma

DIVISION DEUTEROMYCOTA

A group of organisms also called Fungi Imperfecti. Forms included are both parasitic and saprophytic. They produce only asexual reproductive cells. Spores may be borne directly on the mycelium or on special fruiting structures. Some botanists believe species of Fungi Imperfecti may represent alternate stages of Ascomycota or Basidiomycota in which the spore-bearing bodies are produced infrequently, if ever.

CLASS DEUTEROMYCETES

Order Melanconiales./ Family Hyalosporae

Bean Anthracnose
Colletotrichum sp.

Bean anthracnose appears as brownish or purplish, usually sunken spots that later develop into extensive patches. These infect the seeds that in turn infect the succeeding crop. The disease appears on the seed leaves, on the stems, and on the seeds and fruits of beans.

The imperfect fungi to which this plant belongs have life histories that are somewhat in doubt since they lack some of the properties that would place them in one or the other of the major groups that are recognized. Some seem to have relationships with the basidiomycetes and some with the ascomycetes and some may have had or lost the characteristics which would readily classify them with one or the other.

A few imperfect fungi cannot be classified because they seem to exist only as sterile mycelium without normal reproductive bodies one would expect in fungi. Others have merely what seem to be asexual stages of a fungus, but in spite of these shortcomings most of these fungi seem to be able to maintain themselves in too great vigor.

Bean Anthracnose

While many imperfect fungi are saprophytic, living on dead plant material, some are parasitic and may be serious pests as is the bean anthracnose here figured. They form cankers on limbs and branches of trees, spots and rots on fruits, and a number of other unfortunate deformities and injuries.

Bean anthracnose is controlled by selection of resistant varieties, by selection of seeds from clean pods, by burning old infected plants, and by spraying with bordeaux, 5-5-50, particularly when disease occurs early and has not made too much progress before being discovered.

SUBKINGDOM CHLOROTA, DIVISION HEPATOPHYTA

This division introduces us to some of the early land plants, fossils of which appear in Devonian rock strata. Made up of the so-called liverworts, the division contains some 300 genera and nearly 10,000 species. All genera are representative of moist habitats; are aquatic forms. Some botanists include the liverworts with the mosses in one division, the Bryophyta. In the classification scheme accepted for this fieldbook, the division contains two classes: the Hepatopsida or liverworts and the Anthocerotopsida or hornworts. The gametophytes (sex-cell-producing plants) are the more conspicuous phase in the life cycle.

It is believed that the common name, liverwort, probably means liver herb or liver vegetable. The liverworts were thought to have medicinal properties in ancient time as well as in medieval time. Liverworts are still collected and prescribed by herb doctors in some oriental countries.

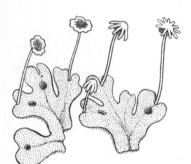

Marchantia

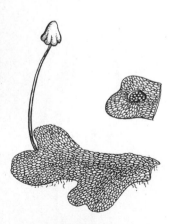

Conocephalum

CLASS HEPATOPSIDA

Order Marchantiales./ Family Marchantiaceae

Marchantia
Marchantia polymorpha
Dull green, forked ribbons, to 5 in. or more long, with prominently sunken midribs above, with upper surface marked off in diamond-shaped areas each with an oval or circular dot in the center. On close examination, the diamond areas are really eight sided and the pores are air holes leading to the interior. Underside of thallus crowded with rootlike structures.

A liverwort of the family Marchantiaceae, which includes *Lunularia* of greenhouses, *Conocephalum*, and *Preissia*. Marchantia is found all over North America and Europe with this species most common in the North and with two other closely related species in the South.

Thallus is either male or female. Male thallus bears erect, flat-topped structures that bear the antheridia that produce the spores that make their way by swimming to the female structures. Female plants bear umbrellalike archegonia beneath branches of which are the eggs that produce the yellow sporophytes. Asexual reproduction.

Plant usually most abundant on newly disturbed clayey or silt-filled soil particularly in burned areas; common on calcareous rocks, in wet places as in gorges. May thrive in crevices between stone flagging. Umbrellalike structures develop even though eggs are not fertilized and reach maturity through June to August. Spores are forced from the capsules of the sporophytes by springlike structures.

Plants are of little economic importance but are most commonly used for illustrating development in the liverworts in classes in biology. Plants lend themselves readily to culture in terrariums. May serve some purpose as soil anchors in areas where other plants could not survive. It is difficult to measure the value of the plants since much of what is to be evaluated, esthetics, is not measurable.

Conocephalum
Conocephalum conicum
Largest of our liverworts with broad flat thalluses, in this case often covering many square feet of surface. Plant body is paler green than is *Marchantia*, with marked areas on the upper surface, larger and more distinctly defined, these often being sufficiently large to be recognized many feet away. When the thallus is broken, it gives off a strongly resinous odor. Considered by some to be aromatic.

A liverwort further distinguished from *Marchantia* by the fact that the air pore in each surface segment is borne on top of a low mound of colorless cells. Like *Marchantia*, *Conocephalum* is found widely distributed over North America and Europe.

Male sperm-bearing antheridia are borne in a wartlike spot on the thallus rather than on a tall flat-topped antheridia-bearing structure as in *Marchantia*. The female archegonia-bearing structure looks like a partly closed umbrella with a long handle. It is watery and does not last long. It usually matures somewhat earlier than does *Marchantia*. Asexual spores are borne in remarkably shallow cups.

Found for the most part on moist earth or on the dripping rocks of gorges and glens, often entirely covering considerable areas to the practical exclusion of other plants. Do not expect to find the spore-producing umbrellas in late summer when *Marchantia* may be in excellent fruiting condition.

Probably of little economic importance though it was once thought to have some medicinal properties. It must serve to some extent to anchor soil as well as to add organic material to small patches of soil that must be more or less sterile. As a source of organic matter on rock surfaces, the decayed plant bodies provide nutrients for next step in plant succession.

Order Jungermanniales./ Family Porellaceae

Scale-moss Liverwort
Porella platyphylloides

Stems to over 3 in. long, branched, prostrate, slender, and hidden by the closely overlapping thin leaves each of which is blunt at the tip, to nearly $\frac{1}{20}$ in. long, but the whole plant with its many branches covering an area sometimes many square feet in extent. Beneath the closely overlapping leaves are underlobes and underleaves that hug the stems.

While *Marchantia* and *Conocephalum* belong to the order Marchantiales or thalloid liverworts, *Porella* belongs to the leafy liverworts of the order Jungermanniales. At least six species of the genus are recognized but are difficult to distinguish one from another. They are widespread in North America and Europe.

Male and female reproductive organs are borne in separate structures on separate plants so that a plant is either male or female. They appear as small buds on the branches and the fruits look like little cups that mature to free the spores in May or June. They are not conspicuous in comparison with the developments to be seen in *Marchantia*.

P. platyphylloides usually lies flat on bark of living trees though the tips may be raised. *P. pinnata* is attached to rocks and is generally submerged. The two species, *platyphylloides* and *pinnata*, are so similar that they are difficult to differentiate as separate species. *P. navicularis* is conspicuously glossy, is found on our northwest coast on trees and logs; and *P. cordaeana* and *P. roellii* are also native of the northwest areas.

Plants are of little importance economically and are of interest to botanists who find in them interesting variations in reproductive habits as well as in branching and leafing habits.

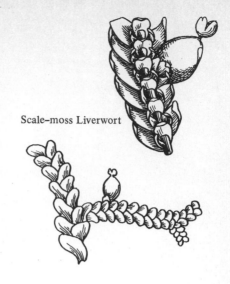

Scale-moss Liverwort

Scale-moss Liverwort
Ptilidium pulchrinum

Leaves without any midrib, in two rows near upper side of the stem, deeply divided into fine lace of threads or rows of cells, but with these joined into a substantial base that is from 6 to 20 cells wide as contrasted with narrower-leaved genera, often somewhat purplish-brown in general color, to $\frac{1}{25}$ in. wide with the leaves spreading when moist and close-pressed when dry. Occurs in dense and brownish mats; some "stems" prostrate and some ascending.

Leafy liverwort belonging to the order Jungermanniales and the family Ptilidiaceae. In this country, the genus is represented by this species and the wider-leaved *P. ciliare,* both of which are circumpolar and extend in America south to Pennsylvania and Illinois, and by *P. californicum,* which ranges on the West Coast from Alaska to California and Idaho.

From the axils of the branches there frequently appear what seem to be white "hairs." At the tips of these are small black capsules in which the spores ripen and are ready to be shed in the spring. The capsule may persist on the plant after the spores have been shed, and the supporting hairs after the capsules have been shed. Male elements short-stalked in the axils of more closely overlapping "leaves." Female elements at end of stalks, terminal, or may be lateral.

Plants are most conspicuous in early spring when they appear most commonly on rotted wood or on tree trunks. They are frequently found mixed in with mosses. Usually, the botanists who are interested in these liverworts use the leaves as a basis for distinguishing the species, the important characters being the number of lobes and the number of hair-tipped lobe divisions.

Plant is probably of no economic importance whatever, but it does appear as an attractive decoration for a tree trunk or a fallen log.

Scale-moss Liverwort

Order Jungermanniales./ Family Lepidoziaceae

Three-toothed Bazzania
Bazzania trilobata

Stem branching, prostrate, and closely crowded with overlapping leaves that seem to form a continuous frill to either side. May form a solid patch to 2 ft across. Leaves are three-toothed at the tip, bent toward the ground when they are dry, are firm and green or brown. Leaf and branch are to $\frac{1}{8}$ in. wide and individual branches are rarely over 5 in. long.

A leafy liverwort of the order Jungermanniales and the family Lepidoziaceae. It is found east of the 95th meridian and in Europe. In Washington State it is represented by *Bazzania tricrenata;* five other species are recorded as being found in North America.

Rarely found in a fruiting condition. Sex organs appear as buds from the sides of the branches. Male organ is a leafy structure while the female is a stalked capsule, or rather produces a stalked capsule that usually reaches maturity sometime late in the summer. One of these plants in fruit is a find many botanists never make.

Found usually on moist shaded logs or rocks; while they usually cling close to the supporting surface through most of their length, they may turn upward at the tips. Structures that appear to be rootlets are borne at intervals along the undersides of the trailing stems. The plant may be found in woodlands or in marshes but is essentially a shade lover.

(Continued)

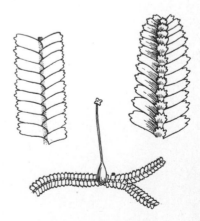

Three-toothed Bazzania

Not of any significant importance economically, and not commonly recognized by botanists except by specialists in the group. Most introductory botany texts may mention the order and may figure one of the plants, but they rarely do more than this. May occur in patches more than 2 ft across.

Order Jungermanniales./ Family Harpanthaceae

Variable Lophocolea
Lophocolea heterophylla
Stems trailing, slender, with rootlike structures beneath. Leaves bright green if growing in exposed sunlight. Underleaves small and inconspicuous. Upper leaves some entire and others deeply two-toothed, hence the name *heterophylla*. In the related *L. bidentata,* the leaves are broadest at base and with two long pointed lobes.

A leafy liverwort. This species common over wide areas in the United States, southern Canada, and Europe. Closely related *L. minor* is about half the size of this species. *L. bidentata* and *L. cuspidata* are found in much the same range.

In *L. heterophylla,* sex organs are borne close to the stem, the female looking like a leafy bud that produces a body bearing a small egg-shaped capsule that sheds spores in May. Sex organs mature in autumn but the sporophytes do not appear in nature until spring, though they may develop in January in the laboratory. *L. bidentata* has male and female organs on separate plants.

Common for the most part on shady banks, on rotten wood or sometimes exposed to strong sunlight. This is the general habitat of most species of the genus. In both *L. bidentata* and *L. cuspidata,* the leaves are broadbased and divided into two long, pointed lobes.

Not of any recognized economic importance. Attractive when well-developed but serves little purpose and does little if any injury to anything. Who can determine the economic value of a mass of this liverwort if one gains only a good feeling from viewing it?

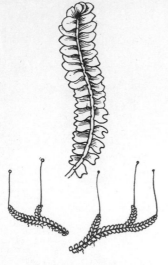

Variable Lophocolea

DIVISION BRYOPHYTA

Some classification systems include the liverworts, hornworts, and mosses in this division. In this fieldbook, we include in Bryophyta only the mosses, comprising some 600 genera and somewhat more than 14,000 species. Bryophytes are relatively small plants that grow in moist places on land. One might expect to find them on rotting logs, on the forest floor, in the vicinity of wet areas such as streams or ponds, or in swamps and marshes. Generally speaking, a moist habitat is a must for Bryophyta. The plant body lacks vascular tissue. Because bryophytes lack vascular tissue, they are condemned to limited growth in height. Probably the bryophytes arose from filamentous green algae. Bryophytes do develop rhizoids, rootlike structures. In the mosses the haploid gametophyte is the dominant stage in the life cycle. What one sees and calls a moss plant is the green, "leafy" gametophyte or sex-cell-producing plant. Usually this sex-cell-producing plant produces a sporophyte or spore-producing stage which is attached to and dependent on the gamete-producing stage which we know and recognize as the moss plant.

Acute-leaved Sphagnum

CLASS SPHAGNOPSIDA

Order Sphagnales./ Family Sphagnaceae

Acute-leaved Sphagnum
Sphagnum capillaceum
S. memorium of some authors.
Stems weak, crowded, and forming often a pure stand. Stems are clothed with two to three layers of empty cells that lack the spiral markings of *S. palustre* and its associates. Branches or branchlets taper to a fine point and may be distinctly purplish, though they may also be red or green. Lower portion of the stem is of course decayed.

There are 10 species in North America closely related to *S. capillaceum* and about 20 similar to the related *S. cuspidatum* that has long narrow branch leaves that are spreading and wavy-margined when dry. Distribution circumpolar in the Northern Hemisphere.

Male organs in this species are usually on red branches, while the female organs are on tassellike branches. The two may be on the same or on separate plants. The stalk of the spore case is of gametophytic origin as is true of the other peat mosses. The spore cases in this species are spherical and the spores are distinctly rusty.

Associated with these plants are the sundews, pitcher plants, cranberry, cotton grass, a number of our most beautiful orchids, bog rosemary, Labrador tea, cassandra, and other interesting plants. Most of these are uniquely equipped to use a minimum of water from beneath though their lower portions are almost always submerged.

This is a bog builder. It is used also in the making of absorbent bandages for use in medicine or for packaging trees and other plants to be shipped in marketing. Dried, living or dead portions may be used as fuel and in some parts of the world are the main source of fuel. They are also used in gardens to cut down weed competition. Peat has been used as a mattress filler. Dried sphagnum is offered for sale in herb shops in parts of the Orient and a tealike preparation is recommended for eye diseases and as a cure for acute hemorrhage.

Boat-leaved Sphagnum
Sphagnum palustre
Stems weak but crowded compactly together to form great mats, sometimes acres in extent, with the upper portion of the plant living and the lower portion in varying degrees of decay. Erect stems are covered with leaves as are the branches, the leaves being broad, spoon- or boat-shaped, with the tips turned back from the stem. Spiral lines on walls of stem scales.

In North America there are six species of *Sphagnum* with the walls of stem scales marked with spiral lines. Spore cases open by a circular lid with the opening unguarded by teeth. In this species the branchlets are stout and in large heads not slender pointed. Common in northern forests of the U.S.

Male and female organs in this species are on separate plants, the male plants being thick, coarse, and yellow, brown, green, or red. The spores are borne in a capsule at the end of a stalk that is gametophyte rather than sporophyte, thus differing from the true mosses and being like the other peat mosses.

Plant is one of the common bog builders. It can take water from the atmosphere by means of the absorbent cells on the branchlets and stems and thus is independent of water from beneath. In bogs and places where this moss grows, the water may be abundant but may also be physiologically useless. Is responsible because of its activities for the low pH (acid condition) of the bog.

Plants are used as fuel either from recent plants or from the peat they produce. They have been used as absorbent materials in surgery, or for holding water in packing around trees and other living plants in the process of being marketed. Peat is built by these plants at the rate of about 1 ft in 100 years. This and other sphagnums making up peat have been used to make a smudge screen against frost. Also, study of pollen from peat helps to identify the character of past vegetation in the vicinity of a bog.

Boat-leaved Sphagnum

Order Bryales./Family Ditrichaceae

Purple Horn-tooth Moss
Ceratodon purpureus
Stems rarely over 3 in. long, erect, densely crowded, branching, leafy. Leaves lancelike, keeled, with irregular margins, with tips commonly somewhat curved backward, bright green in winter but purplish is summer and fall, with margins curled, very narrow and, when dry, very wrinkled, sometimes distinctly reddish-brown.

One of the commonest of the true mosses, particularly in places that become very dry as on roofs, dry walls, at edges of flagging, or on dry lawns. Usually on sunny sites, sterile earth, cliff crevices, or rotten logs. Widespread in North America. This is the common species of the genus. Closely related genus *Ditrichum* has smooth, erect capsule while this one has ridged and more horizontal capsule.

Male and female organs borne on separate plants. From the fertilized egg arises the sporophyte with a stalk almost as long as the stem of the plant, slender and usually conspicuously glossy. Capsule at the end of stalk is dark chestnut or reddish-brown, furrowed and held in an almost horizontal position at an angle with the stalk.

The change in color of these plants from bright green in winter or early spring through brown to purple in summer and fall is rather spectacular. While the soil or surface on which the moss grows may be genuinely dry for a major portion of the year, it must be moistened occasionally if the plant is to survive.

Of little or no economic importance unless it may be considered as a soil maker or anchor. It may serve crudely as an indicator of humidity by the position of the leaves. It is at its best in early spring and does little during the summer months.

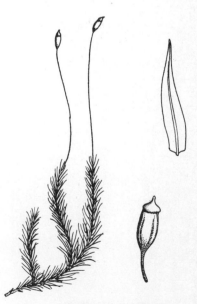

Purple Horn-tooth Moss

Pointed Mnium

DIVISION BRYOPHYTA. CLASS MNIONOPSIDA

Order Bryales./Family Mniaceae

Pointed Mnium
Mnium cuspidatum
Stems and young leaves bright green, becoming much darker with age. Leaves toothed slightly along the margin of the upper half; the midrib continues to form a prominent tip. Leaves near the top of each branch, definitely larger than those lower down. Plants from ¾ to 1½ in. high. Plants spread and form mats which are light to yellowish-green, on ground in woods and fields.

Common particularly in the central and eastern parts of the United States. *M. glabrescens* with elliptical leaves which twist and curl when dry; *M. insigne* a very large species; *M. spinulosum* with strongly elliptical capsule which is often red-brown at neck are common to western North America.

Male and female elements borne in the same cluster. From the fertilized egg develops the sporophyte with its slender, pale stalk, from the end of which droops the neckless capsule from which the spores are freed in spring. In this species, the sterile stems may bend over, root at the point of contact, and reproduce in this manner.

Favors ground that is well-shaded and not too dry, the plant being relatively delicate in comparison with some other mosses. It survives dryness by crisping the leaf margins, thus reducing the area exposed to drying. Related *M. venustum* does not reproduce with its sterile stems as outlined above for this species.

Of no obvious economic importance. It does come up for consideration rather commonly in classes in botany in college and high school levels. It lends itself well to culture in terrariums and is popular in this connection. It may serve some purpose in soil making and in soil anchorage.

CLASS SPHAGNOPSIDA

Order Bryales./Family Bryaceae

Silvery Bryum
Bryum argenteum
Gamete-bearing plants separate, one-sexed; silvery green to whitish; appear wormlike; are glossy, erect shoots. They form short plants, about ½ to 1 in. high, covered with rootlike filaments. Stem turns reddish-brown. Leaves silvery gray, small and ovate, wide at base; midrib of leaf visible, smooth margin. Stems have numerous shining catkinlike branches, each tipped with a brush of hairs. Sporophyte at tip of short shoot. Capsule nodding, cylindric, bright red at maturity.

Plants often densely tufted, sometimes spreading or growing with other mosses; almost always white or silvery; forms compact cushions. Stems almost always erect. Forms sods of tiny silvery shoots.

Occurs on ground, rocks, bricks, throughout the U.S. Considered a weed in many parts of the world. May surprisingly be found on exposed earth; on dry sandy soil, or on cement as at margins of sidewalks or paths, and in rock gardens. Found as a cosmopolitan species from the Arctic to the Antarctic and even in the tropics. However, most abundant locally in temperate regions. Once a student learns the key characteristics for this species, it is difficult to confuse it with others. Some authors classify this species as a moss forming short turf. Worldwide distribution of this species may result from worldwide activities of man. *B. capillare,* widespread in both Northern and Southern Hemispheres may be confused with *B. argenteum.* In western North America as far east as western Montana, we find *B. miniatum,* which is easily confused with other species of *Bryum;* however, *B. miniatum* has ends of leaves which are blunt, is dark green in color, and is most commonly found in wet cliff habitats. Predominantly, it occurs on coastal and montane regions of the Northwest but is also present farther east in the interior. Is a good terrarium plant.

Silvery Bryum

CLASS MNIONOPSIDA

Order Bryales./Family Leucobryaceae

Pin-cushion or Cushion Moss
Leucobryum glaucum
Plants grow as dense, thick, rounded, whitish cushions, spongy disks, or sods, which may reach size to 6 in. in diameter; peaty below. Common on humus in woods. May grow on trees. A very common form; appears from autumn to spring. Capsules rarely seen, ripen in autumn. Grayish or whitish green. Stems erect; capsules erect. Leaves crowded, oblonglanceolate. Appear as conspicuous greenish-white cushions about the roots of trees. Moist cushions are soft and spongy and decidedly green. If dry are easily crumbled to dust; so colorless as to suggest a parasitic or saprophytic existence. Name of moss means "white moss."

Occurs on soil or humus in open woods, in shady places, and at the margin of swamps. Widespread throughout U.S. In fact, genus is universal; some 74 species of which about 9 are known in North America.

A good terrarium plant.

In South, *L. albidum* is common and it is difficult to distinguish between the two species.

Leucobryum

DIVISION BRYOPHYTA. CLASS MNIONOPSIDA

Order Bryales./Family Leskeaceae

Common Fern Moss
Thuidium delicatulum

Stems loosely branched in a featherlike manner, the divisions often being subdivided. Plant appears to be loose, weak, sprawling, and somewhat fernlike. Stem leaves crowded, enlarged at the base. Branch leaves broadly oval, concave at base, long-pointed at tip, with toothed margins. Often forms mats several feet across.

Found from Labrador to British Columbia and south to the Gulf of Mexico. At best in moist shaded places. Closely related *T. recognition* is stiffer, more yellowish, with shorter, broader stem leaves. At least another half dozen common species are also referable to this genus.

Leaves around the female sex organs bear long hairs along the upper margin. Fruits are usually not abundant, do not form readily, and do not appear normally until autumn. Capsule is borne on a long stout stalk and at maturity is conspicuously curved. Lid is conical and spores mature in midwinter. Cap is split up one side.

One of the prettiest and daintiest of mosses, found not only in North America but in South America and Europe as well. It may cover stones in flowing streams or be found on drier ground. It may also be found growing on the roots or trunks of trees.

Merely a beautiful little moss known sometimes as the dainty cedar moss and named *delicatulum* or "dainty" by the great botanist Linnaeus. Ordinarily, it is not a difficult moss to identify. Related *T. abietinum* is cylindrical when dry and with stems only once divided; *T. minutulum* appears chainlike when dry because of curved leaves.

Common Fern Moss

Order Bryales./Family Brachytheciaceae

Rivulet Brachythecium
Brachythecium rivulare

Large, light-green, leafy moss, with stems that are to 2 in. tall and variable, depending on whether the plant is submerged or growing in the air. Leaves rather widely separated in comparison with some close relatives, with the stem leaves larger than the branch leaves, concave, to 1½ mm long, with obscure teeth.

Found commonly on wet rocks, in wet places, or submerged in still or in flowing water that is fresh. Found in Europe, and in North America in Canada and the United States from the Atlantic to the Pacific, south to Missouri and Virginia or North Carolina. At least nine relatively common North American species. *B. albicans*, a European species, is an early colonist on leeward side of sand dunes.

Plants bear either male or female sex organs; after the sperms fertilize the eggs, the sporophyte develops in autumn with long stalks capped by cylindrical spore case that is not beaked but rather broad and spindle-shaped. Spores are shed through the winter or well into it and from these develop a thread that eventually bears new stems.

This plant favors cool swift streams though it may be found elsewhere. On ground is glossy green; submerged and growing in streams, stems become long and slender and tend to appear to float. The related *B. rutabulum* bears male and female sex organs on the same stem and forms large bright green mats while *B. nelsoni* is a form with long pointed leaves that grows in the Wyoming and Colorado region. *B. frigidum* in western North America from Alaska to Mexico east to Rockies.

Probably of little or no direct economic importance to man. It may serve as food and shelter for minute animals that provide basic food for larger animals that have value to man. The plants have an esthetic appeal because of the place in which they grow and because of their fresh greenness. *B. albicans* often crowds out grasses in lawns where too wet and soil is acid. Genus may also serve along stream beds as a collector of organic materials and mineral matter.

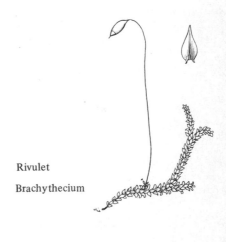

Rivulet

Brachythecium

Order Polytrichales./Family Polytrichaceae

Common Hair-cap Moss
Polytrichum commune

Stems to 1 ft long, erect, crowded. Leaves crowded, to ¾ in. long, with enlarged, silvery, shining base, with margins toothed at least to the middle. In related *P. juniperinum* leaf margins are translucent and rolled over the upper leaf surface. Plants often form pure stands covering areas yards across.

P. commune and *P. juniperinum* have a general distribution over the Northern Hemisphere, with some related species rather limited. *P. piliferum* requires sand or rocks as a base for growing but is found from sea level on the Atlantic Coast to elevations of over 12,000 ft in Colorado. *P. strictum* sometimes as a "single-species community" signals the drying of a bog area.

Male and female organs are borne on separate plants, the clusters bearing the male organs being terminal and swollen into a flowerlike structure. Sporophyte consists of a 4-in. stalk bearing at the end a capsule which is almost cubical, and which at first is erect and then horizontal, capped by a blunt, short-tipped cap and a hoodlike calyptra.

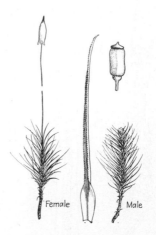

Female Male

Common Hair-cap Moss

(Continued)

Common in woods or at edges of peat bogs or in other relatively infertile places, where they may form clear stands. Prefer wet situations, but can survive some drying. They do not absorb water from the air as do the associated sphagnum mosses. The family is characterized by having from 32 to 64 teeth at the capsule mouth and a light membrane over the capsule mouth.

Of little economic importance though the plants have sometimes been used as a cheap stuffing for pillows and upholstery. They may serve some purpose as soil anchors, as holders of water, and as soil makers. The plants are commonly used for study in botany classes because of their size and abundance. Believed, by some, to have medicinal value if prepared as a tea and taken to help dissolve stones in the kidney or bladder, although no scientific evidence is available to support the contention. Also used, at one period in our history, in decoction form to "strengthen and beautify" ladies' hair. Laplanders use this moss for bedding. In northern England hair-cap mosses have been used to stuff mattresses and upholstery.

Juniper Hair-cap Moss
Polytrichum juniperinum
Conspicuous because of bluish-white bloom that covers the plant. Stems to 4 in. long. Leaves tipped with a red point that is obscurely saw-toothed, with enlarged, sheathing leaf base, but the main margins of the leaves are not notched as they are in the common hair-cap moss. Stem tip often swollen. Probably the most cosmopolitan moss in the world.

Found in dry places where the sand or rocks are basically dry and usually not well occupied by other plants. In coniferous forests at all elevations. Occurs in a mass or may be solitary; more often in sun than in shade. This species is common all over the Northern Hemisphere. The related *P. piliferum* is found from sea level to the 12,500-ft elevation in the Rockies, in full sun and where the rocks are largely silica.

Male and female sex organs borne on separate plants with male at least in conspicuous terminal leafy structures. From fertilized egg develops sporophyte with long red-brown stalks that may be 2½ in. long with rather four-angled, relatively short spore cases at tip. Spore case usually rather erect.

There are at least a half dozen common species of this genus found widely distributed in North America and usually with rather conspicuous differences. In this species, a translucent leaf margin that rolls over the upper surface of the leaf is rather distinctive if the general bloom and other characters are not adequate. Abundance of genus in one location, almost without exception, an acid indicator.

Possibly of little direct economic importance though some species have been used as cheap stuffing for upholstery and the mats undoubtedly do serve to prevent rapid runoff of water from otherwise bare rocks, thus contributing to the prevention of soil erosion and floods.

Order Polytrichales./Family Polytrichaceae

Slender Catherinea
Atrichum angustata
Stems slender, a few inches tall at the most, well-covered with slender leaves. Leaves with midrib-like structure composed of four to seven folds and covering half the leaf. Leaf bases do not sheathe the stem. Under a microscope, the tip margins are finely toothed for most of the upper half, curl conspicuously when dry; dark olive-green when young but reddish-brown when old.

Very common in the northeastern part of the United States particularly on half-shaded, hard, compact soil. Related *A. macmillani* is smaller and more dense. *A. undulatum* with its purple capsule is common in the Great Plains area and *A. crispum,* with smooth leaves, grows on peaty soil from New Jersey to British Columbia and Oregon. *A. selwynii* of western North America found on road and stream banks and in flower gardens.

Male and female organs borne on separate plants. Male rosettes are conspicuously bright red while the females are not conspicuously developed. The developed sporophyte is a dark green or purple capsule, mostly cylindrical and curved, with a beaked lid and one-sided cap and borne on inch-long shining red stalks.

Most conspicuous from late autumn to early spring where patches of plant may stand out conspicuously on the ground in woods and along roadsides in North America, Europe, or Asia. It may form a sort of felt cover for otherwise naked stones.

Apparently of no recognized economic importance though it must make some contribution to the story of soil making and possibly to soil erosion. Plant is sometimes listed in literature as *Catherinea* rather than as *Atrichum,* a difference of opinion significant primarily to systematic bryologists.

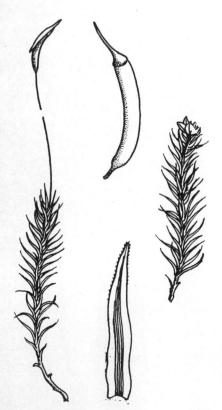

Juniper Hair-cap Moss

Slender Catherinea

DIVISIONS PTEROPHYTA, ARTHROPHYTA, MICROPHYLLOPHYTA, PSILOPHYTA

The following genera and species, (pp. 97–110) would, in an earlier classification scheme, have been considered in a single division, Pteridophyta. In a later scheme, they would have been considered a part of the phylum Tracheophyta and would have been collectively the vascular cryptogams (plants with organized conducting tissues but not producing seeds). We will consider these vascular cryptogams as representatives of four different divisions. They are the division Pterophyta, or true ferns; the division Arthrophyta, or horsetails, pipes, and scouring rushes; the division Microphyllophyta, or club mosses and spike mosses; and the division Psilophyta, or whisk ferns. In this field-book, we will not treat the divisions in phylogenetic order from simplest to most complex. Here we will treat them in an order from the division which is most commonly recognized by the amateur in the field and continue through to the division which is least apt to be so readily recognized.

DIVISION PTEROPHYTA

This division includes the vascular plants (those which have conducting tissues) and encompasses the ferns. There are many fossil forms which lived in Devonian times. Ferns have large, feathery leaves which typically unroll from the tip during growth. They possess sporangia (spore-bearing bodies) on the underside of their leaves or at times on special fronds. Ferns, as a group, are ancient land plants appearing as early as the Devonian. Including fossil and extant forms, we usually find eight orders listed in the Pterophyta, of which only five are extant. One order, Filicales, has the largest number of living ferns; 12 families with some 170 genera with about 9,000 species, or around 90 percent of all ferns.

Ferns have stems, roots, and leaves. In some forms, the stem is much reduced; in others, like the tropical forms, it may be a tall column and hence the name, "tree fern." In almost all ferns, the leaf is the dominant, conspicuous organ.

The sporophyte generation is the conspicuous featherlike, spore-producing plant and may be the only part of the life cycle known to the beginner. The sex-cell-producing generation (gametophyte) is very small and may have been seen but not recognized on flowerpots in greenhouses. The two generations alternate: from sporophyte with spores to gametophyte with sperms and eggs to sporophyte, etc.

Rockcap Fern

CLASS LEPTOSPORANGIOPSIDA
Order Filicales./Family Polypodiaceae

1. Rockcap Fern
Polypodium virginianum

2. Resurrection Fern
P. polypodioides
Fronds to nearly 1 ft long, evergreen, produced in rows, from a brown, densely scaly rootstock; smooth on both sides in (1), but grayish with fine scales beneath in (2). The parts of the blade seem cut almost to the central axis but each part is not separated and borne on a short stem as in the Christmas fern. The general shape of the blade may vary from triangular to oblong or even to long pointed.

From eastern Alberta to Newfoundland and south to Georgia and Arkansas, for (1), with close relatives in Europe which were long thought to be of the same species. (2) ranges from Iowa to Delaware and south into tropical America and in South Africa.

On dry rocky ledges, on boulder piles, on tree trunks, or even on rich woodland soil, doing best on slightly acid soils. (2) covers the trunk and lower branches of trees.

The spore-bearing spots borne on the undersides of the fronds, usually near the tip, are round spots which are not protected as in some other ferns. Sometimes these spots are so crowded as to seem to be continuous. They appear midway between the midrib and margins of the part they occupy.

A decorative fern of gorges and rocky exposures; survives some exposure to sun and may be transplanted but should be left in natural setting since nurserymen supply plants more likely to survive change. Fronds curl in dry weather or in winter but recover with moisture. Known sometimes as "female fern," and a medicine supposed to have purgative powers is extracted from the rootstock. Can be grown in shaded rock garden; also cultured indoors.

Oak Fern

Maidenhair Fern

Bracken, Brake

DIVISION PTEROPHYTA. CLASS LEPTOSPORANGIOPSIDA

Oak Fern
Gymnocarpium dryopteris
Fronds reach a height of 1½ ft, are smooth, except that the main axis may be finely fuzzy; rise in a row from rootstock; are thin, not evergreen. Stipe pale yellow, maturing to a greenish-brown. Blade triangular in general outline, and in three main parts; in this respect, differing from the long beech fern *P. polypodioides*, the broad beech fern or southern beech fern *P. hexagonoptera*, which are with the pinnae connected to the main axis without separate stalks. Confusion in accepted scientific name; may appear in genera: *Gymnocarpium, Polypodium, Phegopteris, Currania*, and others.

From Alaska to Labrador and Newfoundland, south to Virginia and in the northern parts of Europe and Asia.

Rocky woodlands and slopes and at the margins of swamps commonly where the soil is slightly acid but almost always where there is rather deep shade.

The spore-bearing spots are on the underside of the pinnules arranged usually near but not at the margins and appearing more naked and exposed than many similar spots in other ferns. They are not borne on special fronds.

This fern rarely grows in abundance in any one spot and lacks the lacy attractiveness of many other species. It provides a measure of beauty in dense, shaded woodlands at times of the year when other plants of the region are not conspicuous.

Maidenhair Fern
Adiantum pedatum
Height of fronds to 20 in. Rootstocks slender, creeping, black, wiry, and producing fronds in rows. Stipe of frond slender, tough, shining, brownish-black, forked unequally to make a fan-shaped top which is held more or less horizontally. Each branch of stipe bears 5 to 9 branches which bear thin irregularly oblong pinnules which are lobed on upper margin. Whole frond wax-coated, and sheds water. Not persistent in winter. Southern maidenhair, *A. capillus-veneris,* has main stipe unbranched.

Northern maidenhair, from Minnesota to Nova Scotia and south to Georgia, Oklahoma, and Louisiana and in Asia. Southern maidenhair, North Carolina to Texas; grows on limestone rocks.

Wooded, well-drained, rich-soiled slopes, without much preference as to acidity, but doing best where the acid content is low. Should have shade. Color changes due to maturity from green to brown and red-brown.

Spore cases borne in interrupted rows of fruit spots under the downward and inward-bent tip margin of the pinnules, the pinnule margin providing some protection as a shield and limiting slightly the direction in which the spores may be freed. The whole spore-producing section makes a series of cresent-shaped formations along outer edge of each pinnule except near tip.

Easily cultivated and transplanted and for sale by gardeners. Fronds wilt quickly when cut, so are not useful for picked bouquets. Species named by Linnaeus in 1753 from plants from Canada and Virginia. Name may come from hairlike resemblance of stipe or because of use of European Venus's hair fern in hairwash concoctions.

Bracken, Brake
Pteridium aquilinum
Height to 3 ft, with widely spreading fronds that are twice pinnately divided from a central axis. Stipe tough, brown, coarse, erect. Branches twice pinnate. Blade coarse, dull green, with the upper pinnules undivided. Rootstock tough, the diameter of a pencil, and spreading rapidly underground, sometimes as much as 50 ft in a year and producing a row of erect fronds above ground, thus indicating its position. The fronds die back each year.

Newfoundland to Wyoming, Arizona, and Florida, being locally very common. A closely related European bracken is *Pteris aquilina*, long thought identical with the American species. Linnaeus named the European bracken in 1753.

In open woods, stream margins, banks, and slopes but most commonly where it is relatively dry and where the soil is neutral or strongly acid. Its abundance does not seem to be affected by neutral or acid soil.

The margins of the pinnules are incurved all around beneath to cover the sparsely hairy, spore-bearing areas. These fertile margins are continuous and not interrupted as in the maidenhair and some other ferns. The fruiting areas mature in August although new fronds are being produced throughout most of the summer growing season, some of which of course do not reach maturity.

The young fronds may be eaten either raw or cooked, particularly if gathered in spring when they are tender. Fiddleheads covered or partially covered with silvery gray hair and said to be shaped like eagle's

claw. One of first ferns to be killed by frost. Mature leaves of bracken
cause poisoning of livestock. Rootstocks used by Indians in making baskets and fronds as a thatch. Brake is not easily transplanted. Name may
refer to the Greek *pteris* meaning wing. A section of stipe is reputed to
resemble the print of the devil's hoof.

1. Maidenhair Spleenwort
Asplenium trichomanes

2. Blackstem Spleenwort
Asplenium resiliens

3. Ebony Spleenwort
Asplenium platyneuron
Genus, *Trichomanes,* in older texts. (l) and (2), to 8 in. high; and (3) to 2
ft high. Fronds all delicate-appearing but evergreen except the erect
fertile ones in (3), which are only semievergreen. Stipe and long rachis
dark brown in (l) and (3), black in (2). Pinnae almost opposite in (1),
mostly opposite in (2), and mostly alternate in (3). They are pointed at
the base in (1), "eared" in (2), and in line with the rachis in (3).

(1), from Alaska to Nova Scotia and south to Georgia and Arizona. (2),
from Missouri to Pennsylvania and south through Florida, Jamaica, and
Arizona to western South America. (3), Maine to Colorado and south to
Florida and Texas.

(1), on wet rocks and shady crevices or on dry limestone ledges or in
slightly acid soil. (2), on rocks in the shade, usually in slightly alkaline
soil. (3), on rocky ledges or clay, either of acid, alkaline, or neutral nature.

Spore-bearing spots few, short, placed near the center along veins in (1)
and (2); in (3), like short, oblong dashes starting from the midrib and
extending halfway to the margin. In (3), the fertile fronds are erect and
the sterile are prostrate, while in (l) and (2) the fronds are alike whether
fertile or sterile.

Beautiful little ferns that are rare and, near centers of population, are
growing rarer. They do well for a time when transplanted into glass
containers indoors, but usually such plants eventually get destroyed. It is
better to leave the plants in their natural setting, where they are at their
best.

1. Northern Lady Fern
Athyrium filix-foemina

2. Glade Fern
Athyrium pycnocarpon

3. Silver-stripe Fern
Athyrium thelypteroides
(1), to 3½ ft and (2) and (3), to 4 ft tall; difference in size not a definite
criterion but (l) is usually smaller plant. Fronds pinnate, tapering at
base and tip about equally, with pinnae with entire or wavy margins in
(2) but deeply toothed in (1) and (3). Stipe yellow to red in (1) and (2), to
brown in (3), and scaly below. Rootstock short and horizontal in (1) and
(2), slender in (2), producing fronds in clusters in (2) and in short rows, in
(1) and (3), retaining old stipe base in (1). Stipes of (1), pink with pinnae,
pale green in spring but yellow in summer.

(1), Manitoba to Newfoundland, Virginia, and Missouri. (2), Wisconsin
to Maine, Georgia, and Louisiana. (3), Minnesota to Nova Scotia, Missouri with variety in Asia.

Moist shaded woods and thickets in all species, with acid soil favored in
(1) and (3) and neutral soil in (2). Generally does best in cool spots.

Spore-bearing spots under pinnae numerous; in (1), short, crowded, and
hooked; in (2), in long, regular rows; and in (3), shining, and appearing
silver-striped. Spores of (1), yellow-brown. Fertile fronds less spreading
than sterile in (1); spore-bearing spots in (1), at first horseshoe-shaped,
then more linear. Projections at base of stipe good identification.

One of best groups of ferns for replanting about house. (1) gets name
lady fern from resemblance to European lady fern, whose spores in shoes
are supposed to render wearer invisible and also to confer on such persons second sight, useful in finding lost things, these qualities vanishing
when "seed" is lost. Indians used it medicinally to stop urine.

Maidenhair Spleenwort

Northern Lady Fern

Southern Lady Fern

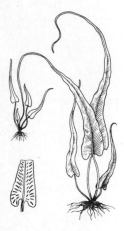

Walking Fern

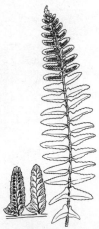

Christmas Fern

Southern Lady Fern
Athyrium asplenoides

Fronds not evergreen, to 30 in. long, broadest just above first pair of segments, once or twice divided, with ultimate segments grown fast to support rather than on independent petioles, with margins coarsely doubly indented and with pointed ends. Fronds with main axis reddish or yellowish. Rootstock long, horizontal, with some old erect bases of older fronds. Stipe nearly equaling blade in length, bears sparse brown scales.

Found where soil is somewhat sour in wet woodlands, swamps, along stream banks and in similar places from Massachusetts to Florida, west to Texas and Missouri, frequently being one of the commonest ferns within this area.

Usually associated with other wetland semisour soil species. It may form clear stands under some circumstances.

Fruiting bodies, longer than broad, are borne halfway between the middle axis of ultimate segments and their margins, incurved, hooklike cases that open from one side and free very dark to black spores. Protecting case may show gland-tipped hairs under the microscope.

This species was first observed about 1803 by the botanist Michaux, who cruised the eastern parts of the United States for plants for France. It was placed in this genus by Amos Eaton, the engineer, geologist, and botanist, about 1817. Can be used as an addition to a woodland garden in its range but spreads rapidly so may not be desirable.

Walking Fern
Camptosorus rhizophyllus

Fronds evergreen, close to ground or only slightly raised; shaped like long, slender spearheads; tips, touching ground take root and produce clusters of similar fronds thus giving rise to common name, walking fern. Fronds to nearly 1 ft long, though more commonly about 4 in. Stipe brown at base. Blades with wavy margins; their veins form rather large areas like a net. Rich, waxy, blue-green.

From Minnesota to southern Quebec, south to central Georgia and eastern Oklahoma.

Commonly on cliffs, and most abundant on limestone outcrops, though may grow on other rocks, in shade, and even on rich humus or tree trunks; usually northerly exposure. Does not normally favor acid soils.

Fruiting bodies borne in long, slender, irregularly placed slits or streaks on the underside of the blades of the fronds, and appearing brown to blackish or rusty depending on degree of maturity reached. Genus name alludes to irregularly arranged sori (fruiting bodies); *kamptos* meaning bent and *sorus* meaning spore-producing body or heap. One of the most easily identified ferns because of these spots and the rooting habit of the frond tips.

Notable fern because of interesting way of spreading and also because it responds readily to transplanting into indoor table gardens in jars and aquariums. It can be purchased from some nurserymen and because of usual rareness should not be collected from the wild state except under favorable conditions.

Christmas Fern
Polystichum acrostichoides

Fronds more or less prostrate, evergreen, from 10 to 30 in. long, dark green above and lighter beneath. Sterile fronds, shorter than those which bear spores; fertile, more erect. Fronds of the year gray-green and unroll from crosiers or fiddleheads at tips. Second growth of fronds may appear in August if preceding month has been unusually rainy. One variety has crisped or ruffled edges that make it exceptionally beautiful. Related holly fern *P. braunii* has blade twice pinnate.

From Nova Scotia to Wisconsin, south to central Florida and eastern Texas with the related holly fern with wider range extending to Alaska and Newfoundland.

In woods and shaded swamps and thickets or in rock crevices irrespective of whether the soil is acid, neutral, or alkaline but rarely where direct sunlight strikes for any considerable period of time.

Fertile fronds bear spores in spots clustered near the central line on the underside of the modified terminal pinnae. Pinnae bearing spores are usually smaller and more widely separated and set at a more acute angle with the main axis than are the sterile pinnae. Spring fronds bear spores in July and the second-growth fronds in September.

One of earliest of ferns to develop each season. Fronds, because of graceful form, deep green color, and evergreen qualities, are used considerably in decorations, particularly in funeral sprays and at Christmas. Much more beautiful and appropriate in natural setting but have a commercial value for cut fronds and for living plants. It has been said by some that the leaflets (pinnae) resemble Christmas stockings, hence its common name; others say it is because of its evergreen nature.

DIVISION PTEROPHYTA. CLASS LEPTOSPORANGIOPSIDA

1. Spinulose Woodfern
Dryopteris spinulosa

2. Evergreen Woodfern
Dryopteris s. var *intermedia*
Only an expert can differentiate between (1) and (2). Sometimes, even they are confused as (1) and varieties (2) *intermedia, fructuosa,* and *americana* interbreed. Fronds to 30 in. long and evergreen. Sometimes not all of fertile frond is evergreen. Rootstocks produce fronds, generally in rows (1), sometimes in clusters (1), (2); color of scales on stipe varies from pale to dark brown in both. Fronds of both dark rich green and ultimately divisions are more or less leathery; some authorities believe (1) less leathery. Subleaflet (1) without prominent lobing, only slightly toothed, teeth appear to bend toward tip; subleaflet (2) with prominent lobes and distinct and divergent teeth.

(1) From Quebec to Minnesota and British Columbia to Virginia, West Virginia, Kentucky, widespread to the West Coast; (2) range similar but perhaps farther south. Not possible to separate (1) from (2) on basis of location or abundance.

Both occur in rich moist woodlands; (1) more frequently in or near wet woodland swamps, (2) generally in drier regions where soil limy or neutral. Both favor shady slopes.

(1) with spore-bearing spots small and near tips of veins; covering of fruit dot (indusium) smooth; (2) spore-bearing spots small but occur away from vein tips, indusium glandular (hairy). (1), (2) spots on underside. Shape of spore-bearing spots similar but some say (1) more kidney-shaped. So much hybridization between (1) and (2) that no one characteristic can be used to separate them.

Beautiful and popular ferns. Unusually hardy and survive transplanting well. Fronds unfortunately cut and sold for ornament though it does not seem that this should be permitted. Apparently no states place this fern on "protected plants" list. Some confusion in classification, (1) may appear as *austriaca* of some authors.

1. Marginal Woodfern, Evergreen Woodfern, Leatherleaf Woodfern
Dryopteris marginalis

2. Goldie's Giant Woodfern
Dryopteris goldiana
Fronds of (1) to 30 in. long; of (2), to 40 in. long; of (1), evergreen; of (2), surviving late into season but not evergreen. Rootstocks produce fronds in clusters in both. Main axis of (1), with shining brown scales; of (2) scales throughout length, especially scaly at base, scales pale tan with darker centers. Both have fronds that are a dark, rich green and are more or less leathery. In (2), there is a tendency for the basal pinnae to be narrowest near their base while this is hardly so in (1).

(1) Found from Nova Scotia to Alabama and west to Oklahoma and Minnesota and in British Columbia as well. (2) Found from Newfoundland to South Carolina and west to Minnesota. (1) is usually a very common species while (2) is rather uncommon.

(1) Hardy evergreen, abundant and common at variety of altitudes, may be on rocky ledges, or in wooded rocky lowlands, along stream banks; a beautiful addition to deep dark ravines, prefers shade and tolerates acid or neutral or basic condition, usually on rich soil; (2) on rich moist woodland soils, prefers cooler inner woods site, in shaded places, frequently on rocky slopes at higher elevations.

In (1), the spore-bearing bodies are borne along the margins of the undersides of the ultimate divisions. In (2), they are borne conspicuously in two rows nearer the center of the midrib. In each case, the spore-bearing areas are essentially round. They may hybridize with each other and with other species of *Dryopteris.*

Beautiful and justly popular ferns that are too hardy for their own good. They are of course cut for ornament but deserve much more protection than they get.

Crested Wood Fern
Dryopteris cristata
Fronds to 40 in. tall, almost but not completely evergreen, borne in clusters from the horizontal rootstock that is covered with pale brown scales. Bluish-green color of this fern stands out in its damp woods location. Main axis of frond sparingly scaly. Lower pinnae definitely widest at base and may be in more compact triangles than indicated in sketch. Fertile fronds wither quickly in fall but those that are sterile may be almost evergreen. Look for wide-spaced pinnae looking like ladder rungs along the stem.

Found from Newfoundland to Virginia and west to Arkansas and Idaho, with range extended east and west to Europe and on into Asia. It is a relatively common plant throughout its range.

(Continued)

Spinulose Woodfern

Marginal Woodfern

Crested Wood Fern

Favors soil that is approaching an acid condition. Commonest in bogs, marshes, wet meadows, thickets or on springy slopes and in gorges.

Spore-bearing areas are borne between the midrib and the margin of the pinnules, usually more or less squarely across branches of the veins. Margins of the pinnules saw-toothed. Spore-bearing bodies round, without hairs evident in some related species.

A fine plant identical with the plants found in Europe. The fronds are probably generally narrower than those here figured and are often characterized by having the divisions twisted so that they are normally held in a horizontal position rather than in the plane of the frond as a whole. *Dryopteris* has a large number of species. Some called oak ferns from species name and from preferred habitat, oak forests. Also called common wood fern. Some species said to have medicinal properties.

Northeastern Marsh Fern
Thelypteris palustris
Fronds to 30 in. long, gray-green in general color, not evergreen, fertile ones being at best in midsummer; borne in a row on slender, horizontal rootstock. Main axis (stipe) somewhat bronzed in its lower areas. Expanded part of frond finely downy. Pinnae deeply cut into rather blunt sections. Veins in the sterile fronds forked; those in the fertile fronds unbranched. Rhizome (underground stem) black, may be wide-creeping, branched.

Found from Newfoundland to Georgia, west to Oklahoma, Florida, Texas, and Manitoba. Also found in northeastern Asia. Closely related to New York fern, and another species which some authors call bog fern.

Does best in soil where acidity is low. Nevertheless, it is found in bogs as well as in marshes; margins of swamps, and wet woodlands where the acidity may vary considerably. One of our very common ferns in its range. May occur in moist places or in shaded locations. Usually in soils which are fertile. Look for it along woodland streams also.

Fertile fronds have their margins curved inward in part over the spore-bearing areas, which are more common on upper leaflets and are concentrated in rows near midveins. Spore-bearing areas are midway between the midrib and the margin of the pinnule and are commonly rather crowded.

Some confusion as to generic name. Sometimes placed in *Dryopteris*. European fern is *Dopalustris pubescens*. That American and European plants were not the same was recognized only comparatively recently. In woods may be associated with woodland orchids to present an attractive array. Also associated with alders, buttonbush, and cattails.

New York Fern
Thelypteris noveboracensis
Height of fronds to over 2 ft. Rootstock slender and producing fronds in a short row. Fronds delicate, yellow-green, thin, with longest pinnae in middle and forming a blade which tapers to each end conspicuously. Pinnae hairy on heavier veins on undersides and veins are mostly unbranched. Stipe relatively short, pale green, and only slightly scaly, slender. Grows in tufts along dark brown rootstock, some three or more leaves form tuft.

From Minnesota to southern Newfoundland, and to Arkansas and Georgia. The related bog fern *T. simulata* extends the range to Prince Edward Island in Canada and south to Alabama. It lacks the conspicuous tapering of the blade at the base.

Moist thickets, swamps, and woods where the soil tends to be slightly acid. Often grows to cover considerable areas to the exclusion of other plants. It occurs mostly in the mountains in the southeastern United States. Generally seen in areas exposed to some sunlight, so not common in deep woods. Usually in drier location than marsh fern, *T. palustris*.

Fertile fronds appear in summer and superficially resemble those that are sterile. Fruiting spots on undersides of pinnules, rather round and borne just back from margins. Protecting portion is covered with fine, glandular hairs.

Most beautiful in its natural setting but, because of its habit of growing in rather clear stands, it is popular with those who collect ferns for decoration for special events. Fronds are not sufficiently hardy to hold for use commercially.

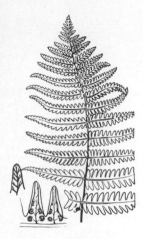

Northeastern Marsh Fern

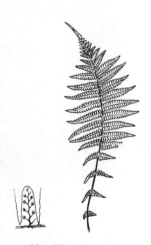

New York Fern

DIVISION PTEROPHYTA. CLASS LEPTOSPORANGIOPSIDA

1. Brittle Fern
Cystopteris protrusa

Bladder Fern

2. Bladder or Bulblet Fern
Cystopteris bulbifera
Fronds from 1 ft in (1) to 4 ft long in (2); rising in clusters from usually short rootstocks but, in (2), often reclining or nearly so. Stipe brown to black near rootstock, then changing to straw-colored or green toward tip, few scales in (1) and yellow in (2); rachis green and smooth in (1), and pinkish in its lower portion in (2). Whole fronds appear to be long and tapering, more particularly in (2), giving always the general appearance of being weak and brittle.

(1), from Alaska, the states of U.S. and northeast to Nova Scotia and North Carolina, Georgia, and Arkansas in its different varieties. (2), from Newfoundland to Manitoba and south to northern Georgia and Arizona; also in Alaska, broken distribution south to Utah.

(1), on slopes and ledges partly sheltered from sun; on soils usually acid; sometimes in rock crevices, frequently locally abundant; forming small mats, sometimes in rich woodland soil in shaded location or on tree stumps. (2), where soil is alkaline or neutral; may appear in swamps, but usually on limestone exposures; very frequently appears on rocky slopes and steep banks; shaded limestone.

Fruiting spots round and on center of lateral veins that branch from central veins on undersides of fronds, in (1). In (2), same general description might hold, but some pinnae bear, in their axils or lobes, small green reproductive bulblets which give plant its common name. These are usually conspicuous in the mid- and late summer months. Bulblets, produced on undersurface, drop off when mature, germinate, and produce new plant.

These ferns ornament many a rock exposure in gorges or hang gracefully over some woodland bank. In hot weather of summer, fronds of (1) turn yellow, then brown, but may recover if wet weather returns. Some specimens of (2) have been said to be fragrant when crushed. Can be grown successfully in a moist, shaded home fern garden but more attractive in their natural locations.

Blunt Lobe Cliff Fern
Woodsia obtusa

Blunt Lobe Cliff Fern

Fronds to 3 ft long, arising from very short rootstocks. Sterile and fertile fronds resemble each other in general but sterile ones are evergreen. Unlike the bladder fern, it has fronds minutely hairy and scaly and veinlets show as small markings at lobe tips. In sun, fronds become yellow. Stipe and rachis pale green to yellow with scales extending up stipe and rachis to glandular hairs above. Stipe jointed above rhizome and breaks at joint.

From southeastern Canada through to Alaska and south through northern United States and in the Rockies. The related rusty cliff fern *W. ilvensis* is found also in northern Asia and Europe. Our commonest and largest *Woodsia*.

Found on well-drained, shady soils, generally where the soil is neutral; often in cracks in walls or even in sand. Common in limestone areas; on shaded, rocky banks and cliffs, also in dry, rocky woods.

Fruiting dots round, large, and on or below minutely toothed lobes of smaller parts of fronds, arranged in irregular rows, usually toward margin, and commonly terminate short veins branching from main vein of lobe. Mature about July and their protecting covers are lobed, spreading back to form small black stars with brown centers.

This species was originally discovered in Pennsylvania. It is a pretty fern when it is fresh and young, but becomes unsightly when mature as fronds become blotched with white. It is at its best in June. It does not transplant well into indoor ferneries under ordinary treatment and should be left where it may have established itself naturally. May be evergreen in north of its range. When growing in shade, is fuller, darker green and appears more delicate.

Hay-scented Fern
Dennstaedtia punctilobula

Hay-scented Fern

Height from 1 to 2½ ft fronds appearing in rows arising from slender rootstocks; older parts of rootstock with reddish-brown hairs. Stipe or frond stem shining brown. Axes of branches pale green. Finest subdivisions thin, paperlike, with downy hairiness and a pale yellow-green color; with appearance of being long and tapering from basal pinnae nearly 5 in. long. Fronds last through summer, and covered with slightly sticky hairs; fragrant because of wax they produce.

Southern highlands of Georgia to Arkansas and north to Minnesota and Nova Scotia, being commonest on rocky slopes and in dry woods.

(Continued)

Soil preferably somewhat acid in nature. Highly sensitive to early frosts and may bleach almost white in autumn. May spread so rapidly as to crowd out all other forms. Difficult to eradicate.

Spores in cases, borne in a recurved toothlet of pinnae of fronds, with one sorus of spore cases usually at upper margin of each lobe. Sterile and fertile fronds superficially much alike. Cuplike indusium bearing spore cases unlike and not easily confused with similar structures in other ferns.

Discovered by Michaux in Canada. Formerly considered in genus *Dicksonia*. Possibly because of fragrant wax is not palatable to cattle and therefore multiplies extensively in suitable pasture lands. Fragrance most conspicuous in ferns grown in dry, sunny spots.

Sensitive Fern
Onoclea sensibilis

Sensitive Fern

Height to 30 in. Sterile fronds broad, in general rather triangular and like coarse feathers, yellowish or sometimes bluish-green, with segments near tip not cut through to midrib. Fronds held more or less vertically or arching gracefully but slightly. Rootstock shallow, much branched and sending up fronds in rows, the sterile appearing early in the season and the fertile usually coming in the autumn. Mass of young plants (fiddleheads) appear pale red in spring.

From Newfoundland to Saskatchewan and south to Florida and Texas with a close relative in Asia. Locally very abundant and conspicuous in muddy ditches, marshes, swamps, seepage areas, under bridges.

Sterile fronds do not survive a frost; die back leaving only erect, beadlike, black-brown fertile fronds or spikes standing.

Fertile fronds produced in late summer persist through winter and produce spores in at least 2 years; may persist battered even more than 2 years. Spores bear chlorophyll and grow at once into sex-bearing stage. Shape of spore-bearing parts of fertile fronds has led to name of beadfern, which is not commonly used, however.

Fern was collected first in America in early eighteenth century and described by Linnaeus in 1753. American variety was named in 1937 by Edgar T. Wherry. Common name may come from fact that it wilts quickly when cut or touched by frost but it is hardy in other ways. Is grown relatively easily. Reported to be poisonous. Generic name reported from unknown ancient origin and adopted by Linnaeus. Not recommended for home fern garden because of its spreading growth.

Ostrich Fern
Matteuccia struthiopteris

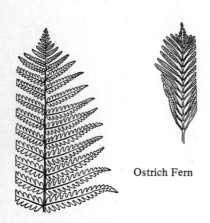

Ostrich Fern

Height of sterile fronds to 8 ft. Stipes; brown and scaly, with whole, sterile fronds, coarse, narrowest at base, and narrowing more gradually toward base than toward tip, in this latter respect differing from cinnamon fern which it otherwise slightly resembles. Rootstocks stout, growing vertically out of ground, but bearing underground, brown, chaffy runners which reach surface at their tips and bear fronds. Sterile fronds surround fertile fronds to form a sort of vase; wither late summer. Name from ostrich-plume-shaped leaves of dark rich green. In winter one finds its lyre-shaped, brown, stiff, erect fertile fronds.

From Alaska to Newfoundland and south to northern Virginia, being much more common in northern part of range or at least more common in United States along northern border.

In swamps, on rubble-covered slopes, in open lowland, and wooded flats where there is an abundance of humus and where the soil is neither strongly alkaline nor acid but essentially neutral.

Spores produced on special fronds borne in center of year's cluster from one rootstock in late summer or autumn and looking plume-like in shape or possibly like a badly withered sterile frond; light green at first, then dark green, then brown. Spores not freed until spring so fertile fronds must survive one winter and perhaps two.

Described in 1803 by Michaux from a plant from Montreal but genus was originally described by Rafinesque. Common name undoubtedly refers to resemblance between sterile fronds and ostrich plumes. One of most beautiful ferns of its range. Can become a successful addition to home fern garden; probably more beautiful in its natural setting, however. Sometimes listed in genus *Pteretis*.

DIVISION PTEROPHYTA. CLASS LEPTOSPORANGIOPSIDA

Chain Fern, Netted Chain Fern, Net-veined Chain Fern
Woodwardia virginica or *Anchistea virginica*

Because it possesses a row of chainlike sori on each side of the midveins of the pinnules, little leaflets, it has been called a chain fern. Medium to large, erect land-based ferns. Underground rootlike portion short, horizontal, or creeping; sometimes erect and more or less branched. Fronds (leaves) of one kind, erect, and branching; deeply cut singly featherlike or doubly featherlike. Stalk of sterile frond very long, double the length of the blade; a very dark and shiny purplish brown in color. Fertile fronds lustrous blackish-brown, appear in autumn; sometimes overtopping the sterile fronds.

Widespread in northeastern and midland United States, in particular in south; chiefly in Atlantic lowlands and then north to Nova Scotia. Also occurs in upland stations from Virginia to Missouri and Michigan; south to Florida, Louisiana, and Arkansas. May be confused in these locations with cinnamon fern, which grows in clusters from individual sources whereas chain ferns occur as close masses growing from subsurface parts. *W. radicans* occurs in northwestern states; probably the best large native decorative fern. In the Northwest it may occur evergreen throughout winter months. *Woodwardia* reported from each of the continental states.

Found in sphagnum bogs, swamps, wet woods; roots generally in water; often a foot or more deep. Prefers acid soil and shady locations. Readily grown in a home fern garden if one has a damp hollow to fill with plants; however, a locale such as this may be impossible to reproduce.

Chain Fern

Order Filicales./Family Osmundaceae

Royal Fern
Osmunda regalis

Fronds to 5 ft long, clustered, and beautiful; divided and in turn divided again; with those that are sterile, the divisions are well separated. Fertile areas are in uppermost part of fronds where units are much more crowded but finer. Axes relatively fine but strong. Fronds generally a pale green and units are practically entire; almost translucent. Up close, the plant's "leaflet" arrangement makes it look like a black locust. Young plants, fiddleheads, wine-colored.

Found from Newfoundland to Florida, west to Texas and Saskatchewan. Also found south into South America and southeast into West Indies. It also occurs in other parts of the world. In our forms, base of each pinnule is more likely to be somewhat heart-shaped than in European forms.

Found in moist woodlands, in marshes, swamps, and sometimes on wet cliffs. It usually favors soils or waters that are somewhat acid though it may survive if planted in neutral areas if other conditions are suitable. One of our more common ferns.

Fertile fruiting fronds appear as dense clusters of light brown structures that on examination are seen to be made up of dense masses of light brown spore cases. These are darker in early stages and light brown when spores are being freed.

A beautiful fern which should be protected. Where transplants are made, they should be made only from dense masses that will reconstruct themselves. Sometimes known as "male fern"; a mucilage from stem is used in treating coughs and diarrhea.

Royal Fern

Interrupted Fern, Clayton's Fern
Osmunda claytoniana

Fronds to 5 ft long, woolly when young, once divided, without the tuft of brown at the base of each section that distinguishes the cinnamon fern. Fertile pinnae in one to five closely placed pairs, with sterile pinnae above and below producing the "interrupted" effect that gives fern its common name. Young plants, fiddleheads, are heavy, thick, woolly, brown in color and appear very early in the spring.

Found in eastern Asia, and in eastern North America from Greenland to Georgia, west to Arkansas and Manitoba. Particularly common along coastal areas. It is also sometimes one of commonest plants in upland wooded areas often making an almost pure stand.

It favors partly acid soils and is at its best in moist woodlands, in open brushlands, or along the margins of swamps and marshes. Common along roadsides and woodland edges. Does well in stony, dry soil. It grows on mountaintops.

Fertile areas are always conspicuous and are at their best in May. After they have shed their spores, the gap in the frond makes the fern an easy one to identify even at a considerable distance. The presence of these middle fertile pinnae is unique among our ferns.

The genus *Osmunda* may refer to the Saxon god Osmunder or Osmund, the Waterman, reputed to have hidden his family in a clump of the related royal or regal ferns. Interrupted ferns are easily transplanted and make good ornamentals for use on the shaded side of a house in poor soil or good loam.

Interrupted Fern

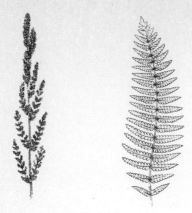

Cinnamon Fern

Water-clover Fern

Rattlesnake Fern

Cinnamon Fern
Osmunda cinnamomea
Fronds to 5 ft long, arising from creeping underground stem in clusters that look like graceful bouquets of wands; generally made up of sterile and fertile fronds. Usually in center of cluster are a few fertile fronds. Sterile fronds easily identified by tufts of brown wool that appear at base of each pinna. Main axis brown, woolly; fronds are at their best in midsummer. Young fronds conspicuously woolly.

Found from eastern Canada through eastern United States, with a closely related variety found in tropical regions to south. Its close relatives include interrupted fern and royal fern.

Found in wet woodlands, stream banks, marsh borders, on wet slopes, or sometimes even on wet cliffs. It is rare where acid content of soil is low and it is not at its best where that content is high.

Fertile fronds are erect and nearly as long as those that are sterile. They bear no sterile pinnae and in this respect differ from interrupted fern with its frond of mixed nature. Spore cases in fertile frond become cinnamon brown before withering away. They spring from short rootstocks in early summer.

This fern transplants easily and is more popular than interrupted fern because of uniformity of fronds. It is hardy in a variety of places but does best where there is reasonable shade. Central parts at base of plant make up rootstock to which many buds are attached. This structure called "heart of Osmund" or "bog onion," is gathered and eaten like raw cabbage; roots are used for rooting and growing orchids. Harvesting the rootstock destroys that plant, of course.

Order Marsileales./Family Marsileaceae

Water-clover Fern, Water Shamrock, Water Clover, Pepperwort
Marsilea quadrifolia
Fronds look like parallel-veined, four-leafed clovers, parts being almost exactly equal and forming a quarter of a circle about 1 in. across. Rootstock slender, tough, and branching to interweave with others forming a dense mat with assistance of rootlike branches.

Found largely wherever introduced, coming to America through Connecticut in 1862 and establishing itself where introduced elsewhere through northeastern United States particularly in New York, Pennsylvania, New Jersey, and Maryland. In some lists from western states, genus given as *Marsilia*. Species described for state of Washington to Montana, southward to California and Wyoming.

It is aquatic or nearly so and floats on still waters or establishes itself on mud flats at edges of ponds, lakes, and streams where soil is neither acid nor alkaline. In still bodies of fresh water it is hardy and persistent once established.

Fruiting bodies appear in late summer and through winter, like small, dark, long peas or beans borne on short branches from near the base of the fronds. These are very hardy and may be dried for years only to resume growth and development when they have access to proper water and temperature.

Plant may have some value in anchorage of soil at margins of waterways and it is usually more beautiful than many plants which grow in same place. May get out of control, however, and spread to territory where it is not wanted but at all events it provides food and some shelter for minute organisms that make their way into open water to be eaten by fish. *M. vesita* has been listed as a weed in parts of Nebraska; also listed as "horse-clover" because horses are said to like it. *Marsilea* grows especially well under cultivation in greenhouses.

CLASS EUSPORANGIOPSIDA

Order Ophioglossales./Family Ophioglossaceae

Rattlesnake Fern, Virginia Grape Fern
Botrychium virginianum
Height 6 in. to 3 ft, with stem of single frond fleshy. Frond part sterile and part fertile. Veins of sterile part not netted. Sterile frond triangular, with three obvious, much divided parts held horizontally. From sterile frond rises fertile frond, often an additional foot. Frond grows from mass of thick, fleshy roots from depth of nearly 1 ft underground, new frond of each year springing from withered base of last year's frond, unfolding in its annual development.

Throughout eastern United States and southern Canada, with close relatives farther north and in the Old World. Closely related grape ferns of many species may extend group still farther, some recognized species forming hybrids with others. Reported in the west from Alaska to Labrador, southward to California, Texas, Florida, and Mexico.

In rich woods, moist thickets, swamps, and similar shady places where the soil tends to be slightly acid. Prefers shade.

Spores bright yellow and mature in June; develop into underground plants which bear sex organs and from which more conspicuous plants arise. May be grown by raising these sexual stages from spores, but it is too difficult a process to be attempted by amateur. Plants may be purchased from nurserymen. Because of depth of root system, does not transplant well from rocky ledges.

This species was originally described by Linnaeus in 1753 from plants brought to him from Virginia in spite of fact that species grows in Europe and Asia. Indians used plant medicinally in treatment of bites. Vernacular name probably from fancied resemblance of clustered sporangia to snake's rattle. In some places in southern U.S. it is known as sang sign because it is believed that the tip of the leaf points to a ginseng plant, a plant with supposed but unsubstantiated medicinal value. Can be readily grown in woodland garden.

DIVISION ARTHROPHYTA

Division takes its name from the Greek *arthros* meaning jointed and *phyton* meaning plant. Known from the Devonian period, about 400 million years ago. Were abundant and dominant during the Paleozoic Era, 180-500 million years ago. Today, one genus, *Equisetum*, survives. Plants in this division may be called articulates, sphenopsids, or arthrophytes. They are vascular plants with jointed stems and a whorled arrangement of the leaves and branches at the nodes. Reproductive structures, sporangia, appear as terminal structures on short lateral branches or terminal cone-shaped structures. Some 25 living species.

CLASS EQUISETINAE

Order Equisetales./Family Equisetaceae

Common Horsetail
Equisetum arvense
Sterile stems appear after fertile ones have died back in late spring. Sterile stems are pale green, with whorls of slender green branches arising from the joints, are to 2 ft high with the branches branched or unbranched. There are 12 to 14 furrows on the rough stems. Sheaths at base of joints are funnel-shaped.

Native of North America as well as of Europe and Asia. Found in America from Newfoundland to Alabama and west to Alaska and California but possibly more common east of the Mississippi than west of it. Related *E. sylvaticum*, wood horsetail, and *E. pratense*, thicket horsetail, also have fertile and sterile stems separated.

Fertile stems arise in season from horizontal underground stem, reach a height of 10 in., appear weak, flesh-colored, or brown and terminate in a conelike spike about 1 in. long from which clouds of spores are freed. These spores make small plants that bear either male or female sex organs.

Found in sterile or sandy soil such as on railroad embankments, in road cuts, or in thickets, woods, open meadows, or other places irrespective of whether soil is acid or alkaline. Roughness of stems due to silica deposited in the stem surface. Ash from burned stems may still be rich in fine silica.

Useful as soil anchors because of branching underground parts. May become a pest in some fields but is usually easily controlled by cultivation. Interesting because of the fact that much plant life of the Carboniferous time when part of our coal was laid down was of this type though considerably larger. Genus name means appropriately "horse bristle," apparently from the taillike appearance of some much-branched species. Considered a weed by many. *Equisetum* causes poisoning in livestock.

Scouring Rush
Equisetum hiemale
Stems to 3 ft tall, arising from a horizontal branching underground system, hollow, evergreen, grooved, with short sheaths at the joints but no branches such as are evident in *E. arvense*. Sheaths at joints, blackish teeth, and disappear early in season. Stems rough and much coarser than in *E. arvense*.

Found native through most of the United States and Canada and so closely related to European species that many question that there is a real difference. Certainly it is slight. Variable in its structural characteristics, and some authors list variable forms as *E. robustum* or *E. praealtum*.
(Continued)

Common Horsetail

Scouring Rush

Fertile portion borne at top of stem and appears like a small conelike spike that is erect and only slightly thicker than diameter of the stem at the point of attachment. Cones mature and free their spores from May to September and of course persist through the year.

Common in springy or dry sandy, open banks or in woods, in noncalcareous or alluvial soils, on railroad embankments, on stream borders and any moist, sterile slope, often making an almost pure stand over a considerable area. It is not evident in soils that are excessively acid or excessively alkaline.

Useful as a soil anchor either at stream edges or on bare steep exposed slopes. Stems have been used as scouring rushes because of the embedded silica that helps remove grease and dirt, or the ashes may be used in making a scouring mixture effective because of the persistent silica. Some few woodworkers use it as a substitute for fine sandpaper. May be referred to as scrubgrass, shave grass, polishing rush, or gun bright, winter-rush. Stems explode in fire, so Indian medicine men used it as a remedy. Scouring rushes may be grown outdoors in beds or in greenhouses for classroom use.

DIVISION MICROPHYLLOPHYTA

These are the club mosses, the spike mosses, and quillworts. In some classification systems this group of plants is given subdivision status: Lycopsida, in the division Tracheophyta, or vascular plants. Plants with true vascularized stems, leaves, and roots. In some forms the spore-producing bodies are on special branches looking like spikes. The sporangia are associated with fertile leaves known as sporophylls. Ancient plants dating back with fossil remains to the Devonian and perhaps even Cambrian. In the Carboniferous period (coal-forming period) this group was represented by some of the largest and most numerous plants. About 11,000 species.

CLASS AGLOSSOPSIDA

Order Lycopodiales./Family Lycopodiaceae

Shining Club Moss
Lycopodium lucidulum
Stems trailing, branching, rooting where in contact with the soil and ascending at the free ends to a height of nearly 1 ft. Leaves rather evenly distributed over the stem but with year's growth usually relatively well-marked, shining dark, rich green about ¼ in. long and arranged all around the stem.

Native of North America and found from Newfoundland to Alabama and west to Missouri and Washington. Not usually common in the lowlands and at its best in Canada, where it is found clear across to British Columbia. A number of closely related species.

Reproduction is by division of the vegetative parts or by the shedding of special reproductive structures which are freed from the upper parts of the branches and are somewhat triangular and waferlike, or it may be by means of spores, which are shed and form underground plants.

In suitable areas this may be frequent in rich woodlands and swamps or in deep humus but more particularly in woodlands in which hemlock is found growing. Spore cases are bright yellow and become mature from August through September. They are borne at the bases of the uppermost leaves.

In the club mosses the conspicuous part of the plant is essentially spore-producing (sporophyte) while in the true mosses the conspicuous part bears real sex organs whose fertilized egg produces the spore-bearing part of the life cycle. An evergreen form, without cone-shaped tips, which sometimes forms fairy rings in woodlands.

Common Club Moss

Common Club Moss, Running Clubmoss, Running Pine
Lycopodium clavatum
Stems run along just under the surface of the ground or at the surface, forking freely and growing to great length, rooting at intervals where contact with the ground is established. Usual height is to about 10 in. Leaves crowded, about ¼ in. long, bristle-tipped, silvery green when new and dark green when old.

Found widely distributed over the world. In North America, found from Labrador to North Carolina and west to Washington and Alaska, with closely related species extending the range. It is found at high altitudes in the tropics in both hemispheres. Fruiting regions are longer than in *L. annotinum*.

Spore cases in cylindrical spikes borne on forked branches which have few and well-separated leaves. Sterile leaves are entire and bristle-tipped, while in *L. annotinum* they lack the long bristles. Spores mature in August and September and are freed in great clouds when ripe.

Found in open fields or woods or mixed with grasses commonly on soil too poor or sterile to support other plants, but usually on soils relatively free of lime. May do well on sand or gravels that would support few competitive plants. Grows particularly well in pine woodlands where soil is loose and acid.

Used in making Christmas wreaths, unfortunately. Spores used in expensive toilet powders, as coatings for pills to prevent them from sticking together, to form a basis of flashlight powders, and in certain kinds of fireworks. Plants collected by campers to fill mattresses but are too slow growing to survive excessive collecting. *Lycopodium* is difficult to grow in the greenhouse. Should not be collected.

Ground Pine, Running Pine, Christmas Green
Lycopodium complanatum

Long horizontal branching stems at or just beneath the surface of the ground bear erect branches. Stem forks freely and erect branches are to 5 in. long, which may grow about 2 in. a season. Leaves are of two kinds and are arranged in four rows to make a flattened frond which fans out widely in one plane. Some confusion as to proper nomenclature among botanists. Most frequently *L. complanatum* is reserved for those plants in the far northern part of the range and *flabelliforme* occurs in the south.

Native of North America and found here from Newfoundland to Georgia and west to the West Coast of the U.S.

Fertile branches are forked into three or more sections and are to 4 in. long, or about equal to the erect sterile part. They are poorly supplied with leaves in the lower parts but the spore-bearing part is closely crowded with spore-bearing leaves that are more or less yellow.

Found in open woodlands or on dry gravelly banks, doing reasonably well in a soil of some acid content but not on soils rich in lime. The spores are windblown and germinate to produce underground sex-organ-bearing structures that are small and rarely known. From these arise the plants we see.

Plants are used somewhat in the making of Christmas decorations and the spores have uses similar to those suggested for *L. clavatum*. For general characters of club mosses see account of *L. lucidulum*. Worthy of protection. Undoubtedly useful to some extent as a soil anchor.

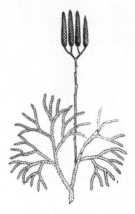

Ground Pine

CLASS GLOSSOPSIDA

Order Selaginellales./Family Selaginellaceae

Meadow Spikemoss
Selaginella apus

Genus name means a diminutive plant resembling a juniper. Large genus with some 700 species; more commonly found in tropics. Abundantly branched, to 10 in. long, appears delicate and weak, an evergreen form. Grows flat, creeps; pale yellow-green. Forms mats but with open spaces in them. Can be easily mistaken for a moss because of appearance; leaves are flat and tiny like some mosses. Leaves in four rows; overlap, narrow, lance-shaped, hairy. Fruiting spike four-sided at end of stem, appears like crowded leaflets on stem.

Found in lawns as a "weed," typical of meadows, muddy stream banks, swamps. Occurs from Quebec and Maine to Wisconsin and south to Florida and Texas.

S. rupestris, rock spikemoss, occurs on thin, dry soil; typically associated with mosses and is more widespread form occurring from Quebec and Maine, south to Florida and Texas and west to West Coast of U.S.

Plant is a beautiful addition to and easily adapted to terrarium life. *S. lepidophylla*, the resurrection plant, of southwest U.S. is a "hygrometer." By absorbing water its branches spread out in a rosette on the ground. When the plant dries, its branches contract and form a ball-shaped mass. Resurrection not really factually accurate as dead plants respond to humidity as do living plants. *Selaginella*, as a genus, has been used as a diuretic by some persons; its spores have been used to coat pills to keep them from sticking together. It is claimed by some to have the power to purify the blood, to stop local bleeding, and to act as a carminative in cases of dysentery. No concrete evidence is presented to support the claims of medicinal value. *Selaginella* species are widely cultivated in greenhouses and provide specimens for class use.

Meadow Spikemoss

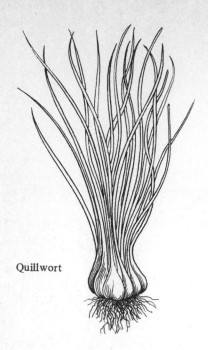

Quillwort

DIVISION MICROPHYLLOPHYTA. CLASS GLOSSOPSIDA

Order Isoetales./Family Isoetaceae

Quillwort
Isoetes engelmanni

Plant appears to be much like a tuft of grass, but the bases of the leaves are swollen and bear masses of spores; the leaves are not sheathed to a stem as in grass and are more delicate than the usual grass leaf, being somewhat four-sided. They may be to 20 in. long in some species but are always in rosettes that arise from a branching root system.

Found in marly ponds, in the mud, and in rivers and streams and lakes from Maine to Pennsylvania and west to Missouri and Illinois. *I. echino-spora* has wider distribution from Alaska to Greenland, south to Washington, Idaho, Utah, Colorado, east to Indiana and New Jersey. At least five species are recognized in the eastern United States region, based largely on microscopic characters of the spores.

Bases of the leaves are really sacs of spore cases with the outer ones bearing larger spores that produce plants bearing the eggs. The bases of the inner leaves bear smaller spores that produce plants producing the sperms, which fertilize the eggs.

While these plants are found in or near marl ponds in which there is much lime, they are usually not found where the soil is strongly acid or strongly alkaline. They may be found on almost dry shores in gravel, or relatively deep in lakes and completely submerged. Plants reach maturity in July to September.

Probably are of little or no economic importance but are tremendously interesting to botanists who consider that they represent a degree of development that is approximately equal to that of the more primitive seed plants. Corms, bulb-like parts, of some quillworts are favorite food of muskrats. Leaves eaten by grazing cattle. Quillwort may be grown in shallow submerged pots in aquariums.

DIVISION PSILOPHYTA

Vascular plants with simple plant bodies. Most are dichotomously branched stems which have no true leaves or true roots and arise from rhizomes which have rootlike projections. Spore-producing bodies at the end of the branches. Possess cylindrical, subsurface gametophytes (sex-cell-producing stage). Division appeared in the Silurian period, some 405 million years ago. Most of the early members of the division lived in the Devonian, some 425 million years ago, and then became extinct. In this division the dominant stage in the life cycle of the plant is the sporophyte. Two extant genera; three species found in tropics and subtropics.

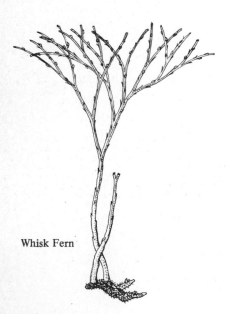

Whisk Fern

CLASS PSILOPSIDA

Order Psilotales./Family Psilotaceae

Whisk Fern
Psilotum nudum

Conspicuous aboveground part of plant is the sporophyte; main stem dichotomously branched several times; to 12+ in. tall. Stems are green. Underground are rhizomes which bear rhizoids. Stem above ground has ridges, frequently five-sided nearer the ground and three-sided toward the tip. No true roots. A very primitive vascular plant. Underground stem does not carry on photosynthesis. Spores produced in three-lobed, globose spore-bearing bodies in notches where short lateral branches arise. Spores colorless, kidney-shaped; will not produce gametophytes in laboratory. In nature, gametophyte is small, perhaps 2 mm in diameter and with many rootlike structures. Gametophyte probably saprophytic; bisexual; usually goes unnoticed; is subterranean. This genus may be a "living fossil" as remnants appear to be present as fossils in Devonian strata. Widespread in tropical and subtropical regions; on soil rich in humus in Texas, Florida, and Louisiana. No economic importance attributed to the plant. May be grown in the greenhouse in a soil rich in organic matter such as leaf mold, with additions of bone meal, and in strong light.

DIVISIONS CYCADOPHYTA, GINKGOPHYTA, GNETOPHYTA, AND CONIFEROPHYTA

As a group, plants in these four divisions make up the gymnosperms, or Gymnospermae, of some classification systems. Gymnosperm refers to production of naked seeds. The habit of producing seeds is a common characteristic of the four divisions listed here as it is with the angiospermous plants, Anthophyta. In some early classification systems the gymnospermous and angiospermous plants were grouped together in the phylum Spermatophyta.

Here we will not be concerned with following an evolutionary sequence in the treatment of the four divisions. Instead, we shall give but short space to three (Cycadophyta, Ginkgophyta, Gnetophyta) and devote considerably more material and discussion to the division more familiar to most of us, namely the Coniferophyta.

DIVISION CYCADOPHYTA

An ancient group of plants. Appeared first in the Permian period, 280 million years ago, and became very abundant in the Mesozoic era, 230 to 135 million years ago. These plants, the cycads, had large palmlike leaves. When one sees pictures of dinosaurs, these palmlike plants are frequently pictured as associated forms. The angiospermous plants replaced the cycads in the Cretaceous period, about 135 million years ago. Today Cycadophyta are represented by nine genera and over a hundred species. Cycads are sometimes termed "living fossils." Only one genus, *Zamia*, is represented in U.S. flora. *Cycas* is an introduced genus and may be found as an ornamental in warmer regions of the U.S. and in greenhouses. The Cycadophyta are characterized by sparsely branched stems and large, pinnately compound leaves.

CLASS CYCADOPSIDA

Order Cycadales./Family Cycadaceae

Sago Palm
Cycas revoluta
Height to 10 ft. Branching or unbranched, with coarse trunk covered by beautiful, recurving, shining, dark green leaves, 2 to 7 ft long whose narrow sections are almost paired opposite each other and end in spinelike tips. Considered as a tree or shrub. Slow grower. A Japanese species.

A common conservatory tree in America, grown indoors in the North and outdoors in the South. It favors open areas, and is grown in subtropical parks and conservatories practically everywhere where suitable temperature may be maintained. *Cycas*, the genus, extends from Japan to Australia. Some Australian members grow to 20 ft in height.

Seed-bearing leaves broad, densely brown-felted, and stiff-tipped. Two or three pairs of seeds are borne at the leaf base. Fruit red, about 1½ in. long and compressed. Seed-bearing cluster of leaves forms a spherical head yielding to 200 seeds about Christmas time. Pollen-bearing leaves make a cylindrical cluster to 20 in. long. The seeds are reported to be poisonous.

Wood of little importance. The plant is subject to a serious blight whose control is not understood. Its antiquity and its fossil relatives make it well worth knowing.

The plant is grown as an ornamental; the seeds are eaten and used as ornaments by the natives. From the pith of the stem is extracted a powdery starch that is excellent in puddings. The durable leaves are crossed on many a human grave. Fossil Cycads occur as fossil forests in Cycad National Monument in the Black Hills of South Dakota.

Sago Palm

Zamia, Coontie, Florida Arrowroot, Comfortroot
Zamia floridana
Tender, ornamental foliage plant. Chosen to represent one of the cycads in this fieldbook because it is frequently found in college greenhouses and public conservatories. Can be obtained from growers or collected in its home range, and can be successfully shipped. A useful specimen because it is small and maintains a good supply of fruiting bodies. At least two species, *Z. floridana* and *Z. umbrosa*, appear in flora of Florida. Genus widespread in the West Indies, Mexico, northern South America.

Plant body a short, upright, conical, turnip-shaped stem. Stem dwarf, may grow to 2 ft. With subsurface fleshy roots. What remains of last year's leaves is visible on upper parts of stem. Dead leaves may remain. Stem has a crown of spirally arranged, leathery, dark green leaves which may reach a length of about 2 ft. Leaves pinnately compound. Cones in center of stem as large as the stem. Sometimes described as an oddity with short trunk but none above ground.

Has a small cluster of ovate, pinnate, fernlike leaves, each with 14 to 20 leathery leaflets which are smooth above and hairy below, with edges rolled under; tapered toward base. In greenhouse condition, will pro-

Zamia

(Continued)

duce new growth of leaves annually.   Sunken stomata on undersurface of leaves only.   Can take considerable mistreatment and neglect, and survive as an ornamental.

Produces a disease, wobbles, in cattle in South America.   Causes permanent injury to limbs resulting in peculiarities of stance.   Starch present in underground portions of the plant.   When used as food, stem is pounded to a pulp and washed in a straining cloth to remove poisons in stem. Seminoles of Florida used starchy pith of stem as flour and called it coontie.

DIVISION GINKGOPHYTA

This division, at least in its extant genus, is characterized by stems which are highly branched bearing small simple leaves.   In some classification systems this division and the conifers are included in one group.   Characterized by stems with little pith and cortex, but abundant xylem. Vascular plants which probably originated in late Paleozoic, perhaps 300 million years ago and then later became dominant in Mesozoic, some 200 million years ago.   *Ginkgo,* the only extant genus, referred to as a living fossil.

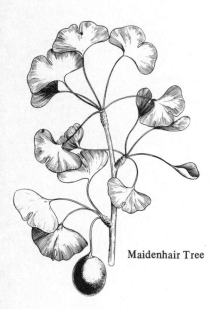

Maidenhair Tree

CLASS GINKGOPSIDA

Order Ginkgoales./Family Ginkgoaceae

Maidenhair Tree
Ginkgo biloba
Height to 80 ft.   Straight, slender, with upward-reaching, relatively few branches.   Twigs, from which the leaves come in clusters or singly, rather coarse.   Leaves to 5 in. long, veins are dichotomous, usually notched at center of outer edge, pale green.

A native of China but known as a fossil from Alaska to England to Spitzbergen before it was found alive.   Has been growing practically unchanged since the mid-Paleozoic age where it appears among the first fossil land plants.   Darwin called it a "living fossil."

Plants either staminate or pistillate.   Staminate catkins slender, stalked and numerous.   Pistillate flowers on long stalks, in pairs, develop seed. The mature seeds have the appearance of small plums.

Grows well in any reasonably good soil; is relatively free from insect and fungus injury, is easily propagated by seed, cutting, grafting, or layering. To some, the fruits are poisonous to touch or eat, but the seeds are edible. Staminate trees are preferred in cultivation because the outer layer of the fruit decays and smells like rancid butter!   Fall color yellow.

Now grown widely as a street and ornamental tree with several horticultural varieties developed to meet certain needs of shape and color. Name, maidenhair tree, is an allusion to similarity between certain leaves of ginkgo and leaflets of maidenhair fern.

DIVISION GNETOPHYTA

Gymnospermous (naked-seeded) plants.   Leaves much reduced to tiny bractlike organs; function only early in development.   Branches largely responsible for photosynthesis.   Perennials, which form secondary wood for a few or many years; branching whorled, and successive whorls alternate. Characteristics of the division are understood best by the professional botanist.

Resemble gymnosperms because of naked ovules and angiosperms in most other characteristics. Some botanists regard division Gnetophyta as a link between gymnosperms and angiosperms, but the ancestry is unknown.   Fossil remains from the Permian, some 280 million years ago.   Three orders, 71 species described.

DIVISION GNETOPHYTA. CLASS GNETOPSIDA

Order Ephedrales./Family Gnetaceae

Mormon Tea Bush, Joint Fir
Ephedra nevadensis
Bushes much branched and densely growing to a height of 5 ft. Branches green. Leaves reduced to length of ⅛ in. in *nevadensis* and to ¼ in. in *trifurca*. Branches look like horsetail and the name *Ephedra* refers to this resemblance.

Grows with greasewood and cacti in desert areas from Texas to California and north to Colorado for the two species listed. There are 40 living species of the genus growing in warmer parts of Europe, northern Africa, and tropical America; as well as in southeast Asia.

Staminate cones of *nevadensis* are six- to eight-flowered while those of *trifurca* bear but one flower. Pistillate cone of the former, two- flowered; of *trifurca*, one-flowered. In each case, the fruit is dry and enclosed in bracts, located at the joints of the stem, with the bract margins in *trifurca* transparent.

Some of the species of the genus are highly ornamental. The plants are not easy to grow where gardens are maintained though they may form a dense thicket in their natural environment.

Of no great significance economically and primarily of interest to botanists. As the name implies, they may have been used as tea substitutes. Also called "Brigham tea." In our arid West, may serve as a sand binder. Most species, especially in loose soils, spread by branching underground stems. Ephedrine used to treat asthma obtained from one Asian species.

Mormon Tea Bush

DIVISION CONIFEROPHYTA

Gymnosperms with simple leaves, often scalelike or needlelike. Within the wood, the xylem is compact; composed mostly of tracheids. Xylem composes bulk of stem with pith and cortex much restricted. Commonly called the cone-bearing plants although some members bear no structures which are truly cones. Made up of three classes: one class, Cordaitopsida, represented by extinct species only; a second class, Coniferopsida with extant representatives in some 50 genera and 500 species. This class includes, among others, many of our favorite "Christmas trees" and lumber and paperpulp producing trees. Generally speaking, these are trees of the Northern Hemisphere although some members grow in the Southern Hemisphere. Commonly, they are called conifers. A third class, Taxopsida, is represented in the U.S. by Yews and Torreyas. Probably appeared first in the late Carboniferous (coal-forming) period about 280–300 million years ago.

CLASS TAXOPSIDA

Order Taxales./Family Taxaceae

California Nutmeg
Torreya californica
Height to 100 ft; with girth to 9 ft but usually much smaller and often only as a shrub. Branches spreading, slightly drooping. Bark grayish-brown rather smooth and thin, with orange tinge on 2-year-old branches. Leaves slender, 1 to 3½ in. long, shining, dark green above, pointed, giving off strong, rather clean odor when broken. Wood and foliage aromatic. Distinguished from other species by its long, flat, rigid leaves.

This species is essentially a California tree. There are four species in North America and Asia. The Japanese *Torreya* is hardiest of all and grows in New England. It has leaves not over 1¼ in. long. Plants favor shaded, sheltered areas with moist soils.

Fruit light green, streaked with brown, about 1 to 1½ in. long, somewhat cherrylike though larger and the flesh is altogether too thin. Propagated by seeds, by cuttings, and sometimes by grafting on closely related forms; plants raised from cutting may be slow-growing. Pit of fruit resembles a nutmeg.

Wood hard, strong, exceptionally durable in contact with the soil. It burns long and well and is smooth and close-grained when worked into furniture, tools, ornaments, and other fancy woodwork. It withstands temperature and humidity changes well and is a good general-purpose wood.

In addition to use for wood in Japan, the trees are grown in America as ornamentals. Brilliant greenness of the leaves coupled with artistic shape of young and old trees makes them generally popular. Found wild only in California; occurs on borders of streams; known as stinking cedar. *Torreya taxifolia* grows in southwestern Georgia and adjacent Florida.

California Nutmeg

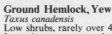

Ground Hemlock, Yew

European Yew

White Pine

Ground Hemlock, Yew
Taxus canadensis
Low shrubs, rarely over 4 ft high and usually sprawling in rather dense formations over ground to exclusion of other plants, though in dense shade this may not be so. Leaves usually under 1 in. long; dark green above, and pale, light green beneath. Twigs rough, dark, and elastic.

Common in evergreen woods from Newfoundland to Virginia and west to Manitoba and Iowa, usually in the shade of other, larger, woody plants. The only native member of the genus in the territory described above and the only species likely to be found in natural areas there. *T. brexifolia*, Pacific yew, or mountain mahogany, or western yew occurs on the Pacific Coast from Alaska to California in wet locations. Wood of this species fine-grained, rose-red, dense, heavy, durable, elastic; used in making bows and in cabinet work. Slow grower; may reach maturity in 300 years. *T. floridana*, Florida yew, grows along the Apalachicola River in Florida.

Flowers in small clusters close to the end of stems, staminate being globular, and pistillate consisting of an erect, seedlike structure that ripens into a beautiful, red, berrylike fruit having a hard nutlike center but is not a true drupe. This fruit ripens about 2 months earlier than that of *T. baccata*, which is planted extensively.

This is a popular shrub for ground cover in shady, steep, stream banks since it serves to hold snow in winter and to hold soil through the year. It also provides excellent year-round cover for small game though its fruits are not wholly useful as food.

European Yew
Taxus baccata
Tree with a height up to 90 ft and a trunk diameter to over 8 ft. Bark deeply furrowed in old trees; reddish, flaky, and smoother in young. Leaves persistent, dark, shining green above and pale yellowish-green with lighter lines beneath; ¾ to 1¼ in. long, with some forms being even shorter; with prominent midrib and tapering to a horny point. Branches droop.

Native of Europe and North Africa east to the Himalayas, but widely established throughout suitable parts of the world as an ornamental. In America, hardy as far north as mid-New England and western New York. In the Far West, the native *T. brevifolia* is found from British Columbia to California.

Fruit ⅓ to ½ in. across, surrounding on the sides the broad, egg-shaped, brown seed that is only about ¼ in. long. Yew trees are usually either staminate or pistillate, with flowers appearing in early spring and being inconspicuous, followed later in year with brilliant fruits. Useful in biology classes to represent naked-seeded fruits, gymnosperms. Foliage, bark, and seeds, green or dry, are toxic to men and all classes of livestock. Bright scarlet fruit attractive to children; pulp not especially poisonous but seed may contain dangerous concentration of the poisonous alkaloid.

Wood hard, strong, durable, uniform, remarkably resistant to weather and suited for maintaining a long fire. Not the "Yule Log."

The tree finds a greater place in America as an ornamental for hedges and for general landscaping than it does as a source of Christmas firewood. There are many horticultural varieties of shade, color, and other characteristics.

CLASS CONIFEROPSIDA

Order Coniferales. / Family Pinaceae (Abietaceae)

White Pine
Pinus strobus
Height to 220 ft. Trunk: in forests, tall, straight, and rarely to 6 ft in diameter; in open, a rather loose, open tree with graceful, partly drooping branch tips. Bark with shallow, broad-topped ridges, to 2 in. thick. Leaves in clusters of four to five needles, each 3 to 5 in. long, soft, blue-green. Twigs smooth or hairy, fragrant, flexible, resinous.

Newfoundland to Manitoba and south through northern United States and on to Georgia in mountains. Great stands of this tree were in states bordering Great Lakes. Replaced to west by *Pinus monticola*, western white pine, which ranges through Columbia River Valley in British Columbia south to California. Essentially a tree of hillsides. *P. flexilis*, the limber pine, also called Rocky Mountain white pine, occurs in the Rocky Mountains from Alberta and Montana to western Texas; found on dry rocky east slopes, summits, tops of ridges, and foothills.

Cones: staminate, bright yellow, appearing in late spring in crowded clusters near twig tips just behind new leaves; pistillate, 5- to 11-in. cones free seeds about second September, though they mature second July. Seeds winged, 26,800 to the pound, with 75 to 90 percent normal germination, narrowed at ends, ¼ in. long, ¼ length of wing, red-brown, with black mottlings.

White Pine

Wood light, 36 percent as strong, 22 percent as hard as white oak; white, uniform, straight-grained, easily worked, and ideal for cabinet work, shingles and walls, masts, and woodenware.   White-pine blister rust, with intermediate host on gooseberry, is a serious fungus pest.   Beetles attack leading shoot, deforming the tree and making it useless for lumber.

One of the most valuable of trees, the basis of much early prosperity of Great Lakes region.   Has been planted rather extensively but less than formerly because of blister rust.   A clean stand is beautiful but a wildlife desert.   Leaves mildly poisonous to cattle.   An old tree is a thing of beauty even though it may be deformed.   Needles, leaves, usually five so children learn to recognize the tree by counting the number of letters (5) in the word, white.

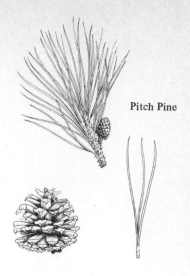

Pitch Pine

Pitch Pine
Pinus rigida
Height to 80 ft; diameter to 3 ft.   Trunk rugged, bearing horizontal, cone-crowded branches, to 4 ft in diameter.   Bark rough, dark, and broken into irregular fissures which expose platelike, dark brown to purplish scales. Leaves in 3s relatively slender and often twisted, 2 to 5 in. long, frequently yellowish-green, scraggly; mostly twisted.

Relatively common on sandy or barren soil from New Brunswick to Lake Ontario and south to northern Georgia and eastern Tennessee.   It is sometimes found in cold, deep swamps but does better on exposed rugged terrain frequently along coasts of either fresh or salt waters.

Cones: staminate, at base of new growth, in clusters, yellow or purple; pistillate, clustered or raised on short, stout stems, light green to purple, developing into stout, persistent cones 1 to $3\frac{1}{2}$ in. long; mature second fall after pollination but remain on tree indefinitely after triangular $\frac{1}{4}$-in.-thick, $\frac{3}{4}$-in.-long winged seeds are shed; 68,200 seed per lb; 70 to 80 percent germination.

Wood brittle, 32 lb per cu ft, light, soft, coarse-grained, durable, light brown to red, not strong, with sapwood that is thick, yellow or white, and relatively soft.   *P. serotina,* pond pine or marsh pine, is found on wetter grounds such as sand flats or even peat swamps.

Used as a tree for poor soils because it is able to resist fire well and yields wood good for firewood, excellent for charcoal, and sometimes suitable for sawing into lumber for coarse construction work.   An old tree with its gnarled horizontal branches and its persistent cones is very beautiful.

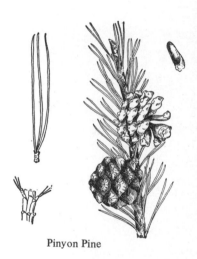

Pinyon Pine, Mexican Pinyon
Pinus cembroides
Tree to 40, but more commonly to 20 ft high; 1 to 2 ft in diameter with horizontal, bushy branches when young, low and round-topped when old. Needles 2 to 3, rigid, dark green, $\frac{3}{4}$ to $1\frac{1}{2}$ in. long.   Branchlets light yellow-brown; sharp-pointed, persistent 3 to 4 years.   Whole plant almost bushlike in northern part of range.

Colorado to New Mexico and Texas, though may be planted and prove hardy as far north as Massachusetts.   One ornamental variety has white leaves mixed in with the green.   Normally found on dry hills and slopes within its range.

Pinyon Pine

Cones almost stemless, greenish-yellow, shining, about $1\frac{1}{2}$ in. long, with scales strongly keeled or ridged.   Seeds about $\frac{1}{2}$ in. long, with a narrow wing that remains attached to scale when separation takes place; sweet and very edible.   Pollination by wind.   Male flowers yellow in crowded clusters; female dark red.

P. edulis normally distributed at altitudes from 5,000 to 9,000 ft; *P. paryana,* another nut pine, or pinyon pine, from 3,500 to 6,000 ft; *P. monophylla,* single-leaf pinyon, from 2,000 to 7,000 ft; and *P. cembroides,* Mexican pinyon, from 4,500 to 7,500 ft.   All are nut pines.

Large oily seeds are an important article of food among Indians and Mexicans.   May be collected by pack rats and the rats' stores robbed by man.   Nuts sometimes found in eastern markets.   European nut pines, "pinocchi" of Italy, are used in cakes and puddings or as delicacies but have strong turpentine flavor.   Wood unimportant, moderately light; hard, brittle although used for fuel, ties, and fence posts.

Shore or Lodgepole Pine, Black Pine, Scrub Pine
Pinus contorta
Height to 150 ft.   Trunk to 3 ft through, or rarely 6 ft.   Branches slender, much forked, light orange when young, and tending to droop.   Crown pyramidal to spire-shaped.   Leaves in 2s, about 2 in. long but may be 3 in.; around $\frac{1}{8}$ in. through, yellow-green, giving a light appearance.   Bark very thin, orange-brown or gray.   May also occur as a stunted bush and any form between extremes described.

Commonest cone-bearing tree in northern Rocky Mountains, forming most forests of that area.   Ranges from the Yukon south, being found as far south as Lower California at 7,000 to 11,000 ft elevations.   The common tree through eastern Washington and Oregon, through Yellowstone and Wyoming.

Shore or Lodge–pole Pine

(Continued)

Cones may remain closed with seeds still vital for as long as 20 years. On the other hand, cones may open and free seeds as soon as they reach mature size. May produce seeds at 10 years of age, 100,000 per lb. Seeds are edible.

Wood light, 30 percent as hard, 40 percent as strong as white oak, weak, soft, close, straight-grained, light yellow, not durable, with thin, lighter colored sapwood. It is at its best in the Sierra Nevada of eastern Washington and Oregon and northern California, at elevations between 8,000 and 9,500 ft.

Wood is used for fuel, in rough construction work in mines, as railroad ties, and sometimes made up into coarse lumber. Shearing strength radially, 672; tangentially, 747 lb per sq in. Serves as a ground cover in areas where other trees do not thrive. Used as an indicator species of Rocky Mountain forests in some lists. Trees along the Pacific Coast may be low and scrubby.

Scotch Pine
Pinus sylvestris
Has many common names; e.g., northern pine, Scotch pine, bois rouge de nord, baltic redwood, yellow pine, Archangel fir, Norway fir, and many others.

Tree. Height 70 to 100 ft commonly; girth 6 to 12 ft, normally smaller. Crown pyramidal when young, irregular when old. Bark reddish, red-brown, with thin areas above; dark, thicker areas below. Branchlets dull, gray to yellow. Leaves in 2s usually twisted, to 3 in. long, bluish-green, slender. Buds resinous, brown, oval.

Native of Europe and Siberia, where it has long been under cultivation. Rather extensively planted as a forest tree in early American forestry practice but now not used to any great extent. A great variety of forms characteristic of different areas within range of species. Occupies an area larger than that of any other pine; from Scotland to the Pacific Coast of Siberia and from Norway to Spain and Arctic Siberia to Mongolia. Also occurs in Mediterranean region.

Cones short-stalked, gray to red-brown, 1½ to 2½ in. long. Seed dark gray, ⅛ in. long, ripening in early autumn, mottled, 68,400 to a pound, with 70 to 80 percent germination; should be collected from October to March and stored in bottles or bags in a cool, dry place until ready for planting in seedbed.

Most important timber tree of Europe. Varieties include slim, tall, red-barked *rigensis*; yellow-coned, slender, pyramidal, broader-leaved *lapponica;* gray-green, thick, stiff-leaved, oblique-coned *engadensis;* columnal, upward-reaching, branched *fastigata*; light, silver-blue-green-leaved *argentea*.

In Europe, it is a timber tree; in America, its greatest use is as a quickly growing cover species and soil anchor. It is relatively free from some common fungus pests but commonly is too irregular in shape to yield timber though it does yield a quick crop of firewood. In eastern U.S. probably most common Christmas tree. Pennsylvania leads states in production of plantation-grown specimens.

Mugho Pine
Pinus mugo
Usually low, spreading, or even prostrate tree, with many branches close to ground, rarely reaching more than 40 ft in height. Many forms are modified by the surrounding topography. Branchlets brown. Bark dark, rough, and peeling off in flakes. Leaves in 2s, crowded, stout, to 3½ in. long, bright green. Described by some authors as one of the dwarf varieties of the Swiss mountain pine.

Native of the mountainous areas of central and southern Europe. Four major varieties are recognized; a compact form, *compacta*; a conic form, *rotundata*; a prostrate form, *mughus*; and an oblique, unsymmetrical form, *rostrata*. Other varieties are based on other characters of the leaves, cones, and color.

Young cones usually tawny-yellow, or dark brown, and shiny; ripen into cones of varying length; in *rostrata*, they are about 2½ in. long, while in some others they are smaller. Varieties and some closely related species frequently cross, so hybrids are abundant and identification is not always easy or accurate.

Possible that prostrate form makes it possible for this species to survive on snow-covered mountains where larger, more erect species could not exist. Except as bizarre ornamentals, the species probably has little commercial value, but landscape gardeners find in it excellent opportunities.

The wood probably would be useful as firewood in areas where other species could not survive, but is commercial value for that purpose would be relatively low.

Scotch Pine

Mugho Pine

Austrian Pine, Black Pine
Pinus nigra

Tree. Height to 150 ft. Pyramidal when young, but flat-topped when old, with upward-bending, coarse branches that are relatively stiff. Branchlets light brown, coarsely roughened. Buds light brown, resinous. Leaves in 2s, stiff, coarse, to 7 in. long, dark green, sometimes slightly twisted, persist about 6 more years. If one is familiar with the pines, he recognizes that the color of the Austrian pine is a darker green than almost all others.

Natives of central and southern Europe and Asia Minor, with many varieties, with rather definite geographic limitations. Variety *austriaca* is the commonly planted ornamental in America. It has slightly shorter leaves than the typical species. Other varieties include Crimean, prostrate, Corsican, and so on.

Cones borne near base of new growth of the year; the staminate leaving a relatively bare area when they are shed. Typical cones to 3 in. long, opening to free winged seeds, with opened scales bent far back, usually yellow-brown but darker in interior; slightly varnished on outside. Attractive for ornamental purposes.

Wood relatively uniform, light-colored, highly resinous, rather fast-growing. Relatively free from disease, and when grown with others of its kind makes a dense, dark background useful in establishing windbreaks and in providing some reasonable shelter for game.

Uses in America largely as an ornamental and as a windbreak but in Europe is an important timber tree. The variety *austriaca* is hardy farther north than other varieties though most of them can survive as far north as Massachusetts. Possibly more beautiful as young plants than when mature. This enhances their use as ornamentals.

Austrian Pine

Norway or Red Pine
Pinus resinosa

Height to 100 ft. Trunk straight, tall, to 3 ft through. Branches stout. Crown broad, open. Bark reddish-brown, shallowly fissured into broad, flat plates. Leaves in clusters of 2s, 4 to 6 in. long, dark green, flexible, persistent 3 to 5 years. Lateral roots stout, rapid-growing, making tree windfirm.

On gravelly ridges, to dry sandy plains. Essentially northern. Nova Scotia and Quebec, south to Pennsylvania, thence west to Minnesota. Locally common. A rather beautiful tree particularly where it grows in pure stands and towers to great heights with uniform straight trunk.

Cones: staminate, ½ in. long, clustered; pistillate, whorled, short-stalked, scarlet. Pollen shed in May. Mature cones conical, closed, with unarmed scales, shedding seeds second autumn. Collect seeds September to October, 61,400 per lb, germination 70 to 80 percent. Wind and squirrels spread seeds.

Remarkably free of insect and fungus enemies. Wood resinous, light, hard, 40 percent as strong as white oak, close-grained, pale red, weighs 30.3 lb per cu ft. Twigs stout, light reddish-brown, not downy, roughened near end of each year's growth by leaf stalks. Seedlings 1¼ in. high at end of 1 year, with six to seven cotyledons.

Valuable timber tree, used in heavy construction, for piles and masts. A good substitute for white pine. Bark sometimes used for tanning leather. Often planted in parks as ornamental and probably is most desirable for ornament of all pitch pines which are northern in their range. Commonly used as a plantation planting in the U.S. in the 1930's.

Red Pine,
Norway Pine

Ponderosa Pine, Western Yellow Pine
Pinus ponderosa

Tree to over 180 ft tall, straight clean trunk commonly to 4 ft in diameter. Branches stout, spreading, often drooping, but then turned up toward tip. Bark yellowish or dark reddish-brown; breaks into large scaly plates which become very large and thick on old trees. These thick plates of bark make a good key character for beginning student of trees. Winter buds conical and to ¾ in. long; scales reddish-brown. Crown short, cone-shaped or flat-topped. Needles 5-11 in. long in bundles of three or two and three; however, rarely one-five, stout; dark to yellow-green, persistent for 3 years. Occurs in western coastal states east through Rocky Mountains south to Mexico, north to southern British Columbia. Occurs in Transition Zone; large altitudinal range from 2,000 to 8,000 ft elevation. Tree of dry places, extremely drought- and fire-resistant, may occur both in mixed stands and as pure stand.

Male flowers yellow, female red; may occur in pairs or in clusters. Cones toward end of branches, occur singly or in clusters, with or without stalks; characteristically spreading scales light reddish-brown in color, ridged, ending in stiff point. Oval-shaped seed about ¼ in. long with wing 1 in. long and to 1 in. wide.

Very important wood of commerce, variable from light, soft, and fine-textured to heavy, hard, coarse-textured. Sapwood white with properties like eastern white pine. Heartwood light brown. Used in rough and

(Continued)

Ponderosa Pine

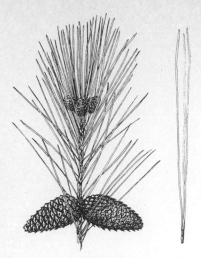

Loblolly Pine

Longleaf Pine

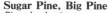

Sugar Pine

finished construction, for paneling, planing-mill products, railroad ties, and mine timbers.

Very slow-growing tree; 350 years or more to reach maturity; produces long taproot; reproduction abundant and fast. Major problems for continued production are bark beetles and fire damage. Also attacked by mistletoe and several fungi. Needles reported as poisonous to cattle, may cause abortion; apparently most damaging in later stages of gestation. Sold as white pine by some lumber dealers.

Loblolly Pine
Pinus taeda

Height to 130 ft and diameter to 4 ft. An important commercial species of the southeastern coastal plain and the piedmont from southern New Jersey south to central Flordia west to eastern Texas, and north in the Mississippi Valley to southeastern Oklahoma, Arkansas, and southern Tennessee. Altitude from sea level to 1,400 ft. Needles 5 to 10 in. long, three in a bundle. Can be confused with shortleaf pine, *Pinus echinata*, which has much shorter needles; longleaf pine, *Pinus palustris*, with longer needles and larger cones; and slash pine, *Pinus elliottii*, which has leaves mostly three but sometimes two in a bundle. Bark reddish-brown; broken by irregular fissures into broad, flat, scaly ridges. Thrives in a variety of soils.

Cones on sides of branches, or sometimes almost at the end of branches, oval-shaped; 3 to 5 in. long; with little or no stalk; scales oblong, about 1 in. long with a ridge along length which ends in stiff spine. Seeds ¼ in. long with wing about 1 in. long.

Wood shares with several other pines the common name, yellow pine. Yellow pine lumber from *P. palustris*, longleaf pine is of superior quality. Loblolly is becoming a useful wood and will probably find greater use in the future. Wood brittle, not durable. Lumber used for general carpentry, boatbuilding, railroad ties; not a major source of pitch. Is used as a source of pulpwood and is commonly planted in pine plantations in the southeastern states along with slash pine.

Longleaf Pine
Pinus palustris

Tree. Height to 120 ft. Trunk tall, straight, tapering, to 3 ft in diameter, with stout, gnarled, twisted branches. Bark to ½ in. thick, light orange-brown, with thin, papery scales, closely appressed except at surface. Leaves in clusters of three, dark green, to 18 in. long, dropping off at end of second year. No other pine has such long needles.

In areas of generally sandy soils from Virginia to Texas, south through Florida, but mostly in a belt about 125 miles wide from southern Virginia to northern Florida, west to eastern Texas and northern Louisiana. Western yellow pine, *P. ponderosa*, a three-needled pine ranging from British Columbia to Mexico and east to South Dakota is also known as long-leaved pine.

New cones appear in early spring in axils of the new leaves; staminate are dark, rose-purple clusters; pistillate appear just below tip of new shoot, in 2s, 3s, or 4s and are also dark purple but develop into 6- to 10-in. cones, with thin, flat scales bearing ½-in., ridged, triangular, thin-shelled seeds, on wings 1¾ in. long and ½ in. wide. Seedling remains in "grass stage" from 1 to 5 years before developing into a tree.

Wood heavy, 40 percent as hard, 57 percent as strong as white oak, durable, pale red to orange, tough, coarse-grained, with winter and summer wood sharply contrasted in color, making wood look streaked. Young and old trees are resistant to fire.

Longleaf pine used in making telephone poles, masts, bridges, railway ties, flooring, buildings, fuel, charcoal, and in general heavy construction. Trees under 20 years old may be used in making paper pulp, rayon, and other cellulose materials. This species probably yields a majority of paints and naval stores used in world. Important source of turpentine.

Sugar Pine, Big Pine
Pinus lambertiana

Described as most majestic of all pines. Takes its common name from a sugary substance which is exuded from trunk of tree and was used by the Indians as a food and a medicine. Grows to 200 ft in height commonly, and may be 10 ft in diameter. Specimens have been cut which were 500 years old. Named by great botanist, Douglas, over a century ago. Young sugar pine looks like western white pine. Young stems and smaller branches smooth, grayish green. Bark of old trees like that of ponderosa pine breaking into broad flat plates; color reddish-brown. Largest pine in range. Huge trunks may be free of limbs for 100 ft. First branches high, very long, and down-curved, producing huge cones; the typical souvenir of the tourist. Cones to 2 ft in length. Leaves deep blue-green, sometimes with whitish tinge, moderately slender, firm, 2 to 4 in. long. Leaves five in a bundle.

Grows between 1,700 and 7,000 ft altitude from Oregon south to Lower California and to western Nevada. Not west of the Cascades; range almost completely confined in Oregon and California. Common on north-and east-facing slopes in Sierra Nevada, and on warmer aspects in most locations. Frequently in ravines and canyons. Associated with ponderosa pine, incense cedar, Douglas fir at lower elevations. Planted as an ornamental in eastern U.S. and in Europe. Fruit, "the big cone," to 24 in. long; very slender compared to length; attached to ends of branches; usually only two cones to a branch. Cones mature at end of second season; fall entire, do not break. Reproduction from seeds is poor and ponderosa pine may take over. Seed ½ in. long, with a wing 1 to 1½ in. long, and ½ in. broad.

Wood is light, 22 to 24 lb per cu ft when dry. Heartwood yellowish to pale brown; sapwood lighter color, yellow. Timber is high-quality, easily worked, takes good polish. Used in building, particularly indoors. Some believe well-cured sugar pine wood is good competitor with eastern white pine for quality. Seeds, pine nuts, were collected by Indians for food.

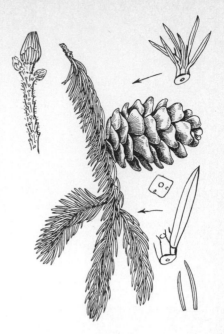

Red Spruce

Red Spruce
Picea rubens
Tree. Height to 100 ft. Tall, narrow, conical, with branches clear to ground when in the open, or absent when crowded in forest. Trunk to 3 ft through. Bark to ½ in. thick, with appressed, irregular, red-brown scales. Leaves about ½ in. long, green, square, sharp-tipped, alternate. Twigs brown, hairy, roughened. Buds red-brown.

On dry hillsides and slopes, often being very common. From St. Lawrence River and Prince Edward Island south to Massachusetts coast, inner New England, and south along Appalachian Highland to West Virginia, Tennessee, and North Carolina, at elevations above 2,500 ft. Sometimes found planted as an ornamental.

Cones: staminate, small, oval, bright red; pistillate, oblong cylinders, developing on short, downward-curving stalks into 2-in. cones, with entire or notched scales; cones shed relatively soon after reaching maturity, unlike black spruce in which they persist for many years. Seeds 131,400 per lb with 59 to 60 percent germination.

Wood light, soft, pale, 40 percent as strong as white oak, close-grained, tinged with red, but with paler sapwood which is usually about 2 in. thick. Among enemies of tree are squirrels, a leaf blister rust, a twig blight, and pecky wood rot. For nurseries, seeds are collected with cones in October and November and stored in a cool dry place.

An important timber tree in Northeast, producing lumber used in flooring, veneer, house construction, sounding boards in musical instruments, paper pulp. Important in airplane construction. Tree yields spruce gum of commerce and twigs yield spruce beer. Too slow-growing for economic reforestation practices.

Norway Spruce

Norway Spruce
Picea abies
Tree. Height to 150 ft. Straight, cone-shaped, with branches in whorls, extending horizontally from the trunk and rising at the tips, but with numerous short branches hanging from them. Twigs hairy or smooth, brown. Buds with overlapping pointed scales. Old bark with large, thick, flaky scales. Leaves to 1 in. long, sharp-pointed, four-sided, stalkless. Said to have, in older specimens, pendulous (drooping) branches on major drooping stems.

Grown along roadsides and near buildings. Especially abundant in Norway but widely cultivated from Maine to Washington, D.C., and west to Kansas. It was the tree most farmers of Civil War times planted about their homes for windbreaks and for decoration after they had cut down the native species.

Cones: staminate, among leaves of last year's growth; pistillate, on upper sides of newer branches, at first erect, but when mature during first year, they droop in great clusters. At maturity, cones may be about 6 to 7 in. long, compact. Seedlings 2 in. tall, with about four medium-broad seed leaves. Pollination May; fertilization, June.

Wood white, soft, brittle, weak, 23.8 lb per cu ft. Over 30 horticultural varieties recognized, these varying in size, form, color as well as in nature of leaves, twigs, branches, and cones; some being mere prostrate plants with little or no erect trunk, while others are tall, stately trees. No taproot and generally shallow-rooted.

Wood used in oars, spars, masts, paper pulp, baskets, rough construction where great strength is not needed. Trees may grow well and quickly in swampy areas. Relatively few enemies, a bud gall and a few fungi being the worst. These are rarely epidemic and some may remain confined to an individual tree for many years. A favorite ornamental planting.

Colorado Blue Spruce

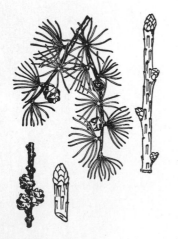

Engelmann Spruce

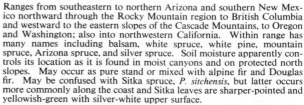

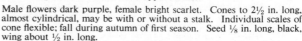

Larch

Colorado Blue Spruce
Picea pungens

Tree. Height to 150 ft. Trunk to 3 ft in diameter, but frequently branched into more than one. Relatively low, with stiff, horizontal branches which curve upward near their tips. Twigs roughened by leaf bases. Leaves square in cross section, stiff, sharply pointed, to 1⅛ in. long, dull blue-green after 3 to 4 years; lighter, earlier.

Favors stream banks but forms important part of cover in the Wyoming region. Ranges from Colorado, Wyoming, and Utah south to northern New Mexico. Planted extensively as an ornamental, particularly in some of its horticultural forms which are grafted. Normally found wild from 6,500 to 11,000 ft elevation.

Cones: staminate, yellow with a tinge of red; pistillate, pale green, with broad scales, developing into 3-in. cones that are found almost exclusively in upper third of tree, are green with a red tinge, and reach full growth by midsummer. Seeds ⅛ in. long, with wings twice that length, produced in abundance every other year. Tree may live to 600 years.

Wood light, weak, soft, pale brown to white, with negligible sapwood, close-grained, and relatively easily worked. Loss of the lower branches as the tree matures makes it unsightly, particularly if it is grown near other trees, but in youth the trees are most attractive in form and also in color.

Wood has no unique uses and is not superior to that of most other spruces, but tree is one of best of ornamental evergreens and has gained a wide popularity which is deserved. Found favor in Europe as well as in eastern United States for this purpose. Growing season is often under 3 months and trees 6 in. through may be 150 years old. A favorite ornamental planting, in particular in U.S. suburbs. Not a good Christmas tree because it loses its needles too soon. Slow-growing spruce, but long-lived; develops moderately deep root system and so good windbreak species.

Engelmann Spruce
Picea engelmanni

Large tree of high altitudes; in North America to 150 ft and with a diameter of 5 ft. Most specimens much smaller. With tapering, pointed narrow crown. Scaly, resinous grayish-brown to purplish-gray to russet bark. Young shoots grayish-yellow with minute scattered hairs. Buds conical to ⅓ in. long; bud scales brown, rounded. Leaves hang on for several years, are arranged so that they expose twig to which they are attached. Leaves four-sided ½ to 1 in. long, soft, ends of needles sharp to touch, light to dark green, with white lines on each side and with disagreeable odor if crushed.

Ranges from southeastern to northern Arizona and southern New Mexico northward through the Rocky Mountain region to British Columbia and westward to the eastern slopes of the Cascade Mountains, to Oregon and Washington; also into northwestern California. Within range has many names including balsam, white spruce, white pine, mountain spruce, Arizona spruce, and silver spruce. Soil moisture apparently controls its location as it is found in moist canyons and on protected north slopes. May occur as pure stand or mixed with alpine fir and Douglas fir. May be confused with Sitka spruce, *P. sitchensis*, but latter occurs more commonly along the coast and Sitka leaves are sharper-pointed and yellowish-green with silver-white upper surface.

Male flowers dark purple, female bright scarlet. Cones to 2½ in. long, almost cylindrical, may be with or without a stalk. Individual scales of cone flexible; fall during autumn of first season. Seed ⅛ in. long, black, wing about ½ in. long.

Bark thick, cinnamon-red to purple-brown, broken into thin loosely attached scales. Wood like that of white spruce; light yellowish-brown to pale reddish-brown, light, soft. Of some importance as lumber and becoming increasingly more important as more favored trees are logged off. Used for telephone poles, railroad ties, mine timbers, general rough wood in construction, fuel, pulpwood. Indians mixed resin with bear grease to make an ointment.

1. American Larch, Tamarack
Larix laricina

2. European Larch
Larix decidua

Height of (1), 60 ft; of (2), 100 ft. Both pyramidal when young, tending to be more open-topped with age. Branches horizontal, ascending near their tips. Leaves shed in winter; borne in clusters, each a needle about 1 in. long, green, turning yellow in fall. Sometimes referred to as a "deciduous evergreen." These, of course, are incompatible terms.

(2) native through northern and central Europe with close relatives in Siberia. (1) Canada, south to Pennsylvania and Illinois and west to Manitoba. In America, larches are found north to 67°; in Siberia, to 72°N. They grow in swamps or on drier lands and are hardy. *L. occidentalis*, western larch, of Pacific Northwest grows at higher elevations on

deep, moist, porous soils of mountain slopes and valleys; *L. lyallu*, subal-pine larch, occurs at even higher elevations in almost same region.

Cones of (1), ½ to ¾ in. long; of (2), ¾ to 1½ in. Those of (1) may appear more globular than those of (2). Seed of (1) minute, brown, winged with 50 to 75 percent germination, shed in autumn and collected for nurseries October to November. Young pistillate flowers of (1) are red. Fastest growth to 40 years; mature at 100 to 200 years.

Wood 30 percent as hard, 60 percent as strong as white oak, durable, heavy, uniform, dark, 32 lb per cu ft. Bark of (1), reddish brown; of (2), gray. Will grow on limestone or on acid soils, wet or dry. Usually propagated by seeds sown in spring but can be started by grafts on seed-lings, by cuttings of ripened wood or even by layering under glass.

As forest tree that grows farthest north, it is unique. One of best woods for heavy construction where exposure to weather is necessary. Bark contains tannin useful in making leather. Roots strong and fibrous. Is victim of a considerable number of insect and fungus pests. Some orna-mental species are popular.

Balsam Fir
Abies balsamea
Tree. Height to 75 ft though at high altitudes it may be merely a low shrub. Trunk may be to 2 ft in diameter. Old bark scaly, and spotted with resinous blisters. Twigs smooth, with needle scars scarcely if at all raised. Leaves flat, blunt, narrow needles, ½ to 1½ in. long, without definite stalks. Buds blunt, resinous. Old bark about ½ in. thick, dull reddish-brown, divided into thin scales.

Varyingly common in damp woods and swamps from Newfoundland and along highlands to Virginia and west to Iowa and north into Canada; and to northwestern Alberta. Sometimes planted as an ornamental but not generally so successful as other species of firs or so hardy as many other conifers that are equally attractive.

Cones erect; the staminate, yellow, tinged with reddish-purple, among the leaves of the preceding year; pistillate, with round, purple scales, with pale green bracts, and developing into 4-in. cones that shed their scales; seeds leave erect cones; seeds about ½ in. long with a wing of same length. Seeds 43,800 to the pound, with 40 to 60 percent germination.

Wood white, weak, soft, brittle, 23.8 lb per cu ft, weathers poorly, coarse-grained, pale brown to white, with brown streaks and a thick, lighter-colored sapwood. Tree seems clean and symmetrical, makes an ideal Christmas decoration because needles are rather long-lasting.

Wood used for making weak, poor packing cases, in the making of paper pulp. Bark yields oil of balsam used in medicine and in the arts. Oil commonly used to cement lenses together, to hold cover glasses on micro-scope slides and for a variety of other purposes. Short-lived tree, 90 to 150 years; with shallow root system; does best on higher sites in combina-tion with spruce, hemlock, and broad-leaved species. Resin supposed to have some curative powers in throat afflictions though not proven. Resin from knots once used as torches. Seeds eaten by ruffed, spruce, and sharp-tailed grouse; twigs eaten by snowshoe hare, white-tailed deer, and moose; bark used by porcupines.

Balsam Fir

Silver Fir, White Fir
Abies concolor
Tree. Height to 250 ft, though inland from the West Coast it rarely exceeds 125 ft. Trunk to 6 ft in diameter. Bark to 6 in. thick; on old trees, divided into broad, rounded ridges, exposing platelike scales be-tween. Twigs relatively smooth and rubbery. Leaves flat, slender, crowded, pale blue, to 3 in. long, blunt, dull green after 2 years. No leaf stem, but circular scar is left when leaf is separated from twig. This is frequently key genus characteristic in taxonomic keys.

From southern Colorado west to the mountain ranges of California ex-tending north to northern Oregon and south to New Mexico, northern Mexico, and Lower California. Tree of moderate altitudes and generally on north slopes. A dominant tree in forests in mountains of southern California; seldom occurs in pure stands, usually with ponderosa and limber pine, Douglas fir, alpine fir, Engelmann spruce, and aspen. Grown as ornamental widely through East and in western Europe.

Cones: staminate, rose to dark red; pistillate, erect, with broad, round scales, developing into mature cones that are to 5 in. long, and only slightly narrowed from the middle to the ends. Cone scales shed, freeing the ½-in. dark brown, shining, rose-winged seeds, the wings finally changing to a gray-brown. 11,000 seeds per lb.

Wood 23 lb per cu ft. Weak, not durable, coarse-grained, pale brown to white, soft and generally poor, but easily made into weak boxes. This is the big tree of parts of its range and is the only fir to be found in Great Basin in New Mexico and Arizona where it grows in surprisingly dry areas.

(Continued)

Silver Fir

A beautiful ornamental tree grown rather extensively for that purpose. Slow grower; a tree 15 in. through may be 150 years old; matures at 250 years. Wood useful only in making such things as cheap packing cases, though it has been used in making butter tubs, and other wooden containers which do not require great strength.

Hemlock
Tsuga canadensis
Tree. Height to 100 ft. Tall, straight, pyramidlike, variable in form. Trunk to 4 ft in diameter, gradually tapering. Bark to ¾ in. thick, deeply cut, with narrow grooves and rounded, somewhat scaly ridges, dark gray or reddish. Buds blunt, brown, 1/16 in. long. Leaves flat, to ⅔ in. long, 1/16 in. wide, with two light lines beneath.

Common, mostly on upland ridges or scattered through broad-leaved forests. Very common gorge tree, not common in open. Nova Scotia to Minnesota and south to Alabama and Tennessee. Replaced to the west by three other species, with black or mountain hemlock of the West being the most different.

Cones: staminate, small, pale yellow; pistillate, pale green at first, but developing into ¾-in. cones which open late in first winter and free minute, winged seeds in which wings are about twice length of seed. Seeds 200,000 per lb with 20 to 30 percent germination.

Wood light, 20 lb per cu ft., 58 percent as strong, 33 percent as hard as white oak, "brash," with unpleasant smell particularly when wet, splits easily and splinters on weathering, does not work easily, light brown with a tinge of red, with sapwood thin and somewhat darker. Cones shed at end of first year and leaves usually at end of their third year.

Wood used for coarse construction, particularly where it will not be exposed to weather. Inner bark has strong astringent qualities, yields most of materials formerly used in tanning leather in Northeast and Canada. Whole forests in Adirondacks were cut down for bark alone. Hemlock oil is extracted from young branches. Tea made from leaves. Poor Christmas tree because it loses its needles as it dries. As a fuel, throws sparks, so less desirable. Seeds taken by woodland songbirds and squirrels; twigs browsed by red squirrel, deer, cottontail rabbit.

Douglas Spruce or Douglas Fir
Pseudotsuga taxifolia
Tree. Height to over 200 ft. Trunk tall, straight, to 12 ft in diameter. Bark on young trees, smooth, thin, gray; on old trees, to 1 ft thick, broken into oblong plates or great rounded ridges, with the outer surfaces showing pressed scales. Buds pointed, ¼ in. long. Leaves flat, pointed, to 1¼ in. long by 1/16 in. wide; persistent for 5 to 8 years or perhaps longer.

In the Rocky Mountains from latitude 55°N, south to northern Mexico but reaches greatest size in British Columbia, Washington, Oregon region where it is the "big tree." It is planted extensively as an ornamental through the East and in temperate Europe. Can survive a wide range of climatic conditions.

Cones: staminate, orange-red, small; pistillate, slender, with a touch of red, developing into mature cones, to 6½ in. long, bearing unique scales, with longer "cover scales." Cones hang down in attractive clusters and reach maturity by midsummer. Seeds ¼ in. long and half as wide, with dark brown wings that are also about ½ in. long. "Shaggy cones" are good identification characteristic.

Wood 25 lb per cu ft, remarkably uniform, 40 percent as hard, 50 percent as strong as white oak, light, yellow to red, with white sapwood that varies greatly in depth. Branchlets pale orange and shining the first season, turning successively red-brown and gray, and fuzzy for 3 to 4 years. Leaves may persist for as long as 16 years. Bark of old trees resists fire reasonably well.

Wood used for all sorts of construction, for buildings, railway ties, telephone poles, piles, and for fuel. Douglas fir is top producer in the timber market. It is the Oregon pine of commerce. Bark is sometimes used as a source of material for tanning leather and, in East, plant is used for hedges, windbreaks, and ornamental purposes. Tea made of leaves.

Order Coniferales./Family Araucariaceae

Norfolk Island Pine
Araucaria heterophylla
Height in native land, to 200 ft, with trunk 10 ft in diameter. Most commonly known in North America as a greenhouse or house plant. General form a central stem with well-separated whorls of horizontal branches that bear needles on secondary branches, or drooping branchlets. Needles about ½ in. long, densely crowded on the stems; of two kinds, on young branches awllike incurved, bright green, up to ½ in. long; leaves of older and fertile shoots dense. Outer bark peels off in thin flakes.

Hemlock

Douglas Fir

Norfolk Island Pine

Native of Norfolk Island between New Caledonia and New Zealand but grown as a house plant almost anywhere. The related *A. imbricata* is a tree grown as an ornamental in America and hardy along our West Coast to Washington and in Florida on the East Coast. There are large evergreen tree species in the genus in South America and Australia.

Cones large, woody globes, about 5 in. in diameter, with a seed under each scale, with the staminate and pistillate cones separate. Greenhouse and house plants are usually raised from seeds, the plants having market value when 2 to 3 layers of branches have been developed. Also grown from cuttings from leader shoots in beheaded plants.

Wood in large trees is used in shipbuilding because of strength, durability, and workability. In spite of the fact that araucarias grow in warmer parts of the world, they do best where night temperatures are not above 60°F and where they are not exposed too long to a direct summer sun.

The use as house plants testifies to their beauty. In Florida and the West Coast areas, araucarias are valuable as ornamentals in landscaping. Member of monkey puzzle family.

Bald Cypress
Taxodium distichum

Height to 150 ft. Trunk to over 12 ft, with great buttresses increasing this dimension; usually hollow in old age. Leaves shed annually, about ¾ in. long, flat, thin, narrow, light green, and borne like the parts of a feather on the branchlets that are sometimes shed with the leaves. A majestic nonevergreen. Sometimes called red cypress, swamp cypress.

In swamps, for most part from Delaware to Florida and west to Texas to elevation of 1,750 ft, and Missouri; southwest Indiana, south Illinois, west Kentucky. Usually where water is strong with lime but sometimes planted successfully on high dry land. 90 percent is at under 100 ft elevation. Related Montezuma cypress of Mexico is planted in southern California and may have a trunk 20 ft in diameter. *Taxodium ascendens*, Pond cypress, is found along the coast of the southeastern United States.

Cones: staminate, in loose, 4-in. purple clusters; pistillate, like roughened 1-in. spheres. Pollen shed in March to May. Seeds ¼ in. long, winged and two to a scale, 5,000 to a pound, 40 percent germination, planted in early spring for nurseries, or cuttings are started in sand saturated with water or in water alone.

Buttress roots serve to provide a broad base for a large tree anchored on a soft soil and to play a role in getting air to submerged parts during time of year when roots are water-covered. Trees in drier soils are more slender pyramids than broad trees growing in swamps.

Wood 64 percent as strong as white oak, 30 percent as hard, close-grained, durable "wood eternal," used for cabinetwork, interior decoration, and other places where strength is not important; 28 lb per cu ft. Basis of important lumber industry. A few fine stands are being preserved in parks. Maximum age 800 years. Second-growth 100-year tree is 100 ft high, with 20-in. trunk. A most valuable lumber tree because of resistance to decay. "Cypress knees" are woody, cylindrical or oval growths which often grow from roots and are exposed in wet and swampy places; may be several feet high. Not useful as a wildlife food plant or as cover.

Big Tree, Giant Sequoia
Sequoia gigantea

Height to 280 ft. Trunk to 35 ft in diameter. Bark darker, richer brown, than in the redwood and known to resist a 7-day fire of logs, to 2 ft thick. Leaves scale like, blue-green, ⅛ to ½ in. long, with longer leaves on the larger twigs, and the longest on the seedlings. Tree might weigh 6,000 tons. *Sequoiadendron* of some older texts.

In Central California, between 5,000 and 8,500 ft, elevation on the west side of the Sierra, usually in groves of which some 28 are considered important. The best-developed groves are to be found on north fork of Tule River in California with extensive forests in King's River country.

Cones 2 to 3¾ in. long by 1½ to 2¼ in. thick, with four to six seeds, each ⅛ to ¼ in. long, under each scale. Seeds shed when mature in the second summer; about 3,000 seeds to the ounce. In nature, only about one seed in a million germinates and a small percentage of these take root. Reproduces only by seed. Seeds fertile to 20 years.

Wood 20 lb per cu ft, weak, brittle, coarse-grained, becoming dark on exposure. Tree known to live between 4,000 and 5,000 years, though it is doubtful if any over 4,000 are now standing. None known to have died of old age and no reason why tree should not live to 10,000 or more years. Annual growing season 30 days to 6 months.

Thought to be oldest living thing in United States but recently bristlecone pines have been found to be older. Some authorities believe that there are older trees of other species in Mexico and in other parts of world.

(Continued)

Bald Cypress

Big Tree

Redwood

Common Juniper

Red Cedar

Big trees have fallen to ax to make commonplace shingles, fences, and toolsheds. Some persons confuse this big tree with redwood, *Sequoia sempervirens*. This should not be done because two species are so different. Compare two descriptions in this fieldbook. Foliage and cones of two genera are distinctly different in size and appearance.

Redwood, Coast Redwood
Sequoia sempervirens
Tree. Height to 364 ft. Diameter, breast high, to nearly 18 ft. Bark lighter in color than in big tree, to over 1 ft thick, widely furrowed and ridged at the base. Needles flat, stiff, borne singly and not scalelike as in big tree; 1/3 to 1 in. long, unequal in length, sharp-pointed; whitish beneath and dark above.

From sea level to 3,000-ft elevations, in a 450-mile strip in fog belt of western California and Oregon, never growing naturally out of this fog belt region extending from Chetco River in southern Oregon to Monterey County, California. Never naturally over 30 miles from Pacific Ocean in this area: usually in groves. Grows on rocky slopes where there is little soil and in moist valleys with a considerable alluvium. In drier parts, occurs in mixed stands; in moist situations, it produces pure stands.

Cones only 3/4 to 1 in. long with four to five seeds under each cone scale. Cones mature in 1 year. Seeds thin and flat. Reproduces either from seed or from stumps of older trees. Redwood 350 ft high, with a 20-ft trunk is about 1,000 years old. Oldest of the trees known is 1,400 years old.

Wood light, 67 percent as strong as white oak, 20 lb per cu ft, light red, durable close to the soil, easily worked and split, close-grained. "Founder's Tree," in Bull Creek Flat, near Dyerville, Calif., is estimated to contain 235,000 board feet of timber. It has a circumference of 72 ft at ground and is 345 ft high.

Fortunately some of finer redwoods are now being preserved in national and other parks in West. Pure forests of smaller trees yield tremendous amounts of lumber. One tree yielded 480,000 board feet excluding waste. Smaller trees frequently raised as ornamentals, both in Europe and in America. Leaves yield a perfume. Burls make ornaments. Attempts by conservationists to preserve areas of the redwoods for posterity should be applauded.

Common Juniper
Juniperus communis
Tree erect or sprawling shrub. Up to 40 ft high, but usually much lower. Leaves needle-shaped, joined in whorls at base and with a broad, whitish band on top, very sharp-pointed; 1/2 to 3/4 in. long, standing out from the stem in young plants and being more scalelike in older plants.

Ranges through North America, Europe, and Asia, circumpolar; in eastern United States sometimes forming dense prickly thickets. Among close relatives are Colorado juniper of the foothills of the Rockies; California or desert juniper, of the desert basins; Sierra juniper of the rocky ridges of the West up to 11,000-ft elevation.

Flowers in axils of leaves, with the pistillate and staminate cones on different plants. Fruit almost stemless, dark blue, from 1/4 to 1/3 in. through, with a bloom that rubs off easily. Berrylike fruit is really a fleshy cone that bears seeds hidden within; flesh, however, is not succulent.

Most commonly found junipers are those used as ornamentals and these are not tall but chosen for their form or color. Among common forms are narrow, columnar Irish juniper with its short leaves; the Swedish juniper that grows to a 40-ft slender spire; the 2-ft-high, spreading mountain juniper and the 4-ft prostrate juniper.

Junipers besides being ornamental provide excellent shelter and some food for game, and are excellent soil anchors on steep slopes. They are propagated by seeds that germinate in the second or third year, though this may be hurried by dipping for under 6 seconds in boiling water. See also red cedar. Oil from wood and leaves of some species used in perfumes and medicines. Bad forage. Provides cover and food for many songbirds and mammals. Cedar waxwings take their common name, cedar, from this tree the berries of which they are particularly fond.

Red Cedar
Juniperus virginiana
Tree. Height to 120 ft. Commonly appears as shrub or pyramid-shaped tree, usually with many branching trunks rather than a single central trunk. Bark shreddy. Trunk to 4 ft in diameter. Leaves mostly opposite, somewhat triangular in outline, with tip free from or pressed to twig, crowded. Twigs: usually four-sided, with covering of leaves.

Common from dry hills or dense swamps ranging from Nova Scotia to Ontario and North Dakota, south to Kansas and Florida. More commonly grows mixed with other species than in a pure stand, though pure stands sometimes do occur in limited areas. Growth is slow, maximum age about 300 years. Seedlings appear second year after seeds are planted.

Staminate and pistillate conelike structures are on different trees, clustered near the tips of the twigs. When young, they appear conelike, but as the pistillate cones mature, they become bluish, berrylike structures, $\frac{1}{4}$ in. through, each with one to three seeds. Pollination in April or May. Mature berries or cones shed after first winter.

Wood heartwood, fragrant, hard, uniform in texture, very durable, 30.7 lb per cu ft, very red but streaked with yellow or brown, takes a brilliant polish; beautiful when varnished. Birds carry seeds. Tree is alternate host to cedar-apple rust and should be eliminated around apple orchards. Bagworms also are pests.

Wood most useful in making chests and certain kinds of furniture, for making cedar closets which are relatively free of moth attacks. Posts are exceptionally durable. Oil from leaves and from wood used in perfumes and in arts. Decoction of fruits is used as tea and medicinally. Berries are consumed by many birds including bobwhite, sharp-tailed grouse, pheasant, and mourning dove, as well as opossum. Bark is good tinder.

True Cypress (Italian)
Cupressus sempervirens

C. sempervirens Italian cypress is an 80-ft tree, with thin, gray bark and erect or horizontal branches. Monterey cypress has dark reddish-brown bark. Leaves of former, $\frac{1}{25}$ in. long; of latter, $\frac{1}{16}$ in. long; of both, closely appressed to the branchlets and usually dark or bright green. Spreading habit like that of cedar.

Italian cypress is a native of southern Europe and western Asia and is famous in some Italian plantings. Classical cypress of the Roman and Greek writers of old. Monterey cypress is native of the Bay of Monterey in California.

Cones of the two cypresses described above between 1 and $1\frac{1}{2}$ in. across, with usually 8 to 12 scales, though there may be two more in the Italian cypress. In nurseries, cypresses are grown from seeds, being transplanted frequently in the early stages to develop a ball-like root system. Otherwise they do not transplant well.

These plants are essentially of interest because of their beauty and their botanically unique characters. Some species are grown as far north as regions of light frosts but California and Gulf States seem to mark northern limits in United States. Hardiest are Arizona cypress and the two mentioned above.
C. arizonica of southwestern U.S. from 4,500 to 8,000 ft elevation, on moist, gravelly, north slopes is used locally for fence posts and mine timbers and has been planted in the southeastern states for Christmas trees. Arizona cypress reaches sufficient size to yield an excellent timber that has the usual properties of its close relatives. Most of the other species do not grow large enough, are too rare, or too subject to frost killing to be of much economic importance. Wood yellow or light brown, moderately hard, close-grained, easily worked, and very durable. Used in building furniture and clothes closets and chests because of workability, color, and fragrance.

1. Atlantic White Cedar
Chamaecyparis thyoides

2. Alaska Cedar, Alaska Cypress, Nootka Cypress, Yellow Cypress
C. nootkatensis

(1), tree to 80 ft high; with reddish-brown bark, fissured into flat, connected ridges; and with inner bark that may be torn into long, strong strips. Branchlets flattened. Leaves with gland on back. (2), tree to 120 ft high; with gray-brown bark, separated on surface into large, loose scales, and irregular fissures. Branchlets not flattened but squarish. Leaf backs glandless.

Six species of genus in North America and eastern Asia. (1) ranges from Maine to Florida and west to Mississippi, being found mostly in low areas. (2) ranges from Alaska to northern California, and in cultivation, in California and through the East; from sea level to 3,000 ft elevations, at higher elevations appearing like a low shrub.

Cones: of (1), about $\frac{1}{4}$ in. across, bluish-purple; of (2), nearly $\frac{1}{2}$ in. across, and dark reddish-brown. One or two winged seeds under each fertile scale, each with wings as broad as the dark, red-brown body; and whole, about $\frac{1}{8}$ in. long. Seeds wind-borne, shed over a relatively long period of time, and produce two-cotyledoned seedlings.

(Continued)

True Cypress

Atlantic White Cedar

(1) favors wet areas; produces a light, soft, weak, close-grained, light brown to reddish, fragrant, uniform wood. (2) produces a hard, close-grained, brittle, durable, bright yellow, fragrant wood with resinous quality, with a very thin, white sapwood. This tree lends itself well to cultivation and is commonly grown as an ornamental.

(1), wood is used in boatbuilding, shingles, interior finish, fence posts, railway ties, cabinetwork, and woodware. (2), wood is used in furniture, shipbuilding, and interior finish. Inner bark of either may be used as fiber of a primitive nature, particularly in making baskets that will hold water.

Incense Cedar
Libocedrus decurrens

Tree. Height 200 ft. Branches erect and relatively short, forming a beautiful, narrow pyramid, with bright green foliage. Leaves oblong, close to bright green, flattened branchlets, free at tips and pointed but most conspicuous because of base that grows to branchlet. Bark bright, cinnamon. Young trees conspicuous because of bright green color.

Ranges from Lower California to Nevada and Oregon but grown in other areas as ornamentals and in arboretums. In general, more common on west than on east mountain slopes; probably related to moisture level. Eight species to be found in western North and South America, in southern and western China, and in Australia. Closely related to genus *Thuja* including white cedar that has more numerous cone scales, with two pairs bearing seeds.

Cones ¾ to 1 in. long, light red-brown, of six woody scales, one pair of which bears two seeds with long wings; staminate and pistillate, on different trees. Seeds planted in spring or plants may be reproduced by cutting made in late summer and rooted under glass. Whips may also be grafted on *Thuja* stock or on Alaska white cedar. Seeds abundant every 2 to 3 years.

Wood light, soft, 60 percent as strong as white oak, unusually durable in contact with soil, close-grained, straight-grained, easily worked, slightly and pleasantly fragrant, 22 lb per cu ft. Hardy farther north than most species. Does best in open, well-drained soils. Because trees lose their lower branches early, young trees are favored in landscaping.

Wood used in shipbuilding, in fence posts, in interior cabinet work, for shingles and sidings of houses and in making pencils. Trees probably are as important to horticulturalists who use them as ornamentals as to foresters who might wish timber. Matures at about 300 years but may live to 500. Tree 200 years old, 90 ft high with 30-in. trunk; 400 years, 110 ft high, 40-in. trunk.

Arbor Vitae, American Arbor Vitae
Thuja occidentalis

Tree. Height to 70 ft. Cylindrical to cone-shaped, but variously shaped. Trunk to 5 ft through. Bark pale, shredding vertically into long strips, free at the ends and curling free from trunk. Leaves flat, opposite, closely pressed to twigs, alternating with pairs of folded needles, also closely pressed to twigs. Arbor vitae, "tree of life," probably named by explorer Jacques Cartier, who used decoction of it to cure scurvy of his men. Cedar swamps provide home and food for deer. Seeds eaten by red squirrels and songbirds. Moose and cottontail rabbits eat twigs and foliage.

Common, particularly in swampy areas, in cool spots from eastern Quebec to Pennsylvania, along the mountains to North Carolina and west to Minnesota and Manitoba. Probably none survive in the Carolinas except in cultivation. Often forms almost impenetrable forests but usually where there is an abundance of groundwater, size of trees being larger in northern part of range.

New cones form in April and May from liver-colored cones, the staminate and pistillate being on different branches of the same tree. Mature cones ½ in. long, with thin, woody scales, and ready by early fall to shed ⅛-in. seeds with thin wings that are as wide as the seed itself. 325,000 seeds per lb.

Wood soft, 33 percent as strong as white oak, very durable, brittle, coarse-grained, 19.7 lb per cu ft, fragrant, yellowish or pale brown, and relatively easily worked for certain purposes. Seedlings about 4 in. tall, with spreading, narrow leaves, in 2s and 4s. Nurseries have produced a tremendous number of varying forms.

Wood used as lumber, in making shingles, spools, boats, railroad ties, and for other woodenware. Bark may be shredded for tinder. In an open fire, the wood "spits" sparks and crackles loudly but is not so dangerous in spreading forest fires as some hardwoods. Extracts from the younger branchlets yield medicinal substances. Maximum age about 300 years.

Incense Cedar

Arbor Vitae. White Cedar

DIVISION ANTHOPHYTA

Commonly known as the flowering plants; the most recent and successful plant colonizers of the earth. Largest in numbers and certainly in diversity; somewhat more than 300,000 species in over 12,000 genera and some 300 families have been described. Found in the widest range of habit and habitat. Occur as herbs, shrubs, trees, vines, floating plants, plants attached to other plants (epiphytes); also, in a few cases, as parasites without chlorophyll. Found in habitats from drylands to wetlands. In some places the Anthophyta would comprise *the* vegetation of an area. They occur as annuals, biennials, and perennials.

The structure of the plant in this division which makes it unique is the flower.

Divided into two classes: In the Dicotyledonae, there are two seed leaves present in the seed and the stem has a stele (the vascular bundle or cylinder composed of pith, xylem, phloem, or pericycle) arranged in a cylinder. There is a functioning cambium or growth layer which produces secondary xylem; secondary phloem, and parenchyma usually in radial rows (in a circle). The other class, Monocotyledonae, has one cotyledon and the stem does not have a cambium.

It would seem that everyone should have some interest in the Anthophyta, as they provide us with a large percentage of our economically important plants; e.g., the source of most wood, fibers, condiments, spices, and raw materials for medicines as well as food.

CLASS DICOTYLEDONAE

Those members of the division Anthophyta which (1) have two cotyledons (seed leaves) in the seed, (2) have stems which are eustelic (having a cylindrical arrangement of xylem and phloem in separate vascular bundles), (3) have a cambium (growth layer), (4) have a special type of leaf venation called net-veined, and (5) have floral parts in fours or fives or multiples of fours or fives are in the class Dicotyledonae. About 250,000 species. The boundaries between the two classes of Anthophyta (Monocotyledonae and Dicotyledonae) may be somewhat blurred. Therefore, we look for clusters of characteristics rather than one single structural characteristic when assigning a member of the division Anthophyta to one or the other of the two classes.

DIVISION ANTHOPHYTA. CLASS DICOTYLEDONAE

Order Casuarinales./Family Casuarinaceae

Beefwood, She-oak, Ironwood
Casuarina equisetifolia
Tree to 150 ft high, with pale green, round, drooping branches, with a generally erect, open appearance. Some species have four-angled branches. Branchlets appear to be jointed like stems of *Equisetum* or horsetail, with sheath teeth at joints, six to eight closely pressed to stem. No true leaves such as are usually found on trees. Also known as Australian pine and horsetail tree.

About 25 species to be found in Australia, East Indies, and New Caledonia as natives. This species planted extensively as shade tree in parts of southern United States, common in Hawaii, Jamaica, Cuba, and similar areas. Since it grows well in brackish and acid soils, it is common along coasts.

Flowers either pistillate or staminate, the staminate appearing in small cylinders at ends of branches, and pistillate appearing as coarsely roughened balls or crude cones, ½ in. in diameter, borne on loosely hanging stems. True fruit is a winged nutlet, borne inside cones.

Wood bright red, giving tree the name beefwood; so dense that it sinks in water or just barely floats. Known as "oak" in Australia, and is valuable for many uses; hard and durable and burns quickly. Propagation normally by seeds or by making cuttings. Sometimes grown in greenhouses because of its unique form, but this is not a common practice.

In South, often planted in rows along roadways and sometimes along streets. Grown in Australia for its wood; along the shores of many tropical areas it serves as an anchorage to hold drifting sands. *C. equisetifolia* is now naturalized in Florida.

Order Salicales./Family Salicaceae

White Willow
Salix alba
Tree to 80 ft high, with short, thick trunk, with coarsely grooved bark. Branches yellow-brown. Branchlets greenish, and permanently silky. Leaves ashy-gray, and permanently silky on both sides, to nearly 5 in. long and to 1 in. wide. Related yellow willow, *S. vitellina*, has yellow branches and leaves that are silky when young but smooth when mature.

White willow is European. While it is grown in parts of America, it does not establish itself so well as does the related yellow willow that has
(Continued)

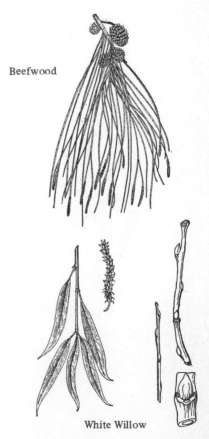

Beefwood

White Willow

become common throughout eastern North America. Both willows favor areas where there is an abundance of available water.

Flowers borne in catkins on short, lateral branchlets, the yellowish scales falling before fruits open to free fluffy seeds. Catkins from 1 to 3 in. long, not recurved, sometimes weak but not drooping. Some consider yellow willow a variety of white willow. Crosses freely with several species of willow.

Wood soft, light brown, easily worked, takes a high polish, tough. Pollination by wind, the staminate and pistillate flowers being on different trees. Seed dissemination by wind, fluffy seeds being carried long distances. Staminate yellow willow is not common in America.

Tree used as an ornamental, as an anchor for controlling stream cutting. Roots may penetrate and clog sewers. Twig trash sometimes troublesome. Some varieties are used in basketry; wood is used in making artificial limbs, charcoal, in wood-burning products, and, by scouts, in making fire by friction. It does not make a lasting fire.

Pussy Willow, Glaucous Willow
Salix discolor

Shrub or small tree. Height to 25 ft. Trunk diameter to 8 in. Trunk usually short, with many ascending branches that form round crown. Twigs at first hairy, then smooth and stout; flexible, red-purple to dark-green; marked by occasional orange-colored lenticels; dark reddish-purple and coated at first; more often shrubby with numerous tall straggling stems. Leaves alternate, smooth, bright green above, smooth, gray beneath when mature, to about 5 in. long.

Found from Nova Scotia to Manitoba south to Delaware and Missouri, being most common at edges of marshlands, along streams, sometimes on moist hillsides, but will grow in dry regions if transplanted when well-developed. Usually grows mixed with other willows and shrubs rather than as a pure stand.

Flowers appear in March, or earlier, on twigs of previous season's growth before leaves appear. Staminate and pistillate catkins on separate plants, catkins being densely flowered and with brown-tipped bracts; develop from the well-known "pussies." Long, shining hairs on flower scales. Fruits, fuzzy capsules.

Twigs may be cut in winter and forced to form "pussies" indoors. Since these have a commercial value with florists, the plant has some use other than as a soil anchor or as an ornamental. It may be easily reproduced by rooting slips in water and transplanting these to suitable ground. Wood weighs about 28 lb per cu ft; soft and weak.

Commercial value lies in its attractiveness an an ornamental, and in use of early spring flower buds for indoor decoration. Frequently grotesque grafts are made to produce unique ornamental trees used in landscaping. This plant frequently the perch in popular pictures of birds singing in the early spring.

Pussy Willow

Black Willow
Salix nigra

Tree to 120 ft high, scrubby, densely branched, with slender twigs and rough, dark, deeply furrowed bark. Twigs with smooth, dark bark, appressed, one-scaled buds, and narrow leaf scars. Leaves alternate, short-stalked, with tapering, curving tips, much longer than broad, to 4 in. or more long; lance-shaped, thin, bright light green.

Common along streams and in wet places but may be found on gravels or sand where there is an abundance of light. Found to elevations of 2,000 ft in New York State, and at sea level on sand barrens. Ranges from New Brunswick to Ontario and the Dakotas, south to Florida, Arizona, California, and Coastal Plain.

Staminate and pistillate flowers in tassels, 1 to 4 in. long, green, and appearing with leaves. Scales pale yellow and not dark-tipped. Nectary yellow, pale when dry as contrasted with a red to black nectary in peach-leaved willow. Fruits smooth capsules, about ⅛ in. long, slender, in loose, drooping tassels, ripening in spring.

Wood light, diffuse-porous, with inconspicuous medullary rays, reddish-brown, 28 lb per cu ft, 22 percent as strong as white oak, has little fuel value. Pollination by wind and probably in part by insects. Seed dissemination by wind, which blows the down-covered seeds when they burst from the fruits. Reaches maturity at 50 to 70 years.

Greatest value is as a soil anchor. Fuel and charcoal are made from wood, and it is used in wood turning in making excelsior, bats, boxes, crates, boats, waterwheels, baskets, and wicker furniture; and in the manufacture of toys and for other purposes where strength is not important as it does not warp, chip, or splinter. Young willow twigs of many species of *Salix* are browse for mammals in northern regions. Tree is always too irregular in shape to have any timber value and wood is too weak for heavy construction purposes.

Black Willow

Crack or Yellow Willow
Salix fragilis

Height to 80 ft. Diameter of trunk to 6 ft. Bark gray, thick, rough. Form tall, slender, profusely branched. Twigs angular, red, or yellow-brown. Bud one-scaled, ⅛ to ¼ in. long. Leaves alternate, smooth, both sides green, to 6 in. long and to 2 in. wide, with 10 to 17 fine teeth to the inch on the margin.

Common where there is an abundant supply of fresh water, being introduced into America from Europe, and well-established through eastern half of United States. Relatively few staminate trees in America but tree has hybridized freely with native species. In the Midwest, with its flatlands, this tree prefers rich damp soil, as it does in other geographic regions where it has been introduced.

Stamens and pistils on separate trees. Staminate tassels 1 to 2 in. long. Pistillate tassels to 3 in. long, and slender. Pollination by wind, in April or May. Dry seed case ⅓ in. long, on a short stalk, matures in early summer, opens to free fluffy seed into wind.

About Civil War time, this tree was sold freely and abundantly to American farmers. It will develop from a stake driven into ground or from cuttings thrust into or dropped on ground. Wood light in weight, salmon-red in color. Attacked freely by twig-cutting, leaf-eating, and wood-boring insects, and by bracket fungi.

Has been grown freely as an ornamental and as a stream anchor. Wood used in making small boats, in making charcoal, but is not useful as a firewood or in heavy construction. It warps badly in curing unless great care is taken. Recorded as being eaten by cottontail rabbit.

Crack or Yellow Willow

1. Purple Willow
Salix purpurea

2. Green Willow
Salix amygdalina

(1), shrub or small tree, with long, erect, uniformly diametered twigs. Leaves to 3 in. long, smooth, paler beneath, sometimes nearly opposite, and with short stalks, or none. (2), shrub or small tree to 30 ft high, with bark in flakes. Twigs long and slender, but coarser than in (1). Leaves to 4 in. long, tapering pointed, light or blue-green beneath.

(1), native of Europe, but escaped and wild in North America, commonly planted for basketry in western New York. True basket willow, *S. viminalis*, is grown in Northwest but is useless in East and South. (2), popular in Europe and in America, particularly along the Mississippi, where it withstands heavier soil and more water than does (1).

Staminate flowers of (1) have two stamens; of (2), three stamens. Bracts of the pistillate flowers dark at tip in (1) and light in (2). Leaf stalks of (1) without glands; of (2), with glands. Pollination is by wind or insects, and seeds are distributed by wind when fruits burst. Normal reproduction is by cuttings, which root freely.

Cuttings made 2 weeks after leaves fall, may be planted in spring, shoots being about 1 year old and 1 ft long, or larger and older, for poorer ground. Usually spaced about 18 to 36 in. apart, are cultivated the first two critical years. Twigs are harvested in December or January, and yield may continue 25 years.

Harvested rods are peeled, dried, packed, and bundled. Willow is grown locally or imported, but centers in America are near large cities.

Purple Willow, Basket Willow

Shining Willow
Salix lucida

Tree or shrub reaching a height of 25 ft, with a trunk diameter of 8 in. Trunk short, with ascending branches that form a broad, symmetrical crown. Bark thin, bitter, reddish-brown. Twigs shining yellow-brown, becoming darker. Buds about ¼ in. long, smooth, pointed. Leaves hairy when young, but shining on both sides; when mature, to 6 in. long.

Found in wet grounds, along banks of streams and other waterways from Newfoundland to Manitoba, south to Pennsylvania, Kentucky, and Nebraska. Often grown as an ornamental where there is sufficient underground water to support it.

Staminate and pistillate flowers on separate plants; appear just before leaves unfurl. Staminate have five stamens, are in densely flowered catkins that are to 1½ in. long. Pistillate in more slender catkins that are to 2 in. long. Fruit a smooth, stalked, straw-colored to pale brown or greenish capsule that frees fluffy seeds.

Wood of no commercial importance, and generally like other willow wood, weighing about 28 lb to the cubic foot but rarely being suitable for timber. Glands on petioles, shining yellow-brown twigs, shining leaves, and five stamens to the flower are most distinguishing characters.

Of value only as an ornamental and as an anchor for soil in low areas where flood damage might result in the washing away of loose, rich soil. Commonly attacked by a variety of insects, some riddling the leaves during the worst seasons.

Shining Willow

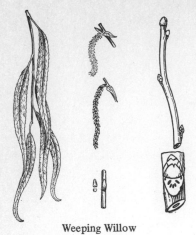

Weeping Willow

White Poplar

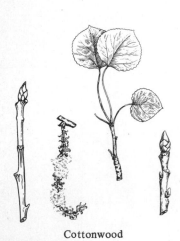

Cottonwood

Weeping Willow
Salix babylonica
Tree to 40 ft high, of rapid growth, with long, slender, purplish or olive-green twigs or young branches drooping from more or less erect trunk and limbs. Leaves to 6 in. long, narrowing at each end to slender points, at first silky but soon smooth, pale beneath. Winter buds slender, sharp-pointed, with only one scale showing.

Native of China but widely cultivated throughout world as an ornamental. Grown frequently in cemeteries as a symbol of sorrow, but a joy to behold. It grows as an escape along riverbanks and lake shores from Connecticut to the west and south.

Flowers appear in catkins, with leaves. Pistillate catkins to 1 in. long, green, slender and developing plump, stemless, fruit capsules that are to $\frac{1}{20}$ in. long, crowded, and burst to free fluffy, wind-borne seeds that often appear in great abundance. Not so hardy as the related *S. blanda*, or Wisconsin weeping willow.

May be propagated by cuttings. Escapes are commonly hybrids of the true weeping willow and some locally hardy species. It crosses freely with the crack willow, but the hybrid lacks the extreme beauty of the typical species. *S. blanda* has broader leaves, less pendulous twigs, longer petioles on the leaves, and longer stalks on the fruits.

Common ornamental willow of cemeteries. Frequently described in poetry of a sorrowful or morbid sort.

White Poplar
Populus alba
Tree to 95 ft high. Trunk to 4 ft in diameter. Twigs slender to stout, usually felty-white, with a woollike down. Buds two-ranked, to $\frac{1}{3}$ in. long, with stipule scars showing. Numerous short spurs. Leaves dark green above, felty-white beneath, coarsely toothed, long-petioled, alternate.

Native of Europe and Asia but extensively grown as an ornamental in America and frequently becomes established as an escape. Closely related to gray poplar, *P. canescens*, and the Chinese white poplar, *P. tomentosa*, the latter having most "restless" leaves and making a sound like falling rain.

Staminate flowers borne in stalkless catkins 2 in. long or less, with 6 to 10 stamens per flower. Pistillate flowers borne in longer catkins, with stigmas slender and lobed. Fruit a thin, slender, two-valved, short-stalked capsule that opens in summer freeing fluffy seeds into the wind.

Wood soft, weak, reddish-yellow, with nearly white sapwood, hard to split, weighs 38 lb per cu ft. Tree frequently spreads by shoots that arise from shallow roots. Easily propagated by cuttings that may be rooted in moist earth or in water. A poplar borer may cause severe injury.

Used essentially as an ornamental shade tree. Wood is poor as fuel but is made into cheap packing cases and flooring and finds some use as rollers that must be light and must not split easily. The Chinese white poplar does not propagate easily by cuttings. It is grown frequently in Chinese temple gardens.

Cottonwood
Populus deltoides
Tree to over 130 ft high, with a trunk diameter to 6 ft or more. Trunk tapers, sometimes straight and unbranched for a considerable distance above ground. Old bark gray and thick; young, thin, greenish-yellow. Twigs with large buds that are resinous and glossy. Leaves to 7 in. long, with petioles flattened near blade.

From Quebec to Rocky Mountains, south to Florida and Tennessee, but widely planted as a street tree in real-estate developments and used as anchors along waterways. Several varieties or races are recognized by botanists. Favors rich, moist soils, borders of lakes, and partially swampy areas.

Flowers appear in March or April, before leaves, with staminate and pistillate flowers in catkins, on different trees. Staminate catkins densely flowered, to 4 in. long. Pistillate catkins remotely flowered, to $3\frac{1}{2}$ in. long. Fruit a dark green, three to four-valved capsule, in catkin, sometimes 1 ft long.

Wood weighs 24 lb per cu ft, diffuse-porous, with pores just visible to naked eye, and indistinct medullary rays. Heartwood dark brown; sapwood, white. Wood soft, warps badly, hard to split. Staminate trees used mostly for street purposes since they do not yield the annoying fluffy seeds. They do clog sewers with roots, however.

Grown as an ornamental in suitable areas; known to grow 5 ft in a year and 40 ft in 10 years. Cuttings placed in wet ground will grow rapidly with little care. Wood has some value as pulpwood and is used extensively in making cheap packing boxes, crates, berry baskets, and tubs. Tree outlawed in some cities.

Lombardy Poplar
Populus nigra var. *italica*

Tree to 150 ft high, with trunk to 6 ft or more in diameter, and with all branches and twigs extending almost straight up, producing a tall, slender, irregularly cylindrical tree. Suckers arise in great numbers from base. Leaves, triangular blades, shining, dark-green, and with petioles flattened near blade.

Originated from a staminate sport of typical *Populus nigra* on plains of Lombardy about 1700 and has since spread widely over world as an ornamental. Thought to be a native of Afghanistan. In the Midwest and in other regions it grows in a variety of soils and exposures. So common is it in areas originally settled by Mormons that it has been called "Mormon tree," since it is characteristic of many of their older settlements in Utah and Rocky Mountain area.

Known only in staminate form although black poplar from which it was developed has both staminate and pistillate flowers. It cannot therefore be reproduced by seeds but depends on suckers for spreading in limited areas and on cuttings for spreading elsewhere. Many trees are reported to have developed from whips.

A rapidly growing tree. Probably true black poplar frequently hybridizes with this. Many forms with minor differences that have been developed so that trees may be found in one form or another in widely different types of climate. Some are so hardy that they can survive severe winter weather.

Essentially an ornamental tree grown as windbreak to shelter farm buildings but also as street tree that occupies little space at base and reaches a great height. Figures prominently in concept of beautiful landscapes of Italy and southern coasts of Europe.

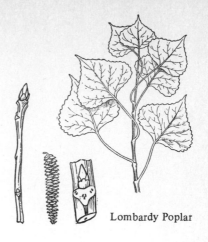

Lombardy Poplar

Trembling Aspen
Populus tremuloides

Tree to 80 ft high, with a trunk to 20 in. through. Crown high, narrow, round-topped. Branches ascending, brittle. Bark on old part, thick, black, deeply fissured; on young, yellow-green to white, with dark blotches under branches. Twigs slender, reddish, shining, with white pith. Leaves to 2 in. long, thin, dark above and light beneath. Roots shallow.

From Newfoundland to Alaska, south to Pennsylvania, Kentucky, Mexico, and California, with probably widest range of any native North American tree. Grows on variety of soils, preferably where dry, does not survive in swamps. Abundant in cutover or burned-over areas mixed with fire cherry and fireweed. Reproduces and forms new stands by root suckering.

Flowers appear in April, with pistillate and staminate on different trees. Staminate catkins drooping, to 2½ in. long, each flower with 6 to 12 stamens. Pistillate catkins when mature may be 4 in. long, drooping, with few flowers. Fruit a two-valved, oblong cylinder, light green, drooping, bursting to free brown seeds with white, fluffy hairs.

Wood weighs 25 lb per cu ft, fine, light brown to white, diffuse-porous, with fine, indistinct medullary rays, not strong or durable. Distinguished from big-toothed aspen by sharper buds free from flourlike crust, and by lighter bark of younger branches.

Wood used in making packing cases, lard pails, buckets, wooden dishes, boxes. Considered weed tree, but with high demand for wood it has found a place as a rapidly growing pulp and cheap wood producer that thrives on a variety of soils under varying conditions. Very important in North Central States. Eaten by ruffed grouse, deer, beaver, mountain sheep, and domestic cattle. *P. tremuloides* and *P. grandidentata*, big-toothed aspen, are most important species of poplars as wildlife foods. Their buds and catkins are eaten by grouse; bark, twigs, and foliage are eaten by rabbits and hoofed browsers, and wood and bark are taken by beavers and porcupines.

Trembling Aspen

Big-toothed Aspen
Populus grandidentata

Tree to 95 ft high, with a trunk diameter up to 2 ft. Crown irregular, with branches usually more horizontal than those of trembling aspen, and with bark of young branches more yellowish. Twigs reddish-yellow, often crusty or pale woolly. Leaves to 4 in. long, dark green above, lighter beneath, rather coarse.

Found from Nova Scotia to Ontario and Minnesota, south to North Carolina and Iowa but planted elsewhere as a quick-growing ornamental shade tree. Does best on rich, moist, or dry, sandy soils, or on abandoned lands where it may form rather pure stands; essentially a forest tree. Reproduces and forms new stands by root suckering.

Staminate and pistillate flowers on different trees; the staminate, in catkins to 4 in. long, and the pistillate, in catkins that at maturity reach a length of 5 in. Fruit two-valved, about ¼ in. long, bursting and freeing brown seeds with mats of white, fluffy hairs.

(Continued)

Big-toothed Aspen

Wood weighs 29 lb per cu ft, is soft, 30 percent as strong as white oak, weak; compact, diffuse-porous, light brown. Bark frequently badly infested with oyster-shell scale that may kill or weaken trees. Frequently associated with shadbush, birch, bird cherry, and scrub oak. Pollination and seed dissemination by wind.

Valuable as a soil anchor and as a species that will quickly cover lumbered-over areas and then give way to more valuable species. Thrives on burned-over areas. Wood used in making boxes and weak crates much as is wood of trembling aspen, though it is a little heavier than that species. A useful pulpwood species. Hybrids of *Populus spp.* have been used extensively in plantings on spoil banks left from open-pit coal mining operations. May be later cut as pulpwood.

Order Myricales./Family Myricaceae

Bayberry, Wax Myrtle
Myrica cerifera
Tree or shrub to 40 ft high, but usually a low, sprawling shrub, forming a dense, stiff, thicketlike area. Leaves narrow and acute, to $\frac{3}{5}$ in. wide as contrasted with obtuse leaves of *M. carolinensis*, whose leaves may be $1\frac{3}{5}$ in. wide, and with entire margins. *M. carolinensis* is true bayberry. Leaves fragrant.

M. cerifera native of swamps and wet woodlands from New Jersey to Florida and the West Indies, west to Arkansas and Texas; one of commonest woody plants forming great thickets in the Everglades, Florida. *M. carolinensis* is native of moist, sandy soils from Nova Scotia to Florida, west to Lake Erie and Louisiana. Commonly sold by nurseries as *M. cerifera*. In New England, the common species is *M. carolinensis*; in western North America, *M. californica*.

Flowers appear in catkins before or with the leaves, those bearing stamens being cylindrical, and those bearing pistils more or less globular. Fruits of *M. cerifera*, under $\frac{1}{12}$ in. in diameter; of *M. carolinensis*, over $\frac{1}{12}$ in. in diameter. Fruits of both, when mature, covered with a pale blue, aromatic wax.

Wood of *M. cerifera* weighs 35 lb per cu ft, is light brown, but has little commercial value. Wax from fruits of both species is removed sometimes by boiling and floating off melted wax, which is then made into candles that burn with a most pleasing fragrance.

In addition to yielding wax used in making candles, these plants are most popular as ornamentals, since the attractive pale blue berries appear in great clumps that persist through winter. They are for sale in florist shops for use as interior decorations in winter but are at their best in the native setting.

Bayberry

Sweet Fern
Myrica peregrina
Shrub to 3 ft high, erect, with spreading, profusely branched top. Leaves slender, with rounded lobes making them resemble some ferns, aromatic, with a prominent midrib persisting on the plant for some time after they have ceased to function, to 6 in. long and to $\frac{1}{2}$ in. wide but usually smaller.

Common on poor soils from New Brunswick and Nova Scotia to Saskatchewan, south to North Carolina and Indiana. In Virginia, found to an elevation of 2,000 ft. Over much of its territory is considered a highland plant. Because of its ability to thrive on poor, dry soils, it finds some favor as an ornamental.

Staminate and pistillate flowers in separate catkins, with both kinds, or only one, on given plant. Staminate catkins clustered at ends of branches, usually under 1 in. long, cylindrical. Pistillate catkins globular, burlike when mature, borne back from tips. Fruit a light brown, shining, somewhat oval nut.

Serves as a good soil anchor and as an excellent shelter for small game, since it provides a dense cover above but adequate open space beneath. Little if any food value for wild animals; however, deer browse on it in the Alleghenies. Game birds and rabbits make use of it. To a very limited amount, buds, foliage, and stems are all used by wildlife. Not a favorite food. Might be classified as an emergency diet for wildlife. Considered by some farmers as an index of poor soil, since it will grow where ordinary farm crops will not. An interesting gall is formed around fruit.

In Revolutionary War times, used in making tea. Plant may be propagated by sowing seeds after they have reached maturity, or by layering, though this species is not so commonly cultivated as are other species of the genus. Leaves are crushed, dried, and used in pillows because of their fragrance.

Sweet Fern

Butternut, White Walnut
Jnglans cinerea

Tree to 100 ft high, with a trunk diameter to 4 ft. Trunk usually relatively short, and often crooked. Crown broad, round-topped, rather open. Twigs coarse, with pith chambered. Terminal bud twice as long as wide. Bark coarse, with long grooves. Leaves to 30 in. long, with 11 to 17 leaflets, sticky. Tree has a scraggly look; often having several dead branches.

From New Brunswick to Quebec, south to Georgia, west to Arkansas and Minnesota. Grows in rich, moist, deep soil, usually along fence rows, at edges of woods, or in abandoned corners. Commonly found at higher elevations than black walnut; survives sour soil conditions better than most nuts. Never found in pure stands.

Stamens and pistils in separate flowers on same plant, former being in catkins near ends of branches, catkins being to 5 in. long. Pistillate flowers in spikes of six to eight flowers. Pollination by wind. Nuts 2 in. long, twice as long as thick, mature first year, germinating following spring if buried about 2 in.

Rarely live over 75 years, being attacked by fungi that sometimes take 20 years to cause death. Fruits weigh 16 to the pound, ordinarily germinate about 75 percent, and should be husked before being stored, should be planted permanently rather than transplanted in early spring; seedlings should be 10 to 18 in. high at end of first year.

Wood, known as "white walnut," weighs 25 lb per cu ft, is soft, 30 percent as strong as white oak, light brown, diffuse-porous, coarse-grained; used in furniture and interior finishings. Sap sweet; added to maple sap, makes sugar. Young fruits pickled in vinegar, sugar, and spice as "pickled oil nuts." Roots used medicinally. Dye from husks and bark used until 1860. Nut meats common in candy and ice cream. May be eaten by squirrels but relative, *J. nigra*, is preferred.

Butternut

English Walnut
Juglans regla

Tree to 70 ft high with a relatively short, straight trunk. Bark light brown or gray, roughened. Twigs coarse, with diaphragms in the pith. Leaves compounded of 5 to 15 leaflets, each 2 to 5 in. long, bright green and almost smooth, not easily confused with the other members of the genus.

Cosmopolitan. Cultivated in America mostly in California and Oregon but grows successfully in the East from Pennsylvania to Georgia. Trees in the more severe climates may survive but do not bear well, partly because of inability to mature flower-bearing parts. Large plantations are maintained in Europe, China, and California.

Staminate and pistillate flowers in separated parts on same tree. Pistillate flowers borne at end of a year's growth. Nuts enclosed in relatively thin husks, and have thin shells. Seedlings usually transplanted at 2 years, the taproot being cut. Many seedlings are budded, or top grafting is practiced.

Wood is a valuable furniture wood, known as French, Turkish, Italian, or Circassian walnut. Fruit known in commerce as English or Persian walnut. Wood durable, lighter in weight and in color than black walnut, and important as a timber tree in India where it reaches a great size. European nuts collected largely from woodlands.

Nuts have high food value. Hot-pressed oil from nuts used in artists' paints, printing ink, and soap. Cold-pressed oil used as human food, oil cake, as stock food. America formerly dependent largely on foreign countries for supply.

English Walnut

Black Walnut
Juglans nigra

Tree to 110 ft tall, with a trunk diameter to 8 ft. Trunk usually straight, clean, unbranched below, and with high, rounded crown. Bark dark, roughened, and broken into squarish units. Terminal bud about as wide as long. Twigs coarse, and with pith in diaphragms. Roots deep. Leaves with 15 to 23 leaflets downy beneath.

Found in rich lowlands and river bottoms where there is good sun and about 150 growing days in a season with an average temperature of 62°F. Best in slightly alkaline soils. From Massachusetts to Flordia, west to Minnesota and Texas, generally in lower, warmer areas than butternut, often in relatively pure stands or small groups.

Stamens and pistils in separate flowers, on same plant appearing before leaves; staminate, in unbranched catkins; pistillate, in two- to five-flowered spikes. Nuts weigh 34 to the pound, with 78 to 80 percent germination, often delayed by several years; should be planted 1½ in. deep, 3 to 6 in. apart. Seedling 3 to 36 in., first year. Matures at 150 years; lives 250 years.

Wood heavy, 38 lb per cu ft, strong, rich, durable, not easily split. Tree relatively free from insect and fungus enemies but has a tent caterpillar and a red butt rot. Relatively fast-growing tree whose nuts are favored

(Continued)

Black Walnut

by squirrels, whose wood is more valuable than most of its associates, and whose bark yields a dye. Unfortunately now becoming a rare species in marketable size.

Nuts are expensive whether bought in the shell or as meats. Wood so valuable it is used largely in making veneers, fine furniture, and gunstocks. Tomatoes and some other crops do not thrive near black walnuts. Can be cultivated with profit in upper Ohio Valley region, Missouri, and southern Iowa. Nut meats common in cookies and candy. A good tree was worth 5 to 12 thousand dollars in 1972, so high as to be almost unbelievable.

Pecan

Pecan
Carya illinoensis
Tree to 150 ft high, with straight, clean trunk. Roots deep. Bark smoother than in most hickories. Twigs coarse, with broad leaf scars. Leaves compound of 9 to 17 leaflets, each tapering gradually to a slender point, minutely downy at first but with maturity becoming smooth.

Does best in rich, low bottomlands such as river bottoms. Ranges from Indiana to Georgia west to Texas and Kansas. Commercial raising of pecans limited to South because tree must have 150 successive warm growing days if nuts are to ripen and tree is to survive.

Pistillate flowers that produce fruits are formed in buds that are not developed until 12 weeks of a growing season have elapsed. Because of this, trees may grow in North, but no nuts will be formed year following a short season. Pollination by wind. Nuts thin-husked and thin-shelled, planted 2 in. deep, 4 in. apart.

Wood hard, durable, 45 lb per cu ft, light brown. Plant winterkills freely in northern part of range. Seedlings and grafts may be bought for starting a plantation. Seedlings for transplanting should be reset every 2 to 3 years to develop proper mass of roots, since this leads to an earlier fruiting habit. Fruit ripens in September.

Nuts of greatest value. Trees may begin to yield 3 to 4 years after being set out. Pecan nuts have a 70 percent fat content, which is higher than that of any other vegetable product. A favorite nut of commerce; especially popular as meats in mixed nuts or as part of Christmas mixed-nuts in the shell. In the South, opossum, wild turkey, and squirrels feed on the nuts.

Shagbark Hickory, Shellbark Hickory
Carya ovata
Tree to 120 ft high, with trunk diameter to 4 ft. Bark gray, stripping off in long, loose, irregular pieces lengthwise of trunk. Form cylindrical. Roots deep. Twigs coarse, with large, loose-scaled, hairy terminal buds. Leaves alternate, compounded of 5 to 7 leaflets, of which basal pair is smallest. Leaves to 14 in. long. Leaflets stalkless. In this species there are tufts of hairs on the teeth of the leaflets. This distinguishes *ovata* from all other species in *Carya*. Crushed leaflet produces an odor unique to hickories. Difficult to differentiate from other species of *Carya* when young; not difficult as old tree because *C. ovata* has definite "shaggy bark."

Shagbark Hickory, Shellbank Hickory

On hillsides and in mixed forests, where soil may be rocky but is well-drained. From St. Lawrence River to Florida, west to Texas and Minnesota, except along southeastern coast. In Midwest, prefers light, well-drained, loamy soil such as those along riverbanks. Requires long, warm summer to mature fruits and wood to bear fruits next year. At least 15 varieties are recognized.

Staminate flowers at base of season's shoot; pistillate flowers, at tip. Pollination by wind. Nuts mature in fall, and are enclosed in a thick husk, which splits to free delicious, hard-shelled nuts weighing about 90 to pound, and germinating from 50 to 75 percent. May live to 300 years but matures much earlier.

Wood exceptionally hard, weighs to 52 lb per cu ft, is hard, tough, 125 percent as strong as white oak, flexible, highly resistant to blows; one of strongest woods and one of best firewoods. Wood ring-porous, with summerwood pores, large, evenly distributed, not in groups or lines. Leaves may bear a mildew. Twigs attacked by a witches' broom disease.

Wood used in making wheel spokes and axles, tool handles, baseball bats, baskets, golf clubs. Nuts have commercial value as food though not so high as walnuts. Nut meats yield excellent salad oil. Nuts favored by squirrels. Sap yields a sweet gum. Bark, mixed with glucose, tastes like maple sugar. Eaten by rabbits and deer. Not a favored lumber species today. It and some other species of *Carya* considered weed species by some foresters.

King Nut, Big Shellbark
Carya laciniosa
Tree to over 100 ft high, with a trunk less robust than that of shellbark. In forests, trunk is long, clean, and slowly tapering. Bark less shaggy than that of shellbark. Twigs stout, usually slightly velvety. Buds larger

than those of shellbark, with outer scales less keeled and more hairy. Leaves to 22 in. long, of 7 to 9 leaflets, dark green above.

Found in rich, wet soils, often in floodlands or in lowlands not commonly the habitat of the shellbark. If found on hillsides, is in area of rich soil. Ranges central New York and Pennsylvania, west to Iowa and Nebraska, south to Tennessee and Arkansas. Rare in mountainous regions of its general range.

Flowers, in general, like those of shellbark, with staminate on wood near base of year's growth, and pistillate near tip of year's wood. Fruit to 2 in. long, globular, with thick husk that splits completely into four parts exposing nut that is white, oblong, slightly flattened, strongly pointed at each end, thick-walled, with sweet, light brown meat.

Wood 50 lb per cu ft, very dark, 116 percent as strong as white oak. Propagated almost wholly by planting nuts, since seedlings do not transplant successfully because of long taproot. Will grow in areas too wet for black walnut, while shagbark will grow in soils too poor for black walnut.

Economic importance of the species centers around the lumber and nuts although in recent years, lumber use is of lesser importance. Distinguish this plant from shagbark by the seven to nine rather than five to seven leaflets, more downy underleaf surfaces, and dull white or yellow, strongly pointed nuts. Wood uses similar to those of the shagbark.

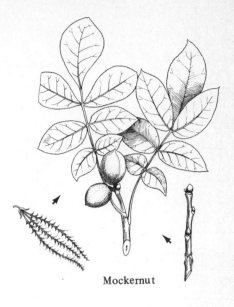

Mockernut

Mockernut
Carya tomentosa

Tree to 90 ft high, but normally between 50 to 75 ft, with trunk diameter up to 3 ft. Crown oblong to round-topped. Trunk often swollen at base. Bark light gray, to ¾ in. thick, not shaggy or smooth, but roughened by irregular furrows separating close, rounded ridges. Leaves to 12 in. long, of seven to nine leaflets, with fuzz that rubs off easily.

Found in rich, moist woodlands where there is abundance of sunshine and water, so usually in fertile valleys at base of slopes facing south. From Massachusetts and Ontario south to Florida, west to Texas and Nebraska, often making up an important part of a mixed hardwood forest. Grows at elevations of 3,500 ft in Virginia. Common on dry, upland soils of the central hardwoods region of the Ohio Valley and southward.

Flowers appear when leaves are half-developed. Staminate flowers in 4- to 5-in. catkins, borne in 3s on a common stalk. Pistillate flowers borne in two- to five-flowered, pale, hairy spikes. Fruits globular, to 2½ in. long with thick, hard husk that splits to middle or base. Nut brownish, four-ridged toward apex, very thick-shelled, sweet.

Wood 51 lb per cu ft, 115 percent as strong as white oak, like that of shellbark but has wider, white sapwood that gives white oak scientific name *alba*. Heartwood is dark brown. Propagated almost wholly by seeds because of deep taproot that is usually injured when seedlings are transplanted. Nuts should be planted about 2 in. deep.

Wood is about the best of hickories for lumber and for fuel so tree should be encouraged where this is possible. Kernel of nut is too small to be considered of economic importance, and leaves form a trash that affects value as a street tree.

Pignut, Broom Hickory
Carya glabra

Tree to 90 ft high. Trunk diameter to 4 ft; usually slender, tapering, clean, long. Crown narrow, usually high, with lower branches often drooping. Twigs slender, smooth, yellow-green to red-brown. Buds oval, blunt-pointed, shedding outer scales in winter. Leaves to 12 in. long, of five to seven leaflets, smooth, dark above.

Pignut Hickory

Found on dry ridges and hillsides, though sometimes found in wet lowlands. From Maine to Ontario and Minnesota, south to Florida, Nebraska, and Texas. This is the hickory of the hilltops within its range. It does well in mixed forests with other broad-leaved trees.

Flowers appear when leaves are half-developed. Staminate flowers in catkins, about 3 to 5 in. long, in 3s on a common stalk. Pistillate flowers in two- to five-flowered spikes on new growth. Fruit to 2 in. long, pear-shaped or oval, reddish-brown, with a thin husk, a thick, bony shell, and a meat that at first tastes sweet but becomes bitter.

Wood 51 lb per cu ft, tough, dark brown, 125 percent as strong as white oak. Tree most easily distinguished in winter by slender, smooth twigs, with small, reddish-brown, oval buds, whose outer pair of scales is smooth or dotted with glands, and which usually fall off before spring. Needs much sunlight to do well.

Of value primarily as a timber and firewood tree, for which its usefulness compares favorably with most other hickories. Nuts inferior to those of other hickories for food, but they are frequently sold mixed with other species and reputation for bitterness is probably exaggerated.

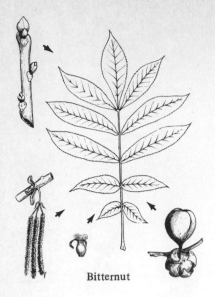

Bitternut

Hazelnut, Filbert

Hop Hornbeam, Ironwood

Bitternut
Carya cordiformis
Tree to 100 ft high.   Trunk to 3 ft in diameter, long, clean, with little taper.   Crown round-topped, usually broadest at or near the top.   Side branches ascending, or with drooping branchlets.   Bark gray, thin, close. Twigs slender, smooth, with distinctly yellowish buds.   Leaves to 10 in. long, of 7 to 11 leaflets, yellow-green above.

Found in low, wet, fertile areas along fencerows and at borders of marshes, woodlands.   Does well on ridges in southern part of range but in north does better in valleys.   From Quebec to Minnesota, south to Florida, Nebraska, and Texas.   Does not occur in pure stands and is usually scattered over an area.

Flowers appear when leaves are about half-developed.   Staminate flowers green, in 4-in. catkins, clustered in 3s on a common stem.   Pistillate flowers in spikes, on new growth, and about ½ in. long.   Fruit to 1½ in. long, with thin husk enclosing a thin-shelled nut that is smooth, short-pointed, and at least as broad as long, with bitter meat.

Wood like that of other hickories, but weaker, more brittle, 47 lb per cu ft, 115 percent as strong as white oak, with less fuel value because of large amount of ash left, but this does not mean that it is an inferior fuel wood as contrasted with most trees commonly associated with it.   May be best in woodlot.   Does not do so well in shaded areas.   Considered a weed tree by some foresters.

Should be developed as ideal fuel wood in areas where other hickories will not thrive.   Unlike other hickories, endures transplanting well and so young seedlings may be placed where they will do best after being started in easily established seedbeds.   Grows more rapidly than the other hickories and will therefore give a quicker return.

Order Fagales./Family Betulaceae

Hazelnut, Filbert
Corylus spp.
Beaked hazel, *C. cornuta* (illustrated), shrub growing to height of 8 ft, with twigs that are not bristly as in American hazel, *C. americana*.   Filbert of commerce, *C. avellana*, tree to 15 ft high, with a number of important varieties such as Barcelona, DuChilly, Alpha, Daviana, and Clackamas. Twig characteristics next to fruits in importance in identification.

Beaked hazel ranges from Nova Scotia to British Columbia, south to Georgia, Kansas, and Oregon; American hazel, from Maine to Saskatchewan, south to Florida and Kansas.   Commercial filberts raised mostly in Washington and Oregon as they winterkill at 0°F and fruit buds are injured below 12°F.   A related tree, *C. colurna*, reaches height of 70 ft.

Staminate flowers in long, pendulous, yellow catkins appearing before the leaves.   Pistillate flowers in short, red-tipped clusters, open for about 4 to 6 weeks.   Pollination by wind.   In orchards, Barcelona is commonly pollinated by white Avellana, Daviana, or DuChilly.   Fruits mature in late summer or fall and average about ½ in. high.

Wood of little value, but trees and shrubs are used in landscape work. Filberts from hybrids of Eurasian species do best on average bottomlands and do not do well on poor soils.   Plants easily pruned, grafted, and improved in production.   Hazel stems were believed by some to be ideal as divining rods for locating water until value was disproved.

Commercial filberts may be made to yield 3,000 lb of nuts per acre, though a normal yield is about 1,000 lb.   Oregon is a major producer of commercial filberts, but most nuts used in America are imported from Europe.   Wood of *C. avellana* is an important hardwood.   Provide good cover for wildlife.   Nuts used by many rodents.   Serves as poor deer browse in the Northeast.   Bark and twigs eaten by several mammals in winter months.

Hop Hornbean, Ironwood
Ostrya virginiana
Tree to height of 60 ft with trunk diameter to 2 ft.   Crown open, high, broad, formed by widely spreading branches.   Bark gray-brown, with loose scales peeling free at ends.   Twigs slender, zigzag, with sharp-pointed buds covered by eight visible, two-ranked scales.   Leaves to 5 in. long, yellow-green above and paler green below, alternate.

On dry, gravelly slopes, but sometimes in moist lowlands.   Usually, in shade and where it is cool.   Never in pure stands.   From Cape Breton Island to Florida, west to Minnesota and Texas.   Four species, two of which are in United States, one being found only in Grand Canyon of the Colorado.

Flowers appear with the leaves; staminate, in loose catkins, about 2 in. long; and pistillate, in short, erect catkins, each in a bladderlike bract. Staminate catkins begin to appear the summer before they mature.   Fruit a small, flat nutlet, in a bladder, making a hoplike arrangement, and shed during the winter, being blown along the snow.

Wood one of hardest, toughest, strongest known, but rarely abundant; weighs 51 lb per cu ft, diffuse-porous, with distinct medullary rays, light brown to white; is considered 30 percent stronger than white oak. In autumn, leaves turn yellow. Tree usually found mixed with other hardwoods as undergrowth for maple, oak, and chestnut.

Used in making tool handles, spokes, axles, and similar wooden equipment. Considered by foresters to be a weed tree, and for decades was destroyed to make place for species that were of more value as lumber. Wood excellent for fires, making a hot fire with coals that persist. Tree attractive, and frequently used as an ornamental. Has little value as a wildlife food.

Blue Beech, Musclewood, Hornbeam
Carpinus caroliniana

Tree to 40 ft high, with a trunk to 2 ft in diameter but commonly much smaller. Trunk short, fluted like "muscles"; close, bluish-gray bark. Crown open and round-topped, the branches ascending but drooping at tips. Twigs at first silky, then smooth, shining, reddish. Buds with to 12 visible scales, four-ranked. Leaves to 4 in. long, deep green above.

Usually found near waterways and hence is known sometimes as "water beech," but may be found on hillsides where there is abundant water available. It is not a swamp tree, however. Ranges from Nova Scotia to Florida and west to Minnesota and Texas in suitable areas. Only one American species but several in Europe and northern and central Asia.

Flowers appear with leaves. Staminate in catkins that droop and are about 1½ in. long, though they first appear the preceding summer as large, long buds. Pistillate in short catkins about ⅔ in. long with conspicuous bright red styles. Fruit a nut about ⅓ in. long with enclosing three-lobed bract. Only native species in birch family in which partially developed male catkin does not appear on the tree in the winter.

Wood diffuse-porous with conspicuous broad and narrow medullary rays, 45 lb per cu ft, light brown with large sapwood. Best distinguishing character aside from trunk and fruits is blunt, downy buds in place of sharp, smooth buds of hop hornbeam. Leaves turn orange or scarlet in autumn rather than yellow.

Tree has some merit as an ornamental but it is a slow grower and produces little volume of wood. Considered as weed species by foresters and is usually cut out to give room for more useful species. Wood makes excellent fuel and is good for tool handles and levers and is split to make binding material by woodsmen. Its seeds are eaten by several kinds of birds and squirrels; its catkins and buds are food for upland game birds.

Black Birch, Sweet Birch
Betula lenta

Tree to 80 ft high. Trunk diameter to 5 ft. Trunk usually continuous, but sometimes divided at a low point. Bark on old parts, dark, broken into large, irregular scales; on young parts, smooth, shining, close. Twigs with numerous short spurs, bearing crowded leaf scars, with wintergreen flavor. Leaves to 5 in. long and 3 in. wide.

Commonly on rich, dry soil on slopes, or even on rocky mountain slopes. Good-sized trees may be seen hanging precariously to edges of cliffs. From Newfoundland to Florida, west through Ontario to Illinois, south to Tennessee. Not found in pure stands but most common in mixed stands of other broad-leaved trees. Most common on cool, moist sites.

Staminate flowers appear with leaves in catkins that lengthen to 4 in. from winter forms ¾ in. long; catkins form in fall preceding maturity. Pistillate flowers in erect, slender catkins, to ¾ in. long, slender, and pale green. Conelike, erect fruit develops to 2 in. long and persists, shedding winged fruits through winter.

Wood 47 lb per cu ft, diffuse-porous, with indistinct rays, dark brown, heavy, strong, with thin, yellow sapwood, superior as firewood and has qualities that permit it to be used as a substitute for mahogany and cherry in making of furniture, takes good polish and holds stains well. Tree a slow grower; attacked by many fungi.

Wood makes superior fuel and furniture. Sap yields some sugar. Oil, from twigs and wood, used as wintergreen oil in medicines and soft drinks. Oil of wintergreen of commerce is now synthetically prepared. May be valuable as oil producer. Excellent "tea" made by steeping shaved bark of twigs in hot water and sweetening. Important food of deer, moose, cottontails, and grouse.

Yellow Birch
Betula lutea

Tree to 100 ft high, with trunk diameter to 4½ ft. Growing in open, has short, thick trunk. Bark pale yellow, peeling off in thin sheets around trunk, but with loose ends hanging free; on old trunks, reddish-brown. Twigs green, then brown, then gray. Buds downier than in black birch. Leaves to 4 in. long, in pairs, but not opposite.

(Continued)

Blue Beech, Musclewood

Black Birch, Sweet Birch

Yellow Birch

Common in moist woodlands, along streams and swamps, usually where there is a rich soil. Sometimes appears in small, almost pure stands. From Newfoundland to Minnesota, south to North Carolina and Tennessee. This species may be one of largest eastern American trees that sheds its leaves. Frequently associated with hemlock.

Staminate flowers appear with leaves after forming in fall, developing from ¾- to 3-in. catkins when pollen is to be shed. Pistillate flowers in short, erect, conelike structures, about ⅔ in. long, with acute scales, that are red and hairy above, green below. Fruit borne in erect, conelike structures, 1½ in. long, that shed winged nutlets.

Wood diffuse-porous, 70 percent as hard, 50 percent as strong as white oak, with indistinct rays, 41 lb per cu ft, compact, but not durable in contact with soil. Heartwood light brown with a red tinge. Sapwood paler. Wood used in making flooring, furniture, interior finish, crates, veneers. Some oil is distilled from it.

Not recommended by foresters for encouragement since most of its associates yield superior fuel and lumber in shorter time, but where it comes up naturally it can well be used. Campers welcome tree because its loose bark makes an excellent tinder for starting fires and wood, when sound, makes good fires with a good bed of coals. Seeds are important late winter food for birds.

White Birch
Betula papyrifera
Tree to 80 ft high, with a trunk diameter to 3 ft. Bark close, chalky white, peeling off in horizontal bands, paper-thin, with a yellow tinge on inner side of peeled-off band. Twigs stoutish except in some varieties, finally smooth, reddish-brown before becoming white. Leaves to 3 in. long and to 2 in. wide, sharp-pointed.

White Birch

Found in rich woodlands, most commonly near waterways, often among hardwoods. From Newfoundland to Alaska, south to Pennsylvania, Michigan, Colorado, and Washington. European white birch grown as ornamental over wider range and farther south. Many varieties including cut-leaf and drooping branched forms.

Flowers appear just before leaves. Staminate, in groups of two to three catkins, each to 4 in. long, but starting in spring at about 1 in. long. Pistillate in clusters, to 1½ in. long, with green scales and red tips. Fruits like cones, hang downward, about 1½ in. long, producing small, winged nutlets.

Wood diffuse-porous, with inconspicuous medullary rays, 37 lb per cu ft, strong, light, hard, brown, tinged with red, with thick, light sapwood. Ornamental European white birch, *B. alba*, is victim of a beetle that bores into trunk and kills tree, so species, common early in century, has about disappeared, but is still planted.

Of much economic importance for dowels and spools. Formerly main sources of bark for Indian canoes. Still used in making woodcraft novelties and rustic furniture. Should not be peeled unless tree is to be destroyed, since beauty of tree is destroyed. Wood is excellent firewood. Eaten by 12 bird species and by deer, rabbits, and stock.

Smooth Alder
Alnus serrulata
Small tree or shrub, to 10 ft high forming dense thickets, in shallow marsh water, or along stream edges. Bark thin, brownish-green, smooth, becoming gray-green or blotched with gray. Buds slightly but plainly stalked. Leaves alternate, smooth or only slighty fuzzy beneath, rarely with impressed nerves. Stipules oval. *A. incana* can be confused with this species. Combination of wedge-shaped leaf bases, nondrooping "cones," and relatively unspeckled bark make *A. serrulata* different.

Smooth Alder

Along streams and close to waterways in northern part of range, usually near sea but in South more widely distributed. From Maine to Florida, and west to Texas and Minnesota. Rarely present along streams that flow over limestone or lime deposits, therefore abundant near bogs and waters full of plant wastes.

Flowers in catkins, appearing before leaves, with staminate and pistillate in separate catkins on same twig. Staminate develop somewhat preceding fall, being in 1-in. catkins in winter and in clusters of two to five. Pistillate in catkins, to ½ in. long, and in groups of two to three. Fruits winged, in ¾ in. woody cones.

Wood diffuse-porous, but with growth rings distinct. Medullary rays variable in width. Sapwood yellow-brown after exposure to air. Pith of twigs distinctly triangular in outline, green in color. Best characters to distinguish it from speckled alder are those given for leaves and fruits.

Probably of considerable importance in anchoring streams to their beds, in providing shelter for trout. A favored food of beaver, which use bark for nourishment and sticks for their dams and dwellings. These dams are important in flood control and in soil building. Deer eat twigs but not a favorite food; even in winter. Several species of songbirds feed on catkins.

Speckled Alder
Alnus incana

Shrub or small tree, sometimes reaching a height of 20 ft, frequently forming a dense tangle. Bark bronze-brown, more lustrous than in smooth alder, and much more spotted with lenticels, usually lacking gray blotches on older bark. Leaves with impressed nerves and usually, but not always, with undersides fuzzy at least along veins.

Common in swamps and along streams, particularly those that are small. From Newfoundland to Saskatchewan, south to Pennsylvania, Iowa, and Nebraska and present but rare along coastal plain where smooth alder is more representative. Distinctly more northern in distribution than is smooth alder. Speckled alder is Eurasian.

Flowers much like those described for smooth alder. While staminate and whole general flower-producing areas are nodding in both smooth and speckled alders, pistillate clusters are erect in smooth alder and nodding in speckled alder. Winged fruit of speckled alder is round, while it is oval in smooth alder.

Woods of smooth alder and speckled alder essentially alike. They cut easily, but regenerate abundantly by means of suckers from the roots or from thickets from shoots arising from branches that may have become prostrate in winter and buried in the spring freshets.

Economic importance of smooth alder and speckled alder is essentially same, and of little direct value, though plants are definitely associated with important changes in the plants, animals, and soils of the areas in which they live. Generally speaking, wildlife food value of alders is low. However, redpolls, siskins, and goldfinches eat their seeds. Other parts of alders are eaten by game birds and browsers. Dense growth of alders provides adequate cover for wildlife, protecting them from adverse weather conditions and their predators.

Speckled Alder

Mountain Alder, River Alder, Barana
Alnus tenuifolia

Tree reaching a height of 30 ft, with trunk diameter up to 8 in. In northern part of range, rarely over 5 ft high. Bark over ¼ in. thick, bright reddish-brown, with small, closely pressed scales on surface. Buds bright red to ⅓ in. long. Leaves to 4 in. long and to 2½ in. wide, smooth, or mealy on underside.

Found along banks of waterways, north to Fraser River in British Columbia, east along Saskatchewan to Prince Albert, south to New Mexico. Common alder of eastern Oregon, Washington, Idaho, and Montana; reaches largest size in Colorado and New Mexico. European alder a tree, *A. vulgaris*. Red alder, *A. rubra*, West Coast lowland species.

Flowers much like other alders. Staminate catkins, in winter light purple, to 1 in. long, becoming 2 in. long when mature. Stamens, four. Pistillate flowers naked in winter, dark brown, to ¼ in. long. Fruit woody, conelike, to ½ in. long. Nutlet nearly circular and surrounded by a thin membrane border. Flowers open in winter or early spring before opening of leaves.

Forms great thickets much as do smaller shrubs and trees that are its close relatives in East. Wood, diffuse-porous. In southern California found at elevations up to 7,000 ft along headwaters of most rivers flowing from California into the Pacific. Wood of related *A. glutinosa* is an important hardwood; that of red alder, *A. rubra*, hard and red.

Chief role that of anchoring banks of streams and helping prevent soil erosion. As with other alders, shade delays melting of snows in highlands; contributes to flood control. Related European alder, grown as ornamental tree and has become established in some areas in eastern United States. Wood of *A. rubra* a good mahogany imitation.

Mountain Alder

Order Fagales./Family Fagaceae

Beech
Fagus grandifolia

Tree to 125 ft high; trunk diameter, to 4½ ft. Bark, close, light gray, smooth, with dark mottlings, relatively thin. Twigs slender, dark yellow to gray, zigzag, with spreading, slender, long, pointed buds. Roots spreading, giving rise to shoots. Leaves to 4 in. long, leathery, long-persistent.

Commonest on rich bottomlands or on fertile highlands. An indicator of rich forest floor flora including many rare and interesting wild flowers. From Nova Scotia to Ontario and Wisconsin, south to Florida and Texas. Favors gravelly slopes and fails to thrive where there is an abundance of limestone or of water that has passed over limestone. Forms relatively clear stands. Related European beech resembles it.

Flowers appear in April, when leaves are ⅓ developed. Staminate flowers in stalked, round head, about 1 in. through; pistillate, in two-flowered clusters, in axils of upper leaves. Trees begin bearing relatively late, sometimes not until 13 years old. Fruit: small, spiny bur, with two pale tan, sweetmeated nuts, germinates fall or spring.

Wood 90 percent as hard, 71 percent as strong as white oak, but does not weather well, weighs 42 lb per cu ft, diffuse-porous, with broad, conspicu-

Beech

(Continued)

ous medullary rays, alternating with small, inconspicuous ones. Highly susceptible to leaf blight, bark canker, and nuts too frequently infested with insect parasites. Trees not easily transplanted.

Wood used commonly in furniture and in tool handles. Makes excellent fuel, and diseased trees are important source of fuel. Nuts were formerly important and delicious food for man and hogs, and wild turkeys fattened on them. Nuts may be crushed; extracted oil useful in cooking or in food for animals. Eaten by 15 bird species, by foxes, squirrels, bears, and deer. This is the tree on which youngsters carve their own and others' initials. These carvings may date, persist, and be evident for several decades.

Chestnut
Castanea dentata
Tree to over 100 ft high, with trunk diameter to exceed 10 ft. Old bark gray, with long, coarse furrows. Bark on younger branches, closer, smoother except where disease-infected. Twigs smooth, greenish-brown, with ¼ in. buds, covered by two to three dark brown scales. Leaves to 8 in. long, and 2 in. wide, with sharp-pointed teeth, along margins. Roots deep.

On hillsides and some flatlands, on relatively poor soils that are well-drained and commonly associated with maple, beech, and similar trees. From southern Ontario to northern Georgia, west to Mississippi and Michigan but practically gone except as saplings and suckers that die about time first nuts are produced. U.S. Forest Service and many cooperating state forest departments are attempting to locate disease-resistant mutations. Some of these efforts have been, to a degree, successful. Many chestnuts are now harvested from the Chinese chestnut, a low-growing shrub.

Flowers appear in late spring, after leaves, staminate in loose, slender, crowded catkins, to 5 in. long; pistillate clusters, near base of staminate. Fruits mature in September and October, with prickly but enclosing two to three plump, russet nuts, pointed at top, and thin-shelled. Nuts germinate spring following planting.

Wood to 28 lb per cu ft, relatively coarse, strong, light, easily split, durable when exposed to the weather. Chestnut blight *Endothia parasitica* has killed most trees in eastern part of range. A leaf blight may also be serious. Trees were formerly budded and grafted to improve quality and quantity of yield.

Both nuts and wood formerly had high commercial value. Nuts were harvested, stored, roasted, boiled. Meal made from nuts was used by Indians and early settlers and wood provided interior trim and heavy construction in many buildings within range.

Chestnut

Chinquapin
Castanea pumila
C. pumila, round-topped tree, sometimes to 50 ft high, with trunk diameter to 3 ft, but often a mere shrub. Bark to 1 in. thick. Twigs slender, woolly when young, smoother when mature, with red-scaled buds. Leaves to 5 in. long, to 2 in. wide. Other chestnuts, European chestnut, *C. sativa*; Japanese chestnut, *C. crenata*; et al.

C. pumila native from New Jersey to Florida, west to Missouri and Texas, where it is found on dry, sandy slopes. European chestnut cultivated in Europe. Japanese chestnut grown in Asia, is possible cross with native chestnuts that may be worth developing. *Castanopsis chrysophylla*, the golden-leaved chestnut, or chinquapin, is from Pacific Coast.

C. pumila bears staminate flowers in spreading catkins in May or June, and pistillate at base of upper catkins. Fruits, burs, with stiff spines, enclosing one, or rarely two, sweet, bright brown nuts. Nuts sprout soon after they fall in autumn. Because of this, chinquapins do not do well in North, since new wood cannot set before severe weather.

Chinquapins do not reach great size of chestnuts so have little value as wood. Wood to 28 lb per cu ft, hard, brown, durable, strong, and splits easily. Nuts may be badly infested with a nut weevil in South, but tree is relatively resistant to chestnut blight, and hybrids between chinquapin and chestnut may solve blight problem.

European chestnuts highly susceptible to blight. Japanese chestnut too coarse to be popular. Chinese chestnut of superior quality, and relatively resistant to blight. Chestnuts used in flour or in stuffing for fowls; too good to be lost to man completely. Chinquapins or Chinese chestnuts or some naturally resistant strain may improve the species.

Chinquapin

White Oak
Quercus alba
Tree to 95 ft tall, with trunk to 6 ft in diameter; tall, in open, two to three times as broad as tall, with gnarled branches. Roots deep. Bark light gray, broken into long, thin, irregular scales. Twigs with star-shaped pith, medium thick, with a bloom first winter, then gray. Buds to ⅛ in. long, dark, reddish-brown, rather blunt. Leaves to 9 by 4 in.

White Oak

In dry upland soils or at best on well-drained, deep, higher bottomlands where there is adequate moisture. Ranges from southern Maine to Minnesota, southward to Florida and Texas, sometimes forming rather extensive and pure forests. Commonly found mixed with other oaks, hickory, basswood, yellow poplar, red gum, black cherry, white ash, and other species.

Staminate flowers in catkins to 3 in. long, yellow. Pistillate flowers bright red, scattered or clustered. Pollination by wind. Fruit an acorn, to 1 in. long, 1/4 enclosed in light, chestnut-brown cup, with thick, warty scales. Acorns ripen in October, 208 per lb, 25 to 90 percent germination; planted 1½ in. deep, to 6 in. apart, grow to 9 in. first year.

Wood strong, ring-porous, durable, 46.3 lb per cu ft, light brown, requires 4,425 lb pressure per sq in. to shear, with thin, light brown sapwood; close-grained, tough, hard. Not a fast grower but may live to 600 years, being less tolerant to enemies in old age. Acorns germinating in fall may freeze unless they have been buried by squirrels.

Important wood for heavy construction, floors, furniture, cabinetwork, tool handles, baskets, railway ties, fences, fuel. Seton believes that spread of this oak is largely dependent on a sustained squirrel population. Acorns were used as food by Indians, and early settlers boiled them and used them as food when necessary.

Bur Oak
Quercus macrocarpa
Tree to 120 ft high, with trunk diameter to 7 ft though normally smaller. Bark darker than in white oak, with more vertical ridging, to 2 in. thick; scales often reddish. Twigs corky at first with pale gray fuzz but usually smooth by first winter. Buds to 1/4 in. long, with light red-brown scales. Leaves to 12 by 6 in., on 1-in. petioles, somewhat rough.

Best on dry clay soils. Extends farther northwest than other eastern oaks. Appears as shrub in Rocky Mountain foothills. Ranges from New Brunswick to Delaware and west to Texas and Wyoming and North Dakota. Grown as an ornamental in many parts of eastern United States and in South Africa. At its best in Ohio River valley. State tree of Iowa.

Staminate flowers in catkins to 6 in. long, with pale, matted hairs. Pistillate on same tree, red tinged, stalked or stalkless. Fruit, a solitary, usually stalked nut, to 2 in. long, broad, egg-shaped, downy at tip and enclosed ½ or more by deep fringed cup, a character that gives tree the name bur oak or mossy-cup oak.

Wood very durable, heavy, strong, hard, close-grained, dark, rich brown, with thin, lighter sapwood, tough. A common associate of such bottomland species as elm, soft maple, pin oak, willow oak, red gum, and cypress and has a high resistance to injury by smoke and gases from industrial and fuel plants.

Grown sometimes as an ornamental and particularly within its range where there is danger of smoke injury. Wood used as is the wood of many other kinds of oaks. It will survive on soils too poor and dry to support some of the better hardwoods and so represents a good product of such soils.

Bur Oak

Rock Chestnut Oak
Quercus prinus
Tree to 70 ft high, with trunk to 3 ft in diameter, tall and columnar, or open branching. Bark thick, brown to black, fissured into deep broad ridges. Roots deep. Twigs stout, orange to brown. Buds clustered at tip, sharp-pointed, with scales light brown, and hairy at apex. Leaves to 9 by 5 in., with coarsely rounded shallow lobes. There are three other chestnut oaks: the yellow chestnut oak, *Q. muelenbergii*, is the common one in the Great Lakes area; *Q. prinoides*, the dwarf chestnut oak, occurs in dry, sandy areas; *Q. michauxii*, the swamp chestnut oak, occurs in the southeastern states.

On dry, poor, rocky soils, sometimes in pure stands but often associated with black oak, chestnut, and pitch pine. Ranges from southern Maine to northern Alabama and Georgia and west to northeastern Mississippi, Illinois, Michigan, and southern Ontario. In southern part of range, is found at elevations of 4,500 ft.

Flowers borne separately on same tree; staminate, in yellow-green catkins to 4 in. long; pistillate, pale or dark red, in dense spikes. Fruit acorn, to 1½ in. long, shiny, egg-shaped, ½ to 1/3 enclosed at base in thin cup whose scales are somewhat grown together; 184 to pound; 25 to 90 percent germination; grows to 10 in. first year.

Wood stout, heavy, 47 lb per cu ft, ring-porous, durable, hard. Bark higher in tannin than in other oaks and is used with tannin from hemlock to offset the red of the hemlock tannin. Unable to compete successfully with other oaks whose wood is of superior quality on the richer soils.

Chief use is as an important source of tannin for use in leather industry and in arts. Wood used as a substitute for white oak, in making railroad ties, fencing, fuel, and heavy construction timbers. Leaves often bear an interesting bullet gall. Old wood commonly parasitized by a white wood rot (*Polyporus*).

Rock Chestnut Oak

Coast Live Oak

Live Oak

Cork Oak

Coast Live Oak
Quercus agrifolia
Forest tree growing to 90 ft. Trunk diameter to 6 ft, the trunk generally dividing low into crooked heavy branches. Leaves stay on for 1 year; 1 to 4 in. long, oval, rather coarse and heavy. Twigs hoary for two seasons. Bark brown to black and deep-furrowed.

Found in dry loams or gravelly soils of open slopes from sea level to 4,500 ft elevation from San Francisco south to Lower California and rarely more than 50 miles from coast. If near salt water and exposed, becomes more shrubby in appearance.

Flowers appear in early spring. Pollination by wind. Fruits maturing in 1 year into acorns 1 in. long, with cup over the lower ¼ or ½ the nut, having silky, closely pressed scales. Seedlings may develop under moderate cover, and maturity may be reached in 150 years.

Trees often found in almost pure stands, alone in the open or mixed with other trees. May be associated with California sycamore, big-cone spruce, and white alder. Trees are too distorted to produce any great amount of clear lumber, but they produce great quantities of acorns.

Acorns of this and of some other oaks, including swamp white oak, chestnut oak, basket oak, and white oak are edible. Indians, especially in West, ground the meat, either before or after roasting, and sometimes removed tannin by soaking in water. From this, flour cakes were made which, according to John Muir, were most valuable food for hiking trips.

Live Oak
Quercus virginiana
Tree. Height, to about 50 ft, but with crown to 150 ft across, trunk being to 4 ft through. Bark dark, reddish-brown, with small surface scales and small furrows. Twigs slender, brown to gray, with buds about $\frac{1}{16}$ in. long, and with white-margined, brown scales. Leaves persistent for 1 year, to 5 in. long and 2½ in. wide, glossy green above.

At best on moist, sandy plains. On dry lands, may be only 1 ft. tall. Does well on flooded lands. Ranges from Virginia to southern Florida and west along Gulf coast through southern Mississippi, southern Louisiana, on to the Rio Grande in Texas, north to central Texas. In Cuba and Mexico. Most important live oak in East.

Flowers staminate, in hairy catkins to 3 in. long; pistillate, in fuzzy spikes, on stems to 3 in. long; stigmas, bright red. Fruit in clusters of three to five; a nut about ¾ in. long, dark brown to black, ⅓ enclosed in cup that is reddish-brown. Tree in Georgia grew to 4½-ft trunk diameter in under 70 years and may live apparently to over 300 years.

Wood one of heavier of native woods, strong, requiring 8,480 lb per sq in. to shear in contrast to hickory, 6,045; ebony, 7,700; black walnut, 4,700; and chestnut, 1,500 lb. Wood tough, durable in various situations. Some trees compound, having a combined buttressed trunk diameter of to 8 ft., with wide, horizontal branches.

Valuable tree for shipbuilding, one of best but now used mostly for tool stock and as ornamental shade tree. Acorns apparently vary in their quality and some of sweeter forms were considered edible by Indians. From them, they produced an oil similar to olive oil. Trees only 1 ft high may bear fruits in some places.

Cork Oak
Quercus suber
Tree. Height to 60 ft, with a trunk diameter of 4 ft, crown, densely spreading. Bark thick, and the most important part of tree to man. Its nature is to be seen in average commercial cork used in bottles. Leaves evergreen, ovate to oblong, sparingly toothed, smooth above, and whitish fuzzy beneath; to 3 in. long.

Native from Atlantic to Asia Minor but especially abundant in Portugal, Spain, Algeria, Tunisia, Corsica, southern France, and Morocco, occupying some 4 million acres of forest lands. Also is cultivated in India and in United States, on plantations in California, New Mexico, and Arizona. Obviously favors warm, dry climate.

Fruit a short-stalked acorn, egg-shaped, to 1¼ in. long, ⅓ to ½ enclosed in a thick-scaled cup whose scale tips are recurved. 60,000 vigorous young cork oaks were planted in America about 1940, and 150,000 when war broke out. First California tree to be stripped yielded 151 lb of ground cork. Tree lives about 500 years. Produces a cork layer which is thick enough and possesses the elastic qualities required for production of commercial cork.

Bark is removed down to inner bark at 20 years, usually working first 6 ft of trunk but sometimes more. New crop is harvested about every 9 years, an average yield per tree varying from 40 to 500 lb, best quality coming from young vigorous trees though first harvest is usually poor.

Some 500 million lb of cork are produced a year, of which United States imports about 1 million lb. Used in making bottle stoppers. Linoleum is made of cork, linseed oil, and burlap. Linotiles are coarser tiles of cork and linseed oil. Cork tiles are made by baking cork under pressure. Valuable insulating agent.

Red Oak
Quercus rubra (borealis)

Tree. Height to 150 ft, with a trunk diameter of to 5 ft; trunk shorter in trees grown in open. Bark on old parts broken by shallow fissures with distinctly flat-topped ridges. Inner bark light red. Twigs smooth, with buds to ⅓ in. long, pointed. Leaves to 9 in. long and to 6 in. wide, dull green, with reddish midribs.

On sandy, porous, clay or gravel soils, in sun. Ranges from Nova Scotia to Minnesota, south to Florida and Texas but planted extensively in Europe and elsewhere for forestry purposes. Common as a shade tree. Red or black oaks have furrowed bark, pointed buds, leaves with bristle-tipped lobes, and fruits maturing the second year.

Flowers appear when leaves are about half-developed. Staminate flowers in slender, hairy catkins to 5 in. long, with four to five stamens. Pistillate on short stalks, develop acorns to 1¼ in. long, flat at base, in a broad, shallow cup, with basal scales apparently in three rows reaching ¾ way up cup.

Wood ring-porous, with conspicuous medullary rays, thin, light sapwood, and heavy, strong, hard, close-grained, light reddish-brown heartwood, weighing 41 lb per cu ft. Tree a relatively fast grower, compared to most oaks used as ornamental shade trees, and remarkably free from serious insect and fungus pests. Acorns harvested in April. 128 to a pound; germination 75 to 95 percent.

In nurseries, young trees reach height of 20 in. first year if planted at depth of 1½ in. 3 to 6 in. apart. Tree may reach height of 18 ft in 10 years; 39 ft in 20 years; 57 ft in 50 years. Excellent shade tree, good fuel tree and good for heavy construction, railroad ties, interior decoration, cooperage, and furniture.

Red Oak

Pin Oak
Quercus palustris

Height to 150 ft. Trunk to 5 ft in diameter. Many short stubby branches or branchlets stick out from branches. Bark gray-brown, smooth at first but eventually with low scaly ridges. Leaves alternate, to 5 in. long and to 5 in. wide, with five to seven narrow, bristle-tipped lobes, smooth above and below except in axils of veins.

Native of North America. Found in bottom lands from Connecticut to Virginia, west to Arkansas and Iowa. In black oak group. Differs from typical black oak by having leaves with smooth undersurfaces except for tufts in vein axils there.

Flowers: staminate and pistillate flowers are borne separately on same tree, staminate being in loose, or-hanging catkins, pistillate in short, rather stiff spikes. Fruit, an acorn about ½ in. long with meat of kernel bitter and nut resting in a thin, saucerlike cup; acorn two to three times as long as cup.

Flowering time from May to June; acorns are borne and ripen second season in September and October. Wood weighs 43 lb per cu ft, hard, coarse-grained, strong, knotty, light brown. Subject to attacks of many twig-gall makers; galls can often be collected in great abundance from the tree.

Name pin oak comes from practice of using the stiff pinlike spurs as pins to fasten timbers together, before nails were generally used. Some authors believe the appearance of the short, tough branchlets suggested the name "pin." Other authors believe that pins were the short tough branches actually used to hold together sections of planks and beams. Tree is commonly grown as an ornamental because it survives transplanting better than many other oaks and also because it is in itself an attractive tree. Closely resembles jack oak, *Q. ellipsoidalis*.

Pin Oak

Scarlet or Spanish Oak
Quercus coccinea

Tree of medium size, usually to 80 ft high but sometimes to 100 ft, with trunk diameter to 4 ft, with small, spreading branches making open crown. Bark light brown to black, with irregular ridges separated by shallow fissures. Buds to ¼ in. long, red-brown and fuzzy above middle. Leaves to 6 in. long and 4 in. wide, with seven to nine deep lobes bearing sharp points.

On dry sandy soils, often in almost pure stands or mixed with oaks and hickories. Native from Androscoggin River valley in Maine to Georgia, west to Wisconsin and Oklahoma, reaching to altitude of 5,000 ft in southern mountains and to 2,500 ft in Blue Ridge, where it is prevailing oak. Grown as an ornamental rather widely in America and in Europe.

Staminate flowers in slender catkins, to 4 in. long. Pistillate flowers bright red, on fuzzy ½ in. stems. Fruit usually borne singly or in pairs; a nut, ½ to 1 in. long, oval, hemispherical, or globose, red-brown, ½ to ⅓ enclosed in a deep, bowllike cup whose scales are shining, red-brown and closely pressed at tips. Flesh of acorn nearly white.

Wood heavy, coarse-grained, ring-porous, strong, light or reddish-brown, with thicker, dark-colored sapwood. Although wood is one of poorer grades of oak, trees are cut and wood mixed with other red oaks in trade,

(Continued)

Scarlet or Spanish Oak

although where there is too much scarlet oak the product may not always bring highest prices.

Used primarily as an ornamental and shade tree since it has marvelous beauty, particularly in fall, and since it responds rather well to horticultural manipulations that modify form for many desired purposes. Hardy, relatively free from attacks of more destructive insect pests, and therefore requires a minimum of care when established.

Black Oak
Quercus velutina
Tree to height of 125 ft; trunk diameter, 4½ ft. Upper branches ascending; lower, horizontal. Bark dark, deep-fissured between thick ridges that are cross-fissured. Inner bark yellow, bitter. Twigs stout, angular, with sharp-tipped, angled buds, with dirty-white to yellow fuzz. Leaves to 6 in. long and 4 in. wide, highly variable, thick.

Black Oak

Found on dry uplands or gravelly plains or in sandy soil, but rarely in rich lowlands. In West, mostly on poor upland soil. Range from Maine to western Minnesota and south to Florida and Texas. Best distinguishing characters are inner bark, cup of fruit, woolly buds, highly variable leaves that are frequently widest toward tip.

Flowers appear when leaves are about ⅓ developed. Staminate, in hairy catkins, to 6 in. long. Pistillate, on short, hairy stalks. Fruit an acorn maturing the second season, to 1 in. long, light reddish-brown, covering about ½ the nut, forming a fringelike margin to the top.

Wood ring-porous, with conspicuous medullary rays, hard, heavy, strong, durable, not tough, checks readily, weighs about 44 lb per cu ft. Heartwood light brown; sapwood, lighter brown. Not an attractive ornamental or a superior timber tree. Acorns mature in October, are collected in March, germinate 75 to 95 percent if planted at depth of 1½ in., 3 to 6 in. apart.

Wood used in cooperage, in furniture, in heavy construction, for interior finish. Speed of growth next to that of red oak. Bark used in tanning industry. Tree sprouts persistently and individuals rarely found over 200 years old. May go several years without bearing acorns. Usually with deep taproot, therefore poor as a transplant.

Shingle Oak
Quercus imbricaria
Tree. Height to 100 ft with trunk diameter to 3½ ft. Trunk seems to be short and crown rounded. Leaves alternate, to 6 in. long and to 2 in. wide, dark, shining green above, pale brownish and downy or hairy beneath, margins slightly wavy; in this respect unlike most other oaks, with short bristle at tip like red oaks.

Shingle Oak

Native of North America and found near streams, mixed with other hardwoods or on moist hillsides from New Jersey to Georgia and west to Arkansas and Nebraska, but not found on Atlantic coastal plain. Belongs to red oak group and a small subgroup that have willowlike or laurellike leaves.

Flowers staminate and pistillate, borne separately on same tree with staminate in catkins, pistillate in short, relatively stiff spikes. Fruit an acorn about ¾ in. long, maturing second autumn, somewhat more than half enclosed in cup that is relatively deep and has reddish-brown scales.

Flowering time April to May; fruiting time, the fall of second season. Wood weighs 47 lb per cu ft, light red-brown, hard, coarse-grained, of comparatively little value except as firewood. More common in southern than in northern part of its range.

Name shingle oak may have come from fact that early botanist Michaux is reported to have found the natives making shingle shakes from its wood. It probably is inferior to many other species for this purpose.

Tanbark Oak
Lithocarpus densiflora
Tree to 150 ft high, with trunk to 5 ft in diameter. Trunk of trees grown in open often grotesque, bearing wide-spreading branches. In close formation may have narrow top. Roots deep taproots, with laterals. Bark of heavy rounded ridges, with deep narrow fissures. Twigs densely fuzzy at first. Leaves persistent, light green above, brown woolly beneath; may have a toothed or entire margin.

Tanbark Oak

Does best in deep, rich, moist, well-drained, sandy or gravelly soils, often in association with Oregon white oak, California black oak, chinquapin, Douglas spruce, redwood. From southern Oregon, south along coast ranges to Santa Ynez Mountains in California, south in Sierra to Mariposa County, from sea level to elevation of 5,000 ft.

Flowers appear in early spring and resemble those of chestnut, with catkins to 4 in. long. Fruit solitary or in pairs, to 1 in. long, an acorn partially or wholly enveloped by cup, maturing at end of second season. Trees begin bearing fruit by thirtieth or fortieth year and mature at 200 to 300 years.

Wood hard, strong, brittle, like chestnut, reddish-brown, with thick, dark sapwood; close-grained. Tree does well in humid area that supports redwood, particularly in San Francisco Bay area. Can exist under shade of other trees throughout its lifetime. When competition is removed, growth and development are phenomenal.

This is the tree used on West Coast as the principal local source for tannin. The wood is used mostly as fuel although it has many of the properties of good timber if the trees were of the type to yield straight stock. The normal development is of average rapidity.

Order Urticales./Family Ulmaceae

Red Elm, Slippery Elm
Ulmus rubra

Tree to 115 ft high, with trunk to 2½ ft through. Branches relatively horizontal and short, without usual droop of American elm; gray, and with buds whose scales show rusty hairs distinctly. Leaves with margins doubly toothed and with surface conspicuously roughened.

Slippery Elm

Does best in rich woodlands where there is good water and drainage. Found from southern Quebec to southeastern South Dakota, south to eastern Texas and Florida. Limestone country seems to be favored by this tree. Grown in some places as an ornamental because it occupies less space than American elm though it lacks its grace.

Flowers bear both stamens and pistils, appear before leaves in spring, each on a short stem. Fruits mature as leaves appear and become ¾-in. wafers, with thin, papery, wing margin that is smooth. Not common in pure stands but grows best with such trees as yellow poplar, basswood, white ash, butternut, and other elms.

Wood heavy, 45 lb per cu ft; hard, strong, tough, durable, dark brown, easily split for an elm, tends to twist out of shape in curing; makes fair but not superior fuel. Bark yields from its inner layers a mucilage used medicinally in gums or lozenges to soothe inflamed tissues, and gives off unique odor.

Tree used as an ornamental. Wood used for fuel and in coarse construction work and for other uses typical of elms. Inner bark yields a fiber useful in making shoe lacings and cordage, also the mucilaginous medicine for sale in drugstores in a ground or processed form. Bark used by Indians in making canoes like birch-bark canoes. Buds are bird food. Slippery elm is seriously affected by the Dutch elm disease and is becoming much less common.

American Elm, White Elm
Ulmus americana

Tree to 100 ft high, with trunk diameter to 5 ft. Usual form a branching vase shape. Bark thick, grayish, roughened, and irregular, with furrows and ridges not continuous. Twigs fine, brown, with alternate egg-shaped, few-scaled buds. Leaves with upper surface somewhat roughened, and base usually uneven. Bark in waferlike layers; alternating brown and white.

American Elm

Does best on rich bottomlands or where springs supply an abundance of subterranean water. From Newfoundland, west through Canada to Rocky Mountains, south to Florida and Texas but planted widely elsewhere as a shade tree. Through East is one of most popular shade trees but may eventually be destroyed by Dutch elm disease.

Flowers appear before leaves, in slender, drooping clusters, each flower with stamens and pistils. Fruit matures shortly after flowering or about as leaves develop; a ½-in. wafer, with a papery margin, notched at apex, hairy on margin; 94,000 to the pound; 50 to 75 percent germination; planted to depth of ½ in., reaches height of 10 in. first year.

Wood heavy, 40 lb per cu ft, strong, tough, coarse-grained, hard to split, with a tendency to bend in curing; rings porous. Bark used as emergency rope and may be peeled off in strips to make shelters, or inner bark may be boiled and stripped off to be used as chair-bottom material. Cankerworms and elm leaf beetles may be serious pests. Dutch elm disease is now destroying American elms throughout their range. No control acceptable to all persons, including recent devotees of the environmental consciousness in the community, is now available. If American elms are to be saved, some other living forms may have to be sacrificed.

Chief use is as an ornamental shade tree, largely because of graceful arching canopy formed over streets when tops meet. Dutch elm disease should be controlled to save this tree. Buds serve as emergency food for some birds including such game birds as pinnated grouse.

Cork Elm, Rock Elm
Ulmus thomasii

Tree reaches height of 100 ft, with a trunk diameter of 3 ft. Trunk divides less than in most elms, often running undivided nearly to top of crown. Bark dry, thin, close, flaky, and, on twigs, found in corky wings or ridges.

Cork Elm

(Continued)

Leaves to 4½ by 2¼ in., coarsely doubly toothed, usually smooth above, and hairy beneath.

Favors gravelly ground such as banks of rivers and hills but where there is ample water supply. Ranges from Quebec to Minnesota, south to New Jersey and Nebraska. Grown to some extent outside this range as an ornamental. Like American elm, it favors areas where there is an abundance of underlying limestone.

Flowers bear both stamens and pistils, appear before leaves in spring, clustered along a central stem instead of at a common point as in other elms. Fruit matures as leaves unfold, as a ¾-in. wafer with encircling hairy wing with margin of seed cavity indistinct as compared with other elms.

Wood tough, heavy, hard, strong, light or reddish-brown and in many respects much like that of slippery elm. Victim of a leaf-spot disease and its leaves are preyed upon by elm leaf beetle so destructive to other elms. Straight form of trunk makes it possible for this tree to develop better timber than other elms.

Used primarily as shade tree; wood used for fuel and for lumber for particular uses such as wagon spokes, tool handles, and some kinds of furniture; also common in cooperage such as barrels, barrel hoops, hubs, ties, and baskets. Fruits gathered in great quantities and eaten by many squirrels.

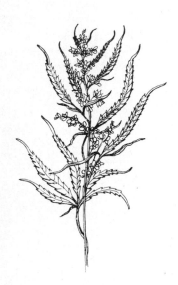

Hackberry

Hackberry
Celtis occidentalis
Tree. Height to 120 ft; trunk diameter to 4 ft. General shape short and bushy, with rounded crown, though it may vary considerably. Bark light gray, with small, warty ridges. Roots relatively shallow. Twigs reddish-brown, slender, zigzag, with buds whose tips are pressed close to twig. Leaves to 4 by 2 in., sharply, finely toothed, with unequal bases.

Favors hillsides and edges of marshes, and often alluvial, floodplain ground. More frequently a plant of periodically inundated areas along streams and occasionally wetted hillsides, but also grown rather commonly as a street tree, particularly in Middle West, where it may rival elms in this respect. Ranges through most of United States except eastern New England and Southwest but more common east of North Dakota and Kansas, where its native western boundary lies.

Flowers may have both stamens and pistils, or either, but both are found on same tree. Fruits are cherrylike but smaller than commercial variety, thin-fleshed, about ⅓ in. in diameter, with a large pit whose surface is rather conspicuously netted. Fruits edible; may not be borne every year, and number varies greatly.

Wood light brown, yellow, heavy but weak, soft, and coarse compared to elms. Does not have durable quality of elms or oaks, therefore of little importance ordinarily though it is sometimes mixed in with elm. A powdery mildew affects leaves, and witches-broom disease affects branchlets, sometimes seriously.

Greatest use is as a shade tree. Fruits are obviously favored by birds, at least 25 kinds of birds being known to feed on them. Because of this, tree may be useful to those who wish to attract birds. Furthermore, bark is close and fine and trees have a neat appearance along sidewalks.

Order Ulticales./Family Moraceae

Soft Hemp, Marihuana
Cannabis sativa
Height to 12 ft, with stalks ½ in. or more in diameter, if grown crowded; or 20 ft high and 2 in. through, if grown in hills. Leaves opposite near base, and alternate near top, composed of five to nine dark green, 2- to 6-in. roughish leaflets, with notched margins. Stamens and pistils borne in flowers on separate plants. Fruit a thick, lens-shaped achene; enclosed in yellow to olive-brown bracts.

A close relative of hops, nettles, mulberries, and similar plants. Has been grown since twenty-eighth century B.C., when Emperor Shen Nung taught its cultivation in China. It came to Europe about 1500 B.C. and to America with arrival of earliest pioneers. Was grown primarily for cordage. Commercial production for fiber, except for World War II years, has steadily decreased. Earlier it was used in twines, oakum, and packing. It endures friction, heat, and moisture and is high in cellulose.

It does best in rich river bottoms. *Cannabis* has spread throughout the U.S. along the major rivers and there is a correlation between *Cannabis* distribution and alluvial stream deposits in areas of the Plains States where intermittent flooding occurs.

Seeds sown in March; using greater amount of seeds per acre on rich than on poor soil, 35 to 50 lb per acre, usually in drills. Mature, yellow-stemmed, taller staminate plants more abundant than pistillate, whose

Marihuana

stems are bushier and pale at maturity. Flowering tops of female plant secrete a varnishlike resin called "hashish" in the West and "charas" in India. Both male and female plants contain psychoactive substance, which is more concentrated in the flowering tops of the female plants.

Harvesting is possible 4 months after planting. Seed weighs 44 lb per bu and has commercial value. Yield of fiber may reach 1 ton to acre.

In North America marihuana consists of a mixture of crushed leaves, flowers, and often twigs of *Cannabis*. It may contain many adulterants when marketed. In 1972, possession of marihuana was a crime punishable by federal law. Effects of smoking or chewing *Cannabis* have been known for over 2,000 years. Effects on individuals are variable; in general the smoker of marihuana claims he feels a period of euphoria and elation followed by a heightened sensitivity to stimulation. Later hallucinations and mental confusion and perhaps comatose sleep may follow. However, none of these conditions can be predicted. The principal active agent in *Cannabis* is thought to be delta-9-tetrahydrocannabinol. Has been reported to be one of the safer drugs for human use. There is evidence that heavy marihuana use is correlated with a loss of interest in conventional goals and the development of a kind of lethargy. More systematic research efforts on the effects of marihuana are needed.

Hop
Humulus lupulus
Coarse, climbing vine 20 or more feet long. Roots perennial, sending forth each spring new vines. Stems angular, rough, weak, coarse, dying back each year. Leaves near base of plant, opposite, rough, rather deeply lobed; near upper part, alternate, entire, and less coarse. Stems, roots, or sometimes seeds used in starting new crop.

Hop

Under cultivation since time of Roman Empire and grown in Europe since the ninth century. Grown extensively in United States, England, and Germany, formerly being an important crop in New York State, Massachusetts, and Vermont, but now grown more extensively in the Northwest in Oregon and Washington.

Staminate and pistillate flowers on different plants; pistillate develop from conelike, hairy catkins into short catkins with thin, concave scales at base of each of which there is an aromatic, resinous, bitter substance that contains resin and the drug lupulin. Set 6 to 7 ft apart, with poles or trellises nearby for vines to climb.

Harvested in late August and early September by cutting vines 1 yd above ground and picking hop cones, which are then immediately dried in kilns for 12 to 20 hours over slats at about 120°F. Bleached to yellow by sulfur fumes, then baled in bundles 20 by 22 by 52 in., weighing 180 to 200 lb, for shipment to market.

Used in medicines, in poultices, as sedative, and for tonic effects but mostly in beer to prevent spoilage by bacterial growth; they impart a characteristic flavor and aroma to the finished product. Chief enemies a red spider, the fungal diseases blue mold and black mold, and a stem borer that was highly destructive to New York plants until skunks were protected by law to destroy borers. Leaves may cause dermatitis.

Osage Orange, Bois d'Arc
Maclura pomifera
Tree to 60 ft high. Trunk diameter to 3 ft. Irregularly cylindrical. Bears stiff, spinelike twigs. Bark dark, rough, longitudinally ridged. Twigs stout, greenish to yellow-brown, with buds pressed to twig. Leaves alternate, with entire margins, shining green, smooth, to 5 by 3 in. Juice milky. Usually found in hedges.

Osage Orange

Grows on various soils, but mostly on good rich bottomlands. Originally from southern Missouri to northern Texas, but planted extensively and established throughout United States. Grown in hedges because of stiff spines and dense tangle formed. Used in barricades to keep human beings and other animals out.

Staminate and pistillate flowers borne separately on different trees. Staminate in loose, pistillate in compact clusters. Fruits, hard structures, buried in spongy mass that assumes size and shape of wrinkled, green orange. Seeds weigh 12,140 to the pound, with 60 to 90 percent germination; planted 1 in. apart and ½ in. deep, grow year seedlings to 10 in.

Wood hard, stiff, strong; to 48 lb per cu ft, bright orange in color, elastic, tough, hardest of our native woods; formerly useful in making bows, axles, woodenware, pulley blocks, paving blocks, and tool handles. In Arkansas, in early nineteenth century, a good Osage bow was worth a horse and blanket. Bark yields a fine, flaxlike fiber.

Yellow, orange-yellow, or golden dye or a base for green dyes is extracted from wood and used as substitute for fustic and aniline dyes in arts and industry. Indians used tree for fiber, dye, and bows. Chips of wood boiled in water with cloth give cloth fixed yellow color. Wood 2 times as hard as white oak and 2½ times as strong.

Paper Mulberry
Broussonetia papyrifera

Tree under cultivation, often shrubby, with branches widespread. Twigs moderately rounded, zigzag, sticky, hairy when young; pith, large, round, with green partition at each joint. Buds with striped scales, cone-shaped. Leaf scars rounded and swollen. Leaves like other mulberries, unlobed, mittenlike or three-lobed, with milky juice. Bark netted.

Native of eastern Asia where three species are under cultivation. *B. papyrifera* is hardy in America as far north as New York State; grown mostly there as ornamental. Fruits are eaten by birds. Not partial to any particular soil, doing well on any good average garden soil.

Rather inconspicuous, greenish-white flowers, with pistillate producing a dense, globelike head of "seeds" surrounded by orange-red pulp. Seeds sown when mature; or wood cuttings made in spring, seasoned in greenhouse in winter; or root cuttings are made. Others are propagated by grafting on desired stock.

Unique in its ability to withstand heat and dust. Makes a favored tree in cities except that those bearing fruits may be untidy in season. Bark peeled and outer bark scraped off with shell. Inner bark then soaked in water and strips pounded together with a mallet over a hard wood log. This may be pounded into tapa cloth.

Probably most interesting use is in connection with making of tapa cloth, particularly valuable for clothing and ornaments in the South Pacific area and in eastern Asia. Some of the cloth is paper-thin and stiff, while other cloth may be thick and yet flexible, like leather. Of course, use as an ornamental must not be overlooked.

Paper Mulberry

Breadfruit
Artocarpus altilis

Tree to 60 ft high, with fragile branches, and often a grotesque form. Juice milky. Leaves from 1 to 3 ft long, leathery, rounded or wedge-shaped at base, and, in upper part, with three to nine long, deep lobes. Leaves usually appear healthy and free from insect or fungus injury. Related jack fruit has leaves entire rather then lobed.

Native of Malayan area, particularly in Polynesia, and widespread in tropics where it has been under cultivation since days of antiquity. Requires a hot, moist climate, with plenty of water but good drainage. Grown in botanical gardens throughout the world where possible.

Staminate flowers in dense, yellow, club-shaped catkin, to 16 in. long. Pistillate in short, compact spike that develops into spiny, globular fruit size of a melon. Some varieties are seedless. An 8-year-old tree in healthy condition may yield to 800 fruits a year, each size of a melon. May be propagated by cutting young shoots.

Carbohydrate content of fruit unusually high. Fruit eaten raw, boiled, baked, roasted, fried, or ground to form a flour used in bread or stored to be used as a paste. Fruits attached to short, thick branches. Natives bake them on hot embers, in pits filled with hot rocks, or sliced and fried.

Probably few plants have a greater yield of useful food for man or beast. Fruits of the related jack fruit may be 2 ft long instead of 6 in. as in the breadfruit, may weigh to 40 lb and are borne on the trunk instead of on the lateral branches or twigs. Fruit pulp of breadfruit brownish-yellow.

Breadfruit

White Mulberry
Morus alba

Tree to 60 ft high. Trunk to 2 ft through, normally smaller. Thick topped, bushy, often escaping cultivation and forming thick copses. Bud scales shorter, more closely pressed to stem than in red mulberry. Buds shorter. Leaves light green, small, smooth above, and sometimes shining and whitish beneath, variously lobed. Bark yellowish, in furrows.

Native of China. Cultivated from earliest times. Introduced into America in colonial times with thought it might stimulate silk industry. Has become naturalized and independent in East and almost a weed tree near some dwellings. About 100 species of mulberries known through world, many of major importance where they are.

Flowers, in general, similar to those of red mulberry. Fruit smaller, more globular, usually white, though some may be black. Most fruits are sweet. Where 1 in. long, may yield a substantial and a favored food delicacy. Russian mulberry more hardy than normal white and is grafted onto white stock, usually in spring months.

Leaves important product outside the fruits. On the whole, probably most important since they provide basic food for silkworms that provide silk of commerce. In silk-growing countries, leaves are gathered and fed to silkworms. Many horticultural varieties of white mulberry exist including a drooping form, a cut-leaf form, and others.

Uses include food for man, for birds, and for pigs, for which purposes white mulberries are commonly grown near homes; food for silkworms, for which they are grown in extensive plantations; for ornamental purposes, and to a limited extent and probably secondarily, for wood, and possible dyes and fibers.

White Mulberry

Red Mulberry
Morus rubra

Tree to 60 ft or less. Trunk diameter to 7 ft; normally much smaller. Crown round-topped, open. Trunk short. Twigs slender to stout, brown, with swollen leaf scars and pointed eggshaped buds. Juice milky. Leaves highly variable, from heart-shaped to deeply lobed once or twice, hairy below, and sandpaperlike above; to 5 by 4 in.

On rich bottomlands or moist hillsides. From southern New England south to eastern Texas and Flordia, with related Mexican mulberry, *M. microphylla*, extending range into New Mexico and Mexico. Not a common tree in most of its range but planted frequently for various reasons. Found in Appalachians to an elevation of 2,000 ft.

Staminate and pistillate, flowers borne on different trees, the staminate in hanging catkins; the pistillate in hanging spikes. Fruit something like a blackberry with the hard achene surrounded by juicy, sweet, fleshy red to purple parts. Fruits of Mexican and Persian mulberries almost black, larger, and more spherical.

Wood light, soft coarse-grained, rather tough, very durable, light orange, with thick, lighter sapwood; useful mostly in boatbuilding, in making cheap wooden boxes, and for light outdoor wooden structures like fences that must withstand weathering. Wood used in making durable wooden pins in shipbuilding.

Tree rarely large enough to yield much lumber, but this has value. Fruits may be eaten by man but are more favored by birds and other forms of wildlife. Some 21 species of birds known to feed on fruit. Commonly planted in bird sanctuaries to attract birds. Squirrels and skunks also eat fruits. Indians made cloth from fiber from bark.

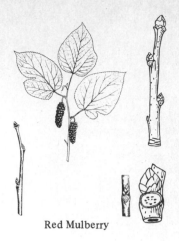

Red Mulberry

India-rubber Plant
Ficus elastica

Tree in tropics reaches height of 100 ft, but in greenhouses rarely exceeds 10 ft. Stems relatively elastic and coarse. Leaves to 12 in. long, with entire margins, coarse, pinnate veins, an abrupt, elongate point at end; dark, shining green above, lighter beneath. Juice conspicuously milky and abundant, when plant is broken.

Native of northern India and the Malayan area but grown widely in the tropics; common potted plants in hotel lobbies. Usually grows in forested area, but most have an abundance of water and tropical temperatures. Related to fig that is in the same genus of plants. Popular indoor plant of temperate regions. Will survive in spite of poor treatment; in particular infrequent watering.

Fruits borne in pairs, in axils of leaves, and at first covered by a hooded shield. When ripe, fruit is greenish-yellow and to ½ in. long. Plants do not mature until at least 50 years old. Since rubber is inferior, it is not of commercial importance.

Stems or roots tapped and milk is allowed to drip onto bamboo mats or to accumulate on trunk. Rubber is then scraped off, cleaned, and dried but there is tremendous waste in methods followed by natives and consequently little inducement for them to use this rubber as a source of income. It is known as "Assam rubber" or "India rubber."

Plants are probably more important as house plants. Trees in native lands, with huge buttresses and conspicuous prop roots, provide a unique appearance, characteristically tropical in effect.

India Rubber Plant

Fig
Ficus carica

Tree to 30 ft high but relatively weak-looking, with weak stems. May also occur as a shrub in certain climates. Leaves thick, somewhat like a mulberry leaf in shape, with five lobes of which two basal ones are the smaller. In this species, veins are arranged palmately while those of other species are pinnate.

Native probably of southern Arabia or of Caria in Asia Minor. Spread early through Mediterranean area. Now grown in all tropical and subtropical areas. In America, may be hardy north to Philadelphia. Found as a pot plant almost anywhere. Frequently mentioned in the Bible. Forms include common, caprifigs, Smyrna, and San Pedro figs.

Common figs bear fruits on old wood with second crop of year in axils of leaves. True fruits are borne on short stalks inside fleshy twigs. Common figs need no pollination and have no seeds. Caprifigs pollinated by gall wasp *Blastophaga grossorum*, yield three annual crops but have no staminate flowers, so use caprifig pollen.

Must grow on heavy clayey soils that remain uniformly moist. May survive temperatures down to 10°F, when tops are usually killed unless bent down. May be grown successfully under glass. San Pedro figs of California yield one nonpollinated crop but second must be pollinated to develop. Wasps in figs do not hurt food value.

Figs dried, preserved, canned, or ground and baked to make fig coffee. In addition to food value, have high laxative quality and some medicinal value. California and Texas leading states in fig production; Turkey, Greece, and Italy lead in Old World. Smyrna fig probably most important of all commercial figs. Leaves named in Bible and used in art as cover for nakedness.

Fig

149

Nettle

Bastard Toadflax

American Misletoe

Order Urticales./Family Urticaceae

Nettle
Urtica dioica

Height to 3 ft; related *U. gracilis,* to 10 ft. Covered with bristling, stinging hairs. Leaves on short stalks, less than half the leaf width, with coarsely cut margins and downy undersurface; heart-shaped at base. Spurge nettle of southern states, *Jatropha stimulosa,* has stinging bristle like hairs.

Related to hemp, hops, and mulberries. Naturalized from Europe, where it has enjoyed some cultivation for fiber since days of early Egyptians. Common stinging nettle of North ranges through southern Canada and from North Carolina to Missouri.

Flowers clustered in loose formations near the bases of the leaves, greenish, and, in some nettles, staminate or pistillate on one plant or on separate plants. Since the plant will thrive on lands that will support few other crop plants, its culture is relatively simple.

Stinging hairs may cause severe injury. Hunting dogs have been poisoned by contact with *U. chamaedryoides.* Has been reported that these dogs show symptoms such as difficulty in breathing, slow and irregular heart beat, and muscular weakness. Stings of some Asiatic nettles are reputed to be strong enough to kill human beings. As weeds, plants may be kept in control by cutting twice a year, grubbing with gloved hands, or salting.

In World War I, fiber was cultivated for making fabrics for tents, wagon covers, clothing. Capable of producing a good, strong, white linen. A yellow dye is obtainable from roots. Young shoots boiled make an excellent spinach substitute. In a machine age, it may again become a valuable source of fiber.

Order Santalales./Family Santalaceae

Bastard Toadflax
Comandra umbellata

Perennial herb, with horizontal, underground rootstock and flowering stems that reach a height of 18 in., much-branched, well-supplied with leaves. Leaves oblong, thin, pale green, darker above than below, to 1½ in. long, with prominent, pale midrib.

Found on dry grounds among growing trees and shrubs or near them. Ranges from central Maine to Wisconsin and south to Georgia. Related varieties occur to the west. Often found in rather dense stands of the herbaceous stems so that if one plant is found, many are likely to be nearby rising from common rootstock.

Flowers pale green to white, bearing both stamens and pistils, in clusters that seem to arise from a common point near top of herbaceous stems. Calyx tube attached to dry, urn-shaped fruit. Flowering period May and June. Fruit nutlike, maturing in midsummer after which plant dies back for season.

Hemiparasite on roots of a wide variety of herbs, shrubs, and trees, even including certain mosses. It takes water and minerals from its host. Doubtful if plant ever does any serious damage to its hosts, though apparently this has not been studied exhaustively.

Interesting because of its hemiparasitic habit but not of commercial importance. There are many species in this genus that are independent.

Order Santalales./Family Loranthaceae

American Mistletoe
Phoradendron flavescens

Shrubby, rubbery, greenish-barked plant that grows from some part of a tree as a parasite. Leaves pale green, easily broken, rather thick, oval, smooth, to about 1 in. long, somewhat succulent. Whole plant may form a sort of loose globe as big around as a barrel, with branchlets spreading in all directions.

About 100 species, of which this one is found on many trees that shed their leaves. Ranges from New Jersey and southern Indiana south to Florida and Texas; also in California. Not cultivated although it has a market value with florists. All mistletoes are American. Most are tropical with but few found in Western states. *P. villosum* is parasitic on oaks from the Pacific Coast states east to Arizona.

Flowers either staminate or pistillate, on different plants. Staminate flowers in small, catkinlike spikes. Pistillate flowers in short spikes, produce pulpy, pale green to white, one-seeded, berries, about ¼ in. in diameter. Berries eaten by birds and seeds planted by them in their droppings at some remote point.

Plants harvested and stored in cool temperatures previous to Christmas season, when they are sold to flower dealers. Once believed to have medicinal properties and those growing on oaks to have magic powers. Make their own sugar and starch with their own chlorophyll but draw on hosts for water and mineral supplies. Has been reported as toxic to animals browsing among oak branches.

Mistletoe may injure many shade and timber trees. Dwarf mistletoe attacks cone bearers, while this species attacks deciduous forms. A hook on a pole or rope is used in collecting mistletoe commercially. Fruits reported as having been fatal to children but not to pigs. License to kiss one under a mistletoe at Christmas has its possibilities and penalties.

1. Desert Mistletoe
Phoradendron californicum

2. Tufted Mistletoe
P. densum

3. Juniper Mistletoe
P. ligatum

4. Colorado River Mistletoe
P. coloradense
Hangs in great, loose balls or festoons from tops of trees or shrubs or erect in tufts, with rubbery greenish stems and opposite leaves. Leaves of (1) small; (2), larger and oval; (3), scalelike and much narrowed at base; (4), thick, larger, and with not evident veins. All seem to be so designed that they may conserve water loss.

Grow as parasites (1) on ironwood, mesquite, paloverde in California and Arizona deserts, (2) on California juniper from southern California to Oregon, (3) on desert and mountain junipers, (4) on mesquites along Colorado River. Another mistletoe of western deserts is variety of eastern form that is conspicuous on cottonwoods along Mohave River (large-leaved).

Staminate flowers sunken in joints or spikes, smell in evening like apple blossoms. Pistillate in separate flowers, developing by November into berrylike fruits that are coral-pink in (1), straw-colored in (2). Berries eaten by birds as source of food and water from November to April and seeds are in droppings. Seedlings attack plants on which bird droppings are left.

Birds known to carry seeds of desert mistletoe include phainopeplas, robins thrashers, desert quail, bluebirds, and others. Some plants like beefwood are able to resist attack of mistletoe by exuding a gum at point of attack.

Birds eat some fruits, many of which are usable as food by man in areas and at times when food might not otherwise be available. Plants obviously harm some forage trees in desert areas but while fruits of eastern mistletoe have been known to be fatal to children, Western Indians used to boil stems of desert mistletoe with fruits and eat them.

Desert Misletoe

Order Aristolochiales./Family Aristolochiaceae

Wild Ginger
Asarum canadense
Height to 6 in., growing from a horizontal, perennial, partly buried rootstock that shows growth and leaf scars conspicuously and branches and persists for years. Leaves about two to a plant, kidney-shaped, thin, long-stalked, 6 or more inches long, with a deep sinus at base of expanded part, with conspicuous veins but not mottled.

Found in rich woodlands where there is moist soil in spring, from New Brunswick to Manitoba, south to North Carolina, Missouri, and Kansas but most abundant in northern part of range. In prairie states, a more pointed-leaved variety is common form. On Pacific Coast, an evergreen species *A. caudatum* takes place of *A. canadense*.

Flowers partly buried in duff of forest floor; purple brown, with globular base on end of slender stalk and expanded at mouth by three sepal lobes that spread 1 in. or more. Flowers April and May in North. Pollination by small flies, possible fungus gnats and flesh flies. Fruit a fleshy capsule that bursts to free many seeds.

Wild Ginger

Plant obviously cannot survive in long direct sunlight and lives most of its year of activity in the spring and early summer months. Rootstock not offensive to taste. Flowers and leaves relatively attractive in a wild garden and easily propagated.

Element in rootstock is mildly purgative and is used to some limited extent in medicine as an irritant. Chippewa Indians called it "namepin" or "sturgeon plant" and used it as an emergency food or as a medicine in case of indigestion. It does not seem to have had wide use by other Indians.

Order Polygonales./Family Polygonaceae

Curled Dock, Yellow Dock
Rumex crispus
Height above ground, to 3 ft, erect, well-branched, slender, tough, smooth, bearing fruit clusters at branch bases. Leaves somewhat narrowly oblong, with conspicuously wavy, crisped margins that are smooth and entire, alternate, rather thick but not succulent or tough. Roots large, red or yellow, with expanded taproot to 1 ft long and 1 in. thick. Perennial.

Curled Dock

(Continued)

151

Found in open dry spaces such as meadows, roadsides, gardens, lawns, fields, pastures, particularly where some neglect is evident. Ranges throughout United States and southern Canada. Naturalized from Europe whence it has spread to other parts of world possibly through being a common impurity of commercial seeds.

Flowers perfect, with six sepals, the outer three expanding in fruit, and inner making a thin envelope around three-sided brown fruit. Flowers individually inconspicuous. Stamens six to each flower, with one one-seeded fruit. Fruits more pointed at free end, polished brown, long-lived under normal conditions. Pollination by wind.

Late summer rosettes may become large in winter, are hardy. Control by late ploughing and frequent cultivation. Leaves bear maggot of leaf miner, the fly *Pegomyia calyptrata,* support the moth *Papaipema nitela,* a pest of corn, tomatoes, and potatoes, the copper butterfly *Chrysophanus thoe,* and a brilliant beetle.

Powdered root used medicinally if harvested just after fruits mature. Fruits eaten by many winter birds. In spring and fall, rosette leaves may be cooked as greens. Commonly cooked with other greens and improved if bacon or salt pork is added. Fruits eaten by over 20 species of ducks.

Smooth Dock

Smooth Dock, Pale Dock
Rumex altissimus
Height to 6 ft or over, rather coarse, smooth, and tough. Leaves sharply pointed at either end, and broadest just below middle, smooth, somewhat shining, with rather obscure veins, and with lower leaves much larger and broader than the higher. Forms a vigorous rosette in spring and fall. Perennial.

Found on rich, loose soil such as is found in river bottoms or in some good gardens, from Connecticut to Nebraska, south through United States to its native Mexico. Through much of Middle West, it is common dock of fields and pastures, much as curled dock is common in East.

Flowers borne on parts of stem above that are almost leafless, in whorls on ends of short drooping stems, these stems being shorter than calyx that expands to enclose small triangular, brown, pointed fruit that looks something like a small brown buckwheat fruit. Fruit paler than fruits of most related docks.

Shallow, horizontal roots running just beneath surface of ground make it possible for this plant to occupy considerable territory if it is kept unchecked. Roots cannot survive considerable exposure to sun. Common control practice is merely that of shallow cultivation. Plants may be pulled by hand if soil is reasonably moist.

Like most docks this is a weed. Raw leaves are not distasteful when young and make good greens if cooked like spinach or similar plants. Seeds of smooth dock are food for birds. Wild mammals of many kinds feed on all parts of this plant.

Broad-leaved Dock

Broad-leaved Dock, Bitterdock
Rumex obtusifolius
Height to 4 ft, with stout, erect stem, unbranched or with a few branches, grooved, dark green. Lower leaves to 14 in. long, with long petioles, with rounded base and blunt or pointed tip. Upper leaves smaller, with less tendency to have a heart-shaped base and with petiole proportionately shorter. Root stout and deep. Perennial.

Naturalized from Europe, well-established through North America in southern Canada and throughout United States. Varies greatly in abundance in different localities. In Middle West not the common dock, being there replaced by smooth dock *R. altissimus.* In East, common large dock is probably *R. crispus.*

Flowers borne in loose whorls at or near tip of stem or at base of upper leaves, on slender drooping stems. Calyx wings pointed below middle, rather large and conspicuous, brown, but only one sepal bears a grain or fruit. Fruit ⅛ in. long, three-angled, glossy brown, with angles standing out conspicuously because of margins.

Many insects find food and shelter on this plant, among them a fly that builds a huge blotch mine, one leaf often being badly infested by many such flies. Root used medicinally but of doubtful value. Tops browsed by deer. Fruits eaten by waterfowl. If plants are cut before flowering in June,. may be kept from multiplying unduly.

Common seed impurity of clovers, which should not be sown with it. Plant can be controlled by avoiding introduction by using clean seed, by heavy cultivation, by using overrun plow, by improving drainage, and by hand pulling of yellow, spindle-shaped root, which comes relatively easily by twisting it from moist ground.

Sheep Sorrel

Sheep, Garden, Field Sorrel
Rumex acetosella
Height to 1 ft, erect, simple, smooth slightly grooved, slender, tough, green or brown, arising from creeping, branching, yellow, tough, fibrous rootstock. Leaves to 5 in. long, usually much shorter, to 1 in. wide,

shaped like an arrowhead; basal ones, long-stalked; grooved along veins, smooth, sour.   Perennial or annual.

Common on fields and roadsides, particularly on acid soils or in sour pockets in soil.   Grows well on ground where there is little water as is case in many waste places.   Ranges throughout North America except in far north.   Established apparently from Europe either as an impurity of commercial seeds or in hay wastes.

Flowers in small, erect clusters, Staminate yellow, pistillate red.   Borne on separate plants.   Pollination by wind or insects.   Flowering time May through September.   Fruits one-seeded, small, brown, obscurely three-angled, usually enclosed in semipersistent calyx.   Seeds may be shaken from plant.   Cycle from seed to seed, a few months.

Considered an indicator of presence of acid soil.   Helps serve as a soil anchor by growing in soil that would not support many other pasture plants.   Seems able to survive drought well and provides some forage in dry times.   Control by cultivation of fields, exposing roots to drying, by enriching soil with fertilizer, or by liming.

New spring or autumn growth yields leaves good raw in salads, or cooked, served with cream sauce or in cream soups providing tart stimulant to flat foods.   Plant may be crowded out by encouragement of rival plants such as grasses or clovers.   Human beings may get dermatitis from leaves.   Animals with low calcium diet may be poisoned by eating leaves.

Knotweed, Knotgrass
Polygonum aviculare
Prostrate, but with stems to 1 yd long, branching, faintly ridged, spreading, with underground parts white, woody, fibrous, with deeply boring taproot.   Leaves bluish-green, to 1 in. long, elliptical, often with grayish bloom, with short stalks or stalkless, relatively dry and tough, with a membranous, thin, gray scale at base.   Annual.   There are a number of very similar, closely related species; some equally common.

Common on bare dry spots along paths, driveways, and barnyards.   Native of Europe, Asia, and America.   Widely established in these continents, mostly where man lives.   Sometimes found in poor lawns where soil has been too compacted for success of clovers and grasses.   Grows commonly in cracks in sidewalks and pavements.

Flowers small, greenish-white, with pinkish margins, borne in axils of leaves; in flower, from June to October in North.   Probably self-pollinated.   Fruits one-seeded, small, dry, dull light brown, sharp-pointed, three-angled, finely ridged, fall from plants.   Seedlings grasslike, first leaves being narrow.

Thought to be troublesome to sheep.   Often harbors a white mildew that gives plant a dusty appearance.   Helps cover dry, unoccupied ground.   To some extent, may hold soil temperature down and control water loss through evaporation.   May also anchor soil that might be eroded.   A few months, from seed to seed.   Seeds have good vitality.

Controlled by hand pulling, hoeing, spraying with usual salt spray used in weed control, or by loosening soil so that other plants may thrive.   Cannot stand such competition.   Seeds commonly found as impurities of many kinds of commercial seeds; listed as food of 10 duck species.   Seeds eaten by some ground-feeding birds.

Knotweed

Water Smartweed
Polygonum amphibium, Persicaria amphibium
Stems to 20 in. long, branched or unbranched, floating on or just below water surface, or sprawling in soft oozes at edges of ponds.   Leaves oblong to elliptic, rather coarse, petioled, to 4 in. long, rounded or narrowed at base, and sometimes hairy.   Sheaths at leaf bases longer than distance between the joints, so stem is largely hidden.

Found in fresh water, mildly alkaline, or mildly acid pools, streams, marshy spots.   From Quebec to Alaska, south to New Jersey, Kentucky, Colorado, and California.   In Adirondacks, found to an elevation of 2,000 ft.   Found also in Europe in similar situations.   A number of varieties are found in America and Eurasia, with highly variable growth habits.

Flowers in compact, flame-shaped clusters at tip of stem, held erect. Since calyx is rose-red, an aggregation of plants may give a pink effect at a distance.   Stamens, five.   Fruit one to a flower, one-seeded, orbicular to oblong, about 1/8 in. long, smooth, shining or granular on surface.   Reproduction by rootstock or seed.   Seeds germinate 26 percent in 45 days.

Plant serves as shelter and food source for many invertebrate freshwater animals that are themselves food for fish.   Mass of plants may serve to anchor soil from wave wash at edge of ooze-bottomed ponds.   Since plants may sometimes be found over extensive areas in almost a pure stand, they add no little beauty to a spot when they are in full bloom.

Plants may clog shallow regions from which waters may leave a reservoir.   Do not help keep muddy beaches clean of vegetation.   Seeds or fruits in diet of shore birds and 15 species of ducks.   Seeds stored at near-freezing temperatures in moist, ground peat moss.   Ruptured seeds germinate about 85 percent.

Water Smartweed

Pink Smartweed

Pink Smartweed
Polygonum lapathifolium

Height to 7 ft under ideal conditions, but normally much shorter, smooth, branching, usually swollen at joints, with sheathing scales at bases of leaves. Leaves to 10 in. long, pointed at each end, with base being blunter; margins entire, net-veined, with midrib rather conspicuous. Annual.

Found in waste places throughout most of temperate America. May have been introduced from Europe, where it is found in great abundance. Also found in Asia. Sometimes becomes a weed particularly in low, moist, fresh, or slightly alkaline situations where average crop plants are not likely to offer serious competition.

Flowers in drooping narrow clusters, to 4 in. long, rather compactly arranged. Calyx, pinkish, white, or greenish, and in five parts. Stamens, six. Fruit a one-seeded achene, one to a flower about 0.1 in. long, flattened, dark brown with a surface that is finely grained or glossy, and with calyx persisting about it.

Wet lands almost essential to success of this plant, so it is not surprising to find it most abundant in newly drained areas, on flood lands, or along water courses. On higher, drier lands, it is never a serious weed. To layman, it looks like a tall, light-flowered lady's thumb smartweed without dark spots on leaves.

Easily kept in control by cultivation of land or by improving drainage. Fruits provide some food for birds of slough areas including ducks (17 species) and upland game birds. Plants might conceivably serve a role in anchoring soil in regions exposed to recurrent floods.

Lady's Thumb Smartweed
Polygonum persicaria

Height to about 2 ft, nodding at tips, smooth, with sheaths at bases of leaves, bristly with hairs. Leaves pointed at each end, roughish, to 6 in. long, without petioles or with very short ones; usually with a dark, heart-shaped V on leaf near center; margins entire or slightly irregular. Annual.

Common in neglected gardens, in waste places usually where soil is looser than that favored by knotweed. Naturalized from Europe, widely established. In America, found from coast to coast but less common to north, being frequently most abundant in fields after a grain crop has been harvested.

Flowers in drooping, usually terminal flower clusters resembling compact spikes; many spring from axils of upper leaves. Calyx pink or purple. Stamens, usually six. Fruit a one-seeded achene like a broad, somewhat lens-shaped egg, or sometimes angular, to almost ⅛ in. long, smooth, shining, and easily shed from cluster. Flowers June to October.

Commonly associated with grains and cultivated lands. Black mark on leaf has some legends, indicating that it is finger mark of a lady, hence name lady's thumb. Reputed to have medicinal value in heart troubles because of heart-shaped "sign" on leaves. Of course, none of these ideas has merit in fact.

Common names include lover's pride, heartsease, redweed, pinkweed, common persicary, redshank, willowweed, blackheart, spotted knotweed, signigicance of each of which is evident. Easily controlled by cultivation, particularly if cut before flowers have set fruits. Seeds found as seed impurities in commercial seed and eaten by wild fowl and upland game birds.

Morning-Glory Smartweed, Black Bindweed
Polygonum convolvulus

Twining, slender stems reaching to 1 yd and climbing on other plants or sprawling on ground. Stems roughish, smooth at joints. Leaves to 3 in. long, heart-shaped, pointed at tip, with slender petioles, slightly hairy, usually a rich green. Root system not extensive, as plant is annual.

Common in waste and cultivated areas throughout most of North America except extreme north and in West Indies. Naturalized from Europe and sometimes locally a troublesome weed. Probably brought to America as an impurity of commercial seeds or in hay trash.

Flowers greenish, hanging in loose clusters from axils of the alternate leaves. Calyx five-parted, closely covering the one-seeded fruit that is dull black, with a wrinkled surface, obscurely or obviously three-sided, pointed at either end and about ⅛ in. long.

While this behaves somewhat like some morning-glories, it does not have typical morning-glory flower or perennial underground parts like those of the morning-glory. Leaves and twining habit provide resemblance that is reflected in the common name. Flowers July to October. Should not be allowed to go to seed.

Control by hand picking, cultivation, and repeated harrowing. Seed common impurity in seeds of grains, particularly of oats and wheat. Vitality of seed probably less than 3 years, so a short rotation including hay for 2 years should check it. Seeds germinate late in spring and seedlings about 3 in. high should be destroyed; eaten by wildfowl.

Lady's Thumb Smartweed

Morning-glory Smartweed

Garden Beet
Beta vulgaris

Stem produced in second year, to 4 ft high, bearing many small flowers. Leaves of the flowering stem, narrow and linear. Basal leaves of different shapes, but usually smooth, commonly green, with reddish veins, usually clustered close to ground. Thickened root may, by finer roots, penetrate 6 ft of soil. Biennial or annual.

Probably a native of Europe; has been known as a cultivated plant since third century. Sugar beets and mangels are common forms of beets, Swiss chard another variety. Some beets grown for ornament; some, for use of leaves as greens. In Swiss chard, midrib of leaves is greatly enlarged.

Flowers numerous, in open-topped, branching arrangement, greenish, with five-parted calyx, five stamens, a sunken ovary that bears three stigmas. Usually, two or more flowers become grown together to form a "fruit" in which true seeds are developed. Each seed enclosed in a thickened calyx. Seeds may retain their ability to germinate for 5 to 6 years.

Seeds planted in boxes for early greens to be transplanted in garden in rows 24 to 30 in. apart; sown in drills in fields, require 4 to 6 lb to acre; produced when plants are subjected to a temperature of 40 to 50°F for 15 days or longer. Roots may expand at or above surface. Hand cultivation expensive, possible only where market high and labor cheap.

Roots with diameter 3 to 4 in. are bunched and sold; larger roots stored through winter at 32°F or in outdoor pits or banks such as were used in earlier days. Also raised for canning. Leading states in acreage of beets are Wisconsin, New York, and Oregon.

Garden Beet

Sugar Beet
Beta vulgaris

Biennial plant, with leaves arising to height of 2 ft from huge vertical root that averages about 1 ft long and to 3 to 4 in. wide at top, tapering to point below. Leaves crinkled, curved much like leaves of spinach, with definite forage value in addition to value gained from root culture. Sugar beet roots are nearly white, although same species as red garden beet.

Probably a native of Europe. Pure sugar first extracted in Germany in 1747. First sugar-beet factory in Silesia in 1799. First profitable factory in United States in California in 1879, with many now established particularly through California, Utah, Colorado, and Michigan areas, with some north into Ontario.

Ordinarily sugar beets are harvested before seeds mature. Seed borne in numerous flowers that are clustered in an open, branching arrangement, but with some flowers grown together to form what seems to be a single fruit. Germination power retained to 5 years. Seed-bearing plants chosen by testing sugar content of different roots and saving best.

Profitable sugar content expected in regions with average temperature of 70°F through June, July, and August. High temperatures cause low sugar yield and low temperatures cause improper ripening and low content. Formerly, yield was about 8 to 10 percent. Now commercial yields of to 18 percent are expected. Rainfall, May to September, should average 4 in. a month.

Processing of sugar is done on ground during season of 60 to 120 days out of year. Since sugar is produced in competition with sugar grown in foreign areas, maintaining market with profit involves many labor and commercial difficulties. Seed required, 5 to 6 lb per acre.

Sugar Beet

Rhubarb
Rheum rhaponticum

Height to 6 ft, with conspicuous joints hollow-stemmed. Leaves to 2 ft long or more, with broad, coarse, thick blades; petioles, sometimes 1½ ft long, and to over 1 in. thick, succulent; when young, break crisply. Roots large, dark brown, branching from a crown. Perennial. Popular variety, Victoria.

Native of Asia, but grown for home and market widely over the world. Most noncity homes have a corner in garden with one or more of these plants. Does best in a soil that has been deeply plowed, is light and well-filled with humus. Normally, such a plant requires little attention.

Flowers numerous, small, greenish-white clusters borne in a rather, loose open formation at top of stem that arises after leaves reach prime. 3 years, from seed to seed. Roots divided every 4 to 5 years and set in rows 4 to 6 ft apart, roots being 1 yd apart; buried 3 to 4 in.

During third season after setting new roots, leaves may be cut for 4 to 5 weeks but after that may be cut for 2 months. Leaf-spot disease affects market value of leaves. Crown rot affects roots; controlled by dusting leaves and treating roots with corrosive sublimate. Relatively free from insect injury. Weeds should be kept down. Mulch desirable.

Petioles harvested, skinned, and eaten raw in salads, or more commonly sliced, stewed, and used as a sauce or in pies tasting not unlike a sour apple pie. Nutritive value is probably essentially in vitamin B though cooking may not preserve this value. Crushed and sweetened makes a delicious drink. Broad part (blade) of leaf should not be eaten as it contains oxalic acid and soluble oxalates which can cause death.

Rhubarb

Buckwheat
Fagopyrum esculentum

Height to 3 ft, erect, strongly grooved when old; with stems smooth except at joints. Leaves to 3 in. long, shaped like a triangle but borne on rather long petioles, with conspicuous veins, with tip somewhat drawn out, with bases of leaves where attached to stem enclosed in rather brittle, small but conspicuous sheaths. Annual.

Escaped in waste places and persisting in fields where it has been cultivated. Known from practically all inhabited parts of United States and in southern Canada. Also found in West Indies, in Europe, of which it is native, and in western Asia, where it is grown rather extensively on poorer soils.

Flowers borne in clusters from stalks arising from bases of upper leaves. Calyx in five whitish parts, with about eight stamens, with eight nectar-bearing yellow glands placed between stamens. Fruit produces a single, triangular, gray achene, on lower portion of which persists some of calyx, to about ⅓ in. long.

Raised as emergency crop where other crops have failed and where some salvage is desired. Also raised along edges of fields as food for game birds. Honey made by bees visiting these flowers may be dark and strong but abundant. Pancakes made from buckwheat flour make a favorite meal for many people.

May cause photosensitization of domestic animals if animal is white or has unpigmented skin, if enough is eaten or if animal is later exposed to bright sunlight. Does not affect animals fed and kept in stalls. Symptoms inflamed eyes, ears, and face, with itching that may appear weeks after feeding. Some human beings get dermatitis from flour or from leaves.

Buckwheat

Order Chenopodiales./Family Chenopodiaceae

Spinach
Spinacia oleracea

Height to 2 ft, stem being usually unbranched or only slightly so. Leaves smooth, green, large, those produced in cool weather forming a dense crown. Leaves of flowering stalk narrow, oblong, variable. Flowering stalk may branch from axils of leaves. Flowers borne on both main stalk and branches. Annual.

Native of southwestern Asia, with one of the four species being widely cultivated. In United States, California holds a leading position in growing spinach. Grown commercially for shipping in 14 states and for local consumption in most states. Texas, California, Pennsylvania, Oklahoma, Arkansas, New Jersey, and New York are important producers of spinach.

Flowers usually either staminate or pistillate, with some bearing both stamens and pistils. Stamen-bearing flowers usually found on leafless spikes; pistil bearers in leaf axils. Fruits enclosed in structures bearing two to four spines, small, generally inconspicuous. Considerable variation in different strains cultivated.

When grown in rows, seeding requires 4 to 6 lb per acre; broadcast, 8 to 10 lb. Rows 10 to 24 in. apart, and plants thinned to 5 in. For Texas, winter crop seed is sown September to January; for fall crop in North in late summer. Diseases include spinach blight or yellows, downy mildew, leaf spot; checked by good fertilizer and care.

Grown for marketing and for manufacture, importance being reflected in acreage; in 1951; spinach for market, 62,758 acres; spinach for processing, 33,148 acres. Importance of salts and vitamins it adds to average diet has been emphasized. "Poyeye" has helped popularize it with many children. In shipping from·California to East, usually requires about ⅔ its weight in ice.

Spinach

Lamb's-quarters, Pigweed
Chenopodium album

Height to 3 ft, well-branched, stems bearing clusters of fruits that persist through winter. Has smooth but powdery appearance, with stems often ridged and grooved. Leaves alternate, to 4 in. long, somewhat diamond-shaped, with entire or shallowly lobed margins; petioles sometimes as long as blades. Root, a deep taproot. Annual.

Waste places and gardens or wherever there is good sunlight, loose soil, and some water. Found throughout United States and Canada; native of Europe and Asia, whence it was probably introduced early in commercial seeds or in hay trash. Not found in extreme north in Canada but wherever man maintains some form of plant agriculture.

Flowers small, greenish in compact clusters close to tips of stems or branches. Calyx five-parted, more or less enveloping fruit when mature and persistent. Petals inconspicuous. Stamens, 6. Pollination of some species by wind; of others, by insects; commonly, by self. One fruit per flower; one seed per fruit, plump, lens-shaped, highly variable.

Rapidly growing weed easily adapting itself to varying conditions. Occupies land growing common crops. Control effected by cultivation, by

Lamb's Quarters; Pigweed

cutting before flowering takes place. Some herbicides are effective in control. However, their effects on the environment should be considered before they are employed. Host for a leaf mining fly *Pegomyia hyoscami*, the skipper butterfly caterpillar *Pholisora catullus*, and many aphids and leaf hoppers, some carrying virus plant diseases.

Useful as a winter lunch counter for many winter birds and as a lifesaver for some winter game birds. Plant not affected by most sprays such as those that destroy mustards. Young growths 6 to 10 in. high are eaten raw by man or are cooked like spinach and eaten. Fruits eaten raw, or dried, ground, cooked as mush, taste somewhat like buckwheat.

Summer Cypress, Belvedere
Kochia scoparia

Height to 5 ft, much-branched, forming a usually symmetric, rather compact, global, pyramidal, or egg-shaped unit. Branches slender, striate, close to main stem, bearing slender leaves that are to 4 in. long, usually a pale, fresh green in summer, turning in late summer or fall to a brilliant red in some forms.

Probably native of China, but introduced in gardens in many parts of world. Some 40 species of *Kochia* in the world including white sage *K. vestita* of our Western salt deserts and *scoparia*, a garden ornamental, native of central Europe. Summer cypress does well in areas that in late summer are relatively dry, such as along paths and roadsides.

Flowers in ball-like clusters, some fertile and others not, those on lateral branches eventually bearing fruits and seeds that are small and enclosed in persistent calyx. Seeds planted indoors in April; seedlings set out in May; or seeds sown outdoors in May in North. Should be thinned so that one plant will be 3 ft from next.

Summer cypress useful in gardens since it continues to stand and be attractive in fall after other plants have been killed by frost, and because of its fresh greenness and regular shape early in season. White sage of desert can survive high salt content in soil at 1 ft depth, and if upper 18 in. is salt-free, plants may do well.

Deserts with white sage may not be expected to yield crops without irrigation but may be worked profitably if irrigation is available to remove undesired alkali. Summer cypress may seed itself on soils that are suitable but may not be easily established in a location too wet or too hard.

Spiny Hop Sage
Grayia spinosa

Shrub reaching a height of 3 ft, with a mealy appearance. Twigs sharp at tips, forming rather formidable spines when mature. Whole plant has a unique, green appearance different from associated plants. Leaves somewhat fleshy, with gray tips but with a tendency to assume a pinkish tint with age.

Common on mesas and flats in desert areas about Mohave Desert at elevations of 2,500 to 7,500 ft, to eastern Washington, Wyoming, and Colorado and in Whitewater Canyon on Colorado desert area. Common locally but with a rather spotty distribution probably because of its physiologic needs.

Flowers small and inconspicuous, as are most of those in this family, but develop into rather attractive fruits that are sometimes a brilliant rose-purple because of surrounding winged calyx which of course persists. Not all plants bear fruits. One fruit to one flower, but flowers and fruits are crowded.

When plants or year's growth are young, soft green herbage provides a welcome and abundant forage on which both sheep and cattle feed readily and on which they fatten well. Later in season, when fruits are ripe, these animals feed on them also with good results. Sometimes fruit harvest is surprisingly abundant.

Values as suggested above, largely as a basis of food for domestic animals living in an area in which forage plants are famous for their absence. Conceivable that plants of this type might have been of lifesaving importance to early settlers who found themselves stranded with their stock in deserts or dry areas.

Shadscale, Saltbush
Atriplex confertifolia

Height to 4 ft, a woody, spiny, much-branched shrub that forms a great, globelike, open plant, with some of its branches well-supplied with oval-shaped entire leaves that are not persistent and with some of the branches naked or nearly so. Root deep and strong, going down to 3 ft or more, depending in part probably on the demands of the situation.

In salt marshes and on exposed limestone, and relatively common in such situations from Idaho to California and east into northern Arizona except in the Colorado desert and of course only where the soil conditions and rainfall are suitable. It is one of the plants that helps define the appearance of the deserts within its range.

(Continued)

Summer Cypress, Belvedere

Spiny Hop Sage

Shadscale

Flowers pistillate or staminate, and relatively inconspicuously hidden in the axils of the leaves but more particularly near the ends of the branches. Fruits enclosed in rounded, wedge-shaped, thick calyx bracts; one-seeded, and when mature, are freed and blown from place to place by wind. Following shedding of leaves, twigs become coarse spines.

Provides some index to nature of soil, as its presence indicates a probable gravelly soil to a depth of 2 to 3 ft that dries in summer and may become salty at a depth of 1 ft or more. Without irrigation, such lands might possibly produce crops; with irrigation and the removal of alkali, they surely can.

Provides some early-season forage before the leaves are shed and before the twigs have developed the offensive and decidedly protective spines. The fact that cattle can survive in shadscale country at all is surprising to those accustomed to seeing them in more favorable surroundings. Shadscale is poisonous if selenium is in soil; causes blind staggers in cattle that eat it. Of fair importance to western wildlife. Seeds are eaten by several kinds of birds and small mammals. The twigs and foliage are taken by antelopes, rabbits, and other browsers. Saltbushes also serve as cover in the relatively open country of the dry plains and deserts.

Glasswort
Salicornia europea

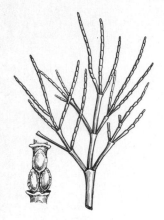

Height to 2 ft, with stem usually erect, much-branched, with branches slender, spreading, upright or rising at tips, with joints two to four times as long as they are thick. Leaves scale like, blunt, under 0.1 in. long, much wider than long. Roots variable, depending largely on nature of subsoil; usually shallow-rooted. Annual.

In salt or alkaline marshes, from Anticosti Island to Georgia, to Manitoba, British Columbia, south through Utah and Kansas but only where there is salt. Found also in Europe and in Asia. Because of high salt requirements, presence of plant is usually associated with lands that are waste from an agricultural viewpoint. Best on tide flats flooded once in 2 to 15 days.

Fruiting spikes are about 1 to 3 in. long but slender, only about ⅛ in. in diameter. In spike, three flowers at each joint, middle one being about twice as high as its neighbors, reaching nearly to top of joint. Stamens two to a flower. Seed compressed, enclosed in a spongy, fruiting calyx. Flowers from July through September.

At a depth of 4 in., soils that support glasswort may have a salt concentration of 2.3 percent of dry weight of soil as contrasted with 1.8 percent for sea blite *Dondia depressa* and 1 percent for *Atriplex*. This indicates an abnormally high resistance to toxic nature of salt as it affects most plants.

In general, soils that will support glasswort must be considered incapable of producing crops either with or without irrigation. Plants often turn vivid red in fall, giving rise to name marsh samphire. Other names include saltwort, pickle plant, English sea grass, chickens' toes, frog grass, crab grass. 28,000 seeds found in one pintail. Geese feed on fleshy branches of glasswort, and in the fall ducks feed on the seeds on the tips of the stems. Several closely related, similar species can be confused with *europea*.

Glasswort

Russian Thistle, Tumbleweed, Saltwort
Salsola kali

Height to 3 ft, with many branching stems extending to form a huge, loose, open globe, sometimes 4 ft or even more through; breaking loose to form a tumbleweed. Leaves alternate, stiff, short, prickle-tipped about ½ in. long, with shorter, sharp-pointed bracts at bases of older leaves. Has general appearance of thistle.

Dry deserts, fields, and waste places, along railroad embankments and in meadows. Native of Asia but widely established in North America from Manitoba to California and east through New York State. In some areas, like western Dakotas in bad years, it may be most conspicuous of all plants as it is a last survivor of drought conditions.

Flowers in axils of leaves, small greenish, whitish, red, or pink, with five-parted calyx which encloses mature fruit, leaving papery margin that helps in wind distribution. Fruits reddish, top-shaped, long-lived, and most abundant. Seedlings grasslike; may give impression that a good grass crop is developing.

Russian Thistle

Green in midsummer when other desert plants are brown. Flowers in July, and may be eaten by cattle. Fruits mature by September, but by then plants are too tough to be eaten. A single plant may produce over 100,000 fruits and a single plant on a railway car truck may plant miles of roadbed.

In depression years, plants are cut green, salted, watered, and fed to cattle, but normally they are destroyed where possible by plowing, cultivating, or burning. Ash has been used an an impure carbonate of soda under name barilla. Amply able to take care of itself under conditions trying for most plants. Plant, accidentally introduced to U.S. in flaxseed, has prospered, spread. Now, in spite of its noxious character, its seeds serve as food for birds and rodents, and hoofed browsers feed on young plants.

Greasewood
Sarcobatus vermiculatus

Low, branching, but conspicuous shrub, with somewhat spiny branchlets
and bright green, rather thick leaves whose unbroken margins make them
look somewhat like branchlets. Rather like gray shadscale. Roots penetrate to depth of 6 to 7 ft; much larger than might seem necessary for such
a small plant. Stem to 3 in. through.

Found in arid deserts, but usually where there is usable water within
reach of long roots. Widely distributed, most conspicuous in alkaline
soils from Alberta to North Dakota south to Texas, California, and into
Mexico. May be dominant plant, in fact, exclusive plant, on alkaline
flats. Favors regions of clay flats or where there are salty areas and may
form rather pure stands in good conditions.

Flowers rather inconspicuously clustered at base of some of leaves; staminate, in cylindric spikes to 1 in. long; pistillate, producing one hard fruit
with a calyx wing that may expand to ½ in. wide. Flowering period
June and July. Fruit matures on plant from September through October;
windblown.

Typical of undrained depressions in salt deserts and associated with
pickle weed *Allenrolfea* and salt grass *Distichlis*. Loses little water and
gets an adequate supply because of extensive root system. Geologists
consider it an indicator of a nearby supply of fresh groundwater even
though salt may be present at a depth of 1 ft.

Cattlemen consider greasewood a valuable fall and spring browse even
though it contains oxalates of sodium and potassium. Thousands of
sheep have been killed by it in early spring in New Mexico, Nevada, and
Oregon. Symptoms of poisoning are weakness, weak pulse, poor respiration. Tender twigs are cut, washed, boiled, and eaten with butter by
man. Hopi Indians use the wood for fuel and for making planting sticks.

Greasewood

Order Chenopodiales./Family Amaranthaceae

Redroot Pigweed, Green Amaranthus
Amaranthus retroflexus

Height to 9 ft, erect, branching profusely, coarse, rough and, in winter,
with fruit-bearing clusters that persist. Commonest at branch tips.
Leaves alternate, to 6 in. long, rough, dull green, long-petioled, with wavy
margins; not persisting on winter stalks. Roots comparatively small, extending directly downward, red, with few side roots.

Waste places, gardens, and cultivated fields where ground is not already
occupied by sod throughout North America except in extreme north.
Came from tropical America by way of Southwest. Widely established
wherever agricultural seeds are planted in suitable territory. May act as
a crude tumbleweed but *A. graecizans* is better example.

Flowers inconspicuous, three-bracted, five-stamened, greenish. Bracts
pointed, stiff, longer than flowers which appear hidden among them.
Pollination by wind or by self from July through September. Each
flower one-fruited and one-seeded. May mature seeds in a few weeks or
months. Seeds may remain viable for many years.

Amaranth Pigweed

Does not thrive in shade but competes too successfully with economic
plants in sun. Host in July and August of stalk borer *Papaipema nitela*
and of larva of skipper butterfly *Pholisora catullus*. Some protection
against being eaten by cattle because of tough nature and stiff bracts.
Fruits naked, shining lenses.

Provides abundant food and shelter for winter birds. Young growth
cooked like spinach; good with strong-flavored greens. Small, shining
fruits, parched or ground and used by Indians as food. Young plants,
eaten raw as salad. Control by cultivation, burning, and preventing formation of new seed. Chinaman's greens, careless weed are other common names.

Iresine
Iresine herbsti

Height to 6 ft in tropics, grown mostly for foliage; in most gardens, not
producing erect, flower-bearing stalk that looks much like top of amaranth pigweed. Leaves to 5 in. across, with deep notch in tip, normally
about 2 in. long, nearly round in general effect, purplish-red, with yellowish or greenish veins.

Native of South America, where it is probably perennial. Grown widely
in greenhouses and in formal gardens and named in honor of Herbst, a
propagator in Royal Botanic Gardens at Kew, England. Some relatives
are almost shrubby in nature. *I. lindeni*, commonest other garden species, has pointed leaves, without terminal notch.

Flowers not common in gardens, small, hidden by bracts as in amaranths.
In practice, plants propagated by cuttings made in fall, kept over winter
in cool greenhouse, in February given more heat, and in March given
more heat and water and cut back. Cuttings root readily if properly
conditioned; set 6 in. apart for mass bedding.

Iresine

General culture of iresines, like that of coleus such as is found in almost
any houseplant group. Great variety of horticultural forms made by
(Continued)

Pokeweed

Bougainvillaea

Wild Four–o'clock

cuttings produces bronze, blood-red, greenish, yellowish, and other types. Frost is fatal to foliage of these plants outdoors.

While these plants are primarily of importance as house or bedding foliage plants, one species *I. paniculata* grows wild in dry soil from Ohio to Kansas and south to Florida and Texas. About 20 species native of warmer parts of the country, particularly in the Southwest.

Order Chenopodiales./Family Phytolaccaceae

Poke, Skoke, Inkberry
Phytolacca americana
Height to 12 ft, smooth, sparingly branched, strong-smelling, weak, with stem interior divided by disks that form lens-shaped cavities, easily broken down. Leaves pointed at both ends, to 12 in. long, with petioles to 4 in. long, smooth, deep green, somewhat drooping. Root sometimes to 6 in. in diameter, spreading. Perennial.

In open or in woods, where it is dry or wet, from Maine to Minnesota through Ontario, south to Florida and through Arkansas to Mexico, also in Bermuda. Commonly grows in groups of a few plants, rarely forming any extensive stand, and possibly is most common at borders of woodlands though this should not be considered a rule.

Flowers borne along short stems that arise from bases of upper leaves. Calyx of four to five persistent round sepals. Stamens 10, inserted at base of sepals. Fruit juicy, black berry, bearing 5 to 15 seeds. Flowering time July through September, fruits remaining on even after leaves have been destroyed by frost. Seeds probably bird-borne.

Contains more than one active poisonous substance, the exact identities of which are unknown. One should be careful in mistaking roots for those of horseradish, *Armoracia rusticana*, as they look alike. Humans are poisoned by eating roots or berries, which act as a slow emetic, causing vomiting about 2 hours after eating, also purging, spasms, convulsions, and sometimes death. Juice from berry provides a primitive and relatively permanent dye.

Young shoots may be cooked and eaten safely and taste much like asparagus. Only shoots under 4 in. long should be used. Cut shoots are boiled, the first water being then discarded; then, cooked again and served like asparagus. Root poisonous. Berries are not edible and should be avoided at all times as food.

Order Chenopodiales./Family Nyctaginaceae

Bougainvillaea
Bougainvillaea glabra
Height to 10 ft or more when allowed to grow unhindered. Sometimes may develop a trunk to 1 ft or more in diameter, or may grow as a sprawling, woody plant over buildings, walls, or banks. Common porch and arbor vine of the tropics and warm countries and tremendously popular in Hawaii. Leaves oblong lanceolate, smooth, bright green.

Native of Brazil, but grown widely as an ornamental in the warmer and tropic parts of the world and to some extent grown in pots and kept as a dwarf form in the temperate regions in greenhouses. About 10 species, all native of South America, and a number of varieties based on flower color and other characters.

Flowers enclosed by great, showy bracts that are purple or more commonly magenta-colored; some, a deep rose and distinctly, rather conspicuously veined. Commonly grown by cutting young shoots a few inches long and rooting them in a few weeks in sand at 65 to 70°F. Fruit a five ribbed nutlet.

Withstands drought remarkably well. May be pruned severely without seriously affecting surviving plant. Cuttings made in 6 to 12 in. lengths in April, May, or June. Stock is taken from field in early autumn to be grown in pots indoors for early blooming effects. Plants used in bedding, prostrate on ground.

This plant lends much color to arbors and fences in California and Hawaii and obliterates many obnoxious signs. Named after French navigator De Bougainville, who died in 1811. The name has new significance after World War II. *B. spectabilis*, a closely related species, is reported as the most commonly cultivated form in Hawaii. Ideal plants for poor gardeners; tolerates much mistreatment.

Wild Four-o'clock
Mirabilis nyctaginea
Height to over 4 ft, with four-sided stem repeatedly forked in upper portions, nearly smooth. Leaves opposite, entire, petioled, heart-shaped blade, to 4 in. long and to 3 in wide, acute-tipped, but uppermost leaves mere bracts on stem, smooth, relatively dark green. Root thick, fleshy, vertical taproot. Perennial.

Native from Manitoba to Illinois, through Minnesota, south to Texas and Louisiana, being most common on dry soils such as are of wasteland type. Found in isolated areas to east of range given. Some five species in eastern part of United States. Garden four-o'clock is *Mirabilis jalapa* of same family. They require a sunny location and a light well-drained soil.

Flower bears three to five stamens and pistils together in same flower, in expanded structure on hairy, slender stems that spring in clusters in forked terminal areas. Fruits hard nutlets, five-ribbed, angular, brown, about ⅛ m. long, borne apparently from colored "involucre" that persists, appears corollalike, and is rose or purple.

Weed that propagates itself by seed and by root. Roots easily destroyed if plants are cut in dry weather. Spreads freely from seed and so, to keep in control, plant should be cut close to ground to injure roots and to prevent seed formation. Garden four-o'clock may be grown from seed, or roots may be lifted in fall as dahlias are.

Wild four-o'clock an interesting and common weed through Middle West. In garden four-o'clock, flower part is a colored calyx and flower clusters are surrounded by a group of bracts in some related species. Erroneously supposed that jalap, a purgative, came from root of garden plant. It comes rather from *Exogonium jalapa.*

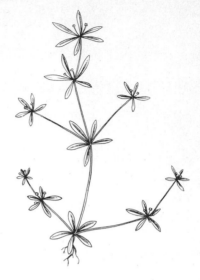

Carpetweed

Order Chenopodiales./Family Aizoaceae

Carpetweed, Indian Chickweed, Whorled Chickweed
Mollugo verticillata
Prostrate, sprawling weed, spreading in all directions from a common central root system to form a patch sometimes 20 in. across and often making an almost conventional design by regular branching system. Leaves borne in whorls of five to six at branching joints, narrow, entire, to ½ in. long, narrowest at base. Root rather deep. Annual.

Common on sandy riverbanks, in sandy gardens, roadsides, and waste places in mid and late summer from New Brunswick and Ontario to Minnesota, on to Washington, south to Florida, Texas, and Mexico. Native of warmer parts of America; probably spreading, though not likely through being a commercial seed impurity. Also in Africa.

Flowers borne in axils of some leaves, particularly of whorls near branch tips, under 0.1 in. across, with sepals oblong and slightly shorter than mature capsule that develops as fruit. Seeds somewhat kidney-shaped, smooth, shining, red, and almost dust size. Flowering period June through September.

Apparently plant does best in loose, dry soil that will not support many garden plants. Does not appear in early spring, but later when conditions more nearly approximate the climate of the South and West.

As a weed, is not serious since it is kept in control by reasonably clean cultivation in late season or by hand hoeing of isolated plants. It bears common names of devil's grip, Indian chickweed, whorled chickweed as well as carpetweed, so must be well-recognized or it would not have gained these names.

Order Caryophyllales./Family Caryophyllaceae

Common Chickweed
Stellaria media
Height to 16 in., much-branched, weak, sprawling or rising at ends, smooth stems, except for lines of pale hairs along stems and branches. Appears as pale, fresh green, tufted mass of weak vegetation. Leaves opposite, to 1½ in. long, the lower on rather long petioles; the upper with none. Blades of leaves sometimes heart-shaped. Roots relatively weak. Annual.

In waste places, meadows, gardens, along foundations of buildings, or in cracks of neglected paths and roadways throughout most of North America. Naturalized from Europe, native of Asia; almost universally distributed through temperate parts of world. Found in bloom in northern states in almost any month of the year.

Flowers to ⅓ in. across, with pale green sepals and deeply two-parted, whitish petals, the sepals being longer. Stamens three to seven and about as long as pistil. Pollination by insects or self. Fruit a capsule longer than the persistent calyx, yielding many rough, wrinkled, reddish, somewhat flattened seeds that are blown by wind rather easily as sand.

Common Chickweed

Grows most vigorously during cool weather, when there is ample moisture; can bear mature seeds in dead of a northern winter in reasonably protected place. Can be kept in control by simply raking tops off, by shallow cultivation, or by hand weeding. Spray of iron sulfate (100 lb to a barrel of water) solution is most effective in destroying plant.

Plants eaten by pigs, chickens, and even by human beings. Leaves or whole plant eaten raw as a salad or cooked like spinach, in which case spinach taste is closely duplicated. Eaten in Europe much more commonly than in America. Handful of chickweed mixed with some mild mustard makes an acceptable emergency salad on almost any field trip.

Mouse-ear Chickweed

Mouse-ear Chickweed
Cerastium vulgatum
Height to 18 in., branches being erect or rising from a sprawling base, sticky, fuzzy, tufted, and rather weak but tougher than in common chickweed. Leaves opposite, the upper to ½ in. long, the lower longer; blunt, hairy (like a mouse ear). Roots at joints to form dense masses, particularly in lawns and dry meadows. Biennial or perennial.

Native of Europe, North America, northern Asia, and widely established in practically all of our states, in fields, woods, pastures, and other places where it may, in some cases, become a bad weed. Particularly in lawns it may crowd out more desirable grasses and clovers. It may grow even on rocky banks where there is little soil.

Flowers few to several, borne at or near tips of branches on individual stems in loose formation. Sepals, four to five. Petals five, each deeply cleft, giving effect of two, white. Stamens usually ten. Pistil about as long as stamens. Fruit, one-celled capsule, to ⅖ in. long, bearing many kidney-shaped, roughened, flattened, chestnut-brown seeds.

Cannot survive cultivation practices. In lawns, stems may be broken loose and removed by simply raking and permitting resultant bare spot to be occupied by other more useful plants. Spray of iron sulfate or even of dilute sodium chloride is often sufficient to eliminate plant, as spray sticks readily to hairy leaves and stems.

Chickens may feed on this plant but it is not ordinarily so recognized as an emergency food for man as is common chickweed. While it is a recognized weed, it is easily eliminated and so can hardly be called a serious pest. A half dozen species of genus to be found growing wild in East; many more elsewhere in world.

Corn Cockle, Purple Cockle, Corn Campion, Corn Mullein
Agrostemma githago
Height to over 3 ft, unbranched or bearing a few branches, erect, densely covered with whitish, appressed, somewhat sticky hairs. Leaves linear or nearly so, sharply pointed at either end, erect, to 4 in. long, ¼ in. wide, the lower ones being narrowed much at base, with a rather conspicuous midrib. Annual, or sometimes biennial.

In waste places, and more particularly in grainfields, of varying abundance, in certain localities being very abundant. Native probably of Europe and of northern Asia but widely introduced wherever small grains are grown. Often found as winter annual in fields of winter wheat and rye surviving winter with these plants. Not common on roadsides.

Flowers showy, to 3 in. broad, purplish-red, often spotted with black but paler in spots. Calyx three to four times length of corolla tube, exceeding petals, slender, leaflike, dropping off in fruit. Flowering time July through September. Fruit a capsule containing many triangular, kidney-shaped, black seeds, with coarse tubercles on them.

Control by planting clean grain or by harrowing grain crop just before it emerges from ground, and again when it is 3 in. high or, where there is bad infestation, letting land lie fallow through summer. Seeds rarely hold their vitality more than 1 year and if plowed under are lost. Best cure is to use only clean seed. In lawns, can be controlled by close mowing or spraying with recommended amounts of solutions of iron sulfate or dilute sodium chlorate.

Feeds made from screenings containing any considerable number of corn cockle seeds may be dangerous to poultry and stock as seeds are definitely poisonous, the poison being the glucoside githagin which makes suds, causes vomiting, nausea, vertigo, and diarrhea. ¼ lb. of ground seed to 100 lb. live weight of animal may be fatal.

White Campion
Lychnis alba
Height to 2 ft, with stems erect or loosely branching, sticky, with mass of fine hairs. Leaves opposite, to 3 in. long, pointed at ends and tapering at bases or joining around stem with opposite leaf. Upper leaves without petioles. Basal leaves may form a rosette arising directly from perennial, rather substantial, vertical taproot.

Found in meadows, waste places, gardens, along paths, on railroad embankments, from Nova Scotia through Ontario to Michigan and even on to Pacific Northwest, south through New York and Pennsylvania. Native of Europe. Commonest as a weed in United States in East on rich, well-drained soils that have been free from cultivation.

Flowers few, in loose clusters or solitary; some staminate; others, on other plants, pistillate. Sepals five, joined into tube enclosing five petals that are pink or white, much longer than calyx, ten stamens, and one pistil. Fruit, dry capsule, that opens at end freeing many kidney-shaped, gray to tan seeds that bear coarse tubercles.

Flowers open in evening rather than at midday. Plant is a weed but not serious one since it yields to competition of any clean cultivated crop or to plowing. Scattered plants may be pulled by hand and plants that survive should be cut before they mature seeds to avoid a repeated appearance. Short crop rotation generally fatal.

Corn Cockle

White Campion

Many common names testify to fact that this plant has wide distribution and is well-known. It is known as bull-rattle, cow rattle, white robin, snakeflower, thunderflower, and cuckoo flower. It resembles a night-flowering catchfly but has five styles rather than three.

Night-flowering Catchfly
Silene noctiflora
Height to 3 ft branching or unbranched, with rather coarse stems that are sticky with fine, clammy hairs. Lower leaves to 5 in. long, narrowed at base into a long petiole, sticky, hairy; upper leaves without petioles, and joined with opposite at base forming a cup. Root system not elaborate. Annual or winter annual.

Native of Europe but introduced widely wherever small grains are grown; well-established in eastern North America and in Pacific Northwest; also found south to Florida and north into Nova Scotia, usually in waste places. Favors rich, gravelly soils where new seedings have been established.

Flowers few, in clusters at tips of branches, white or pinkish, to 1 in. across, with calyx to 1.3 in. long, tubular, conspicuously veined, and loosely enclosing capsule that develops as fruit. Seeds freed from openings in end of capsule, many, kidney-shaped, gray, with dark-brown lines and blunt tubercles. Moth-pollinated.

Control by cutting plants, by mowing before seeds have had time to set, by sowing clean seeds, by using a year after fall plowing for growing of a clean cultivated crop, by harrowing young grain, and by hoeing or hand pulling. Flowers fragrant, opening at dusk and remaining open until next morning.

Has many common names such as sticky cockle, clammy cockle, night-flowering catchfly, all of which emphasize the stickiness of the plant and the frequency with which insects of many kinds find themselves stuck on plant. Stated that stickiness prevents crawling insects such as ants from reaching flowers to steal pollen.

Soapwort, Bouncing Bet
Saponaria officinalis
Height to 2 ft, erect, relatively stout, sparingly branched, smooth, leafy, rather pale green, coarse, but easily broken. Leaves opposite, entire, joined at bases with opposite leaf, to 3 in. long, with some rather conspicuous parallel veins, pointed at tips and narrowed at base. Root system a short rootstock. Perennial.

Native of Europe, but introduced early into America and now widely established in North America, being most abundant in East and in areas on Pacific Coast. In East, is almost certain to have been established in flower gardens of early settlers and to have persisted after original dwellings disappeared.

Flowers showy, bluish- or pinkish-white, or white, about 1 in. broad, bearing both stamens and pistils. Sepals five, joined to form long tube enclosing tube of corolla. Stamens, ten. Pistil with two styles. Fruit a dry capsule that opens at end to free numerous black, flattened, short, kidney-shaped, tubercle-roughened seeds. Flowers July to September.

Yields a soapy material that, mixed with water, has a cleansing effect, the principle being the glucoside saponin. It is probable that taken internally this glucoside might have poisonous effects. Control is by means of spraying with sodium chlorate, by cutting the plants frequently close to the ground, by plowing, or by deep cultivation.

Was grown originally probably as a garden flower but has escaped and established itself widely. Known as fuller's herb, wild sweet william, lady of the gate, woods phlox, mock gillyflower, sheepweed, old maid's weed, chimney weed, Boston weed, scourweed, Sweet Betty, and hedge-weed, giving evidence of its wide use and popularity.

Sweet William
Dianthus barbatus
Height to 20 in., with four-angled stems; branched or not, smooth, and rather brittle. Leaves broad, flat, and with five conspicuous nerves, joined at base with leaf on opposite side to form a slight cup about three to four times as long as wide, with upper leaves the smaller, and around 4 in. long. Biennial, with substantial root system.

Native of the area from Russia to China, south to Pyrenees Mountains but widely established where it has escaped from culture in American gardens. Often persists in abandoned sites of early pioneer homes or may appear spontaneously elsewhere.

Flowers many, crowded into rather flat- or rounded-topped clusters, with five petals bearing a small beard at throat of tube and being rose, red, white, purple, or variegated with these colors; petals, with outer edges toothed and with beard. Fruit one-celled, from which the flattish seeds are shed.

(Continued)

Night-flowering Catchfly

Soapwort, Bouncing Bet

Sweet William

Seeds may be sown in August and seedlings that develop before frost should be sheltered through winter for flowers next season, or plant may be propagated by cutting perennial underground structures. Perennial pinks can survive winter but fall-planted annuals will need protection to survive.

One of the common old homestead garden flowers still found in considerable abundance and reasonably established at the sites of pioneer homes in the East. In some places, they may have spread to occupy a considerable territory almost as a pure stand, judging from the flower display in season.

Carnation

Carnation
Dianthus caryophyllus
Height to 2 ft, with stems sometimes branching, and with swollen joints from which the relatively narrow, smooth, backward-curving leaves appear in pairs. Many kinds, some being bushy and hardy enough to survive being outdoors, while others are delicate and require care in greenhouses to survive.

Under cultivation for over 2,000 years. Name *Dianthus* given it by Theophrastus, some 23 centuries ago in reference to its "divine" fragrance and *Caryophyllus* referring to clove-like quality of fragrance. In Old English, plant is called "gillyflower," varieties being developed 500 years ago in England, France, Germany, and Holland.

English call flowers of one color "selfs"; those with ground color of white or yellow streaked with scarlet, rose, or purple, "flakes"; those with same ground color but marked with two or three colors, "bizarres"; those with a pure white or yellow ground color but with a colored band on margins, "picotees." This nondistinctive classification is probably as good as any.

Culture: new cuttings 3 in. long are made in January and kept at about 50°F; in February, all but one bud removed, still kept cool; in March temperature raised to 60°F in day, 48° at night; in April, stock is set out, and by May flowers may be harvested. From June to August plants develop vigor and are put back in benches. From September to November they may yield crop, and in December rest.

One of most universally and prolongedly popular flowers at any season of year. Makes the perfect, dignified buttonhole bouquet or mainstay for the flowers in the center of the table. Culture is a science, which pays well those who learn how to make it succeed. The state flower of Ohio.

Garden Pink

Garden Pinks
Dianthus sp.
Flowers of considerable variety including sweet williams and carnations already discussed and clove pinks, China pinks, and maiden pinks. Clove pinks, *D. plumarius*, are low, usually under 1 ft in height. Maiden pinks, *D. deltoides*, have short, linear leaves that are smooth or roughish, and minute flowers. China pinks, *D. chinensis*, usually are tall, single-flowered, broad-leaved.

Clove pinks native of Siberia on to Austria but widely established in gardens, formal and otherwise. Maiden pinks native of Scotland, Norway, and Japan and usually grown outdoors in massed beds. China pinks native of China and Japan and grown more for their individual beauty than for mass effect.

Flowers of clove pinks, pink, white, or red, fragrant, solitary or in 2s or 3s, with calyx lobes, short and broad, with pressed tips; of maiden pinks, red, with a crimson center, ½ to ¾ in. across, with calyx bracts, narrow, pointed, and spreading; of China pinks, solitary, pink or lilac, with calyx lobes, broad to leafy and spreading.

Pinks resemble phlox flowers in many respects but differ from them primarily because petals of flowers are not grown together into a tube as in phlox, although they may appear so.

Generally speaking, pinks are hardy plants that in most cases can survive in gardens with little care, or if annual are able to yield an abundance of bloom with a minimum of care and expense. Since they are naturally found in more or less dry places, they can survive in locations and under conditions not suitable to many garden flowers.

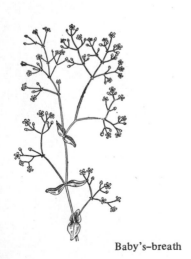

Baby's-breath

Baby's-breath
Gypsophila paniculata
Height to 3 ft or more erect, profusely branched, with stem fine and delicate-looking. Leaves narrow, sharp-pointed, smooth, with the longer to the bottom, reaching a length of 4 in.; the upper, shorter. Some varieties have stiff wiry stems. Does not wilt readily under normal treatment. Perennial. Ordinarily hardy.

Native of Europe and of northern Asia but established by escape from cultivation in various parts of North America. When grown outdoors, favors open, sunny, relatively dry spots such as are characteristic of many rock gardens. Of the 50 to 60 recognized species, most of them come from western or southern Asia, eastern Europe, or northern Africa.

Flowers small, and at many ends of widely branching top, numerous and white, with stem supporting individual flower two to three times as long as calyx of flower itself. Some double varieties have been known but they are not easily grown. Plants may be grafted on roots of established plants.

Name implies that plant favors soils with lime in them but common name refers to effect of the whole plant as like appearance of one's breath on a cold day. This species is grown in greenhouses for sale as cut flowers to "lighten up" a bouquet, or is allowed to grow in gardens to fill in bare spots.

Planted in masses in gardens, the flowers give a dainty appearance to a spot that might otherwise appear bare. The cut tops seem to be favorite cut flowers for weddings, funerals, and for banquet centerpieces, when used with other, more stolid species.

Order Caryophyllales./Family Portulacaceae

Garden Portulaca
Portulaca grandiflora
Sprawling, or slightly ascending stems that are succulent, sparingly branched, and to 1 ft long. Leaves alternate, round in cross section, about 1 in. long, 0.1 in. thick, and more clustered near tips of branches. Scattered hairs or hair tufts may be found on stems or in leaf axils. Root, a central taproot.

Found in gardens where showy flowers are desired and where soil is sufficiently dry and sandy. Introduced from South America but widely established as indicated by the many common names that refer to a locality. "Mexican rose" is more appropriate than "Kentucky moss" as a significant place-name.

Flowers to 2 in. broad, brilliantly pink, white, yellow, or red, but open only in the bright sunshine. Petals five, and slightly cleft at tips. Stamens many. Fruit like other portulacas, a capsule that opens by a lid at the top to free many seeds. In this species, seeds are shining gray and small.

Term *moss* commonly used in describing this plant of course refers to remote resemblance of the plant to a moss when flowers are not present. Common name "rose" refers to resemblance of flower to a rose. Neither a rose nor a moss. Limitation of flowering to hours of bright sunlight interests many people.

One of common garden flowers, its seeds are readily available. Showy flowers pictured are intriguing but results are often disappointing because plants cannot stand competition and must have a relatively dry loose soil in which to do best. Known as wax pinks, French pussley, and showy portulaca. Varieties available with flowers of a variety of colors; yellow, purple, scarlet, pink.

Garden Portulaca

Pussley, Purslane
Portulaca oleracea
Sprawling plant, with stems to 10 in. or more long, radiating from a central root system that penetrates the soil rather deeply. Stem smooth, succulent, easily broken, much-branched and green. Leaves to nearly 1 in. long, fleshy, rounded at the tip, alternate, clustered at ends of branches, easily broken from stem. Annual.

Common in gardens and waste places, in bare spots on lawns and in dooryards. Native of Europe. Naturalized and cultivated in Europe and is considered a useful food plant in some parts of the world.

Flowers inconspicuous, yellow, solitary in axils of leaves, to ¼ in. across, with broad, green sepals and five petals. Pollination largely by bees and by butterflies. Flowering time from July through September. Fruit small capsule that opens by lid, shedding many fine dustlike, flattened, wrinkled seeds. Seed to seed in few months.

Frost kills plant normally. Flowers are open only for few hours in early morning. Plant is able to survive pulling if parts of it can fall on moist ground where they will root. Plant harbors a white mold but can be destroyed by repeated hoeing while plant is still in seedling stage and by taking care to destroy hoed plants. Good pig food. Seeds eaten by several different songbirds. Mammals, e.g., rabbits and chipmunks, eat the seeds and other parts.

Plant may be eaten raw, or cooked by steaming or boiling and adding a little salt. Makes good salad greens if collected reasonably young and eaten with other plants with a stronger taste. Mixed with mustard and lamb's-quarters, it may be really good. Sold regularly in food markets in China, India, France, and Mexico.

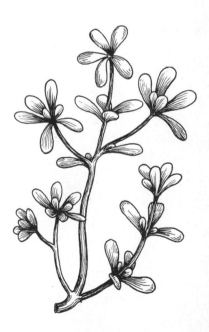

Purslane

Bitterroot

Bitterroot
Lewisia rediviva
Practically stemless plants. Leaves arise directly for most part from top of coarse root, narrow, oblong, nearly round in cross section, smooth. Flower-bearing stems little if any longer than leaves. Root coarse, vertical, loaded with starch, and with a bark which is very bitter but which slips easily from the root at flowering time.

Found in open fields from Montana to Arizona, and to some extent north and west of these centers of abundance; e.g. reported to range from British Columbia to California. Is not limited to natural range as it is commonly grown in rock gardens and similar places as a low ornamental that can survive difficult climatic conditions. Over a dozen species, of which this is by far most important to man.

Flowers conspicuous, rose or white, borne singly, with to eight persistent, bractlike sepals, and to 16 spectacular petals. Stamens many. Fruit, capsule that breaks by means of a line running around tip and opening somewhat like a lid, freeing many seeds from within. May be reproduced from root or from seeds.

White, inner part of root is starchy, mucilaginous, and at first bitter, but bitterness is removed by boiling, after which root is edible. Indians collected and dried it, first chipping off bitter bark. It could then be stored and when boiled regained its food value and mucilaginous qualities. Collections were made in spring months.

State flower of Montana. Long a favorite food with Indians. Oregon Indians called it "spatlum" and it is after this plant that the Bitterroot Mountains were named. Mucilaginous qualities of root make it possible for plant to survive in areas which may be dry for considerable periods of time.

Carolina Spring Beauty

Carolina Spring Beauty
Claytonia caroliniana
Height rarely to 1 ft. Less likely to sprawl than is Virginia spring beauty. Basal leaves to 3 in. long, ¾ in. wide, being conspicuously proportionately wider than in Virginia species. Upper leaves borne on petioles to ½ in. long. Root broad, deeply buried, with smaller roots coming from many places.

In damp woods, from Nova Scotia to Saskatchewan, south to Connecticut, in highlands to North Carolina, and in west to Ohio and Missouri. In Virginia, found up to 5,000-ft elevation but is rare near seacoast. A related *C. lanceolata*, with shorter leaves is found from the Rocky Mountains to the Pacific coast.

Flowers similar to those of Virginia spring beauty but usually fewer and more likely to be smaller. Flowering period from March through May. Flowers open only when exposed to sunlight. They close relatively quickly if plant is plucked. Pollination by insects seeking nectar and pollen.

Common pollinators are beelike flies, bumblebees, and numerous species of butterflies.

Both spring beauties can be purchased from those who supply stock for wild gardens. A few bulbs given care may in a few years produce many. Spring beauties should not be picked, since picking the leaves as well as the flowers deprives the plant of survival.

Virginia Spring Beauty

Virginia Spring Beauty
Claytonia virginica
Height rarely to 1 ft, rarely branched, erect or somewhat sprawling, succulent, smooth, easily broken, bearing two to three leaves, the upper being opposite and shorter than the basal that are to 7 in. long and ½ in. wide. Plant not active long during year but arises from a deep tuberous root, shaped something like a chestnut.

Found in early spring in moist woods from Nova Scotia to Saskatchewan and Montana, south to Georgia and Texas. Usually grows associated with trees of a mixed forest rather than in a pure-stand type of forest. Not common near evergreen trees. As this species occurs further and further south it becomes more and more a plant of open grassy areas; does survive clearing, and flourishes in open, grassy areas.

Flowers white or pink, or with flesh-colored markings, to 0.4 in. across, in loose, open clusters at top of plant, on individual stalks to 1.5 in. long. Petals five distinct, somewhat notched at tip. Sepals, two, persistent. Stamens five, fastened to base of petals. Fruit, capsule that is shorter than sepals. Seeds many and small.

Even in winter thaws, this plant may be making progress to be ready to bear flowers as soon as light conditions are suitable. One of first of spring flowers. Unfortunately is attractive enough to be collected by flower pickers but is not a good bouquet flower since it wilts quickly and soon looks bedraggled.

Bulbs are starchy and may be eaten raw or boiled. However, the flowers are so attractive as they are that the bulb should not be collected for food except in an emergency. Even then the amount of food gained may hardly justify the energy expended in its collection.

Order Ranales./Family Ceratophyllaceae

Hornwort
Ceratophyllum demersum

Submerged, aquatic plant, with weak stems sometimes 8 ft or more long, with no roots at any time, with leaves in whorls of from 5 to 12, slender, forked two to three times; with segments, stiff, rigid, and with toothed margins, to ¾ in. long, easily broken from stem, conspicuously crowded at growing tip giving rise to common name coontail.

In ponds and slow streams, particularly where water is hard, throughout North America except in far north and in Cuba. Variations which are sometimes conspicuous probably are not sufficient to warrant establishment of many species so only one is recognized, variations chiefly being in length and arrangement of leaves on stem. Known in Europe.

Staminate flowers having from 10 to 20 stamens, usually found at different joints along stem though sometimes they may be separate flowers at same joint. Fruit to ¼ in. long, varying in appearance from being winged or spined or warty; one seed per fruit, some classifying plants with spiny fruits in a variety *echinatum*. Anthers float free in pollination.

Provides food and admirable shelter for small animals that are eaten by useful fishes. Known that muskrats eat plants and some ducks eat either fruits or tips of plants themselves. Plants may be so abundant as to clog waterways seriously. Also may assist in pollution. In fall, tips break, sink, act as winter buds. Require little light.

Ducks known to feed on fruits or plants include black duck, baldpate, bufflehead, canvasback, goldeneye, mallard, pintail, redhead, ringneck, ruddy, bluebill, shoveler, scoter, wood duck, blue-winged teal, and green-winged teal. Used sometimes as an aquarium plant in spite of its odor and rapid growth.

Hornwort

Order Ranales./Family Nymphaeaceae

Cow Lily, Spatterdock
Nuphar advena

Leaves oval, with blades to 12 in. long and 9 in. wide, on petioles that are spongy and 1 ft or more long; floating, or sometimes erect or submerged, growing from end of a thick, spongy rootstock, from 3 to 4 or more inches through; sometimes 1 yd or more long, with weak, white, smaller side structures, and conspicuous triangular and circular scars.

Found in fresh or acid (to pH5) ponds and slow streams or at their edges or to depths of 6 ft from Labrador to Nova Scotia, west to Rocky Mountains, south to Florida, Texas, and Utah. In few cases, is grown in artificial pools but not so commonly as those with more attractive and fragrant flowers. Six species in United States; 25 in world.

Flowers to 3½ in. in diameter, like flattened globes, with six sepals; petals fleshy and under ½ in. long. Flowers appear yellow, with a purple tinge mostly on outside. Stamens, many, in five to seven rows. Disk of pistil yellow or red, with 12 to 24 parts. Fruit egg-shaped, to 2 in. long and 1 in. thick, with many seeds inside. Seeds should be stored wet.

Thick rootstocks, eaten raw by deer and men, taste sweet. May be boiled with meat or roasted. Indians used to collect them in fall or steal them from caches in muskrat houses, according to some records. Deer eat leaves and flowers and beaver eat submerged parts as do porcupines. Sometimes plants crowd out other water plants. Ducks eat seeds.

In Northwest, closely related *N. polysepalum* yields seeds that are important food to Klamath Indians. Seeds are extracted after pods have been dried, then parched to loosen covering and after covering is removed, contents are again parched and increase in size like popcorn. Seeds stored or ground as flour and used in breadmaking.

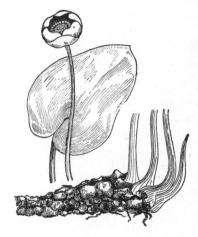

Spatterdock, Cow Lily

Sweet White Water Lily
Nymphaea odorata

Leaves with round blades that float and have a single break from margin to center. Blades to 1 ft across, smooth, green, shining above, covered with algae and small animals usually beneath. Petioles reach from buried rootstock to water surface and have four main air channels. Underground tubers 3 to 4 in. thick, fleshy, with tuberous offshoots.

In ponds and slow streams, where water level does not change radically, from Newfoundland to Manitoba, south to Florida, Louisiana, and Kansas, with related species extending range west to Idaho, with plant established as an ornamental in pools in many parts of the world. May spread by means of tuberous offshoots that establish new plants.

Flowers white, with a center of yellow stamens. Petals sometimes pink. Flowers to 6 in. across, remarkably fragrant, and in absence of sun, enclosed by green sepals. Fruit globelike or somewhat flattened, of 12 to 35 cells, with seeds enveloped. Flowers float on water surface or held above it. Fruits ripen beneath water surface. Seeds float.

Some botanists recognize two species; *N. odorata* that is fragrant and has flowers open from 7 A.M. to 1 P.M. with flowers seldom over 4½ in. across with no purple on the petiole; and *N. tuberosa* with less fragrant flowers that may be over 4½ in. across and open from 8 A.M. to 3 P.M., and with four to five purple petiole streaks.

White Water Lily

(Continued)

167

Tubers well-filled with starch, with food value. Fruits of some relatives gathered by natives, allowed to decay to free seeds that are then washed, dried, stored, parched, and eaten. Seeds contain starch, oil, and protein and are considered very nutritious. Tubers are almost solid starch and treated as though they were potatoes. Ducks eat seeds. Moose make water-lily pads a principal item of their diets. Muskrats eat the plants.

Cabomba, Carolina Water Shield
Cabomba caroliniana
Stem slender, underwater, to several feet long, branching, weak, with a gelatinous slime. Leaves opposite or whorled, with threadlike divisions, arranged like a fan, petioled, to 2 in. broad, with floating leaves, alternate or opposite, with blades narrowly oblong and to 1 in. long.

Found in ponds and slow streams. From Missouri, Illinois, and Michigan south to North Carolina, Florida, and Texas and possibly introduced in western Massachusetts. Found in almost any state wherever aquarium supplies are sold.

Flowers borne on long stems from bases of uppermost leaves, to ¾ in. across, white or yellow, but not particularly conspicuous. Plant is reproduced by merely breaking stem and letting new growth take place.

Used as food, shelter, and ornament in household aquarium, the food and shelter of course being for occupants of aquarium. Seems to be relatively free of offensive odors that may come from hornwort, looks less brash, and leaves less trash possibly because it is eaten by aquarium fish.

Stems may be anchored in aquariums by burying ends in sand but plant seems to do reasonably well without such help. Not common for aquarium plants to produce oval floating leaves and flowers, so this may be evidence that it is merely existing rather than living as it might in its native waterway.

Water Shield
Brasenia schreberi
Leaves with blades to 4 in. across and 2 in. wide, almost circular and without slit common in water lilies. Leaves and petioles covered with a clear mucilage. Petioles flexible and long enough to reach from buried slender rootstock to water surface; attached to middle of blade. Rootstock branches rather freely.

In ponds and slow streams, from Nova Scotia to Florida, west through Manitoba, Nebraska, and Texas, also along Pacific Coast from Washington to California, and in Asia, Africa, and Australia. Does not seem to be so abundant as water lilies that have a more limited distribution but locally may be common. Found also in Cuba and Mexico.

Flowers inconspicuous, to 1 in. across, on stout stems reaching from rootstock to water surface, with three to four sepals and three to four petals that are purple or mauve. Fruits a cluster of club-shaped pods to ⅓ in. long, enclosed in calyx when mature and rarely conspicuous at any time. Flowering time June to August.

Japanese eat young mucilage-covered leaves and stems as a salad, with vinegar. Indians of California commonly collected tuberous roots that are full of starch and ate them as a salad. Rootstocks in condition to be collected for use as food from autumn through spring months. Starch tubers may be baked, boiled, or eaten raw.

One of prettiest of water plants if one does not demand attractive flowers. Many common names such as deerfood, frogleaf, little water lily, water target, and water shield. Reported as food of black duck, canvasback, bluebill, mallard, pintail, redhead, ruddy, and other ducks.

American Lotus
Nelumbo lutea
Leaves huge shields, almost circular in outline and to 2 ft across, appear raised above the surface of water and arise from rootstock that may be 50 ft long and bears tuberous enlargements that in fall are well-filled with starch, but these may be well-buried in mud under water and not easily accessible.

In ponds, mud-bottomed marshes, and lake margins from tropical America and the Gulf States north to the Great Lakes region, from northern Minnesota and Nebraska through southern Ontario to Massachusetts and southern Connecticut but highly localized in North and often exterminated there.

Flowers like pale yellow water lilies, to 10 in. across. Sepals and petals numerous and in many rows. Stamens numerous and shed early. Pistils, 25 to 35 to a flower and embedded in a swollen receptacle in which the fruits rattle around when maturity is reached, each ½-in., hard, globular seed or fruit remaining in its own separate cavity.

Crisp tubers and growing tips of rootstock are baked for food and taste much like sweet potatoes. Seeds are size of acorns and when half-ripe taste like chestnuts. Ripe seeds are roasted and starchy interior is eaten

Cabomba

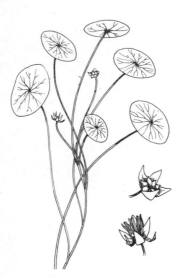

Water Shield

Lotus

dry, baked, boiled, or made into bread.   Seeds may be collected even in winter.   Leaf stalks and young leaves are also edible.

Related lotus of China and Egypt grown extensively for food, seeds being a regular food in China; rootstocks systematically cultivated.   Because of rarity and beauty, plants in North should not be considered a food source but in South might be developed.   Common names include sacred bean, water chinquapin, and nelumbo.   Superior breeding site for malarial mosquito.

Order Ranales./Family Ranunculaceae

Water Buttercup, White Water Crowfoot
Ranunculus aquatilis
Submerged, with weak, flexible stems, over 1 ft long, branching.   Leaves to 2 in. long, fan-shaped, divided again and again into almost threadlike segments, unable to stand erect out of water, with no flat, floating leaves such as are found in the yellow water crowfoot *R. delphinifolius.*

Common in shallow water from Labrador and Nova Scotia to North Carolina, west to Lower California and Alaska.   A number of varieties are recognized.   Placed by some in genus *Batrachium.*   Most abundant apparently in the Northeast where it is recommended as a wildfowl food plant.   Found also in Europe and Asia in one or more forms.

Flowers solitary, white, ½ to ¾ in. across, on weak stems to 2 in. long. Flowers float at water surface.   Many stamens and many pistils per flower.   Fruits to ⅛ in. across and borne on hairy receptacle held near water surface.   Blooms from June through September.   Apparently many closely related species with minor differences.

Definitely aquatic, though can survive some shore conditions.   Can survive 1,500 parts of alkali per million of water and 1 percent or less of sodium chloride.   May form dense mass of growth even in relatively swiftly moving water such as is found in brooks.

Of minor importance as food for waterfowl and other game species. Leaves and fruits reported as found in stomachs of waterfowl.   Plant studied to determine its importance to wildlife.   Propagated by transfer of growing material or by sowing of fruits.   May be found in water to 4 ft deep that is moderately acid.

Water Buttercup

Bitter Buttercup
Ranunculus acris
Height to 3 ft.   Erect, branching, particularly in upper areas.   Upper leaves without stems or with but short ones, three-parted.   Lower leaves on long stalks with blades cut or divided into from three to seven parts radiating from a common point.   Stems hollow, hairy, and branched. Roots clustered and fibrous.   Plant bitter.

Found in fields and roadsides particularly where there is abundant moisture.   Generally common.   Ranges from Newfoundland to British Columbia, southward to Virginia and Missouri, also in Bermuda.   Naturalized from Europe and possibly introduced by seed as it occurs as a seed impurity.

Flowers yellow cups, about 1 in. broad, shallow with five yellow petals, two to three times length of sepals and many times broader, borne on long, slender stems.   Stamens very numerous.   Pistil many to a flower, one-seeded, compressed and short-beaked.   Flowers from May through October.   Insect-pollinated.

Green parts contain volatile poisonous substance that vanishes in hay. Poison affects horses, sheep, but most particularly cattle, causing inflation of intestinal membranes and producing bitter and sometimes reddish milk from dairy cattle.   Because of bitter taste, plants not normally eaten if better forage is available.   Seeds are eaten by birds and some small mammals.

Since cattle avoid this plant, it survives to produce seeds unless these are cut or otherwise destroyed.   Will not survive cultivation or drainage so control is by cultivation, cutting, draining, or encouraging rivals by good fertilization.   Hay not dangerous since poison vanishes with drying.

Bitter Buttercup

Swamp Buttercup
Ranunculus septentrionalis
Height to 3 ft, branching.   Smooth or fuzzy, usually the latter.   Older branches may droop to ground and take root at joints or form long runners.   Leaves unequally three-parted, with divisions commonly stalked; long-stemmed (sometimes 1 ft or more); divisions wedge-based.   Roots fibrous and relatively weak.

Commonly in swamps and wet areas and rather commonly in shade. Ranges from Georgia to New Brunswick and west through Manitoba and Kansas with at least one established race in Minnesota-Missouri area with somewhat sticky, bristly covering on the plant.

(Continued)

Swamp Buttercup

Flowers to 1 in. or more across, bright yellow, not crowded, with five petals broadest near free end and themselves twice length of spreading sepals. Stamens and pistils, many per flower, with head of fruits to ⅓ in. in diameter. Fruits with broad wing margins and with a beak about the length of fruit body. Flowers April through June.

Pollination is largely by the beelike flies belonging to the Bombylidae and the small bees of the family Andrenidae. Fire and drought may eliminate the species from certain areas though it can withstand flood rather well.

The plant is easily controlled as a weed by drainage of the soil and by only moderate cultivation. Murie has reported it as a food favored by moose in the Isle Royale area and its fruits may possibly be considered of value for some birds.

Early Meadow Rue
Thalictrum dioicum
Height to over 2 ft. Slender, but reasonably tough, branched, smooth, pale green stems. Leaves two or more times compounded, of thin, pale green, widely spreading leaflets usually obscurely three to nine lobed. Leaves relatively few. Roots fibrous and not yellow.

Found in relatively open but protected woodlands from Maine to Alabama, west to Missouri and Saskatchewan ascending in North Carolina to 4,500 ft. elevation. It is also recorded from Labrador. At least eight closely related species in the eastern part of the United States, some favoring more open habitats than does this species.

Pistillate and staminate flowers on separate plants; pistillate inconspicuously pale green, with rather long tips to pistils, develop into clusters of ribbed, egg-shaped fruits; staminate appear as delicate clusters of lace, composed of slender anthers. Pollination probably by wind. Flowering period April and May.

Genus is represented by plants mostly of temperate zone with at least 100 known species, of which 11 are found in United States. Common Eastern species include this woodland species and a more open country *T. polygamum* or tall meadow rue. Best-known Southern species is well called maid of the mist. A Pacific Coast species does best along streams.

Garden species of *Thalictrum* have been developed from European and Asiatic species. Among these is the feathered columbine. *T. minus* is an attractive little rock garden plant, about 10 in. tall, with beautiful fernlike leaves.

Rue Anemone
Anemonella thalictroides
Height to 9 in. Smooth, delicate. Basal leaves resemble those of meadow rue and appear after the flowering stalk has developed and compounded into 3s. Leaflets all thin and pale green or olive and on long stems. Root system a cluster of small tuberous roots, generally elliptic in shape.

Found in woods that are usually thin; ranging from New Hampshire and Massachusetts south to Florida and west to Ontario, Minnesota, and Kansas. Only one species in the genus, found only in eastern North America.

Flowers borne in clusters of two to three at the top of early-appearing stems and arising from a point common with the involucre leaves. Sepals 5 to 10, white and petallike. Petals, none. Stamens, numerous. Pistils, to about 15 per flower and developing into 8 to 10 ribbed, oval fruits.

Flowering period from March through June. Name refers to resemblance of plant to meadow rue and to anemone, the flowers being like latter except that they are clustered, and leaves being like former in general appearance. Pollination by bees and beelike flies.

Plant is sold for planting in wild gardens but since stock is generally collected from wild this does not seem to be a good practice. Roots may be eaten as "potatoes" and are high in starch content but the returns in food cannot make up for loss of beauty of the plant.

Blunt-leaved Hepatica
Hepatica americana
Height to about 6 in. Rather leathery, hairy, satiny leaves and flower stems arising from a common point close to the ground. Leaves on long stalks, with three blunt lobes, to 2½ in. broad, when mature resting on the ground, may be green under winter's snows. Roots fibrous and profusely branching.

Common within range, on dry or open wooded banks, in gravelly soil that may be acid or neutral, unlike related *H. acutiloba* that favors lime soils. Ranges from Nova Scotia to Manitoba and south to Missouri and northern Florida and in Alaska. To 2,600 ft elevation in Virginia.

Flowers blue, purple, or white, saucer-shaped, to 1 in. broad, flowering from December through May in some seasons. Petals absent, being re-

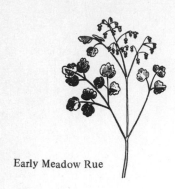

Early Meadow Rue

Rue Anemone

Blunt–leaved Hepatica

placed by to 15 colored sepals, with green involucre acting like calyx. Stamens numerous, as are the separate, one-seeded, hairy, pointed fruits. Pollination by bees and flies.

Tannin is extracted from full-grown leaves with alcohol and believed by some to have slight astringent value but not normally used as medicine. In past, was believed to be of value in "stirring a torpid liver" on assumption that since leaves looked somewhat like the liver, they could correct liver troubles.

State flower of Missouri and of Minnesota. A most popular spring flower that survives picking rather well. Known as three-leaved liver-wort, herb trinity, ivyflower, golden trefoil, squirrel cup, mouse-ear and other names, indicating its popularity. Survives some transplanting to suitable gardens.

Wind Flower

Windflower, Wood Anemone
Anemone quinquefolia
Height to 9 in. Nearly smooth. Basal leaves long-stemmed and appearing after the flower-bearing stem, five-parted, with divisions pointed at base and dentate at free ends. Leaves on flower stem usually three-, five-parted, springing from a common point. Root system rather slender, horizontal rootstock.

Common in low open woodlands from Nova Scotia to Georgia, west to Tennessee, Minnesota, and Ontario rising in Virginia to elevations of 3,500 ft. Resembles slightly the European windflower *A. nemorosa* which has escaped into territory normally occupied by our species.

Flowers single (as contrasted with many in rue anemone), to 1 in. broad, with four to nine whitish or purple sepals and no petals. Stamens, many. Pistils rather numerous and crowded into a coarsely bristly head, each part being hooked and slightly fuzzy. Pollination by bees and early bee-like flies.

Flowers from April through June. Plant rarely is as abundant as many other spring flowers and flowers too few or leaves too many to make it popular for picking. It transplants easily and is relatively hardy in wildflower gardens.

The popularity of the plant is attested by its many common names such as nimble weed, Mayflower, wild cucumber, wood flower, and so on. It is obviously more slender in most ways than its European namesake and relative.

Thimbleweed

Tall Anemone, Thimbleweed
Anemone virginiana
Height to 16 in. Densely greenish-gray, hairy. Hardly erect. Leaves much-divided and like a fuzzy bitter buttercup leaf though coarser and somewhat smaller, the lobes being slender. Rootstock thick, horizontal, bearing basal leaves and flower-bearing stalks.

Occurs as patches or clones in dry, open, prairie lands from Illinois to Texas and west to Nebraska and British Columbia, also in Europe and in northern Asia, though the European and Asiatic forms usually have narrower leaf parts than do those of America. 18 North American relatives known.

Flowers gorgeous, solitary, blue, lavender, or white, with light centers, without petals, the sepals being conspicuous and making a flower 1 in. or more across. Stamens, many. Pistils form a conelike structure that breaks to free long feathery individual fruits heaviest at base.

Flowers March and April. Perennial. Fresh plant produces a virile, poisonous, bitter alkaloid anemonine that may cause severe inflammation and irritation to sheep that may graze the young plants or flowers in open pasture.

Its beauty should make it a popular wild-flower-garden plant but it apparently demands certain conditions if it is to thrive. Its popularity as a spring flower is testified by its many common names such as wild crocus, prairie smoke, badger, gosling, headache plant, rock lily, and April fool.

Pasque flower

Pasqueflower
Anemone patens
Height to 1½ ft, rarely more than 1 ft. Stem stout, rather tough, branching. Basal leaves long-stemmed, three-parted, broader than long, with divisions wedge-shaped at bases and variously cleft. Leaves not arising from the base, on stems to 2 in. long and much like basal leaves. Roots fibrous.

Common along dry roadsides, in dry woodlands, and in some waste places from Nova Scotia to South Carolina, west to Arkansas, Kansas, and Alberta, with closely related species extending the range considerably. There are of course a number of related species that are ornamentals.

Flowers borne singly at the ends of relatively long erect stalks. Sepals five, greenish-white or white, and blunt or pointed at the tips. Petals, *(Continued)*

Virgin's Bower

Cowslip

Goldthread

none. Stamens, numerous. Pistils, many, forming a thimble-shaped head which splits to expose rather fluffy individual fruits.

Flowering period June through August. Pollination by bumblebees, honeybees, other small bees, and by syrphus flies. Other wild anemones in relatively same range include the long-fruited *A. cylindrica*; the Canada anemone with fruits that are not fluffy, and others.

The name anemone means flower, shaken in the wind. The plants are of little economic importance but of considerable esthetic beauty, although they are a bit too coarse to compete with their woodland relatives or the pasqueflower of the open prairies. Anemones contain ranunculin, which breaks down to a substance which is toxic when eaten by grazing animals. *A. patens* may cause plant "hairballs" in the digestive tract of sheep.

Virgin's Bower
Clematis virginiana
Climbing vine often many yards long and sprawling over shrubs and low trees, walls, or fences. Leaves nearly smooth, of three leaflets, broad at base and acute or deeply lobed at tip. Stem more or less woody. Leaves and leaflets coil about support, assisting in climbing habit. Buds blunt, somewhat sunken.

Common along woodland borders, over shrubs and fences from Nova Scotia to Georgia, west to Tennessee and Manitoba. It often covers riverbanks with an intertwining mass of stems each well over 10 ft long. Related wild species include the purple virgin's bower *C. verticillata* and leather flower *C. viorna*.

Flowers, ½ in. to 1 in. across, white or greenish-white, in loose, relatively open clusters, the pistillate and the polygamo-staminate on separate plants. Sepals four, greenish-white. Petals, none. Stamens, many. Pistils develop into slender structures with long, curved, feathery tips that together make fluffy balls.

Flowers from July to September. Pollination by syrphus flies, small bees, and beelike flies attracted by abundant nectar. From October on, the fruits make the plant attractively conspicuous and give the plant the name old-man's-beard. Also called woodbine, devil's-hair, wild hops, and devil's-darning-needle.

Sometimes cultivated as an ornamental. Handling of the leaves may cause a dermatitis to some persons. European species closely related to this one also cause dermatitis. Plant is easily propagated by pulling up sections of wild plants and setting these out.

Marsh Marigold, Cowslip
Caltha palustris
Height to 2 ft. Stem erect, smooth, mellow, branched above, rich dark green. Leaves long-stemmed, kidney-shaped to broadly heart-shaped, from 2 to 8 in. broad, smooth, bright dark green, with conspicuous veins, lower leaves longer-stemmed than upper. Underground root system soft and spongy with fine rootlets.

Found in marshes and other wet spots, frequently forming spectacular displays. Ranges from Newfoundland to Saskatchewan, south to South Carolina and Nebraska, with two other American species found in much the same area, one more or less aquatic. Native of North America. *C. biflora* is the Pacific Coast relative found in comparable locations.

Flowers to 1½ in. broad, shaped like a shallow cup, bright yellow, blooming in May and June. Petals none, the sepals appearing like petals and dropping off relatively early. Stamens, numerous. Pistils 3 to 12 or more, splitting down one side, to 1 in. long, freeing many seeds. Perennial. Pollinated by syrphus flies and bees.

Green parts contain poisonous alkaloid jervine and glucoside helleborin, both of which are destroyed by drying or cooking plant. Sheep, cattle, and horses may be poisoned by eating green tips. No cases of livestock loss are recorded in continental U.S. Name means cup of marsh. Other common names include capers, soldier's buttons, boots, meadow boots, drunkards, crazy bet.

While green plants eaten by cattle may cause stoppage of milk, diarrhea, and stomach inflation, cooked greens are a standard spring green for human beings and are delicious. Buds are picked, parboiled in salt solution, and pickled as capers. In preparing greens, it is suggested that first water be poured off and not used.

Goldthread
Coptis groenlandica
Flowering stalk to 5 in. but usually lower. Leaves evergreen, rich, dark green, shining, of three leaflets, each wedge-shaped, sharply toothed, and obscurely three-lobed. Root slender, branching, horizontal, well-buried, brilliant yellow and bitter. Whole plant a thing of beauty.

Found in rich mossy woodlands and swamps where acidity is relatively high and in bogs. Ranges from Newfoundland to Maryland and Tennes-

see and west through Tennessee, Iowa, Minnesota, and British Columbia to Alaska, reaching an elevation of 3,500 ft in the Adirondacks in New York. About nine related species.

Flower one to a flowering stalk, though sometimes two are present, to nearly 1 in. across. Sepals five to seven petallike, dropping off early. Petals smaller, equal in number, slender. Stamens, numerous. Fruits in three to seven parts, about ¾ in. long, spreading at the tip. Golden anthers are conspicuous in white flowers. Golden sterile anthers function as nectaries, "honey leaves."

Pollination effected probably by fungus gnats and by a small beetle *Anaspis*. Flowering May through August. Those who know this plant always like to discover again beautiful yellow thread of the root system, no matter if it has been seen hundreds of times before.

The plants provide medicinal stock for the old-time herb doctor, possibly on the assumption that something as beautiful and as bitter as the root must have some medicinal value. The plants grow well in indoor gardens in terrariums and are popular for this reason.

Peony
Paeonia anomala
Height to 3 ft. Flowering stem usually one-flowered. Leaves twice-divided, with segments slender, pointed, smooth beneath and sparingly hairy above, abundant, forming a rather compact clump. Roots tuberous and not spreading by stolons near surface as does *P. tenuifolia*. Herbaceous.

Native of northern Europe and Asia. *P. officinalis* with crimson to white flowers and *P. albiflora* with white or pink flowers have leaves whose segments are not narrow. *P. anomala* with crimson flowers and *P. tenuifolia* with crimson to purple flowers have leaves with narrow segments, the latter not so narrowed and pointed at end as the former.

Flowers to 6 in. across, showy. Sepals, five, persistent. Petals, 5 to 10, but much more numerous in cultivated forms. Stamens many, not conspicuous. Carpels on a fleshy disk containing large seeds. Coloration and multiplication of parts have been accomplished in cultivation. In this species, fruits are reddish, in variety *insignis,* fuzzy.

Roots planted to depth of about 2 in. to avoid frost heaving and to ensure flowering. Transplanting should be avoided but may be done in September into rich soil; roots should not be in contact with manure. Tops should be collected and burned in autumn to avoid disease. Bad bud rot, sometimes serious.

Disease may be controlled by bordeaux spray in early spring and continued once a week for a month. Fertilizer may be a dressing of bone meal and sheep manure in spring. Mulch may prevent frost from heaving roots out of ground but usually this is not necessary in old established plants. The "piney" is a symbol of early American garden.

Wild Columbine
Aquilegia canadensis
Height to 2 ft or more. Open, branching habit, usually smooth. Leaves two to three times compound, 4 to 8 in. broad, with leaflets to 2 in. broad, those at end being without stalks. Lower leaves on long stems. Lower side of leaflets paler than upper surfaces. Root system long, coarse, and tough.

Common in rich woodlands and in loose soil in gorges and on wet cliffs. Sometimes rather abundant even in fields. Ranges from Nova Scotia to Northwest Territory area and south to Texas and Florida and in Rocky Mountains. Found at 5,000-ft elevation in Virginia.

Flowers scarlet and yellow, nodding, 1 to 2 in. long, with nearly straight spurs, each to ¾ in. long and tipped with nectar gland. Sepals, five, drop off early. Stamens numerous, often with tips turned back. Pistils, five, each splitting down side to free many seeds. Flowers April through June. Pollinated by moths and butterflies.

Has many common names such as rock lily, jack-in-trousers, bells, meeting-house, clucky, and honeysuckle. It is of course not at all related to the woody plant known as honeysuckle. A number of related garden species, some of them with blue flowers and usually with larger flowers.

Flowers are so beautiful and wilt so quickly that there is no reason why plants should be collected for bouquets particularly since plants are usually badly torn by ruthless collectors.

Garden Columbine, Rocky Mountain Columbine
Aquilegia caerulea
Height to 3 ft. Smooth below, but slightly fuzzy above. Basal leaves divided into 2s, smooth above and somewhat fuzzy beneath, or rather with a grayish bloom on the lower surface. Substantial root system of the usual herbaceous perennial.

This species native of Rocky Mountains. To 50 species found in north temperate regions, of which this is probably basis of most of blue garden
(Continued)

Peony

Wild Columbine

Garden Columbine

columbines. Other types are known as chrysantha and the Canadian to garden trade and are kinds most frequently recommended for garden use.

Flowers to over 2 in. across, blue and white with spurs not curved inward or hooked at tips and with flowers in this species erect and not nodding. Sepals to 1½ in. long, deep purple. Petals white, with pale blue spurs, to about 2 in. long. Stamens not protruding from flower. Fruits about 1 in. long, curved.

Seeds sown 1 year develop into plants that should bloom next and should increase in size from then on. Since plants cross, it is best to destroy those that are not desirable so that future seedlings from stock may be as wished. Recommended that plants be bought and in this way good stock be assured.

State flower of Colorado. Other relatively common garden species include those from Japan (incurved, spurred) *A. flabellata* and *A. glandulosa, A. siberica* of Siberia, common *A. vulgaris* of Europe and Siberia, and straight-spurred *A. formosa* of our West Coast and Siberia, *A. skinneri* of Mexico, and *A. chrysantha* of our West.

Larkspur

Larkspur
Delphinium sp.
Four common cultivated larkspurs are recognized: annual rocket, *D. ajacis;* annual branching, *D. consolida;* perennial rocket or candle; and perennial branching or bouquet. Wild tall larkspurs grow to height of 8½ ft, while low wild larkspurs rarely exceed 1 ft. The perennials may have stout root systems.

Native species in eastern and in western North America. Low larkspurs are found from sea level to 10,000-ft elevation while tall larkspurs are best above that elevation. Low grow well in open exposed places while tall species favor moist habitats. Garden larkspurs come from many ancestries. Over 250 species.

Flowers showy, commonly blue but may be white, pink, or other colors. Sepals, five, colored, one prolonged to rear into a long spur close to spurs of upper pair of petals. Stamens many. Carpels of pistil, from one to five, splitting when ripe to free seeds. Chinese larkspurs among perennials that flower first year.

Seeds long known to be poisonous. Plants, if grazed by cattle, may in many species be poisonous. While plants are most attractive for bouquets and as garden flowers, cattlemen recognize them as among their worst enemies. Strange to say, plants that are poisonous to cattle may be harmless to sheep that may graze them.

In common garden parlance, the name larkspur is applied to annual delphiniums while delphinium is reserved for the perennials. Many varieties of colors and of single or double forms have been developed for gardens. Seeds sown early develop into plants that should be thinned to 1 ft apart.

Locoweed

Locoweed, Poison Larkspur
Delphinium glaucum
Other locoweeds include *Oxytropis lambertii, Astragalus mollissimus,* and *Astragalus diphysus.* Best known are *Astragalus* spp., which are legumes. Larkspurs poisonous to cattle are rivaled only by the above. This is a tall larkspur with almost smooth stems and is by some considered a variety of *D. scopulorum,* the other most common variety being *subalpinum.*

Found mostly in higher areas ranging from Alberta in Canada south through Washington and Oregon to Nevada and California. Closely related species, of similar effect, extend range of poisonous larkspurs east through Colorado, Wyoming, Idaho, and Utah for tall species and to Pennsylvania, Virginia, and Georgia for low species.

Flowers essentially as those described for the garden larkspur. Pollination in most delphiniums is by bumblebees, honeybees, and beelike flies with tongues suited for collecting hidden nectar. Seeds of difficult species vary considerably in size but most are considered to be poisonous.

Poisoning of cattle caused by grazing just after snows have left. Sheep may be poisoned by overfeeding on some young plants. Poison is due to alkaloid delphinine and other alkaloids. The alkaloid deltaline is reported in one species.

Symptoms of larkspur poisoning are staggering gait, loss of appetite, rigid extension of limbs after falling, constipation, nausea, and bloating. In general, the practice of control has been to graze the sheep where species are not injurious to them.

Wild Monkshood

Wild Monkshood
Aconitum uncinatum
Height to 3 ft, climbing or turning upward. Leafy, slender. Leaves thick, alternate, broader than long, to 4 in. wide, with three to five deep clefts almost to the center, smooth or nearly so. Root substantial, normally poisonous if eaten.

Found in woodlands from southern Pennsylvania to Georgia and west to Wisconsin and Kentucky, reaching an elevation in Virginia of 3,000 ft. Related species extend range of genus into Canada and to West Coast; among these being introduced, *A. napellus* and yellow-flowered *A. lutescens* found in the West.

Flowers blue, with a conelike, slightly beaked hood, to 1 in. broad or larger. Sepals, five. Upper petals, two. Pistils, three to five, with pods being several-seeded and to nearly ½ in. long, beaked and spreading outward at tips. Drug aconite comes mostly from European plant.

Flowers, seeds, and particularly roots yield aconite depending in part on climatic conditions. Poisons are alkaloids aconine and acontine that affect horses and sheep, causing muscular weakness, bloating, impaired breathing, altered pupils, and difficult but urgent swallowing.

Cure for poisoning includes digitalin or atropine injections or inhaling of ammonia or camphor fumes or both. Seeds sown in fall in well-drained loam yield roots ready for harvest by plowing the second season after the flowering stems have died. Harvesting is in fall.

Black Snakeroot

Black Snakeroot
Cimicifuga racemosa
Height to 8 ft. Leaves most abundant at top. Leaves thrice-divided and in three main parts, broad, forming a great, flat, open, fanlike structure. Leaflets sharply toothed, thickish, nearly smooth, light green. Rootstock large and thick.

Found in woodlands from Maine to Georgia, west to Missouri, Indiana, and Ontario rising in North Carolina to 4,000-ft elevation. Related heart-leaved snakeroot has a more limited range within that of this species.

Flowers borne along graceful frond that is to 3 ft or even more long. Flowers appear as feathery tufts of white; stamens numerous. Fruit berrylike and purple. Flowers have a most offensive odor in contrast with their general attractiveness.

Flowers appear in June through August. Scientific name means drive bug away, the *cimic* referring to *Cimex* the bedbug. Whether plant is an effective deterrent to these may be doubtful. Pollination effected by green flesh flies attracted no doubt by odor.

In spite of its offensive odor, plant is often grown as an ornamental because of beauty of graceful tall flowering stalk. It does reasonably well in a wild-flower garden, but is best placed well back from the path where it may be seen but not sniffed.

White Baneberry

White Baneberry
Actaea pachypoda
Height to 2 ft. Bushy, branching. Leaves thrice-divided and of three major parts with the lobes of the white species more sharp and more numerous than in the red, *Actaea rubra,* widely spreading and relatively few. Rootstock substantial and virulently poisonous if eaten.

White baneberry found in woods from Nova Scotia to Georgia, west to Missouri and Minnesota, rising in Virginia to 5,000-ft elevation. Red baneberry ranges from Nova Scotia to Pennsylvania and west to Nebraska and South Dakota. *A. arguta* and other species extend range to west.

Flowers borne on short stems from top of flowering stalk. Petals shorter than stamens. Fruits white, berrylike spheres, with dark-purple ends resembling doll's eyes. In red baneberry, berry is red; in white, supporting stems may be red though this is not always the case. Flowers April through June.

Pollination by small bees especially of genus *Halictus.* Berries may provide food for caterpillar of one of azure butterflies. Flowers provide pollen, not nectar to visiting pollinators. Poison affects heart and is known to be fatal to children who eat berries.

Rootstock poison is violently purgative and emetic. Eating a few berries may cause increased pulse, colic, dizziness, and general sickness. Common names of white baneberry include necklaceweed, snakeroot, and white beads, while red is known as poisonberry, rattlesnakewort, herb Christopher, and grapewort.

Goldenseal

Goldenseal, Orangeroot
Hydrastis canadensis
Height about 1 ft. Leaves, a single leaf from base and a pair of smaller leaves on flowering stem. Basal leaf with long petiole, to 8 in. wide, five-to nine-lobed, lobes being broad but sharply toothed. Stem leaves borne at or near top, upper being just below flower. Rootstock thick and yellow.

(Continued)

175

Found in woods from Connecticut to Georgia, west to Missouri, Kansas, Minnesota, and western Ontario, reaching an elevation in Virginia of 2,500 ft. Rather extensively cultivated as a drug plant and as such is found outside normal range even to Pacific Northwest. A shade lover.

Flowers solitary, at end of flowering stem, greenish-white, to about ⅓ in. across. Sepals, three, dropping off soon. Petals, none. Stamens, numerous. Carpels, over 10, each supplied with one to two seeds and becoming grown into a fleshy, red berry, at maturity to ⅔ in. through.

Poisonous or medicinal elements are alkaloids berberine, hydrastine, and canadine, of which hydrastine is most active. Poisons cause ulcers and catarrhal mucous inflammations. Grown under shade in woods well-supplied with leaf mold, with bone meal, cottonseed meal, and potash added as fertilizer.

Seeds are sown in fall but must not be allowed to dry. Started in sand layers after fruit pulp is removed. Seedlings are transplanted into lath houses and roots are ready for harvest in 4 years. Goldenseal is becoming very rare in nature over much of its range. It was collected as the source of the drug hydrastine, used to inhibit uterine bleeding.

Order Ranales./Family Magnoliaceae

Cucumber Tree, Magnolia
Magnolia acuminata
Height to 100 ft or more. Diameter of trunk to 4½ ft. Trunk straight, and only slightly ridged. Twigs relatively coarse and weak, with alternate buds. Leaves to 10 in. long and to 6 in. wide, pointed at tip, with unbroken margin. Roots deep, spreading.

In moist, fertile, loose, deep soils growing with other hardwoods such as white oaks, ash, black gum, beech, tulip, poplar, and hickory. Ranges from New York and Ontario south to Georgia and west to Mississippi, Arkansas, Missouri, and Illinois, being found in Virginia up to 4,200-ft elevation.

Flowers greenish-yellow, to 2 in. high, with petals much longer than sepals that quickly drop off. Stamens, many. Pistils form a cone-shaped or cucumberlike fruit to 4 in. long and 1 in. through that becomes rose-colored when ripe and frees ½-in. red seeds suspended on slender threads.

Wood weighs 29 lb per cu ft. Heartwood yellow-brown. Sapwood lighter in color. Growth relatively rapid as tree reaches maturity in 80 to 120 years. Wood has much the same uses as that of tulip poplar. Tree not sufficiently abundant to be commercially important.

Related magnolias are well-known for brilliant or large showy flowers and this of course places them among ornamentals. Some have flowers appearing before leaves, which adds to their attractiveness. Trees are clean and therefore relatively popular as shade trees.

Tulip Poplar, Yellow Poplar
Liriodendron tulipifera
Height to 200 ft but average about 100 ft. Trunk diameter to 12 ft, but average about 4. Bark ash-gray, uniformly furrowed. Twigs coarse, relatively weak, with diaphragmed pith. Buds large, look like a duckbill. Leaves smooth, dark green, to 6 in. long, with two basal and two terminal lobes. Roots deep, spreading. For the first 50 to 100 years tulip poplar exhibits a tall, straight growth habit. This makes it a valuable ornamental and a source of merchantable lumber.

In relatively dry woodlands, usually on hillsides. In wild, with black cherry, white pine, white oak, chestnut, hickory, black walnut, basswood, butternut, and magnolia, not commonly in pure stands. Ranges from Rhode Island to Florida, west to Mississippi and Michigan. Known from Cretaceous times.

Flowers greenish-yellow, to orange within, to 2 in. high. Petals, six, to 2 in. long and erect or curving upward, with three sepals recurving downward. Stamens many, with slender anthers. Pistils form erect, relatively dry cone to 3 in. long and nearly 1 in. through from which winged units are freed.

Wood uniform, greenish-gray to pale yellow, or brown with sapwood nearly white; weighs about 26 lb per cu ft. About 2 to 15% of seeds fertile. 300-year-old trees are known. Trees may be transplanted in spring if roots are not allowed to dry. Intolerant of oaks.

Desirable ornamental if grown with other trees away from strong winds. Wood known as "whitewood" and a valuable cabinet wood; also used in boatbuilding, exterior finishing of houses, shingles, woodenware. 12,000 to 14,000 seeds per lb. Seeds eaten by bobwhite, white-tailed deer, red squirrel, cottontail rabbit. Recommended for growth in ravines as future timber and as soil anchorage.

Cucumber Tree

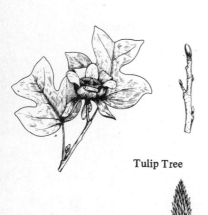

Tulip Tree

DIVISION ANTHOPHYTA. CLASS DICOTYLEDONAE

Order Ranales./ Family Annonaceae

Papaw
Asimina triloba
Tree to 45 ft high, with trunk to 10 in. through. Often grows more as a shrub than as a tree. Young shoots and leaves dark fuzzy, becoming smooth when mature. Leaves to 12 in. long, with petioles to 6 in. long, rounded or pointed at base, pointed at tip and with unbroken margins, alternate. Buds hairy, without bud scales.

Relatively common along streams from southwestern Ontario and western New York south through Pennsylvania and western New Jersey to Florida and west to Texas, Kansas, and Michigan. About seven species related to this one in eastern and southeastern North America.

Flowers on shoots of preceding year, appear with leaves, to 1½ in. across, dark purple. Petals, six, in two series, the outer being much larger and more spreading. Sepals to ½ in. long. Stamens, numerous. Fruit a fleshy berry, to 7 in. long, 2 in. thick, brown, sweet, and edible, borne in clusters.

Wood to 24 lb per cu ft, not of great commercial importance. Seeds 1,200 per lb, with 50 percent germination and about 250 plants to be expected from 1 lb of good seed. Few insect enemies, though a moth injures flowers and fruits in some areas rather seriously. Flowers in March and April, fruits in October.

Known as fetid shrub and custard apple. Fruits best after they have been frost-bitten but even then not liked by everyone. Some persons develop an unpleasant dermatitis from handling fruit. Recorded that members of Lewis and Clark expedition were saved from food shortage by abundant papaws found on return journey.

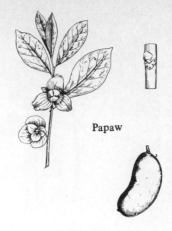

Papaw

Order Ranales./ Family Menispermaceae

Moonseed
Menispermum canadense
A slender, twisting, climbing vine, to 12 ft long, smooth or slighty fuzzy. Leaves alternate, heart-shaped, to 8 in. wide, to 4 in. long, with unbroken margins, a short, pointed tip or usually with three to seven lobes. Stems woody, slender, without stipules or stipule scars. May spiral around woody plants, eventually strangling them.

Found in woodlands and along streams from western Quebec south to Georgia, west to Nebraska, Arkansas, and Manitoba ascending in Virginia to 2,600-ft elevation. Family is most abundant in tropical regions; three small genera represented in flora of eastern United States.

Flowers white, small, to ⅙ in. across, in loose, open clusters. Sepals four to eight in two series and longer than the six to eight petals. Stamens, 12 to 24. Pistils develop into fleshy, bluish, globular, stone-centered fruit, to ⅓ in. across and with stone which is shaped like a large crescent or ring. Fruits resemble small grapes but should not be treated as such.

Flowers from June through July. Sharp edges of pits when eaten may cause mechanical injury to intestines. Probably some poisonous properties to fruits just as bitter poisonous alkaloids are known to be found in good-sized rootstock. Resemblance of fruit to grapes is unfortunate.

Plant may be considered as a rather attractive wild climber but doubtful if it should be encouraged in gardens or where children might mistake the fruits for grapes.

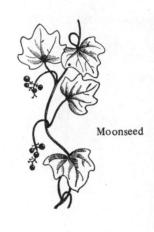

Moonseed

Order Ranales./ Family Berberidaceae

Mandrake, Mayapple
Podophyllum peltatum
Height to 18 in. Erect, rather brittle stem bears one to three large, umbrellalike leaves that are five- to nine-lobed with clefts sometimes almost reaching center and being nearly 1 ft across; much lighter green beneath, smooth or nearly so. Rootstocks stout, horizontal, bearing clusters of radiating roots.

Common in moist woodlands, ranging from Quebec and southern Ontario to Florida and west to Texas, Kansas, and Minnesota, being found up to 2,500-ft level in Virginia. Varies greatly in abundance through range. In some places, is conspicuous along railroad rights of way but probably equally so along other woodland borders.

Flower a single, beautiful, white, saucer-shaped flower borne at top of flowering stalk usually between two leaves. Flower pendant; in fact, people often overlook it because it hangs downward underneath the leaves. It is considered ill-smelling by some. Petals six to nine, stamens, twice as many as pistils; bearing abundant, yellow pollen. Fruit a yellowish or brown succulent berry that is to 2 in. long. Flowering time May.

Pollination by bumblebees and other bees though some believe it may be self-pollinated. Rootstocks may cause severe dermatitis if handled by some. They contain poisonous, resinous podophyllin that is used medicinally as a purgative but that used in excess may be fatal.

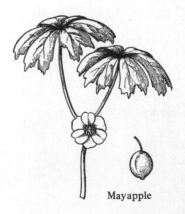

Mayapple

(Continued)

Plant has some commercial value as a drug producer. DeWitt Clinton wrote of probable sale of ripe fruits in markets of New York after opening of Erie Canal, "Clinton's Ditch," but that one dream of his did not come true. Ripe fruits may be used in preserves but green fruits should be avoided always.

Blue Cohosh
Caulophyllum thalictroides

Height to 3 ft. Smooth, slender, blue-green herb, the erect stem bearing a large thrice-compounded leaf that in early season is conspicuously bluegreen to purple. Stem arises from coarse, knotty, rather thick rootstock that bears scars of previous years' growth.

Favors moist rich woodlands with deep soil in woodlands. Ranges from New Brunswick to South Carolina, west to Missouri, Nebraska, and Manitoba, being found up to 5,000-ft elevation in North Carolina. A native of region it now occupies. Usually associated with plants of mixed hardwood forests.

Flowers to ½ in. across, greenish-purple, borne in terminal clusters. Petals, six, and smaller than and opposite to sepals. Stamens, six. Pistil matures before stamens, and is short. Pollination by bumblebees and other bees. Berrylike seeds look like blueberries, to ⅓ in. through and borne in loose, open clusters. Flowers in April and May.

The rootstock contains a bitter, poisonous alkaloid, methyloytisine, and certain glucosides. Because of bitter taste of leaves, plant is probably avoided by cattle and other grazers. No one should try to eat berries, which appear attractive enough but are highly irritating to mucous membranes.

Some people are susceptible to properties in leaves and may develop a dermatitis from such experience. Well-known and has such common names as papoose root, blue ginseng, squawroot, and blueberry. The scientific name of the genus means stem leaf.

Blue Cohosh

Common European Barberry
Berberis vulgaris

Shrub to 8 ft high, with arching branches and drooping ends. Twigs gray with rough bark, armed with three-parted spines at bases of alternate leaves. Leaves to 2 in. long, with margins bristly notched, clustered on older portions of shoots. Roots relatively deep, tough, and yellowish-wooded.

Naturalized from Europe and Asia, widely established as an ornamental and as an escape through southern Canada and the United States, being common in thickets and in waste places, sometimes grown as a hedge because of the spines that provide an excellent barrier.

Flowers in drooping, many-flowered clusters that are to 2 in. long, individual flowers being yellow, to ⅓ in. long, with disagreeable odor and with six petals and six stamens, the last being sensitive to touch at the base. Fruit a scarlet, oblong, juicy berry that is sour but edible.

Flowering time May and June, fruiting time from September on. Plant is intermediate host for wheat rust, bearing aeciospores in characteristic cups on underside of leaves. From these spores may develop structures that cause wheat rust on wheat. Tremendous efforts, through eradication campaigns, have been and are being conducted to remove this plant from much of its range. Fruits eaten by grouse, pheasant, quail, and other birds.

Popular as an ornamental in regions where wheat growing is not important. Berries make excellent preserve or jelly or may be made into delicious pies. Young leaves have pleasing sour taste and juice mixed with other fruit juices makes a good drink. Juice alone with sugar makes a good jelly. Yellow dye from root bark.

Common European Barberry

Japanese Barberry
Berberis thunbergii

Shrub, to around 4 ft high, with dark, tough stems well-armed with spines that are sharp, stiff, and more or less simple. Leaves dark green, with unbroken margins, to 2 in. long but normally smaller, with relatively constricted bases. Roots tough but relatively shallow, withstand abuse of frequent and careless transplanting well.

Does best on well-drained, sunny hillside but grown most frequently along borders of buildings, paths, and the like. Native of Japan but has become widely naturalized in many parts of United States and may be found at considerable distances from communities. Found in many cultivated varieties.

Flowers borne singly rather than in clusters as in European barberry. Flowers yellow, shallow cups, with a broad pistil in center surrounded by stamens that respond readily to touch and reach over, touching the pistil. A person can demonstrate this by touching the center of an open flower with the point of a pencil. Fruit an oblong, scarlet berry.

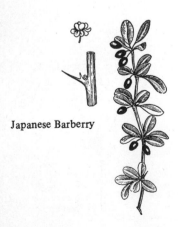
Japanese Barberry

Competes poorly with grasses and may succumb to drought.   Branches root freely where they touch ground.   Plant not attacked by wheat-rust fungus.   About 25,000 to 28,000 seeds per lb with a germination of 80 percent, 1 lb of seed yielding normally about 4,000 usable plants.   Berries that are hard in fall become softer through the winter months.

One of most popular and hardy of low ornamentals.   Fruits known to be eaten by pheasant, grouse, bobwhite, and other species of birds and by cottontail rabbit.   Provides good cover and food for many useful small forms of wildlife and roots serve well as soil anchor for controlling erosion.

Oregon Grape

Oregon Grape
Mahonia aquifolium
Height to 6 ft.   Evergreen shrub, with compound, hollylike, evergreen leaves whose five to nine leaflets are to 4 in. long, dark, shining green above, and lighter beneath with coarse, bristle-tipped scallops along the margins.   Branches are unarmed.

Ranges from British Columbia, south through California, growing along roadsides and as undergrowth at edges of wooded areas.   Also grown as ornamental widely, with many horticultural forms developed, some with variegated leaves.   Some 45 species native of North America, Central America, and eastern Asia.

Flowers, many, in a cluster arising from axils of leaves, clusters being erect in this species and relatively compact.   Sepals, nine.   Petals, six with basal nectaries yellow.   Fruit a dark blue, relatively small berry.   Flowers in evidence in May and fruits in late summer and fall.

Favors humid soil, sheltered from strong winds and direct sunlight.   Usually transplants easily.   Commercial propagation by means of seeds sown soon after they mature, or stratified in sand and sown in spring, or by cuttings of half-ripened wood in greenhouses or by means of abundant suckers in some species.

Plant is popular as an outdoor ornamental in regions where weather conditions are not too severe for its survival.   Superficially it gives effect of a blue-berried holly though it is not closely related to holly species.   It is the state flower of Oregon.   Serves as alternate host to wheat rust.

Order Ranales./Family Lauraceae

Sassafras
Sassafras albidum
Tree reaches height of 125 ft.   Today, usually found as a short tree or even shrublike.   Forms small colonies or clones.   Twigs rubbery, often bright green or reddish brown and, when mature, shining.   Bark of young stems thin, reddish-brown; old trunk, to 1½ in. thick, with deep, irregular fissures.   Leaves highly variable, of one, two, or three lobes, to 6 in. long and 4 in. wide and red, yellow, or orange in fall.   *S. officinale* of some authors.

Sassafras

Found in dry, sandy soil from Maine to Florida and west to Texas, Iowa, Michigan, and Ontario but frequently suffering badly from winterkilling in the northern part of the range.   May form considerable thickets and frequently produces suckers.

Flowers to ¼ in. broad, in loose, open clusters, with stamens equal to calyx segments and ordinarily opposite them.   Fruits appear at ends of thickened stems and are bright blue but when immature surrounded by remains of scarlet calyx, to ⅓ in. long.

Wood soft, brittle, weak, brown, coarse, but light-colored in sapwood, weighs 31 lb per cu ft.   Bark of roots particularly rich in aromatic oil that gives plant its character.   Wood may be used for fence posts, rails, and in cooperage but bark has a more popular use.

Bark stripped from young roots may be used fresh by boiling to make delicious tea or may be dried to be used later.   With sugar flavoring, it is delicious.   Oil is used in flavoring candies, tobacco, gum, soaps, perfumes, and medicines and is one of sources of artificial heliotrope.   Oil most commonly found in all bark and in pith.   Planted to a limited degree as an ornamental in the Eastern States.

Spicebush
Lindera benzoin
Shrub to 15 ft high, nearly smooth, with greenish, spotted, rubbery twigs that are highly aromatic.   Leaves alternate, entire, slightly hairy beneath, relatively thin, to 5 in. long and all parts fragrant when crushed.   Whole plant appears to branch grotesquely but attractively.

Found in moist woodlands, thickets, and by streams or sometimes planted as ornamental ranging from Maine to North Carolina and west to Tennessee, Kansas, Michigan, and Ontario ascending in Virginia to 2,500-ft elevation.   Not closely related to plant *Styrax* that produces benzoin of commerce.

Spicebush

(Continued)

Flowers yellow, appearing before leaves, fragrant, to $\frac{1}{10}$ in. across but in clusters that give a larger effect. Some flowers staminate, some pistillate; on separate bushes or some with both stamens and pistils. Sepals, six, dropping off early. Stamens, nine. Fruit oval, red, stoned, fleshy, aromatic.

Bark and fruit most commonly used as source of aromatic, tonic, astringent oil that has some reputed medicinal properties; certainly it is pleasing to sense of smell. Flowers appear in March to May and fruits from August through September. In autumn, plant has attractive, clear yellow leaves.

Woodsmen welcome plant when building fires since it is one of few plants whose wood will burn well while still green. It bears common names of snapwood, benjamin bush, wild allspice, and spicewood as well as spicebush. There are at least a half dozen other species known from different parts of the world.

Bloodroot

Order Papaverales./Family Papaveraceae

Bloodroot
Sanguinaria canadensis
Height to about 1 ft. Smooth, with leaves apparently coming from ground. Petioles to 14 in. long. Leaf blade to 1 ft broad and 7 in. wide, five- to nine-lobed, with conspicuous cleft at free end. Leaves curved with underside outermost. Rootstock thick, juicy, red, horizontal.

Found in rich woodlands from Nova Scotia to Florida and west to Alabama, Arkansas, Nebraska, and Manitoba, ascending to 2,500-ft elevation in Virginia, sometimes surviving after woodlands have been cut but ordinarily disappearing when too exposed to sun.

Flowers borne singly at end of stem that exceeds leaves, white, to 1½ in. wide, rarely with two flowers to a stem. Sepals, two. Petals, 8 to 16, shed easily and early, closing or opening on disturbance. Stamens, numerous, yellow. Fruit single-celled, narrow, of two parts to 1 in. long. Flowers April and May.

Red juice contains alkaloid sanquinarine which is toxic, causing vomiting, a burning sensation in mucous membranes, faintness, vertigo. Some commercial value for the drug but it is negligible.

One of most popular spring flowers but also most disappointing because while flowers look attractive before they are picked, they almost immediately close or lose their petals and must then be discarded. Plants may be set out in wild-flower gardens and in proper soil succeed in giving showy white flowers early in season.

Celandine

Order Papaverales./Family Papaveraceae

Celandine
Chelidonium majus
Height to 2 ft. Weak herb with a pungent yellow juice, sparingly hairy. Leaves to 8 in. long, thin, almost twice compounded, with segments usually rounded or lobed. Leaf bases expanded and sometimes partly clasping stem. Veins in leaves conspicuous and relatively large.

Naturalized from Europe but established well from Maine to North Carolina, west through Pennsylvania and Ontario. Common along roadsides and steep embankments, in rather shaded damp waste places, or sometimes in woods or at their edges.

Flowers to ¾ in. across, in loose open clusters that arise from axils of leaves on slender yellow stems. Sepals, two. Petals, four. Stamens, many. Fruit an elongate, almost linear but coarsely roughened capsule that frees many shining smooth crested seeds. Flowers from April through September.

Juice and other parts of plant contain alkaloids chelidonine, protopine, and chelerythrine and may be used with care as a purgative or as a diuretic. Juice may poison skin of some individuals so children should be discouraged from painting themselves with it.

In spite of poisonous properties, plant is useful as an ornamental. It requires little care and once established maintains itself. Grows in areas not suitable to some other species. Known as tetterwort, killwort, wartwort, and swallowwort.

Oriental Poppy

Oriental Poppy
Papaver orientale
Herbaceous perennial that reaches a height of to 4 ft, appears to be stout, and is well covered with stiff hairs. Leaves to 1 ft long, deeply cut into sharply toothed or lobed segments that are relatively narrow and uniform and usually pointed at free end.

A native of Mediterranean region east to Persia. Widely grown in gardens as ornamental. Other ornamental poppies include Iceland poppies *P. nudicaule,* corn poppy *P. rhoeas,* and the opium poppy *P. somniferum.* All are to be found in gardens in various parts of the world.

Flower to over 6 in. across. Petals, six, broad at free end, narrowed at base, scarlet with black area at base. Stamens, numerous. Fruit a capsule to 1 in. long, with a row of openings along upper edge from which are shaken enormous numbers of small seeds. Some poppies annual but most biennial or perennial.

Opium poppy, which differs from Oriental in having clasping stem leaves, is source of opium of commerce, which is collected from milky juice found in unripe fruits. It is source of many alkaloids such as morphine, codeine papaverine. Misuse of opium and its derivatives has resulted in addiction in human beings. This has prompted control legislation at state, federal, and international levels.

Opium juice is usually sedative. This plant has probably done much to shape world history, particularly in the Orient. Of use only in the garden, since flowers wilt quickly when cut, and the season is normally short. Massed in a garden with flowers of a single color effect is hard to beat.

California Poppy

California Poppy
Eschscholzia californica
Height to 2 ft. Erect, spreading, or sprawling to form rather compact mats. Leaves long-petioled, alternate, divided into slender parts, those on stems being smallest and most slender. Plant pale green and smooth.

Native of California and Oregon but found wild from Lower California to Columbia River and east through Nevada, Utah, New Mexico, and Arizona. Escaped and naturalized in central Europe and grown rather widely as a hardy garden ornamental.

Flowers yellow to deep orange shallow saucers spreading to 3 in. wide when open and responding conspicuously to sun. Petals four, inserted on receptacle with numerous stamens. Fruit a slender capsule to 4 in. long, that splits into two parts, freeing seeds.

A perennial grown largely as an annual, this species often blooming even after first killing frost. If protected in severe winters, plants may survive in North and give early flowers in spring. It self-sows and once established may prove reliable as a border plant where bright showiness is sought.

Chief use is as an ornamental. It is state flower of California. Plant was named after the Russian naturalist Eschscholtz, who found it in 1815 in great profusion on sand where San Francisco now stands.

Dutchman's-breeches

Order Papaverales./Family Fumariaceae

Dutchman's-breeches
Dicentra cucullaria
Height to 10 in. Leaves and flower stalks arise from a common point close to the ground. Plant delicate, immaculately smooth, and pale graygreen. Leaves slender-stalked with finely divided parts that may be almost linear. Underground parts bulbous but not like yellow tubers of squirrel corn.

Often very common in rich woodlands where there is deep leaf mold, but mostly over lime soils, surviving on more sterile soils than squirrel corn or growing with it. Ranges from Nova Scotia to Minnesota, south to Missouri, Kansas, and North Carolina. In Virginia found to elevation of 4,500 ft.

Flowers white, nodding on ends of drooping stalks, with two spreading trouser-leg-like spurs, usually with point of spurs uppermost. Flowers in April and May about 1 week before squirrel corn. Sepals, two, scalelike. Petals make the "breeches." Stamens, six, in two sets opposite outer petals. One pistil. Seeds, 10 to 20.

Has common name of little blue staggers, possibly referring to poisonous properties of leaves that include toxic alkaloid cucullarine. More concentrated in this species than in squirrel corn. May be fatal to cattle, effects appearing to 2 days after eating and including trembling, labored breathing, frothing, convulsion, glassy eyes.

A most attractive spring flower, harmless to the touch but not to be eaten. Normally not eaten by cattle if other more suitable food is available and does not last late into the season. Pollination by long-tongued bumblebees, shorter-tongued honeybee visitors merely eating pollen but not reaching nectar. Worth protecting as a wild flower.

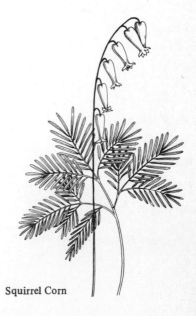

Squirrel Corn

Squirrel Corn
Dicentra canadensis

Herbaceous perennial. Leaves almost identical with those of Dutchman's-breeches though possibly with more white bloom on undersides. Leaves all basal and arising from buried rootstock that bears numerous clustered spherical orange tubers quite different from those found in Dutchman's-breeches.

(Continued)

181

Bleeding Hearts

Found in rich woodlands where soil is deep, from Nova Scotia to Virginia and west to Tennessee, Missouri, Nebraska, Minnesota, and Ontario, or slightly more restricted than Dutchman's-breeches. They may grow side by side in same woodland and often form considerable patches.

Flowers nodding, on short stems branching from a rather erect slender stalk that reaches a height of 1 ft, to ¾ in. long, broadest at base but without spreading spurs characteristic of Dutchman's-breeches, white or greenish-white, slightly fragrant. Inner petals more crested than those of Dutchman's-breeches.

Like Dutchman's-breeches leaves contain poisonous alkaloids though not in such great concentration. Since this poison is also found in tubers, one should not experiment with eating them. Cattle may eat leaves when other forage is not available and become violently though not fatally ill.

Plant is attractive spring flower and worthy of protection. Best in its natural setting and so might well be left unpicked. However, picking only the flowers has little bad effect on the general economy of the plant.

Bleeding Heart
Dicentra spectabilis
Height to 2 ft. Stems leafy, the leaves being alternate, apparently doubly compound or deeply cut, light transparent green and showing silvery under water; usually in shape of a broad fan. Substantial underground parts.

Native of Japan, where it was discovered in middle of nineteenth century by Robert Fortune; introduced into English gardens about 1847. Related to our native squirrel corn and Dutchman's-breeches, as should be evident from appearance of flowers and leaves. About 15 species of *Dicentra* in Asia and America.

Flowers like conventional red hearts, hence the name. In this species arranged along a single bending stem. Sepals, two, small and scalelike. Petals, four, with two outer spurred. Stamens, six, in sets of three. Pistil, one, ripening into a two-valved capsule that frees crested seeds when it breaks.

Prefers rich soil but does best when it is light. Propagation may be by separating the underground parts or by seed. Requires a good supply of moisture and if this is available may flower through summer. Young shoots normally appear in May but may be forced under glass to begin development earlier.

A most popular plant in old-fashioned flower garden. Children delight in pulling flower apart and giving parts names based on their shape, such as rabbits for the red outer petals, a harp, a pair of glasses, and a bottle—the pistil in the center.

Sweet Alyssum

Order Papaverales./Family Cruciferae

Sweet Alyssum, Madwort
Alyssum maritimum, Lobularia maritimum
Stems sprawling, to height sometimes of 1 ft, weak and rising at tips but so numerous as to form a relatively compact mass. Leaves slender to narrow paddle-shaped, generally pale green and poorly covered with grayish hairs, giving plant gray-green appearance. Roots fibrous.

Native of Europe but widely grown in America and other parts of world where it favors loose rich well-drained soil. Also does well on compact soil. Related to golden tuft *A. saxatile* or madwort, known also as golden alyssum, with yellow flowers that appear as mats in spring.

Flowers pink, white or pale yellow, with four petals which are little longer than four green persistent sepals. Fruit small and many-seeded. *A. saxatile* grows taller and forms larger head of flowers. Plants lend themselves readily to separation; hardy under such treatment.

Seeds sown in early spring indoors or out. Little plants transplanted to stand 6 to 8 in. apart so that they will make matted rows of white or yellow flowers. If cut back after blooming, will show new growth and again bear flowers. Seeds may be sown at different times.

One of most popular and easily grown border plants. Related *A. saxatile* is perennial and popular where early yellow mat effect is desired along foundations of houses. Seeds sown in August should produce flowers in early spring. Once believed useful cure for hydrophobia. Also, successful rock-garden plant. See generic name *Alyssum, a* meaning "not" and *lyssa* meaning "rage"; i.e., "without rage," or serene, bland.

Shepherd's Purse, Shepherd's-Bag, Pepper Plant
Capsella bursa-pastoris
Herb. Height to nearly 2 ft, with slender, tough, sparsely leaved, openly branched stems commonly curving upward from leafy rosette. Leaves alternate, varying from slender upper clasping forms to rather deep-toothed basal ones. Root a deeply penetrating taproot.

Found in gardens, along roadsides, in waste places, usually where there is sun and where it is not too wet, often where ground is firmly beaten.

Shepard's Purse

Naturalized from Europe and widely established through America and in other parts of the world.

Flowers small and inconspicuous. Stamens and pistils about equal in length, permitting self-pollination when insects fail to visit. Flowers appear through most of year but are less conspicuous than three-cornered flattened fruits. Fruits two-celled and crowded with reddish seeds. Plant may produce 2,000 seeds. Seeds have typical mustard-family taste when eaten.

Can withstand severe drying and cold and because of rapid growth quickly fills unoccupied ground. Seeds may remain vital and dormant although buried many years. Seeds or fruits may be eaten by birds. Whole plant may be eaten by chickens and by other animals. Many rosettes may survive winter.

Plant is common salad green but inferior to many others easily collected. Leaves may be boiled or eaten raw. Tender blanched leaves taste like cabbage and young plants make excellent sandwich filler. Hoeing cuts taproot. May be controlled by spray of iron sulfate or copper sulfate; plant cannot withstand much competition. Also called Case Weed, Pick-purse.

Pepper-grass

Peppergrass
Lepidium densiflorum
Stems to 18 in. high, well-branched and spreading from a compound erect stalk. Leaves alternate, slender, usually toothed and nearly scentless but somewhat peppery to the taste. Root system a slender vertical deeply penetrating taproot with branches of minor size. Annual or winter annual. There are several other species, introduced and native, with very similar properties.

Common along roadsides, in gardens, lawns, and waste places frequently where ground is hard-packed as on playgrounds. Most conspicuous in fall but may be found green throughout year. Ranges from Maine to Ontario and California, south to District of Columbia and Texas. Native of Eurasia and introduced comparatively recently.

Flowers inconspicuous, with petals usually lacking. Stamens, two. Pistil about length of stamens, which makes self-pollination a simple and effective matter. Seeds orange-brown, common as a seed impurity, expelled from fruit by an explosion. Flowers June through October. Seed to seed in a few weeks.

Not commonly found in shade or where there is too much moisture. Somewhat sticky seeds adhere to wet animals and take up water quickly when planted, thus getting a start on other plants. Low rosettes of many leaves may preserve some plants through winter. Acts as host plant in July and August for black cherry aphid *Mysus cerasi*.

Plant used occasionally as a salad. Seeds are used as a poor substitute for canary seed for feeding caged birds but do not induce desired behavior in such birds. Plants serve to slight extent as soil anchor in areas not occupied easily by most plants. Controlled by hand-pulling individual plants and early and repeated cultivation.

Radish

Radish
Raphanus sativus
Height of flowering stalk to 3 ft. Leaves stalked, roughened with sharp hairs, or smooth, with lateral segments paired, and terminal segment largest of all. Root thick, fleshy, variously colored depending upon variety, with a sharp taste which makes it delicious if it is grown rapidly. Annual or biennial.

Native of Europe and Asia, and under cultivation. Prized in days of the Pharaohs and ancient Greeks. Grown commercially in United States, particularly in California, Mississippi, and other Southern states. Grown in most gardens for home consumption. Best in light soils.

Flowers white to lilac or blue, with four petals appearing after flowering stalk has developed and after root has stored its maximum of food reserve. Fruit about 3 in. long, rather spongy, contains 1 to 6 seeds; has rather long slender beak and does not split to free seeds naturally. Seed, round.

Roots may be ready for harvest 3 weeks from time seeds are sown. Roots vary greatly in color, shape, size, season of maturity, and texture of the flesh. Varieties are generally classified according to the time the roots require to attain maturity. Radishes are grown in most home gardens, in particular; early spring gardens. Best practice calls for several plantings at intervals of 10 days, so new fresh roots will be coming to maturity for home use. Winter varieties should be planted in late summer and may be stored for use in winter. When planted for commercial crop, rows are 12 in. apart; 12 lb of seed per acre.

Where root maggots and aphids can be controlled and a quick growth can be guaranteed, radishes may make a good cash crop. They are tied in bunches of 6 to 12, shipped like spinach, while winter varieties are shipped like turnips with tops removed before they are put into storage. Slow-growing plants are tough and strong.

White Mustard

Field Mustard

Rutabaga

White Mustard
Brassica alba
Height to 6 ft. Stout annual, rather sparsely covered with hairs. Leaves oval in outline but usually deeply divided, even to midrib, into 1 to 3 pairs of notched, angled lobes. Stem leaves long-petioled. Upper leaves without petioles. Stem branched freely in upper parts. Known in some locations as charlock or senvre or kedlock.

Found in waste places widely distributed either as an established weed or as an escape from cultivation. Introduced from Europe and is a native of Asia. Essentially a plant of temperate parts of world. Not common in Southern Hemisphere.

Flowers yellow, to ¾ in. across, borne on rather stout stems, developing fruits that are slender, nearly round in cross section, tipped with long flat swordlike beak that may be as long as or longer than rest of fruit. Seeds found in a single row in cells and not winged or margined.

Seeds yellow on outside and white within; contain, among other things, a mucilage, some fixed oils, a glucoside (sinalbin), and proteins. White mustard seed has a oil content of from 26 to 30 percent, as contrasted with 30 to 40 percent in black mustard. White mustard is not major source of table mustard.

Leaves used for greens. Ground seeds mixed with water and glucoside break down through activities of enzyme to yield nonvolatile sulfur compound. Used mostly as a medicine or condiment, oil as counterirritant (mustard plaster), lubricant, or illuminant. Popular mustard of England.

Field Mustard
Brassica kaber (arvensis)
Height to 2 ft. Erect, annual, with spreading branches. Stems and leaves well-covered with bristly stiff hairs that may be scattered in some spots. Leaves highly variable but likely to lack extensive lobing and divisions found in most other closely related mustards; more lobed than cut to midrib.

Native of Europe and widely distributed as a common weed that has become naturalized in fields and in waste places. Grows too abundantly mixed in with more desirable plants and may have been introduced through use of poor seed.

Flowers yellow, to ⅔ in. across. Fruit smooth or bristly, spreading or turning upward, with constrictions obvious around seeds, to ⅔ in. long and 1/12 in. wide, tipped with flattened, drawn-out, sometimes one-seeded beak that is to ½ in. long. Conspicuous nerves on outside of fruit. Flowers from July through August.

Common mustard of grainfields. Disliked by most grain growers because of competition plant offers. Hairy surface makes it possible to use some sprays as a control and if these are properly applied they may be selective.

General control measures are careful inspection of seed to avoid its inclusion, spring weeding with harrow when drilled grain is about 4 in. high and sun is hot enough to kill injured plants, and hand pulling where this is possible, particularly before new seed crop is produced and freed.

Rutabaga
Brassica napobrassica
Smooth biennial, much like turnip but with a more elongate root, denser flesh, and tubers tapering downward from lower half or third. Flesh usually yellow but sometimes white. Leaf-bearing crown with leaves like those of rape except that stem leaves are like those below but smaller.

Grown as a crop in most temperate zones where conditions are suitable. In Canada and England, known as "Swedish turnip" or "turnip-rooted cabbage." In northern United States grown most commonly as fall crop but may be grown also in spring.

Flowers much like those of cabbage except that stems bearing fruits are stout and spreading, rather than ascending, and beak of fruit rarely exceeds ⅜ in. in length, instead of being slender. In fact, whole fruit of rutabaga is shorter and more compact.

For fall crop, plant seeds in midsummer preferably in deep fertile loam. In South, sow from July through September in rows 18 in. apart, thinning to about 6 in. apart to allow for normal root development. Will withstand slight freezing but should be harvested before hard freeze comes. Stores successfully in cool cellar.

Of excellent food value, containing 11 percent dry matter, 9.3 percent digestible nutrients, 1.3 percent protein, 1 percent digestible protein, and potassium, nitrogen, calcium, and phosphorus in smaller amounts and in about that order. In some regions rutabagas are grown regularly for feeding to livestock.

Rape, Colza
Brassica napus

Height to 3½ ft, branching, purple at base, with leaves to over 1 ft long and half as wide, mostly smooth though sometimes with stiff hairs sparingly placed. Lower leaves lobed and contracted to base into winged petiole. Upper leaves clasping stem with bases. Root, slender.

Common variety is dwarf. Annual type is grown for seed in Europe, India, and China. Biennial is commonly grown for forage or for human consumption in Europe and in America. Sometimes occurs as an escape from cultivation.

Flowers light yellow, to ¾ in. across, on slender spreading stems. Fruit to 4 in. long, with slender beak. Seed may be sown in Canada by June 15. In South, fall sowing is best. Seeds planted ½ in. deep, broadcast, drilled in rows, or mixed with oats or other crops. Becomes bitter as plants mature.

When broadcast, 4 lb of seed needed per acre; when drilled in 24-in. rows, 2 lb will do. Sown with grain when grain is to 2 in. high and with corn at time of last cultivation. 1 acre can yield to 50 tons of green feed from rape but 10 tons is a better average.

Rape grown for seed for oil in Europe, India, and China. Good field of pasture rape can support 12 sheep per acre for 1 month or 24 hogs for same time. Cattle fed too much rape may bloat, and swine may have similar trouble. Salt should be available when rape is fed, and dairy cattle should be fed rape after milking rather than before.

Rape, Colza

Black Mustard
Brassica nigra

Height to 10 ft. Branches freely and widely, stems being rather coarse, almost smooth or with stiff hairs. Leaves highly variable, usually with stiff bristly hairs, lower ones not greatly developed but terminal lobe usually large, with margins notched or toothed. All leaves well-petioled.

Naturalized from Europe and escaped as a common weed except in Northwest. Native originally of Asia and grown rather extensively in California, England, Germany, Holland, Austria, and Italy, possibly the best coming from Trieste area.

Flowers bright yellow, to ⅓ in. across, in twiglike open clusters crowned by unopened buds. Fruit under 1 in. long, tipped with a conic beak and closely pressed to supporting stem, four-sided. Seeds dark brown outside, yellow inside.

Chemical content much as in white mustard, but essential oil is most powerful and can blister the skin and injure membranes of eyes and nose. Oil content of black mustard seed is 30 to 40 percent, or more than that of white mustard.

Leaves used for greens. Glucoside (sinigrin) from ground seeds yields volatile oil that gives fragrance and flavor to mustard used on table. Black mustard favored over white on European continent. Used in medicine, in soap making, in preparing sardines, pickles, salad dressing. Prepared for table use by adding salt and vinegar.

Black Mustard

Cabbage
Brussica oleracea cv. 'Capitata'

Leaves in center form compact head, leaves themselves up to 1 ft across and nearly circular; the outer coarse, green, and more or less flat; the inner paler, more tender, and making the head. Varieties include Brussels sprouts, cauliflower, broccoli, and sprouting broccoli. Root relatively deep. Biennial or perennial.

Grows wild on chalky seacoasts of England, in Denmark, in northwestern France, and in Greece. It was introduced into cultivation in European gardens in ninth century. Known to history back to 2500 B.C. Favors a cool, moist climate, so in South is grown in winter. Has little soil preference but does best in sandy loams.

Flowers in long cluster, whitish-yellow; flowers sometimes produced only in plants that have been subjected to low temperatures. Plants not cooled known to produce three heads in 1 year and six heads in 2 years. These plants then taken to a cool area produced flowers and seeds as would normally have happened. 2 months' cooling best.

In North, seeds sown in hotbeds in January, transplanted to 2 in. apart, then transplanted into field. For late crop in North, seed sown 6 to 8 weeks before transplanting. 1 lb of seed provides for 4 acres if seedlings grown in greenhouse. Should be "hardened" to withstand temperature to 17°F for short duration. Storage is most successful at 32°F.

Probably most important crop of genus *Brassica*. Grown in market gardens throughout United States, but winter production has increased greatly in South. States leading in acreage in prewar years were Texas, New York, Wisconsin, Pennsylvania, and California. Average price per ton is determined by supply and demand. Shrinkage varies from 7 to 27 percent with temperature, ventilation, and time.

Cabbage

Kale

Cauliflower, Broccoli

Brussels Sprouts

Turnip

Kale, Borecole
Brassuca oleracea 'Acephala'
Leaves thick but not crowded into compact heads as in cabbage or with other modified parts. Some kinds have cut or frilled margins while others are more regular in shape. General color blue-green. Kales probably more nearly like original wild form than are most other varieties. Giant kales 9 ft high.

Found wild today on cliffs of southeastern England, so were probably developed from similar plants. Now grown widely where climate is suitable. Popular varieties include Scotch curled and Siberian. Many general types include tree kales, collards, and curled kales.

Flowers are, in essence, like those of cabbage and other close relatives. Seeds sown broadcast or in rows to 18 in. apart. For market use, plants are thinned to 6 in. apart, thinnings generally being used for home consumption. Clean cultivation essential to success in raising good kale.

Grown more in other countries than in United States. Good resisters of heat, cold, and drought. Frost improves flavor. Kale will grow in any type of soil but application of 20 tons manure to acre or 1,500 lb of fertilizer of 4 percent nitrogen, 8 percent phosphoric acid, and 4 percent potash should give excellent results.

Kale is grown as winter crop in South and as late fall crop in North. In England, stout stems of tree kales are used for rafters and for canes. Kales are also known as borecole, cow kales, kitchen kales, and so on.

Cauliflower, Broccoli
Brassica oleracea cv. 'Botrytis'
Low, with broad dense head grown over by leaves. Part that is eaten is really thick short stalks of undeveloped flowers and their supporting bracts. In a highly desirable cauliflower center is white. In broccoli heads are smaller, leaves larger, and whole plant greener.

In France and England broccoli is favored over cauliflower, which is more popular in America, where Long Island and California have been centers of cauliflower culture. Asparagus, or sprouting, broccoli differs from regular broccoli in not producing a solid head.

Cauliflower seed is grown in Europe and must be handled with care. For early crop, plants are started under glass, and after frost danger has passed set out where they will have no further disturbance and where moisture, temperature, and other conditions are favorable.

Best soil is a rich loam that does not dry easily. Therefore, sand is not ordinarily desirable. Rotted stable manure is excellent fertilizer or a commercial fertilizer of 4 percent nitrogen, 8 percent phosphoric acid, and 6 percent potash should be used at about 1 ton per acre depending on original fertility.

Plants should be in position for reaching maturity 4 months before killing frost. Cheesecloth covers protect plants from a cabbage maggot fly that may be very destructive. Considered more delicate and easy to digest than cabbage.

Brussels Sprouts
Brassica oleracea cv. 'Gemnifera'
Height to 3 ft, stems being erect and unbranched. Leaves short and about as broad as long, often lobed near base. Buds, or small edible "heads," to about 1 in. diameter and borne in axils of leaves at lower end of stem. Stem, stout. Leaves best developed at crown.

As scientific name implies, this is merely a cultural variety of species to which cabbage belongs. Relationship is obvious. Not cultivated in America to extent they are in some other parts of world. Popularity here is localized but apparently increasing. Long Island has long been an important center.

Seeds sown in late spring or early summer and transplanted to permanent bed when about 6 in. tall, 18 to 24 in. apart, in rows 2½ to 3 ft apart. When small heads begin to crowd lower leaves, they are cut off and this continues on up stem as plant develops. A good yield from a plant is about 1 qt.

In mild regions, plants may be left in open all winter but where weather is too severe plants are lifted with earth clinging to roots and set in moist sand in a cool cellar, there to continue some production. Brussels sprouts are more delicate than cabbage but are cool-area plants nevertheless.

For serving, Brussels sprouts are cleaned of larger outer leaves, washed, placed in boiling salted water, cooked quicky in uncovered dish, and served hot with butter or other dressing. Market not so strong as for more generally liked and more easily produced relatives, such as cabbage and turnips.

Turnip
Brassica rapa
Leaves clustered around base, with clasping bases, or along flower stalk. Those near base commonly narrow in general outline and softly prickly. Root, a taproot, usually white-fleshed and not produced into slender

crown at top but rather conspicuously flattened from above and often colored above ground.   Biennial.

Grows wild in Russia and Siberia and may have originated there. Known to have been introduced into America in 1609 in Virginia.   Essentially a cool-season crop, being raised in North in summer and in South in winter or as an early-season crop in North occasionally.

Flowers small, under ½ in. long, yellow; those with open flowers usually closely crowded and topping those not yet open.   Fruits about 1½ to 2½ in. long, with a long slender beak at free end, like little spheres; germinate quickly as is case with many relatives of this plant.

Seeds sown where plant is to mature; if planted early in season, plant may be grown as an annual.   Do not do well in hot weather.   Seeds planted ½ in. deep in rows 12 to 24 in. apart at rate of 2 lb to acre; plants thinned to stand from 2 to 6 in. apart in rows.   Removed plants are used as greens. Young turnips are bunched and sold or mature roots stored.

Make an excellent food for persons without good storage facilities, because they grow easily, store well, have a good food value and require relatively little care.   In North, popular varieties are Purple Top Globe, White Milan and White Flat Dutch; in South, Seven Top.

Kohlrabi

Kohlrabi
Brassica caulorapa
Conspicuous because of swollen tuber that appears above ground as a sphere 4 in. or more in diameter, with a typical cabbage root beneath and long-petioled leaves growing from and above tuber.   Leaves to 10 in. long, of which ⅓ is petiole and remainder blade, rounded, with a few lobes at base.

Popular garden plant grown in America for stock consumption but in Europe largely for consumption by human beings.   Species differs primarily from *B. oleracea* in tuberous stem and small-bladed leaves.   Some persons consider it as var. *gongylodes* of *B. oleracea.*

Plants started in open or in hotbeds.   Rows should be 12 to 30 in. apart, with plants 6 to 8 in. apart in a row.   Usual practice is to start plants at 2-week intervals so as to have a steady group of plants reaching maturity through season.   Cultivation is essential in early stages.

Naturally a cool-weather plant requiring rich, well-drained loam.   Early quick-maturing varieties are recommended for table use and later forms for feeding to stock.   Where growth is slow, tuberous portion is likely to be too tough for most persons to enjoy it.

Preparation for table use calls for peeling the tuber, dicing it into ½-in. cubes, then cooking it and serving it much as cauliflower is prepared. For marketing, plants are sold in bunches of 3 to 5, with leaves usually left on.

Hedge Mustard

Hedge Mustard, Tumbling Mustard
Sisymbrium altissimum
Height to 6 ft.   Branches spreading, with tips curved upward and stiffest toward base.   Fuzzy, with simple hairs, pale green.   Leaves alternate, deeply cut but with end lobe by far largest and with some of lateral segments sometimes reduced to slender strips.

Common as a weed throughout northern United States and southern Canada but most troublesome in Northwest and on Pacific Coast.   Introduced originally from Europe and now grows in almost any good grain soil, in waste places, or along roadsides.   Most evident in late summer and winter.

Flowers pale yellow, about ¼ in. across, borne on stems of about same length.   Fruits develop into slender, diverging, cylindrical pods about as thick as stems that bear them and break into two sections, each with many seeds that are usually flattened on one side, brown, to ⅟₂₅ in. long. Fruit to 4 in. long.

Good cultivation will keep this weed in control.   If tops are left to mature, whole plant breaks loose and tumbles as a tumbling weed.   Fruit tips act as springs to help keep plant going and break to free new seeds which are broadcast in process.   Frequently found piled in fence rows.

Seeds of some hedge mustards are used as flavoring but apparently not seeds of this species.   Hand pulling, disking, and harrowing a few times will usually keep weed under control but it should not be allowed to mature fruits.

Rocket, Dame's Violet

Rocket, Dame's Violet
Hesperis matronalis
Height to 3 ft.   Erect, sparingly branched in upper parts.   Lower leaves to 8 in. long, tapering at base into a petiole, slightly fuzzy on both sides, in general lance-shaped with shallow teeth along margin.   Upper leaves smaller than lower and less likely to be petioled.   Biennial or perennial. Undoubled form resembles somewhat the rocket and is more common. Double form is illustrated.

(Continued)

187

Stock, Gilliflower

Water-cress

Horseradish

An escape from gardens found widely distributed along roadsides and in waste places from Nova Scotia to Pennsylvania, west to Iowa and Ontario; native of Europe and Asia. Seems hardy and well able to maintain itself with a minimum of care. Often found in garden spots of abandoned homes.

Flowers to 1 in. across, purple, pink, or white and fragrant with blades of petals widely spreading to make a most attractive and showy flower. Pods to 4 in. long, spreading or ascending, constricted about seeds so that their location is obvious in a ripe fruit. Flowers from May through August.

This is common old-fashioned garden plant. Does best in relatively moist deep soil, either in shade where it becomes a slender plant or in sun where it may be more substantial. Can grow successfully with other plants crowding it rather closely. Pollination apparently by insects.

Popularity is attested by its many common names such as dame's violet, dame's gillyflower, night-scented gillyflower, rogue's gillyflower, winter gillyflower, sweet rocket, rocket, summer lilac, and damask violet. Scientific name refers to fragrance of flowers in evening.

Stock, Gillyflower
Matthiola incana
Height to 2 ft. More or less woody particularly at base, densely felty. Leaves to 4 in. long, with narrow petiole-like base, blunt at free end, entire. Upper leaves smaller. Whole plant except flowers seems to be grayish and woolly. Biennial or perennial.

Native of area from Greece to southwestern Asia. Standard plant for old-fashioned gardens and has long been popular. Related evening stock *M. bicornis* is more open in its general shape and has pods terminating in two long horns, which do not appear on this plant.

Flowers in terminal open clusters, single or double, canary-yellow, white, rose, salmon, orange, blood-red, purple, light blue, or pink, about 1 in. long, fragrant, variable. Fruits to 2 in. long, thick, stout, and ending in a two-lobed tip. Considerable variation in time for maturity in different varieties.

Ten-weeks, or independent, stock is annual that blooms in about 10 weeks from time seeds are sown. Normally sown in February or March, transplanted to 1 ft apart. Various types include some that bloom through summer and autumn because of flowering side branches and some that produce a single large flower group.

Name gillyflower frequently applied to this common garden flower probably is a distortion of Julyflower. Ten-weeks stock of course refers to seed-to-seed period required in annual variety. Fragrance makes plant popular even with those who might not be impressed by its beauty.

Water Cress
Nasturtium officinale
Grows in more or less tangled masses of stems and leaves in shallow water, usually over mud. Stems weak, easily broken. Leaves of 3 to 9 segments, of which terminal is largest; attractive green, semisucculent, rather crisp. Roots appear at nodes of stem.

Commonly found in springs and spring-fed streams, sometimes at very edge of rather rapidly flowing water. Naturalized from Europe but thoroughly and widely established in America from Nova Scotia to Virginia west to California and Manitoba. Also found in Asia, South America, and West Indies.

Flowers borne at stem tips in rather open clusters, on short stems, to ⅙ in. broad, white, with petals twice as long as sepals. Fruits to over 1 in. long, spreading or curving upward on stems that may be equal to their length and with seeds arranged in two rows. Rather stout tip to fruit.

While plant is usually found in clear spring water, that water may be badly contaminated, so unless sure of the source, cress should be disinfected with some wash such as chlorazene.

This is one of best plants along trout streams for production of such troutfood organisms as fresh-water shrimps, water sow bugs, and the like. It deserves popularity as a salad plant. It should not be confused with the garden nasturtium (*Tropaeolum*) or with water hemlock, which may grow in same place and slightly resembles it but is seriously poisonous.

Horseradish
Armoracia rusticana
Height to 3 ft. Husky branched plants. Basal leaves on thick petioles, to 1 ft long, with blade as long again, rough, with crinkled edges or variously modified along margins. Upper leaves narrow, without petioles, with or without toothed margins. Root a solid coarse deep taproot, usually thick with horizontal branches.

Common in ditches, damp spots, and wet waste places particularly in northeastern United States and Canada, where it probably escaped from

gardens after being introduced from Europe. Also cultivated like field
crop where conditions are suitable. Does not favor heavy clay or light
sandy soils. Potash fertilizers are best. Grows best in the north temperate regions of the U.S.; states like Massachusetts, Montana, Washington.

Flowers borne in loose clusters at ends of stems and from leaf axils,
white, showy, to ⅓ in. broad, appearing in summer. Fruits oblong to
globular pods, with short beaks, borne on ends of upward-turning stems.
Can be most easily propagated by planting surplus root systems of harvested plants.

Roots, planted in furrow, 2 to 2½ ft apart, horizontally or vertically,
preferably where fertilizer has been placed deep. May be grown with
cabbage as two plants have enemies that can be controlled similarly.
Digging should be done in spring or fall as summer roots are bitter. Ideal
roots are deep, straight, and uniform.

Roots should be ground soon after digging as stored roots deteriorate
rapidly. To grated root are usually added vinegar and sugar or even
mustard to make a standard condiment, considered essential by Europeans, particularly the British, in eating roast beef, broiled fish, or oysters.

Winter Cress

Winter Cress
Barbarea vulgaris
Height to 2 ft. Erect and somewhat branched upward. Lower leaves to
5 in. long, apparently of one large terminal lobe and one to four pairs of
usually opposite lateral ones, most of which are narrowest at base. Lower leaves well-petioled; upper without petioles or with petioles greatly
reduced.

Common weed in gardens, fields, and waste places. Naturalized from
Europe. Well-established from Labrador to Virginia and in many places
west to Pacific Coast though these stations are not necessarily united with
each other. Perennial mustard. Commonly turns early spring fields
bright yellow. There are a number of closely related species with similar
habits.

Flowers bright yellow, to ⅓ in. across, with youngest at top of developing
flower cluster, borne on relatively stout individual stems. Fruits spreading or upward-reaching pods, somewhat four-angled, to about 1 in. long,
roughly six times as long as supporting stems.

Grows splendidly through warm periods in winter and so appears as
bright green rosette in gardens in early spring. Can be eaten on St.
Barbara's day in early December when young stems and leaves are tender
and about as bitter as best of dandelion greens. Bitterness is a bit too
strong for most persons.

As a potherb, this plant should be cooked in two to three waters, fresh
water helping to eliminate bitterness. Best to put plants directly into
boiling water after they have been thoroughly washed. Salt may well be
added and greens should be eaten soon after they have been prepared.
Also used as a constituent of green salads.

Honesty, Moonwort

Honesty, Moonwort
Lunularia annua
Height to 4 ft. Rather stout, somewhat branched. Leaves: earliest, opposite; later may be alternate; heart-shaped or triangular in general outline, and with well-developed petioles, margins distinctly toothed, veins
rather conspicuous. Root system annual and not extensive.

Introduced from Europe but sometimes established as an escape where it
is not too seriously disturbed. Sometimes found growing in cultivated
crops from southern Ontario through Connecticut, eastern Pennsylvania,
and New York. Most commonly seen as a garden flower particularly in
old-fashioned gardens.

Flowers purple, to 1 in. or more across, showy, in loose, relatively few-
flowered terminal or axillary clusters. Fruits to 2 in. long and 1 in. wide
but very thin, flat, and rounded at both ends to make a waferlike oval or
circular structure. Fruits turn to a straw color and are showy. Seeds
relatively few.

Related satinflower resembles this one but fruits are pointed at ends.
Plant is perennial rather than biennial or annual. In practice, honesty is
sown in late summer and rosettes form before winter comes. From these
come flowers next year. Flowering time May and June.

Grown as an ornamental garden flower. Dead tops collected and used in
making dry bouquets. Sometimes dried tops are sprayed with colored
paints. Known as "matrimony plant" and "pennyflower."

Pepperwort, Toothwort

Pepperwort, Toothwort
Dentaria laciniata
Height to 15 in. Erect, smooth or otherwise. Basal leaves develop after
flowering and are much like 2 to 3 stem leaves that appear opposite or
whorled and below flowers. Leaves to 5 in. across, usually deeply cut to
five parts, with sections themselves deeply cut and in this species narrow.
Tubers deep, white.

(Continued)

Spring Cress

Pitcher Plant

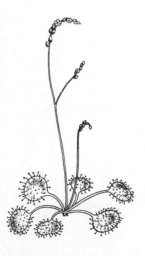

Round–leaved Sundew

Common in rich, moist, deep-soiled woodlands from Quebec to Florida, west to Louisiana, Kansas, and Minnesota. At least six species closely related to this one are found in approximately same range. In this species, tubers are rather spindle-shaped while in some others they are more continuous and less easily broken.

Flowers white or pinkish, to 1¾ in. broad, with backward-curving petals. Petals to ⅔ in. long. Sepals about half as long as petals. Flowers rather attractive. Pollination probably by honeybees, other small bees, and syrphus flies that feed on nectar produced. Fruit slender, to 1½ in. long, curving upward.

Tubers make a mildly flavored horseradish substitute which by many is favored over horseradish. Amount produced by plant is very small, good for an occasional nibble while on a woodland hike. If end of tuber system away from growing stem is eaten, plant need not be injured.

Some favor mixing root with vinegar but others like to slice it thin and serve it as part of a sandwich or mixed with some relatively tasteless wild plant food. Certainly it needs to be toned down to be appreciated to the full.

Spring Cress
Cardamine bulbosa
Height to 2 ft but usually much less. Branched or unbranched. Basal leaves oval or round on ends of long petioles, blade being to 1½ in. long. Upper leaves with or without petioles and to 2 in. long. Rootstock bears tubers. Perennial.

Common in wet meadows or at woodland margins from Nova Scotia to Florida, west to Texas, Minnesota, and Ontario, ascending in Virginia to 3,000-ft elevation. Frequently found near springs or where water is oozing from rocks in gorges and similar places.

Flowers white, to over 1 in. broad, with petals three to four times as long as sepals, white, and rather crowded along upper end of stem, with buds and blooms at very tip and fruits below. Fruits slender, to 1 in. long and narrowest at tip to conspicuously slender end. Seeds oval.

Rather inconspicuous wild flower that blooms from April to June, bears fruits later in season. Not sufficiently attractive to be sought by wildflower pickers or sufficiently aggressive to establish itself as a pernicious weed. Therefore, it is likely to survive where others might be destroyed.

Tops make reasonably good salad and rootstocks may be eaten at any time of year. Rootstock tastes likes a mild horseradish and may be used as such. Flavor is present also in young green stems and leaves. Older plant parts may be too strong to be palatable.

Order Sarraceniales./Family Sarraceniaceae

Pitcher Plant
Sarracenia purpurea
Height to 2 ft. Leaves spring from base and curve upward to height of 1 ft, are inflated and commonly well filled with liquid, may be purple- or green-veined, narrowed at base and broadly winged, with open upper end covered on one side with hairs pointing toward opening.

Found in peat bogs and similar acid situations from Labrador to Florida and west to Kentucky, Iowa, and the Canadian Rockies, with related *S. flava* extending range from Virginia to Florida and west to Louisiana. About eight species in North America.

Flowers large and borne at end of scape that reaches a height of 2 ft, nodding, deep purple or yellow, nearly spherical, to 2 in. or more across, with narrow-middled five petals curving over yellow pistil tip. Fruit, five-celled. Seeds, numerous.

Leaves are commonly well filled with liquid in which many drowned insects may be found. The plant produces enzymes which digest proteins. In spite of this capacity to digest insects, the larvae of at least one mosquito live comfortably in the water of *S. purpurea*.

Always of interest to nature enthusiasts. If given reasonable protection, it can recover from depredation. Physiological problem set by its leaves, an interesting gall frequently found on flower, and other characteristics make plant well worth knowing better. Official flower of Newfoundland.

Order Sarraceniales./Family Droseraceae

Round-leaved Sundew
Drosera rotundifolia
Height to 10 in. Leaves held relatively close to ground, expanded in terminal portion to round blade that bears on upper surface numerous sticky-tipped structures whose tips are capable of movement toward center.

Found in peat bogs and similar wet acid places or in wet sand from Newfoundland and Labrador to Florida, west to Alabama, California, and Alaska, though California stations are in mountainous areas.   In Catskill area in New York plant ascends to 2,500-ft elevation.   Also found in Europe and Asia.

Flowers to ⅙ in. long, borne along slender erect flowering stem that may be branched once and may bear to 25 flowers, each on a short stem about as long as flowers are wide.   Flowers open mostly in bright sunshine. Petals white to red, longer than sepals.   Fruit a many-seeded capsule.

Another so-called "carnivorous" plant.   Leaves capture insects and digest their bodies thus getting needed nitrogen.

Popular plant for field botanists, some of whom collect it for herbarium use or for experimentation.   Plant has such common names as lustwort, red rot, youthwort, rosa solis, eyebright, and dew plant.   Australia has more species of sundews than any other continent.   Shiny drops of secreted mucilagelike substance appearing on each hairlike structure thought to have suggested the plant's common name.

Mossy Stonecrop

Order Rosales./Family Crassulaceae

Mossy Stonecrop
Sedum acre
Spreading, branching mat, up to 3 in. high and of indefinite length because of rooting at joints.   Leaves alternate, to ⅙ in. long, overlapping, sometimes in six rows, light yellowish-green, somewhat succulent, thick, oval, without petioles.   Root system, a branching underground arrangement.

Escaped from cultivation on rocky exposures along roadsides from Nova Scotia to Virginia, west through Ontario.   Native of northern Asia and established in Europe.   Obtainable from most rock-garden supply houses and easily established in suitable environments.   At least 30 genera and 60 species in family.

Flowers about ⅓ in. broad, in clusters of a few without individual stems. Petals bright yellow, narrow, pointed and to four times as long as sepals; central flower of a cluster usually five-parted and others four-parted. Fruits spreading, slender-tipped, and to ⅙ in. long.

Representative of many interesting rock-garden plants or of plants that are grown indoors in pots for their live-forever qualities.

Group has little economic importance except possibly as holders of soil and moisture in crevices between rock exposures or for sale as rock-garden plants.   Yellow flowers of this species and the ease with which it grows make it universally popular.

Live–forever

Live-forever, Garden Orpine
Sedum purpureum
Height to 1½ ft.   Stems erect from horizontal underground system, stout, smooth, unbranched.   Leaves alternate, broadly oval, thickened, succulent, with or without coarsely toothed margins, to 2 in. long, narrowed at base or sometimes with short petioles; lower definitely larger than upper.

Found among rocks, banks, and particularly in old gardens or in areas adjacent to them where plants have escaped and established themselves; usually where there is some moisture.   From Quebec to Maryland west to Michigan and Ontario and in gardens elsewhere.   Native of western Asia but naturalized in Europe, whence it was probably introduced.

Flowers in densely crowded clusters that are to 3 in. long, with individual flowers about ⅓ in. across.   Petals twice as long as sepals.   Stamens, 10. Fruits, to ⅙ in. long, tipped with a slender point.   Blooms little, spreading itself more by freely branching underground parts.

Children may press and blow up the broken, succulent leaves to make "balloons."

An excellent old-fashioned garden plant that can be grown by anyone anxious to make a new garden quickly, but once established it may be difficult to eliminate.

Order Rosales./Family Saxifragaceae

Early Saxifrage
Saxifraga virginiensis
Height to 1 ft.   Leaves basal, in rosettes, to 3 in. long or longer, with conspicuous midribs, narrowly margined petioles, and toothed or only slightly irregular margins.   Leaves appear in well-developed rosettes in fall ready to produce flowers in early spring.

Found in rocky crevices in gorges, by streams, or even in rocky woodlands.   From New Brunswick to Georgia west to Tennessee and Minnesota, up to 3,500-ft elevation in Virginia.   At least a dozen species to be found in northeastern United States, some of which range from Gulf to Arctic and west to Pacific.

Flowers borne at ends of branching stems in more or less flat-topped clusters of a few flowers, white, erect, to ¼ in. broad, with petals to about
(Continued)

Early Saxifrage

Strawberry Geranium

Foamflower

Bishop's–cap, Miterwort

twice length of sepals, there being five of each. There are 10 or some-times 15 conspicuously yellow stamens. Parts of pistil spread apart at tips.

One of popular spring flowers of gorge banks that is fortunately too small to be considered worth plucking and therefore survives destruction at hands of flower collectors. Pollination from March to May may be ef-fected by early bees or by some early butterflies.

Some relatives of this species such as *S. pennsylvanica* are reputed to have roots of medicinal value but it is doubtful if this is of great importance.

Strawberry Geranium
Saxifraga sarmentosa
Height to 25 in. Leaves all basal, between round and heart-shaped, with very long petioles, with upper surfaces white-veined; red beneath, with upstanding hairs. Whole plant loosely hairy. Has running habit of ordi-nary strawberry plants, weak rootstock, fibrous roots. Perennial.

Native of Asia, more particularly of China but widely grown as old-fashioned window plant. May escape and establish its own colonies in milder parts of world but is not hardy to North. Of half-dozen saxifrages cultivated for their flowers, this is probably most popular.

Flowers many, white, arranged along upper parts of erect and branching flower stocks. Petals unequal, two being much larger than other and upper three, longest being to ½ in. long or about four times as long as shortest, with acute tips, sometimes red spotted or pale pink. Stamens, 10.

Requires an abundance of sun and does best at temperatures between 60 to 70°F, with a moderate amount of water and a soil composed of ⅔ ordinary garden soil and ⅓ well-rotted manure. Pollination may be ef-fected by butterflies. Few insect enemies or fungus pests.

An attractive plant grown not only for its pleasing small lacelike flowers but for beauty of its foilage. Spreading stolons make it easy to propagate without relying on flowering process to produce seeds.

Foamflower, False Miterwort
Tiarella cordifolia
Height to 1 ft. Leaves irregularly heart-shaped, with coarsely rounded toothed margins, long-petioled, five- to seven-lobed, with blade to 4 in. long, inclined to be smoother than those of closely similar bishop's-cap leaves, or with hairs on underside more definitely confined to ribs. Leaves rise from rootstock.

Common in rich woodlands where there is good moisture, ranging from Nova Scotia to Georgia, west to Indiana, Minnesota, and Ontario. In Virginia, found up to elevation of 5,600 ft. Six species in North America, of which five are in the West. Others range on to Japan and Himalayas.

Flowers borne at upper parts of erect stem; appear like little white bursts of bloom, the five white petals being exceeded in length by 10 conspicu-ous stamens whose anthers are yellow, with long white central pistil. Fruit divided at tip into two parts like a tiara.

One of popular early spring flowers too attractive for its own good. Ap-pears rather later than such flowers as bloodroot and hepatica and is at its best about time leaves of shade trees come into full shade-producing con-dition. Pollination probably by visits of bees, butterflies, and syrphus flies.

Lends itself readily to transplanting in wild gardens and is commonly found there. Listed in some wild-plant catalogs but doubtful whether plants are grown from seed or collected from wilds. Such a fine wood-land flower that it is worthy of every protection. Known as coolwort, false miterwort, gem-fruit.

Bishop's-cap, Miterwort
Mitella diphylla
Height to 18 in. Leaves mostly basal except for two on flowering stem. Basal leaves long-petioled, irregularly heart-shaped or three- to five-lobed, with blades to 2 in. long and with hairs scattered irregularly over both surfaces. Upper leaves in general similar but with shorter petioles and opposite.

Common in rich, moist woodlands from Quebec to North Carolina and west to Missouri, Minnesota, and Quebec and found up to 2,600-ft eleva-tion in Virginia. Four species in North America and Asia, of which this is by far most common in eastern North America, some others being decidedly limited in range.

Flowers borne along upper portion of erect flowering stalk that has two characteristic small leaves. Flowers white with lacy petals, with 10 sta-mens not so conspicuous as in foamflower. Fruits flattish, broad, and opening in upper parts, shaped somewhat like a bishop's miter.

Fortunately, this flower is so small that its beauty is not appreciated except by those willing to give it close examination. Were it large it would undoubtedly be so popular that it would be quickly eliminated from much of its range. Appears in midspring usually a little before foamflower and before tree leaves have developed fully.

This fine little plant is reputed to have some medicinal value. Common name currant-leaf is as appropriate as its common name bishop's-cap or scientific name *diphylla*. Equally appropriate is the common name fairy cup or fringe cup. Also called false sanicle.

Hydrangea
Hydrangea paniculata

Shrub or small tree reaching height of to 30 ft, forming dense round head. Leaves opposite, to 5 in. long, with toothed margins and drawn-out tips, somewhat fuzzy especially on veins on underside, with petioles but no stipules. May be rooted at joints.

Native of Japan and China but grown widely in suitable places particularly where there is enough moisture in a rich soil and where there is enough sun. *H. paniculata* var. *grandiflora* is one of hardiest of hydrangeas and therefore one of more popular in northern part of range. *H. opuloides* is not hardy in North.

Flowers in terminal, compact panicles often with marginal flowers sterile and exceptionally showy. Tiny inconspicuous fertile flowers in center have four to five petals and sepals, usually 10 stamens, and a pistil that develops into a fruit that splits freeing many small seeds. Flowering time August and September. Flower buds form in autumn.

Tree form may be developed by cutting back all but strongest shoot of many that may start from a vigorous plant growing in rich soil where there is enough sun and moisture. Some of best blooms are produced by a vigorous pruning of most of plant in fall or early spring, but such plants may need support for huge flower clusters.

One of truly showy plants of front yards in suitable climate. Related greenhouse or southern species have a ready sale as house plants available from florists. Best cuttings are made in February from plants forced for Easter, rooted in sand at 70°F in March, potted in May, hotbedded in August, and benched in December.

Hydrangea

Mock Orange
Philadelphus coronarius

Shrub. Height to 10 ft. Branching but with long wandlike shoots that make a graceful hedge plant. Leaves opposite, to 4 in. long, on short petioles, smooth above and slightly fuzzy beneath, with rather distant shallow teeth along margin, three nerved, rounded or narrowed at base, more pointed at free end. Common name, syringa, should be dropped as it is the scientific name of the lilac.

Common ornamental shrub, native of central and southern Europe but widely planted in America. Escaped from cultivation in many places in eastern United States. Grows well on steep slopes or in good garden soil. Some 30 species found in North America, Asia, and Europe of which some reach a height of 20 ft.

Flowers numerous, at ends of branches, in clusters, to 1½ in. across, creamy white, very fragrant, with four petals three to four times as long as sepals whose pointed lobes are longer than calyx cup is deep. Stamens very numerous and possibly half as long as petals. Fruit a dry capsule, flat-topped, splitting lengthwise.

Pruning should be done after flowering since flowers appear on wood of previous year. Propagation by layerings, by cuttings if desirable hybrid qualities should be maintained, or by seeds if hybrids are desired.

One of the favorite ornamentals for suburban and rural homes. Somehow we associate the flowers with graduation time since they are so commonly used for floral decoration then. The related *P. lewisii* is the state flower of Idaho.

Mock Orange

Common Currant
Ribes sativum

Shrub. Height to 6 ft, but usually lower when well-managed. Young growth rather densely fuzzy. Leaves three- to five-lobed, to 3 in. across blade, well-petioled, with margins distinctly toothed and veins conspicuous, usually heart-shaped at base. Twigs without spines at joints.

Native of western Europe. Established rather widely as cultivated plants or as escapes. Some 150 species in genus, these being found mostly in cooler and temperate parts of world. *R. nigrum*, black currant; *R. sativum*, red currant; *R. rubrum*, red currant; and *R. alpinum*, mountain currant grown for fruit. Other species, like *R. odoratum*, grown for flowers.

Flowers commonly in clusters of more than 4 and sometimes 10 or more, drooping, greenish-yellow, with a purple ring inside the stamens and outside the style. Fruit, berries that are usually red but sometimes white or
(Continued)

Common Currant

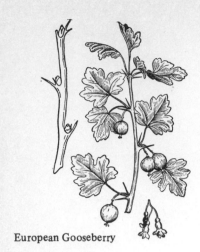

European Gooseberry

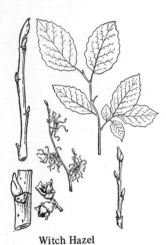

Witch Hazel

Sweetgum

even striped, succulent, tart, but definitely edible. Ornamental golden currant has beautiful yellow flowers.

Best near water in moist atmosphere. Can endure shade. Host to blister-rust fungus of white pine; therefore unpopular in some places. Propagated by cuttings taken in fall, put in sand for winter, set out in spring. Fruits borne on stem spurs 2 to 5 years old, black currant on 1-year-old wood. Stems that have borne should be pruned.

Fruits of currants were collected from wild species by Indians, stored and eaten later. Yield of cultivated plants may be to 400 bushels to acre. Fruits keep surprisingly well. Eaten fresh with sugar or more commonly made into currant jelly or jam.

English Gooseberry
Ribes grossularia
Shrub. Height to about 3 ft. Stocky, bushy plant whose stems are erect, ascending, or reclining. Usually with three stout spines at joints though sometimes even better armed. Leaves stiff, with nearly round blades that are to 2½ in. broad, with three to five blunt lobes and margins somewhat toothed, with inturned edges.

Native of Europe, north Africa, and southwestern Asia. Common species cultivated for fruit are *R. hirtellum* with its smooth fruit and *R. grossularia* with its somewhat fuzzy fruit. *R. hesperium* with green-red flowers and *R. speciosum* with red flowers are cultivated for ornament.

Flowers in loose clusters of one to two, rather than in groups of many as in currants. Flowers of this species greenish. Four- or five-parted, with fuzzy calyx tube about as long as calyx lobes. Fruit not smooth and shining as in currants or as in *R. hirtellum,* or prickly as in *R. hesperium* or *R. speciosum.*

Propagation largely by layering. Plant is cut back in fall, producing vigorous July shoots that are covered at base by several inches of soil and rooted by fall. They are then separated from parent plant and grown in nursery one to two years before being set in field. Host of dangerous white-pine blister-rust fungus.

Fruit produced on 1-year-old wood and 1-year-old spurs on older wood. Branches commonly pruned after fruiting 3 years. Usually about three shoots of an age for a 3-year cycle are kept for a plant. Fruits can remain ripe on bushes after ripening and are eaten raw or made into pies, jams, or jellies.

Order Rosales./Family Hamamelidaceae

Witch Hazel
Hamamelis virginiana
Shrub or small tree. Height to 25 ft. Twigs slightly roughened, yellowish to brown, not stiff, with alternate buds and a peculiar bud at end larger than others. Leaves short-petioled, to 5 in. long, with irregularly shallow-toothed margins, unequally heart-shaped at base and conspicuously veined.

Native of America, ranging from Nova Scotia to Florida and west to Texas, Minnesota, and Ontario. Tolerant to shade of other woody plants. Two species in genus.

Flowers in bright yellow, flaglike clusters, to ¾ in. long, borne in axils of leaves, appearing late in season when leaves are falling. Flowers pollinated in fall yield seeds following fall by expelling them violently from compact, somewhat nutlike fruits that are to ⅓ in. long.

Leaves commonly affected by aphid gall that makes tentlike structure; fruits, by gall that makes burlike structure. Wood heavy, hard, 43 lb per cu ft, excellent for making brooms and toothbrushes in campcraft. Believed, wholly erroneously, to be of value as divining rods for locating water underground.

Leaves and twigs source of witch hazel that is commonly used as an astringent and as an additive in rubbing alcohol. Element of medicinal nature is extracted in alcohol; its fragrance is familiar to most persons. Plants wilt slowly and were therefore recommended for use by the Army as a green camouflage. Branches of Witch Hazel placed on a campfire produce the familiar odor.

Sweet Gum, Bilsted
Liquidambar styraciflua
Tree. Height to 140 ft. Bark, rough. Twigs and branches with conspicuous corky ridges. Twigs with large, angled pith, smooth or slightly fuzzy. Trunk to 5 ft in diameter, straight, with relatively slender branches. Leaves to 9 in. wide, alternate, of three to seven lobes forming a star, smooth and dark above. Sap resinous.

Found in low woodlands from Connecticut to Florida, west to Mexico, Missouri, and Illinois. In lower Mississippi areas regularly flooded each year, is great forest tree. Definitely favors an abundance of water.

Flowers: staminate and pistillate in separate flowers on same plant. Staminate flowers lack corolla and calyx. Pistillate flowers lack corolla, develop into burlike structure, to 1½ in. in diameter that remains persistent through winter with sections opening to shed ½ in. winged seeds in autumn.

Wood to 37 lb per cu ft, hard, not strong but takes an excellent finish and used extensively as veneers for interior woodwork. Gum collected as resinous substance used as substitute for storax gum of commerce. 128,000 seeds per lb with 50 to 75 percent germination. Flowers appear in April and May.

In autumn, this is one of most beautiful of trees, as its leaves turn a gorgeous red. Twelve species of birds including bobwhite and wild turkey have been known to eat seeds or fruits; it ranks thirteenth in importance as quail food in Southeast. Gray squirrel, chipmunks, and marsh rabbits also known to feed on plant.

Order Rosales./Family Platanaceae

Sycamore, Plane Tree
Platanus occidentalis

Tree. Height to 110 ft. Trunk to 11 ft in diameter, rising in a straight column to 80 ft, almost of uniform diameter. Limbs erratic branching, making rather open head to 100 ft across. Bark conspicuously mottled and scaling off in sheets. Buds alternate, surrounded by leaf scar. Leaves to 9 in. wide, somewhat star-shaped.

Found in wet woodlands and along streams from Maine to Florida, west to Texas, Kansas, Minnesota, and Ontario. Family is small, containing only one genus and eight species, all native of North Temperate Zone. European sycamore or a hybrid is planted extensively as a street tree because of its relatively narrow shape. To 2,500-ft elevation in Appalachians.

Flowers very small, in ball-like clusters; staminate, with few stamens; pistillate with about same number of pistils. No calyx or corolla apparent. Fruit a ball-like structure to 1 in. in diameter hanging through winter at end of long swinging stem and rupturing to free wind-scattered nutlets. May live 600 years.

Wood weighs 35 lb per cu ft, reddish-brown, hard, difficult to split, weak. Probably largest tree in eastern North America though not tallest. Germination poor except in moist, open, mineral soil. Seeds harvested in fall should be frozen in dry place, planted in spring to ¼ in deep.

Wood used for butcher's blocks because of its hard splitting qualities. Also used in making tobacco boxes and to some small extent as veneer because of beauty of medullary rays in radial section. Much of it made into furniture but it has a tendency to warp sometimes; now used to some extent as interior finish in houses. In eastern Texas, a good shade tree.

Sycamore, Plane Tree

Order Rosales./Family Rosaceae

Ninebark
Physocarpus opulifolius

Shrub. Height to 10 ft. Branches recurve gracefully. Bark peels off in long sheets, often with many layers showing. Twigs and leaves smooth. Leaves alternate, petioled, bluntly three-lobed or sharply so, commonly with heart-shaped base, with medium toothed margin. Blade to over 2 in. long.

Common on riverbanks and in gorges in the wild from Quebec to Georgia and west to Tennessee and Michigan. More commonly found growing as an ornamental in some landscaped walk border or against a wall or building foundation. A dozen North American species and one native of Manchuria.

Flowers in terminal, somewhat flat-topped or globular clusters to over 2 in. across, with each flower from ¼ to ½ in. across, whitish or sometimes pinkish, with many stamens and one to five pistils developing in each flower. Fruit two-valved, yielding few seeds per unit. Units of fruit are to nearly ⅓ in. long. Flowers June and July.

Favors moist but well-drained site either in sun or shade. Fruits available as food for some kinds of wildlife in September and October but are inferior to most others at that time. About 1,600,000 seeds per lb but propagation is largely by layering and separation of existing plants.

Used primarily as an ornamental and as a hedge or barrier. Recommended by some soil-erosion technicians as a soil anchor for use on hillsides. Provides good cover for wild birds. Although known to have been eaten by at least three species, it nevertheless is not an important food source for wild birds.

Ninebark

195

Meadowsweet

Meadowsweet
Spiraea latifolia
Shrub. Height to 6 ft. Erect, branched or not, with purplish-red nearly smooth stems. Leaves alternate, smooth or nearly so, coarsely saw-toothed, to around 2 in. long and 1½ in. wide but much larger on young vigorous shoots or plants, paler on the underside.

Found in rocky, moist woodlands or open fields for the most part east of the Alleghenies from Newfoundland to Virginia but also west in the northern part to Saskatchewan. Sometimes planted as an ornamental. About 75 species of the genus native of the Northern Hemisphere, of which 4 are described briefly below.

Flowers in dense terminal, somewhat wand-shaped clusters. Individual flowers white or pink, to ½ in. broad. *S. prunifolia* has flowers borne on old wood in leafless, flat-topped clusters; *S. salicifolia* has pink flowers; *S. alba* has white flowers and a fuzzy flower cluster; *S. tomentosa* has leaves tawny beneath and whole plant more or less fuzzy.

Among the most popular of ornamental shrubs offering variety of color, flowering time, and shape of branches and leaves. *S. vanhouttei* with many white flowers in small clusters on drooping branches may be most common, followed by more delicate *S. thunbergii* with slender smooth leaves. *S. brumalda* bears pink blossoms in July and August.

The common names of some of the above include in order as named bridal wreath, willow-leaved spiraea, meadow-sweet, hardhack or stee-plebush, Van Houtt's spiraea, Thunberg's spiraea, and Brumald's spiraea. The meadow-sweet here emphasized may occur as a troublesome weed in New England.

Indian Physic

Indian Physic, Bowman's Root
Gillenia trifoliata
Herb. Height to 4 ft. Erect, branching, smooth, with slender reddish stems. Leaves alternate, compounded of three leaflets that are to 3 in. long, pointed at ends, with sharply toothed margins and small stipules at base. Leaves sometimes merely lobed instead of compound, particularly the upper. Root system perennial.

Native of woodlands from Ontario to Georgia, west to Missouri and Michigan ascending in North Carolina to 4,500-ft elevation. Is success-fully cultivated but not widely adopted. Two species in eastern North America suitable for use as ornamentals.

Flowers in few-flowered open clusters on slender stems. Petals, five, narrow, spreading, white or pinkish, to ½ in. long. Calyx cylindrical, persistent, five-lobed, reddish, and narrowed at the throat. Stamens, many. Fruit a somewhat fuzzy, podlike structure slightly longer than calyx, five pistils to a flower. Flowering time May through July.

Plants are hardy and easy to cultivate in ordinary garden soil. Usually propagation is by using the seeds or by dividing the root system. Of the two species, in *G. trifoliata* the plant is smooth and the pods are fuzzy; in *G. stipulata* the plant is fuzzy and the pods are smooth.

G. trifoliata is commonly known as bowman's root, false ipecac, Indian hippo, western dropwort, and Indian physic while *G. stipulata* is known also as Indian physic and as American ipecac. *G. stipulata* ranges through central and southern United States.

Apple

Apple
Pyrus malus
Tree. Height to 20 ft. Trunk to 3 ft in diameter. Bark rough, scaly, and dark. Twigs rather coarse, somewhat flexible, with fruiting spurs crowded with leaf scars. Leaves to 4 in. long, darkest above, coarse-veined, alternate, short-petioled. Roots both shallow and deep.

Native of Europe and Asia but cultivated in temperate regions where seasons are suitable. Can exist in a variety of well-watered, well-drained soils. Special varieties (500) developed for eating in different seasons, for use in pies, cider, evaporation, and so on. Crab apples are grown par-ticularly for use in making jellies.

Flowers borne on fruiting spurs, opening as leaves develop. Sepals, five. Petals, five, white or pink, to 3 in. across. Pollination mostly by bees or by flies and necessary for crop development. Fruit too well known to need description, maturing from midsummer to snow depending on vari-ety.

Propagation by slip grafting on seedling stock and by slip and bud graft-ing on developed trees. Attacked by scale insects, codling moths, a cur-culio, many aphids, red bugs, tent caterpillars, meadow mice, deer, and many fungi. Spray program must recognize necessity of pollinating in-sects. Red cedar host for apple-scab fungus.

A major fruit crop of United States. Fruits rich in vitamins B and C. Storage at 29°F prolongs keeping qualities, and control of carbon dioxide in storage area helps. Orchard may bear in 4 years and yield to 15 bu per tree by tenth year and to 50 bu per tree or 500 per acre at maturity. Wood weighs 50 lb per cu ft, excellent firewood.

Pear
Pyrus communis
Tree. Height to 60 ft. Pyramidal, long-lived, less open than apple. Trunk to 3 ft in diameter. Twigs commonly with thornlike spurs that bear short-pointed leaves to 4 in. long, with finely cut margins, coarse veins, and dark upper surfaces, with petioles sometimes as long as or longer than the blades. Roots relatively deep.

Native of Europe and Asia but established sometimes as an escape from Maine to Pennsylvania in woodlands and cultivated widely throughout the world in temperate regions. Asian pear is a native of China. Most cultivated pears are hybrids of European pear *P. communis* and Asiatic pear *P. serotina*. Hybrids include Keiffer, LeConte, and Garber pears.

Flowers in few- to several-flowered clusters, white, to 2 in. across, borne on ends of short twigs of preceding year. Petals, five. Sepals, five. Some varieties self-sterile and do best when pollinated by some other variety. Among these are Bartlett, Kieffer, Sheldon. Best to plant varieties that have overlapping flowering time.

Commercial propagation by seeds, by importing seedlings, and by grafting seedlings as in apple. No more than four rows of a variety recommended for one orchard. Fruits usually hand-picked and stored with care. Bartlett pears may be held 3 months at 30°F. Codling moth and fire blight among serious pear pests.

Fruit rich in sugars and eaten by many animals. Vitamin B present in good quantity. Some 30 commercial varieties in America, some producing to 400 bushels per acre. The South is too hot ordinarily for good pear production. Dwarf varieties have been developed that simplify harvest problems.

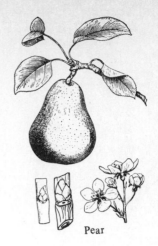

Pear

Mountain Ash
Sorbus aucuparia
Tree. Height to 50 ft. Round-topped, relatively open. Bark gray-brown, relatively close in young trees. Twigs and buds fuzzy but not sticky. Leaves compound, of 9 to 15 leaflets, each to 2½ in. long, short-pointed or blunt, sharply and finely toothed along the margin and not individually petioled.

Escaped or planted along roads or even in waste places. Native of Europe and Asia but now widely established and frequently escaped and persistent. About 10 species, of which 3 are native of North America, *S. americana*, and *S. decora* in the East and *S. scopulina* ranging from Labrador to Alaska and south to Pennsylvania and Utah.

Flowers about ⅓ in. across, in flat or round-topped, rather compact clusters that are to 6 in. across. Sepals united into a five-lobed cup. Petals, five, spreading, with slender bases. Stamens numerous. Odor unpleasant. Fruit globular, bright-red, persistent, usually under ⅓ in. in diameter. Whole cluster produces an impressive effect.

Wood weighs about 35 lb per cu ft, soft, weak, light brown. Seeds stratified in sand 90 days at 32 to 11°F. May be sown in spring. Seeds weigh 83,000 to 105,000 per lb. Mountain ashes are susceptible to San Jose scale and to borer that enters the trunk if base is not kept clear of trash.

Fruits of European species eaten by many birds but less favored than those of American species, eaten by ruffed grouse, blue grouse, sharp-tailed grouse, fisher, marten. Twigs eaten by moose and white-tailed deer. Fruits make good jelly if cleaned with hot water, cooked with 1 cup water per lb of fruit and 1 cup of sugar to 1 cup of juice before cooking to jelly.

Mountain Ash

Japanese Quince
Chaenomeles japonica
Shrub. Height to 6 ft. Profusely branching, with spiny branches. Leaves dark-green above and lighter beneath, alternate, roundish, coarsely toothed, to 3 in. long, forming a rather compact mass of vegetation stimulated to compactness by pruning.

Native of Japan. Widely planted as a hedge or individual ornamental plant. Related *C. sinensis* is not hardy in the North. Four distinct species, all coming from China or Japan and all having some merit as ornamentals.

Flowers in clusters of two to six, appearing with or just before leaves, some scarlet and a few forms pink or white. Flowers are to 2 in. across; since they appear on many parts of the plant they make an effective showing during flowering period, normally in March and April. Fruit ovoid, 2 to 4 in. long, green to yellow, fragrant.

Propagation by seeds that are stratified in sand in the fall and sown in spring, or by grafting on greenhouse stock or on the root system of the common quince, which may be more hardy than the Japanese ornamental varieties. May also be propagated from cuttings using half-ripened wood, or branches may be layered to take root at joints.

One of most popular of spring ornamental woody plants, serving dual purposes as flowers and as a thorny, practically impenetrable hedge. *C. sinensis* cannot be expected to survive north of Philadelphia. Fruit edible.

Japanese Quince

Quince

Agrimony

Shadbush or Serviceberry

Quince
Cydonia oblonga
Tree. Height to 20 ft. Bark very dark to almost black. Branches conspicuously crooked, spineless, slender. Leaves to 4 in. long, blunt-tipped, entire-margined, fuzzy on the undersides, alternate, with rather conspicuous stripules at base.

Native of central and eastern Asia and naturalized and established in the Mediterranean region. One closely related species comes from Turkestan. Flowering quince was formerly considered in same genus but is now in *Chaenomeles.*

Flowers borne on growth of year, singly and appearing with foliage, white or pinkish, to 2½ in. across. Calyx lobes are shed. Petals, five. Stamens, 20. Styles of pistil, five. Fruit appears without stem at ends of stout twigs, globular or pear-shaped, fuzzy, to 4 in. through, yellow, sour, five-celled.

Should be grown in heavy, relatively dry soils. Propagation by stratification of seeds planting them in spring, by cuttings of wood that is from 1 to 4 years old, by layering the branches, by budding, or by grafting. Growth slow. Flowers may appear in fall when plant is in fruit as well as in spring.

Grown primarily for fruits used in preserves and jams. Plant could serve as an ornamental because of its grotesque habit of growth and because of the beauty of foliage, flowers, and fruits.

Agrimony
Agrimonia gryposepala
Herb. Height to 4 ft. Leaves compounded of five to nine leaflets exclusive of small secondary ones. Leaflets more or less elliptic to oblong. Part bearing flowers covered with minute down of a sticky quality with longer, widely spreading hairs interspersed. Relatively slender root system.

Found in woodlands, thickets, and waste places from Nova Scotia to North Carolina and west to California and Minnesota. Some 15 species of the genus native of the Americas and ranging south into the Andes of South America.

Flowers yellow, to ½ in. across, arranged along sides of an erect hairy stem. Petals, five. Calyx tube, five-cleft. Stamens, 10 to 15. Fruit definitely top-shaped with deep, vertical grooves along sides of lower half and a ring of upward-stretching hooked appendages at widest part.

Probably best known because of the sticktight qualities of the fruits, which enable them to attach themselves to clothing, particularly wool socks, and become firmly embedded. Probably plant spreads itself largely by the rides stolen by the barbed fruits.

This is a perennial of dry soil areas and while fields may sometimes become badly infested, the plant yields to ordinary cultivation practices. It can hardly survive a season in which a cultivated crop is maintained.

Serviceberry, Juneberry, Shadbush
Amelanchier sp.
Shrub or small tree. Height to 60 ft. Trunk to 2 ft in diameter. Bark pale steel-gray, faintly and beautifully streaked longitudinally. Twigs slender, with buds somewhat like those of beech but smaller and thicker at base. Leaves to 3 in. long or longer on young shoots, sparingly silky when young but smooth with maturity.

Common in dry woodlands, chiefly on hillsides. Ranges from Nova Scotia to Florida, west to Louisiana, Arkansas, and Quebec. Other closely related species extend the range considerably and may only be separated from this one with difficulty. At least 25 species of which a dozen occur in North America and one in Mexico.

Flowers appear just ahead of leaves. Petals white, to four times as long as sepals, each to ¾ in. long; slender and not stiff. Flowers from March through May. Fruits small, red or purple, spherical, deliciously sweet and nutritious, ripen from June through August.

Wood hard, brown, weighing 49 lb per cu ft, used for tool handles and as the "lancewood" of fish poles. Seeds germinate at about 70 percent; about 10,000 usable plants can be expected from 1 lb of seeds numbering about 50,000. Seeds normally distributed by birds.

Fruits known to be eaten by 27 species of birds including ruffed grouse, bobwhite, and mourning dove, while the leaves or bark are eaten by white-tailed deer, cottontails, and other game. Plants used in erosion control and as ornamentals. Fruits used by man as are blueberries. Named shadbush because plant bloomed at time shad ran up rivers.

Hawthorn
Crataegus sp.

Usually relatively short, stout trees reaching to a height of possibly 20 ft, with a trunk diameter of less than 1 ft, with slender rather crooked branches, often well-armed with spines of variable size. Leaves alternate, variously shaped, usually double-toothed. Usually have long taproots.

Commonest in pasture lands and along fence rows. Some 1,300 species in the north temperate regions with the eastern United States being particularly rich in species. "Species splitters" among botanists recognize few limits to the numbers that might be eventually described.

Flowers somewhat like small apple blossoms but many have a most disagreeable odor, usually clustered at ends of twigs, white or whitish, ½ to 1 in. across and appearing in late spring or early summer. Fruits with relatively large, hard, separable centers, usually red and variously flavored.

Wood of most species, hard, close-grained, reddish-brown, weighing about 45 lb per cu ft and used in turning napkin rings, tool handles, canes, and rulers. 1 lb of seeds numbers about 6,000 to 40,000 with 40 percent germination and may yield 2,500 usable plants.

Provides good food and cover for many kinds of wildlife, making up nearly 10 percent of food of grouse in northeastern United States. Also eaten by pheasants, quail, deer, cottontails, and cattle. Considered a source of honey. Juiciest fruits may be made into acceptable jelly for human consumption. Trees give soil anchorage. A most troublesome shrubby weed in old pastures in the northeastern United States.

Hawthorn

Firethorn
Pyracantha coccinea

Shrub. Height to 12 to 15 ft. Twigs armed with slender spines to 1 in. long. Leaves evergreen, to 2 in. long, narrowed at base and short-petioled, blunt-ended, somewhat shining, darkgreen on upper surface, smooth on both sides, forming when healthy a dense mass of foliage.

Native of southern Europe and western Asia but grown in America as ornamental and escaped from cultivation and established independently in isolated areas from Massachusetts to Texas. About 40 species in the genus native of the Old World, none in Japan.

Flowers in many-flowered, somewhat flat-topped clusters usually at or near twig tips, about ¼ in. across, with five-lobed calyx, five petals, and many stamens. Fruit, bright red, with two bony seeds in each cell instead of the usual one found in *Crataegus.* Blooms in May and matures fruit by October.

Does well in any good garden soil. Evergreen species are propagated by half-ripened cuttings kept under glass and started about August, or by layering in fall, or by grafting on mountain ash, quince, or hawthorn. Foliage of some species is brilliant in autumn. Seeds may be sown in fall or stratified for spring planting.

Grown as an ornamental for brilliant or evergreen foliage or for flowers or for black or reddish fruits. Fruit is generally attractive to birds, particularly robins, and the plant provides some wildlife cover where it is established.

Scarlet Firethorn

Wood Strawberry
Fragaria vesca

Height to under 12 in. Leaves arise from a basal tuft with runners giving rise to other leaf tufts. Leaves long-petioled, of three leaflets, thin, sharply toothed, with terminal one the larger and the petioles and underside of leaves well-supplied with hairs.

Naturalized from Europe in eastern and middle Atlantic states but native to the North and found from Connecticut to New York, west to Kentucky and Ohio as native plants. In the closely related variety *americana* the hairs on the petioles are more closely appressed and less abundant.

Flowers white, borne on flowering stalks reaching to sometimes to nearly 1 ft, with a few flowers in an open cluster. Calyx. Five-lobed. Petals, five, white. Whole flower to nearly 1 in. across. Stamens numerous, with yellow anthers. Fruit red or sometimes white, with "seeds" borne scattered on a smooth or almost even surface.

Pollination by insects. Natural propagation probably as much by runners as by fruits. The determination of species, subspecies, and varieties is a problem for specialists. To most of us, this plant is either long, delicate, small-seeded woodland strawberry or something so close to it that it makes little difference.

The fruits of this strawberry are so delicate that it is almost impossible to collect a "mess" of them. They practically melt in one's mouth. The plants do not bear the abundance of fruits nor are the plants themselves so abundant as the related field strawberry.

Wood Strawberry

Wild Field Strawberry

Cultivated Strawberry

Barren Strawberry

Wild Field Strawberry
Fragaria virginiana
Height usually to under 8 in. Leaves arise from tufts on runners, with petioles to 6 in. long and three rather large leaflets that are coarser than those of the woodland strawberry. Leaflets rather coarsely toothed, blunt-tipped, and hairy, with lateral ones uneven at the bases.

Commonest of the wild strawberries in its range and found often in abundance in dry soil in fields and pastures from Newfoundland to Florida and west to Oklahoma and South Dakota, there being several races within this range and several closely related species in territory adjacent to it. These races roughly represent Canada, New England, Virginia, Illinois, and so on.

Flowers white, in relatively few-flowered, open clusters at the end of erect flowering stems, to nearly 1 in. across. Sepals, five, green and in some erect lobes. Petals, five, white. Stamens, numerous. Fruit an irregularly globular mass of red flesh with the "seeds" embedded in pits over the rather roughened surface.

Izaak Walton says of the strawberry that "doubtless God could have made a better berry but doubtless God never did." It is certain that those who have eaten wild strawberries straight, in shortcake, or in a sherbet doubt if God could have made a more tasty fruit.

Undoubtedly, the fruits provide some food for wild birds and mammals as well as for man and the plants act as soil anchors with their binding runners and leaf tufts scattered here and there.

Cultivated Strawberry
Fragaria ananassa
Perennial, with leaves growing in clumps from scaly perennial more or less woody runners. Leaves of three leaflets each well-toothed, rather glossy above, bluish-white beneath, and, in this species, overtopping the flower-bearing stalks.

Garden strawberry is probably a hybrid of many species including particularly *F. ananassa* of the Pacific Coast region from Alaska to California and from Peru to Patagonia, *F. virginiana* the field strawberry of the East, *F. moschata* the hautbois of Europe, and *F. vesca* the alpine and perpetual strawberry of Europe.

Flowers white or reddish in some forms, in open clusters with or without stamens, with five-lobed calyx and five petals. Whole flower to over 1 in. across. Fruit an enlarged, red pulpy juicy structure with "seeds" embedded in the surface. When imperfect varieties are planted, one row of perfect flowers is set to every four.

Propagation is usually by runners set in rich, well-drained warm soil free from alkali in rows 18 in. apart, in rows 3½ ft apart (8,300 plants per acre), or in hills 3½ ft apart (3,630 plants per acre); cultivated to free weeds first summer; mulched with straw in winter; and in season, picked over every 2 to 3 days.

Yields of 3,000 to 4,000 qt per acre are common and 9,000 qt per acre have been produced. Sold as fresh or frozen fruit or made into syrup, jams, candies, conserves, ice creams, and so on. Important enemies include weather, a nematode, birds, slugs, a weevil, a crown borer, and many fungi.

Barren Strawberry
Waldsteinia fragarioides
Tufts of three-foliate leaves arising from a rather stout creeping rootstock. Leaves long-petioled, with leaflets blunt at the free end, rather narrowly wedge-shaped at the base, with toothed margins, darker green above than below, leaflets being to 2 in. long.

Often growing in rather dense stands in woodlands and on shaded hillsides. Ranges from New Brunswick to Georgia, west to Indiana, Minnesota, and Ontario. Native of America, five species being found in the north temperate parts of the world. This species found in Oregon as well as in the general area indicated.

Flowers like yellow strawberry blossoms, three to eight on a flowering stalk, in loose, open, but erect clusters, each flower being to ⅔ in. broad. Calyx five-lobed. Petals, five. Stamens, many. Fruit not fleshy like a strawberry but made up of four to six finely fuzzy "seeds." Flowering period May to June.

The plant is named in honor of the German botanist Franz Adam von Waldstein-Wartenburg, who lived from 1759-1823. This species has leaves that look so much like those of a strawberry to an amateur, that its common name is easily understood, but to botanists so many differences are apparent that it is hard to understand such confusion.

The plant has little economic importance. Since the flowers fall to pieces almost as soon as they are picked, they make inferior wild-flower bouquets. The fruits may be eaten by some forms of wildlife but they perform no outstanding service in this connection.

Old Field Cinquefoil
Potentilla canadensis

Sprawling herb with runners to over 2 ft long bearing tufts of leaves and flowering stalks at intervals. Leaves of five leaflets, long-petioled, somewhat hairy. Leaflets rather blunt at the apex and acute at the base, to 1 in. long, with saw-toothed margins and rather distinct veins, lighter in color on the underside.

Found on lawns and in dry soil such as pastures from New Brunswick to Georgia, west to Texas and Minnesota, being found up to 6,300-ft elevation in North Carolina. Over 300 species of *Potentilla*, all native of the north temperate regions, of which nearly two dozen are found in the northeastern United States.

Flowers yellow, to ½ in. across, with many stamens, five broad petals, and five acute calyx lobes that are about as long as petals but much narrower. Fruit appears as a bunch of dry "seeds" without fleshy structure found in related strawberry. Some potentillas have large showy flowers while others have inconspicuous ones.

This species does best where competition is low on poor, relatively dry soil. It does not stand competition and may be taken as an index of poor fertility. It survives on acid soil and some consider it an indicator of sour soil. Where this plant is found growing in abundance, one should not expect good results of forest trees set out for reforestation.

Control is by enrichment of the soil and encouragement of competition. Plowing and sowing rye in infested land may serve to eliminate it. It probably serves some value as an anchor for soil that might otherwise be exposed to the elements and be eroded away.

Cinquefoil

White Avens
Geum canadense

Herb. Height to 2½ ft. Branched in the upper parts, softly hairy throughout. Leaves of three to five, segments, with the terminal lobe of the basal leaves being much larger than any of the lateral segments. Basal leaves well petioled, and stem leaves with short petioles or none.

Found in shady waste places. Ranges from Nova Scotia to Georgia, and west to Louisiana, Kansas, South Dakota, and Minnesota. Some 40 species of the genus found in north temperate regions, in southern South America, and one in South Africa. At least a half-dozen in eastern North America.

Flowers white, with five reflexed calyx lobes and five relatively inconspicuous white petals making a flower about 1 in. across. Many stamens. Bases on which hooked fruits are borne, are densely hairy after fruits have been shed. Fruits provided with a long style terminating in a hook and making a sticktight.

Pollinated no doubt by insects. While some of the genus can well be classified as weeds, it is doubtful if this one can. The purple avens *G. rivale* is sometimes called chocolate root because the root may be used as a substitute for cocoa.

A number of the geums are known as weed seed impurities but most of them yield readily to cultivation and are not so persistent that they can be considered serious. The fruits of most of them are characteristic hooked sticktights found in clothing or in the fur of some animals.

White Avens

Blackberry
Rubus allegheniensis

Height to 8 ft. Old canes are purplish and bear numerous straight stout prickles. Leaves compound, of three to five leaflets with terminal one the largest, rather soft and velvety beneath, with the margins shallowly toothed. Root system extensive, survives separation readily.

Relatively common in dry soil from Nova Scotia to North Carolina and west to the Allegheny Mountains, with other closely related species extending the range and with cultivation practices spreading the plants through most of United States. Valuable varieties include Agawam, Kittatinny, Snyder, Early Harvest, and Taylor.

Flowers borne on rather long-stemmed clusters with supporting stems covered with sticky hairs. Flowers white, to 1½ in. across, with petals broader toward tip. Fruit, shiny black, juicy globes with hard centers grown together and not appearing like a cap when plucked.

New canes set 3 to 4 ft apart, in rows to 9 ft apart. Young shoots 2 ft long are clipped back to few inches to induce branching. When developed, lateral branches are pruned to 8 to 20 in., the upper being kept the shorter. Canes that have borne fruit are cut back to ground and next crop borne on new canes that have developed.

A most important commercial crop particularly where superior hybrids have been developed. The fruits keep well and can be shipped considerable distances in safety. They make excellent juices for flavoring, superior pies, jams, and jellies and are at their best taken fresh from the shrubs when dead ripe.

Blackberry

Flowering Raspberry

Red Raspberry

Moss Rose

Flowering Raspberry, Thimbleberry
Rubus odoratus

Shrub. Height to five ft. Erect, branching canes free from prickles, with bark that peels off in long strips, with younger twigs somewhat bristly. Leaves long-petioled, not compound, with three- to five-lobed blade, to nearly 1 ft broad, with heart-shaped base and finely cut margins.

Found at the edges of woodlands particularly in areas where soil is shallow. Ranges from Nova Scotia to Georgia, west to Tennessee, Michigan, and Ontario, with a number of races represented within the range. The closely related salmonberry *R. parviflorus* has smaller white flowers.

Flowers showy, to 2 in. across, in rather open clusters at the ends of branches, purple or sometimes white, with yellow anthers on the many stamens and rather long though slender calyx lobes. Fruits like flattened hemispheres composed of a number of flesh-covered, stonelike "seeds"; have a poor flavor.

Persons differ in judgment of the edible qualities of the fruits, some contending that they are good while others disagree vigorously. The large fruits can be used to piece out more delicious species if the substitution is not too generously observed.

Well over 60 species of birds known to eat *Rubus* fruits, the fruit ranking eighteenth in importance as quail food in the Southeast. It is also eaten by bear, squirrels, deer, porcupines, rabbits, elk, moose, hares, beaver, skunks, chipmunks, and other animals.

Red Raspberry
Rubus idaeus

Shrub. Erect, to 6 ft high, well-armed with hooked prickles on the stems. A well-managed plant has about nine vigorous canes on it. Stems root where they contact the soil. Leaves alternate, compound, of three to five leaflets of which the terminal is the largest, with petioles well-covered with whitish hairs.

R. occidentalis, blackcap raspberry, is smaller fruited than *idaeus*, and has curved canes.

R. idaeus has hybridized and produced asexually reproducing clones which keep their characteristics. At times, some of these hybrids have been mistakenly described as species.

Flowers to ½ in. across or more, white, erect, with somewhat fuzzy sepals. Fruit red, and when picked comes free of receptacle as a hollow cap. Varieties of the European include the Antwerp and Amsterdam; of the American, the Marlboro, June, and Surprise; of the blackcap, the Ohio, Plum and Palmer.

American red raspberry propagated by transplanting suckers; the black and purple by tip layering. Cool summers are best. Set plants 2 ft apart in rows 6 ft apart. Train to two wires. Remove old wood after the fruiting, keeping about five to seven new canes to develop each year. Do not pinch back red raspberry in summer ordinarily.

Well-managed purple raspberries may yield more than any other with black next and red last. Good crops may be borne for to 15 years though 8 to 10 is the average for red and 6 to 8 the average for black or purple.

Moss Rose, French Rose
Rosa gallica

Shrub. Height to 4 ft. Canes and other woody parts with crowded, stiff prickles of unequal length intermixed with weak bristles giving a "mossy" effect. Leaves alternate, thick, compound, of three to five leaflets, to 6 in. long, the leaflets being to 2 in. long, dull above and fuzzy beneath.

Native of Europe and western Asia. An old popular garden rose that persists around abandoned homes or escapes and becomes well established. Closely related to the cabbage rose *R. centifolia*, native of the Caucasus and one of the most ancient of cultivated roses.

Flowers single or double, usually erect (nodding in cabbage rose), solitary or in open groups of two to three pink to deep red, to 3 in. across; with petals spreading and sepals leaflike with ragged edges. Fruit nearly spherical or somewhat elongated, to ½ in. through, dull red and without the sepals.

Chosen here as an old-fashioned garden rose, useful either for its flowers or for its foliage. It flowers through summer and into autumn, and with its large flowers and dense dark foliage provides excellent landscaping service. It is easily propagated by division or by cuttings.

Druggists recognize oil of rose, rose water, and syrup of rose as products of roses, largely of their aromatic petals. Some of these are used in candies, some in medicines to improve taste, some to improve fragrance.

Meadow Rose
Rosa blanda

Shrub. Height to 4 ft. Canes erect and armed with only a few prickles or with none. Leaves compounded of five to seven leaflets, the leaflets being to 1½ in. long, rather sharply saw-toothed along the margins, short-stalked, and pale beneath. Stipules rather broad.

Found commonly in rocky places from Newfoundland to New Jersey and west to Missouri and Ontario, with related species extending range through most of the United States and southern Canada. *R. pratincola* is the common wild rose of the Middle West and has densely prickled stems.

Flower showy, to 3 in. broad, solitary or in few-flowered clusters, pink, single, with many yellow stamens in the center. Fruit nearly globular and almost ½ in. through, almost smooth, and with the mature fruits showing the sepals erect rather than spread as they may be at first.

The related *R. pratincola* may spread considerably by means of underground structures that may be over 1 ft underground and radiate 5 ft or more from the central system. *R. blanda* is a rather early wild rose blooming in June and July while others may bloom through to the autumn months.

Provide beauty, game cover, and game food and may become a weed that yields readily to cultivation. Indians ate the fruits raw. Jellies are now made from them, usually after frost has come. High in vitamin C. Known food of grouse, quail, prairie chicken, deer, opossum, coyote, bear, pheasant, mountain sheep, and wild turkey.

Meadow Rose

Tea or Garden Rose
Rosa odorata

Shrub. Height variable to 3 ft or more. Stems smooth, with prickles that are relatively few, scattered, hooked, and stout. Leaves compounded, of five to seven leaflets, partially evergreen, with the leaflets to 3 in. long, dark shining green above and lighter and smooth beneath, with saw-toothed margins. Root system substantial.

R. odorata is native of western China but is cultivated in many forms and has been hybridized freely so that the garden rose most of us know is really a combination of many kinds of roses or many variations. Most popular tea roses of gardens and florists are probably of this species.

Flowers borne on the shoots of the year, one to three, white, pink, salmon, or yellow, to 3 in. across, exceptionally fragrant. The variety *gigantea* has flowers that are 6 in. across. In typical China rose, sepals are entire (not shown in illustration) and fruits are almost globular and red.

Roses may be produced by seeds but the most common practice is to layer the canes, make cuttings, or graft desired cuttings on hardy stock. For the amateur the layering process is the simplest and safest. A mature cane may be bent to the ground and a portion of it buried in earth. When it has become rooted, it may be separated.

Roses are important commercially, making the backbone of the florist business. Sentimentalists have ascribed a vocabulary to different kinds.

Garden Rose

Choke Cherry
Prunus virginiana

Shrub or small tree. Height usually not over 10 ft. Bark gray, close, with a disagreeable odor and a bitter taste. Leaves thinner than in black cherry, and with marginal teeth spreading rather than incurving, less rounded and more slender, sometimes with secondary teeth showing on the major teeth; egg-shaped, oval, sharp-toothed leaves with hairless midribs or midveins.

Along riverbanks and wet hillsides or along abandoned fence rows. Ranges from Newfoundland to Georgia, west to Texas and Manitoba. A close relative, the Rocky Mountain wild cherry, continues the range on to the Pacific Coast, differing in that the fruit is sweet and the leaves thick, more like those of black cherry.

Flowers white, to nearly ½ in. across, borne along a drooping central axis that terminates leafy shoots. Petal about the length of the stamens. Sepals persistent rather than dropping off as they do in the black cherry. Fruit astringent, deep red or yellow, with a stone that is not corrugated like that of black cherry.

Wood of little economic importance since the plant never reaches a large size. Leaves, like those of the black cherry, may cause serious poisoning to cattle if eaten after they have wilted. The poisoning is rapid and affects the animal's brain as well as the heart and lungs.

Plant used as a soil anchor in erosion control because of habit of developing long roots that sprout at intervals to form new tops. Always occurs as clusters or stands, resulting from single mother plant, reproducing underground by vegetative propagation. From 3,000 to 5,000 seeds per lb, which may be collected from July through September. Fruits or plants known to be eaten by bear, rabbits, quail, pheasants, and grouse. Fruits make good jams.

Choke Cherry

Wild Black Cherry

Pin or Wild Cherry

Plum

Black Cherry
Prunus serotina

Tree. Height to 100 ft. Trunk diameter to 5 ft. Twigs with bitter aromatic bark. Bark of trunk dark, peeling off in squarish flakes irregularly. Leaves alternate, to 6 in. long, stiff and firm when developed, shining above, light green beneath where also most smooth; almost always has midrib on underside of leaf fringed with brown hair. Does not have spur branches along main branches.

In open places and woodlands. Closely related to the choke cherry, the teeth of whose leaves are much less coarse. Ranges from Nova Scotia to Florida and west to Texas, Kansas, and South Dakota with closely related species extending the range considerably. More likely to favor dry spots than is choke cherry.

Flowers borne on short stems along a central axis, the whole cluster erect at first but eventually drooping and being found at the end of leafy branches, relatively small, about ½ in. across, white, with clusters to 6 in. long. Calyx persists with the stamens. Fruit nearly globular, black, ½ in. through, thin-skinned.

Wood hard, beautiful reddish-brown, strong, capable of taking a beautiful polish, weighing 36 lb per cu ft. Wilted leaves may seriously poison cattle, with those from vigorous young shoots being the most dangerous and those cut in spring and summer worse than those cut later in the season.

Wood very valuable in cabinetmaking, being one of the best of native woods. Fresh ripe fruit of food value to man and may be eaten raw, cooked, or made into preserves or jellies. An important food of wildlife including songbirds, ruffed and sharp-tailed grouse, prairie chicken, pheasant, bobwhite, raccoon, black bear, red fox, white-tailed deer, cottontail rabbit, and gray squirrel. Domestic stock poisoned by wilted leaves may have staggers, convulsions, difficult breathing, and may die within an hour.

Pin or Wild Red Cherry or Fire Cherry
Prunus pensylvanica

Tree. Height to 40 ft. Relatively open, branched. Trunk to 20 in. in diameter. Bark of young twigs smooth, thin, red-brown; of old trunks, to ½ in. thick, breaking into horizontal bands, bitter. Leaves to 4 in. long and to 1¼ in. wide, with margins bearing fine incurving tipped teeth, often with glands at the tip; clusters of buds at or near ends of red twig tips.

Common in rich soils, in mixed forests, appearing abundantly on burned-over areas; does well even on sand. Ranges from Newfoundland to the shores of Hudson Bay to the eastern slopes of the coastal range of the mountains in British Columbia, south through Iowa, Tennessee, and North Carolina and somewhat farther south in the Appalachians.

Flowers bear both stamens and pistils, appear when the leaves are half-grown, on slender inch-long stems in groups of four to five, arising from a common point; in clusters that are umbrellalike. Fruit ripens from July through September as a ½-in. pale red-skinned, large-stoned, thin-fleshed, sour cherry that seems to be a favored food of many species of birds; 23 species known to feed on it.

Wood light, soft, light brown, 31 lb per cu ft, close-grained, uniform. Seeds 8,000 per lb and available for harvest from August through October. Tree short-lived but an excellent hedgerow plant though it may be host to some apple pests. Is known to be emergency food for beaver where poplar is lacking.

Leaves that have been wilted are reported as being poisonous to livestock. Used frequently as grafting stock for commercial sour cherry where poor soils are present. Fruit a bit too sour for human consumption when fresh but may be made into excellent jelly or cooked with other fruits. Moose, bear, deer, rabbits, prairie chicken, quail, grouse, and pheasants known to feed on it. Said to be called bird cherry because its fruits attract so many birds; fire cherry because it is an early pioneer in burned-over areas.

1. Domestic Plum
Prunus domestica

2. Japanese Plum
Prunus salicina

3. American Prune Plum
Prunus americana

Large open trees, with young twigs of (1) mostly fuzzy, of (2) mostly smooth, of (3) either smooth or fuzzy. Leaves alternate; of (2), with finely cut margins; of (3), with coarsely cut edges; of (1), to 4 in. long, coarsely and irregularly cut along the edges, thinly hairy above and fuzzy beneath. There are species of plums other than (1), (2), (3) above which are productive of plums.

A cultivated tree. (1) native of Europe and grown most extensively in the Western states; (2) native of Asia and introduced more widely than

(1); (3) native of America with its greatest popularity centering in the Central and Southern states. All may occur as escapes near areas where they have been cultivated.

Flowers few in a cluster, white or creamy, to 1 in. across, fragrant, growing from short spurs, with the supporting stems hairy or minutely fuzzy. Fruit highly variable in different varieties but usually not depressed about the stone, firm, commonly bluish-purple and relatively tough-skinned.

Pollination by bees, syrphus flies, and other insects or by self if insects fail to visit. Commercial propagation commonly by budding on a variety of stocks depending on soil and other requirements. Hardiness is procured by grafting desired varieties on American stock. Pests include a curculio and a brown rot.

Over 30 varieties recognized. Properly cared for, a plum tree in the East may yield 1 to 4 bu, and more in the West. The common prune is (1); dried, it makes an important article of commerce. For home use, fruits are hand-picked. Fruits are rich in sugar and may be canned or preserved.

Apricot

Common Apricot
Prunus armeniaca
Relatively small, round-topped tree. Bark distinctly reddish and much like that of a peach. Twigs smooth. Leaves to 3½ in. long, alternate, abruptly short-pointed; margins with blunt-tipped teeth, smooth above and fuzzy on the veins beneath, with glands on the petioles.

Native of China or of Japan. Three species recognized: *P. armeniaca,* the common apricot of Europe; *P. mune,* the Japanese apricot; and *P. dasycarpa,* the purple or black apricot. Crossed with the plum, it produces the plumcot. Russian apricot is a small race of the common apricot.

Flowers pinkish-white, usually borne singly or in small groups much in advance of the leaves, to 1 in. across and with a short stem or none at all. Fruit highly variable, rounded or flattened, smooth or, when young, with a felty covering, yellow with a reddish overcast. Flesh firm and free from ridged stone.

Not so hardy as a peach and cannot survive heavy, soaked soil. Does best where there is a late spring in the East. Propagation is by seed or more largely by grafting desired tops on rooted seedlings, preferably on peach or plum stock rather than on apricot. It bears every other year. Susceptible to curculio insect injury and brown rot disease; hardier than peach trees but flower buds may be killed by late frosts.

Fruit used largely as a dessert, either fresh, dried, or preserved in syrup. Brandy is also made from the apricot. Because of the beauty of the bloom, some apricots are grown as ornamentals and trained against walls, preferably against those that do not face south.

Almond

Almond
Prunus amygdalis
Tree. Height to 25 ft. With light-colored, smooth branchlets, and with old bark bitter, gray, and close. Leaves alternate, shining, firm, with margins finely uneven, to 4 in. long, taper-pointed, and the petiole usually with glands on the upper side.

Probably originally from Asia or the Mediterranean region but now cultivated widely and in suitable territory established as an escape. In America, most frequently cultivated on the Pacific Coast where it has been grown for more than half a century. More tolerant of drought than most fruit trees but will survive standing water over root.

One of the earliest flowering of the fruit trees, in California blossoming in February. Bitter almonds are white-flowered; sweet almonds, pink-flowered. Propagation mostly by budding and grafting, trees being kept compact in warm regions and open in cool. Over 25 varieties grown in California.

Wood hard and compact but not of value compared with the nut. In dry weather, the hulls normally open, freeing the nuts. Sulfur used in removing stains. Bitter almonds are raised for the extraction of prussic acid. Sweet almonds include thin- and hard-shelled varieties; in the latter ¾ weight may be edible meat.

Before World War II, United States imported 11,441,000 lb of almonds mostly from Italy, with Spain and France following. Used largely as food. Sweet almonds yield 44 to 55 percent oil; bitter, 35 to 45 percent oil. Oil is used in cosmetics and with drugs. Rod of Aaron supposed to be from almond tree so Jews carry almond rods in certain festivals.

Peach

Peach
Prunus persica
Tree small, low, weak, relatively short-lived, with colored, smooth, close bark and rather rubbery smooth twigs. Leaves to 9 in. long, smooth, shining above and lighter in color beneath, alternate, taper-tipped, com-
(Continued)

205

monly with glands on the petioles, and with tendency to fold toward the upper surface and to be twisted.

Native of China.   Grown in United States and Canada where the temperature does not go below −15°F.   Many varieties of *P. persica* have been developed, some of which are Golden Jubilee, an early yellow; Carman, an early white; Halehaven, a midseason yellow; Champion, a midseason white; Afterglow, a late yellow, and Belle of Georgia, a late white.

Flowers beautiful, solitary, pinkish, to 2 in. across, appearing before leaves.   In some, petals are smaller, even down to ¼ in. long.   Fruit has hard central stone surrounded by a delicious flesh and thin furry skin, with stone deeply pitted, hard and clinging to or free from flesh.

Propagation commonly by grafting onto suitable seedling stock.   Pollination by insects or sometimes by self.   Fruits vary in flavor, lasting quality, and size and for shipment are generally picked before ripe but are best for canning when tree-ripened.   Nectarine is a peach variety with smooth plumlike skin.

Tree may yield to 4 bu of fruit and 1 acre may produce to 250 bu a year, but financial return varies greatly because of market conditions, storm, weather injury, and insect or fungous pests.   Tree may begin bearing the third year and continue until many years old, the yields being greater every other year.

1. Sour Cherry
Prunus cerasus

2. Sweet Cherry
Prunus avium

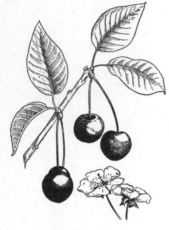

Sweet Cherry

Tree.   (1), small, with close gray bark, producing suckers freely from root.   (2), large, one-trunked tree not producing suckers.   Leaves (1), stiff, short-pointed, without developed glands on petiole, to 4 in. long; (2), limp, long-pointed, with one, two, or more petiole glands, to 6 in. long.

Both are cultivated and natives of Eurasia and may exist as escapes.   (1) includes cherries known as pie cherries of such varieties as Amarelle and Morello while (2) has been developed to produce at least three types, the sour Duke *regalis,* the soft-fleshed Heart *juliana* and the firm-fleshed Bigarreau *duracina.*

Flowers white, about 1 in. across, appearing in (1) slightly in advance of and, in (2), with the appearance of leaves.   Fruit of (1), globular, sour, soft-fleshed, and globular-stoned, red; of (2), globular or heart-shaped, yellow or red, sweet or sometimes bitter and more firm-fleshed, with variations suggested above.

Growing of sweet cherries and sour cherries for profit on a one-crop basis is a gamble because while sour cherries are grown easily, their market is poor; and while the market for sweet cherries is good their culture is rather difficult.

Fruits are sold fresh, canned, made into fruit juices for flavoring drinks, or are used in making brandy.   Cultivated cherries are usually pruned to head 2 to 3 ft from the ground to permit easy harvest.   Sweet cherries for sale as fresh fruit are picked before fully ripe and the sour when they are ready for use.

Japanese Flowering Cherry
Prunus serrulata

Japanese
Flowering
Cherry

Tree.   Height to 30 ft.   With pale gray, close bark.   Branches spreading, to make a thick, relatively close head.   Leaves with rather long, toothed marginal outline, green or somewhat reddish when they begin development.

This species native of Japan and introduced into Europe about 1870. There are many varieties of *P. serrulata* of which *lannesiana* is one.   Also, there are many closely related species. All come from Japan or China or eastern Asia.

Flowers of all borne on branched, short stems, with the bud from which the flower cluster comes forming an enclosing cup to ½ in. long.   Flowers of *lannesiana* are pink and fragrant, while those of *serrulata* are white, of *sieboldi* pink, and of *yedoensis* pink or white.   Fruits not important when compared with others.

Flowering cherries vary considerably in their ability to withstand severe winter weather, some being hardy even over the Canadian border.   Probably most famous of Japanese flowering cherries are those at Washington, D.C., that may bloom at Easter.

These plants are grown primarily as ornamentals.   In Japan, the common species in the temple grounds is *P. yedoensis,* its flowering occurring at about the time of certain of the national festivals.

DIVISION ANTHOPHYTA. CLASS DICOTYLEDONAE

Order Rosales./Family Leguminosae

Cat's-claw, Una de Gato
Acacia greggii

Tree. Height to 30 ft. Trunk diameter to 1 ft. With many spreading branches that are angled lengthwise and armed with stout recurving spines at bases of leaves. Bark to ⅛ in. thick, furrowed and peeling in narrow scales. Leaves to 3 in. long, as shown, with leaflets to ¼ in. long, in four to five pairs opposite each other.

Found on dry gravelly places and steep hillsides or canyon walls from western Texas to southern California, north to Colorado; more common at the lower altitudes. Some 300 species of *Acacia* with several in the United States but most of them in Africa and Australia, only a few found in temperate zones.

Flowers on slender stems, in heads that are to 1 in. through, yellow, with two to three clusters together toward the branch tips, fragrant. Calyx half as long as petals. Fruit, pods that in mid-August are light green, turning to red and showy, to 4 in. long and to ¾ in. wide, contracted between the seeds.

Wood heavy, strong, close-grained, hard, brown or red; with yellow, thin sapwood. Tree drought-resistant. Seed germination about 60 percent, 1 lb of seeds yielding about 200 usable plants. One of the best bee plants, yielding an abundance of useful nectar. Cattle feed on it and the plant withstands heavy grazing. Poor shade plant.

Spines give the names cat's-claw, devil's claw, and tear blanket. Seeds used by Indians for food. A gum similar to gum arabic exudes from the plant. Named after Josiah Gregg, a frontier author and "doctor." Excellent cover for jack rabbits, nesting site for verdin, roost for Gambel's quail. Good soil anchor. Wood used locally in various ways.

Cat's Claw

Mesquite
Prosopis juliflora

Tree or shrub. Height to 20 ft. Trunk diameter to 1 ft. Gracefully drooping branches make a round-topped plant. A pair of sharp spines at the axils of the leaves. Leaves to 10 in. long, twice pinnate, with long slender petioles, each with slender leaflets to 2 in. long and to ⅛ in. wide, smooth, dark green, sharply pointed.

May be almost buried in sand with only a few feet visible while the roots extend downward for 50 ft to water. A dryland plant ranging from Texas to California, into Mexico and north to Kansas with related species extending range. About fifteen species, of which two occur in our Southwest and the others in tropical or subtropical areas.

Flowers in densely flowered spikes to 5 in. long, appearing from April to June; much visited by bees. Petals to four times as long as sepals. Fruit a slender, pointed pod to 8 in. long and ½ in. wide, constricted between the seeds. Pods ripen in September or October and are sweet-meated.

Indians eat fruits, pounding whole pods to powder and extracting woody part. This is mixed with water and baked in a basket under sand in hot sun for several hours to make a sweet cake. Indians wove cloth from the bark. Plant is attacked by a bark-boring beetle and by another that bores into the wood.

Cattle graze on the plant with no ill effects to either. It provides food, cover, or nesting sites for jack rabbits, Gambel's quail, white-winged doves, ground squirrels, porcupines, raccoons, coyotes, deer, skunks, phainopeplas, mockingbirds, road runners, robins, thrashers, and others. It is a common host for desert mistletoe.

Mesquite

Screwbean, Tornilla
Prosopis pubescens

Tree. Height to 30 ft. Trunk diameter to 1 ft. Branches round in cross section, smooth, light red-brown when mature and armed with strong spines to ½ in. long. Distinguished by the gray-barked immature twigs from the darker-barked twigs of mesquite. Leaves to 3 in. long, compounded of oblong leaflets to ⅔ in. long and ⅛ in. wide.

Found along gravelly and sandy flats or in loam along canals and rivers from the Rio Grande Valley in western Mexico through northern Mexico to southern California and north to Nevada, Utah, and Colorado. Closely related to the mesquite.

Flowers yellow, opening first in early spring; found in clusters in axils of leaves in cylindrical spikes and to 3 in. long. Petals three to four times as long as sepals. Fruit ripens through summer and falls off in autumn, characterized by being twisted into a narrow spiral of to 20 turns, to 2 in. long, and with seeds to 1/12 in. long.

Wood hard and durable in contact with soil. Plant provides a good browse for cattle and for some forms of wild animals. The green or ripe beans are eaten by cattle; Indians eat the pods raw or grind the ripe ones to make a meal from which a cake is baked. Bobwhite, road runners, and Gambel's quail use the thickets of screwbean for cover.

Indians and white men alike use the wood for fuel and for certain kinds of construction. This species reaches its greatest size in the lower Colorado Valley and in the valley of the Gila River in Arizona.

Screwbean

207

Kentucky Coffee Tree

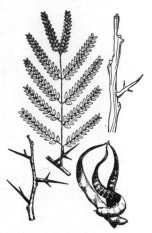

Honey Locust

Wild Senna

Kentucky Coffee Tree
Gymnocladus dioica

Tree. Height to 110 ft. Trunk diameter to 3 ft. Usually with a narrow, rounded top but often dividing from 10 to 15 ft above ground. Twigs coarse, blunt, to ⅓ in. through at tip first year, dark-brown to gray. Recognized by its very distinctive bark with its curlicue patterns of flat ridges and scales. Leaves to 3 ft long and 2 ft wide, divided into five to nine parts, subdivided to six to fourteen leaflets.

Found usually on rich bottomlands or planted as an ornamental. Ranges from southern Ontario to Pennsylvania and west to Oklahoma, Nebraska, and South Dakota but not common anywhere. This is the only species in the genus. Genus closely related to that of the honey locust.

Flowers showy, white, at ends of twigs clustered along a central axis. On pistillate tree, clusters may be to 1 ft long, while those on the staminate are much shorter. Fruit a coarse pod to 10 in. long and to 2 in. wide, containing seeds that are to ¾ in. long and embedded in a dark, sweet pulp through the winter.

Wood, 43 lb per cu ft, hard, strong, coarse, dark red, very durable in contact with soil, with thin light-colored sapwood, rarely attacked by insects. Seeds, 200 to 300 per lb, with high vitality and slow germination. Leaves and sprouts may be poisonous to cattle.

Wood used much as fence posts where it is available, also in cabinetwork and in general construction. Tree used as an ornamental or as an oddity. Long's expedition used the roasted seeds as a coffee substitute in 1820 in Missouri, and a decoction of the pulp of unripe fruits is reputed to have some medicinal value.

Honey Locust
Gleditsia triacanthos

Tree. Height to 140 ft. Trunk to 6 ft through. Bark to ¾ in. thick and in narrow, irregular, vertical ridges. Coarse three-forked spines are found on the branchlets or persisting even on older wood. Leaves doubly compound, or singly compound, to 8 in. long, of 18 to 28 leaflets, dark green above, to 1½ in. long, pale beneath.

In woods, along hedgerows or even planted as an ornamental. Ranges from Ontario to Georgia, west to Texas, Kansas, and Michigan. It has become naturalized as an escape in many areas outside the normal range, growing well in practically all temperate climates in the world.

Flowers in clusters, from axils of leaves of previous year; the pistillate in slender clusters to 3½ in. long; the staminate in short fuzzy clusters to 2½ in. long. Fruit a flattened, twisted pod, to 18 in. long, with thickened margins, mahogany brown, in clusters of two to three with oval seeds to ⅓ in. long in pulp.

Wood 42 lb per cu ft, very durable, bright, yellowish-brown, with darker streaks and with pale, relatively thin sapwood of to a dozen annual layers' thickness. Pollination by insects. Seeds may be collected from September to February, being shed by rolling fruits. Seeds, 3,000 per lb; yield about 1,000 plants.

Wood used for fence rails, wheel hubs, general construction. Trees planted as ornamentals and as hedge. Seeds stratified in sand and soaked in hot water before spring planting. Cattle enjoy eating the fruits and deer, hares, cottontails, squirrels, starling, and quail are known to feed on the plant. Sweet pulp makes a good relish. Cultivated as an ornamental and shade tree in temperate climates. One cultivated variety, Inermis, is now very commonly grown because of its lack of spines.

Wild Senna
Cassia hebecarpa

Height to 8 ft. Sparingly branched. Root perennial. Leaves compounded of 5 to 10 pairs of leaflets, each more or less blunt and to 2 in. long and ½ in. wide, rounded at base, somewhat hairy, with a slender clublike gland at or near the base of the main petiole. The whole plant is only slightly branched.

Found in rich wet soil or swamps. Ranges from New England to North Carolina and west to Minnesota and Louisiana. Among the more important close relatives are *C. chamaecrista,* the partridge pea, a rather common upland weed, and *C. nictitans,* the wild sensitive plant ranging from New England to Florida and Arizona.

Flowers in clusters of many flowers, in axils of leaves in upper part of plant, to ¾ in. across, of five petals almost equal to each other. Stamens, ten, of which upper three have no anthers. Fruit a slender, flat, at first hairy pod whose segments are as long as they are broad, to 4 in. long and ¼ in. wide. Seeds flat.

The cassia of the drugstore comes from *Cinnamomum cassia,* a tree of Egypt and India and other tropical areas, and is used in confections and as a laxative. The drug senna comes from several species of *Cassia* native of Egypt and Arabia. The leaves are collected, dried, baled, and shipped for use.

The plant is most attractive. Its flowers are a golden yellow except for the contrasting chocolate brown of the anthers. Flowering time from July through August.

Redbud, Judas Tree
Cercis canadensis

Tree. Height to 50 ft. Trunk to 1 ft through. Branching distinctly grotesque. Twigs and branches dark, irregular, relatively slender. Leaves long-petioled, with a distinctly heart-shaped blade whose margin is entire and whose veins are conspicuous. Blade to 6 in. broad and about the same length.

Found along stream borders, on rich lands appearing as a dense forest undergrowth or slope cover. Particularly conspicuous along the mid-Mississippi region. Ranges from Ontario to Florida and west to Texas and Nebraska. Two related species extend range to Central and Western states. Seven species from North America, Europe, and Asia.

Flowers appear before leaves; are borne in dense clusters, all along the stems and upper trunk, pinkish to purple, about ⅓ in. long, pealike. Fruit a pod, flat, slender, pointed at each end, to 3 in. long and to ½ in. wide, falling in the autumn and freeing ¼-in. seeds.

Wood weak, hard, 40 lb per cu ft, dark reddish-brown, close-grained, with thin, light-colored sapwood to about 10 annual rings thickness. Seeds about 25,000 per lb, with a germination of 80 percent yielding some 2,000 usable plants per lb, if treated with hot water before planting in the spring.

A beautiful ornamental and a good cover plant. It does not have nitrogen-fixing nodules. Found in slightly acid, calcareous, sandy, or loamy soils. Eaten by deer and by three kinds of birds including bobwhite. Neither it nor its relatives have any forage value for livestock. Supposed to blush for Judas's betrayal of Christ.

Red Bud

Scotch Broom
Cytisus scoparius

Shrub. Height to 10 ft. Much-branched, nearly smooth, with long straight branches that are angled and erect. Leaves of three leaflets, the lower being petioled. Leaflets blunt, to ½ in. long, with a compressed layer of fine fuzz and a short pointed tip. Upper leaves without petioles and smaller than lower.

Established in waste places or planted as an ornamental. Ranges from Nova Scotia to Virginia, west to California and Vancouver Island. Probably native of Europe and naturalized widely in America and elsewhere. May become somewhat of a weed crowding out all normal competitors. Some 45 species native of Europe, western Asia, and northern Africa.

Flowers bright yellow, to 1 in. long, in long terminal leafy clusters. Calyx and supporting stem smooth. Anthers of stamens alternately large and small. Fruit a flat pod, to 2 in. long, flattened on sides and with a coiled style often remaining attached to free end, browning to black. Flowers in May and June.

Drought-resistant and will survive in salty soils or will grow on sand or clay. Considered as in part parasitic on oak roots but this is not a necessary relationship. Seeds 65,000 per lb. Propagated by seeds soaked in hot water before planting in spring or by cuttings merely stuck into the ground.

An ornamental or a weed, depending upon the use made of the soil in which it grows. Has been recommended for restoration of soil. Can be controlled by burning over fields. Recognized as a valuable soil anchor, particularly in the West. Leaves considered poisonous to livestock but are usually avoided. Eaten by bobwhite, cottontail, and California quail.

Scotch Broom

Yellowwood
Cladrastis lutea

Tree. Height to 60 ft. Trunk diameter to 4 ft. Often divided rather close to ground, with spreading branches and hanging, zigzag branchlets. Leaves to 1 ft long, of 5 to 11 leaflets each to 4 in. long, with terminal one usually shorter than others, turning bright yellow in autumn.

Found on limy or neutral soils. Ranges from North Carolina west to Kentucky, Tennessee, and Missouri but planted rather widely as an ornamental in New England and in southern and western Europe. Only this one species in the genus and it is closely related to a genus found in Manchuria.

Flowers appear in mid-June, in clusters measuring to over 1 ft long and 6 in. wide, usually flowering every other year. Corolla white, about 1 in. long. Stamens, 10, distinct. Fruit a short-stalked smooth pod to 4 in. long and ½ in. wide, containing from 6 to 20 seeds and ripening in September, falling soon thereafter.

Wood hard, 39 lb per cu ft, strong, yellow but changing to brown when exposed, with sapwood nearly white and thin. Ordinary propagation by seeds that are harvested in fall, stratified in sand through winter, and sown in spring. Seeds should be collected in September and October though some may be found in winter.

Essentially an ornamental tree grown for the showy flower clusters. Wood has some commercial value for use in making gunstocks. It is used as fuel and yields a clear beautiful yellow dye.

Yellowwood

Wild Lupine

Blue Lupine

Bluebonnet

Wild Lupine
Lupinus perennis
Height to 2 ft. Erect, tufted at the base, branched, rather stout and finely fuzzy in the upper parts. Roots deeply penetrating and dug with difficulty. Leaves long-petioled, to 3 in. broad, with the slender petiole bearing 7 to 11 leaflets that are each to 1½ in. long and to ½ in. wide, dark green and widest toward tip.

Sometimes abundant in dry sandy soils particularly where there is good exposure to the sun. Ranges from Maine to Florida, west to Louisiana, Missouri, Minnesota, and Ontario. Closely related, longer podded *L. parviflorus* extends range to New Mexico and Washington, while others extend range over most of United States and adjacent territory.

Flowers alternating along terminal flower stalks in rather loose, open clusters, the whole cluster being sometimes nearly 1 ft long. Flowers blue, pink, or sometimes white, and to ¾ in. long. Fruit a pod to 1½ in. long, slender, fuzzy, breaking into coils and freeing four to six seeds. Flowering time June through July.

This plant grows in some of the poorest of sandy soils and does not thrive if transplanted to what might seem to be a more desirable environment. It grows from subterranean rootstocks that of course are not easily dug up because of their depth. The plant flowers at about Memorial Day, in the Northeast, and has been abused by flower pickers seeking bouquets.

A massed bed of this plant in full bloom would make anyone stop and look at it with pleasure. The flower is known in different parts of the country by different names, varying from the correct wild lupine to old-maid's-bonnet, Quaker bonnet, and sundial, the last referring no doubt to the response of the plant to the positions of the sun. *L. taxensis*, a close relative of *L. subcarnosus*, is the state flower of Texas. Many species of lupines are cultivated in flower gardens. *Lupinus* is chiefly a North American genus.

Blue Lupine
Lupinus hirsutus
Height to 2 ft. but often low, sometimes higher. Branches freely toward the top. Leaves long-petioled, composed of seven to nine leaflets, each to 1½ in. long, somewhat brown, hairy, and widest toward free end. Leaflets hairy on both sides, thus differing from reddish-purple-flowered *L. hirsutissimus* with its stinging hairs.

Native of southern Europe but rather extensively grown for fodder or for garden purposes. Races of this species show color variation in the flowers from the typical blue to blue-red, red, or white flowers. The related yellow-flowered *L. luteus* is, like this species, an annual and has much the same general uses.

Flowers scattered along an erect terminal flower stalk, alternate above and whorled below, blue with usually a white tip on the keel, to ¾ in. long, appearing in July and August. Pod large and hairy and developed in late season. Seeds bean-shaped, to ½ in. long, grayish or brown and rough. Probably should not be considered as edible.

While this species is grown for fodder and for ornament, many related species are poisonous, particularly if the fruits are eaten. Of these, possibly the perennials are more likely to be suspected but this is not always true. The alkaloids lupinine and spartenine are found in *L. luteus*. Lupine poisoning may be by leaves or by fruits.

This species is planted like any animal forage crop, for cutting, grazing, or plowing under as a green fertilizer. As a garden flower, it is grown as almost any common garden annual is except that it needs a fairly long season. Seed required for sowing 1 acre, 1½ to 2 bu.

Bluebonnet
Lupinus subcarnosus
Height usually under 1 ft. Silky in some areas. Leaves long-petioled with five leaflets each to 1½ in. long arising from the petiole end; smooth above, hairy beneath, the petiole exceeding the length of the leaflets, and with narrow, pointed stipules at their bases. Annual.

Common in dry open spaces in its native range, which is small and practically limited to Texas. Included here because this is the official state flower of Texas. More than 100 species of the genus in the world, of which 70 occur in North America, most of them being in the western parts.

Flowers in a short, terminal, rather open cluster, somewhat scattered, blue with a whitish or yellowish spot in the center of the standard, about ½ in. long. Calyx hairy and with the upper lip much shorter than the lower. Fruit a pod to 1½ in. long, hairy, and containing the seeds that are from ⅛ to 3/16 in. across and mottled strongly.

Western lupines, of which this is one, would make a book in themselves. They vary from treelike *L. arboreus* and shrubby *L. densiflorus* of California to somewhat woody *L. diffusus*, deer cabbage of the East, to the great group of perennials, some of which like *L. argenteus* of the West may cause serious poisoning of sheep, cattle, and horses.

Animals poisoned by lupine may froth at the mouth, have labored breathing, become nervous, have convulsions, and die. One record shows that 1,150 sheep out of a flock of 2,500 grazed in late season died of lupine poisoning. Bluebonnets are protected for their great beauty.

Red Clover
Trifolium pratense
Height to 2 ft. Clumps of ascending stems from a deep taproot. Biennial. Leaves long-petioled, with three leaflets, the terminal stalkless, commonly somewhat hairy; each leaflet to 2 in. long, with or without V marking on upper side. Leaves and stems may be smooth. Stipules long-pointed. New growth comes from the crown of taproot, not from runner.

Native of Europe and Asia but widely cultivated through the United States and southern Canada. Probably was first cultivated in Spain in fifteenth and sixteenth centuries; brought from there to Flanders; in 1633, to England; probably to America about 1747. It is not popular in the Great Plains or Rocky Mountain areas but is excellent in the Northwest.

Flowers in egg-shaped terminal clusters, remaining erect as individuals after maturing. Heads about 1 in. long. Red flowers about ½ in. long. Calyx hairy, with teeth shorter than the corolla. Honeybee's ¼ in. proboscis too short to reach bottom of ⅖ in. tube, but bumblebee's proboscis is long enough to effect pollination. However, bumblebees are seldom present in sufficient numbers to ensure a good seed crop.

Seed may be sown alone in spring at 8 to 10 lb of seed per acre or mixed with other plants such as timothy or small grain, such as oats, flax, or winter wheat. In South, seeding may be done in late summer. After grain crop is harvested, clover may reach height of 6 in. before winter and a good crop may be cut second spring to be followed by another later in the season.

Red clover is one of the best of the clovers raised for hay. Difficult to get seed set because of insect problems and because of weevils that may destroy the set seed. Popular varieties include June Red, Mammoth Red, Medium Red, of which Mammoth is likely to bloom late and avoid some early pests. Taproot system of this species is robust and almost as deep as that of alfalfa. Red clover does not have the drought resistance of alfalfa.

Red Clover

Alsike Clover
Trifolium hybridum
Partly reclining, smooth plant, from leaf axils appear successively new flowering branches. It does not lie prostrate nor root at joints as does white clover. The underside of leaflets is dull, not shiny as in white or hairy as in red clover. Leaflets unmarked, toothed, pale, shorter than stalks. Stipules broad, with branching veins.

Native of northern Europe and first cultivated in Sweden. Brought to America in 1839. Now grown extensively over United States and Canada but especially from Maine to Minnesota, south to Ohio River. Used as a hay, pasture, or forage plant but also as a soil enricher because of its nitrogen-fixing qualities.

Flowers white to pink and turning back on the stem when mature. Appear in globular heads. Calyx much shorter than white corolla. Flower-bearing branches arise successively from each leaf axil, so that the youngest flowers are at the ends of the stems. Like white clover, the plant is a good feeding source for bees and these are necessary for pollination. Seed yield may be greater and more certain than in red clover. Each pod yields from two to four seeds.

Alsike Clover

Alsike seed is often sown with other clover seed or with grass mixtures with excellent results. For seed crop, from 8 to 15 lb of seed are required per acre, though 4 to 5 may be adequate as there are about 700,000 seeds per lb. One crop may be expected the second season and this may be followed by limited grazing. Best in cool moist climate.

A good hay crop for soils too wet or too sour for red clover. Seed crops may yield to 4 bu per acre. Roughly, digestible food ingredients in 100 lb of clover feed are, in percent: Protein: red, 7.38; alsike, 8.15; white, 11.46; crimson, 10.49. Carbohydrate: red, 38.15; alsike, 41.7; white, 41.8; crimson, 38.1. Fat: red, 1.81; alsike, 1.36; white, 1.48; crimson, 1.29.

White Clover
Trifolium repens
Prostrate, with no upright stalks such as are found in red, alsike, and crimson clover. Leaves of three small leaflets, with the free end somewhat notched and the margins definitely finely notched; undersides of leaflets smooth and glossy; petioles sometimes very long; without mucronate tips. Stems sprawling, rooting at joints; sometimes 1 yd long and a dozen to a plant. Stipules small.

Various races from different parts of the world. Found in some form in almost all temperate parts of the world. Wild white clover is perennial,

White Clover

(Continued)

Crimson Clover

native of North America; Dutch white clover is probably native of Europe; and Ladino white clover is probably from southwestern Europe. Ladino does best on nearly neutral soils and favors much moisture.

Flowers normally in white globes but sometimes pinkish, to ½ in. long, with corolla two to three times as long as the calyx, whose teeth are shorter than its tube.   In Ladino flowers are much larger than in other white clovers.   Flowering time from May through December and when mature flowers bend back.   Pods contain two to several yellow seeds.

Except to secure a seed crop, white clover should not be planted alone.   1 lb of white clover added to other clover acre allotments or 2 lb added to grass mixtures are usually suitable.   Normally, 4 to 6 lb per acre is best for a clear clover sowing.   Seed should not be covered more than ¼ in., rolling usually providing this.

Valuable for hay, for green manure, for enriching soil, and for pasture, with varieties different in service.   Dutch white may be only annual requiring new seeding.   Wild white stand may last 3 to 6 years or even longer.   Ladino white, the largest and probably best, may last 4 to 6 years with one seeding.   White clovers improve yield and quality of associated grasses.   Often used in mixture with lawn grass seeds.   Accepted by some people as the true "shamrock," the plant selected by Saint Patrick to illustrate the doctrine of the Trinity.

Crimson Clover
Trifolium incarnatum
Height to 3 ft.   Sparingly branched, covered with soft fuzz, generally erect.   Leaves compounded of three leaflets, all arising from same point; each to 1¼ in. long, usually blunt or slightly indented at free end, with margins finely toothed and terminal leaflet, stalkless.   Stipules at base of leaves blunt, to 1 in. broad, thin; with toothed margins.

Found in fields and waste places or grown for ornament or as a forage or green manure plant.   Introduced from Europe and established sparingly from Canada to the Gulf, sometimes temporarily elsewhere.   Mainly used southward, in the Gulf States and the Carolinas.   Of some 300 species of clovers found through temperate Europe, Asia, North America, Africa, and South America, only 8 are regularly cultivated.

Flowers in rather long pointed conelike heads that are to 2½ in. long, with the heads single.   Flowers crimson and very showy, to ½ in. long, with the corolla equaling or exceeding the calyx lobes, the calyx being conspicuously hairy and the hairs being stiff at maturity.

Seed sown from August to October on prepared seedbed, 10 to 15 lb per acre, germinates promptly and if water is present a quick stand is established.   100 lb of good unhulled seed equals about 60 lb of hulled.   Usually seed is saved for local use but some sections such as Maryland and Tennessee produce it for sale elsewhere.

Chief use of crimson clover is for plowing under as green manure.   Mature clover cut and used for hay may have hairy heads that collect in balls in stomachs or intestines of domestic animals and may cause death.   Was introduced into United States in 1818.   Can be considered ornamental.

Fenugreek
Trigonella foenum-graecum
Height to 2 ft.   Stem erect, usually unbranched, fuzzy, heavily scented. Leaves of three leaflets like most clovers but with the indentations near tip obscure.   Leaflets ¾ to 1 in. long, with veins commonly running out into the teeth.   Stipules close to petiole.   Root system rather deep for an annual.

Fenugreek

Native of western Asia and naturalized in Mediterranean region; little grown in America but commonly grown in southern Europe.   Grown for flavoring in India.   Some 70 related species.   Name Trigonella means little triangle and probably refers to shape of leaflets.   Best on well-drained loams; poor on clay, sand, wet, or sour soils.

Flowers white, blue, or yellow, in an obscure head.   Petals free from stamens and standard rather oblong.   Keel shorter than the wings.   Stamens in two groups, of nine and one.   Fruit a pod, to 6 in. long, slender, round in cross section or flattened, curved and with a beak, with about 16 seeds.   Unlike alfalfa, pod is straight.

Requires potash and phosphoric acid for seed production; nitrogen for forage.   Needs deep plowing and thorough harrowing.   Sow broadcast, 10 to 20 lb to acre or drill 7 to 10 lb per acre, in rows 18 in. apart.   Thin at 2 in.   Clean culture to maturity yields forage in 4 to 5 months 950 lb per acre.   Seeds flavor stock foods.

Seeds used as human food; in Egypt, with flour in bread; in Greece, boiled or raw, with honey; in India, with condiments; in Orient, to fatten women.   Seeds not used as internal medicine but may incite thirst and stimulate drinking.   Not found commonly in U.S., but is used in the coastal areas of southern California as a winter green manure crop.

Birdsfoot Trefoil
Lotus corniculatus
Stems to over 2 ft long, sprawling, ascending, slender, rather smooth, with many from a common root that may be exceptionally long and penetrating. Leaves of five leaflets, three of which are like a clover at tip and two of which are near base of a central axis. Leaflets pointed ovals, dark green, to about ½ in. long. Perennial.

Closely related to true clovers. About 90 species, having as a group a wide distribution over world and of course considerable variation. This species is native of Europe and Asia but is widely established in North America. Some races are much more valuable as forage than others.

Flowers in clusters of 3 to 12, springing from a more or less common point, sweet-pea-like, yellow or tinged with red, to about ½ in. long, the cluster on a stem to 6 in. long. Lobes of the calyx as long as the tube. Fruit, slender pods, to about 1 in. long, often forming what looks like a bird's foot, hence the common name.

Probably originally entered America as a weed in ship's ballast. Established as a weed along Atlantic Coast and was considered primarily as a weed. Then it began to be used as an ornamental in rock gardens and wild gardens and finally was recognized here as an important soil builder and forage plant.

It has long been grown as a forage plant in Europe but now in America it is recognized in some types of soil as a competitor of the clovers for producing a maximum of nourishing food for cattle while contributing a maximum in enriching the soil. Proper seed for particular places must be used to get good results.

Birdsfoot Trefoil

White Sweet Clover
Melilotus abla
Height to 10 ft. Erect, freely branching, smooth or nearly so, rather tough when mature. Leaves of three leaflets, of which two lower seem to be stemless, fragrant when crushed or dried. Leaflets narrowed at base, with whole margin shallowly cut, each to 1½ in. long, may fold so that surface exposed is reduced, with mucronate tips. Root a deeply penetrating taproot.

Native of Europe and Asia but widely established in America and common as a weed or grown as a crop throughout most of the United States. Related yellow-flowered *M. officinalis* seems to be less abundant though it is more common in the Middle West than in the East. Its leaflets are less inclined to be blunt.

Flowers in delicate slender "spires" at ends of branches, erect, numerous, and to 4 in. long. Flowers themselves are white, to ¼ in. long, fragrant, and with mature ones at base of spire. Fruit a pod about ⅛ in. long, bearing a single oval yellow seed somewhat like a flattened pea. Annual or biennial.

Does best on soils rich in lime but succeeds on poor soils. Excellent as a bee forage plant and the root systems serve well as soil anchorage in erosion-control practices. About 20 lb of seed are required for sowing 1 acre. Outdoorsmen frequently strip seeds from weed tops and sprinkle them on waste bare land to help give game cover and soil anchor.

Where grown as a hay crop, it may yield very well the second year or may provide good early grazing if cattle are accustomed to eating it. Known commercially as Bokhara clover. It harbors nitrogen-fixing bacteria and so enriches the soil as well as holds it. Poets speak of "the spiring of sweet melilot." White sweet clover flowers attract many more butterflies than yellow sweet clover. Seeds of white sweet clover similar in size and shape to alfalfa seeds can be separated easily. Alfalfa seeds fluoresce, white sweet clover seeds do not. "Sweet" principle in plant is coumarin.

White Sweet Clover

Black Medic, Yellow Trefoil
Medicago lupulina
Sprawling by means of prostrate stems that are to 2 ft or more long and many from a central root system. Roots at the joints. Leaves petioled, dark bluish-green, of three leaflets, each with conspicuous veins and shallowly toothed margins near the tip. Leaflets to ⅔ in. long, narrowed or rounded at the base, capable of folding. Stipules with toothed margins.

Native of Europe and Asia; naturalized widely in all temperate areas but not in extreme cold north and not so abundant in tropics. Not easily confused with its close relative alfalfa or with less closely related but larger-flowered yellow clover. Often covers dry waste spots to exclusion of all other plants.

Flowers in small compact almost spherical heads that are to ⅝ in. long. Flowers to 1/12 in. long, yellow and crowded. Fruit nearly smooth, curved into a partial spire, conspicuously veined, one-seeded and decidedly black, giving rise to common name black medic. Flowering time from July through September.

Plant does some service as a soil anchor and as an enricher of soil because of its nitrogen-fixing nodules. Its seeds have been used as deliberate adulterant of red clover but with new recognition of grazing value of

(Continued)

Black Medic

213

Alfalfa

Gorse, Furze, Whin

Prairie Turnip

better varieties, particularly in the South, this has discontinued.   Legislation and inspection of course have assisted in this.

Probably serves best if sown in grass mixtures suitable to local situation. Then it can help feed the grass and be kept from clinging too close to the ground; thus being available to grazing animals.   Since it is either a biennial or an annual, it is important that one be familiar with the seed one is buying.   It is known as "nonesuch."

Alfalfa
Medicago sativa
Height to 3 ft.   Erect or partly sprawling, branched, more or less smooth, arising from a crown.   Leaves compounded of three somewhat petioled leaflets, each to 1 in. long, with slightly toothed margins particularly toward free end, with narrowed bases and mucronate tips.   Root system a long, coarse taproot with numerous side branches.

Native of Europe but naturalized through most temperate parts of the world.   Established in waste places from New England to Virginia, west to Minnesota, Kansas, and even on to Pacific Coast.   Some 50 species of the genus, about a dozen of which are of some commercial importance.

Flowers in dense, short, terminal clusters, violet or purple, with petals about ¼ in. long and with yellow stamens capable of being "tripped" by a visiting insect and thus effecting pollination.   Fruit pod, elongated into two to three spirals containing several olive-green seeds.   Pod slightly fuzzy.   Seeds have peculiar knob near one end.

Seed sown in new fields may need inoculation with bacteria.   Sow 12 to 20 lb per acre in deep well-drained loam, with ample potash and phosphorus, usually about July, on weed-free soil to a depth of ½ to 2 in., with or without nurse crop.   If new crop turns yellow, clip back to 6 in. high. Good stands yield from 1 to 10 cuttings a year for many years.

Excellent forage plant, soil restorer, soil anchor, bee plant that is fairly free from insect and fungous pests.   Hay should be cut just as plants come into bloom and should be protected from rain.   For cattle, alfalfa provides excellent food for blood, bones, and nerves, good food for muscles, and fair food for energy.   It is a good source for vitamin A and fair for vitamin B.

Gorse, Furze, Whin
Ulex europaens
Height to 6 ft.   Stiff, tangled, striped shrub that is fuzzy when young and smoother when old.   Leaves reduced to mere scales or narrow and fuzzy, to ½ in. long and prickly, making branches and whole plant a mass of spines and an almost impenetrable tangle.   Underground system suitable for soil anchorage.   Appears evergreen because of mass of green young branches.

Native of western and southern Europe but established somewhat in North America, particularly along Atlantic Coast from Massachusetts to Virginia but also on West Coast on Vancouver Island and in California. Some 20 species recognized in eastern Europe, of which this is probably the best known.

Flowers in clusters of to three or singly in axils of leaves but crowded toward ends of branches, bright yellow and fragrant, to ¾ in. long, with calyx only slightly shorter than corolla; individual stalks of flowers very short.   Fruit a small, flattened, few-seeded pod, dark brown, fuzzy, over ½ in. long.   Gorse may become a serious pest, crowding out its perhaps more desirable associated plants.

Flowering time, May through July, when plant may be attractive and serve as a source of nectar for bees.   A double-flowered, sparsely spined variety is known that has more appeal as an ornamental than the typical plant.   A tangled mass of gorse provides an excellent cover for many small animals.

Found on sand, clay, loam, and other soils and makes a superior soil anchor since it can survive burial in sand and can exist in salty water.   It is not accepted by stock as fodder, probably because of spines.   Makes excellent fuel of almost explosive qualities.   Plant figures in many nursery stories of European origin.

Prairie Turnip, Pomme Blanche
Psoralea esculenta
Height to 1½ ft.   Stem erect, stout, slightly branched.   Whole plant, covered with a dense felt of whitish hairs.   Large spindle- or turnip-shaped root or cluster of roots, about size of a hen's egg, whitish inside but with thick tough brown outer covering.   Leaves of five short-stalked leaflets, each to 2 in. long and to ¾ in. wide.

Native of the plains areas from Manitoba through North Dakota, Nebraska, and Missouri to Texas, west to Montana.   About 120 species in the genus widely distributed over the world, with at least two dozen in United States of which this is probably the most interesting though not necessarily the most common.

Wild Indigo

Flowers in spikes to 3 in. long and half as thick, bluish, with individual flowers with bluish petals about ¾ in. long, almost equaled by calyx bracts. Fruit a small pod to ⅓ in. long, smooth, wrinkled, and hidden by calyx tube. Flowering time, June. Fruiting time July and later.

Reported to have played an important part in solving food problem of early travelers in range of the plant. Indians harvested the roots, stored them, and ate them raw, ground, baked, boiled, roasted, or made into cakes. Digging done usually in August after the top had died down.

Cultivation of the plant for food has been investigated, but it cannot compete with potato which can be grown in the same range. Fresh root flesh tastes something like a flat chestnut. Common names include Cree potato, tipsin, prairie potato, Missouri breadroot, prairie apple, pomme de prairie.

Wild Indigo
Baptisia tinctoria
Height to 4 ft. Branched, somewhat juicy, erect, smooth, rather pale green herb. Leaves with three leaflets, each bluntly rounded at the free end and broadest toward tip, entire, turning black with age, to 1½ in. long and about ½ as wide. Root system substantial and perennial.

Found in dry soil from Maine to Florida and west to Louisiana, Minnesota, and Ontario. About two dozen related species found in eastern and southern North America of which this one and blue-flowered *B. australis* are sometimes grown for ornamental purposes. Other species have flowers of different colors.

Flowers in few-flowered clusters at ends of branches, on short individual stems. Flowers bright yellow, to ½ in. long, obscurely pealike. The many flower clusters on a single plant are conspicuous. Fruit an oval pod, long-pointed at each end, to ½ in. long and much longer than supporting calyx tube.

Flowers appear in June and through to September; visited by syrphus flies, butterflies, and honeybees, leaf cutter and other bees, of which *Halictus* is possibly an important pollinator. Does best on high, dry, loose sandy soil, often existing free from much competition. Root has a rather distinctive and permanent dye.

Root is used sometimes as a source of dye material. Roots and herbage are used as a source of some medicines. Herbage may cause some poisoning to animals if they graze on it. Probably its most interesting use is as an ornamental in wild-flower gardens.

Bastard Indigo

Bastard Indigo
Amorpha fruticosa
Shrub. Height to 20 ft. Smooth or somewhat fuzzy, branched but with tip pointing upward for the most part. Leaves compound, of 11 to 25 thin leaflets to 2 in. long and nearly 1 in. wide, entire and grayish-green, rather conspicuously veined, short-stalked.

Ranges from Pennsylvania to Florida, west to Saskatchewan, Colorado, and Chihuahua. Escaped from cultivation elsewhere. Some 15 related species found in North America and Mexico with *A. canescens* being frequently rather conspicuous in the South.

Flowers in slender, spikelike clusters to 6 in. long. Individual flowers to ⅓ in. long, violet-purple, pealike, with corolla to three times as long as the supporting calyx tube. Stamens extend beyond corolla and may appear conspicuously yellow against the blue. Fruit a smooth, two-seeded, ⅓-in. pod.

Plant commonly grows along streams or on embankments. It makes an attractive ornamental particularly when it is in bloom. About 60,000 seeds per lb; germination of 65 percent give 10,000 usable plants per lb of seeds. Plant does well on limy soils, favoring sun but surviving in shade.

Plant may be slightly poisonous to stock that graze on it. Bobwhite are known to feed on the fruits and the plant gives good cover for game. Recommended as a soil anchor for use along stream banks and popular as an ornamental.

Purple Prairie Clover
Petalostemum purpureum
Height to 3 ft. Erect, branching at tip, smooth or slightly fuzzy, a clean green. Leaves clustered, short-petioled, of three to five leaflets, each narrow and to ¾ in. long and 1/10 in. wide, pointed at tip and narrowed at base, short-stalked, with terminal three leaflets coming from an almost common base.

Found in plains and prairie county. Ranges from Indiana to Texas and west to Colorado, Saskatchewan, and Manitoba. About 50 closely related species found in United States, Canada, and Mexico, some having white flowers while others are pink, purple, or blue and with foliage either smooth or fuzzy.

(Continued)

Purple Prairie Clover

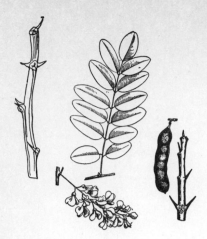

Black Locust

Wisteria

Milk Vetch

Flowers in rather stout, compact, cylindrical spikes that are to 2 in. long and about ½ in. thick. Individual flowers violet or purple and to ⅙ in. long, pealike but with keel and wings rather uniformly oblong. Calyx densely gray and silky, providing an attractive background for the brighter petals.

Roots usually long and deeply penetrating, providing the plants with moisture when many nearby plants may have become withered. It probably never becomes a weed and is so attractive when it is in bloom in July and August that many persons may wish to transplant it. It does not survive cultivation.

Plant may be grown as an ornamental in gardens and is frequently popular in this way. It deserves more protection than it gets in its natural setting. The related white-flowered prairie clover *P. candidum* is almost as attractive as this species.

Black Locust
Robinia pseudoacacia
Tree. Height to over 80 ft. Tall, cylindrical, irregularly branched. Trunk to 4 ft through. Bark dark reddish-brown with squarish scales. Roots shallow, tough, yellowish. Twigs slender, with buds hidden in swellings and with short stiff thorns. Leaves alternate, compounded of 7 to 10 small oval green margined leaflets. May be known as acacia or yellow locust.

Common on rocky high grounds. Ranges from New York to Georgia, west to the Ozarks but planted and naturalized widely east of Rocky Mountains and to some extent in Great Basin. Introduced into Germany in 1601. Now probably the most widely distributed of native North American trees. Up to 3,500-ft level in the Appalachians.

Flowers white, pealike, in loose open clusters, with the standard of the flower large, rounded, turned backward, and scarcely longer than the wings and keel. Fruit a flat brown dry smooth pod, to 4 in. long, and ½ in. wide, containing four to eight seeds, each about ³/₁₆ in. long, dark brown, and like flattened hard peas.

Wood 46 lb per cu ft hard, strong, greenish-brown with yellow sapwood, durable, close-grained, swells and contracts little in changing moisture content. Roots host to nitrogen-fixing bacteria so tree enriches the soil. May grow to 4 ft a year in youth. Has many galls and fungous diseases.

Used as a soil anchor and enricher. Hardy street tree that yields some trash. Fragrance of flowers popular. Grown in managed forests for posts, railroad ties, telephone-pole arms because of speed of growth and durability of wood. Roots, leaves, and bark poisonous if eaten. Seeds eaten by bobwhite, deer, and rabbits.

Wisteria
Wisteria sp.
Japanese *(floribunda)*, with seven to nine pairs of leaflets; Chinese *(sinensis)*, with fewer than seven pairs of leaflets and flowers over 1 in. across; silky *(venusta)*, with silky leaves; American *(frutescens)*, with flowers less than 1 in. across. Most have stout woody climbing trunks. Probably Japanese is the most popular.

Do well in dry sandy soil but better on deep rich ground. Common names indicate parts of world from which different species come, the silky wisteria coming from Korea and Japan. American wisteria ranges from Virginia to Florida and west to Texas with a smaller-flowered species *(macrostachys)* from Illinois south.

Flowers of *W. floribunda* violet to blue, pealike, in long, drooping clusters with a hairy calyx whose upper two teeth are broad and flat. *W. sinensis* blooms later in the season; it may form great drooping clusters from the ends of the branches, the clusters often being over 1 ft. long. There are double varieties but they are not hardy.

Usually rather difficult to transplant so that a vigorous growth can start soon. Highly temperamental about blooming and some plants never seem to bloom no matter what treatment is followed. May be started by cuttings of wood, by root cutting, by top grafting established roots. When grown from seeds do not normally "come true" in all respects.

One of the most popular of the ornamental woody plants grown against buildings or alongside porches. May be vigorous enough to climb to top of building four to five stories high and hardy enough to grow almost anywhere in the United States. Pruning back to short spurs every year is recommended for abundance of bloom.

Milk Vetch
Astragalus distortus
Height to 15 in. Much-branched and with many stems arising from a common deeply penetrating root system. Mostly smooth but with some scattered hairs. Leaves compounded of 11 to 25 leaflets, each to ½ in. long, rounded at tip, entire and narrowed at base, usually opposite each other and not greatly crowded.

Common in dry soil from West Virginia to Mississippi and west to Texas and Iowa.   Some 1,500 species of which the species *mollisimus, lentiginosus, diphysus, earlei, wootonu, nothoxys, thurberi, argillophilus,* and probably others plus members of the genus *Oxytropis.*

Flowers purple, to ½ in. long, in short loose spikelike clusters that reach about as high as the leaves and are few-flowered.   Calyx tubular, with teeth about equal.   Fruit a pod that is not stalked in calyx tube, slightly inflated, strongly curved, grooved on one side, slender and to 1½ in. long. Flowers appear in March through July.

The genus is most important because of locoweed danger to cattle, being particularly bad on soil containing selenium.   Dried green parts of plants may be fatal to horses and cattle, making them drag their feet, become dull and listless and finally die.   Gum tragacanth used in medicine and in calico printing comes from *A. gummifera* of Eurasia.

Horses suffering from locoweed poisoning are treated with 22 cc of Fowler's solution daily in grain or in water while cattle are given ⅕ g of strychnine hypodermically in the shoulder daily for 30 days.   Plants of a given species may be poisonous if grown over selenium soils and harmless over other soils.

Tick Trefoil

Tick Trefoil
Desmodium glutinosum
Height to 3 ft.   Openly branched.   Stems slender, erect, smooth or slightly fuzzy.   Leaves clustered near the upper parts, with petioles to 6 in. long; composed of three leaflets, each to 6 in. long, nearly round but with pointed tips, entire margins, green on both sides.   Perennial from a substantial root system.

Relatively common in dry open woods.   Ranges from Quebec to Florida, west to Louisiana, Oklahoma, Kansas, South Dakota, and Ontario. About 200 species in the genus native for the most part of North America, South America, Africa, and Australia with about 40 in the southern part of the United States.

Flowers large, purple, with two-lipped calyx and pealike petals, 10 stamens fastened together at their bases; flower cluster, open.   Fruit looks like a pea, with each seed surrounded by a conspicuous lobe of the pod, the sections breaking apart and sticking to objects by means of a covering of fine stickers.

Blooms through June to September.   Pollination effected by honeybees and similar small bees, many being in the genus *Halictus.* Seeds disseminated by becoming attached to clothing or to wool of animals and being carried some distance until they drop off or are otherwise removed. Fruits may tangle long wool rather badly.

Probably of little economic importance but rather well known because of portions of fruits that cling to the clothing of those who wander through the woods in fall or early winter.   A common genus with many species, all with hooked hairs on the pod; name from Greek *desmos,* a bond or connection; probably having reference to the jointed pod.

Hairy Vetch

Hairy Vetch
Vicia villosa
Height to 4 ft. but usually sprawling, crawling, or climbing.   Finely fuzzy all over stems and leaves.   Leaflets 8 to 24, opposite, narrow, to nearly 1 in. long about ⅕ as wide, whole leaf ending in paired, twisted, tendrillike structures.   Annual or biennial.

Cultivated for fodder and sometimes established as an escape from such treatment.   Native of Europe and Asia.   Closely related *V. cracca* or cow vetch is a perennial that is smoother and native from Newfoundland to New Jersey and west to Iowa, Washington, and British Columbia. About 130 species in the genus widely distributed.   With few exceptions, *Vicia* spp. are semiviny plants.

Flowers about ¾ in. long, violet-blue or sometimes white, in one-sided clusters arising from the axils of the leaves, pealike.   Stamens in groups of nine and one.   Fruit a pod, to 1 in. or less long, pale green, smooth, containing two to eight small black smooth nearly globular seeds.

Vetches grown for fodder and for enrichment of the soil.   When sown for hay, for winter cover crop, or for soil improvement, 1 to 1½ bu of seeds is sown per acre, usually in mixtures with oats or beardless wheat.   Sown for green manure, the combination is usually with rye.   Often sown in orchards to help soils.   Some vetches are winter annuals in regions having mild winter temperatures.

The green seeds of some vetches are eaten in Europe as we eat green peas.   In South vetches are sown on Johnson grass sod and cut the following summer, the vetch reseeding itself by the time Johnson grass is ready to be cut and eventually improving yield as well as soil.   *Coronilla varia,* crownvetch, has been developed as an excellent soil holder; used extensively to impede erosion on roadcuts in northeastern United States.

217

Japanese
Clover

Lespedeza, Japanese Clover, Bush Clover
Lespedeza striata
Height to over 18 in., sprawling or reaching upward. Tufted, much-branched, with some closely pressed fuzz. Leaves of three leaflets about equal to each other, rather blunt-tipped, each to ¾ in. long and to ⅓ in. wide, narrowed at base and almost stemless; with margins entire, partly hairy.

Found in fields and meadows from Pennsylvania to Florida, west to Kansas and Texas. Naturalized from China and Japan and Korea but rather widely established. Of some 40 species native of Asia, Australia, and North America, nearly half are found in North America. This species is one of most commonly cultivated.

Flowers in clusters of one to three, relatively inconspicuous, with or without petals, in axils of leaves, pink or purplish, with about ½ in. corolla. Calyx smaller than the pod, which is oval, pointed, one-seeded, and to ⅒ in. long. Related *L. japonica* has pure white flowers. Our species flowers through July and August.

Does well on sandy soils and serves to enrich these more or less sterile soils by adding nitrogen and humus. Best grown with Bermuda grass in the South. Since it recovers quickly from close grazing once it has become established, it is a good forage plant. Sow at rate of 15 to 25 lb per acre.

On good lands *Lespedeza* will produce excellent hay but it may be necessary to reseed either by sowing seed, by leaving uncut strips, or by cutting early and leaving recovered crop to produce seeds for succeeding year. Known as "hoopkoop plant." Many species of *Lespedeza* are important argicultural legumes today. Some, e.g., *L. sericea*, are used in wildlife plantings. They have also become important erosion-control plants in some southern states.

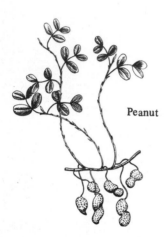

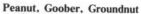
Peanut

Peanut, Goober, Groundnut
Arachis hypogaea
Height to 20 in. Two plant types recognized: a vine and the bunch-nut type. Stems branched, slightly hairy. Leaves of four leaflets, each to 2½ in. long, blunt or short-pointed, entire, without tendrils, rather smooth and green. Root system spreading; underground stems bear flowers at intervals.

Native of Brazil but carried to Old World by Portuguese explorers and brought back to Virginia by African slaves. One of the South's most important crops. Not grown extensively on a commercial basis much north of Washington, D.C. It does best on sandy and loamy soils. It is given for forage and soil building to the North.

Flowers yellow, in close spikes of one to three flowers which appear at first above ground where immature fruit is enclosed in calyx. Flowers then thrust underground by plant and there matured. Fruits are almost too well-known to need description but consist of two dry shells usually enclosing two oblong seeds.

Plants cannot survive frosts, and tops to be fed should be harvested before frost. Plants grown in crop rotation to enrich soil, as forage for hogs, food for man, and as source of fats and oils of commercial value. 1 lb of peanuts yields 2700 calories of food, as contrasted with 900 for 1 lb of beef.

At least 20 kinds recognized. Peanut oil cold-pressed from seeds is used as salad oil, as olive-oil substitute for packing sardines, while seeds pressed with heat are used in soaps, lubricants, and as illuminants. The cake residue is one of best of stock foods. Some plastics are basically peanut oil.

Groundnut, Wild Bean
Apios americana
Height to 10 ft, climbing over other plants and twining extensively. Juice milky. Leaves of five to seven or sometimes three leaflets, each with rounded base and rather entire margins, to 3 in. long and pointed at tips. Root system a necklacelike series of tubers on a long winding axis, each to 3 in. through.

Relatively common in thickets and damp places from New Brunswick to Florida, west to Texas, Kansas, Nebraska, Minnesota, and Ontario. About five species in the genus of which two are from China and one in the Himalaya region. Of the two species in eastern North America, *A tuberosa* alone has any wide range outside Kentucky and Tennessee.

Flowers brownish-purple, pealike, strongly scented, to ½ in. long, pollinated by honeybees and similar insects, borne in rather loose clusters along a central axis arising from axil of a leaf, held more or less erect. Stamens 10, with 9 in one group. Fruit a straight pointed 3-in. pod, like a slender bean.

Reproduction essentially by tubers or by seeds. Tubers may be eaten raw or peeled, scalded, roasted, or boiled like potatoes or cooked with other foods. They may be dried and stored for use in winter. Seeds may be prepared for use like peas. Many prefer tubers and seeds to cultivated potatoes and peas. Generic name *Apios* from *apion*, pear, refers to the shape of the tuberous roots.

Groundnut, Wild Bean

Beautiful enough to serve as an ornamental for foliage and flowers. Was used by early settlers as a basis for bread. An important food plant of the American Indians. Asa Gray considered it one of the most important native American tubers. Pilgrims used tubers as food the first winter. Whittier in "The Barefoot Boy" speaks of "Where the groundnut trails its vine."

Hog Peanut

Hog Peanut
Amphicarpa bracteata
Slender vine, to 8 ft long, twining about plant supports to form tangled masses. Stem clothed in fine brownish hairs. Leaves of three leaflets, each to 2 in. long with rounded bases and pointed tips, thin and rather light green. Root system perennial, bearing subterranean flowers near the surface. *A. monoica* of some authors.

Common in rich damp rather open woodlands. Ranges from New Brunswick to Florida, west to Louisiana, Nebraska, and Manitoba. Only two of the seven species found in America, the others being natives of Asia and the Himalaya regions. It may have been encouraged by early pioneers but cannot compete with other plants.

Flowers of two kinds. Those above ground are pealike, in loose hanging clusters, purplish-white, to ½ in. long, producing pods to 1 in. long that split into spirals, freeing the seeds. Subterranean flowers, rudimentary, borne near base of stem, with few stamens but producing a pear-shaped fruit with one large seed.

Flowers appear through August and September. Underground fruits may be collected from mid-September through the winter. Seeds of both kinds of flowers eaten, though the underground ones are favored; they taste better than raw peanuts. Of course, underground flowers are self-pollinated.

Seeds of flowers above ground used like peas, or may be boiled into soups. Seeds of underground flowers good raw but better roasted or boiled. May be collected without destroying root system. Pioneers robbed caches of underground fruits made by squirrels. Hogs dig fruits for food. Quail feed on the peanuts.

Soybean

Soybean, Soja Bean
Glycine max, Soja max
Height to 6 ft. Stems erect, stiff, hairy, with some of the side shoots more or less vinelike. Leaves compounded of three leaflets, each to 6 in. long, entire, about ⅔ as wide as long and borne on long stems. Whole plant can form a rather dense tangle of considerable depth.

Native of China, India, and Japan but now extensively grown in the United States. In United States, it is grown along northern border of land suitable for growing cowpeas, namely, north of a line from Kansas to Maryland. It does well on poor soils.

Flowers inconspicuous, white or purple. Calyx hairy, with the two upper teeth more or less united and the standard of the corolla broad. Stamens usually fastened together. Fruit a pod that hangs on a short stalk, to 3 in. long and ½ in. broad, brown, hairy and with two to four globular black, brown, green, or yellow seeds.

Does best on warm well-drained loam, planted in drills 2½ to 3 ft apart or in hills to 20 in. apart, cultivated frequently in early stages until ground is shaded, then cut for hay or for silage, cut and threshed for seed, or plowed under for green manure. Often grown in orchards for enrichment of soil. Considered a feed-forage-green-manure crop.

Seeds have a variety of uses including food for man or beast, roasted or baked or ground into meal or made into soups. Industry uses great quantities for plastics, paints, and other materials. Often sown with sorghum or cowpeas. Contains little starch. Second in importance to rice as a food crop, especially in China and Japan. Not used for human consumption as a vegetable in U.S. In fields where harvested, old stubble and grain not harvested may serve as food and suitable cover in winter. A critical element in the "plant to domestic herbivore to man" food-chain.

Kudza

Kudzu Vine
Pueraria thunbergiana
Woody vine. Perennial. With three to four vines per crown and roots penetrating to depth of 12 ft. Vine may be over 100 ft long for 1 year's growth. Leaves of three leaflets, with leaflets not lobed deeply, with entire hairy margins. Roots large, tuberous, and starchy. Vine may grow 1 ft a day on good soil in good weather.

Native of China and Japan but introduced rather widely and established through the South and as far north as Philadelphia, though it may winter-kill badly in northern limits. The related *P. tuberosa* and *P. phaseoloides* are cultivated variously in the South, the latter being introduced into the United States in 1911.

(Continued)

Flowers pea-shaped, purple, borne late in the season in the axils of the leaves, in spikes that are not particularly showy. As the flowers are borne on the older parts of the plant, they may not appear abundantly in the northern part of the range. Flowers smell like Concord grapes. They are from ½ to ¾ in. long, fragrant and produce large, flat, hairy pods bearing many seeds.

Propagation by cuttings of roots or leaves or by seeds. 500 crowns may be enough for 1 acre planted 1 to every 85 sq ft. Bare ground at 140°F may be kept as cool as 89°F if covered with kudzu, with corresponding reduction in loss of moisture. Roots fleshy and yield a good grade of starch; bark contains a fiber suitable for weaving into cloth.

May yield 3½ tons of hay a year. Dried leaves make good fodder, fair breakfast food, and good chicken feed. Leaves produce a rich litter equal to forest litter. Excellent soil anchor, and ideal for restoring worn-out soils, particularly in exhausted parts of the South, as it adds to nitrogen content of soil. Has a good ornamental value.

Cowpea
Vigna sinensis

Height to 18 in. or trailing as a vine depending on the variety. Vines may be several feet long. Leaves of three leaflets, each to 6 in. long, entire or faintly angled, with the older ones commonly short-pointed and nearly as wide as long. In general, the herbage looks like that of the ordinary garden bean.

Probably native of Asia, having been under cultivation in southeastern Asia for better than 2,000 years. It has been widely introduced in warmer parts of the world coming to the West Indies in the seventeenth century and to the United States in the eighteenth. Still an important crop particularly in China and India.

Flowers pealike, greenish-yellow, on long stalks. Stamens 10, in groups of 9 and 1 and usually opening in early morning, closing by noon and falling in the afternoon. Fruit a slender pod to 1 ft long, not flabby or inflated when green, hanging and bearing seeds that are to ½ in. long and of many colors and shapes.

Seeds about size of a navy bean are fed to cattle and poultry or even used by man as a coffee substitute. Plant cannot survive frost. Seeds planted in drills at 5 pk per acre; in rows, at 3 pk; and broadcast, at 8 pk. When grown for hay, commonly harvested in September and cured 2 to 3 days as cut.

One of better plants for restoring nitrogen to soil. Hay is rich in protein. Where plants are allowed to mature seeds, hay or straw is less valuable as food for cattle. Cowpeas may be pastured easily. A common rotation is cotton, corn with cowpeas, winter grain planted after corn, and then cowpeas.

Florida Velvet Bean
Stizolobium deeringianum

Vine with length to 100 ft or more and with exceptional twining ability, somewhat white and velvety. Leaves compounded of three leaflets, of which the terminal one may be smallest; the largest, to 6 in. long and ⅔ as wide, with wider portion away from leaf center; entire and definitely velvety beneath.

Native of Asia and probably of Malaya but widely planted and thriving along our Gulf Coast and in Florida, where it may be raised for forage. Grows reasonably well as far north as Virginia and Kentucky, but north of that the season is not long enough for it to reach maturity. Of about a dozen species, three are of economic importance.

Flowers from 5 to 30, borne in a long hanging cluster with each flower purple, to 1½ in. long, with a whitish calyx having a broadly triangular upper lip. Fruit a pod to 3 in. long, stiff, ridged, with a black velvety covering containing three to five somewhat flattened but plump speckled, streaked, or dark seeds.

Requires about 110 to 130 days from planting to reach maturity but will produce vines sufficiently robust for use as fodder or green manure in a shorter time. Commonly planted with corn, which provides a support for the vines, sometimes on a basis of two rows of corn and one row of velvet beans, although this decreases the corn yield.

Vines too coarse for cutting as ordinary hay, but a field grown to corn and velvet beans may be used as pasture and grazed with considerable profit to the cattle. Vines provide good soil anchorage and serve to restore nitrogen to the soil. The related *S. pruriens* is known as "cowitch" because of stinging hairs on pods.

Cowpea

Velvet Bean

Lentil
Lens culinaris

Height to 1½ ft. Profusely branched, lightly fuzzy. Leaves compounded of two to many pairs of alternate or opposite leaflets, totaling to about 14, each to ½ in. long, slender or narrowly oval. Leaves usually end in short tendrils one to each leaf. Roots bear nodules of importance in soil building.

Native of southeastern Europe, where it is grown much more commonly than in America. About a half dozen species in the Mediterranean region and in southwestern Asia of which only this one is of any great economic importance. They favor dry sandy soils to the richer loams.

Flowers in clusters of one to three on slender stems, small, being about ¼ in. long and the calyx sometimes enclosing the white or pale blue corollas. Fruits are short broad pods which contain two flattened seeds round in outline, convex on both faces, and greenish-brown or some other dark color.

Seeds sown in drills in March, in lines 1½ to 2½ ft apart. Plants require little attention once they are planted. Seeds are frequently preyed on by a weevil. Herbage makes good fodder for cattle or may be plowed in for green manure. Seeds do not keep well after being removed from pods.

Name lens refers to shape of seeds. Seed used mostly in lentil soup; cheaper than beans or peas and about equally nutritious. Esau is reported to have sold his birthright to Jacob for a "mess of pottage" made from red tentils.

Lentil

Sweet Pea
Lathyrus odoratus

Height to 4 ft or more. Weak-stemmed, climbing by means of tendrils at ends of leaves. Stems roughly hairy, flattened into wings. Leaves of two narrow oval leaflets, each of which is to 2 in. long and ½ in. or more broad. Branched tendrils at end of leaves.

Native of Italy. Over 100 species in the genus, some being annuals and some perennials, most of them found native in Northern Hemisphere but some in South America. About two dozen more or less standard garden flowers in the group that are grown for cut flowers or for ornament.

Flowers one to four, in clusters that are held erect on stout stems reaching above the leaves. Flowers to 1 in. long; sometimes to 2 in. broad, frequently and originally purple, now found in many colors. Fruit a pod about 2 in. long bearing several nearly globular gray-brown seeds.

Annuals require a rather rich deep soil planted early and so arranged that soil about roots is kept relatively cool. To help in this, seeds are often planted in trenches under wire; as plants grow older the earth is filled in around the roots. Dwarf sweet pea known as "cupid" does not climb.

Justly one of most popular of garden and hothouse flowers because of daintiness of flowers and excellence of fragrance. Flowers should be cut rather than plucked from their rather strong stems, to avoid injury to plants. Perennial sweet peas have long fleshy roots and are long-lived. Because of a climate which is right for seed production of sweet peas, California has become an important source of seeds. *L. odoratus* var. *nanellus,* known as cupid, dwarf, or bedding sweet pea; a low, compact, nonclimbing plant with a variety of colors has been developed.

Sweet Pea

Garden Pea
Pisum sativum

Height to 6 ft. Climbing vine. Smooth and rather dark green. Leaves compounded of paired leaflets and ending in a branched tendril, with large leaflike stipules at base of leaves. Leaflets to 2 in. long and about ⅔ as wide, entire. Whole plant more or less succulent.

Native of Europe and Asia. Now widely cultivated in temperate countries. Field peas *P sativum* var. *arvense* have bluish or lilac flowers and are raised for angular seeds. Sugar peas *P. sativum* var. *saccharatum* are raised for their juicy, somewhat fleshy, edible pods. The several varieties of peas may be grouped as having smooth and wrinkled seed, as well as into bush and tall types.

Flowers few, usually about two in a cluster, on ends of rather long stems arising from axils of leaves, white. Fruit a pod that easily splits into two parts, is to 4 in. long, and contains 2 to 10 nearly globular sweet green seeds that become wrinkled or smooth when ripe.

Peas for canning are sown in fields in drills in rows of 12 to 30 in. apart. About 50 to 100 seeds in 1 oz, and ½ pt should plant a row to 80 ft long, which will give enough peas for a small family for 10 days. Best to plant at intervals of 1 week to keep a new crop coming into maturity, planting first very early.

Canned peas are a staple food for households almost the world over and their production makes a major agricultural industry. Now, peas are an important frozen-foods item. Like other legumes, peas are particularly high in protein. Field peas that become too tough for use as "green peas" are sold as "split peas." Marrowfat peas are popular because of their size and sweetness, but usually smaller peas are sweeter.

Garden Pea

Lima Bean

Lima Bean
Phaseolus limensis or lunatus
Pole lima beans may be vines to over 10 ft long. Velvety, with thick coarse leaves that have three leaflets each, possibly to 5 in. long and often 3 in. wide with thick texture, long points, and rounded bases. Related sieva bean *P. lunatus* is of bush type and is smoother, with thinner leaves and smaller leaflets.

Native of tropical America. Most extensively cultivated in California. There are small-seeded lima beans, and large-seeded limas, and bush and pole forms. Lima beans are grown mostly in the frost-free parts of the South and shipped to Northern markets for sale. Potato lima beans are of the small-seeded type.

Flowers of the lima bean with slender calyx bracts about ⅓ as long as the calyx; of the sieva bean, with oval, strongly veined calyx bracts. Fruit a broad, flattened, thick, heavy pod to 3 in. long and 1 in. wide with thick margins, containing large plump white seeds.

Lima beans require 120 days free from frost to mature successfully and should not be planted until the soil is well warmed, so are unreliable as a northern crop. The bush types of course require more space than the pole types.

Seeds are rich in protein and starch, are good tissue builders and energy developers. They equal lean meat in many qualities, and if pork is added, provide an excellent balanced ration. Because proteins in beans are in thick-walled cells, prolonged cooking is preferred method of preparation. Marketed dried or canned or frozen. Georgia, Florida, New Jersey, and South Carolina are leading producers.

Field Bean
Phaseolus vulgaris
Types include pole or climbing and bush, dwarf, or bunch forms. True dwarfs include the red kidney bean and the short-vined navy bean. Climbing field beans are not common or popular in gardens. Annuals. Leaves compound, with rounded leaflets, entire and to 6 in long.

Probably native of South America. Domesticated and used by Indians north into North America before arrival of white man. Beans have been grown in Europe and elsewhere since the discovery of America. Field beans include white and red kidneys; white, red, and yellow-eyed marrows; white mediums; and navy or pea beans. 200 species.

Flowers much as in the garden bean. Seeds planted after frost danger, 1 to 2 in. deep, 4 to 6 in. apart in rows or hills 2 ft apart, at 70 lb per acre. May yield harvestable crop in 60 to 80 days, as contrasted with 100 to 130 days for usual field bean. Fruits are handpicked when ready for use.

Anthracnose is a bad fungous pest, and plants are subject to a bad bacterial wilt. Insect pests include a weevil that attacks the fruits and seeds and a leaf beetle. Food value of green string beans is about 190 calories per lb with 7 percent carbohydrate, 2 percent protein, and a little fat, or much lower than in fresh or dried field beans.

Green beans are good for vitamin A, good for vitamin B, and excellent for vitamin C, in these respects being better than field beans usually are. Snap beans and string beans provide basis for a big industry in fresh foods and canning, as well as the packaged, frozen form.

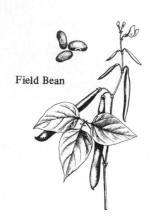

Field Bean

Garden Bean
Phaseolus vulgaris
Types include pole or climbing and bunch or dwarf, with the climbers more popular in garden than in field varieties. Stems of climbers may be many feet long. Leaves compound, with leaflets to 6 in. long, pointed and rounded at base. Garden beans grown for many purposes, each demanding a different type of fruit.

Like the field bean, garden bean is native of America and one of some 200 species and of many varieties. Popular dwarf varieties are Dwarf Horticultural and Goddard, and popular pole type is Horticultural. Popular green-podded snap bean is Stringless Greenpod. Popular bush wax bean is Golden Wax bean. Common canning form is Kinghorn Special or Tendergreen or similar varieties. Of great economic importance in Latin America.

Flowers white, cream, red, or violet, with prominent broad calyx bracts, equaling the calyx. Seeds planted in rows to 36 in. apart at ½ bu per acre with navy beans, after soil is warm and frost danger has passed. Should not be cultivated when leaves are wet. Plants are harvested, stacked, dried, and threshed and seeds are stored.

Dried navy beans yield 1560 calories per lb, 22 percent protein, 60 percent carbohydrate, 2 percent fat as contrasted with 720 calories, 30 percent carbohydrate, and 9 percent protein from freshly shelled beans. Plants make excellent forage and trash is fed to stock. Fruits mature poorly in soil that is rich in nitrogen or too wet. Many insect and fungous enemies.

Beans rank high as source of protein and starch in man's food and serve well as tissue builders and as source of energy. Baked beans are a good substitute for lean meat. Served with salt pork to add fat, they make a well-balanced diet. Preparation requires long and adequate cooking.

Garden Bean

Order Geraniales./Family Linaceae

Flax
Linum usitatissimum

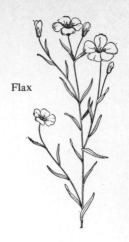

Flax

Height to 4 ft. Stems under ⅛ in. in diameter, commonly divided at or near base into two or more nearly equal branches that are themselves unbranched until near the top. Leaves narrow, simple, to 1½ in. long. Best fiber comes from long, slender unbranched stalks and from inner bark of unbranched stem.

Under cultivation since time of early Egyptians and earlier Chinese while a closely related species, *L. perenne*, was used in Europe in Stone Age. Flax is now cultivated throughout Europe except in the Balkans, and in Siberia, China, Japan, Australia, East Africa, Chile, Canada, and the United States.

Flaxseed, weighing 56 lb per bu is sown for fiber at 1½ to 2 bu per acre; for seed, at 2 to 3 pk per acre preferably on fertile clay loam. Seed broadcast or drilled, and soil is then rolled so that it is buried to a depth of ½ to 1 in. Crop matures in 75 to 100 days. Flowers blue, ¾ in. broad, delicate.

Fineness depends on thickness of stand and uniformity of growth, so drought or other weather conditions may affect quality. Fiber flax is pulled by hand, an acre being average day's work. Seeds threshed out. Stems retted in water 10 to 20 days, or in ground 3 to 6 weeks, then wood is broken and fiber scutched, baled, and marketed.

Yield may vary up to 700 lb of fiber per acre. Used in making linen. Grown for seed, matures in 100 days. From a bushel of seeds, about 2½ gal of linseed oil can be extracted, leaving behind a mucilaginous cake which is used, with caution, as food for cattle. Very excellent paper for Bible paper, cigarette papers, and time-resistant writing paper are made from flax fibers. Used as food and for extracted oils. Cattle may be poisoned by soaked linseed cake.

Order Geraniales./Family Oxalidaceae

Yellow Wood Sorrel
Oxalis corniculata

Yellow Wood Sorrel

Height to 6 in. Branched above, erect, smooth or sparingly hairy, arising from a buried horizontal rootstock that gives rise to a number of erect stems. Leaves many, long-petioled, of three more or less heart-shaped leaflets that droop or fold variously, each leaflet being to ½ in. wide and about the same length.

Native of warm and tropical parts of Old and New World. Found wild as far north as Ontario. Common in greenhouses throughout the world as a not too aggressive weed. Does best on dry or moist but not too wet soil. 300 to 500 species in the genus *Oxalis*, which makes up 99.9 percent of the whole family.

Flowers in loose open clusters of two to six, each on a slender stem joined with the stems of other flowers. Flowers pale yellow, to ½ in. broad; produce somewhat cylindrical yet pointed fruits that split along vertical lateral lines expelling the many small brown seeds. Fruits usually on stems that have become reflexed.

Flowering time from February through November or through year in greenhouses and in portected areas. Pollination usually by syrphus flies, small bees, or even by some small butterflies though self-pollination is possible. Leaves contain oxalic acid, as family and genus names imply.

As a field weed, easily controlled by persistent cultivation and crop rotation. Leaves eaten by hikers or used sparingly in salads, though it is unwise to eat them in too large quantities. Underground runners are much tougher than stems above ground and persist when weeds are pulled by hand.

Wood Sorrel
Oxalis acetosella

Wood Sorrel

Height to 6 in., arising from a sprawling, branched or unbranched, scaly rootstock that roots along its length. Erect stems branched little if any, with sparse brownish hairs. Leaves three to six per stem, to four to five in. long, with three broad heart-shaped leaflets at the end, each being wilder than long and conspicuously veined.

Native of North America, Asia, Europe, and Africa and found most commonly in damp, acid, cool woodlands, being relatively common from Nova Scotia to North Carolina, west to Saskatchewan and north shore of Lake Superior. Its rather fragile daintiness seems in keeping with the nature of the environment in which it grows.

Flowers nearly 1 in. across, pinkish or white with pink or red stripes, color being deepest in the center of the flower where there is usually a small solid dark area. Tips of the five petals are rather conspicuously notched and there is usually only one flower to a branch. Fruit to ⅙ in. long, smooth, with one to two seeds per cavity.

Some flowers do not open but mature fruits underground near the base of the stem. Pollination may be effected by small bees and by syrphus flies.

(Continued)

Nasturtium

Wild Geranium

Carolina Crane's-bill

Flowering time from May through July, depending to some extent on area and season. Plant yields what druggists call "salt of lemons" though this does not make it of commercial importance.

A frail, beautiful low plant with many names including cuckoo-meat, sour trefoil, shamrock, stubwort, sleeping clover, ladies'-clover, and sleeping beauty. It cannot be easily confused with the more common yellow sorrels, which are more inclined to grow in open places. May be eaten sparingly as a salad.

Order Geraniales./Family Tropaeolaceae

Nasturtium
Tropaeolum majus
Strong, climbing or twining, succulent smooth pale green stems, often a number of feet long. Leaves with round blades borne at ends of long, usually upward-curving petioles, with margins somewhat angled and frequently with evidence of a spiral leaf miner. Roots fibrous.

Native of Peru, Colombia, and Brazil but commonly cultivated in greenhouses throughout United States or in open in summer months. Usually grown as a mass garden border plant. It has been in cultivation in Europe since 1685. Some 45 species of vines or weak spreading plants native from Mexico to Peru.

Flowers solitary, showy, to 2½ in. across, yellow, maroon, orange, white, variegated, blotched or otherwise colored, with a long spur to the rear that ends in a nectar-bearing area. Petals rounded. Sepals, five. Stamens, eight. Ovary, three-celled, each flower producing not more than three seeds. Annual.

Large seeds germinate in about 8 days, live 3 to 4 years. Plants bloom 2 months after seeds are sown and continue until killed by frost. Pollination by insects. Sow seed in rows about 1 ft apart after danger of frost is past and thin plants to 9 to 12 in. apart. Keep weeds down and keep soil loosened. Too much fertilizer tends to increase leaf growth and lessen flower production. Flowers should be picked daily.

A useful ornamental for garden or commercial greenhouse. Plant lice may be controlled by nicotine solution, cabbage worm by lead arsenate spray, leaf miners by picking and destroying affected leaves. Leaves often used in salads and young seeds are sometimes preserved as pickles.

Order Geraniales./Family Geraniaceae

Wild Geranium, Crane's-bill
Geranium maculatum
Height to 2 ft. Branched or unbranched, particularly above, erect, fuzzy, arising from a thick coarse rough rootstock. Basal leaves with long petioles and blades more or less round but deeply cut or almost divided into three to five parts that may be obscure. Stem leaves, two, opposite, short-petioled.

Found in woodlands or on shaded rocky hillsides from Newfoundland to Georgia, west to Alabama, Nebraska, and Manitoba. Sometimes found in rather exposed places among weeds along roadsides but probably persisting from a cutover woodland area. Nearly 200 species of geranium, with about 60 being found in North America.

Flowers rose-pink to magenta or purple, in loose, open clusters of relatively few flowers, 1½ in. across, with the five sepals pointed-tipped and the five petals woolly at the base. Pistil develops a fruit to 1½ in. long that splits vertically from the base, hurling seeds by a unique sling technique. Seeds with curving lines on surface.

Flowering time from April through July. Pollination is effected by honeybees, other small bees, and commoner syrphus flies. Self-pollination practically impossible because stamens mature much before pistils. Flowers wilt quickly when picked and therefore make poor bouquets.

Common names include stork's-bill from the appearance of the fruit, chocolate flower and shameface from the color of the flower, rockweed from the site favored, and alumroot or alumbloom from the nature of the rootstock.

Carolina Crane's-bill
Geranium carolinianum
Height to 15 in. Erect fuzzy-stemmed, grayish, branched, and relatively stout from a coarse rootstock. Leaves long-petioled, with blades round or kidney-shaped in general outline but deeply and conspicuously cut into about five segments, each of which in turn is rather deeply cut to form a coarse lacelike appearance.

Found on poor soils in woodlands or in the open and often among rocks from Nova Scotia to Florida, west to British Columbia and Mexico as well as in Bermuda and Jamaica. This is one of the more common wild geraniums of the South, replacing the larger flowered *G. maculatum* of the north.

Flowers pale pinkish or magenta to white, in more or less compact clusters at ends of branches. Each flower to ½ in. across, with petals 1 to 1½ times the length of sepals. Stamens, 10, all fertile. Pistil, fuzzy below, developing into a fruit that is to 1 in. long, sharp-pointed, and bearing fine curving marked and pitted seeds.

Flowering time from April through August. Pollination may be brought about by visits of small bees or syrphus flies or it may be self-pollinated, since pistils and stamens may be mature at the same time and may contact each other. It cannot easily be confused with the larger, coarser *G. maculatum*.

Probably of little economic importance but in regions of abundance may be reasonably well-known. Apparently has relatively few common names. Distinguishing characters are the small flowers rather crowded into clusters, the branched nature of the plant, and the inch-long fruit that is fuzzy below.

Herb Robert

Order Geraniales./Family Geraniaceae

Herb Robert
Geranium robertianum
Height to 1½ ft. Weak, sprawling to erect, freely branched to form a rather ball-like mass, strongly scented, covered with somewhat sticky hairs. Leaves thin, finely cut and divided, with sections blunt-tipped and in three to five somewhat obscure parts, the divisions frequently being twice-divided; borne in pairs. May turn bright red in fall.

Found in rather open rocky woodlands, in crevices in gorges or sometimes in sand from Nova Scotia to Pennsylvania and west to Missouri and Manitoba. Also found in Europe, Asia, and northern Africa. Rarely found in exposed sunny places and does best where moisture is good.

Flowers to ½ in. across, reddish-purple, with the petals entire about twice the length of the five sepals and narrowed at base. Fruit nearly 1 in. long with a long slender point splitting from base to free smooth seeds suddenly a considerable distance. Fruit almost smooth.

Flowering period from June through October, making it one of later blooming of larger geraniums. Some botanists place it in genus *Robertiella*, in part on basis that leaves are divided rather than parted and that parts of pistil are shed rather than persistent after seeds are shed.

This is a conspicuous late summer and fall woodland plant, blooming close to the ground when white snakeroot is blooming well above the forest floor. Particularly attractive late in the season when whole plant may appear a flaming red.

Dove's-foot Crane's-bill

Dove's-foot Crane's-bill
Geranium molle
Usually has weak sprawling stems to 18 in. long, branching, hairy. Leaves long petioled, with blade nearly round in general outline but cleft nearly to the middle into as many as eleven lobes, each of which is further divided into three to five shallow lobes at the tip. Hairs of the plant relatively soft.

Native of Europe but widely established in America, being found in Pacific Northwest of Washington and Vancouver and in East from Maine to Pennsylvania west to Ohio and Ontario. It probably will spread rather widely as its seeds may be found in seeds of commercial lawn grass that may be shipped almost anywhere.

Flowers dark purple, to nearly ½ in. broad. Sepals, five, blunt and not bristle-tipped but very hairy. Stamens, 10. Fruit, with cross wrinkles and beak nearly ½ in. long, with some hairiness, splitting to free smooth seeds that lack pits and wavy markings though there may be straight lines on them.

This rather pretty little weed is often found growing among lawn grasses, low enough to be passed over by a lawn mower and yet not sufficiently aggressive to crowd grass out over any considerable area. Not so conspicuous in the early season as later, when it mars the uniformity of a grass lawn.

Many common names have been given for this geranium, among them being dove's-foot crane's-bill, culverfoot, pigeonfoot, probably referring to the general appearance of the fruiting area, and starlights, referring to the flowers. More recently developed chemical weed killers effective on broad-leaved plants can be used, with precautions, to control plant in lawns.

Alfilaria, Stork's-bill

Alfilaria, Stork's-bill
Erodium cicutarium
Height to 1 ft. Branched and tufted, densely fuzzy, somewhat sticky. Leaves to 7 in. or more long, finely and profusely divided and subdivided, the upper leaves being without petioles and the lower with rather long petioles; whole plant arises from substantial underground structure.
(Continued)

225

Native of the Old World where it is a common weed.   Found in America on lawns, in fields, and in waste places from Nova Scotia to Pennsylvania west to Oregon and Texas, being more abundant in western part of this range than in the eastern.   It may be spread through agency of seeds. Flowers borne in rather small, compact to 12-flowered clusters at ends of stems arising from stem tips or leaf axils.   Flowers purple to pink, to nearly ½ in. across.   Stamens five instead of the usual ten found in the geraniums, but with some sterile filaments.   Fruit a slender beak to 1½ in. long, breaking at the tip.

Flowering time April through September.   Annual or biennial.   Reproduces largely by seeds but persists by means of root system, which must be destroyed if an established plant is to be eliminated.   May successfully crowd out lawn grass or the less aggressive field plants.

Control is by digging out roots once a plant has shown itself or by shallow cultivation, inducing seeds to germinate before a crop is planted and then following with clean cultivation of crop.   Common names include wild musk, pin clover, pin grass, pinweed, and heron's-bill, mostly based on character of fruit.

Geranium; Zonal or Fish Geranium
Pelargonium hortorum
Height may reach 15 ft or more but in potted plants is usually about to 30 in.   Stem stout, branched, and becomes woody if old unless killed by frost.   Roots freely at joints.   Leaves with round or kidney-shaped blades, to 5 in. across, with coarsely scalloped edges, soft-haired and pungently fragrant.   Roots fibrous.

Native of South Africa but grown widely through world as a hothouse or a bedding plant.   Among more popular varieties are white, red, the brick red, the light pink, and the salmon.   In mild climates the plants are perennial.

Flowers normally appear in late winter but may be induced at any time of year; borne many in a cluster on a long stalk arising from axil of upper leaves.   Open flowers turn downward.   Single varieties have five petals and six sepals but double varieties have many petals.   Fruit long, slender, of five parts.

Requires soil high in phosphate and potash.   In beds, apply 20 percent superphosphate at 10 lb and 1 lb potassium chloride to 100 sq ft.   Cuttings set in June 1 ft apart.   Too much nitrogen makes plants too succulent.   Best soil temperature 50 to 70°F.   Buds show about 4 weeks before blooms are at their best.   Normally avoid watering the leaves.

Among the most popular pot plant and bedding plant species.   May be propagated from stems or leaves since either roots readily in water. Plants may be cut back and used 2 to 3 winters.   Never use fresh manure as a fertilizer for geraniums.   A serious disease is leaf spot caused by overwatering and poor ventilation.

Rose Geranium
Pelargonium graveolens
Normally seen as a pot plant, or in greenhouses or flower beds.   Usually under 3 ft high.   Branches freely and when old the stem becomes woody. Roots fibrous.   Leaves with blades nearly heart-shaped, and with five to seven deep lobes which have smaller lobes; long-petioled, rose-scented. Perennial except where killed by frosts.

Native of South Africa but widely grown under cultivation wherever man desires an ornamental flower or a decorative foliage plant.   *P. graveolens* commonly cultivated in France and Spain and *P. odoratissimum* in Florida, Texas, and California for commercial purposes indicated below.

Flowers, in general, similar to *P. hortorum* but narrower, with the petals rose or pink and with the two upper and larger of the five petals commonly purple-veined.   Fruit long, slender, and five-valved.   Flowers and leaves should not be wet and general culture is similar to that outlined for *P. hortorum.*

Stem cuttings are commonly started in sand in summer, covered with glass until rooted, then shifted to soil.   Requires much sun, good drainage, and little nitrogen.   Common diseases include leaf spot and blackleg; worst insect pest is Mexican mealy bug, the last being controlled with Lethane 440.

Geranium oil extracted from *Pelargonium* is a common adulterant or substitute for oil of roses used in making soaps and perfumes.   The oil is extracted by distillation for the most part from the leaves, and industry associated with its production may develop into something substantial in the United States.   The species *graveolens* has been one of the parents of numerous cultivars and hybrids in the U.S.

House Geranium

Rose Gernaium

Order Geraniales./Family Zygophyllaceae

Creosote Bush
Larrea divaricata
Height to 10 ft or stunted to 2 ft. Leaves evergreen, fragrant, particularly when wet, conspicuously dark green with a protecting varnish of resin making a conspicuous contrast when associated with the almost black stems that are slightly swollen at the joints. *L. tridentata* of some authors.

Found on light desert soil marking line between upper and lower division of Lower Sonoran life zone rather definitely. One of most widely spread and conspicuous of desert plants of North America. In Mohave Desert, it is found in pure stands over wide areas and is low in the Panamint Valley.

Flowers large, yellow, appearing suddenly with first appearance of the rains in April and May; and with almost equal and surprising suddenness they produce the furry white fruit balls that drop in later summer, at which time plant becomes essentially dormant until new spring rains come. Desert rodents eat the seeds. Of some value to wildlife.

Sweet scent of creosote bush adds flavor to food cooked over fire made of it. Leaves yield a medicine that, in Mexico at least, is used in treating wounds, burns, rheumatism, tuberculosis, and stomach troubles.

Galls on the bush may be caused by the gall midge *Asphondylia*. These appear as conspicuous walnut-sized swellings. Used by man in land classification as an index that, unless irrigated, land bearing the plant has little or no value for grazing livestock. Associated with gray burro bush, *Franseria dumosa*.

Creosote Bush

Order Geraniales./Family Rutaceae

Prickly Ash
Zanthoxylum fraxineum
Shrub or tree, reaching a height of 25 ft and trunk diameter of 6 in. Branches armed with prickles at bases of leaves and with orange buds in winter. Leaves alternate, of to 11 leaflets, each being dark green above and lighter beneath, to 2 in. long, pointed at tip and almost entire.

Found in woods, wet swamps, and thickets, from Quebec to Virginia and west to Kansas and South Dakota with the related *Z. clava-herculis* extending range to Florida and Texas. At least 150 species of the genus to be found rather widely distributed in temperate and tropical parts of the Old and New Worlds.

Flowers greenish, minute, in loose ball-like clusters borne at axils of leaves, flowers themselves lacking a calyx and having four to five petals that spread to make a flower only about $\frac{1}{10}$ in. across. Flowers borne before the leaves appear, on wood that was developed previous year. Fruits two-seeded, black, and smelling like citrus when crushed.

Wood light, brown, soft, weighing 35 lb per cu ft. Leaves, like those of the related oranges, are aromatic when crushed but plant has no particular use either for ornament or for the production of useful oils or other products. Flowers in April and May.

Not an economically important species. It may add something to the impenetrability of swampy lands and the spines on the twigs may not be pleasant when they strike bare skin, but there do not seem to be any poisonous or medicinal properties to give the plant importance. Known as toothache tree, suterberry, and angelica tree. Is a host of caterpillar of the giant swallowtail, *Papilio cresphontes*.

Prickly Ash

Hop Tree, Wafer Ash
Ptelea trifoliata
Tree or shrub. Reaches a height of to 25 ft with a maximum trunk diameter of 6 in. Twigs and branches spineless but bark bitter and strongly scented. Leaves alternate, composed of three leaflets, which when young are downy and pointed, to 5 in. long, with the side leaflets oblique.

Found in woods from Connecticut to Florida, west through Ontario to Minnesota, south to Mexico, with cultivated forms being grown rather widely outside the normal range. Three species to be found in the United States, of which this apparently has the widest range.

Flowers about $\frac{1}{2}$ in. broad, in rather loose terminal clusters, with a disagreeable odor, greenish-white, unisexual or with both stamens and pistils; whole cluster usually rather well-hidden by leaves. Fruit a thin waferlike structure to $\frac{3}{4}$ in. across, conspicuously veined, with small swollen oval center.

Wood light brown, weighing to 43 lb per cu ft. Not of economic importance. The bitter fruits are sometimes used as substitutes for hops but the plant is most commonly encouraged as an ornamental: some with attractive, conspicuously smooth, yellowish leaves.

The common name quinine tree probably refers to the bitter taste of the bark. Other names suggestive of one characteristic or another are ague bark, swamp dogwood, prairie grub, pickaway anise, and wingseed. One cultivated form is very fuzzy while others are glisteningly smooth. Sometimes planted as an ornamental of parks and gardens. A host of caterpillar of the giant swallowtail, *Papilio cresphontes*.

Hop Tree, Wafer Ash

227

Gas Plant

Gas Plant
Dictamnus albus
Height to 4 ft. Stem stout, bushy, forming into clumps. Leaves glossy green, alternate, composed of 9 to 11 egg-shaped unevenly margined leaflets, the whole plant having a strong, rather unpleasant odor. Roots exceptionally hard and heavy but may be broken to be used in forming new plants.

Native of southern Europe and northern China; while it is rather widely grown in gardens of the world it is not exceptionally common in United States. Several color races have been selected for cultivation but they are all probably merely varieties of a single species. Related to the oranges.

Flowers white, pink, pale purple, or other colors, borne in long showy clusters well above foliage. Petals, five. Stamens, ten. Ovary, deeply five-lobed. Whole flower about 1 in. long. Both flowers and leaves fragrant. Plant slow-growing, requiring to 3 years from seed to seed but living to through three generations of men.

Flowers appear in June and July. Insect-pollinated. Propagation by root division or by sowing seeds when ripe in fall, covering 1 in. deep, and transplanting the 2-year seedlings. Should bloom the third year and many years thereafter. Seeds hard to germinate. Plants not easily affected by droughts or frosts or by diseases.

Attractive as a border plant in a garden; once established may be considered permanent for a number of lifetimes. Name derived from fact that a volatile oil given off in hot, still, muggy weather may cause a gas flash if a lighted match is held near a flower. Dermatitis caused by contact with seed pods is persistent.

Lemon

Lemon
Citrus limon
Relatively small tree that bears short stout stiff thorns and is relatively smooth. Leaves to 4 in. long, rather blunt, with short petioles bearing narrow margins, with a conspicuous joint between petiole and blade. Under cultivation, trees are usually kept to a maximum height of 20 ft.

Native of Asia but introduced widely throughout the world where the temperature does not go below 29°F for any considerable period. Apparently succeeds best in the United States in California. Three common varieties recognized. Margined rather than winged petiole separates lemon from orange and grapefruit.

Flowers solitary or clustered in axils of leaves, pink-tinted outside and in the bud but white within, to ⅔ in. long, with 20 or more stamens. Pollination by insects. Fruits oval, nippled at each end, sour, of 8 to 10 segments, with relatively thick spongy skin, to 2½ in. in diameter.

Commonly grown on sour orange stock by grafting or budding. Fruit picked and packed by gloved hands throughout year but more particularly in winter months when they reach a maximum size. Lemon hardier than lime, less hardy than orange, and less resistant to fungi than orange or grapefruit.

Vitamin A absent or nearly so in the fruit juice; B, present in considerable amounts; C, in large amount. Useful in flavoring many foods, drinks, and candies and to some extent medicinally. One of most important staple citrus fruits; too sensitive to cold to be raised over long periods in most of Florida.

Grapefruit

Grapefruit
Citrus paradisi
Tree to 30 ft tall commonly; may grow to 50 ft. Trimmed to a rather strong low tree, with slender rather dull spines, which are sometimes lacking. Twigs finely furry, greenish. Leaves to 6 in. long, with broadly winged petioles, those on young shoots tapering to a blunt point; without toothed margin. Root system well-developed. Wood uniform. Bark rather close.

Native of Asia, but planted widely particularly in Florida, California, Texas, and Louisiana, on light well-drained sandy fertile soils. The Duncan is a common and popular California variety while "pinks" are favored in Texas. Different races sometimes emphasize different popular names as grapefruit, pomelo, and shaddock.

Flowers solitary or in axillary clusters with large white recurving petals and 20 to 25 yellow stamens, exceptionally fragrant, medium-sized. Fruit almost globular, to 6 in. in diameter, with about 12 segments, pale yellow with rather thick rind, sour, or when tree-ripened may be sweet. Related pomelo fruits weigh to 20 lb. Fruits are often in grapelike bunches.

Propagated by seedlings, budding, or grafting on seedling stock. Begins bearing at 3 years and reaches full bearing stage at about 10 years. Fruits held late on trees may sprout seeds and injure fruit flavor. More susceptible to cold than most other citrus fruits and grows better on poor soil than the oranges.

Fruit rich in vitamins B and C; lower in calcium, protein, and most minerals than oranges or lemons. Introduced commercially into United States in 1809 and now one of major citrus crops of the country, particularly in Florida. Many varieties including pink-fleshed Foster and Thompson have been developed.

Sour or Seville Orange
Citrus aurantium

Tree. To 30 ft high, with many blunt spines that are nevertheless sharp. Leaves to 4 in. long, blunt at tip usually but sometimes pointed, with petiole conspicuously and broadly winged, lustrous green, and in general waxy in appearance. Bark relatively smooth and close.

Native of southeastern Asia. Brought to Spain and there cultivated some thousands of years before the sweet orange. Stranger yet, it was brought to America before the sweet orange and was possibly the first fruit tree brought to America by the early settlers and explorers. It was spread by the Indians.

Flowers several in a cluster or single in axils of leaves, of medium size, white both in bud and when open, fragrant, with 20 or more stamens. Fruit globular or slightly flattened, about 3 in. in diameter, rough, with sour pulp and bitter dividing membranes and at maturity with hollow core.

Grown as an ornamental and as a base for grafting sweet oranges in United States since it can survive some conditions the sweet orange cannot. Name Seville orange indicates Spanish origin of fruits.

Essential oil is extracted from rind and used in perfumery and in liqueur curaçao. Fruits used as food and in making orangeade and marmalade. Peels candied and used as a condiment. Vitamin and food values are much the same as those listed under sweet orange.

Orange

Sweet Orange
Citrus sinensis

A tree of medium height, to 40 ft. Has a compact round-topped shape above trunk line. Spiny branches; especially when young. Spines flexible, some branches lack spines. Leaves are oval-shaped, pointed at the tips, rounded at the base. Petiole slightly winged and jointed at junction with leaf blade. Flowers medium-sized, white, and white in the bud; sweet-scented; occur singly or in clusters on new growth.

Native of southern Asia. Introduced into Europe in fifteenth century. Columbus brought it to America. Probably moved to the West Coast by Franciscan monks. Several different types have been developed including varieties which originated in Florida as hybrids or mutations.

Grown extensively around the world wherever climate is suitable. In the U. S. many millions of boxes are grown each year chiefly in Florida and California. Large quantities are processed as orange juice, some frozen. Also grown in Texas and other Gulf Coast states as well as in Arizona. Now enough different varieties are grown to supply a year-around market; Hamlin is a nearly seedless early variety; Parson Brown is a somewhat yellow or yellow-orange very early variety; Pineapple is a deep orange color with reddish shading and a midseason variety; Valencia is a deep golden-yellow standard late orange with few seeds. The navel orange is seedless and in the tip of the fruit a second smaller fruit is formed. It is California's principal variety. All oranges are good sources of vitamin C.

Sweet Orange

Tangerine, Mandarin
Citrus reticulata

Good-sized tree. Leaves of mandarin variety narrower than those of tangerine and oranges and wings on petioles narrower than those on orange leaves. Leaves to 2½ in. long, entire or finely broken margin, with mere suggestion of petiole margin. Mandarin tree smaller than tangerine.

Native of southeastern Asia but widely introduced through the world in general areas where oranges are grown. Closely related to Satsuma and King oranges. Terms, tangerine and mandarin, may be used interchangeably; however Mandarin is more inclusive term and Tangerine refers to those varieties of *C. reticulata* having deeper orange or, in fact, scarlet rinds.

Flowers small, white, fragrant, with 18 to 24 stamens. Fruit reddish flattened spheres, to 3 in. in diameter, from which rinds may be peeled easily and cleanly, giving the name kid-glove orange, and in which sections are easily separated. Core hollow in variety *unshiu* and the few seeds are not beaked.

All oranges are susceptible to many insect and fungous pests including scale insects, thrips, mealy bugs, mites, gum diseases, and rots. Protection against frost effected by fires. Sprays, beneficial insects, proper fertilization, and pruning help with other problems. Appearance is important in marketing, and so culture is a serious problem.

Food value comparable to that of oranges. Tangerines do not keep well in markets and the abundance of seeds is against their popularity. Frost damage makes oranges dry. Canning of oranges and tangerines not developed to the extent that grapefruit canning has progressed.

Tangerine, Mandarin

Kumquat

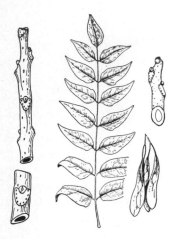

Tree of Heaven

Field Milkwort

Kumquat
Fortunella margarita
Oval kumquat, *F. margarita*, a small tree, to 12 ft high, with or without thorns. Leaves to 4 in. long and to 2 in. broad, tapering evenly at each end, paler beneath, with petiole somewhat margined. Round kumquat, *F. japonica*, has relatively broader evergreen leaves that are not commonly over 4 in. long.

Native of China whence it was brought to England by a naturalist named Robert Fortune, of the Royal Horticultural Society of London, in the middle of the nineteenth century. In America, kumquats are hardy farther north than most citrus fruits, being found into Georgia.

In *F. margarita* flowers are solitary or in small clusters in axils of leaves, rarely over ½ in. across and borne on short stems. Fruit to 1½ in. long, orange-yellow, with bitter skin, acid pulp, and abundant oil glands. *F. japonica* has globular fruit, to 1 in. through, with milder skin and bright orange color.

In addition, *F. crassifolia* differs radically in that it has a sweet rind and a juiceless pulp. Grown for ornamental purposes as well as for their fruits, the edible qualities of which vary greatly with the species and with taste of prospective consumer.

Kumquats are smallest of citrus fruits. Fruits eaten whole, raw or preserved; seeds not usually extracted since they are too small to be offensive. Evergreen quality of foliage, fragrance of flowers, and hardiness add to ornamental value of the plant.

Order Geraniales./Family Simaroubaceae

Tree of Heaven
Ailanthus altissima
Tree. Height to 90 ft. Twigs coarse, rather smooth, brown, with large leaf scars with an upper notch half surrounding the bud. Leaves to 3 ft long, smooth, petioled, composed of to 41 mostly opposite leaflets, each to 5 in. long, that are pointed at the tip, with obscure teeth at base, and in some with a few conspicuous glands.

Native of China but well-naturalized in eastern North America. Common in courtyards of some of our larger cities, where it seems to persist in spite of all obstacles and where it provides a welcome shade but an unwelcome clutter and sometimes odor. Ten related species and many varieties.

Flowers: staminate, small, greenish, ill-scented, with five petals and ten stamens, borne on tree other than that bearing pistils, though pistillate flowers may bear some stamens also. Fruits to 2 in. long, like a small oval seed surrounded by a long twisted wing. In some varieties fruits are bright red. Fruits borne in fall.

Wood coarse-grained, weak, soft, pale yellow, not durable. Foul-smelling staminate flowers make it undesirable to allow such trees to develop in populated areas. Tree ordinarily undesirable as shade tree because of objectionable suckers but so vigorous that it survives in cities where no other tree could exist.

Was introduced into America as ornamental from China. May form tree with a trunk to 40 in. through. Apparently it can grow almost completely surrounded by pavement. Dermatitis from leaves is possible. Attractive to ants, which feed on sweet sap collected on leaves. Food plant of larva of beautiful cynthia moth, *Philosamia*.

Order Geraniales./Family Polygalaceae

Field Milkwort
Polygala sanguinea
Height to 15 in. Stems slender, branched above, somewhat angled, smooth, well-supplied with leaves, not arising from a perennial rootstock. Leaves slender, to 1¼ in. long, blunt or pointed but some much narrower than others, rather evenly distributed over the stem.

Found in open fields and meadows from Nova Scotia to North Carolina, west to Minnesota, Louisiana, and Kansas. Usually does best in moist sandy ground that is loose. Sometimes locally very common but in other areas most uncommon. It may be common some years and uncommon others in a given place.

Flowers in heads that form small globes or are sometimes more elongate but always crowded; rose-pink, purplish, or sometimes white or green. Wings of the flower longer than pod and rather broad for their length. An attachment to seed is in two parts, each of which is almost as long as seed itself. Colored part, the calyx.

Flowering time from June through September.

Such common names as strawberry tassel, pink milkwort, purple milkwort, and others indicate that the plant has attracted attention. Apparently it has no important uses, and it yields so quickly to cultivation that it is not a pest as a weed.

Seneca Snakeroot
Polygala senega

Height to 18 in. Several erect nearly smooth, slender but rather tough stems arise from a woody horizontal rootstock. Leaves alternate, smooth, scalelike at base, but upper, to 2 in. long and to ¾ in. wide, without petioles and rather uniformly distributed over upper leaf-bearing stems.

Found in rocky woodlands or on steep sides of ravines from New Brunswick to North Carolina, west through Hudson Bay area to Alberta and south to Arkansas, being limited in southeastern portion of this range to mountainous areas.

Flowers borne on rather heavily flowered but slender long spikes at upper ends of stems, greenish-white to white, to about ⅛ in. long, with wings conspicuously concave and crest of corolla short and with few lobes. Seeds hairy, slightly longer than lobes of an attachment that persists.

Flowers from May through June. This polygala bears no underground flowers such as are described for fringed polygala. Perennial like fringed polygala and unlike annual pink milkwort. Apparently there are more annual than perennial polygalas in our range.

Listed in a number of books reporting quack remedies but apparently has no recognized medicinal properties. Known as rattlesnake snakeroot, Senega snakeroot, or mountain flax, indicating that it has attracted sufficient attention to be considered a common plant.

Seneca Snakeroot

Fringed Polygala
Polygala paucifolia

Height to 7 in. Erect or ascending, branching, arising from a perennial horizontal rootstock that is to 18 in. long. Leaves more or less clustered at top or upper portions of erect stems, with uppermost the larger, being to 1½ in. long and nearly 1 in. wide, dark green; the lower leaves scale-like.

Found in rich or rocky woodlands or on light soils from Anticosti Island and New Brunswick to Georgia and west to Illinois, Minnesota, and Saskatchewan. In Virginia it is found to an elevation of 2,500 ft. Some 450 species in the genus, of which about 40 are in eastern and southern United States.

Flowers of two kinds. Large showy flowers, one to four, arising from leaf axils of upper leaves, rose-purple to wine-colored to white, with wings to nearly 1 in. long. Some small, inconspicuous flowers are borne underground on short lateral branches and do not open though they produce fertile seeds (cleistogamous).

Leaves may be confused with those of wintergreen since Fringed Polygala and wintergreen have a somewhat similar color and shape though of course Fringed Polygala lacks the fragrance and flavor of wintergreen. This gives the plant the name flowering wintergreen. Flowers borne May and June; pollinated by honeybees and other small bees such as *Andrenidae* and *Halictus.*

An attractive woodland flower. Pollinating bees alight on a wing of the flower; their weight depresses a part of flower forcing rigid stamens and pistils to come free and contact the visitor. Visitors of lighter weight are unable to expose the essential organs so easily.

Fringed Polygala

Order Geraniales./Family Euphorbiaceae

Spotted Spurge, Eyebane
Euphorbia maculata

Sprawling weed, with stems to 1 ft long, freely branched. Covered with fine hairs, exuding milky juice freely when broken. Leaves opposite, from ¼ to 1 in. long, finely toothed, with a central purple-brown spot; covered with fine hairs. Root system rather deeply penetrating, branched but not coarse.

Found along roadsides and in dry paths and waste places generally where soil is relatively poor. Ranges through North America except in far North and a few other places. Native of North America but established widely in other parts of the world. Over 4,000 species in the family and over 1,000 in the genus.

Flowers whitish, in inconspicuous clusters in axils of leaves. Both staminate and pistillate flowers are found on the same plant from June through October. Pollination by insects; in some flowers where pistils and stamens are present, self-pollination takes place. Fruits, angled pods containing three four-angled gray pitted seeds.

Relatively free from insect and fungous pests and avoided by grazing animals probably because of milky juice. An annual reproduced by seeds that may lie dormant more than a year. Seeds probably present in most waste soil. Plants yield to competition and so rich soil is not a common habitat. Juice contains the poison euphorbon.

Persistent hoeing will keep plant down. Serves as cover and somewhat as soil anchor in exposed waste places. Called wartweed but is more
(Continued)

Spotted Spurge

likely to cause wartlike spots than to cure warts. Juice may cause severe blisters, loss of hair, general illness, and act as purgative. Cattle may die of scours after eating some euphorbias. Cases of lamb poisoning causing death have been reported. Seeds of spurges are popular with a number of upland game birds.

Cypress Spurge

Cypress Spurge
Euphorbia cyparissias
Stems to 1 ft high, smooth bright green, arising in large closely clustered patches from substantial horizontal perennial rootstocks; more or less branched in upper parts. Leaves slender to threadlike, the uppermost being in whorls around flower-bearing clusters, to 1 in. long; the lower, alternate and without petioles. Like other euphorbias, exudes a milky sap from the broken stem.

Escapes from gardens; established widely in cemeteries of old-fashioned type and in abandoned homestead sites. Ranges from Massachusetts to Virginia, west to Colorado but originated in Europe where it is grown commonly as an ornamental. It is most evident in America in the northeastern states.

Flowers borne in clusters at top of stems in groups whose stems often come from a common point (umbels) or arising from axils of upper leaves, usually bearing broadly oval but short leaflike structures beneath a small flower cluster. Seed case to 1/6 in. in diameter, bearing smooth oblong gray seeds.

Grown as an ornamental but rather unwisely so, as it may sometimes be eaten by cattle and may be seriously poisonous. Flowers borne from May through September, conspicuous because of yellowish supporting bracts. Quack doctors made a purgative from the roots.

Cattle eating hay containing this plant may be seriously poisoned, become weak, collapse, have scours, and die after mouth and throat become greatly swollen and eyes inflamed. Many common names include welcome-to-our-house, Irish moss, tree moss, quacksalver's spurge, cypress, kiss-me-Dick, garden spurge, balsam, and graveyard weed.

Snow-on-the-mountain

Snow-on-the-mountain
Euphorbia marginata
Stem stout, erect, to 3 ft tall but usually much shorter, bright green, smooth or somewhat fuzzy, repeatedly two-forked to form a rather compact mass. Leaves many, light green to whitish or variegated, to 3 in. long, broadly oval or more pointed at the base; entire margins.

Native of region of dry soils from Minnesota to Colorado, south to Texas but introduced to the East and frequently established as an escape or as a weed where waste soil has been dumped. Grown rather commonly in some sections as a bedding plant in gardens and on steep banks where it may exclude other plants almost wholly.

Flowers small and relatively inconspicuous though the supporting bracts may be rather obvious. Flower cluster composed of three forked rays and margins of supporting bracts are conspicuous. Fruit like a flattened globe, to 1/4 in. through, fuzzy, with three rounded lobes containing round ash-colored seeds.

Flowers appear from May through October. Since plant is unusually free of insect and fungous pests, patches look uniformly healthy. Annual so can be kept in control with reasonable care. This is probably the euphorbia that most frequently causes dermatitis in humans who grow the plant as garden borders.

As in most euphorbia poisoning, white milky juice causes blisters and inflammation of the skin but this varies greatly with the individual and with the condition of the plant. Because of its beauty, it will probably be grown as a garden plant in spite of its poisonous properties.

Poinsettia

Poinsettia
Euphorbia pulcherrima
Shrub or herb. In proper climate reaches a height of 10 ft or more. Leaves elliptical to narrower, to 6 in. long, entire or lobed, fuzzy beneath, milky-juiced and those just below flower cluster a bright vermilion red. These upper leaves are usually narrower and more entire than lower leaves.

Native of Central America and tropical Mexico. Known mostly in the North as a greenhouse plant popular at Christmastime. In some of the many forms the showy upper leaves are white, yellow, or pink rather than vermilion. Related Mexican fire plant has upper leaves blotched with red and white.

Flowers in small relatively inconspicuous green clusters at ends of stems or in center of cluster of showy leaves, with individual groups about 1/4 in. across, bearing a large yellow gland on one side, and without the petallike attachments found in some euphorbias.

After flowering, poinsettias are placed under benches 2 to 3 months and allowed to dry at temperature of 50 to 60°F. In mid-April, they are potted in new rich soil and cut back. Cuttings may be made from old plants after binding areas with wet moss. Soils should be rich, porous, and dry.

Cuttings made in morning, dropped in cool water 5 minutes, and planted in sand root in 16 days and can be transplanted in 4 weeks. Stock plants are fed urea, 1 oz to 5 gal of water, or ammonium sulfate 1 oz to 2 gal every 2 weeks. Ideal temperature is 60 to 65°F.

Croton

Croton
Codiaeum variegatum

Tree or shrub. Height to over 6 ft. Smooth, alternate, simple, entire, rather thick leaves that have petioles and are commonly prettily variegated in dark green and a yellowish-green or, in some varieties, a conspicuous red and green. Some varieties have leaves that are twisted spirally and otherwise.

About six common species native of Malaya and the Pacific Islands, with this one commonly grown in warmer climates as an ornamental or in greenhouses where outside survival is impossible. Number of varieties produced by horticulturalists is almost endless.

Flowers: staminate, in open clusters that are to 10 in. long; pistillate, borne usually separately on same plant. Fruit a whitish or light green capsule to ⅓ in. in diameter. Pistillate flowers without petals; staminate bear 20 to 30 stamens.

Requires night temperature of 70 to 75°F. Cuttings are rooted from October to June with soil temperature to 80°F, or rooting may be stimulated by tying wet moss about the stem for about 3 weeks and then cutting and resetting the rooted portion. Plants require much sunlight.

Tung–oil Tree

Unfortunately mealy bugs and other hothouse pests thrive on crotons. These may be checked by tobacco and other common insecticides. Seeds may be ripened under glass and used for propagation. Important varieties include Delicatissimum with green, yellow, and red narrow leaves; Bravo with yellow, red, and green broad leaves; and Rex with yellow, green, and red twisted leaves.

Tung-oil Tree
Aleurites fordii

Tree. To 25 ft high, with conspicuously smooth branches and rather like an orchard tree in general appearance under cultivation. Leaves somewhat heart-shaped at the base or three-lobed, to 5 in. long, with a loose fuzz on the underside that vanishes with age.

Native of western and central China but introduced widely in suitable territory for cultivation, now grown in considerable acreage through Gulf Coast territory. Can thrive on relatively dry soils that are unsuitable for some other crops. Probably at best in the United States in Florida.

Flowers appear before leaves, in open clusters that appear reddish-white. Petals 1 in. or more long. Fruit three- to five-celled, while the fruit of the related candlenut is two-celled. Fruit to 3 in. in diameter, smooth, with rough seeds appearing when dry like a black walnut.

Few plants have revolutionized an industry more than has this tree that produces the tung oil of commerce; efficient because it makes a hard elastic varnish that dries more quickly than other oils.

Plants begin to yield fruits when from 4 to 10 years old. The oil is only slightly affected by salt water and is therefore useful in painting ships. Like candlenut, the oil cake is poisonous, unsuitable for food but useful as a fertilizer. Plant known as China wood oil tree as well as tung.

Castor Bean

Castor Bean
Ricinus communis

Tree. Reaches height of 40 ft in suitable climate or a height of 15 ft as an annual herbaceous plant. Leaves like great stars, thick, to 3 ft across, lobed, smooth or with spiny structures on underside and with divisions of leaf reaching more than halfway to midrib.

Probably native of Africa but now widely cultivated in tropical and semitropical lands. Escaped from cultivation and sometimes established. Found in waste places from New Jersey to Florida, west to Texas. Great numbers of forms based chiefly on color of leaves, markings of the seeds, and nature of the fruit. Generic name from *ricinus,* a tick, which the seeds of this plant are supposed to resemble.

Flowers, staminate and pistillate borne on same plant, with staminate below pistillate. Flowers in rather open terminal clusters having numerous stamens with branched filaments. Fruit breaks into two to three sections freeing smooth, attractively marked seeds. Fruits conspicuously coarse-spined.

(Continued)

Smoke Tree

Staghorn Sumac

Smooth Sumac

Plant grown as an ornamental in United States but in Oklahoma it is grown for the seeds, which contain from 25 to 40 percent of a thick colorless or greenish oil.   Extracted oil is boiled with water and filtered to remove mucilage and proteins; the poisonous residue cake is used for fertilizer.   Considered a blood poison (ricin).   Since seeds are very poisonous if eaten, tree may well be considered an attractive nuisance if grown as an ornamental.

Castor oil used medicinally as a purgative.   Seeds put in rodent burrows and when eaten often kill rodent pests.   Oil has been used commercially as a lubricant for airplane engines, in increasing the flexibility of leather belts as on motorcycles, as a base for soap, and as an illuminant.

Order Sapindales. / Family Anacardiaceae

Smoke Tree
Cotinus coggygria
Small tree.   Height to 40 ft.   Trunk diameter to 15 in.   Well-branched, usually branching close to ground.   Leaves alternate, entire, to 6 in. long and to 2 in. wide, usually blunt at tip and narrowed at base, with blade running down onto petiole.

Found on rocky hillsides or frequently in plantings on lawns and in gardens.   Native from Missouri and Oklahoma, south and east to Tennessee and Alabama but widely planted and sometimes established over much of the United States.   Some authors place this plant in the genus *Rhus. C. ovatus* is the native smoke tree.   *C. coggygria* is the Old World species.

Flowers small, to ⅛ in. across, green or sometimes purplish, borne in loose open clusters at tips of branches, clusters being to 8 in. or even more long, densely fuzzy and at a distance having the appearance of smoke.   Fruit relatively few but each about ⅛ in. long.

Wood weighs to 40 lb per cu ft, weak, soft, orange-yellow, capable of yielding a rich orange dye.   Flowering time April through May.   Fruits or fruiting clusters may persist well through summer.   Propagated by cuttings or by seeds.

Grown essentially as an ornamental.   Dye not used commercially but is popular with amateur naturecraft workers.   Common names include yellowwood, chittamwood, American smoke tree.   The European species has smaller, fuzzier leaves, more rounded at the base and more leathery in texture.

Staghorn Sumac
Rhus typhina
Tree.   Height to 40 ft, with trunk diameter to 12 in.   Often shrubby.   Branched because new year's growth begins from bud back from tip.   Twigs coarse, furry, somewhat sticky.   Leaves alternate, to 2 ft long, compound, of 11 to 31 paired leaflets, each to 5 in. long, dark green but turn red and yellow in fall.

Native of America.   Ranges from Nova Scotia and New Brunswick to Georgia, west to North Dakota and Alabama, being found on dry, rocky, thin, relatively poor soils such as do not support most trees.   May form almost pure stands on dry talus slopes and is considered a weed tree by foresters.

Flowers borne in terminal compact clusters appearing in June or July after the leaves; yellowish-green and usually with the staminate in larger, more open clusters on different trees from those bearing the pistillate flowers.   Fruits globular, about ⅛ in. through, turning from green to red and in almost compact mass when mature.

Wood soft, greenish-gold or brown, weighing 27 lb per cu ft, burns quickly but leaves little ash.   Bark rich in tannin.   Pith large and more or less orange.   Buds more or less hidden by leaf scars.   Sour fruits quench thirst on a hike or make a delicious lemonade if boiled in water and sweetened with sugar.

Wood used in naturecraft for making attractive picture frames needing no finishing with varnish or paint.   Plants sometimes grown as ornamentals, particularly a cut-leaf variety, but the clumps soon look bedraggled and lose popularity.   However, they do give a quick growth.   Fruits eaten by many birds and wild animals.   In earlier times, young stems were cut, the pith was removed, and the "pipe" so formed was used as a spile to collect maple sap from holes bored in sugar maple trees.

Smooth Sumac
Rhus glabra
Tree or shrub.   Height to 25 ft.   Trunk diameter to 3 in.   Branches freely and close to ground to form a flat-topped crown, branching being similar to that of staghorn sumac.   Leaves alternate, to 3 ft long, compound, of 11 to 31 leaflets, smooth.   Twigs coarse and smooth, without fuzz of staghorn.

Native of North America, ranging on poor soils of pastures, hillsides, and waste ground that is usually well-drained, from Maine to British Columbia, south to Florida and Arizona, making it one of the more widely distributed of the woody plants of the country. A common hybrid, *Rhus X pulvinata,* results from the cross of *R. typhina,* staghorn sumac, and *R. glabra,* smooth sumac.

Flowers appear in June and July, after leaves, with staminate and pistillate in separate flowers on different trees, staminate being in more open clusters. Fruit a depressed globe, to ¼ in. in diameter, covered with close, somewhat sticky fuzz that is sour. Inner part of fruit smooth, orange-brown.

Wood weak and, in general, like that of the staghorn sumac. Not considered a poisonous plant and freely grown as an ornamental particularly in the cut-leaf form. Because of the absence of "fur" on twigs, plant is more likely to be clean than is staghorn sumac, which may become filthy in smoky areas.

Landscape architects may consider this one of best plants for quick mass planting in Northeast. Easily propagated by seeds or from suckers arising near base. Cuttings may also be used. Known as shoe-make, vinegar tree, senhalanac, and Pennsylvania sumac. Dried fruits used medicinally as an astringent and as a gargle.

Poison Ivy, Poison Oak
Rhus toxicodendron

Poison Ivy

Woody shrub or vine, the vine climbing by aerial rootlets that cling readily to trees. May have a stem diameter to 4 in. Root system shallow but widely spread. Leaves alternate, compound, of three leaflets borne on a long petiole and each to 4 in. long, entire or notched, dark waxy green above and lighter, more fuzzy beneath. *Toxicodendron radicans* of some authors.

Found in dry fields and rocky exposures as well as on rich croplands or woodlands sometimes high up in trees, but at its best possibly along fencerows and stone walls. Ranges from Nova Scotia to Florida, west to British Columbia and Mexico as well as in Bermuda and the Bahamas. Closely related to poison oak and poison sumac.

Flowers borne on slender open axillary clusters with a calyx of five yellowish-green sepals; five separate stamens, and a one-celled ovary that ripens into a white berrylike structure, to ¼ in. in diameter, persisting through winter into spring. Stone in the fruit gray and to ⅙ in. in diameter.

A vicious plant because of its contact poison, which may be carried by smoke. Toxin is a colorless or milky substance; volatile oil present in almost every part of the plant. Poison causes inflammation, spreading blisters, and scabs usually beginning 12 to 24 hours after contact. It may cause severe suffering and temporary blindness.

Treatment of poison: wash several times with hot water and soap. Even with many washings, it is difficult to remove all poisonous material. Several moderately effective patent lotions and creams are now available in drugstores. In severe cases, physicians may administer ACTH or cortisone as a treatment. No completely successful known cure available. If severe, ask assistance of a physician.

Poison Sumac
Rhus vernix

Poison Sumac

Tree or shrub. Height to 25 ft. Trunk diameter to 6 in. Rather coarse, open plant with gray bark. Leaves alternate, to 15 in. or more long, compound, of 7 to 13 thin oval to pointed leaflets, each to 4 in. long and to 1½ in. wide. Twigs rather coarse and with small buds. Root system spreading. Whole plant very poisonous.

Limited fortunately to swampy lands but ranges from Maine to Florida, west to Minnesota, Missouri, and Louisiana. About 150 species of the genus and some 400 in the family. Eight species of *Rhus* are considered economically important, including the tree which yields the base of lacquer.

Flowers in loose open clusters borne in axils of leaves, slender, the clusters being to 8 in. long, and individual greenish flowers to 1/12 in. broad. Fruit gray, globular, with a pulpy covering over a hard center, smooth, to ⅙ in. in diameter, and persistent through winter.

Wood weighs 27 lb per cu ft, soft, yellow-brown. Poisonous properties similar to those of poison ivy but if anything more virulent. Fortunately, the habitat of plant in swamps makes it less likely to be contacted, but brilliant coloring of leaves makes it attractive to amateur collectors, often with disastrous results. Frequently poison sumac affects the person about the face and head. As a result, the poisoning is more obvious and sometimes worse than poison ivy. Poison ivy, although it does climb on trees, is more commonly found along the ground. Therefore, poison ivy more frequently affects the feet, ankles, and legs.

(Continued)

Treatment and symptoms of poison of poison sumac and of poison ivy are the same. Poison sumac is also known as thunderwood, poisonwood, swamp dogwood, poison dogwood, poison ash, poison tree, poison elder, and swamp sumac. Most of these names appropriately emphasize the plant's poisonous properties.

Pepper Tree

California Pepper Tree
Schinus molle
Tree. To over 20 ft high, with conspicuously pendulous, slender branches giving top a rounded appearance. Leaves alternate, smooth, to over 1 ft long, compound, of 30 or more slender leaflets, each to 2 in. long and to ½ in. wide, evergreen. Bark scaly.

Native of tropical regions of America, with this species from Peru but about 17 other species known from tropical America. Possibly most commonly grown ornamental tree in southern and central California. Known as mastic tree for its resemblance to the real mastic tree.

Flowers borne in much-branching open clusters, yellowish-white, frequently sold in flower shops for their decorative beauty. Fruits rose-colored, almost spherical, to nearly ½ in. in diameter and fully as attractive as the flowers. Fruits of related Christmas berry tree are smaller and bright red.

In California, this tree is at its best in warmer parts of state. More satisfactory as an ornamental than as a shade or street tree because its branches frequently extend almost if not quite to the ground. Staminate and pistillate flowers on separate trees, so not all trees will bear fruits.

This tree should be familiar to all motion-picture fans since its artistic beauty makes it a tree favored by producers even if setting is in New England or Canada where the trees never grow. It, eucalyptus, and some palms make characteristic California landscape. Harbors the black scale, a menace to citrus orchards. Attractive to many species of songbirds which eat the fruits.

Cashew

Cashew
Anacardium occidentale
Tree. Height to 40 ft. Juice milky, but hardening into a gum when exposed to air, much like gum arabic. Leaves alternate, oval, rounded, sometimes slightly notched at tip, to 8 in. long and to 4 in. wide, evergreen but very susceptible to frosts, borne on short petioles.

Native of tropical America particularly in West Indies but may survive climatic conditions in southern Florida. Widely cultivated in tropical regions of world. Grown in greenhouses in part as a curiosity. Eight species native of American tropics but only this one is cultivated.

Flowers rose-tinted, in loose, open clusters, fragrant, with staminate flowers on same plant that bears both stamens and pistils in the same flower. Stamens to 10. Petals, 5. Fruit to 3 in. long, white, yellow, or red, sweetish-sour, edible cashew apple that is crowned by the 1-in. curled nut.

Nut enclosed in two-layer, leathery, oily, acrid covering. Oil (cardol) driven from cashew apple by heat is highly irritant to skin and eyes and may cause dermatitis. Roasted nuts delicious, may yield, if crushed, a valuable cooking oil equal in some respects to olive oil. Trees grown from cuttings of mature, leaf-bearing wood.

United States imports cashews largely from British Indies. A fermented, distilled beverage (kaju) is made from cashew apples in Brazil and West Indies, sometimes bottled. Gum from juice is used in varnish to resist termites and other insects. Ripe fruits are eaten raw in tropics. Vast Brazilian stands are untouched commercially.

Pistachio

Pistachio
Pistacia vera
Tree. Height to 40 ft, with spreading branches. Leaves at first fuzzy, then smooth, alternate, compound, to 7 to 15 stemless leaflets, borne most abundantly and with greatest leaflet number on young shoots. One cultivated variety has only three to five leaflets. Bark of twigs brown.

Native of Mediterranean and western Asiatic regions, with a close relative *P. texana* found from Texas into northern Mexico. Eight to nine species, most of which are found in warmer regions from China through Mediterranean area.

Flowers brownish-green, in loose open clusters, with staminate and pistillate flowers on separate trees, without petals. The staminate, with five-cleft calyx and five stamens; pistillate, with three- to four-cleft calyx. Fruit a wrinkled, reddish, 1-in., double-shelled nutlike structure, with outer shell red and inner brittle.

Kernel has a pale green or creamy yellow rich sweet oil that has exceptional flavoring possibilities. Related species of importance include the Cyprus turpentine tree *P. terebinthus* and the mastic tree *P. lentiscus*, which produces a gum useful in varnishes and in a form of chewing gum.

Pistachios are not cultivated in United States, our imported products coming mostly from the British Indies, which yield about ⅔ of our needs, while ¼ come from Syria and the remainder from Iran, Italy, and elsewhere. Turkey produces the largest crop; the soft outer flesh is removed and the thin-walled hinged pit is cured in brine or dried; then marketed. The "shell" is sometimes dyed. Reputed to have been sent by Jacob into Egypt and has been grown for centuries in Palestine.

Mango
Mangifera indica

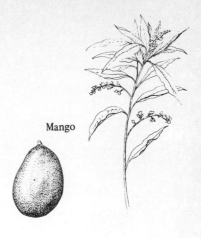

Mango

Evergreen, tropical tree. May attain a spread of 120 ft; to 90 ft high. About 30 species from tropical Asia. Native of Burma and Malaya. Only *M. indica* important in the U.S. Cultivated out of doors only in Florida and warmest part of California. Name *Mangifera* from Hindu name for fruit, *mango*, and *fero*, to bear. Several species produce delicious mango fruit of the tropics. Leaves alternate, entire, with distinct petiole. Individual plant bears staminate and pistillate and perfect flowers. Flowers in branching cluster at ends of stems. Sepals four to five and deciduous, petals four to five.

Fruit fleshy with a stonelike seed. Mango, like its relatives poison ivy and poison sumac, contains a toxin which causes dermatitis. Poisonous principle is in the stalk, which holds the mango fruit so mangoes should always be peeled before using. Mango is "the apple of the tropics." Wood is used for boat and canoe making, for houses, and for boxes. Wood is gray, rather soft and easily worked.

It is said that Gautama Buddha contemplated under a grove of mangoes. Hawaii and the Philippines are leading producers through improved horticultural practices. Widely used as human food especially in the East. Most frequently eaten raw but are also used in making preserves, jellies, or pickles, wine, and glucose. Kernel of the seed may be roasted and eaten like chestnuts.

Mangoes are sometimes picked green and then shipped. Must be handled as carefully as bananas in shipment to avoid damage. Mangoes can be grown as attractive houseplants. Their foliage is dark green and glossy. Plant should be kept wet and at temperature of about 70°F for best growth indoors. Fresh seeds of mangoes, planted, germinate in about 3 weeks. Dry seeds usually do not germinate. When plant develops dense root system, it should be given liquid fertilizer.

American Holly
Ilex opaca

American Holly

Tree. Height to 100 ft. Trunk to 4 ft in diameter. Branches short and slender. New branchlets stout and at first covered with brown fuzz; after first season smooth. Bark gray, rough, ½ in. thick. Buds sharp, to ¼ in. long. Leaves evergreen, to 4 in. long and 1½ in. wide, dark green, leathery, with spine-tipped scallops or rarely entire.

In moist woodlands from Massachusetts to Florida, west through Pennsylvania to Indiana, Missouri, and Texas, being found in North Carolina at an elevation of 3,000 ft. Probably at its largest and best in the rich bottomlands of Louisiana. Related English holly is planted rather widely where suitable. Some large trees are found in Pacific Northwest.

Staminate flowers found in clusters to 1 in. long and on different trees from those bearing the pistillate. Pistillate flowers more scattered, usually only 1 to 2 in a cluster though a number of clusters may be close together. Fruit fleshy, with stone center, sometimes to nearly ½ in. through, spherical, red or rarely yellow. Trees are either male or female. To produce red berries, both kinds of trees must be planted near one another.

Wood hard, white, uniform, weighing to 36 lb per cu ft, tough but weak, turns brown with exposure or age and used extensively in wood turning, cabinetwork, interior finishing, and otherwise. Tree highly resistant to salt spray so is suitable for growing for ornament along the coast. A relatively slow grower.

Tree most popular for the fruits, borne of course only on pistillate tree, the fruits being used mostly for Christmas decoration with the leaves. So popular is this use that unless trees are protected, they are almost certain to be despoiled. Flowers appear in the Northeast in May and June. Plant propagated by cuttings. Berries reported to be poisonous. Many songbirds take the fruit. Evergreen, so provides a year-round cover for some forms of wildlife. Has been successfully planted using seeds in wet locations.

Black Alder, Winterberry
Ilex verticillata

Black Alder

Bush or shrub. Height to 25 ft. Twigs brown and smooth or slightly fuzzy. Old bark darker. Leaves to 3 in. long and to 1 in. wide, sharply pointed at tip, with shallow-toothed margins and relatively short petioles, thick, leathery, smooth above and slightly fuzzy beneath, not evergreen, turning black in fall.

(Continued)

Found in swamps, low grounds, and other wet places. From Connecticut to Florida, west to Wisconsin and Missouri, with varieties extending range north to Nova Scotia; related species extend range south and west. Over 250 species of *Ilex,* mostly in America but also in Asia, Australia, and Africa; 5 in United States.

Flowers: staminate, in clusters of to 10 flowers borne on short stems in leaf axils; pistillate, in clusters of 1 to 3 borne close to stem at bases of leaves, almost as though they formed a ring around the stem. Fruit bright red, to ¼ in. through, containing smooth nutlets.

While leaves of related, *Ilex vomitoria,* yaupon, and *I. glabra,* gallberry, yield a fair substitute for tea, in part because of reported caffeine in leaves, this substance is apparently lacking in leaves of black alder. Nevertheless it is suggested as a tea substitute. Berries reported to be poisonous.

Plant popular as a winter ornamental, and fruit-bearing twigs find a too-ready market in florists shops; is also cultivated for planting as an ornamental but its moisture requirements limit its usefulness in this respect. Known as false alder, striped alder, white alder, and black alder.

Order Sapindales./Family Staphyleaceae

Bladdernut
Staphylea trifolia

Shrub or small tree. Height to 15 ft. Branches profusely. Trunk to 6 in. in diameter. Bark smooth and striped. Leaves opposite, of three leaflets, or rarely five, each to 2½ in. long, pointed-tipped; lateral leaflets without stalks but whole leaf rather long-petioled; smooth when mature but partly fuzzy when young. Twigs are slender and may have wartlike lenticels.

Found in wet woodlands and brushy territory from Quebec to South Carolina west to Minnesota, Missouri, and Kansas. Some eleven species known, all native of temperate climates in Northern Hemisphere; and one, *S. bolanderi,* found native in California. 22 species in the family, giving group a wide distribution. Chiefly Asiatic.

Flowers in rather loose, semihanging clusters, borne from between a pair of leaves; flower stems, slightly longer than flowers themselves. Fruit spectacularly inflated into a bladder of three or rarely four parts, each of which contains a globular seed, the bladder being to 2 in. long.

Will grow in almost any soil but at best in rich moist shaded area. Propagated by using seeds, cuttings, or by layering the growing branches. Cuttings forced in greenhouses so that they form dependable root systems. Pinkish flowers appear in May; conspicuous nodding fruits, in fall and winter.

Grown primarily as an ornamental for its clean pale green foliage, its intriguing, bladderlike fruits, and rather attractive flowers. Hardy in northern United States but not so rugged as some of the more commonly used ornamentals.

Order Sapindales./Family Celastraceae

Bittersweet
Celastrus scandens

Climbing and twining vine with a length of 30 ft or more. Twigs yellowish-green to brown, frequently winterkilling at the tips. Leaves alternate, more or less two-ranked, to 4 in. long and to 2 in. wide, smooth on both sides, with petioles to ¾ in. long and margins obscurely uneven or with shallow curved teeth, pointed at tip.

Found in rich soil, along roadsides, or in woods from Quebec to North Carolina, west to Manitoba, Kansas, and New Mexico, being found at high elevations in southern part of range. While there are some 30 species of the genus, only this one is in North America though a few are found in tropical America.

Flowers, staminate may be on one plant; other plants may bear both stamens and pistils; greenish, relatively inconspicuous, in clusters that are to 4 in. long, at the ends of branchlets. Petals longer than lobes of calyx but flowers very small. Fruit a yellow capsule, to ½ in. through, that bursts to expose a showy, attractive, red seed cluster.

Conspicuous in fall and winter and even into spring because of red-centered fruits on fertile plants. Propagation by seeds, or more simply by cuttings that root readily. Such cuttings should be from plants that bear fruits if plants are grown for ornamental purposes. Horses have been poisoned by eating leaves.

Some forms grown for ornament have variegated foliage. Fruits collected on twigs in fall, made into bunches and find a ready sale in flower shops for semipermanent winter bouquets indoors or for a substitute for holly at Christmas time. Twining nature of plant may cause it to kill useful forest trees by strangulation. Bittersweet is not classified as an important plant for wildlife. Some few upland game birds, songbirds, and fur and game mammals eat the seeds, buds, and leaves.

Bittersweet

Bladdernut

Burning Bush, Wahoo
Euonymus atropurpureus

Tree or shrub. Height to 25 ft. Trunk usually 4 to 6 in. in diameter. Twigs greenish or green-blue, conspicuously but bluntly four-angled, often conspicuously long and straight. Leaves opposite, to 5 in. long and to 2½ in. wide, pointed at tip, sparingly fuzzy beneath, thin, with ½-in. petioles and margins, with shallowly curved teeth that are very small.

Grown as an ornamental but native of area from Ontario to Florida, west to Montana, Oklahoma, and Nebraska. Persists in spite of abandonment and is relatively common about abandoned homesites in some parts of its range. Some 120 species native of North Temperate Zone, of which 2 are California species.

Flowers purple, to ½ in. across, in clusters the parts of which branch repeatedly in 3s. Petals, usually four. Flowers appear from May to middle of June. Fruit a smooth, deeply three-lobed or parted capsule that is ½ in. or more through and bursts to expose a showy, attractive red seed cluster.

Hardy. Seems to do well in variety of soils with propagation commonly being by seeds or by cuttings. Seeds collected in fall, stratified in sand, and sown in spring. Wood cuttings made in fall from matured wood, rooted under glass in winter, planted in spring. Wood white, 41 lb per cu ft.

Name, burning bush, from spectacular colors of the foliage. An attractive and popular ornamental. Bears the name arrowwood because it is assumed Indians used straight branches as arrow shafts. These were sometimes stimulated by cutting the tree or shrub back or burning it and then using the vigorous straight suckers that came on afterwards.

Burning Bush

Japanese Euonymus
Euonymus fortunei

Low shrub or more commonly a climbing, clinging vine that fastens itself to walls of buildings and may reach to a height of 20 ft or more, providing a green cover for otherwise bare walls. Branches round, warty. Leaves green, rounded at base, white-veined, to 2 in. long, evergreen.

Native of Japan but widely grown as a cover and ornamental through temperate parts of the Northern Hemisphere. A large number of varieties based upon size and color of leaves, general growth habit, and other factors have been developed by horticulturalists. Better than a dozen species are considered as stock ornamental woody plants. Many varieties of *E. fortunei* are available; cv. 'Vegetus' is a spreading shrub to 5 ft, or climbing if supported; cv. 'Radicans' from Japan and South Korea has smaller less pointed leaves which are sharply serrate. Varieties may be difficult to distinguish. Consult a reputable dealer for recommendations.

Flowers pale greenish white, in clusters generally of a few, the usual number of parts four; borne on slender stems making a rather open cluster. Fruit smooth, pink, globular capsule that matures usually in October. Flowering time June or July.

Propagation, like that of most woody plants of this type, usually by cuttings of mature wood rooted through winter in greenhouses and planted following spring. Because of evergreen leaves and climbing habit, this is a popular ornamental on school buildings.

Japanese Euonymus

Order Sapindales./Family Aceraceae

Norway Maple
Acer platanoides

Tree. Height to 70 ft. Head round and well-spread. Bark finely checkered, ridged, relatively close, dark, uniform. Leaves opposite, to 5 in. long and of equal width, with five lobes much as in sugar maple but with a slightly milky juice in petiole; underside somewhat shining green and smooth.

Native of Europe and western Asia but widely planted as a superior ornamental in America where it frequently excels sugar maple as a street tree in popularity. A variety *schwedleri* has bright red leaves when young but these change to dark green with maturity. Horticulturists have developed many interesting forms.

Flowers yellowish-green, in erect, smooth, spherical, rather open clusters, appearing at about the same time as the leaves do or a little before. Sepals and petals distinct. Fruits paired, equal, maturing in September, with wings widely spread. 40 to 70 percent germination.

Many garden forms are based on shape and color of leaves, as suggested by names *albovariegatum* (white blotched) and *rubrum* (red) as well as *dissectum, globosum,* and *laciniatum* suggesting differences in shape. *A. platanoides aureomarginatum* suggests a unique color and shape form.

Leaves turn pale yellow in fall rather than red as do sugar maples. Attractive fruits, rather pretty flower clusters, smooth even dark bark, and general cleanness of this species entitle it to the popularity it enjoys as a shade tree. Has few insect and funguous enemies and is tolerant of city conditions.

Norway Maple

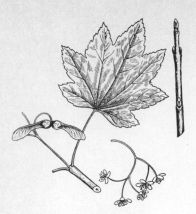

Vine Maple

Mountain Maple

Broad–leaved Maple

Vine Maple
Acer circinatum
Tree. Height 30 ft with a trunk to 1 ft through, or just as commonly a prostrate vinelike plant that branches freely in many directions. Bark thin, smooth, red-brown, with many shallow fissures. Leaves rather circular, but with seven to nine pointed lobes, with conspicuous veins, rose or red in spring and fall.

Found along stream banks from British Columbia south through western Washington and Oregon to the Sacramento River country in California. In Oregon and northeastern California, found up to 4,000-ft elevation, but at its best in rich lowlands.

Flowers appear with leaves, in loose open clusters of to 20 flowers that hang, with staminate and pistillate flowers in same cluster. Sepals red, purple, and much longer than the broad greenish-white petals. Fruit with thin wings, double, spread at right angles, red or rose, to 1½ in. long, ripe in fall.

Wood heavy, close-grained, hard, not strong, light brown to white with much lighter sapwood. Winter buds blunt, to ⅛ in. long, with bright red scales. Hardy in East as an ornamental, north into Massachusetts; easily transplanted or grown from cutting of the vinelike plant.

Popular as an ornamental because of red flowers, rose-colored fruits, and attractively shaped leaves that are delicate light green in summer and a brilliant orange or scarlet in fall. Wood used as fuel, as tool handles, and because of twisted shapes is popular in making fishing-net bows.

Mountain Maple
Acer spicatum
Bushy tree. Height to 35 ft. Trunk diameter to 8 in. Bark green and not striped. Often appears merely as a weak bent shrub in a thicket formation. Leaves opposite, to 5 in. long, rather coarsely notched along margins but with three distinct pointed lobes, conspicuous veins, and thin texture.

Found in damp rocky woodlands or on sides of gorges, apparently surviving in shade or sun but doing best where neither is too strong. Ranges from Newfoundland and Labrador to Georgia, west to Manitoba, Minnesota, Iowa, and Tennessee. Found at 5,000-ft elevation in North Carolina. In southern part of range, is found in mountain areas.

Flowers appear after leaves are fully grown, greenish-yellow, borne in narrow, erect, long-stalked, terminal clusters with fertile or pistil-bearing flowers near base of cluster. Stamens, seven to eight, free. Petals usually five, pointed, narrow, longer than sepals. Fruit red, compressed, leathery, drooping, each to ½ in. long.

Wood weighs to 33 lb per cu ft, soft, coarse-grained, relatively weak, light brown with red tinge and with lighter sapwood. Winter buds bright red, slender, pointed becoming to 1 in. long at times. This species can survive strong sunlight better than the striped maple can.

Of little economic importance though fruits may provide some bird food. Wood makes fair campfire fuel and is sometimes cut for a regular fuel supply. Tree often grown as an ornamental in gardens and parks where conditions are suitable.

Broad-leaved or Oregon Maple
Acer macrophyllum
Tree. Height to 100 ft, with tall straight trunk with a diameter of to 3 ft and branches that tend to droop or hang at the ends. Twigs pale green at first but become red or bright green by winter. Leaves opposite, deeply five- or three-lobed, to 1 ft long, with a petiole to 1 ft long, dark green and shining above.

Native of western North America but not hardy in the northern parts above 55° latitude. Ranges from Alaska to Oregon, and in the mountains south to the border, being between the 2,000- and 3,000-ft level in Sierra Nevada and rarely above 3,000 ft farther south. At its best in rich soils of southern Oregon.

Flowers appear after leaves, bright yellow, fragrant, to ¼ in. long, on slender drooping stems nearly 1 in. long. Sepals, petallike. Staminate and pistillate flowers separate. Fruit grown by July, mature by late fall, double, with wings to 1½ in. long and to ½ in. wide. Seeds ¼ in. long, brown.

Wood not strong, light brown, soft, close-grained, red-tinged, with thick, nearly white sapwood, extending through 60 to 80 annual rings in larger trees. Grown primarily more as a shade tree, in part because of its beautiful large leaves.

Wood used for fuel, for cheap furniture, flooring, and general interior finishing, also for tool handles such as broom handles or sometimes even for cheap ax handles. It is possibly the most valuable wood produced by any of the deciduous trees of the western part of the country.

Sycamore Maple
Acer pseudo-platanus

Tree. Height to 70 ft. Head large and spreading. Bark rather thick, gray, well-furrowed longitudinally. Twigs rather coarse, straight, large-budded. Leaves opposite, five-lobed, to 6 in. long and of similar width; margins with coarsely curved teeth; veins rather conspicuous, dark green above and light beneath.

Native of Europe and western Asia but planted extensively in the temperate parts of America. Does well in exposed positions where its form may become broad-crowned with rather divergent branches for a maple.

Flowers appear with or after the leaves, in rather long hanging loose clusters, with rather long hairy stamens. Fruit paired, smooth, with widely spread wings, shed in autumn, larger than in sugar maple and somewhat smaller than in larger-fruited silver maples.

Horticultural varieties with suggestive names include *bicolor* (yellow and white), *tricolor* (purple, yellow, and green), *quadricolor* (green, purple, white, and yellow), *albo-variegatum* (white-spotted), *purpurascens* (purple), *erythrocarpum* (red-fruited), *nervosum* (conspicuously veined), and *villosum* (fuzzy beneath).

Because of irregular shape and coarser bark and leaves, this species lacks the popularity as a shade tree enjoyed by other species of a generally similar nature. In its native Europe, it is justly considered an important hardwood, taking the place there of our sugar maple.

Sycamore Maple

Striped Maple, Moosewood
Acer pensylvanicum

Tree. To 30 ft high. Usually slender, straight, and beautiful. Trunk to 10 in. through. Bark smooth and attractively streaked vertically with dark green and gray or white. Roots shallow. Twigs slender, weak, usually straight, red and green, with two-scaled buds. Leaves opposite, like track of goose's foot, rounded at base, finely notched along margin.

Common in mixed woodland borders or near streams but always in the shade of other trees. From Nova Scotia to Minnesota, south to Georgia in country that provides typical cover for moose and deer. Not found among giant trees of other species or in pure stands of its own kind. Largest in Big Smoky Mountains.

Flowers borne in May and June, with staminate and pistillate flowers separate but on the same plant, in loose drooping clusters. Pollination probably by insects. Fruits paired samaras that may remain on tree through first winter, being finally broken free and carried by wind. Seedlings develop year following the flowering. Fruit red in July, ½ in. long.

Wood of little importance commercially and generally weak; weighing 33 lb per cu ft; soft, pale, close-grained. Leaves turn to orange or scarlet in fall. Common name goosefoot maple refers to resemblance of leaves to shape of goose's foot.

Since it cannot survive strong light, it is of value as an ornamental mixed with other trees. Bark, fruits, leaves, and flowers make it popular with man. Twigs furnish a favored browse for deer and moose and give plant the name moosewood. Leaves of this plant are very large for a maple.

Striped Maple

Silver Maple, Soft Maple
Acer saccharinum

Tree. Height to 120 ft., with trunk diameter to 5 ft. Main trunk usually short, dividing at least by 10 to 15 ft and eventually with drooping or pendulous slender twigs. Leaves opposite, deeply five-lobed, with narrow segments that are often narrowest at base, dark green above, silvery beneath, to 6 in. long.

Native of America. Ranges from New Brunswick to Florida, west to South Dakota, Kansas, Oklahoma, and Louisiana but planted frequently outside its natural range and sometimes established as an escape. Favors more moist situations than red maple does and is found on swamp borders and along streams in wild country.

Flowers appear early in season, long before leaves, and mature their fruits early. Staminate and pistillate clusters separated either on one tree or on different trees. Some flowers may bear both stamens and pistils. Fruit smooth, unequal, double, winged, with wings to 2 in. long and rather widely spread.

Wood is "soft maple," weighs to 32 lb per cu ft, light-colored, hard, strong but weaker than sugar maple. Sap sweet, used somewhat for maple sugar but amount flowing is negligible and sugar content much less than that of sugar maple. Chief value is in the ornamental qualities of the tree.

Cultivated kinds, cultivars, include cv. 'Laciniata' (cut-leaved), cv. 'Pendula' (drooping-twigged), cv. 'Heterophyllum' (unequally divided), cv. 'Tripartitum' (three-parted leaves), cv. 'Albovariegatum' (white or pink-spotted), cv. 'Crispum' (crimped), and cv. 'Lutescens' (yellow). Known also as soft, river, creek, swamp maple. Seeds 21,000 per lb, 25 to 50 percent germination.

Silver Maple

Red Maple

Sugar Maple

Box Elder

Red Maple, Soft Maple
Acer rubrum

Tree. Height to 120 ft. Trunk diameter to 4½ ft. Bark gray, closer than in silver maple but possibly rougher than in sugar maple. Makes compact, rounded head. Leaves opposite, three- to five-lobed, with irregularly cut margins, bright green above, pale green or whitish beneath, to 6 in. long and nearly as wide.

Native of North America. Found in swamps and low grounds, associated with black ash, red gum, pepperidge, and cypress or often on higher ground. Ranges from Newfoundland to Florida, west to Manitoba, Nebraska, Oklahoma, and Texas. Found at 4,000-ft elevation in Virginia. A number of varieties recognized.

Flowers appear much ahead of leaves, giving a bright yellow or red cast to tree; axillary. Petals, narrow. Stamens, five to eight. Staminate and pistillate flowers in separate clusters on same or on different trees, with stamens often in the pistillate flowers. Fruits double, winged, smooth, incurved, each to 1 in. long, with wing to ½ in. wide.

Wood weighs 38 lb per cu ft, light reddish-brown, medium hard, medium heavy, not strong, close-grained, light brown with rose cast, with thick pale sapwood, sold in trade as "soft maple" with silver maple; rather good campfire fuel. *Closely related to Carolina maple.* Winter twigs lack odor of silver maple when crushed.

Makes a fair ornamental or street tree but best planted at edge of a forest for color effect in spring. Wood used in gunstocks, cheap furniture, flooring, woodenware, crates, and wood turning. Bears common names of swamp maple, white maple, soft maple, shoe-peg maple, and others. Seeds number 18,400 per lb and have 25 to 60 percent germination.

Sugar Maple, Rock Maple
Acer saccharum

Tree to 135 ft high. Egg-shaped to cylindrical. Trunk to 5 ft in diameter. Bark relatively close but with vertical ridges, irregular in shape and depth, grayish. Roots shallow. Twigs russet, straight, slender, with opposite, pointed, many-scaled buds. Leaves long-stalked, thin, with blade wider than long, somewhat five-lobed, red or yellow in autumn.

Common on rocky highlands from southern Newfoundland, west to southeastern Manitoba, south through eastern North Dakota and Iowa to northeastern Texas, east through northern Gulf States to piedmont in Virginia and north. Grows frequently in relatively pure stands but may be found mixed with beech, birch, and ash with large trees of these kinds represented. Commonly planted in northeastern United States as a shade and ornamental tree.

Flowers appear with leaves, small, yellow, in groups near ends of twigs, borne on slender drooping stalks, with both stamens and pistils in same flowers. No petals. Fruits, paired winged samaras with 1-in. wings, about parallel to each other or only slightly spreading. Mature in fall. Seedlings develop the following spring, with two narrow seed leaves showing first. May live 300 years.

Wood heavy, hard, uniform, close-grained, light-colored, easily polished, durable, with thin sapwood, sometimes with curly grain yielding bird's-eye maple. Seeds 7,000 per lb, with 30 to 50 percent germination, planted at once in fall at depth of ½ in. In 1 year reach to 9 in. Tree with 15-in. trunk has ⅓ acre leaf surface that needs 100 tons of water a season.

Wood used in furniture, flooring, interior finish, pulley blocks, broom and tool handles, and as fuel. Wood ashes high in potash and used by pioneers in making soap. Sap contains 2½ percent sugar and is boiled to make maple syrup and maple sugar. Average tree may yield ½ gal of syrup or 4 lb of sugar, though large trees in good season may yield 15 times this amount. 40 gal sap equals 1 gal syrup or 8 lb sugar.

Box Elder, Ash-leaved Maple
Acer negundo

Tree. Height to 75 ft. Trunk diameter to 4 ft. Trunk short, usually more or less deformed. Crown broad and rounded. Stump sprouts very readily. Leaves opposite, to 15 in. long, compound, of three to nine leaflets, each to 5 in. long and to 3 in. wide. Sometimes doubly compound in part. Leaves smooth.

Usually favors wet areas along waterways but may be planted and will thrive in drier situations. Ranges from Maine to Florida, west to Manitoba, western Minnesota, Kansas, Texas, and Mexico though it is planted extensively outside this range. Closely related to western ash-leaved maple *A. interior* of Rocky Mountains.

Flowers appear in April and May, with or before leaves, on growth of preceding season, yellow-green with tasseled staminate on one tree and pistillate on another, the latter being borne in narrow drooping clusters that elongate in fruit. Fruit smooth, paired, winged structures, one-seeded, to 2 in. long, persisting.

Wood weighs 27 lb per cu ft, light, weak, soft, white, close-grained but capable of being worked easily into structures that do not require any considerable strain. Trees frequently winter-kill and assume an unkempt appearance but they grow rapidly and so have been popular as shade trees in new real-estate developments.

Inferior shade tree because of relatively short life and considerable trash forming beneath. Staminate trees drop less trash than the pistillate. Wood used somewhat as fuel, in making crates, cheap furniture, and coopering. Some horticultural varieties have been developed. Young growth frequently confused with poison ivy. Box elder has opposite leaves while poison ivy is alternate. Is frequently attacked by the box elder bug *Leptocoris trivittatus,* which is very familiar around cities and in plantations of this maple.

Soapberry

Order Sapindales./Family Sapindaceae

Soapberry
Sapindus drummondii
Tree. Height to 50 ft. Trunk diameter to 2 ft. Old bark with well-fissured surface. Leaves to 1½ ft long, compound, of 7 to 19 leaflets, each entire-pointed particularly at tip and to 4 in. long; usually slightly curved because one side is shorter than the other.

Locally common on hillsides or river valleys from Missouri to Kansas, south through Louisana, Texas, and Arizona into Mexico. The related soapberry of commerce *S. saponaria* can be grown successfully in southern Florida and in southern California. Its leaves are evergreen, while this tree has deciduous leaves.

Flowers appear in May or June, borne in open clusters that are to 8 in. long with the individual flowers white and to ⅙ in. across, with many in a single cluster while the stems are fuzzy and many-angled. Sepals much shorter than petals, which are rounded at tip. Fruit is ripe in fall, a yellow globe ½ in. through, turning black.

Wood weighs 59 lb per cu ft, heavy, hard, close-grained, light brown with a yellow tinge, with still lighter sapwood that is about 30 annual rings thick. Wood splits easily into thin strips, each representing an annual growth in thickness. Fruit soapy. Plant does well on moist clay or on dry limestone soils.

Wood used rather extensively in making baskets for harvesting cotton and for frames for packsaddles. Berries used as a source of soap, though handling them may cause a dermatitis because of contained saponins. The cultivated soapberry flowers in fall and bears fruits in spring.

Gate Tree, Shower Tree, Varnish Tree

Gate, Shower, Varnish Tree
Koelreuteria paniculata
Tree to 30 ft high. Bark finely furrowed. Head round and rather compact. Twigs coarse, brown-barked. Leaves alternate, once or twice compound, to 18 in. long, of 7 to 15 leaflets, each to 3½ in. long, coarsely veined, dark green above, coarsely indented along margin, smooth, attractive.

Native of China, Korea, and Japan. Cultivated in United States particularly from Kansas to south and in region of southern Indiana, though it survives well in New York and to the west where there are hot dry winds. Related to soapberry *Sapindus saponaria,* there being about 1,000 species in family, mostly trees. Prefers a position exposed to full sun and a well-drained loamy soil.

Flowers about ½ in. long, in many-flowered clusters, forming a loose open cluster that hangs downward, cluster being to 1½ ft long while individual flowers are about ½ in long. Flowers yellow, appear in July and August or sometimes earlier. Fruit a 2-in. papery attractive inflated bladder.

Tree may winter-kill in northern New York, but in southern Indiana it may grow uninjured to mature size. Grown as a street tree for its most attractive flowers and fruits and because of its general beauty. Common names varnish tree and China tree properly belong to other plants.

Known in some places as "gate tree" because when New Harmony settlement was made in southern Indiana, Thomas Say and others set trees at gates. This was once a custom so that some campuses of the country have these trees growing at the gates. Also called golden rain tree.

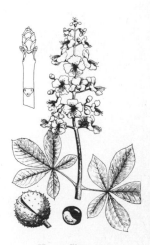

Horse Chestnut

Order Sapindales./Family Hippocastanaceae

Horse Chestnut
Aesculus hippocastanum
Tree to 75 ft high. Trunk to 3 ft in diameter, frequently continuous to top of tree but usually branching. General shape conical, lower branches drooping but turning up at tips. Bark dull brown with irregular scale plates. Twigs coarse, with large brown to black sticky buds. Leaves opposite, palmately compound, of to seven leaflets, each to 10 in. long.

Originally native of Greece but planted widely and established in many places where it has been introduced as a shade and ornamental tree.

(Continued)

243

Ohio Buckeye

Touch–me–not

Closely related to Ohio buckeye and to red horse chestnut of Pacific Northwest. Also a dwarf shrubby form *A. parviflora* frequently used in formal plantings.

Flowers showy white with a red tinge, to nearly 1 in. across, in open panicles to 1 ft long. Petals, five. Flowers borne at ends of twigs causing twigs to branch the next year. Fruit a weak short-spined bur, green, containing a few large brown large-scarred seeds having a high percentage of germination.

Wood light, soft, close-grained, whitish or yellowish, not strong. Seeds mature in October or November; must be buried and frozen before they germinate. In spring, planted 1 in. deep and 5 to 8 in. apart, reaching a height of to 18 in. as seedlings first year. A severe blight of leaves. Winter-killing often deforms or injures trees.

Wood used by wood turners and wood carvers because of fine uniform quality and because it can be cut easily. Seeds appear to be edible but are bitter and are considered by some to be poisonous. A flour paste made from seeds is reported to be poisonous to insects. Alcohol has been extracted from fermented seeds. Great popularity is as a shade tree.

Ohio Buckeye
Aesculus glabra
Tree. Height to 70 ft. Trunk diameter to 2 ft. Branches small and spreading. Twigs rather coarse. Winter buds not sticky as in horse chestnut, to ⅔ in. long and more pointed. Leaves opposite, compound, of five to seven leaflets, each to 6 in. long and arranged like the fingers on the hand, mostly smooth and with a petiole to 6 in. long.

Found along river bottoms and stream banks from southern Pennsylvania to northern Alabama west to Iowa, Nebraska, and Missouri. Ohio buckeye has many horticultural and other varieties and is cultivated in eastern United States and Europe. Hardy into northern Massachusetts. Related California buckeye and others extend range considerably.

Flowers in loose, open, terminal clusters that are to 6 in. long while the individual flowers are yellow, to 1½ in. long or twice the length of their supporting stems; appear in April or May. Stamens conspicuously incurved. Fruit to 1½ in. in diameter; prickly when young but smoother at maturity, freeing large brown seeds.

Wood weighs to 28 lb per cu ft, white, soft, not strong, close-grained, often with brown blemishes, decays easily unless kept dry, with thin dark sapwood that is 10 to 12 annual layers thick. Pollination probably by bees. There are shrubby forms, large-flowered forms, and others well-known to horticulturists.

Wood was used in making paper pulp, artificial limbs, and woodenware; sometimes sawed into rather inferior lumber. An extract of the bark has been used as an irritant for the cerebrospinal system. Probably the most universal use, however, is as an ornamental tree or shrub. Unofficial state tree of Ohio, the Buckeye State.

Order Sapindales./Family Balsaminaceae

Touch-me-not, Jewelweed
Impatiens pallida
Herb. Height rarely to 6 ft. Annual, smooth, with semitranslucent stems showing the bundles within, hollow, semisucculent, and showing swellings at the joints. Leaves thin, pale green, alternate, smooth above and below, semisucculent, blunt, to 3½ in. long, with coarsely toothed margins and petioles to 4 in. long.

Found in moist ground, commonly in the shaded areas such as at woodland borders, from Nova Scotia to Georgia, west to Saskatchewan and Kansas. Closely related to spotted touch-me-not with orange and yellow flowers, somewhat smaller and possibly somewhat more common in many parts of range. Also related to greenhouse balsam.

Flowers pale yellow or sometimes dotted with red-brown, to 1¼ in. long, suspended at end of slender stem arising from a leaf axil, like a funnel-shaped sack somewhat closed at larger end, with a curved spur at smaller end in which there is nectar. Fruit a slender green capsule about 1 in. long that bursts spontaneously or upon being touched, freeing dark brown seeds.

Flowers from July through September. Pollination effected probably by bumblebees and honeybees powerful enough to force way into semiclosed flowers and reach nectar sac, which in this species is about ⅓ length of flower. Stem commonly contains larva of a borer that eats its way through the joints.

An attractive plant found abundantly at woodland borders. Seeds delicious tidbits, tasting much like butternuts. Stamens develop as a rule before pistils. Plants are considered fairly emetic and poisonous if eaten by stock in fresh condition. *I. capensis,* the orange-flowered species, is common in the eastern U.S.

Order Rhamnales./Family Rhamnaceae

Buckthorn
Rhamnus cathartica

Shrub. Height to 20 ft. Lateral twigs may end in short stiff stout thorns. Bark, dark. Leaves alternate to opposite, smooth, to 2½ in. long and 1 in. wide, oval, well-petioled, blunt or acute at tip, with margins with shallow rounded teeth, with rather conspicuously paired veins, deciduous, dark green above and lighter beneath.

Native of Europe and western and northern Asia but widely planted in America as a hedge plant and often escaped and established in many places. About 100 species of the genus in the world, found largely in temperate and the warmer regions. Does well on dry soil. Related *R. frangula* yields high-grade charcoal used in making gunpowder.

Flowers of two kinds, with staminate and pistillate on separate plants, clustered in axils of leaves, greenish, appearing shortly after leaves, with narrow, relatively inconspicuous petals. Stamens four to a flower. Fruit a small globular black cherrylike structure, to ⅓ in. in diameter, with three to four grooved nutlets.

Relatively common hedge and fencerow plant, the seeds being planted by birds. Propagated mostly by seeds that are collected in fall, stratified in sand for winter, and sowed in place in spring. Related evergreen species propagated by cuttings of mature wood. Fruit yields the dye Chinese green.

Bark, leaves, and berries have strong purgative qualities but because of bitter taste they are rarely eaten. Dried bark of related *R. purshiana* from a western shrub is described elsewhere. Related *Karwinskia humboltiana* or coyotillo has berries that if eaten may cause a paralysis of hind legs of domestic cattle. In the Western States species of *Rhamnus* provide, in their black fleshy fruits, important food to the pileated woodpecker, mockingbird, catbird, crested mynah, and thrushes. Deer and bighorn sheep browse on the plants in California.

Buckthorn

Cascara, Bearberry, Coffee-tree
Rhamnus purshiana

Tree. Height to 60 ft. Trunk diameter to 20 in. Bark close, mottled, usually under ¼ in. thick, dark or light brown, often tinged with red and broken into small thin scales on surface. Winter buds are naked, hoary or fuzzy. Leaves alternate, almost entire, to 2 in. wide and with prominent midrib and branching veins.

Native of Northwest. Ranges from southern British Columbia, northern Washington, and the Puget Sound region to Idaho, Montana, and south to central California and southern slope of Grand Canyon in Arizona. Western North America has monopoly on drug. Early Mexican and Spanish priests in California learned of its use from Indians. Commonly found on rich bottomlands and the sides of canyons, usually in coniferous forests.

Flowers relatively inconspicuous, hung on slender stems from axils of leaves. Corolla minute, of five petals. Fruit like small black two-to three-seeded cherries to ⅓ in. through, with a thin juicy flesh; with nurlets flattened on inner face and with a thin gray shell. Inner surface of thin seed coat is bright orange.

Since 1877 cascara has had official listing as a helpful drug. Bark peeled from woody parts down to 2 in. through, though that of older trees is superior. Bark cured for 2 years is best. Trees peeled should be left with live bark near base and cut to encourage new sucker shoots. If under 5 in. through, they should not be cut.

Cascara

Drug cascara sagrada produces the constituent frangulin, a mixture of frangulic acid, emodin, and a bitter resin, the latter being modified in some fluid extracts by magnesium oxide. A 3-in. tree may yield only 5 lb of bark.

New Jersey Tea
Ceanothus americanus

Shrub. Height a few feet. Densely branching, with parts erect or ascending. Root system deep, red, with especially well-developed taproot. Leaves alternate, to 3 in. long and to 1 in. wide, obviously with three main veins, with petioles to ½ in. long, pointed at tip, rounded at base and with finely toothed margins. Leaves used as a tea substitute during the American Revolution, hence its name.

Found in dry woodlands or exposed rocky areas. Ranges from Maine to Florida, west to Manitoba and Texas reaching an elevation of 4,200 ft in North Carolina. Some 55 species in United States and northern Mexico including. *C. velutinus* or snowbush of western U.S., a valuable browse species for hoofed mammals such as deer, elk, mountain goat, and mountain sheep, and *C. thyrsiflorus*, which provides an evergreen cover on sides of canyons from central California to Oregon either as tree-sized specimens or as chaparral.

Flowers in dense clusters, with a number of clusters at ends of branches or arising from leaf axils, small, white, on short stems, to ½ in. long and arising more or less from a common point. Fruit like a flattened sphere that becomes black and is to ⅙ in. high and wider than this; dry, producing three smooth nutlets.

New Jersey Tea

(Continued)

245

Woodbine

Concord Grape

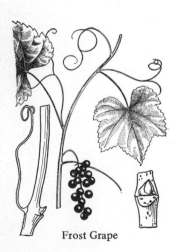

Frost Grape

Root system penetrates deeply into clay, sand, limestone, sandstone, and other soils and provides a superior soil anchor with the dense brushy tops serving to anchor windblown material. Most of the *Ceanothus* species seem able to recover after a fire, probably in part because of substantial root system.

Leaves and twigs collected when they are just reaching full size may be dried quickly and used as a tea substitute. Many western species are trees of considerable size and provide browse for cattle, deer, rabbits, porcupine, beaver, and other mammals and food and cover for quail and other birds. Red root yields an excellent dye.

Order Rhamnales./Family Vitaceae

Woodbine, Virginia Creeper
Parthenocissus quinquefelia
Woody vine. Climbing 20 or more feet by means of tendrils that bear at the end of five to eight branches some adhesive disks that grow firmly to wood, brick, or stone. Leaves alternate, of three to seven but usually five leaflets, each to 6 in. long and coming from a common point at end of a rather long petiole. Leaflet margins rather coarsely toothed.

Found on fences, walls, buildings, climbing trees, or in thickets or on the ground from Quebec to Florida and west to Saskatchewan, Texas, and Mexico and grown rather extensively as a hardy climbing ornamental. Found wild in Cuba and the Bahamas. About a dozen species native of eastern North America and Asia.

Flowers inconspicuous, yellow-green to white, with pistils and stamens in same flower or with pistils and stamens in separate flowers on one plant, borne in an open broad cluster that expands to bear by fall attractive two to three seeded blue berries to ½ in. in diameter, with their stems becoming red.

Found on sand, loam, clay, and a variety of soils. Resists grazing rather well. Seeds run 12,000 to 19,000 per lb. Easiest propagation usually by cuttings, stems rooting easily in proper soil or layering naturally and establishing a plantation over a considerable area.

Leaves turn a beautiful scarlet in fall. This five-leaved ivy is not to be confused with poisonous three-leaved ivy, since while this may sometimes bear three leaflets the poison ivy never bears more than three. Berries or other parts eaten by turkeys, pheasants, quail, grouse, prairie chicken, deer, chipmunks, and other birds and mammals.

Concord Grape
Vitis labruscana
Heavy, woody vine, climbing by tendrils and rooting freely with contact with ground. Leaves as broad as long, with blade to 8 in. across, alternate, with three obscure lobes at apex, dark green above and gray felty beneath, with a petiole nearly as long as blade's diameter.

Derived from native fox grape *V. labrusca* whose range is from New England to Georgia, west to Indiana. Other grapes from same source include the Worden, Hartford, and Vergennes while the Catawbas, Niagaras, and Isabellas result from crosses between *V. labrusca* and *V. vinifera*. Labruscan leaves are usually more felty beneath.

Flowers relatively inconspicuous, borne at about time leaves appear and at base of current year's growth. Flowers like small greenish cups, actually or functionally either staminate or pistillate. Fruits well-known, soft, blue, juicy berries, with many seeds embedded in a juicy sweet pulp.

Pruning commonly cuts canes to 8 to 10 buds each, with three to four canes to a vine and with most 2- and 3-year-old wood cut away. In East, commonly trained on a two-wire trellis. Vines may produce to 60 bunches weighing to 20 lb. Fruit rots and mildews controlled in part by bordeaux mixture, and grape rootworm by lead arsenate.

Fruits have considerable quantities of vitamins B and C. A food portion of 4.8 oz equals a 100-calorie portion. Yield may be as much as 5 tons per acre. Sold fresh, for extraction for juice, either unfermented or fermented into wine. In some forms, as in the raisin grapes, fruits are dried. Many wild grapes supply important game food.

Riverside or Frost Grape, Chicken Grape
Vitis vulpina
Climbing, trailing, woody vine. More or less smooth throughout. Bark scales off in long strips. Branches rounded, angled, green, with pith interrupted and with tendrils. Leaves alternate, shining, three to seven-lobed, with notches angular and points acute but with terminal lobe largest.

Common by streams and woodland borders from New Brunswick to Maryland, west to Manitoba, Kansas, Colorado, Texas, and West Virginia. About 50 species of wild grapes known, of which half are American and of these about ⅔ are native of eastern part. Species vary greatly in leaf shape, fruit, bark, and other characters, and may cross freely.

Flowers: staminate, with stamens that are erect or nearly so; those with both stamens and pistils, with downward-curved stamens borne on medium-large, rather open clusters, fragant. Fruits, bluish-black to nearly ½ in. in diameter, with two to four seeds, with a bloom, acid and very juicy, ripening from September through November. Flowers in May and June.

Grows well in well-drained, loose, sandy soils; poorly in lime or clay soils. Roots long, thin, hard, fibrous. Plant resistant to *Phylloxera* and to cold but not to heat and drought. Propagated easily by cuttings, 85 percent of which usually root readily, easily grafted. Seeds germinate quickly and readily and weigh 14,500 per lb.

Commonly used in Europe as a base for grafting *V. vinifera* and other domestic stock. Fruit constitutes more than half of total fruit food of pheasants in Michigan and four percent of their total food. Also eaten by grouse, deer, cottontail, skunk, fox, wild turkey, mourning dove, quail, cedar waxwing, and other forms of wildlife.

Order Malvales./Family Tiliaceae

Basswood, American Linden
Tilia americana
Tree. To 80 ft high. With trunk to 4 ft in diameter. Crown dense and rounded. Trunk straight and slightly tapering, with dark gray, somewhat scaly, deeply and regularly fissured bark. Leaves alternate, simple, unequally heart-shaped, with incurved marginal teeth; texture, firm. Roots lateral, deep, spreading, without taproot.

A close relative of Old World linden. Family has about 370 species. Found from New Brunswick to Manitoba, south to Georgia and eastern Texas and is usually rather abundant within its range though it does not occur in pure stands. It favors rich woods and loamy bottom lands.

Stamens and pistils in fragrant, yellowish-white flowers borne in flat drooping clusters in June and July. Pollination by insects, particularly bees, results in woody, spherical, pea-sized fruits, attached singly or in groups, on a wing, and distributed in October or fall by the wind. Seedlings 1 to 2 in. high first year; with two four- to five-lobed palmate cotyledons.

Rarely attacked by fungi but sometimes by a canker of bark and a leaf spot and by an insect pest, the basswood leaf roller. Propagated by seed, stratified when ripe and planted in spring. Dry seed may germinate after 2 years. Seeds 5,000 per lb, germinate 10 to 15 percent. Grows better in a mixed forest than alone and does not resist wind injury well or severe ice storms.

Inner bark yields a cordage fiber; whole bark used by early farmers for tying logs together. Inner bark has some edible qualities, and buds are good to eat. Wood valuable as "whitewood," light, soft, moderately strong, fine-grained, white, weighs 28 lb per cu ft. Flowers with help of bees yield a superior honey. Staple food of cottontail and white-tailed deer.

Basswood, Linden

Indian Jute
Corchorus capsularis
Height 15 ft. Stems straight, branching only near top. Stem diameter to ½ in. Leaves shaped somewhat like an arrowhead, to 4 in. long, with basal teeth sharply pointed. In *C. capsularis* leaves are bitter and seed pods spherical; in related *C. olitorius*, Desi or Nalta jute, leaves are not bitter and seed pods are cylindrical. Annual.

More than 20 races of Indian jute known. Requires a rich loose, sandy soil, with good drainage and climatic conditions such as are to be found in Burma, Bengal, southern China, northeastern India, and the southern islands of Japan. Does best on high lands but is grown on lands that are flooded in late summer season.

Flowers small, yellow, often solitary, mallowlike, with five petals, five sepals, and ten or more stamens, which are free of each other. Seeds sown by hand, 10 to 15 lb to the acre, March to May. Cultivated when plants are 6 in. high, and hoed thereafter. Harvested from flowering time to maturity, the earlier the better commercially. In both Pakistan and India, the foliage of jute is prepared as a potherb.

Plants cut, then held under water at about 80°F for 10 to 20 days, after which fiber is extracted by running to three stalks at a time through the fingers, a woman being able to strip to 40 lb a day, for which the wage has been the waste fiber to be used as fuel. Yield about 1,300 to 1,800 lb per acre. Exported in 400-lb bales. Raw fiber prepared by cheap labor in Calcutta is shipped to ports in Scotland and elsewhere to be spun into yarn.

In normal times, India grows about 3 million acres of jute. Normal world yield between 1 and 2 million tons. Used in making burlap sacks, coarse cordage, twines, and some coarse papers. In commerce, the material is known as Bengal gunny, the short fibers being used in papermaking and the long fibers being twisted or spun into cordage and similar materials.

Indian Jute

Musk Mallow

Round–leaved Mallow

Hollyhock

Order Malvales./Family Malvaceae

Musk Mallow
Malva moschata

Height to 2 ft or even more. Branching, covered with long fine hairs or smooth and erect. Leaves alternate, to 4 in. wide. Basal leaves with five to nine broad rounded lobes and long petioles. Stem leaves deeply cut into repeatedly cut divisions. Root system substantial, with a deep branched or unbranched taproot and finer roots.

Naturalized from Europe but escaped from cultivation or as a weed impurity. Now well-established sometimes as a weed from Nova Scotia to Virginia, west to British Columbia and Oregon. Sometimes planted in wild or semiwild gardens as persistent attractive plant. Closely related to round-leaved mallow or cheeses.

Flowers to 2 in. broad, borne in showy clusters at tip of plant or in axils of upper leaves, white, pink, or sometimes somewhat bluish, with delicate petals that are five to eight times as long as green sepals. Fruit composed of to 20 segments, rounded on back, densely hairy, and enclosed in calyx when mature.

Flowers in June through August and produces seeds succeeding months. Survives in midst of tall grass and may decrease value of hay where it is an established weed. Control is by keeping tops cut before seeds have time to form and digging out perennial and rather large root systems.

Flowers grown in wild gardens for ornament and are used to a slight extent for bouquets in spite of the fact that they are inclined to wilt quickly after they have been picked. Delicate musk odor of flowers gives them added attractiveness.

Round-leaved or Common Mallow
Malva neglecta

Height to 2 ft. Usually a sprawling plant covering a square yard or more. Stems round, smooth, branched, green. Leaves with blades round or heart-shaped, with toothed or scalloped edges and to 6-in. petioles. Blades to 3 in. across, five- to nine-lobed. Root system deeply penetrating. Whole plant tough. Stem and root not easily separated.

Common in waste places, on poor lawns, and in neglected gardens usually at best in relatively dry areas. Ranges through North America except in extreme north. Native of Europe and Asia, but widely established in other parts of world. Persistent once it has become established anywhere. 30 species in the genus, mostly Old World.

Flowers five-parted, with pale lavender or white delicate petals. Stamens fastened together into a tube and at first enclosing the blue-tipped pistils that eventually emerge from the top of the stamen tube. Pollination effected by insects or by self. Fruits a disklike collection of cases, each containing one flat seed.

Flowers from May through October. Fruits from June through November though flowers and fruits may be found at any time of the year. Seeds long-lived and may persist in soil many years. Deep roots help plant survive drought. Mucilage in seeds helps them get a start in germination and makes them desired by children who chew the "cheeses."

Plant reported to have medicinal properties of an obscure nature. Tender shoots eaten as salad in France. Greeks, Romans, Egyptians, and Mexicans use the whole plant as a potherb like spinach. Young plants properly cooked are good as greens. Two flies live as leaf miners. Fruits eaten by birds and by a few insects.

Hollyhock
Althaea rosea

Height to 9 ft. With erect unbranched spirelike hairy stems. Leaves deep green above and lighter beneath, alternate, rough, lobed, long-petioled, rather thick, with wavy margins and relatively conspicuous veins, often spotted with disease marks. Root system heavy, deeply penetrating, and often branched, with short side-crown buds.

Native of China. Carried to Europe by the Crusaders; to America by the Pilgrims; now commonly established throughout United States and through much of temperate world. Sun-loving but tolerates partial shade. Lower leaves demand light. Grows best in well-drained fertile soils. About 15 species in the genus. Closely related to marsh mallow.

Flowers nearly stemless but arranged in a wandlike formation along main stem; large, single or double; to 3 in. or more across, many-colored. Many stamens fastened into a tube at base through which pistils thrust. Fruit a disklike collection of cases, each with many seeds. Blooms July to September, each flower lasting three to four days. Perennial or biennial.

Propagation by seeds or by division. Seeds germinate in 5 days and may live 4 to 5 years dry. Self-sowing maintains production but plants hybridize freely. Sow seeds in fall, cover lightly in deeply dug soil, transplant in spring, mulch in fall. Insect pests include rose chafer, stalk borer, abutilon moth, leaf roller. Should be treated as a biennial when rust disease is troublesome.

Excellent ornamental perennial or biennial for growing near walls and borders. Popular with amateur plant breeders. Occupies little space horizontally but much vertically. Insects controlled by picking or lead arsenate spray; destructive rusts, by bordeaux spray or sulfur dusting; red spiders, with a water spray. Some varieties resistant.

Marsh Mallow
Althaea officinalis
Height to 4 ft. Erect, downy, leafy stems. Leaves heart-shaped or oval, three-lobed or undivided, densely fuzzy, alternate, conspicuously veined and inclined to curl rather than to lie flat. Root system substantial, deeply penetrating, brown-barked but lighter within and not too tough.

Native of Europe but escaped in America and sometimes established at edges of marshes such as the salt marshes of New England. Now considered established from Massachusetts to Pennsylvania and reported in Michigan, Arkansas, and District of Columbia. Little cultivated in this country.

Flowers pink, about 1½ in. across, borne terminally and in axils of leaves, showy because of colored, relatively thin petals, usually in clusters of a few though clusters are well-distributed over upper parts of plant. Fruit composed of to 20 sections, fuzzy. Seeds like flattened thin sections of an apple.

Brown bark of root contains 1 percent asparagin, 25 percent bassorin (mucilage), 10 percent pectin, and 8 percent sugar when dried. Roots collected when 2 years old. Mucilaginous properties demonstrated in marshmallow confection so well-known.

Used in medicine as a demulcent. Common names include "mortification root" and "sweatweed." Leaves sometimes used as poultices but mucilage of the roots finds best use as a soothing agent for inflamed tissues. Plant is sometimes grown as an ornamental because of its attractive flowers.

Marsh Mallow

Cotton
Gossypium hirsutum
Height to 8 ft. Much-branched but with one more or less central stem. General shape of plant like a cone. Leaves variable, to 6 in. long and 5 in. wide, heart-shaped, three- to seven-lobed, and coarse-veined. Roots a main taproot, with finely divided lateral roots that fill the soil. Sometimes called herbaceous shrub. Annual. Name from Latin *gossypion*, cotton tree.

American cottons include upland cotton, *G. hirsutum*, Sea-island cotton, *G. barbadense*, Peruvian cotton, *G. peruvianum*. Asiatic cottons include Bengal cotton *G. neglectum*. Upland cotton constitutes about 99 percent of all cotton grown in the United States and is of course leading crop in the Southern states.

Flowers on short stems, creamy white at opening, changing to red by second day. Sea-island cotton has yellow flowers. The bolls which develop from flowers may be 2½ in. across, and a good plant may bear 50 bolls. Seeds 1 to 3 bu per acre, planted 2 weeks after last killing frost produce plants that in September to November may yield mature cotton.

Requires about 200 days, with a mean temperature in summer of around 77°F and rainfall, after harvest, of about 10 to 22 in. May and June are critical months, and a rainy fall may ruin a harvest. Seed hairs vary in length, strength, and fineness in different cottons. Cotton boll weevil has been most important insect pest.

World's most important fiber crop, having been used since 1000 B.C. In United States next in importance as an agricultural crop to wheat, potatoes, and corn but leads in areas between 30 and 40° north or south of the equator. Cottonseed valuable in industry for food and oil. 2 million acres of cotton in United States yield about 10 million bales.

Cotton

Flowering Maple
Abutilon hybridum
Shrubby. Stem erect, stiff, rather coarse. Leaves alternate, somewhat five-lobed, resembling some maples superficially, coarse, prominently veined, long-petioled but varying greatly from lobed to unlobed, smooth to fuzzy, large to small, and so on. Root system fibrous and branched.

A hybrid of many species including such garden varieties as snowball, fireball, caprice, Savitzii, golden fleece, and others. Some 100 species of herbs and shrubs in the genus, many native of South America. Commonly grown in greenhouses or coolhouses or as summer border plants or sometimes as house plants.

Flowers red, purple, pinkish, yellow, white, or mixed in color; open, drooping from slender stems at tip of plant or its branches, relatively light-sensitive. Stamens many, fastened together at base, at first enclosing pistils but later freeing them through tip. Seeds sown in spring yield large blooms in autumn.

Flowering Maple

(Continued)

249

May be propagated by slip taken in spring or autumn as well as by seeds. Plants taken into greenhouse in September may be cut back and grown at 55°F but bottom heat is necessary for satisfactory development. Mealy bugs, red spiders, thrips, and aphids are bad pests controlled by sprays, fumigation, and picking.

High nitrogen content of soil produces heavy foliage and few flowers so low fertility is favored where plants are grown as houseplants or for window boxes. Shrubby potted plants have a fair popularity for sick and will bloom all winter if given care but of course cannot survive severe weather outdoors in the North.

Velvet Leaf, Indian Mallow
Abutilon theophrasti
Height to 6 ft. Erect, branched above, rather stout, densely velvety. Leaves with a strong odor, velvety, with a heart-shaped blade that may be to 1 ft across, nearly entire, with a drawn-out point at the tip or blunt, conspicuously veined, with petioles about as long as blade is wide. Deeply penetrating roots.

Locally common in waste places, vacant lots, cornfields, barnyards, and similar localities through most of United States and southern Canada particularly in warmer areas. Some 100 species of genus to be found in tropical areas, but this is most common one to reach into temperate zones.

Flowers orange-yellow, to ¾ in. across, drooping or erect bells, solitary, on rather stout stems that are shorter than the petioles. Stamens united in a column enclosing pistil tips. Fruit an interesting, many-chambered capsule that frees seeds from tip. Seeds gray-brown, thinner on one side.

Flowering time from August through October. Propagated by seeds only. Seeds known to retain their living qualities for 60 years, so once soil has been well-supplied with these seeds, it may continue to yield plants almost indefinitely. Control of course centers around destroying the first plant before it bears seeds.

This is sometimes a serious weed in crop fields but it succumbs to cultivation and pulls easily. It was considered as a possible source of fiber to substitute for manila and sisal but cannot rival these plants. Known as butter print, mormon weed, Indian hemp, American jute, cottonweed, pieprint, piemarker, and buttonweed.

Flower of an Hour
Hibiscus trionum
Height to 2 ft. Erect or rather weakly so, branching freely, covered with rather fine spreading hairs. Leaves long-petioled, with blades rather circular in general outline but deeply lobed or divided one or more times into three to seven lobes, with the middle one the longer. Root system rather deeply penetrating. Annual.

Found in rather open dry waste places. Native of southern Europe. Now rather well established in fields and near grain plantings from Nova Scotia to Florida, west to South Dakota and Kansas. Possibly more abundant in western part of this range. Closely related to rose of Sharon, rose mallow, and similar plants.

Flowers attractive, showy, pale yellow with dark-purple centers, to 2½ in. across. May fade from a deep sulfur-yellow to a pale weak yellow or petals may be purple-tipped at outer edge. Calyx five-angled, almost globular, hairy, conspicuously veined, and like "spun glass." Fruit a globular capsule containing many roughened dark seeds.

Flowers from July through September. Does best on gravelly or limy soils. Seeds retain life long, either buried in soil or may survive in silage. Since reproduction is by seeds alone, every effort should be made to prevent the plant going to seed, because once a seed supply is established in soil, the plant may persist for years.

Control is by means of hoeing, pulling, and cultivating as late as crop will permit. Plant entitled to place in flower garden but should be kept there as it does not improve fields where it may become established. Known as modesty, shoofly, Venice mallow, black-eyed Susan, devil's-head-in-a-bush, flower of an hour.

Rose of Sharon
Hibiscus syriacus
Tree or shrub. Height to 12 ft. Branches freely. Has dark bark and is nearly smooth throughout. Leaves alternate, somewhat triangular, dark green, conspicuously three-ribbed, to 3 in. long; lower ones inclined to be three-lobed, margins rather coarsely but shallowly toothed and notched. Leaves give plant beauty as an ornamental even without flowers.

Native of eastern Asia, cultivated under many names depending on time and color of flowering, shape and coloring of leaves, and other characters.

Velvet Leaf

Flower of an Hour

Rose of Sharon

Relatively common as a lawn or hedge ornamental as far north as the Canadian border and into Ontario, though not all varieties are equally hardy.

Flowers to 3 in. or even more across, single or double, showy, short-stemmed, commonly rose or purple though other colors are known, usually darkest near base, borne in axils of leaves on young wood of year but late in the season. Fruit about 1 in. long, short-beaked and splitting into five valves to free the seeds.

Flowering time summer and early fall. Propagation by seeds or by cuttings of wood that has ripened by fall, these cuttings being rooted in greenhouses through winter for spring planting. Plants are also grafted rather easily, this making it possible to retain unique varieties that might not persist through the seed stage.

A generally popular ornamental closely related to hibiscus so commonly planted along roadsides of tropical countries. A number of perennial herbs such as swamp rose mallow *H. moscheutos* that are hardy in North and many other shrubs, trees, and herbs that have attractive flowers and other pleasing qualities. National flower of the Republic of Korea.

Spiny Sida
Sida spinosa

Spiny Sida

Height to 2 ft. Erect herb that branches freely and is covered with fine soft hairs. Leaves alternate, with blade and petiole totaling to 2 in. long and blade to nearly 1 in. wide with curved teeth along margins and with soft hairs such as are found on stem. Petiole somewhat shorter than blade.

Locally common in waste places and, unfortunately, in cultivated fields from Maine to Florida west to Iowa, Kansas, and Texas, being possibly more common in western part of this range; probably spreading. Has wide distribution in tropical America whence it is supposed by some to have moved north.

Flowers borne in axils of leaves usually singly, yellow, to ⅓ in. across and thus relatively inconspicuous, on short stems. Calyx encloses much of the flower and is conspicuous because of its rather sharp teeth that are about equal to end of petals. Fruit composed of five sections, each of which frees a red-brown seed.

Blooms June through September. Annual. Propagates solely by seeds that are long-lived, in soil or out. For this reason, control requires cutting of plant before it has had time to sow a crop of seeds which may continue to produce plants for a number of years. In South, it may be a serious weed in certain areas.

Seeds rather commonly found as impurities of commercial seeds. Control measures are largely limited to cultivation and hand pulling though sprays are useful, in part because fuzzy covering of the plant holds them until they may become effective.

Order Malvales./Family Bombacaceae

Silk-cotton Tree, Kapok
Ceiba pentandra

Silk-cotton Tree, Kapok

Tree. Height to over 100 ft. Trunk enormous, with giant buttresses that stick out to 30 ft in various directions and extend upward for considerable distance. Branches more or less horizontal and sometimes spiny. Leaves compounded of about seven leaflets, each to 6 in. long and arising from a common point. Root system relatively shallow but widespread.

Family includes baobab tree *Adansonia*, of Africa, India, and South America, whose trunk is probably larger than that of any other tree. Also in the family is the balsa *Ochroma* and the durian *Durio*. *Ceiba pentandra* is a native of the tropics of Asia, Africa, and America and is grown sparingly in southern Florida and in southern California.

Flowers with corolla to 3 in. long, with the white or rose petals distinctly hairy on the outside and with the calyx cup-shaped. Fruit important, to 8 in. long, shaped somewhat like a plump banana but composed of five sections that break at maturity to free the many seeds, whose woolly covering is probably hidden in pillows in most homes in America.

Tree begins to yield fruits when only 15 ft high. A mature tree may bear to 1,000 pods yielding a total of about 10 lb of cottony floss that is white, yellowish, or brown and water-resistant but cannot be spun into threads. Wood soft, white, brittle. Seeds contain to 23 percent of a fatty oil extracted for soap and for food.

Kapok floss, of which the United States imports annually some 10,000 tons, is used for life preservers, sleeping bags, mattresses, pillows, and upholstery. Java, the Philippines, and Ceylon are the chief sources of supply. Natives of Jamaica believe that the duppies or "little folk" live in the huge buttresses of the trees, where they are seen only by the believers.

Cocoa

Tea

St. John's Wort

Order Malvales./Family Sterculiaceae

Cocoa
Theobroma cacao
Tree. Height to over 25 ft. Twigs rather coarse and fuzzy. Branches wide-spreading, particularly in wild plants. Leaves evergreen, to over 12 in. long, leathery, conspicuously veined, more or less entire, with short petioles and a rather short point at the tip, usually hanging rather conspicuously downward.

Native of Central America and northern South America, being grown rather extensively there and in the West Indies. Lends itself easily to cultivation. A close relative is the cola or goora nut, from which a popular soft drink is prepared and which lacks petals such as are present in cocoa plant. Over 40 varieties of cocoa.

Flowers borne in clusters directly on bark of trunk and main branches, each on slender stems to over ½ in. long. Flowers have rose-colored calyx, yellowish petals, and are to ¾ in. across. Fruit an elliptic red, purple, or brown pod with hard thick walls, to 1 ft long and 4 in. through, with flat 1-in. seeds.

Seeds washed or fermented to remove mucilaginous coat and pinkish or white pulp that adheres to them. Single fruit may yield to 60 seeds. Trees begin to bear at 4 years and may yield to 50 years. Fruits ripen in about 4 months. Extracted seeds are roasted at 257 to 284°F and ground to a paste. Oil is removed by pressure.

Seeds contain under 1 percent alkaloid theobromine and a trace of caffein, 30 to 50 percent fatty oil, 15 percent starch, and 15 percent protein. Sweet chocolate made by adding sugar and spices; milk chocolate, by adding milk and spices; cocoa, by removing ⅔ of fatty oil by pressure; cocoa butter from the oil used in cosmetics. Wastes used as fertilizers and cattle food.

Order Parietales./Family Theaceae

Tea
Camellia sinensis or *Thea sinensis*
Tree or shrub. Height to over 30 ft. Mostly smooth. Leaves alternate, evergreen, to 5 in. long, fuzzy beneath, when young, blunt or acute, somewhat toothed on margins, short-petioled, with pointed tips. Leaves quite different in given varieties, varying from dark to light green and considerably in shape and size.

Native of China and India but widely cultivated in suitable territory. About 14 species in tropical and subtropical Asia. Grown only rarely as curiosities in America. Closely related to camellias grown for their attractive flowers. Loblolly bay or tan bay of Virginia a close relative.

Flowers white, fragrant, showy, of five petals, to 1½ in. across, with many stamens and with a many-celled pistil that is short and hairy and breaks, freeing one large seed from each cell. Seeds sown in fall grow into seedlings that are transplanted following spring, developing to yield first crop in 4 years; full crop in 10 years.

Requires 60 to 200 in. of rainfall annually. Can be grown from sea level to 5,000-ft altitude. Known to live 200 years. Pickings may be made from 4 to 30 times a year. Acre may yield to 1,000 lb a year over period of more than 50 years. Tea contains 2 to 5 percent alkaline theine and 13 to 18 percent tannin.

Grade of tea may depend on age of leaves. Young buds yield golden tips; of black teas, the best quality of buds and small foliage become orange pekoe; next larger and coarser is pekoe, then souchong (coarsest leaves) and last "pekoe dust." Leaves for black tea are fermented after being covered and kept warm, while green teas are cured in sun or by rolling with hands or machines.

Order Parietales./Family Hypericaceae

Saint-John's-wort
Hypericum perforatum
Herb. Perennial. Height to nearly 3 ft. Stem erect, tough, slender, wiry, much-branched toward top; many arising from common base and its runners. Stems flattened somewhat. Leaves opposite, entire, showing minute translucent dots when held to light, to 1 in. long, speckled. Root system tough, strong, deep.

Common weed in fields, waste places, and pastures where it is eliminated with difficulty once it is well-established. Naturalized from Europe. Well-established in Canada and the United States except in extreme northern and southern portions. Some 200 species in the genus, with wide geographical distribution.

Flowers yellow, to 1 in. broad, crowded at top of plant, erect, with five petals and a great tuft of black-dotted stamens more or less separated into three groups. Fruits persistent capsules filled with enormous numbers of small cylindrical dark tan seeds that shake free with the swaying of the plant.

Juices sour and blistering. Plant is therefore usually unpalatable and not eaten by cattle. It then survives to multiply. If cattle are forced to eat it because of lack of other fodder, a poisoning effect may result particularly in unpigmented-skinned cattle that must remain exposed to the sun.

Plant can best be eliminated by persistent hand pulling or by frequent plowing and close cultivation sufficient to destroy the runners. Hay is unsatisfactory as food for cattle. Poisoning due to photosensitizing agents, hypericin and hypericum red. Affected cattle and sheep develop high temperature, rapid pulse, diarrhea, and dermatitis.

Order Parietales./Family Violaceae

Pansy
Viola tricolor
Herb. Height to about 6 in. Sprawling, smooth, weak-stemmed, branched, green, erect at tips. Leaves petioled, with broad, heart-shaped blades on basal leaves and more oval blades on those arising from upper part of stem, with conspicuously large divided stipules at bases of leaves.

One of most popular annual garden flowers, long known under cultivation. Derived from violets. Escaped from cultivation, flowers become small and lose showy coloring of garden plant. Some writers think it should have a species name separate from *V. tricolor*. More than 300 species of violets.

Flowers erect, showy, mostly blue, white, or yellow, with petals making a broad "face," with a spur to rear that is about twice as long as calyx parts. Some flowers remain close to ground and do not open, yet produce fertile seeds (cleistogamous). Fruit a capsule that splits to free large numbers of small brownish seeds.

For spring blooms, seeds are sown in cold frames in August, then transplanted to 3 in. apart. Protecting mulch of straw on branches allows plants needed air and prevents crushing. If this is removed gradually in spring, flowers should appear early. Flowers should be picked and seeds should not be ripened if blooming period is to be long.

Flowers picked in early morning before visiting insects have effected cross-pollination last longer than those picked after pollination has taken place. For summer and fall flowers, seeds may be sown indoors in boxes from February to June. A moist pansy bed fertilized with well-rotted manure produces superior blooms.

Pansy

Marsh Blue Violet
Viola cucullata
Annual. Weak, sprawling, green, the sprawling stems branching to yield tufts of leaves on relatively short stems. Leaves petioled, with heart-shaped blades that are rich green and sometimes to 3½ in. broad when mature, with rather conspicuous veins and margins with shallow, rounded teeth. Stout rootstock but no runners.

Native in America. Found in wet places where soil is appropriate from Quebec to Ontario, south to higher parts of Georgia. Closely related species extend range westward. Garden pansy is a violet. Sweet violet of the garden and greenhouse is *V. odorata*.

Flowers borne on stems that are longer than the leaves but in this species there is no true erect stem; violet-blue, darker colored at throat or sometimes white, with a conspicuous beard at throat but with spurred petal smooth and somewhat shorter than those at the side. Some flowers do not open to mature seeds.

Pollination effected mostly by bumblebees that touch stigma of pistil before they reach pollen-bearing stamens. While this species blooms normally in spring, violets may be kept blooming year round if each day they have an amount of light similar to that of their normal flowering time.

Popular spring bouquet flowers. Flowers may be picked with little harm to plant unless whole leaf cluster is torn free. Since plant is not an annual, it is not dependent solely on seeds to maintain an existence from year to year. Violet is state flower of Illinois, Rhode Island, Wisconsin, and New Jersey. Commercial perfume extracted from *V. odorata*.

Marsh Blue Violet

Order Parietales./Family Violaceae

Sweet White Violet
Viola blanda
Herb. Height to under 6 in. Stem long and slender, producing slender runners from ends of which arise tufts of leaves and flowers. Runners are branched, thus enabling plant to occupy considerable territory. Leaves alternate, on long petioles that may be tinged with red, with heart-shaped blades, usually under 2 in. broad.

Native of America. Found mostly in cool, shady, moist places such as ravines in clumps of rich soil. Ranges from western Quebec and New England to northern Georgia, west to Minnesota, with closely related species extending range particularly to the west.

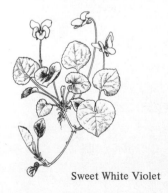

Sweet White Violet

(Continued)

Flower cream-white, small, fragrant, without a beard on the petals, with upper pair rather longer and narrower than is typical of violets and also often bent backward or twisted as shown in illustration. Some flowers do not open to produce fertile seeds but remain close to ground. Fruit a purple capsule bearing brown seeds.

Flowering time April and May, usually a little later than earliest violets. Pollination effected by honeybees and other bees such as *Halictus* that are attracted probably by fragrance and conspicuous character of flowers, even though they are small for a violet.

An attractive spring flower that does not yield well to transplanting probably because transplanting is usually to an unsuitable site. *V. cucullata* seems to establish itself readily in a variety of places but not this species. Violets were cultivated as potherbs in Egypt. Shoots are eaten for salad in France and Italy. Pythagoras thought of violets as spinach.

Begonia, Wax Plant

Order Parietales./Family Begoniaceae

Begonia, Wax Plant
Begonia semperflorens
Height to 18 in. Stem above ground soft, not much branched, green or reddish. Leaves to 4 in. long, nearly oval, oblique at base, with toothed margins, a large stipule at base, with hairs between teeth in leaf margin. Roots relatively fine and fibrous.

Native of Brazil. Such varieties as Chatelaine, Henry Martin, Vernon, Carrierei, and Erdfordii have been evolved from this species. Group is commonly known as the ever-blooming begonias in contrast with the rex begonias, the tuberous-rooted begonias, and the fibrous-rooted begonias.

Flowers rose-red to pink or sometimes white, on few-flowered clusters arising from leaf axils, not conspicuously drooping. Pistillate flowers have two sepals and two petals, while staminate have the same number but narrower petals. Stamens, many, yellow. Pistil two- to three-celled, with wings and a twisted stigma.

Cuttings may be started in spring or old plants may be repotted in summer if it is desired that they be grown indoors for more than a year. They do best at temperatures between 60 and 70°F in soil with a moderate amount of water but with good drainage. They need good light but little direct sun.

A popular houseplant commonly reproduced by cuttings of almost any vegetative part but more particularly stems or leaves since these root easily and quickly establish their independence. Soil to be at best should be porous and well-mixed with sand and leaf mold.

Rex Begonia

Rex Begonia
Begonia rex
No upright stem. Rootstock lies more or less horizontally near surface of ground and bears coarse attractive leaves. Leaves to 1 ft or more long with blade held perpendicularly, from long hairy succulent petiole, red beneath, silvery and green above or varicolored but usually conspicuous.

Native of Assam. Common cultivated varieties are probably *B. rex-cultorum* derived from many variations of *B. rex*. This is really a group of begonias, most of which are at their best when grown indoors or in the shade if not in their native land.

Flowers few, in loose clusters that reach above leaves, pale rose or pink, with two sepals that are color of petals but narrower in staminate flowers. Petals in staminate flower, two; in pistillate, two or more of the same size. Fruit two- to three-celled, with wings and a twisted stigma.

Requires moderate sunshine, favoring the morning sun. Does best at a temperature of 60 to 70°F with a moderate amount of water applied to roots rather than to leaves. Can be reproduced easily by cutting a leaf in half and sticking the leaf vertically in moist sand.

Planted outdoors, rex begonias do best in quiet, sheltered spots where there is a rich leaf mold. This is essentially a foliage houseplant. Red spiders, mites, thrips, aphids, mealy bugs, and nematodes attack begonias not adequately cared for, many of which are encouraged if the leaves become too wet.

Lemon Vine, Leaf Cactus

Order Cactales./Family Cactaceae

Lemon Vine, Leaf Cactus
Pereskia aculeata
Shrub that becomes a vine, with climbing woody stems. Stems to 30 ft long and branching. Related *P. grandifolia* a shrub or tree but not a vine. Spines on lower part of stem, one, two, or three in a group, slender and straight; axillary spines, usually in 2s and recurved. Leaves to 3 in. long, short-petioled.

Widely established in tropical America. Normal range through West Indies, on east and north coasts of South America, and also grown in Florida and Mexico. Plants in Washington and New York greenhouses bloom rather regularly. Not grown more commonly as an ornamental largely because of its offensive odor.

Flowers white, showy, fragrant, sometimes pale yellow or pinkish, to $1\frac{1}{2}$ in. across in small clusters. Stamens numerous. Lower part of the pistil with scale, leaves or spines on it. Fruit light yellow like a lemon, to $\frac{3}{4}$ in. in diameter; when mature, smooth and juicy. Seeds black and somewhat flattened.

Species has been in cultivation in Kew Gardens, England, since 1760; and in other places longer. In Argentine it is called sacharosa but this is not a correct use of this common name. Also known as Barbados gooseberry and West Indian gooseberry. Flowers fragrant but odor of vegetation is most offensive to some persons.

Cultivated in some parts of the world for fruits, which have market value, grown as hedge plant and as cover for walls and for some kinds of buildings. Related *P. grandifolia* common in greenhouse collections and can be distinguished by solitary straight spines on young growth in place of recurved pairs.

Cochineal Cactus

Cochineal Cactus
Nopalea cochinellifera
Height to 12 ft. Trunk diameter to 8 in. Branches spreading or ascending, with oblong joints, spineless, or the older ones with minute spines; bright green, particularly when young. Leaves small, awl-shaped, and falling off early. Spine clusters bear many small spines and an occasional larger one.

Probably native of Mexico and/or Guatemala but is found widely scattered in tropical and semitropical countries where it was cultivated. Spaniards found it under cultivation by Mexicans in 1518 and transplanted stock to Spain, whence it spread to India, Africa, the Canary Islands, and elsewhere.

Flowers usually abundant, appearing from tops of joints, about $2\frac{1}{2}$ in. long, with scarlet petals and sepals, the petals being somewhat longer. Stamens pinkish, many and extending about $\frac{1}{2}$ in. beyond the petals and sepals. Fruit red, about 2 in. long, rarely maturing under ordinary cultivation as in greenhouses.

Plants are set in rows about 4 ft apart. Minute cochineal insects, *Dactylopius coccus,* are placed on the joints or branches, where they multiply, and in about 4 months are collected by brushing off into bags. Two to three collections may be made in a year. Cochineal is a scarlet, brilliant dye, now mostly supplanted by aniline dyes.

Small cochineal insects killed by heat from stoves give the natural silver-gray cochineal, while those killed by hot water provide what is called black cochineal. Color, dye, produced by insects, was used in lipsticks and other cosmetics. Can withstand temperatures as low as 30°F. Because of beauty of flowers, members of *Nopalea* made attractive additions to a cactus collection.

Indian Fig, Prickly Pear
Opuntia ficus-indica
Tree or bush. Height to 15 ft or even more. Stem jointed, with units to 15 in. long, oblong or elliptic, thickened, usually spineless, and with a bluish bloom over otherwise smooth surface. Main trunk woody, rather cylindrical. Spines drop off early and are very short.

Indian Fig, Prickly Pear

Native of Mexico (?) but widely grown in warmer parts of world. Sometimes maintained under cultivation and sometimes escaped and established as somewhat troublesome weed. Over 250 species of *Opuntia,* all native of America; about 90 in western United States. *O. humifusa* ranges from Massachusetts to Florida and west to Kentucky.

Flowers yellow, to 4 in. across, with showy corolla and calyx blending apparently one into the other, with little difference between sepals and petals. Stamens much shorter than petals. Fruit over 3 in. long, red, edible, somewhat top-shaped, bristly, reddish-fleshed.

Forms free from prickles are grown as stock food and introduced from the Mediterranean area. Joints are broken and planted in well-drained, light soil 8 ft apart in furrows 12 ft apart. These begin bearing in 3 years and bear regularly thereafter.

Fine bristles that cover the fruit can be removed by rubbing with a leaf or cloth. Earlier varieties mature in June and the later in November. To prepare fruit for eating, remove thin slices from each end, slit the skin from end to end, and unwrap the peel from the edible pulp inside.

255

Hedge Cactus

Old–man Cactus

Organ–pipe Cactus

Hedge Cactus
Cereus peruvianus
Treelike or somewhat sprawling, to a height of 50 ft. Branches green, to 8 in. in diameter, sometimes smooth, with to nine longitudinal ribs bearing clusters of 5 to 10 sharp brown to black spines, each to 1¼ in. long. Night-blooming cereus (*Hylocereus*) is a climber, with three thin ribs and spine clusters of to three small spines.

Native of southeastern South America, but widely planted and adjusted to existence in tropical America and other tropical parts of the world, and grown in greenhouses in temperate regions. Over 100 species of *Cereus*, all native of South America, though some are now considered as belonging to different genera.

Flowers about 6 in. long, white, with a thick tube, particularly abundant on lower part of stem, superficially like a water lily except that they open at night instead of in bright sunlight. Fruit globular, slightly fuzzy, about 1½ in. in diameter with black rough seeds; inside fleshy and orange-yellow.

When grown in house or in greenhouse, care should be taken to provide suitable soil conditions and to place plant where temperature conditions are appropriate. These call for good drainage, an average temperature of around 70°F, and a soil that is not too rich, such as can be attained by mixing sand and garden soil. Genus introduced as ornamental into most tropical and subtropical countries.

The fruits of many members of the genus are edible, but great use for plant is a hedge and for its beautiful flowers, in spite of fact that these are only about ½ as wide as flowers of *Hylocereus*. Some species of genus are sprawling vinelike climbers while others are prostrate. In the tropics some hedge cactus plants grow as epiphytes.

Old-man Cactus
Cephalocereus senilis
Erect columns, usually unbranched and rising to a height of over 5 ft, with truck diameter to 1 ft. Branching, if present, is more likely at top but may be at base. Ribs 20 to 30, of large wartlike units, with a head of long gray bristles and with basal spines to 1 ft long. Top of plant "white-haired," hence name "old-man cactus."

Native of Mexico, where it is found wild in Hidalgo and Guanajuato, particularly on the limestone hills of eastern Hidalgo where it may be the most conspicuous plant on the landscape. At least 48 species recognized in the genus, but this is probably the best known and most widely distributed.

Flowers numerous, to 4 in. long, red outside and rose within, with the tube bearing few scales. Fruit egg-shaped, to 2 in. long, bearing at the top the base of the flower which has a few scales and hairs still attached. Seeds black.

Young plants covered with long, white silky hairs, giving the plant its name. Large plants rarely seen, though small plants are one of popular cactuses for house plants. Little wood tissue, and largest may be cut down with a penknife with ease.

Great quantities of the small plants have been shipped to Europe and to other parts of the world for sale to cactus hobbyists. Bases of old plants have weak gray bristles that are 1 ft long, but such plants are not often seen except in the native areas.

Organ-pipe Cactus
Lemaireocereus marginatus
Height to 25 ft. Stems usually erect and unbranched. Ribs 5 to 6, usually sharper in the younger plants and much blunted with age. Spine clusters closely crowded, their wool forming a dense cushion along ridge of each rib. Spines 5 to 8 in a cluster, with the center one conspicuous. Also listed in genera: *Cereus, Stenocereus, Armatocereus*.

Native of Mexico and found wild in Hidalgo, Queretaro, and Guanajuato. Widely established in Mexico, Jamaica, Cuba, and similar areas, where it is grown deliberately or occurs as an escape from cultivation. At least 10 species of the genus recognized, all native of Mexico or of southern California.

Flowers funnel-shaped, to under 1 in. long, including the supporting ovary. Tube and ovary rough and scaly, often with bunches of wool and small spines at edge of scales. Fruit globular, over 1 in. in diameter, covered with wool and spines that drop off at maturity, not particularly fleshy. Seeds numerous, black and shining.

Figures in lore of areas in which it grows, in part because it usually separates native dwellings. Properly cared for, it can provide an impenetrable hedge that is not unattractive; because of its spines, it is an effective barrier even though there are apparent openings at intervals.

Around estates and the lands of the poor, this plant takes the place of the old stump fence or stone wall in rural northeastern United States or of the wire fence of today. It has a type of beauty not possessed by any of these devices. In its native area, it is sometimes called "organo" in reference, of course, to the organ-pipe-like stems.

Snake Cereus

Nyctocereus serpentinus or *Cereus serpentinus*

Stems grow in clusters that are at first erect but eventually clamber or hang and reach a length of nearly 10 ft, with a diameter of not over 2 in. Ribs 10 to 13, low and rounded.   Small areas on surface crowded, felted, with sharp or bristlelike spines that are to 1 in. long.   Tips of spines usually darker than remainder.

Native of Mexico, probably from near the eastern coast.   Not now known from the wild state but is widely cultivated and has escaped from cultivation in areas.   About five species of the genus supposedly native of Mexico and Central America, of which this species is most widely cultivated.

Flowers borne at upper ends, sometimes literally terminal, white, to 7 in. long and to 3 in. wide, with tube and support exceptionally bristly, funnel-shaped.   Pistil and stamens of about same length.   Fruit red, covered with spines that drop off easily, to 1¼ in. long.   Seeds black, rather large for such a small cactus fruit.

Known as night-blooming cereus, as are a number of other cactuses here considered.   May be propagated by making cuttings from new growing areas.   These plants can make an old rock pile attractive since they climb in and over the units and eventually burst into bloom at night.   It is known in Mexico as "junco" or "junco espinosa."

The scientific name, of course, means night cereus and because of this might claim for the genus priority on the common name night-blooming cereus.   However, its common name snake cereus is even more appropriate and probably is the name by which it is known by most persons. Fragrance of blossoms is attractive.   Used in cactus gardens and window decorations.

Snake Cereus

Saguaro

Carnegiea gigantea

Height to 40 ft.   Composed of erect cylindrical columns to 2 ft in diameter and unbranched, or up to 12 branches, the branches parallel to main trunk.   With to 24 vertical ribs, each blunt and to over 1 in. high.   Spines of two kinds, those above being yellow-brown, the lower ones stouter and to 3 in. long.

Found in Arizona, southeastern California, and Sonora, Mexico. Known to science since 1848.   Some related Mexican species in other genera are larger and would weigh more; of these, *Lemaireocereus weberi* is considerably taller and stouter.

Flowers to about 4 in. long and sometimes as broad as they are long, supported by a green-scaled, white-felted tube about ⅗ in. long and about 1 in. wide at the throat.   Stamens white (one flower had 3,482).   Pistil white or cream-colored, to over 2 in. long.   Fruit a red or purple, edible, 2- to 3- by 1-in. berry, with to 2,000 seeds.

Fruits eaten by Indians.   In a national monument of 2,000 acres of desert land 20 miles east of Tucson, Ariz., on a rocky hillside, these plants will be forever protected.   Largest individuals are considered to be 200 years old.   Plants 4 in. high are 10 years old; 3 ft, 30 years; after that they grow 4 in. a year.

Primitive peoples used heavy rods from stems for building construction; fruits and seeds, for food and drink.   In wet seasons, plant may be 98 percent water, most of which is lost in May and June.   Woodpeckers burrow into the trunks and pygmy owls nest in abandoned woodpecker holes.   Wounds in rainy season may become badly infected with bacteria. The beautiful blossom of the "sage of the desert" is the state flower of Arizona.   Fruits are edible and highly prized and are eaten raw or converted into candy.

Saguaro

Night-blooming Cereus

Hylocereus undulatus

Climbing.   Stems adhere to walls, trees, and similar supports by aerial roots, often to 40 ft long, green, with three ribs, thin and with a horny margin.   Spines short but effective, arranged in groups about 1½ in. apart and with one to four spines to a group.   Whole plant may twist and wind around its support and branch freely.

Native of West Indies, Central America, and Mexico but established in the tropics and subtropics generally as an ornamental and in greenhouses the world over as an interesting decorative plant.   Listed in literature as *H. tricostatus* and as *H. triangularis*.   Other genera considered popularly as cereus are *Selinicereus, Aporocactus,* and *Echinocereus.*

Flowers about 1 ft long, white, with yellowish-green outer segments turning backward and numerous center ones erect.   Abundant stamens are cream-colored.   Whole flower looks like an enormous white water lily superficially and is fully as fragrant.   Pollination by night by moths. Fruit red, scaly, except when young when it is smooth, edible, oblong, to 4½ in. through.

Flowers appear at night remaining open until dawn.   May be picked and kept open in a refrigerator for some time.   It flowers in homes and greenhouses as far north as New York.   Of the 16 species of the genus, the one

(Continued)

Night–blooming Cereus

here listed is undoubtedly the most popular as an ornamental or as a wall cover.

One is fortunate indeed who has the opportunity to observe a hedge of this plant in full bloom.   Commercial reproduction is largely by cuttings and rooting the slips in sand.

Night-blooming Cereus
Selenicereus pteranthus
Trailing or climbing plants but relatively stout, with stems to 1½ in. or more in diameter.   Stems blue-green to purple, conspicuously four- to six-angled, with ribs on younger branches to ⅙ in. high.   Spines in clusters one to four, short, dark, rarely over ⅛ in. long but most effective.

Native of Mexico but known mostly from conservatories, where it has long been a popular species.   Probably commonest of conservatory-grown night-blooming cereus plants.   16 species in the genus, ranging from southern Texas to South America, one extending down to the Argentine.   *S. grandiflorus*, from Texas southwards to Argentina and the West Indies is a most popular species worthy of culture and easy to grow.

Flowers to 1 ft long and 15 in. across white, very fragrant, the tube and throat about half the length of flower and swollen in upper part.   Lower cluster of stamens attached to petal tube for about 3 in.   Fruit red, globular, about 2½ in. in diameter, covered with long white silky hairs or bristles.

Flowers open in night and produce remarkable effect because of their beauty and fragrance.   Some species are reported to have medicinal value, but beauty and fragrance of flowers should have a therapeutic effect on most confirmed of pessimists.   Not surprising that flowering time is announced in newspapers.

A plant that can be grown in a 10-in. pot and produce flowers that may be 15 in. across is bound to hold interest of almost anyone.   Related *S. grandiflorus* yields a heart tonic extracted from the green branches and originally discovered in Naples in 1889.   The medicinal agent is probably an alkaloid.

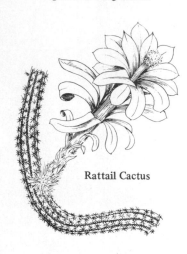

Night-blooming Cereus

Rattail Cactus
Aporocactus flagelliformis
Slender, vinelike creeper that twists and turns over its supporting tree or wall or sometimes hangs suspended from such a support.   Stems weak and rarely over ¾ in. in diameter, with 10 to 12 low, inconspicuous, somewhat warty ridges and well-supplied with clusters of many brown spines.   Has aerial roots.

Native of Mexico and Central and South America.   Reported to have been found growing wild on trees on the coast of Jamaica but has not been known wild from that island in recent times.   Best known in temperate regions from greenhouses and sometimes as a house plant.   Five species of the genus.

Flowers to 3 in. long, pink to crimson, with outer segments more or less bent backward, the inner spreading only slightly, and all these parts relatively narrow.   Stamens in somewhat of a tube terminated by the yellow anthers.   Fruit to ¼ in. through, red, bristly, globular, with yellow pulp.

Flowers remain open or go through opening process for 3 to 4 days.   In Mexico, the dried flowers are sold as a household remedy under the name of "flor de cuerno," and this is sometimes found in drug markets under the same name.   A common window plant in Mexico.

Apparently was introduced into Mexico from Peru about 1690, but it is not known in a wild state anywhere at present time.   Mexicans frequently plant it in the open end of a cow's horn and hang this on the outside of the house.   Flowers from mid-December to mid-January. Easily grown.

Rattail Cactus

Hedgehog Cactus, Strawberry Cactus
Echinocereus polyacanthus
Height a few inches, 5-in. plants bearing flowers.   Stems cylindrical but narrower toward top, making whole thing somewhat like a typical plump cucumber.   Stems often grouped to make a mass.   Ribs 9 to 13.   Radial spines stout, 8 to 12, lower one longest, being to about 1 in. long; upper, ½ in. white to red and dark-tipped.   Sometimes listed as *E. triglochidiatus* var. *polyacanthus.*

Found native from Chihuahua and Durango in Mexico to western New Mexico and southeastern Arizona.   Described in 1848 by Engelmann, but confused in its classification for some years.   Now found rather commonly in cactus collections together with the purple-flowered *E. pectinatus*, the true hedgehog cactus *E. engelmanni*, and the rainbow cactus *E. rigidissimus.*

Flowers scarlet to salmon, 2½ in. long, lateral but held erect, with a funnel-shaped base and widespread showy parts.   Flower tube yellow; spines on flower base are well intermixed with a cobweblike wool.   Fruit spherical, about 1 in. long, spiny, greenish-red, uncommon.   Some authors claim fruit is unknown.

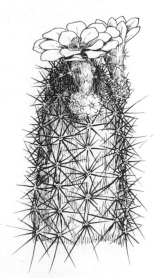

Hedgehog Cactus

Fruits of some of members of genus are edible, spines that cover them being easily removed when fruits have become mature and skin is unusually thin.   Seeds black and bear small tubercles in most members of genus.   Some 60 species known in genus of which nearly ½ are grown as ornamentals.

Some members of the genus have sprawling stems.   In this species the small scarlet flowers separate it from many of other common species whose flowers are crimson or purple but not scarlet.   In *E. pectinatus* there are several central spines in spine group; in *E. rigidissimus* there are none.

Mescal Buttons, Peyote
Lophophora williamsii
Globular.   To 3 in. in diameter, arising from a coarse taproot extending to a depth of 4 in. or more.   Ribs 6 to 13, nearly vertical or irregular and indistinct and composed of a series of tubercles each of which is crowned with a delicate bunch of spines; dull bluish-green.

Found from central Mexico through southern Texas.   Looks so much like mushroom that it has been considered to be one by some, and bears common name of "sacred mushroom" as well as a host of other names. There seems to be only one species in spite of fact that some writers consider there are two.

Flowers found at top of plant, white, to 1 in. across when fully opened, surrounded by a mass of relatively long hairs, the outer flower parts being greenish on the back and somewhat swollen at the tips.   Stamens and pistil much shorter than surrounding parts.   Fruit under 1 in. long, naked, pink.   Seeds black.

Yields the narcotic anhalonin, although the narcotic effect may be caused by resins.   Drug causes persons using it, in drinks or otherwise, to lose all sense of time, as does hashish from *Cannabis indica*.   Drug users also have remarkable visions.   Its use dates to pre-Columbian times and is forbidden by law.

Indians used the plant in "breaking fevers" and in religious rites.   Plants are cut and dried to make "mescal buttons."   Indian names include xicori, pellote, peyote, peyotl, hiculi, camaba, seni, huatari.   A spree with the drink is followed by a long period of wakefulness.

Mule Cactus, Fishhook Cactus, Candy Barrel Cactus
Ferocactus wislizenii or *Echinocactus wislizenii*
Height: to over 6 ft.   At first almost spherical but when mature forms a cylinder.   Usually unbranched but if injured may bear several heads of branches.   Ribs to 25 or more, about 1 in. high.   Spine groups with one central strongly hooked spine surrounded by many smaller ones and by brown felty areas.   Spines may be light pink or gray-pink with translucent tips.

Found growing wild from El Paso, Tex., through southern New Mexico and Chihuahua to Arizona and Sonora, and possibly on to Lower California.   Reported, probably erroneously, as from Utah.   Some 30 species of the genus, all globular or cylindrical and all with well-developed straight or hooked spines.

Flowers to 2½ in. long, yellow, supported by green-scaled tubular base. Some flowers may be reddish or orange, borne only on younger growth just above spine clusters.   Stamens very numerous, borne in throat of flower and much shorter than showy parts of flower.   Fruit to 1½ in. long, yellow, oblong.   Seeds dull black.

Grown as an ornamental to some extent.   Related *F. glaucescens* is smaller and has no curved spines.   Usually, picture showing a Mexican drinking water from a "barrel cactus" represents a plant of this genus. Doubtful if the plants can be considered a source of water except in case of dire need.

While the interior may consist of a moist pulp, it is not filled with clear cool water, as some publicity agents would have us believe.   Plants have a unique beauty that should justify their continued protection.   Small plants are grown indoors by hobbyists quite commonly.   *F. ancanthodes* is the barrel cactus, and *F. johnsonii* the devil's big toe.

Golden Cactus, Hedgehog Cactus
Echinocactus grusonii
Large balls.   Sometimes to nearly 1 yd in diameter, unbranched.   Ribs from 21 to 37, thin and high.   Spines vary in color from golden-yellow when young to pale or even white, and finally to dirty brown; with the radial spines to over 1 in. long and the usually four central spines up to 2 in. long.   Plants usually grow singly.

Native of Mexico and is found wild from San Luis Potosi to Hidalgo. Also found as natives in Arizona, Texas, and California.   A popular plant in collections maintained by hobbyists and by greenhouse operators.   Genus has at least nine species, of which this is probably best known.   Other species lack the bright yellow spines that give the group its common name and its popularity.

(Continued)

Mescal Buttons

Mule Cactus

Golden Cactus

Flowers red and yellow, borne at the top and center, to 2½ in. long and opening fully only in bright sunlight. Stamens yellow, numerous, forming a cylinder. Pistil yellow, divided at the top into 12 lobes. Fruits spherical, bearing pointed scales, with an abundance of wool in their axils, to ¾ in. long. Seeds blackish, smooth and shining.

Woolly crown; woolly, thin-skinned fruit and smooth seeds are characteristic of the genus; the golden spines, of the species. Large specimens are not commonly seen. Plants may bloom at 6-month intervals through the year, beginning about mid-May. Flowers will open in 3 days under good conditions.

New plants are obtained from seeds, or a large plant is cut off at top, stimulating development of buds that are removed and started as new plants. Flowers sunken rather deeply in stem and surrounded by felt cushion so that to be collected they must be actually dug out of surrounding tissue. Beginning of scientific name from *Echinas* meaning a hedgehog, which refers to its spiny appearance.

Orchid Cactus
Epiphyllum ackermannii
Stems many, to over 3 ft long, somewhat recurved, with branches that are for most part under 1 ft long, unarmed, with middle and side ribs, with short bristles on younger and lower portions. The flat two-edged branches have waved or shallow-toothed margins as in some leaves of other plants.

This is not known in wild state, though most other members of genus come from Mexico, Central America, and northern South America. By some writers, this is considered a hybrid and not a natural species. It was originally described from material sent by Ackerman from Mexico. It may represent *Epiphyllum* and *Heliocereus*.

Flowers to 6 in. across, blooming in the daytime, flaming red or scarlet outside and carmine within, with a greenish-yellow throat and very short tube. Tip of pistil, pink. Flowers closely resemble those of *Heliocereus*. Lower part of pistil is more or less bristly.

Culture of plant is usually by cuttings and rooting of the sections in suitable earth. Some prefer to cover the young shoots with glass until they get a start, and bottles with the base broken off may serve this purpose well.

One of at least a half-dozen genera that have long been popular as houseplants and in warmer parts of the country as outdoor ornamentals. Genus name *Epiphyllum* means "on leaf" and no doubt refers to the fact that the flowers appear on what seems to be the leaf, though it is really the stem.

Christmas Cactus, Crab Cactus
Zygocactus truncatus
Plant hangs in large bunches from trees and similar supports. Stem flat, with joints 1 to 2 in. long and ¾ to 1 in. wide, thick, green, soft, blunt at tip of each joint; with upper part of each joint more or less curved inward like a blunt horn. Whole stem relatively weak and somewhat succulent.

Native of Brazil. Many of cultivated forms found in houses and greenhouses are hybrids of different species of the genus, or hybrids of these with forms of *Cereus* or of *Epiphyllum*, the last of which is normally found growing on trees as does typical Christmas cactus.

Flowers solitary, growing from ends of young joints, showy, magenta red, 2½ to 3 in. long, with calyx and corolla alike and composed of many curled-back segments. Stamens many, with long pink filaments. Fruit pear-shaped, red, to nearly ½ in. in diameter.

Flowering time late winter. Commonly reproduced by stem cuttings. Ideal soil about ⅔ garden soil and ⅓ sand, with good drainage. Should be placed in a sunny window where it will not be disturbed, provided with water sparingly except at blooming time, and kept at temperature between 60 and 70°F to be at best.

After houseplant has bloomed, water should be withheld for some time. Plant should be kept in same pot for years without changing. Plant responds to treatment and unless used sensibly may be disappointing instead of an outstanding delight.

Pincushion Cactus
Mammillaria longimamma
Form rather large clumps of tuberclelike structures, each to about 2 in. long and terminated by a circle of about 6 to 12 spines, each about 1 in. long, with whitish hairs when young but naked with age. Spines arranged like spokes of a wheel and sharp, white or pale yellow. Whole plant takes a general dome shape. *Mammillaria* may appear as small or low plants.

Orchid Cactus

Christmas Cactus

Pincushion Cactus

Native of central Mexico. Cultivated to some extent for unique ornamental qualities. At least 150 species known, of which nearly ⅓ are considered of importance as ornamentals by hobbyists.

Flowers lemon-yellow, with many pointed petallike parts surrounding shorter stamens and pistils that are crowded to center. Whole flower to over 2 in. across, borne in the woolly axils between the tubercles. Fruit nearly smooth and berrylike.

These bright green spiny cushions are relatively common in greenhouses. In the southern part of the United States, may be grown outdoors where soil conditions are suitable. Some species of the genus may be found in flower from March through November.

Species commonly grown for flowers include *M. albicans, M. bocasana, M. camptotricha, M. elegans, M. elongata, M. hahniana, M. hemisphaerica. M. microhelia, M. parkinsonii, M. parbella, M. rhodantha,* and *M. trichacantha. M. macdougalii* known as pincushion cactus, *M. fragillis* as thimble cactus, *M. dioica* as candy cactus, and *M. plumosa* as feather bed cactus. Very popular plants owing to their dwarf habits plus several other characteristics which make them attractive.

Order Myrtales./Family Thymelaceae

Leatherwood
Dirca palustris
Shrub to 6 ft high. Twigs conspicuously jointed, smooth, somewhat yellowish-green to very dark. Leaves egg-shaped, entire, rounded at base, somewhat pointed at tip, to 3 in. long, fuzzy when young but smooth when mature. Bud scales with brown hairs that drop off, three to four showing. General shape of shrub almost globular if in open. Nodes are encircled by leaf scars.

Found in damp rich woodlands or sometimes in the open but usually only where there is good wet soil. Ranges from New Brunswick to Florida and west to Minnesota, Missouri, and Tennessee. Sometimes grown as ornamental but not a superior plant for that purpose. Two species in the genus and over 400 in the family.

Flowers develop just before leaves and apparently surrounded by newly developing leaves, yellow, under ½ in. long, with pistil longer than the eight stamens borne on the corolla. Every other stamen longer than its neighbors. Fruit fleshy, with a stone center, to ½ in. long, reddish, appearing in summer.

Bark contains substances that can cause severe blisters and irritations to skin of some persons. Taken internally, bark can cause severe vomiting. In this connection it has some medicinal uses. Berries reported to have narcotic property. Seeds sown in fall or spring. Plant remarkably free of insect and fungous pests.

Moose and deer sometimes use it as forage. Wood not at all leathery as name implies but bark is exceptionally tough. Indians and pioneers used it as an emergency fiber for thongs and other cordage. Sometimes found being used as an ornamental. Worthy of such recognition. Known also as wicopy, wickup, and leatherbark. Bark once used by Indians for making bowstrings, fishlines, and baskets.

Leatherwood

Order Myrtales./Family Lythraceae

Cigar Flower
Cuphea ignea
Stem much-branched, slender, smooth, often red on one side, soft. Leaves abundant, to 1½ in. long and ¾ in. wide, sharply pointed at free end, nearly smooth, entire, opposite, narrowed at base into a distinct petiole, slightly scaly. Root system fibrous.

Native of Mexico, now rather well-established as a popular houseplant that is usually most attractive in midwinter. Over 200 species of the genus, some being herbs like this species, others shrubs. Some are hardy enough to survive outdoors in southern United States.

Flowers solitary, red, cigar-shaped, borne in leaf axils and hung on a slender stem, spurred. Calyx tube about 1 in. long, flaring at end, six-pointed, red, with a dark ring around it at end. No corolla. Stamens 11, inserted near end of calyx tube. Fruit long and slender, ¾ in. long, backward-spurred.

Commonly propagated by stem cuttings started in summer in soil and kept in shade for at least 1 week. Favors abundance of sun, in soil composed of ⅔ garden soil and ⅓ well-rotted manure, at temperature between 60 and 70°F. Requires an abundance of moisture.

Plant useful apparently only as an ornamental. Has few fungous or insect enemies. May be grown from seeds but cuttings may shorten time to maturity.

Cigar Plant

Order Myrtales./Family Lecythidaceae

Brazil Nut
Bertholletia excelsa

Tree. Height to over 150 ft. An evergreen tree with fruits also known as "para nuts." Trunk to over 4 ft in diameter. Bark smooth. Branches high. Leaves alternate, to over 2 ft long and about 6 in. wide, leathery, bright green, with wavy margins, somewhat oblong. The tall tree is rather conspicuous for way in which it branches most abundantly near top.

Native of Brazil, where it is found in great forests largely along banks of Amazon and the Rio Negro. Not grown in open anywhere in United States but its nuts are known in practically every home in America. Some 18 genera and 230 species in family in South America, western Africa, and Malaya.

Flowers cream-colored, with calyx parts united but tearing into two parts when flower opens, these two parts dropping off; borne in open clusters. Petals, six. Stamens, many, united, uppermost without anthers. Fruit about 6 in. across, with hard, thick shell, containing 18 to 24 hard, three-sided nuts.

Tree not hardy enough to survive outdoors in United States and not attractive enough for growth as an ornamental anywhere. Oil extracted from meats by pressure; finds various uses particularly by artists and watchmakers. Bark used in calking ships particularly at Para, Brazil.

Common trade name of castana describes the nuts of a number of species of the genus. Chief point of exportation is Para. Natives collect fruits and break them to free the nuts. United States normally imports from 20 to 30 million lb a year. Known as monkeypots in some places.

Brazil Nut

Order Myrtales./Family Myrtaceae

Eucalyptus, Blue Gum
Eucalyptus globulus

Tree. Height to over 300 ft. Bark peels off in great longitudinal strips or hangs swinging, leaving trunk smooth and bluish-gray. Leaves opposite, on young shoots, thick and leathery, pointed at end, to 1 ft long, often somewhat whitened, with a mealy covering, relatively broad.

Native of Australia. Naturalized in California, where it is a common street tree and is grown for various purposes. Over 300 species coming from Australia and Malayan region. Many of these have been adapted to cultivation in the warmer temperate regions of the world. 14 species were introduced into California in 1856.

Flowers large, to 1½ in. across, solitary or in groups of not over three. Calyx tube hard, warty, with a blue-white wax covering with the lid in the center, shorter than tube itself. Fruit stalks flattened. Fruit, a capsule with lid made of united calyx and corolla. Stamens, many. Flowering time from December through May.

Wood easily stained, not durable in contact with soil, checks badly in logs when being cured, strong, rapidly growing. Tree endures minimum temperature of 25°F and high temperatures but young trees cannot resist drought well. Flowers yield much nectar and attract honeybees but the honey is not popular because of its flavor.

Oil distilled from leaves of *E. globulus* and *E. dives* is used in treating colds, malaria, and other fevers. Imported timber trees are karri, *E. diversicolor*, and jarrah, *E. marginata*. Mallet bark, *E. occidentalis*, yields to 50 percent tannin. Gum kino, used in medicine, comes from *E. rostrata*. Trees use much water and reduce mosquito hazard.

Eucalyptus

Order Myrtales./Family Onagraceae

Fireweed, Great Willow Herb
Epilobium angustifolium

Herb. Height to 8 ft. Stem stout, smooth, or sometimes finely downy, branched or unbranched, erect. Leaves alternate, short-petioled, to 6 in. long and to 1 in. wide, dark green above and pale green beneath, narrowed at each end, pointed, ends of cross veins not coming to edge of leaf but joining next vein forward.

Native of Eurasia but thoroughly established in America from Greenland to North Carolina and west to California and Alaska. Found in Rocky Mountains south into Arizona. Usually most abundant in burned-over areas either in low ground or in the mountains. About 80 species in the genus, some 400 in the family; widely distributed.

Flowers showy, to 1½ in. across, usually purple but sometimes white, in long terminal spikelike clusters, with lower ones appearing longer. Petals, four, entire, broad. Sepals, four, narrow. Stamens, eight, shorter than pistil. Fruit a capsule to 3 in. long, slender, breaking to free whitish plumed seeds.

Flowers in June, July, and August. Bears fruits and seeds through August and September. Flowers give a hillside the appearance of being in flame; seeds, the appearance of a great cloud of smoke. Lower flowers mature first so sometimes a plant may bear both flowers and seeds at same time.

Fireweed

Plant well known, as is indicated by many common names such as purple rocket, Persian willow, rosebay, French willow herb, bay willow herb, firetop, spied willow herb, great willow herb, Indian wickup, blooming sally, blooming willow, flowering willow, and herb wickopy. Shoots eaten like asparagus; leaves, as tea; and pith, in soup. Known as willow herbs because of the shape of their leaves and fireweeds because of their ability to encroach after fire has destroyed an area.

Evening Primrose
Oenothera biennis
Herb. Height to over 6 ft. Branched or only slightly so, erect, rather coarse, rough or more rarely smooth, biennial. Leaves alternate, to 6 in. long and to 1½ in. wide, pointed at tip with clasping base; margins obscurely, shallowly, and sparingly notched. Lowest leaves may be with short petioles.

Native of America. Found on dry soil and in waste places from Labrador to Florida, west to Minnesota and Texas with closely related species extending range to Pacific Coast, with other closely related species in the Old World. Related *O. lamarkiana* has been grown as an ornamental in European gardens for over 100 years.

Flowers open at dusk, to 2½ in. broad, clean pale yellow, with an elongate tube at base. Sepals, four. Petals, four, broad and spreading. Stamens, eight, of equal length, about length of pistil. Fruit a capsule that opens at top to free small brown seeds, to 1½ in. long and to ¼ in. thick.

Under short days of winter plant forms a rosette; under constant illumination it grows into a tall plant. Under normal conditions, it forms a rosette first season and tall flowering stalk second. Rosette characterized by pinkish cast to light-colored veins. De Vries used *O. lamarkiana* in developing his mutation theory of evolution.

Roots of the rosette in spring used as salad or cooked, but it must be collected early or it will be too peppery. Best to cook it in two waters to remove the strong taste. Properly cooked, tastes like salsify or parsnips. As a weed, it can be eliminated by simple hoeing.

Evening Primrose

Order Myrtales./Family Haloragidaceae

Spiked Water Milfoil
Myriophyllum exalbescens
Stems weak and a number of feet long, depending much on depth of water. Leaves whorled, in 4s or 5s, with a central narrow rib from which equally narrow branches develop like parts of a feather, these portions being narrow enough to be threadlike. Leaves about the flowers entire and whorled. Perennial aquatic.

Native of North America but also found in waters of Europe and Asia. In North America, ranges from Newfoundland to Maryland, west to California and Alaska. This species has not been correctly recorded from Florida. Some 20 species in the genus with wide distribution in the world.

Flowers held above water, in spikes surrounded by a whorl of very short leaves. Spike to 3 in. long. Petals, four, drop off early. Stamens, eight, relatively inconspicuous. Fruit, small, to under 1/10 in. long, with the parts rounded on the back, with a wide groove between ridges.

Supplies shelter and food for many insects that are food for fishes. Ducks feed on fruits and a few eat the foliage but it is not an important wildlife food. Muskrats eat whole plant sparingly and moose have been known to eat it. Related *Proserpinaca* is a valuable duck food.

Myriophyllum is often sold in aquarium supply shops as an aquarium plant because of its general beauty. Differences between species are usually not easily recognized from an examination of the leaves and even with the fruits the differences are sometimes slight.

Water Milfoil

Order Umbellales./Family Araliaceae

Wild Sarsaparilla
Aralia nudicaulis
Apparently stemless, as leaves seem to arise directly from root. There is, however, a short stem present. Leaves rise to a height of over 1 ft, petioles alone being that length. Leaves are compounded, with three main sections, each with three to five slender stalked leaflets that may be to 5 in. long. Leaflet margins saw-toothed. Root, substantial and long.

Found commonly in moist woods from Newfoundland to Georgia, west to Manitoba, Idaho, Colorado, and Missouri. Not easily confused with other aralias of same range largely because it is apparently stemless and nearly smooth.

Flowers in three umbels, usually raised higher than highest leaf tip on a stem independent of single leaf that apparently starts at the root, greenish,
(Continued)

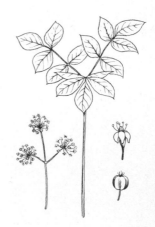

Sarsaparilla

263

Ginseng

English Ivy

Poison Hemlock

to ¹/₁₀ in. broad or slightly larger. Petals small, and conspicuously bent backwards. Stamens five, greenish, conspicuous. Fruit a purple-black berry, ¼ in. through, borne in clusters.

Flowering time May and June, the fruit following for next 2 months. Root is strongly aromatic, used as a substitute for true sarsaparilla, *Smilax officinalis*, forming basis for some of flavoring in root beer. Rather strong if eaten directly.

Indians formerly used root as an emergency food particularly when they were on forced marches, or long journeys. The plant is known as shotbush, rabbit's-root, wild licorice, and false sarsaparilla. It has some medicinal properties. The true sarsaparilla grows in Honduras.

Ginseng
Panax quinquefolius
Perennial herb. Height to 15 in. Single erect relatively slender stem bears about three leaves at crest, each composed of five thin pointed leaflets that arise from a common point at the end of petioles that are to 4 in. long. Leaf looks much like that of a small horse chestnut. Flower cluster arises from between leaves.

Native of rich woodlands from Quebec to Alabama, west to Minnesota, Missouri, and Nebraska. Has been grown extensively. Root so valuable that wild stock has practically vanished from existence. Seven species in the genus native of eastern North America and of Asia.

Flowers in an umbel containing from a half dozen to over five dozen, each flower under ¹/₁₀ in. broad, yellowish-green, with some staminate that smell like lily of the valley and others with stamens and pistils. Whole cluster makes a single globular mass. Fruit a bright ruby-red berry, to nearly ½ in. through, containing two to three seeds.

Reproduction by division of root or by seeds. Seeds ripen in September; may germinate first season or second. Roots may develop to sale size in 3 to 4 years. Plants are grown under artificial shade. Roots are large and believed by Chinese to have remarkable medicinal properties.

Roots divided so that they resemble a man with legs are considered most valuable and the name ginseng is from the Chinese jin-chen meaning manlike. Apparently, medicinal properties are more imagined than real. There are a variety of remedies and health foods and fads which are made from ginseng root. Some persons believe ginseng has powers to treat all types of human problems including dread diseases. This may reflect on generic name *Panax*, which is of the same root as panacea, i.e., all-healing.

English Ivy
Hedera helix
Creeping and climbing vines that cling to walls with assistance of vigorous abundant holdfasts. Leaves evergreen, three- to five-lobed, dark green above and much lighter beneath, with lighter colored veins, to 5 in. across, rather stiff, with margins that are essentially entire.

Native of the area from the Canary Islands to Asia but grown extensively as a covering for brick walls of buildings, being particularly popular on college buildings. Many races and varieties have been developed during years the plant has been under cultivation, and a half dozen species native from Europe to Japan.

Flowers imperfect, greenish, in umbels at ends of branches. Calyx five-parted or five-toothed. Petals, five, small. Stamens, five. Ovary, five-celled. Fruit black or yellow, globular, to ¼ in. in diameter, containing relatively few seeds embedded in flesh, the usual number being under five.

Leaves and berries contain poisonous glucoside hederin. Children who eat the berries or cattle that eat the leaves may be poisoned by them. Some persons suffer from a severe dermatitis after handling the leaves or stems, or merely picking up the drying clippings after the plant has been pruned.

Commonest use is as a wall cover or as a houseplant. The so-called "ivy league" of colleges is composed of the longer established institutions of the Northeast whose buildings characteristically bear English ivy and possibly some English traditions. Plant may be reproduced by seeds or by simple cuttings.

Order Umbellales./Family Umbelliferae

Poison Hemlock
Conium maculatum
Herb. Perennial. Erect. Height to 5 ft. Profusely branched in upper portions. Lower leaves with long petioles and, like upper petioleless leaves, at least twice compound, with leaflets thin with toothed margins. Base of petioles somewhat inflated and enclosing the stem.

Naturalized from Europe and well-established in wet waste places from
Nova Scotia to Delaware, west through Michigan and on to California
though the range is not continuous.   Found in Mexico, in South America, and other parts of the world.

Flowers borne in umbels to 3 in. broad, with some tendency to be flat-
topped, with each umbel supported by a ray to 1½ in. long which arises
from a point with other umbels.   Flowers white but under ¹⁄₁₀ in. broad.
Fruits a pair of nutlets, each with five ribs that are prominent when dry.

Poison is particularly concentrated in fruits at fruiting time.   Contains at
least five distinct yet closely related alkaloids.   Poisonous property
known for ages.   Death results from impairment of breathing; however,
ingestion of plant is not always lethal.

Seeds may be easily mistaken for anise but should not be.   This plant was
used by the Greeks as a means of quick death for disposing of prisoners
before the refinements of modern warfare had been discovered.   This is
probably the hemlock Socrates drank.

Water Hemlock
Cicuta maculata
Herb.   Perennial.   Height to 6 ft.   Smooth hollow conspicuously jointed
stems, often purple at the joints.   Leaves alternate, two to three times
compound, with a conspicuous sheathing petiole base, to 1 ft long or
more.   Upper leaves smaller than lower.   Roots several, tuberlike, fleshy
and have appearance of being edible.

Native of North America ranging from New Brunswick to Florida, west
to Manitoba and New Mexico.   Common in swamps and low grounds
where it is wetter than habitat of poison hemlock.   Eight species in the
genus in North America including Mexico, of which four are definitely
western.

Flowers, many in each of a number of umbels making up a compound
umbel, white, with five white petals and five stamens.   Fruit slightly over
¹⁄₁₀ in. long, composed of two segments grown together, making an almost
globular structure except for being flattened and for prominent corky ribs
between, which are solitary oil tubes.

Rootstock shows platelike cross partitions when cut lengthwise.   If cut
crosswise, exudes aromatic yellow oil.   Poisonous property in the roots,
leaves, and fruits is cicutoxin.   Rootstocks richest in poison.   Tops may
be eaten safely by cattle in hay.   Root size of walnut can kill a cow in
spring.

Poisoning by water hemlock is fatal, usually without convulsions; water
hemlock poisoning involves violent convulsions following labored
breathing, mouth frothing, loss of sight, vomiting, increased pulse, diarrhea, and great pain.   Induced vomiting or purgatives to eject the poison
seems to be best treatment.

Water Hemlock

Caraway
Carum carvi
Perennial or biennial herb.   Height to 3 ft.   Stems slender, erect, smooth,
furrowed, hollow.   Leaves compounded like the parts of a feather into
many narrow segments, twice or thrice compounded, smooth, light green.
Bases of petioles widely dilated and clasp the stem.   Bracts at base of
flower cluster, three, narrow.   Roots thick, spindle-shaped, tuberous.

Native of Old World but well-naturalized in many parts of America.
Not uncommon as a weed in waste places from Newfoundland to Penn-
sylvania, west to South Dakota and Colorado.   Probably named from
Caria, a place in Asia Minor.   Some 50 species native of temperate and
warmer parts of the world.   Main supply source of caraway seeds is
Holland.

Flowers small, pink, or more commonly white, appearing in May and
June, in compound umbels to 2½ in. broad and of to 10 major rays, each
to 2 in. long.   Fruits about ⅛ in. long, brown, with alternating light and
dark brown ridges and grooves, with two more or less connected.   Cut
across, fruits show one to two oil tubes to each groove.

Fruit yields a volatile oil, about 5 percent of which is used in making
commercial tincture of cardamom.   Fruits find commercial outlet as flavoring for breads, cakes and as added flavor for candies.   Plants easy to
cultivate, being raised much as are carrots.   In Europe, plant is known as
kümmel.

As a medicinal agent, relieves gas in stomach.   Also serves as stimulant,
as a flavor for perfumery and liqueurs, and as a common home medicine
that may be used safely without expert advice in treating mild cases of
colic.   Usual dose, 1 tsp of fruits cut small to 1 cup of boiling water, a
cupful being drunk slowly each day.

Caraway

Parsley

Parsley
Petroselinum crispum
Herb. Biennial or short-lived perennial. Height to 30 in. Much-branched, bright green, smooth, not too tough. Leaves compounded into 3s repeatedly with the last divisions deeply cut. Uppermost leaves smaller and in some cases entire. Leaves a deep rich bright green, sometimes crisped, and with a good taste.

Native of the Old World but much cultivated in gardens. Not infrequently escaped and established by itself about farm buildings, particularly in the Northeast. Probably native of Mediterranean region. Five species in the genus, all native of Europe, and many cultivated varieties. Moss Curled is a desirable variety for garnishing but the plain or single leaf variety is favored for flavoring food. A cool-season crop; seed is planted at about the same time as carrots.

Flowers greenish-yellow, in flat-topped umbels, of which the whole group may be 2½ in. across. Smaller umbels are grouped into a larger umbel whose rays or supports may be about 1 in. long. Petals inconspicuous and with incurved points. Fruit smooth, though ribs may be prominent when the fruit is dry, to ⅕ in. long.

Seeds may be started in early spring in good soil in the open or in hotbed. Germination slow. Plants should be thinned so that they are 6 to 8 in. apart and should then be cultivated as one would an ordinary root crop. Leaves gathered fresh and used as soon as possible. For winter use, roots are planted indoors. Will withstand considerable frost.

Under cultivation for 2,000 years. One of the commonest plants used for garnishing fish, meats, and omelets and common ingredient of stuffing for fowl. The tops are sometimes boiled as a potherb.

Fennel
Foeniculum vulgare
Height to 5 ft. Perennial. Stem erect, branched, more or less downy, moderately slender. Leaves three to four times compounded, in a featherlike arrangement, with narrow threadlike segments and with the bases of the long broad petioles clasping the stems. Whole plant is aromatic when it is crushed.

Native of southern Europe but rather widely and well established in America. Most common wild in America from Connecticut to Virginia, west to Missouri and Louisiana but grown in gardens almost everywhere. Commercial fennel comes for the most part from Germany. Grown by Chinese, Egyptians, Romans.

Flowers yellow, in large open, loose compound umbels, with to 25 major rays that are stout, smooth, and to 3 in. long and smaller slender rays to ⅓ in. long. Ripe fruits about ¼ in. long, slender, oblong, smooth, round in cross section, flattened from the back, with prominent dorsal ribs and single oil tubes between ribs in cross section.

Cultivation is simple and like that of carrots. Name refers to haylike odor. Medicinal property a volatile oil found chiefly in nearly ripened fruit where it constitutes about 2 to 6 percent of the material. All parts of plant may be used. Fruits common in flavors for candy, perfume, soap, medicine, and liquor. Leaf stalks used as a vegetable.

Oil from fruits used chiefly to settle intestinal disturbances, such as gas on stomach; extracted commonly by steeping them in boiling water. Finely divided leaves are used for fish sauces and garnishing. Thrives in any ordinary well-cultivated soil. Young leaves of plant are used as a salad and as a garnish. Fruits and leaves are used as a condiment. Fruits may be used medicinally.

Fennel

Order Umbellales./Family Umbelliferae

Coriander
Coriandrum sativum
Annual or perennial. Herb. Height to 3 ft but usually less. Slender, erect, branching, smooth, strong-smelling. Leaves compounded into three parts or with parts arranged somewhat like those of a feather, with the uppermost leaves divided into fine slender parts which appear to be almost threadlike.

Native of Mediterranean region but most extensively grown in Morocco, India, and Europe. Found on waste ground from Massachusetts to North Carolina and to the Western states. Was mentioned in Egyptian, Hebrew, Sanskrit, and Roman literature. Name connected with Greek word for bug.

Flowers small, white or pinkish, in compound umbels to 2 in. broad, with slender rays. Outer flowers of cluster enlarged like ray flowers. Flowers appear in July. Fruits globular, yellow-brown, and when fresh have an unpleasant odor that is replaced by a fragrance when fruits are dried.

Dried fruits yield a volatile oil, coriandrol, at rate of ½ to 1 percent. Fruits used extensively for flavoring sweet dishes, particularly in India and in Europe. Essence of fruits considered better than the fruits themselves for flavoring purposes. Fruits are also candied, making the pleasing confection known as sugarplums.

Coriander

Oil used as a medicine acts primarily in reducing gas in intestinal tract. Used in flavoring whisky, gin, and other liquors but probably the service rendered is not solely in adding flavor but more importantly in controlling digestive and intestinal disturbances caused by the liquor. Of considerable commercial importance.

Celery
Apium graveolens var. *dulce*

Perennial or biennial herb. Height to 3 ft. Stems much-branched, more or less grooved and jointed. Leaves compounded, in two to three pairs usually with an extra terminal one and each part compounded. We eat the petiole principally. Root leaves crowded, on a short stem before flowering time comes.

Native of marshy places from Sweden to Algeria, Egypt, Abyssinia, and in Asia to the Caucasus, Baluchistan, and mountainous India. Has been found growing wild, probably as an escape, in California, Tierra del Fuego, and New Zealand. Many varieties suitable for different climates, conditions, and appetites have been developed.

Flowers white, small, borne on stems of irregular length, among leaves and springing from joints in stem. Fruits small, compressed, prominently two-angled, with large oil tubes showing in section. Seeds commonly soaked before planting, particularly for a late crop, and should not be dried too long in storage. In recent years, there has been an increasing use of unblanched (green) celery, which is higher in vitamin A. Blanching means that light is excluded from the plant.

Seeds commonly sown 8 to 10 weeks before transplanting into field. In North, transplant at beginning of hot weather and store when sufficiently developed. In South, sow seeds in summer and shade to conserve moisture. 2 oz of seed enough for 1 acre where grown in single rows 3 ft apart. Transplanting is done about 1 month after sowing.

One of most important of field crops. Storage should be at 32°F, the celery itself averaging 2 to 3° warmer than the room. Storage may be by burying in root cellar, by field trenching, or by mulching. Commercial storage plants can hold celery safely over 2 months. Was first grown for food in France in early seventeenth century. California, Florida, and Michigan are leading states in production of celery.

Anise
Pimpinella anisum

Herb. Perennial. Height to 2 ft. Erect, smooth, somewhat branched, rather slender. Leaves three times compounded, the basal ones with long petioles, or simple, with coarse teeth and broad lobes. Upper leaves have finer parts than lower. Root spindle-shaped, woody, with finer branches, from main axis.

Native of Egypt but now cultivated extensively in Asia Minor, India, and South America. Escaped from cultivation in warmer parts of Europe and in America over wide areas. Genus name refers to compound nature of leaves. Mentioned in earliest Greek, Hebrew, and Roman literature. According to Matthew 23:23 Pharisees gave tithes in anise.

Flowers small, yellow to white, in large loose open compound umbels. Fruits mature in late summer or in autumn, are not bristly or scaly, are about 1/8 in. long, gray-brown, slightly downy, compressed from sides, with evident ribs. Oil from the fruit fragrant and stimulating to the taste, being somewhat sweetish.

Fruits provide an element used for flavoring bread and "licorice" candy and even for making medicines palatable. Between 1 and 3½ percent of the fruit is the volatile oil that may be extracted in part by water or almost wholly with the use of alcohol. Culture of the plant is simple and like that of carrots.

Distilled oil is used in preventing nausea and in treatment of colic in children. The stimulation of taste makes it popular with manufacturers of beverages who like to maintain sales demand. Oil is used in perfumery, candy, and sprays. The liqueur anisette gets its name from anise. Plant was used as a potherb in England about 1542 and was heavily taxed.

Parsnip
Pastinaca sativa

Stems with flowers reach a height of to 5 ft, are hollow, grooved, branched, and relatively stout. Biennial or perennial. Leaves compounded into three to four pairs of leaflets, with an odd leaflet at end; leaves appearing, on the whole, to be rather coarse. Upper leaves much smaller than lower. Large root may be 20 in. long.

Native of Europe. Under cultivation before beginning of Christian era. Now grown widely in the cooler temperate climates of world. Grows best in deep moist soil. When escaped from gardens, it loses its thick roots, sometimes becoming an annual. Under these circumstances may become a troublesome weed.

(Continued)

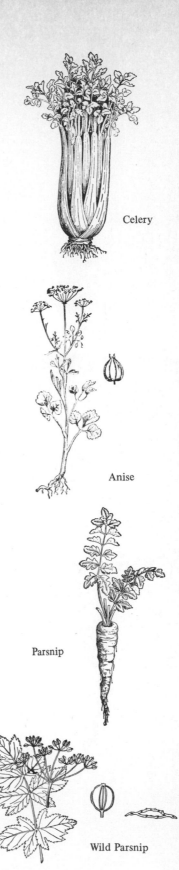

Celery

Anise

Parsnip

Wild Parsnip

Flowers yellow or red, in compound umbels that are to 6 in. broad and composed of to 15 rather slender rays, each to 2 in. long, with final rays slender and to 2 in. long. Fruits developed the second season with exhaustion of large root, which the first season may reach a diameter of to 4 in. Seeds may live 2 years.

Seeds planted at rate of 4 to 8 lb per acre for stock, or 1 oz for every 200 ft of row, or 3 to 4 lb per acre in rows 15 in. apart when hand-cultivated. Radish seeds often sown in row with parsnip since they mature early without interfering with parsnip development. Parsnips delicate when young and do not withstand weeds. Hollow Crown is a common variety and has tapering roots which are 2 or 3 in. in diameter at the crown and 12 to 15 in. long.

Grown extensively as commercial crop. Deep roots make harvest expensive. Quality improves with storage, those stored 2 weeks at 34°F. equaling others left in ground over 2 months, the change being from starch to sugar such as sucrose. Wild parsnip may cause serious dermatitis to wet skin.

Wild Carrot, Queen Anne's Lace
Daucus carota

Biennial.   Height to 3 ft.   Erect, bristly, branched above, with rather slender but nonetheless strong stem.   Leaves lower and basal, two to three times compounded, segments being relatively slender but not linear and smaller units being conspicuously saw-toothed.   Upper leaves smaller and less finely divided.

Native of Asia but naturalized from Europe.   Now commonly established as a weed in fields, pastures, and waste places.   Found from coast to coast in North America but may be commoner in East.   25 species in the genus.   From this species has been developed the valuable cultivated carrot.

Flowers borne in compound umbels that total to 4 in. across and open or close according to humidity, with crowded rays, outer ones being to 2 in. long; rays of smaller umbels, slender and to ⅕ in. long.   Flowers white except that the central one in whole umbel may be purple to almost black.   Fruits with bristly back, to ⅕ in. long.

Flowering time from June through September.   Fruits may be found at almost any time of year since tops persist above snow and through early summer.   Pollination effected by butterflies, moths, bees, flies, and other insects if this can be judged by visits made.   Insects possibly attracted by strong odor of plant.

Queen Anne's lace considered a beautiful plant by some flower lovers, but same plant known as wild carrot is considered a despicable and pernicious weed by farmers.   Seeds persist long in the soil and so tops should not be allowed to come to fruit.   Cultivation, hoeing and crop rotation will ordinarily keep plant in control.   Wet leaves cause dermatitis.

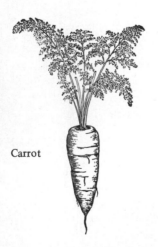

Wild Carrot

Carrot

Carrot
Daucus carota var. *sativa*

Height to 3 ft.   Annual or biennial with top much-branched and more or less bristly hairy.   Leaves on long petioles, finely divided so that finer segments seem almost linear.   Root thick, coarse, orange-yellow, with substantial taproot that penetrates deeply and becomes swollen at end of first season, losing size with flowering.

Native of Africa, Europe, and Asia, widely spread.   Now widely naturalized in America.   Some species of carrots are native of North America. This variety was cultivated by the ancients but has proved popular as a food plant in America and Europe.   It does best in deep, loose, loamy, rich soils, with large yields of early crops coming from sandy soils.

Flowers white or yellow, very numerous, bearing stamens and pistils. Fruits small, flattened or concave on one side, with rows of weakened spines on other, with oil glands under these secondary ribs.   3 lb of seed needed for planting an acre that is to be hand-cultivated in rows 18 in. apart; 4 to 6 lb where sown for stock feed.

Seeds covered to depth of ½ to ¾ in., the lighter soils favoring deeper burying.   Quick-maturing varieties may be harvested from seeds sown 2 months before a killing frost; others require to 5 months.   Carrots ¾ in. through are harvested and sold in bunches.   They keep well for 6 months at 32°F, deterioration being largely due to sugar loss in respiration.

United States crop of commercially grown carrots is on the increase. Food value of root is high.   10 tons of carrots will remove from the soil 100 lb of potash, 32 lb of nitrogen, 18 lb of phosphoric acid.   Yield in acid soils is low or negligible.   Such soils should not be used for growing the plant commercially.

Flowering Dogwood
Cornus florida

Tree or large shrub. Reaches height of to 40 ft, with rough "alligator" bark that is close and relatively uniform. Leaves opposite, dark green above and lighter beneath, smooth or finely hairy above, to 6 in. long, narrowed at base and pointed at tip. Twigs green and rubbery. Root system spreading.

Found in woods and on hillsides, usually among other woody plants, on sandy or gravelly soils but mostly where soil is somewhat acid. Ranges from Maine to Ontario, south to Florida, Kentucky, Minnesota, and Kansas, reaching an elevation of 4,000 ft in Virginia, with close relatives west to the Pacific Coast.

Flowers in small compact clusters, in center of a cluster of four showy greenish, white, or pinkish bracts that are ordinarily considered to be the petals. Flowers appear from April through June and make the plant most attractive. Fruits appear later and through to winter as scarlet structures to ½ in. long crowned by the persistent calyx.

Name cornus means horn, which refers to nature of the wood. Wood brown, weighing to 50 lb per cu ft. Leaves in autumn often turn brilliant red. Fruits sometimes persist through winter making plant attractive throughout year because of flowers, leaves, fruit, and bark. Has few obnoxious insect pests.

Commonly planted as an ornamental but not always with success. Fruits collected in September yield 4,200 seeds per lb with good germination, may be stored by burying for 2 years, to be planted in spring at ¼ in. depth. Has common name of boxwood, false boxwood, cornelian tree, white cornel, nature's mistake, and Indian arrowwood.

Powdered bark can be made into a toothpaste and was used as a quinine substitute. At one time, hard wood was used in making shuttles, tool handles, and golf clubs. Important food of many game mammals and game birds. Dogwood usually is considered a name equal to dagwood and daggerwood.

Dwarf Cornel, Bunchberry
Cornus canadensis

Woody at base but herbaceous at top of stem that is erect, to 9 in. tall, scaly and arising from a slender creeping underground rootstock. Leaves opposite, short-petioled, uppermost crowded into what seems to be a whorl, somewhat downy or smooth, acute at each end, entire, to 3 in. long, conspicuously veined.

Native of North America and found in cool low woodlands from Newfoundland to New Jersey and west to Minnesota, Colorado, California, and Alaska. Obviously closely related to flowering dogwood, which is a tall tree in some parts of its range.

Flowers borne in a greenish-white cluster surrounded with four to six petallike white pointed leaves that appear to make up the flower petals, the whole cluster borne on a slender stem to 1½ in. long. Fruit a bright red globular berry, about ¼ in. in diameter, containing a smooth globular stone a bit longer than it is broad.

Berries considered edible but are tasteless. It has been suggested that with the addition of some flavoring such as lemon juice a palatable dish might be made from the berries but it would seem that many superior fruits could be found wherever lemon juice was available.

Plant is attractive and finds a place in some wild gardens. Because of its beauty, it deserves protection from thoughtless pickers of wild flowers. Ripe berries have been used to catch minnows for bait when other initial baits were not available.

Alternate-leaved Dogwood
Cornus alternifolia

Shrub or tree. Height to 30 ft. Trunk diameter to 8 in. Unique branching resembling that of a deer's antlers is caused by unequal length of areas between leaves. Leaves alternate, slender-petioled, clustered near ends of year's growth, to 4 in. long, smooth above and pale downy beneath. Twigs with white pith.

Native of North America. Found in shady woodlands, from Nova Scotia to Georgia, west to Minnesota, Missouri, and Alabama, reaching an elevation of 4,000 ft in Virginia. Some 20 species in the genus native in the North Temperate Zone, Peru, and Mexico, of which most have opposite leaves although this one does not.

Flowers in downy flat-topped clusters to 4 in. across. Petals white or creamy, slender. Flowers with number of petals equaling number of stamens; petals, to ⅛ in. long, reflexed. Fruit borne on red stems in loose open cluster, almost spherical, dark blue-black or sometimes yellow, ⅓ in. through, one- to two-seeded.

Wood hard reddish-brown, weighs 42 lb per cu ft, with light sapwood, to 20 to 30 annual layers' thickness. Plant is subject to a twig blight that turns twigs yellow and may kill the whole plant. Flowers May to July. Fruits may remain on until fall unless eaten.

(Continued)

Flowering Dogwood

Dwarf Cornel

Alternate–leaved Dogwood

Panicled Dogwood

11 species of birds, including ruffed grouse, known to feed on berries. Cottontails and white-tailed deer have been observed feeding on plant. Sometimes grown as an ornamental in the East, but because of twig blight is hardly a wise selection for permanent ornament. Known as umbrella tree, pigeonberry, blue dogwood, purple dogwood, and green osier.

Gray Dogwood, Panicled Dogwood
Cornus racemosa
Shrub. Height to 15 ft. Twigs and bark smooth, gray. Leaves opposite, to 4 in. long, wedge-shaped or blunt at base, relatively long-pointed at tip; when full grown, finely downy above and below, often pale below, with slender petioles, entire; when first emerge from bud leaves are pale green with red tinge and slightly downy. Known by a combination of two characteristics; light gray branchlets and clusters of white fruits attached to red stems.

Native of North America. Found along stream borders, by thickets, at woodland borders, or in ornamental plantings from Maine to North Carolina, west to Minnesota and Nebraska. Has been recommended for wholesale planting for soil control and so its range has been greatly increased as it is relatively hardy.

Flowers in loose open clusters, not necessarily flat-topped, to 2½ in. across, but the individual flowers are small, white, and borne on smooth rays. Fruit white, to ¼ in. through, with a furrowed stone that is slightly broader than high and essentially globular. Known by a combination of two characteristics: light gray branchlets and clusters of white fruits attached to red stems.

Flowers May and June. Fruits appear August through November, though they may persist through winter. Easily propagated by cuttings or by seeds, of which there are 3,500 to the pound, uncleaned. Plant endures city smoke well and occurs on clay, sand, or gravel soils, dry or moist, in the sun or shade.

Fruits eaten by pheasant, bobwhite, ruffed grouse, and sharp-tailed grouse, making up about 5 percent of the total annual fruit food of pheasants and being one of most important pheasant foods in southern Michigan even into May. Plants also eaten by cottontail rabbit. A popular hedge plant grown for ornamental purposes. State flower of Virginia.

Red Osier Dogwood
Cornus stolonifera
Shrub. Height to 10 ft. Usually sends many branches erect from prostrate branches to make a dense thicket. Twigs smooth, bright reddish-purple, or downy when young. Leaves to 5 in. long, slender-petioled, smooth on both sides or downy above or below, whitish or pale beneath and darker above. Root system relatively shallow.

Red Osier Dogwood

Native of North America. Found in wet or moist soil or planted in hedges and borders from Newfoundland to Virginia, west to Yukon Territory and California, being found up to 2,400-ft elevation in New York. Related silky dogwood *C. amomum* has silky leaves and twigs, and panicled or gray dogwood *C. racemosa* has grayish twigs.

Flowers in flat-topped clusters to 2 in. across, with supporting twigs downy and white; relatively few flowers in a cluster, with no conspicuous bracts such as are found in dwarf cornel or flowering dogwood. Fruit white to blue, globular, to ⅓ in. through, with stone that varies greatly in general size. Known by combination of characteristics: red twigs, white pith, underside of leaves white, fruits white. Red osier blooms very early as compared to other *Cornus* spp.

Does best in moist but well-drained area, in sun or in shade. Flowers May through July. Survives in alkaline soil, new erect shoots arising from branches that touch the ground. About 17,300 seeds per lb. Seeds may be collected at almost any time of year though most abundantly in late summer and fall.

Known to supply food to ruffed grouse, bobwhite, sharp-tailed grouse, Hungarian partridge, elk, deer, moose, snowshoe hare, and mule deer. Used extensively as an ornamental and easily propagated by cuttings either in fall or spring or by layering and then cutting off the rooted stock.

Sour Gum, Pepperidge
Nyssa sylvatica
Tree. Height to 120 ft. Trunk diameter to 5 ft. Bark to 1½ in. thick, light brown or tinged with red, deeply fissured. Leaves crowded at twig ends in lateral branches or remote on vigorous shoots, alternate, entire, with slightly thickened margins, to 5 in. long and to 3 in. wide. Roots thick and hard.

Found in swamps or at their borders in poorly drained soil, and, in southern part of range, on mountain slopes. Ranges from Maine to Florida, west to Michigan and Texas, with closely related species within much the

Sour Gum

same range or extending it only slightly. Seven species in the genus native of eastern North America and Asia.

Flowers appear in early spring when leaves are about half grown, in many-flowered, rather compact clusters with staminate with more flowers and in heads that droop, while pistillate have two to a few flowers in heads that do not droop. Stamens in staminate flower shorter than petals in pistillate. Fruit dark blue, to ⅔ in. long.

Wood hard, heavy, weighing to 40 lb per cu ft, tough, not durable, light yellow to white, with thick dark sapwood to 100 annual layers thick. Has few insect pests. Tree resists fire well but not wind. Seeds 3,000 to 4,000 per lb, from stone fruit appearing in September and October but persisting through winter.

Wood was used for wheel hubs, ox yokes, shoe soles, wharf piles. A beautiful tree in autumn, when it is the first to turn a brilliant red. Grown sometimes as an ornamental. Fruits eaten by ruffed grouse, bobwhite, wild turkey, pheasant, prairie chicken, black bear, white-tailed deer, marsh rabbit, and other game species. Fruit makes excellent preserves.

Order Ericales./Family Ericaceae

Pipsissewa
Chimaphila umbellata

Pipsissewa

Evergreen, woody, with erect stem rising from a sprawling stem to a height of 1 ft. Stem rather tough and not much branched in erect portion. Leaves thick shining dark-green, to 2½ in. long and to 1 in. wide, not mottled, broader toward tip, with fine, sharp, small saw teeth along edge.

Native of North America, Asia, and Europe. Found in America in dry deep woodlands from Nova Scotia to British Columbia, south to California, Mexico, and Georgia, being possibly more abundant in the West than in the East. Most common under evergreen trees such as pines, hemlocks, and spruces.

Flowers white, pink, or flesh-colored, waxy, dainty, with five-lobed corolla commonly marked with a deep pinkish ring, lobes turning back as flower matures. Stamens, 10, with downy or hairy filaments. Fruit a capsule, to ⅓ in. through, globular, brown, dry, maturing in late summer.

Flowers June through August; visited by many kinds of bees and small flies that effect pollination, among the genera being *Andrena* and *Halictus*. Delicate scent of flowers and their pale color no doubt serve as an attraction. Leaves have a delicate flavor and are nibbled by woodsmen.

A plant for shaded parts of a woodland garden. Use rich woodland soil which is sandy and acid for best results. They may be planted in early fall or in the spring. There are cultivated varieties. A common ingredient in root beer. Plant is used in Christmas decorations. Known as love-in-winter, bitter wintergreen, ground holly, noble pine, pine tulip, king's cure, and bittersweet.

Indian Pipe
Monotropa uniflora

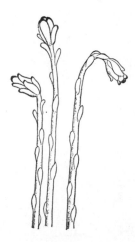

Indian Pipe

Height to 10 in. When fresh is pinkish white but turns black at maturity; finally loses its succulence and becomes brown and relatively dry. Stems arise in clusters from spreading, brittle, matted masses of roots. Only leaves are rudimentary, scalelike, scattered along erect stem bearing the flower.

Found in rich moist woodlands where shade is heavy, from Anticosti Island south to Florida, west to Washington and California thus including practically all of United States. In North Carolina it reaches an elevation of 4,200 ft. Ranges south into Mexico and is found in Japan and in Himalayas. There are two species, one ranging into Colombia.

Flowers borne singly at top of flowering stalk, at first drooping to form a "pipe" but later becoming erect when the fruit is formed. Sometimes two flowers to a stalk. Flower to 1 in. long, with four to six petals and ten stamens. Fruit erect, obtuse-angled capsule, nearly ½ in. through.

Lives chiefly on decaying woody plant material, so needs no green tissue to use in manufacture of food for itself. Can also take its food from roots of living plants on which it is parasitic. Entire root system compacted into an almost solid clump of fungus mycelium so that roots themselves do not come into contact with soil but receive their nourishment from the fungi, which in turn get theirs from the decaying wood.

Botanists do not always agree as to dependence of this plant on fungi that constitute its substitute for a root system but it is certain Indian pipe could not grow in wild without assistance of fungi. Possibly the fungi may receive something in return from the Indian pipe. Related false beechdrops *M. hypopitys* is fragrant and bears many flowers; considered edible by some persons.

Labrador Tea

Pink Azalea

Great Laurel

Labrador Tea
Ledum groenlandicum
Shrub. Height to 4 ft but usually much lower. Twigs densely fuzzy. Leaves to 2 in. long and to ⅔ in. wide, dark green; roughened above, with incurving edges bending toward the brown, wool-covered undersurface, alternate, usually more abundant in upper portions of stem, evergreen.

Relatively common in peat bogs. Sometimes found in swamps from Greenland and Labrador, south to Pennsylvania and west to British Columbia and Washington. It survives in well-drained or in water-soaked soil, in sun or in shade. Five species in the genus extend range from coast to coast and into subarctic regions.

Flowers to nearly ½ in. across, in many-flowered terminal clusters at top of plant, each borne on brown downy stems that are to 1 in. long and recurved when fruits are formed. Stamens, five to seven. Fruit a nodding, oblong capsule about ¼ in. long, bearing elongate seeds. Flowering time May through August. Fruit available in August.

At its best in sour peat bogs where it often forms dominant plant along edges. Ecologists use unique form of leaf to illustrate what they call "roll-leaf" form of a plant to be found growing in water that may be physiologically useless because of its chemical content.

Early explorers called it "muskeg tea" and made a substitute for tea from its leaves. Linnaeus stated that the leaves were frequently mixed with stored corn to keep mice away. It appears in the list of useful medicinal plants issued by quack herb doctors but not in accepted medical literature. It is reported to be a poisonous browse for grazing animals, as is *L. glandulosum* and *L. columbianum* of our western states. Often associated with leatherleaf, and bog rosemary in northern bogs.

Pink Azalea, Pinxter Flower
Rhododendron nudiflorum
Shrub. To 6 ft high, well-branched above, with conspicuous terminal buds from which many new smooth or stiff-haired twigs arise with the new year's growth. Leaves more or less pointed at each end, to 4 in. long, with hairy margins; more hairy when young than when old. Root system substantial and relatively tough.

Relatively common in sour soils, open or thick woodlands, over sand or gravel or even in swamps and bogs. Ranges from Massachusetts and New Hampshire south along mountains to Florida and Texas, reaching an elevation in Virginia of 3,000 ft. Relatively commonly planted as ornamental about homes.

Flowers pink or white, expanding from a short, narrow tube into a showy, somewhat two-lipped, nearly 2-in. broad bloom. Blooms April through to June. Stamens conspicuously long, spreading, with pistil tip curving upward. Fruit an erect, rather slender capsule to ¾ in. long, splitting to free seeds. Pollination by honeybees and moths.

May be transplanted in spring but not in fall. Should not be placed where they will get full glare of sun. Soil should be well-watered and sour rather than limed. Horticulturists graft plants freely. In winter, young plants may be temporarily protected by a leaf mulch but mature plants should be hardy in their range.

Valuable wild and ornamental shrub protected from despoliation by law in some states. Should be more thoroughly protected. Plucked flowers do not survive long and so are not satisfactory material for bouquets. An edible, juicy gall may be found on twigs or leaves by end of May that is delicious eaten fresh or may be pickled in vinegar for later use.

Great Laurel
Rhododendron maximum
Height to 40 ft. Trunk diameter to 1 ft. Leaves to 7 in. long and to 2½ in. wide, evergreen, tough, thick, smooth beneath, dark green on both sides, drooping in winter, somewhat curled in bitter cold weather, on stout petioles to 1 in. long, rather pointed at end. Root system sturdy but compact.

In woods and along streams or sometimes in bogs, usually where there is muck, gravel, or damp rocky soil, usually sour but not commonly with limy soils or exposures. Ranges from Nova Scotia to Quebec and Ontario, south through Ohio and New England to Alabama and Georgia with related species to west.

Flowers showy, to 2 in. broad, borne on sticky stems, white, purple, or rose-colored, with yellowish or whitish spots within, somewhat waxy in appearance. Calyx nearly as long as petals but hidden by them. Fruit a somewhat fuzzy capsule, to ½ in. high. Flowers appear from June through July. Pollination probably by bees.

Wood one of hardest and strongest, weighing 39 lb per cu ft, light brown. Green parts contain the poison andromedotoxin that may be injurious or fatal to sheep and other animals that eat leaves when more desirable forage is not present. Honey from bees feeding on nectar reported to be poisonous.

Rhododendron of this or related species is the state flower of West Virginia and of Washington. Some 100 species, of which this is most popular as a landscaping plant. Transplants well, forms a dense year-round bank, and is attractive because of leaves or flowers. Poisonous properties not commonly troublesome. Deer eat small quantities safely. May serve as year-round cover for some forms of wildlife.

Mountain Laurel
Kalmia latifolia
Shrub. To 20 ft high, forming dense thickets, with stiff round twigs. Sometimes grows to tree size, to 40 ft high. Leaves alternate or sometimes a few opposite, smooth dark green, evergreen pointed at each end, to 5 in. long and to 1½ in. wide, with margins somewhat curved in toward underside. Root system tough and persistent.

Common in rocky and sandy soils, sometimes forming extensive pure stands. Frequently grows where other plants could not survive. Frequently found in abundance in cleared areas under power lines. Ranges from New Brunswick to Ontario, Indiana, and Kentucky, south to Florida and Louisiana but more abundant in middle of this range.

Flowers beautiful, pink or white, to ¾ in. across, in compact clusters at end of twigs, erect, on densely glandular slender stems. Stamens, about twice as many as petals and close to them, surround longer up-curving pistil. Stamens bear web-entangled pollen that attaches itself to visiting insects.

Wood very hard, fine, brown, weighing 44 lb per cu ft, used in making wooden implements that need to be carved. Green parts contain the poison andromedotoxin, most commonly eaten by sheep causing intense salivation and even paralysis but an animal ill for 2 days may yet recover.

A legitimate crop if harvested wisely for decoration where abundant. Used also in making wooden tools. Has been called spoonwood, broadleaved kalmia, ivy bush, big-leaved ivy, poison laurel, calico bush, and American laurel. Eaten sparingly by deer and grouse. It is state flower of Connecticut and of Pennsylvania.

Bog Rosemary
Andromeda glaucophylla
Shrub. Height to 3 ft. Sparsely branched. Leaves rather crowded, slender, usually acute at tip and narrowed at base, sour, alternate, to 2½ in. long and to ⅓ in. wide, with margins curved sharply inward and downward, dark green above and almost white beneath, short-petioled, evergreen.

Native of North America, Asia, and northern Europe. Found in bogs from Newfoundland to Alaska, south to British Columbia, Michigan, and northern Pennsylvania and New Jersey, with several close relatives extending range to south. Only a few species in the genus; only three native of United States.

Flowers in few-flowered clusters borne at branch tips, nodding; pink or white in corolla and pale red in calyx. Stamens, 10. Although flowers nod, fruit becomes erect as it ripens. Fruit a reddish or brownish dry capsule about as long as it is wide. Flowers May through July.

Grows best in wet peat bogs exposed to sun, reproducing and spreading largely by creeping rootstocks. Rarely attacked by any insects or by fungi. Leaves contain the poison andromedotoxin, but this is rarely troublesome since leaves are too bitter to be eaten in abundance. Most dangerous in spring.

Fruits known to be eaten by ptarmigan, and leaves on occasion by stock. Does not survive as an ornamental in usual situations. Of interest to plant ecologists and physiologists because of nature of leaves in relation to nature of available water supply. Leaf covering probably limits loss of water reducing water needs of plant.

Leatherleaf, Cassandra
Chamaedaphne calyculata
Shrub. Height to 4 ft. Profusely branched, with slender twigs and branches. Leaves leathery, oblong, thick, covered with fine round scales that rub off when young, to 1½ in. long and less than half that width usually, evergreen, alternate, decreasing in size from bottom to top of plant.

Found in bogs and swamps in North America, in Europe, and in northern Asia. Ranges from Newfoundland to Georgia, west to British Columbia, doing well in moist or even in well-drained soils, but only on acid peat soils and preferably in a place exposed to sun. Only one species in the genus in North America.

Flowers borne along tips of uppermost branches, with small leaves at base of each, bell-shaped, hanging from one side of branch, white, with five acute sepals in calyx and ten stamens. Fruit a five-celled capsule bearing many flat seeds. Flowering time April through May or June.

(Continued)

Mountain Laurel

Bog Rosemary

Leatherleaf

Sourwood

Trailing Arbutus

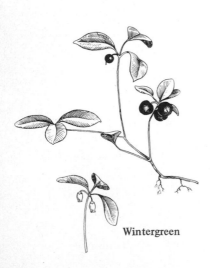

Wintergreen

Propagated by dividing mature plants or by growing seeds sown on sand and sphagnum moss under glass in winter or spring and transplanting them when sufficiently large. According to some authors leaves may contain the poison andromedotoxin but others recommend their use in making a tea substitute.

Fruits form staple winter food for sharp-tailed grouse. Plant is eaten freely by cottontail rabbits and snowshoe hares. Can hardly be considered as on ornamental since its soil requirements are such that it would be hard to grow in ordinary situations. Serve to consolidate floating peat bogs in which cranberries grow.

Sourwood, Sorrel Tree
Oxydendrum arboreum
Tree. Height to 60 ft. Trunk diameter to 1½ ft. Bark smooth on twigs and branches but deeply fissured on trunk. Leaves alternate slender, essentially smooth, to 7 in. long, commonly somewhat wrinkled, entire, green on both sides, finely veined in a netlike manner, slender-stalked.

Native of North America and found ranging from Pennsylvania to Florida, west to Louisiana and Indiana but grown as an ornamental rather considerably outside normal range. Grows wild in woodlands associated with other species. In its southern range is more or less confined to higher lands.

Flowers borne in branching one-sided clusters at ends of twigs, whole clusters being to 10 in. or more long and individual flowers being about ⅓ in. long; like little bells constricted at free end, with shallow cuplike calyx at base. Fruit a capsule to ½ in. long, ripening in September and October.

Flowering period June through August. Wood hard, reddish-brown, weighing 46 lb per cu ft. Tree survives in sun or shade in dry or well-drained soils, is rarely attacked by insects or by disease, turns a brilliant scarlet in autumn, is propagated by seeds sown in fall or spring under glass on sour soil.

Wood has occasional commercial uses but plant is best known as an ornamental. Sour leaves are commonly chewed by hikers as a thirst quencher. Flowers yield a nectar that is collected by bees and makes a good honey. Leaves are eaten by white-tailed deer as a forage. Known as elk tree, titi, and by other names.

Trailing Arbutus
Epigaea repens
Trailing woody vine. To 15 ft long; slender with hairy branches that are tough, branching, and close to ground. Leaves alternate, thick, dark green above, coarsely veined, heart-shaped or rounded at base, mostly smooth above and somewhat hairy beneath, to 3 in. long and 1½ in. wide, on short petioles. Roots tough.

Found in pastures, rocky woods, and sandy regions, but most particularly in or under evergreen woodlands, often in large patches more or less completely covering ground. Ranges from Newfoundland to Saskatchewan, south to Florida, Kentucky, and Wisconsin and sometimes established in wild gardens.

Flowers in clusters of few to several, intensely fragrant, pink or white, to ¾ in. long and nearly as broad with the five petals expanding and longer than the sepals. Fruit breaks into five sections that spread apart exposing a white fleshy interior. Flowers March to May. Pollination most commonly by early flying queen bumblebees. Flowers of three types.

Leaves reputed to have an astringent tonic though this is not always recognized as important. A related species is found in Japan. A taller plant that is not the trailing arburus goes by name of arbutus. Genus name *Epigaea* means in Greek on the earth and refers to trailing habit. Propagation is by division in October and by seeds.

State flower of Massachusetts. Known as shadflower, gravel pink, winter pink, mountain pink, Mayflower, and ground laurel. Is protected against collectors by law in some states. Should not be destroyed nor should its sale be encouraged or legalized. Flowers eaten to relieve thirst; taste spicy.

Wintergreen, Checkerberry
Gaultheria procumbens
Stems woody, creeping, freely branching and bearing erect portions that reach a height of to 6 in. Smooth throughout, fragrant. Leaves clustered mostly at top of erect branches, dark green, shining on upper surface and lighter beneath, to 2 in. long, rather leathery but tender and delicious when young.

Native of North America where it is found in woodlands, particularly of coniferous type, from Newfoundland to Georgia, west to Manitoba and West Virginia. About 100 species in the genus ranging through North America, through the Andes in South America and in Asia. Salal of Pacific Coast is in the group.

Flowers usually solitary in axils of leaves, borne on recurved stems that are to ⅓ in. long. Corolla white, waxy, five-toothed, to ¼ in. long. Stamens, 10, borne on inside of corolla. Fruit a bright red berrylike mealy, fragrant, 5-celled structure, ripening in autumn and increasing in size through winter.

Found on sandy and other soils, in sun or shade, in mats. Produces oil at rate of 1 lb of oil to 1 ton of leaves. Similar oil obtained from birch and made synthetically. May be propagated by seeds sown under glass in winter of spring or by division. About 2,800 fresh berries per lb.

Leaves are delicious in spring, and fruits in winter. Commercial oil of wintergreen or of checkerberry comes mostly from birch or from synthesis. Plant eaten by bobwhite, ruffed grouse, sharp-tailed grouse, pheasant, mountain sheep, white-tailed deer, spruce grouse, and chipmunk. Mixture of blueberries and wintergreen is marvelous.

Salal
Gaultherla shallon
Shrub. Height to 2 ft. Spreading with somewhat hairy branches. Leaves evergreen, alternate, round or heart-shaped at base, to 4 in. long, dark green, with margins somewhat broken, smooth when mature, with petiole to ¼ in. long, acute at free end. Branchlets fuzzy.

Native of western North America, ranging from Alaska to southern California, growing in dry or well-drained areas either in sun or in shade but needing a humid atmosphere to reach maximum development. About 100 species of the genus, best known in the East being wintergreen or checkerberry.

Flowers in slender clusters to 5 in. long, nodding, borne from tip of branches or from leaf axils. Calyx white or whitish, fuzzy, with triangular lobes. Corolla fuzzy, white or pink, to ⅓ in. long. Fruit purple at first; when mature, black or indigo, hairy, about ⅓ in. across.

Flowers borne in May and June. Fruits July through December. In dry sunny situations plants are dwarfed; but in humid regions on sandy soil near sea may form a dense mat. May be propagated by sowing seeds in winter or spring under glass and transplanting seedlings.

Known as salal or shallon. Used primarily as an evergreen ornamental shrub but of importance as an emergency food. Eaten regularly by Indians. Not palatable for livestock but is important food for Roosevelt elk, an emergency food for Olympic wapiti, and food for mountain beaver, deer, bear, grouse, pigeons, and chipmunks.

Bearberry
Arctostaphylos uva-ursi
Trailing or sprawling, branched, shrubby plant with branches to 2 ft long but usually sprawling close to ground. Twigs finely fuzzy. Leaves evergreen, entire, smooth or finely fuzzy toward base, somewhat leathery, to 1 in. long and to ½ in. wide, with fine crooked veins and short petioles, or sometimes apparently none.

Native of North America. Ranges from Labrador to Virginia, west to California, north through Arctic America to Alaska. Also found in northern Europe and in Asia. Some 40 species in the genus, of which this one is widely spread and others are better known in western North America. For example, *A. pungens* and *A. pringlei* are characteristic plants of the chaparral association. In the west members of *Arctostaphylos* vary from low-spreading shrubs to small trees.

Flowers borne in small, few-flowered clusters at ends of branches. Corolla white, narrowest at free end that shows five shallow lobes, and bears, inside at the base, ten or sometimes eight stamens shorter than the pistil, which is, in turn, usually shorter than the corolla. Fruit red, smooth, rather tasteless, nearly ½ in. through, with five nutlets.

Succeeds on dry or well-drained soil, in sun or in shade. Flowers May and June. Bears fruits through year, but mostly from August through March. Propagated by cuttings taken in late summer and rooted under glass, or by seeds sown in similar situations. May serve somewhat as a soil anchor on otherwise sterile soil.

Raw berries mealy but nourishing. Cooked berries better. Reported to have some medicinal value, since it contains arbutin, ercolin, ursone, tannic acid, and gallic acid used as urinary antiseptic, astringent, and diuretic. Eaten by grouse, turkey, deer, bear, and sheep but considered of little value as a browse for domestic livestock of any sort.

Manzanita
Arctostaphylos manzanita
Shrub. Upright. Height to 13 ft. Branches widely spreading, with branchlets finely fuzzy or smooth. Leaves evergreen, entire, smooth or nearly so except when young, dull green, more or less blunt or with a short point, to 2 in. long, petioled, rather broadly egg-shaped.

Ranges from Oregon through Calfornia, particularly in the humid regions to the west of the mountains and near the sea. Related woolly
(Continued)

Salal

Bearberry

Manzanita

275

manzanita *A. tomentosa* has wider range extending down into Arizona and New Mexico. Its leaves are not smooth. Nearly 40 species of *Arctostaphylos* spoken of collectively as manzanitas.

Flowers in dense clusters over 1 in. long and for most part smooth, white or pink, with corolla to ⅓ in. long, showy in spring and fragrant. Fruit to ½ in. through, like a sphere, flattened from the top, reddish-brown and smooth.

A relatively common chaparral shrub. With in its range, conspicuous because of branching and reddish branches. While most reproduce by root shoots this species is easily killed by fire and does not crown-sprout. Relatively slow-growing, doing best on dry, well-drained soil in a humid atmosphere but exposed to sun.

A most important honey plant within its range. Fruits available the year round. May make an important food for game birds such as sharp-tailed grouse and dusky grouse. Seeds sown in spring after stratification in moist sand at from 41 to 50°F. Of little value as forage for cattle but will be eaten by goats and some sheep. Valuable food and cover for wild birds and mammals.

Madrona
Arbutus menziesii

Tree. Height to 125 ft. Trunk usually tall, straight, and to 5 ft in diameter. Branches stout, spreading or upright. Slender branchlets light red, pea green, or orange, smooth when young, becoming reddish-brown by first winter. Leaves dark green above and white beneath, evergreen, to 5 in. long and 3 in. wide, thick. Often shrubby rather than treelike.

From British Columbia through Washington and Oregon into northern California, with *A. arizonica* extending range into Mexico and *A. texana* extending it into Texas. *A. menziesii* is occasionally grown in gardens in western and in southern Europe; is probably best known species generally. Does well on a variety of soils. About 12 species in all.

Flowers in relatively few-flowered clusters arranged along terminal stems beyond the leaves, each flower being about ⅓ in. long and whole cluster being to 6 in. long. In *A. menziesii* and in *A. arizonica,* the ovary is smooth; in *A. texana,* it is fuzzy. Fruit orange-red, fleshy, to ½ in. long, with several angled dark brown seeds.

Flowers March and May. Fruits available from July through January. About 30 seeds per berry and 1,000 useful plants from 1 lb of seed. Difficult to transplant. Wood heavy, hard, close-grained, light brown shaded with red, with lighter-colored sapwood of to 12 annual rings thickness. Stumps sprout abundantly.

Wood used commercially for furniture and charcoal. Bark used as source of tanning material. Nectar of flowers an excellent source of honey. Tree a thing of beauty. Fruits eaten by doves, pigeons, turkeys, raccoons, and ring-tailed cats. Leaves browsed by deer, goats, sheep, and cattle. Plant serves as a soil anchor.

Huckleberry
Gaylussacia baccata

Shrub. Height to 3 ft. Erect, with gray forked branches that are usually fuzzy when young. Leaves entire; when young, very resinous; alternate, smooth or nearly so, rather stiff, to 2 in. long rather uniformly green on both sides, not evergreen, short-petioled.

Native of North America ranging from Newfoundland to Florida, west to Manitoba, with closely related species extending range considerably. In America, about 40 species in genus, most of which are valuable because of their delicious fruits. Of several kinds in cultivation, *G. brachycera,* the box huckleberry of eastern U.S. is to 18 in. high, with short stems bearing dark green, evergreen leaves. However, *G. baccata,* also known as *G. resinosa,* is one of the prime sources of commercial huckleberries.

Flowers relatively few to a cluster, clusters being at ends of short spurs arising from ends of branchlets or from below ends, in one-sided clusters, pink or reddish, with five-angled corolla that is much longer than calyx and to ¼ in. long. Fruit black or blue, with or without a bloom but sometimes white, to ¼ in. through.

Flowers May and June. Bears fruits in July and August. Fruits rather seedy but very sweet. Soils favored are sandy or distinctly acid. Some persons consider fruits to be poisonous, but there is no basis for this except that berries resemble those of the puckery but not poisonous chokeberry, *Pyrus arbutifolia.*

Leaves feel sticky when pinched, owing in part to yellow-brown resin or varnish they contain. Fruits are much more tasty than more abundant and smaller blueberries though they contain more conspicuous seeds that make them somewhat unpleasant. Known as crackers, black snap, and black huckleberry. Excellent food for wildlife.

Madrona

Huckleberry

High-bush Blueberry
Vaccinium corymbosum

Shrub. Profusely branched. Height to 15 ft. Twigs round in cross section, finely warty, commonly but not always smooth. Leaves to 3 in. long and to 1½ in. wide, alternate, entire, not evergreen, green and smooth above and lighter green and sometimes slightly fuzzy beneath, short-petioled.

Native of North America. Ranges from Quebec to Virginia, west to Louisiana and Minnesota, with closely related species extending range on to the Pacific Coast and with other closely related species extending range to the north. Found in swamps and in woodlands. Grown principally as a cultivated crop in New Jersey, Michigan, and North Carolina. Needs an acid soil to be successful.

Flowers borne at time leaves appear, in short clusters of relatively few flowers, the flowers being equal to or longer than immediate stems that bear them; white, or sometimes faintly pink, cylindrical, with throat slightly narrower, to ½ in. long and ¼ in. through. Fruit blue, with a bloom, to ⅓ or more inches through.

Flowers May and June. Fruits July through August, generally recognized as a later form than *V. pennsylvainicum*. Produces abundant suckers. Plants are easily divided and survive transplanting well. May be easily propagated by means of cuttings and therefore lends itself to cultivation practices.

Valuable commercial blueberry used for late-season market. Fruits sold fresh, preserved for use in pies, or made into jams and jellies. Wild fruits provide a superior food for most species of game birds and particularly for ruffed grouse, ring-necked pheasant, and mourning dove; also for cottontail rabbits.

High–bush Blueberry

Cranberry
Vaccinium oxycoccus

Slender, creeping, woody, vinelike plant that roots at joints and has stems to 1½ ft long and a number of branches. Leaves alternate, evergreen, thick, entire, dark green above and light beneath, to ¾ in. long and nearly ½ as wide, with margins curved downward and inward. Root system extensive and, like stems, long and spreading.

Native of Old World and of New World. Found in Asia and Europe as well as in North America. Found in cold peat bog formations from Newfoundland to North Carolina, west to Washington and Alaska but grown under cultivation in some parts outside natural range. Differs from American cranberry in having spherical rather than oblong fruits.

Flowers nodding pink bells, up to six on a branch, arising from terminal shoots, about ⅓ in. across, with parts divided to near base and recurved conspicuously, with filaments of stamens ½ length of anthers. Fruit red spheres, to ½ in. in diameter in wild form or larger under cultivation; tart, sometimes spotted.

Flowers May through July. Bears fruits that are at best through August and September but may persist through winter, keeping well in spite of severe weather. Common market cranberry is *V. macrocarpon*. Principal areas of commercial production are Massachusetts (Cape Cod), New Jersey, Wisconsin, Oregon, Washington, and Nova Scotia. Development of a cranberry bog is very expensive. Berries may require much sugar but a teaspoonful of salt may equal a cupful of sugar in counteracting sourness.

Highly valuable in making cranberry sauce, a Thanksgiving essential, cranberry pie, and in jellies, jams, and drinks. Mixed with gelatin or white of egg, makes an excellent desert or may be made into a mock cherry pie. The industry of raising and marketing cranberries is an important one. Known as sourberry, moss melons, crowberry, moorberry.

Cranberry

Order Primulales./Family Primulaceae

Early Garden Primrose
Primula polyantha

Herb. Hardy, usually in rosette form. Height to about 1 ft though usually less. Leaves narrowed into a long-winged petiole, rather conspicuously wrinkled, with uneven margins, about 6 in. long and about ⅓ as wide, bluntly pointed at free end, prominently veined.

Probably a hybrid of a number of species including *P. acaulis*, *P. veris*, and *P. elatior* or *P. vulgaris*. By some, the whole group of garden primroses of which this one is called *P. variabilis*. Since it is a hybrid, its geographical origins are probably widespread. This is grown commonly in borders, in greenhouses, and in houses.

Flowers borne in clusters like umbels at top of flowering stem though in some garden primroses single flowers are borne close to ground. Flowers single or double, maroon, orange, bronze, white, red, yellow, or of mixed colors, sometimes with center differently colored from outer areas.

(Continued)

Primrose

277

Usually propagated by sowing seeds as soon as they are ripe or by division of plant itself. Seeds sown in greenhouse from February to May may yield flowers by Christmas. Best soil is mixture of half leaf mold and half sand. Plants should be kept cool. Some species, like *P. obconica,* may cause a dermatitis to some persons.

Seedlings with three leaves transplanted into 2½-in. pots with loam and rotted manure, with crowns at ground surface. In August, reset in 6-in. pots with rich fertilizer. In September, plants returned to greenhouse and kept at 45°F, to be raised when flowering needs to be stimulated. *P. malacoides* is grown to be sold as a cut flower.

Whorled Loosestrife
Lysimachia quadriflora
Herb. Height to 3 ft. Smooth, slender, little branched, erect. Leaves in whorls of three to seven, but usually of four to five, or sometimes a few are simply opposite, without petioles, to 4 in. long and to 1½ in. wide, usually bearing small black dots, entire, rahter prominently veined, pointed at each end or more blunt at base.

Native of North America. Ranges from New Brunswick to Georgia, west to Tennessee and Wisconsin. Found for most part in shady thickets or in relatively open woodlands. About 70 species of the genus found mostly in Northern Hemisphere but some are found in Africa and Australia.

Flowers borne in axils of leaves, on rather long slender erect or drooping stems, to 1½ in. long. Flowers to ½ in. broad, yellow, with dark spots or streaks, smooth. Sepals pointed, slender, with filaments fastened together at base. Fruit a capsule that is about length of calyx lobes.

Flowers appear from June through August. Pollination effected by honey bees, bumblebees, and bees of the genus *Macropis.* Their visits are necessary to produce pollination since stamens are too short to reach pistil otherwise. Plants do well on sandy soil or on moist ground.

Of no considerable importance economically, but a rather attractive flower when growing in its natural setting. It has little beauty if picked and apparently no medicinal or food values have been recognized.

Moneywort
Lysimachia nummularia
Sprawling herb. With stems to 2 ft long that frequently take root at joints where they touch ground, smooth, light green. Leaves opposite, the pairs being about equal, to 1 in. long, with short petioles, blunt at both ends or heart-shaped at base, with some small black dots scattered over surface.

Found on wet grounds in shade or in sun. From Newfoundland to Virginia, west to Illinois and Michigan or farther. Native of Europe but now well-established here. Most of other lysimachias are erect, slender plants not easily confused with this prostrate sprawler.

Flowers borne in axils of leaves, individual, yellow, deeply five-lobed, with calyx half as long as corolla. Stamens fastened together at bases. Fruit much shorter than calyx lobes and practically hidden by them at maturity. Many small seeds to each fruit.

Flowering time June through September or later. Fruiting time still later of course. Plant easily established by division or by cuttings and maintains itself without trouble in suitable environments. It persists in gardens, lawns, or on waste soil particularly if there is sufficient moisture.

Popular in hanging baskets or as a quick filler in garden space but must be kept under control. It can be eliminated by digging it out, by raking, by close mowing, or by spraying with a solution of sodium chlorate. Possible it does serve as a soil anchor. It may survive complete submergence for a considerable time.

Fringed Loosestrife
Lysimachia ciliata
Herb with erect, slender stems, much-branched, mostly smooth but with rows of hairs on certain parts. Height to 4 ft. Leaves thin, membranous, opposite, to 6 in. long and to 3 in. wide, on petioles to ½ in. long, these being definitely hairy, often with clusters of small leaves at bases of normally sized leaves.

Ranges from Nova Scotia to Georgia, west to Arizona and British Columbia, with related species extending the range. At least five species native of North America. This species ascends to elevations of 6,300 ft in North Carolina. Habitat is swamps and forests for the most part.

Flowers erect, spreading or nodding, singly from axils of leaves, yellow, to 1 in. across. Corolla of five parts. Calyx of five parts that are shorter and narrower than petals. Stamens, five, with distinct filaments and slender anthers. Fruit a capsule, conspicuously longer than calyx lobes that persist. Seeds angular.

Whorled Loosestrife

Moneywort

Fringed Loosestrife

Name refers to presence of some stamens in flowers that are without anthers and to hairy petioles. Flowering time June through August. A rather conspicuous terra-cotta-colored ring about the center of the otherwise pure yellow corolla. Plant is apparently of no importance medicinally.

An interesting, inconspicuous plant that rarely if ever could become obnoxious as a weed. Attractive flowers do not survive picking, and are not large when compared with extent of foliage.

Order Ebenales./Family Ebenaceae

Persimmon
Diospyros virginiana
Tree. Height to 100 ft. Trunk diameter to 2 ft. Bark to 1 in. thick, dark brown with a tinge of red or gray, deeply checkered into squares. Winter buds to ⅛ in. long, with lustrous scales. Leaves alternate, entire, taper-pointed, petioled, to 5 in. long, dark green above and pale beneath, smooth when mature. Known by a combination of characteristics; leaves without teeth, dark-colored buds, and regularly furrowed bark.

Native of North America. Found in fields and woodlands. Ranges from Connecticut to Florida, west to Kansas and Texas. One other tree of the genus in North America, *D. texana* or chapote, ranges from Texas to southern California. About 160 species in the genus, mostly native of Asia.

Flowers borne singly or in small clusters in axils of leaves, mostly four-parted; with greenish-yellow corolla and with stamens and pistils in separate flowers, the pistillate usually being solitary and the staminate in clusters. Pistillate flowers to ½ in. long and about twice the size of staminate. Fruit to 1 in. through, reddish-yellow.

Flowers May and June. Fruits September through November; become sweet when completely mature after a frost but are astringent when they are green. Wood hard and brown, weighing to 40 lb per cu ft, with nearly black heartwood. Leaves reported to be high in vitamin C.

Wood was used in wood turning, shoe lasts, and for shuttles. Fruits eaten raw when ripe, or made into jellies, syrups, coffee substitutes, tea, vinegar, or beer. Coffee is made by roasting the seeds, but Indians made flour for bread from the seeds. Known as Jove's fruit, possumwood, winter plum, date plum, lotus tree, and seeded plum. Fruit eaten by nearly all birds and mammals.

Persimmon

Order Gentianales./Family Oleaceae

Privet
Ligustrum vulgare
Shrub. Height to over 15 ft. Freely branched, particularly if pruning has been severe. May form a dense impenetrable hedge. Twigs dark and finely fuzzy. Leaves opposite, smooth, dark green above and lighter beneath, short-petioled, to 3 in. or more long, with margins practically entire. Root system spreading and sometimes giving rise to shoots.

Native of Europe, Africa, and Asia. Naturalized in many parts of the world and not uncommon in remote places where plants may have been started by seeds carried by birds. One of commonest of hedge plants in eastern United States, where a well-trimmed privet hedge is evidence of pride in the home.

Flowers borne in rather dense terminal clusters, to 3 in. long, somewhat pyramid-shaped. Individual flowers white or greenish-white, small, with a funnel-shaped corolla that is longer than calyx, which is four-parted and with two stamens. Fruit a blue-black, two-celled, one- to two-seeded berry that persists through winter.

Flowering time June through July. Fruits first available about September. About 13,000 seeds per lb. Propagation by means of seeds, by division of root system, or by simple cuttings thrust into ground and rooting if the soil is wet enough. Plant does well in dry well-drained soil, perferably in the sun.

Primarily of importance as a hedge plant and ornamental and because of its hardiness will withstand considerable neglect or abuse. Fruits eaten by a number of wild birds and are reported in southern Michigan as an important food for pheasants. Horses have been poisoned by browsing on leaves. European children have been killed by eating fruits. Poisoning has been but rarely reported.

Privet

White Ash
Fraxinus americana
Height to 120 ft. Tall, open, straight, with relatively coarse branches standing more or less horizontally. Trunk to 6 ft in diameter. Roots shallow. Twigs coarse, smooth, straight, with broad, crescent leaf scars, and bundle scars in curved line. Buds round, brown to black. Leaves to 12 in. long, compound, of to nine stalked leaflets, whitish beneath. An ash with hairless twigs plus deeply cut leaf scars is usually white ash.
(Continued)

White Ash

Found on rich hillsides, in mixed forests rather than in pure stands. From Minnesota to Nova Scotia, south to Florida, eastern Texas, and Kansas but planted outside normal range for one reason or another. Commonly associated with maples, oaks, and birches. Does not thrive in low, wet lands on which black ash does so well.

Flowers appear before leaves, stamens appearing as close clusters on one tree while pistils are in loose open clusters on another. Pollination by wind. Fruits winged samaras, from ¾ to 2 in. long, with seed at one end; wing almost wholly terminal and seed portion almost round in cross section. Fruits 8,000 per lb, with 35 to 50 percent germination.

Wood light brown, tough, strong, heavy, hard, with thick, lighter-colored sapwood; used in making tool handles, oars, furniture, boats, carriages, and the like. Fruits collected in September to December are buried in sand, planted in spring at ½-in. depth. Seedlings reach to 9 in. in 1 year, to 6 ft in third season. Stumps sprout to 5 ft first year.

Fine timber tree where conditions are suitable and often trunks rise straight and uniform to surprising heights. Tree is sometimes used as a shade tree but compound leaves produce an untidy trash that makes it unpopular. Wood makes superior fuel and commercial wood. Good campfire fuel; similar to oak and hickory.

Red Ash
Fraxinus pennsylvanica

Red Ash

Tree. Height to 60 ft. Trunk diameter to 2 ft. Compact, irregular head, with slender twigs that are round in cross section, conspicuously fuzzy, and with semicircular leaf scars. Leaves compound, of five to nine leaflets, each to 6 in. long and to 2 in. wide, green on both sides, petioled, and sharp-pointed. An ash growing in a wet location and having narrow toothed leaflets and with leaf scars nearly straight on their upper edges and with hairy new twigs and leaves is red ash.

Native of North America. Found mostly in swamps and wet places from New Brunswick to Florida, west to Minnesota and Texas. Grown somewhat outside its normal range as an ornamental or for other reasons. About 50 species of the genus found mostly in north temperate regions; over 20 in North America.

Flowers: staminate and pistillate on separate trees, appearing late in spring about as leaves begin to appear. Each staminate flower has two stamens; each pistillate flower appears like a deeply divided cup. Fruit borne in an open, smooth cluster, winged, with wing extending to about middle of fruit body.

Wood weighs to 44 lb per cu ft, strong, brown, tough, and split in spring into splendid material for making baskets and cane seats. Grows rapidly when young but slower when mature, with about 11,000 seeds per lb having 50 percent germination; green ash yields 3,000 usable plants per pound of seed.

Wood used extensively in basketry but on whole less important than that of white ash. Rarely attacked by fungi or by insects. Seedlings reach height of 4 in. first year. Plants in one condition or another provide food for a number of birds and deer, rabbits, and hares. Known as water ash, river ash, swamp ash. Green ash is a glabrous form, *subintegerrima*, of *F. pennsylvanica*.

Black Ash
Fraxinus nigra

Black Ash

Tree. Height to 100 ft. Trunk diameter to 2 ft. Bark ash gray, without deep ridges but with thin scales. Twigs smooth, dull, with black buds and circular or semicircular leaf scars. Leaves opposite, compound, of 7 to 11 leaflets, each to 6 in. long and to 1½ in. wide, stalkless and with rounded bases.

Native of North America. Found in wet woods and swamps and planted along streets. From Newfoundland to Virginia, west to Arkansas and Manitoba. Differs from other ashes of its territory in presence of black buds, leaflets without petioles, and because fruit wing extends all around the "seeds."

Flowers: staminate and pistillate on separate trees, or sometimes on same tree, purple, calyx absent, appear in May or about same time as leaves. Fruit winged, to 1½ in. long, with wing running all way round "seed" but at one end just the same, carried by wind and shed in fall.

Wood weighs to 39 lb per cu ft, soft, weak, dark brown. Root system conspicuously shallow. Tree easily overturned by wind, easily damaged by fire but avoided by fungi and insects. Seeds weigh 3,000 per lb and have low vitality. Spring cankerworm may attack leaves; ash wood borer, the lumber.

Used as a shade tree and ornamental. Wood used in interior finish, in cabinetwork and in making barrel hoops and baskets. Also generally makes a good firewood. Does not succeed as a hedge plant. Cottontail rabbits feed on bark and leaves and evening grosbeaks on fruits in winter. Also known as hoop or basket ash.

Golden Bell, Forsythia
Forsythia viridissima

Shrub. Height to 10 ft. Branches erect or in graceful arches. Twigs remarkably uniform in diameter for long distances, ridged, yellow-brown. Buds clustered at joints. Leaves opposite, to 6 in. long and to about ½ that width, tapering at each end, entire or irregularly notched above middle. Twigs with conspicuous lenticels which look like corky warts.

Native of China. Widely grown as an ornamental in temperate parts of world. One of most popular of hedge plants. Of four species that originate in China, Japan, and southeastern Europe, *F. suspensa* has hollow branches, this species has generally erect branches, and *F. intermedia* has conspicuously arched branches.

Flowers bright greenish-yellow, with one to five in axils of leaves, with a four-parted calyx and a deeply four-parted corolla that in this species is less likely to open widely than in *F. intermedia,* with two stamens fastened to inside of corolla tube. Fruit, a woody capsule that splits to free many winged seeds.

Flowers appear before leaves; are to 1 in. long and spectacular. Easily propagated by using cuttings of fresh green wood, or by layering drooping branches, or by use of seeds. If undisturbed, drooping branches may themselves take root and plant may thus occupy considerable space.

One of most valuable and popular of hedge plants because of beauty of flowers in early spring, dense thicket formed by branches, and clean appearance of dark green leaves. Winter twigs bearing flower buds may be made to produce flowers by bringing them into the house early. May flower in fall. Most cultivated forsythias are hybrids of *F. intermedia,* a cross between *F. viridissima* and *F. suspensa.*

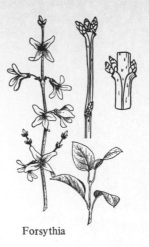

Forsythia

Lilac
Syringa vulgaris

Tree or shrub. Height to 20 ft or more. Rather profusely branched, smooth, with light gray bark. Usually two buds at ends of twigs. Leaves opposite, to 4 in. long and ⅔ that width, somewhat egg-shaped but pointed at tip, usually bright green, thin, not evergreen, heart-shaped at base.

Native of eastern Europe but widely grown as an ornamental and so hardy that it survives in homesteads long after other evidence may have vanished. One of commonest of ornamentals in old-fashioned gardens. May escape from cultivation. About 30 species native of Europe and Asia, but widely grown.

Flowers lilac or white and well-known, most fragrant, borne in rather compact clusters that may be 8 in. or more long, with individual flowers about ½ in. long. Calyx and corolla both in four parts. Stamens two, fastened to inside of corolla tube. Fruit a dry, somewhat woody capsule, to ¾ in. long, splitting to free seeds.

Flowers April or May; may persist until early June. May be forced by bringing plants into greenhouse and holding at temperatures of about 60°F for a few days, then to 88°F, increasing gradually while clusters develop, then lowering to 66° when first flowers open. Many forcing tricks developed by gardeners.

Valuable plant for greenhouse, for use as cut flowers or as houseplants, or possibly more valuable as a "friendly" sort of hedge ornamental, particularly if lower areas can be filled in with some low species. Leaves in summer are commonly covered with a powdery mildew that does no particular harm. State flower of New Hampshire.

Lilac

Fringe Tree
Chionanthus virginicus

Tree or shrub. Height to 30 ft or more. Trunk diameter to 10 in. Branches stout, ashy to brown, forming an oblong head. Bark to ½ in. thick, with small thin brown scales tinged with red. Leaves opposite, to 8 in. long and to 4 in. wide, thick, firm, dark green above and paler beneath, yellow in fall.

Native of North America. Found ranging from Pennsylvania to Florida, west to Texas, Oklahoma, and Arkansas. Three species in the genus native of North America and China, of which this one may be rather commonly planted as an ornamental particularly in eastern United States and western Europe.

Flowers white, appear when leaves are about ⅓ grown, in long drooping open clusters that are to 6 in. long and lengthen in early summer to make the fringe. Each flower to about 1 in. long, slightly fragrant; some flowers with both stamens and pistils, but sometimes these elements are on different trees.

Fruits ripen in September, are like dark blue to black stone fruits, to ¾ in. long, with a bloom that may be rubbed off. Plants do best in moist sandy soil in sun. Propagation by grafting onto ash seedlings, by layering, by cuttings, or by seeds sown in fall in layers of sand and transplanted as seedlings.

(Continued)

Fringe Tree

Fringed Gentian

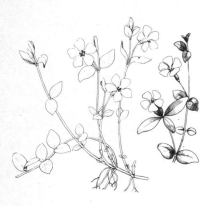

Myrtle

Spreading Dogbane

An ornamental with limited and local popularity.   About 2,000 seeds per lb.   They are not listed as being commonly eaten by wildlife.   They have been found in the stomachs of pileated woodpeckers.   May thrive on soils sourer than those usual for most trees and shrubs.

Order Gentianales./Family Gentianaceae

Fringed Gentian
Gentiana crinita
Annual or biennial.   Height to 3½ ft.   Erect, relatively slender, somewhat four-angled stems.   Leaves opposite or approximately so, lower ones being widest toward tip, and upper, widest near base; smooth, entire, to 2 in. long; upper leaves with clasping bases.   *G. procera*, narrow-leaved fringed gentian, is similar but has very narrow leaves and the lobes of the petals are toothed rather than lobed.

Found about springs or in moist meadows and woodlands.   From Quebec to Georgia, west to Iowa and Dakota.   Considered one of the rarest of the desirable wild flowers.   Some 400 species of this genus occur in the North Temperate and Arctic Zones.

Flowers, several, borne at tips of terminal branches.   Calyx unequally four-cleft, with sections as long as tube of corolla.   Corolla a delicate blue or white, to over 2 in. long, spreading at tip, with petal lobes beautifully fringed all around the edge.   Fruit spindle-shaped.   Seeds numerous, scaly.

Reproduction by seeds only, therefore promiscuous and extensive collection of the flowers easily leads to plant's extinction.   Remaining stations in an area should be closed to collectors.   Fortunately, the plant can persist once it is established but it is very difficult to establish.   Seeds do not survive drying.

The flowers are open only when the sun is shining and close tightly in the shade.   For this reason if for no other they are at their best in their natural setting.   Will not adapt to planting in a wild-flower garden.

Order Gentianales./Family Apocynaceae

Periwinkle, Myrtle
Vinca minor
Trailing stems.   From 6 in. to 2 ft long.   Perennial.   Leaves opposite, evergreen, green on both sides, entire, short-petioled, to 2½ in. long and to 1 in. wide, firm, narrowed at base, distinctly glossy.   Stems root rather freely at joints to form a dense mat that usually excludes competition.   *V. major* is similar, but larger, with erect stems up to 3 ft tall and flowers 2 in. across.

Introduced from Europe but firmly established in America where it was grown by early settlers.   Often at its best in old cemeteries or on abandoned homesites.   Grown as a border plant in gardens and found from New England to Georgia and westward.

Flowers usually borne singly in axils of leaves, not too abundant on a given stem, but massed may appear to be so, blue, with wide spreading lobes making whole flower to over 1 in. across.   Calyx lobes deeply parted, smooth, green.   Stamens with anthers connected.   Fruits slender, few-seeded, splitting freely when mature.

Easily propagated by division of horizontal rooted stems.   These are merely buried lightly in soil.   A clump of plants every foot or so in newly graded, relatively shaded soil may spread to cover the whole surface to the exclusion of other species, particularly if weeding is carried on for the first year or so.

Justly popular as a ground cover for steep banks and for areas where shade is too deep to support many other species.   Flowers February through May.   Shining leaves present throughout the year.   Ability to form a pure stand makes it attractive and a good soil anchor.   Used in · erosion control.

Spreading Dogbane
Apocynum androsaemifolium
Height to 4 ft.   Branched, freely spreading, smooth, tough-stemmed.   Leaves to 4 in. long and to 2½ in. wide, opposite, smooth above but pale and somewhat fuzzy beneath, with a short blunt point at tip, milky-juiced.   Rootstock substantial, horizontal, buried, perennial, spreading.

Found in waste fields, thickets, and flood plains.   From Anticosti Island to Georgia, west to Arizona and British Columbia, reaching an elevation of 3,500 ft in Virginia.   Related *A. cannabinum*, Indian hemp, has more closed flowers and narrower, more distinctly petioled leaves.   11 species, mostly found in temperate North America.

Flowers pink, like bells opened suddenly and widely, somewhat angled, to ⅓ in. long, with calyx tube considerably shorter than tube of corolla as contrasted with longer calyx in *A. cannabinum*, where it about equals corolla tube.   Fruit a pair of slender inward-curving pods that burst to free many seeds.

Flowers June through July. Pollinated by bees and butterflies. Visited intensively by the metallic red and green dogbane beetle, *Chrysochus auratus.* Considered a poisonous plant but not a very attractive food to grazing forms.

Poisonous principle may be the glucoside cymarin, a poisonous resin, and probably other elements. A crude fiber made from the bark. Milky juice has been considered a possible source of rubber. Young plants eaten by cattle are emetic and cathartic in their effect. Known sometimes as "Indian hemp."

Common Milkweed

Order Gentianales./Family Asclepiadaceae

Common Milkweed
Asclepias syriaca
Height to 5 ft. Stem stout, usually unbranched, tough-barked, often fuzzy in upper parts. Leaves opposite, to 9 in. long and to 4 in. wide, dark green above and lighter beneath, with main veins widely spreading and conspicuous, petioles stout. Root system to 15 ft. long, just under surface; proliferates new plants, forming colonies by vegetative propagation.

Native. Found in fields, along roadsides, in waste places, and sometimes in gardens. From New Brunswick to North Carolina and west to Kansas and Saskatchewan. A few varieties are recognized, based on differences in leaves, in flower clusters, and in flowers.

Flowers borne in many-flowered ball-like clusters in axils of upper leaves in umbels; from pale brown to lilac, dull crimson-pink to lilac-pink, with a greenish five-parted calyx, rather fragrant, abundantly visited by bees, flies, and butterflies, with anthers free to be torn loose by visiting insects.

Survives dry wasteland. Young shoots eaten raw or cooked, although considered to be poisonous by some persons, are delicious when eaten with spearmint as a sandwich. Cattle and sheep have been poisoned by eating mature plants. Pods stout, to 4 in. long, open to free many parachuted seeds sometimes throughout the winter.

Bark makes an excellent fiber substitute, fluff of seeds used in making life preservers in World War II, milky juice used in search for rubber source, root used in medicine of a homely type, root system serves as soil anchor. Plant is host of beautiful monarch butterfly. Can be kept under control by cultivation and by salting.

Great Bindweed

Order Polemoniales./Family Convolvulaceae

Great Bindweed, Wild Morning-glory
Convolvulus sepium
Stems trailing, twining and climbing, often to length of over 10 ft, sometimes slightly fuzzy or more commonly smooth. Leaves alternate, gray-green, darker above, slender petioled, with triangular or arrow-shaped blade to 5 in. long. Root system of slender light-colored rootstocks.

Native. Found in fields and thickets from Newfoundland to North Carolina, west to New Mexico and British Columbia. Some 200 species found in temperate and tropical areas, with 30 to 40 in the United States. Small bindweed, *C. arvensis,* and others are bad weeds in some areas.

Flowers borne singly in axils of leaves, on stalks about equal in length to the flower that is about 2 in. long, white or pink, with white stripes, funnel-shaped, with widely spreading edge. Calyx has two large enclosing bracts that are absent in smaller flowered *C. arvensis.* Fruit a dry spherical capsule.

Flowers may be cross-pollinated by honeybees or bumblebees or may be self-pollinated. They usually close before noon much as do morning-glories. Flowers June through August. *C. arvensis* has root system that may reach 15 ft underground and is therefore difficult to destroy by ordinary cultivation methods.

Ordinary control of bindweeds is by frequent cultivation at 6-day intervals, cutting the plants to depth of 6 in. followed by planting of dense shade crop like millet, sorghum, or alfalfa; or may be covered with paper or heavy straw mulch; or sprayed with arsenical sprays, sodium chlorate, or other weed sprays. Any chemical herbicide used should be employed with precautions.

Morning Glory

Common Morning Glory
Ipomoea purpurea
Annual. Long, trailing stems binding their supports and reaching to 10 ft in height, relatively slender and weak. Leaves to 5 in. long, dark green above and lighter beneath, broadly heart-shaped at base, short-pointed at tip, entire, with petioles about equal to blades in length.

Native of tropical America but grown extensively to north and generally in gardens and in temperate regions where it may reseed itself and become weakly established. Over 400 known species from various tropical parts of the world. Many races and varieties, some horticultural in nature owing to crossings.

(Continued)

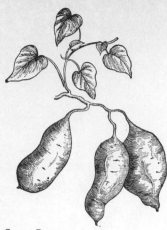

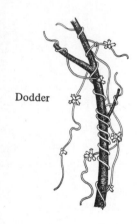

Sweet Potato

Dodder

Moss Pink

Flowers showy, wide-lipped funnels, variously and usually brilliantly colored, opening in morning sun, with tube usually lighter-colored than other parts, to 3 in. long, with stamens and pistils usually shorter than tube and stigma sticking above anthers. Commercial varieties may have blue, purple, crimson, or rose flowers. Fruit, a dry capsule.

Many double-flowered forms have been developed. Used frequently in making studies in genetics because of conspicuous life cycle and fact that plants may be grown in pots easily in house or greenhouse.

A popular, old-fashioned-garden flower. Figure shown does not show typical *I. purpurea* leaves, these being broadly heart-shaped rather than triangular as shown. Related *I. pandurata*, man-of-the-earth, yields roots weighing to 30 lb and edible when roasted. *I. purga* of the tropics is poisonous.

Sweet Potato
Ipomoea batatas
Stems many feet long, trailing, with thin hairs, rooting freely at joints, milky-juiced. Leaves highly variable in different parts of same plant, some being heart-shaped, some oval, and some triangular, but all rather coarse and dark green. Root system, deep-penetrating reddish tubers.

Native of South America. Probably cultivated by native Americans thousands of years before Columbus landed. Known in China before the Christian era. Carried to Europe from the West Indies in 1526 and was grown in Virginia in 1650. Requires long warm growing season with moderate rainfall, particularly at first.

Flowers like those of the morning-glory with funnels of rose or blue, with darker instead of lighter centers, about 2 in. long, with the sepals of the green enclosing calyx each about ½ in. long. Seeds angular, smooth, dark, enclosed in a fragile spherical fruit that breaks readily.

Commercially cultivated by slips, by cuttings of vines, or by separation of roots. Requires 6 bushels of "seed" to set an acre though quantity varies with nature of slips. Large roots produce fewer plants per bushel of seeds than smaller roots do. Roots should be disinfected before planting to control rot.

Georgia leads in production of sweet potatoes with 80 percent of the United States crop usually produced in Georgia, Alabama, Louisiana, North Carolina, South Carolina, Mississippi, Tennessee, Texas, and Virginia. Improved storage and commercial drying techniques make possible a wider use of this valuable food so subject to decay. In parts of the U.S. certain varieties of sweet potatoes that have a soft, moist texture are called yams. Yam usually refers to the similar underground part of *Dioscorea*, a monocot. Ornamental sweet potato "vines" can be grown by suspending one half of a root of sweet potato in a jar of water, leaving the remainder of the root above the surface.

Dodder
Cuscuta gronovii
Annual. Stems slender, yellow to orange, climbing, with haustoria that tap bundles of plants on which dodder grows, these being used not only as supports but as means of getting food. Plant is leafless and not green; therefore, must be parasitic. Contact with ground is lost at maturity.

Found growing on many kinds of plants, both herbs and woody plants. From Nova Scotia to Florida, west to Texas, Montana, and Manitoba. About 100 species of dodder, of which some 30 are found in United States either as natives or as introduced species. This species is rather widespread.

Flowers July and August, in dense clusters on short stems, with bell-shaped corolla only about ¹⁄₁₀ in. long, with its rounded lobes about equal in length to that of tube, these alternating with short stamens that appear at corolla clefts. Fruit a dry capsule, ¹⁄₁₀ in. through.

Mealy brown seeds, somewhat spherical. When they germinate, they develop a seedling that eventually comes in contact with a plant that may serve as a host. Then contact with the ground ceases. *C. planiflora* is a common western species introduced from Europe.

Many species attack particular plants as dodders of alfalfa, clover, hemp, onion, thyme, and hazel, though some of these may not be too particular. Hand pulling, burning of infested plants, and heavy mulches are control measures. Strict laws govern occurrence of dodder seed as impurities of imported commercial seeds.

Order Polemoniales./Family Polemoniaceae

Moss Pink
Phlox subulata
Stems in tufts, from trailing, prostrate stocks, the erect stems being rarely much over 6 in. tall, and often rather profusely branched in upper regions. Leaves to nearly 1 in. long, but only about ¹⁄₁₀ in. wide at most,

pointed, persistent, slender, entire, dark or yellowish-green, spreading, in clusters at joints.

Native. Found on dry, sandy, rocky, or poor soil from New York to Florida, west to Michigan and Kentucky. Frequently grown in mass beds in flower gardens with considerable success. In West Virginia, is found to elevation of 3,500 ft. Between 40 and 50 species of phlox are found native in America and Asia.

Flowers pink, purple, or white with a darker center, with five spreading lobes arising from top of a tube which is about as long as the lobes. Calyx green, with pointed lobes about ½ total length and with whole calyx shorter than corolla tube. Fruit capsule about ¹⁄₁₀ in. long.

An excellent ground cover for poor soil, being at its best in spring when erosion factors are at their prime. One of most popular and relatively easily grown of bedding garden flowers. Flowers April through June. Pollinated probably by bumblebees and butterflies.

A dozen or more garden varieties of this species, based on habit and color. Apparently the wild plants are buried in other vegetation in the late summer when they are not so conspicuous.

Perennial Phlox

Perennial Summer Phlox
Phlox maculata
Height to 3 ft. Stem slender, erect, smooth or slightly fuzzy, branched or unbranched, often marked with purple dots. Leaves opposite, smooth, firm, to 5 in. long, widest near rounded base, pointed at tip, the lower being sometimes almost linear. Perennial.

Native of moist woodlands and along streams. From Connecticut to Florida, west to Mississippi and Minnesota. Commonly grown in gardens everywhere, particularly in northern part of range where it is a usual summer garden flower. Regular garden phlox is *P. paniculata*, which has less pointed calyx teeth and blunter leaves; found wild from New York to Georgia and Arkansas and blooms from July to September. Different varieties show a wide range of color: white, rose, crimson, scarlet, lavender, and purple.

Flowers borne at tops of branches in clusters of a few, but with many clusters in a given area and each cluster developed along short lateral stems. Corolla white, blue or purple, with a tube that suddenly and widely expands. Stamens usually short and enclosed in corolla tube. Pistil longer. Fruit a capsule to ½ in. long.

Common phloxes include annual *P. drummondii*, the early wild *P. divaricata*, the tall summer perennials *P. maculata* and *P. paniculata*, the rock-garden *P. multiflora*, and the moss pinks *P. subulata*, all of which are popular in their particular time and places. The garden species usually need room and water.

Heliotrope

Perennial phloxes do not normally reach their full blooming vigor until some 3 years after they have been transplanted. Old clusters may be divided every 2 to 3 years but this delays full blooming, so the general practice is to leave some plants for blooming while others are divided for new plantings.

Order Polemoniales./Family Boraginaceae

Common Garden Heliotrope
Heliotropium arborescens
Small trees, vines, or shrubs in their native land. In America, best known as low shrubby plants of the garden or greenhouse. Leaves oval to oblong, with conspicuous lighter veins. In *H. peruvianum*, leaves are not narrowed at the base as they are in *H. corymbosum*. Leaves usually dark green and roughened. Roots strong and fibrous.

Native of Peru, with upward of 250 related species in the genus, all of which are native in the warmer regions. Now widely distributed, and many are popular as border plants or in greenhouses. When grown outdoors, this species favors strong sunlight and normally rich soil.

Flowers about ⅛ in. long, violet or purple, with most pleasant vanilla scent. Some forms are white-flowered. Flowers arranged in compact rather long recurved formation (cyme). Stamens short and included in tube of corolla. In *H. peruvianum* corolla is little longer than calyx; in *H. corymbosum*, it is twice as long.

May be raised from seed to blooming in 1 year if started indoors. Usually trained to trellis. While they need sunlight when grown indoors, they do not do well if temperature rises much above 45°F, in winter months. Many enemies, but temperature most serious control. Attacked by most greenhouse pests.

Heliotropium means turning to the sun, indicating sensitivity to sunlight. Popular old-fashioned-garden flower. Fragrance compensates for usual scraggly appearance of the plant. Flowers used commonly as long-lasting buttonhole bouquet.

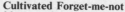

Forget–me–not

Cultivated Forget-me-not
Myosotis sylvatica
Erect or ascending stems, reaching a height of to 2 ft but normally much lower. Green-stemmed, slightly fuzzy, much-branched. Leaves with little or no petiole, rather blunt at free end and tapering at base, and, like stem, apparently hairy.

Native of Europe and northern Asia, but grown in America commonly. Frequently established in gardens and along waysides as escapes from cultivation. Some 30 or more species, of which at least a half dozen are native of eastern United States, with fewer in western and southern parts. *M. scorpoides* reaches height of 9 in.; blooms through spring and summer, is suitable for planting at waterside or in bog garden.

Flowers borne in terminal clusters on small stems that are longer than deeply cleft calyx; whole cluster rather conspicuously loose and open. Corolla pale blue with a yellowish center, to about ⅓ in. across. In some varieties, flowers are pink or white. Fruits, four, margined nutlets.

Common garden plant used to fill in bare spaces and sometimes troublesome to keep under control. Favors shade or wet places at borders of shade. Many wild species do rather well in swamps partly submerged in water. Seeds may be sown in August and young plants given winter cover.

Among the recognized garden varieties are white gem, distinction, grandiflora, Victoria, perfection, and so on. Plants may be forced in greenhouses and set out in mats, or whole mats may be transplanted after the flowering time has passed.

Virginia Bluebell
Mertensia virginica
Height to over 2 ft. Stems ascending, branched or unbranched, rather stout, smooth. Leaves alternate, smooth, pale green, to 5 in. long and 1½ in. wide, usually more narrowed at base than at free end, with rather conspicuous veins that join inside margin, entire, rather distinctly petioled.

Virginia Bluebell

Native of North America, where it is found in wet woodlands and at stream margins. From Ontario to South Carolina, west to Kansas, Nebraska, and Minnesota. Sometimes grown in a wild-flower garden. Some 40 species of the genus (all of which are native to the Northern Hemisphere) in North America, Europe, and Asia. Some of most important western North American kinds are *M. laevigata, M. longiflora, M. nutans, M. oblongifolia,* and *M. platyphylla.* Generally these western species are best left in their natural setting as they do not do well in cultivation.

Flowers beautiful blue bells, with relatively slender tubes that may be held erect at first but later droop, borne in loose terminal clusters, blue to purple, and showy, to l in. long, with the calyx short and sepals about ¹/₁₀ in. long. Fruit dull, rounded, roughened nutlets.

Plants in gardens should remain undisturbed for years, not being dug up and divided as is practice with so many garden plants. Best propagated by seeds that are sown almost as soon as they are ripe. Leaves of the year disappear shortly after the flowering period, which is from March through May.

Justly popular wild-flower-garden plant. Unusual as smooth member of borage family, most species of which are extremely rough. At best when grown in masses, particularly when found in a mass growing in their natural environment. Reported by some to have a slight medicinal value but not considered authentically valuable.

Order Polemoniales./Family Verbenaceae

Garden Verbena
Verbena hybrida
Height to 2 ft. Stems may arise from a more or less creeping base. Entire plant grayish, with long stiff spreading hairs. Leaves opposite, with margins with rounded or slightly lobed teeth, petioled, dark, with rather conspicuous veins providing attractive foliage. Roots usually tough and fibrous.

Verbena

Native of Brazil. Introduced into United States in 1840. About 100 species in the genus, chiefly native of tropical America, but some are weeds in fields and waste places on either dry or wet ground north into Canada. *V. hybrida* favors rich, well-drained loam and does best when exposed to full sunlight.

Flowers in flat heads, on long floral stems, pink, blue, white, purple, or mixed, tubular, with right-angled border, slightly fragrant. Stamens, four. Ovary, one-celled. Fruit, dry, ribbed nutlets. Closely related *V. chamaedry-folia* has scarlet flowers and appears greenish rather than gray at a distance. Annual.

Seeds germinate in about 8 days but may live 3 years. Seedlings raised indoors in hotbeds in March, or seeds sown outdoors after danger of frost. Transplant to 1 ft apart. Cultivate lightly, pinch back plants to improve them. Pick flowers to prolong blooming. Common enemies, thrips and aphids, controlled by nicotine sulfate.

Good for beds, window boxes, or use as cut flowers. Bright colors, fragrance, and durability make plant popular both indoors and out. Late-blooming habit helps popularity. A leaf tier pest is controlled with arsenate of lead; a red spider, with water spray, and verbena bud moth by picking off affected buds.

White Verbena
Verbena urticaefolia
Height to 5 ft. Stems slender, erect, tough, branched, four-sided, finely hairy. Leaves opposite, to 5 in. long and to 1 in. across, with veins joining inside saw-toothed margin, thin, rough, rather long-pointed, rounded at base, with relatively long petioles, particularly in lower ones.

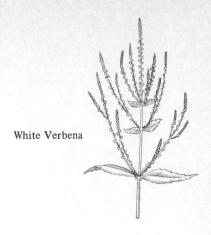

White Verbena

Native of North America. Found in wet fields and waste places from New Brunswick to Florida, west to Texas and North Dakota. About 100 species of the genus native of the Americas, with one in Mediterranean region and nearly 30 in North America. Some are grown as ornamentals while others are serious pests.

Flowers borne on slender erect spikes at tips of branches, arranged with spaces between flowers rather than crowded as in blue verbena *V. hastata* or the bracted verbena. Flowers white, about $\frac{1}{10}$ in. across, with spikes to 6 in. long. Sometimes flowers are blue or purple. Fruit oblong, brown, to $\frac{1}{10}$ in. long.

Flowering time June through September but dead tops persist through winter, providing food for birds and a persistent source of seed infection. Substantial underground parts help plants survive severe conditions but they do not survive frequent burning nor are they drought- or shade-resistant.

Control is by burning or plowing. However, in some lowlands the underground parts may provide some desirable soil anchorage. The fruits of the blue vervain are gathered by Indians of California, roasted and eaten as a bitter flour. In the Midwest blue vervain is an "indicator" plant of poor pasture management.

Lantana
Lantana camara
Height to 6 ft, or in tropics to 20 ft, where it forms a dense impenetrable thicket. Stems sometimes with short, hooked prickles; woody when old. Leaves opposite, to 6 in. long, rather thick, rough particularly above, fuzzy beneath, short-petioled, with a strong aromatic odor. Roots coarse and fibrous.

Lantana

Native of tropical America but ranging wild north to Texas and southern Georgia. In most cases, plant is known from pot plants. For instance, *L. montevedensis* or *(delicatissima)* is excellent for growing in hanging baskets. Some 50 species, most of which are American but some are found in the Old World. In areas such as Hawaii, lantanas along paths may be impenetrable.

Flowers in round flat-topped clusters with a green bract at base of each. Flowers to $\frac{1}{2}$ in. across; heads, to 2 in. across. Usually pink or yellow at first but changing to orange or scarlet. Bud folded like an envelope. Stamens, four. Pistil, two-celled.

Propagation by seeds or by stem cuttings. Started usually in early summer in soil that should be not too rich. Should be exposed to the sun and temperature should be from 60 to 70°F. Only moderate amount of water needed. Few enemies of plant when grown in house but many including some galls affect wild plants.

Best known in temperate regions as a houseplant that can withstand neglect and yet appear to be healthy and vigorous. Old plants taken from garden in late summer may be cut back vigorously, potted, and brought into the house to provide a welcome winter-blooming houseplant.

Order Polemoniales./Family Labiatae

Horehound
Marrubium vulgare
Height to 3 ft. Stems ascending, white, hoary, aromatic, square in cross section. Leaves opposite, rough, conspicuously veined above and woolly beneath, roundish oval, with a narrowed petiole $\frac{1}{2}$ to 2 in. long, bitter-juiced. Roots perennial and fibrous.

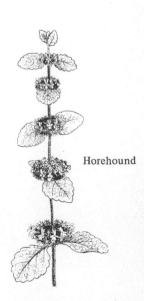

Horehound

Native of Europe but well-naturalized in America, where it has long been established as a weed from Maine to North Carolina, west to California and British Columbia, being most abundant along Pacific Coast. Cultivated in America and abroad, particularly in Europe and Asia. Favors dry sandy wastelands.

Flowers crowded in dense clusters at bases of leaves, relatively small, white, with calyx bearing 10 teeth, alternating large and small and each ending in a hooked spine. Stamens, four. Corolla two-lipped, the upper

(Continued)

287

being erect and commonly notched. Fruit deeply four-lobed, making nutlets.

Propagated by seeds or by division. Seeds sown in spring in cold frames are, when seedlings, transplanted to stand 6 to 18 in. apart in rows, the closer rows yielding finer stems and better quality. Harvesting is just before August-September flowering, yielding, when good, 1 ton of dry herb per acre.

United States normally imports about 100,000 lb annually for use in hot-water infusions and in lozenges for treatments of colds, rheumatism, dyspepsia, and other troubles. In large doses, it acts as a purgative and in this connection is used in treatment for worms. Not recognized as an official drug. Used much in candy. May become a persistent weed as an escape.

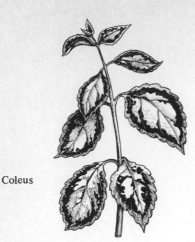

Coleus

Coleus
Coleus blumei
Herb or semishrub. Height to 3 ft. Branched or unbranched stems that are weak and square in cross section. Leaves of various shapes and colors, opposite, nearly regularly toothed, the teeth being rounded; variously and conspicuously colored with yellow, purple or red. Roots fibrous.

Native of Java, though averge greenhouse plant may have been developed from many stocks. Common crisped-leaved form is variety *verschaffeltii;* has more brilliantly colored and denser foliage. Related forms come from India and from Africa. Important varieties have red and yellow or yellow and green leaves. About 90 species.

Flowers in long terminal blue or whitish clusters like spikes. Calyx small and green, five-toothed, persistent after corolla is shed. Corolla two-lipped with the lower the longer. Stamens four, united at base with round pollen sacs. Pistil with two two-lobed carpels. Nutlets smooth.

Commonly propagated by cuttings made in late summer, rooted in sand that is kept moist through winter, then transplanted. Since plant is sought for its foliage rather than for its flowers, forcing techniques are not so important. Does best at a temperature of 60 to 75°F, where there is abundant moisture.

Common potted plant or border plant that does well also in window boxes if properly cared for. Best to grow new plants each year since cutback old plants are scraggly. Requires direct sunlight for best colors. Mealy-bug enemies controlled by nicotine fumigation, or spray with fish-oil-coal-tar-nicotine spray.

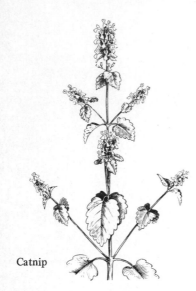

Catnip

Catnip
Nepeta cataria
Height to 3 ft. Stems erect, spreading, branched, grayish-green and lighter than most mints. Leaves opposite; dark above and light beneath, to 3 in. long, with rather deeply scalloped or notched margins, sometimes clustered close to ground, rather long-petioled, not stiff, rather thick. Substantial root system.

Naturalized from Europe. Established from New Brunswick to Georgia, west through Utah to Oregon; also in Cuba. Native of Asia as well as of Europe. Some 150 species of the genus mostly native of Europe and Asia but this is probably one of the better known.

Flowers in relatively small coarse spikelike clusters at ends of branches. Clusters to 4 in. long. Flowers white to purple or pale blue, with purple markings on dots. Corolla to ½ in. long, with main lobe a little longer than calyx. Insect-pollinated. Fruit four brown nutlets, with paired white spots.

Leaves and tops used in drug trade. Propagation by division or by seeds. Seeds sown in rows in late fall or spring; 4-in. plants thinned to 1 ft apart. Shallow cultivation encourages growth. Tops harvested when in full bloom and cured quickly to preserve color, with a normal yield of 1 ton of dried tops per acre. Mauve catmint, *N. mussinii* or *faassenii,* is a favorite for perennial borders or edging.

Free from insect and fungous pests. Oil extracted and used by trappers. Tops used medicinally as a mild tonic. Fruits found as impurities of commercial seeds. Not a persistent weed pest, since cultivation controls it adequately. Cats are affected by the plant. They eat the leaves and rub against the plant. Leaves are collected, dried, and packaged in cloth bags which are used by cats as playthings.

Gill–over–the–ground
Creeping Charley

Gill-over-the-ground
Glecoma hederacea
Sprawling stems. May reach length of 18 in., with short branches turning upward. Stems root freely, usually at joints. Leaves opposite, to 1½ in. in diameter, with scalloped heart-shaped or round blades on petioles as long as or longer than blades, green on both sides, with distinct veins and strong odor.

Native of America. Found through United States and southern Canada. Close relation to catnip is not superficially obvious nor is it recognized by all botanists. It does well in all sorts of waste places but particularly where there is a good supply of moisture and little sun.

Flowers borne in axils of leaves in few-flowered clusters to nearly 1 in. long, with corolla tube to three times as long as calyx. Corolla blue and white or with dots that are darker. Upper stamens much longer than lower pair. Apparently pollinated by bees. Fruits, oval.

Flowering time March through June. Plant sometimes becomes a pest in lawns and flower gardens where moisture and light conditions are right. On rare occasions, it may establish itself in croplands but usually such lands are not suitable for the plant.

Control effected in lawns by close mowing after raking up the tops, or by sprays of iron sulfate, sodium chlorate, or the newer weed-killer sprays. Infested fields may be cleared by plowing. Known as creeping Charlie, robin runaway, crow-victuals, tunhoof, snakeroot, hedge-maids, gill-ale, and cat's foot.

Heal–all

Heal-all, Selfheal
Prunella vulgaris
Height to 2 ft. Stems erect, ascending or sometimes sprawling; branched or unbranched; smooth or finely fuzzy; four-angled, rather stout. Leaves opposite, to 4 in. long, petioled, with larger leaves usually above, with coarsely but shallowly toothed margins. Stems commonly root at joints.

Native of Europe but established throughout most of North America, where it was also probably native. Also native of Asia. Only one of five species of the genus that is established in North America. May become established in patches in lawns and crowd out all other vegetation.

Flowers borne in rather compact spikes at ends of branches, spikes being to 1 in. or more long and to ½ as thick when in flower, or to 4 in. long in fruit. Flower to ½ in. long, purple, violet, or sometimes white, with corolla about twice length of green or purplish calyx. Fruits brown, pointed.

In past, was considered to be of some medicinal value but not now. Infested lawns may have to be dug up and reseeded but in fields ordinary cultivation and plowing keep it in check. Sprays of iron sulfate or similar substances are effective as controls.

The many common names of the plant attest its conspicuous cosmopolitan nature. It is known as heart of the earth, sicklewort, bluecurds, dragonhead, brownwort, carpenter's weed, thimble flower, hookheal, and hookweed, as well as selfheal and heal-all.

Motherwort

Motherwort
Leonurus cardiaca
Height to 5 ft. Stem relatively slender, usually well-branched, conspicuously four-angled, erect or sprawling at base. Leaves opposite, long-petioled, relatively thin, 3- to 5-cleft with pointed segments, to 4 in. wide, dark green even in winter when hardy rosettes persist. Strongly scented at all times.

Naturalized from Europe but native also of Asia. Widely established in waste places from Nova Scotia to North Carolina, west to Kansas and Montana or farther. Three of the ten Eurasian species are established in North America but this is the most typical and abundant.

Flowers in axils of leaves but usually shorter than petioles. Corolla pink, white, or purple, to ½ in. long, with a ring of hairs within tube, with lower lip mottled, white, and woolly on outside. Stamens protrude from corolla throat. Calyx persistent, with stiff teeth. Pollination by bees. Fruit, four-angled nutlets.

Requires water but comparatively little sun. Underground rootstock permits persistence but this may be destroyed easily by plowing or by cultivation. Green currant aphids live in summer on motherwort as an intermediate host; because of this, the plant may well be destroyed.

Once believed to have considerable medicinal value but now believed to be worthless in that respect. It is also known as lion's-ear and cowthwort.

Garden Sage

Sage
Salvia officinalis
Shrub or semishrub. Height to 1 ft. More or less white woolly. Leaves opposite, entire, long-petioled; the lower white with abundant wool beneath, to 2 in. long, roughened with rather conspicuous veins. Plant appears to be an irregular mass of ascending stems springing from a common root base.

Native of southern Europe, where, in Dalmatia, it is raised extensively. Throughout Mediterranean region it has long been under cultivation. Old in America. Does well in any rich soil. Related to such sagebrush-

(Continued)

Scarlet Sage

es as *S. carnosa,* Great Basin blue sage; *S. columbariae,* China; and *S. mohavensis,* Mohave sage of the Southwest.

Flowers purplish, blue, or white, with corolla about ¾ in. long and with an interior hairy ring; arranged in whorls about stem, somewhat bell-shaped with thin but conspicuous calyx; bloom during summer months. Some varieties have yellow, red, or white leaves. Fruit four-parted. Seeds germinate readily without acid treatment.

Plant in well-drained, fertile soil, in early spring, in rows 2 ft apart; thin to 12 in. apart, or set as cuttings that had winter mulch cover. Collect leaves at maximum growth, dry them away from sun, but stir to prevent mold. Fair crop first year, but better succeeding years; yields to 1 ton dried tops per acre.

Volatile oil contains pinene, cineol, and salvoil, all having warm bitter sharp taste, used in flavoring flat foods. Gum and resin also produced. Oil obtained by water distillation by alcohol extraction. Used as stimulant or tonic or in infusion; standard portion is 1 tsp of leaves to 1 cup of boiling water per day.

Scarlet Sage
Salvia splendens
Height to 3 ft. Stems shrubby or almost so, branched, light green, smooth or with fine hairs. Leaves opposite, to 3½ in. long, pointed, with conspicuous veins, rather well-petioled, with brilliantly colored floral leaves and bracts of the flowering portion, and with both leaf surfaces smooth.

Native of Brazil. Grown extensively even in temperate regions as an annual. Most commonly seen in North in mass planting in gardens in late summer, in window boxes, or sometimes in sun parlors. Some 500 species in the genus found in warmer parts of both hemispheres.

Flowers borne in terminal, rather open, somewhat wandlike clusters, with calyx and corolla a brilliant scarlet and corolla to 1½ in. long, with two to six flowers in a whorl. Actually, calyx is brilliant red and persists and corolla is relatively unimportant in the showy color. Some varieties have purple, crimson, or white flowers but commonest color is scarlet. Stamens uniquely hung from the middle, thus effecting clever pollination.

Does best when grown in strong sunlight and may display full bloom from July through September. Ordinarily seeds are sown indoors in early spring so that seedlings are ready to transplant after last frost. Does not do well in soil rich in nitrogen or when season is too wet or dark. Late plants may be potted for display a few months.

One of most popular of mass ornamentals but must be protected from aphids and spiders. Flowers are illustrated in many botany texts because of unique stamens and plant is usually available for classroom study in the fall. Over 50 species and many more varieties are recognized by horticulturists as valuable ornamentals.

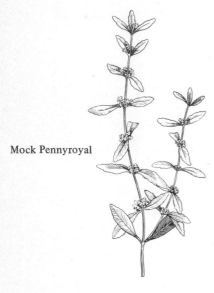

Mock Pennyroyal

Mock Pennyroyal
Hedeoma pulegioides
Height to 18 in. Stem slender, tough, branching, ascending from sprawling bases, finely fuzzy, fragrant. Leaves opposite, thin, petioled, to 1½ in. long and to ¾ in. wide, with upper much smaller, blunt at tip and narrower at base, rather pale green but in winter dropping off and leaving slender stems. Annual.

Native of North America. Found from Nova Scotia to Florida, west to Nebraska and the Dakotas, being found in dry fields and on exposed banks. Over a dozen species native of North America. The pennyroyal of commerce is *Mentha pulegium,* a native of Europe and Asia.

Flowers in whorls at bases of leaves; rarely as long as leaves, more or less erect but not in compact clusters. Calyx fuzzy, swollen at base, with five teeth, two of which are larger than others. Corolla purplish, to about ⅓ in. long. Stamens with or without anthers but more often without. Fruits, small nutlets.

Grows well on any dry, sandy or gravelly soils. Seeds sown in rows in fall and covered to not more than ¼ in. depth. Seedlings appear in early spring. Weeds should be kept out. The plants should be cut in early summer when in full bloom, dried quickly in shade, and after all large stems have been removed, should be packed. Oil is distilled by steam from plants without first drying.

Under cultivation, 1 acre may yield to 1,200 lb. Oil is used in soaps and medicines and as a supposed flea and mosquito repellent.

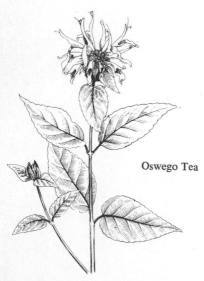

Oswego Tea

Oswego Tea, Bee Balm
Monarda didyma
Height to 3 ft. Stem stout, erect, branching, square in cross section, smooth or fuzzy. Leaves opposite, thin, petioled, conspicuously dark green to blue-green, rather long-pointed at tip and blunter toward base, to

6 in. long and to 3 in. wide, with longest petiole to 1 in. long. Upper leaves and stems often tinged with shades from bronze to dark red. Substantial perennial roots.

Native of North America. Found from Quebec to Georgia, west to Tennessee and Michigan growing in moist places along roadsides or streams; in North Carolina found up to 5,200-ft elevation. About a dozen species of the genus to be found in North America including Mexico, with this best known in the East.

Flowers usually in only one terminal whorl to a branch or to a plant. Whorls supported by green bracts. Calyx green, smooth or fuzzy, with teeth about as long as the corolla tube. Corolla brilliant scarlet, to 2 in. long, with two stamens whose anthers extend beyond the upper lip although they are exceeded by the pistil. Very showy when flower head is in bloom.

A beautiful wild flower worthy of all protection but adjusting well to transplantation in suitable soil in gardens. Though native to East, it is raised as an ornamental over a much wider range. Easy to transplant.

Considered a good honey plant, with honeybees apparently seeking it wherever possible. As an ornamental it is popular for its late summer or early autumn bloom. A variety with large heads of salmon pink flowers; in the Middle West this species is supplanted largely by *M. fistulosa,* the wild bergamot, with pale blue or lavender flowers.

Spearmint

Spearmint
Mentha spicata
Height to 2 ft. Stem arises from underground stems, erect, nearly smooth, square, leafy throughout with ascending branches. Leaves without petioles or with very short ones, normally under 2½ in. long, with saw-toothed margins, opposite, conspicuously ribbed, more or less green throughout, or lighter beneath.

Native of Europe and of Asia. Widely naturalized in North America. Particularly common in wet places about old gardens in eastern United States and in Middle West where it is often cultivated. Known in Biblical times: in Matthew 23:23 the Pharisees gave tithe in mint. About 30 species in the genus in North Temperate Zone. Spreads by underground stems and readily escapes to grow wild.

Flowers in whorls on slender interrupted spikes, central spike to 4 in. long, longer than lateral ones. Flower clusters narrower and more pointed than in peppermint. Corolla pale blue, smooth, sometimes purple. Calyx smooth, with teeth about ½ as long as whole calyx.

Grows well in fertile moist soil from seed or cuttings. Tops harvested before leaves begin to fall or after first flowers appear. Oil extracted from whole plant is volatile; has ability to reduce gas in stomach.

Greatest use is as flavoring, particularly in chewing gum, but also used in mint juleps, in sauces particularly for use with lamb, and in delicious jellies. Applied externally, oil may cause a flushed skin but is really harmless. It is an official drug plant, of commercial and medicinal importance in many cases. When found after an extended hike, it is a real pleasure to chew the young leaves.

Peppermint
Mentha piperita
Height to 3 ft. Stems erect, more or less crowded, branched, smooth, aromatic. Leaves without petioles, to 3 in. long, smooth or slightly fuzzy along veins on underside, opposite, rather dark green, spotted with little oil globules, pointed at free ends.

Native of Europe. Widely naturalized in America where it grows in relatively thick stands in water-soaked or well-watered ground. Has been cultivated as a crop in America for more than 100 years. Related Japanese mint *M. arvensis* var. *piperascens* is cultivated in Japan as a source of menthol.

Peppermint

Flowers in thick spikes to 3 in. long, central one often being shortest. Corolla purple or rarely white, smooth, mintlike. Calyx tubular, five-toothed, resin-dotted. Flowers July through September, but rarely conspicuous.

Propagation by roots and runners set 3 in. deep continuously in rows 3 ft apart, on any good cornland. Does well on muckland or on loam, with 12 tons per acre of well-rotted manure. May yield for a number of years 3 tons of dry material per acre.

Oil used in candy, soap, perfume, gin, and in external and internal medicines to cure rheumatism, toothache, neuralgia, stomach troubles, and colds. An officially recognized drug. It is a pleasure to chew the young fresh leaves and savor the flavor of peppermint. The mint used in making mint juleps for which there are many different recipes and for which our Southern states are famous.

Order Polemoniales./Family Solanaceae

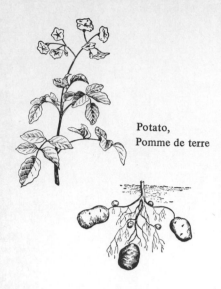

Potato,
Pomme de terre

Potato, Pomme de Terre

Solanum tuberosum

Height to 3 ft. Stem usually weak, pungent, somewhat sticky, with characteristic odor. Leaves to 10 in. long, compound, with an odd number of leaflets; larger pairs often with smaller pairs between. Underground tuber is the stem we eat, the eyes being rudimentary buds. Biennial.

Native of temperate part of Andes in South America but widely introduced throughout habitable parts of the world and serving as a staple food. Does best where there is a cool growing season, this being more important than soil. England, Germany, New Brunswick, Maine, Montana, and Idaho are important potato-producing centers. There are special varieties of potatoes developed for baking, boiling, and other means of food preparation. For market purposes there are five basic types: round white, long russet group, round russet, round red group, and long white group.

Flowers white to bluish, somewhat star-shaped, borne in rather long drooping supports, from 1 to 1¼ in. across. Fruit, infrequently produced, a yellowish-green ¾-in. globular berry with two to three cells, each containing a number of seeds. "Seed" potato of commerce is a piece of the tuber, not a true seed.

Seed potatoes grown in North are commonly used in South. Northern farmers grow their own. When cash is not available for purchase of seeds, they are grown by planting second crop and getting low-yield crop suitable for seed but not for market. Cut tubers into sections and plant pieces.

In 1846, a potato blight in Ireland caused a terrible famine. In World War II, Germans fed tops to stock and ate the tubers themselves or made them into flour or used the starch in manufacturing. The plant is attacked by serious fungus blights and by the Colorado potato beetle.

Jerusalem Cherry

Jerusalem Cherry

Solanum pseudo-capsicum

Shrub. Height to 4 ft. Branches freely, smooth and generally erect, varying greatly with the conditions under which it is grown. Leaves narrow, to 4 in. long, narrowed at base and more blunt at free end, entire or with wavy margins, smooth on both sides but upper surface more shining than lower.

Native of Old World. Widely introduced and grown as a houseplant or sometimes in suitable climates as an outdoor ornamental. It is related to the potato, like more than 1,000 other species of the same genus.

Flowers white, somewhat like those of the potato, to ½ in. across, borne separately or in small numbers from sides of stems rather than from tips. Fruits brilliant yellow or orange and about size of a large cherry, long-persistent, making the heavy-bearing plant an attractive ornamental.

A relatively common ornamental house plant that can be grown from seeds or from slips. The attractive fruits unfortunately contain the poisonous alkaloid, solanine, and may cause a severe poisoning if eaten by human beings.

A popular houseplant particularly where it is desired that the attractiveness be considerably prolonged.

Eggplant

Solanum melongena

Height to 4 ft. Bushy, gray fuzzy, sometimes spined stems, much-branched. Leaves to 15 in. long, not compound, thick and heavy, pointed at tip; with obscure lobes or angles, alternate, thick, coarse, somewhat gray, hairy, or even with weak spines. Root system spreading.

Native of India but established widely as a commercial food plant. Popular variety is New York Improved or *S. melongena* cv. *'Esculentum'*. Varieties have been developed to meet many needs. States which lead in production are Florida, Texas, Louisiana, and New Jersey.

Flowers borne opposite leaves, singly, nodding, to 2 in. across, violet in color and much like the better-known flowers of potato or tomato. Fruit a large blue-black, egg-shaped berry that may be to 1 ft long, shining and sometimes yellow, white or striped, with pulpy interior in which are embedded many seeds.

Grows best in warm, sandy, well-drained loam. Seeds started in hotbeds, from 8 to 10 weeks before being transplanted after danger of frost is past. Plant seeds at depth of 1 in. and transplant to 2 ft apart in rows to 2 ft apart. Fruit matures in about 100 days after transplanting. Fruit is often commonly peeled, sliced, and fried in deep fat; also used as the basis of many different casseroles. A tender plant which requires a long, warm growing season.

Fruit believed to be a fair source of vitamin B. Fruits picked before completely mature. Pests include a wilt and fruit rot, controlled by rotation and by treating seed with corrosive sublimate, and a flea beetle and the potato beetle, controlled by screening or by bordeaux and lead arsenate sprays.

Eggplant

Tomato
Lycopersicon esculentum

Length of stems and branches to 6 ft. A spreading, sprawling, or somewhat climbing annual; if perennial, the young plants are erect at least at first. Leaves strongly scented, compound, to 18 in. long, with individual leaflet to 3 in. long and with incurved margins. Roots freely from where stem touches earth. Most authorities recognize four cultivars: *'Commune'* or common tomato; *'Grandiflorum'* or large-leaved tomato; *'Validum'* or upright tomato; *'Pyriforme'* or pear tomato.

Native of South America but now grown outdoors throughout the world where climatic conditions are suitable or grown under glass where the necessary conditions do not exist outdoors. Among more popular varieties in the East are the early red Bonny Best or the Earliana. Many varieties developed to thrive in different conditions.

Flowers in clusters of three to seven, each to nearly 1 in. across, borne on jointed stalks, deeply cleft into 5 parts, yellow, with recurving petals, nodding and with persistent calyx. Fruits red or yellow spheres or flattened spheres, to 3 in. or more through, juicy, well-supplied with seeds.

Thrive best where night temperature does not go below 60°F. Overwatering of plants under glass a common fault. Pollination effected, in greenhouse, by shaking plants once a day. Seeds, started in hotbeds in late March, produce seedlings that are set in June 4 ft apart in rows to 4 ft apart, yielding fruits within 90 days of resetting time.

Fruits rich in vitamin A and more so in vitamins B and C. 1 lb of fresh tomatoes equals 103 food calories, ½ oz of carbohydrate, and 1 oz of protein. A plant may yield 6 lb of fruit or 12 tons to the acre. 1 bu weighs to 60 lb. Sold fresh, green, or canned, either whole or as juice.

Tomato

Red Pepper
Capsicum frutescens

Shrub. Height to 8 ft. In North is grown as an annual. Wood hard. Trunk to 3 in. in diameter. Whole plant tender when young. Leaves to 5 in. long, usually pointed at each end but highly variable in shape, some being small and less than 1 in. long.

Native of Central and South America, with one related species from Japan. Many varieties include *fasciculatum,* red cluster pepper; *longum,* long pepper; *conoides,* cone pepper; *grossum,* sweet pepper; and *cerasiforme,* cherry pepper. Of course, the plant is best known under cultivation.

Flowers small, to ½ in. broad, usually five-lobed, developing into the well-known thick-rinded, variable-shaped berry. Large fruits are produced by pinching off growing tips of plant. A seed packet normally contains about 200 seeds, or more than enough for a family garden.

Relatively free from insect pests in the field but in greenhouses is subject to attacks of red spiders and aphids. Some persons are allergic to red pepper and develop burning sensation if skin is exposed; application of milk will relieve burning sensation. Paprika comes from long-pointed varieties. Bell peppers mildest, used as stuffed peppers.

Food value of 13 oz is about 100 calories. Merit lies more in stimulation of appetite for other foods than in flavor of the pepper itself. Not to be confused with condiment, black pepper, which comes from another family of plants. Pimiento, or Spanish pepper, is a mild, thick-fleshed type grown particularly in Georgia.

Red Pepper

Belladonna
Atropa belladonna

Height to 6 ft. Erect, well-branched in upper parts, somewhat hairy. Leaves alternate, entire, on short petioles, often opposite in upper parts, to 6 in. long, dull green, pointed at tip. Root thick, fleshy, creeping horizontally, perennial, with red sap.

Native of Europe and Asia where it grows in waste places and among ruins. Has become naturalized in some places in eastern United States. Four species of genus native of Old World. Widely cultivated in United States, Europe, and India. Red sap used in cosmetics.

Flowers in leaf axils, singly or in pairs, nodding or erect. Corolla purple or dull red, to 1 in. long. Flowers May through August. Fruit matures in September, about ½ in. across, crowded with seeds, shining black, nearly globular, with a suggestion of being two-lobed, with purple-violet juice.

Seeds treated 45 seconds with concentrated sulfuric acid, then washed before planting in flats. Transplanted to rows 3 ft apart or 4,000 per acre (20,000 seeds per oz). Requires frequent cultivation and spraying with pyrethrum or arsenate sprays to keep down cutworms. Yields hyoscyamine, atropine, and other solanaceous alkaloids. Toxins present in all parts of plants; especially in seeds, roots, and leaves.

Known for centuries as a poison. Drug hyoscyamine used to dilate pupils of eyes, to stimulate the heart, and externally to allay pain. Used in treating whooping cough, neuralgia, scarlet fever, rheumatism, and convulsive disease, but too dangerous to be used except by physicians.

Belladonna

Henbane

Ground Cherry

Tobacco

Henbane
Hyoscyamus niger

Height to 2½ ft. Usually little over 6 in. Leaves large, to 8 in. long, oblong, commonly crowded, the lowermost with petioles clasping the stem and the upper without; midrib with long hairs, pale green. Root long, thick, brown outside and white inside, tough, fetid like tobacco.

Introduced from Europe but established in the Northern states and in Canada, particularly in the Northwest. Locally common in some areas of the northern Rocky Mountain states. Rare, otherwise. A weed in waste places or a cultivated plant. Common about cemeteries, foundations of old houses, and other abandoned places.

Flowers funnelform, almost or wholly without stems, in a one-sided terminal cluster, yellow or greenish-yellow with purple veins and purple throat, five-lobed. Fruits enclosed in calyx, two-celled, many-seeded capsules that open with a transverse lid. About 35,000 egg-shaped, brown seeds per oz. Biennial or annual.

Treat seeds 15 seconds with concentrated sulfuric acid, then wash quickly to help germination. Leaves, fruits, and stem harvested when stems are not over ¼ in. through. Leaf drug collected from second-year biennials. Plant yields hyoscyamine, hyoscine, and atropine.

Any part of plant may be poisonous to eat. Drugs from it are used like belladonna as a sedative, in treatment of convulsive diseases, in treatment of pain, to help nervous sleep, and for other purposes. Too potent to be used except on physician's advice. Known to be poisonous to poultry.

Ground Cherry
Physalis heterophylla

Height to 3 ft. Erect or sprawling, with erect tips. Stem rather sticky and hairy, weak and green, arising from a slender, long, underground rootstock, branching or unbranched. Leaves alternate, to 4 in. or more long, coarsely but shallowly and bluntly toothed at margin, with veins joined inside the leaf margin.

Native of North America. Found on rich garden soil or similar places from New Brunswick to Florida, west to Texas, Colorado, and Saskatchewan. Closely related species extend range to other continents with 30 in the United States, 2 in Europe, 6 in India and Australia, and others in South America.

Flowers borne singly in axils of leaves, greenish-yellow with a brown or bluish center, to nearly 1 in. across, with calyx not half petal length. In fruit calyx is inflated to enclose completely and loosely the yellow, many-seeded berry.

Known as "husk tomatoes," the members of this genus are eaten raw, preserved, or cooked. Some are considered of enough importance to be cultivated for sale in markets or for home consumption. Unless ripe, they may have a strong unpleasant taste, but ripe and cooked properly, they closely resemble tomatoes.

Domestic animals have been poisoned by eating large amounts of the tops or unripe fruits. Vegetation normally so distasteful that poisoning only rarely occurs. The South American *P. peruviana*, with longer-tipped, nonsticky leaves is most commonly cultivated. Related *P. alkekengi* is known as Chinese lantern plant and winter cherry. Orange fruits are cut late in season for use indoors as decoration. "Fruits" are really swollen calyxes of the white flowers which open in summer.

Tobacco
Nicotiana tabacum

Height to 8 ft. or even more. Branched or unbranched, erect stems bear enormous leaves that may be to 2 ft long and at least half as wide, whose edges may be somewhat curled downward but whose texture varies greatly with purpose for which plant is raised. Annual. Plant may be somewhat woody at base.

Native of tropical America. Now widely cultivated over the world, being grown as far north as the northern border states in the East. Of the 60 species, some are grown only for ornament and some are trees or shrubs. One is native of Australia, others of the Pacific islands. Cultivated in Virginia by white man since 1612.

Flowers borne in loose open clusters at tops of branches, opening in full sunlight, to 2 in. long. Corolla funnelform, woolly on outside, with somewhat swollen throat; the limb pink to red, varying to white. Plant contains alkaloid, nicotine, which can be extracted and is a violent poison. Calyx rose-colored and much smaller than corolla. Fruit to about ¾ in. long, nearly equaling the calyx, a capsule.

Plant may yield 1 million seeds, of which 1 tsp will sow a bed 100 ft square in fine soil, which when transplanted will plant an acre if weeds are kept down, water is properly applied, and diseases are kept out. Transplanted about May in rows 3 to 4 ft apart, topped to leave 12 to 20 leaves and suckers kept down.

Kentucky among most important tobacco states. Varieties include cigar-leaf, white Burley, heavy or export, and bright yellow. Harvested leaves are specially dried and processed for particular purposes.   Evidence from research demonstrates that the use of tobacco, especially in the form of cigarettes, is a health hazard.   Smoking has been linked to cancer of the respiratory tract and to cardiovascular problems.

Jimson Weed
Datura stramonium
Height to 5 ft.   Stem stout with spreading branches, purple to green, smooth or fuzzy, strongly scented.   Leaves alternate, unevenly toothed, smooth, thin, green, narrowed at base, to 8 in. long, with petioles to 4 in. long, strongly scented, rather conspicuously veined.   Annual.

Native of tropical regions but found established in waste places from Nova Scotia to Florida, west across the United States especially to the south.   About a dozen species of the genus found over wide stretches of the world.   Many races differing in color and shape of flowers and fruits.

Flowers on short stems, in axils of branches.   Corolla funnelform, to 6 in. long, with a spreading border to over 2 in. broad, white or violet or mixed or purple.   Calyx about half length of corolla.   Stamens, five, separate.   Fruit, spiny, to 2 in. long, egg-shaped; a capsule bearing large seeds.

All parts of plant are poisonous to cattle, horses, sheep, and human beings, even contact with leaves sometimes producing a severe dermatitis.   Poisons are several solanaceous alkaloids, principal among them being atropine, hyoscyamine, and hypscine in the leaves, fruits, roots, and seeds, and hyoscine in the roots.   Plant is sometimes grown as an ornamental, but is dangerous.

Poison causes headache, vertigo, thirst, nausea, loss of sight and of coordination, mania, convulsions, and death.   Named from Jamestown, where early settlers became crazed after eating it.   Other common names include devil's-trumpet, fireweed, dewtry, Peru apple, devil's-apple, mad apple, and Jamestown lily.

Jimson Weed

Garden Petunia
Petunia hybrida
Height to 3 ft. or more.   Stems branched, sticky, covered with fine hairs; weak and must be supported.   Leaves: broad, heart-shaped, covered with soft hairs, the upper without petioles and the lower with short ones, conspicuously veined, somewhat sticky, alternate below and opposite above.

Native of Argentine but widely cultivated for growth outdoors or in.   Of the dozen or more species, all apparently come from South America.   Many have established themselves as escapes from old gardens in regions where climate is suitable.   Common petunia is product of crossing many kinds.

Flowers to 5 in. across, funnelform, fringed, unfringed, or double, with color varying from white to red or purple, with stripes or bars or with special throat markings.   Calyx much smaller than the corolla, with slender pointed sepals about equal to the fruit capsule.

Seeds planted in fine-powdered soil and lightly covered, thinned or transplanted so that plants are to 18 in. apart.   Plants should bloom from 2 to 2½ months after seeds are planted in the open in suitable conditions but are susceptible to early frost so probably should be started indoors.   May be propagated by cuttings.

Valuable indoor and outdoor ornamental.   Modern cultivars are of two types: one having large, double or single blooms and exhibiting a wide range of different colors; and a second with compact small flowers and bearing many flowers per plant.   Completely double flowers are propagated by slips taken in September and October.   Houseplants need night temperature down to 45 to 50°F.

Petunia

Mullein, Velvet Plant
Verbascum thapsus
Height to 8 ft.   Stem branched or unbranched, woolly, fluted, coarse and brittle, sometimes somewhat sticky.   Leaves alternate, densely woolly, light green, to 1 ft long and half as wide, the basal ones with broadly margined petioles; the stem leaves with petioles that extend down stem as flutings.   Commonly biennial.

Native of Europe and of Asia.   Firmly naturalized in America and established as a weed from Nova Scotia to Florida west to California and north.   Some 125 species of the genus native of the Old World, of which a few are established in America.   Apparently, no native American species of the genus.

Flowers borne in long compact spikes to over 1 ft long.   Flowers yellow, crowded, to over 1 in. across, appearing from July through September.   Calyx buried in the spike.   Stamens unequal, with three upper ones short and hairy and the two lower longer, smooth, with larger anthers.   Fruit a capsule to ⅓ in. high, longer than calyx.

Pollination by self or by insects.   Many small black thrips are found among the hairs in fall.   Plant is a sun lover that can prosper on bare
(Continued)

Mullein, Velvet Plant

Butter–and–eggs

Kenilworth Ivy

Snapdragon

lands not suitable for competitors and is left untouched by grazers because of felt of hairs. Hairs may cause a dermatitis to some persons. Fall rosettes close to ground are common.

Seeds and plant listed by quacks as having some medicinal value but this has not been recognized. It is considered of some ornamental value. The seeds are found as a relatively common impurity of commercial seeds. Deep-penetrating root system helps it survive drought conditions well.

Butter-and-eggs
Linaria vulgaris
Height to 3 ft but usually less. Slender erect stems are smooth, leaf-crowded, and pale green, branched or unbranched, arising from short perennial rootstocks. Leaves narrow, entire, mostly alternate, to 1½ in. long and without distinct petioles, with a fine fuzz that may rub off easily. Perennial.

Native of Europe and of Asia but established in America from Newfoundland to Georgia, west to California and Oregon. Also found south into South America. It lives mostly on dry waste soil and may form clear stands crowding out other species. Over 100 species in the genus, mostly of north temperate regions.

Flowers borne in compact clusters at ends of branches, making a spike-like formation but for fact that flowers are short-stalked. Flowers to 1¼ in. long, with spur of the erect corolla darker. Corolla orange and white or yellow and white. Calyx short. Fruit dry capsule containing black wafer seeds.

Flowers July through October. Pollination by bumblebees and some beetles. Black wafer seeds are blown freely by the wind, being freed on through the winter months. Plant is beautiful enough to be an ornamental but too independent of care to be popular.

When it becomes established as a weed, the recommended procedure is to plant the area to open cultivated crops and cultivate late into the season; then follow this by a close annual crop that may smother the weed. It often becomes established near cemeteries with such plants as periwinkle and cypress spurge, other mat formers.

Kenilworth Ivy
Cymbalaria muralis
Length to about 12 ft. Stems trailing, weak, rooting freely at joints, relatively thin, smooth, branched. Leaves mostly alternate, to 1½ in. broad, with rounded blades, distinct lobes, petioles longer than blades, smooth, purple-tinted beneath and a rich green on top.

Native of Europe. Established in wet places from Ontario to New Jersey, west through Pennsylvania but grown rather commonly as a greenhouse or sunroom ornamental in pots. Old World species in the genus but none apparently to be found native in America.

Flowers solitary, in axils of leaves, to about ½ in. long, blue or lilac; in general, resemble a diminutive snapdragon or butter-and-eggs flower. Corolla irregularly two-lipped, with three lobes in one and two in the other. A small calyx and four stamens. Fruit a capsule containing several seeds.

Propagated by seeds or by cuttings, seeds being normally planted in medium garden soil in good light and kept at a temperature close to 60 to 70°F to be at best. A relatively easy plant to grow, rather free of insect and fungous pests although aphids may attack it sometimes.

Essentially useful as an ornamental either for growing indoors in sun parlors or greenhouses or outdoors in window boxes or on walls where a crawling plant is to be desired. Apparently, it has no importance as a weed, poison, or producer of medicine. There is a white-flowered form which has variegated leaves.

Snapdragon
Antirrhinum majus
Height to 3 ft or more. Commonly branched or unbranched stems, downy in the upper areas but smooth below, slender, arising from a perennial root system. Leaves entire, opposite or alternate, narrow, to 3 in. long, dark green, either pointed at both ends or essentially oblong with pointed ends, smooth, short-petioled.

Native of southern European countries but escaped from gardens and sometimes established as a weed in the East. Cultivated widely in gardens and to some extent as a greenhouse plant. Related lesser snapdragon, *A. orontium,* also is an escape from cultivation. About 40 species native of Europe, Asia, and North America.

Flowers to 1½ in. long, in rather open clusters at ends of branches, red, purple, white, or mixed but commonly reddish-purple in the escaped forms; sometimes borne in axils of leaves, with a small calyx and a corolla with a wide tube and a wider spread two-lobed lip. Stamens, four. Fruit a capsule to ½ in. long.

Normally perennial but grown as an annual in the North. Seeds planted early either in open or under glass may produce plants that flower in August. These may be potted and held over winter or covered with a deep mulch. They survive in same sort of conditions that favor geraniums and similar houseplants.

Often planted in masses to make beds or as borders though for latter purpose they are higher than is usually favored. Many cultivated varieties offering low plants with almost any desired selection of color in flowers. Common names include tiger's-mouth, rabbit's-mouth, calf's-snout, toad's-mouth, dragon's-mouth, dog's-mouth, liver snap.

Monkey Flower
Mimulus ringens
Height to 3 ft. Stem branched or unbranched, four-angled and four-winged, erect, smooth, bright green. Leaves opposite, clasping at base, to 4 in. long and to 1 in. wide, with conspicuous veins joined inside the shallowly notched leaf margin, clean green, long-pointed at free end. Perennial.

Monkey Flower

Native of North America. Found from Nova Scotia to Virginia, west to Texas and Manitoba reaching an elevation of to 3,000 ft in Virginia. Cultivated to some extent and a number of varieties have been developed for ornamental purposes. About 40 species in the genus native of North America.

Flowers borne in axils of uppermost leaves, on slender stems that in fruit reach to 2 in. long, or two to four times calyx length. Calyx nearly as long as corolla tube and with pointed tips. Corolla tubular, with irregular flaring brim, to 1 in. long, blue, white, pink, or violet. Fruit a $\frac{2}{3}$-in. capsule.

Garden species is *M. luteus,* with yellow dark-spotted corollas to 2 in. long. Seeds of this species sown indoors in January in fine soil may yield plants that will flower by late summer. They favor cool shady wet spots and may be kept through winter if night temperature goes down to around 45°F.

Common name refers to assumed resemblance of corolla to face of a monkey, an assumption that is not always easy to justify. Many horticultural varieties of the garden species; the wild species here listed is commonly used in wet wild-flower gardens.

Foxglove
Digitalis purpurea
Height to 5 ft. Stem stout, rigid, erect, usually unbranched, slightly fuzzy, green. Leaves alternate, with relatively long petioles, and margins with rounded saw-teeth, lower being relatively long petioled, and to 10 in. long, while upper lack petioles and are smaller. Perennial or more commonly biennial.

Foxglove

Native of Europe but established in North America as an escape from cultivation. From southeastern Canada to New Jersey, west to California and Washington, but not extensively in many places. Cultivated in many parts of the world, including the United States. 20 species of the genus native of Europe and Asia.

Flowers in clusters along tips of branches, on short slender individual stems, more or less from one side, hanging bell-like with inflated tube and small flaring unequally lobed lip, to 2 in. long, purple to pale blue or white. Calyx usually under $\frac{1}{3}$ length of corolla. Fruit a capsule splitting to free seeds.

Hard to raise. Best in ordinary well-drained garden soils. Best to sow seeds in February, mixed in fine sand, distributed evenly over well-protected flats in greenhouses, then set 1 ft apart in rows in fields. Blooms form second year, so plants need protective mulch first winter. Leaves contain the glucosides digitoxin, digitonin, and digitalin.

Leaves of either year valuable for medicine; an aglycone is the active substance. Medicine prepared has twofold function: strengthening the force of the contraction of the heart and indirectly prolonging the duration of diastole. Poisoning from excessive use may cause nausea, loss of appetite, dullness, increased pulse, contracted pupils, and intestinal disturbance.

Thyme-leaved Speedwell
Veronica serpyllifolia
Height to 10 in. Stems sprawling, slender, branching, rising at tips or erect for most part. Leaves opposite, the lower petioled and the upper without petioles, with wavy or entire margins, to ½ in. long, oval or oblong in general shape, smooth or finely roughened. Perennial.

Native of North America and of Europe. Found from Labrador to Georgia, west to California and Alaska reaching to elevations of 2,600 ft in the Catskill regions of New York. Some 200 species of the genus found throughout world with better than a dozen native of North America.

Thyme–leaved
Speedwell

(Continued)

Lousewort, Wood Betony

Bladderwort

Flowers in loose open clusters at ends of branches, with each flower borne on a separate little stem as long as or longer than the calyx, which is in turn nearly as long as the corolla. Corolla blue with darker stripes, or white, to ⅓ in. across. Fruit broader than high. Seeds, flat.

Flowering time April through September. Sometimes a weed in lawns, pastures, or other places where cultivation is not common, but it yields quickly to competition or to cultivation. Hoeing, crop rotation, and sprays of ammonium sulfate, sodium chlorate, or other weed killers are effective.

Two of the related species, *V. officinalis, V. chamaedrys,* have found use as a substitute for tea. Another, *V. americana,* has been recommended as a dietary preventative for scurvy and some species have been used as potherbs; all are high in vitamin C.

Lousewort, Wood Betony
Pedicularis canadensis
Height to 1½ ft. Stems in tufts, smooth, or smooth above and fuzzy below, usually unbranched, leafy. Leaves alternate, many of them arising from crown of underground portion, to 5 in. long, mostly slender-petioled, with margins notched halfway to midrib.

Native of North America. Found in woods and relatively dry thickets, often on poor soil, from Nova Scotia to Florida, west to northern Mexico, Colorado, and Manitoba, with related species extending range. About 30 species in the genus in North America; about 125 in the world, all in Northern Hemisphere.

Flowers in crowded rather dense spikes at tips of stems, with the spike lengthening to 8 in. sometimes when in fruit. Corolla to nearly 1 in. long, with overhanging upper lip longer than three-lobed lower lip and tube much longer than calyx. Flower is extremely variable in color: pinks, reds, browns, yellows, whites. Fruit about ⅔ in. long.

Flowering time April through June, when the plant is visited by such pollinating insects as bumblebees and honeybees. Some members of genus are used commonly as ornamentals in rock gardens and in wild-flower gardens but some may be parasitic on certain other species of plants.

Plant is named *pedicularis* beccause it was formerly reputed to cause sheep that fed on it to become infested with lice. It is rather doubtful whether this association is justified.

Order Polemoniales./Family Lentibulariaceae

Greater Bladderwort
Utricularia vulgaris
Stems to 3 ft long, somewhat branched, without roots, floating horizontally under the surface of fresh but stagnant water. Leaves alternate, to 2 in. long, with sections repeatedly forked into 2s or 3s and bearing near bases bladders that are to ¹⁄₁₀ in. long.

Native in one variety or another from Eurasia through North America, this variety being found from Newfoundland to Maryland, west to Lower California and the Yukon. American variety has longer stems than the European. About 75 species of the genus, of worldwide distribution.

Flowers borne on erect stems that rise above surface of water, the individual flower stalks being to nearly 1 in. long and the whole cluster to possibly 2 ft high. Corolla yellow, to nearly 1 in. long with the lower lip longer and broader than the upper. Fruit, a capsule.

Bladders are unique structures that capture minute water animals and apparently digest them. Such animals as water fleas are known to be captured when they thrust their antennae into the bladder opening, after which they are held until the animals die. Hairlike structures and spines are arranged in a definite pattern around the mouth, opening, of the bladder. Normally, the bladder is compressed. When a small water organism touches one of the hairs, the bladder inflates. Water plus the water organism is sucked into the bladder. A transparent trapdoor closes the opening and the organism is digested. Special hairs in the bladder absorb the digested materials.

Plants are eaten by a number of the larger animals that feed in marshes, such as moose and deer, and by some aquatic birds. Winter is survived by the production of winter buds that start new plants the next year, or by the production of seeds.

Order Polemoniales./Family Orobanchaceae

Beechdrops
Epifagus virginiana
Stem to 2 ft high, branched or unbranched, with a scaly base and few scales along the erect stem arising from the fibrous and brittle root that contacts the beech tree host's root system. Tough, brown-stained, almost woody, medium slender, and looking more or less dead.

Beechdrops

Native of North America. Found from Nova Scotia to Florida, west to Louisiana and Michigan or Wisconsin, or about same range as that of host plant, the beech. Only one species in the genus though there are four few-specied genera in family in eastern North America. About 200 species in whole family.

Flowers more or less tubular, to nearly ½ in. long and about ¹⁄₁₀ in. thick, with the calyx under ⅓ the length of the relatively closed-mouthed striped purple and white corolla. Stamens about length of corolla although style of pistil is a little longer. Fruit a capsule to ¼ in. long. Lower flowers to ¹⁄₁₀ in. long.

Flowering time August through October, when flowers are visited by bees that cross-pollinate the upper flowers. Lower flowers are self-pollinated (cleistogamous) and lack purple and white markings of upper insect-pollinated units. As suggested above, plants are parasitic on root of beech.

Of little damage to the host plants and probably of no great economic importance. The plant is listed by quack doctors as having some medicinal properties but this is doubtful.

Order Polemoniales./Family Bignoniaceae

Catalpa, Indian Bean
Catalpa bignonioides
Tree. Height to 65 ft with trunk diameter to 4 ft. Trunk short, breaking into relatively few coarse branches that make a symmetrical crown. Leaves opposite or whorled, to 6 in. or more long and nearly as wide, heart-shaped at base, pointed at free end, on petioles to 6 in. long, pale downy, with an unpleasant odor when broken.

Native of the Gulf States. Now extensively cultivated into Northeast as far as central New England and west to Rocky Mountains. Also found in Europe. A number of ornamental varieties. Closely related *C. speciosa* has flowers with purple on the interior, not spotted as are the interiors of this species. In any case, trash from tree makes its value decrease.

Flowers in huge open erect terminal cluster, with individual flowers white, mottled with yellow and purple inside, or lined with purple, to 2 in. long, very showy with inner part of tube with yellow blotches and parallel purple ridges. Stamens, two, slightly longer than corolla tube. Fruit a cylindrical capsule, podlike in shape. Illustration of fruit and leaf not drawn to same scale. Leaves are 5 to 8 in. long and 4 to 5 in. broad; fruits are 8 to 20 in. long and about ¼ in. in diameter.

Wood light, weak, soft, coarse-grained, durable in contact with soil, gray-brown with a bluish tinge and thin whitish sapwood. Tree is a quick grower but winter-kills freely in the northern part of its range. Flowering time is from June through July with the fruits persisting through winter, breaking and shedding waferlike seeds.

Used as an ornamental or as a windbreak where other trees often will not survive. Wood used as fence posts, poles, and railroad ties largely because of durability in contact with the soil. Foliage and flowers are definitely attractive, but in winter the tree looks bedraggled.

Order Polemoniales./Family Acanthaceae

Water Willow
Justicia americana
Height to 6 ft though a portion of this may be sprawling. Stem branched or unbranched, woody, smooth, slender. Leaves opposite, entire, to 6 in. long and to ¾ in. wide, narrowed near base to form short petiole in lower leaves and with drawn-out tip.

Native of North America. Found growing in wet places or in edge of waterways from Quebec to Georgia, west to Texas and Ontario, with closely related species extending range westward and southward. Some 100 species of the genus, mostly native of tropical America, Asia, and Africa.

Flowers borne in relatively few-flowered heads, at ends of slender stems arising from axils of leaves, with white or violet corollas to ½ in. long, with tube longer than lips and base of lower lip definitely roughened. Fruit a capsule to ½ in. long, compressed, longer than calyx.

Flowering time May through September. Ecologists recognize a service performed by the plant in anchoring soft soil shorelines from erosion by waves or by stream currents. Trash accumulating among the stems helps build soil.

In some places, planting of this species has been recommended for fixing shorelines that might otherwise be destroyed or shifted. Apparently the plant is eaten occasionally by browsing animals that feed at the water's edge.

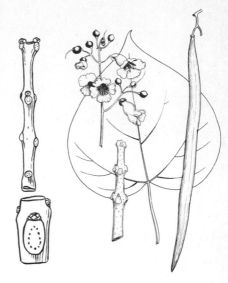

Catalpa

Water Willow

Lopseed

Order Polemoniales./Family Phrymaceae

Lopseed
Phryma leptostachya

Height to 3 ft. Herbaceous. Stem erect, four-angled, slender, branched, with swollen joints, dark green. Leaves opposite, to 6 in. or more long, thin, green, the lower on long petioles and the upper sometimes without them; margins of blade with shallow rounded teeth; veins joined inside margins.

Native of North America. Found in woodlands and brushy spots from New Brunswick to Florida, west to Kansas and Manitoba; also in Bermuda and eastern Asia. Apparently, in spite of wide distribution, only one species in genus and one genus in family, an almost unique situation.

Flowers borne along sides of erect tips of branches, rather well-separated from each other and from the leafy portion, mostly opposite, about ¼ in. long, extending out from stems at right angles when in flower but depressed to the stem when in fruit, with the calyx surrounding the fruit.

Flowering time June through September. Fruit at best in late summer and fall after it has become mature, when it breaks free from the stalk violently when touched. In this way, it may be thrown for some distance. Calyx may serve in a weak way to attach freed fruit to wool of animals.

Not of any apparent economic importance, nor is it considered of a medicinal or poisonous nature. An interesting plant to investigate when the fruits are mature, merely to see how far the fruits may be hurled when freed.

Order Plantaginales./Family Plantaginaceae

Common Plantain
Plantago major

Stemless. Leaves arise in rosettes from head of fibrous and many-branched root system, with five to seven lengthwise veins and with conspicuously long-channeled petioles to 10 in. long, with entire or coarsely toothed margins, firm, usually flat on the ground unless there are competitive plants. Perennial.

Native of Europe. Also apparently of America, where it is now found through most of North American continent and in West Indies. Also found in Asia. Grows in almost any place where man maintains lawn grass.

Flowers borne along an erect flowering stem, crowded to make a long "rattail" spike that may be to 18 in. long. Pistils mature before the stamens so flowers of one part of spike may, with the help of wind, gravity, or insects, pollinate other flowers of the same spike. Fruit a capsule containing to 18 sticky seeds.

Flowers May through September. Fruits June through December. Serves as food for larvae of a beetle; five flies use leaves as mines. Seeds eaten by insect larvae and by birds. Plant 81 percent water, 11 percent nitrogen, 2 percent fiber, and 3 percent protein and fat. Sticky seeds help in distribution. Plant cannot survive competition.

Used in China as a potherb and often used in America as spring greens, or dried leaves used as a tea substitute. May be controlled by cutting, by salting, by digging, or by the better weed sprays. Does not survive cultivation nor does it do well in very rich soils. Seeds collected and fed to caged canaries by some owners. Favorite food of rabbits. Many rodents eat plantain seeds. Can be, and is, easily confused with closely related native American *Plantago rugelii,* which is becoming a serious weed. The native form is more common.

Common Plantain

Narrow-leaved Plantain
Plantago lanceolata

Stem above ground short or apparently absent, with leaves rising directly from root system which is a short rootstock, with contributing fibrous roots. Leaves narrow, with entire margins and usually five more or less conspicuous parallel ribs, to 1 ft long and to 1 in. wide, usually more or less pointed at each end.

Common on lawns and in waste places of wide areas in the world. Native of Europe and of Asia. Found in America from New Brunswick to Florida, west to California and British Columbia though in the west it is replaced in abundance by closely related species. Some 200 species in the genus, some being serious weeds.

Flowers borne in short, dense spikes at end of leafless flowering stems that rise to a height of over 1 ft. Pistils mature first at upper end of spike, with the long-filamented stamens following and waving yellow anthers freely. Pollination mostly by wind but sometimes by insects. Flower produces a one-fruited, two-seeded capsule.

Serves as host for larvae of some tiger moths and the buckeye butterfly. In summer an apple pest, *Aphis sorbi,* lives on this too common weed. Survives drought or flood and since seeds retain vitality in soil many years may become a persistent pest even where care is observed. Tops should not be allowed to seed.

Cultivation, hoeing, and the better weed sprays may be effective in keeping it under control. Formerly believed to have medicinal value but not

Narrow-leaved Plantain

now so used. Tops provide food for seed-eating birds including pheasants and quail. Perennial. Known as rib grass, ripple grass, blackjack, jackstraw, English plantain, buckhorn, and rattailed plantain. Seeds collected by some landowners and fed to their caged songbirds such as canaries.

Bracted Plantain

Bracted Plantain
Plantago aristata
Stemless except for flower-bearing stem that reaches a height of 1½ ft. Leaves slender, much like those of narrow-leaved plantain, to 1½ ft. long, and to ⅓ in. wide, often conspicuously three-ribbed, entire, narrowed into fine petioles, smooth or downy, gray-green. Annual.

Native of North America. Found in dry meadows, grasslands, waste places, and sometimes in lawns from Maine to Georgia, west to New Mexico and British Columbia, being more common in Middle West than in East. Related to other species extending the range.

Flowers borne in spikes at tops of erect flower stalks and interspersed with bracts, which may be nearly as long as the 6-in., terminal cylindrical spike or 10 times the length of small inconspicuous flowers. Fruits, capsules bearing two seeds, with convex and concave surfaces.

Flowers May through October. While it may become a weed in some localities, it may usually be controlled by using a green manure crop that may be plowed under, and following this with clean-cultivated crops for at least two seasons until most seeds have been destroyed. Plants should not be allowed to seed.

A conspicuous weed in waste places in the Middle West but not so resistant to control as the narrow-leaved and the common plantains. Has such common names as clover choker, western buckhorn, western ripple grass, large-bracted plantain, bottle brush, and Indian wheat.

Order Rubiales./Family Rubiaceae

Woodruff
Asperula odorata
Height to 8 in. Stems square, erect or ascending, arising from a slender creeping rootstock. Leaves usually in whorls of eight, with finely toothed or roughened margins, pointed at each end, fragrant when dried, smelling much like hay, to 1½ in. long. Some species smooth, some rough. This species is perennial.

Woodruff

Native of Europe and the Orient but grown as an ornamental through most of the temperate regions of the world. Do best in moist soil and in shade. About 80 species in the genus, of which some have been classified as weeds.

Flowers borne in terminal clusters of several flowers each, white or pink, to ¹⁄₁₀ in. long, funnelform but with a widely spreading limb, four-lobed. Stamens, four, inserted on the corolla. Fruit, two-celled, bristly, with short bristles, about ¹⁄₁₀ in. across, rough, borne on short individual stems.

Flowering time May and June. Propagation by division of plants or by means of seeds. Where they are grown in dry sunny places, they become stunted and die back during the summer while in moist, shady places they may thrive into the fall months. Plants are grown in gardens as carpets and as borders; especially in sunny or semishady places. They are at their best if planted in the fall or spring. Propagated by separating the plants into rooted pieces in early spring.

Known by the Germans as Waldmeister, and used by them in May wine, or *Maitrank,* or in their summer drinks. Since the fragrance of the vegetation is long-lasting, it is commonly sprinkled on stored cloth material to give it a pleasing fragrance. Also used rather commonly in home medicines though it seems doubtful if it is really valuable.

Bedstraw
Galium aparine
Length to 5 ft. Stems weak, sprawling over the ground, bushes, or walls, with fine backward-bending but weak short stiff hairs borne on angles of stem. Leaves in whorls, with six to eight at a joint, slender, pointed at each end and thickest near tip, to 3 in. long and to ½ in. wide, with roughened margins and midribs. Annual.

Bedstraw

Native of Europe and possibly also of America. Found from New Brunswick to Florida, west to Texas and the Dakotas. May occur as a weed over wide stretches of territory. Over 250 species in the genus found in various parts of the world.

Flowers in clusters of one to three in axils of upper leaves, on flowering stems to 1 in. long, small and relatively inconspicuous, with a four-lobed corolla and four stamens. Fruit to ¼ in. across, covered with short hooked bristles nearly as long as thickness of each half of the fruit, grayish-brown, warty in appearance.

(Continued)

Flowering time May through September. Commonest near woodlands but not necessarily in them, thriving on almost any rich sandy or alluvial soil or on bare spots in fields and meadows. Fruits apparently distributed by adhering to passing animals. Shoots will cling to passerby. The reflexed hairs attach it. Some species are considered of ornamental value. A weak stemmed plant which may form dense patches in the woods.

Reputed to have certain mystic medicinal values that are not now recognized. Known by at least 60 common names, among them being cleavers, goose grass, gripgrass, loveman, scratchweed, poor robin, burhead, beggar lice, hedgebur, gosling grass, pigtails, and sweethearts.

Coffee

Coffee
Coffea arabica
Height to 15 ft. A small tree or shrub with a main trunk that soon branches freely close to ground, the branches being horizontal, opposite or in whorls of three. Leaves evergreen, dark glossy green, to 6 in. long and to 2 in. wide with a ½-in. point, opposite, thin, elliptic. Perennial.

Native of tropical Africa whence it was introduced into Arabia some 500 years ago. Reached Ceylon and Java about 1700, West Indies about 1720, and Brazil in 1770. From Arabia, it reached Venice in 1615, Paris in 1645, and London in 1650. Libyan coffee is *C. liberica* and Congo coffee *C. robusta.*

Flowers starlike, fragrant, with a delicate odor, pure white, with corolla units to ¾ in. long, this being longer than corolla tube. Fruit a deep red, two-seeded berry about ½ in. long. Seeds planted in beds and seedlings are set at 6-ft intervals. Plants begin to bear in third year, give best yield in fifth year, bear about 30 years.

Young plants are grown in shade. Picked berries are dried in sun or run through pulping machine and dried by artificial heat. Parchment is removed by polishing machines and seeds are bagged. Roasted coffee contains 0.75 to 1.5 percent caffein, the volatile aroma-producing oil caffeol, and glucose, dextrin, proteins, and a fatty oil.

United States is greatest coffee-consuming nation. Production is limited by climate, Arabian coffee needing 50 in. annual rainfall and preferring to 120 in. in areas between 25° N and 25°S. Can be grown from sea level to altitudes of 6,000 ft. Brazil apparently leads world in production of this important crop. Constituents of coffee that are of chief significance are the flavor substances, the bitter substances, and the caffein. Extracts of coffee beans contain 14 amino acids which have been identified from green coffee beans as well as several organic acids including acetic, pyruvic, caffeic, chlorogenic, malic, citric, and tartaric, and a trace of formic. Predominant acid is chlorogenic and the one in lowest concentration is acetic. The pH of coffee brew reaches its minimum with a light brew. Caffein, an alkaloid, contributes to coffee's stimulating property as well as its bitterness. Most of the flavor substances are volatile and are changed by heat.

Partridge Berry

Partridge Berry
Mitchella repens
Length to 1 ft. Stems somewhat woody, trailing close to ground, branching freely, smooth or with a slight down, rooting freely at joints, slender, tough. Leaves opposite, short-petioled, with round to heart-shaped blades to nearly 1 in. long but usually much smaller, dark green, shining, evergreen, conspicuously light-veined.

Native of North America. Found in cool woodlands from Nova Scotia to Florida, west to Texas and Minnesota, reaching an elevation of to 5,000 ft in Virginia. Two species in the genus of which this is the only one in America; other is Japanese.

Flowers borne in pairs at ends of branches, with corolla white, to ½ in. long, tubular, with a flaring mouth divided evenly into four segments somewhat shorter than tube. Calyx much shorter than corolla. Stamens, four, grown to corolla. Pistil, longer than corolla tube. Fruit a red berry to ⅓ in. through. Fruit is a "double berry," actually the fusion product of the ovaries of two flowers. This results in the observation that partridge berry has "two eyes."

Flowering time April through June, with sometimes a late fall flowering period. Fruits persist through winter unless they are eaten by wildlife or by human beings. In spite of frequency with which fruits of this plant are mentioned as being of medicinal value in quack literature, it apparently has no such value.

Excellent ornamental for growing in terrariums indoors. Fruits are edible but lack flavor of wintergreen and are too few and too small to be of any considerable importance. Scientific name *Mitchella* refers to early Virginia botanist John Mitchell, who was a frequent correspondent of Linnaeus.

Innocence, Bluets
Houstonia caerulea

Height to 7 in.   Stems erect, arising in tufts from slender perennial creeping rootstocks, usually unbranched except sometimes near top.   Leaves
crowded at base or paired up stem, about ½ in. long, pointed at each end
or with a short obscure petiole, often slightly downy or hairy.

Native of North America.   Found in meadows, where soil is poor, or on
wet exposed rocks from Nova Scotia to Georgia, west to Missouri and
Michigan.   Over two dozen species native of North America including
Mexico, of which this is possibly the best known.

Flowers borne in pairs at ends of branches, erect, with a small calyx
much shorter than tube of corolla that spreads into a flat four-parted top.
Corolla beautiful pale blue and white or violet with yellow center.   Tube
about length of split lobes.   Fruit a capsule about ⅒ in. long and slightly
broader than long.

Flowering time from April through July, though flowers may be produced throughout summer under proper weather conditions.   Pollination
by bees and smaller butterflies such as clouded sulfur and painted lady.
Used generously in rock gardens and for borders in flower gardens.
Does well in some shade.

It also is known as eyebright, quaker-ladies, brighteyes, Venus's-pride,
little washerwoman and star-of-Bethlehem, all names indicating its popularity largely because of its beauty.   Successful dwarf rock-garden plants.

Bluets, Innocence

Order Rubiales./Family Caprifoliaceae

American Fly Honeysuckle
Lonicera canadensis

Height to 5 ft.   Shrub.   With smooth, light-colored twigs that branch
rather freely.   Leaves opposite, bright green on both sides and attractive,
thin, to 4 in. long and to half that width, with short petioles, rounded at
base and pointed at tip, downy when young but smooth when mature,
with hairy margins.

Native of North America.   Found in woodlands where there is water.
From Nova Scotia to Maryland, west to Indiana, Minnesota, and Saskatchewan.   Cultivated as an ornamental and grown outside natural
range.   Over 150 species in the genus, native mostly of north temperate
areas.   About 25 in the United States.

Flowers borne in pairs from axils of leaves, the flowers being borne on a
single flowering stalk that is about as long as greenish-yellow, ¾-in.-long
corolla.   Bracts at bases of flowers small.   Open lobes of corolla short.
Fruit, reddish, oval berries about ¼ in. in diameter.

Flowering time April through May.   Fruits available for collecting from
June through September.   Plants require moisture and shade to succeed.
Propagation by seeds, by division, or by cuttings.   Most species are not
favored food of cattle, but white-tailed deer are known to feed on this one.

This species is grown somewhat as an ornamental in proper places but
other species are more suitable for growing in sun and where soil is drier.
Known sometimes as "medaddy-bush."   Other popular species include
L. dioica, smooth honeysuckle. *L. japonica*, Japanese honeysuckle, is of
importance to wildlife.   Where it occurs, it provides food and cover for
wildlife.   Unfortunately its habit of growth sometimes crowds out
equally or more desirable species; it is a very serious pest.   Japanese
honeysuckle is spreading northward.   Tartarian honeysuckle, *Lonicera
tartarica*, is an important plant which is now spreading and has been used
repeatedly as a successful wildlife planting.

Fly Honeysuckle

Coralberry, Indian Currant
Symphoricarpos orbiculatus

Height to 7 ft.   Branches slender, dark, upright or drooping and wandlike, downy, densely covered with leaves, purplish.   Leaves opposite, to
1½ in. long, oval, short-petioled, blunt or acute, not evergreen, smooth or
downy beneath, shed late in atutmn.

Native of North America.   Found from New York to Georgia, west to
Texas and South Dakota being usually found on exposed rocky banks or
along streams.   10 species native of North America including Mexico.

Flowers borne in many-flowered, short, dense clusters at axils of leaves
which in fruit and flower are shorter than the leaves.   Corolla bell-
shaped, pink, downy inside, to ⅕ in. long, with stamens included in the
bell and the lobes relatively shallow.   Fruit a purplish-red berry, round,
to ⅕ in. through.

Flowering time June through July.   Fruits available September through
to June.   About 407,000 seeds per lb.   Propagated by seed or by hard- or
softwood cuttings made below the joints and stimulated by potassium
permanganate.   Leaves contain saponin but in small amounts, and poisoning of stock is rare.

Commonly grown as an easily propagated ornamental that spreads once
it is established.   Provides excellent cover for small wildlife.   Fruits are
eaten by grouse, pheasant, quail, and deer.   Provides an important
(Continued)

Coralberry

Twinflower

Horse Gentian, Wild Coffee

Highbush Cranberry

browse in Utah, Nevada, and southern Idaho. Also important as a soil anchor by streams or on steep banks. In the Bluegrass country, this plant becomes exceedingly common, covering large areas as a shrub.

Twinflower
Linnaea borealis
Stems trailing, to length of 2 ft, slender, branched, downy, rather tough, close to ground or under forest duff. Leaves opposite, to ⅔ in. long and about the same width, sometimes wider than long, on short petioles, evergreen, borne on short erect spurs arising from the trailing prostrate main stem.

Native of North America, Asia, and Europe. Found in cold, wet, or dry woods from Newfoundland to New Jersey, west to Colorado, Vancouver, and Alaska. American form is considered a variety of *L. borealis* by some authorities. Three to four species in the genus to be found in north temperate regions of the world.

Flowers borne in pairs at top of slender, erect, threadlike stems to nearly 1 in. long. Corolla bell-shaped, nodding, to ½ in. long, fragrant, whitish, flesh-colored, or tinged with rose-purple and hairy inside. Calyx small. Fruit supported by a pair of scales that may be grown to ovary.

Flowers appear June through August. Plants do well in dry well-drained soil, in sun or in shade or where it is moist. Plant often forms a good ground cover in areas not occupied by other plants. It occasionally flowers in late autumn and sometimes there are four rather than two flowers in group.

Named after Linnaeus, the taxonomist who is usually portrayed with a sprig of twinflower in his buttonhole. It is a delight to the sight and to the nose of man. It is eaten by grouse and deer.

Horse Gentian, Wild Coffee
Triosteum perfoliatum
Height to 4 ft. Stem erect, not woody, stout, finely downy, greenish. Leaves opposite, joined at bases around stem, entire, soft-downy beneath, slightly hairy above, to 9 in. long and to 4 in. wide, rather thick but flexible, dark green, widest at about middle and pointed at free end.

Native of North America. Found in rich soils on hillsides or in lowlands. From Massachusetts to Alabama, west to Kansas and Nebraska. Of the six known species of the genus, one is native of the Himalayan region, two of Japan, and three of North America, of which this is possibly the most widely distributed. Frequently found on less disturbed sites and on edges.

Flowers borne in axils of leaves, usually with one to each leaf in upper regions. Corolla green in lower portions and brownish-purple above, to nearly 1 in. long and rather conspicuously sticky. Calyx about length of corolla, but of slender sections. Fruit an orange to yellow berry, downy, with three nutlets.

Flowering time April through May. Fruiting time August through September. The pioneer American botanist Muhlenberg is reported to have dried and roasted the berries and claimed they made an excellent coffee substitute, hence the name wild coffee. Plant does not seem to have wide use in this way.

Reported by many quacks to have medicinal value but this is not considered a valid claim. That it has been widely recognized is shown by many common names such as feverfew, wild ginseng, horse ginseng, tinker's weed, white gentian, wild ipecac, and wood ipecac.

Highbush Cranberry
Viburnum opulus
Height to 12 ft. Stem branched, erect, smooth, relatively coarse and weak. Leaves opposite, broader than long, rather obviously three-lobed, with coarse saw-toothed edges, three-ribbed, rounded at base, to 3 in. or more long, with petioles to 1 in. long and bearing glands at the top.

Native of North America, Asia, and Europe. Found on low wet lands from Newfoundland to New Jersey, west to Oregon and British Columbia. Chosen as a representative of the subgenus *Opulus*, which includes the squashberry *V. pauciflorum*, with scaly winter buds. Some 100 species in the genus; some 20 in North America.

Flowers borne in open clusters at ends of branches, clusters being to 4 in. in diameter; outer flowers being more conspicuous and to 1 in. in diameter, having greatly enlarged corollas but being without either stamens or pistils. Inner flowers produce red, sour, ½-in. translucent fruit.

Flowering time from June through July. Fruiting time September through October, though they may persist until May. 16,000 seeds per lb used in propagation, or cuttings are made. Fruits of American variety are made into delicious, beautiful jelly preferred to the bitter jelly of the European wayfaring tree.

Known in cultivation as snowball tree, white dogwood, dog rowan, whitten wood, gaiter tree, cherrywood, red elder, witch hopple, pincushion tree, squawbush, May-rose, cramp bark, and other names.   Fruits eaten by grouse and pheasants but leaves and twigs do not figure as important browse.

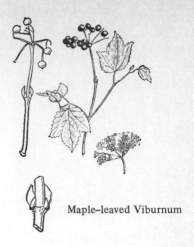

Maple–leaved Viburnum

Maple-leaved Viburnum
Viburnum acerifolium
Height to 6 ft. or rarely more.   Shrub.   Bark gray, smooth, with slightly downy, slender straight twigs.   Leaves opposite, somewhat in shape of typical maple leaf, being rather three-lobed, with heart-shaped base and saw-toothed margins, downy when young but smooth when mature; to 5 in. broad.

Native of North America and found in dry woods or where soil is thin from New Brunswick to Georgia, west to Alabama, Michigan, Minnesota, and Ontario.   Chosen here as a representative of subgenus *Euviburnum,* which in East is represented by at least five species including *V. dentatum* and *V. rafinesquianum.*

Flowers borne in open clusters to 3 in. across.   All flowers bear both stamens and pistils and are essentially alike, with stamens much longer than tube of corolla, which flares widely above relatively shallow tube. Fruit a stone fruit that is at first crimson but turns to black.   Stone with two ridges on each side.

Flowering time May through June.   Fruits borne normally in September and October but may persist until following July.   About 4,500 fresh berries per lb.   Plant a slow grower but endures smoke of cities well.   It may form thickets and when cut or burned back may form straight shoots useful as arrow shafts.

Indians used wood for arrow shafts.   Fruits are not fleshy enough to be of food value to man.   Ruffed grouse, white-tailed deer, and cottontail rabbits are known to use the plant as food and it undoubtedly provides ideal cover for many species of small wildlife.

Arrowwood
Viburnum dentatum
Height to 15 ft. but usually much less.   Shrub with smooth gray bark and twigs with sometimes remarkably straight shoots.   Leaves opposite, oval, but with margins evenly saw-toothed, with conspicuous veins, to 3 in. broad and somewhat longer, with petioles to 1 in. long, smooth or downy.

Native of North America.   Found in moist woods and thickets from New Brunswick to Georgia, west to Minnesota and Ontario.   Closely related to maple-leaved viburnum but lacks the three-ribbed maple-shaped leaf. This plant survives in moist or dry soil in the shade or in the sun.

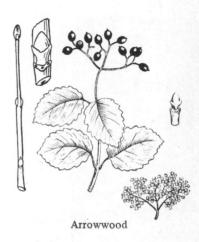

Arrowwood

Flowers in clusters to 3 in. across, all bearing stamens and pistils, capable of producing fruits, cluster being more round-topped than with some other common viburnums.   Stamens protrude beyond tube of small corolla.   Fruit a blue-black stone fruit, ¼ in. through; with stone grooved on one side and rounded on other.

Flowering time April through June.   Fruits available for harvest from October through December.   About 17,900 seeds per lb used in nursery practice, or plant is propagated by cutting.   It, like the maple-leaved viburnum, survives city smoke and so is popular as a city ornamental.

Plant also provides good cover for wildlife.   Fruits known to serve as food for ruffed grouse and other birds and for chipmunks.   If properly handled, makes a good full head of leaves and is therefore attractive. Straight new shoots good arrow-shaft material.

Nannyberry
Viburnum lentago
Height to 30 ft but usually lower.   Small tree or shrub.   Trunk diameter to 10 in.   Twigs relatively slender and winter buds smooth, slender, and pointed.   Leaves opposite, to 4 in. long, smooth on both sides, with finely saw-toothed margins, with slender wavy-margined petioles to 1 in. long.

Native of North America and found in rich-soiled areas from Quebec and Hudson Bay area, south to Georgia and west to Kansas, Colorado, and Manitoba.   Species given here as representative of subgenus *Tinus* which includes black haw *V. prunifolium,* wild raisin *V. cassinoides,* and a few others.

Flowers in many-flowered clusters that may be over 5 in. broad, the individual flowers being small with five-lobed corollas and stamens that protrude much beyond the corolla itself.   Fruit, blue-black, sweet and edible, to ½ in. long with circular flat stone.

Flowering time from May through June.   Fruit persists from August through September or later. 3,000 to 5,000 berries per lb and 4,300 seeds per lb.   Commonly propagated by seeds, by suckers, or by cuttings. Wood weighs about 45 lb per cu ft and is orange-brown and hard.   Plant withstands city smoke.

Nannyberry

(Continued)

Black-berried Elder

Red-berried Elder

Teasel

A good ornamental. Pulp of fresh ripe fruit is delicious though stone is distasteful. Usually some tart fruit is added to make the best jelly. Fruit eaten by ruffed grouse, pheasant, raccoon, skunk, fox, gray squirrel, cottontail rabbit, and, to some extent, by sheep and goats, as well as by man.

Black-berried Elder, Common Elder
Sambucus canadensis
Height to 10 ft. Stems coarse, light-colored when young, easily broken, white-pithed, with large lenticels, large buds, and a pungent odor when broken. Leaves opposite, of 5 to 11 but usually 7 leaflets, each to 5 in. long and with saw-toothed margins, paired, with the odd one at the end.

Native of North America. Found along roadsides, in moist spots in hedgerows or at edges of forests or marshes, in sun or out. From Nova Scotia to Florida, west to Texas and Manitoba as well as in West Indies. In North Carolina, found to elevation of 4,000 ft. About 25 species, of which half occur in North America.

Flowers borne in open, flattish-topped clusters that are to 6 to 8 in. across. Individual flowers white, to $\frac{1}{10}$ in. across, with five stamens, each about length of lobes of corolla and commonly alternating with those lobes. Fruit a deep purple to black stone fruit, with sweet pulp, to $\frac{1}{4}$ in. through, with roughened nutlets.

Flowers from June through July. Fruits available from August through October. Plant bears 4 years after seeds are sown, 175,000 to 468,000 seeds per lb. Grows well on coal-stripped lands where some other plants will not grow. Fresh leaves, fruits, flowers, and roots yield a purgative destroyed with cooking.

Fruits are used to make excellent pies, wine, pancakes. Common food of grouse, quail, pheasant, dove, wild turkey, red squirrel, white-tailed deer. Leaves eaten freely by livestock. Stems used by pioneers in making spiles for collecting maple sap. Gives some cover as well as food to wildlife. Bark may be considered as poisonous. Inflorescences make magnificent fritters.

Red-berried Elder
Sambucus pubens
Height to 12 ft. Stems and twigs commonly covered with down. Twigs coarse, easily broken, showing a reddish pith in younger parts. Buds large, with a bluish tinge. Leaves opposite, compound, of five to seven leaflets, rounded at base and pointed at free end, with fine saw-toothed margins. Leaflets to 5 in. long, stalked.

Native of North America. Found along roadsides and in rocky places from Newfoundland to Georgia, west to California and Alaska. In Virginia, it reaches an elevation of to 5,000 ft. Closely related to black-berried elder but these are not closely related to the black alder *Ilex*.

Flowers borne in somewhat cone-shaped but rather open clusters, not flat-topped as in black-berried elder. Clusters considerably longer than broad. Flowers white but become brown when they dry. Fruit a stone fruit that is red rather than blue or black, to $\frac{1}{4}$ in. through, finely roughened nutlets. Flowers about 90 days earlier than common elder, and fruits are ready when the flowers of common elder appear.

Flowering time April through May. Fruits available June through November, or later into January. 48,300 seeds per lb. While red berries are generally considered poisonous, evidence is not conclusive.

A beautiful plant when in fruit so that a poet can speak of "the torching of red-berried elder" in a wholly justified expression of enthusiasm. Reputed poisonous qualities of fruit diminish its popularity. Elders are important summer food of songbirds. Game birds, rodents, and some kinds of browsers feed on the fruit, foliage and young twigs of *Sambucus.*

Order Rubiales./Family Dipsacaceae

Teasel
Dipsacus sylvestris
Height to 5 ft. or more. Stem stiff, erect, branching in upper areas, covered with conspicuous coarse prickles, somewhat angled, arising from a biennial root system that penetrates soil deeply. Leaves opposite, joined at base to make a cup, coarse, with prickles and toothed margins. Fall rosettes of leaves well-formed.

Native of Asia and of Europe. Naturalized in America. Well established from Maine to North Carolina, west to Michigan. Closely related to *D. fullonum,* the fuller's teasel used in carding wool in early days. Some 15 species in the genus native of the Old World; 2 established in America and 1 cultivated.

Flowers borne in terminal spikes that are essentially cone-shaped and conspicuous because of bristles. Flowers mature first in middle of spike and work up and down simultaneously from the middle. Corolla blue, lavender, or purple, to $\frac{1}{2}$ in. long, four-parted, with calyx tube fastened to it. Fruits dry, four-sided, oblong, grooved on each face.

Flowering time July through September. Pollination by insects or by self. Crawling insects are unable to pass water pools at leaf bases but fruits are badly parasitized by other insects. Juice disagreeable to taste of cattle. Since it is biennial plant, can usually be controlled by plowing and clean cultivation.

A beautiful plant if one can look at it without prejudice. In fuller's teasel, the bristles are stiff enough to raise the surface of woolen cloth. The heads are split and mounted on belts or rollers that move over the cloth.

Order Cucurbitales./Family Cucurbitaceae

Field Pumpkin
Cucurbita pepo
Herbaceous vine. Annual. Stem prostrate, armed with stiff, harsh-feeling hairs, rather succulent and easily crushed. Leaves rigid, erect, stiff, to 1 ft in diameter, with deep lobes separated by deep notches, unlike leaves of summer and winter squash, with hollow petioles that are harsh to the touch. Fruiting stalks five-sided, swollen above.

Native of America. Found among relics of cliff dwellers. Some contend they are of Asiatic origin. Undoubtedly cultivated in America as early as 1500 B.C. Plants now growing wild in Texas resemble them. Summer squash is var. *condensa;* yellow-flowered gourd, var. *ovifera,* the latter being grown for ornament.

Flowers yellow, staminate flowers being without green globular base possessed by pistillate flowers. Corolla a great, flaring funnel, very narrow at base. Calyx with short, narrow lobes, generally not conspicuous. Fruit to 1 ft or more in diameter, depressed sphere, with longitudinal ribs, yellow, white-seeded, with a medium-hard rind.

Seeds sown in rich soil, with good moisture, six seeds per hill, and hills 6 ft apart, at depth of 1 in., thinned to three plants per hill and soil kept loose about plant. Downy mildew causes angular leaf spots controlled by bordeaux 3-6-50 when vines begin to run; powdery mildew making white leaf coatings controlled by later bordeaux application.

Sometimes planted as supplementary crop with corn. Made into pies either fresh or canned or frozen. Fair source of vitamin B and one of the essentials of a successful Thanksgiving dinner. Also popular with youngsters as the basic material for jack-o'-lanterns for use at Halloween celebrations. Stored successfully at temperatures between 50 and 60°F. Raised for cattle food.

Summer Squash
Cucurbita pepo melopepo
Common name bush pumpkin implies lack of the long, running vines of the typical pumpkin; varietal name further implies a compact plant. Otherwise it is very similar to the ordinary pie pumpkin in shape of leaves. Ordinarily not a tendril bearer. Annual.

Native of America and developed in many forms, with many different shapes and qualities. Among the types recognized are the cocozelle, zucchini, straightneck, crookneck, White Bush Scallop. Most frequently, however, name summer squash is associated with yellow form of crookneck.

Flower much like that of pumpkin, yellow, funnelform, with stem jointed where it joins fruit and expanded at that point. Staminate flowers without swollen base characteristic of pistillate flowers. Pollinated by insects. Cross-pollination between type and variety is easily possible but undesirable.

Sow in garden soil, in hills about 4 ft apart, six seeds per hill, at depth of 1 in. after danger of frost has passed. Thin seedlings to three plants per hill and keep soil loose about plants. Fruits are picked when ready to use and not allowed to ripen on vine or after. Pests similar to those of pumpkin.

Pumpkins and squash should not be planted close together as fruits resulting from crosses are usually not desirable. Summer squash commonly cooked when fruits are young and before rind has started to harden. Summer squash not so easily stored as winter squash. Yield a fair amount of vitamin B. Widely known varieties of squash and of pumpkin belong to *Cucurbita moschata* though it is not grown in some parts of the United States.

Yellow-flowered Gourd
Cucurbita pepo ovifera
Vines long and running as in typical pumpkin, slender, often climbing, tendril-bearing. Leaves much like small pumpkin leaves but more deeply lobed or scalloped.

Native of North America. A variety of typical pumpkin, with vines and leaves much like typical species. Many subvarieties based on size, shape, and color of fruit and on other characteristics. So varied are these gourds that some interesting studies in genetics have been made from them.

(Continued)

Field Pumpkin

Summer Squash

Gourd

Hubbard Squash

Watermelon

Cucumber

Flowers similar to those of pumpkin and summer squash but smaller than either. Flowers apparently require sun to succeed and so should be grown in exposed places. Fruits much smaller than either pumpkin or summer squash, with a brittle hard rind; when dried, they can withstand much abuse.

To succeed, these plants must have an abundance of water during the growing period. Seeds should not be planted until all danger of frost has passed and usually they are not planted in hills but near some tall support. Gourds are grown principally for their hard shells used as ornaments, cups, pails, birdhouses, and other things.

Some related gourds have unique uses, as the dishcloth gourd whose young edible fruit is baked or boiled and eaten while the fibrous center serves as a dishcloth or bath sponge. Many gourds have holes bored in them, their interiors removed, and pebbles or shot inserted so they may be shaken as rattles for primitive music.

Winter Squash
Cucurbita maxima
Vines extending to 20 ft or more, soft and not harsh as in pumpkin. Leaves with lobes rounded rather than pointed as in summer squash. White spots are supposed to be absent from leaves of winter squash though this is also often true of the pumpkins and summer squash. Leaves not rigid. Annual.

Native of Americas. Includes many radically different varieties such as mammoth Chile with enormous fruits sometimes exhibited as pumpkins, Boston marrow squash, Hubbard squash, and turban squashes. An essential botanical difference between this plant and pumpkin is that it lacks swollen end of supporting fruit stem.

Flowers with broad backward-bending lobes spreading from a tube that has more parallel sides than in pumpkin or may be swollen at base. Fruit stalks soft and yield easily to fingernail, unlike hard stalks of pumpkin and summer squash. Seeds white or brown, with slanting scar, and margin colored like rest of seed.

Seeds planted in hills, often with other crops such as corn, and thus require no extra land or cultivation. Hills may be 6 ft apart, with seed six to a hill and seedlings thinned to three per hill. Plants sometimes started in boxes and transplanted. They do not require a long season and do well on a variety of soils.

Hubbard and other winter squash are usually baked and soft inner rind or pulp eaten direct from shell. Plant subject to many insect and fungous enemies, among worst of which is a wilt. Winter squash store better than other similar fruits and sweeten with storage because of change of starches to sugar.

Watermelon
Citrullus vulgaris
Stem a long, running vine with branched tendrils, hairy but not rough as in pumpkin and summer squash. Leaves with three to four pairs of lobes, with heart-shaped base, to 7 in. long, smooth or hairy but not rough. Annual. Vines may be so extensive as to cover a field with their branches.

Native of tropical Africa where it grows voluntarily in some places. Has been under cultivation for centuries in India and Egypt; is shown in early Egyptian paintings and described in Sanskrit. Of four known species, three are African and one Asiatic.

Flowers small for such a large fruit, being about 1½ in. across, of five lobes divided nearly to base, light yellow, usually borne singly in leaf axils. Fruit weighs to 50 lb or more, striped with light and dark green, with hard but sensitive rind, red watery flesh, and black or white seeds. Stamens and pistils in separate flowers.

Seeds planted in hills 5 to 7 ft apart, 1 in. deep, four to ten per hill, after danger of frost has passed. Seedlings thinned to two to three per hill. Requires deep rich soil and 90 to 100 days to mature fruit. Attacked by anthracnose, powdery mildew, and striped cucumber beetle. Plants often sprayed and netted in protection.

An excellent fruit for eating fresh, in conserves, or otherwise. The cultivar *citroides,* the citron, has a white, more solid flesh and a high pectin content and is used in jams, jellies, and preserves. Watermelons constitute an important agricultural crop in the South; Texas, Georgia, Florida, South Carolina have important centers of production. They stand shipment rather well when harvested properly.

Cucumber
Cucumis sativus
Vine with stems angled, trailing or climbing, tender, rough and hairy, with simple unbranched tendrils that help vines climb on trellises or over other plants. Leaves to 6 in. long, somewhat three-lobed, about as wide as long, very rough to touch, easily crushed, rather long-petioled. Roots fibrous. Annual.

Native of southern Asia. Introduced throughout the world where climate is suitable or grown anywhere in greenhouses. English forcing cucumber is a var. *anglicus* and the bur gherkin and muskmelon are related species *anguria* and *melo,* respectively. About 30 species and many varieties known to the genus.

Flowers either staminate or pistillate, to about 1½ in. across, with corolla lobes acute, with many staminate flowers often closely clustered but the pistillate commonly borne singly. Seed to seed in 65 to 90 days. Fruit somewhat prickly, oblong or elongated, yellow when mature, with small flat white seeds and wet pulp.

Seeds planted in hills to 6 ft apart. Seedlings thinned to three plants per hill. Soil kept loosened. In greenhouses, pollination by hand may be necessary. In fields and gardens, fruits picked green, not allowed to mature on vines. 1 oz of seed enough for 50 hills; one plant may yield 100 cucumbers weighing 50 lb per bu.

An important food plant of garden and greenhouse. Possibly fair as source of vitamin B but yields little A or C. 1 lb yields 68 food calories and 0.39 oz of carbohydrate; 1 acre may yield to 100 bu of pickling cucumbers. Fruits made into pickles or preserved when ripe or eaten fresh and green. Florida, South Carolina, New York, New Jersey, and California some of leading states for growing cucumbers.

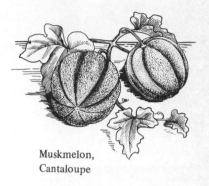

Muskmelon, Cantaloupe
Cucumis melo
Vine. Climbing, sprawling, and trailing, with ridged or angled stems covered with soft hairs. Leaves to 5 in. across, nearly round, not distinctly lobed, but with margins shallowly and broadly toothed; hairy and somewhat roughened, petioled, definitely blunt at free end. Annual.

Native probably of central Asia. Grown widely in fields and gardens where conditions are suitable. Three major American groups include nutmeg or netted melons of var. *reticulatus,* winter or cassaba melons var. *inodorus,* and cantaloupe melon var. *cantalupensis.* Name cantaloupe erroneously applied to netted melons.

Muskmelon,
Cantaloupe

Flowers about 1 in. across; stamens grouped in more than one axillary flower, pistillate usually borne separately in an axil. Corolla margins like rounded teeth. In greenhouses, pollination is by hand. Fruits of many varieties with fragrant yellow, green, or flesh-colored inner rind, musky in odor. Mature in 90 to 100 days.

Grown usually in relatively dry areas or in greenhouses, 1 vine yielding three to four mature fruits. Seeds planted in hills to 6 ft apart, thinned to three to four per hill, cultivated like other melons. Little danger of cross-pollination with others. Striped cucumber beetle, worst pest, controlled by sprays and netting.

1 lb of pulp yields 180 food calories; only about ½ purchased weight is useful. Fair as source of vitamin B. Used chiefly as appetizer. Now being quick-frozen for sale at any time of year. By forcing young plants in greenhouses, crop for November harvest may be planted in July. Should be dry during ripening season. Most important melon-producing states are California, Arizona, Texas, Indiana, and Maryland.

Prickly or Wild or One-seeded Cucumber
Echinocystis lobata
Climbing, tendril-bearing vine, sometimes to 25 ft long but weak, slender, smooth, branched, angular, grooved, and sometimes with hairs at joints or near them. Leaves borne on petioles, to 6 in. long, thin, three- to seven-lobed to form an obscure irregular star with toothed, pointed lobes.

Found in moist waste places such as wet lowlands climbing over other vegetation, sometimes forming a considerable mass. Ranges from New Brunswick to Virginia, west to Manitoba, Kansas, and Texas. Found sometimes in wild gardens where it may have been introduced as a semi-ornamental.

Prickly Cucumber

Stamens and pistils in separate flowers on the same plant, the former abundant, in rather long, loose open clusters; the latter, borne either singly or in pairs. Fruit to 2 in. long and about half as thick, spongy, covered with weak spines, capable of expelling seeds. Calyx in six parts and lobed.

Aside from the rather attractive nature of the plant, the most interesting feature is probably the expulsion of the seeds from the fruits. Seeds flat, pointed at one end, bluntly rounded at other, brown and few in number. Flowering time from June through September, followed immediately by fruiting.

Of little economic importance since the fruits are not edible and the plants are not outstandingly attractive, nor are they destructive of the plants on which they may rest for support. Common names include creeping Jennie and wild cucumber.

Order Campanulales./ Family Campanulaceae

Canterbury Bells
Campanula medium
Height to 4 ft. Stem erect, hairy, relatively stout, covered with fine bristly hairs. Leaves, those of base to 10 in. long, with long, tapering petiole-like base; those of upper stem to 5 in. long, with clasping bases and wavy margins. Biennial.

Native of southern Europe but grown rather widely in many parts of world as a garden ornamental. Some 250 species of genus found for most part native in Northern Hemisphere. Of course many varieties have been developed with varying colors of flowers and time of flowering. One colored variety, *calycanthema,* is known as cup-and-saucer Canterbury bell, which has an enlarged, colored, spreading, petallike calyx.

Flowers borne in a long terminal stalk, singly or in 2s along stalk. Corolla like an inflated bell or vase, with tube to 1 in. in diameter and lobes spread wide or bent backward, violet, blue, pink, white, or other colors. Calyx with its lobes about ¼ corolla length. Fruit a five-celled capsule.

Seeds sown in open to produce flowers succeeding year or started indoors in early spring so that seedlings may be transplanted by early May to flower both first and second year. Seedlings from summer-planted seeds may be taken up in fall to be potted for winter blooming indoors.

This species is possibly most popular of this group of plants. Among many varieties are some with calyxes colored like corollas or with corollas broadly flaring or with corollas more truly bell-shaped than is case with typical species. Unfortunate that plant is not perennial.

Canterbury Bells

Order Campanulales./ Family Lobeliaceae

Cardinal Flower
Lobelia cardinalis
Height to 4½ ft. Stem erect, not commonly branched, smooth or with some down, well-supplied with leaves. Leaves alternate, thin, green, smooth or somewhat downy, conspicuously veined, to 6 in. long and to 1½ in. wide, with clasping bases and wavy margins. Perennial offshoots appear at season's end.

Native of North America. Found in wet or moist soils from New Brunswick to Florida, west to Texas, Colorado, and Ontario. Some 250 species in the genus found widely distributed over the world but with about 30 to be found in United States. Unfortunately, being gradually eliminated.

Flowers borne largely from near top of stems, numerous, and on short individual stems, with corollas to 1½ in. long, with lower lip of three relatively distinct lobes and the tube itself about 1 in. long, brilliant scarlet or rarely white. Bracts at base of flowers usually glandular.

Flowering time July through September, and when abundant plant provides a spectacular sight. Pollination is largely by hummingbirds or more rarely by bumblebees, latter having difficulty in clinging to weak, split lower lip of corolla, but former able to probe flower without alighting.

This flower should receive protection against promiscuous collectors at all times. Many hardy and tender annuals and perennials are *lobelias.* Lobelias are best grown as annuals. Of tall-growing lobelias, the best known is *L. cardinalis. L. siphilitica,* blue lobelia, is also popular. Species name *cardinalis,* meaning cardinal, refers to the shape of the corolla, like the cardinal's miter. It has nothing to do with red, the color of cardinal, the bird.

Cardinal Flower

Order Asterales./ Family Compositae

Joe-pye Weed
Eupatorium purpureum
Height to 10 ft. Stem erect, smooth or downy, branched or unbranched, green or bluish, round or ridged. Leaves opposite, or more commonly in whorls of three to six, rather long-petioled, with saw-toothed margins, to 1 ft. long and to 3 in. wide, smooth or downy.

Native of North America. Found at edges of marshes or in other wet places, and in wet wastelands from New Brunswick to Florida, west to Texas and Manitoba. Belongs to tribe *Eupatorieae,* whose heads are in disks and whose flowers are all perfect and tubular. Over 500 species in the genus.

Flowers in numerous, more or less cylindrical heads with pinkish or bluish outside bracts and with enclosed flowers pinkish, bluish, purple, or sometimes white. Fruit with about five angles, crowned with rough bristles. Normal reproduction by means of wind-distributed fruits.

Flowering time August through September. Fruits found on dead tops of plants on into winter. As a weed, plant is not a serious pest since it yields readily to cultivation or to drainage. Usual procedure is frequent mowing where control is necessary.

Joe–pye Weed

An attractive marsh plant that figured in early pioneer medicine but has no recognized merit in the modern drug trade. Many common names attest its general familiarity as motherwort, niggerweed, skunkweed, gravelroot, boneset, purple boneset, kidneyroot, and king of the meadow. Some people like to use this species in a cultivated herbaceous border, in a wild garden, or along the margin of a pond or stream.

Thoroughwort, Boneset
Eupatorium perfoliatum

Height to 5 ft. Stem branched in upper areas, pale green, hairy, stout, ridged, erect. Leaves opposite, joined at base so that they seem to be pierced by stem, hence *perfoliatum,* widest at base, tapering evenly to tip, to 8 in. long and to 1½ in. wide, with finely toothed margin, pale green.

Native of North America. Found commonly in wet meadows, marshes, and wastelands. From Nova Scotia to Florida, west to Texas, Nebraska, and Manitoba. This species and others here treated are members of subgenus *Eupatorium,* characterized by having bracts of head enclosure (involucre) unequal.

Flowers in closely arranged heads, with 10 to 40 flowers in each head and heads to ¼ in. long. Bracts enclosing head are unequal in length, are pointed and are in 2 to 3 series. Corolla tubular, white or blue. Fruit five-angled, black or brown, with yellowish dots, crowned with hairy parachute.

Boneset

Flowering time July through September. Fruits may be shed from mid-fall on into winter. Plant perennial but not persistent and cannot survive drainage, repeated cultivation, or hoeing. Leaves contain the glucoside eupatorin used in medicine as a tonic or stimulant or, in large doses, as an emetic.

Plant had widespread use medicinally in early days and is still recognized as having some merit. Dried leaves and tops yield fluid extract, of which 30 gr is considered an average dose. Raised as a drug plant it yields 1 ton of dry herb per acre. Other names include agueweed, Indian sage, crosswort. Common name boneset refers to supposed ability of this plant to be cure-all for many human ailments. The name may have been related to the fusion of the leaves indicating that the use of the plant might be used to get fractured and broken bones to come together again.

White Snakeroot
Eupatorium rugosum (urticaefolium)

Height to 4 ft. Stem erect, smooth, somewhat sticky, much-branched, relatively slender. Leaves opposite, much like those of nettle, hence *urticaefolium,* to 6 in. long and to 3 in. wide, with petioles to 2½ in. long, conspicuously veined, with saw-toothed margins, turning conspicuously dark to black with the first frost.

Native of North America. Found in woodlands where the soil is rich. From New Brunswick to Florida, west to Texas, Nebraska, and Minnesota. Plant is one of the conspicuous plants of many woodlands in late summer and fall.

White Snakeroot

Flowers in heads, arranged in a flat-topped formation, with from 8 to 30 flowers in a head, each bearing pistils and stamens. Corolla conspicuously white so that en masse plants may give woodlands appearance of being smoke-filled. Bracts of involucre that encloses each head are about equal to each other. Perennial.

Flowering time July through November. Leaves and stems contain poisonous alcohol tremetol, which is soluble in milk fat. Cattle eating this may yield poisonous milk, causing milk sickness to both cattle and man. The cattle tremble, are depressed, constipated, and nauseated, have labored respiration, and stand with difficulty.

Milk sickness in man may cause weakness, nausea, vomiting, stomach pains, thirstiness, dry skin, constipation, slow pulse, slow respiration, low body temperature, weakness, collapse, and in some cases death. Cattle eat plants usually only when other forage is not available.

Tall Ironweed
Vernonia altissima

Height to 10 ft. Stem erect, smooth or practically so, freely branched in flower-bearing portion, relatively slender. Leaves alternate, smooth on both sides, to 1 ft long and to nearly 2 in. wide, with shallow saw-toothed margins and rather conspicuous veins, pointed at each end, with suggestion of petiole at base.

Native of North America. Found in moist, rich soils, mostly in open from New York to Florida, west to Louisiana, Kentucky, and Michigan, with range extended farther west by closely related species. Over 500 species of genus of wide distribution in temperate areas but most abundant in South America.

(Continued)

Tall Ironweed

Flowers borne in heads of 15 to 30 flowers, all tubular, all bearing stamens and pistils, to ¼ in. across, with bracts of involucre, enclosing heads of varied length, with shortest to outside. Fruit dry, crowned with parachute of bristles arranged in 2 rows, with outer row shorter than inner. Bristles, purplish.

Flowering time July through September. In full fruit, shows rather conspicuous purplish top. Ironweeds of one species or another are rather conspicuous in late summer and fall prairie flora through much of the Middle West. They are not so evident in the East.

Classified as a weed but not as a serious one since it yields readily to control measures such as plowing and cultivation. Plant is recognized as being a valuable nectar producer by beekeepers who have resented its destruction along roads and railroad right-of-ways. In this respect, it is like its close relatives the eupatoriums. A standard indicator of overgrazing. Ironweed is very popular with butterflies, especially the small summer and fall marsh skippers. The flowers serve as an excellent collecting bait.

Blazing Star

Tall Blazing Star
Liatris spp.
Height to 5 ft. Stem erect, rather stout, usually unbranched even in the flowering portion. Leaves deep rich green, alternate, hoary, narrow, to 6 in. long and to ½ in. wide, entire, with upper even proportionately more narrow than lower, usually somewhat resinous. Root system cormlike and substantial. A large and complex group of plants all of which are apt to be encountered in dry open country.

Native of North America. Found in open prairie country widely spread from Maine to Florida, west to the Rocky Mountains and into southern Canada. It favors dry soil. The genus is often referred to as *Lacinaria*. About 35 species in the genus native of eastern and central North America.

Flowers borne in not too crowded heads along upper portions of erect stem, heads being hemispheric, to 1 in. across, composed of 15 to 45 flowers, with brilliant magenta-purple corollas. Bracts of enclosing involucre may be in 5 to 6 series and parachute that crowns fruit of finely barbed bristles.

Flowering time from August through September. Any attempt to describe prairie flora for this period without reference to one or another of the species of this genus is inadequate. Some of the species are grown as garden ornamentals. Of these, *L. elegans* and *L. pycnostachya* are most popular. Division of underground parts is a common propagation practice.

Difficult to understand why this plant is not more widely grown as an ornamental. Has many common names such as gay feather, blue feather, rattlesnake master, devil's-bite, button snakeroot, blue blazing star, and so on. Albinos are quite common and showy. Many of these plants are beautiful in open prairies.

Gumweed

Gumweed, Gumplant, Tarweed, Rosinweed
Grindelia squarrosa
Height to 2 ft. Stem erect, branched, smooth, commonly with reddish cast. Leaves alternate, with clasping bases, rigid, roughened, to 1½ in. long and to ½ in. wide, with sharply toothed margins, widest in upper half, with the lower commonly narrowed at base into petiole. Root system a substantial vertical rootstock with branches.

Native of North America. Found in open dry pastures and meadows from New York to Pennsylvania, west ot Texas, Arizona, and Mexico and to Manitoba. Not abundant in the eastern portion of this range but a common prairie plant through Middle West.

Flowers borne in heads that usually have distinct tarry or gluey feeling outside where bracts of involucre are narrow and pointed with free ends spread back. Flowers in heads are of two kinds, with outer ray flowers bearing pistils only and inner ray flowers with both stamens and pistils. Inner and outer ray flowers may be sterile. Ray flowers to 1 in. long, or absent.

Flowering time June through September, so it is a long-season plant. Fruits bear two to three rather stiff awns at top rather than any abundant parachute. Plant may be considered a seriously poisonous forage for grazing animals in regions where there is selenium in the soil. In other regions, it is harmless.

Plant is considered by range managers a good index of overgrazing. Management calls for hoeing or cultivating the plant before fruits form. It yields rather well to plowing and cultivation, so a crop rotation with open cultivated plants in the cycle serves as a good check. This of course may be impractical in range management.

Canadian Goldenrod
Solidago canadensis
Height to 5 ft. Stem erect, slender, smooth or downy above, somewhat angled or ridged. Leaves definitely three-nerved, with branching veins, to 5 in. long and ½ in. wide, with shallow saw-toothed margins, narrowed at either end or the lower ones with petioles.

Native of North America. Found from Newfoundland to Virginia, west to Tennessee and Saskatchewan. Not ordinarily found bordering salt water. Goldenrods, asters, gumweeds, and others here considered belong to the tribe *Astereae* whose flower heads have flat smooth styles.

Flowers borne in small heads, erect on arching to horizontal upper branches, composed of four to six yellow short ray flowers with bracts of enveloping involucre pointed at tip, thin and narrow. Fruit crowned by many slender, equal, hairlike bristles.

Goldenrods are the state flowers of Nebraska, North Carolina, Alabama, and Kentucky but it is doubtful which species is meant in some cases. Goldenrods have been imported into London with the thought that they could heal wounds.

It has been proposed that since goldenrods of one sort or another are to be found conspicuously and widely distributed over America, they should be our national flower. Unfortunately, this might be taken to mean that we worship gold and the idea has not won favor. They are considered to cause hay fever. Parts of the plant may be eaten by birds and small mammals.

Canadian Goldenrod

Late Goldenrod
Solidago gigantea
Stem erect, to 8 ft high, smooth or sometimes with a bloom, somewhat irregularly ridged. Leaves with three main veins, with sharply but very shallowly cut margins, to 6 in. long and to 1½ in. wide, the lower with petioles, smooth on both sides or finely downy on the underside.

Native of North America. Found from New Brunswick to Georgia, west to Texas, Utah, and British Columbia, reaching an elevation of 2,300 ft in Virginia. It favors moist meadowlands. Of about 125 species, almost all are natives of North America with only 3 or 4 outside our continent.

Flowers borne in relatively small heads, arranged usually along upper side of arching upper branches to form wandlike structures. Heads are to ¼ in. high and the bracts of the enclosing involucre are blunt and rather thin. About 7 to 15 flowers in each head.

State flower of Nebraska. This species does not survive cultivation although it is perennial and persistent if undisturbed. Formerly, goldenrods were thought to yield wound-healing juices but this is not considered a valid assumption at present.

Some goldenrods have a place as ornamentals in wild-flower gardens. Two species of 24 eastern species have been found to contain sufficient latex in the juices of their leaves to warrant their improvement as a possible source of rubber. Solidago may cause hay fever in some persons.

Late Goldenrod

Silverrod, White Goldenrod, Silverweed
Solidago bicolor
Height to 4 ft. Stem usually unbranched but not always so, erect, often densely hairy but sometimes almost smooth, rather stout for a goldenrod. Leaves alternate, to 4 in. long and to 4 in. wide, widest at free end with base of lower leaves definitely narrowed into a winged petiole, usually downy.

Native of North America. Found in dry soil, frequently in the shade of woodlands or in some thicket, not so common in open sunshine. From mouth of the St. Lawrence to Georgia, west to Tennessee and Minnesota. In North Carolina, found to 6,300 ft elevation.

Flowers borne in clusters of heads arranged in alternate groups along upper portion of stem. Heads to ¼ in. high, with ray flowers white instead of usual yellow of a goldenrod and with bracts of involucre that encloses the head whitish and helping to give the name silverrod. Fruits smooth.

Blooms from July through September. Rarely appears in lists of wild flowers through it has a beauty of its own. Not a weed in strictest sense nor does it apparently have medicinal or commercial possibilities. Its common names include bellyache weed, from the assumption, probably false, that it either caused or cured that malady.

It is just as dangerous to assume that all goldenrods cause hay fever as it is to suppose that none of them do. They do bloom at the time when less conspicuous but more virulent ragweed is at its height and they are blamed for much hay fever that other plants may cause.

Silverrod

Flossflower

Flossflower
Ageratum houstonianum
Stems relatively weak, to about 1 ft high or sometimes higher and forming an open loose mass. Leaves blunt or rounded at base. In *A. houstonianum,* leaves are heart-shaped at base. Leaves mostly opposite and with distinct petioles. Roots fibrous and much-branched.

Native of Mexico. Hardy ageratum belongs to a related genus. Of about 30 species, native herbs of tropical America, only 2 appear commonly in greenhouses or gardens. Favor loose relatively rich well-drained soils and withstand heat well.

Flowers tubular, blue, rarely pink or white, and in heads ¼ in. across or less. In *A. conzoides,* the heads are commonly under ¼ in. across. Flowers fragrant, long-lasting, always attractive. Flowers all summer, or until frost comes. If protected, it is perennial.

Propagated by cutting to get early blooms or by seeds planted indoors in late March or outdoors after danger of frost has passed. In greenhouse, blooms profusely through winter. Flowers must be kept picked if blooming is to be prolonged. Worst pests are aphids, red spider, and thrips. Difficult to keep varieties true, so it is a gamble what may be expected from most seed. Tall and short strains seem to be more permanent.

Seeds sown in February produce marketable plants in 4-in. pots by May at 50°F. Pinching off ends develops bushiness. One of most popular plants for growing in hanging baskets and window boxes, where they are grown with geraniums, wandering Jew, and coleus.

European or Garden Daisy

European or Garden Daisy
Bellis perennis
Practically stemless. Appears as flat compact tufts of leaves, all of which of course are basal. Leaves to 2 in. long, obviously narrowed at the base into petioles, downy to hairy, widest toward free end, with margins only slightly broken. Root system perennial and substantial.

Native of Europe but established rather widely in America, appearing usually as patches in lawns either in shade or sun. Found here most commonly in the Northeast from Nova Scotia to Pennsylvania, west in broken areas to California and British Columbia. Also native in Asia.

Flowers borne in crowded heads on ends of short scapes or stems; one head to a scape, to 1 in. broad. Ray flowers many; either white, purple, or pink, while bracts of involucre that bind flowers of head are usually purple. Flowering period April through November.

Probably introduced with grass seed. Once established, may spread over small patches excluding all other plants. Rarely however do they take over a whole lawn. At least 11 well-defined races have been identified in this daisy but they do not justify varietal status.

This is the true English daisy, not closely related to ours. It has many common names such as ewe-gowan, May gowan, childing daisy, herb Margaret, March daisy, bairnwort, boneflower, white daisy, and lawn daisy. Attractive border plant.

China Aster

China Aster, Annual Aster
Callistephus chinensis
Stems erect, to 2½ ft high, branching, with slightly sticky, short, stiff hairs. Leaves broadly triangular to egg-shaped, deeply and irregularly toothed, the upper being narrower than the lower and without stalks common on lower. Roots fibrous.

Native of China and Japan. Cultivated in Europe more than 200 years; in America more than 100 years. Grows best in rich, sandy loam either in full sunshine or in part shade. Well-decayed manure makes best fertilizer. Favors temperate climate.

Flowers in terminal heads, showy, single or double, large or small, in tints of pink, blue, and white. Early-flowering varieties sown in July, benched in late August, mature in January if 5 hours of additional light is added each day by 50-watt lamps spaced at 4 ft, and raised 18 in.

Sow in March; transplant to get early bloom; or sow outdoors in May, for September bloom. Cover seeds ½ in. deep. Thin plants to 18 in. apart, cultivate lightly and frequently, add lime if soil is sour, keep fertilizer from direct contact with roots. Seeds germinate in 8 days; live 3 years. Spray with kerosene emulsion for tarnished plant bug. Mix nicotine dust in soil to control root lice. Knock off blister beetles into can of kerosene. Dig and burn plants attacked by yellows, wilt, or rust. Yellows not common under glass but leaf hoppers may be serious pest here.

Valuable and popular cut-flower plant grown best in greenhouses. Keeps best after cut if cut ends are in 2 percent sugar solution instead of in water. Chief types are Ostrich Plume, to 18 in.; Giant Comet, 15 to 18 in.; Peony-flowered, 24 in.; California Giant, to 36 in.; also miniature forms, Miniature Pompon and Lilliput strains.

Large-leaved Aster
Aster macrophyllus

Height to 3 ft or sometimes more. Stems coarse, rough, angular, reddish and rarely straight, arising from a long thick rootstock. Leaves basal, like huge pointed hearts, on long petioles, with three to four to a stem, to 1 ft or more long, rough above and generally harsh to the touch.

Native of North America. Found in relatively open woodlands or shady dry areas. From southeastern Canada to North Carolina, west to Minnesota. A number of different races have sometimes been given varietal standing by different writers but others group them all under this species.

Flowers in relatively large heads that are usually widely separated, to ½ in. high, with about 16 pale lavender, violet, or almost white ray flowers over ½ in. long surrounding the smaller, tubular, perfect, yellow disk flowers of the center.

Flowering time August into early September, making plant a relatively early species. Leaves appear early in spring and remain late; since they are so large, they often make a conspicuous part of woodland ground flora. When completely mature, disk may be reddish-brown.

Young leaves of this aster are sometimes used as herbs comparable to dandelion and spinach. They become too tough to be generally popular and should ordinarily find favor only when they are young and fresh. Even then, they can hardly rival other sources of food usually available.

Large–leaved Aster

New England Aster
Aster novae-angliae

Height to 8 ft. Stem branched, stout, with abundant sticky stiff hairs well covered with leaves. Leaves entire, to 1 ft long and nearly half as wide, thin, a rich dark green, more or less uniform in width, pointed at free tip. Root system perennial, horizontal, and substantial.

Native of North America. Found often most abundantly along edges of swamps or roadsides or in neglected fields. From Quebec to South Carolina, west to Alabama, Colorado, and Saskatchewan. Some 250 species in genus, of which more may be native of North America than of other continents.

Flowers borne in heads at ends of or near ends of uppermost branches, erect for most part, with 30 to 50 brilliantly colored ray flowers that are slender, to ¾ in. long, usually bright violet-purple, pink, red, white, or rose, with pistils but no stamens, and tubular, yellow disk flowers that have both stamens and pistils.

Time of flowering August through October, at which time they may give their surroundings a color effect that may well rival maples at their best. Fruits freed to winds from September through November. May be starved by repeated cutting. Cultivation immediately eliminates it as a possible weed.

Too beautiful to be destroyed if area it occupies is not needed by other plants. While it yields no important medicinal elements and it is not considered a good forage for domestic stock, it adds beauty to its surroundings and provides welcome cover to many small wild game animals.

New England Aster

Wild Heath Aster
Aster pilosus

Height to 3 ft. Stem widely branched, slender, tough, smooth or downy, making a bushlike top to the plant. Leaves: basal or stem ones, to 3 in. long and to ¼ in. wide or with base a long-drawn-out petiole and free end relatively blunt; uppermost leaves narrow and relatively short and dry.

Native of North America. Found in one or more forms from southeastern Canada and Maine to Florida, west to Missouri and Minnesota. As a rule asters of this group do not produce any valuable forage to the territory in which they may grow.

Flowers in numerous small heads to about ½ in. broad, with from 15 to 25 rays that are white or tinged with rose and are pistillate; with smaller number of yellow, tubular disk flowers that bear both stamens and pistils and produce fertile seeds. Parachute that bears mature fruit is white.

Flowering time September through December, making plant one of late-blooming asters. Not uncommon to find good flowers of this plant in perfect bloom in protected areas even after first snowfall. Some aster relatives of the West, such as *A. parryi* and the alkali aster *A. glabriuscula*, are believed to poison sheep.

Some asters provide good grazing for elk, deer, horses, and sheep while others are of practically no importance. A few as suggested above are poisonous. On the whole, asters are merely good to cover land with a degree of beauty all their own.

Wild Heath Aster

Robin's Plantain

Daisy Fleabane

Horseweed

Robin's Plantain
Erigeron pulchellus
Height to 2 ft but usually lower. Stem usually unbranched except in flowering portion, densely covered with hairs arising from a perennial rootstock or from prostrate stems that help spread plant. Leaves mostly basal, to 3 in. long and to 2 in. wide, short-petioled, hairy, and usually almost entire.

Native of North America. Found from Nova Scotia to Florida, west to Louisiana, Kansas, and Minnesota. *E. philadelphicus*, perennial, with leaves almost entire. Unlike *E. annuus* and *E. ramosus*, which are usually annuals. Genus name from *eri* meaning early, and *geron* meaning old man; having reference to the hoary growth or down on some members of genus.

Flowers borne in relatively large beautiful heads that are to 1½ in. across while heads of *E. philadelphicus* are rarely to 1 in. across. Ray flowers abundant, narrow, purple, violet, or blue and contrast with greenish-yellow disk flowers that form a flat center of the head.

Flowering time April through June, while that of Philadelphia fleabane is from April through August. This may provide a simple superficial way of separating the two at least in late summer. Propagated by seeds and by stem offshoots that develop at base of clusters.

A beautiful flower that might find a place in a flower garden if it were not considered as a weed. Since it yields readily to almost any control measure, it is difficult to consider it at any time as a serious weed pest. Hoeing the crowns is an effective means of destroying a thing of beauty.

Whitetop, Daisy Fleabane
Erigeron annuus
Height to 5 ft. Stem erect, stiff, slightly ridged, greatly branched in upper areas, sparingly hairy, with hairs spreading. Leaves alternate; the lower, to 6 in. long and to 3 in. wide, with a petiole and rather coarse marginal teeth; the upper usually without petioles, narrower, often nearly entire.

Native of North America, established from there in Europe, Bermuda, and probably other parts of world. Found from Nova Scotia to Georgia, west to Missouri and Alaska. It grows in common roadsides and waste places usually thought of as ideal for what we call weeds.

Flowers borne in numerous heads at tips of uppermost branches, heads being "daisylike," about ½ in. across, with ray flowers essentially uniform in width throughout their length as with other erigerons; unlike asters in this respect. Rays white or tinged with purple. Annual. Parachute on fruits, double.

Flowering time May through November. Fruits from June to end of year. Sheep reputed to favor white-top over good grass hay and if turned into a field will destroy the fall rosettes that may have developed from early seedings. Ordinarily plants should be destroyed at first blooming.

Fields badly infested with this plant may be cleaned by plowing and sowing to winter wheat, following this with a clean-cultivated crop. Known as lace buttons, tall whiteweed, and sweet scabious. In spite of fact that sheep favor it as a food, farmers do not recognize it as a valuable forage plant.

Horseweed
Erigeron canadensis
Height to 10 ft. Stem densely covered with rather stiff hairs or smooth, much-branched at top with branches commonly turning upward. Leaves alternate, the lower to 4 in. long, blunt, broadest near tip, entire or obscurely notched, while upper leaves are much narrower. Annual. This plant may be given scientific name *Conyza canadensis*.

Native of North America and found over all of North America except in extreme north. Widely distributed in Asia, Europe, South America, and West Indies. Over 150 species of the genus, of wide geographic distribution. Common in fields and waste places generally.

Flowers borne in small, rather widely separated heads, numerous, about ⅕ in. across, enclosed by a rather compact involucre of slender, smooth, sharp-ended bracts. Ray flowers short, white, numerous but relatively inconspicuous, with flat part shorter than tube that bears it.

Flowering time June through November. A drug from this plant produces a dermatitis, smarting eyes, sore throat, and prostration. It is a diuretic, tonic, and astringent, used in treating chronic diarrhea and certain types of hemorrhage. Dermatitis is most commonly observed in those who handle the dry plant as hay.

Name horseweed probably comes from fact that plant is abundant in horse pastures. Others contend that farmers formerly rubbed their horses down with it, but it may affect some animals as does poison ivy. Controlled by repeated cutting, by burning to destroy the fruits, or by growing an open cultivated crop. Has other common names: fleabane, bitterweed, hog-weed, mare's-tail, blood stanch.

Plantain-leaved Everlasting
Antennaria plantaginifolia
Height to 20 in. Stems sprawling from a crown, with ascending leaf-bearing tips, more or less erect, crowned with flowering heads, either stout or slender, hairy or not. Leaves three-ribbed, to 3 in. long and to 1½ in. wide, pale green above, silvery beneath. Perennial.

Native of North America. Found on dry soil particularly in open hilly woodlands from Labrador to Georgia, west to Texas, Nebraska, and Minnesota. About 50 species in genus found in North Temperate Zone and well-represented in South America.

Flowers in heads, crowded to make a powder-puff effect at top of erect flowering stems. Plants have either flowers bearing stamens or flowers bearing pistils. Each head is about ¼ in. in diameter, with those bearing pistils appearing pink to crimson and those bearing stamens appearing white. Fruiting heads appear gray.

Flowering time April through June. Fruits found in varying amounts through latter part of year. Apparently plants of this genus are not eaten by grazing animals when other more desirable plants are available. Therfore overgrazing removes competitors and allows antennarias to take over extensive territory.

Where this plant becomes established as a weed in pastures, plowing, disking, and cultivating are usual control measures. However, these squat plants may serve a useful function as soil anchors where soil is thin and might be washed away were no other cover provided. Some relatives may form indigestible cud balls in cattle.

Plantain-leaved Everlasting

Pearly Everlasting
Anaphalis margaritacea
Height to 3 ft. Stem branched in upper areas, densely covered with a white, woolly mass, and well-supplied with leaves. Leaves to 5 in. long and to ⅓ in. wide, woolly beneath and somewhat downy above, green, with margins frequently turned inward and lower leaves the shorter.

Native of North America. Found on dry soil from Newfoundland to North Carolina, west to Oregon and Alaska. Established in Europe and northern Asia. Of the three dozen species, most are native of North Temperate Zone. This one is sufficiently attractive to be listed as a wild flower.

Flowers in numerous heads grouped into a compact flat-topped cluster that may be 8 in. across. Individual heads may be ⅓ in. across with bracts of enclosing involucre; with outer apparently shortest and pearl white. Tubular pistillate, and slender staminate flowers are usually in same head. White daisylike flowers appear in summer. If cultivated, must be planted in a sunny place in autumn or spring.

Flowering time from July through September. Sometimes grown as ornament in wild-flower gardens. Of related species known as cudweeds, a number are listed as weeds but few if any are serious pests. Most can be controlled either by drainage or by cultivation of the land.

The name everlasting is applied to a number of species of this genus and of other genera, for example, *Helichrysum*, the strawflower, which appears commonly in bouquets sold at roadside stands in the West. Flowers of *Anaphalis* spp. are grown for sale by florists as flowers dyed in a variety of colors and used as winter decoration under the name of Immortelle.

Pearly Everlasting

Low Cudweed
Gnaphalium uliginosum
Height to 8 in. Stems sprawling on ground or turning upward at tip, or erect, well-branched close to ground, covered with relatively closely pressed white wool. Leaves to 2 in. long, narrow, well-covered with white wool, pointed. Root system shallow or with deep penetrating taproot. Annual. Name *Gnaphalium* derived from *gnaphalon*, soft down, and has reference to woolly covering of plants.

Native of North America and of Europe. Found usually in damp soils from Newfoundland to Virginia, west to Oregon and British Columbia. Well over 100 species of the genus covering a wide geographic area. Sometimes, owing to various conditions, they may apparently crowd out all other plants.

Flowers grouped in relatively small heads at end of erect stems and often partially enclosed in leaves. Flowers all tubular, with outer flowers with pistils and no stamens, and larger, relatively few inner flowers with both pistils and stamens. Enclosing bracts whitish.

Flowering period July through to October. Fruits found from midsummer to end of year. Favors wet lowlands, the sides of streams, and waste places and must have water to be at its best. Consequently it is often easily controlled by drainage practices.

Sometimes plant is grown in borders in wild-flower gardens and even in other gardens. Hoeing will keep it within bounds in such situations. Where it seems to have crowded out other plants, it may be merely surviving where others could not stand existing conditions.

Low Cudweed

Elecampane

Rosinweed

Common Ragweed

Elecampane
Inula helenium
Height to 6 ft. Stem coarse, branched or unbranched, usually rising from a large rootstock, densely downy to hairy in upper parts. Leaves to 20 in. long and to 8 in. wide, downy, rough above; the larger lower ones with long petioles, the upper with none but with clasping bases.

Introduced from Europe and well-established in America. Also native of Asia. Found in dry fields and meadows from Nova Scotia to North Carolina, west to Missouri and Minnesota. Of nearly 100 species of the genus to be found in Europe and Asia, this is only one that seems to have become established in America. Rare in Pacific Northwest. Not uncommon as a roadside or field weed in Midwestern States.

Flowers borne in large heads at or near tips of uppermost branches. Heads to 4 in. across and to 1 in. high, with outer ray flowers pistillate and tubular center flowers with both stamens and pistils. Rays, numerous. Bracts of enclosing involucre almost leaflike. Fruit a dry, smooth, four-sided, one-seeded achene.

Flowering time from July through September, with fruits being borne from August through October. Fruits usually insect-infested. Cultivated best in deep clay soil. Usually propagated by divisions of old roots planted in fall. Set 18 in. apart, in rows 3 ft apart, or may flower from seed second year. Second-year roots dug in fall, dried, and sliced.

1 acre may yield to 1 ton of roots. Roots used as a drug; or fresh roots are candied in sugar and used as a medicinal confection. Leaves reported to be edible but this is questioned seriously. Also given common names: horseheal, horse-elder, yellow starwort, elf-dock, elf-wort.

Rosinweed, Compass Plant
Silphium laciniatum
Height to 12 ft. Stems branched or unbranched, distinctly resinous, rough and harsh, stiff, arising from a thick, deeply penetrating root. Leaves alternate, long-petioled, to over 1 ft long, deeply cut, the upper with clasping bases; sometimes almost entire.

Native of North America. Found on open prairie country and along dry roadsides from Ohio to Alabama, west to Texas and the Dakotas. A dozen species of the genus to be found in North America. Indian cup, *S. perfoliatum*, is a wet-soil species but most are at their best in dry lands. Does not succumb easily to man's activities; so persists along roadsides and on spaces not disturbed by man's activities.

Flowers in several heads at ends of branches. Heads to 5 in. across, with individual ray flowers to 2 in. long and from 20 to 30 in number. Rays notched at tip, yellow, pistillate and produce fruits. Yellow disk flowers tubular, with stamens and pistils but do not produce fertile seeds.

Flowering time July through September, fruiting time August through October. Perennial. Propagation commonly by seeds. Leaves frequently arrange themselves in a plane that indicates north and south but this is not wholly reliable.

Plant is a relatively common weed but it yields to cultivation. It has been reported that the Omaha Indians used the tarry juices that come from scarred areas and some persons break the plants to induce flow of this substance.

Common Ragweed
Ambrosia artemisiifolia
Height to 6 ft but usually much less. Stem profusely branched, downy, smooth, or intermediate. Leaves thin, the upper alternate, the lower opposite for most part, twice-divided into a lacelike arrangement, to 4 in. long, deep green above and paler beneath. Root system deeply penetrating. Annual.

Native of North America. Found as a pernicious weed from Nova Scotia to Florida, west to Mexico and British Columbia, though Canadian line marks almost the northern limit of serious abundance. Of the 15 species of the genus, most are natives of North America, where some are serious weeds.

Flowers borne in two kinds of clusters. Conspicuous staminate flowers borne in terminal clusters that are to 6 in. long, with head drooping, with both heads and clusters numerous. Pistillate flowers in axils of upper leaves, concealed, the one flower producing a top-shaped fruit with a pointed crown, to ½ in. long.

Flowering time July through October. Fruits maintain a ready fertility for 5 years or more so fruit production should be stopped. Some plants yield only pistillate flowers. This plant harbors a borer in the stalk with a wasplike parasite that destroys Oriental fruit moth, a serious peach-orchard pest.

While the plant is one of major sources of pollen that causes hay fever, nevertheless fruits form a major winter food for grouse, quail, and wild turkeys. Has been listed as a part of the diet of birds and mammals. Seeds may pass through animal undigested. Related *A. psilostachya* is more western in its range and is perennial.

Giant Ragweed
Ambrosia trifida
Height to 18 ft. Stem coarse, erect, branched, rough, scaly, smooth or with fine stiff hairs. Leaves opposite, petioled, of three to five lobes, coarse, dark green; the lower ones to over 1 ft wide and the upper smaller and sometimes not lobed; veins joined inside leaf margin. Annual.

Native of North America. Found usually in rich moist lowlands while common ragweed is at best in drier uplands. Ranges from Quebec to Florida, west to New Mexico, Colorado, and Manitoba. One of more conspicuous wasteland, lowland plants of Middle West, where this species is almost certain to be found in field borders.

Flowers staminate, in terminal or axillary clusters to 10 in. long with the heads small and clustered along main axis; pistillate, clustered in axils of upper leaves and yield top-shaped fruits to ⅓ in. long and crowned with one large and a number of small tubercles.

Flowering time July through October, at which time amount of pollen freed from a patch must be enormous. It, like the pollen of the common ragweed, causes hay fever. Since plant grows in lands that normally are not cultivated, it usually continues contaminating an area without interference.

One of worst hay-fever weeds but also one whose fruits are eaten by some game birds as well as being parasitized by a number of insects. Forms escape cover for pheasants and rabbits. Known as buffalo weed, horse cane, bitterweed, richweed, and wild hemp. Because of the large size of seeds and their hard coat, this species is of little importance as "bird-food-seed." Ragweeds take over in grain fields after crop is harvested.

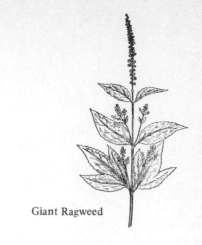

Giant Ragweed

Cocklebur
Xanthium strumarium
Height to 2 ft. Stem rough, light green to light straw-colored, coarse, branched, angled, sometimes with red spots. Leaves alternate, simple, long-petioled, rough and downy, to 5 to 6 in. long, lobed or only slightly so. Root system deep. Annual. May be listed as *orientale, pennsylvanicum, canadense,* or *commune.*

Native of North America. Found through southern Canada to Mexico, being one of worst weeds of Mississippi Valley area. More than a dozen species in the genus, with representation over a wide geographic area. Most of them favor sandy ground where water is available at no great depth.

Flowers: staminate and pistillate in separate heads on the same plant; the staminate in short ball-like heads at ends of branches; the pistillate developing into coarse-spined burs crowned by two inwardly curving spurs which contain two fruits, flattened and generally oval in outline.

Flowering time August through October. One fruit may germinate the first year and the other the second year. Seeds and young plants contain the poisonous glucoside xanthostrumarin, which decreases as plant matures but if eaten may cause depression, vomiting, weak pulse, lowered temperature, spasms, and often death. Recent research would indicate hydroquinone is another poisonous principle in *X. strumarium.* Seeds contain toxic substance but rarely eaten. Poisonous principle distributed in seedling as seed germinates.

May be a serious pest in crops. Plants should be hoed and cultivated out of existence, controlled by modern weed sprays or by rotating crops to include clover and grass, which compete successfully and eliminate the cocklebur. If poisoning has resulted from eating the fruits, milk may be drunk to counteract it. Cocklebur has a very precise reaction to day length in coming to flowering, and is therefore a popular research object with plant physiologists. It is a short-day plant, and a classical experimental organism in some plant science courses.

Cocklebur

Oxeye, False Sunflower
Heliopsis helianthoides
Height to 5 ft. Stem coarse, smooth, erect, branched, somewhat ridged. Leaves opposite or in whorls of three, to 6 in. long and to nearly 3 in. wide, petioled, with even-sized, saw-toothed margins, with blade rounded at base and pointed at tip, with relatively conspicuous veins. Perennial.

Native of North America. Found in open meadows and waste places from southeastern Canada to Florida, west to Tennessee and the Dakotas. All six species of the genus are native of North America. Two commoner species in East are separated in part on smoothness, this one being smooth while *H. scabra* is rough.

Flowers in heads, to nearly 3 in. across terminal or axillary in uppermost branches, with ray flowers that are to 1 in. long, pistillate and developing into fruits. Disk flowers tubular, with stamens and pistils. Bracts of enclosing involucre rather large, with those to outside larger than those toward center.

Flowering time July through September. Unlike sunflowers and coneflowers, ray flowers persist on top of fruit that yields a fertile seed. Easily mistaken for a sunflower. Is apparently of no medicinal or economic importance, nor is it poisonous under any circumstances.

(Continued)

False Sunflower

Coneflower

Black-eyed Susan

Purple Coneflower

This species is not ordinarily listed as a weed, and since it yields readily to cultivation, it cannot be considered as competitive with farmers' interests.

Coneflower, Golden Glow
Rudbeckia laciniata

Height to 12 ft. Stems smooth, rather slender, much-branched, stiff, somewhat ridged. Leaves finely five to seven divided, with division in turn deeply toothed along margins. Whole leaf may be over 1 ft wide, on relatively long petiole, rather thin, downy, at least above. Stem leaves smaller. Perennial.

Native of North America. Found often abundantly in thickets from Quebec to Florida, west to Arizona, Colorado, Idaho, and Manitoba. Variety *hortensia* is the golden glow of flower gardens. The related *R. speciosa* is a 2-ft plant known as "showy coneflower" and relatively common in flower gardens. Tends to be a lowland organism; frequently in edges of marshes, swamps, and low woods.

Flowers in heads at ends of branches, to 4 in. across, with 6 to 10 ray flowers that may be to nearly 2 in. long, drooping, bright yellow, and with central cone of disk flowers shaped like a chocolate drop, greenish-yellow and about twice as long as thick. Golden glow has "double" flowers, or more ray flowers.

Flowering time from July through September. Plant has been suspected of being poisonous to hogs, with poisoning similar to that caused by belladonna, but this has not been proved. Golden glow of the garden and coneflower of the field are both beautiful.

Stems of garden golden glow often break when they are in full bloom, so a rigid support for each major clump should be provided. Both golden glow and coneflower have red aphids as serious pests in great numbers. Where the flowers are to be picked, the plants should be sprayed with aphid-control liquids. Propagation of perennial cultivated kinds of *Rudbeckia* is carried out by division of plants in autumn or spring. They may also be raised from seeds sown in spring under glass.

Black-eyed Susan
Rudbeckia hirta

Height to 3 ft. Stems erect, stiff, rather brittle, simple or sparingly branched, roughened, with fine, stiff hairs, with many stems sometimes forming tufts. Leaves thick, almost entire or only obscurely notched along margins; the lower ones with long petioles, to 7 in. long and to 2 in. wide; the upper, smaller.

Native of North America. Found commonly in open dry meadows and pasture lands, or in waste places where there is good sun. Found from Ontario to Florida west to Texas, Colorado, and Manitoba. Related western coneflower, *R. occidentalis* provides some forage and erosion control in the western areas of the country.

Flowers in attractive heads at ends of stems or branches. Heads to 4 in. broad, with 20 to 40 neutral orange-yellow ray flowers to nearly 2 in. long, notched or not at tip, with compactly crowded dark-brown tubular disk flowers that produce fertile seeds. Fruit has no parachute.

Flowering time May through September. Dead tops bearing some seeds may be found through the succeeding winter. In Colorado and Wyoming, plant is reported to provide some forage for sheep and cattle but it may be merely that nothing else will survive where it does.

An attractive flower that is sometimes grown in wild-flower gardens, more often is considered as a weed by farmers, and as suggested above may be welcomed as a lunch by a starving sheep or cow. The western coneflower lacks the ray flowers so characteristic of the coneflowers and black-eyed Susans. Maryland state flower.

Purple Coneflower
Echinacea purpurea

Height to 5 ft. Stem stout, simple or sparingly branched, covered with fine stiff hairs or only slightly so. Leaves of *B. purpurea*, to 8 in. long and to 3 in. wide, long-petioled, with blade with shallowly scalloped margins; of *B. angustifolia*, as shown in the figure. Perennial, with thick dark rootstocks.

Native of North America. Commonly found in deep moist rich soils. From Pennsylvania to Georgia, west to Louisiana, Arkansas, and Michigan, with narrower-leaved species extending range much farther west. Probably more than three dozen species of the genus native of North America.

Flowers in huge solitary heads at tips of branches, with 10 to 20 rays sometimes as much as 3 in. long, drooping or spreading, purple, red, crimson or pale purple. Ray flowers neutral. Disk flowers tubular, bearing both stamens and pistils. Fruits, husky, four-sided achenes.

Flowering time July through October. Fruiting time August through November. In some parts of the country, this plant may be a pest, but it yields rather readily to hand pulling or to cultivation. If it needs control, care should be taken to prevent formation of a new crop of seeds.

This plant is so attractive that while Middle Westerners think of it as a weed, Easterners plant it in their flower gardens and are pleased when it becomes established. Certainly it has beauty in excess of many plants considered worthy garden plants. Two cultivated varieties, Taplow Crimson and the King, grow to 3 ft high, bloom in July and August, and are useful for cutting.

Garden Sunflower
Helianthus annuus
Height to over 15 ft. Stem coarse, somewhat branched or unbranched, rough-scaled, somewhat hollow. Leaves alternate, at least in upper areas, petioled, with three conspicuous veins, with a blade to 1 ft or more across and somewhat longer, somewhat heart-shaped at base and pointed at tip. Lower leaves may be opposite. Annual.

Native of North America or of Peru. Found wild in gardens and in rich soils. From Minnesota to Texas, west to Pacific Coast. Reported as being found north into Saskatchewan. Has occurred as an escape in widely separated parts of the world. May establish itself where it is cultivated.

Flowers in heads at ends of branches. Heads may have a diameter to 2 ft, and in wild plants to 6 in. Ray flowers yellow streamers, mostly entire, relatively few. Disk flowers brown to purple. Fruits flattened oval, usually striped with light and dark gray, widest near upper end.

Flowering time July through September. Root system of a sunflower may occupy over 1 cu yd and 1 acre of sunflowers may lose nearly 400,000 gal of water through transpiration. Fruits sought by many seed-eating birds, particularly goldfinches, who may ruin a commercial crop.

Raised commercially for their fruits, which are excellent food for poultry and highly favored for bird-feeding stations. Fruits are high in oil. Plant is cultivated in Peru and in many other parts of the world for the production of sunflower oil used as food, in soap, and on leather. Was cultivated in American Southwest before A.D. 1000. Garden sunflower may be North America's main contribution to economic plants of the world. The Russian cultivar of this species presents an extraordinary example of gigantism and produces a very large head.

Garden Sunflower

Jerusalem Artichoke
Helianthus tuberosus
Height to 12 ft. Stems coarse, branched in upper areas, downy or with short, stiff hairs arising from fleshy white thickened horizontal rootstocks that bear tubers as offshoots. Leaves opposite, rather stiff, rough above and downy beneath, petioled, to 8 in. long and to 3 in. wide. Perennial. Also known as girasol and topinambour.

Native of North America. Found in loose moist soil or in good garden soil from Nova Scotia to Georgia, west to Arkansas, Kansas, and Manitoba. Occurs in cultivation in a number of areas. At least 70 species of Helianthus to be found in the Americas.

Flowers borne in heads at tips of uppermost branches. Heads to nearly 4 in. across, with from 10 to 20 yellow ray flowers that are notched at tip, well over 1 in. long, and neutral. Yellow, tubular disk flowers, perfect, bearing both stamens and pistils. Fruits, downy.

Flowering time September though October. Tubers mature at end of fall and may be collected then or in spring. Cultivated by the Indians for centuries and introduced into Europe in 1616. Has been cultivated more extensively in Europe than in its native America and is known as "Canadian potato."

The plant is also grown as a forage crop and as a competitor of weeds. Tubers are eaten raw or cooked; flesh is watery and sweet with a flavor palatable to some tastes. Food value of tuber is not high as the starch is not readily digested.

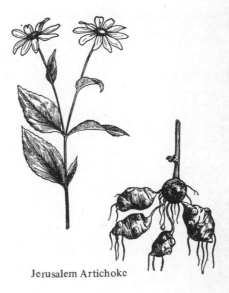

Jerusalem Artichoke

Wild Sunflower
Helianthus decapetalus
Height to 5 ft. Stem branched, slender, smooth or almost so, arising from a substantial, sometimes thickened rootstock. Leaves alternate above; opposite below, though some of those on flowering branches may be opposite; thin, to 8 in. long and to 3 in. wide, rough above and downy beneath; the lower with petioles.

Native of North America. Found from Quebec to Georgia, west to Missouri and Michigan. Many closely related species that extend range of native sunflowers to the west. In fact, Kansas is known as the Sunflower State and the sunflower is naturally its state flower.

(Continued)

Wild Sunflower

Zinnia

Sweetbush

Dahlia

Flowers in good-sized terminal heads, with ray flowers numbering from 8 to 15 and being much longer than the disk is wide. Whole head sometimes more than 3 in. across. Disk flowers tubular, yellow, and rather compactly crowded together. Fruits, smooth dry dark achenes.

Flowering time August through September. Fruits of most wild sunflowers gathered by Indians and used as food either raw, crushed, roasted, or in mixtures. Sunflower seeds toasted and salted are in many respects as good as or better than salted peanuts. Unfortunately, their shells must be removed.

Sunflower fruits when properly roasted may make a drink that has some of the qualities of inferior coffee. This species is not listed as a bad weed but others are not so free from criticism.

Zinnia
Zinnia elegans
Stems stiff, erect, hairy, and to 3 ft high. Leaves opposite, compound, lobed, more or less clasping at bases, with roughened surfaces which are somewhat sticky and which show prominent veins. Roots fibrous, relatively strong.

Single form, native of Mexico; double form, of French origin, where it first appeared in a garden in 1856. Introduced through United States and other temperate parts of world. Best in rich, well-drained, loose loam and favors sunny areas.

Flowers in solitary, terminal heads, from 2 to 4½ in. across. Ranges from crimson to scarlet, salmon, rose, purple, orange, yellow, and white. *Z. haageana* has orange flowers. *Z. multiflora* has slender linear red or purple rays. Pollination mostly by moths and butterflies. One kind, Youth and Old Age, of which there are many showy varieties, is commonly used for summer beds and border plantings.

Annual. Propagated from seeds which germinate in about 5 days and live to 4 years. Seeds sown indoors in March, transplanted after frosts cease, thinned in rows to 12 in. apart. Cultivate lightly; water thoroughly in dry weather. Prune smaller heads to produce larger remaining heads. Enemies include leaf rollers controlled by lead arsenate spray, tarnished plant bug controlled with nicotine sulfate spray, cut worms controlled with poison baits, and European corn borers controlled by burning infected plants. Hummingbirds visit flowers.

Good in beds and borders, dwarf *Z. haageana* being better for edgings. Flower heads should be removed before seeds are formed if blooming is to be continued. Flowers attractive; keep well after being cut, some lasting many days. Some are almost shrubby. Indiana's state flower.

Sweetbush
Bebbia juncea
Height to about 5 ft. Stems woody, branching freely to form an almost globular head, to 5 ft in diameter but in the dry season looking practically dead, whitish, thus adding to dead appearance. A shrub with rushlike appearance. Root system deeply penetrating. Leaves alternate, entire, absent for much of year. Perennial.

Native of deserts of southwestern areas. Found in Mohave and Colorado deserts, being most abundant and best developed on broad sandy washes of Colorado desert. Genus does not have any common representatives to the east but is closely related to *Coreopsis.*

Flowers in yellow fragrant heads. Appear in great numbers whenever rains furnish enough water to roots to start development and may practically hide stems, which are leafless at the time or practically so. In full bloom, plant is one of most attractive desert plants.

Bebbia or sweetbush is a good honey plant. When in bloom, is visited by great numbers of bees. Sweet flowers are eaten by a number of animals. Lizards known as "chuckwallas" feed upon them greedily. Apparently, it offers no particular indication of ecological condition of its surroundings.

An attractive flower of the desert that may intrigue the tourist. Apparently it is of no great significance to man aside from its beauty, fragrance, and indirect contribution to honey production.

Dahlia
Dahlia pinnata
Height to 4 ft. Stems smooth, considerably branched. Leaves green above, grayish beneath, with upper divided, and in turn divided again, opposite or whorled, with bases of petioles almost grown together. Some related dahlias are woody and perennial while others are annual.

Native of Mexico and extensively developed in United States and elsewhere, mostly as spectacular ornamentals for their huge blooms. Roughly, two major groups of dahlias, the tree and the bush dahlias, of which the latter are most commonly seen in gardens.

Flowers of *D. pinnata,* in heads to 8 in. across or sometimes much over this, nodding or horizontal. So-called "single dahlias" have a single row of ray flowers and have yellow centers. Semidouble forms are called "duplex dahlias," while peony dahlias resemble peonies (see illustration).

Seed of some dahlias sown in spring may yield flowers in fall; but usually propagation is by division of roots. Need sandy soil where nights are cool, where there is little nitrogen as fertilizer, since this stimulates leaves rather than blooms. Roots planted 2 to 3 ft apart, to 4 in. deep, and shoots cut back to one.

Dahlias are good as cut flowers if they are dipped in hot water that is allowed to cool before flower is placed in a container of water. Roots should be dug in fall before frost danger; stems should be cut off and roots stored where cool and not too dry or too wet.

Cosmos
Cosmos bipinnatus
Stems to 6 ft high, smooth or nearly so and relatively slender. Leaves opposite, compound, or lobed and cut into slender parts so that they look lacelike, but this is not the case with all species. Roots fibrous and relatively strong. Plant as a whole most attractive. Also called Mexican aster.

Native of Mexico. Only earlier flowering forms hardy enough to bloom well in northern United States. Best in well-drained sandy loam which is not too rich and which is exposed to full sunlight for relatively long periods. If soil is too rich, plants fail to bloom.

Flower heads solitary or in loose clusters on long stems; single or double; about 3 in. wide, with ray flowers pink, red, or white and central disk yellow. In *C. sulphureus,* rays are yellow; in *C. diversifolius,* disk is red. Floral bracts oval and unequal.

Annual. Propagated from seeds which germinate in about 5 days and live about 3 years. Early varieties sown in April are transplanted when weather permits, thinned to 18 in. apart, cultivated until blooming begins in September. Blooming continues until frost. In South, plant self-sows. Spotted cucumber beetles bad cosmos pests, controlled by spraying with arsenate of lead. European corn borer and a stem blight also attack the plant. Injured plants should be burned to prevent spread of trouble. Since plant may break, it should be staked.

Excellent plant to use in masses and supply attractive cut flowers which do not wilt readily and may last several days. Form may be controlled by pinching off parts not wanted to fill desired space. In Korea, cosmos is planted by school children along roadways each spring.

Stick-tight, Beggar Ticks
Bidens frondosa
Height to 5 ft. Stem erect, purplish, stiff, branched, smooth or nearly so. Leaves opposite, usually smooth, thin, to 4 in. long and 1 in. wide, the lower divided into three to five segments, with end one particularly long-pointed, long-petioled with a groove on upper side of petiole. Annual. Known also as devil's boot-jack, and pitchforks. Fall hikers in eastern woodland borders have most certainly carried home some *Bidens* hitchhikers attached to their clothing.

Native of North America, where it is found in gardens, meadows, pastures, and roadsides where it is wet. From Nova Scotia to Florida, west to California and British Columbia. Introduced and has become established in southern Europe. 75 widely distributed species in the genus.

Flowers in numerous heads, at tips of upper branches, each head to ½ in. long and to nearly 1 in. wide. Ray flowers may be absent or inconspicuous. Disk flowers orange, and produce flat black two-awned fruits. Bracts of the enclosing involucre usually rather leaflike.

Flowering time July through September. Fruiting time August through to December. Since the plant requires abundant moisture, drainage provides an effective control measure if repeated cutting is not sufficient. Plants should be cut to prevent reseeding. Fruits spread by sticking to animals.

Mechanical injury may be caused to sheep and other animals by the stiff-awned fruits that may make their way to the skin and cause serious irritation. In Africa, the natives make leaves into a sort of greens, but this does not sound appealing to those who have smelled the plant.

French Weed
Galinsoga ciliata
Height to 3 ft. Stem erect or more commonly sprawling, slightly downy or hairy branched, pale green, slender, weak. Leaves opposite, to 3 in. long and well over half as broad, three-nerved, petioled particularly in the lower leaves, with the margins scallop-toothed. Upper leaves sometimes without petioles. Annual.

(Continued)

Cosmos

Beggar Ticks

French Weed

Native of tropical America but established from Maine to Florida, west to California, Mexico, and Oregon. Has been introduced into and has become established as weed in Europe. Of five species in the genus, all are native of tropical and subtropical America. Usually, this is the species found in North.

Flowers in heads, borne on stems singly, in axils of upper leaves, to ⅓ in. across but commonly smaller. Ray flowers white, with three teeth at free end, pistillate, and producers of fertile fruits. Disk flowers yellow; produce fertile fruits; bear both stamens and pistils. Fruits, dark.

Flowering time June through October. Fruiting time September to December. Northern range determined largely by length of period between killing frosts. Since it is an annual, control measures might well center around prevention of formation of mature fruits.

A harmless weed of waste places such as along paths and at edges of buildings. When it comes into competition with any cultivated plant, it retires with good grace at first sign of a hoe or cultivator. Were its flowers larger, they would be considered as ornamental. Edible as greens. A pest in gardens which could easily become an important potherb if Americans developed an interest in it as have peoples of some Southeast Asian countries.

French Marigold
Tagetes patula

Bushy annual. With stems to 1½ ft high, compact, relatively stout. Leaves dark green, compound, and divided into about 12 long-toothed segments. Glands on leaves give off a distinctive odor. Leaves of *T. lucida* not compound. Roots fibrous and relatively profuse.

Native of Mexico, but greatly modified by cultivation. Grown widely in United States and Canada. About 20 species ranging from New Mexico and Arizona to Argentina. This species prefers sunny area where there is light loam and moderate amount of moisture.

Flowers in heads, single or double, yellow or orange and marked with brown or maroon. Heads about 1½ in. across. *T. erecta*, African marigold, reaches height of 2 to 4 ft and bears large, single, or double orange or yellow flowers. *T. signata var. pumila*, Mexican marigold, to 12 in. tall, has gold-colored flowers. *T. lucida*, sweet-scented marigold, to 12 in. tall, has golden or orange-yellow flowers.

Propagate from seeds which germinate in 5 days and live to 4 years. Seedlings transplant well and thrive better if transplanted frequently. Should be thinned to 15 to 18 in. apart and cultivation should be light but frequent. Will blossom from June to frost under proper treatment and may self-seed. Enemies include yellow woolly caterpillars controlled by hand picking, leaf tiers controlled by lead arsenate spray, tarnished plant bugs controlled by nicotine sulfate spray. In greenhouse cultivation, requirements are light soil, good ventilation and temperature of 55°F. Needs room.

Desirable plant for border edges or in mass plantings. Also popular with commercial florists as commercial early flower for spring sales. Flowers arrange themselves and keep well when cut. For commercial sales, seeds are sown in January; seedlings are usually sold in lots of 12 in shallow containers.

French Marigold

Sneezeweed, Bitterweed
Helenium autumnale

Height to 6 ft. Stem angled or winged, branching in upper areas, pale green, slender, smooth or only slightly downy. Leaves alternate, coarsely toothed, firm, to 5 in. long and to 2 in. wide, with blade pointed at each end and petioles forming narrow wings that extend down stem. Perennial.

Native of North America. Found in wet areas such as swamps and low meadowlands or in flooded regions, often in gravel. From Quebec to Florida, west to Arizona, Nevada, Oregon, and Manitoba, reaching the 2,600-ft elevation in Virginia. Of 24 species in genus, all are native of North and Central America.

Flowers in conspicuous terminal heads to 2 in. or more across, with 10 to 20 drooping yellow ray flowers to over 1 in. long, with three well-defined teeth at the free end, pistillate, and productive of fertile fruits. Disk flowers darker, yellow when completely matured, formed into a spherical head.

Flowering time August through October. Fruiting time September through November. A bitter substance is found all through plant, but particularly in mature flower heads, which if eaten create a strong appetite for more. Powdered flower heads are used medicinally for the purpose of stimulating sneezing.

Sneezeweed

Cattle, sheep, and horses with an appetite for the plant may have accelerated pulse, difficult breathing, staggering gait, sensitiveness to touch, spasms, and convulsions that may end in death. A general reputation for toxicity of *Helenium* spp. exists; however, evidence is not verified in all cases. Poisonous principle assumed to be dugaldin, which acts like the poison, aconite. Control of the plant as a weed is by cutting, cultivation, and drainage of the site.

Pot Marigold
Calendula officinalis
Height to over 2 ft. Stem relatively stout, straight, somewhat hairy. Leaves alternate, simple, to 4 to 5 in. long and about half that wide, with bases that are more or less clasping, entire, or only very obscurely toothed with leaves or bracts of flowering branches smaller than lower ones. Annual.

Native of southern Europe but grown as an ornamental in many parts of the world. Flower culturists have improved size, color, and form of flowers. About 15 species known in the genus native either of Europe or of territory not far from it. Does well either in open or in greenhouses, under cultivation. Grows best in full sunshine in rather dry location. Seeds may be planted in fall or spring of year before flowers are desired.

Flowers borne in relatively large heads at ends of branches. In old varieties, flower heads were single, but now they are double for most part. Typical heads are to 2 in. across, with the flat rays varying from pale yellow to deep orange and closing the head at night. Flowers almost continuously.

Seeds may be planted in open ground or cold frame and in good soil and sun. Flowers should develop by early fall; plants should self-seed but in this case fancy varieties will revert to original single type. These are the "winking Mary-buds" of Shakespeare's *Cymbeline.* Summer-sown plants are potted for winter.

Leaves and heads used as potherbs. Has been considered a remedy for sore teeth and as a "comforter of the heart and spirits," also used as an emetic and in the treatment of warts. Flower heads have been used, dried, in cookery to flavor soups and stews and to color butter. Varieties include the rich yellow Chrysantha, and Orange Chrysantha; Orange King, which is the earliest to bloom and is double orange; the lemon-yellow Lemon Queen; Radio with quill-like orange petals; Golden Beam with quill-like golden petals; and Prince of Orange, which is deep orange with a black center.

Gaillardia, Blanketflower
Gaillardia aristata
Erect stems to height of 3 ft, more or less short-haired, relatively slender and stiff. Leaves alternate, gray-green, longer than wide, entire or cleft; lower ones paddle-shaped; in some species leaves almost entirely basal. Roots stout and fibrous.

Native of United States, from Minnesota and Manitoba, west and southwest, but now spreading east. 12 related American species, mostly from Far West. Grows well in sandy soil with good drainage. Favors strong sunlight. Can endure partial shade and resists cold.

Flowers in terminal heads 2 to 4 in. across; daisylike, yellow or yellow-orange. Related annuals *G. amblyodon* and *G. pulchella* have flowers normally red; lobes of disk flowers of former are obtuse while those of latter are acute. Pollination by insects.

Propagated by seeds, division, or by stem or root cuttings. May bloom first year from seed but is better second. Seeds germinate in 20 days, live about 2 years; should be sown early. Named varieties should be grown from cuttings to assure desired characters. There are some annual forms. Not often seriously injured either by insects or by fungi. Hardy in severe weather. If flowers are kept picked, blooming will continue from June to November, long after many other common species have ceased. Typical plant of western plains.

Vivid, sprawling, hardy plants that deserve popularity they enjoy. Flowers cut before they are open last longer than those cut when mature. In gardens, perennials are favored because they do not need to be purchased or planted new each year.

German Chamomile
Matricaria chamomilla
Height to 2 ft. Stem erect, smooth, much-branched. Leaves two to three times cut, like parts of a feather, with short narrow segments, light or dark green; generally appear like loose fragile lace, making whole plant highly ornamental. Leaves to 1 in. broad. Annual.

Native of Europe but established as escape in New York and Pennsylvania. Cultivated as an herb plant for its drug in many parts of world. Only about 4 of the 20 species of genus have been established in eastern North America; others are Eurasian and African. Cultivated mostly in Germany and Hungary.

Flowers in daisylike heads, with 10 to 20 rather slender and spreading white ray flowers; central disk flowers arranged in a conelike manner. When flowers are broken from receptacle, they leave no remaining scales. Fruits without downy parachutes, faintly three- to five-ribbed.

(Continued)

Pot Marigold

Gaillardia

German Chamomile

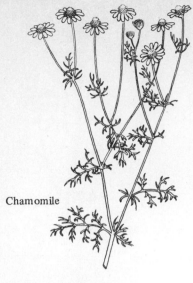

Chamomile

Yarrow

Oxeye Daisy

Flowering time June through August or September. Seeds sown in spring yield flowers in 8 weeks, so crop may be grown between others. Flowers collected in full bloom yield a strong unique aromatic odor and a bitter taste. Extracted oil is thick deep blue, then brown, and known as "matricaria" or "anthemic acid."

Matricaria oil extracted by boiling has been used to reduce gas in stomach, as a stimulant or tonic, to soothe children who are teething, or for earache or other localized pains. Relatively harmless as a home remedy. Used as a hair rinse to add luster to the hair. 1 acre can yield 400 lb of dried flowers. *M. inodora*, a garden variety, to 2 ft tall, with double white flowers thrives in any average soil. *M. matricarioides* is the very abundant weed called pineapple weed.

Chamomile, Mayweed
Anthemis cotula
Stem to 2 ft high, well-branched to form a sprawling or spherical mass, smooth or downy, with a bitter taste and a strong unpleasant odor. Leaves to 2 in. long and to 1 in. wide, finely twice-divided to form a lacelike effect, weak, abundant, alternate, with upper leaves commonly smaller than lower. Annual. Also known as dogfennel.

Native of Europe. Firmly established in America. An ever-present weed of barnyard, roadside, and waste places. It finds a home in America, Asia, Africa, or Australia and establishes itself firmly. It lives on a variety of soils, with a great variety of plant associates.

Flowers in heads, something like small daisies, about 1 in. across, with 10 to 20 ray flowers, white, and usually with three teeth at free end, with or without pistils, usually bent somewhat backward. Disk flowers tubular, yellow, and producers of fertile fruits. Fruits somewhat cylindrical, dark, with long rows of tubercles.

Flowering time June through October. Fruiting time from July through to December. May be controlled as a weed by heavy cultivation and spring harrowing. Capable of causing a burning sensation to wet skin. Cattle that eat it may give bad-flavored milk. May be a pernicious weed in gardens.

Related field or corn chamomile *A. arvensis,* with straw-colored fruits and without ill-scented juices, and yellow-flowered yellow chamomile *A. tinctoria* are controlled in much same way as this species. Fruits of all three are relatively common seed impurities in commercial seeds. Roman chamomile *A. nobilis* raised as a drug plant. It is eaten by several kinds of birds, e.g., green-backed goldfinch; Oregon junco; golden-crowned, Lincoln's, savannah, and white-crowned sparrows; and the brown towhee.

Yarrow
Achillea millefolium
Height to 2 ft. Stems usually unbranched except at top, stiff, erect, rather slender, smooth or with webby hairs. Leaves alternate, lower reaching a length of to 10 in., deep green, strongly scented, twice-divided and finely toothed to give a lacelike appearance. Root system horizontal, substantial. Perennial.

Native of North America. Some persons contend it was naturalized from Europe and Asia, where it is also found. Of the more than 75 species recognized, most are native of the Old World.

Flowers borne in relatively small heads arranged in flat-topped clusters. Individual heads to ⅓ in. across and more or less downy. Ray flowers number from four to six per head, are white or pink and, like the yellow disk flowers produce fertile fruits that are pale gray or straw-colored thin oblong wafers, easily windblown.

Flowering time June through November. Fruiting time August to end of year. Plant does well either in severely hot or cold climates. Achilles is purported to have used the plant to heal the wounds of soldiers in the siege of Troy. It still has some slight medicinal uses. Cattle eating it may yield bad-flavored milk.

As a drug, it has a small demand. A very common weed throughout U.S. This species and *A. lanulosa,* the principal representative on the Western rangelands, are both used to a limited extent; leaves are eaten by upland game birds as well as fur and game mammals, other small mammals, and hoofed browsers. Leaves have been used in a tea but it is too bitter to be recommended. Where the plant is a weed, cattle have usually avoided eating it because of its taste. Cultivation keeps it in easy control.

Oxeye Daisy, Marguerite
Chrysanthemum leucanthemum
Height to 3 ft. Stem erect, green, somewhat ridged, tufted, slender, nearly smooth, usually unbranched, arising from a short thick horizontal perennial rootstock. Leaves to 3 in. long, dark lustrous green, lower ones with relatively long petioles, rather conspicuously toothed along margins. Perennial.

Native of Europe but too well established in America on pastures, mead-ows, and waste places through almost all of North America though less abundant in South and West.   It is also native of Asia and is widely established in other continents.   About 100 species in the genus, with few in America.

Flowers borne in heads at ends of branches, heads being to 2 in. across. Ray flowers number 20 to 30 and are flat, white, and often with two to three teeth in the end.   Disk flowers small, yellow and densely crowded into an almost flat disk, sometimes depressed in the center.

Flowering time May through November.   Fruiting time June through to December.   Control possible largely by cultivation and by cutting tops before seeds have time to mature.   Unfortunately, seeds may be distrib-uted in stable manure.   Salting plants helps control them but is a slow process.

Young leaves are eaten as a raw salad but if too old they are strong and unpleasant.   Leaves may also be cooked as a potherb.   Salting rosettes stimulates cattle to eat the plants as well as helping kill plants by direct action.   State flower of North Carolina.

Max Daisy
Chrysanthemum maximum
Height to 30 in.   Stems grow in clumps, branched somewhat but more commonly unbranched, somewhat downy but becoming smooth.   Leaves to 1 ft long, lower ones petioled, upper without petioles, usually under 1 in. broad.   Leaves dark green and clean-looking.   Short-lived perennial.

Max Daisy

Native of Pyrenees in southwestern Europe but grown as relatively hardy garden plants or as greenhouse plants wherever ornamentals are grown in reasonably temperate climates.   Related marguerite daisy *C. frutescens* has leaves cut almost to midrib while those of this species are not so deeply cut.

Flowers in heads at ends of branches, to 3 in. or more across, with narrow blunt rays white, usually about 1 in. long; in cultivated varieties with heads to 4 in. across they may be 1½ in. long.   In some forms, rays are held out stiffly while in others they droop or are even somewhat incurved.

Flowering time through most of summer months and in some forms on into fall.   True Shasta daisy is an early-flowering form and in North it may be reasonably hardy if given some winter protection with a good mulch.   Requires a sunny location, deep rich well-drained soil, and 1 ft of space between plants.

Raised primarily as ornamentals.   Single forms may be raised easily by seed but double forms are propagated largely by division of perennial underground part.   Plants make effective mass plantings for a relatively long period of time and require a minimum of care and attention.   *C. maximum* and its varieties thrive in ordinary, well-cultivated garden soil. Flower stems should be supported.

Florists' Chrysanthemum
Chrysanthemum morifolium
Stems erect, to 4 ft high, in hardy varieties much-branched, ridged or fluted, generally heavily downy, with a strong odor.   Leaves to 6 in. long, usually with two prominent lobes, relatively short-petioled, with wrin-kled margins and variety of marginal smaller lobings.   Perennial, but grown as annual in greenhouse management.

Florist's Chrysanthemum

Common "mum" of the greenhouse is probably a hybrid of *Chrysanthe-mum morifolium* and *C. indicum* from China and from Japan, though very recently Korean mum *C. coreanum* has won considerable popularity. Hardy outdoor chrysanthemums with small heads may have ancestry very similar to the large greenhouse varieties.

Flowers in great heads of considerable variety including single forms, pompon forms, Japanese forms with ray flowers doubled and either curved inward, outward, or backward; hairy forms that look featherlike; anemone forms that are more nearly like single forms; and also cushion mums that make a mound to 1 ft high, sometimes called azaleamums.

Heads may be composed of modified ray flowers or of modified tubular disk flowers, the latter producing an anemone type of head.   Chrysanthe-mums have been under cultivation for over 2,000 years.   Best soil is a fibrous loam with well-rotted manure.   Lime should be avoided.   Cut-tings with 2 to 3 joints are used.   Rooted in March, should flower in fall.

Ideal conditions call for night temperatures below 50°F, with care taken to syringe flowers a number of times a day to keep down red mites and to reduce demands on roots.   There are many chrysanthemums, and a buy-er should consult the floriculture department of his state college of agri-culture before purchasing flowers, seeds, or plants.   Chrysanthemum flowers and plant parts are eaten by Japanese as part of tempora meal.

Tansy

Sagebrush

Wormwood

Tansy
Tanacetum vulgare

Height to 3 ft.   Stem unbranched up to the flowering portion, then rather profusely branched, smooth or somewhat downy, stiff, rather stout, strongly scented.   Leaves to 1 ft long, deep dark green, smooth, alternate, featherlike, with divisions similarly divided, petioled, with upper leaves smaller than lower.   Perennial.

Native of Europe.   Well-established in North America as an escape or otherwise from Nova Scotia to Georgia, west to Nevada and Oregon. Particularly common in the East around abandoned homestead sites.   Of the 30 species in the genus, 2 are established in eastern United States, in California; the rest are Eurasian for most part.

Flowers borne in many heads at ends of fine branches at top of plant arranged to form a flat-topped, rather compact mass.   Individual heads to ½ in. across, yellow with most flowers tubular, though there may be a few ribbon-shaped marginal flowers in the heads.   Ray flowers without stamens.   Fruit a five-angled achene.

Flowering time July through September.   Fruiting time August through October.   Leaves and stems contain the oil tanacetin, which is poisonous to man and to domestic animals.   Raised commercially for a small market, the plants being cut in late summer when in full flower.   Propagation by seeds or by division, with sets 18 in. apart.

1 acre may yield to 1 ton of dry leaves.   Michigan has distilled around 1 ton of oil a year with average oil yield per acre about 20 lb.   Has been used in tea.   Poisoning causes convulsions, violent spasms, dilated pupils, weak pulse, and frothing.   Abortion in cattle has been linked to ingestion of this herb.   It is purported to have antihelminthic properties in man.

Order Asterales./Family Compositae

Sagebrush
Artemisia tridentata

Height to 10 ft.   Stem branching freely, shrubby, with silvery-gray hairs in abundance, tough and arising from a deeply penetrating tough root system.   Leaves to 1 in. long and ½ in. wide, narrow, wedge-shaped but with three to five blunt teeth, broadest near free end.   Whole plant more or less fragrant when broken.

Native of North America.   Found on dry plains from Nebraska to Colorado, Utah, and Montana, west to California and British Columbia. Commonest in high northern desert and sagebrush areas where it is sometimes found in almost pure stands.   Some 225 species in the genus in North America, South America, and Eurasia.

Flowers borne in heads, in crowded clusters in axils of leaves, and at ends of stems with terminal clusters often many times divided.   Each head only about ⅛ in. through, with all flowers bearing stamens and pistils and all alike.   Ripe heads fall free and are blown by winds for long distances.

Flowering time July to September.   Fruiting time August to December. Survives drought by shedding most of its leaves.   Its presence indicates light-textured soil, low runoff, and no salt for first foot of soil at least. Purple, spongelike galls in summer are made by gall midge *Diarthronomyia*.

Foliage eaten by sheep and goats; and provides principal food of sage grouse; rich in proteins and fats.   Sagebrush furnishes cover for many desert animals.   Range cattle also use sagebrush as forage.   Tea made from bitter leaves is used in treatment of colds, sore eyes, and as a hair tonic.   Ripe fruits were ground for meal by Cahuilla Indians.   Wood excellent for a quick hot fire.   Nevada's state flower.

Wormwood
Artemisia absinthium

Height to 4 ft.   Stem branched, finely downy, rather tough and reasonably slender at least in upper areas.   Leaves to 5 in. long, covered with fine white silky hairs when young turning to soft gray-green when mature, thrice-divided, with segments finely divided and lower leaves petioled.

Native of Europe and escaped in America where it was grown in gardens. Now established in many waste places from Newfoundland to North Carolina, west to North Dakota and Montana.   Principal centers for cultivation in this country are Wisconsin and Michigan.   Related of course to the common sagebrush.

Flowers borne in loose open clusters to 1 ft in length while individual heads are less than ⅛ in. through, drooping, not usually crowded, yellow, numerous.   Central flowers of a head have both stamens and pistils. Marginal flowers have pistils only, or stamens and pistils.

Flowering time July to October.   Fruiting time August through November.   Plants do best under cultivation in deep moist soil, seeds being broadcast in fall after a grain harvest or started in seedbeds or from cuttings, set 18 in. apart in rows to 3 ft apart.   Planting may yield crops to 3 years.

Oil used in manufacture of absinthe, for which there is little demand.   1 acre may yield 2,000 lb of dry plants or 40 lb of oil but a more normal

yield is 20 lb of oil. Oil distilled by steam. Little profit in raising plant.
Milk of cattle eating plant may be badly tainted. *A. filifolia,* sand sagebrush, produces a disease known as "sage sickness" in horses. Not a
serious disease. *A. dracunculoides,* linear-leaved wormwood, to 4 ft tall,
with simple linear leaves to 3 in. long, is a late-flowering species from
Illinois west to the Pacific.

Coltsfoot
Tussilago farfara
Height to 1½ ft. Stem appears only at flowering time and bears relatively closely pressed leaflike structures (phyllodia) that are slightly more
crowded at top. Leaves arise from thick, rather juicy, branched and
spreading more or less horizontal rootstocks. Leaves to 7 in. across,
long-petioled, densely hairy beneath.

Native of Europe. Well-established in North America, being particularly common on dry banks such as railroad embankments. Ranges from
Nova Scotia to Pennsylvania, west to Minnesota. Only one recognized
species in the genus and it is native of both Europe and Asia.

Flowers appear in early spring before leaves and have usually freed their
fruits by time true leaves have reached any considerable development.
Flowers borne in single terminal heads about 1 in. across and vaguely
resembling a dandelion head. Ray flowers fertile. Disk flowers sterile
even though ray flowers lack stamens possessed by disk flowers.

Flowering time April to June. Fruiting time May to July. Fruits beautiful objects when viewed under a lens. Hairs on undersides of leaves are
collected and used medicinally as an emetic. If plant is considered as a
weed, it may be controlled by soil management. It does well only in
moist clay soils so drainage and enrichment control it.

Branching root systems serve a useful function in anchoring soil on steep
slopes that might otherwise become badly eroded. Since these are easily
destroyed by cutting or by cultivation, this suggests another common and
simple means of control.

Coltsfoot

Fireweed, Pilewort
Erechtites hieracifolia
Height to 8 ft. Stem erect, smooth or downy, conspicuously grooved,
branched with the branches turning upward. Leaves thin, to 8 in. long,
pale green; the lower with clasping petioles, the upper without any; with
margins deeply toothed to more than halfway to the midrib. Annual.
Takes its name from being one of first plants to pioneer burned-over
areas.

Native of North America, the West Indies, and South America. Widely
established from Newfoundland to Florida, west to Texas, Mexico, and
Saskatchewan. The only American species of the 12 in the genus, the
others being native of Asia and the vicinity.

Flowers borne in more or less erect cylindrical heads at ends of branches,
each head being to nearly 1 in. long and to ⅓ in. through, with area at
base rather conspicuously swollen. Flowers all fertile, greenish-white,
and only slightly longer than the enclosing involucre. Fruit parachute,
white, shiny.

Flowering time July through September. Fruiting time August through
October. May become a weed in gardens and waste places. Once it has
established a good crop of seeds in soil, it is persistent. Plants break too
easily for hand pulling. Hoeing or cultivation seems about the only
check.

It has been listed as a medicinal herb, the whole plant being pulled and
dried just before blooming. It turns black on drying. Juices are rank to
smell and offensive to taste.

Fireweed, Pilewort

Common Groundsel
Senecio vulgaris
Height to 1½ ft. Stem smooth or slightly downy, ridged, hollow, well-
branched, easily broken, erect. Root system fibrous and well-branched.
Leaves to 6 in. long, with margins roughly toothed, to halfway to midrib,
with many smaller teeth all along the margin. Lower leaves petioled.
Annual.

Native of Europe. Well-established in North America. From Newfoundland to Georgia, west to Pacific Coast. Does not usually form any
dense clean stands as do some other weeds. Over 1,200 species of the
genus of wide geographic distribution and, of course, of considerable
variation.

Flowers borne in more or less erect heads somewhat crowded at ends of
branches, each head being to ½ in. high to ⅓ in. through. No ray flowers. Flowers all yellow. Bracts of involucre slender, some black-tipped.

Flowering time April through October. Fruiting time May through November. Fruits wind-borne, with assistance of a generous fine white
parachute of hairs. Where it is established as a weed, cultivation, hoeing,
(Continued)

Common Groundsel

and use of modern weed-killer sprays are recommended procedure. Seeds should not be allowed to form.

Apparently weed does not have conspicuously important medicinal uses, nor is it so troublesome as to be taken seriously in most places. Not unattractive. White fruits could be considered worth seeing if one did not realize that they might produce weeds that would compete with our interests. Some species of *Senecio* are poisonous plants but *S. vulgaris* has not been implicated specifically.

German Ivy

German Ivy
Senecio mikanioides
Stem long and twining, thin, smooth, and green arising from a mass of thin fibrous much-branched roots. Leaves five- to seven-angled, lobed or nearly triangular, with sharp angles and a notched base, alternate, smooth, not toothed, apparently more abundant near base of the plant, pale green. Annual.

Native of South Africa. Found mostly as a greenhouse plant and in that capacity widely distributed. One of two species of *Senecio* that climbs, other being *S. scandens* whose flower heads have rays while in this species they are rayless.

Flowers, 12 to 15 in a head, with several heads in a cluster, none being ray flowers, greenish-yellow, with heads borne in close groups on branches springing from axils or from tips of branches. Bracts of enclosing involucre in each head are shorter than flowers themselves.

Flowering time indefinite because of greenhouse conditions. Plant propagated either by seeds as an annual, or by cuttings that root readily. Cuttings started in damp sandy soil and transplanted to medium garden soil. Plants need little sun; do best at temperature of 60 to 70°F. A common green aphid thrives on it.

An attractive ornamental for hanging gardens in sunrooms and in greenhouses, easily propagated. Easily cared for if it is sprayed occasionally with any good aphid control such as soapy water with some nicotine dust in it, or a nicotine fumigation process is used. As a climber it can be propagated by cuttings in the spring. Tips of shoots, 2 in. long, removed and inserted in moist sand; all of which are propagated under glass cover until rooted and ready for potting.

Burdock

Burdock
Arctium lappa
Height to nearly 10 ft. Stem coarse, ridged, dark green, strongly scented, much-branched, with soft pith when young, rather tough when old. Leaves alternate, relatively thin above, but lower leaves and rosettes may be coarse, long-petioled, to 1½ ft long. Root system deeply penetrating. Biennial.

Native of Europe but thoroughly naturalized in America. This species and *A. minor* together occupy most of United States. This species may be a troublesome weed in waste places in the East and in the Middle West. Of the six or more species known for the genus, all are natives of Europe and Asia.

Flowers in heads, clustered at ends of branches. In this species, heads may be to 1½ in. across while in *A. minor* they are rarely half that width. Flowers all tubular; disk flowers with pink corollas are each capable of producing a fertile fruit. Anthers of stamens purple. Stigmas and pollen white.

Flowering time July through October. Fruiting time September through winter. First year spent in growing a rosette; second, in growing top and flowers. Under cultivation, seeds are sown 1 in. deep in fall. Seedlings thinned to 6 in. apart. Roots harvested at end of first year's growth.

1 acre may yield 1 ton of dried roots. Pith of young stems removed from outer portion tastes much like rhubarb and makes good greens. Roots and stems may be peeled, boiled, and eaten. Almost anyone who has hiked in the fields in fall has taken home a few fruits attached to clothing. Fruits have effective hooked bracts by which they are attached not only to clothing but to fur and hair of animals.

Bachelor's-button, Cornflower

Bachelor's-button, Cornflower
Centaurea cyanus
Height to 2½ ft. Stems slender, branched, well-supplied with leaves, sometimes woolly when young. Leaves slender and almost grasslike, to 6 in. long, with lower ones sometimes obscurely toothed; the upper entire, alternate, grayish-green, and usually somewhat woolly. Annual.

Native of Europe. Probably introduced into America as an ornamental. Now escaped from cultivation and safely established in fields and waste places. From Quebec to Virginia and extending west particularly to the Pacific Northwest where it is most abundant. The 350 or more known species of the genus are all natives of the Old World.

Flowers borne in single heads at ends of branches, to 1½ in. wide and almost as high, though there are still larger cultivated forms. Flowers blue, purple, pinkish or white; flowers in margins of heads have large, almost fringed corollas, much larger than those of central flowers. Fruit uniquely lopsided at the base.

Flowering time July through September. Fruiting time August through October. Fruits retain their vitality for many years so it is important that plants be prevented from fruiting. They may be a really serious pest in grainfields of the Northwest. It is a long fight to eliminate the plants where once established.

Flowers used as rather long-lasting buttonhole bouquets. The name *Centaurea* refers to mythical centaurs who were reported to ascribe a medicinal value to the plant, which is not now so recognized. It is known as French pink, witchbells, brushes, hurt sickle, bluebonnets, and knapweed. *C. maculosa,* purple star thistle or spotted knapweed, is an exceedingly common weed in the northeastern states, north central states, and the Pacific Northwest.

Canada Thistle

Canada Thistle
Cirsium arvense
Height to over 4 ft. Stem straight, nearly smooth, somewhat woody, slender, grooved, green. Leaves alternate with crimped margins, well-supplied with vicious spines that break off easily, with clasping base; the basal to 8 in. or more long and petioled; the upper, smaller. In contrast to most thistles this is a patch-forming plant. The colony is formed by extensive proliferating roots.

Native of Europe. Thoroughly established in America, where it is found in any neglected place, in gardens, lawns, and grainfields as well as in meadows and pastures. Ranges from Newfoundland to Virginia, west to Utah and British Columbia. One of most persistent plants.

Flowers borne in many terminal medium-sized heads or in axillary heads, heads being about 1 in. through and 1 in. high, with fragrant purple flowers that may be pistillate or staminate. Corollas of flowers longer than bracts of enclosing involucre and so protrude.

Flowering time June through August. Fruiting time July through September. Fruits carried by wind because of their delicate feathery parachutes. So unpopular is this weed that it has long been legislated against. It apparently is not law-abiding and continues to prosper in spite of grubbing, cultivating, and spraying.

Strawflower

Modern weed-killer sprays control the plant better than earlier ones have. Crop rotation with clover on occasion tends to discourage the plant. Goldfinches feed their young on the fruits in the milk stage and line their nests with the down. Peeled young stems boiled in salt water are delicious. Also called creeping thistle, small-flowered thistle, perennial thistle, and green thistle. A noxious weed by any name.

Strawflower
Helichrysum bracteatum
Stout stems to 3 ft tall, sometimes somewhat branched, minutely roughened, with scales that may rub off. Leaves numerous, up to 5 in. long, narrower near base and smooth or nearly so, dry and retaining form well when deprived of water. Roots relatively tough and fibrous.

Native of Australia but has established itself in gardens and in hearts of many Americans. Around 300 related species native of Europe, Asia, Africa, and Australia. Grown mostly as an "everlasting." It prefers rich loamy soil but does well where there is plenty of sunshine.

Flowers in terminal, solitary, disklike heads, 1 to 2½ in. across. Red, white, yellow, brown, pink, or red, the colored portion being not the corollas of the flowers but the bracts which subtend the flowers. In related *H. petiolatum,* heads are clustered rather than solitary.

Seeds may be sown indoors or out and transplanted when danger of frost has passed. Seeds germinate in 5 days; live to 3 years. Plants should be thinned to 12 in. apart and cultivated frequently. Blooming should continue from July until frosts come. Many seeds are produced in each head. Common pests include aphids and tarnished plant bugs, controlled by a nicotine sulfate spray. Chief limitations to its more general growth are climatic. Favors moderate amount of moisture and does better in warmer climates.

Probably largest and best of everlastings. Flowers should be cut when half-open, in evening, for winter bouquets. Should be fastened in bunches and allowed to dry hanging downward in a warm room. May be arranged after they have dried.

Globe Artichoke

Globe Artichoke
Cynara scolymus
From 4 ft high, coarse, thistlelike, except that in this species leaves are only slightly spiny and bracts of head are unarmed while in cardoon, *C. cardunculus,* leaves and bract tips are spiny and the latter plant may grow over 6 ft tall. Leaves large and variously lobed. Rootstocks produce offshoots after crown of year dies. Perennial.

(Continued)

Native in southern Europe and northern Africa and grown extensively, particularly in France. Popularity in America varies greatly. Has been cultivated in the South for over 100 years. Enjoys a wide range of soils, but does best on fertile, deep, well-drained areas. Must be protected in winter from heavy freezing by a cover, coal ashes proving popular.

Flowers borne at ends of main stem and of lateral branches, in heads to 3 to 4 in. through, with a fleshy base and thick bracts or scales. Flowers of cardoon, purple. Plants grown from seed do not "come true" so usual reproduction is by underground crowns. Fruits thick, smooth, four-angled, seedlike.

Young plants selected when about 12 in. high as offshoots of parent plant, avoiding injury to parent root system. Roots set 6 to 8 in. deep and 6 ft apart, with 8 ft between rows, and cultivated. Freezing during growing may be fatal; heat may cause buds to open, ruining marketability. Greatest production in plants 2 to 3 years old; plantings renewed every 4 years. Manure is an unsatisfactory mulch as it causes rotting.

Large buds most desirable, so grading and packing is done, heads being packed in paper-lined boxes. Long-distance shipping in refrigerator cars with usual icing common for lettuce and similar vegetables. In desirable marketable heads, stem is cut about 1 in. below base of head. Old stems are cut off after last bearing, burned, used as silage or buried in field for green fertilizer. We eat the soft, fleshy receptacle of the flower head and the thickened bases of the flower scales. These are sometimes eaten raw but more usually boiled. Some persons eat the blanched stems and leaves.

Chicory

Chicory
Cichorium intybus
Height to 4 ft. Stems sparingly hairy, erect, relatively slender, hollow, well-branched, green but becoming purplish, then red, then brown; woody and tough when mature. Leaves somewhat dandelionlike, to 6 in. long, with deeply toothed margins; lower with long petioles, upper without. Perennial.

Native of Europe but firmly established in fields, waste places, gardens, and roadsides from Nova Scotia to North Carolina, west to California and Manitoba. Of the eight species in the genus, all are native of the Old World. Only this one is well-established in America.

Flowers in showy heads, arranged along stem in upper areas. Heads to 1½ in. broad, usually somewhat grouped, with as many as four being close together, but only one in a cluster seems to be open at a time. Flowers all alike, blue, rarely white or pink, with notched tips, producing fertile fruits.

Flowering time July through October, usually in the morning. Fruiting time August through to end of year. Young leaves used as a slightly bitter salad plant. If tops are repeatedly cut and roots well-watered, a delicate green may be produced.

Roots dug just after flowering ceases, ground, roasted, and used as a coffee substitute or even as a deliberate adulterant of coffee. Used routinely in coffee in Louisiana and nearby states. Some persons prefer it to real coffee. It was used as a means of extending coffee supply in the Civil War and in the World Wars. Sometimes called blue sailors.

Salsify, Goat's Beard
Tragopogon porrifolius
Height to 4 ft. Leaves keeled like a boat, tapering to clasping, rather broad bases; in fact, leaves look almost grasslike. Plants smooth, dark green, with long relatively slender taproots somewhat resembling those of parsnips. Biennial that requires a long first season to succeed.

Native of southern Europe and escaped in various parts of the world including North America. Also found in north Africa and central and southern Asia in closely related species, most of which are cultivated to some degree. Known to have been cultivated since 1600 in southern Europe. Mammoth Sandwich Island is the most common variety grown.

Flowers purple or yellow, in large heads which close to form pointed cylinders, open in the morning but closed by noon. Fruits sticklike, crowned with a feathery yellowish down that serves as a parachute to help in distribution by wind. Flowers enclosed by green case except when it is expanded.

Salsify

Seeds usually sown in drills from 12 to 15 in. apart, and plants are thinned to be about 2 in. apart in the row. Long season necessary makes it usual practice for roots to be dug in winter but some are dug earlier and stored, as are beets, turnips, parsnips, and similar root crops.

Because of "oyster" flavor, these plants have some popularity, but they never hold an important place in a commercial market. Their value probably warrants more general use. Roots marketed in bunches, usually tied at each end with tufts of leaves showing at the end. About 10 roots make a standard bunch.

Dandelion
Taraxacum officinale
Height of flower stalk to 1½ ft. A stemless plant, leaves rising directly from top of coarse bitter thick root that may be well over 10 in. long. Leaves to 10 in. long and to 2½ in. wide, narrowed at base into petioles, but with deeply toothed margins, with many teeth curving backward. Perennial.

Dandelion

Native of Eurasia. Thoroughly established in North America in lawns and waste places. Found in Southern Hemisphere as well as in Northern. In North, there is a closely related red-seeded dandelion that is much less common than this species.

Flowers yellow, all bright yellow straps, with the ends notched or toothed. All with stamens and pistils. Would seem that pollination would be effected either by self or by insects but pollen is sterile and there is no true pollination or fertilization. This does not matter, however. Seeds of dandelions contain little buds of the parent plant and do not require fertilization. Thus dandelions are asexual, and the successful strains are uniform.

Flowering time January to December. Fruiting time the same. Cultivated for a drug. Probably one of the most successful weeds if we consider dispersal as a criterion for success.

As a weed, this plant is a bother to those who like uniform green lawns. Can be easily eliminated by the newest of the weed-killer sprays but treatment must be repeated because new crops are sown by wind each year from untreated adjacent territory. Leaves make an excellent potherb. A wine is prepared by fermentation of the flowers.

Sow Thistle
Sonchus oleraceus
Height to 10 ft. Stem only slightly branched, weak, erect, somewhat juicy, smooth, green, hollow between joints, with milky juice, angled, arising from a deep white taproot. Leaves alternate, deeply toothed or lobed with smaller teeth around the margin, the lower with petioles and with bases clasping the stem, provided with weak spines. Annual.

Native of Europe. Well-established through most of North America except extreme north. Also established in Central and South America. Too common a weed in cultivated lands, particularly in gardens as well as in waste places. Will grow in lawns that are not mowed frequently. About 45 Old World species in the genus.

Flowers borne in rather crowded heads at top of plant. Individual heads over 1 in. across, with a broad base on which are crowded 50 or more pale yellow flowers, all ray flowers with notched tips. Fruits red or brown, carried by wind by a dirty white parachute.

Flowering time June through September. Fruiting time July through October. Since plant is annual, one of best control measures is to keep tops cut back each year before they have time to fruit, so that any seed supply in soil may eventually be destroyed. Fall plowing and cultivating help control also.

Sow Thistle

While juices of plant are definitely bitter, young leaves may be made into a pleasant-tasting green if boiled, the water removed, and the material well salted. It is more popular as a food in Europe than it is ever likely to be in America.

Garden Lettuce
Lactuca sativa
Stem to 4 ft high, well-supplied with leaves, branching above. Basal leaves crowded, to 10 in. long, forming a more or less compact ball depending on variety, and varying in texture, shape, and color. Root relatively deep. Varieties include three classes: butter, including Boston lettuce; crisp, including Iceberg; and Cos. Annual.

Native of Europe and Asia. Has been under cultivation for 2,500 years, being mentioned by writers in 500 B.C. Grown in almost every home garden. Favors a cool growing season, and to do best has rather definite requirements for water.

Flowers borne in erect, yellow heads, open in forenoon and close in afternoon; produce a flattened seedlike fruit, blown by wind with help of downy parachute. Fruit either straw-colored, brown, or black. 12 to 16 flowers or fruits per head, fewer than in related sow thistle. High temperatures cause premature seeding and lessen commercial value.

Seeds do not germinate well at above 80°F so may be iced 4 days before planting. Seeds 2 lb per acre, when rows are 14 to 18 in. apart and thinning is to be to 10 to 18 in. between plants. Thinning should not be delayed. Cultivation avoided when plants are well-grown and should be shallow. Can be stored at 32°F, if not harvested after rain, for from 3 to 4 weeks.

Garden Lettuce

Value of lettuce raised in United States greater than for any other vegetables except tomatoes and potatoes, the leading states being California and Arizona, which produce about 80 percent of the commercial crop, followed by New York, New Jersey, and Texas.

Wild Lettuce

Wild Lettuce
Lactuca scariola

Height to 7 ft but usually shorter. Stems erect, with short side branches, green, smooth, relatively weak, smooth or with weak bristles at base, with milky juice. Leaves alternate, light green, with waving margins and prickles along the margins along the midrib below, clasping at base. Annual.

Native of Europe. Widely established in North America from coast to coast in United States except in the southern areas. Found on all sorts of soil, usually in the open and where there is a good sun exposure. Many races and varieties differing largely in leaf characters. In most, leaves are deeply toothed.

Flowers borne in open clusters of heads at or near top of plant. Heads under ½ in. through; contain only from 6 to 12 flowers, or fewer than in sow thistles. Heads accordingly much narrower. Uppermost flowers open first and as the season advances lower flowers mature. Fruits ridged, brown, with parachutes.

Flowering time July through October. Fruiting time August through November. Plants, when young, used as a salad substitute, and may easily be accepted as a desirable potherb. Since plant is an annual, it can be kept under control by preventing tops from forming fruits.

Cultivation, hoeing, and use of the modern weed-killer sprays are usually sufficiently effective to keep this plant under control. This variety is generally more common than species with its more deeply cut leaf margins.

Orange Hawkweed
Hieracium aurantiacum

Height to 20 in. Stem usually without leaves but bearing heads of flowers at top, not profusely branched, well-covered with hairs, slender. Leaves mostly in tufts at base of stems, entire or obscurely toothed, hairy, conspicuously veined, to 5 in. long and to 1 in. wide. Sprawling stems take root. Perennial. Also known as devil's paintbrush.

Native of Europe. Altogether too well established in certain parts of North America. Known well from New Brunswick to Pennsylvania, west to Corn Belt. Related species occupy same range and extend it west. Flowers of related species usually bright yellow, not the beautiful reddish-orange of this species. Over 300 species of the genus native of North Temperate Zone and of higher lands of South America.

Flowers borne in few or several crowded heads at top of erect flowering stems. Each head is to 1 in. broad and composed of brilliant orange-red flowers that open or close with intensity of sun. Rays toothed at tips. Involucre that encloses heads composed of two to three rows of hairy bracts.

Flowering time June through September. Fruiting time July through October. A real weed pest on poor farmlands. May occupy soil to exclusion of other plants because soil is too poor to provide competition from normal plants of the area.

Control may be by means of destructive measures such as sprays, salting, and the like, but a better procedure seems to be to use fertilizer and stimulate rival plants that if well-nourished will quickly crowd out this offensive but beautiful plant. In a sense, presence of plant is a sign of poor soil management.

Orange Hawkweed

CLASS MONOCOTYLEDONAE

Those members of the division Anthophyta which (1) have a single cotyledon (seed leaf) in the seed, (2) have a stem with scattered vascular bundles, (3) have leaves with parallel veins, and (4) have floral parts in threes or multiples of three are in the class Monocotyledonae.

Unfortunately the boundary between this class, Monocotyledonae, of Anthophyta and the second class, Dicotyledonae, is somewhat blurred. Therefore, combinations of structural characteristics rather than one single characteristic become clusters of "key" characteristics when we separate members of the division into classes. Represented by palms, lilies, orchids, grasses and other plants. About 50,000 species.

Order Pandanales./Family Typhaceae

Common Cattail
Typha latifolia
Height to 6 ft or more. Leaves alternate, narrow, ribbon-shaped to nearly 1 in. wide, or as narrow as ¼ in., sheathing the stem, filled with large air cells. Rootstock horizontal, branching, 2 to 3 in. thick, white within, well-stocked with starch, and forming a firm floor about at usual water line. Some authors list *T. domingensis* as a species in marshes at low altitudes from California, east to Atlantic and south through tropical areas.

Found in marshlands throughout temperate North America, Europe, and Asia with next species continuing the range into tropical areas. Often forms extensive, clear stands giving characteristic appearance to surroundings. About 17 species in tropical and temperate areas.

Flower stalk to height of over 8 ft; ends in two flowering spikes, uppermost is staminate, with loose stamens that are shed in early summer; lower, pistillate. Pistillate and staminate areas not separated by bare portion as in next species, *angustifolia.*

Cattail marshes provide superior shelter for waterfowl but little food. As a result, in marsh management it is often necessary to cut bayous that increase the shoreline accessible to birds that must get their food in more open waters. Geese and muskrats use the underground stem a great deal. Carp sometimes suck away sediments from roots and expose them for the muskrats. Long-billed marsh wrens, redwings and yellow-headed blackbirds nest in cattail marshes, and muskrats use the plants in constructing their houses.

General use similar to that outlined in next species, *angustifolia,* though this one provides superior leaves for calking barrels and for making rush-bottomed furniture. Fluffy fruits of both species are used for stuffing pillows and insulating blankets. Young stems, up to 18 in., minus rind are delicious. Young fruits eaten as "Cossack asparagus." Roots eaten as salad.

Common Cattail

Narrow-leaved Cattail
Typha angustifolia
Height to 6 ft. Leaves erect, flat, smooth, green ribbons, about ½ in. broad as contrasted with inch-wide leaves of broad-leaved cattail. Underground a stout, horizontal, perennial rootstock well-stocked with starch, branching and giving rise to leaves and flower stalks.

Native of Eurasia and other north temperate parts of world. Found from southern Maine to North Carolina and west to the Pacific Coast, but most abundant near coast. Has more restricted range than broad-leaved cattail and is rarely so abundant. Both grow in shallow waters or marshlands.

Flowers appear in early summer, either pistillate or staminate. The two are borne on same erect stalk with staminate at top and separated slightly from pistillate, a character which helps distinguish the two species. Pollen wind-carried. Fruits wind-carried when spikes break during winter or following spring. Seedlings grasslike.

Erect leaves expose maximum surface to sun when it is at angle; minimum surface when it is strongest or directly above. Creeping rootstocks submerged for considerable portion of year, particularly when fire may sweep exposed upper portions. Underground parts provide good food for muskrats, and upper parts summer protection.

Leaves of this and of broad-leaved cattail may be cut in late summer, dried either on racks or in shooks, and twisted or braided into cords, which may be used in making rush-bottomed furniture. Staminate heads, harvested at their prime, yield a flourlike nutritious powder, submerged rootstocks, a starchy food, and plants, flood control. Cattails may appear in almost any wet place and frequently are first invaders of a newly excavated pond. May be but a small number of plants visible, yet an extensive root system may have developed.

Narrow-leaved Cattail

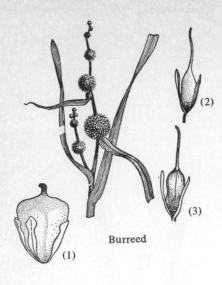

Burreed

(1) (2) (3)

Order Pandanales./Family Sparganiaceae

Burreed
Sparganium sp.
Up to 5 ft high, of rosettes of long, erect, limp, or partly surface-floating, flat, keeled or three-sided leaves, with fruiting stems either erect or limp. Identified largely by fruit clusters. Do not generally produce beds but appear along margins of waterway. Important species: (1), *S. eurycarpum;* (2), *americanum;* (3), *chlorocarpum;* (4), *angustifolium;* (5), *multipedunculatum.*

Found in marshes, along the margins of still or slowly moving bodies of water. Various species occupy all southern Canada and United States, except states bordering Mexico but including California. (1), widest spread.

Fruiting stem bears at and near tip spheres of staminate flowers that shed pollen into wind for lower spheres of pistillate flowers. Mature fruit is necessary for identification of most species. *S. eurycarpum* has two stigmas. Perennial.

Plant serves as anchor for soil in wave-beaten or river-flooded flats and as food for many birds, mammals, and insects. Some of plants may be tuber-bearing and supply extra food through these. Transplanted by seeds or rootstocks.

All species provide superior cover and favored food for muskrats, preferred food for deer, and good food and cover for waterfowl, the last eating nutlets and tubers. Of the species mentioned, all are probably important to wildlife.

Order Najadales./Family Najadaceae

Floating-leaved Pondweed
Potamogeton natans
Stems unbranched or only slightly branched. Leaves of two kinds: floating type has the blade base slightly heart-shaped, to 4 in. long and 2 in. wide, with petioles; 0.1 in. thick and stipules to 4 in. long; submerged leaves slender, soon die and are shed, with late-season ones even smaller.

Widely distributed over world in temperate and subtropical regions in ponds and slowly moving streams. Tolerant of acid. Best at depths of 1 to 5 ft. Many related species with floating leaves.

Flowers of most pondweeds are borne in slender spikes that extend well above the water surface in quiet weather. Some if not all flowers may be pollinated by wind. Spikes and their fruits are drawn below the surface at maturity. Short branches or "winter buds" of some species are produced in axils of leaves. These fall to bottom, take root, and produce new plant.

Important as a duck food since rootstocks and nutlets are available late in season in great abundance. Because these nutlets are held late, this species outranks some others in importance in managing ponds for wildlife production. Propagated by seeds and rootstocks.

Unique shape has appealed to artists, and its leaves and weak stems provide an almost standard motif for artists interested in representing the environment in which the plants live. It may have some food and fertilizer value. Cuttings for stocking are dropped in clay balls every 6 ft., or 3 to 5 bu an acre. Important food for several species of waterfowl, marsh birds, shorebirds. Muskrats use it. Seeds and tender leaves taken.

Floating-leveled Pondweed

Sago Pondweed
Potamogeton pectinatus
Stem almost threadlike and repeatedly branching into two. Leaves exceedingly narrow and bristlelike with points tapering but sides parallel for most part. Stipules joined to base of leaf making a sheath for stem. Tubers borne on horizontal stem at base of plant.

Does best in water 2½ to 5 or more ft deep, frequently found in shallow water, over sandy mud, in fresh or up to even 44 percent seawater. Most important aquatic in alkaline lakes of the West. Ranges from Quebec to British Columbia, south to Florida and Mexico. Whole plant is normally submerged.

Fruit with short beak, borne in spikes of 2 to 6 rather well-separated whorls; each fruit about ⅛ in. long, compressed, rounded on back, poorly ridged on sides. Tubers reasonably abundant on horizontal part of plant that is just beneath soil surface.

This species is listed as most important of all the pondweeds for encouragement of ducks, which feed freely and abundantly on nutlets that constitute an important part of fruit and on tubers at base of plant. Its wide range also makes it of importance to migrants.

Can be reproduced by transplanting underground rootstock and tuber-like buds; 1,200 to an acre or slips at 6 ft intervals. If dried in windrows, seeds may remain virile for as long as a year. Tubers harvested in spring with dip nets can be shipped wet for transplanting if water is changed and temperature is kept down. Tubers and seeds have been planted in many localities to improve feeding places for ducks. Introductions have probably extended range of this species beyond normal range. Stems, leaves, rootstocks, seeds taken by wildlife.

Sago Pondweed

DIVISION ANTHOPHYTA. CLASS MONOCOTYLEDONAE

Pondweed
Potamogeton crispus

Stem compressed and weak. Leaves with margins, finely toothed, with sides not parallel, but usually wavy-margined. Common gall-like winter buds, the well-developed stout beak of the fruit, the oblong crisp leaves plus the short, often coronalike stipules make identification of *crispus* easier than most other potamogetons. Related whitestem muskie weed, *P. praelongus,* has boat-shaped leaf tip, and persistent stipules. Bassweed or clasping stem pondweed, *P. Richardsonii,* has greatly reduced stipules.

P. crispus found in hard, or brackish water, often where it is polluted; introduced from Europe. Common from Massachusetts to Ontario and Virginia. *P. Richardsonii,* from Quebec to British Columbia and south to New England and Nevada. *P. praelongus,* from Nova Scotia to British Columbia, New Jersey, and Mexico. In California *crispus* is found in slow-running streams or canals at low altitudes and has been reported in the San Joaquin and Sacramento Valleys as well as in the San Francisco Bay region.

Chief means of propagation of *P. crispus* is by burlike winter buds formed by hardened short branches grown into leaf bases. Fruits in spikes, ripen in June and July, but whole flower cluster is relatively inconspicuous. When fresh, fruits have a soft portion.

Eaten by ducks but may become a bad weed in ponds where water is polluted. Closely related *P. Richardsonii* is reported more frequently as a duck food but this may be due to nature of its range. Winter buds of most of the pondweeds make a staple food for ducks.

Provides food and shelter for game of various sorts. While such game fish as bass and muskellunge do not feed on the weeds, they find in them shelter and smaller food which does eat them. Ducks may eat these species because they are abundant. Most do best in clear water 6 to 8 ft deep.

Curly Pondweed

Widgeon Grass, Ditch Grass
Ruppia maritima

Stems threadlike, forking, bearing even more threadlike alternate leaves, that sheath stems at base. *R. occidentalis* is a large form found in alkaline and saline lakes in the interior. *R. occidentalis* may be a form of the *maritima* with longer leaves.

Found in shallow bays along entire coast to a point where salinity is only 2 percent normal sea water. It can live in fresh water or in a salinity half again that of normal sea water. *R. maritima* and *R. occidentalis* occupy almost any saline or alkaline waterways in the United States. *R. maritima* more common in the western United States.

Normal propagation is by seeds or vegetative portions. Seeds of *R. maritima* much like those of Sago pondweed but are smaller, black, pointed, and borne in slender-stalked clusters. Cuttings do not normally begin to root unless they are in contact with soil.

Considered excellent duck food because of seeds and vegetative portions, both of which are eaten, but it is important in this respect only in areas where water is brackish or alkaline. Since it is more tolerant to these conditions than many other aquatics, it is uniquely useful.

Seeds of this and related plants are planted in spring or after normal crops are harvested in late summer or fall. Fall planting is usually preferred because it uses stock that has not been stored so long and is therefore more likely to be virile. Spring planting avoids destruction by migrating birds.

Widgeon Grass

Horned Pondweed, Grass Wrack
Zannichellia palustris

Stems numerous, threadlike, from extensive, creeping rootstocks. Leaves are opposite and frequently bunched closely together and are narrower, longer, and less crowded than in *Naias.* Stipules sheathing and membranelike. Placed in family Zanichelliaceae by some authors.

Found from the north edge of the Gulf of St. Lawrence west to the Yukon Territory and south into Mexico, being abundant all over the United States. Does best on good soil in fresh spring-fed water and in water up to 40 percent normal seawater salinity. Common, especially in brackish and subsaline habitats, ponds, streams, and ditches. Cosmopolitan through the world's temperate and tropical regions.

Flowers unisexual, adjacent in same axil of leaf. Fruits flattened and usually toothed down one side, borne in axils of the leaves in groups of 2 to 5 and show a distinct, hornlike form; are short-stalked or stalkless, with a body up to 0.1 in. long, with one variety with a slightly longer fruit.

Fruits and much of plant parts are eaten readily and with profit by most waterfowl. It has been reported in the diet of black duck, gadwall, mallard, pintail, redhead, ringneck, bluebill, shoveler, blue-winged teal, cinnamon teal, and widgeon.

In wildlife management, plants are propagated by distributing seeds or vegetative parts of plants into new territory. It is not difficult to find or to transplant.

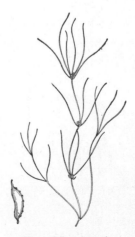

Horned Pondweed

Grass Wrack, Eelgrass
Zostera marina

Stem jointed, creeping, wholly submerged, sheathed by bases of long, slender, blunt-ended, ribbonlike, three- to five-nerved, flat leaves whose margins are not notched or irregular like those of the otherwise generally similar leaves of *Vallisneria,* even though those notched are fine and often overlooked. Tapelike marine plant. Placed in family Potamogetonaceae by some writers.

Usually submerged in 2 to 6 ft of water, in brackish water up to 25 percent that of normal seawater, and where the water is relatively cool. It is found in one variety or another along both coasts of the United States but not the Gulf Coast, and north to South Carolina. Occurs in shallow water in sheltered bays and coves in intertidal zone; often found washed ashore on beaches. Widely distributed in both hemispheres.

Reproduction in nature or under management is normally either by seeds or by rootstocks. Flowers are either staminate or pistillate and only one kind is found on a given plant.

Seeds and vegetative portions provide superior food for waterfowl in brackish and saline waters. Along Atlantic Coast, it has been attacked by a fungus that has greatly reduced its abundance. Great losses along Atlantic Coast in 1930s now almost recovered, plant has made a good comeback. Measures are being taken to supplant it there with a West Coast variety that may be disease-resistant.

Importance of this plant in salty waters is such that efforts will be made to save it from complete destruction or to supplant it with other species or varieties that may be equally useful. Important to marine fisheries as it provides "cover" along shoreline and food for forms on which fish feed. Probably the "grass" fed to horses of Julius Caesar when he invaded Africa.

Grass Wrack

Bushy or Northern Naiad
Najas flexilis

Stems very slender, often branched into two parts. Leaves opposite, slender to threadlike, and commonly crowded into whorls formed usually at a fork in the stem; slightly wider at base but leaf tapers before it widens at base.

Commonest representative of genus in northern half of United States and southern Canada, extending north to central Hudson Bay and Yukon Territory and south to Florida and south central California, except for most of Texas, Colorado, Utah, Nevada, and south. Most abundant in Great Lakes area.

Both staminate and pistillate flowers may be on same plant or may be on different plants. Fruit to ⅛ in. long, narrowly oblong, with seeds that are slender ovals and smooth and shining; located in axils of leaves rather than in clusters as in pondweeds. New plants may be propagated by separating larger plants and rooting cuttings.

Naiads can grow at depths of 20 ft or more, apparently with less light than most other seed plants of water, in fresh or brackish water. This species is most commonly found growing over sandy bottoms with wild celery. Most naiads are at their best in water 1 to 4 ft deep.

This is probably most useful of all duck foods, nutlets and vegetative parts being eaten regularly and developing in such a way that large flocks of waterfowl may be maintained without injury to the forage. Seeds and leaves used by marsh birds and shorebirds.

Naiad

Order Alismales./Family Alismaceae

Wapato, Water Nut, Swamp Potato
1. *Sagittaria latifolia*
2. *Sagittaria graminea*

Height to over 3 ft. Leaves like broad spearheads, coarse, dark green, varying greatly with depth of water (1); slender, spear-like (2); although much variation exists and so no strict generalization can be made. Tubers size of a potato, twice as long as broad, with slender point (1); smaller (2); difference may not be consistent since great variation exists.

(1), at freshwater margins from southern Canada throughout most of United States and sometimes almost completely submerged. (2), mostly submerged in shallow mud-bottomed marshes, streams, and ponds from Newfoundland to Florida, Saskatchewan to Texas. *S. latifolia* much more widespread.

Flowers produced in summer, on erect stems, white, with parts in whorls of three with three petals and three sepals surrounding crowded nutlets whose beaks in (1) are horizontal and incurved and in (2) incurved and remarkably short. Both are flattened and partially winged.

Mostly perennial, milky-juiced, succulent plants with variable leaf shapes. Water loss through leaves is great and therefore in some reservoirs plants are eliminated to help save water. Tubers of (1) may rival size of ordinary cultivated potatoes. (Illustration continues at top of p. 339.)

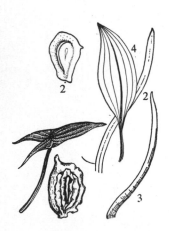

Both fruits and tubers are eaten by waterfowl and the tubers by muskrats.
16 species of ducks and geese are known to feed on the plants, which are
commonly set out where it is desired that ducks should be attracted.
Common names also include Indian onion, wild onion, duck potato, and
arrowhead.

3. Dwarf Wapato
Sagittaria teres

4. Delta Duck Potato
Sagittaria platyphylla
Plants only a few inches high with leaves usually slender throughout their
length (3), or plants to over 1 ft high with leaves on slender petioles with
expanded blade at tip (4). Tubers of (3), small spheres about the size of
BB shot; of (4), to nearly 1 in. through.

Found in mud flats in very shallow water from Massachusetts to Florida
(3) and in marshes that are either fresh or with brackish water from
Alabama to Texas and north to Missouri, Kansas, and Tennessee (4).

In each of these species, flowers are either pistillate or staminate and both
are borne on same plant. In (4), stems bearing fruits bend backward at
maturity, but not in (3). Fruits of (3) have a very short, erect beak while
in (4) beak is not erect.

Some persons consider shape of leaves of (4) indicative of saltiness of
water in which plants grow. In (3), waterfowl commonly feed on whole
plants while in (4) only the tubers may be selected. In neither case are
the fruits important food.

Tubers of (4) have been found in stomachs of gadwalls, ring-necked
ducks, pintails, mallards, canvasbacks, and pheasants. One canvasback
alone had 36 tubers in its stomach and gizzard, some as much as 1 in. in
diameter. Obviously, this is an important duck food, particularly in the
South.

Considerable confusion in classification to species. Some authors treat
S. graminea and *S. platyphylla* as one. Use of a good key to flowering
aquatic plants is the best way to separate members into species since
differences are minute.

(1), (2) Wapato, Water Nut, Swamp Potato
(3) Dwarf Wapato   (4) Delta Duck Potato

Water Plantain
Alisma subcordatum
Height to over 2 ft. Leaves in rosettes, with long petioles for, on the
average, over half the length and broad-pointed, egg-shaped, coarse-
veined blades or if submerged, no blades or these simply ribbonlike. Related species, (2) *A. gramineum;* sometimes *A. plantago-aquatica.* Only
latter species recognized by some authors. Appears as part of the flora of
our western states. (2) commonly narrower bladed.

Found in marshes, ponds, streams, and wet spots, from British Columbia
to California, Florida to Nova Scotia, with *A. gramineum* widespread in
Europe, Asia, and Africa.

Flowers small, on ends of slender, branching, stiff structures. In *A. plantago-aquatica,* petals are white; in *gramineum,* pink. Nutlets crowded in
fruits with two ridges and a groove down back; in (1), three ridges and in
(2), two grooves. Perennial and maintained through winter and adverse
conditions by a stout underground corm, which can divide and produce
new units.

Must have water in abundance at some time to grow and may survive
submergence for a long time.

Not an important plant in marshes or for game species. Small quantities
of nutlets have been reported as entering diet of wild ducks and of pheasants, but never in significant quantities. Ducks concerned include 12
species.

Water Plantain

Order Hydrocharitales./Family Hydrocharitaceae

Waterweed, Elodea
Anacharis canadensis
This is the *Elodea* of older classifications. Stems weak, long, not profusely branched but crowded throughout most of length with leaves about
½ in. long, with minutely irregular margins; not broadened at base as in
naiads. Will grow unrooted or will root on ground contact.

Found across the continent, from Gulf of St. Lawrence to southern British Columbia, except northern Maine, and south to central California
and South Carolina. Generally favors fresh water over mucky bottoms
at various depths.

Pistillate flowers borne on long, slender stems that reach to water surface.
Some flowers with stamens and pistils. Fruit small and oblong, with few

Waterweed, Elodea

(Continued)

Wild Celery, Tape Grass, Eel Grass

Southern Cane

Bamboo

seeds. Commonest method of reproduction, breaking of plants and starting of new ones from fragments. Fruits never abundant.

Plants grow so readily and root so quickly that they may become really water weeds. In England and Europe, often serious water pests. Ducks feed on fruits and leaves, one redhead having been found with 600 fruits filing half its stomach.

Plants are sometimes fed to domestic and captive ducks as a greenstuff since they grow so quickly, but should not be introduced where more desirable species exist. Leaves used in studying cell structure and protoplasm movement in biology classes. Members of the genus frequently available from stores for purchase as aquarium plants.

Wild Celery, Tape Grass
Vallisneria americana

Stems buried in the mud but producing tufts and clusters of leaves like slender, pale green ribbons. Common names, tape grass and eel grass, suggest leaf type. Stem roots freely at joints and new buds are delicious.

Found from Nova Scotia to North Dakota and south to Florida and Louisiana, but not in all parts of lower Mississippi Valley. Common in spring-fed lakes and streams, submerged in water from 1 to 5 or more ft deep, and in water of salinity up to 28 percent normal seawater in growing season.

Staminate flowers break loose and rise to surface, floating to pistillate flowers that float at surface at end of slender, long stem. After pollination, pistils and maturing fruit are drawn down by a spiral development of supporting stem.

In North, mature fruits found from September to November; in South, they mature in late December. Fruits, leaves, and more particularly underground buds provide one of best foods for ducks and for those who may eat the ducks that have fed on wild celery.

Wild celery is transplanted in wildlife management by using seeds, winter buds, or portions of rootstocks. Plants are eaten by fish, muskrats, and all sorts of waterfowl and are probably not exceeded in their food value by any other aquatic plant. They are propagated relatively easily. Interesting to note that canvasback duck *Aythya valisineria* feeds on wild celery.

Order Graminales./Family Gramineae

Southern Cane, Wild Bamboo
Arundinaria gigantea

Height to 30 ft, with numerous, short, spreading branches. Arise from stout, underground rhizomes. Erect stems only sparingly branched the first year, but very leafy and branched the second. Leaves 4 to 8 in. long and to 1½ in. wide, smooth or fuzzy, with persistent sheaths. In some classification systems, listed in subfamily Bambusoideae.

Native in the United States, being found from Virginia to southern Ohio and Illinois, south to Florida and Texas and being cultivated as an ornamental elsewhere. The plants sometime form almost pure stands, making the well-known cane brakes of the South that often are extensive in lowlands. Genus contains about 100 species found principally in southeastern Asia and adjacent islands, Japan to Madagascar.

Flowering branchlets are on main, erect stems appearing second year or later. Spikelets of 8 to 12 flowers, to 3 in. long, springing in clusters commonly from forks from which side branches come. Stems that have borne fruits soon die.

Favors rich lowlands along rivers; in damp woods, wet ground, and swamps; with plants flowering simultaneously over great areas at irregular but frequent intervals. Erect stems or culms may be as much as 3 in. through, hard and strong, and make an impenetrable barrier to traffic since they grow crowded closely together.

Young shoots make excellent forage for cattle. Leaves and seeds favored by cattle. Shoots frequently canned and may be purchased in groceries. Stems used in fish poles, pipes, baskets, mats, trinkets, fences, floors, rafts, and to some extent in buildings. Members of the genus serve to anchor soil because they send down a great number of fibrous roots.

Japanese Bamboo
Arundinaria japonica

Height to 10 ft. Erect stems arising from underground rhizomes are greenish or golden, stiff and tough, and bear single branches from the axils of the leaves. Leaves to 10 in. or more long and to 2 in. wide, smooth, shining above, and whitish or fuzzy beneath, with conspicuous sheaths.

As the name implies, this is a native of Japan but it is grown extensively in parks and gardens in the South as an ornamental. It is particularly popular in cities because of its general hardiness.

Flowers borne in rather open, spikelike clusters that spring from the axils of the leaves. In closely related *A. Simonii,* also from Japan, flowers appear very frequently and erect stems do not then die down as is the case with many other arundinarias, such as the native southern cane.

Bamboos are among the most rapidly growing of all plants, some species increasing as much at 18 in. in height in a single day. Some species, not American, may reach a height of 100 ft or more and may form almost impenetrable jungles because of the close-growing, tremendously hard stems.

This species is of interest to us because of its hardiness and its ornamental beauty. Some species yield materials useful in making of paper and textiles. Uniformity, lightness, and strength of stems make them ideal for construction of temporary houses, bridges, and rafts. Bamboo is a most important "tree" in the Orient. Where westerners use wood or steel, orientals substitute bamboo. Used as poles, tree props, fishing poles, fish-net handles, garden stakes, scaffolding, and in fence building.

Switch Cane

Switch Cane
Arundinaria tecta
Height to 12 ft but more commonly not over 6. Arise from stout, horizontal rhizomes. Erect stems slender, stiff, and bear leaves that are rough and to 6 in. long. In one variety, the leaves are shed each fall, the leaves turning yellow and new ones appearing each spring.

Found in swamps and moist soil from Maryland to southern Indiana and to the south and west through southern Oklahoma and Louisiana. This and the southern cane are the only native bamboos of the United States, but others are found frequently grown as ornamentals in various parts.

Flowers are borne in clusters near tips of leafless or almost leafless shoots that grow from the base of the plant or seem to arise independently from hidden or creeping rhizomes. Reproduction is either by seeds or by breaking up of underground parts.

Shoots, leaves, and younger stems are all eaten by cattle and hogs. In fact, hogs grub out and eat great quantities of underground portions of the plant. Dense crowding of stems provides an excellent shelter for many kinds of game and so plants are valuable for both food and shelter. Valuable livestock forage on the humid, forested ranges of southeastern United States.

These plants are too small for uses commonly made of other bamboos, but small canes, trinkets, and some kinds of furniture are not infrequently made of erect stems. Branching rhizomes provide a good soil anchor in flooded places and so prevent soil loss. Hogs may eradicate the plants.

Flint Corn
Zea mays var. *indurata*
Height to 12 ft or more, with stems; coarse, succulent. Leaves large, mature relatively early, and reasonably hardy and strong. Root systems sometimes deep.

Native of America. Flint corn is grown for grain farther north than are the different kinds of dent corn. It requires a deep, rich, sandy loam, with humus and sufficient water for best results. Flint corn is preferred for cultivation in some of our northern states and in Canada because cold soil does not inhibit its germination so much as that of dent corn. Is also favored in tropical lowlands because it is resistant to attacks by weevils.

Annual. Flowers in general, as in other corns. Ears long, slender, generally yellow-brown when mature. Kernels smooth, hard on top, closely crowded on relatively slender cob. Seed to harvest, about 90 to 100 days.

Kernels planted in hills or rows. If in hills, about three to five are planted in hills about 3½ ft apart, to a depth of 1 to 2 in. If planted in rows, there should be about one plant to every foot of the row. With proper management procedures and applications of fertilizer, it is possible to place corn plants much closer. Planting should be made after danger of frost. Cultivation as with other kinds of corn. In the East, stalks are harvested for ensilage.

As food for man, kernel is fair as a muscle builder (protein); excellent as a blood builder (iron) and as a maker of energy (carbohydrate); fair for bones and nerves (calcium and phosphorus). Fair for vitamin A and good for vitamin B. In West may be eaten by cattle in fields. Ears are picked when mature and stored.

Flint Corn

Corn

Sweet Corn
Zea mays var. *rugosa*

Height to around 8 ft, intermediate between pop corn and dent corn. Leaves clasping at the base, curving downward. Roots fibrous, with special prop roots near ground.

Native of America. Common trade varieties, golden bantam and Crosby evergreen. For early crops, favors well-drained sandy loams; for late crops, a richer, water-holding type. In sweet corn sugars are not changed to starches at as early a stage as in dent corn. So kernel, achene, remains sweet and can be boiled, roasted, or frozen while still "green," not ripe. Because the food reserves are not starches at this stage, the kernel is susceptible to wrinkling and shrinking.

Annual. Staminate flowers in tassels at top of plant. Pollen windblown to "silk" or tip of pistils in "ear" lower on plant. Mature ear 5 to 8 in. long. Mature kernel yellow, wrinkled, horny, more or less translucent. Immature kernel soft, milky, sweet, edible. Seed to harvest, 90 to 100 days. Kernel an achene.

Kernels planted in hills 3 ft apart, three to four per hill, 1 to 2 in. deep, after frost danger. Planting at 2-week intervals prolongs harvest. Cultivate at least four times when soil is dry; first deep, then shallow after roots develop. Animal enemies include grasshoppers, earworms (in South), wireworms; plant enemy, corn smut.

Sweet corn is a fair energy builder for man and cattle; fair source of calcium and phosphorus for bones and nerves. Used to make alcohol, syrups, gums, starches, and oils from fruits. Ton of corn yields about 90 gal of alcohol for explosives, fuel, and other uses. Paper from stalks, explosives from pith, special charcoal from cobs.

Pop Corn
Zea mays var. *everta*

Height to 6 to 7 ft, but usually the shortest of the corns. Leaves smaller than in the other kinds of corn. Roots in general similar to type already described.

Native of America but now grown farther north than was the original corn plant. Soil requirements in general similar to those of sweet corn. Product largely of Iowa.

Annual. Flowers in general similar to those of sweet corn. Number of ears to a stalk may be a hereditary character. Mature ear 5 to 8 in. long, relatively slender. Kernel, when ripe, hard, smooth, sharp-tipped, and crowded into many rows. Hard surface essential to good "popping"; hard starch and water are held within the membrane. When heated, water turns to steam, starch expands with steam, bursts the membrane, and makes "popcorn." Seed to harvest, 90 to 100 days.

Kernels are planted in hills or rows, 3 ft apart. About five kernels to a hill, to depth of 1 to 2 in., after frost danger. Cultivate as with other kinds of corn. Pop corn should not be harvested until the kernels are mature but before danger of frost. Keep in well-aired, cool, dry place. Animal enemies much the same as for sweet corn.

As food for man, pop corn is essentially a producer of starches that provide the usual carbohydrate parts of a diet. Pop corn is not normally fed to cattle, although the vegetative part of the plant has some silage value. Cutworms that work at night cut plants off close to the ground. Plant enemy, corn smut.

Dent Corn, Field Corn
Zea mays var. *indentata*

Height commonly well over 12 ft, with coarse, succulent leaves and stems that are excellent for ensilage and good strong, sometimes deep root systems that can combat drought.

Native of America. This is the important corn of the Corn Belt and the backbone of the hog and cattle industry. It does best on the deep, rich soils of the long-season Middle West.

Annual. Flowers in general like other corn flowers. Ears relatively short and thick, with cob proportionately small but actually large. Kernel indented on the top, yellow, white, or red. Requires 145 days or more from seed to harvest.

Kernels are planted in hills or rows. If in hills, 3½ ft apart. About 4 to 7 qt will plant an acre. Planted in rows about 1 ft apart. Planting after frost danger. Can be planted 10 to 14 days before average date of last killing frost. Can be planted in greater concentration per unit of area if good management procedures and heavy applications of fertilizer are made. Seedbed tillage should be kept at a minimum to maintain soil structure and decrease soil compaction. For about a quarter of a century, the average yield per acre in the United States was between 25 and 30 bu. Depends upon man to develop disease-resistant strains.

It is not sound economics on a nationwide basis to feed corn to hogs, but it pays the farmers of the Corn Belt. As food for man dent corn is

(Continued)

Sweet Corn

Pop Corn

Dent Corn

essentially like flint corn, although the volume of dent corn is much the larger. Harvested green and chopped and stored as silage. Corn harvested for silage will yield about one-third more feed nutrients per acre than corn harvested for grain. Male-sterile corn, grown for high stalk sugar content, produces no grain and usually yields considerably less dry matter than grain-producing corn hybrids. In America, dent corn is largely yellow as a result of the discovery that yellow form has more carotene and so is more nourishing. White dent corn is used to make corn meal, mush, and hominy grits in the South while yellow dent corn is preferred in the North to make corn meal for mush, Johnnie cake, and corn bread. Also used farther south to make tortillas. Cornstalks are used as litter and bedding for cattle in barns; for making insulating boards, strawboard, wood alcohol, and in preparation of a substitute for hard rubber. Cobs are used chopped for litter for barns and made into bowls for pipes for smoking. Development of high-yielding hybrids has increased total possible yield to over 200 bu per acre under intensive fertilizer and management practices and proper soil conditions.

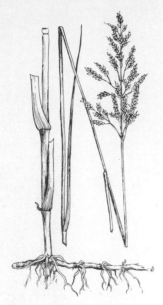

Johnson Grass
Sorghum halepensis
Height to 7 ft. With tough, persistent, creeping, sheathing rootstocks. Stems ½ in. thick, smooth and leafy. Leaves smooth, with roughened edges, conspicuous midribs, 1 ft or more long and ½ to 1 in. wide. Generally vigorous in appearance. Except for perennial rootstock, plant resembles sorghum. Now considered a pernicious, perennial weed in some locations where it has escaped from cultivation; spreads by seeds and rhizomes. Important as a hay crop in many parts of the South.

Native of southern Europe and Asia but now all too firmly established in Southern states and elsewhere. Introduced into United States about 1830 by Governor William Johnson of Alabama and Governor Means of South Carolina, with thought that it might provide a superior pasture and hay grass.

Flowering panicle open, spreading, to 2 ft long, with two to three branches at a joint, and more or less drooping. Spikelets in pairs at joints, or in 3s at ends, each containing one flower that produces a fruit. Grain free, closely resembles sorghum grain. Spikelets usually awnless. Blooms in June and July.

Thrives on great variety of soils, rich or poor, wet or dry. Does not grow during drought but grows rapidly immediately after rains. Rarely persists where ground freezes to 6 in. Yields two to four crops of hay a year which, if cut young, is palatable and nutritious. Does not stand grazing well. Sow 1 to 1½ bu per acre.

Eradication by prevention of seeding, by pasturing a year or two, then plowing to depth of 3 to 4 in. to expose roots to frost, by disking in early summer and then planting to a crop such as cotton that is cultivated, or cowpeas that smother with dense growth. There are chemical weed killers which act on Johnson grass but should be used with precautions. Plant may be poisonous to cattle in dry weather. Hogs relish rootstocks.

Johnson Grass

Sudan Grass

Sudan Grass
Sorghum vulgare var. *sudanensis*
Height to 10 ft. Leaves many, up to ½ in. wide. In general much like Johnson grass but without perennial, spreading, tough rootstocks that make it a pest. Narrower leaves provide an easy superficial identification character. When grown in cultivated rows, Sudan is planted like other sorghums using 10 lb of seed to the acre and in cultivated rows. Yields seeds at rate of 200 to 300 lb per acre.

First coming from South Africa and becoming established as a crop grass in the limited rainfall of our area north to central Kansas. The related Rhodes grass of central Africa does well in the dry soil of the southern part of the Gulf States. Prefers a warm summer climate of relatively low humidity and is sensitive to frost. Recovers quickly after drought if rains occur.

Fruiting and flowering cluster of Sudan grass about 1 ft long and half as wide; that of Tunis grass, 2 ft or more long and slender. In both, spikelets bear awns for the most part and both are annuals. In Sudan grass, spikelets are usually brown; in Tunis grass, green.

Sudan grass has ornamental possibilities, but its greatest value lies in its forage properties. Being an annual, it yields well to crop rotation practices and while the amount of forage produced may not equal that of Johnson grass, the quality is better and the plant may be kept in control more easily. Sudan grass growth after pasturing or mowing and three or four cuttings a year can be obtained in the southern Great Plains region. Crop of hay is produced in 60 days.

It is possible that this plant may prove of increasing value as a crop in areas where water is not sufficiently abundant to support more common crop plants. It probably does not play an important part in wildlife management but does give a quick hay crop when this is needed. Probably planted on an area of more than 3 million acres annually. Has in recent years become a choice food of waterfowl and there are problems when ducks decide to feed in farmers' Sudan grass fields. Food for many wildlife forms.

Broom Corn
Sorghum vulgare var. *technicum*

Two varieties include standard, with a height up to 15 ft; and dwarf, with a height up to 7 ft. Stems solid and tough. Root system, in general, similar to that of corn, but more likely to be deep-rooted. Not used as forage.

Native of tropical Asia and Africa. Grown extensively in Europe, for the most part in Italy, Austria, Hungary and Germany. Known in Europe for over 300 years. About seven types of broom corn are grown in North America. It is thought that Benjamin Franklin first planted broom corn on the North American continent. In America, standard is commonly grown in Illinois and adjacent states; dwarf, in Oklahoma, Kansas and Texas. Requires a fertile soil and abundant moisture. Grown in United States since 1700s.

In standard, the "brush" or flower panicle is 18 to 30 in. long, slender and flexible; in dwarf, 1 to 2 ft long and much stiffer and broader. Rays or branches of panicle or brush are naked below, stiff, and arise from almost a common point but branch at the ends. Spikelets awned.

Standard is planted in rows 3½ ft apart, with the plants 3 in. apart; dwarf, in rows 3 ft apart with plants 2 in. apart. From 3 to 5 lb of seed are used to plant an acre. In harvesting standard, tops of two rows are bent toward each other to form a "table" 2½ ft up. A bushel of broom corn seed weighs about 45 lb, and an acre should yield 20 to 30 bu of seed.

Raised for the brush, useful in making brooms. A crop of dwarf should yield 400 to 500 lb to the acre. It is sold in bales of 300 to 400 lb of material, cured in layers to 3 in. deep after the seeds have been threshed out.

Broom Corn

Sweet Sorghum
Sorghum vulgare var. *saccharatum*

Height to 15 ft. Leaves numerous, and broader than in most of its close relatives. Stem coarse, pithy, and well-supplied with a sweet juice; conspicuously jointed. Root system much like that of corn, but may be more branching to form a continuous, shallow, underground mat. At early stages of growth looks like corn. Also known locally as sorgo.

Native of Africa and southern Asia; introduced into America probably about 1875 for serious culture. Among leading cultivated forms are amber and orange. It has been developed from same species that produced broom corn, durra, and kaffir corn. It is important in the South and Southwest. Grown in Egypt before 2200 B.C.

Flowers are borne in terminal, loose, drooping panicles, with spikelets showing protruding red or yellowish-red fruits from between dark red or black enclosing scales. Amber matures early; orange, later; gooseneck and redtop, later. Black amber matures in 85 to 90 days; red amber, 90 to 100; orange sorgo, 110 to 115.

Seed is planted in drills much like corn; when it is in rows, must be closer. Broadcast seeding requires 75 to 100 lb (1½ to 2 bu) to the acre; close rows, uncultivated, 50 to 75 lb; wide rows cultivated, 8 to 20 lb. It may be planted after corn and yield of forage is higher than with corn.

Raised for fodder, for ensilage, as a smother crop, for a fermented drink, for production of syrup. Sugar production from sorghum is possible but too expensive. As pasturage, is dangerous after frost as poisonous properties may develop. These disappear when plant is cured for storage as fodder. Sought as winter food by ducks in some localities. Attractive as a food not only for waterfowl but also for songbirds, shorebirds, and some small mammals. Problems of conflict of interest between landowners and wildlife forms erupt locally.

Sweet Sorghum

Kaffir Corn
Sorghum vulgare cv. *'Caffrorum'*

Some grain sorghums may grow to height of 15 ft. Kaffirs are stout, with somewhat juicy stems, crowded leaves, with juice almost sour. Milo sorghum is less leafy than kaffir. Feterita, another variety, reaches a height of 14 ft and is either slender or stout, by some classed as a durra. (Hegari is the variety shown.) There have been, over the years, so many varieties developed from *Sorghum vulgare*, it is difficult to differentiate between them.

Kaffir and most grain sorghums are native of Africa, but kaffir is probably most commonly grown variety in United States. Grown in India, China, Africa, and in America in Great Plains area between Rocky Mountains and 98th meridian. Texas, Oklahoma, Kansas, and Nebraska favor crop.

Mature panicles are compact, dense, crowded with spikelets and with the large grains. Varieties of kaffir include red, white, pink, and blackhull.

(Continued)

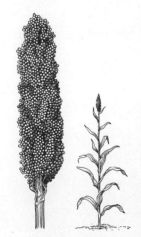

Kaffir Corn

345

Others have pink grains. Spikelets loosely hairy, awnless, with supporting scale about ½ as long as grain that is broad to spherical. Annual.
Pink matures in 110 days.

Kaffir was brought to Great Plains area of United States about 1875.
Commonly drilled into soil. Provides excellent ensilage and a fair quality of forage. Area devoted to growing grain sorghums of United States
approximates that devoted to rye. From 4 to 6 lb of seeds are needed to
acre for grain; 50 to 70 lb for fodder. Seed heads attractive to many birds
which may be considered pests by some farm owners.

Grains used commonly for feeding stock, poultry, and to some extent
humans, grains being fed whole or crushed. Breakfast foods, pancakes,
and bread of kaffir locally common. Stalks properly cured make excellent fodder. Grain 16.8 percent water, 6.6 percent protein, 3.8 percent
fat, 70.6 percent carbohydrate, 2.2 percent ash.

Durra
Sorghum vulgare var. *durra*
Durra includes milo and feterita, according to some authorities. Stems
slender to mid-stout. Leaves 8 to 10. Smaller in all respects than kaffir,
although feterita may be almost as high. Pith dry and not sweet. Some
durras are mature when no more than 2 ft high. Leaves commonly break
off before maturing. Over the years there has been much crossing of
varieties, so identification is very difficult; some authors say impossible.

Durra is native of the Nile region in Africa, from which we get the varieties most commonly grown. Yellow milo, Jerusalem corn, and feterita
are all kinds of durra. Closely related shallu from India and tropical
Africa, and kaoliang from China and Manchuria, have not established
themselves in this country.

Panicle of durra is compactly oval, erect or goose-necked (bent over).
Grains are flat or flattened, large, unlike those of kaffir and are commonly
white or reddish-brown and covered by greenish, strongly nerved scale at
base. Stems ripen with or before the grain, 72 to 82 days. Total growing
period 102 to 150 days.

Durra is smaller than kaffir, less valuable as forage, and earlier but is
better adapted to growing on dry areas, particularly where the growing
season is short. Plants are drilled so that they will be about 4 to 6 in.
apart if grown for grain or forage, with 4 to 6 lb of seed per acre.

Durra grows slowly until settled warm weather, when it matures quickly.
It is harvested by cutting and shocking, and shocked material is threshed.
Durra makes good poultry food and, in some areas, fields attract ducks
for waste grains left. All forage sorghum types are about equal to corn
and barley in feed value for cattle. There are bird-resistant types of
sorghums, identified by dark seed coat. In making up rations for poultry,
they should account for no more than 10 percent of the ration. In general, sorghums yield better than corn in dry years and on droughty soils.

Sugar Cane
Saccharum officinarum
Height to 15 ft. Stems solid and heavy as in corn, with leaves springing
singly from solid joints. Leaves stiff, to 2 in. broad, 1 ft or more long,
smooth on both surfaces, sharp on edges, with prominent midrib and long
tapering point. Sheath of leaf overlaps and is hairy at top.

Probably native of southeastern Asia. Important sugar-producing areas
now are India, Java, Cuba, Hawaii, Puerto Rico, as well as Africa, Australia, and South America. In United States, sugar is produced most
abundantly in Florida and Louisiana. It cannot survive killing frosts.
Superior sugar canes may be result of chance mutations or accidental
crossings of wild species. Wild forms known but not for a long time.

Flower cluster is an open panicle, fluffy, 1 to 2 ft long, little spikelets
being conspicuous because of their white, downy tufts of hairs. In agricultural practice, reproduction is by cuttings usually from tops of plants.
Sometimes underground parts are divided and used to start new plants.
Perennial.

Fields are plowed deeply, even to depth of 2 ft. Stable manure best
fertilizer if potash and phosphorus are returned from sugar-mill trash.
Stem sections 4 tons to an acre, are laid every 2 ft in rows 4 to 6 ft apart
and covered with earth and fertilizer. Clean cultivation is necessary.
Crop matures in a year. Controlled burning sometimes used to destroy
leaves on plants in fields before harvest. In harvesting, cane is cut at its
base and the leaves are removed. Later canes are loaded and transported
by truck, cart, or railway cars to nearest raw sugar factory. Canes sometimes floated through flumes to sugar cane factory.

One of most important food plants grown between 33°N and S of equator. Sugar content lost rapidly in first 24 hours, so must be removed near
the growing field. Rollers remove 75 percent of sugar first time. Remainder is removed by further rolling and spraying with hot water. Sugar most efficient energy-producing food available to man. Fed upon by
cotton rat. A by-product of sugar cane refining is a material for making
paper.

Durra

Sugar Cane

Big Bluestem, Bluejoint, Turkeyfoot, Beardgrass
Andropogon gerardi

Height to 6 ft. Rather stout stems arise from underground, short rhizomes, in large tufts and may be branched near summit. Leaf to ⅖ in.
wide, with margins very rough, and surfaces often densely soft and hairy,
this being particularly true in lower portions of plant. Name bluejoint
refers to stems. Vigorous, coarse bunchgrass. Major distribution in and
associated with that region termed the Tall Grass Prairies.

Native plant ranging from Quebec and Maine to Saskatchewan and Montana, south to Florida and Mexico and particularly in Wyoming, Utah,
and Arizona. Favors dry soils, prairies, and open woods and grows frequently mixed with other species of grasses.

Flowers borne crowded at tip of erect stems that branch often into three
parts and resemble a turkeyfoot, thus giving rise to the name. These are
usually purplish and each section is to 4 in. long. Stamen-bearing spikelets are slightly longer than those bearing fruit. Perennial. Flowering
stalks are stout, coarse, and solid in contrast to most other grasses which
have hollow stalks. Growth of plant starts in late spring and continues
throughout the summer.

A native plant that is unquestionably a valuable forage grass in prairie
states and in adjacent Mississippi Valley. Not extensively cultivated but
has an underground system that is persistent and permits grazing without
plant being destroyed. Sometimes recommended in grass mixtures.
Successful plantings of big bluestem have been made on many soil types;
may be seeded alone or with associated species. Seedings should be
made on a well-prepared, firm seedbed, free from weeds, 15 lb of seed to
the acre.

This species has won favor as a soil anchor against wind and water erosion. Extensive root system penetrates deeply. This grass grows well on
most soil types. It makes good-quality "wild hay." Big bluestem was
one of the principal grasses which helped to produce the great agricultural black soils of our Midwest cornbelt. Unfortunately, little of this
native prairie is left.

Big Bluestem

Little Bluestem
Andropogon scoparius

Plants green or reddish purple; from 1 to 3 ft tall. Vigorous, long-lived,
native bunchgrass. Upper part of plant freely branching; leaves smooth,
wide, flat; blades to ¼ in. wide and 4 to 8 in. long.

A plant characteristic of the dry prairies and the Plains States. Widespread, however, over most of the U.S. in prairielike situations, old fields,
rocky slopes, and open woods. Adapted to a wide variety of soil types.
Occurs from Quebec and Maine to Alberta and Idaho south to Florida
and Arizona. Most prevalent in Flint Hills of Kansas and Oklahoma.
In eastern part of range, in particular through the Tall Grass Prairies,
little bluestem is associated with big bluestem, *A. gerardi*. Little bluestem
is more drought-resistant.

Spikelets in pairs arranged on one main stem which is undivided. As
wildlife food, seeds are used by songbirds and upland game birds, and
many mammals use leaves, stems, and seeds. Supplies dependable grazing and cured forage in short-grass prairies. Not considered of major
economic value because grazing animals do not prefer it as the plants
mature. Good forage grasses until they become woody. Used in crop
rotations and in seeding for abandoned cultivated land. Seedings made
on firm seedbed, free from weeds, and at a rate of 15 lb per acre. Pound
of seed contains about 250,000 seeds.

Little
Bluestem

Crab Grass
Digitaria sanguinalis

Height to 4 ft, arising from sprawling stems that root freely at the joints.
Leaf blade 3 to 6 in. long and ¼ to ½ in. wide, rough, more or less hairy
particularly in the basal region, dark green. A number of varieties and
closely related species give many modifications.

Common in fields and gardens, lawns, pastures, and waste places
throughout the United States at low and medium altitudes, being most
common in East and South. Found generally in temperate and tropical
parts of world. Native of Europe and cultivated in Germany and Poland.
Probably reached U.S. as impurity in seeds.

Flowers borne in slender, crowded branches at tip of flowering stalk,
often appearing like fingers of the hand and giving the common name
finger grass. Flowers from July to October and is propagated by long-
lived seeds or by plant parts. Seeds from August to October. Annual.

Appears quickly in fields after usual grass crop has been harvested; this is
recognized as valuable in the South since forage is nutritious and is easily
produced. In gardens, particularly in moist seasons, plant may become a
pest. Fruits appear as impurities in commercial seeds. Serious weed in
lawns but is most successful during dry conditions; in fact, may take over
from other grasses in later summer; forms large patches in lawns.

If plants are prevented from seeding a few years and clean weeding is
kept up, crab grass will disappear from a garden. Forage value of plant
(Continued)

Crab Grass

should not be overlooked when it is being condemned as a garden weed.
Seeds are cooked in milk like sago in Europe and eaten by man as a
highly nutritious food. Chemical herbicides now available which control
crab grass, but should be used with precaution. Outstanding in usefulness to songbirds and upland game birds. A number of our songbirds
obtain a considerable part of their food from crab grasses.

Proso Millet
Panicum miliaceum
Height to over 4 ft. Stem slender, either smooth or hairy, usually unbranched, in an open or compact cluster with tips usually bent to one
side. Leaves narrow, somewhat hairy with basal sheaths with weak hairs
and beards at upper edge. Unlike perennial Para grass, *P. barbinode,* this
species is annual.

Probably native of East Indies. Now grown rather widely over world in
suitable situations. In North America, had popularity in Northwest and
more recently in Middle West; now chiefly in Colorado and the Dakotas.
Found as an escape on waste ground. May be the most ancient of cultivated crops. The Romans knew it as *milium.* Evidence shows that it
was cultivated by early lake dwellers of Europe. It has many weed-grass
relatives, there being some 400 species in genus.

Flowers borne in rather open clusters somewhat like stiff, erect, corn
tassels. Each spikelet bears one fertile floret, beneath it two empty
glumes or scales and a neutral one that bears stamens. Fruit to $\frac{1}{16}$ in.
across, smooth, white or straw-colored, shining. Colors vary in different
varieties.

Grown in America, mostly as a forage grain that is highly nutritious. In
addition to carbohydrates of plant, grains contain to 10 percent protein
and to 4 percent fat. Makes an excellent forage for fattening hogs in field,
often being offered as a substitute for sorghum and maize in planning hog
diets. Is also used as a forage crop and for birdseed as well as chick feed.
A very limited crop, the seeds of which do not mature at the same time.
So the crop is mowed, windrowed, and allowed to dry before threshing.

Little or no value as a hay producer. A palatable bread has been prepared from fresh fruits. Leading countries in raising this crop are Russia, China, Japan, India, and southern European countries. Known variously as Proso millet, millet, French millet, hog millet, Russian millet,
Indian millet, and broomcorn millet.

Tickle Grass, Old Witch
Panicum capillare
Height to 2 ft. Stems erect or spreading from base, smooth or densely
covered with rather long hairs. Leaves to 10 in. long and nearly 1 in.
wide, with densely hairy sheaths and with shorter hairs on both surfaces
of leaf blades. Stems break readily at joints.

Common, particularly on sandy soils in fields and waste places from
Nova Scotia to British Columbia and south to Florida and Mexico. A
native plant often appearing as a tumbleweed when tops break loose and
are blown across fields or snow. One of the commonest weedy grasses of
fields and disturbed soils.

Flowering cluster, a large, open, spreading group of threadlike branches
often well over 1 ft across; when mature, branches are stiff and brittle.
Grains small, grayish-brown, shining, borne singly in little spikelets that
are about 0.1 in. long. Annual.

Plant of doubtful value. In rich soil, it does not stand competition with
other plants and gives in quickly to cultivation so it is not a serious pest.
Seeds eaten by many birds. Found as seed impurity of commercial seeds.

Easily controlled by preventing development of seed, by hand pulling,
and by ordinary cultivation practices. The related *P. dichotomiflorum*
and *P. virgatum* occasionally serve as food for wild waterfowl.

Barnyard Grass
Echinochloa crus-galli
Height to 5 ft. Stems branching profusely from base and ascending,
stout, rather succulent. Leaves and sheaths smooth. Related *E. walteri*
in Southeast may reach a height of 9 ft. Plants break easily, leaving
reproductive portions in the ground.

E. crus-galli found from the mouth of the Gulf of St. Lawrence to southern British Columbia, and south to Mexico common on wastelands and
along the borders of freshwater areas. Probably introduced from Europe. A variety *mitis* abounds along marshes in northern states and others are recognized. *E. walteri* may be found in brackish areas.

Usually, flower cluster appears to be bristly because of the awns, but
awns are lacking in some varieties. Flower cluster may appear purple or
pale green, erect or drooping. Fruits appear to be hard and firm, more
pointed at one end than the other, and are able to retain their vitality for
a long period.

Proso Millet

Tickle Grass, Old Witch

Barnyard Grass

Under name of wild millet is recognized as one of best duck-food plants of shallow marshes and moist soils. Attracts songbirds, upland game birds, and some forms even attract muskrats. Variety *frumentacea*, known as "billion-dollar grass" or "Japanese millet," is occasionally cultivated as bhasti. A serious weed of rice plantings in California.

Seeds common as impurities in commercial seeds. Plants are unusually valuable in wildlife management because they produce abundant seeds, are hardy, grow with little care, and are generally cheap. Planted in drills at 1 to 2 pk an acre. Grass is 76 percent water, 0.7 percent fat, 1.5 percent protein, 12 percent fiber and ash.

Foxtail Millet
Setaria italica

Height to 5 ft. Stems smooth, conspicuously jointed, somewhat branched at top. Leaves many, broad, flat, long, pointed to over 1 ft long, to 1 in. wide, rough, with large sheathed bases. Root system relatively small.

Worldwide in cultivation. Foxtail millet or German millet was said to be one of five species of plants that Chinese Emperor sowed each spring in a public ceremony established 2700 B.C. Other common millets are Hungarian and Japanese millets.

Flowering panicle compact, interrupted or lobed, heavy, usually purple, yellow, or green, to 1 ft long, to 2 in. wide. Fruits about $\frac{1}{16}$ in. long, smooth, yellow, brown, red, or nearly black, egg-shaped but somewhat flattened on one side. Annual. Matures 6 to 10 weeks after planting.

In United States, millets are grown mostly for forage or in a crop rotation where a quick, green vegetation is needed; grown mostly in the Great Plains and southwestern states. Varieties include black-seeded Hungarian grass; red- or orange-fruited Siberian or Turkestan millet; and yellow-fruited, German, or golden wonder millet.

In India, Japan, Korea, and China, millet is raised extensively for human consumption, but in America its place is that of a quick crop to replace an early crop failure, or for forage, or for grains for poultry. Excellent for planting at field margins and, in suitable waste places, as food for certain upland game birds. Mixed with wheat flour, ground millet makes excellent bread. Stalks used as hay. Seeds used in poultry feed.

Green Foxtail, Bristlegrass
Setaria viridis

Height to 3 ft. Erect, usually unbranched, springing from tufts close to the ground. Leaves to 1 ft long and to 0.4 in. wide, roughened on the margins, with smooth sheaths, dark green. Vigorous, quick-growing plant with erect stems curving near the ground.

Native of Europe but widely established throughout world in almost any kind of soil but particularly in poorly cultivated lands and waste places. In America, found commonly everywhere but in far north. Found mostly in late summer but tops remain erect through the winter.

Flower clusters spikelike, but made of many spikelets, to 4 in. long, nearly 1 in. through. Bristles 2 to 6 at base of each spikelet, barbed upward. Grain about $\frac{1}{12}$ in. long, doubly convex, commonly brownish or greenish, smaller than yellow foxtail. Blooms July to September. Annual.

Fruits commonly found in clover seed. Plant has some food value, being 14 percent water, 7 percent ash, 7 percent protein, 3 percent fat, 19 percent fiber, 50 percent nitrogen-free extract, this analysis being of freshly dried hay. Sometimes a bad pest in lawns, where it is found more commonly than yellow foxtail. One of commonest weeds of cornfields and other areas of disturbed soil.

Extermination is by hoe-cutting or hand-grubbing of plants in lawn or garden. Sometimes fields are burned over to destroy foxtail and if the pest is to be controlled, seed production should be prevented as much as possible. Very popular wildlife food plant.

Yellow Foxtail, Pigeon Grass
Setaria glauca

Height to 4 ft, and arising from branching bases. Stalks commonly compressed at base, sprawl over ground before rising. Leaves to 6 in. or more long and to ½ in. wide, flat, smooth, and usually hanging twisted, with loose sheaths, lower of which may be red. Roots clustered and fibrous.

Found in cultivated lands all over world, being particularly common in cultivated areas and waste places where soil is loose. Naturalized from Europe and probably introduced with commercial seeds since it matures its fruits at same time many valuable species do. Like *S. viridis*, a common weed of cornfields and other disturbed soil.

(Continued)

Foxtail Millet

Green Foxtail

Yellow Foxtail

Flowering clusters spikelike, but composed of many spikelets, these being closely crowded, one-sided and subtended by 6 to 10 upwardly barbed bristles that are brownish-yellow and much longer than the spikelets. Grains about 1⅕ in. long, yellow, brown, or pale green. Annual.

Cattle will eat young grass but old plants are worthless as fodder. Green plant is 81 percent water, 10 percent fiber and ash, 2 percent protein, 7 percent nitrogen-free material. Once seeds are established in soil, their vitality is great and plants may appear regularly.

Game birds and birds that remain north in winter find fruits above snow a most acceptable food and will clean plants when they are to be found. For this reason, plant has a place in wildlife management. To control pigeon grass in cultivated lands, cultivate late, graze with sheep, and prevent seeding.

Sandbur

Sandbur
Cenchrus tribuloides
Stems to 2 ft long, branching and rising at tips to 10 in. or more. Leaves to 5 in. long, with loose, flattened sheaths, with hairy margins, with fringed throat and smooth, flat, sometimes curled blades. Shorter stems sometimes erect, not sprawling.

Common from Maine and Ontario to Florida and southern California as well as in other parts of world. It is not found in all areas in this range, but in South and Middle West it is particularly troublesome. As name implies, it is common on sandy soil, shores, and dry waste places although it sometimes is found in lawns.

Flowers borne in 8 to 20 clusters, of two to six flowers that are almost enclosed by prickly bracts that make burs about ¼ in. in diameter, thickly set with strong prickles stiff enough to penetrate shoe leather, barbed backward, and most difficult to remove. Spines break off freely. Annual.

One of pests of prairie states for those who like to roam in waste places with or without protection of strong leather. Burs often get into wool of sheep and other animals and are almost impossible to remove. They may make their way to the skin and cause serious irritation.

Little if any good use for plant. Too well able to take care of itself. However, roots are shallow and do not stand exposure well, so if there is shallow cultivation of an infested spot the plant may soon disappear. Animals carry seeds from place to place in burs. Infested areas are commonly burned over. If abundant, may furnish some feed for livestock when the plant is young. Value to wildlife is slight; however, a few songbirds eat the seeds.

Rice

Rice
Oryza sativa
Height to 4 ft. Erect, growing from stools. Fruiting stem angled, smooth, almost entirely enclosed by sheaths of leaves that are smooth and conspicuously nerved. Leaf blades long and about ½ in. wide, more or less roughened. One seed will produce several of the erect stems. Over 2,400 varieties recognized. Wet- and dryland forms. Is man-dependent; cannot survive without cultivation.

Has been under cultivation in China at least 4,000 years. As it grows wild in southern China, it is probably native of the area. Introduced into western Asia, then Europe. Brought to America for growing in 1694 near Charleston, S.C. Average summer temperature above 77°F.

Flowers in compact panicles. Hull usually yellowish-brown. Inner grain white and hard. In hull, rice is known as "paddy." Without hull, called "cleaned rice." Carolina gold rice, commonly grown in Carolinas, has golden-yellow hulls; Japan and Honduras rices grown in Texas and Louisiana have yellow-brown hulls.

Polished rice has about 12.3 percent water, 79 percent carbohydrate, 8 percent protein, 0.3 percent fat, 0.4 percent ash, and 102 calories per 100 g. Unpolished rice has a higher food value. Seed sown, 1 to 3 bu per acre in drills, mid-April to mid-May, water is not applied until rice is 8 in. high. Water to 3 to 6 in. is then flooded in until crop is mature.

Probably eaten as staple by more people than any other grain and is their principal source of nourishment, but low in protein.

United States imports more rice than it raises but is not a major rice-consuming country. Rice bran is fed to cattle. Rice straw is made into rice boards and straw hats, woven into rope, and woven to form fabric for bags, but rice paper comes from mulberry or bamboo. Polished rice lacks vitamins necessary for preventing the disease beri-beri, which in some stages may be cured by eating unpolished grain. Much research done to produce varieties which increase yields. Short-grain rice, poor in yield, thought to be most delicious by some.

Wild Rice
Zizania aquatica
Height to 9 ft. Leaves to 2 in. wide, to 40 in. long, flat, green, curving
backward from base that sheaths stem. Roots relatively short, easily
pulled. Two varieties recognized, variety *interior,* of the Middle West,
being about two times height of the eastern *angustifolium.*

Found from mouth of St. Lawrence to central Manitoba, south to Kansas
and Virginia and around coast to Louisiana. Common in shallow water,
on muddy shores, or in water to 3 ft deep and in salinity up to 3.5 percent
that of normal seawater. Water salty enough to be tasted is not suitable
for wild rice. Found also in ditches, ponds, streams, and marshes.

Flowers borne in loose open clusters, at top of plant, uppermost bearing
pollen and lower ones bearing valuable fruits. Fruits enclosed in thin,
papery, minutely roughened scales. Flowering takes place in July and
August in northern part of range. Annual.

One of most commonly cultivated of wild foods for game. Usually is
planted by broadcasting fruits in fall or in early spring. Ducks that have
fed on wild rice and wild celery are supposed to be superior food for man.
Whole grain 9.5 percent water, 12.9 percent protein, 1 percent fat, 75.2
percent carbohydrate, 1.4 percent ash.

One of more important of foods for most plant-eating wildlife. Fruits
and plant parts eaten, but fruits are most important. They attract waterfowl of all sorts, songbirds, upland game birds, muskrats, deer, and moose
over a wide area for a considerable season. An important food item,
menomin, of the American Indian who threshed the standing plants into
canoes. Early French explorers called it wild oat. An expensive delicacy available in some stores today.

Wild Rice

Reed Canary Grass
Phalaris arundinacea
Tall grass reaching a height of 4 to 7 ft; coarse, sod-forming cool-season
plant. In thin or volunteer stands, it occurs in clumps which measure
several feet in diameter. Heavy sods produced from solid seedings.
Leafy stems stout and so do not lodge. Seed-bearing structures purplish.

Adapted to much of the northern half of the U.S. and southern Canada;
in U.S. from Maine to Tennessee and from Washington to central California. Occurs where grasses will grow, and also in wet places. Flowers
borne in dense cylindric clusters, 3 to 6 in. long. Seeds shatter promptly
when ripe. Seeds, waxy, gray to gray-black and about ⅛ in. long. Perennial.

Considerable difference of opinion among experts concerning its palatability for livestock and also concerning its nutritive value. Seeding with
legumes not successful because Reed Canary grass is too tall and so
crowds out the associated legume. However, experiments with some
legumes (bird's-foot trefoil, alfalfa, alsike, and ladino clover) have been
successful.

Plants form dense colonies in marshes and in other wet places, like
ditches. Now planted as an important erosion control plant on farm
waterways. When new plantings are desired, plants are cut, chopped in
short lengths, and forced into the mud by various means. Particularly
useful in Midwest, where deep loose soils are common as in loess of Iowa.
Some forms have white-striped leaves and are grown as ornamentals.
Plants cut for forage and/or hay. Has been seeded on poorly drained
soils subject to flooding and silting. Can be under water for periods of
time without excessive damage. Reed Canary Grass is unchallenged in
its value to heal and control gullies; even if shoots are buried under 6 in.
of sediment, they recover.

Reed Canary Grass

Canary Grass
Phalaris canariensis
Height to 3 ft. Erect fruiting stems smooth, usually unbranched. Leaves
with sheaths, loose, rough, shorter than space between joints; with blades
to 1 ft long and nearly ½ in. wide, rough and flat. Root system not
elaborate. Same common name given to *P. arundinacea,* reed canary
grass.

Found from Nova Scotia to Alaska, south to Virginia, Kansas, Wyoming,
and California, sometimes farther south; introduced into Mediterranean
region. Found mostly in gardens, on roadsides, and in waste places.
Grown to some extent and has established itself widely through being an
impurity of some commercial seeds.

Flowering head a dense spike about 1 in. long and ½ in. thick, composed
of one-flowered spikelets that crowd each other, overlapping closely and
showing keeled scales that are whitish with green veins. Flowering time
from July to August. Fruits produced in August and September. Annual.

Probably known to more people because of shining fruits that are fed to
canaries as "canary seed." Conspicuous grass but can hardly be considered a common weed except in limited areas. However, it sometimes
gets established as a weed and may need control measures to keep it from
spreading.

Canary Grass

(Continued)

351

If seed production is prevented, plant will disappear quickly in most areas where it is found. Vegetative part is worthless as hay or as forage for livestock. Flour from seed is used in making weaver's glue for sizing cotton.

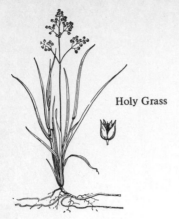

Holy Grass

Holy Grass, Vanilla Grass, Indian Grass
Hierochloe odorata
Height to 2 ft, erect stems arising from a creeping rootstock that winds, divides, and forms a mat. Erect stems unbranched, slender, and smooth. Leaves of flowering stems short, smooth or slightly roughened. Late leaves flat, rough, deep green, and longer than earlier leaves.

Ranges from Newfoundland to Alaska, south to Pennsylvania, Colorado, and Oregon and also a native of Europe and Asia. Found in moist meadows, bogs, and prairies. Commonest in northern part of range along coast and Great Lakes area usually in moist meadows and prairies, though it sometimes becomes a pest in managed pastures that are wet.

Flowers borne in open panicles, to 5 in. long, in May and July. Individual spikelets about ⅕ in. long, brownish, enclosed by two scalelike structures. Those bearing stamens are awnless; those bearing grain, hairy at tip. Much reproduction is by spreading rhizomes. Perennial.

After fruits have matured by first of June, a new growth appears that is more vigorous than early vegetation. This growth provides a reserve of food that is stored underground and makes possible early flowering and fruiting the following year. Vanillalike fragrance is best at flowering time.

Useless as hay or forage. Dried leaves and tops are made into "sweet grass" baskets, put into pillows and dresser drawers because of their fragrance and are strewed at church entrances in northern Europe to give off fragrance, whence the name holy grass. Eradication by spring plowing or by burning ripened grass.

Mexican Dropseed

Mexican Dropseed
Muhlenbergia frondosa
Height to 3 ft. Stems smooth, commonly branching at the base and rising to erect position from the ground, with roots at joints that touch the ground. Leaves to 6 in. long and under ¼ in. wide, smaller on branches and more crowded there. Roots clustered and relatively shallow.

Found in moist fields and meadows, in ditches, and on banks of waterways from Newfoundland to British Columbia, south to Florida and Mexico covering thus whole of United States. Also found native in Europe and Asia. Seems to be more common in eastern parts of United States than in western.

Flowers borne in slender, loose, weak clusters, whole panicle sometimes up to 6 in. long and hanging almost as though it were wet. Spikelets small, about 1/12 in. long, and about equal in length to brown seeds they produce. Embryo end of seed is darkest. Perennial.

When young, leaves are eaten readily by livestock but old plants are so wiry they are rejected. Fresh hay of a 2-ft plant is 73 percent water, 10 percent nitrogen-free content, 10 percent ash and fiber, 5 percent protein, 1 percent fat. Water content may vary from 50 to 84 percent, the latter being in the young plants.

Dropseed may provide food and shelter for some kinds of game and at times may be useful forage for cattle in places where shade and amount of water are too great for other plants to survive, but usually better plants can be found. It is a good soil anchor and pioneers in this along streams. Cultivation destroys it.

Redtop

Redtop, Bent Grass
Agrostis alba
Height to 4½ ft. In creeping bent, *A. palustris,* there are long, rooting runners at base and erect stems sag at base. In black bent, *A nigra,* runners are short, stout, and leafy; in typical redtop, *A alba,* rootstocks are horizontal and fruiting stalks are erect. Leaves to ⅖ in. wide. Spreads by creeping stems.

Native of Europe, but generally naturalized in northern parts of United States and in Canada. Brought to America by the first colonists. Under humid conditions, it is tolerant of a wide variety of soil and moisture conditions; will withstand considerable drought. It favors meadows and open fields where pasturing is common. Different varieties are sufficient to provide grasses suitable for being grazed or for being cut as hay.

Flowering panicle to 1 ft long, open, freely branching, of slender parts, starting in whorls from a central stem, greenish, purplish, brown, or red, and slightly rough. Scales outside fruits nearly equal in length, the shorter being over half as long as the other. Spikelets one-flowered. Perennial.

In seeding, sown at rate of 12 to 15 lb to the acre. Blooms about 6 weeks
after Kentucky bluegrass. Grain is about $\frac{1}{25}$ in. long and is enclosed in
scales about half again as long.

This pasture grass is less valuable than timothy or bluegrass but it forms
a sod more quickly than bluegrass and lives on soils too heavy, wet, or
acid for many other grasses to survive. Some rank redtop hay as second
in value to timothy. Its palatability is less, though its nutritive value is
about equal. Used in lawn grass mixture for golf courses, meadows,
terrace outlets, and roadsides.

Bent Grass

Rhode Island Bent Grass, Velvet Bent
Agrostis canina
Height to 2 ft. Usually seen cut so short that 2-ft height seems untrue.
No true rhizomes, fruiting stems rising directly. Erect stems and leaves
fine, smooth, weak, with leaf margins often incurled. Lower leaves al-
most hairlike and others very narrow.

Commonly found on putting greens of golf courses and on lawns. Gets
common name velvet bent from the smooth close sward it may make.
Native of America from Newfoundland to Quebec and south to Delaware
and Michigan but has been widely introduced throughout world where
golf is played.

Flowers borne in loose, open panicle, to 4 in. long. Scales enclosing the
single grain are to $\frac{1}{10}$ in long. Outermost scales are to $\frac{1}{4}$ again as long as
inner and the customary inner scales found in closely related forms are
minute or lacking.

Useless as a source of hay but seed has a high market value because of its
special use in developing golf courses. Some related bent grasses may
produce vegetation and fruits that have been found in the stomachs of
ducks, but this role cannot be important.

Of the bent grasses, the following have special uses: colonial bent, *A
tenuis,* on pastures and lawns; creeping bent, *A. palustris,* on golf greens;
velvet bent, *A canina,* on putting greens; redtop, *A. alba* on meadows,
pastures, and lawns. Some are suited for landing fields at airports.

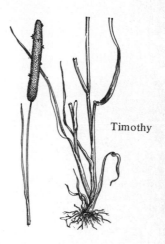

Timothy

Timothy, Herd's Grass
Phleum pratense
Height to 6 ft; more commonly 20 to 40 in. tall. Fruiting stems unbranched, erect, smooth, arising from a thickened bulblike base. Leaves
arising from crown or from fruiting stems, flat, about 1 ft long and $\frac{1}{2}$ in.
wide, rough; with long, smooth, streaked sheaths. Leaves of erect stems
more numerous than others. Root system does not spread to form a sod.

Found from Europe through North America where pastures and hay-
fields are managed, but in United States most important in states of
Northwest and Middle West. Does best on clay loams where there is
good moisture and better in moist climates than in dry ones. Does not do
so well on acid soils. In America since 1700.

Fruiting head a slender spike of closely crowded spikelets, the whole
being to 6 in. long and to $\frac{1}{3}$ in. through. Spikelets single-flowered; gray-
ing-silver fruit is about $\frac{1}{12}$ in. long and is usually enclosed in a loose,
papery set of scales.

Probably best of all hay-crop species. Seed weighs 42 to 50 lb to the
bushel, legal weight usually being 45 lb. Seed sown at 8 to 15 lb to acre,
or with clover, at 7 to 10 lb. Without nurse crop, seeding may be done in
July or August. Crop usually cut in July; cures quickly and cleanly.
Timothy in mixture with a legume makes a better growth than when
grown alone.

Hay usually has 6 percent protein, 45 percent carbohydrates, 2.5 percent
fats, and 29 percent crude fiber of which only about $\frac{1}{2}$ is considered
digestible. Timothy is a bit lower in protein than other grasses but is
otherwise equal. When harvested for seed, yields from 2 to 3 bu to the
acre. From 1910 to 1960 there was a steady decrease in production of
timothy hay in the U.S. which was related to decreasing number of
horses. After 1960 the decrease in production was even greater because
other grasses such as brome grass and orchard grass became more popular. Some wildlife forms such as upland game birds, songbirds, some
small mammals, and deer feed on timothy.

Oats

Oat
Avena sativa
Height to 5 ft. A single fruit produces from three to seven erect, jointed,
hollow, fruiting stems from a base of fibrous roots. Leaves numerous, 6
to 12 in. long and about $1\frac{1}{2}$ in. wide, blue-green, rough, flat, with sheaths
that are loose and extend practically from one joint to next.

Probably native of central and western Asia. Although early Greeks and
Romans used oats for feeding domestic animals, little evidence that they
were used as long as most other cereals. Early American colonists
(Continued)

Wild Oat

Poverty Grass

brought oats. A red-rust-proof oat of South was developed from African *A. sterilis.* Grown north to 69° in Alaska.

Fruiting head an open panicle, to 1 ft long, bearing from 40 to 75 spikelets, each spikelet usually consisting of two or more flowers, of which two bear fruits. In some varieties, only one fruit is borne in a spikelet; in others, three are produced. Flowers open for only a few hours, pollination taking place before flowers open.

In most oats, grain weighs 60 to 75 percent of total. Early oats ripen in 90 to 100 days; late, in 115 to 130 days. In late April, oats are drilled or broadcast, 2 to 3 bu an acre, this being sometimes harrowed in two to three times. Yields average around 30 bu an acre but may reach 200 in Northwest. Hessian fly and a sawfly are worst insect pests.

As food for man, are fair for protein, excellent for minerals and for carbohydrate, and fair for calcium and phosphorus, the bone and nerve builders. Fair for vitamin A and good for vitamin B. About 70 percent of American-raised oats are used on home farm. Rusts and smuts are serious fungus pests. Wildlife such as water birds, marsh birds, shorebirds, upland game birds, songbirds, and small mammals use oats as food. Oats rank fourth among our cereal grains, behind wheat, rice, and corn. In the Corn Belt, oats are frequently grown in rotation with corn and clover.

Wild Oat
Avena fatua
Height to 5 ft. Erect, growing in tufts from a common cluster of fibrous roots. Smooth. Leaves green, about ½ in. wide and to 1 ft long. In general, the plant vegetatively closely resembles the cultivated oat.

Common throughout United States but most troublesome in grain-growing areas of Canada and from Minnesota to Oregon and California. Grows in almost any kind of soil. Commonly found in fields planted to cereals and other grasses and to flax. Matures earlier than some of these and thus seeds itself.

Flowering panicles loose and open, much like those of cultivated oat. Flowerlets become much more easily separated from heads in wild oat; stiff twisted awns are nearly twice as long as spikelets that bear them. A grain will germinate and develop a plant when buried 5 in. deep.

Wild oats occupy land used better by cultivated species. Twisted awns cause serious irritation in noses, mouths, and digestive tracts of animals that eat them; seeds can survive going through animal's digestive system. Twisted awns cling to sheep's wool and grain sacks.

Wild oats can be controlled by sowing only clean seed, by following an infested crop with corn or with a clover-timothy meadow. When wild oats appear in hay, every effort should be made to clean it out, for protection of the animals that eat the hay. Wild-oat hay cut before fruits mature is nutritious for cattle. In the Pacific States wild oats are common weeds of roadsides and agricultural land. As such, they are used by many wildlife forms.

Poverty Grass, Poverty Oats, Poverty Oatgrass
Danthonia spicata
Height to 2½ ft but normally much lower. Flowering stems round, with leaves that are smooth or sparingly haired. Leaves springing from base are usually twisted and curled at slender tips, forming a cottony cushion close to the ground. Leaves gray-green, about 4 to 5 in. long.

Found through eastern United States and Canada, west to Dakotas and south to Gulf of Mexico, being found often as a clean stand in "worn-out" soils whether sandy, rocky or clayey. Presence of plant may be used as an index of soil exhaustion, since it does not do well on rich soils.

Flowering stalks somewhat like oats but few-flowered, with branches short, with spikelets about ½ in. long, and with twisted awns longer than scales that support them. Flowering panicle about 2 in. long. Seeds usually ripen and fall before hay grasses are ready for cutting.

Interesting to outdoor persons in part because of behavior of leaves in varying degrees of humidity. When mature and dry grass may be collected and piled either loose or in a tick to make an excellent emergency mattress in areas where stones are exceptionally hard for those who sleep out.

Best control of this weed, if this is necessary, is to enrich soil so that it cannot survive competition with plants that may then be supported by it. This implies that soil must also be in condition to hold moisture for other plants as well. One of few native forage grasses. Forage value very low.

Bermuda Grass, Scutch Grass
Cynodon dactylon
Height to 1 ft. Flowering stems erect or sprawling, slightly flattened, much-branched, stiff and wiry when mature. Leaves smooth, flat, light green, stiff, numerous, to 4 in. long, with smooth sheaths crowned with a ring of white hairs. Sheaths overlap and are crowded at base.

Bermuda Grass

Native of India. Introduced from Europe and established in warmer
parts of United States from Maryland west to Pacific Coast and south. It
winter-kills as far north as Virginia but can be grown during usual season. If roots are exposed to freezing, plant cannot ordinarily survive.

Flowering stalks end frequently in a five-branched fingerlike arrangement much like that of crab grass. Spikes four to five each to 2 in. long.
Spikelets to 0.1 in. long, closely pressed to axis of spike. Perennial in
suitable climates. Rarely reproduces by seeds but mostly by rooting runners.

Prefers warm sandy areas, surviving droughts that would kill most other
grasses. Because of this, used in South on exposed lawns and golf
courses where the heat may be severe. Because of its creeping habit, it
may become a pest in fields requiring cultivation. Sometimes serious as
a pest in cotton. Hard to destroy.

Best methods of eradication call for use of a smother crop which for a
time will cut it off from all light. Millet, cowpeas, or sorghum planted
close is often used since they can compete successfully and will also produce a useful crop. Has been one of the most important pasture grasses
of the Southern States. As a plant to prevent soil erosion, it is unsur-
passed. Will tolerate flooding for long periods. Seeds and parts of the
plant eaten by a few species of waterfowl, marsh birds, and shorebirds, as
well as some small mammals. Not an important wildlife food plant.

Wheat
Triticum aestivum
Height to 4 ft. Flowering stems erect, unbranched, hollow except at
joints, to ⅛ in. in diameter. Leaves to 15 in. long by ½ in. wide, taper-
pointed, with a long loose sheath that is finely hairy or smooth. Roots
usually near surface, but in favorable soils may penetrate to 7 ft; grow in
whorls.

Cultivated since first records of history and known to have been grown in
China at least in 3000 B.C. Mentioned in first book of the Bible, and was
grown in western Asia, Europe, and northern Africa long before it was
introduced into America. It grows to within 200 miles of the Arctic
Circle; on mountains at equator. Several types of modern wheat appear
to be descended from successful crosses among wild species of true
wheats, *Triticum,* and goat grasses, *Aegilops,* and perhaps *Agropyron.*

Flowering spikes compact, composed of numerous spikelets, with two or
more flowers in each spikelet. Spikes to 4 in. long, cylindrical except in
some varieties. Fruit an oblong grain about ⅟₁ in. long that frees itself
readily from the enclosing scales and is usually fuzzy at the top. Self-
fertilization is the rule. Annual.

About 60 percent of American wheat is winter wheat, most of which is of
turkey red variety. 1½ to 2 bu are drilled per acre to depth of 1 to 2 in.
to mature in about 100 days of growing weather. Fields are harrowed in
drilled direction. Worst enemies are rusts and smuts of wheat, chinch
bugs, and Hessian flies.

The "staff of life" as food for man is fair for protein, calcium, and phosphorus, and vitamin A; good for vitamin B; and excellent for minerals
and carbohydrates. An average yield per acre is about 14 bu, but yields
to 30 bu have been reached. Very important crop in southern Canada.
About a fifth of land in cultivation is devoted to growing wheat. Wheat
straw is used as livestock feed, animal bedding, compost, mulch for strawberries and other plants, and fertilizer. Very important wildlife food crop
for water birds, marsh birds, and shorebirds, upland game birds, songbirds, as well as fur and game mammals.

Rye
Secale cereale
Height to 6 ft. Flowering stems erect, unbranched, smooth, except near
top, slender. Leaves to ½ in. wide, long-pointed, soft, nearly smooth,
blue-green, many curved. Roots fibrous, well-matted. Tallness of rye in
fact and in appearance is always characteristic.

Comparatively recent among cultivated plants. Not known to early
Greeks and Romans. Probably originated in western Asia and southeastern Europe. It will generally grow on poorer soils than those required by other cereals. It does best on loam soils. Pennsylvania, Wisconsin, and Michigan are rye states.

Flowers borne in terminal spike to 6 in. long, well-awned, narrow, closely
flowered, with spikelets with two seed-producing flowerlets and possibly
one that is not. Flowers more likely to cross-pollinate than most cereals.
Fruits about 1³⁄₁₀ in. long, light brown, narrow, pointed, with narrow
groove on face, smooth.

Planted in drills or broadcast at 5 to 6 pk per acre, planted to a depth of
2 to 3 in., seldom cultivated. Planted in fall or spring but matures in
about 100 growing days of spring. If planted for fall pasture, seeds are
sown in August in North or in September in South; if for grain, September in North, October in South.

(Continued)

Wheat

Rye

Quack Grass

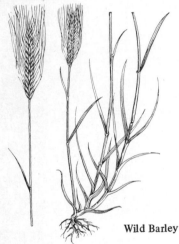

Wild Barley

Barley

Average production per acre in good states is 20 bu, although 17 is considered excellent in others. Used to make "black bread" in poorer countries in northern Europe, as grain to feed hogs and horses in America, and in making alcohol for whisky. About ⅔ goes to animal food and ⅓ to flour and alcohol. Is used by wildlife as food but limited acreage affects its usefulness. Seeds may be infected with ergot, a fungus which causes abortion in cattle. Seeds infected with ergot poisonous to man if eaten.

Quack Grass
Agropyron repens
Height to 3 ft. Flowering stems erect, or curving from base, arising from long, creeping, yellowish rootstocks that persist for a long time buried in soil. Leaves with blades, relatively thin, flat, mostly sparsely covered with short hairs on upper surface, to ⅓ in. wide. Spikelets with side to axis.

Found in waste places from Newfoundland to Skagway in Alaska and south to North Carolina, Arkansas, New Mexico, and California. Introduced into Eurasia. Common in waste places, among cultivated crops, particularly grasses, and most common in northern part of range within United States. Cosmopolitan weed.

Flowering cluster spikelike, to 6 in. long, with main axis roughened on angles, spikelets mostly four- to six-flowered, to ⅗ in. long. Flowers appear in June; mature fruits in July. Cutting underground parts of plant does little damage but rather increases number of plants since each section establishes itself.

Considered by farmers worst of grass weeds because it offers too keen competition with other plants and once established is difficult to eliminate. Often badly infected with a stem rust and with ergot. Quack grass can make a good hay, yielding two crops a year, and cattle relish the new growth as pasturage.

Best control is by putting on such a heavy mulch that underground portions are starved out. Late fall plowing followed by persistent harrowing in fall and spring may help, followed by a crop that will be hoed, followed by burning trash after harvest. Quack grass is a superior soil anchor against erosion of waste hillsides.

Wild Barley, Foxtail Barley, Squirreltail Grass, Foxtail Grass
Hordeum jubatum
Height to 30 in. Flowering stalks unbranched, erect or inclined, slender, tufted. Leaf blades to 1½ in. wide, rough, somewhat grayish-green, to 5 in. long, flat. Sheaths shorter than space between joints. Some stems may lie close to ground for a considerable distance.

Found in waste places, roadsides, fields, meadows, and open pastures from Newfoundland and Labrador to Alaska and south through Maryland, Illinois, Missouri, Texas, California, and Mexico. Has been introduced through the Eastern States but is a native of America.

Flowering spikes nodding, to 5 in. long, with spreading awns making them about as thick as long. Spikelets in 3s, on opposite sides of main, flattened, jointed axis. Each spikelet that bears a seed has seven long barbed awns; those not producing fruits have three such awns. When still fresh, the awns often have a purplish or pink sheen. Perennial, biennial, or annual.

Single clump may produce to 2,000 awned "seeds" that may be borne great distances in wool of sheep, by wind, or by water in irrigation ditches. Awns become brittle when ripe, penetrate lining of mouths of cattle that eat them and cause "lumpy jaw," get into eyes and cause blindness, or pierce alimentary canal and cause death. Wool grower further penalized because of skin condition as a result of irritation of awns.

Even though grass may have some value as forage in young stages, dangers from mature plants are too great to encourage plant except where more desirable kinds will not grow. Extermination is by repeated early mowing, cultivation, or burning over of infested areas.

Barley
Hordeum vulgare
Height to 3 ft. Flowering stems arising in clumps up to 15 to 20, unbranched, smooth or roughened under the flowers. Leaves short, long-tapering, pointed, to ¾ in. broad; sheaths, loose and smooth. Leaves appear broader than in other grains and more conspicuously gray-green.

One of oldest cultivated plants. Carvings on Egyptian tombs show it. Mentioned in earlier books of the Bible. Probably cultivated as early as wheat and earlier than oats or rye; probably originated in Asia Minor from a form now wild there, *H. spontaneum.* Was brought to earliest Massachusetts and Virginia colonies. Stray plants may appear but barley is never found as a truly wild plant.

Head of flowers much like wheat, to 4 in. long, densely flowered, erect, or
nodding, with many stout, erect, very long beards that extend far beyond
end of spike, some being 6 in. long. Fruit elliptic, about ¼ in. long,
short-pointed, smooth, furrowed length of face, usually enclosed in hull.

Types include two-rowed, four-rowed, and six-rowed, bearded, hooded,
beardless, hulled, hull-less; winter and spring forms. Winter barley less
hardy than winter wheat but more so than winter oats. Winter varieties
are mostly six-rowed. Almost any barley can survive winter in South.

Planted 6 to 10 pk an acre in drills in late April; matures in 90 to 100 days
after being harrowed two or three times in direction of drill. About ½ of
crop used in making malt by partly germinating seed. In Mississippi
Valley, important in hog foods. As human food, fair for protein, phosphorus, and calcium, excellent for mineral. Most of barley raised in U.S.
now is used for feeding dairy and beef cattle, sheep, and pigs. Barley also
used in making barley malt, which contains enzymes which change starch
to fermentable sugars. These fermentable sugars are useful in the brew-
ing and distilling industries. Problem occurs where waterfowl and barley appear together as the birds enjoy both the young seedlings and the
grain. Owners who take the grain with mechanical harvesters become
concerned because the waterfowl eat the grain before the harvest.

Wild Rye

Wild Rye, Canada Wild Rye
Elymus canadensis
Height to 4½ ft. Erect, tufted, green, or gray-green. Flowering stems
slender, unbranched, drooping at tip. Leaves flat, roughened above, to ⅕
in. wide, with sheath smooth, or rarely finely hairy. Roots fibrous and
rather widely spreading. Wild ryes make up a group of coarse bunch-
grasses with rough and relatively unpalatable foliage.

Found in moist, open, or shaded ground, or in open woods, on river
banks, or low ground from Quebec to southern Alaska, south to North
Carolina, Missouri, Texas, Arizona, and northern California, with varieties extending the range considerably.

Flowering spike nodding, to 10 in. long, with spikelets commonly in 3s
and 4s, slightly spreading, and conspicuous with loosely spreading awns.
Scales about fruits are narrow, and mostly with two to four nerves, with
awns about as long or longer than scale itself. Perennial. Produces
seeds from July through September.

A relatively large group of these rye grasses are to be found growing wild
in many places in United States. Serve no great need but add beauty
often by their nodding tops that persist through winter months. Has
been experimentally planted for forage in the Midwest. Hay is palatable
if harvested before heads appear. Becomes ergotized, so may be dangerous for feeding to cattle.

Bottle Brush

To botanists, interesting because of closely related species with distinct
characteristics, which include erect Virginia wild rye *E. virginicus;* the
slender, finer, *E. striatus,* and others. May yield some small amount of
food and protection to wildlife.

Bottle Brush
Hystrix patula
Height to 3½ ft. Flowering stem erect. Spikes to 6 in. long; very open,
bristly; blades flat, rarely hairy. Some authors place this grass in genus
Elymus, which it resembles to some extent.

Commonly found in moist, rocky woodland areas.

Occurs from Nova Scotia to North Dakota, south to Georgia and Arkansas. *H. californica* occurs on West Coast; is limited to woodland sites and
shady ravines in California coast ranges. In this grass, spikelets not so
bristly, close together on the stem and not spreading; erect.

Spikelets two- to four-flowered, one to four at each node of a continuous
stem. The pair of bracts, glumes, at base of spikelet may be tiny or
missing; if present bristly. Bract of flower tapering into a long, flexible
awn. Spike looks like a bottle brush with few but rather regularly spaced
bristles; hence, its common name.

An interesting, though somewhat fragile, addition to a dry flower bouquet. Of no special interest to wildlife enthusiasts although achenes
probably eaten by some birds and young plants eaten. Probably best left
in its natural setting where it makes an attractive contrast because of its
unique appearance.

Perennial Rye Grass

Perennial Rye Grass
Lolium perenne
Height to 30 in. Flowering stems erect, or sagging at base, unbranched,
smooth. Leaves flat, smooth, 2 to 5 in. long, to 1⅗ in. wide, dark green,
with sheaths definitely shorter than space between joints. Best field characteristic is glossy nature of whole plant. Native of all temperate Asia
and North Africa.

(Continued)

Found in meadows and wastelands, from Newfoundland to Alaska, south to Virginia and California and even farther south. It was introduced from Europe and is cultivated sometimes in meadows, pastures, and lawns. Sometimes called English rye grass. Short-lived perennial. Most successfully cultivated on the Pacific Coast and in the central and southern parts of the Atlantic States. Useful in permanent pastures.

Flowering spike 3 to 8 in. long, with spikelets set with their edges to the main axis and their tips well away from axis. Spikelets have lowermost scale slightly shorter than spikelet, being in this respect unlike related poison darnel, in which this scale is longer than spikelet.

Requires moist ground and a cool, moist climate. Its seeds closely resemble those of meadow fescue and are sometimes used as an adulterant of that more valuable grass. It does best in northern parts of British Isles. Thomas Jefferson reported the grass to be a good producer in Virginia as early as 1782. If it is allowed to go to seed, it is unusually exhaustive to soil.

Planted sometimes in grass mixtures where a field varies greatly, with thought that it may grow where others of mixture will not. It can be controlled by early cutting. Though it is a perennial, it is not ordinarily long-lived and does not stand competition well. In California, seeds are reportedly eaten by several kinds of birds and by the pocket mouse.

Fescue Grass

Fescue Grass
Festuca elatior
Height to 4 ft. Loosely tufted, and often from short creeping rootstocks. Flowering stalks erect, smooth, with nodding flower cluster at the top. Leaves with flat blades, to ½ in. wide, roughened above. Sometimes called English bluegrass.

Native of Europe. Widely established in America throughout United States and Canada but does best in the cooler areas. It is cultivated for meadows and pastures. Also called *F. pratensis.* Found widely dispersed in meadows, pastures, roadsides, and waste places in northern states.

Flowering panicle erect or nodding, usually much branched but becoming contracted after flowering, to 8 in. long. Spikelets to ½ in. long, with six to eight flowers, rarely with short awns. Empty scale is about ¼ in. long, smooth, faintly streaked. Produces seeds in June and July. *F. aurundinacea,* tall fescue, more commonly used in U.S.; especially as a grass constituent throughout western intermountain region. Also common in West, Northwest, and Southeastern states.

In Europe and particularly in England, fescue is considered one of more important grasses. In America, it is most commonly grown in mixtures. In the timothy-growing region in America, fescue can never compete with that grass.

Seeds or fruits are occasionally found mixed with commercial brome grass or rye grass but they are considered inferior to these by most farmers. To seedsmen, standard fescue should be 95 percent pure fescue, and 85 percent of the seeds must germinate. Forage value good.

Reed

Reed
Phragmites communis
Height to 12 ft. Growing from stout, horizontal rootstocks that interlock firmly and crowd out other species. Leaves flat, to 2 in. wide, to 2 ft long, smooth and with overlapping sheaths at the base. A single rootstock may extend over 30 ft.

Found in wet places, at edges of marshes or ditches, across continent, and in Europe and Asia, but most commonly in or near brackish water. Minor differences between reeds of Europe and America not to be considered here. Reeds grow in same kinds of places as wild rice. Fairly common weed along irrigation ditches in parts of the Far West states.

Reproduction commonly by means of spreading rootstocks. In fact, seeds are only rarely perfected. Whole flowering part of plant is conspicuous because of long, silky hairs on flowering spikelets that are individually about ½ in. long but are crowded into a large mass.

In part, because plant matures few seeds, this is not considered a desirable species in wildlife management. Widespread distribution in part due to the fact that its seeds are high fliers, airborne, assisted by their light weight and long silky hairs. While wild rice pulls easily because of weak root system, it is almost impossible to pull up reeds. This makes the reed useful in flood control where the rice would be useless.

Reeds would be planted only for shelter by a wildlife manager. Instead, they are often destroyed by crushers that cut up the undergrown system, since mowing does not seem to destroy them. A mass of reeds has a unique beauty when in bloom. Fossil rhizomes of reed have been found in Europe, thus making it one of the few grasses known from geologic past.

Canada Bluegrass
Poa compressa

Height to 1½ ft. Flowering stems sprawl on ground before becoming erect, strongly flattened, usually not crowded. Leaves and whole plant conspicuously blue-green. Leaf blades short, not over ¼ in. wide. Perennial. With extensive, creeping rootstocks.

Found from Newfoundland to Alaska, south to Georgia, Alabama, New Mexico, and California and introduced from Europe. Grows fairly well on poor soil where more valuable species cannot survive. Found mostly in open fields, along roadsides, in waste places, in open woods, and meadows.

Flowering·panicle to 3 in. long, usually one-sided, much more sparse than that of Kentucky bluegrass. Usually short branches are in pairs bearing spikelets to the base. Spikelets crowded, to ¼ in. long, with three to six flowers. Seeds of Canadian smaller and lighter than those of Kentucky bluegrass. Characteristics which readily separate this from Kentucky are: *P. compressa* has compressed stems which long remain green, a single shoot at the end of each rhizome, and close narrow panicles. Perennial. Hardy.

Standard commercial Canada bluegrass seed must be 90 percent pure and must germinate 45 percent. Of considerable value as forage species on farms with poor soil of marginal and submarginal type, and particularly on heavy clayey soils unsuitable for Kentucky bluegrass. Finds most favor in eastern Canada, New York, and New England.

Without grasses of this type, many farms would not be profitable. Helps build soils and make them eventually suitable for more valuable grass species. Excellent as a soil anchor, for flood and soil control. While volume of forage produced is not great, nutritive value is relatively high. For lawns and golf links and similar purposes it can be used where dry and under these conditions is superior to Kentucky bluegrass. Seeds eaten by songbirds and rodents, leaves eaten by rabbits.

Canada Bluegrass

Kentucky Bluegrass
Poa pratensis

Height to 3 ft. Flowering stems tufted, erect, slightly compressed. Leaves with boat-shaped tips, those on flowering stems rarely over 6 in. long and ¼ in. broad, but the basal leaves are larger. Blades soft, flat, or folded. Roots rather shallow, but underground spread to form a dense sod.

Either a native of region from Pennsylvania west to the Mississippi or introduced from Europe at an early date. Essentially a grass of open country but lives in open woods, as well as in meadows and humid pastures. Grows at altitudes higher than those acceptable to most cultivated grasses. Adapted to more or less well-drained loams and heavier types of soil of medium to better productivity.

Flowering panicle pyramid-shaped, open, lower branches coming off usually in a whorl of five, with usually a long central one and shortest pair on outside. Spikelets three- to five-flowered, with a length of to ¼ in. Grain is enclosed in two scales which are from ⅙ to ¹⁄₁₀ inch long, being longer than that of Canada bluegrass.

Grown alone may yield to 900 lb of valuable forage per acre. Wild white clover grown alone may yield to 3,000 lb per acre. A mixture of this clover and Kentucky bluegrass may yield about 5,000 lb of forage even more nutritious than either of other two yields. Does best on soils with lime.

Kentucky bluegrass usually grown in mixtures; usually planted in early fall at 4 to 6 lb per acre in mixtures that include other grasses; mixed with legumes twice as much Kentucky blue seed per acre is used. For turf purposes, 20 to 100 lb per acre is recommended. Seed sowed on firm seedbed and covered lightly. Weighs 24 lb to the bushel. As a mixture, usually not over 12 lb per acre are used. Grown for seed, a good yield may be to 100 lb of cleaned seed. This grass is perennial and one of best of forage grasses. Attractive as a wildlife food to the same extent as Canada bluegrass.

Kentucky Bluegrass

Chess, Cheat
Bromus secalinus

Height to 3 ft. Flowering stems erect, rather slender, but sturdy, smooth. Leaves 3 to 10 in. long, flat, slightly hairy above but smooth beneath. In related soft cheat *B. hordaceus,* whole plant is softly hairy.

Introduced from its native Europe and more or less too generally established throughout United States. Commonest in grainfields and waste places. In fact, its range might be defined as any place where ordinary cultivated grains will grow successfully. It is especially troublesome in the Southwest.

Flowering panicle rather open, and nodding, to 5 in. long, with three to five lower branches, unequal and drooping. Spikelets to ⅘ in. long, ³⁄₁₀ in. wide, with awned scales forming a somewhat flattened cluster of flowerlets. Flowers appear in June and July; fruits in July and August. Annual.

Chess, Cheat

(Continued)

Brome Grass

Orchard Grass

Zoysia Grass

Makes such a vigorous growth in winter grainfields that where stand is thin, cultivated grass has no chance. Leaves have some slight value as hay. Nevertheless, it is grown in some places where soil is poor. Plant about 80 percent water, 11 percent fiber and ash, 4.5 percent nitrogen-free content, 2.5 percent protein, 2 percent fat.

When young, the plant has considerable value but its competition with more desirable species when mature makes it highly undesirable in most cases. It is eliminated easily in fields by crop rotation and by using clean seed. Seeds of brome grasses are large and are eaten by a number of different kinds of birds and rodents.

Brome Grass, Smooth Brome
Bromus inermis
Height to 4 ft. Flowering stems arise from creeping rootstocks. Leaf blades smooth, or nearly so, to $2/5$ in. wide, to 1 ft long, long-pointed. Leaf sheaths closed nearly to top. Flowering stems smooth, unbranched, well-supplied with leaves.

Introduced, probably from Hungary, into California as early as 1884. A hay grass especially popular from Minnesota and Kansas to eastern Oregon and Washington and occasionally east through Michigan and Ohio, where it runs wild now and then. Introduced along roadsides and on mountain ranges particularly in northern and western parts of United States. Most widely utilized of cultivated brome grasses. Effective range corresponds roughly to the Corn Belt.

Flowering panicle about 10 in. long, loose and open with spikelets on ends of loose branches. Spikelets about 1 in. long, $1/4$ in. wide, composed of 6 to 10 flowers, narrow, purplish, usually without awns though not always so. Fruit flattened, boat-shaped, $1/3$ in. long and $1/12$ in. wide, enclosed in scales.

Thought by some that brome grass may someday replace Kentucky bluegrass as a hay and pasture grass, particularly in the Northwest. Others do not agree. Has long been cultivated in southern and central Russia. Most palatable to cattle. Can be grown successfully on a variety of soil types, including sandy loams. A sod-forming perennial.

Excellent in improving worn-out lands. Usual seeding is 15 to 20 lb per acre or in mixtures at 6 to 10 lb. As a seed crop, yield should be between 400 and 500 lb to the acre. Unfortunately, brome grass seed too commonly has quack grass seed mixed with it. Its value was demonstrated following the drought years of the 1930s as a grass to replace depleted bluegrass and native pastures. Most effective in combination with a legume like alfalfa.

Orchard Grass
Dactylis glomerata
Height to 3 ft. Grows in tufts or bunches, not from a creeping rootstock. Roots deep, penetrating to a depth of at least 2 ft. Fruiting stalks smooth, unbranched, angled. Leaves many, flat, with long sheaths, rough, to 2 ft long and $1/2$ in. wide, rather conspicuously light gray-green. Common name from its shade tolerance and consequent occurrence in orchards and other shady places.

Native of Europe. Probably in cultivation in the U.S. since 1760. Found in fields, meadows, and waste places from southeastern Alaska to Newfoundland, south to Florida and central California; also found in Asia. Commonly cultivated in England, and known as cocksfoot. Not found in arid and semiarid regions. North Carolina, Tennessee, Kentucky, and Arkansas important centers. Only fairly drought-resistant.

Flowers produced in June, about when red clover flowers, in one-sided panicles, spikelets containing three to four flowers and spikelets being in dense clusters, whole cluster being to about 4 in. long. Grain enclosed in a scale which is about $1/8$ in. long while grain itself is about $1/10$ in. long. Perennial. Bunchgrass.

Seed weighs 14 to 22 lb to a bushel; is sown for hay at 35 lb to the acre or, when mixed, to as low as 6 lb. Most seed is produced near Lexington, Ky.; removed by threshing about 2 weeks after hay is cured. Hay should be cut when plant is in bloom. Remarkably free of disease.

Good for early pasture and excellent for hay. For livestock, food value is fair for proteins, good for minerals, fair for calcium and phosphorus, and fair for carbohydrates. Does not stand pasturage well and a stand that lasts over 4 years with close grazing is exceptional. Seeds eaten by birds.

Zoysia Grass, Manilagrass
Zoysia matrella
Leafy perennial, grows low; develops and spreads by many wiry underground stems. Forms a dense turf. Produces harsh, fine-tipped leaves. *Zoysia* species native of southeastern Asia and New Zealand.

Flower stalks slender, a few-flowered spike; spikelets single, arise alone at nodes along zigzag-shaped stem.

Papyrus, Umbrella Plant

Does not survive well above 40°N. Is fairly successful in shade if in moist, warm climate. Grows slowly. Perennial. Introduced into U.S. as a lawn grass and has escaped to an extent.

Z. tenuifolia, mascarenegrass, is the smallest and finest zoysia used in the U.S. and is of least importance. Grows to 2 in. tall and develops only a shallow root system. Used in the South and as far west as California. *Z. japonica,* Japanese lawn grass, has broad coarse leaves similar to redtop and produces a dense cover. Winter-hardy to Boston. Often grown for lawns in southern states. Produces good, deep lawn mat; however, turns brown and becomes dormant earlier than some homeowners would prefer. Seldom produces flowers and hence seeds so usually transplanted by means of small "plugs" of the sod. Does not survive heavy frost in the fall and remains without green color until after last frost in the spring. In the District of Columbia remains green from mid-April to late October. Some survival of plants of one zoysia grass as far north as Rhode Island. When planting an area, one sq yd of turf can be cut into plugs to plant 750 to 1,000 sq ft of lawn. Frequently homeowner plants seeds of another grass between plugs to fill in bare spots during that time it takes for zoysia to spread over total lawn surface area.

Order Graminales./Family Cyperaceae

Papyrus, Paper Reed
Cyperus papyrus
Height to 15 ft. Flowering stalks strong, erect, smooth, roughly three-angled, arising from a stocky, woody rhizone that lies horizontally just below surface of ground. Leaves all spring from base, though sheaths on flowering stem resemble leaves. Leaves are mere sheaths at base. Family usually known by its solid three-sided stem.

Native of southern Europe, Syria, and Africa but often grown in aquariums and park or hothouse ponds for unique artistic effect. Used as lawn ornaments in California and survives winter in milder parts of South. Of the many varieties, this is one most associated with Palestine.

Tall fruiting rays have great clumps of drooping rays to 20 in. long, that make what appears to be an elevated cushion or stool. Spikelets from 1 to 1½ in. long, with many smaller, spreading spikelets. Fruit nut-like, three-cornered, and gray.

Need an abundance of sunlight and air. Highly susceptible to attacks by aphids that deform them. Need an abundance of water but cannot stand a continued jet. Plantings from seeds are made in fall or spring, and seedlings may be well-developed first season.

Plant is famous as source of paper for ancient Egyptians. In fact, name papyrus and paper are practically synonymous. In America, only important use is decorative, young and old plants each being valuable. Probably used in papermaking as early as 2400 B.C. According to the King James Version of the Bible, this is the plant from which reeds were taken to build the "ark of bullrushes" in which the baby Moses was hidden.

Chufa

Chufa, Nut Grass
Cyperus esculentus
Height to nearly 3 ft., stout, with fruiting stems and leaves about equal in length. Leaves spring in clusters from horizontal, twisting stems that lie just beneath ground and bear walnut-shaped tubers at their ends.

Found in Europe, Asia, and America from Nova Scotia to Minnesota, from southern Nevada north along Rockies to Alaska and south to Mexico. Also reported in wet soils in the Imperial Valley, Central Valley, and coast ranges and in coastal southern California. Grows in low, wet grounds along waterways and frequently spreads into cultivated fields.

Flowers borne in light chestnut or straw-colored clusters about ½ in. long. Many fruits are borne in a single cluster. Most effective reproduction is by tubers that are thrust in all directions from parent plant.

As high as 94 percent of food found in gullets of pintail and mallards has been nutlike tubers of chufa, some containing 300 or more. Tubers are often known as "ground almonds" and may constitute the best duck food in areas that are flooded.

Plant may be a pest in cultivated lands. Where used for encouragement of wildlife, tubers are planted at a depth of about 1 in. at 1 bu an acre, usually between April 1 and June 15 in areas where floodwaters do not disappear before July.

Spike Rush
Eleocharis sp.
Eleocharis palustris, common spike rush, grows to a height of 5 ft, is round-stemmed, and fruiting stalks terminate in a spike about ½ in. long. Stems to ⅕ in. in diameter; there are numerous, horizontal, underground stems from which erect clusters rise at intervals. Leaves reduced to bladeless sheaths.

(Continued)

Spike Rush

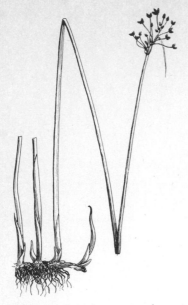

Great American Bulrush

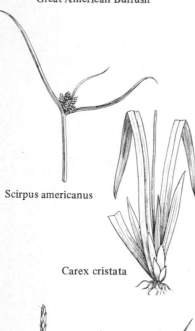

Scirpus americanus

Carex cristata

Carex stipata

Sedges

E. palustris ranges from Labrador to British Columbia and southern Alaska, south to Florida and into Mexico, thus covering the entire United States. Most species grow at the edge of water and some grow submerged. Also appear in vernal pools and ditches.

Small spike at end of erect stems contains fruits. In *E. palustris,* spike is like a pointed cylinder with scales loose, reddish-brown with white margins and a greenish back. Underground stems also serve to spread plant and in general reproduction. Perennial.

Common spike rushes provide food for ducks and geese over a wide territory. Blue geese and Canada geese may graze on it extensively in the Hudson Bay area, eating both nutlets and vegetative parts. Marsh birds, shore birds and ducks eat the seeds (nutlets). Cottontail rabbits and muskrats feed on the plant.

Where artificial propagation is desirable for raising food for wildfowl, usual procedure is to cut and plant rootstocks. Data on germination of seeds are not available. Most species are of fresh water but salt water supports at least a half dozen species including *E. palustris.* Much confusion as to classification. Some authors divide members of this species into several separate species.

Great American Bulrush, Soft-stem Bulrush
Scirpus validus
Height to 8 ft. Fruiting stems arise from stout, horizontal, scaly rootstocks that form dense interlocking mats at about low-water mark. Sheaths at base soft, light green, relatively short, with ragged, almost transparent margins.

Found at pond and stream margins, often in several feet of water, over a wide territory, on this and other continents. Some 150 species over world, including hardstem bulrush, *S. acutus,* river bulrush, *S. fluviatilis,* and the wool grasses including *S. cyperinus.*

Reproduces considerably by spreading of underground or underwater rootstocks that may be broken by flood and washed to a new site. Nutlets dull black, nearly $\frac{1}{10}$ in. long, $\frac{2}{3}$ as wide as long, broadly egg-shaped but flattened on one side, borne in clusters, on loose stems, to 2½ in. long.

Bulrushes provide food and shelter for many forms of wildlife of lowlands. *S. fluviatilis* bears 2 to 3 in. tubers; *S. validus,* fleshy rootstocks that are edible and nutritious for ducks and muskrats. Nutlets of *S. fluviatilis* may be a major duck food at certain times.

Bulrushes serve to anchor loose soil, and to build new land that may be valuable agriculturally. Fleshy rootstocks raw or bruised and boiled, cooked with cornmeal, or dried and made into sweet flour, or thick submerged bases of white, crisp, juicy, young shoots are good food for man at any time.

Sedges
Carex sp.
Crested sedge, *Carex cristata,* and awl-fruited sedge, *C. stipata,* are illustrated. Former grows to height of 3 ft, fruiting stalks being longer than leaves. Leaves soft, flat, to ¼ in. wide, and spring from numerous sterile shoots that form close to ground. *C. stipata* to 3½ ft tall, with flowering stalks strongly three-angled and often winged along angles. Leaves flat, soft, to ⅗ in. wide, and spring from sterile shoots close to ground. Most sedges have solid, usually angled stems that bear fruits. Leaf sheaths not split, and usually leaves are coarser and rougher than those of most grasses.

C. cristata found in wet swales and wet woodlands, from Massachusetts and Vermont south to Pennsylvania and Missouri and west to British Columbia. Probably some 900 species of *Carex* known to science. Most of these are to be found in wet areas. Usually occur in more or less pure stands of a given species and lend a characteristic effect to landscape. Many are a yellowish-green, while others are a distinctly blue-green cast. Some species live in high and dry lands, but they are in minority and not so likely to establish pure stands. Plantain-leaved sedge *S. plantaginea* is common in eastern woodlands.

In *Carex,* nutlets or fruits are enclosed in a papery sac that may be loose or tight. Stamens and pistils borne in separate flowers either in separate spikes, in different parts of same spike, or mixed inconspicuously in same spike. Pollination largely by wind or, in some cases, by insects or by gravity. Sac that encloses fruit is possibly most important structure used in classifying different species. Many sedges reproduce by spread or separation of ordinary vegetative parts. Crested sedge is an example where flowers are in separate spikes.

Beaver, moose, elk, deer, muskrat, bobwhite, sharp-tailed grouse, upland game birds, and waterfowl generally have been reported to feed on sedges. Roots and sprouts are relished by muskrats and other mammals, while the nutlets are more important parts eaten by birds. Most commonly reported species to figure in diet of wildlife is *Carex stricta.*

Sedges do not figure as an important forage crop for cattle either as green feed or as hay.   They yield rather readily to competition with cultivated plants and are not normally considered as weeds in spite of their frequent abundance.

Sedges are most commonly of importance to man as soil anchors and land builders in lowlands.   They prevent undue erosion in flooded lands and provide excellent cover as well as some food for game birds and animals that might otherwise compete with man for more economic species.   At least two species are cultivated for their ornamental beauty and are grown in greenhouses or in pots as house plants.   These are *C. morrowii* and *C. comans,* former being a native of Japan and latter of New Zealand.

Order Palmales./Family Palmaceae

Date Palm
Phoenix dactylifera
Height to over 100 ft.   Trunk shaggy, rarely perfectly straight or completely erect.   Leaves often yards long, more or less erect, loose, open, featherlike, stiff, spine-tipped.   Suckers rise freely from and near base, eventually developing roots of their own and making separation possible.

Native of North Africa and Arabia but planted widely through tropical world, particularly in southern Europe and Asia and to some extent in southwestern United States.   Family Palmaceae includes about 140 genera and over 1,200 species, some extending range into warmer temperate regions.

Staminate and pistillate flowers on separate trees, pollen normally being spread by wind.   One staminate tree is sufficient for 100 pistillate trees. Pollen can be gathered and dusted on pistillate flowers immediately or stored for over a year for later periods of use.   Flowers on some species of *Phoenix* are cut off and the sweet sap is collected and dehydrated to sugar or syrup.   Fruits ripen over long period, some being improved by picking and storing in warm moist place.   Seeds or suckers planted at any time in suitable regions.

Favors hot climate but not humid air; can endure more alkali than other profitable tropical fruits and can survive 10°F for short time but must have hot climate during fruit ripening.   Survives condition in Salton Sea, Salt River Valley, and Lower Colorado Valley in United States, and similar conditions elsewhere.

Food value high as nutriment.   Has served man as food through ages. Seeds roasted and used as coffee substitute or pressed for oil and pomace used as stock food.   Sugar made from sap.   Valuable in tropics for providing shade for other plants.   Individual fruit clusters may weigh to 30 lb.   Individual tree may bear fruits for 100 years.   80 percent of the world crop of dates comes from Iraq.

Coconut Palm
Cocos nucifera
Height to 100 ft.   Tree with leaves pinnately compound and to 20 ft long, springing from a common head or restricted area.   Trunk tough, fibrous, often curved, and diminishing in diameter to head.   Bark rough and fibrous, broken irregularly but often in musclelike corrugations.

Probably originally from America but widely spread by man early in history.   Requires mean annual temperature above 72°F, and unless underground water is available must have at least 40 in. of rainfall a year. Found in California and Florida in United States, in tropics but generally near seacoast.

Flowers borne at head of tree, at 4 years of age in some but more commonly from 7 to 10 years.   Flowers clublike spathes, white or yellow; staminate having six stamens and pistillate producing one-seeded fruits. Fruits may float unaffected in salt water for great time and distance giving wide distribution.   White "meat" of the coconut is the thick layer of endosperm and very small embryo of the seed.   The "coconut milk" is the milky fluid or sap found in the cavity of the endosperm.   Palm tree produces about ten coconuts when 5 to 7 years old and produces 60 to 70 "nuts" at full growth.   May bear coconuts for 60 to 80 years.

May tolerate salt water and withstand drought, though latter may affect crop for a year or more.   Tough fibers of trunk, up to 20 percent elastic, used in making coir.   Subject to a number of insect and fungus pests including a scale, a beetle, rats, and certain crustacea.   Nut sizes vary from 3,300 to 7,100 necessary to produce 1 ton of copra.

Liquor from flower cluster is fermented to a toddy, sugar, vinegar, or yeast.   Copra, most important product, used by man and beast as soap, lubricant, and so on.   Copra is dried coconut meats.

Date Palm

Coconut Palm

Tall Palmetto

Jack-in-the-pulpit

Arrow Arum

DIVISION ANTHOPHYTA. CLASS MONOCOTYLEDONAE

Palmetto
Sabal palmetto
Height to 80 ft. Stem erect, covered in upper portion with leaf bases. Leaves to 8 ft long, with long, slender base and broad fan-shaped blade, blade being shorter than petiole or base. Leaves arise from a terminal bud, youngest being uppermost and older eventually dropping to stem.

Found from North Carolina to Florida and Bahamas. Forms dense thicket often close to ground where tops have been repeatedly destroyed by fire or otherwise. Favors relatively dry land but grows in sandy stretches close to salt or fresh water. At least 18 species in United States, Mexico, and Central America. Planted as a "street tree" in the southern states.

Flower cluster shorter than leaves and produces a number of black, somewhat cherrylike fruits, 1/3 to 1/2 in. in diameter. Some palmettos have twisted, underground stems that assist in multiplying the plant and in helping it survive fire and other disasters.

Fanleaf palms include a variety of plants. The California fan palm is *Washingtonia.* In the Southeast are the folded-leaved scrub palmetto, the 60-ft cabbage palm of Florida, the saw-edge-leaved saw palmetto, the flat-leaved, low, fibrous-margined dwarf palmetto, and others not here considered.

Palms and palmettos provide excellent cover for wildlife of many sorts. Some make excellent thatching for houses. Raffia of commerce, betel nut of Asia, a sugar, a wine, a vegetable ivory, an oil come from different palms and palmettos. None probably exceed in importance date palms and coconut palm already considered. Unfortunately land developers are destroying palmetto palms with their overly ambitious bulldozers at an alarming rate. Also there is a great demand on Palm Sunday for the unopened palm leaves as a religious symbol. Perhaps acceptance of an artificial biodegradable "palm leaf" might be in order.

Order Arales./Family Araceae

Jack-in-the-pulpit
Arisaema triphyllum
Height to 3 ft. Leaves one to two, each of three parts, almost like leaflets, with netted veins and unbroken margins, each part to 7 in. long and 3½ in. wide. Leaves arise from a deeply buried, turnip-shaped, solid, bulb-like stem base. Roots spring from base of this structure.

Found in rich, moist woods and thickets, from Nova Scotia through Ontario to Minnesota, south to Florida and Louisiana; up to altitudes of 5,000 ft in North Carolina. Often found in crevices of rich earth on sides of shaded gorges. Closely related to green dragon, *A. dracontium,* which ranges to south and southwest.

Usually, flowers in a given cluster are either staminate or pistillate. Flowers on spadix or "jack" in "pulpit" (spathe). Spathe hooded, marked with green or purple or brown. Fruit a scarlet berry containing a few seeds, fruit being exposed when mature. Perennial.

Pollination effected by beetles and small flies that visit the "hood," flies commonly being of family Mycetophilidae. Seeds carried by animals or merely drop to ground. Fruit or "turnip," eaten or bitten raw, causes a severe stinging pain in mouth for some time. Flowers April to June.

Fresh turnip reported to have some medicinal properties as an emetic; mixture made of this and of fruit is used to kill insects. Turnip, boiled or baked, then peeled, dried, pounded to flour, heated again, and allowed to stand becomes mild and edible and was an Indian food. Not to be eaten raw. Both leaves and bright red fruit eaten by wildlife to a limited extent.

Arrow Arum, Tuckahoe
Peltandra virginica
Height to 30 in. Leaves bright green, to 30 in. long, 8 in. wide, with blade like a spearhead, rather thick, strongly veined. Root system a thick tuft of fibers and rootstock from which leaves arise without any apparent stem between.

Found in lake and stream margins or in shallow water, from Maine to Ontario and Michigan, south to Florida, Louisiana, and Missouri. Common in low, flooded lands where there is an abundance of loose, rich silt in some parts of the range. Known as "Virginia wake-robin." Blooms May through summer.

Pistillate flowers cover about 1/4 length of enclosed spike on which they are borne, remaining 3/4 being covered by staminate flowers. Enclosing envelope (spathe) to 8 in. long at maturity, thrust under water or close to ground. Fruit green berries, with one to three seeds each.

Seeds have been reported by wildlife researchers as being eaten by waterfowl, sometimes assuming importance, but they and plant are rarely eaten by muskrats. In addition to species of arums here considered, there are many others, including the peculiar *Monstera* often seen in banks and in greenhouses.

Starchy rootstock is acrid and poisonous when fresh but may be eaten safely after it has been roasted or boiled a long time. May be made into meal or flour. Berries and fleshy stalk are boiled and were a luxury to Indians. Seeds that have been cooked thoroughly and long may be eaten as beans. Berries are choice food for wood duck.

Skunk Cabbage
Symplocarpus foetidus
Height to 3 ft or more. Leaves large, to nearly 2 ft long and almost same width arising from a stem that elongates as season progresses but in early stages makes plant resemble a cabbage. Rootstocks thick, going deep into soil and bearing whorls of fleshy, fibrous roots.

Common in bogs and moist, open, or wooded lands from Nova Scotia to western Ontario and Minnesota and south to North Carolina and Iowa. Flowers appear early in spring, even thrusting way into icebound soil and frequently being found under snow. Leaves appear after flowers of year have been pollinated.

Flowers bear both stamens and pistils, flowers being crowded onto a spherelike structure enclosed in a broad slipper-shaped hood that is green and red or brown. Pollination probably by carrion flies. Seeds borne one to a flower, sunken in pulpy fruit that is size of a man's fist.

Name comes from resemblance of leaves collectively to cabbage and of odor of juice to offensive scent produced by skunks. Young roots and flesh of fruit have medicinal properties as an irritant, producing vomiting and even temporary blindness.

Young leaves and shoots may be cooked until tender, discarding one or two waters, which makes them not too offensive as greens. Rootstocks cooked long and thoroughly are palatable, but plant should be used as a food only in an emergency. Remotely resembles poisonous false hellebore. Large seeds eaten by some wildlife forms.

Wild Calla, Arum Lily
Calla palustris
Height to under 1 ft. Leaves with heart-shaped blades, with entire margins, thick, to 4 in. wide, shining, rich green, on long petioles sheathed at base and arising from a thick, horizontal rootstock that is covered with scales and bears fibrous roots at joints.

Found in bogs and shallow waters from Nova Scotia to Hudson Bay, south to Virginia and Iowa. Also grows wild in Europe and in Asia. Known as swamp robin, water lily, or female or water dragon. It is almost always a shade lover and may grow in almost pure stands over a limited area.

Flowers appear in May and June, fruits in July and August. Flowers, at least lower ones, have both stamens and pistils, borne on rather short spike about 1 in. long, backed by a greenish-white spathe about 1 in. wide and 2½ in. long. Fruits red, few-seeded berries, in a large head at maturity.

Pollination helped by water snails as well as by insects. Finding a plant as beautiful as wild calla in a cool, dark bog is a memorable experience.

Of little or no economic importance. While this little flower looks much like the calla lily of Easter fame, the two do not even belong to the same genus. The small wild calla probably yields to cultivation only for adding to beauty of a landscaped pool.

Calla Lily
Zantedeschia aethiopica
Height to 2½ ft or more. Stocky plant, with smooth, shining green leaves. Blades many, shaped like a spearhead, spongy, and arising from a thick underground rhizome. Often bases of leaves are bristly.

Eight species native of tropical and southern Africa and growing commonly in wet, rich lowlands. This species native of the Nile; rose calla, of Natal. Other species, some with yellow flowers and some with the spathe black-throated.

White spathe that opens to one side, flares widely, and tapers to a free tip is well-known. Inside is spadix, covered for most part with many small, naked, crowded flowers, sometimes with stamen-bearing flowers mixed with those that produce one- to three-celled berrylike fruits.

Thrives best in good light, at a minimum temperature of 55°F. In commercial planting, roots are kept dormant in winter in storage at around 45°F, then potted and kept dry until roots start in about 2 weeks. After this, plant may be given water, a higher temperature, and may be expected to bloom in 10 to 12 weeks. Liquid fertilizer is usually used.

Quite a calla industry in California, where plants may sometimes be grown safely in the open. Main source of revenue is bulbs, grown for shipment and sale to greenhouses, which will develop plant sold in flower shops. Commercially, plants do not thrive in hot, gravelly, or stony soil.

Skunk Cabbage

Wild Calla

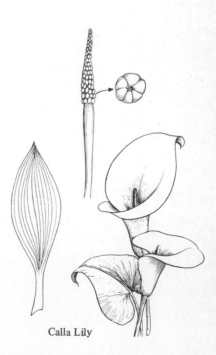

Calla Lily

Golden Club

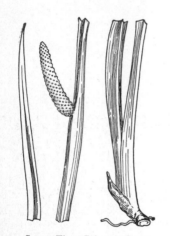

Sweet Flag, Calamus Root

(1) Duckweed

Golden Club
Orontium aquaticum
Height to 2 ft. Leaves erect or floating, depending on depth of water. Sometimes nearly 3 ft long, with blade about 1 ft long. Blade dull green above, paler beneath, to 5 in. wide, usually narrowed about equally at base and tip. Rootstocks thick and under mud.

Found in ponds, bogs, and swamps, from Massachusetts to Pennsylvania, south to Florida and Louisiana but mostly near coast. Found up to a 2,000-ft elevation in Pennsylvania. May form a solid mass acres in extent and may be a pest to boat navigation.

Flowers borne on a stem to 2 ft long, upper half bearing crowded flowers to form a club up to 4 in. in diameter that becomes much thicker in fruit. Pollinated by small insects and snails. Spathe reduced to a scale or bract only 4 in. or less long. Flowers appear in April and May and fruits a month or so later.

Named from Syrian river, Orontes. Size and shape of leaves affected by amount and nature of surrounding water. Grown rather commonly as an ornamental in landscaped pools where it is started by cuttings of rootstocks; rather easily grown.

Bulbous rootstocks may be eaten as food only after repeated boiling and changing water or after roasting to remove acrid juice that makes them otherwise unpalatable. Seeds were gathered by Indians, boiled repeatedly, and then eaten like peas.

Sweet Flag, Calamus Root
Acorus calamus
Height to 6 ft. Leaves slender, linear, erect, sharp-pointed, usually under 1 in. wide, with a stiff midrib throughout length, sheathing each other, forming a two-ranked arrangement. Horizontal, long, branching, aromatic rootstocks bear rather coarse secondary roots.

Found near waterways as along lake shores or streams or in swamps, from Nova Scotia to Ontario and Minnesota, south to Louisiana and Kansas. Also found wild in Europe and Asia. Has many common names including sweet myrtle, myrtle flag, sea sedge, beewort, and sedge cane.

Flowers borne in a spikelike spadix, extending at an angle from erect scape or flowering stem, with spathe merely continuation of structure below spadix. Flowers contain both stamens and pistils. Fruit, when mature, is dry but gelatinous inside and bears few seeds.

Thrives best in moist soil but may be grown in shallow water or even on dry land. A variety *variegatus* with deep yellow stripes is most commonly cultivated, usually by division of underground parts in spring or in autumn.

Pungent, aromatic rootstocks may be bought at drugstores either in natural condition or ground. Used as candy and reputed to have medicinal value. Used in flavorings, and in powdered form mixed in making of sachet powder. Distilled in oil, used in perfumes.

Order Arales./Family Lemnaceae

1. Duckweed
Lemna trisulca

2. Lesser Duckweed
Lemna minor

3. Large Duckweed
Spirodela polyrhiza

4. Watermeal
Wolffia sp.
Lemna trisulca forms irregular net of branching fronds, each to ⅖ in. long, green, often without rootlets, usually obscurely three-nerved, offshoots usually remaining connected. *L. minor* appears as one or more egg-shaped, thin, floating plants, each section to ⅕ in. across, with only one root; green above and below.

Both species are found floating on top of quiet, stagnant waters or washed on nearby shores. *L. trisulca* ranges from Nova Scotia to New Jersey, south to Texas and west to Pacific and in tropical and temperate zones. *L. minor* ranges commonly through most of North America.

Reproduction mainly by division of vegetative plants in each of these species. In *L. trisulca*, newly formed sections tend to remain together while in *L. minor* they separate quickly. Flowers are inconspicuous and are borne in a sac at margin of frond; essentially two stamens and one pistil per unit.

Remarkable for quickness with which they can multiply and cover a considerable surface of water. Either may form a dense mat almost completely covering water beneath. Useful in aquariums in biology class as examples of a quick vegetative reproduction.

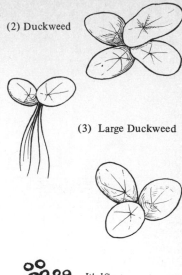

(2) Duckweed

(3) Large Duckweed

Different species are essentially alike in wildlife management, the duck-weeds having been reported as serving as food for ducks, geese, muskrats, pheasants, beaver, and other small animals, with two species here considered being most frequently mentioned. Ducks take in many small animals when they feed on duckweeds. May be one of the principal foods of ducks when seeds of other plants may no longer be available in quantity.

3. Large Duckweed
Spirodela polyrhiza

4. Watermeal
Wolffia sp.

These are the largest and the smallest of the duckweed group. *Spirodela* is easily recognized as a floating oval or group of ovals, each about ⅓ in. long, green above and purple beneath, with a tuft of many roots beneath. *Wolffia* is so small as to resemble small green dots, not over ¹⁄₂₅ in. long. It is rootless.

Wolfia
(4) Water meal

Both *Spirodela* and *Wolffia* are found floating on surface of stagnant waters, often mixed in with *Lemna*. *Spirodela* is common from Nova Scotia to British Columbia and south to North Carolina, Alabama, Texas, and California. *Wolffia* ranges from Ontario to Gulf of Mexico and west to California.

Reproduction is for the most part by means of vegetative division, which under favorable conditions may be very rapid. Flowers of *Spirodela* are marginal, while those of *Wolffia* are on upper surface or burst through it. They are of course microscopic in each case.

Wolffia is interesting to biologists because it is probably the smallest of all flowering plants; although it may be abundant, it may never be noticed.

Both *Spirodela* and *Wolffia* have been reported as good duck food, with *Spirodela* sometimes constituting a substantial portion of ducks' diet and also often eaten by pheasants.

Order Xyridales./Family Xyridaceae

Pickerelweed, Tuckahoe
Pontederia cordata

Height to 4 ft. Erect, bright green. Leaves mostly coming from near base from a stout rootstock, thick, dark green, smooth, with blades broadly triangular, spearhead-shaped, and borne on long petiole to 10 in. or more long and to 6 in. wide.

Found in shallow ponds and streams or at water's edge, in rivers and bayous, from Nova Scotia to Minnesota, south to the Gulf of Mexico and Florida and Texas; south to Argentina. There are several races, based partly on width of leaves.

Flowers borne in a coarse spadix or spikelike structure supported by a small, leaflike spathe. Spadix bears flowers about 4 in. long. Flowers crowded, weak blue or deep violet-blue, with distinct, yellow-green spot. Stamens, six, three being long. Fruit bladderlike, containing only one seed.

Seeds occasionally of value as duck and deer food and plant is good muskrat food. Leaves lose a large amount of water through transpiration and so plant may be undesirable around water reservoirs. It is, however, most attractive and will probably always find favor.

Scientific name refers to a professor of botany at Padua about 1730. Common name refers to fact that plant grows at edge of waters in which pickerel are likely to be found.

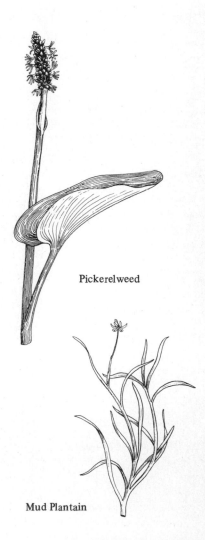

Pickerelweed

Order Xyridales./Family Pontederiaceae

Mud Plantain
Heteranthera dubia

Length to many feet. Usually submerged aquatic, with a slender, forked, flattened stem, joints sometimes being 3 ft apart. Roots at the joints. Leaves slender, flat, very long, finely parallel-veined. Often has color of dying grass.

Found in still or flowing fresh water from Quebec and western New England to California and south to Cuba and Mexico. While it is usually out in water deep enough to cover it, it sometimes is found growing as a low plant on muddy shores that may be temporarily submerged.

Flower light yellow, inconspicuous, borne singly apparently at the end of a long threadlike tube. Colored part in six divisions. Stamens longer than pistil they surround. Fruit a one-celled, many-seeded capsule. Sometimes two flowers are borne together.

Another aquatic that provides some food and much shelter to fishes and other animals of its environment. In some localities it is reported to

(Continued)

Mud Plantain

serve as a good duck food, but it cannot compete with many better species in this respect. Related species have more beautiful flowers.

Of little direct importance to man. Rarely sufficiently abundant to interfere with water transportation, as does its close relative water hyacinth, and not so beautiful as the pickerelweed. It probably is more useful as food for aquatics than either of these relatives. It does well in alkaline water, tolerates temperatures to 50°F; propagates freely vegetatively from cut stems.

Water Hyacinth

Water Hyacinth
Eichornia crassipes
Height to 1 ft or more above water in which plant floats. Underwater leaves, if on young stems, slender. Leaves at water surface have bases, are air-filled, and serve as floating bladders. Other leaves may be more or less erect with swollen bases. Roots extend into water or soil.

About six species, mostly in South America but one from Africa. *E. crassipes* is a pest in Florida and along Gulf Coast and is found as far north as Virginia and Missouri; west to California. Is grown as a curiosity in ornamental pools in the North. Brazil is center of abundance. Not winter-hardy in northern states.

Flowers borne on erect stem about 1 ft high, supported by wavy-margined sheath partway up stem. About eight, violet, six-parted flowers, upper lobe of colored part being marked with dark-blue spot, with a bright yellow spot in middle. One variety has yellow flowers; another, rose-lilac.

Reproduces largely by extension of main plant body. This breaks, floats away, and starts new colonies; process is continued until whole surface of water may be so closely packed that boats can get through only with great difficulty. Not valuable as food for wildlife.

Known in South as "million-dollar weed," not because it is worth that much but because it is estimated that it might cost that much to keep it under control in valuable waterways. Plants grow rapidly. Even in short season in North may cover a considerable space in a small pond. A suitable species for large aquariums in that its roots provide cover and shelter for young fish and insect larvae.

Order Xyridales./Family Bromeliadaceae

Pineapple

Pineapple
Ananas comosus
Height to 4 ft. Erect, short, stout stem, bearing coarse, spiny-edged, thick, closely crowded, sword-shaped leaves. *Crowns,* leaves coming from top of fruit. *Slips,* small plants formed just below the fruit. *Suckers* develop from axils of leaves close to ground. *Rattoons,* buds developing underground.

Raised chiefly in Hawaii, Cuba, Puerto Rico, and formerly, to limited extent, in Florida. Has been grown outside the tropics, but without commercial success. Hawaii produces 75 percent of world's canned pineapple and most of West Coast's fresh fruit; Puerto Rico produces most of East's fresh fruit.

Plants are raised from seed only for breeding purposes, as about 12 years are required to mature a plant from seed. Rattoons are usually left to replace harvested crop. Suckers producing fruit in June may produce five suckers by mid-September. Slips require a year longer than suckers to produce a crop. Barrel-shaped pineapple of commerce represents entire inflorescence. The pineapple consists of a thickened central stem and numerous small berries which form from flowers. These berries come together to form a multiple fruit.

Some 8,000 to 15,000 plants needed to set an acre. May require 18 months to yield first crop of 50 to 350 crates, but second crop may be double the first and third crop may be even larger. Yield may continue 8 to 10 years. Fruit harvested week before it is ripe. Soil must be well-drained but water must be available.

Canning fruit and juices and shipping fresh fruits is big business. Fruit high in sugars of high nutritive and palatable nature but only about ½ vitamin A value in canned pineapple as found in fresh strawberries. Fruits weight to 32 lb. Pina fiber comes from leaves.

Spanish Moss

Spanish Moss
Tillandsia usneoides
Length several feet. Usually found hanging in irregular profusion from branches of trees. Stems slender, hoary gray. Leaves scattered, 1 to 3 in. long, narrow and threadlike. Both leaves and stems are flexible and wave freely in wind from support above. Roots attached to support above.

Found from Virginia to Texas and south, being common in most of its range and usually very conspicuous. Any picture of a southern woodland in moist regions will show Spanish moss. It extends south to southern Brazil.

Flowers solitary, borne in axils of leaves, inconspicuous, with yellow petals bent backward at ends. Reproduction may be by vegetative division of parent plant or by means of seeds that are borne in an inconspicuous capsule.

Moisture, warmth, and sufficient elevated support are necessary for survival of this plant. It does occur to some extent in regions of killing frost and snow. A number of insects, including thrips, are to be found dependent on the plant. Takes moisture from air or precipitation, mineral elements from dust it collects.

Man uses Spanish moss not only for decoration of his trees in proper territory but for packing material, upholstery stuffing, and, in outdoor living, for making emergency mattresses. Used by a few birds as nesting material.

Order Xyridales./Family Commelinaceae

Wandering Jew
Zebrina pendula
Common wandering Jew is either *Zebrina,* which is a sprawling or hanging plant with leaves that are striped white and green above and blue beneath, or *Tradescantia fluminensis,* which is a prostrate herb, rooting at joints and whose leaves are generally broader. Roots fibrous.

Zebrina is a native of Mexico but is grown commonly in greenhouses and window gardens almost anywhere; in some parts of the tropics is an escape. *T. fluminensis* is a native of Brazil, Uruguay, and Paraguay and is also a common greenhouse and window box plant.

Flowers of both species usually have six fertile stamens. In *Zebrina,* petals are united into a tube at base at least; while in *Tradescantia,* petals are free of each other or nearly so. In *Zebrina,* calyx tube is whitish and corolla white with red or purple segments. Commonest reproduction by cuttings.

Interesting to teachers of biology because epidermis of leaf peels off beautifully to show stomates, and because hairs found on the stamens show circulation of protoplasm in an excellent manner. They also survive careless treatment well and grow rapidly. Best temperature, 60 to 70°F.

If these are grown as houseplants, they may be "thickened up" by pinching off ends of stems, forcing abnormal branching. Cuttings may be rooted in water before planting in rich, moist, or water-soaked soil. Plants will grow for a long time in water.

Spiderwort
Tradescantia sp.
T. virginiana, a relatively common species, has green, flat, linear leaves, uppermost of which are somewhat downy. Bracts beneath flowers leaflike, long, usually extend at right angles from main stem, or upward. Leaves all keeled. Stem unbranched, to 1½ ft high.

More than 30 species of genus, of which *T. virginiana* and wandering Jew, already discussed, are sometimes cultivated. *T. virginiana* ranges from Maine to Virginia and west to Rocky Mountains. More common in western part of its range.

Flowers in clusters at tip of stem. Sepals thickly covered with hairs. Petals a rich, purplish-blue, to ⅘ in. long, therefore conspicuous. Flowers do not last long and wilt quickly if picked for a bouquet. Pollinated by queen bumblebees who eat pollen.

Tradescantias in general are considered hardy hothouse plants and wild forms seem able to survive difficult conditions. Railroad embankments in Middle West are often blue with their flowers in season. All are easily propagated by cuttings.

Aside from being attractive wild flowers and garden plants, spiderworts are useful in supplying material in biology classes for studying circulation of protoplasm. This is seen easily in hairs found in centers of flowers of many species. Species figured is *T. reflexa.*

Dayflower
Commelina coelestis
Height to 18 in. Stems diffusely branching, weak, more or less succulent. Leaves alternate, clasping stem at base, broad, but pointed at tip, parallel-veined, smooth, arising from fibrous or tuberous root that may or may not figure in commercial propagation. Virginia dayflower more slender, 3 ft high.

(Continued)

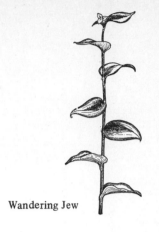

Wandering Jew

Spiderwort

Dayflower

Native of mountains of Mexico, with a considerable number of natural and horticultural varieties. About 100 species of commelinas, or about ⅓ of all of spiderwort family. Two of these are grown outdoors in pots as ornamentals. Virginia dayflower, New York to Texas.

Flowers in groups of four to ten from axils of upper leaves, or one to two flowers from axils of lower leaves. Flowers deep blue, although there are white or blue-white variations. Three sepals, three petals, and six stamens of which only three are fertile. Fruit a three-celled capsule.

Dayflowers are propagated by seeds, by cuttings of tuberous root, or by cuttings of stems. Cuttings are made in March, hardened, and then set out. Tubers may be divided when setting out in spring. Seeds may be planted in cold frame in April. Roots should be taken up and stored in sand through winter.

Cultivated commelinas all hardy, some surviving intact while others may die in severe weather; used as garden flowers or as houseplants because of attractive foliage, growth habit, and flowers; used as a potherb in Europe.

Order Liliales./Family Juncaceae

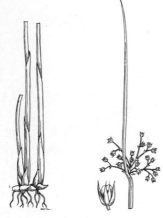

Rush

Juncus effusus
Height to 4 ft. Fruiting stalks erect, arising from creeping, horizontal rootstocks, bearing fibrous roots, and often appearing in tussocks or hummocks. Leaves small, and reduced mostly to sheathing scales. Name said to be derived from *jungo,* to join, a reference to the use of the stems and narrow leaves for binding purposes.

Native of North America and Europe, being found almost anywhere in North America where soil and water conditions are suitable. Favors areas such as low meadows and pastures, edges of ditches and waterways, and rich lowlands that may become overflowed. Particularly troublesome in Puget Sound area.

Flowers contain both pistils and stamens and have three petals and three sepals each. Fruit a greenish-brown capsule containing an enormous number of minute seeds each to about 1/50 in. long. Plants may be reproduced by spreading and eventual separation of underground parts or by seeds.

Presence of this plant indicates too much moisture in soil for most cultivated crops. Related path rush of woods may be used as an indication of location of a path. Pith of stems beautiful under a microscope and is used as an example of unique cells. This is Japanese mat rush used in "matting."

Where this plant has to be destroyed as a weed, plants are grubbed, piled, and buried or burned with impression filled or leveled by cultivation. Cultivation of crops on an infested field for 2 years should eliminate plant. Muskrats feed on roots and bases, as do moose. Ducks find shelter in rushes.

Rush

Order Liliales./Family Liliaceae

False Hellebore

Veratrum viride
Height to 8 ft. Stem stout, unbranched, covered with leaves. Short rootstock has many coarse, fibrous roots.

Found in wet areas, open or wooded, from New Brunswick and Quebec, through southern Canada and northern United States, south to Oregon and mountains of Georgia. In Adirondacks in New York it is found to an elevation of 4,000 ft. Known also as swamp hellebore, itchweed, tickleweed, earthgall.

Flowers borne in a panicle which is to 2 ft long, many-flowered, branched, with lower branches drooping. Flowers yellow-green, to 1 in. across. Fruit to 1 in. long and ½ in. thick, smooth, containing many seeds, each to about ⅓ in. long. Flowers mature in May, June, and July. Perennial.

All parts of plant are poisonous if eaten. Several plant alkaloids have been extracted from this species. Since it resembles edible skunk cabbage, two plants should be differentiated. Has a bad taste and so would not be eaten by animals normally.

Yields quickly to cultivation; will not survive drainage that would precede agricultural practices. Plants that persist can be removed by grubbing. European species, *album,* is source of a drug which functions as a heart stimulant. Some species produce an insecticide powder.

False Hellebore

DIVISION ANTHOPHYTA. CLASS MONOCOTYLEDONAE

Bellwort
Uvularia grandiflora
Height to 20 in. Stem above ground, erect or nodding, forked, smooth, round in cross section, arises from a short rootstock with fleshy roots, and bears leaves that are 2 to 5 in. long, pierced by the stem, oblong to oval, smooth above but fuzzy beneath when young.

Found from Quebec to Ontario and Minnesota, and south through Georgia, Tennessee, and Kansas, with related species extending the range. Closely related *Oakesia* has leaves that are not pierced by stem. *U. perfoliata* that is smooth throughout is sometimes called wild oat.

Flowers yellow, drooping bells, to 1½ in. long, with three distinct yellow sepals, three distinct yellow petals, six distinct stamens, each with long, narrow pollen sacs and a three-lobed, three-celled, capsulelike fruit about ½ in. long. Flowering time April and May.

Bellworts not hardy after shade-producing woods in which they normally live are removed. They sometimes form considerable colonies but rarely if ever occur in as pure stands as many other early spring woodland plants. A favorite spring flower of anyone who has observed it in its rich woods habitat.

Young shoots reported to be as good as asparagus and the roots are supposed to be edible when cooked. Recommended that they be eaten only as an emergency food, however.

Bellwort

Wild Leek
Allium tricoccum
Height to 1½ ft. Stems arise from a deeply buried, pointed bulb that is to 2 in. long. A short, fibrous rootstock may connect a number of bulbs. Leaves erect, to 1 ft long and to 2 in. wide, appearing early but drying up before flowering time. Leaves narrowed at each end.

Found from New Brunswick to Minnesota, and south to North Carolina, Tennessee, and Iowa. Often found in extensive beds of almost pure stands in rich, damp woodlands. It also grows well in gravelly or sandy pastures. It is a native plant. *A. stellatum*, prairie wild onion, has slender flat leaves and grows to 16 in. tall; with numerous pink flowers; found on rocky banks and sandy prairies of tall grass prairies region.

Reproduction by bulbs or seeds. Flowers appear after leaves of year have wilted, in June and July; on top of stem or scape, to 15 in. high. Group of flowers at first enclosed by two bracts. Flowers white, to ¼ in. long. Fruit, a capsule, to ¼ in. broad. Seeds smooth, black spheres. Pollination by bees.

Sometimes persists as weed in open, sandy fields. Has leaves with a strong leek or onion odor that if eaten by cattle may affect taste of milk and butter produced. Because of this, it is essential that plants in pastures be kept in control if not destroyed.

Occasionally, leaves are eaten by persons. They give the breath a never-to-be-forgotten fragrance that is much worse second- than firsthand. Used sparingly, they add flavor to otherwise tasteless foods. Wild onions can be planted with ferns in woodland gardens. Sometimes served like asparagus. Can be stored for winter use if dug, placed in cold frame or in cellar where soil can be kept moist.

Wild Leek

Onion
Allium cepa
Height of flower stalk to 4 ft, hollow, swollen below middle, and much taller than leaves. First-year leaves basal, hollow, developing bulb in late season. Some varieties produce new bulblets from bases of larger ones; some produce bulblets instead of seeds in flower; some produce regular flowers and seeds. Mostly biennial.

Native of Asia, probably from Palestine to India; mentioned in Bible and reported as one of things Israelites longed for when lost in wilderness. It has been cultivated in America since 1629. In acreage, leading states raising onions commercially are Texas, New York, Michigan, California, Colorado, New Jersey, Massachusetts, Indiana, Idaho.

Flowers, many, lilac or white, in large, loose cluster springing from spreading bracts. Stamens six; pistil three-celled, with seeds black, spherical. May be grown directly from seeds, by using "sets" from tops of some varieties, by bottom sets springing from underground bulb, or by sets grown from seeds in hothouses. In North, commonly grown from seeds. Those grown from sets should be used while young as they would otherwise most likely produce inferior bulbs. 1 lb of small onion sets will plant 100 ft.

In South, seed is planted in September at rate of 30 lb per acre, but 17 lb of seed per acre of hotbed should produce enough sets to plant 10 acres in rows 14 in. apart. Transplanting generally gives better results than direct growing. Large sets, to 1 in. in diameter, may produce a yield of 94 bu to the acre, while small sets under ½ in. may yield but 9 bu.

(Continued)

Onion

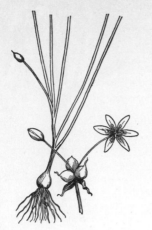

Wild Onion

Garlic

Tiger Lily

Most important American bulb crop. Food value is high and popularity of onion as food is well established and sustained, so there is a dependable market. When harvested, plants are pulled; bulb is allowed to dry and cure for several days in the sun. Dirt is then brushed off and onion is ready for storage in cool, dry place. Some favorite varieties are Ebenezer, Early Yellow Globe, Southport Red Globe, Sweet Spanish, and White Portugal or Silverskin.

Wild Onion, Meadow Garlic
Allium canadense
Height to 2 ft. Bulb, usually one to two in a place, to 1 in. high, covered with a fibrous net. Leaves linear, remaining active through time of flowering, arising from basal bulb, flat or flattish, to 1½ in. wide and to 1½ ft long. All dark green, and with strong onion taste.

Found in meadows and brushlands, from New Brunswick to Minnesota, south to Florida, Louisiana, Texas, and Colorado. In Virginia may be found at an elevation of 2,500 ft. Essentially an early summer plant.

Reproduces by bulbs, bulblets, and seeds. Bulbs divide underground. Bulblets appear at end of erect stem or scape where flowers would be found; are more common than flowers. Flowers white or pinkish, with stamens just enclosed by surrounding parts. Pollination by bees, butterflies, flies, and moths.

Since this plant if eaten by cattle affects milk and butter produced, it is necessary to keep it under control or eliminate it in pasture lands. Ingestion of large amounts of plants of this species has caused death in cattle. It does not withstand competition of cultivated crops and so can be controlled relatively easily. Bulblets are interesting to botanists.

Indians formerly harvested these plants for use through the year. Green tops may be eaten raw or may be used to improve the flower of soups, meats, and salads. They may be cooked in several changes of water if the flavor is too strong to be acceptable to the eater.

Garlic
Allium sativum
Height to 1 ft or less. Leaves narrow, or keeled, not round in cross section as in the onion, under 1 in. wide, long-pointed. Bulb composed of several parts enclosed in a silky whitish or flesh-colored sac. Unique in group and is common product in grocery stores so it should be well-known.

Native of southern Europe, but grown in gardens rather widely, particularly where peoples of Mediterranean extraction are to be found. Europeans have grown it for more than 2000 years. Popular varieties are Italian, Tahiti, Creole, and Mexican. California is most important producing state. Soil favored is an ordinary garden soil of loose, relatively rich loam.

In gardens, reproduction is primarily by planting the little bulbs, or "cloves," or "bulbils." If these are set out in spring, they mature in early autumn; if in the fall, they mature in spring or summer. Bulbils replace flowers, which are purple.

If soil is too rich, plants may produce too much of relatively useless top and too few valuable bulbs. If top growths are broken down when they reach maturity, this situation may be helped. Cloves are set in drills from 4 to 6 in. apart.

Garlic is for sale in stores in strings of bulbs or in separate cloves. It is used in flavoring meats, soups, and salads, the amount depending on choice of eater and artistry of cook. It can "break" a meal as easily as "make" one. Garlic salt is made from dehydrated cloves. Garlic has antiseptic and bactericidal properties.

Tiger Lily
Lilium tigrinum
Height to 4 ft. Stem purplish-brown, covered with cobwebby substance. Leaves 75 to 100, crowded, each 4 to 6 in. long, more or less clasping at base, narrow, straight or curved, and with bulblets in their axils. Bulb to 4 in. in diameter, spherical, white or yellow, or tinged with purple or red.

Native of China and Japan but naturalized as an escape in Maine and New York. Several natural and cultural varieties, one of them having double flowers, some being paler in color of flowers; some being larger in every way.

Flowers horizontal, or slightly declined or drooping, opening wide, without a tube remaining, orange-red or salmon-red, with black spots, and with apparently long pistils and stamens. From 1 to 15 flowers on a plant, each 3 to 5 in. in diameter and with conspicuous red anthers on the stamens. Lower buds open first.

One of easiest of all lilies to grow. Hardy, survives neglect, and yet is as beautiful as most of its relatives. Grown in old-fashioned gardens and shown in pictures as typical of the fence in a New England home. Bulbils from leaves produce flowers in 3 to 4 years. Plant fall or spring, 10 in. deep in rich, loose soil.

Bulbs planted in June should produce flowers about Oct. 15. Those potted in July may give flowers at Thanksgiving time. Recommended that bulbs be planted 6 to 8 in. deep in late autumn or in the spring. Potted tiger lilies need much fresh air and comparatively little water to thrive. Bulbs formerly imported from Japan seemed to be among the best. Popular forms of the tiger lily are splendens, which has salmon-red flowers with purple spots, is 4 to 5 ft high, and flowers in August; fortunei, a giant 5- to 7-ft tall plant, which produces 20 to 30 large flowers on each stalk and is salmon-orange spotted with purple and flowers in August and September; and planescens, the double-flowered tiger lily with orange-red flowers which appear in August and September on shorter stems, 3 to 4 ft high.

Wild Yellow Lily
Lilium canadense
Height to 5 ft. Stem slender, smooth, light green, with leaves in whorls of four to ten, there being four to eight whorls, leaves being to 4 in. long and to ¾ in. wide. Usually no leaves on lower part of stem. Bulbs borne an inch or so apart, on a stout rhizome, each to 1½ in. in diameter.

Found in meadows and open woodlands, from New Brunswick to Ontario and Minnesota, south to Missouri, Georgia, and Alabama. Probably fewer than 100 species of genus *Lilium,* but most of them are beautiful graceful flowers, too attractive for their own good in the wild. Common also in bogs. Thrives on acid soils, but does equally well on neutral soil.

Flowers nodding on ends of stems arising from uppermost whorls of leaves. From one to ten flowers per plant, each to 3 in. long or wide, bright orange-yellow, with purple-brown spots, with red anthers on stamens. Pollination mostly by honeybee and leaf-cutting bee. Blooms from late June through July. Perennial.

Probably commonest of wild lilies growing within its range, and, because of its beautiful, nodding flowers, always popular. In related Turk's-cap lily lower leaves are whorled but others are crowded along stem, and outer flower parts bent farther back.

Fleshy bulbs of these lilies were used by Indians to thicken soups and were eaten cooked. However, now no reason for destroying these beautiful flowers merely for sake of eating them. Can be grown successfully in cultivation; does well in sun or partial shade, or in moist well-drained situation. Plant the bulbs 6 to 8 in. deep in a compost of sandy peat and leaf mold or lime-free loam.

Madonna Lily, Annunciation Lily
Lilium candidum
Height to 4 ft. Stem erect, deep green, smooth, bearing from 60 to 100 leaves that are to 9 in. long and to 2½ in. wide, lower ones being larger and appearing horizontally in fall; shortest being just below flowers. Bulb globular, white to yellow, to 4 in. in diameter.

Native of southern Europe and southwestern Asia, from Corsica to Caucasus Mountains and Iran. This lily is widely grown by florists and is one of the commoner flowers sold in pots at Easter. Believed to be lily referred to in the Bible. Commonest outdoor white lily cultivated.

Flowers held horizontally or nearly so; trumpet-shaped, widening upward; three to twenty, each to 9 in. long, to 2½ in. wide, with yellow anthers; tube, pure, waxy white. Fragrance delicate and pleasing. Anthers golden. Seeds germinate in 21 days, live about 2 years but require several years to reach flowering stage. Blooms during June and early July.

For Easter, bulbs are potted in December in heavy, fibrous, slightly acid soil. Individual bulbs set in 5- to 6-in. pots. First months 54 to 56°F, then raised gradually to 60°, this bringing flowers in 13 weeks. Usually, 6 weeks required after buds show; though at 75° this may be shortened to 2 weeks. August and September are the best times for transplanting. Several forms of Madonna lily are in cultivation.

L. candidum is used for Easter when season is late; *L. longiflorum,* when it is earlier. *L. candidum* is popular with florists for Mother's Day and for Memorial Day. *L. regale* may be brought to flower in 65 to 70 days with first 10 days at 70° and the remainder at 60°F. Plant lice and a rust are worst pests.

Wild Yellow Lily

Madonna Lily

Day Lily

Tulip

Sego Lily,
Mariposa Lily

DIVISION ANTHOPHYTA. CLASS MONOCOTYLEDONAE

Day Lily
Hemerocallis fulva

Height to 6 ft. Leaves basal, to 1 in. wide and to over 2 ft long, sword-shaped, erect, but not reaching higher than flower-bearing stem. Roots coarse, fibrous, producing new shoots that form a pure stand even against competition. Generic name from *hemera,* a day, plus *kallos,* beauty; an allusion to the short life of the bloom.

Found from France to Japan and probably was introduced into Europe early. In America, found commonly from New Brunswick to Virginia and Tennessee as an escape and probably generally wherever flower gardens are maintained. Now established either in cultivation or as an escape in the 48 mainland states. Needs no winter protection.

Flowers short-lived, fading and closing at end of a day; borne at top of erect flowering stems, with 6 to 12 flowers on a single stem, not fragrant, orange-red 3 to 5 in. long, with a slender 1-in. tube. Fruit oblong, three-celled, three-angled, and many-seeded. Perennial. Plants flower from May to September. An increased stock of plants may be prepared by digging clumps in spring or fall and separating the bulbs and replanting.

In general, orange day lily resembles yellow day lily except for color of flowers. Yellow day lilies have narrower leaves. All are unusually free from insect and fungus pests and survive conditions of cold, drought, and flood. Divide every 4 to 5 years.

These lilies mark last evidence of many an early homestead and their presence in quantity usually means that nearby is the foundation of an old home. They are used for quick effects in new gardens but are sometimes difficult to keep in control or to eradicate.

Tulip
Tulipa gesneriana

Height to 2½ ft. Leaves, three or more, on lower part of stem, to 4 in. broad and 1 ft long, smooth, parallel-veined, with entire margins, uniformly green and usually with edges curved upward in early stages at least. Leaf surfaces vary in different varieties.

Native of Russia and Asia. Developed for garden purposes extensively in Holland and Denmark and imported into America for a long time. Possibly was first grown in Europe in Austria in 1554, in England in 1577. Now one of most popular spring garden flowers. Some 60 species and many more varieties.

Flowers, one to a plant, erect, 2 in. or more high, and, when open, twice that width, not fragrant. Typically, flowers are scarlet with a blackish-purple center, but may be yellow, or white, or almost black. Anthers purple or yellow. Fruit a capsule that produces many seeds. Six "petals" are actually three sepals and three petals in two whorls.

Best known of fancy varieties are Darwin tulip, a tall, late-flowering variety with deep colors; and parrot tulip, with segments of flower crisped or wrinkled around edges. In open, tulips are planted in fall, mulched with straw, and should bloom by spring.

Tulip bulbs may be potted in the fall and stored in a cool place to get roots established. This should take 8 to 12 weeks at 45°F. Then, forcing at 70 to 80°F for 3 weeks should produce blooms, or at higher temperatures, blooms may be forced in 10 to 14 days. Lowering temperature holds flowers. In some localities, rodents feed on bulbs. To prevent this, wire mesh is used beneath the bulbs and to the side and above the plantings. Early in the spring the surface layer of wire mesh is removed. Tulip bulbs left in the ground should be protected with compost or manure in places where winters are severe.

Sego Lily, Mariposa Lily, Globe Tulip
Calochortus nuttallii

Height to 2 ft. Stem stiff, erect, and arising from a bulb base. On erect stem, usually only one leaf. Leaves ashy green and surprisingly small in comparison to size of flower. A vigorous underground system in which nourishment is accumulated and used at time of flowering. *C. splendens,* lilac mariposa, is found on dry stony hills, often in low brush called chaparral in the coast ranges of California. Stems of this species are branched with a flower at end of each branch.

Found from the Dakotas and Nebraska, west and south to New Mexico and California. Other species have a more limited range and are less well-known. Especially characteristic of sagebrush deserts such as are common in Great Basin.

Flowers to 2 in. long, white, with greenish-white or a lilac tinge, and with a purple spot or band above the yellow base. From one to five flowers on a single stem. Mariposa lily, *C. venustus,* common in some Western national parks, is a close relative.

Most members of genus can withstand severe cold but do not prosper where there is much freezing and thawing at short intervals. Globe tulips of gardens are members of this genus as are meadow tulips and butterfly tulips. Sego lily is the state flower of Utah.

Siberian Squill

Underground portion has been used as food, and early Mormons in Utah found it valuable in this respect. As garden flowers or houseplants, blooms can be forced in the spring by potting the plants in the fall and keeping them growing slowly at a temperature of 50°F.

Siberian Squill
Scilla sibirica
Height to 6 in. Flowering stalks, one to five rising from buried, scale-coated bulb. Leaves rather succulent, to 5 to 6 in. long and ¾ in. broad, blunt. *S. peruviana* reaches to height of 1 ft; English bluebell, *S. nonscripta,* to 1 ft; Spanish bluebell, *S. hispanica,* to 20 in.; *S. amoena,* 6 in.

S. sibirica comes from Russia and southwestern Asia; *S. peruviana,* from the Mediterranean region; *S. nonscripta,* from Great Britain and vicinity; *S. hispanica,* from Spain and Portugal. Star hyacinth, *S. amoena,* comes from mid- and southern Europe, as in Germany and Italy.

Flowers of *Scilla* have only one nerve in each of the conspicuous segments instead of three as in *Camassia.* They appear early, with the leaves. *S. peruviana* has over 50 flowers, while others rarely have to 20. *S. sibirica* has three or fewer flowers to a flowering stem; *S. amoena,* 4 to 6; and *S. nonscripta,* at least 12. Most effective as a decorative plant if planted in masses; also effective as a border planting for paths.

Squills are grown mostly for their early appearance and their hardy qualities. Siberian squill blooms just for a few days; English bluebell may bloom from April to June. For most part, they need not be disturbed for years though a good top dressing in fall may do no harm. Bulbs of *Scilla* are quite large and when planted outdoors should be set at a depth of at least 6 in.

Scillas may be raised or forced in greenhouses by putting a half dozen bulbs in a 5-in. pot with soil, leaving them cool until growth starts, and giving them little water. When growth has started, temperature may be raised and amount of water increased.

Camass
Camassia quamash
Height to 3 ft. Stem erect and arising from a coated bulb, with leaves arising from base. Leaves under ⅔ in. wide, but to 2 ft or even more long. Some forms are much lower; some are slender, while others are stout. Bulbs ordinarily produce no offsets.

Camass

Native of North America, where there are five to six species ranging from British Columbia and central California to Arkansas and Texas. *C. quamash* found from California to Utah, and north to British Columbia. Favors meadows that are very wet in winter and spring and very dry in summer. *C. esculenta,* wild hyacinth, with blue or whitish flowers grows along streams and in open fields in eastern half of the U.S.

Flowers, 10 to 40 on a single stalk, with white to blue, showy parts, each segment being three- to five-nerved rather than one-nerved as in *Scilla.* Pedicels or individual flower stems, shorter than flowers, and showy parts of flower not usually evenly arranged.

Camassia may be cultivated in ordinary garden soil, although it normally thrives best in heavy soil that has much moisture in early season. Planted 3 to 4 in. apart, 6 in. deep early in fall. Seeds develop into plants bearing flowers in 3 to 4 years. Excellent as cut flowers.

Camass bulbs formerly constituted a major portion of plant foods of the Vancouver Indians. Bulbs collected and stored are either boiled, roasted fried, or cooked in pies with other food. A reasonably good molasses may be made from camass bulbs. Collecting is done in spring and summer. Known to the Indians as quamash, now its species name.

Squill, Sea Onion
Urginea maritima
Height to 3 ft. Flowers appear before leaves. Leaves come from underground bulb that is unusually large, being to 6 in. through and rather fleshy, smooth, to 1½ ft long and to 4 in. wide, somewhat shining; reaches a weight of 15 lb.

From the Canary Islands to Syria, and in South Africa. Is grown extensively in homes in southern Europe and can survive long periods without water. About 75 related species, ranging from Europe to India and through tropical Africa to South Africa.

Flowers bloom in summer or autumn in England, appearing sometimes as early as July but always before leaves. From 50 to 100 flowers on a common stem, each ½ in. across, whitish, with greenish-purple areas present. Seeds strongly compressed, winged, borne 10 to 12 in each section of fruit.

Sea onions are not hardy outdoors in the North and are not grown much in America. Confused with *Ornithogalum caudatum* under common name of sea onion, and with *Scilla* under common name of squill.

(Continued)

Sea Onion, Squill

Grape Hyacinth

Hyacinth

Bowstring Hemp

Bulbs contain to 22 percent sugar. Used in making croup medicine, as cathartic, as an emetic. Also used in rat poison that seems to have a fatal effect only on rats because poison cannot be eliminated by vomiting among species of rats. Flowers should have greenhouse popularity, since they last long.

Grape Hyacinth
Muscari botryoides
Flower stem to 1 ft high, in center of cluster of six to eight leaves that spring from a small, egg-shaped bulb and are 1 ft or more long and ⅓ in. wide, dark rich green and grasslike. Name refers to resemblance of flowers to a bunch of grapes and similar fragrance.

Native of southern Europe but planted widely in flower gardens and naturalized in many nearby places. Seems to do well in any kind of soil but best in ordinary garden soil of a rich, loose loam, in the sun or not. Most frequently planted in borders in clumps.

Flowers appear in early spring, outdoors or indoors, may be made to flower in late winter. There are 20 to 40 tiny blue urn-shaped flowers to a stalk, each with six petallike parts, about ⅛ in. long, the uppermost sterile. Commonest method of reproduction is by division of bulbs. The most brilliantly colored of all grape hyacinths is *M. armeniacum*, Heavenly Blue, which may produce grapelike clusters of bright blue flowers on stems 8 to 9 in. tall. There are blue, white, and pink forms of *M. botryoides*.

Indoors, plant six to eight in a flat 6-in. pot; cover with ½ in. of soil of ⅓ sand, ⅓ rotten manure, and ⅓ garden soil. Set this away for 2 weeks, in the dark, at a temperature of 60 to 70°F until started, then bring into light and warmth. Outdoors, bulbs are set 1 to 2 in. apart in fall for early spring flowers. In gardens, bulbs are set in masses rather than singly.

Popular, hardy garden flower of paths and rock gardens. Garden-set plants should be reset every 3 years to avoid undue crowding of new bulbs that appear as offsets from old.

Hyacinth
Hyacinthus orientalis
Flowering stem to 1 ft high, smooth, hollow, succulent, arising from a large 2-in. sheathed, nearly spherical bulb. Leaves 8 to 12 in. long by ½ to 1½ in. wide, many-nerved, lengthening after flowers mature. Roots fibrous, developing from base of bulb.

Native of Asia Minor, Greece, and northern Italy, but developed for gardening mostly in the Netherlands and similar western Europe territory. Roman hyacinth is smaller, early, and comes from southern France. All favor deep, rich, well-watered soils.

Flowers in many colors, double or single, bell-shaped, drooping with a swollen base and spreading segments, 30 to 40 to a stalk, to 1½ in. across, unusually fragrant; with three colored sepals, three similarly colored petals, and six stamens, of which three are longer than the other three. Perennial.

Favored varieties include for single flowers the blue Baron Van Thuyll, the white Alba Maxima, the red Florence Nightingale, and the yellow King of the Yellows. Double flowers include the yellow Bouquet d'Orange, the blue Charles Dickens, the white Prince of Waterloo, and the red Bouquet Tendre.

Bulbs received in September may be potted one bulb to a 4-in. pot, with soil ⅓ garden soil, ⅓ manure, and ⅓ sand, about ½ in. below surface, watered and set under straw or kept at 45 to 50°F. May be forced in 2 to 4 weeks at 75 to 80°F with good moisture, then cooled down to 50°F for longevity. Bulbs can be planted outdoors in autumn between the middle of September and the middle of October. Should be planted so that tip of bulb is 3 to 5 in. below the surface of the soil and set from 6 to 8 in. apart. Grown in France for making perfume.

Bowstring Hemp
Sansevieria sp.
Leaves 8 ft long, to 3 in. wide, springing directly from root, since plant has no stem above ground. Leaves rounded on back, with dark green, longitudinal lines on back and lighter green, transverse lines on inner face. A number of varieties differ by color and banding of leaves.

Some 50 species, for the most part native of Africa and Asia, although many are suitable for cultivation in North America and are to be found as houseplants, since they survive abuse and lend themselves to limited floor space and exposure to the sun. Named after a Prince of Sanseviero, who was born in Naples in 1710.

Flowers whitish to yellow, in an open cluster. Tubes swollen at base, with six stamens attached to throat. Fruit is a berry, with one to three seeds. Occasionally plant will flower as a household plant, in which case its fragrance is most welcome. Closely related species include cylindrical-leaved *S. cylindrica* and flat-leaved *S. thyrsiflora*.

In India, Queensland, Java, and South China, plant is grown for its fiber, which is soft, pliant, silky, very strong, and easily extracted by machines. It is uniformly good throughout length of leaves. 1 ton of leaves will yield 50 lb of fiber, with 1 acre of land yielding 13½ tons of fresh leaves. Propagation is simple and by roots.

Known in many homes as "snake plant." Also known as "murta" or "moorva." Probably commonest potted plant in hotel lobbies of United States, and next to geranium, ranks high as a houseplant. Most commonly cultivated in the Bahamas. Slightly more difficult to handle as a commercial crop than sisal is.

Asparagus

Garden Asparagus
Asparagus officinalis
From 4 to 10 ft tall, with spring shoots succulent and thick, and summer growth slender and tough. Leaves reduced to scales that are pressed closely to stems, with slender stems assuming role of leaves in mature plant. Stems smooth, much-branched. Roots of many cordlike fleshy parts spring from a more or less common center. Perennial.

Native from Britain to central Asia and Africa, where it grows wild on coasts and in sandy regions. Known as a food in Europe for 2,000 years; grown in U.S. since colonial days. Escapes cultivation in America and may be found in a variety of places. Does best on deep, loose soil; sandy loam best for early crops; and should at all times be well-drained since plant cannot long survive submergence. Usually planted 8 in. deep.

Plants bear either staminate or pistillate flowers, rarely those with both stamens and pistils. Flowers bell-shaped, borne in groups of one to four, in axils of leaflike stems. Staminate flowers yellow-green, ¼ in. long; pistillate, smaller, and with three longish stigmas. Fruit a berry, under ½ in. through, turning from green to red and bearing about three seeds.

If soils are acid, liming may be necessary. Doubtful if use of salt in controlling weeds is desirable. Seeds germinate in 2 to 6 weeks at 86°F; may be soaked to 3 days. Staminate plants yield 25 percent more crop, but pistillate yield larger spears, a year-old staminate plant yielding 26 spears, first spears appearing, in East, at 2 years. Should produce for 20 years with good care. Enemies, rust and beetles.

May yield over 2 tons per acre. Food value of shoots is decreased with removal of summer tops. Grading, washing, packing, bunching increase price. Fresh shoots, preferably green; those to be canned, often blanched. Fresh shoots toughen rapidly first 24 hours, unless cooled before shipping at ordinary storage temperatures. Sold packaged as frozen form. For long-distance markets, asparagus spears are packed in bunches, placed in crates, and shipped under refrigeration.

Asparagus Fern

Asparagus Fern
Asparagus plumosus
Height to over 4 ft. Stem thin, twisting, smooth, with a few short spines, drooping. Leaves flat, green, leaflike branchlets, 1 in. long and ¼ in. wide in clusters around slender stem. Root system with short white tubers.

Native of South Africa. Many of its varieties are purely horticultural and have been developed by greenhouse methods. There is a dwarf form, *nanus;* a compact form, *compactus;* a robust form a climbing form, and so on.

Flowers in loose clusters of one to four, at the ends of twigs, pinkish or whitish, with calyxlike corolla, and each composed of three segments that are united into a bell-like form. Fruit a berry, ¼ in. across, bright red, and with one to three seeds that germinate in 3 weeks.

Commonest method of reproduction, by dividing tubers so that each section has one or more "eyes." Plants are set in dark in sandy soil and given liquid manure. At 60 to 70°F they grow rapidly. Seeds should be soaked for 24 hours before being planted, and in a year should develop large plants. Cuttings about 4 in. long can be taken from side shoots of old plants in April or May, planted in pots filled with a mixture of wet sandy loam and leaf mold, and covered with a bell jar until they root.

Plants are started in summer; ends are pinched off to thicken vegetation. Ammonium sulfate at rate of 1 oz to 2 gal of water given twice a week stimulates growth. Must be kept in shade in warm hot months. Greenhouse white fly is bad pest. Always a ready market for this plant.

Lily of the Valley

Lily of the Valley
Convallaria majalis
Height to 1 ft, but usually less. A slender rootstock branches profusely and gives rise to leaf clusters that rise in 2s. Each leaf long, elliptical, with the base lengthened to a petiole, smooth, not toothed or cut on margin, parallel-veined, uniformly dark green.

(Continued)

Native of Europe and Asia, and in the mountains from Virginia to South Carolina. Escaped from cultivation in many places and appears as a survival in abandoned homesites, particularly in the East. Does best in shade in relatively dry spots and often forms pure stands.

Flowers attractive white bells, nearly ½ in. across, borne commonly more to one side of the erect or drooping flower stalk, each stalk bearing a few to many uncrowded flowers. Fruit a red, few-seeded berry. Petals like the sepals, and there are three of each. A most delicate fragrance.

If plants are potted, they are set in equal parts of sand, garden soil, and humus, using 8 to 10 pips in a 6-in. pot. Potted plants are kept in dark for first 2 weeks and are then brought into full sun and a temperature of 65 to 70°F. Pots are probably best set in water. Pips can be planted outdoors in autumn; when established they spread rapidly and become crowded. When crowded, they do not flower so abundantly, so the plants should be lifted from the bed and separated and replanted in larger areas.

Flowers characteristic of old homesteads of longer settled parts of country and cherished by some as belonging to bridal bouquet. In greenhouses, may be forced to bloom at almost any time but commonest in January and February. Plants outdoors usually bloom in June. Yield the poisons convallarin and convallamarin, both glucosides.

Plantain Lily, Funkia
Hosta plantaginea
Height to 3½ ft. Flower stems tower above spreading foliage. Leaves to 10 in. long and to 6 in. broad, long-petioled, with prominent ribs, strong side ribs in oval blades of leaves, and a short sharp point at end. Leaves may be large ovate to lance-shaped, green or variegated. Known to many as funkia. Root strong and cordlike.

Native of China and Japan but introduced and well-established widely in the United States and Canada as a hardy plant for the flower garden. Group might well be represented also by the blue-flowered plantain lily, *Hosta caerulea*, whose leaves are greener and more shining. Seeds flat, black, winged at end.

Flowers to 4 in. long, trumpet-shaped, showy, waxy white, fragrant, appearing above foliage. Showy segments, six. Stamens, six and long. Flowering period through August into September, with pollination effected by insects. Blooms second year from seeds and is perennial.

Prefers rich, deep, moist soil in partial shade; easily sunburned or destroyed by frost. Seeds sown as soon as ripe. More common to divide root system in fall or early spring. Water through dry weather. Mulch lightly with well-decayed manure after ground freezes. Remove mulch in spring after frost danger has passed. Relatively free from insect pests.

One of more desirable of old-fashioned garden flowers, the appearance and fragrance of which somehow mean midsummer to all who have sought the odor of orange issuing from the delightfully attractive though short-lived flowers. Unusually fast maturing from seed, when compared with many of its relatives. Attractive, perennial border plant which does well in moist, shady place.

Aspidistra, Parlor Palm
Aspidistra elatior
Height to 20 in. Leaves arise from long rhizome that is ringed with finer roots. No tall flower stalk. Leaves, many, to 24 in. long, to 4 in. wide, narrowed at both ends with a smooth, unbroken margin, parallel-veined; sometimes variegated and most attractive.

Native of China, Java, and Japan, with three to four species represented in that area and up into the Himalayas. It needs little sun, stands neglect as a houseplant well, and is generally popular in hotels and in greenhouses. Will survive outdoors in shady places where winters are mild.

Flowers borne singly, at surface of ground, beneath leaves, and often are not seen even though they are produced. Three sepals and three petals, all colored alike, a purple-brown, and all united into a tube. Fruit, four-celled.

In common practice, plant is propagated by separating underground parts. It does best at temperatures between 60 and 75°F. It needs only a moderate amount of water and has few insect or fungus pests. Aphids sometimes develop but they are easily removed.

The plant's ability to withstand fumes from gas stoves, to survive dry air, dust, and small amount of sunlight of homes and hotels makes it a popular houseplant. *A. variegata* must have poor soil to do well. Sometimes called "cast-iron plant" from ability to withstand hard usage.

Plantain Lily, Funkia

Aspidistra

St. Bernard's Lily
Anthericum liliago

Height to 3 ft. Flowering stem unbranched and usually leafless. Leaves, many, to 1 ft long, and very slender. Underground parts, a tuberlike rhizome, bearing numerous, smaller, branching roots; underground or ground surface branches or stolons help in reproduction.

Native of southern Europe and northern Africa. Widely grown as outdoor ornamental largely because of ease with which it is maintained. More than 50 different species in genus and 100 in closely related genera found in Europe, Africa, and America but mostly in Africa.

Flowers star-shaped, whitish, not showy, about ¾ in. across, borne loosely on long, woody stalks. Petals and sepals alike totaling six, usually with greenish tips. Fruit round and three-celled. Seeds, three-angled.

Needs little light. Does best where temperature remains around 60 to 70°F. Has few insect or fungus pests, so needs little care. Little water needed, but should be available in generous quantities when plant is in full bloom. If grown outdoors, needs winter protection with a mulch or something similar.

A relatively common garden plant. New plants are usually made either by bringing one of stolons to ground, letting it take root, and then cutting it off, or by cutting off a bunch of small leaves at end of a natural stolon and giving it independence.

St. Bernard Lily

Adder's Tongue, Trout Lily, Dog-tooth Violet
Erythronium americanum

Height to about 10 in. Leaves arise from deeply buried bulb. Young plants bear one leaf; mature plants, two. Leaves to 1 ft long, green, or green with purplish-brown markings, pointed at each end, with unbroken margins, smooth, usually curving toward upper surface.

Found from New Brunswick to Florida, west to Arkansas and Minnesota, with close relatives extending range; one found on West Coast. Favors moist woodlands, near streams, and may persist for some years in open after forests have been cut. Some species prefer open country to woodlands. Prefers a neutral or very nearly neutral soil.

Flowers produced seventh year after seed has been shed; first 2 years, little showing, and, in intervening years, only one leaf per plant. Many offshoots are formed from bulb and help in multiplying the plant. Flowers yellow, nodding, with segments to 1½ in. long.

E. denscanis, the dog-tooth violet of Europe, as well as the native species, may be planted in wild gardens and in rock gardens. It is smaller than this species. Pollination by bees such as queen bumblebees, by cabbage butterflies, sulfur butterflies, and by some flies of the early season.

Bulbs edible, either raw or boiled. Beauty of flowers is such that bulbs should be eaten only in emergency when other food is not available. If flowers must be picked, leaves should be left, since loss of leaves and flower in a single year may destroy a plant that took 7 or more years to mature. Bulbs, which are smooth and egg-shaped, should be purchased from a nursery since in their natural state they probably are at a depth of 6 to 15 in. and so too much damage may be done in digging them. Suitable for groups in rock gardens and for planting in woodland gardens. In gardens, well-drained soil should be provided.

Adder's Tongue,
Trout Lily

Clintonia
Clintonia borealis

Height to 15 in. Leaves rising from base only, commonly two to five, but usually three, oval, thin, to 3½ in. wide and to 10 in. long, blunt-pointed. Rootstock slender, branching, and giving rise to erect leaves and their clusters.

Found from Newfoundland to Manitoba south to North Carolina and Wisconsin, and to an elevation of 4,500 ft in Virginia. Usually in moist woodlands and brushy bogs, frequently in acid soil. There may be many plants in a restricted area if they have not been destroyed.

Flowers drooping, greenish-yellow bells, in loose clusters of three to six, on slender, upright stems, reaching to height of 15 in. Each flower to nearly 1 in. long, and relatively inconspicuous largely because of color. Fruit a blue, many-seeded berry, to ⅓ in. in diameter. Flowers in May and June.

A plant of the wild places in much of its range. A shade lover. Has been called such names as beartongue, cowtongue, dogberry, wild lily of the valley, and Clinton's lily. Its scientific name is from Governor DeWitt Clinton of New York, an early naturalist.

This is a plant that poet and nature lover instinctively like. The poet speaks of "the rare blue of Clintonia berries" and says more to the nature lovers than all the academic description to be found in a botanical manual. It is reputed to have some medicinal properties. Chipmunks eat the berries. If planted in a woodland garden, they require a shady location and a soil which is loose, such as a mixture of loam, peat, and leaf mold.

Clintonia

False Lily-of-the-Valley

False Solomon's Seal

Solomon's Seal

False Lily of the Valley, Beadruby, Canada Mayflower
Maianthemum canadense

Height to 7 in. Stem erect, arising from a slender rootstock, zigzag, and bearing one to three leaves, but usually two. Leaves to 3 in. long, broadly egg-shaped, but pointed at free end and heart-shaped at base, base sometimes enclosing stem. Stem and leaves smooth or finely hairy. Name *Maianthemum* is Greek for Mayflower; from *maios,* May, and *anthemon,* blossom.

Found from Newfoundland to Northwest Territory and south to North Carolina, Tennessee, Iowa, and the Dakotas, reaching an elevation of 5,000 ft in Virginia. It favors moist woodlands and brushlands, often where soil is poor but more commonly where it is loose and deep. There is a species found in the Pacific Northwest.

Flowers, small, with segments only about $\frac{1}{16}$ in. long. Showy part of flower (perianth) in four parts. Stamens, four. Ovary and fruit, two-celled. To 12 flowers on a common axis. Fruits pale red, speckled berries appearing in summer. Flowers appear from May to July.

The name means Mayflower. Name of beadruby, refers to the fruit, and one leaf refers to the fact that young plants bear only one leaf. It is a dainty flower, fortunately too small and insignificant to invite the average flower collector to add it to a bouquet.

This plant is used sometimes in table garden terrariums and occasionally in shaded rock gardens where a touch of green or some red jewel fruits are needed. Plant lends itself to care in such a garden but is not so hardy as other plants similarly used. Suitable for growing in shady places or shrubberies as well as in woodland gardens. Should be planted in early fall. Pieces of slender rootstock are planted 3 in. apart and 1 in. deep.

False Solomon's Seal
Smilacina racemosa

Height to 3 ft. Stem slender, arched, slightly or strongly angled, finely haired above, arising from a thick, fleshy, horizontal, rough rootstock, and bearing numerous oval or pointed, elliptic leaves, without petioles, usually to 6 in. long and to 3 in. wide, conspicuously parallel-veined.

Ranges from Nova Scotia to British Columbia, south to Georgia, Missouri, and Arizona, reaching an elevation of 2,500 ft in Virginia. Native not only to North America but also to the Himalayas and eastern Asia, including Japan. Found in moist, open woods, usually where there is a good rich, rather deep soil, but sometimes on thin soil of hillsides where there is sufficient moisture.

Flowers borne in a many-branched cluster at tip of stem, crowded but not into a compact head or spike. Flowers to $\frac{1}{8}$ in. broad, whitish, appear in May to July. Fruit to $\frac{1}{4}$ in. in diameter, a red, fragrant berry commonly speckled with purple. Persists to fall.

Name false Solomon's seal comes from superficial resemblance of plant to Solomon's seal. In this plant, however, flowers are borne at tip of stem instead of along it, and underground part is rough and lacks regular leaf scars that give Solomon's seal its name.

Berries edible either raw or cooked. Root has been used in home remedies but it is not recognized officially as having great medicinal properties. Plant is a thing of beauty, either in flower or in fruit and helps make a woodland a place of joy for the nature lover. May be planted in a woodland garden in a shaded, protected location. Soil must be moist and not too heavy. Can be grown in greenhouse but probably best left in its natural woodland location.

Solomon's Seal
Polygonatum biflorum

Height to 3 ft or more. Stem wandlike, slender, smooth, zigzag in the upper parts, arising from a horizontal, thick rootstock, that bears on upper surface seallike scars of stems of previous years, and on lower side, rather coarse, sparse roots. Leaves to 4 in. long and to 2 in. wide.

Found from New Brunswick to Ontario, to western Michigan south to Florida and Tennessee, with closely related species extending the range. Found mostly in damp, open woodlands where soil is rich and either deep or well supplied with moisture in spring months.

Flowers borne in clusters of one to four, hanging on short stems from axils of leaves, each one to $\frac{1}{2}$ in. long, greenish, and relatively inconspicuous. Fruits dark blue berries, to $\frac{1}{4}$ in. in diameter, persisting until late in season unless eaten by some animal.

Name springs from seallike impressions made on rootstock by scars of earlier stems. Rootstock branches rather freely, and it is interesting to trace the development of a group of plants by uncovering the rootstocks and counting the years' growth.

Early spring plant has been eaten, either boiled like asparagus or raw. Indians have eaten starchy roots and were said to cultivate the plant for this purpose, although this is doubtful. Early French colonists in America were said to have eaten the roots apparently only to avoid starvation. Solomon's seal makes an excellent pot plant for a cool greenhouse or window garden; however, probably most appreciated in its natural woodland habitat.

Indian Cucumber
Medeola virginiana
Height to 2½ ft. Stem erect, arising from a horizontal, whitish, edible rootstock that is to·3 in. long. Stems relatively slender, stiff, and bearing whorls of leaves at remote intervals. Leaves to 5 in. long, to 2 in. wide, those of lower whorl being much larger than those of upper.

Found from Nova Scotia to Minnesota, south to Florida and Tennessee. Only one species in genus and this is confined to eastern North America. May be relatively abundant but never too much so. Found in rich, open woodlands, where there is usually a deep, loose soil.

Flowers, sometimes described as "spiderlike," borne on short, inch-long stems from axils of leaves of uppermost whorls, and are held, either erect or drooping, in flower or in fruit. Exceptionally long styles, and segments of conspicuous part are nearly ½ in. long. Fruit a dark purple berry, ½ in. in diameter.

This perennial plant is too pretty to destroy for the delicious morsel cucumber-shaped and cucumber-flavored rootstock it provides. Difficult to see how anyone could eat rootstock without destroying plant, and for that reason its use as food should not be encouraged.

Indians are said to have eaten rootstock regularly. If you must try eating it, never collect a plant unless there are three others in a 3-ft radius, so plants will not be destroyed.

Indian Cucumber

Trillium
Trillium grandiflorum
Height to 18 in. Stem stout, erect, arising from a deeply buried, irregular bulb, easily broken, and bearing at top a whorl of three leaves, each to 6 in. or more long, and nearly half as wide, with rather prominent branching veins. Stem and leaves smooth and dark green.

Ranges from Quebec to Minnesota, south to North Carolina and Missouri and found at elevations up to 5,000 ft in Virginia. Favors rich, moist woodlands with a deep, loose soil, and will not long survive sun coming with destruction of woods. It will grow in shade by houses. Does well in neutral or nearly neutral soils.

Flowers erect or nodding, composed of three green sepals, each to 2 in. long, and three petals each to 3 in. or more long, that are white at first but change to blue or pinkish with maturity. Stamens have anthers about ½ in. long. Fruit, a berry 1 in. long, red or black.

One of most beautiful woodland flowers of late spring. Has many equally attractive relatives but none have pleasant odors. Wake-robin or red trillium, *T. erectum,* has a flower with a most offensive odor. The painted trillium, *T. undulatum,* with red and white flowers, is popular with wild-garden enthusiasts. *T. chloropetalum* ranges from Washington to California and was named from a specimen which was the greenish-flowered form. Generally the petals are maroon or greenish-yellow or even white.

Leaves considered by some outdoorsmen to be edible but plant is too attractive to be destroyed for food. It makes a hopeless bouquet because it lacks fragrance and wilts quickly. No excuse for collecting it for bouquets and thus ruining a beautiful woodland. "The wake-robin's trinity sign."

White Trillium

Twisted Stalk
Streptopus roseus
Height to 2½ ft. Stem erect, or nodding, arising from a stout, fibrous, root-covered rootstock, or rooting from semiprostrate stems. Stem may be branched, somewhat fuzzy, ridged, and twisted. Leaves rather evenly spaced along stem, to 4½ in. long, pointed at free end and egg-shaped toward base.

Found from Newfoundland to Manitoba and Oregon, south to Michigan and Georgia, occurring in Virginia at an elevation of to 5,600 ft. Found in same kind of moist woods in which Solomon's seal, trillium, and Indian cucumber are found, where there is a rich, deep, well-watered soil, at least in spring.

Flowers borne in axils of leaves, with one to two to an axil; sometimes described as six-pointed bells. They are on short stems to 1 in. long, are rose-purple in color, with slender parts of calyx and corolla to ½ in. long, and usually hang loosely downward. Fruit, a cherry-red, translucent berry. Perennial.

Pollination probably by visits of bumblebees and beelike flies. A closely related plant has greenish-white flowers, and leaves whose bases closely

Twisted Stalk

(Continued)

clasp the stem. Both are conspicuously branched and by this character
quickly separated from Solomon's seal and spikenard.

This is one of attractive woodland flowers. It has no value as part of a
bouquet and since plucking the top may mean death of the whole plant, it
should be left undisturbed where possible. Is a successful addition to a
woodland garden where it thrives with little care.

Greenbrier

Greenbrier
Smilax rotundifolia
Stem woody, winding, and climbing, covered with prickles that are stout,
straight, or sometimes slightly backward-curved or sometimes absent.
Stem round, green, arises from a long, tuberous rootstock. Leaves on
petioles, to 6 in. long, oval but pointed at free end, with three to five
prominent veins. Blades to 6 in. long.

Found from Nova Scotia to Minnesota, south to Florida and Texas,
growing in rich lowlands in woods and thickets, or more particularly
along borders of woodlands. This assists in making dense thickets almost
impenetrable.

Flowers on short stems, springing from a common point in clusters of 6 to
25, whole group supported by a common stem. Sepals and petals together total six similar structures, with recurving tips. Fruit a berry, to
¼ in. in diameter, black, one- to three-seeded, matures first year.

In much of its range, this and other species of *Smilax* are the only woody
monocotyledons. This should not be confused with the smilax of the
florist. There are approximately a dozen species of *Smilax* in the northeastern United States, none of them being of any great economic importance. *S. herbacea,* carrion flower, is fairly common on riverbanks and in
wet places in the eastern half of the U.S. The name of the plant is
descriptive of its odor and also indicates that its flowers are pollinated by
carrion flies.

A related species, *Smilax officinalis* of Honduras; another, *S. medica* of
Mexico; and a third, *S. ornatus* of Jamaica have roots which when dried
yield the sarsaparilla of commerce. Sarsaparilla is rarely used except in
combination with such oils as wintergreen. *Smilax* spp. provides protective covering for wildlife and some winter food. Berries are eaten by
many birds, e.g., catbird, fish crow, mockingbird, and ruffed grouse.

Joshua Tree

Joshua Tree
Yucca brevifolia
Y. brevifolia, height to 40 ft; 2 to 3 ft in diameter. Spread to 20 ft.
Branching with disklike bases, from which spread small, tough roots.
Leaves crowded near ends of stout branching stems and persistent. *Y.
gloriosa,* Spanish dagger, height to 8 ft. Leaves to 2½ ft by 2 in., with stiff,
red tips. Bark 1 to 1½ in. thick, deeply divided into oblong plates frequently 2 ft long. Wood, light, soft, spongy, difficult to work, light brown
to white.

Y. brevifolia found in deserts where there is at least an 8 to 10 in. annual
rainfall, sometimes forming forests. Found in Nevada, Utah, Arizona,
and parts of California. *Y. gloriosa* found along Atlantic Coast from
South Carolina to Florida, in dry sandy areas. Most abundant and of its
largest size on the foothills of the desert slope of the Tehachapi Mountains.

Joshua tree may or may not flower each year depending on temperature
and rainfall. Flowers appear from March until the beginning of May.
Formation of flowers stops growth in that direction and causes branching. Wind and animals distribute seeds. Spanish dagger has lilylike
flowers that are greenish-white to red, 3 to 4 in. across, appear in summer,
hang in clusters.

Joshua tree is parasitized by boring beetle, *Scyphophorus yuccae;* by butterfly larvae, *Megathymus yuccae;* and by other insects. Night lizard,
Xantusia vigilis, is entirely dependent on it for food and shelter. At least
25 species of desert birds known to nest in it.

Near Palm Springs is a Joshua Tree National Monument of several
square miles designed to save these unique trees from destruction. Indians used fine red roots in basketry and used seeds as food. Of Spanish
dagger, there are at least a half dozen varieties recognized by gardeners,
usually on leaf form and color.

Spanish Bayonet

Spanish Bayonet, Yucca
Yucca filamentosa
Adam's needle, height to 12 ft but no stem except flower stalk. Leaves
not rough-edged, 2½ ft long and 1 in. wide, sharp-pointed, with long
curly threads along edges. Spanish bayonet, height to 25 ft. Trunk to 3
ft. Leaves 2½ ft long and 2½ in. wide.

Adam's needle native from South Carolina to Mississippi and Florida,
but planted extensively through much of the United States including

New York, Minnesota, and similar Northern states. Spanish bayonet native of West Indies, and around the Gulf north to Virginia and south into Mexico. *Y. glauca,* bear grass to Texans, occurs in the southwestern U.S. primarily but is found throughout the dry grasslands region and is often planted as an ornamental.

Flowers of Adam's needle borne in tall, loose, open arrangement, hanging like white bells, each to 2 in. long, waxy. Fruit a capsule with rounded angles and frequently with insect borings showing. Spanish bayonet, flowers to 4 in. across, white, with purple tips, bell-shaped. Fruit black to purple, to 4 in. long, fleshy.

Pollinated by Pronuba moth that thrusts the pollen into the pistil where it will be effective, and lays her eggs in the ovary that will ripen food for her young. Number of seeds ripened by pollination act of moth is in excess of number eaten by moth larvae. Pollination may be easily affected by man.

Wood of many yuccas is favored by boy scouts for equipment to make fire by friction. The plants have a deserved popularity as ornamentals. Juice of some yuccas is fermented for making of an alcoholic beverage similar to pulqué. The fiber "palma istle" comes from immature yucca leaves. Some yuccas are called soapwoods or soap plants because the juice from the stem can be worked into a lather.

Dracaena
Dracaena fragrans
Treelike, to 20 ft high, branched or not. Leaves weak, shining green, to 3 ft long and to 4 in. wide, recurved, narrowed at base, and gradually narrowed at free end into a sharp point. Many varieties based in part on different color patterns in leaves, some highly decorative, others plain. Supposedly takes its name from *Drakaina,* a female dragon, and this alludes to red juice in stems of some species, in particular *D. draco.*

Native of Guinea, or more definitely, Upper Guinea, but known throughout the world as a potted plant in hotel lobbies, in tropical gardens where the climate is suitable, and at weddings, banquets, and funerals where a green effect is desired in a hurry.

Flowers to ⅔ in. long, clustered in open panicles that are to 1 ft long. Flowers yellow, with narrow šepals and petals. Fruit a small, orange-red berry. Flowers not often seen and so plants are usually classified by characters apparent in leaves. Most dracaenas are long-lived.

At least 14 species and many more varieties of dracaenas are cultivated by greenhouse folk. One of most interesting plants in its natural environment is dragon tree, *D. draco* of Canary Islands, and the Teneriffe dragon tree, which is 70 ft high and famous as one of oldest of trees.

Of great value in interior decoration. *D. cinnabari* of East Asia exudes a red resin known as "dragon's blood" that was used extensively in the eighteenth century to stain dyes and varnishes used in making violins. A similar resin is sometimes obtained from American Dracaenas.

Dracaena

Order Liliales./Family Amaryllidaceae

Daffodil
Narcissus pseudo-narcissus
Height to 18 in. Grows from a deep bulb that is to 2 in. or more in diameter. Flowering stem smooth, semisucculent, easily broken. Leaves, four to six, all starting from base; flat, smooth, to 18 in. long and to 1 in. broad, sometimes with a whitish bloom.

Native from western Europe to Sweden to England, Spain, and Austria. More popular varieties include Golden Spur, Empress, and Ajax. Favors deep, loose, very wet soil but will grow in grass meadows either in open or in shade of nearby trees.

Flower, one, on end of flowering stem, with a trumpet to 2 in. long and a frill of petals that are relatively narrow. Flower pale yellow, held horizontally, pointing somewhat upward, or drooping. In the large-crowned narcissus trumpet or crown is as long as or longer than segments. Really a member of a group designated as Trumpet Narcissi according to the classification system developed by the Royal Horticultural Society.

For blooming indoors, bulbs are planted one to a 4-in. pot, in a mixture of ⅓ sand, ⅓ garden soil, and ⅓ manure, and kept in dark until rooted first, at around 45°F; then, after rooting, temperature is raised to 65° and water is supplied in abundance. Placed in sunshine when flowers are about to open.

For outdoor planting, bulbs should be covered to a depth of 4 to 5 in. If planted in borders, it is advisable to lift the bulbs and transplant every 3 to 4 years. In a woodland, hillside garden, the bulbs should be planted at greater distances apart than in border plantings. A hillside woodland with daffodils blooming in the spring is a beautiful sight.

Daffodil

Polyanthus Narcissus

Jonquil

Snowdrop

DIVISION ANTHOPHYTA. CLASS MONOCOTYLEDONAE

Polyanthus Narcissus
Narcissus tazetta

Height to 20 in. or more. Flowering stalk distinctly flattened, both leaves and flower stalk rising from a bulb that is often to 2 in. in diameter. Leaves four to six, to 1½ ft long and to 1 in. broad, parallel-veined, flat and somewhat grasslike, but more succulent. Flowers few to several, usually 4 to 8, horizontal or a little drooping, white with light yellow crown, fragrant, small (1 to 1½ in. across); tube-shaped toward stem. Name probably old European in origin, perhaps related to Italian *tarza*, meaning a cup.

Native from Canary Islands to China and Japan. Widely established as a house and garden plant. Grown in homes and schools from one end of America to another, wherever there are 10-cent stores to supply intriguing bulbs. Related poet's narcissus comes from France and Greece. About 30 recognized species.

In *N. incomparabilis,* crown is about half length of segments, while in other grass-leaved narcissi it is shorter, except in daffodil. *N. tazetta* bears four or more flowers on a single stem, crowns not crisped. In *N. biflorus,* usually two flowers on a stem, crisped crowns. In *N. poeticus,* only one flower to a stem with crisped, red-edged crowns.

N. tazetta may be potted to flower before Christmas. In winter, about 6 weeks are required to bring a bulb to flower; but in spring it may be done in 3 weeks. *N. poeticus,* poet's narcissus, and others are potted late in September in a fibrous, medium loam, a 6-in. pan holding six bulbs, then "heeled in" for early winter.

Heeled poet's narcissus if kept at 50°F should develop roots by January, then forced at 50 to 55°F until buds are free, then raised to higher temperatures, but nights should be 60° or less. This will produce flowers in 3 to 4 weeks. Christmas blooms should be forced from mid-November, if potted in September.

Jonquil
Narcissus jonquilla

Height to 1½ ft. A deep bulb produces two to four leaves, that look rushlike rather than grasslike, are almost round in cross section, and with a narrow channel on one face. Leaves to 18 in. long, glossy, dark green, and semisucculent, easily broken. Flowering stem not flat as in narcissus. Name jonquil probably shortened *jonquilla,* from Latin *juncus,* a rush.

Native of southern Europe and Algeria, but hardy in such states as New York, where it has established itself in or near old homesteads and persisted long after other signs of human habitation may have disappeared. Favors loose, deep soil, where there is much moisture in spring.

Crown of flowers small, much less than half as long as segments back of it. Two to six flowers on a single stalk, held horizontally or drooping. They are yellow throughout, most fragrant, and with a slender rear tube to about 1 in. long. Crown not over ⅛ in. long.

Popular varieties include General Pershing, and most fragrant Regulosi. Related species include Campernelle jonquil, *N. odorus,* whose leaves are not quite round in cross section; the hoop petticoat daffodil, *N. bulbocodium,* in which crown of flower is as long or longer than segments behind.

All of narcissus group produce beautiful flowers, popular with houseplant lovers and with outdoor flower gardeners. They bring results rather quickly after initial care has been given and this makes them popular with amateurs. They are relatively free from insect and fungus pests. *Narcissus* bulbs are poisonous to man, causing gastroenteritis plus associated symptoms.

Snowdrop, Fair Maid of February
Galanthus nivalis

To 10 in. high. In snowdrop, stalk supporting flowers is solid; in snowflake, *Leucojum,* it is hollow. Leaves in snowdrop, only two to three, in snowflake, many. In *G. nivalis,* leaves are to 9 in. long and to ¼ in. wide. Bulb to 1 in. thick, with sheath slit down one side.

Native from central and southern Europe, to the Caucasus, with different varieties in different areas and of course with some different horticultural varieties. At least a dozen species of snowdrops. Generally favor cool, moist, shady areas, often forming beds in and near lawns. Plants will increase slowly over the years until large clumps are formed from an original planting of a few bulbs.

Flowers single, on a slender, nodding, flowering stem, with outer segments white, to 1 in. long, blunt, veiny, and inner segments about half as long, white with green in notches. One flower for about every 10 plants. Blooms in spring but a related species blooms in the fall.

First of showy flowers of year. At latitude of Philadelphia, it may flower in January or February. Since they may be produced in great abundance at a time when there is no competition, they are usually most welcome. By midsummer, they have flowered and leaves have almost completely disappeared. Bulbs are set about 3 to 4 in. deep, preferably in loose soil, and in masses. Planted in rock gardens, in lawn borders; as for edging for shrubbery.

Bulbs are usually cheap. Plants seem to be popular in some areas and less so in others. Giant snowdrop, *G. elwesii*, has flowers to 1½ in. long and ¾ in. wide.

Sisal Hemp, Henequen
Agave rigida
Leaves 8 to 10 ft long, average 6 ft by 4½ in. wide, by ½ in. thick, with a tough, green skin and pulpy interior mixed with fibers; about 3½ to 5 percent of leaf is fiber. Weight of leaf 1 to 2 lb. Stem erect, coarse, and, in cultivation, conspicuous, with diamond-shaped leaf scars, cylindrical, barrel-shaped.

About 300 other species in genus including century plant or blue aloe *A. americana*, small-leaved aloe *A. angustifolia*, and others. Sisal hemp also comes from *A. morrisi* and *A. vivipari*. Mauritius hemp comes from *Furcroya gigantea*. *A. rigida* is cultivated in Bahamas, Cuba, Mexico, Florida, and northeast India. Native of Yucatan. For those who have driven in the desert country of Mexico and the American Southwest, a familiar sight is the tall 8- to 12-ft flowering stalk with its cluster of flowers arising from a base of spearlike leaves. These *Agave* plants die after flowering.

Shoots planted 4 ft apart, in rows 12 ft apart. In 4 to 5 years, or when leaves assume horizontal, first leaves may be cut. Plants allowed to bear 25 to 27 leaves and yield 12 to 15 a year cut at intervals. If leaves are not cut when mature, plant may send up 8 to 12 ft flower stalk and die. Properly handled, a plant may yield for 14 years in rocky soil, sometimes 1,500 leaves.

A bed of semidecomposed coral rock with hot, dry air is best environment. In native Yucatan, a plant may yield 20 crops over a span of 25 years. If one plant in an area is allowed to produce its flower stalk, nearby plants may, strangely enough, do the same and die, so a plantation requires care to have continued yield. 600 to 1,200 plants to an acre.

Leaves are worked when green and fiber is removed by machines, but juice attacks wrought iron. Yellow leaves yield inferior fiber. Fiber whiter, flatter, and less pliable than manila. It is used in making ropes, nets, hammocks, and cordage. Some sisal is extracted by steeping leaves in water 10 days after drying for 4 days and then beating fiber out.

Sisal Hemp, Henegquen

Order Liliales./Family Dioscoreaceae

Yam
Dioscorea alata
Climbing vine, with winged or fluted stems. Leaves heart-shaped, but rather long-pointed at free end. Chinese yam *D. batatas* may have a stem 30 ft long. Tuber of *D. alata* may be 8 ft long and weigh 100 lb, though it is usually smaller. It also has aerial tubers. *D. batatas* has 2-ft tubers.

D. alata, native of South Sea Islands and India but is cultivated widely throughout the tropics. It is becoming more important in America. It should have a deep soil but can withstand drought unusually well. *D. batatas* will survive New York winters. Stems of Chinese yam are round, not winged.

Reproduction is essentially by tubers that may continue to increase in size for years. In Chinese yam, small tubers are produced in leaf axils that if planted will produce in 2 years large edible yams. Flowers of yams either pistillate or staminate, with a plant producing one or the other. Seeds, winged.

In commercial practice, yams are grown from tubers, seeds, or stem cuttings, and there are many varieties of the different species. One of the most interesting is a yam from Hawaii that produces large edible, aerial tubers. It is grown along the Gulf of Mexico.

Yams are commonly baked, boiled, or ground into flour, dried, stored, and used in the making of starchy foods of many sorts. In the Southern States, they are raised chiefly as food for stock, but they are relished by many humans and, properly cooked, are delicious. Name yam is used also for sweet potato, *Ipomoea batatus,* a member of the morning-glory family, Convolvulaceae.

Yam

Order Liliales./Family Iridaceae

German Iris
Iris germanica
Height to 3 ft. Stems branching, usually with two flowers at top, one on a short, and one on a longer lower branch. Rhizome stout, creeping, and, for most part, exposed, with numerous roots coming mostly from lower side. Leaves, broad, stiff.

(Continued)

German Iris

385

Blue Flag

Gladiolus

Native of central and southern Europe; German origin implied by the name is definitely disputed. Many varieties of the species have been developed and it hybridizes with some others. It prefers rich, well-drained, limestone soils, endures half-shade and much drought, and suffers with heavy winter mulch.

Flowers large, showy, purple, yellow, white, or of intermediate or mixed hues. Conspicuous parts of three upright standards and three drooping "falls." Above the "brush," on the falls, are stamens covered by a sterile part of style of pistil, stigma being beyond end of stamen.

Rhizomes are best divided in July and August. They are planted horizontal to surface, about 8 in. apart, and deep enough to be half-covered with earth pressed close around divisions. Water is needed during the blooming period, and weed competition should be kept down for best results.

Excellent garden and path flower. To control crown and pod borers, spray with lead arsenate and destroy infected part, wash contaminated rhizomes with weak potassium permanganate. Dust parts affected by leaf blotch with sulfur, and burn parts affected by sclerotic rot. Most tall bearded irises of gardens are hybrids of *Iris pallida variegata* crossed with plants from Asia Minor.

Blue Flag
Iris versicolor

Height to 3 ft. Stems round in cross section, often branched, and usually leafy. Leaves erect, shorter than stem, to 1 in. wide, like pointed swords, slightly curved, parallel-veined. Rhizome horizontal, thick, fleshy, and root-covered, roots being fibrous and light-colored. Perennial.

Found from Newfoundland to Manitoba, south to Florida and Arkansas, mostly in and at edges of marshes, wet meadows, and in open wet brushlands where sun can reach the ground. Not usually in pure stands as are some other plants of same general environment. Western blue flag, *I. missouriensis,* is an attractive western form with range from the Prairie States west to southern California and north to eastern Washington. Flowers are usually two on a stem, pale blue to violet in color.

Flowers several, usually violet-blue, with yellow, white, and green, the segments being generally more slender and smaller than in most of cultivated species and varieties. Fruit a three-lobed, oblong capsule, to ¾ in. long, with seeds in two rows in each cell, and each to ½ in. broad.

Rootstock contains irisin which causes indigestion in humans or cattle, but this is rare because roots have an offensive taste. Skin diseases may be caused from handling iris roots. It has been assumed that death results from calves' eating the roots.

Flower attractive. Leaves provide some shelter for aquatic game. Supplies little food to any animal and because of this is inferior to other species that give both food and shelter. Orris root of commerce comes from root of *Iris florentina* and is used as perfume and in flavoring. It smells like violets.

Gladiolus
Gladiolus sp.

Height to 5 to 6 ft. Thick, flattened, scaly corms of stems, with papery, scaly covering, produce erect stems that bear linear, sword-shaped, slightly curved leaves, parallel-veined, with rather prominent ribs and regular margins. Roots thick and fibrous.

Most important species came from South Africa but some came from Mediterranean region and some from Asia Minor. At least 200 natural species and thousands of horticultural varieties. Hardy plants that do best in rich, well-drained, limestone soils cannot stand drought and shade well. Can be grown successfully in all 48 states of the U.S. mainland. Some varieties of merit are: large-flowered type—Apple Blossom, Golden Arrow, Purple Birma, and Royal Steward; miniatures—Burnt Orange, Golden Trills, Peter Pan, and Starlet.

Flowers large, showy, somewhat trumpet-shaped, but broadly opened, purple, yellow, white, red, or of intermediate hues and mixtures, generally with 6 to 30 flowers on a spike, closely crowded and maturing from bottom up. Blooms from 60 to 120 days after corms are planted. Pollination by insects and birds.

A tender perennial. Plant corms 4 in. deep, 2 to 3 in. apart, in rows 2½ to 3 ft. apart, and cormlets about half as deep and far apart. Water thoroughly once each week during blooming period. Cut spikes, leaving 3 leaves to renew the corm. Dig corms in fall, and store away from frost.

Important garden and cut-flower plant. Attacked by thrips and aphids but controlled by nicotine sulfate. Also attacked by a blight, borers, a neck rot, and a scab so avoid planting infected stock or planting in infected soil.

DIVISION ANTHOPHYTA. CLASS MONOCOTYLEDONAE

Blue-eyed Grass, Satin Flower

Sisyrinchium montanum

S. angustifolium or *montanum* has stems bearing flowers. These reach a height of 2 ft, are stiff, pale, and covered with a gloss. Leaves vary from being longer than stem to half its length, and are to ⅙ in. wide, flat, sword-shaped, rather stiff, and with fine marginal irregularities.

Found in fields and meadows, from Newfoundland to British Columbia south to Virginia, Nebraska, Colorado, and Utah, with closely related species extending range considerably. Some of species are difficult to separate from others.

Flowers to ½ in. long, deep blue with some yellow, borne in small clusters, on short stems, at base of a spathe, which appears essentially as a continuation of stem that bears it and flowers. Fruits spherical, to ¼ in. high, brown, purplish, or white, and often held erect.

Names star grass, pigroot, blue-eyed grass, blue-eyed Mary, grassflower, and others indicate that the plant has attracted attention widely and favorably. It never grows in sufficient abundance to be a severe pest and it rarely competes at all with important crops.

Men are interested in this plant because of its beauty and possibly because it is difficult to interpret the different species properly. Parts of the plant reported eaten by wild turkey, ruffed grouse, and mountain quail. Generic name reported by the Greek Theophrastus. Hardy varieties make ideal border plantings where they do best in a cool, moist location. They should be planted using a compost of loam and peat; can be planted in spring or fall. Plants are easily increased by dividing them in the spring. The chief hardy kinds are *S. californicum,* which is bright yellow; *S. douglasi (grandiflorum),* which is purple, and its white variety *album;* and *S. angustifolium,* which has pale blue flowers.

Blue-eyed Grass

Freesia

Freesia refracta

Height to 18 in. A weak plant, standing only partly erect, remotely branched, and with comparatively few leaves that are reduced, slender, about 3 to 10 in. long, to ½ in. broad, flat, parallel-veined, and sheathing stem at base. Stems arise from an egg-shaped, bulblike corm.

Native of South Africa in region of Cape of Good Hope, but grown extensively by florists the world over, being offered to trade mostly at Christmastime. In century and a half that plant has been manipulated by florists, it has changed considerably in form and color.

Flowers, five to six in a cluster, held in a horizontal row, at or near tip of stem, fragrant, pale yellow, funnel-shaped, with tubes to 1½ in. long, tube being sharply restricted below middle. Sepals, three, lower one being with a dark yellow spot. Petals, three. Stamens, three, with long, white anthers. May be obtained in a wide range of colors—mauve, lavender, blue, yellow, orange, pink, and carmine-rose.

Planted six to eight in a flat, 6-in. pot, about 1 in. deep, in ⅓ manure, ⅓ sand, ⅓ garden soil, in August, should yield flowers in 10 to 12 weeks if temperature is the ideal between 50 and 70°F, if there is moderate sunshine and enough moisture to keep the soil moist at all times. Corms may be kept dry during summer.

Freesia

Grown easily even by amateurs who will follow rules and have suitable equipment and patience. Since flowers can be produced in from 6 to 7 months from seeding or in 10 weeks from setting bulbs, a generous plantation for home window box or garden can be built up quickly.

Crocus

Crocus sp.

Height to 8 to 10 in. From flat-topped, rough-fiber-covered corm arise leaves. Leaves grasslike, dark green, with a white stripe down middle, parallel-veined, enclosed in a papery sheath at first, and to 10 in. long. Roots, from the base of the corm; fibrous.

Native of Europe and Asia, mostly from Italy to southern Asia. Common species include fall-blooming *C. sativus,* or saffron crocus, and at least four spring-blooming species. Among these latter are yellow-flowered, *C. moesiacus* and *C. susianus,* cloth of gold crocus, and the blue, white, or yellow, *C. biflorus* and *C. vernus.*

In most species, flowers appear first thing in spring. Usually goblet-shaped, 3 to 5 in. long, with slender tubes, with three sepals, three petals, three stamens and three parts to ovary. Reproduction most commonly effected by offshoots from corm, though seeds may be produced and used.

For indoor growing, plant six to seven corms in a 4-in. pot or flat, in late summer or fall, in a mixture of equal parts sand, garden soil, and rotten manure at a temperature of 45°F at first, in the dark, then after a few weeks raise temperature to 60° and then to 70°. Seeds produced at ground surface easily overlooked.

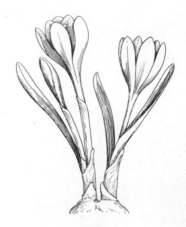

Crocus

One of most popular and effective of early spring flowers, but to get best results, bed should be dug up at least once in 3 years and corms separated and replanted. Plants do best if there is an abundance of moisture at

(Continued)

Manilla Hemp

Banana

Plantain

time of blooming. Corms set out outdoors in October should bloom early following spring. Can be planted in a variety of locations; e.g., along a path, on a bank, in front of shrub plantings; almost anyplace they can be left undisturbed after they have flowered. They flower each spring, year after year.

Order Musales./Family Musaceae

Manilla Hemp, Abaca
Musa textilis
Height to 20 ft. General appearance almost identical with banana. Leaves often spotted, with the fiber found in the sheathing leaf bases that form the 20-ft trunks. Like banana, plant is not a tree, nor is it a woody plant. Inner leaves more valuable. A cut stem weighs from 20 to 80 lb. Perennial.

Close relative to banana and to many ornamentals. Grows best in Philippines, especially on Luzon and to the south, where there is a warm, moist climate, a deep, rich, well-drained soil, and 60 in. or more of annual rainfall. Attempts to introduce abaca into American tropics have failed, partly because of diseases.

Fruits inedible, with seeds the size of BB shot and borne on drooping spikes. Reproduces largely by sucker shoots. Field is cleared and burned, suckers planted, and competitive weeds removed. Plants mature in about 3 years, when they should be cut. Young leaves give weak fiber; older leaves, harsh, brittle fiber.

Leaf bases contain 90 percent water. Requires 5 acres of plants to yield 1 ton of fiber. The shoots cut are replaced by developing suckers. Grades of manila fiber include "current," "fair," "brown." Knotted abaca is hand-sorted and tied fiber.

One of most valuable Philippine exports. Knotted abaca is twisted into skeins and then braided by hand or by simple machines in Japan. Hemp fiber, which for many years was used in the manufacture of rope, now is used mostly in the production of small yarns, twines, and canvases, and to some extent in making a special paper; is yet used in making marine cables because it is strong, durable, and resistant to both fresh and salt water.

Banana
Musa paradislaca sapientum
Height to 30 ft or more. Not a tree or a woody plant, but a perennial herb. Trunklike structure bears a crown of erect or arching paddle-shaped leaves, and arises from sucker or huge underground bud about size of a football. Suckers often produced in abundance at base of main stem but should be removed to increase vigor of bearing plants.

Native of West Africa; introduced into East Africa and India later and now grown extensively in the tropics around the world. Dwarf Chinese banana, *M. cavendishii,* grown for ornament along coast of Gulf of Mexico and in southern California as well as in greenhouses of botanical gardens. First introduced into United States in 1804. First full cargo of 1,500 bunches in New York in 1830.

Commercial bananas do not produce seeds. Flowers appear in erect clusters that bend downward as they mature. Plant or erect stem bears fruit but once, usually from 10 to 12 months after shoot development or transplanting begins. Because of developing suckers, plant may bear several years if free from pests. Staminate flowers drop off.

Light frost will kill leaves and prevent fruit formation or development, but new suckers will arise and fruits may be borne the succeeding year. In a hothouse, plant may not bear fruit until shoot is 2 to 3 years old. Not economical to raise the plants under glass for their fruits.

Fruits are picked green and may be eaten raw. Vitamins B and C present. Banana flour is made by drying ripe bananas. Plant provides an abundance of cheap, highly nutritious food and is one of most important of all foods in the tropics. A "bunch" of bananas is called a hand. The long stalk with many green hands of bananas is cut with a large knife. They are shipped green and ripen after they reach their destination. The large banana plant is cut down during the harvesting.

Plantain
Musa paradisiaca
Height to 30 ft. Stem herbaceous but rigid. Leaves erect or ascending, brighter green than in bananas, to 9 ft long and to 2 ft broad, usually rounded at base, and with petiole to 1½ ft long. Underground swollen areas suggested for banana are also found in plantain. Known as "Adam's fig."

Native of India, but cultivated extensively in the tropics particularly in West Indies and Central America. This is the species of which banana is considered by some to be a subspecies. The subspecies has many varieties including the ladyfinger, the Dacca, the Chotda, the Lacatan, and the red banana.

Plantain has staminate flowers that are persistent while those of banana drop off. There are about a dozen yellow-white staminate flowers to a cluster and to 80 pistillate in a bunch. Fruit to 1 ft long, yellowish-green when ripe, and with firm flesh.

No record of plantains ever having been reproduced by seeds by man. In spite of the fact that fruit is inedible when raw, it has superior nutritive values when cooked. Because of ease with which cooked plantain or plantain flour is digested, it is frequently recommended for use by invalids and children.

While a visitor to the tropics will always have the opportunity of eating plantain, he will probably have to develop an appetite for it but afterward the dish is likely to be a favorite. Baked plantain, yams, papayas, mangoe, and plant-ripened pineapples are among the things remembered by those who visit tropics.

Order Marantales./Family Zinbigeraceae

Ginger
Zingiber officinale
Height to over 3 ft. Reedlike. Leaves to 8 in. long, to ¾ in. wide, slender, without petioles, arising from a coarse, irregular, tuberous rhizome, with flower-bearing stem not unlike leaf-bearing stems. Both leaves and rhizomes strongly aromatic.

Native of tropical Asia, but widely cultivated throughout tropics of the world. Introduced into southern Florida and grown somewhat in greenhouses. This is not the wild ginger of the North *(Asarum)* or the flowering ginger *(Hedychium coronarium)* used so much in leis in the Hawaiian Islands.

Flowers borne on dense spikes, to 2 in. long, pale green with yellow margins, may be spotted with purple or yellow, lip being usually purple, and segments being about ¾ in. long. Fruit an oblong capsule that breaks open irregularly.

In Florida, ginger is grown on rich soil, in shade. Probably nothing in climate of southern Florida and southern California to interfere with commercial cultivation of ginger there. Rhizomes contain starch, an essential oil, gums, and oleoresin, the content varying with the variety.

Medicinal ginger comes from dried rhizome; ginger for flavoring, from the green root; candied ginger, from succulent, young roots. Ginger is used in culinary preparations (soups, curries, puddings, pickles, gingerbread, cookies), and for flavoring beverages such as ginger ale and ginger beer. The flavoring agent is an oleoresin (gingerin) which is responsible for the pungent taste, and an essential oil which imparts the aroma. Candied roots are peeled and then preserved in a syrup. Leaves of ginger are supposed to roll upward; those of the garland flower, downward.

Ginger

Order Marantales./Family Marantaceae

Zebra Plant
Calathea zebrina
Height to 3 ft. Leaves spring from a short base bearing up to 20 leaves. Leaf blades to 2 ft long and to 1 ft broad, rich green above, marked in alternating yellow-green and dark olive-green bars, with underside gray-green when young and purple-red when mature. Petiole to 2 ft long.

Native of Brazil but one of commonest of foliage potted plants grown in greenhouses. Requires more warmth than prevails in some homes and must be therefore kept in a warm greenhouse. Over 100 species known and about half of these are cultivated for some purpose.

Flowers in a spike that is almost globular. Flowers of *Calathea* larger than those of true *Maranta*. *Calathea* lacks zigzag stems and branching flower stalks of *Maranta*. However, these plants are known mostly from their foliage and are reproduced mostly by dividing underground parts.

Ideal temperature for these plants is not below 65°F at night and up to 90 to 95° in the day, with a high humidity at most times. They require a well-drained soil but a generous amount of moisture at all times. In greenhouses, they should be sprayed regularly.

Outdoors, where conditions are right, the plant grows and multiplies rapidly. It can be grown outdoors in some parts of Florida successfully but should be transplanted regularly to prosper. Repotting is done annually to prevent plant from becoming pot-bound and thereby decreasing growth. In summer, pots of zebra plants must be kept in the shade and moist but not wet. In winter, plant is watered only when soil has dried.

Zebra Plant

Canna

Fringed Habenaria

Cattleya

DIVISION ANTHOPHYTA. CLASS MONOCOTYLEDONAE

Order Scitaminales./Family Cannaceae

Canna
Canna indica
Height to nearly 5 ft. Such related species as *C. edulis* may grow to 10 ft high. Leaves green, about twice as long as wide, smooth throughout, with entire margins, and with bases clasping stem. Roots heavy, tuberous, branching, with many finer, spreading roots.

Native of tropical America, though thought by Linnaeus to be a native of Asia and Africa. About 50 species native of Western Hemisphere, some of which have become naturalized outside their normal range. Favors deep, loose, well-watered soil. When brought from West Indies in 1830, was grown for leaves only.

Flowers normally red or orange, with 1¼-in. petals, mostly borne in 2s, erect, relatively narrow, with lips orange spotted with red. Seeds will not normally germinate unless filed to let in water, may be soaked in warm water a day before being sown. Hard seeds give name Indian shot.

Roots dug in autumn, stored in cellar with earth left clinging. May be started indoors in March after being divided so that each piece has one bud or eye. Potted plants rarely set out before May in northern states, should then be set 16 to 18 in. apart; are tender, subject to drought, frost, or too much water.

Most attractive of foliage plants. Purplish-leaved species, *C. edulis* and *C. warscewiczii;* the former bears edible tubers that yield Queensland arrowroot starch as well as being attractive foliage plant. *C. indica* is old-fashioned green canna, so common in old gardens and in lawn corners. This plant in earlier times became associated with mass plantings near railroad stations and in city parks. As a result it lost favor. Recently, with the development of new varieties, it has gained favor again for home plantings.

Order Orchidales./Family Orchidaceae

Large Purple-fringed Orchid
Habenaria grandiflora
Height to about 5 ft. Lower leaves oval, to slender lanceolate, becoming reduced above to mere bracts. Lower leaves to 10 in. long and to 3 in. wide. Stem ridged and clasped by bases of leaves. Roots branching from base of stem.

From Newfoundland to Ontario, south to North Carolina. A close kin to *H. psychodes* that is smaller and ranges over the *H. fimbriata* range and west to Minnesota and Tennessee. A score of habenarias in northeastern United States, seven on Pacific Coast, and about fifty in all.

Flowers in spikes as shown, spikes being to 1 ft long and to 2½ in. in thickness with flowers densely crowded. Lip of petals to about ½ in. long, magenta-pink or white. Flowers fragrant, with a spur to 1½ in. long, spur being exceptionally slender.

Pollination by visits of moths and butterflies, on whose heads pear-shaped masses of pollen become glued when insects have sought nectar for their food. This orchid favors rich meadows or woodlands.

Of little economic importance, but a common popular orchid that should be left in its glory rather than picked for temporary admiration. The habenarias in general are called the rein orchids. A worldwide group numbering about 500 species. The long-spurred rein orchid is often found in the South, floating with masses of water hyacinths.

Cattleya
Cattleya trianaei
Structures known as "pseudo bulbs" lie between leaves and roots. In this species these are erect, club-shaped, to 1 ft long, and bear one leaf that is to 8 in. long, relatively thick, broadly paddle-shaped and smooth. Roots in part visible and spongy, and of a heavily branching type.

Native of Colombia but is grown widely in greenhouses and outdoors in suitable climates as an ornamental. Some 50 species of *Cattleya,* this one sometimes being considered a variety of *C. labiata,* which species includes many of the important cultivated forms.

Flowers to 6 in. across, white, delicate rose or amethyst-blue in spots. Sepals much narrower than petals that are blunt and with crisped edges. Lip narrow, long, usually rose-colored. Throat yellow, and often streaked. Front lobe shorter than tube.

During spring, winter, and fall, where cattleyas are raised in greenhouses, night temperatures should be between 50 and 55°F, but at all times they should have as much fresh air as is possible. In hot weather, they should be watered in the evening with a delicate spray.

This is possibly the most popular group of commercial orchids available from florists. There are many varieties of the species and many related species, but this one is common for the amateur and for the professional orchid raiser. There are some 40 different species and hundreds of different horticultural varieties of *Cattleya,* one of which probably made up the last orchid corsage you saw.

Vanilla
Vanilla fragrans

Stem climbing, leafy, fleshy, tall, bearing many thick leaves to 8 in. long and to over 2 in. broad, with indistinct, parallel veins. Both leaves and stems somewhat fleshy. One closely related variety grown for ornament has variegated and striped leaves.

Native of Mexico and Central America, but grown to some extent in other parts of the tropics, with greenhouse varieties grown rather extensively outside the tropics as an ornament and as a curiosity.

Flowers, 20 or more, in an open cluster, yellow-green, to 2½ in. or even more long, with lips narrow, trumpet-shaped, and shorter than other parts. Fruit a long, fleshy, bean-shaped capsule that opens little if any for shedding seeds.

Commercial propagation is effected by cuttings. Plants are trained on posts and flowers are artificially pollinated by man. Pods not aromatic until they are cured. Unripe fruits are picked, exposed to sunlight during morning, then "sweated" under blankets in afternoon, and put in airtight boxes for the night.

Spaniards found the Aztecs using vanilla to flavor chocolate. Flavor is due to a crystallin substance vanillin, caused by an enzyme working on a glucoside in the pods. Vanillin can be synthesized cheaply from petroleum or from papermill wastes, but there is still a strong demand for the natural product.

Vanilla

Showy Lady's Slipper
Cypripedium reginae

Height to 2 ft. Stem rather densely covered with somewhat sticky hairs, stout, leafy clear to the top. Leaves to 7 in. long, to 4 in. wide, elliptic, acute at free end, with conspicuous, parallel veins. Roots coarsely fibrous.

From Newfoundland to Minnesota, south to Georgia. Found mostly in swamps and wet woodlands, often in considerable numbers. In the same general area are found other leafy-stemmed Cypripediums such as small yellow *parviflorum;* large yellow *parviflorum* var. *pubescens,* and small white *candidum.* Approximately 50 species of *Cypripedium* grow across world in the Northern Hemisphere. There are 11 species and several varieties in North America.

C. hirsutum has one to three flowers, each with a conspicuous white or pink-tinted lip, that appears inflated and is to 2 in. long. Sepals longer, and broader than the petals. This is easily one of the most beautiful of all wild flowers. Fruit a capsule. Seeds minute.

This species is always known as *C. reginae,* which name well describes dignity of the whole plant, while name *hirsutum* describes nature of leaves and stems. Hairs found there yield a fatty acid that is poisonous to many people, causing blisters and inflammation in 8 to 10 hours.

The beauty of this plant is such that it should be left where it is. If people generally knew that they might get something similar to ivy poisoning from it, they might leave it alone. If also grows in the same places where poison sumac grows.

Showy Lady's Slipper

Moccasin Flower
Cypripedium acaule

Leaves two, lie almost flat on ground, to 8 in. long and to 3 in. wide, dark green, with conspicuous, parallel veins. Roots fleshy, fibrous, and much-branched. Flower-bearing stem may reach a height of 15 in. and may bear a small, leaflike bract on it.

Found from Newfoundland to Manitoba and Minnesota south to North Carolina and Tennessee. In Virginia, ranges to altitudes of 4,500 ft. It bears a number of common names such as Noah's ark, squirrel's shoes, camel's foot, nerveroot, old goose, Indian moccasin, and two lips. *C. parviflorum* var. *pubescens,* large yellow showy lady's slipper, has wider range across the U.S. from Maine to Washington south to Georgia and Oregon and is the most common of all of our lady slippers.

Flower borne singly at top of a tall stalk or scape; conspicuous largely because of its inflated lip, which is 2 in. or more long, pink or sometimes white, and marked usually with dark lines. Sepals greenish-purple, spreading, to 2 in. long, with lateral ones united.

Favors sandy or gravelly woodlands, usually where soil is thin and likely to be slightly acid. Leaves may be found through most of year with flower or fruit body on scape that rises between them. Flowers appear in May and June.

This is state flower of Minnesota, and in New York State and some other states it is illegal to pick the flower on state-owned lands, or to transplant or injure the plant at any time when it is growing on such lands.

Moccasin Flower

Showy Orchis

DIVISION ANTHOPHYTA. CLASS MONOCOTYLEDONAE

Showy Orchis
Orchis spectabilis

Flowering stems reach to a height of 1 ft, are angled, coarse, fleshy, and too easily broken. Leaves two, borne close to the ground, often resting on it; to 8 in. long and to 4 in. wide, a beautiful, clear green, smooth, and clammy to the touch, parallel-veined. Roots coarsely fleshy, fibrous.

Found from New Brunswick to the Dakotas, south through Georgia, Kentucky, Missouri, and Nebraska. In Virginia, it is found at altitudes of to 4,000 ft. The plant is usually found in areas to which the plant-collecting botanist has no access. In Eastern States this species is typical of mixed deciduous forests of rich beech and sugar maple stands.

Flowers borne in a spike of three to six flowers. Flowers about 1 in. long, violet, purple and white, with lip whitish, entire, and about as long as petals. A spur about ¾ in. long may be conspicuous. Fruit a conspicuously angled capsule, about 1 in. long.

This plant favors rich, wet woods and is often found at lower edges of slopes from steep hills, where soil is deep and water abundant at flowering time in May and June. Whole plant is much too attractive for its own prosperity in densely populated regions.

Probably no great economic importance to plant, but those who know it best wish it had the legal protection afforded many of its fellows.

The primary divisions of the animal kingdom are the phyla. In some cases, phyla are divided into subphyla. Phyla and subphyla are, in turn, made up of classes. The classes are groups of orders. Some orders are subdivided into suborders but all contain one family or more. The families are divided into genera, which may in turn be divided into subgenera. In each genus there is one or more species, of which there may be two or more subspecies. Some zoologists use the terms *variety* and *subspecies* interchangeably, but modern usage inclines toward restricting the term *subspecies* to populations that differ from each other in hereditary qualities, while varieties represent differences brought about by environmental factors such as temperature, humidity, and differences in food. *Variety* has many different meanings and may refer to differences that are not genetic. In recent times, we have heard persons speak of *ecotypes* or *ecophenotypes,* terms which allude to differing forms of the same species which have resulted in response to environmental differences. Subspecies, then, are geographic races and in many cases are doubtless incipient species; that is, species now in actual process of differentiation. They are separated by barriers that act to reduce or prevent interbreeding between the populations on the two sides. Most of our species of birds and mammals are divided into such geographical subspecies. In many cases the Rocky Mountains form a barrier between our subspecies. The lower Mississippi effectively separates two subspecies of the ribbon snake. The Isthmus of Panama, of comparatively recent origin geologically, has brought about differentiation of certain fishes on its two sides. The great arid regions of our Western States constitute another important barrier; in addition, these deserts have a characteristic animal population. Breeds are limited groups resulting from selective breeding controlled by man. Strains are even more limited than breeds. Attempts by systematic zoologists to agree on the definition of these terms have led in the past to endless but not necessarily futile discussion.

There are probably several million species of animals, of which only about 1⅕ million or so are known to science and slightly less than 1,000 are considered here. Taxonomists divide the animals into large groups called phyla (phylum, singular). There may be as few as 18 or as many as 31 phyla reported. Some of the phyla are listed below. An asterisk indicates that a representative member of the phylum is included in this fieldbook.

The common animals we notice in the field belong to relatively few groups and are unevenly distributed among these groups. If the number of woodchucks that could survive on a 10-acre lot were put end to end, they would probably reach less than 100 ft. The microscopic nematode worms that might live unseen and comfortably in the same area if placed end to end might well reach around the earth. Most of us on a field trip see the woodchucks. Few if any would see the nematodes. We notice the color, odor, luminescence, taste, or feel of some of the smaller animals, and for this reason a few are mentioned in this fieldbook.

Agriculture, medicine, commerce, conservation, health, and government are all dependent on an understanding of the interrelationships that exist between different animals, plants, and types of environments. The understanding necessary for continued successful civilization may in large part be developed from work in this field.

Phylum	Estimated number of species	Phylum	Estimated number of species
Protozoa*	29,000		
Porifera*	5,000	Echiuroidea*	60
Coelenterata (Cnidaria)*	5,300	Annelida*	8,500
Ctenophora	80	Arthropoda*	838,000
Platyhelminthes*	12,700	Phoronida*	15
Nemertinea*	550	Bryozoa*	3,500
Acanthocephala*	500	Brachiopoda	250
Aschelminthes*	12,500	Hemichordata*	80
Priapulida*	8	Echinodermata*	6,000
Mollusca*	85,000	Chaetognatha*	50
Sipunculoidea*	250	Chordata*	43,000

*A member of the phylum is described in the fieldbook.

PHYLUM PROTOZOA

Some 29,000 species of Protozoa are know to science. The only other large groups of animals are the arthropods, the mollusks, and the chordates.

The Protozoa are minute one-celled animals, aquatic or parasitic, without special body or sense organs but with specialized structures having particular functions. The animals live singly or in colonies and may move with or without the help of whiplike structures called *cilia* or *flagella*. Respiration and excretion are carried on through the body surface and solid food may be similarly engulfed. The outer covering, when present, is not made of cellulose as the cell walls of plants are. Some of the primitive Protozoa bear chlorophyll; they are claimed as plants by botanists and as animals by zoologists since they have some of the characters of each. For example, *Euglena* sp. is listed here in Protozoa and earlier in the fieldbook in the division of plants called Euglenophycophyta. In like manner we might have listed *Synura* sp., *Dinobryon* sp., and *Uroglena* sp. in the plant division Chrysophycophyta, as well as here. They were first studied by Leeuwenhoek with his microscope in 1674, and the name Protozoa was first used by Goldfuss in 1820.

Three classes are represented here, although it is not easy to recognize the differences since a colony of animals, each of which has a few flagella, may appear like a single animal with many cilia. *Paramecium* belongs to the class Ciliata while *Amoeba* belongs to the class Sarcodina, and *Euglena* sp., *Synura* sp., *Dinobryon* sp., and *Uroglena* sp. belong to the class Flagellata.

Animals of a fourth class, the Sporozoa, usually possess no organs of locomotion. However, they are important and cause many diseases of other animals. Of these, *Plasmodium,* the cause of malaria, is carried by certain mosquitoes and *Babesia,* the cause of Texas fever, is carried by ticks.

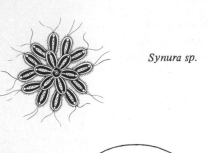

Synura sp.

Few persons interested only superficially in field natural history will ever see individual *Synura, Dinobryon,* or *Uroglena* although they recognize their presence. They foul stagnant waters such as swimming pools. With a simple microscope, many of these forms, and a host of others, become visible. Places like mud puddles and pond water contain a world of fantastic diversity that would interest many people. *Synura* is found in stagnant fresh waters. It makes such waters smell like ripe cucumbers. An individual is about $\frac{1}{1000}$ in. long. Water abounding with this animal has a bitter, spicy taste and is undesirable for drinking purposes. The odor or taste is due to the freeing of aromatic oils so strong that one can detect by odor the presence of 1 part of the oil in 25 million parts of water. Copper sulfate in the water will keep the animals down and if used in small quantities will not harm the water for drinking purposes.

Dinobryon, also found in fresh water, gives the water a fishlike odor and commonly fouls waters in reservoirs and ponds.

Uroglena, found in drinking water, gives flavor like that of cod-liver oil. An individual is about $\frac{1}{125,000}$ in. long.

The green color appearing in stagnant water in late summer may be due to a number of protozoa, among them *Euglena*. Some relatives give a red color to snow in various parts of the world and some give a yellowish cast to sea water. *Noctiluca* may color the sea red or yellow by day and may glow with intense phosphorescent light at night, particularly in the wakes of vessels or animals moving over warm seas. In the Mediterranean there are about 4 billion protozoans to every quart of water, some being found at depths of over 3,000 ft. For the most part these creatures serve a useful function as basic food for larger organisms or in breaking down undesirable compounds.

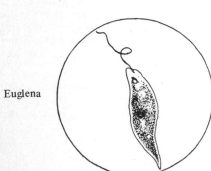

Euglena

Many microscopic living things inhabit the soil. In 1 acre of rich soil there should be about 250 lb of bacteria and an equal weight of microscopic animals and fungi. A single gram of such soil might well contain close to 10,000 individual animals. Representative of these is *Amoeba*. At least six genera and many more species of amoebae are found is fresh water and in salt water. Many of these are within bodies of animals, one being the cause of amoebic dysentery in human beings. This disease might easily be a critical factor in deciding a war or in maintaining a successful peacetime economy. Other amoebae are highly useful in preserving the fertility of the soil by assisting in the breakdown of certain organic compounds.

Of the Ciliata, we show *Paramecium*, commonly studied in beginning biology classes. There are some 2,500 species of ciliates, sometimes spoken of as Infusoria because they are commonly found in hay infusions.

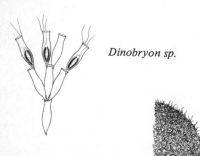

Dinobryon sp.

Uroglena sp.

Paramecium

Amoeba

PHYLUM PORIFERA

The Porifera are aquatic animals that live attached to the bottom. Sponges exhibit radial symmetry; generally have a cylindrical body either branching or irregular. They possess an internal skeleton of spicules and/or fibers (spongin); have a surface with many pores which connect to canals and chambers lined with flagellated collar cells. There are three classes: Calcarea, calcareous sponges, with calcareous spicules, marine forms which occur in shallow water; Hexactinellida, glass sponges, with siliceous spicules, marine forms which occur at relatively great depths to 300 ft; and Demospongiae, with skeleton that may be siliceous or of spongin or both, or none, both marine and fresh-water forms.

CLASS CALCAREA

Order Heterocoela./Family Grantiidae

Simple Sponges
Scypha coronata
Length about ¾ in. Like a slender goblet or vase, with central cavity connected with the outside by numerous canals through which water is kept in motion by whiplike flagella. Mass supported by spicules of harder material. Attached to ocean bottom. Sometimes listed as genus *Sycon.*

Related classes include larger commercial forms. Found attached to rocks in shallow salt water, but some are found in fresh waters on submerged objects. Some 400 species of Calcarea, nearly 10 times that number in other groups, and many others as fossils. They date back to Pre-Cambrian times.

Reproduction by budding, formation of gemmules, which survive drying when parent colony dies, or by sexual reproduction. Animal produces both male and female cells in jellylike middle layer. Eggs become fertilized and after some development are freed as free-swimming larvae which finally settle down and start a new colony.

Feeding by extracting food from stream of water forced through canals in walls and out through opening at top. A good-sized sponge may force a gallon of water an hour, day in and day out, through itself. Subject to number of diseases and in some parts of world have practically vanished.

Too small to be of great economic importance, but interesting to zoologists because of relative simplicity. They may be listed in materials furnished by biological supply houses and may find a limited market for this purpose.

Simple Sponges

Fresh-water Sponge

CLASS DEMOSPONGIAE

Order Haplosclerina./Family Spongillidae

Fresh-water Sponges
Spongilla fragilis
Usually yellow or brown but sometimes forms green patches of spongy material on objects in standing or in running water. *S. lacustris,* the commonest fresh-water sponge, is branching and usually green, while *S. fragilis,* next most common, does not branch.

S. lacustris found typically in sunny, running water in late fall. *S. fragilis,* in standing or running water where there is no direct sunlight for long periods of time. Relatively common *Ephydatia fluviatilis* found either in still or in running water. Some members of family are found in brackish water and some in water 200 ft deep.

Fresh-water sponges usually die and disintegrate in autumn, forming gemmules, which have hard coverings and live over until next spring. Gemmules may survive drying for over a year and may survive extremely cold weather. Young leave parent body as swimming larvae.

While common fresh-water sponges appear as shown, some are cushion-shaped (as *Ephidatiamülleri).* They vary greatly in color and in site selected for growth.

Food, minute animals. Sponges may themselves be food for larger animals, although they are of little economic importance.

Order Keratosa./Family Spongiidae

Bath Sponges
Euspongia officinalis
Skeletons of these animals at one time well-known in homes and in commerce. Common sponge once used on automobiles was horse sponge, *Hippospongia equina,* while the more delicate household sponge was the bath sponge, *Euspongia officinalis.* Fibers are of spongin, often with sand embedded; they enclose cavities of variable size; cavities large in the horse sponge, and smaller in others. Sponge, as a household item, has been largely replaced by a manufactured item, of cellulose, which resembles the sponge in its properties as a household cleaning item.

(Continued)

Sheep's-wool Sponge

Marine animals. Best bath sponges are from Asia Minor and are highly elastic and light yellow; but they are also found in the Bahamas, West Indies, and Australia. Florida and Bahamas produce sheep's-wool sponge, *H. gossypina;* and Florida, the West Indies, and the Mediterranean produce horse sponge.

Young bath sponge may have a conic form and one opening; but with maturity it becomes broader with many openings. Sponges may live at least 50 years. Commercially attempts have been made to plant them by attaching small portions to sunken concrete blocks. Sponges can regenerate lost parts and may reproduce sexually or by budding.

Feeding habits are much like those of other sponges.

Collected by diving or hooking in summer months. Dredging destroys young animals and is poor conservation practive. Animals are allowed to die out of water and are then rotted in water, beaten, squeezed, and washed under water; then dried, bleached, trimmed, and packed. Sponge, as a household item, has been largely replaced by a manufactured item made of cellulose, which resembles the sponge in its properties as a household cleaning item.

PHYLUM COELENTERATA (CNIDARIA)

Coelenterates are some of the lower many-celled animals and the lowest animals which possess tissues. They exhibit radial symmetry. They occur as individuals or as colonies of two types: a *polyp* with a tubular body having a closed end which may be attached and the other end with a central mouth usually surrounded by soft tentacles; or a *medusa,* with a gelatinous umbrella-shaped body with a margin of tentacles and a central mouth on the underside and in the middle of the umbrella. Tentacles are used for taking the food. Coelenterates are of some economic importance; for example, some coral is used as jewelry; some are pests to bathers who suffer from their tentacles' stinging cells. Some corals form the 1,500-mile Great Barrier Reef off Australia which is an offshore protective shield to shoreline erosion. Efforts of some magnitude are being developed to control *Acanthaster planci,* a starfish, which is destroying the coral reefs. Some Pacific island natives eat jellyfish.

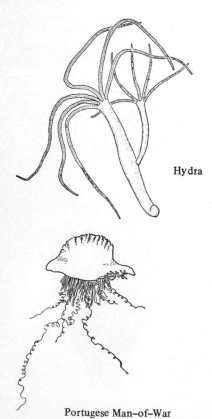

Hydra

Portugese Man-of-War

CLASS HYDROZOA

Order Hydroida

Hydra
Hydra littoralis
Body about ⅔ in. long, with tentacles to 1⅓ in. long. Pinkish or greenish-orange. Attached at base but free at oral end. In six common species, four have individuals either male or female, while two have both sexes in the same individual.

Relatively common in fresh water, attached to bottom or to objects on it. Five American species described. Related green hydra *Chlorohydra* bears green algae in outer layer. Brown hydra *Pelmatohydra* has lower portion of stalk narrowed. Former common in America and Europe, latter most numerous in central states.

Reproduction by budding or sexually, latter commonest in fall. Egg (embryo), surrounded by sticky shell, may drop to bottom or stick to objects. Tentacles of many forms, of which three types have the following functions; one holds prey with nettling cells that paralyze it, one coils and helps in locomotion, and one holds animals fast.

Food, small animals captured by tentacles, sometimes including fish fry; in hatcheries they may be pests. Are themselves food for larger animals. Commonly studied in beginning biology classes as representatives of primitive structure, reproduction, defense, food getting, and so on.

Order Siphonophora

Portuguese Man-of-war
Physalia pelagica
Appear like blue elongate bladders with mass of long streamers beneath. Appear on surface of sea or washed on shore and are about 4 to 5 in. long, with wrinkled crest. Tentacles beneath may be 50 ft long and of many types, some relatively short and straight, others long, slender, and frequently coiled.

Often common in Gulf Stream from Florida to Bay of Fundy. Pacific Coast animal is *Velella lata,* which looks like a dark blue oval elliptic raft, with a clear triangular sail above and short tentacles beneath. Pacific Coast form belongs to order Chondrophora, while Atlantic Coast form, with about 30 other species, is of the suborder Cystonectae.

PHYLUM COELENTERATA (CNIDARIA). CLASS HYDROZOA

Colony of specialized individuals. Some are unbrellalike and produce eggs and sperms which are involved in sexual reproduction but are not freed from colony. Others serve in moving or in capturing food. Stinging tentacles can sting a human worse than many wasps. These sting fishes and other animals and bring food to colony for consumption.

Dangerous to swimmers or beachcombers who may handle them. Commonly noticed from ships particularly in Gulf Steam. Bladders may be inflated or deflated at will, much to amusement of ships' passengers who watch them from decks. Of little direct economic importance but interesting to zoologists. There is disagreement as to classification of these organisms.

CLASS SCYPHOZOA [Jellyfish]

In most of these, there is alternation of n and 2n number of chromosomes from generation to generation or an alternation of sexual and asexual methods of reproduction in which the free-swimming medusa stage is usually more conspicuous than the hydroid stage. Hydroid stage may be less than ½ in. long while medusoid stage may be 6 ft across top of "umbrella," with tentacles over 100 ft long. All are marine; relatively common on shores and offshore.

At least five orders, of which Stauro-medusae have no free-swimming medusoid stage, Coronatae have constriction about middle of umbrella, Rhizostomae have no tentacles on margin or subumbrella, Cubomedusae have four long marginal tentacles or tentacle groups, and Semaeostomeae have eight or more marginal or subumbrella tentacles.

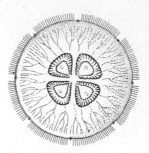

Order Semaeostomeae

Of 3 families, Pelagiidae have long marginal tentacles, Ulmaridae have short marginal tentacles, and Cynaeidae have long subumbrella tentacles but no marginals. May grow from 1-in. diameter in April to full size in July.

Common Jellyfish

Family Ulmaridae

Common Jellyfish
Aurelia aurita
Disk may be 10 to 35 in. (in some species 7 ft) across. Color white or bluish with pink sex organs. Body flat, with four large pockets, oral arms, that curve toward center and eight marginal indentations, 10 genera in family. Eyes and nerve centers in notches on margins. About $\frac{1}{20}$ of body weight is solid. Some species flash lights in wakes of ships.

Found in Atlantic and Pacific Oceans, breeding in summer and sometimes lasting through winter. In Britain it may survive being frozen in ice, but this is fatal in Florida. Specimens from Halifax die at 29.4°C, which is temperature at which some Florida individuals are most active. Some are eaten. Sting.

PHYLUM COELENTERATA (CNIDARIA). CLASS ANTHOZOA

In this class, no medusa generation is present. Body attached sometimes permanently to foot. Mouth a slit at opposite end surrounded, in same species, by hundreds of hollow tentacles. All members of the subclass Alcyonaria have eight tentacles. Most of the class have horny skeletons and most animals have only one sex. All of the 6,000 living species are marine.

[Subclass Alcyonaria]
Order Gorgonacea
The order, Pennatulacea, sea pens and sea feathers, are capable of independent movement while others are not. Gorgonacea, including red corals and sea fans, are stationary and have a central axis. In latter, family Corallidae include red corals which are hard and branch profusely, while Gorgoniidae or sea fans are more erect and usually branch in one plane. Red corals are the coral of commerce used in jewelry and otherwise.

Sea Fan

Sea Fan
Gorgonia flabellum

Colony like a fan or flattened tree, up to 20 in. high with meshes up to ¼ in. across. Yellow or red, though living animals are themselves yellowish-white. Horny material is a protein, lower in sulfur content than true horn such as is found in vertebrates.

Found in shallow seas in West Indies and South Atlantic. Sea fans sold for ornaments and serve as bases for other marine animals. Substance gorgonin, which forms central support, is replaced by solid calcium carbonate in red corals. Lend unique beauty to warm seas by color. Some are phosphorescent.

Commercial red corals at best in Mediterranean and off Japan. Horny relatives of sea fans are known from Tertiary times and calcareous relatives of red corals from Cretaceous, though red corals are not themselves known as fossils.

[Subclass Zoantharia]

Subclass includes about 1,000 species of sea anemones and 2,500 species of stony corals. Sea anemones are abundant along seacoasts. In warmer waters corals may form high reefs. Forms in this subclass are widespread; e.g., some occur in polar regions, others in deep waters. Characterized as anthozoans which never have eight single septa or eight branchlike tentacles like Alcyonaria.

Order Actiniaria

About 1,000 sea anemones belong to this order and other orders of Zoantharia, some free-swimming but more commonly attached by a broad sucker foot. Some of the most beautiful members of the animal kingdom. Quite diverse in form. Oral disk, at top of animal, may bear one or more whorls of tentacles over entire surface. Many feed on small particles. Descriptions of sea anemones appear in most beginning biology texts.

Sea Anemone

Family Sagartiidae

Sea Anemone
1. Sargartia sp.
2. Metridium dianthus

(1) *Sagartia leucolena* with body to 2½ in. long and to ⅓ in. through; with translucent fleshy color and 96 retractile tentacles in four rows. (2) *Metridium dianthus* with length to 4 in. and width to 3 in.; varies from brown to yellow and is one of the largest and commonest Atlantic Coast sea anemones.

Animals serve as food for a few larger forms of marine life, but some are avoided because of their stinging tentacles. Members of one genus, *Adamsia,* become fixed to backs of hermit crabs or other crustaceans and are moved about by them; provide some protection for their bearer and in return may share his meal.

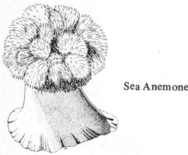

Sea Anemone

Order Zoanthidea
Small, anemonelike. No skeleton. Many attach themselves to hermit crabs, sponges, hydroids, and other marine objects. Most abundant in warm tropical waters but widely distributed. Not illustrated in this fieldbook.

Cerianthus sp.

Order Ceriantharia
Anemonelike anthozoans with greatly elongate bodies adapted for living in sand burrows. Bodies cylindric, muscular, and smooth. Forms tube by secreting mucus in which sand grains and detritus collect. Animal lives on bottom with only tentacles and oral disk extended. *Cerianthus,* the most familiar genus, is common in locations with sandy bottoms in both American and European waters; may be found in shallow or deep water.

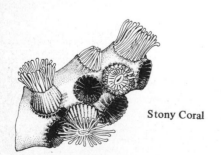

Stony Coral

Order Madreporaria
Made up of true or stony corals. Chiefly responsible for coral reefs and islands of warm coral seas. Generally occur in colonies, but may be solitary. Always attached to firm substratum. Possess hard calcareous covering. Prefer depths from 100 ft to 350 ft, but exceptions occur. Most abundant source of reef-forming corals in Great Barrier Reef, Australia, where more than 200 species have been reported. Polyps of animal expand at night or during dark overcast days and feed on the abundant plankton of the warm seas.

PHYLUM COELENTERATA (CNIDARIA). CLASS ANTHOZOA

Family Astrangidae
Astrangia danae
Beautiful small coral; grows in shallow water, in colonies up to 3 in. across. Encrusts on rocks and shells in sheltered locations. Common in cooler waters. Found on Atlantic Coast of U.S. from the Carolinas north. Coral skeletons often beautifully marked. Fossil forms of this and other corals used in dating deposits. Not a reef-building coral. Can be kept in aquariums of clean sea water, fed small animals or bits of meat.

PHYLUM PLATYHELMINTHES [Flatworms]

These soft flat worms lack a body cavity, a distinct head, or paired appendages for the most part. In some, even a mouth is lacking and there is no circulatory or respiratory system. Three classes: free-living Turbellaria, which have bodies with external cilia; and parasitic Trematoda and Cestoidea. Trematoda are flukeworms, with a mouth and intestine, but small and unsegmented. Cestoidea are tapeworms, with no mouth or intestine but with segmented bodies. Over 12,000 species.

CLASS CESTOIDEA

Order Cyclophyllidea./Family Taeniidae

Beef Tapeworm
Taenia saginata
To 30 ft or more long, with over 1,000 segments each under ⅓ in. long (except end ones which may be nearly 1 in. long) and containing reproductive organs. Head end smallest, with hooks useful in holding worm to lining of host's alimentary tract. Both sex organs in a single segment.

Terminal reproductive segments become freed in dung of host. From these escape hooked embryos that may be eaten by cattle. Develops in flesh of cattle and goes into resting stage. If this muscle is eaten poorly cooked or raw by man, the creature develops to maturity in man's alimentary canal.

Commonest human tapeworm. It robs human beings of their food and may make them subject to disease, though worm itself rarely causes fatal injury. Control is effected by eating only clean well-cooked meat or by freeing animal from its hold and then flushing it out with a physic. Cattle may be restrained from eating grass growing in dung and thus avoid infection.

Beef Tapeworm

CLASS TREMATODA

Order Digenea./Family Fasciolidae

Sheep Liver Fluke
Fasciola hepatica
Cause of disease commonly called "liver rot." Large thin worm (to almost 2 in. long); lives in bile ducts of many mammalian hosts, commonly in sheep and cattle, infrequently in man. Commonly illustrated in introductory biology textbooks. Costly to sheep production because sheep infested become subject to other diseases. Disease thrives only on ranges and pastures with marshy areas where alternate host, a snail, occurs. Common in many parts of the U.S. and Europe, and in other parts of the world.

Body leaf-shaped, rounded at head end and bluntly pointed at tail end. Life cycle complex, egg shed in feces of sheep develops into free-swimming form which enters snail; develops within snail and emerges into water, travels to grass blade; attaches itself and then is injested by another host (sheep). Effective controls not available.

Sheep Liver Fluke

CLASS TURBELLARIA

Order Tricladida./Family Planariidae

Planaria
Dugesia tigrina
Common inhabitants of fresh waters in temperate zones. Frequently easily collected, sometimes in large numbers under stones along shores of ponds and streams. Commonly found in springs and spring-fed waters. Can be "baited in" by using raw beef on a thread dangled in the water. In nature, probably feed on small crustaceans among other things. Easily kept in laboratory in covered jars. Should be kept cool. Can be fed thin strips of liver, earthworms, or clams. Mouth is on ventral surface and is at the end of a tubular pharynx which can be extended to capture food.

Planaria

(Continued)

Have triangular heads with conspicuous auricles, "ear-shaped extensions," and two eyespots. Body tapered, brown to black; commonly spotted above. May have a light stripe along midline of dorsal surface. One tenth in. to 1 in. long.

Occurs throughout U.S. Exists in both sexual and asexual phases. If sexual reproduction takes place, cocoons in the form of brown to black spheres are formed.

Planarians avoid strong light and remain under objects during the day. They crawl actively after dark. Head region responds to weak mechanical or chemical stimuli. Have been used in regeneration experiments in which the head end is split several times. Generally "new heads" are regenerated in laboratory studies.

PHYLUM ASCHELMINTHES

A complex group of animals which are classified many different ways. Bilaterally symmetrical, wormlike, mostly cylindrical in shape but sometimes flattened. Not segmented. Fresh-water, marine, or terrestrial; free-living, commensal, or parasitic. 12,500 species. Five classes: [1] Nematomorpha, here represented by hair snakes, which look like a living hair from a horse's tail and are parasitic larvae in arthropods and free-living as adults; [2] Nematoda, unsegmented roundworms; many are free-living in the soil or water; others are parasitic on plants and animals. Most are small, minute; some grow to a meter in length. In this fieldbook nematodes are represented by two human parasites, hookworm and ascaris. The vinegar eel sometimes found in vinegar which is not pasteurized, and *Trichinella,* the cause of human trichinosis, are nematodes. [3] Rotifera, minute to microscopic animals, are attractive because of their color and movements. Class name Rotifera comes from many beating cilia on the anterior end of body which appear like a wheel when in motion. Free-living in fresh water such as ponds and lakes, even in ditches and eaves troughs. *Hydatina,* a fresh-water rotifer common in field collections, is discussed in this fieldbook; [4] Gastrotricha which in size and habits resembles some ciliate protozoans, occurs in fresh and salt waters, common in bottom samples taken by aquatic biologists and associated with algae; [5] Kinorhyncha, sometimes listed in Echinodera, are small, to 1 mm in length, marine worms of the bottom sand and mud in both shallow and deep waters. They feed on small algae and detritus.

Hookworm

CLASS NEMATODA

Order Strongylida. / Family Ancylostomidae

Hookworm
Necator americanus
Length of some, female ⅔ in., male ⅓ in. White. Head end narrower than body, curved backward. Mouth large, with oblique opening above and with cutting teeth or plates and two lips on lower margin; upper portion separated by conical tooth which projects into it.

Eggs pass out of host in dung. Young worms live in water or moist earth and infect new host by piercing skin of feet, or by being ingested in drinking water or on infected substances (such as unclean garden vegetables). Young bore into blood vessels, are carried to heart and lungs, then to throat; later become attached to walls of small intestine.

Probably about 2 million residents of United States are infected with hookworm; more in other countries. Going barefoot, eating infected food, or drinking infected water are common sources. Human wastes should be disposed of safely. Oil of chenopodium, tetrachlorethyl, or carbon tetrachloride are used as drugs to drive worms from infected body.

PHYLUM ASCHELMINTHES. CLASS NEMATODA

Order Rhabditida. / Family Ascaridae

Ascaris
Ascaris lumbricoides
Length of female to 16 in., male smaller, 6 to 10 in.; diameter of female to ¼ in.; male smaller. Living specimens white or yellow or pink. Preserved specimens common in biology laboratories are gray. Body slender, almost circular in cross section, tapers toward both ends. Mouth end rounded with three flaplike lips, tail end somewhat pointed.

Adults live in intestine of man. Other vertebrates, in particular pigs, are also infected with *Ascaris* but usually species other than *lumbricoides*. Reproduction takes place in intestine of the host. One female may lay 200,000 eggs a day. Eggs discharged with fecal wastes and develop in moist earth. Eggs taken into next host usually on plant foods; move to intestine where they hatch. Young worms penetrate the tissue of the intestinal wall and get into the venous circulation of the abdomen; travel usually to the liver, then the lung, to the trachea, then down the throat to the stomach and back to the intestine.

Path of worm's migration through the body described known for some lower vertebrates and probably similar in man.

Ascaris is common where human wastes are not treated and where food is not properly cooked. In oriental countries it appears to be more common in rural areas. Medical treatment, antihelminthetic, is available and infestation in man can be controlled.

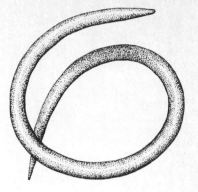

Ascaris

CLASS NEMATOMORPHA

Order Gordioidea. / Family Gordiidae

Hairworm, Hair Snake
Gordius lineatus
Some hairworms are over 2 ft long, though this species is shorter. Male has V-shaped ridge at end; this is lacking in female. Commonly found in water or on wet vegetation or inside insects and other animals, often coiled into a "Gordian knot," which gives animal its name. Food absorbed through body wall.

Eggs laid in water in strings, in some species 8 ft long and numbering to 6 million; hatch into larvae; these enter some water insects and develop into resting stage. If host is eaten by fish, bird, or insect, next stage develops into typical worm, which lives in alimentary tract of host until mature, then leaves host, enters water, mates, and lays eggs.

Interesting example of parasitic worm whose life history is not understood by layman, who will readily believe fanciful theory rather than accept discovered fact. Some who will not accept the theory of evolution readily believe that a horsehair can turn into this worm even though attempts to prove it always fail.

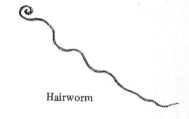

Hairworm

CLASS ROTIFERA

Order Enoplida, Monogononta. / Family Brachionidae

Rotifer
Hydatina senta
Known as fresh-water rotifers and wheel animalcules. Common in fresh water. Microscopic, yet one of the larger rotifers. Classified in a variety of ways by different authors; sometimes in a class, Rotifera, or in a phylum, Rotifera. In fact they have been classified in many phyla from the protozoans to the arthropods. This genus also listed as *Epiphanes*. Free-living. Body pear-shaped or cone-shaped; with a "head" region and a narrow taillike foot. On the head region, there is a so-called corona which is rimmed with cilia, hairlike projections. These cilia beat and look like a whirling wheel, therefore the name, rotifer, wheel animalcules. The beating cilia bring food into the animal and also are used in locomotion. If one looks carefully at the corona, it may appear to be made up of two continuously moving wheels. Female produces three types of eggs: "summer eggs," which develop without fertilization into females, parthenogenic "male" eggs, and thick-covered resting eggs which have been fertilized and develop into females only.

Rotifers are probably important links in food chains of fish; some species are of interest because of their ability to withstand desiccation, and other species because there are apparently only females. Anyone examining a drop of fresh water containing living things under a microscope is almost certain to find rotifers, a fascinating group of animals.

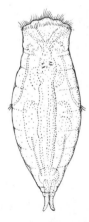

Rotifer

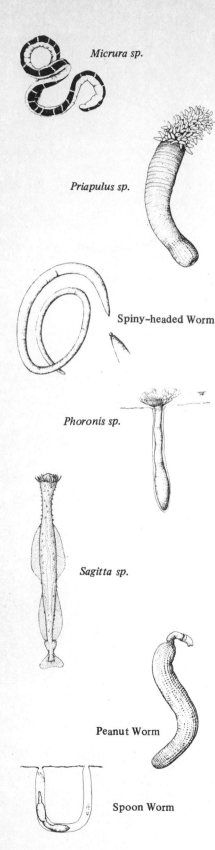

Micrura sp.

Priapulus sp.

Spiny-headed Worm

Phoronis sp.

Sagitta sp.

Peanut Worm

Spoon Worm

PHYLUM NEMERTINEA [Ribbon Worms]

Some 550 species. Slender worms with soft, flat, unsegmented bodies. Length from ⅕ in. to 80 in. Color varies; red, brown, yellow, green, white; some uniformly colored and others striped or banded. Mostly marine, living among algae and stones between high and low tide region. Some few species inhabit deep waters.

Genus *Micrura,* small, flat and soft nemertean with a copulatory appendage on tail end. One species, *leidyi,* may be to 6 in. long; red or purple on back, usually with a lighter median line and lighter below. Common from New Jersey to Cape Ann, in the sand near low-water mark. Breeds in summer.

PHYLUM PRIAPULIDA, Priapulids

Priapulus sp.
Eight known species described; common genus is *Priapulus.* Priapulids are wormlike creatures, cylindrical; may be yellow or brown and to 3¼ in. long. They inhabit mud or sand bottoms from Boston and the Belgian coast north; and from Patagonia to Antarctica. Found in crevices in rocks or between external structures of sessile animals such as mollusks. Predaceous, feed on soft-bodied invertebrates.

PHYLUM ACANTHOCEPHALA [Spiny-headed Worms]

Macracanthorhynchus hirundinaceus
Some 500 species described. Worm-shaped parasites of peculiar structure and function which live as larvae in arthropods and as adults in the intestines of vertebrates. The distinctive feature of acanthocephelans is an anterior cylindrical proboscis which bears rows of recurved spines serving to attach the worm to the gut of its host. *Macracanthorhynchus hirundinaceus* is common in pigs and may occur in man. Its intermediate host is the "white grub" of the June beetle.

PHYLUM PHORONIDA [Phoronids]

Phoronis sp.
Fifteen species; all marine, in temperate regions. Live in self-made tube which may be buried in the sand or attached to objects such as shells, rocks, or piling in shallow water. Plankton feeders; food particles stick to mucus-covered tentacles. Never emerge from tube. May be red, orange, or green and, if abundant, give color to shallow sea bottom. *Phoronis* of Pacific Coast grows to 8 in.

PHYLUM CHAETOGNATHA [Arrowworms]

About 50 species; length 1 in. to 4 in.; marine, torpedo-shaped, transparent animals of plankton; with fins on sides and tail fin. Move swiftly. Most abundant in warm, shallow seas but occur in cold polar waters. Mouth bordered by bristles adapted for seizing prey. Has paired compound eyes. Hermaphroditic. Newly hatched young are miniature adults. Voracious feeders. Eat single-celled plants and animals, larvae of crustaceans, and other small marine animals. *Sagitta,* arrowworm or glass worm, will eat herring as large as itself; bristles used to capture and hold prey. Eaten by larger plankton feeders of the sea.

PHYLUM SIPUNCULOIDEA

[Peanut Worms]
250 species, vary in length from ¹/₁₀ in. to 24 in.; slender or chunky, unsegmented yellow to gray-colored worms; marine; may be found in sand or mud or empty shells on seashore. Found from high-tide mark to great depths. Has a fringe of tentacles at one end which can be withdrawn when disturbed; then takes on club shape or peanut shape; hence the common name. Feeding habits not well understood but probably take soft-bodied organisms and plankton which are caught on its tentacles.

PHYLUM ECHIUROIDEA [Echiurids, Spoon Worms]

About 60 species. Gray or reddish or yellowish worms; live in mud or sand burrows; usually in shallow water. Recent studies in the North Pacific have discovered many new species at great depths. Found in warm or temperate seas. Commonly digs U-shaped burrow. Varied lengths from ³/₁₀ in. to as much as 5 ft.

Urechis lives along the California coast; builds a U-shaped burrow. Feeds by producing mucus in funnel-shaped collar at one end of burrow; then pumps water through the funnel; particles are trapped in the mucus. The animal then ingests the mucus funnel and its contents. Some larger echiurids are used as fish bait in Europe; some are eaten by rays and bony fish.

PHYLUM BRYOZOA [Moss Animals]

Some 3,500 marine and 35 fresh-water species of living Bryozoa, and thousands of fossil species. Commonly occur in colonies of thousands of animals. Each animal usually small, cylindrical, and partly surrounded with a hard or limy wall which persists after death.

CLASS GYMNOLAEMATA

Order Cheilostomata./Family Bugulidae

Bugula
Bugula turrita
Branching, limelike tufts, with branches in spirals, the whole forming dense clusters 1 ft or more across. Lower stem often yellow or orange; upper parts, white or yellow. Some of the small, cylindrical animal cases may have attached to them a specialized individual which, under magnification, looks like a bird's head and beak (avicularium). The avicularium helps keep the colony free of debris and traps small organisms.

Common between low-water marks and to depth of 100 ft, covering rocks and piles from North Carolina to Maine, with many other kinds found throughout the world.

Young partly developed within parent body, freed through opening, swim about near surface, then settle and start a new colony which develops by budding. Two ends of digestive tract are near each other. Tentacles capture food and help in water circulation. Special cells, with birdlike beaks, capture minute living things for food. No locomotion after colony is fixed. Intercommunication between colony members possible although there is no nervous system. No special sense organs.

Have slight value as food to some marine animals. Have been considered as seaweeds, then as corals and hydroids, and now as a degenerated group of animals. True nature recognized about 1830. One can collect young *Bugula* colonies by submerging glass slides in the sea for 1 to 2 weeks.

Bugula, Moss Animal

CLASS PHYLACTOLAEMATA

Order Cyclostomata./Family Plumatellidae

Fresh Water Moss Animal or Fresh Water Bryozoan
Plumatella polymorpha
Colony made up of brownish or transparent cylindrical branching tubes. With 30 to 50 tentacles. These, as a colony, may trail like a vine or grow in a clump. Some stand erect. Top of the tube, sometimes referred to as the head, is white. May be hundreds or thousands of individuals in one clump on surface of a submerged stone. May cover a square foot in area. Genus *Plumatella* represents the commonest freshwater bryozoan.

Prefers place with low light intensity; so frequently in either dark or shaded submerged freshwater location; common on undersurface of floating or attached objects. Ponds, streams, on water pipes, in slow-moving waterways.

Widely distributed in the eastern U.S. and west to Montana.

Single individual, zooid, may produce one or more new individuals by budding. Buds are produced throughout the summer in large numbers. In winter, colonies may die leaving their tubes filled with old buds which have hard brown bivalve cases; very tiny, microscopic. When conditions for germination occur, the old buds start new colonies. Also reproduces sexually.

Individual animals are visible with a hand lens. May be collected, placed in a dish or pan for observation. If the dish is not disturbed, the animals will be seen to constantly rotate their headlike cluster of tentacles. Tap the container or disturb the animals in any way and they will retreat into their tubes. Tentacles appear like a horseshoe-shaped wreath.

Food, microscopic organisms, chiefly diatoms. May become troublesome pest in sources of drinking water and have been, in a limited number of instances, removed by using shovels. Not known to be favored food of any animal. Probably ingested at times as part of plankton.

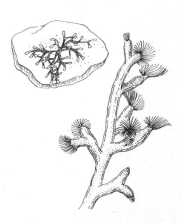

Fresh Water Moss Animal

PHYLUM ANNELIDA. [Annelid Worms]

Some 6,500 species of living annelid worms. Have elongate segmented bodies with distinct heads, digestive tract, and commonly paired unjointed appendages. Three classes, some of which have common well-known forms. Hirudinea or leeches have suckers at either end and bodies with external rings. No other classes have suckers. Remaining two classes with setae. The Polychaeta have distinct body segmentation and are marine. The Oligochaeta, including the earthworm, are fresh water or terrestrial, and have definite body segmentation.

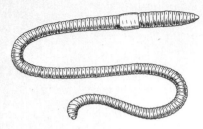

Earthworm

Leech

CLASS OLIGOCHAETA

Order Megadrili./Family Lumbricidae

Earthworm
Lumbricus terrestris
Up to 11 in. long, with pointed head end and flattened tail end. Sex organs: those of male open in fifteenth segment; of female, in fourteenth. Bristles on underside together with secreted mucus enable animal to climb even on smooth glass. Body can be retracted into burrow quickly when necessary.

Animal burrows underground but feeds outside, mostly at night on decaying organic material. Favors damp soil and is found in Europe, America, and other temperate regions where conditions are suitable.

Two individuals mate by placing sperm in receptacles in each other, each individual lying with head toward partner's tail. After mating a secreted cocoon is shed over head, taking with it eggs of one and sperms of other. Eggs hatch 2 to 3 weeks after fertilization in cocoons; young worms like adults in general appearance. Animals are sensitive to soil vibrations and may be brought to surface by vibrations of post driven into soil.

Darwin contended that 1 acre may contain 63,000 earthworms which in a year may bring 18 tons of soil to surface and in 20 years build a new 3-in. layer. Do not "rain" down as is sometimes supposed; rather are forced out of their burrows which become filled with water.

CLASS HIRUNDINEA

Order Rhynchobdellida./Family Glossiphoniidae

Leech
Placohdella parasitica
Length to over 2 in. Greenish-black with suckers at each end. Hang on with one end while feeling about with other. Individuals bear both sex organs, the male opening between eleventh and twelfth segments; the female on the twelfth.

Commonest American leech, usually found on turtles, stones, and other submerged objects. May be more active at night in search of animal food.

Two individuals mate, eggs of one being fertilized by sperms of other. No cocoon formed in this species, though some make cocoons as earthworms do. Young may remain attached to parent, which they resemble in shape although usually more greenish in color. In species which lay eggs in a cocoon, this is formed on the ninth, tenth, and eleventh segments. European medicinal leech belonging to order Gnathobdellida may be 8 in. long and has been introduced into some streams in eastern United States.

Medicinal leeches were formerly used by physicians for bloodletting. Animals common in temperate zones are pests, which may or may not be serious enemies of living things.

PHYLUM ECHINODERMATA

Nearly 6,000 described species in this group, all marine or brackish-water animals. Bodies radially symmetrical, usually five-rayed. The five classes include armed Crinoidea, Asteroidea, and Ophiuroidea, and armless Echinoidea and Holothuroidea. In Crinoidea, or sea lilies, mouth opens upward. In Asteroidea, or starfish, there is a deep groove along each arm, which is lacking in Ophiuroidea or brittle stars. In armless groups, Echinoidea, or sea urchins, have firm body walls while the Holothuroidea or sea cucumbers, have soft body walls.

CLASS ASTEROIDEA

[Sea Stars]
Order Forcipulata./Family Asteriidae

Starfish
Asterias forbesi
Usually five-armed. Width up to 6 in. Greenish-black, with orange plate near base of two arms, and red eyespot at end of each arm. Upper surface with coarse spines; lower, with rows of tubular walking feet. Mouth under central disk. Sexes usually distinct. Can right self if placed on back. Sensitive to light, disturbance, or chemical change.

Starfish

If cut in two, parts may regenerate. Eggs laid in June, hatch into transparent larvae which swim at surface, feed on animals until about ⅓ in. long, then in a few hours take starfish form and settle to bottom. May reach 2½ in. size in 1 year and breed. Young may eat 50 young clams in 6 days, and large animal may exert 2½ lb pressure on clam it is attempting to open. May thrust own stomach into clam it can only partly open.

Found from shore to depths of 180 ft with larger species on northern Pacific coasts. This species ranges from Maine to Mexico, with related species far-flung. One of greatest enemies on beds of oysters and clams, destroying annually millions of dollars' worth of these valuable food animals.

CLASS OPHIUROIDEA

[Brittle Stars]
Order Ophiurae./Family Ophiodermatidae

Brittle Star
Ophioderma brevispinum
Central disk, five sides, thin, about ½ in. across, with five slender arms each about 1½ to 2½ in. long extending from corners. Green or brownish-gray. Arms apparently segmented and often broken at tips. Disk rough. Sexes separate. No eyespots evident on arms, but animal retreats from shadow so must be light-sensitive. Lost parts easily restored.

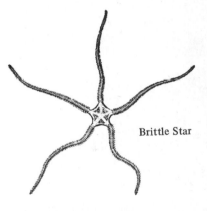

Brittle Star

Egg sacs each with two openings in pairs on lower sides between arms. Water circulates through egg sacs, possibly to supply oxygen. Eggs freed when fertilized to become free-swimming (pluteus) forms which finally transform into adult form and settle to bottom for rest of life. Closely related basket starfish walks along bottom on tips of branches of arms. About 1,000 species of living brittle stars.

Found from shore to depths of over 700 ft; particularly in deep water. This species ranges from Cape Cod south, with other brittle stars extending range to Europe, the Pacific Coast, and the tropics. Interesting to collectors because of difficulty of finding a perfect specimen. Dates back to Devonian ancestor.

CLASS ECHINOIDEA

[Sea Urchins]
Order Arbacioida./Family Arabaciidae (described), Echinidae (illustrated)

Sea Urchin
Arbacia punctulata (described); *Echinus* sp. (illustrated)
Body up to 1¾ in. through, protected by skeleton of lime plates under skin; covered with stout movable ¾ in. spines, usually arranged in five broad areas, separated by narrow unprotected areas. Mouth on underside with five white teeth. Some species have poisonous spines. Five double rows of sucker-bearing tube feet.

Sea Urchin

Eggs number 20 million per individual; expelled from four pores near summit of body; developed into clear larvae with reddish spots and eight rods. Swim for several weeks; then in a few hours change into minute sea urchins which settle to bottom. Strong-toothed mouth grinds animal and vegetable food, even making sand finer by grinding. Teeth and jaws form a flattened sphere shell on beach known as "Aristotle's lantern." Spines usually break off at death.

On sea bottom, from shore to depths of 700 ft; ranges from Cape Cod to Yucatan, but other species make range worldwide. Internal parts, particularly eggs, used as food by man; Marseilles fish market handles 100,000 dozen a year. Animals also produce a dye and provide food for many marine animals.

Sand Dollar

Sea Cucumber (*Stichopus*)

Sea Cucumber (*Cucumaria*)

PHYLUM ECHINODERMATA. CLASS ECHINOIDEA

Order Clypeasteroida./Family Scutellidae

Sand Dollar
Echinarachnius parma
Flattened disks, more or less circular, less than 3 in. in diameter, covered with minute spines and bearing what appears to be a five-parted flower marking in center. Mouth in center. Purple to gray. On beach, fine spines may be worn away, leaving flat dollarlike thin disk.

Common on Atlantic Coast north of New Jersey and on Pacific Coast south of Puget Sound, with other species extending range widely. Found in sea from tide lines to great depths, but usually on a sandy bottom. Able to recognize whether right side up and can turn over on a sandy but not on a hard bottom. Moving spines resemble waving grain.

Sexes distinct. Eggs and sperms freed into sea water by parents and fertilization takes place independent of them. After free-swimming larval stage, known as the "pluteus," young assume shape of parents, settle to bottom, and continue life there. Intestinal opening in this species is on margin. Sand, bearing food, is taken in with help of spines.

Of little importance as food to other animals. Skin and spines ground and mixed with water have been used in making an indelible ink. Fossil species closely resemble living forms, so they have been long-persistent. Hardy and thrive rather well in salt water aquariums.

CLASS HOLOTHURIOIDEA

Order Aspidochirta./Family Holothuroidea

Sea Cucumber
Stichopus californicus
Length to nearly 18 in. Body flat beneath, with three rows of black feet; upper surface covered with nipplelike projections, usually in three rows. Color reddish-brown. At one end 18 to 20 tentacles, yellow-tipped and somewhat like flowers on stalks, with mouth in center of this group. In one related species (*S. johnsoni*), feet are red-tipped.

Common along Pacific Coast. Sea cucumbers are found in most seas at various depths, moving over bottom slowly or resting partly buried in sand and mud, or sometimes among rocks. Atlantic Coast genus, found along Florida coast and in West Indies, is *Holothuria,* up to 30 oral tentacles instead of 20 in *Stichopus.*

Sea cucumbers commonly free eggs and sperms into water where eggs become fertilized and, after free-swimming period, larvae settle to bottom and assume adult activities. In some, fertilization takes place within body of mother; in one, young develop within her body and are freed through opening broken in her body wall. In all sexes are distinct. More sensitive than starfish; tentacles show greater ability to select suitable food, which includes plants and animals. Unduly disturbed, a sea cucumber may eviscerate most of its internal organs. These can be completely regenerated.

Dried, known as *bêche-de-mer.* Used as food by man, being cooked in sea water, dried, then boiled in fresh water or sold as trepang to Chinese who use them in making soups. Greatest use as food centers around Malay Peninsula. Cannot be used as substitutes for cucumbers, as their name implies, since they resemble these fruits in shape only. Have a rather thick leathery or rubbery skin which is not easily utilized as food. Animals interesting to zoologists, not only because of unique regenerative ability but because of ability to receive and respond to a variety of external stimuli.

Order Cucumariidae./Family Cucumariidae

Sea Cucumber
Cucumaria frondosa
In this order, tentacles are irregularly branched like those of a tree. In one genus, *Psolus,* lower surface is flattened, providing a crawling surface. In other genera lower surface is not flattened. In *Cucumaria,* feet are in rows, while in *Thyone* they are scattered over surface. This species is about 10 in. long and 4 in. through.

Common from low-water mark to a depth of 1,200 ft along Maine coast and in Europe. Reddish-brown above and lighter beneath. Other species and related genera extend range greatly. Some 200 species in family.

PHYLUM MOLLUSCA

Some 85,000 species described in this group; estimated 45,000 living species. Bilaterally symmetrical; shell and soft parts coiled in some species. Soft-bodied, covered by a thick or thin mantle that usually secretes a limy shell. Usually with an anterior head and a muscular foot used in locomotion. Respiration usually by gills. Some species found on land but more commonly in salt or fresh water.

CLASS AMPHINEURA

With plates resembling set of false teeth or section of tiled floor. Central plates described as "butterfly shells."

Order Polyplacophora./Family Mopaliidae

Mossy Chiton
Mopalia muscosa   1
Length to about 2 in., apparently composed of jointed sections bordered by mosslike hairs. Outside color commonly obscured by growths. Interior, bluish-green. Removed with difficulty from rock and may fall to pieces when dried. Ranges from Lower California to British Columbia.

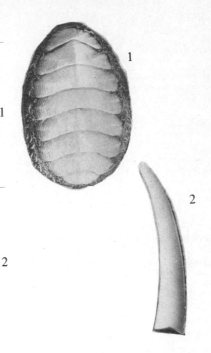

CLASS SCAPHOPODA

Family Dentaliidae
Shells and animals suggest both pelecypods and gastropods. Like pelecypods, they lack true head and are bilaterally symmetric. Like gastropods, they have a single shell. Some are to 5 in. long.

Tooth, Tusk, or Precious Tusk Shell
Dentalium pretiosum   2
Shells resemble small slightly curved hollow delicate white teeth or tusks; to 1 in. or more long; larger open end is the upper, from which animal extends foot in digging sand; from smaller lower end extend tentacles that capture food. Indians used shells for money and made them into beads. Live mostly in sand. Collected by comblike dredges. From Forrester Island, Alaska, to San Diego, Calif.

CLASS GASTROPODA

Gastropods have a single shell, commonly coiled but sometimes greatly reduced, with an opening, the aperture, that may be closed by an operculum. Muscular foot may be drawn into protection of shell. Eyes may be present either on tentacles or at their bases. Food rasped with long filelike tongue that may be extended greatly. They may breathe by lungs or by gills, and may hear by means of special organs, otocysts, on foot. React variously to light. May locate food by smell.

Order Archaeogastropoda./Family Acmaeidae

White-cap Limpet
Acmaea mitra   3
Length about 1 in. Shell like a small cone or tent, with smooth white surface. Animal may move, but when disturbed clamps firmly to support and can hardly be removed with knife. Knife thrust quickly under before disturbance frees animal easily. Eaten by man. More commonly used as fish bait. Shells used as earrings. Aleutian Islands to lower northern California.

Order Archaeogastropoda./Family Haliotidae

Green Abalone
Haliotus fulgens   4
Length to nearly 1 ft. Width ¾ length. Spiral almost obscure. Shell like broad shallow bowl with row of small oval holes along one side. Outside rough, coarse, frequently covered with other organisms. Inside layer of mother-of-pearl or nacre rich in rainbow tints, and glistening. Animal remains attached to rock, breathing through holes. New holes appear with age and old ones fill. Flesh has high commercial value. Found from Point Conception, California, to Magdalena Bay, Lower California.

Order Achhaeogastropoda./Family Fissurellidae

5 Rough Keyhole Limpet
Diodora aspera
Length to over 2 in. but usually smaller. Conical, with elliptical base and with ridges radiating from hole at crest of cone. Outside dark purple to gray-brown. Inside pearly white. Cling to rocks. Cook's Inlet, Alaska, to northern Lower California.

Order Archaeogastropoda./Family Trochidae

6 Puppet Margarites
Margarites pupillus
Length about ⅓ in. Yellow-brown. Whorls four, plainly marked with spiral ridges. Openings nearly circular. Nunivak Island in Bering Sea, to San Pedro, Calif., where it is found in deep water.

7 Black Top Shell
Tegula funebralis
Length to just over 1 in. Purple to black outside. Greenish-white and pearly as outer layers wear or break. Normally four whorls, with upper often worn off or broken. May carpet rocks exposed to violent wave action. Eaten freely by fish. Vancouver Island, B. C., to Cerros Island, Lower California.

Order Archaeogastropoda./Family Trochidae

8 Ringed Top Shell
Calliostoma annulatum
Length to about 1 in. Beautifully and finely engraved pointed cone, with spiral grooves usually marked in deep purple. Edge of lip fine and thin. Live on seaweeds in deep water. On bright days may be near surface. Forrester Island, Alaska, to northern Lower California.

Order Archaeogastropoda./Family Turbinidae

9 Wavy Turbine Shell
Astraea undosa
Length to 4 in. Covered with brown fibrous skin. Whorls wrinkled and ornamented. Operculum horny within. Vegetarina. Point Conception, Calif., to Cerros Island, Lower California.

Order Archaeogastropoda./Family Epitoniidae

10 Angular Wentletrap, Angled Staircase Shell
Epitonium angulatum
Length to ¾ in. With 6 to 11 whorls, there being nine ribs to each whorl. Mouth opening almost circular. Many pure white. Longitudinal ribs represent periods of rest or slow growth. Give off purple fluid when disturbed. Prey on other animals. Eat beef hungrily. New York to Texas.

Family Naticidae

11 Little Moon Shell
Natica canrena
Length about 1 in. Brown with longitudinal zigzag streaks. About five whorls, smooth and regular, with limy operculum. No eyes. Has huge foot that envelops food. Prey is another shellfish. Soft parts eaten out after a hole has been drilled through shell. May eat dead fish and act as scavenger. Often found in sand, burrowing for bivalves. Common from North Carolina to West Indies.

12 Moon Shell, Sandcollar
Polinices duplicatus
Length about 2 in. Width same. Shell solid, with five or more whorls and a prominent spire. Opening oval and oblique. Horny operculum closes opening in shell. Lip thin, sharp. Interior pearly or chestnut. Outside ashy-gray to brown. Found in shallow waters. Massachusetts Bay to Gulf of Mexico.

13 Lewis' Moon Snail
Polinices lewisii
Sometimes to 6 in. through. More normally 3 to 4 in. Yellowish-white, smooth-surfaced, with rather large aperture closed by broad horny brown operculum. Strictly carnivorous, preying on other shelled animals through whose shells a hole is drilled, permitting access to soft flesh inside. Unusually large foot. Moves freely in search of food. Duncan Bay, B.C., to northern Lower Calif.

Order Archaeogastropoda./Family Calyptraeidae
Shells have shelf on inside of one half.

Flat Slipper Shell 14
Crepidula plana
Length to 1⅔ in. Width just over ½ in. Shell frail, flat with white polished interior and milky white outside, frequently curved to conform to shape of support. Female reported to be 15 times size of male. Common in all warm seas, often attached to shells or seaweeds from low-water mark to depth of 3,000 ft. Prince Edward Island to Texas, West Indies.

Arched Slipper Shell, Quarter-deck, Boat Shell 15
Crepidula fornicata
Length to 2 in. Brown or white. Obliquely oval, with a white strong solid diaphragm. Eats seaweed or other mollusks. Found attached to shells or other animals. Has commercial value as base for oyster beds. Common in shallow water. Prince Edward Island to Texas.

Hooked Slipper Shell 16
Crepidula adunca
Length to ⅘ in. Shaped like slipper or small boat, but recurved at tip. Surface brown. "Deck" white. In young animals point at end is sharp and less curved. Queen Charlotte Islands, B.C., to northern Lower California.

Order Archaeogastropoda./Family Littorinidae
Shell spiraled, more or less golbular or top-shaped, with round opening and horny operculum or opening cover. Widely distributed.

Periwinkle 17
Littorina littorea
Length up to 1 in. Width ⅔ length. Shell solid, roughened, yellow, black, brown, or red, with dark bands and somewhat glossy. Inside white or brown. Shell thick, with 6 to 7 whorls, apex acute. Males smaller than females. Head projects, has conical tentacles with eyes at bases. Foot divided longitudinally so that as animal moves it swings alternately from side to side. Tongue two times length of animal. In shallow water or on rocks. Eats plant material. Eggs laid in masses on rocks and weeds. Eaten by man in Europe and in America. Also used as fish bait. Common in European waters. Introduced on our Atlantic Coast and now abundant from Labrador to Cape May, N.J.

Checkered Littorine 18
Littorina scutulata
Length to just over ½ in. Greenish to brownish-gray marked with lighter white bands or spots. Inside of shell conspicuously purple at opening. Clings to rocks between tide marks. Kodiak Island, Alaska, to Turtle Bay in Lower California and Socorro Island.

Order Archaeogastropoda./Family Vermetidae

Worm Shell 19
Vermicularia fargoi
Length to 10 in. Shell develops regular or irregular open spiral. Yellow, brown, or white, with thin membranelike portion with which opening may be closed. Animal retreats deep into shell. Interior partitioned by cross walls. Foot not well developed. Head long, with two conelike tentacles with eyes at bases and two other tentacles at sides of mouth. Essentially stationary. In crowded tangled masses in shallow water, in sponges, on coral, among weeds, or to depth of 1,000 ft. Florida to Texas.

Order Archaeogastropoda./Family Cerithiidae

California Horn Shell 20
Cerithidea californica
Length to slightly over 1 in. Black outside, glossy within; 10 strongly ribbed whorls and relatively large circular aperture that is closed by thin brown operculum. In mud of tidal flats. Undisturbed by exposure to air. Bolinas Bay, California to central Lower California.

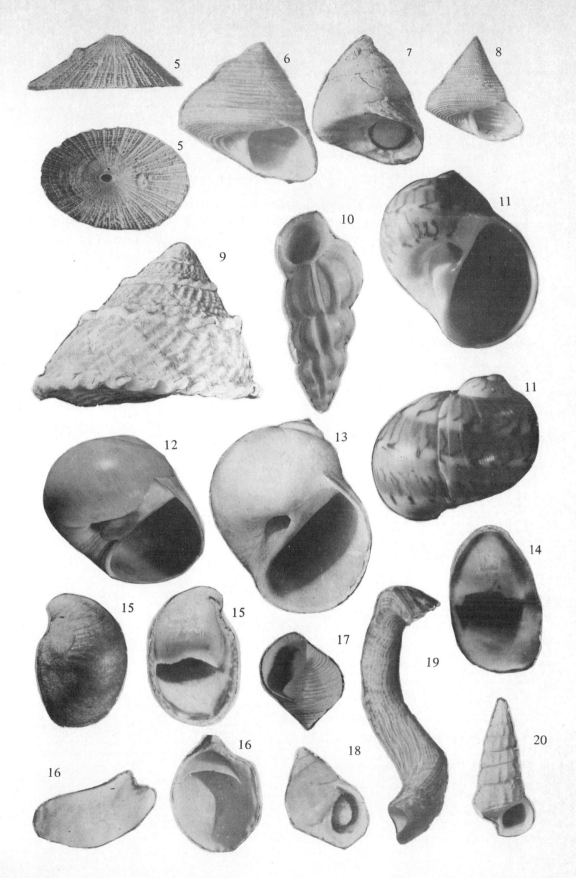

Order Archaeogastropoda./Family Strombidae
Large, solid-shelled, conical-spired animals. Shells with expanded lips, deeply notched. Opening long, narrow, and may be closed by operculum. Eyes large, at ends of pair of long stalks.

Fighting Stromb, Conch 21
Strombus alatus
Shell to 4 in. long, orange to brown or purple, with thin skin that covers animal but wears off in age. Actively attacks animals by series of jumps from side to side. If upset, animal rights self by somersault motion. In shallow seas. South Carolina to Florida and west to Texas.

Order Archaeogastropoda./Family Ficidae
Shells small-spired, with large openings in which animal is not protected by operculum, at least as adult.

Fig Shell, Paper Fig 22
Ficus communis
Shell brown with dark interior, to 3½ in. long. Foot large. Siphon long and narrow. Cape Hatteras to Gulf of Mexico, to Mexico; West Indies.

PHYLUM MOLLUSCA. CLASS GASTROPODA

Order Neogastropoda./Family Muricidae
Ancients extracted royal purple dye from members of this family that includes borers *(Eupleura)* and rock shells *(Murex)*. Foot, long. Opening uniform in outline.

Festive Rock Shell 23
Pteropurpura festivus
Length to over 3 in. Shell elaborately ornamented by coarse frills, ridges, and sculpturings. Three growth ridges to a whorl. White, gray, or dingy, but when young sometimes brilliant scarlet. Clings to sunken weeds and trash. Best in warmer waters. Santa Barbara, Calif., to Magdalena Bay, Lower California.

Drill, Oyster Drill 24
Eupleura caudata
Length to 1½ in. Width 3½ times length. Shell brown, gray, white, or red-brown. Two growth ridges to a whorl instead of three as in *Murex*. Foot yellow, remainder white. Lip, thick. About seven angular whorls at maturity. Common at depths of 6 to 50 ft. Massachusetts to Florida.

Oyster Drill 25
Urosalpinx cinerea
Length rarely to 1 in. Growth ridges, nine to a whorl. Yellow to gray but brown within. Foot small, with yellow border. Eyes small and black. Bores neat round holes through shells of oysters and other shellfish and sucks out soft parts. Eggs laid in parchment cases, each containing to a dozen eggs and attached in rows to rocks. Single female may lay 100 such cases in few weeks, usually placing them just below low-water mark. Probably oyster's worst enemy. Prince Edward Island to Florida. Also at San Francisco.

Leafy Hornmouth 26
Pterorytis foliata
Length to over 3½ in. Breadth to over 2½ in. With three broad conspicuous winglike, shinglelike structures and strong spiral ridges that spread in fanlike form. Dull white, but usually stained. Sitka, Alaska, to San Diego, Calif.

Sculptured Rock Shell 27
Ocenebra interfossa
Length to 1 in. Breadth to ½ in. Spindle-shaped with spirals marked by deep, well-defined grooves. Yellow, gray, brown, and dull outside but white inside. Appears coarse and rugged. Clings to rocks. Relatively common. Semidi Islands, Alaska, to Santo Tomas, Lower California.

Order Neogastropoda./Family Thaisidae
Short-spurred shells with no growth ridges. Many produce a crimson dye.

Little Rock Purple, Dog Winkle, Horse Winkle, Sting Winkle, Whelk 28
Thais lapillus
Length to 1¾ in. Reddish, yellow, white, or banded. Feeds on oysters, mussels, and other mollusks, whose shells it drills until its tongue can be thrust in to remove soft parts. Enemies include starfish, hermit crabs, fish. Eggs resemble pink rice on tiny stalks attached to rock surface, each group containing 29 to 40 young. One individual capable of producing more than 200 such groups. Eggs laid any time of year in capsules like slender eggs attached at one end and open at other. Injures valuable shellfish. On rocks, commonly in shallow water. Newfoundland to New York and in northern Europe.

Wrinkled Thais, Purple Dog Winkle 29
Thais lamellosa
Length to 2 in. Exterior variable from smooth to wrinkled or frilled. Some are plain white, while others are conspicuously painted with rich brown bands. Usually conspicuously colored. From waterline to depths of 150 ft or more. Port Clarence, Bering Strait, to Japan Sea and east to Aleutian Islands, south to Santa Barbara, Calif.

Unicorn Shell 30
Acanthina punctulata
Length to about 1 to 1½ in. Mature spire has four whorls. Small opening bounded by series of toothlike knobs. Appears like waterworn granite. Usually on rocks among seaweeds. Monterey, Calif., to San Tomas, Lower California, with variations and subspecies widening range considerably, particularly to south.

Order Neogastropoda./Family Nassariidae
Shells more or less egg-shaped, with lip-lining enamel spreading around opening; carnivorous.

Lash Nassa, Basket Shell 31
Nassarius vibex
Length ¼ in. Chestnut and white, in bands or washes that are brilliant in fresh shells. Shell with six whorls. Foot, forked behind. Long Island to Gulf of Mexico, West Indies.

Worn-out Basket Shell 32
Nassarius obsoletus
Length 1 in. or less. Width ½ length. Shell brown or muddy. Opening dark brown with white bands. Body mottled gray. Animal commonly sand-covered. Eggs laid in spring on an empty shell; each sac spiny, transparent, stalked. Scavengers, or may bore shells of other shellfish, including own kind. Common where water is brackish at mouths of streams in shallow water. Probably commonest north Atlantic Coast shell. Gulf of St. Lawrence to Florida.

Dog Whelk 33
Nassarius tegula
Length to ¾ in. With stout, strong, evenly tapering, pointed spire and relatively small opening. Dark gray, with rather conspicuous bumps forming interrupted ridges from one end to other. Mud flats. Santa Barbara to San Ignacio Lagoon, Lower California.

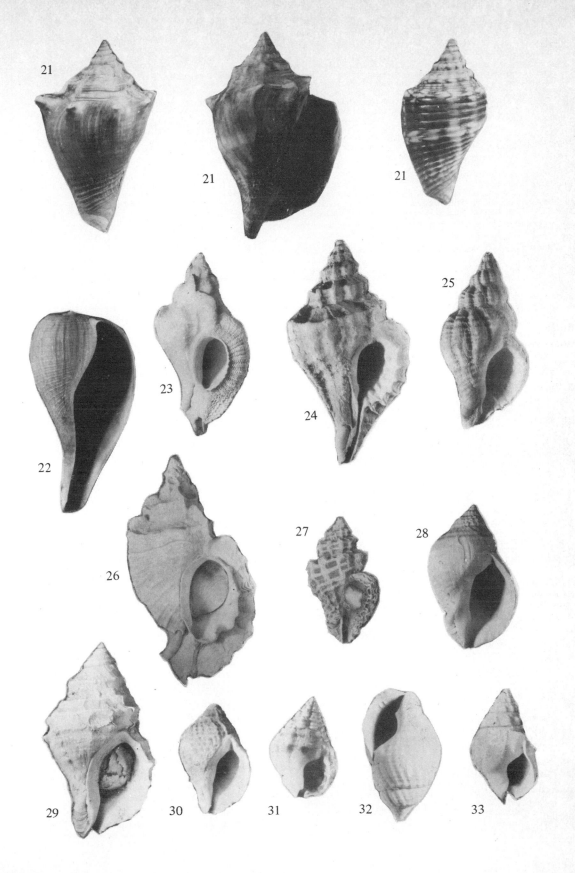

21

21

21

22

23

24

25

26

27

28

29

30

31

32

33

411

Channeled Dog Whelk 34
Nassarius fossatus
Length to 1¼ in. Diameter to ¾ in. In older animals, lip much thickened in outer areas. Pale ash-gray but with enamel showing at lip as bright orange. In mature shell, seven whorls marked by conspicuously spiraling ridges crossed by more obscure coarse longitudinal ridges. Long breathing tube extends into clear water as animal feeds in deep mud. Vancouver Island, B.C., to San Ignacio, Lower California.

Order Neogastropoda./Family Columbellidae

Frieze-covered Dove Shell, Eelgrass Shell 35
Mitrella carinata
Length to less than ½ in. About size of wheat grain, chestnut brown, polished, glistening and marked with spots, stripes, or dots. Spire conical. Lip thickened. At roots of eelgrass in enormous numbers. Forrester Island, Alaska, to Salina Cruz, Calif.

Columbian, Wrinkled Amphissa 36
Amphissa columbina
Length to nearly 1 in. Light yellow-brown. Closely related to Joseph's-coat amphissa, which is about ⅓ in. long and variable in color. Found in mud where sea is about 100 ft deep. Chiachi Islands, Alaska, to San Pedro, Calif.

Order Neogastropoda./Family Buccinidae
Shell with unusually large opening that ends in a wide notch or canal through which the siphon extends. Except in *Busycon*, shells lack color.

Waved Whelk 37
Buccinum undatum
Length to 6 in. Width ⅔ length. Gray outside, yellow around opening, and usually white within. Several hundred eggs laid in half-pea-shaped sac; about 500 sacs fastened together in a mass. Newly hatched young feed on own kind, some 2 months being spent in egg sac, usually during winter months. Steals bait from fish hooks. Forms large part of diet of cod and other fish. Used as bait by fishermen. In Europe, fried and eaten in soups by man. Low-water mark to deep water. Labrador to New Jersey and in Europe.

Order Neogastropoda./Family Neptuneidae

Tabled Whelk 38
Neptunea tabulata
Length to 5 in. Diameter to 2 in. Coarse beautiful yellowish-white with flat-topped whorls making a spiral table that gives common name. Outer lip thin and smooth. Inner lip hard and crusted. At depths of 150 to 1,200 ft. British Columbia to San Diego, Calif.

Short Distaff 39
Colus stimpsoni
Length 3 to 6 in. Generally covered with velvet. In water 6 to 2,700 ft deep, from Labrador to North Carolina, favoring deeper water in South.

Order Neogastropoda./Family Busyconidae

Left-handed Whelk 40
Busycon sinistrum
Length to 10 in. Fawn-colored with blue-brown stripes; when young, tan. Spiral turns to left. In India, such shells are sacred, so many are shipped from Florida to serve in ceremonies in Indian temples. Shell lining, shining brown. Animal, black. Eggs in cases, attached in row to long cord. Burrow in sand for food, which is largely other mollusks. Cape Hatteras to Dry Tortugas, west Florida; Yucatan.

Channeled Whelk 41
Busycotypus canaliculatum
Length to 9 in. Brown, with numerous revolving lines. Channel follows joint of whorls. Egg sacs much as in knobbed whelk, but with narrow instead of double ridged edge. Indians made wampum of shells. Shells used as containers and as cutting tools. Shallow water. Cape Cod, Mass., to St. Augustine, Fla.

Knobbed Whelk 42
Busycotypus aruanum
Length to 11 in. Width about ½ length. Usually smaller on Long Island. Gray to brown. Shell red within. Lip, thin. Knobs on outer whorl give shell its name. Food, soft parts of oysters and other shellfish procured by drilling small round *beveled* hole. Eggs laid during warmer months in disk-shaped sacs, with double ridge around edge, fastened on long spiral ribbon to a stone. Egg ribbons may be 1 yd long. From ribbons emerge young whelks shaped like adults. Sandy beaches near oyster beds and elsewhere. Cape Cod, Mass., to Cape Canaveral, Fla.

Crown Melongena 43
Melongena corona
Length to 5 in. Shell pear-shaped, with curved flattened spines that give it a crownlike appearance. Spines arranged in two to three series. Surface polished. Preys on other mollusks, sticking long snout into soft parts and rasping muscle free. In shallow brackish water, usually under mud. Florida and West Indies; Gulf States.

Order Neogastropoda./Family Fasciolariidae

Giant Band Shell 44
Fasciolaria gigantea
Length to 2 ft. Shell solid, heavy, with 10 whorls. Strong ribs with less distinct ones between them. Reddish-brown epidermis over shell. Soft parts, conspicuously red. In shallow waters of tidal pools or open shores. Florida to Gulf of Mexico.

Tulip Band Shell 45
Fasciolaria tulipa
Length to 8 in. Shell somewhat tulip-shaped, with nine well-rounded whorls, with sculptured region near line between whorls that is lacking in *F. distans*. Plain orange, dark mahogany, or other colors. Feeds on dead mollusks and crabs, as scavenger. North Carolina to Antilles.

Pale Tulip Sell 46
Fasciolaria hunteria
Length to 3½ in. Shell gray, with shadowy white design or marked with 14 dark brown bands. North Carolina to Gulf States.

Order Neogastropoda./Family Cymatiidae

Oregon Triton 47
Fusitriton oregonensis
Length to about 4 in. Shell covered with shaggy, brown, hairy material in life. Inside of shell pure white. From line of floating ice in Bering Sea, near Pribilof Islands, to Sea of Okhotsk and Japan, and to San Nicholas Island and San Diego, Calif.

Order Neogastropoda./Family Cypraeidae
Cowry shells were important as media of exchange in primitive South Sea civilizations. Animals now sold to tourists or used as bait for octopuses that have food value. Ordinarily not easy to collect. Known to all South Sea travelers.

Chestnut Cowry 48
Cypraea spadicea
Length rarely to 2 in. Shell with brown center, with bluish lips beneath and about 20 whitish "teeth." Center of back lighter than adjacent area. Young with thin short-spired cones and large aperture. Rarely found. Monterey, Calif., to central Lower California.

Order Neogastropoda./Family Triviidae

Solander's Trivia, Sea Button 49
Trivia solandri
Length to nearly 1 in. Shell black, with horizontal groove crossed by conspicuous ridges at right angles, the groove exposing white area that underlies colored surface. Dark chocolate brown to light pink variations found. Palos Verdes, Calif., to Panama.

Order Neogastropoda./Family Olividae
In *Oliva* there is no operculum protecting the opening. In *Olivella*, there is an operculum.

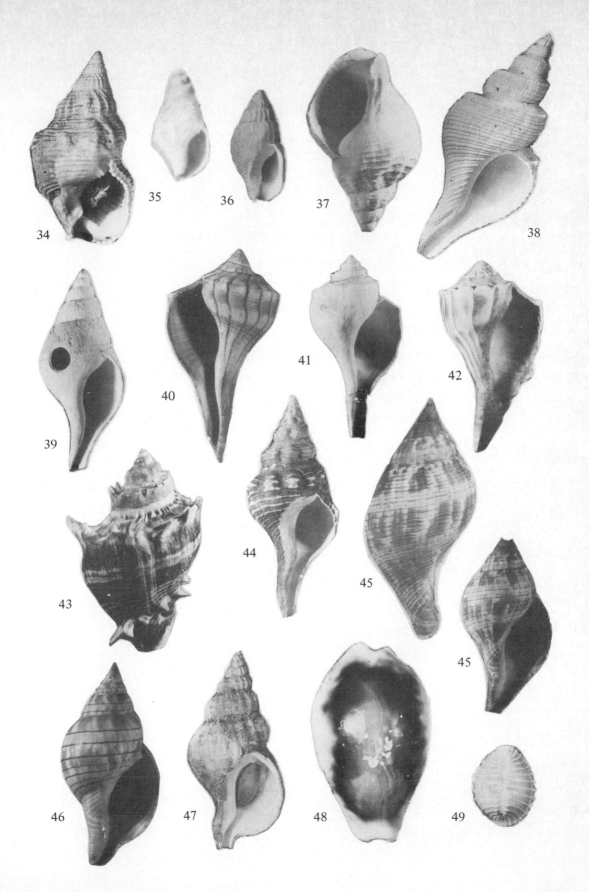

34
35
36
37
38
39
40
41
42
43
44
45
45
46
47
48
49

413

Lettered Olive Shell, Panama Shell   50
Oliva sayana
Length to 2½ in. Shell usually with two bands of irregular markings and zigzag ornaments, tapering at both ends. Shell covered by animal in life, so that animal looks like piece of fat pork. Animals live in colonies, often buried in sand. North Carolina to Florida to Gulf States.

Purple Olive Shell   51
Olivella biplicata
Length to 1 in. Shell about size and shape of olive pit but smooth, polished, and beautifully marked. White to dark slate, usually with purple markings. Usually found just under surface of sand, with siphon extending to open water above. Vancouver Island, B.C., to Magdalena Bay, Lower California.

Order Neogastropoda./Family Terebridae
Shells are shaped like augers. Some species have poisonous glands.

Auger Shell   52
Terebra dislocata
Length to 1⅖ in. Animal has long proboscis with which it squeezes, suffocates, poisons, and sucks its victim to death. Movement normally sluggish. Egg cases on shells in May. Virginia to Texas; West Indies.

Order Neogastropoda./Family Conidae
Name cone is almost adequate description. Shells heavy and chinalike, with narrow opening. Some can remove inner whorls to increase available room. Some have poison glands and strike at their enemies or prey. Some have great value to collectors.

Florida Cone Shell   53
Conus floridanus
Length to 2 inches. Shell yellow-white blotched with brown, with indistinct white band at shoulder and at center. North Carolina to Florida.

Order Neogastropoda./Family Cancellariidae

Cross-barred or Nutmeg Shell   54
Cancellaria reticulata
Length to 2½ in. White with brown bands. Surface finely ribbed and grooved. Slow and shy. Explores with forepart of foot and eats plants. Cape Hatteras to West Indies.

Order Opisthobranchiata./Family Bullaridae

Gould's Bubble Shell   55
Bullaria gouldiana
Length to 2 in. Shell thin, polished, mottled brown or white, slate or yellowish, of eggshell appearance and almost as delicate. Spire is pushed inward. In and on mud flats. Santa Barbara, Calif., to Gulf of California.

Order Stylommatophora./Family Polygyridae

White-lipped Land Snail 56
Triodopsis albolabris

Size to 1¼ in. in diameter. Broader than high, with 5½ low whorls making a low spire. White around lip of opening. Without teeth at opening. Animal varies from cream to gray, black or brown on head. Eyes on ends of the two longer "tentacles" on head.

Genus occurs in eastern United States and Canada east of 100th meridian, with subgenus in West. Few related species; common and widely distributed. This species prefers forests of oak, hickory, elm, and walnut and is found among forest debris.

56

Both sexes represented in one individual, though two individuals mate reciprocally. Spherical eggs with shells are laid in a moist situation and hatch into small snails difficult to identify because lip and other distinguishing characters are not adequately developed. Eggs laid in May hatch in 20 to 30 days.

Maturity reached in about 2 years. Animals avoid direct exposure to sun. Feed upon plant material rasped free by tongue. Can detect food by scent. Eyes on long feelers permit seeing over obstructions. Deposit mucous trail. Some land snails are carnivorous. Nature of shell varies with availability of lime.

Essentially scavengers. Preyed on freely by shrews and other animals that often leave caches of them. Some land snails are carriers of flukes and other parasites dangerous to man, but these are limited in their distribution. These parasites may develop further in reptile, bird, or mammal.

56

Order Stylommatophora./Family Succineidae

Land Snail 57
Succinea ovalis

Length to almost 1 in. in subspecies *optima,* in which spire is longer than in typical form and shell is yellow or amber rather than green. Shell relatively thick and marked with coarse sculpture. Animal too large to withdraw completely within shell, is blunt before and tapering behind, with two short and two shorter "feelers."

Shells, commonly known as "amber shells" in this species, found under leaves in forest debris or high on tree trunks, where they may await rainy season before continuing foraging. This species has been found 15 ft above ground. One genus in family, represented in Illinois by five species and more races.

Individuals represent both sexes, mating being reciprocal with external organs of both sexes joined in one opening in an individual. Life history probably much like that of white-lipped land snail, with shelled, relatively inconspicuous eggs laid in moist debris. Growth undoubtedly relatively slow.

57

Animal may vary in color during season, one with a dark body in spring becoming yellowish in summer and fall. Food essentially plant material rasped by tongue. Trail of mucus is left behind as animal moves about. This species favors shady upland wooded areas, while smaller species are commoner in lowlands.

Probably of little direct economic importance though it may be considered as a scavenger and as source of food for some animal eaters such as shrews, moles, some mice, some birds, and other creatures. Tree-climbing habit is found in many beautiful tropical and subtropical snails, many of which have been destroyed by rats introduced as in Hawaii. Found generally throughout U.S.

Order Stylommatophora./Family Limacidae

Field Gray Slug 58
Deroceras reticulatum

Shell a mere thin plate, white-bordered. Animal with body 1 to 2 in. long but average about 1½ in., narrow, gray, black, white, yellow, amber, brown and spotted or blotched with irregular black areas. Foot yellowish and with copious milky mucus. Head with two pairs of feelers, the longer pair bearing eyes at tips.

Slug introduced from Europe, where it is common, appearing early in last century at Boston, New York, and Philadelphia, but now widely distributed over eastern United States and represented in western part by larger forms. Smaller *D. laeve* lacks white bordered breathing pore of this species and has prominent tubercles over upper surface of body.

58

Both sexes in one individual with reciprocal mating. Eggs 500 to 800 per slug, ¹⁄₁₆ to ⅛ in. long, white, laid in moist soil among roots, in groups of about 50, hatch in 3 weeks to over winter. Wintering slugs may lay eggs in May that by October may have developed into slugs sufficiently mature to mate and lay eggs or may winter over as adults.

Bad pests in wet seasons in gardens. Fine ashes placed about plants cause animals to secrete mucus to point of exhaustion and death; or poisoned baits of stale beer or Paris green on cabbage or lettuce in protected
(Continued)

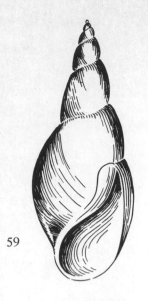

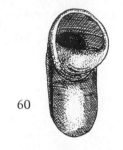

59

60

60

61

spots may eliminate individuals. Can detect food by sense of smell and therefore commonly forage at night. Trails of *D. laeve* glisten on sidewalks in morning.

In some areas in wet seasons these slugs may destroy a vegetable garden or ruin the marketability of an otherwise acceptable crop by eating into the center of heads of lettuce or cabbage. Make excellent experimental animals for studying their senses of smell since they possess olfactory receptors, and sight and their resistance to conditions of varying humidity.

Order Basommatophora./Family Lymnaeidae.

Right-handed Pond Snail 59
Stagnicola palustris elodes
Shell to 1⅕ in. long and ½ in. through, composed of about six whorls in a relatively acute spire, with large opening from which animal protrudes to the right when shell is placed so that point is away from observer. Shell dark, pale brown to black; may vary in thickness in different waters.

Found in stagnant ponds and still fresh waters where lime is abundant or not. Over 200 species in genus, of which 65 are American. This species is circumpolar, being found in Europe and subspecies *elodes* in northern America. Other important species are long-spired *Lymnaea stagnalis jugularis* and fragile *Pseudosuccinea columella*.

Many spherical eggs are laid in clear jelly mass on some support under water, and as young develop they may be observed easily with a hand lens as they twist and turn in their shells within jelly mass. Growth gradual, without conspicuous stages once egg stage and jelly mass have been passed.

Food probably largely plant materials, living or dead, although they have been known to attack living animals. These snails are generally less active than are those of genus *Physa*. Some snails are known to carry disease organisms, and much research is being done to improve our understanding in this field.

Animals serve as scavengers, help keep the glass in aquariums clean, and provide food for many organisms of food value to man. Some fish live exclusively on mollusks.

Disc Pond Snail 60
Helisoma trivolvis
Shell a flat coil, to nearly 1 in. wide and to ⅖ in. high, with coil steadily increasing in diameter and opening relatively wide in mature animal. Usually with about four whorls, yellowish or brown, of varying opacity, with an opening that has a sharp lip and V-shaped angle above. Sinistral when young.

Commonest species found in North America east of Rockies. Different species found mostly in temperate North America, where there are some 25 species in all. This species common in streams and muddy-bottomed ponds, commonly among water plants. In related western *H. tenue* of Pacific slope, growth lines are not raised.

Both sexes represented in a single individual, with mating reciprocal between two individuals. Eggs laid in small jelly masses, hatch in 2 to 3 weeks, depending largely on temperature. Stages of development gradual, young differing only in that spire seems to be as in *Physa*.

Not an active pond snail. Food plant and animal matter rasped by mouth and eaten. Animal may serve some function as a scavenger. Can retreat within shell and secures reasonable safety from most enemies, of which there are many. Position in plants gives additional safety.

Not apparently of importance to man, though it may be a reasonably common food for some species of fish that man eats or enjoys catching. Probably host of a number of parasites that may spend other stages in other animals.

Left-handed Pond Snail 61
Physa heterostropha
Shell to ⅗ in. long and to ⅓ in. through, with opening to the left if point is away from observer; usually four-whorled, the first being very large and others very small, with an acute tip. Opening about ¾ length of whole shell. Yellowish-brown to black. Smoothly gliding, active pond snail.

Very common in stagnant ponds and other still waters. Common in eastern states and perhaps found in Central states, with close relatives extending range throughout the United States. Over 150 species in genus of wide distribution, with some 22 found in United States.

Individuals represent both sexes, though two individuals mate reciprocally. Numerous eggs deposited in clear jelly masses on supports of plants or stones under water; hatch in 2 to 3 weeks. Growth is gradual, this species maturing in about 1 year, though related species may require 2 years to mature.

Food plant and animal matter rasped by tongue and eaten. Active little snails, excellent acrobats of an aquarium, dropping by mucous strings from plants or floating to new grazing grounds in an amazing manner, twisting shell suddenly or exploring with two feelers to know their homes better.

Excellent pond and aquarium scavengers and a common food for many fish and other aquatic animals. Possibly a factor in pollution. Filamentous tentacles, foot pointed at rear, single-pieced jaw, are among indentification characters.

CLASS BIVALVIA

Order Veneroida./Family Unionidae

Fat Mucket, Mussel   62
Lampsilis radiata

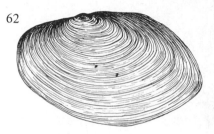

62

Shells of species reach 4 in., rounded in front, oval behind. Light or dark green, with eccentric ridges or growth lines around hinged areas as center. Male longer and lighter, female shorter and heavier. A single fleshy spade-shaped foot.

Common in freshwater lakes, streams, and ponds where water moves little if at all and where bottom is well covered with loose mud. Common genera include *Anodonta,* in which hinge lacks the teeth that in other genera align the valves and pivot opening shells; and *Unio,* the type genus.

Many minute eggs that have been fertilized by male become lodged in gills of female and develop there from August to May. Young larvae freed in May attach to bass, develop about 20 days, reaching length of $1/_{100}$ in. only to leave fish and spend about 2 years to reach mature size of to 2 to 3 in.

Scavengers, feeding on organic matter brought into animal with water. Different species of freshwater mussels or muckets have different fish species on which they are parasitic in larval stage, during which time they are carried considerable distances before beginning independent life. Sensitive to disturbances.

63

Harvested by dredging. May contain pearls. Not ordinarily edible. Shells formerly source of pearl-button material, with best button material from animals dependent on such apparently worthless fish as gars. Collecting controlled by law in some areas. Waste shells of some fertilizer value.

Order Veneroida./Family Veneridae

Clam, Quahog, Venus   63
Mercenaria mercenaria

Shells two, $5\frac{1}{2}$ by $4\frac{1}{2}$ by 3 in. or thereabouts; dirty white with prominent eccentric ridges on each shell; inner surface of shell dull white, with lower margin purple or violet; shell margin slightly uneven. Hinge with three spreading teeth to each valve. Sexes distinct in this species.

Common on sandy and muddy sea bottoms from tide levels to depths of 50 ft or more, but commonest in shallow bays even in brackish water. On coast, from Nova Scotia to Yucatan, but commonest from Cape Cod to South Carolina with related genus extending range considerably. Introduced on West Coast; Humboldt Bay, California.

Numerous small eggs, expelled into water by female, are fertilized at time by male. In about 10 hours, hatch into shell-less, free-swimming stage that lasts about 1 day. Shell-forming stage of 6 to 12 days follows and animal settles to bottom. Attached until $\frac{1}{2}$ in. long, then burrows. May live about 25 to 40 years. Immature called cherrystone.

63

Growth rate about 1 in. a year. May live to 4-5 yr. Wide muscular foot permits burrowing in mud or in sand. Water carrying food enters through short blunt yellow siphon and leaves through opening nearest shell hinge. Microscopic food collected between gills and forced to mouth by cilia, undesirable material being ejected. Sensitive to disturbance.

This species annually yields hundreds of thousands of dollars worth of food. Shells were used by Indians in making wampum, with blue portion considered the more valuable. Soft-shelled clams (see page 424) may yield twice value of this species. Many people make their living by digging and dredging clams.

Order Veneroida./Family Sphaeriidae

Fresh-water Clam 64
Sphaerium solidulum

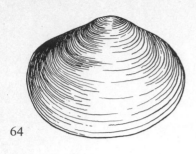

Shells two, nearly equal, each about ½ in. long and nearly as broad, relatively thin and weak, shining yellow-brown with rounded eccentric ridges on surface sometimes apparently translucent but not always so.

Fairly common in fresh water streams and ponds, in sand or in mud, or on plants if in a stream. Common in Central States while related smaller *S. striantinum* is common east of Rockies. About 20 American species; 75 in world, with genus widespread over world.

Comparatively few eggs are laid; these are nursed in gills of female. There larvae develop until relatively large and are freed to independence without parasitic stage described for *Lampsilis*. Breeds from April through midsummer. When winter approaches, animal buries self in mud by means of long extensible foot.

May serve somewhat as scavenger and may be of direct importance as food for fish and for other eaters of aquatic animals. Bottom of suitable streams or ponds may be conspicuously white-flecked with empty shells. Nature of shells may vary with availability of lime in surrounding water.

Of little if any direct economic importance to man, though it is a food organism of species that man eats. Interesting because of the brooding of few young in gills of adult, in contrast to freeing of many young to a parasitic stage as practiced in muckets.

Order Pterioida./Family Ostreidae

Oyster 65
Crassostrea virginica

Length to 18 in., usually 4 to 5 inches. Two shells, the upper usually flat and the lower more convex; attached to bottom by left shell; exposed shell irregular, thick, or in folded layers; inside of shell often blue mixed with white. No foot present in adult. Sexes distinct, or both present in one individual.

Gulf of St. Lawrence to Gulf of Mexico; West Indies; introduced on West Coast, British Columbia to Morro Bay, Calif. "Bluepoints" and "lynn-havens" are special forms from locality of name. Found in shallow brackish water, commonly on hard bottoms and more commonly near mouths of rivers or bays or for some distance up such arms of ocean.

Females produce about 9 million ⅟₅₀₀-in. eggs. Fertilized eggs hatch in 5 hours to free-swimming forms. At 32 hours shells appear; in 6 days shells enclose soft parts; in 3 weeks become established as spat and become fixed; growth increases; may be ¼ in. across in 10 days, 2 in. in 82 days, sexually mature in 2 years. Sexes indistinguishable externally.

Food, minute plant and animal matter drawn in with water used in breathing; waving hairs on gills cause current. Heart in freshly opened oysters appears as bulb beating close to inner side of great muscle that closes shell. Sensitive to heat, to sound or vibration, to touch and to chemical change, shells closing when animal is irritated.

Important marine animal of annual market value in the millions. Rarely contains pearls. May shelter small soft-bodied oyster crabs, *Pinnotheres ostreum*, males of which are hard-shelled, and free-swimming. Sanitary practices demand pollution control where oysters are harvested commercially, and management is essential for a maximum yield under any conditions.

Order Mytiloida./Family Mytilidae

Mussel (Marine) 66
Mytilus edulis

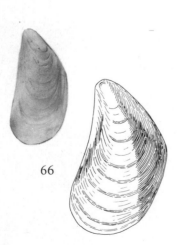

Length to 4 in. Width to ⅓ length. Shells about equal, pointed in front, round behind, smooth; violet to dark brown, pearly within and with violet margins. Attachment, a fibrous byssus secreted by glands at posterior end of foot. Gills, rows of filamentous structures.

Widely distributed, Arctic to Cape San Lucas, Lower California, Arctic to South Carolina. Common in European waters. In sand or mud, attached to rock and piles or to each other between tide lines or in shallow salty or brackish waters. Species in family vary in size, color, and edibility.

Sexes distinct; sex glands distributed through animal. Tiny yellowish eggs found almost anywhere, as in foot, gills, or other parts. A California species may lay to 100,000 eggs a year, these developing within year to animals to 3½ in. long. Become attached early by network of strong byssal threads and resist wave action easily.

Moves by means of foot and by byssus, gluelike threads it puts out. Often mantle tip protudes from separated shells. This attracts fish that may come too close and be captured by closing shells, as shells snap quickly when animal is disturbed. Sensitive to touch and to salinity. Protection is by burrowing, by exposed nature of habitats, and by shells.

Species in common use in France for food, there cultivated and harvested for profit. Also sold in markets on East and West Coasts of America, but

67

used more in chowder.   Has unique color when cooked.   Since it thrives in polluted water, its abundance may be a snare and a delusion.   Animals unable to close shells should always be avoided as food for man.

Order Pterioida./Family Pectinidae

Scallop, Atlantic Bay Scallop 67
Aequipecten irradians
Shells two, one flatter than the other, with a straight hinge, without teeth; about 20 radiating, conspicuous ribs on shell and abundant lines of growth; up to 1 ft across but commonly much smaller.   Edges of mantle may protrude from shell, exposing 30 to 40 bright blue eyes that may glow.   Between eyes are delicate feelers sensitive to disturbance.   Color varies from white, orange, purplish to brownish with several conspicuous radiating orange bands.

Found from Nova Scotia to Texas.   Common pecten of Atlantic Coast, abundant on eastern Long Island Sound, in shallow water of saltwater bays where bottom is covered with sand or seaweed, on mud flats, or even in water hundreds of feet deep.

Breeds in early summer and grows so rapidly that by winter young may reach a length of 1 in.   After a free-swimming stage that develops from fertilized egg, animals attach by secreted stems to submerged support and develop to width of 1½ in.; then change to free roaming life with usual life span in the neighborhood of 4 years.

Adults move in zigzags by flapping shells, much as butterflies move with wings.   A shell flap is accompanied by a jet of water and may move animal several feet.   Has definite seasonal migration and is somewhat gregarious.   Food, minute living things taken in with water used in breathing.   Sufficiently sensitive to take to flight with a passing shadow.

Basis of industry of several hundred thousand dollars in the East.   Animals captured by dredging but abundance varies greatly.   Muscles are eaten most commonly but other parts are also edible.   Shell is used in many art designs and in the Middle Ages one species had religious significance.   Shell is now the trademark of an important oil company.

67

Order Arcoida./Family Arcidae
Comblike teeth, in rows on hinge.   Foot, large.   Mantle edges with row of eyes.   Shells relatively heavy with thick velvety or hairy covering or periostracum, which wears off variously in dead shells.

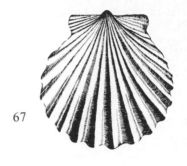

68

Black Widow, or Ponderous Ark 68
Noetia ponderosa
Length, 2½ in.   Heavy, swollen, with about 32 flattened radiating ribs; yellowish-white with dark shaggy covering.   Interior yellowish-white, but white and glossy at edge.   Hatteras to Key West, west to Gulf of Mexico; Texas.

Ark Shell 69
Anadara ovalis
Largest at Cape Cod.   Length to 2 in.   Left valve ridges, flatter and narrower.   Also known as "bloody clam," Cape Cod to West Indies.

Ark Shell 70
Anadara transversa
Length 1½ to 2 in.   Width about same.   Height, ⅔ length.   Brown. Ribs, about 35, deeply cut.   Shallow water, Massachusetts to Texas, especially on Nantucket, west coast of Florida, and Gulf.

69

Order Arcoida./Family Pectinidae
Mantle hangs like finely fringed curtain inside each shell, with conspicuous row of black eye dots along its base.   Notch under ear of shell marks opening through which byssus attachment may be spun.

70

Pink Scallop 71
Chlamys bericius
Length to 2½ in. Pink. With unequal ears. Thin ribs bearing slender spines. Under valve, white; upper, pink-banded, with fewer ribs. Deep water. Port Althorp, Alaska, to San Diego, Calif. Off San Juan Island, Wash., at depths of 150 to 200 ft.

Pacific Scallop 72
Acquipecten circularis aequisulcatus
Length to 3 in. Typical form, round scallop, with bulging shell. Lives in sandy or muddy water. Deep water, or at shoreline at low tide, lying on right side or attempting to escape. While whole soft parts are edible, only adductor muscle is sold and darker parts are rejected. From Monterey to Cape San Lucas in Lower California.

Purple-hinged or Rock Scallop 73
Hinnites multirugosus
Diameter to nearly 2 in. Height to 4 in. Length to 8 in. Free-swimming when young, with shells nearly alike. With age, becomes attached by one shell that conforms to rock. Near hinge inside always deep purple. Disturbed, an adult may clap its shell, forcing water stream several feet into air. Queen Charlotte Island, B.C., to Abreojos Point, Lower California. Shallow water, sheltered or rocky areas. One of the best edible scallops; large muscle sliced.

Order Mytiloida./Family Anomiidae
Left or lower shell pierced; animal attaches firmly through hole to some object; fixed shells lose locomotion muscle. Upper shells convex. Both shells let light through easily and are thin and pearly.

Plain Jingle Shell 74
Anomia simplex
Diameter 1 to 2 in. Largest of genus on Atlantic Coast. Oval or circular but variable, fitting self to object of attachment. When fixed, may bore foot into cover. Outer surface dark, fragile; when worn, dead shells show golden or greenish mother-of-pearl. Shallow water, particularly on oyster beds. Nova Scotia to West Indies, Gulf of Mexico.

Rock Oyster, Jingle Shell, Pearly Monia 75
Pododesmus cepia
Length to 4 in. Attached, one shell much smaller than other, with hole through which muscle scar is visible. Rough outside but pearly inside, with purple and green tints. Flesh bright orange, edible. At Puget Sound, at 100 to 250 ft down. Found attached to abalone. British Columbia to Gulf of California.

Family Mytilidae
Commonly between tide lines, attached separately, grouped in nests or sunk in burrows in wood or earth. Two shells equal and generally long.

Hooked Mussel 76
Brachiodontes recurvus (M. hamatus)
Length 1 to 2 in. Shell thick, dark-colored; surface densely striped. Rhode Island to Texas and West Indies. Particularly abundant in Florida.

Ribbed Mussel 77
Geukensia demissa
Length 2 to 4 in. Narrow, yellow-green, triangular, fat, dingy-looking, brittle, compressed behind, plaited and with finely radiating lines or ribs, apparently varnished. Lining silvery white. Attachment long and strong. Makes nests of shells or burrows. Probably commonest eastern mussel. In tidewaters or mud flats at stream mouths, usually near high-water mark. Often among reeds in burrows in banks. Being common and at its best in polluted waters, it should not be eaten. From Nova Scotia to Georgia; Texas. Smaller in northern range. Introduced in San Francisco Bay. Used for poultry food.

Straight Horse Mussel 78
Modiolus recta
Length to 4 in. Brown rather than black, either conspicuously bearded or ridged. Umbo not at extreme end. Usually solitary, attached, partly buried in mud or gravel. Edible but not commercially important. From British Columbia to Magdalena Bay, Lower California.

Hooked Pea-pod Shell 79
Botula falcata
Length to 3 in. Shell thin, flexible, pearly white inside, dark chestnut outside, with conspicuous transverse wrinkles. Bores holes in solid rock and attaches with strong byssus. More common in sheltered shallow water than on exposed wave-beaten rocks. From Coos Bay, Ore., to Lower California.

Order Pholadomyoida./Family Pandoridae
Shells pearly, irregular, thin, usually with right shell flat and mate convex. Two spreading teeth in right shell form grooves for tooth in left.

Pandora 80
Pandora gouldiana
Length to 1½ in. Height ⅔ in. Width ⅕ in. Delicate clear white, rough, very thin with iridescent interior. In oyster beds, in sand or mud, burrowing freely. Common to depth of 180 ft, from Prince Edward Island to New Jersey.

Order Hippuritoida./Family Chamidae

Agate Chama, Rock Oyster 81
Chama pellucida
Length to 2 in. Outside irregular, rough, frilled, hard, translucent, white, rosy or otherwise. Inside lined with opaque white layer frilled at margin. Living shells usually too firmly attached to be removed entire. Looks like nubbin of rock or like limpet. Young formerly considered separate species. On pilings and breakwater rocks from Oregon to Chile and Galapagos Islands.

Order Veneroida./Family Cardiidae
Shell margins conspicuously toothed.

Heart Shell, Basket Cockle 82
Clinocardium nuttallii (Cardium corbis)
Length to 4 in. Shell brittle, easily broken, conspicuously grooved outside, white inside. Active. Makes shallow burrow with long strong foot. No long siphon tube. Edible but not commercially important. Bering Sea to Japan and San Diego, on tide flats. Common West Coast species.

Morton Cockle, Duck Clam 83
Laevicardium mortoni
Length to 1 in. Width about same. Height ⅔ length. Shell thin, small, smooth. Pale fawn color, sometimes with brown spots on outside and bright yellow with purple blotch within. Often brilliant in muddy water, but color fades. Shallow waters (to 30 ft) on sand flats. Massachusetts to Gulf of Mexico. Common south of Cape Cod. A small cockle.

Great Cockle 84
Dinocardium robustum
Length to 4 in. Height to 5¼ in. Heart-shaped. Posterior somewhat flattened. With 35 to 37 regularly arranged broad radiating ribs and crenulated margins. Yellow-brown with irregular rows of chestnut or purple spots. A most conspicuous Florida beach shell. Found from Cape May to Florida, Gulf of Mexico to British Honduras and Central America. Subspecies, West Florida.

Order Veneroida./Family Veneridae
Shell valves equal and strong. Foot, strong. Of variable beauty but of wide distribution.

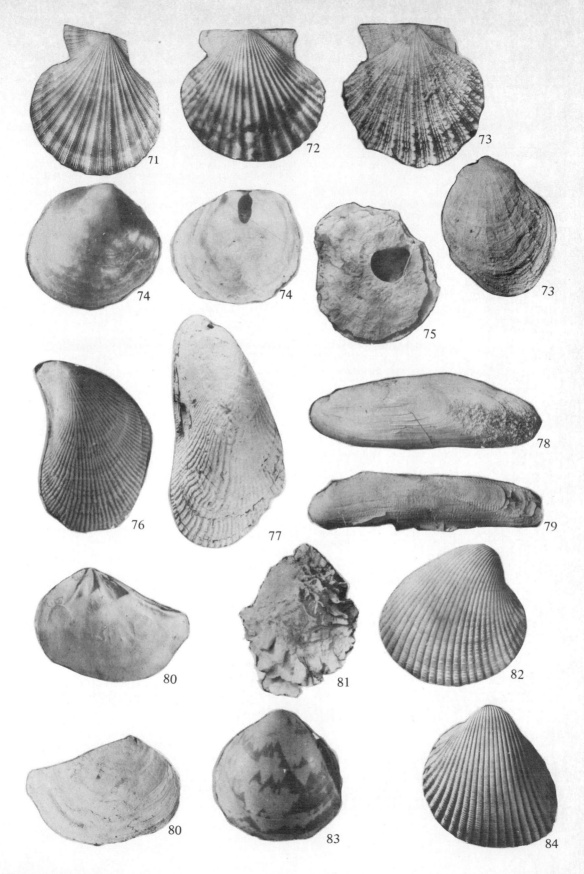

Elegant Dosinia 85
Dosinia elegans
Flat, almost disklike. Length to 2½ in. Surface distinctly marked with concentric raised ridges. Prefers warm offshore waters. Sometimes abundant. Cape Hatteras to Yucatan and in West Indies.

Hard-shelled Clam 86
Pitar morrhuana
Length to 2 in. Shell plump, thin but hard, chalky, with concentric scratches. Prince Edward Island to Cape Hatteras.

Spotted Calliata 87
Macrocallista maculata
Length to 3 in. Beautiful. Surface marked with violet-brown patches or waves and a shining horny skin. Inside white. Flesh edible but peppery. From Cape Hatteras to West Indies, and on to Brazil.

White Amiantis, Sea Cockle 88
Amiantis callosa
Length to 4 in. Pure white. Edges thin. No ribs, but many often paired concentric ridges. Less triangular than Pismo clam. Beaches at extremely low tide. Edible but not commercially important. From Santa Barbara, Calif., to Lower California.

Pismo Clam 89
Tivela stultorum
Length to 7 in. Weight to 4 lb 3 oz. Not legal to take those under 4¾ in. across, which weigh about 1⅓ lb. Shell apparently varnished, pale brown, often with faint concentric purplish markings. Foot short and like a plowshare. Spawns from July to September and young are free-swimming for a short time. Takes 7 to 11 years to reach legal size. Individuals known to have lived 25 years. Only on open surf-beaten sandy beaches, almost invariably above low-water mark. Second only to abalone in weight of crop harvested in California, but rivaled by introduced soft-shelled clam. Half Moon Bay, Calif., to Socorro Island, Mex.

Cross-barred Venus 90
Chione cancellata
Length to 1½ in. Heart-shaped to triangular but with narrow raised ridges crossing on surface. Dirty white to yellow-brown outside; inside white, violet, purple, or orange. Commonest Venus-like Florida shell. From Cape Hatteras to Florida, Texas and West Indies.

Ribbed Rock Venus, Hard-shelled Clam, Littleneck Clam 91
Prototbaca staminea
Length rarely over 3 in. Valves so deeply arched as to make animal appear round. Longer than true cockle. Foot flattened, strong but not long. In bays and gravel but not in mud. Esteemed food. Commercial importance. Commander and Aleutian Islands, to Kamchatka and Japan and to Puget Sound and Socotto Island, Mex.

Rock Cockle 92
Prototbaca laciniata
Shell covered with radiating ribs, crossed by concentric ribs of equal prominence and often with small spines at the intersections. From Monterey, Calif., to Lower California.

Butter Clam, Washington Clam, Money Shell 93
Saxodomus nuttalli
Length to 5 in. Shells with rough eccentric ridges; when young, with brown markings on beaks and purple trace inside. Name money shell refers to use of shells as money by Indians. Delicious. Found between tide marks in sand and gravel. Humboldt Bay from Monterey Bay, Calif. to Lower California.

Thin Copper Shell 94
Compsomyax subdiaphana
Length to 2 in. Shell thin, white, glistening, apparently swollen. From Sannak Island, Alaska, to Lower California.

Gem Shell 95
Gemma gemma
Like small yellow peas, to ⅛ in. through and about ½ as wide. Pink, yellow, white, or violet-tinged or colorless. Surface smooth and shining. In sand, between tide marks or in shallow water. Often abundant. Nova Scotia to Florida, Texas, West Indies. Introduced in Puget Sound. Young carried a long time by parent and when freed active and independent.

Order Veneroida./Family Petricolidae

Heart-shaped Rock Dweller 96
Petricola carditoides
Length to 2 in. Dingy white, oval, with radiating ribs. Variable, from long and narrow to flat and oval. Bores into soft rock, shell conforming to shape of hole. On sandy and muddy beaches. From Vancouver Island to Magdalena Bay, Lower California.

Order Veneroida./Family Tellinidae

Bodega Tellen 97
Tellina bodegensis
Length to 2 in. Shell creamy white with polished surface etched with fine concentric lines. Found on outer sandy beaches. Strong, active. Edible but not commercially abundant. From Queen Charlotte Islands off British Columbia to Gulf of California.

Bent-nosed Macoma 98
Macoma nasuta
Length to 2½ in. Whitish or brown stained with mud. Shells light, thin-edged. Siphons long, indicating deep-burrowing habit. In mud in sheltered bays or on sand beaches. Survives brackish water well. Common in Indian midden heaps. Popular in Chinese market in San Francisco, Kodiak Island and Cook's Inlet, Alaska to Scammon Lagoon, Lower California.

Order Veneroida./Family Donacidae
Wedge-shaped shells of clean sandy ocean beaches. Foot, long. Gills, variable. Mantle, open, fringed.

Wedge or Pompano Shell, Coquina 99
Donax variabilis
Length to 1 in. Beautiful blue-white with purple or reddish bands. Variable in color and pattern. Common on clean sand beaches. Buries self rapidly. In spring tossed alive on beach in great numbers. Excellent soup base. Shells made into "birds" and "flowers." Commonest East Coast wedge shell. From Cape Hatteras to Texas.

Wedge Shell 100
Donax californica
Length under 1 in. Shell attractive, light, thin and like related Gould's wedge shell often well-colored. Just beneath sand surface. Good for soups. Santa Barbara, Calif., to Magdalena Bay, Lower California.

Order Veneroida./Family Diplodontidae

Round Diplodonta 101
Diplodonta orbella
Length about 1 in. Shell white, often hidden by cemented sand. Bering Sea to Gulf of California.

Short Razor Clam 102
Tagelus plebius
Length to 4 in. Height 1/3 length. Width 1/4 length. Shell thick, almost cylindrical. Foot large, retractile into shells. Siphons two tubes, each longer than shell; each may maintain separate hole through sand, giving burrow two exits. Cape Cod to Forida and Texas.

California Razor Clam, Jackknife Clam 103
Tagelus californianus
Length to over 4 in. Shell thin, gray, dull. Lives in vertical burrow in soft sandy mud. Favored food of ducks and eaten by human beings. Sold as bait but not commercially important. Humboldt Bay, Calif., to Panama.

Order Veneroida./Family Solenidae
Valves of shell gape at each end and much elongated.

Sword Razor Clam 104
Ensis directus
Length to 6 in. Width 1/8 length. Yellow or green. Right shell with one projecting tooth and long ridgelike tooth back of it. Left shell with two teeth and double ridge. Burrows rapidly in sand at low-water mark by digging motion of club-shaped foot. Animal larger than thin shell. Laborador to Florida, Texas.

Sea Clam, Razor Shell 105
Siliqua patula
Length to 6 in. Shell thin, brittle, smooth inside and out. White outside, zoned with violet; inside tinted either with pink or violet. Animal too big for shell. Moves rapidly through sand and can bury self in 6 seconds. Disturbed, it may squirt water high in air. Delicious. Favorite in South where sufficiently abundant to have commercial importance. Canned in Alaska and in Washington. Alaska to Pismo, Calif.

Order Veneroida./Family Mactridae
Shells with heavy, thick skin. Siphons united and fringe-tipped. Foot flattened. Mantle open in front.

Surf, Giant, or Hen Clam 106
Spisula solidissima
Length to 7 in. Height 4/5 length. Width 2/5 length. Shell large, solid, brown or white. Largest Atlantic Coast bivalve. Can leap a foot in escape. May be caught by thrusting stick between open valves. Excellent food. Caught in shallow water. Labrador to Carolinas. Subspecies, *similis* from Massachusetts to Gulf of Mexico, most abundant south of Cape Hatteras.

Triangular Clam 107
Mulinia lateralis
Length to 3/5 in. Height to 1/2 in. Width 2/5 in. Whitish, triangular, apparently smooth but really wrinkled. Smoother in South. Bare, white, with brownish skin. Muddy bottoms, near river mouths. Often abundant, particularly in Long Island Sound. New Brunswick to Texas.

Washington Clam, Summer Clam, Gaper 108
Tresus nutallii
Length to 8 in. Weight to 4 lb. Shell thin, easily broken, not including all of animal. Foot small. Siphons long and stout. Shell may lie buried 3 ft deep usually below midtide. May squirt water 3 ft high when disturbed. Siphons tough, but flesh is delicious. Indians dried siphons for later use in soups and chowders. Called "geoduck" but this name should correctly be applied to *Panope generosa* of family Saxicavidae. Typical form ranges from Puget Sound to Scammon's Lagoon, Lower California. *T. capax* from Kodiak Island to Monterey, Calif.

Order Myoida./Family Myidae
Soft-shelled clams. One shell has spoon-shaped tooth that fits into corresponding opening in other.

Sand or Soft-shelled Clam, Nannynose 109
Mya arenaria
Length to about 4 in. Width over 1/2 length. Open at each end. Shell chalky white with brownish cover. Skin wrinkled, thin, dirty brown. Common between tide lines, in shallow water and mud flats. Burrows to 1 ft depth, leaving small opening through which water squirts. European. Greenland to Cape Hatteras. Victoria, B.C., to Monterey, Calif.

California Soft-shelled Clam 110
Cryptomya californica
Length about 1 in. Shell white, thin, almost smooth, slightly ashy, and slightly gaping. Gulf of Alaska to N. Peru.

Checkered Soft-shelled Clam 111
Platydon cancellatus
Length to 3 in. White, gray, ashy, with greatly bulged valves. Surface with fine, irregular growth lines overlaid with brownish-gray cover. Near bay entrances where current is brisk. Burrows into firm hard clay but not in rock or shifting sands or mud. Edible but not commercially important. Queen Charlotte Island to San Diego, Calif.

Order Pholadomyoida./Family Pholadidae
Shells gape at both ends, with toothlike sculptures in front and no hinge or ligament.

Angel's Wing Shell 112
Cyrtopleura costata
Length to 8 in. Width and height each about 1/3 length. Shell white. Animal yellow. Shells meet only near tips. Burrows in colonies down several feet in sandy mud or clay, or in wood or rocks. Massachusetts to the West Indies. Commonly sold for food in Cuba.

Little Piddock 113
Martesia cuneiformis
Shell small. Length to 4/5 in. Closed and divided by oblique serrated canal. May burrow into timber or soft rock. North Carolina, West Indies to Brazil.

California Piddock 114
Parapholas californica
Length to 5 in. Upper end of shell with scales. Shells white, delicate. Burrows into hard clay, using waste to build stout, conelike channel to protect siphons that reach up to open water. Edible but not commercially important. Santa Cruz, Calif., to Lower California.

Order Pholadomyoida./Family Hiatellidae

Geoduck, Giant Panope 115
Panope generosa
Largest of burrowing clams, shells to 7 in. long, partly enclosing soft animal that may weigh 6 1/2 lb. Huge siphons cannot be pulled inside shell; latter white, with conspicuous growth lines. Each valve has a prominent tooth. Considered a delicacy. Ranges from Forrester Island, Alaska, to Scammon's Lagoon, Lower California. Common name geoduck pronounced "gooey duck."

102

105

103

106

104

107

108

109

110

111

112

114

113

115

115

PHYLUM MOLLUSCA. CLASS CEPHALOPODA

Cephalopods are marine, usually without an external shell, with a circle of long arms about the mouth, with from 8 to over 90 arms, and head distinct from body.

Nautilus

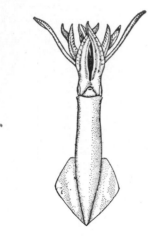

Squid, Sea Arrow

Octopus, Devilfish

Order Tetrabranchia./Family Nautilidae

Nautilus
Nautilus pompilius
Shell to 10 in. in diameter, coiled, with waving markings of alternating white and yellow-brown. Inside of shell pearly and divided toward center of coil by cross walls, cells left being filled with a gas. About 90 tentacles, in four groups of 12 to 13 about mouth, two groups of 17 larger tentacles, and two thicker ones. Living species of *Nautilus*, 4; fossil species, over 300. Living forms dwell on sea bottom in Pacific and Indian Oceans, not floating on surface as sometimes suggested. Favors rather deep water but usually near shores. Group dates back to Cambrian, reached maximum development in Silurian and Devonian. This is not paper nautilus.

Generally lives on the bottom in shallow water. Gas-filled chambers said to make it buoyant and counteract weight of shell. Each section, chamber, once enclosed the soft parts of the animal. Chambers, compartments, coiled in one plane. As the animal grew larger the smaller chamber was abandoned and a new wall was produced. Animal lives in outermost compartment. Partitions concave toward mouth of the shell.

Pursues and catches food in part because of semibuoyant shell. Tentacles lack suckers but hold prey firmly. Food is entirely animal matter caught by pursuit by squirting water through funnel. Eye, a deep pit with small opening which admits light to sensitive surface as in pinhole camera. Shell used much in arts. Animal known to many schoolchildren through Holmes's poem. Paleontologists are interested in animal as a primitive form which has long persisted. Lacks ink, sac common to squids and devilfish.

Order Dibranchia./Suborder Decapoda./Family Loliginidae

Squid, Sea Arrow
Loligo pealei
Length to 8 in. Width to 2 in. Body a flattened cylinder, pointed at fin end, with fin over ½ length of trunk. Dark gray with reddish spots. Tentacle-bearing arms partly retractile, with four rows of suckers. Arms, 10. Eye, with a cornea, conspicuous. Giant squids may be over 50 ft long. *L. pealei* very common in Atlantic Ocean from Maine to South Carolina. Related species, 31; 3 on our Atlantic and 2 on our Pacific Coast. Related cuttlefish is common in the Mediterranean. Giant squids, the largest mollusks, are sometimes found on Nantucket beaches.

Sexes separate. In some forms males introduce sperm capsules into female with a special arm and fertilization is internal. Eggs embedded in jelly and attached to objects on sea bottom. Young squid able to swim and capture food when it becomes free, after few weeks. Probably has reached ¼ in. when freed, 1 in. by 3 months, 2 in. by 5 months, and full grown in 12 months. Swims by quick expelling of water or, if in danger, of ink, and by use of fins, with body being held horizontally. Common in schools; able to change color with remarkable speed. Sensitive to light. Food is held by tentacles and torn by sharp jaws.

Certain species are eaten by man. Squids form basic food of sperm whales and of many other marine animals. Harvested by tons as bait for codfish and other marine species and for use as fertilizer. Some are dried and shipped as food, 3 tons of wet squid producing 1 ton of dried. Famed sea serpents are undoubtedly giant squids. In some Oriental countries squid is an important source of protein in man's diet and is considered a delicacy.

Order Dibranchia./Suborder Octopoda./Family Octopodidae

Octopus, Devilfish
Octopus bairdi
Body to 3 in. long and 1½ in. wide, with eight tentacle-bearing arms to 40 in. long, webbed ⅓ their length. Bluish-white speckled with brown. West Coast species, *O. bimaculatus,* has a pair of large blue spots midway between eyes and third pair of arms from front. Common in deep water of Atlantic north of Cape Cod and in Europe. On West Coast, *O. bimaculatus* is found from Lower California to Alaska. Four Atlantic Coast species and three Pacific, with 50 in world. Paper nautilus is a close relative.

Sexes separate. Third arm on right in male modified for placing sperm capsule in mantle of female during mating. As in squids, eggs are attached to objects on sea bottom and hatch into well-developed young, which grow relatively rapidly where food is abundant. Can walk forward on bottom or swim backward rapidly by ejecting water. Can change color to pink, brown, yellow, or black to match environment. Hides in crevices, coming out to capture prey with long arms. Prey drawn to mouth and paralyzed by poison of salivary glands. Food, mostly crabs, but it attacks man and is a ferocious fighter.

Eaten by men in many parts of world and greatly enjoyed by Orientals and Italians, who use arms in soups or preserved in oil. Considerable sport in capturing them by spearing with the help of flares.

PHYLUM ARTHROPODA

These animals, including over 840,000 species, have segmented external skeletons that are shed during periods of growth, and segmented appendages. The phylum is commonly divided into two subphyla; the Chelicerata, including members of the following classes: (1) Eurypterida, consisting of animals extinct since the Paleozoic era; (2) Pynogonida, rare in some areas but common in others; the sea spiders, marine animals; (3) Merostomata, the ancient king crab or horseshoe crab; and (4) Arachnida with the spiders, ticks, mites, daddy longlegs, scorpions, and their relatives; and the Mandibulata, including members of the classes: (1) Crustacea with the crayfish, lobsters, shrimps, and crabs; (2) Chilopoda or centipedes (commonly referred to as hundred-legged worms); (3) Diplopoda or millipedes (commonly referred to as thousand-legged worms); (4) Insecta, in some lists called Hexapoda, and two less well-known classes, the Pauropoda and the Symphyla. Of every six known species of animals, five belong to this phylum.

CLASS CRUSTACEA

These gill breathers have two pairs of antennae. Commonly, the head and thorax sections are fused into a cephalothorax that in many cases is covered by a continuous shield, the carapace. The 26,000 species are divided into eight subclasses.

[Subclass Branchiopoda]
The 800 branchiopods have leaflike appendages on the thorax, with which they breathe and swim.

Order Anostraca./Family Chirocephalidae

Fairy Shrimp
Eubranchipus vernalis
Length about 1 in. Semitransparent or pinkish-green. Male with larger claspers on head. Swims on back by waving appendages. Relatives include claw shrimps and tadpole shrimps.

After mating, spherical eggs are carried in sacs by female. Resting during summer is in egg stage. Young hatch the following fall into females which, after a number of stages, lay eggs. At approach of resting stage, males may develop; these mate with females and fertilize resting eggs. Males develop from unfertilized eggs. Winter eggs larger than summer eggs.

Common in freshwater pools from fall to spring, with different species ranging over United States, southern Canada, and Mexico. Serve a slight function as food for larger animals, some of which are food for man.

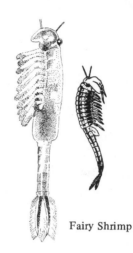

Fairy Shrimp

[Subclass Branchiura]
(Elongate and segmented, or modified as parasites. Thoracic appendages are cylindrical.)

Order Arguloida./Family Arguloidae

Fish Louse
Argulus versicolor
Body strongly depressed. A piercing organ in front of mouth. A pair of compound eyes, and a single eye on median line. Movable joint between fourth and fifth thoracic segments, but cephalothorax covered with a more or less round shell or carapace. Length 1/5 in.

Pairs of antennae very small, first pair provided with a large claw for attachment to the host. Some of mouthparts are modified to form two large suckers. Has a spine in front of mouth cone with which it punctures skin of host. This species is common on pike and pickerel in lakes of New England. Related species are found on other fish. Their presence on salmon may indicate that latter are fresh-run from the sea.

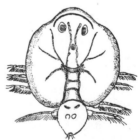

Fish Louse

[Subclass Branchiopoda]

Order Diplostraca./Family Daphniidae

Water Flea
Daphnia longispina
Length about 1/12 in., just visible as individuals to naked untrained eye. Body oval, with tail spine almost as long as body. Swimming antennae obvious. Head apparently bent forward under a helmet.

Sexes distinct. Females carry young in brood sacs. May go through winter as special eggs. A pool or aquarium with temperatures from 70 to 80°F will reach a maximum population in about 21 days, so cultures are normally used and renewed each 3 weeks. Do not thrive in copper or zinc containers or where these metals form a part of the aquarium that has contact with the water.

Water Flea

(Continued)

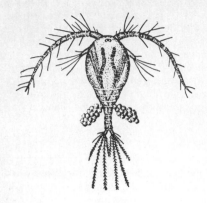

Cyclops

Ostracod

Barnacle

Live year round in fresh water through America and Europe where dirty water and algal and bacterial food are present. Important as fish food, in studies of population cycles, in genetic studies, and as food for aquarium animals. One aquarium stocked with *Daphnia* will feed animals in many others.

[Subclass Copepoda]
Order Eucopepoda./Family Cyclopidae

Cyclops
Cyclops viridis
Body $\frac{1}{15}$ to $\frac{1}{5}$ in. long, but commonly not easily found in field. Usually greenish, and first noticed because of erratic motion caused by swimming strokes of antennae. First antenna 17-jointed and short; second, four-segmented. First four pairs of legs three segmented; fifth pair two segmented. A single conspicuous red eye.

Commonest American species, widespread in ponds here and in Europe. Variable, the larger forms being the less common. Some relatives are found in the sea, sometimes in such great abundance that they color it red for miles around.

Female lays summer eggs which hatch quickly. Later, more hardy winter eggs are laid, and these see animals through severe conditions. Female carries as many as 200 eggs in her two conspicuous egg sacs suspended at middle of body. After early stages in development, young resemble adults. Complete life cycle in 7 days.

Food microscopic plants and animals, captured in journeys through water. Sensitive to light, heat, and disturbance by jarring. Desirable in aquariums for keeping down protozoan population and keeping water clear, as well as serving as food for aquatic animals. More fertile with plant food than on a strict animal diet.

Important basic food for aquatic animals, including some whales. May be fed in aquariums with protozoa cultures made by steeping timothy hay in hot water for an hour, filtering liquor into open dishes, and inoculating with cultures made by keeping fresh horse manure in jars of tap water to which a little pond water has been added. A manure jar should be productive for 2 months. One species of *Cyclops* may be host of parasitic fish tapeworm which may be transferred to man if he eats raw or insufficiently cooked fish. Other *Cyclops* species serve as intermediate hosts of Guinea worm of India, Arabia, and Africa.

[Subclass Ostracoda]
Order Podocopa./Family Cypridae

Ostracod
Cypria sp.
Like minute swimming clams, with swimming structures extending from one end between the "shells," which are technically valves. Seven pairs of appendages. Two pairs of antennae are used for swimming and for orientation. Length about $\frac{1}{25}$ to $\frac{1}{12}$ in.; shell about twice as long as wide. In some, shell is almost transparent, in others not.

Common at bottom of lakes, sluggish streams, and pools where there are dead plants and animals which have passed stages of active decay but where decay is continuing slowly. Some 2,000 species, living in salt or fresh water in America, Europe, and elsewhere in world. Most are marine.

Unisexual. Most lay eggs, which may be attached to refuse or carried within shell. In some genera, no males have been discovered, eggs developing without fertilization. Some have two shells when born; others do not. Young vary in degree of development at time of hatching.

Food largely microscopic animals, though some species live as parasites on larger animals like crayfish. Laboratory cultures may be maintained indefinitely if natural conditions can be approximated, but different species have specific requirements for light, temperature, food, and salinity. May be cultured from individuals or from egg masses found on stones.

Thin slices of potato or pieces of decayed lettuce in partly covered dishes of water provide reasonably good aquariums in which ostracods may be reared. They make excellent food for the earlier stages of fish, salamanders, and other aquatic animals, whose life histories should be known for wise conservation practices.

[Subclass Cirripedia]
Order Thoracica./Family Balanidae

Barnacle
Balanus tintinnabulum
Rock barnacles are stalkless, with six-plated shell joined by thinner shells to enclose soft body. Method of overlapping of plates is significant in classifying groups. In the goose barnacles the five-pieced shell protecting appendages, sometimes referred to incorrectly as feet and body, is borne at end of a stalk from $\frac{1}{2}$ to 12 in. long.

Barnacles are common on ships, piers, and logs, sometimes fouling them badly. Barnacles are common on rocks between high- and low-tide marks, are worldwide in distribution, except that they are not found in fresh water.

Sexes usually not distinct. Eggs carried outside fold of skin beneath shell. Young free-swimming "nauplius" stages swim near surface of sea; molt to two-valved, "shelled," "cypris" stage and swim about as butterflies fly; finally come to rest on support, where they remain for rest of lives and shed swimming legs.

Lies on its back inside shell, with legs extended from shell. Legs force stream of water, bearing food, to mouth. Legs move with clocklike regularity, giving a mass of barnacles a different appearance from a nonfeeding group. Highly sensitive to jar. In some species, males are small animals without mouths or intestines, living within female.

Some species have food value for man and for fish. Ship barnacles often cover bottoms of ships, retarding their speed though doing little actual damage to ship itself. Their removal causes much expense and delay.

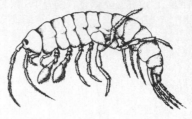

Scud

[Subclass Malacostraca]
Order Amphipoda./Family Gammaridae

Scud, Amphipod, Sand Hopper
Gammarus fasciatus
Up to ⅝ in. long, with high arched back, rather conspicuous antennae, and a brood patch carried beneath the larger female. In *Hyalella* first antennae are shorter than second; in *Gammarus* both are the same. Greenish-white to brown. Rather conspicuous eyespot.

One of commonest of smaller freshwater shrimplike crustaceans, ranging through America to Arctic Ocean, to Europe, and elsewhere. It thrives in aquatic vegetation, in swift or still water. Many species live in fresh or to salt water.

Mates April to November. Lays average of 22 eggs every 11 days, incubating them in brood sac about 11 days. Young are like adults except in size. Mature in 39 days, laying about 6 eggs first time. Adult may lay as many as 46. Mated pairs swim together for a long time, male smaller. Do not normally breed at temperatures below 64°F.

Swims backward, on back, by swimming legs. May eat own kind, or living or dead plants, or animals. Can see well and apparently has excellent sense of touch. Each generation produces large number of young. Hides in protecting vegetation.

One of most valuable organisms for converting aquatic plants and small animals into suitable food for fish. Important food source in fish hatcheries and in nature. Serves as a scavenger. Too small to be directly significant as food for man.

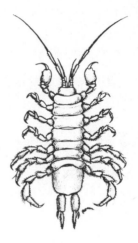

Water Sow Bug, Isopod

Order Isopoda./Family Asellidae

Water Sow Bugs
Asellus communis
Length to ⅗ in.; breadth ⅕ in. First pair of legs suitable for holding, others for walking; last three pairs longest. A single shieldlike plate on abdomen. First pair of antennae short, second pair long. Sexes distinct.

Common among freshwater vegetation and in oozes on bottom of pools and stagnant ponds. This species common through eastern United States, being there commonest aquatic isopod. Related species extend range of genus through America and to other parts of world.

Breeds during warmer months, mating and laying eggs every 5 to 6 weeks. Eggs carried by mother in a conspicuous whitish brood sac under front part of body. Young remain here about a month, until they reach stage where they can maintain independent existence.

Food dead or living plants or animals. In some cases mother may even eat her own young, or young eat brothers and sisters. Move by walking but may burrow in mud, leaving gill openings to rear protected by abdominal plate, thus preventing suffocation.

Serves role of scavenger, but may also serve as food for many common aquatic animals. Abundance in areas where decay is active indicates that it can live in water with relatively low oxygen content.

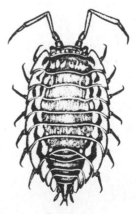

Sow Bug

Order Isopoda./Family Oniscidae

Sow Bug
Oniscus asellus
Length to ⅗ in. Width ⅓ in. Deep slate-gray spotted with white, or light gray with white along edges of back. Related pill bugs, family Armadillidae, are able to coil themselves into compact spheres.

Common under stones and boards, close to soil or in rotting wood, under bark, and in similar places where there is some dampness, though this is
(Continued)

not always obvious. Common in eastern and central states and in Europe, with related forms extending range widely over world.

As in water sow bugs, eggs are carried by female in ventral brood sac formed by flat projections of thoracic legs. Young are like adults in general shape and appearance, there being no metamorphosis like that common in many other crustaceans.

Food organic matter, plant or animal, which abounds in places where they are found. They rarely live exposed to direct sunlight or great heat, even though humidity and food conditions might seem suitable for their needs otherwise.

Serves essentially as scavenger and to some extent as food for enemies. Rarely develops abundance sufficient to be considered as pest but may eat young plants; controlled with Paris green bait.

[Subclass Malacostraca]
Order Decapoda./Family Cragonidae

Shrimp
Crago septemspinosus
Length 2 in. Second pair of antenna long. Claws on second pair of legs like structures back of mouth, with pincers. In swimming, special abdominal appendages (pleopods) are used, this not being case in crabs and lobsters next discussed. In shrimps abdomen is well-developed and body more compressed than in crabs and lobsters.

Sand shrimps commonly found on and in sand at low-water mark and in shallow water on to depths of 300 ft. This species ranges from Labrador to South Carolina and in Europe and on Pacific Coast. Other species replace it on other shores, so that relatives are to be found on most seashores of the world.

After the mating, eggs are laid. Young stages in some species spent in egg. Some, from South, go through series of four larval stages, which are, in sequence, nauplius, protozoa, zoea, and mysis; in some species all four stages are passed within egg. Breathing by gills hidden in sides; swimming by flexing abdomen, which ends with tail fin and has swimmerets beneath, last help in carrying egg; walking by legs of middle of body.

Food captured by pincers, as in crabs and lobsters. May change color to match surroundings. Sensitive to light change. Hides when disturbed.

Industry important, particularly on Gulf Coast, Pacific Coast, and along Atlantic Coast of southern states. Valuable bait and basic food for larger animals. Consistently a leading money-maker for the American fishing industry.

Order Decapoda./Family Homaridae

Lobster
Homarus americanus
Length to 2 ft. Weight to over 28 lb, average about 2 lb. Dark green above, with darker spots, and yellowish beneath. First pair of claws large and used as pincers. First appendage of abdomen of male hooked, of female normal. 15-in. female may carry 100,000 $\frac{1}{16}$-in. eggs under abdomen, 10 to 11 months. Young, transparent.

In shallow salt water in summer, in deeper water in winter. Very young swim at surface 6 to 8 weeks. Common along coast of Atlantic from Labrador to North Carolina. Individual may range 12 miles in 3 days. Dig holes into which they retreat backwards.

Mate in May. Eggs extruded in June, develop on female to second May. Young swim at surface 2 months. Average molts: first, aged 3 days; second, 6 days, $\frac{2}{5}$ in.; third, 11 days, $\frac{3}{7}$ in.; fourth, 26 days, $\frac{1}{2}$ in.; fifth, 41 days, $\frac{3}{5}$ in.; sixth, 46 days, $\frac{2}{3}$ in., goes to bottom; seventh, 56 days, $\frac{3}{4}$ in.; eighth, 70 days; ninth, 80 days, 1 in.; tenth, 90 days, $1\frac{1}{8}$ in.; twelfth, 173 days, $1\frac{1}{4}$ in.; fifteenth, 282 days; twenty-fifth, 5 years, $10\frac{1}{2}$ in., $1\frac{3}{4}$ lb. Mates. Frees first young at 6 years and every other year thereafter.

Sensitive to light, pressure, and chemical nature of water. Prefers water temperature about 55°F. Uses antennae as sense organs.

Probably our most valuable crustacean. In past, over 100 million captured a year. Useful as scavengers and as food for valuable fish. Conservation demands that no animals be taken until they have freed the first generation of young or are over 12 in. long. Laws regulate size of lobster taken. Caught in baited "pots," wooden traps. Lobster fisheries now in trouble. Demand far exceeds supply.

Order Decapoda./Family Astacidae

Crayfish
Cambarus bartoni
Length to 5 in. Two large claws in front, four pairs of walking legs, long and short antennae, and jointed tails. Male with first appendage of abdomen hooked; female with this not hooked. Atlantic and Mississippi ge-

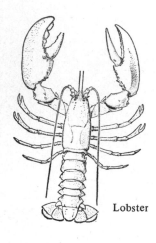

Shrimp

Lobster

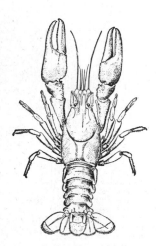

Crayfish

nus *Cambarus,* with 17 pairs of gills; Pacific slope genus *Astacus,* with 18 pairs. About 100 species in all, 70 of which are *Cambarus.*

In fresh water, in streams or ponds, hiding under stones in small, excavated caverns, or among water plants. Some terrestrial, living in holes sometimes with turret entrances on "crawfish land" where water is near surface. Some species of *Astacus* found in Europe and Asia.

2 years from egg to egg. Mate March 6;* 200 or more spherical eggs laid March 24 and held under female's abdomen 2 months; eggs hatch, May 18; first larva, May 18 to 20; second May 20 to 26; third May 26 to June 13, desert mother; fourth larva, June 13 to 30; fifth July 1 to 5; sixth ⅗ in. long, July 6 to 17; by October, 2 in. long; by November, 2 to 3 in. long. Mates following spring and may lay eggs. May live 6 to 7 years. This species may mate and spawn through year; *C. diogenes* mates in fall, spawns in spring.

Food variety of plants and animals which may be caught and held with claws. Can regenerate lost limbs. Sheds outside skeleton as size increases. Then gets sand into pits on side of head which serve as balance organs. Can move in four directions along the bottom; forward, backward, and to either side.

Excellent scavengers. Superior food for many valuable fishes. May destroy fish eggs. Used as food in many large Eastern cities. Popular laboratory animal in biology and zoology classes where it is usually studied in its most uninteresting condition, dead and pickled. 333 tons harvested annually.

* Initial date selected at random.

Hermit Crab

Order Decapoda./Family Paguridae

Hermit Crab
Pagurus sp.
Small crabs, usually about 1 in. long, found occupying spiral shells of other animals, with soft abdomen hidden in shell and claws protecting shell opening. Great variation in species and in kind of shell occupied. Abdominal appendages of female suitable for egg carrying. Eyes on long stalks between antenna bases.

Some species found in shells on land, some even living in trees throughout their lives. Common species include *P. hirsutiusculus,* Alaska to Lower California; *P. longicarpus,* Maine to South Carolina, in shallow water; *P. pollicaris,* Maine to Florida, in deeper water. Over 100 species.

Coral-colored eggs, about size of mustard seeds, laid January to March. Early in year, zoea stage with short frontal spine appears as a free swimmer. Then comes glaucothoe (megalops) stage, in which animal is symmetrical and abdomen has five pairs of two branched appendages; this lasts 4 to 5 days. Then follows shell-occupying period. As crab grows larger, it seizes larger shells of dead mollusks or other hermit crabs and makes a rapid transfer.

Great fighters, readily attacking and killing own kind but retreating to protection of shell before more powerful enemies. Corals, sea anemones, and the like rest on shells. Walking is by second and third pairs of legs; fighting and food getting by first. Sensitive to light, touch, water pressure, position, and other factors.

Serves as scavenger, as food for fishes, and as means of locomotion for organisms which live on the shell. Frequently covered by the hydroid *Hydractinia echinata,* which enlarges the empty shell of the crab by building up its free edge.

Blue Crab

Order Decapoda./Family Portunidae

Blue Crab
Callinectes sapidus
Width of shell 2½ times length. Good-sized male, 2½ by 6½ in.; female smaller, with smaller claws. Apron under male, inverted T; of virgin female, equilateral triangle with concave sides; of fertile female, a plump triangle. Gray or blue-green. In family Cancridae, genus *Cancer* (illustrated), first antennae are folded longitudinally; in Portunidae, transversely or obliquely.

Blue crabs found in shallow salt or brackish water, chiefly on mud bottoms near river mouths where vegetation is abundant, from Cape Cod to east coast of South America, or sometimes in adjacent fresh water. Edible crab of California is *Cancer magister,* found along entire Pacific Coast.

Eggs light orange to brown, 1 to 5 million; in mass carried by female, hatch into free-swimming zoea, followed by quieter megalops, which develop into crab forms. Summer hatch transforms by winter. Molts weekly, later monthly, increasing each molt about ⅓. Matures third summer. Adults do not molt. Males carry females before final molt.
(Continued)

431

Females mate once. Males defend virgin females from rivals, may carry mates 2 days before mating and 2 days while mating. Females produce two lots of eggs each summer, the first several weeks after mating.

In summer, found close to shore; in winter, in deeper waters. Larger males favor deeper waters. Soft-shelled, newly molted individuals hide. Build no permanent homes. Swimming movements so rapid animals appear like a flash. Run sideways. Sensitive to light, touch, chemical nature, and temperature. Food plant and animal matter, particularly decaying animal matter.

Valuable scavenger and important human food. Caught in nets or on lines baited with dead animal matter. Sold fresh, canned, or cooked. Provides some sport, and for many persons, a meager livelihood. Commercial value about ½ million dollars a year.

Fiddler Crab

Order Decapoda./Family Ocypodidae

Fiddler Crab
Uca sp.
Shell width up to 1⅔ in.; length to 1 in. Light brown to purple and dark brown. In male, one claw, usually right, is much larger than other; in female pincers are about equal in size. Eyestalks exceptionally long and slender. Among larger on Atlantic Coast is *U. minax.* Among smaller is *U. musica* (shell ⅓ in. long), which makes a noise.

Found usually in droves at water's edge, or in burrows to 2 ft deep or, as in *U. minax,* some distance from salt water and in fresh water. *U. pugnax* found from Cape Cod to Louisiana; *U. pugilator* from Cape Cod to Texas; *U. crenulata* (illustrated), in southern California; and *U. musica,* from Vancouver Island to Mexico.

Male courts female with elaborate motion of huge claw, often leading her into tunnel. Eggs carried by female until young are able to shift for themselves, then freed gently into water, usually on an incoming tide. Free-swimming stages precede adult forms which are more sedentary; more or less similar to those of closely related crabs already described.

Food mostly small organic materials rolled into little pellets and often carried into burrow for use. Female uses both claws in food getting; male the smaller claw only. Runs sideways, male usually holding large claw in threatening attitude. Obviously able to sense soil vibration well.

Of some value as food for such enemies as birds and fishes. Often trapped for use as fish bait. May burrow through dikes, causing damage by the resultant leaks. Essentially scavengers and always interesting to watch.

Alaska King Crab

Order Decapoda./Family Lithodidae

Alaska King Crab, Lithode Crab
Paralithodes camtschatica
Moderately sized body with extremely long legs. Specimens may measure 6 ft though usually 3½ to 4 ft from tip to tip of legs. Weight to 22 lb, more generally to 14 lb in Bering Sea. Carapace, shell, of large specimens, to 11.5 in. for a male and to 8.5 in. in female. *P. platypus* and *P. breviceps* also a part of king crab fishery.

Occur in cold waters of Sea of Japan, Bering Sea, Okhotsk Sea, and south along the Alaskan water to British Columbia. A creature of the continental shelf.

During mating, the male "handshakes" the female. Fertilization is external on swimmerets of female. Eggs incubate for a year before hatching.

Food: clams, sea urchins, salmon, plankton.

Prior to World War II only a small commercial Alaska king crab fishery was carried on in Alaska. The catch was some 10,000 lb. After the war, Alaska king crab meat became popular. Supplies of the animals dwindled, prices rose sharply, and regulations as to numbers taken and size limitations had to be agreed upon. The Japanese have always been important exploiters of the Bering Sea and Alaskan coastline king crab resource. Their canneries are floating.

Crabs are caught in large gill nets, some 150 by 80 ft, floated by glass buoys, 5 in. in diameter. Fishing is carried out at 20 to 50 fathoms. Season from April to September with peak of production in May and June according to Japanese; April to November with peak in fall and winter according to American fishery.

Meat from one leg may be equivalent to two 1-lb cans of crab meat. For many years king crab from floating canneries (Russian and Japanese) provided about 90 percent of the canned crab meat in the United States. Becoming scarce and worthy of protection.

PHYLUM ARTHROPODA. CLASS MEROSTOMATA

Order Xiphosura./Family Limulidae

Horseshoe or King Crab
Limulus polyphemus
Up to 20 in. long, with slender tail and broad flat forepart consisting of abdomen and still broader foresection. Relatively flat. Males much smaller than females. Two large separated eyes and smaller pair close together. Back of abdomen flattened; bears along outer edges a row of movable spines.

In shallow sea water; along shore, in sand, at breeding time; below low-water mark, unless washed up by a storm. Atlantic Coast, from Maine to Mexico, with related species along eastern coast of Asia. One genus with five species. Survivors of group of animals once abundant and wide-spread.

Mate in early summer, when large females come to shore bearing one or more slender males, each clinging to tail of animal ahead. Eggs laid in depression in sand prepared by female. fertilized by male, and left for waves to cover; hatch in midsummer. Young crabs which may cover beach, finally make way to sea and disappear until well grown. Young lack spinelike tail. When molting, shell splits on front edge.

Walk with four pairs of front feet. Swim with hind feet. With help of shell, borrow in sand for small animals as food. Air breathers, even though they live in sea. Protected by shell and hiding ability. Not aggressive. Feed mostly at night.

Commonly fed to hogs and chickens or used as fertilizer. Some people consider them good food, but generally they are of little economic importance, in spite of their generous size.

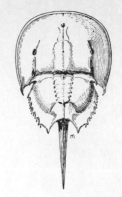

Horseshoe or King Crab

PHYLUM ARTHROPODA. CLASS ARACHNIDA

(The abdomen of this class lacks locomotor appendages. Some 36,000 species in 11 orders)

Order Scorpionida./Family Centruridae

Scorpion .
Centrurus sp.
Members of this order usually have a long segmented "postabdomen." In this family, there is usually a spine under stinger at end of postabdomen. Length of species shown about 3 in. Pair of claws in front helps hold food. Flexible tail brings stinger into play when needed. Some species reddish, some yellow, some green.

About 300 species, 25 of which are found in United States, from North Dakota west and southeast. Common in drier areas. Not common northeast of line from North Dakota to Carolinas, but increasingly common southwest of that line. Found in open, in wooded areas, in buildings, and in a great variety of places. Some tropical species to 10 in. long.

Male courts female, mates with her, is commonly eaten by her after mating is over if not before. No eggs laid; incubated in enlarged female. Young up to 60, may live for week or more clinging to back of mother before leaving to maintain independent existence. Two adults rarely together unless mating or one is eating other.

Sting of no United States scorpion known to be fatal to a healthy adult human being, though it may in some cases cause vomiting, convulsions, and death to children. Tropical species much more dangerous. Not affected by own poison so cannot commit suicide. Active, mostly at night, seeking shelter in shoes, clothing, of crevices at dawn.

Dangerous to man but more dangerous to insects. Their prey is primarily insects and spiders, which they hold, then paralyze, then eat. In scorpion country clothing should be shaken before it is put on.

Scorpion

Order Phalangida./Family Phalangiidae

Daddy Longlegs, Harvestman
Leiobunum vittatum
Spiderlike with body ⅕ in. long, and eight legs varying in length from 1 to 2 in.; second pair longest. Body brown, legs black. Male smaller-bodied and more brilliantly colored than female. Abdomen segmented (unlike spiders), and with effective stink glands.

In fields and meadows, or sometimes crowded in considerable numbers in holes in logs and in brush piles, but usually close to ground and most commonly more or less solitary. About 60 American species, comprising 2 families and 15 genera.

Eggs laid in fall through a long ovipositor which is thrust underground or into debris; hatch in spring. Young white, with black eyes at first, but otherwise much like adults, whose color they soon develop. By fall young have matured, then they mate, lay eggs, and most individuals die, wintering being in egg stage.

Food small insects and other animals, or decaying or dead matter. More or less nocturnal, though may be active on bright sunny days. Build no
(Continued)

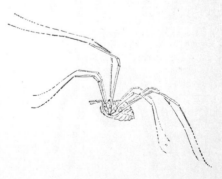

Daddy Longlegs, Harvestman

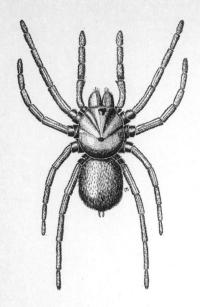

Trap–door Spider

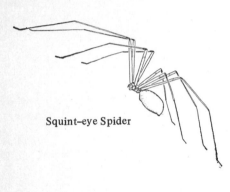

Squint–eye Spider

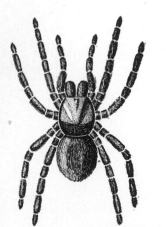

Tarantula, Bird Spider

web or nest. Move relatively slowly, getting some protection from noxious odor. May shed legs and effect escape, though this may provide poor protection against persistent enemy.

Interesting to children who have been told that they will point in direction cows are to be found. Since they point in all directions, this is a reasonably safe but useless generalization. General role probably that of scavenger. Of little economic importance.

Order Araneida [Spiders]./Family Ctenizidae

Trap-door Spider
Bothriocyrtum californicum

Large. Body of adult female 1⅙ in. long. Legs long and stout. Color of whole spider rich dark chocolate-brown, with legs slightly darker. Third claw well developed with rake for hole digging.

Half dozen American genera of trap-door spiders, some with several species, closely related to tarantulas. Fairly common throughout southern and western United States, but more common farther south. Lives almost wholly on or under ground, in burrows which may be 6 in. or more deep.

Life history habits in general similar to those outlined for tarantulas. Young remain in nest burrow with mother for about 8 months, leaving home burrow after winter rains cease and establishing own burrows. New layer of web added to door each year. May live over 7 years.

Food larely insects and other small invertebrate animals which can be captured with quick forays from nest. Life of trap-door spider seems ruled by fear, particularly of parasitic wasp enemies, of which there are many. Highly sensitive to jar and to light. Can hold cover of nest shut from beneath.

Useful as destroyer of insects and similar animals captured at burrow entrance while door is held open by abdomen. Some tunnels have two doors; some branched, all lined with waterproof silk; some camouflaged at entrance. Excellent terrarium and garden pets; harmless.

Order Araneida./Family Pholcidae

Squint-eye Spider, Long-legged Spider
Pholcus phalangioides

Body about ¼ in. long; longest legs just over 2 in. long. Body conspicuously long and slender for a spider, pale brown in color. Might easily be confused with daddy longlegs but for presence on web and general body characters which make it a true spider.

Closely related to house spider, though not in same family. American species of genus two; related American genera six. Common in cellars, under porches, in houses, both in Europe and in America. Closely related species limited, for most part, to Florida region. Favors dark warm places.

Eggs laid in inconspicuous egg sac that female carries about with her, rather than leaving it hanging in web as does house spider. Egg sac thin, seen only when looked for carefully. Young develop in egg sac and, when released, live for a while on mother's web.

Builds a large irregular web which is loose and open. Spider hangs head down on web, except when alarmed. When prey enters web, spider jumps around on web, shaking it violently and entangling victim, which is then killed. Food almost exclusively small insects caught on web.

Useful as destroyer of insects and interesting because of its behavior on loosely hung web. Of course it clutters up neglected places with its webs, but this does no great harm, and insects captured cannot do harm they might do otherwise. Lives comfortably in dark box if fed.

Order Araneida./Family Theraphosidae

Tarantula, Bird Spider
Dugesiella californicum

Largest of American spiders, some species measuring 2 in. in length of body alone, with legs correspondingly large. Body and legs conspicuously velvety. Legs stout but able to carry spider rapidly. One South American tarantula has legs that spread over 7 in.; another has body 3½ in. long.

Closely related to trap-door spiders; about 40 species in family but does not belong to same family as Italian spiders whose behavior gave name tarantella to a dance. Lives on or near ground, hiding in holes under stones and in logs and debris. Common in southwestern United States.

Mates in fall. Following summer female puts 300 to 600 eggs in thick-walled cocoon, which she protects. Young grow slowly first half year, rapidly next 4 to 5 years, molting about twice a year, living as long as 16 years. Only 2 to 3 may mature out of an initial 5,000 young.

Not poisonous in spite of contrary convictions. Bites painful but not serious because jaws are suitable for crushing its food of insects. Some species can kill birds. When annoyed, assumes defense attitude. Can hear camera click. Bite only when forced to do so. Largely nocturnal. Common prey of wasp called tarantula hawk.

Valuable as insect destroyer, of no serious injury to man's interest, so really worthy of protection. Relies on sense of touch. Probably cannot see difference between light and darkness but seems to be "charmed" by bright light at night. Makes excellent terrarium pet and thrives if given reasonable care.

Order Araneida./Family Uloboridae

Triangle Spider
Hyptiotes cavatus
Male $\frac{1}{12}$ in. long. Female $\frac{1}{6}$ in. long. Generally inconspicuous. Back of female with four hair-bearing humps, which in male are less conspicuous. At rest, closely resembles tree bud, for which it is commonly mistaken until it moves on being disturbed.

One species is best known in United States, but there undoubtedly are others as yet undiscovered; ranges to West Coast. Found on dead twigs, commonly in gorges and similar places where air may be relatively quiet. Usually on a twig next web. Frequently inhabits pines.

Egg sacs not easily noticed as they are placed on twig and resemble bark or silk. They are oval and flat and have an outer covering over $\frac{1}{4}$-in. sac. Covering is gray like bark, or dirty white due to fact that portions are made of black silk mixed with white. Young hatch in brood sac.

Makes triangular web of four threads with cross threads, one corner terminating in single long thread. Spider takes up slack on this long thread, and when prey enters web, releases it to snap web and tangle prey in meshes. Rests feet uppermost beneath this trip cord, slack lying loosely between front and hind feet.

May be kept in a small bottle, where it will build its web, lay its eggs and, if food is avilable, live with reasonable comfort. Too small for general appreciation. Included here because of unique method of capturing prey and because of remarkable resemblance to its immediate environment. Protective coloration almost perfect.

Order Araneida./Family Agelenidae

Funnel-web or Grass Spider
Agelenopsis naevia
Medium-sized, varying in size and color but with two longitudinal black lines bordering gray lines on the back. Male $\frac{1}{2}$ in. or less long. Female up to $\frac{3}{4}$ in. long. Legs do not vary greatly in length. Median gray band down back serves as quick rough identification character.

Several species are commonly found in United States; *Agelenopsis* also represented elsewhere in world. Commonly found on or near their webs in grass, on lawns, in meadows, most conspicuous when webs are covered with dust or dew. Probably one of most universally common spiders.

Live only 1 year, winter being spent in egg stage. Adults die soon after eggs are laid. Eggs in irregularly shaped flat sacs, under rubbish and usually protected by female until her death. Male wanders in late summer, visiting webs of females, seeking mating, but may be eaten for his trouble.

May stay on one web all summer if undisturbed, web being enlarged with season. Trip thread common between supports above web. Spider runs on top of web (unlike most spiders), and hides in funnel which has a rear escape from which prey may be pounced upon. Webs of young seen by May.

Useful as a destroyer of insects that may be injurious to crops. Most excellent subject for study in fall biology because of abundance, food-getting habits, physics of nest, sensitivity, locomotion, and protection. May be attacked by parasitic wasps. Webs used by some birds in nest making.

Order Araneida./Family Micryphantidae

Dwarf Spider
Erigone autumnalis
Body about $\frac{1}{20}$ in. long, easily recognized by generally light color and bright yellow head. Male has claspers used in holding female in mating. Related species are mostly larger and darker in color. Legs moderately long. Movement relatively slow. Does not frighten easily.

A very large family of spiders most of which are of small size; difficult to identify. Four American genera. Found for most part on ground, in grass, among leaves, moss, and other small plants where they spin their small webs during summer months. In fall found high on some post or exposed plant.

(Continued)

Triangle Spider

Funnel-web Spider

Dwarf Spider

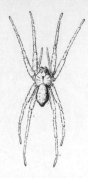

Nursery Web Spider

Wolf Spider

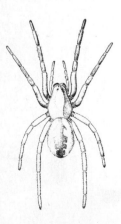

Turret Spider

Little known of life history. One member of family builds dome web about 3 in. across; others build small webs like plain sheets. Some favor houses; others, fields; and others, woodlands; but all are most commonly inconspicuous; webs usually noticed only when dew-covered.

About time first frosts appear, great numbers of erigonids climb to elevations and spin long streamers of web when air is rising, which serve as balloons to journey great distances. Sometimes streamers cover a field, causing "gossamer," "flying summer," or "old-woman's summer" effects.

So conspicuous are gossamer threads when covered with dew or frost that everyone has noticed them in season.

Order Araneida./Family Pisauridae

Nursey Web Spider
Dolomedes tenebrosus
Large, long-legged, the sexes differing so greatly that they have been described as separate species. Male with conspicuous yellow band around forepart of body. Female with large abdomen bearing median yellow band, with three to six pairs of small yellow or white spots. Relatively active, with remarkable locomotor abilities.

One of the largest spiders of the family. By some, grouped with wolf spiders, the Lycosidae, but in *Dolomedes,* upper row of eyes is less curved and smaller. *Dolomedes* commonly found on, under, or near water, and because of skill in diving is sometimes called diving dolomedes.

Eggs laid in egg sac almost equal in size to that of abdomen of female; may number to 300. Young freed by mother when sac becomes mottled; may be guarded in web nusery, or in web under stones. Young feed on each other at first; not fed by mother. Web made just as young are freed from sac.

Feed on water insects and other small animals caught largely by pursuit. Normally wary; hide skillfully under water or vegetation. Webs built apparently only for protection of young. A ¾-in. spider has been known to capture and drag from water a fish 3½ in. long weighing four times as much as spider.

Probably serves some role in maintaining balance among small animals of waterways. Most interesting because of ability to swim under water, to build and use rafts from which to dash after prey, and because of webs which so obviously serve as a nursery. Harmless and should be studied more.

Order Araneida./Family Lycosidae

Wolf Spider
Lycosa helluo
Medium-sized. Male about ½ in. long. Female ¾ in. long. Color gray, yellow, grayish-brown, or brown, with a narrow yellow middle stripe down the back. Legs shorter and stouter than those of water spider *Dolomedes,* although some species of *Lycosa* have relatively slender legs.

In same general group with water spiders and turret spiders, a group including a majority of larger spiders which run on ground without making nets. Commonly found on or under stones or in holes underground; when pursued, runs rapidly, even going onto or under water to escape many enemies.

Eggs laid in spherical sacs, each consisting of two pieces. Sacs carried by female near rear of body. Female frees young from sac and they may crawl over her body, taking a ride with her wherever she goes. Young soon leave parent and shift for themselves. Female builds a silk-lined nest for shelter in early summer.

Food small animals, mostly insects caught by pursuit. May be active during day but most active at night when its food is more active and its enemies, such as many wasps, are less active. Fight between wolf spider and wasp, which uses spider as food supply for its young, is well worth watching.

Probably of considerable importance in controlling insect enemies of plants of fields and gardens. Difficult to conceive as harmful to man's interests. Makes excellent and interesting terrarium pet for home, camp, or school and should be better known.

Turret Spider
Geolycosa pikei
About ⅞ in. long, with stout body and strong legs. Male more slender than female. Reddish-brown obscured by gray hairs, with broad light band down back which becomes narrower toward abdomen. Body hung low; female's abdomen conspicuously larger than that of male.

In same general group with wolf spider, which it resembles slightly. Roughly 114 species in America in family Lycosidae, all of which live on or near ground. This species digs burrows underground, entrance being marked by little turrets of sand. Aso known as burrowing wolf spider.

In late May, male courts female; usually eaten by her after mating.   In June, female retires to burrow, closes it, spins and protects cherry-pit-sized cocoon containing eggs.   Young emerge from cocoon in July, live on mother's back, molt twice by August, leave to dig own burrow.   Mature in size by third fall.   Sexually mature the next spring.

Preys on small animals caught by pursuit.   First year, burrows few inches deep; second year, 10 in. or more; mature spider burrows to 18 in., often ending in enlarged cell in which winter is spent.   Entrance turret may be reinforced with plant material.   May go months without food without ill effects.

Useful enemy of insects that may injure plants.   Worth studying because of great sensitivity to jar, light, odor, sound, and moisture.   Can be frightened by footsteps 12 ft. from burrow entrance.   Will come to entrance if beam of light is centered on it.   Rushes to entrance of burrow at slightest disturbance.

Black Widow Spider

Order Araneida./Family Theridiidae

Black Widow Spider
Latrodectus mactans
Black, with hourglass-shaped red spot beneath abdomen, latter large and almost spherical.   Body of female ½ in. long, over twice size of male which is less than ¼ in. long.   Legs slender, front and hind pairs longer and more slender than middle pair.

From New Hampshire to Patagonia and west to Pacific Coast but more common in southern and western United States.   Favors damp dark areas such as are found in cellars, under boards, in stone walls, and frequently in outhouses.   Closely related to harmless house spider.

In June or July female makes 3 to 4 ½-in. cocoons, each of which may contain about 300 eggs.   Young hatch, feed on each other.   In few weeks survivors emerge from cocoons and begin eating spiders, insects, and other small animals.   May mature and breed by fall, and females may live through winter.

Not ferocious, normally must be forced to bite.   Bite causes serious discomfort.   While death has been reported in 10 percent of those bitten, no record of death occurring to a healthy adult human.   Pain in general not accompanied by great local swelling; slight temperature lasts 3 days.

Entitled to worst reputation enjoyed by any spider.   If bitten, use ligature, cut wound, apply weak ammonia and carbonate of potash.   Since bite tends to paralyze intestines, cathartics should be used also.   Name black widow probably resulted from habit of female eating male, but this is common in other spiders as well.

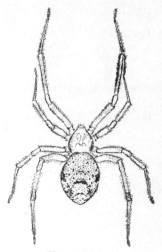

House Spider

House Spider
Theridion tepidariorum
Female with body ¼ in. long.   Male with more slender ⅙ in. body.   Legs nearly three times length of body.   Exceedingly variable but possessing several dark chevrons above end of abdomen and on underside.   General appearance dirty and dusty.

Over 300 species of genus have been described, about 40 of which are found in North America.   *T. tepidariorum* is most common species.   Commonly found in and about houses, particularly in dark unused places such as under porches, in attics, and like.

Males court female, but may be eaten if unwelcome or after mating.   Eggs 50 to 200, laid in spheres hung in webs 6 to 8 weeks after mating, hatch in 1 week.   In next 8 weeks young pass through five molts; leave cocoon in second stage, web of parent in third.   Nine cocoons may be made in one season.

One mating fertilizes eggs of many cocoons.   At first young feed on each other, then on scraps in mother's web.   Finally leave to make own webs.   Webs irregular; adults may or may not have place in which to hide and await prey.   Fourth pair of legs has comb of fine teeth which help fling silk on prey.

Ability to adapt itself to varied climates has made species almost worldwide.   Useful as destroyer of house insects such as flies, moths, and mosquitoes.   Webs collect dust.   Should be studied living in every biology class in preference to too much book reading.

Order Araneida./Family Araneidae

Black and Yellow Garden Spider
Argiope aurantia
Large, black and yellow.   Male about ¼ in. long.   Female about 1 in. long, with front legs as long as body.   Colored conspicuously, with rear of body black, and with bright orange or yellow spots that form a band along sides.

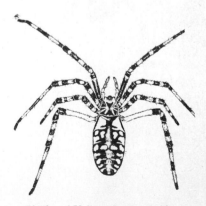

Black and Yellow Garden Spider

(Continued)

Rather closely related to silk spiders, from whose silk cloth has been made. Appears frequently in literature as *Miranda*. About 120 American species. Favors marshes, gardens, or tall plants protected from wind. Webs sometimes found on shrubs.

Male and female build similar nests. Male visits female's nest in fall; after mating, female lays eggs in paper cocoon size of hickory nut, which is firmly fastened to vegetation. Eggs hatch in midwinter but young remain in cocoon until May. Mature by midsummer, disappear by October.

Remarkable web, sometimes 10 ft across, with ladderlike structure near center and shelter web where adult hides. May add secondary web near center. Can kill large grasshoppers that land in web, by enmeshing in silk and then biting. If frightened, drops to ground and hides in vegetation. Web remarkably strong.

Probably useful as insect destroyer. Physics of web worth studying. New web usually made each night. Not all portions of web are sticky, as may be discovered easily. Spider is commonly parasitized by ichneumon flies, which may themselves be parasitized.

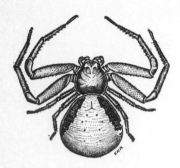

Flower Spider, Goldenrod Spider

Order Araneida./Family Thomisidae

Flower Spider, Goldenrod Spider
Misumena vatia
Medium-sized, with front legs spreading three times body length. Male $\frac{1}{8}$ to $\frac{1}{6}$ in. long. Female $\frac{1}{3}$ to $\frac{1}{2}$ in. long. Color yellow or white, depending upon flower upon which it lives; usually with pink or red markings. Male usually darker, particularly on sides.

A dozen species, distributed through United States, a number from Far West. Commonly found on flowers or resting in throat of some bell-shaped flower. In spring seems to favor white flowers; in fall, yellow; may change from white to yellow in 10 to 11 days. Not common in winter. Family name, crab spider.

Eggs placed by female in flattened silken sac which is composed of two equal parts. Female protects sac and young, sac usually fastened to some leaf or other plant part and hidden on some incurled surface. Young feed on each other until they leave sac to shift for selves.

Makes no web but lies in wait, usually in throat of flower, to capture insect visitor, then hangs on with hind legs while forelegs are used in capturing prey, even though this means a struggle when prey is larger than spider itself. Most active in warm sunny seasons.

Possibly useful as insect destroyer but always interesting as garden or terrarium pet. May wander from flower to flower, but if hiding places are limited, may usually be found easily. If undisturbed may stay for days in a single flower.

Zebra Spider

Order Araneida./Family Salticidae

Zebra Spider
Salticus scenicus
Relatively small. Female slightly larger than male, reaching length of $\frac{1}{4}$ in. Male conspicuous even to naked eye, because of larger mandibles or jaws which hold female in mating. Gray, mottled, hairy, with numerous white bands and spots. General appearance flat or low. Front eyes touch each other.

Only common species in United States, but widely distributed even into Europe. Three other species not so common. Found on walls of rooms, on window casings, on fences, barns, wooden houses, commonly where it is sunny or where wood has been weathered. Not offensive in spite of its abundance.

Male courts female by dancing before her, spreading his legs out at sides and up over his back and head. Eggs laid early in season in cocoons, often several in single nest; hatch early. Young guarded by mother until independent or well through summer, then leave mother's nest.

Although this spider spins silk, it makes no web for capturing prey. It moves rapidly in any direction; when necessary, leaps into air to capture its prey, "letting out a line" of silk until it is ready to stop, then climbs back up silk to place it left. Hides in silken nests in winter.

Amusing and harmless resident of rooms in which we live. Obviously favors dryness, sunny spots, and warmth; active in daytime rather than at night. Excellent subject for classroom study. Should be better known by all students and housewives. Makes common cobwebs that hang from ceilings.

Order Acarina./Family Sarcoptidae

Itch Mite
Sarcoptes scabiei
Almost too small to be seen by naked eye. Mite is white. In female, first two pairs of legs end in stalked suckers; in male, all four pairs have suckers; fourth pair in female tipped with long bristle. Size about $\frac{1}{50}$ in. long and $\frac{1}{75}$ in. wide.

Other related species also cause such skin diseases as scab and mange. About 100 known species in group, of which 15 are American. This species parasitic on man and hogs. Common where cleanliness is not observed, with female laying her eggs under skin of host.

Female makes irregular burrows under skin, laying oval eggs, singly, in rows of about 22 to 24. In about 7 days eggs hatch into little mites whose explorations cause further irritation. Time from eggs to eggs about 4 weeks, though cold weather may extend this period considerably. May live away from host.

Position under skin makes it practically impossible to remove pests mechanically, and attempts to do so only spread infection. Others may be infected by contact with clothing, towels, bed clothing, or anything handled by infected person. No significance in the name "7-year itch," as infection need not be tolerated.

Infected person should be bathed and rubbed thoroughly in hot water and green soap, followed by bath in hot water and by applications of sulfur ointments which should remain on body several hours. Infections should be handled promptly particularly where persons come in contact with others in public places.

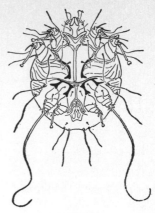

Itch Mite

Order Acarina./Family Dermanyssidae

Chicken Mite
Dermanyssus gallinae
Flat, reddish, pear-shaped, about $\frac{1}{35}$ in. long and slightly more than half as wide, with hind legs not reaching to rear end of body. Female with long piercing mouthparts. Males with pincerlike mouthparts. Body not conspicuously constricted at any place. Young with three pairs of legs like insects.

Several species in genus, but 24 American genera in family. This genus found almost exclusively on birds, though this species may live, at least temporarily, on humans, dogs, cats, horses, or other domestic animals. Relationship is of course as parasite on host.

Elliptical pearly white eggs, laid from 3 to 7 at a time, with a total of 25 to 35 during 8 periods. Eggs laid in cracks near roosts or in nests; hatch in 2 days into small mites, which, after third molt, become adult. Time from egg to egg may be 7 days; individuals may live 4 months.

Crawls about roosts and birds, sucking blood, most commonly at night. Hides easily because of smallness, flatness, and other qualities. In family are some mites which live independent of hosts; some which are parasitic on insects as well as on warm-blooded animals; some which use insects primarily for transportation.

May be kept in control by scrupulous cleaning of roosts and nests, using crude petroleum mixed with kerosene. If too abundant, may kill chickens or so weaken them as to injure their productivity. May kill young of wild birds which use same nest succeeding seasons.

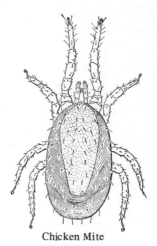

Chicken Mite

Order Acarina./Family Ixodidae

Cattle Tick
Boophilus annulatus
Female length, $\frac{1}{2}$ in.; body elliptic to rectangular, yellowish or slate-colored. Male length, $\frac{1}{10}$ in.; body oval, brown. Eyes present but often indistinct. Young, $\frac{1}{32}$-in. "seed ticks." Plate around spiracle, round in this genus; comma-shaped in genus *Dermacentor*, which includes common wood tick and spotted-fever bearer.

On cattle or ground, particularly through Southern States. Commoner in fields where cattle or other hosts have lived constantly. Numbers can be kept down by dipping cattle periodically. Disease caused by tick parasite may prove fatal to cattle or may merely prevent them from producing flesh, young, or milk.

Mature female leaves cattle host, drops to ground, lays over 5,000 eggs which hatch in 15 days to over winter. Young ticks climb vegetation, attach selves to passing cattle and, in this species, pass all molts on cattle. May infect cattle with the sporozoan *Babesia bigemina*, which causes Texas fever.

Infected adults can pass sporozoan through eggs to young, which may infect another host; hence Texas fever is not contagious between cattle but is spread by ticks. Ticks of all sorts should be removed from humans, cattle, and other hosts to avoid possible infection.

(Continued)

Cattle Tick

Controlled by pasture rotation or by dipping cattle in crude petroleum or arsenical mixture. Keeping cattle out of field for few months usually starves all ticks and makes field again suitable for pasturage. Usually four lots used, moving cattle before new eggs hatch. Ticks should not be crushed but can be forced to free themselves by applying match or alcohol to exposed end. Loss may reach 60 million dollars a year.

American Dog Tick
Dermacentor variabilis

Body oval, chestnut-gray, brown, or blue-gray, tough-skinned. Four life stages—egg, larva, nymph, and adult. Adult with eight large legs and measures about 3/16 in. without blood. Male with hard, white-marked shield over all of back; female with only one-half body covered. Fully engorged female may be 1/2 in. long, becomes gray-blue in color; skin of body appears like pliable, heavy plastic. Adults may live to 3 years but more normally less than 3 months. Mating takes place while female is attached to host. She drops to ground after loosing mouthparts. 4,000 to 6,500 eggs laid, hatch in about 7 to 14 days. Larvae six-legged and 1/40 in. long; attach to passing small animals; feed, then change to eight-legged nymph; feed again, drop to ground, transform to adult. Adults seek larger animals. This tick may take 2 years or even longer to complete its life cycle. Close relative of Rocky Mountain wood or spotted fever tick, *D. andersoni,* in the West; American dog tick distributed to eastern North America.

Control measures never complete but helpful. In tick country, check body twice a day, especially hair on head. Keep dogs from roaming in high-grass areas. Dust dogs with rotenone to kill collected ticks. Spraying dwellings with 0.5 percent toxaphene or 0.03 percent lindane as directed by supplier will help rid premises of ticks. If a tick is removed from a person, a good antiseptic, tincture of iodine, should be applied to wound. Because of disease-carrying capacity, tick infection should not be taken lightly. Ticks can be removed by causing them to release their attachment by applying a few drops of chloroform, gasoline, or turpentine on the body. Firm grip with tweezers or fingers and a steady pull will prove successful.

American Dog Tick

Order Acarina./Family Hydryphantidae

Fresh-water Mite
Hydryphantes ruber

Body 1/10 in. long, almost round in outline, with swimming hairs on last three pairs of legs. Body soft, with chitin plate around median eye on top surface; in this species, no other chitin plates above. Red or brown or intermediate in color, sometimes rather brilliant, sometimes dark.

Common, particularly in spring in and near woodland ponds, or frequently in some stage on insects flying about. Some 17 species in genus and 700 American species in family, with many variations. Eyes and plates are characters used in distinguishing many.

Eggs may be laid on plants, stones, or animals. At first, young have 6 legs; pass through complete metamorphosis before reaching maturity. Larval stages leave water and are parasitic on aerial insects; adults return to water to breed. Related clover mite and red spider may live wholly out of water.

Usually burrow into animal or other material which they eat, but frequently conspicuous on surface because of red color. Some are specific as to host while others are not so particular.

These mites are of little economic importance normally, but on cultivated plants may be serious pests. Flowers of sulfur or kerosene emulsion used as spray provides most common control practice. Some are bad pests in greenhouses and on houseplants, as well as on garden truck crops. Although they "breathe" air and are able to travel rapidly over water surface film, they can stay submerged for a relatively long period. Some water mites are active throughout the year, even beneath ice.

Fresh-water Mite

Order Acarina./Family Trombiculidae

Chigger
Trombicula irritans

Minute harvest mite with greater constriction forward of middle of body than most others of group. Larva is six-legged and has five hairs on dorsal plate. In immature and pestiferous stage, it is small enough to make its way through meshes in ordinary cloth used in clothing.

Several hundred species, in none of which are the adults parasitic. Adults spend their time on the ground and in the soil. This species occurs from New York to Minnesota and Kansas and through South, but close relatives extend ranges, particularly westward in South. Found on vegetation and then on animals.

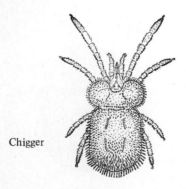

Chigger

In northern part of range, larvae appear in June and disappear by October. They attach themselves to animal hosts that brush against plants on which they lie in wait. They do not burrow under skin but attach themselves and cause intense itching and pink or red spots. Usually leave host in 1 to 7 days.

Larvae parasitic on man, mice, rats, rabbits, prairie chickens, quail, toads, box turtles, and snakes. Humans attacked severely may run a slight temperature and develop certain nervous diseases due to inability to sleep. Chiggers in Far East transmit scrub typhus, a rickettsial disease, fatal to 33 percent of victims.

A great pest attacking man and beast as chiggers of Europe do. Repellents including dibutyl phthalate and benzyl benzoate are very effective. For those who prefer simpler remedies, a good dusting with flowers of sulfur on those portions of the body where clothing is tight-fitting, e.g., beltline, within socks, will be effective.

CLASS CHILOPODA

Order Scutigeromorpha./Family Scutigeridae

House Centipede
Scutigera coleoptrata
Members of class Chilopoda have reproductive pore near rear end, while Diplopoda have it forward. Chilopods also have only one pair of legs to a segment, instead of two. This species has a body about 1 in. long which is light brown with three dark longitudinal stripes. Last pair of legs nearly 2 in. long.

Common in United States from New England through central and western states, but more common in South. Favors damp places such as cellars. While members of this family have 15 pairs of legs, those in related family Geophilidae may have 173 pairs. About 1,000 species of centipedes, among largest being a tropical form 1 ft long.

In Geophilidae and Scolopendridae, young have all their legs at time of birth; in Lithobiidae and Scutigeridae, they have only seven pairs at this stage in development. Centipedes not normally social animals. Species here considered most delicate and difficult to collect whole.

Food largely insects such as cockroaches captured and killed by poison forced through opening in special poison-bearing claws. Will kill many small animals but is not dangerously harmful to man, though it may be painful. Treat by bathing in ammonia solution and retarding blood flow.

Some tropical species are dangerous, but most temperate zone species are highly useful as insect destroyers. The small ¼-in. garden centipede may be a pest of asparagus and garden flowers. It can be controlled by winter flooding of field for 3 to 4 weeks. Too frequently killed by homeowners who think it is a harmful pest.

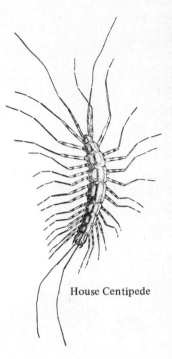

House Centipede

CLASS DIPLOPODA

Order Julida./Family Julidae

Millipede, Thousand-legs
Julus sp.
Members of class Diplopoda have two pairs of legs on most body segments. Ventral plates of *Nemasoma* free; body segments number 35 to 45 in. ½-in. *N. minutum.* In *Julus,* only two forward ventral plates are free. 30 to 35 segments in ½-in. *J. virgatus;* 52 segments in ⅘-in. *J. venustus.* Mating legs of male hidden.

Common in dark moist places such as meadows and gardens, in decaying food materials. Some 125 American species in 8 different families, with 50 species in family here considered. Members of family lack sucking mouthparts, have over 30 body segments, have stink glands, and all legs on seventh segment are of mating type.

Mating effected with assistance of legs on seventh segment of male. Eggs laid in cluster in damp earth. When young first hatch, in about 3 weeks, they have only three pairs of legs; as they go through successive molts number increases until commoner species have 30 or more pairs.

Food usually decaying plant material but, in wet weather particularly, may feed on roots of living plants. Protection effected by curling so that hard plates of back cover leg-bearing surface. Stink glands along sides of body, some bearing prussic acid, make animals obnoxious to prospective enemies.

Milipede

PHYLUM ARTHROPODA. CLASS INSECTA

Insects have bodies divided into head, thorax, and abdomen. In adults, head bears one pair of antennae and thorax commonly three pairs of legs and one or two pairs of wings; or sometimes no wings. Insects are air breathers when adult. More than 750,000 known kinds.

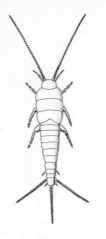

Silver Fish

Springtail

Snow Flea

Order Thysanura [Bristletails]

Wingless, with 11 abdominal segments and usually two to three slender segmented caudal appendages. Chewing mouthparts.

Order Thysanura./Family Lepismatidae

Silver Fish

Lepisma saccharina

About ⅓ in. long. Glistening, scaly, silvery white with yellowish tinge; two conspicuous appendages in front and three behind. Active runners. Related firebrat *Thermobia domestica* more dusky. Bodies of both flattened from top to bottom.

Found on floors, tables, and shelves, particularly where there is starched material such as clothing or books or glued areas. Favor dry areas and live where little seems available for food until books begin to shed their bindings and labels and clothing to show holes.

As in succeeding orders young and adults resemble each other, differing chiefly in size. Metamorphosis such as is common in higher orders of insects is accompanied by much less pronounced changes.

Food either plant or animal material, although some authorities contend that only plant material is eaten. It probably varies with different kinds of Thysanura. Some species that live in dark places have lost their eyes. They also differ considerably in requirements of temperature and humidity.

May cause serious damage to books, clothing, carpets, and to museum specimens, particularly if storage rooms are damp. In case of serious infestations, contact County Cooperative Extension Service as to recommendations for appropriate control measures.

Order Collembola [Springtails]

In these wingless insects, young resemble adults. Six abdominal segments. On first are pair of organs useful in walking on smooth surfaces. On third are usually short appendages that hold springing organs borne on fourth. Springing organ hurls creature into air. None over ⅕ in. long.

Order Collembola./Family Sminthuridae

Springtail, Garden Flea

Sminthurus minutus

Body shorter and more compact than in snow fleas, and with a more obvious constriction between first and second thirds. Jumping ability well-developed, as is ability to cling to smooth surfaces. Appear comical when seen under microscope.

These relatives of snow fleas live in drier areas than other Collembola. One species, *Sminthurus hortensis*, is common on leaves of many garden plants, such as turnips, cucumbers, and cabbages, particularly when they are young.

Young resemble adults, as in other members of order. Little seems to be known about their general life history and habits.

Food plant material. Amount of activity is modified considerably by amount of moisture available. These animals have tracheae, the openings of which are on sides of neck. They do not breathe through body surface as snow fleas do.

Where these insects are sufficiently abundant to be pests, they may be controlled by bordeaux spray mixed with pyrethrum powder. This must be applied thoroughly to all surfaces, as insects may congregate in great numbers in relatively inaccessible places.

Order Collembola./Family Poduridae

Snow Flea

Achorutes nivicolus

Snow fleas belong to suborder Arthropleona, all members of which have long rather slender bodies in which segments are conspicuously distinct. In some members of family, leaping organ is absent.

These insects live in, on, or close to water. *Achorutes nivicolus* is often common in great numbers on surface of snow, where it appears like jumping grayish dot. On fungi there are frequently great numbers of *A. armatus*. On surface of standing fresh water is related *Podura aquatica*; and along tide marks, *Anurida maritima*.

Young like adults, as in other members of order, but little is known about their general life history. They seem to be seasonally abundant, have obvious environmental preferences so far as amount of sunlight is concerned.

Food probably mostly plant material. Since there are no true breathing organs, animals doubtless breathe through surface of body. This may explain their sensitiveness to varying climatic conditions.

Normally these insects are of little economic importance, though some of their relatives may be pests of certain garden plants or may swarm around maple-sugar buckets and possibly do small amount of damage.

Order Orthoptera [Roaches, grasshoppers, and crickets]

Insects which usually have two pairs of wings; front pair long, narrow, thick, and with many veins; hind pair membranous, broad with many veins and folded fanlike under the front pair. Chewing mouthparts.

Order Orthoptera./Family Blattidae

American Cockroach
Periplaneta americana
Length to 1½ in.; one of our largest cockroaches. Wings long and well-developed. Antennae may reach beyond end of extended hind legs. Brown; wings usually lighter in color than body.

Among most primitive of insect groups, abundant in Carboniferous period. This species widespread in United States, most abundant in central and southern portions; competes for disfavor with German cockroach or Croton bug, *Blattella germanica*, which came into prominence in New York City when Croton Dam was built. Oriental cockroach, *Blatta orientalis*, common in Europe and in eastern and southern United States.

Female lays reddish-brown disks of eggs numbering between 200 and 300. Eggs are stuck to floors and cracks, and in good temperatures hatch in about 70 days into young cockroaches that resemble their parents except that they have shorter wings. It may take a year in this species for development from egg to adult, but other species have a shorter cycle.

Food any organic material, particularly foodstuffs, bookbindings, or, in some cases, even nails of human beings. Filth and partly digested foods are expelled onto good foods, spoiling them. Generally cockroaches avoid light, favor moist areas, and are likely to thrive in large numbers of their kind.

Serious enemies of foodstuffs and destroyers of books, carpets, paper, and sometimes clothing. Because they may eat same foods as man, have been used as experimental animals in studying effects of different diets on development through many generations. Serve as scavengers. Control by fumigation, cleanliness, traps, and powdered arsenic baits.

Order Orthoptera./Family Mantidae

Praying Mantis
Mantis religiosa
About 2 in. long. Green or brown. Forelegs used in grasping prey; held often as though in prayer, with forepart of body bent upward, head twisting from side to side. Antennae erect and short. Hind legs used in walking and leaping. Four well-developed wings permit slow, extended flight.

European mantid, native of Europe, well-established in eastern United States. Known as rearhorses, mule killers, devil's horses, and soothsayers. Native, smaller mantid of eastern United States is *Stagmomantis carolina*. About 20 species in country. Chinese mantid *Paratenodera sinensis*, has established itself in Eastern States. Mantids probably entered U.S. as egg masses on nursery stock or packing.

Large straw-colored egg masses, of to 1,000 eggs, are laid in fall on boards, weeds, or buildings. They hatch in May or June into little mantids that resemble parents except for absence of fully developed wings. Young grow slowly, acquiring wings and maturity in August. Egg masses often 1 in. long, often hatched indoors. Young are cannibalistic and eat their brothers and sisters as they emerge.

Food largely flies, grasshoppers, and other insects. Female frequently eats her smaller mate after breeding act, or even before if she is hungry enough. Prey is grasped and held by spine-laden forelegs, eaten at leisure with a most comical display of apparent indifference.

All exceptionally useful destroyers of other insects and should be encouraged in gardens, where they may play an important role in insect control. Larger species widely introduced. Egg masses are collected and sold for establishment of colonies.

American Cockroach

German Cockroach

Praying Mantis

443

Order Orthoptera./Family Phasmidae

Walkingstick
Diapheromera femorata
Length, overall, to 4 in. or more, with diameter of body only about ⅛ in. at most. Six legs are in widely separated pairs and may be held so that whole resembles a green or dead stick, since color may be green or brown. Legs and antennae about equal in length. Eyes small and to front.

Relatively common in United States, except in more southern states. Distributed from Texas, Rocky Mountains, and Manitoba to northern Florida, Maine, and central Ontario. About 20 species found in country and group is mainly tropical, where it is represented by over 600 species; some over 16 in. long; one is the famous tropical leaf insect, *Pulchriphyllium scythe*.

Eggs white with a black stripe; dropped to ground from trees in fall, sounding like rain, hatch in spring. Young resemble adults in shape but not in size. Full size may be attained in 6 weeks after hatching. In some species, eggs do not hatch until second spring after they are laid.

Food largely foliage of trees of forest. Sometimes destruction of leaves is complete and hence serious. Activity is greatest at night. No noises are made. Since there are no wings, locomotion is by walking, though some species leap and others run. Individuals may remain practically motionless during daylight.

May be serious pests to trees but entirely harmless to man. When abundant as pests, they are sometimes controlled by use of a sprayed stomach poison. Food is chewed, not sucked. Superficially resemble water scorpions of freshwater ponds, but these suck their food. In spite of protective coloration, is eaten by common grackle and some 15 other species of birds as well as some lizards and rodents and mantids.

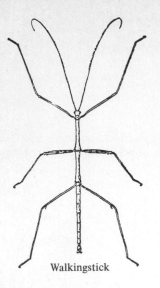

Walkingstick

Order Orthoptera./Family Tettigoniidae

Two members of this family are shown. For most part, these grasshoppers are green, have long hind legs useful in jumping, long slender antennae, with organs of hearing on the forelegs and, except in the mole crickets and sand crickets, females have long curved bladelike ovipositors with which they lay eggs in ground or in plant leaves or stems. Important subfamilies include false katydids, including *Scudderia*; true katydids; meadow grasshoppers, including *Orchelimum*; cone-headed grasshoppers; shield-backed grasshoppers.

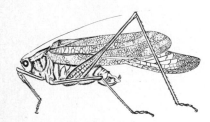

Forked-tailed Bush Katydid

Forked-tailed Bush Katydid
Scudderia furcata
Length about 2 in. overall. Forewings nearly uniform in breadth throughout length, differing in this respect from angular-winged katydid, *Microcentrum*, which has a short, abruptly up-curving ovipositor, and the oblong-winged and the round-winged katydids, *Amblycorypha*, which have long, curved ovipositors.

Scudderia, in different species, is found throughout United States and most of Canada, but mostly east of Great Plains, living on bushes. True katydid, *Pterophylla*, is found in trees, usually in restricted colonies, but throughout United States east of Rockies.

Scudderia lays flat eggs, ⅓ to ¼ in. long, in rows on leaves in late summer. These hatch in following season and young resemble adults except for absence of wings. No pupal stage.

Food leaves of plants on which insect is found. *Scudderia furcata* gives a soft, high, oft-repeated *zeep, zeep, zeep*, from bush or tree; *S. curvicauda* calls *bzrwi* in day and a shorter *tchw* at night; *Amblycorypha* gives a shrill *shrie-e-e-k*; *Microcentrum* calls a high, repeated *tzeet-tzeet-tzeet-tzeet-tzek-tzek-tzuk-tzuk*.

Economic importance not great.

Meadow Grasshopper
Orchelimum vulgare
Length about 2 in. from tip of wing to tip of head. Slender, pale green, with long slender antennae. Head slender, pointed. Eyes relatively small and near front of head. *Conocephalus* is smaller and more slender and commonly with straighter ovipositor.

Found in moist pastures and meadows, usually among grassy plants, with *Orchelimum vulgare* ranging from Rocky Mountains to Atlantic Coast and generally common. Not commonly found high above the ground, even though suitable food might be available.

Eggs laid in plant tissues with the aid of curved ovipositor; hatch in spring. Young resemble adults in general form but do not develop wings until maturity is reached by early summer. No pupal stage.

Meadow Grasshopper

Food mostly grasses. Front wings of males differ from those of female.
In male, left front wing overlaps right and by means of scraper causes
rasping sound, due to vibrating membrane. These sound-producing organs are on wings close to body. Ears of cone-headed grasshoppers appear as two vertical slits on forelegs.

Order Orthoptera./Family Gryllacrididae

Cave Cricket
Ceuthophilus gracilipes
Wingless. Pale brown or dirty white, with brown spots. With high
arched back and remarkably long hind legs and antennae. Body short
and thickset for a grasshopper; in female, terminates in curved swordlike
ovipositor. Lacks the swayback of related sand cricket.

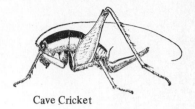

Cave Cricket

Common in cellars and moist caves, or in woods, gorges, or wells. Usually where there is little light. At least 12 species of cave crickets or
camel crickets in the United States. Related sand crickets are found on
the Pacific Coast.

Since these crickets have no wings, they can make no sound. Eggs laid in
late summer, probably in ground, may hatch in fall but more commonly
in spring, into young nymphs that resemble parents.

Food meat, fruit, vegetables, or almost anything organic and available.
While cave crickets are not ordinarily considered of great economic importance, related Jerusalem cricket and western cricket, *Anabrus purpurascens*, may be serious pests to crops in West, sometimes reaching plague
proportions and eating even their own kind.

Order Orthoptera./Family Gryllidae

Field Cricket
Gryllus assimilis
Length overall of these crickets about 1 in. Female (illustrated) with
long slender ovipositor, which is lacking in male. Smaller brown cricket,
Nemobius, is possibly more common. It is ⅜ in. long; brown, with three
darker abdominal stripes.

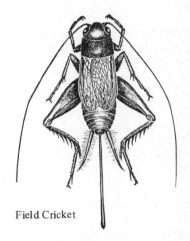

Field Cricket

Field crickets are found in dwellings, fields, pastures, gardens. Cricket
on the hearth or house cricket, *Acheta domestica*, is slender pale yellowish
brown with ⅗-in. body, native to Europe but has been established in
America. Family includes also mole crickets.

In late summer or fall, female lays several hundred eggs, singly or in
masses, in holes in ground. Eggs hatch in fall or spring. Those hatching
from eggs in July may mature by mid-August. Immature crickets resemble adults but lack developed wings. No pupal stage.

Food plant or animal matter, but mostly plants. Males begin singing by
wing rubbing about June and continue through August, purpose apparently being to defy other males. Chirp loudest when weather is brightest
and warmest. Males will fight each other if confined, chirp pitch rising.

May destroy grass, clothing, food, and other materials. In great numbers,
may be extremely damaging to young seedling plants by cutting stems at
ground level; by stripping grains from cereal grasses; may eat holes in
fruits of garden and truck crops such as tomatoes, strawberries, cucumbers, and peas. May serve as vector of disease-producing fungi. For
recommendations concerning control measures, consult your County Cooperative Extension Service.

Snowy Tree Cricket
Oecanthus niveus
Slender, pale green. About ½ in. long, with long antennae. Female
with wings closely wrapped about the body. Male with more slender
body and with wings broader and relatively free and, of course, without
long ovipositor. Male secretes at base of wings a liquid that female
seems to enjoy eating.

Snowy Tree Cricket

Commonly found singly or in pairs on trunks of trees or shrubs in late
summer, with several species covering United States and Canada. Common species include the snowy, the narrow-winged, the black-horned or
striped, the four-spotted, and the broad-winged, all of which are good
singers.

Male sings. Female may approach and feed on liquid on back. Mate.
Eggs deposited on bark or stems such as raspberry canes, 40 to 50 together; hatch in early summer into little crickets that resemble parents
except for wings. Reach maturity by midsummer. Killed with advent of
cold weather.

Generally most active at night. Males fly readily and give calls in unison
with other males. Add 37 to number of times males call in 15 seconds to
(Continued)

get approximate Fahrenheit temperature. Calling is done during night of warmer months and is described as "slumbrous breathing." Amply protected from enemies by coloration.

Young and old feed largely on plant lice and so are essentially useful to man. One species, *O. nigricornis*, may injure raspberry canes by laying eggs too closely together on stems, but most other species do little if any damage and much good, so are worthy of protection because of their pleasing music and useful food habits.

Order Orthoptera./Family Acrididae

Lesser Migratory Locust
Melanoplus mexicanus
Length about 1 in. Yellow or tan, with dark bars across hind legs. Forewings grayish, extending beyond end of body, with a few dark spots near middle. Female with abdomen as illustrated. Antennae relatively short. Hind legs suitable for jumping great distances, even without help of wings.

Wide distribution in North America. Found in fields and meadows. Often serious pest in East. Commonly found with red-legged grasshopper, *Melanoplus femur rubrum*. The Rocky Mountain locust, formerly known as *M. spretus*, is merely the long-winged migrating phase of *M. mexicanus*, which until recently has been known as *M. atlanis*. Prefers well-drained light soils and sparse vegetation. Adults may gather in swarms and migrate hundreds of miles and destroy crops and range plants.

Mates and in fall female lays eggs underground, in clusters of 12 to 80, in 2-in. burrows. Eggs about $3/16$ in. long, like taper-ended cylinders; hatch in spring into nymphs that shed skins five times before attaining maturity and wings. No pupal stage.

Food almost any plant material, often serious pests to crops. Some grasshoppers give weak sounds. Carolina locust of roadside displays yellow-bordered, black hind wings, and the cracker locust gives a clicking sound, possibly in courtship. Eardrums or membranes are on sides of body behind wing bases.

Late fall plowing may destroy eggs in ground. Rolling of fields may kill young before they have developed wings. Hopperdozers dragged across fields, and ditches dug through or around them, provide some control. Baits may be used. Possibly some use of dried grasshopper bodies as chick or fish food may make capture profitable and offset losses caused.

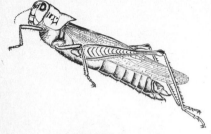

Lesser Migratory Locust

Order Isoptera [Termites]

Made up of small, soft-bodied and frequently pale-colored insects. Members exhibit social organization and caste differentiation. Antennae usually short, either thread-or beadlike. Tarsi four-segmented and cerci usually short. May be with or without wings. Winged forms with two pairs of wings similar in shape and size; somewhat long and narrow or at least as long as the body. At rest, wings are held flat over abdomen. Chewing mouthparts. Simple metamorphosis.

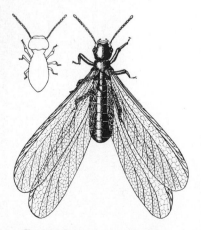

Eastern Subterranean Termite

Order Isoptera./Family Termitidae

Eastern Subterranean Termite
Reticulitermes flavipes
Social, colonial. Four castes: first, reproductive caste; both sexes winged, black or chestnut, functional eyes, with wings over half their length beyond end of body. Second, reproductive caste; both sexes pale, eyes partly pigmented, wings short wing buds. Third, reproductive caste; dirty white, blind, rarely mature sexually, workers. Soldiers wingless, rarely lay eggs.

Most species tropical, though some live in temperate regions. In United States live in hidden nests, usually in wood of buildings or furniture, though some may build in living trees. Some build covered ways from nests to food or other necessities. Some in Africa build mounds 12 ft high.

First reproductive caste sheds wings after mating flight and female acts as egg-laying queen. Should she die, her place taken by one or more second reproductive caste queens with small egg capacity, or eggs may be laid by queens of a third reproductive caste, or even by female soldiers or workers. Nymphs cared for by workers. Colony defended by soldiers.

Food of this species, mostly wood and other plant material. Differentiation in members of castes comes early; interpretation of what establishes an individual in one caste or another differs with authorities. Certainly early stages of third reproductive caste and of workers appear to be almost identical.

Very destructive to wooden things; great care must be taken to control establishment of colonies. May eat interior out of a wooden table, leaving only thinnest shell that may collapse suddenly. Same damage may be done to wooden foundations of buildings. Wood in contact with earth must be protected from attack.

Order Neuroptera [Antlions and Dobsonflies]

Insects with four membranous wings, both pairs of wings about the same size. Usually wings are held rooflike over body of insect when at rest. Wings with many veins. Antennae long, many-segmented; threadlike, clubbed. Chewing mouthparts. Complete metamorphosis.

Order Neuroptera./Family Corydalidae

Dobsonfly, Hellgrammite
Corydalus cornutus
Dobson flies have wingspread of over 5 in. Adult males have long slender crossed jaws, while females have shorter stouter jaws. Larvae aquatic, strong-jawed, breathing through gill tufts at base of abdominal segments. Fearsome-looking in most stages. Fluttery flight; generally found near streams.

Larvae live in moving water and adults fly near it, often attracted by lights. At least six species of family found in United States. Smoky alder fly is about 1 in. long; other common members of family usually larger; one, *Nigronia*, has conspicuous white-banded black wings.

Mates outside water. Eggs laid in white blotch, sometimes 1 in. across, on objects over water, about 2,000 to a mass. Hatch into larvae that drop to water and develop for about 35 months; then leave water to pupate for 1 month in cell, under cover, on bank. Because of long aquatic larval stage, must live in permanent streams.

Food almost exclusively aquatic animals caught and held by strong jaws that are able to inflict painful bites on humans. Hold on tenaciously. Protection effected by aquatic habitat and by position under stones. Fish are greatest enemies, and hellgrammites are enemies to young fish.

One of best of bass baits in larval stage and, as such, is worthy of protection. Larvae may be destructive in fish hatcheries, but are rarely found there. Because of appearance and congregation around lights, they frighten people. Many are killed needlessly. Popular subjects of study in entomology classes because of size and primitive nature of larvae. Larva shown is *Chauliodes*, a fishfly.

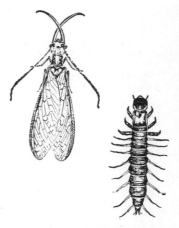

Dobsonfly, Hellgrammite

Order Neuroptera./Family Chrysopidae

Goldeneyed Lacewing
Chrysopa oculata
About ⅕ in. long; antennae are long and slender. Pale green, delicate, golden-eyed, lacy-winged, with some black markings on head. Abdomen slender. Odor most offensive when insect is held close to nose. Rather awkward fliers. Larvae as shown, active.

About 60 species of lacewings live in United States, most of them being found on plants in fields, gardens, or orchards. Adults are attracted by light and are commonly found on screens and on radiators and windshields of automobiles. Eggs and larvae occur on leaves and stems of plants.

Eggs placed on plants at tips of erect, slender stalks. Larva hatches in about 5 days, crawls down stalk seeking animal food and its brothers and sisters safe on nearby stalks. Larval stage about 2 weeks, with skin shed twice. Cocoon papery, thick, egg-shaped, about ⅙ in. long. In summer, pupates about 26 days; may winter in cocoons.

Food largely plant lice, on which larva feeds voraciously. Adult may live through winter, but rarely. Egg laying begins shortly after emergence and mating. One female known to lay 617 eggs during short life as adult. Apparently detects food by feeling rather than by smelling or seeing. Odor may be protective quality.

Highly useful as destroyer of plant lice, both in larval and adult stages. Single larva may destroy 200 aphids in 20 days. Called "green flies" though they are not true flies. Should be protected and encouraged in every way because of food habits.

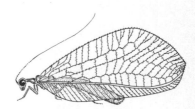

Lacewing, Goldeneye

Larva

Eggs

447

Order Neuroptera./Family Mermeleontidae

Antlion, Doodlebug
Myrmeleon immaculatus

Adult appears somewhat like a damsel fly but has fair-sized knobbed antennae. Length about 1½ in. Wings four, fore pair slightly larger than rear. Abdomen long, slender, or about ⅘ total length of insect. Rather awkward fliers. Larvae like small ovals, with extraordinarily long jaws.

Widely distributed in United States in sandy areas, but more particularly in dry parts of South. About 60 species in United States, belonging to 11 genera. About 338 Neuroptera in United States and about 4,670 in world, with several known as fossils.

May require 1 to 3 years to reach maturity. Larva or doodlebug lives at base of self-made pit in sand, into which prey may fall, only to be caught by larva at base and eaten. Larva's jaws are hollow and prey's juices are sucked in. A globular silk sand-covered cocoon size of a large pea covers pupa.

Food largely ants caught in pits. Once an ant enters pit, larva at base keeps it tumbling by hurling sand at it and undermining its footing. Legs and large body of larva serve to anchor it against struggles of ant. Larva moves backward more readily than forward.

Useful enemies of harmful ants and most interesting insects in larval stage. Children like to bait doodlebug larvae with grass stems and watch larvae hurl sand when sufficiently disturbed. No antlion is harmful.

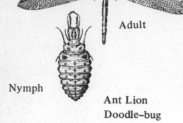

Adult

Nymph

Ant Lion
Doodle~bug

Order Ephemerida or Ephemeroptera [Mayflies]

Small to medium-sized, elongated, soft-bodied insects found near water. Front wings large, triangular in shape and with many veins. Hind wings small and rounded; in some species, missing. Wings, at rest, held together above body. Abdomen with 2 to 3 hairlike tails. Antennae rather inconspicuous, small and bristlelike. Simple metamorphosis. Mouthparts vestigial.

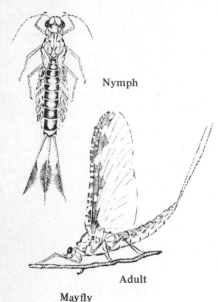

Nymph

Adult

Mayfly

Order Ephemerida./Family Ephemeridae

Mayfly
Callibaetis sp.

Adult thin, delicate, with wings held over back when at rest. Usually with three but sometimes with two long delicate "tails"; front pair of legs held conspicuously forward. No pupal stage. Nymphs aquatic, with 2 to 3 tails; gills on sides of fore part of abdomen, but variable in different species.

Some 550 known species of Mayflies in United States; at least 1,500 in world; many fossil species. Appear in great numbers in and near bodies of fresh water, particularly about rivers and streams, nymphs living under water.

Adult Mayflies live only a day or two, mating in flight. Females drop eggs or place them under water, 500 to 1,000 per female. Eggs hatch and develop through sometimes 21 molts, taking from 6 weeks to 3 years to complete a generation. Unlike other insects, may molt once during adult life, even though this stage may last only a few hours.

Nymphs feed chiefly on plant material, which they chew. Adults eat little or nothing, as their mouth parts are mostly useless. Possibly some Mayfly nymphs are carnivorous, but these would be in minority. Some nymphs burrow in mud; some cling to stones; some are almost free-swimming.

Unquestionably highly valuable as food for fish, serving splendidly in turning aquatic plants into animal food for larger animals. Adults may seem to be pests because of numbers but do little damage. Might possibly be dried and converted into fish or bird food.

Order Odonata [Dragonflies, Damselflies]

Insects with two pairs of similar elongate, membranous, many-veined wings. Wings, at rest, usually held outstretched in suborder Anisoptera, the dragonflies; or together above the body in suborder Zygoptera, the damselflies. Compound eyes, large; antennae very short; prothorax small. Chewing mouthparts. Metamorphosis simple, incomplete.

Order Odonata./Family Coenagrionidae

Damselfly

Ischnura sp.

Adults slender-bodied, with four netted wings that fold in a vertical plane over the back when at rest. Head large and movable, with large bulging round eyes. Mouth parts suitable for chewing. Male frequently more brilliantly colored than female. Nymph aquatic, with three taillike plate-like gills, and with a distinct movement from side to side.

Three American families: the Lestidae and Coenagrionidae, with narrow wings, and the more beautiful and brilliantly colored Agrionidae. Adults relatively slow-flying when compared with true dragonflies, as wings are moved in a more flapping style. Found as nymphs in ponds and streams and as adults near these bodies of water.

Male holds neck of female with claspers at end of abdomen and brings end of female to male sex organs on underside of forepart of abdomen. Female may lay 200 to 300 eggs; these are ovate, $1/25$ in. long, and commonly inserted in tissue of water plants. Eggs hatch in 3 weeks to several months, and in going through 10 to 14 molts, spend 225 to 624 days.

Food of nymphs probably any aquatic animal that can be overcome, and of adults any insect that can be captured in flight, even though its own flight is relatively poor. Protection effected largely by hiding, though escape is usually rather easy. Summer is their adult season; sunny days their choice.

May be thought of as fish food, particularly in nymph stage, though they are competitors of fish for small animals that make up food of both. Nymphs are interesting aquarium animals for school or home, where their enmity for mosquito wrigglers may be effectively demonstrated.

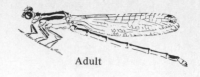

Adult

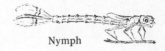

Nymph

Damselfly

Order Odonata./Family Libellulidae

Ten-spot Dragonfly

Libellula sp.

Adults four-winged, with wings held horizontally at right angles to body when at rest. Wings gauzelike with fine net of veins. Abdomen long and slender. Legs medium-sized. Head large, freely movable. Eyes large and conspicuously bulging. Mouth parts, chewing. Nymph aquatic, with hinged shovellike lower jaw that can be extended remarkably. No external gills in nymph stage. No pupal stage.

Between 5,000 and 10,000 dragonflies known; one fossil form has wing-spread of more than 2 ft. Darners fly high; skimmers, lower; ten-spot (illustrated) at 5 ft; amber-wing, at 6 ft; and argias, at 2 ft.

Male has claspers at tip of abdomen and accessory reproductive organs on second abdominal segment; may hold female by neck. Female drops or places eggs in water or inserts them in mud or plant tissue. Incubation usually long, though some may hatch in 2 weeks. Usually with this species one brood a year, beginning in June. Transition into adult form beautiful to see. Some dragonflies of temperate regions may have two annual broods and some require 2 years to mature.

Food mostly larvae of other water insects but may eat tadpoles, small fish, or any other animal matter small enough to be captured. Adults feed on insects caught on wing, studies showing that most of these are probably injurious. May roost in considerable numbers together overnight but normally are not gregarious. Most are sun lovers.

Except in fish hatcheries, are probably useful insects. Always harmless to man in spite of such names as "snake doctor" and "devil's-darning-needle." As mosquito destroyers should be encouraged, and as insects of beauty are to be admired.

Adult

Dragonfly

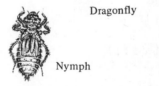

Nymph

Order Plecoptera [Stoneflies]

Elongate, flattened, soft-bodied insects, usually found near water, in particular streams. Have four membranous wings. Wings, at rest, held flat over abdomen. Chewing mouthparts. Metamorphosis simple, nymphs usually aquatic.

Order Plecoptera./Family Perlidae

Stonefly

Perla sp.

Chewing mouthparts, in some adults, do not function. Two diverging "tails" at rear of abdomen, reaching about to end of wings. Nymphs aquatic, with gills most commonly in pads just behind base of legs and with two tails. Adults fly reasonably well.

Ten American families of the order known. Species known, 1,550, mostly confined to flowing streams, with adults of different species emerging even when streams are full of ice. Adults found on land or objects near streams, or flying near streams. Nymphs cling to stones under water, where they crawl about in search of food.

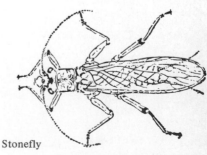

Stonefly

(Continued)

449

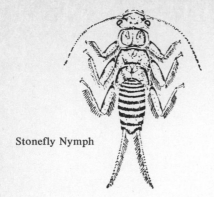

Stonefly Nymph

Adults mate soon after emerging from water. Females drop eggs in masses in water; 200 to 800 eggs per female; in some species, several thousand eggs per female. Eggs hatch into nymphs that may take 1 to 3 years to reach adult stage. No pupal stage.

Food of nymphs, small forms of animal life captured while prowling around under water, under submerged stones and trash. Adults may eat nothing; die soon after eggs are deposited. Adults attracted to lights. Earliest species to emerge in East is *Capnia pygmaea*, which looks like a snow fly and sometimes covers posts near streams in great black swarms.

Useful as food for fishes, adults making superior trout bait, particularly if moved upstream from point at which they are emerging. A western species, *Teniopteryx pacifica*, is a serious pest to fruit-tree buds in adult stage, but other stoneflies do little damage. Nymphs might be harmful in trout hatcheries.

Order Thysanoptera [Thrips]

Slender, tiny pale to black insects. Wings, if present, four, long and narrow; few or no veins and bordered by long hairs. Sucking mouthparts, largely plant feeders.

Flower Thrips

Order Thysanoptera./Family Thripidae

Flower Thrips
Frankliniella tritici

About $\frac{1}{25}$ in. long. Pale yellow to brown, with two pairs of long, narrow, few-veined wings fringed with long hairs and hence featherlike. Abdomen with 10 segments, the last conelike in female and rounded in male. Mouthparts formed into cone attached to rear of head beneath and used in sucking. Pupal stage moderately active.

Most persons notice thrips as little, slender, active insects in flowers like daisies and dandelions, though they live on many kinds of plants and on many parts of these plants. Important as pests are onion thrips, flower thrips, pear thrips, citrus thrips, and others. Vary in length from $\frac{1}{50}$ to $\frac{1}{2}$ in. Flower thrips attack strawberry blossoms and fruit does not form. In California it is a particular pest of alfalfa.

Eggs microscopic, often inserted into leaf tissue, hatch in about 4 days and then go through varying molts. Some species give birth to living young. In South cycle of this species may be completed in 12 days, while in North it may take 3 weeks or more. Young resemble adults in general form but lack wings and are paler. Wing pads appear in last two molts preceding adult stage.

Food, plant juices sucked through peculiar mouth; some prey on small insects and mites; others eat fungus spores; some take in juices of rotting wood. In South, flower thrips may be active through winter while in North it may winter as adult or nymph in protected places in grass of fields close to the ground. In general, thrips show a preference for feeding away from direct sunlight. More than 3,000 species of thrips known in the world; 600 species in the U.S.

Many species are not important economically, but others are serious pests. One or two transmit virus diseases of plants. Consult County Cooperative Extension Service for suggestions of recommended control measures.

Order Mallophaga [Chewing Lice]

Small, wingless ectoparasites of mammals and birds. Chewing mouthparts. Metamorphosis simple.

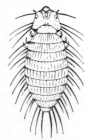

Shaft Louse

Order Mallophaga./Family Menoponidae

Shaft Louse
Menopon gallinae

Bird lice usually are from $\frac{1}{10}$ to $\frac{1}{25}$ in. long, with large heads, hard flat bodies, wingless, and always with biting mouth parts. Color usually whitish. *M. gallinae*, on chickens, is $\frac{3}{100}$ in. long, with short legs and relatively long backward-bent hairs.

Most members of this order are found on birds, though some live on domestic mammals. Two large suborders of Mallophaga are recognized. Important kinds are *Trinoton*, goose louse; *Esthiopterum*, duck louse; *Columbicola*, pigeon louse; *Liteurus*, head louse of chickens and turkeys; guinea and peacock louse of the genus *Goniocotes*. 1,500 known bird lice.

Sexes distinct; easily distinguished in some speicies. Eggs laid on feathers, hatch into nymphs that resemble adults except in size. Go through a number of molts and, in few weeks, reach maturity without pupal stage. In head louse of chickens, eggs hatch in 4 to 5 days, maturity reached in 3 weeks.

Bird lice confined to particular hosts. At least seven kinds on common domestic fowls. Turkey may be host to at least four kinds. Geese, pigeons, ducks, and guinea fowls each have many. Lice feed mostly on feathers, which they bite. Host irritated more by clinging feet than by biting. Most commonly found about the vent and wings of the host. May be 35,000 or more individuals on one chicken. For suggested control measures, consult your County Cooperative Extension Service.

May cause severe injury to birds.

Order Anoplura [Sucking Lice]
Small, flattened ectoparasites of mammals. Sucking mouthparts, short antennae, head small, metamorphosis simple.

Order Anoplura./Family Pediculidae

Human Body Louse
Pediculus humanus
Anoplura are small wingless insects with sucking mouthparts, with or without eyes, and with large abdomens and small heads. Head louse to $\frac{1}{10}$ in. long, male being $\frac{2}{3}$ size of female. Pale gray with blackish margins, head louse browner, tougher, and smaller than body louse.

Human lice worldwide in distribution and found only on man, rarely on apes. Pediculidae have eyes and live on man; tubular-headed Haematomyzidae on elephants; short, spined Echinophtheriidae on marine mammals, and blind Haematopinidae, on rodents and ungulates. "Crab" is pubic louse, *Pthirus pubis.* Some separate head louse from body louse as subspecies; the former occurs on the head and attaches itself to hairs; the latter occurs on the body and attaches itself to clothing.

Female mates 24 to 36 hours after reaching maturity and frequently through adulthood; may lay to 300 eggs during month she lives, ceasing only a day or so before death, usually laying 10 eggs a day. Eggs or "nits" conspicuous on dark hair, hatch in 6 to 30 days into nymphs that reach maturity in about 18 days after about three molts. Nymphs resemble adults in general.

Food blood of host, about 1 mg per meal. Will starve in 3 days at normal temperatures of 86°F or in 5 days at 75.2° but may survive week at lower temperatures. Die in 5 minutes at 124.9°F; in 10 to 30 minutes at 121.1°F; and in 45 to 60 minutes at 115°F. Eggs destroyed in 5 minutes at 128.3°F.

Bite or presence may cause infection with *Rickettsia prowazeki*, cause of typhus that killed 3,000 in England in 1 year; with *Rickettsia quintana*, cause of trench fever; and with *Spirochaeta recurrentis*, cause of relapsing fever. DDT has been used effectively to control human body lice.

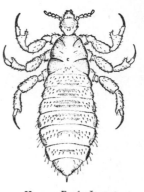

Human Body Louse

Order Anoplura./Family Haematopinidae

Hog Louse
Haematopinus suis
Length to $\frac{1}{4}$ in. Conspicuous black marginal band around body, with six long legs armed at tips with grasping hooks. Blind.

Hog louse lives on hogs, particularly on neck; sometimes on man, hiding in hair and under scales of skin. 29 genera in family that includes over half of Anoplura. Common forms include sucking louse of horses, *H. asini;* short-nosed cattle louse, *H. eurysternus;* goat louse, *Linognathus stenopsis;* and other lice of sheep, dogs, mice, rats, rabbits, and man.

After mating, female lays her white $\frac{1}{10}$ in. eggs one at a time on hairs. These hatch in 12 to 20 days and develop through three molts in 29 to 33 days to adult form that mates and repeats cycle. Mating and egg laying begins about third day after maturity is reached and may continue through most of adulthood.

Food, blood of host taken in from 1 to 4 meals a day. Blood sucked with little irritation to host at time of feeding but considerable irritation later. Louse crawls about in search of food, possibly moving at about same speed as head louse of man, which can move 9 in. in a minute.

Control of infestations of lice on domestic animals necessary. Since animals lick infested areas, poisons cannot be used. For recommendations as to appropriate control measures, consult your County Cooperative Extension Service.

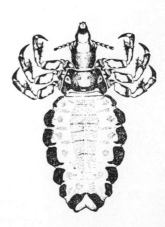

Hog Louse

Order Hemiptera [True Bugs]

Hemiptera vary in size from minute forms to those 5 in. long. Most have flattened backs on which there are usually four wings in the adult. Wings with basal part usually horny and tips membranous, the hind wings entirely membranous. All suck their food and have incomplete metamorphosis. The term *bug* is properly applied only to members of this order; 23,000 species known in the world.

A generalized key character for many forms is the letter X on the back of the insect, formed by folded wings.

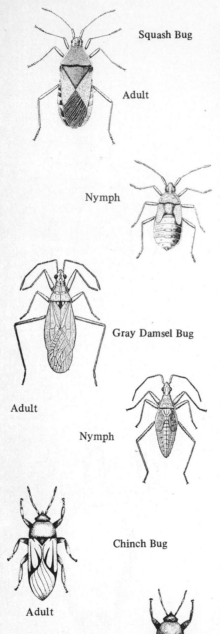

Squash Bug

Adult

Nymph

Gray Damsel Bug

Adult

Nymph

Chinch Bug

Adult

Nymph

Order Hemiptera./Family Coreidae

Squash Bug
Anasa tristis

Length about ¾ in. Adult dark brown but finely mottled with gray or light brown. Wings of adult reach almost exactly to tip of abdomen; wing pads of nymph increase in size as insect approaches maturity. Antennae about equal in length to forelegs and to proboscis.

Common throughout United States on plants of squash, pumpkin, melon, cucumber, and related species, spending winter under rubbish and in protected places. Pest species in South closely related to species here considered. Adults appear in spring when host-plant leaves are developing.

Eggs laid in clusters of varying number, on underside of leaves of host plant; oval, convex, resin-brown, laid so they do not touch each other; hatch in about 10 days into reddish or greenish nymphs that change to black, green, or gray, molting five times to reach maturity in 4 to 5 weeks. Development may be more rapid in South.

Food sucked and, where numbers are large, a whole crop may be completely destroyed. If frost kills host plants before insects reach maturity, they may attack fruits but normally fruits are injured only through destruction of leaves that provide them with nourishment.

Control is by removal of trash that might provide shelter, stimulation of vigorous growth of plants to overcome damage, protection of seedlings by cheesecloth covers; or by using spray of 1 part of 40 percent nicotine sulfate to 400 parts of water, on nymphs. This will not control adults. Consult County Cooperative Extension Service for recommended control measures.

Order Hemiptera./Family Nabidae

Gray Damsel Bug
Nabis ferus

Length ¼ in. Narrow-shouldered, slender, with small head and slender legs. Antennae about length of front legs. Pale grayish-brown or yellowish, with numerous brown dots; wing veins brown. Wings of adult reach to end of abdomen, finely veined.

Common insect from Atlantic to Pacific, most commonly in leafy parts of nonwoody plants relatively close to ground. It favors tall grasses and is a common sight in haymows where hay has been freshly cut. Also found in gardens and cornfields.

Winters as adult, hibernating under matted vegetation. In spring, leaves hiding and flies to feeding ground. Eggs laid in late June; female inserts them into plant tissue, laying about 60 eggs in 3 weeks; hatch in 6 or more days; nymphs resemble adults. In about 25 days and five molts, adult stage is reached.

Food probably exclusively other insects, captured by sudden dashes followed by a piercing thrust with beak; prey's body juices are sucked out. Aphids, leafhoppers, plant bugs, corn-ear worms are among important kinds of insects fed upon.

Food habits entirely useful, but some plants may be injured by slits made by female during egg laying. This injury is more than offset by destruction of plant-feeding insects.

Order Hemiptera./Family Lygaeidae

Chinch Bug
Blissus leucopterus

Length under ¼ in. Body black or dark gray with conspicuous white wings, each with a single black spot. Chinch bugs of Mississippi Valley have long full-sized wings, while those of Atlantic states and Great Lakes areas have short wings that only partly cover abdomen.

Native of tropical America that has migrated northward through Mississippi Valley, Atlantic Coast, and Pacific Coast to St. Lawrence River, Manitoba and Washington. Most abundant and destructive in Mississippi Valley.

Long-winged form winters as adult in grass tufts and corn shocks. In spring, flies to grainfields. Each female lays several hundred eggs on ground, at base of grain plants or on roots underground. These hatch in 2 weeks into nymphs that suck plant juices during 40 days of development to adult. Nymphs at first yellow, then orange, then vermilion, with adults darker yet.

Food plant juices, with result that plants are killed. When food organisms are destroyed before insects become adult, nymphs march to new territory. Favorite foods include corn, millet, kaffir corn, and similar grains. Short-winged forms cannot fly and so must migrate by marching.

Dry weather is favorable and wet weather unfavorable for this, one of six worst American insect pests. Control by burning of fields, which may destroy 50 to 75 percent of adults. Consult County Cooperative Extension Service for recommended control measures.

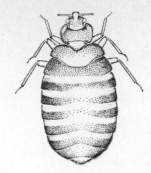

Bedbug

Order Hemiptera./Family Cimicidae

Bedbug
Cimex lectularius
Length about ⅕ in. Flat, reddish-brown, wingless or with only wing stubs showing in adult stage. Possessed of an objectionable and strong odor. Antennae four-jointed. Beak three-pointed. All bedbugs are parasitic, but each kind is limited to few hosts.

Found in houses or poultry houses, in cracks in walls, floors, furniture, or in beds. American species four north of Mexico, including this species parasitic on man, mice, and poultry; *C. pilosellus*, a long-haired species on bats; *Haematosiphon inodorus*, a long-haired species on poultry; and *Aeciacus vicarius*, in swallows' nests.

Female may lay to 200 white oval eggs, bearing a rim, in batches of 6 to 50, in cracks and crevices. Eggs hatch in 6 to 10 days; after five molts, in about 69 days, with an average of 8.75 meals, adult stage is reached. Completes cycle in about 7 weeks, with about four annual generations in warm houses.

Food is blood, preferably of man, but other animals may be used; without food, starvation may come in a year. Bite is poisonous to many people but not to others. Immunity may be developed. Known to transmit relapsing fever, plague, kala azar, and possibly leprosy by contaminated mouthparts. Bite generally not felt immediately, but itching, burning, and swelling soon follow. Symptoms are from venom injected into host. Usually thrive under crowded conditions where many persons live in close quarters. Consult County Cooperative Extension Service to learn recommended control measures.

Meadow Plant Bug

Order Hemiptera./Family Miridae

Meadow Plant Bug
Leptoterna dolobrata
Length to ⅖ in. Wings narrow and folded on back when at rest. Legs long and slender. Antennae about as long as forelegs. Gray or yellowish-gray with darker markings, and with antennae conspicuously black. As in chinch bugs, long- and short-winged forms.

Particularly common among cultivated grasses. Introduced from Europe and now widely established over eastern United States from Atlantic to Minnesota and Kentucky. Closely related to more widely distributed tarnished plant bug.

Winters in egg stage. Eggs laid in fall in grass stems below point of cutting, about 1 day after mating. Female may lay 20 in a half hour, and a total of 70; may hatch next spring, go through five nymphal stages, averaging 6 to 7 days each, for a total of 30 to 35. Nymph resembles adult except for wing development.

Food of adult and nymph plant juices, chiefly of important grasses sucked through piercing mouthparts. Orchard grass, timothy, and fescue grasses seem to be favorite foods; since these are valuable, insect is an enemy to man's interests.

Control by crop rotation, by early and close cutting of hay, by burning over grassy fields, particularly where hay has not been cut, by fall pasturing by sheep, horses, or other close-cropping species. Best to do burning in winter so that humus will not be destroyed.

Tarnished Plant Bug
Lygus ineolaris
Length ⅕ to ¼ in. Brown mottled with red, black, and yellow, with conspicuous white V on back between wing bases. Wings of adult almost reach to end of abdomen. Shorter and broader than meadow plant bug.

Found on over 50 different kinds of plants in Europe and America, including corn, apple, dahlia, aster, wheat, oats, and so on. Family has
(Continued)

Tarnished Plant Bug

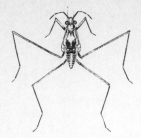

Nymph

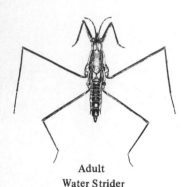

Adult
Water Strider

Back Swimmer

been called Capsidae, and all in it feed on plant juices; many cause plant deformities. In many western states *L. hesperus* represents a serious pest in alfalfa, especially in alfalfa seed fields.

Eggs usually laid on flower of a composite plant, particularly horseweed, alternating feeding and egg laying for weeks. Eggs hatch in 7 to 10 days into nymphs that, during five molts, change from a pale yellow insect $\frac{1}{25}$ in. long, to green spotted with red, to brown, and finally, in 4 to 6 weeks, to adult. Two annual generations.

Food plant juices, sucked, puncture ruining market value of apples, pears, quince, peach, plum, currant, strawberry, and grape and affecting cabbage, potato, turnips, beets, beans, many grains, and garden flowers. Winter is spent in adult stage, in leaves and trash.

Damage may be severe. Control is by flypaper barriers on trees, kerosene emulsion sprays, nicotine sulfate or soap sprays in early morning, covering crop plants with cheesecloth, and destroying weeds in borderlands. Consult County Cooperative Extension Service for recommended treatment of this pest.

Order Hemiptera./Family Gerridae

Water Strider, Jesus Bug
Gerris sp.
Body less than $\frac{1}{2}$ in. long; legs long, with fore pair suitable for grasping; second and third pairs suitable for driving, on water surface. Middle pair merely rests on tips on water. Some species have short wings and some species have wingless individuals. Abundant in northern U.S. and in Canada from Manitoba and to the east.

On surface of fresh or salt water, often crowded together in flocks. More common on quiet water but sometimes found far at sea. *Halobates* found at sea.

Mates from early spring through summer, male sometimes cutting off wings of female. Eggs laid on supports at water surface; hatch in about 2 weeks by a longitudinal split. Young jump about actively. Nymph like adult except for wings, with five stages each lasting about 1 week.

Food insects captured at or near surface, or sometimes by jumping into air. May feed on snails near water's edge. Some can swim under water, surviving submergence for several hours, while others drown quickly under water. Most are highly nervous in captivity. Protection effected by escape or flight.

Serve chiefly as scavengers of water's surface. Make poor aquarium pets because of nervous nature but may be fed dead flies or other insects dropped wounded on water surface. Winter is spent on or under sticks or stones. Of little economic importance. To observe "marks" on water surface made by legs of *Gerris* and in the correct light to produce shadows is indeed an interesting experience.

Order Hemiptera./Family Notonectidae

Back Swimmer
Notonecta sp.
Length usually about $\frac{1}{2}$ in. Body boat-shaped. Swims with back downward. Back lighter colored than under parts; usually black and cream-colored. Each pair of legs specialized as to function; first pair for seizing prey, middle pair for holding things, and hind pair functions like oars. Hind legs of *Notonecta* and *Buenoa* flattened and suitable for swimming. Eyes large. Furrow on underside enclosed by hairs, forming air chamber into which air is forced by hind legs. Night fliers; leave water, land belly side down, and then fly.

Notonecta undulata general over United States, wherever water is suitable. *Notonecta*, including the largest and most common species, may float at pond surface, while *Buenoa*, smaller and more slender, commonly swims slowly beneath surface. Related genus, *Plea*, contains one species, a small insect under $\frac{1}{8}$ in. long that feeds on crustacea.

Mates beneath water, in spring. Eggs white, $\frac{1}{12}$ in. long, glued to submerged stems, hatch in about 3 weeks; young resemble adults but of course have shorter wings. Some males, when adult, have structures that they rub to produce high-pitched sounds. This may be associated with reproduction.

Food chiefly small aquatic animals caught by pursuit and killed, or paralyzed by stinging. Some secrete a milky juice from thorax. Stings so severe that they may hurt man badly, so insects should be handled with care. In captivity, may be fed mosquito wrigglers. High mortality during molting.

Probably of little economic importance, but may serve as checks on multiplication of such aquatic pests as mosquitoes. May also kill young fish and serve as fish food. Always interesting as pets in aquariums. May have nematode enemies that kill off many if they are too crowded.

Order Hemiptera./Family Nepidae

Waterscorpion
Ranatra sp.

Length overall, sometimes around 3 in. Slender and sticklike, with slender legs, the fore part suitable for grasping prey. Other legs used in walking and in poor swimming. Filaments at rear of body form a breathing tube through which air may be drawn while insect feeds under water. Sucking mouth parts. Mostly plain brown in color. Considered a "scorpion" because forelegs resemble large claws of a scorpion and the long breathing tube at end of the body resembles a scorpion's tail.

Food either in deep water, in trash, or at surface, often bask in sun, entirely dry. *Ranatra* slender, common, and widely distributed. *Nepa* $\frac{3}{5}$ in. long, flat, and less common. *Curicta* intermediate in shape and size, and found in south central states.

Mating a prolonged process. Eggs laid under water in decayed vegetation or on live plants, inserted in holes; hatch in 2 to 3 weeks; young resemble adults but go through five molts; about 40 days from egg to adult. Female a most prolific egg layer. Winters in adult stage. May chirp by jerking rigid limbs and a shoulder rasp.

Food other animals caught and held by forelegs while juices are sucked out. Prey caught by lying in wait. Scorpions are predaceous upon eggs and other stages of aquatic insects as well as other small invertebrates; take small fishes and fish eggs. When first brought into strong light, avoids it, but soon reacts by going toward it. May hibernate during times of year when oxygen content of water is below normal.

Not of economic importance but may serve to check multiplication of small forms of water life. May be reared in captivity by feeding on cockroach nymphs. Interesting to naturalists because of unique breathing, locomotion, food capture, and reproduction.

Waterscorpion

Order Hemiptera./Family Belostomatidae

Giant Water Bug
Lethocerus americanus

Length 2, 3, or more in. Broad, flat, brown. With conspicuous strong grasping forelegs, short piercing beak, strong middle and hind legs suitable for swimming, prominent beady eyes, and wings that fold compactly to cover the abdomen above. Powerful insects and able fliers.

Genus *Lethocerus* includes large members of the family, with five species known in United States and Canada, relatively well-distributed. *Benacus* includes another large-sized similar insect. *Belostoma* and *Abedus* are usually smaller.

Egg laying and mating alternate with feeding during much of adult stage. Eggs laid in trash, hatch in 1 to 2 weeks; nymphs resemble parents except for wing development and go through five nymphal stages to reach maturity. Breeding period may last from May to August, during which time 150 eggs may be laid.

Food any animal found in its environment that can be captured. Prey held by strong forelegs while beak is used to pierce and suck out juices. Fish, frogs, other insects of many sorts serve as food. Adults are attracted by strong lights at certain time, hence name electric-light bugs.

May be real pests in fish hatcheries, where they destroy eggs, young, and even adult fish and consume food that these fish might need to live. Might themselves serve as food but not enough to compensate for loss. Known to kill fish four times their own size. Are eaten by ducks and herons. One large form of giant water bug is cooked and considered a delicacy in China. Bite of *L. americanus* can be painful for man for several reasons including the fact that a digestive enzyme is injected into tissues.

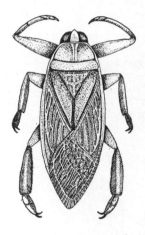

Giant Water Bug

Smaller Giant Water Bug
Belostoma sp.

Length about 1 in. Much like electric-light bug except for size. Brown. Grasping forelegs less conspicuous than in giant water bug. Wings of adults folded on back. Basal segment of beak longer than second, while in giant water bug it is shorter.

Found in waterways and about lights but feeds mostly in waterways. Flies readily from pond to pond in adult stage. Some relatives in Brazil and Guiana are at least 4 in. long. Genus *Abedus* has a strong keel on underside of forepart of body, which is lacking in genus *Belostoma*.

Mates mostly in water. When mating is over, male finds himself with fertilized eggs glued to his back by female. He carries these about during 1- to 2-week incubation period. Young nymphs free themselves and leave father bearing the shells. Molts, five. Nymphs resemble adults.

Like giant water bug, these insects feed on any animal they can overcome, but mostly on those to be found in water. They do not go out on land to seek prey. Feed readily in captivity, sometimes almost immediately after having been caught. Attracted by light.

May serve to some extent as food for fishes but may also do considerable damage to young fish; serve as direct competitors for food with many important kinds of fishes. Normally not of serious economic importance.

Adult

Smaller Giant Water Bug

Nymph

Order Hemiptera./Family Corixidae

Water Boatman
Corixa sp.

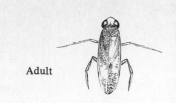

Adult

Length to nearly ¾ in. Body shaped like a blunt-ended boat, with hind pair of legs suited for swimming with an oarlike motion, and front pair of legs suited for scraping oozes that make up food. Males with last four segments under abdomen not symmetrical. Swim right side up, unlike back swimmers. Eyes often red. Dorsal surface of body flattened with narrow dark crosslines.

Found in lakes and streams of fresh water, moving or stagnant; widely distributed through world. A few species are found along seashore in brackish pools above the high-tide mark.

Name coris, from which family name was probably developed, means bug; given group possibly because they smell like bedbugs.

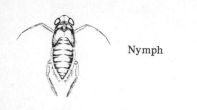

Nymph

Males may chirp by rubbing front feet against beak and opposite leg. Eggs yellow, top-shaped, laid on submerged plant trash, or, in one species, on living crayfish. Young show eyespots in 4 days, with five molts; nymph more nearly like adult. Winters as adult, though young have been found in January.

Food exclusively plant oozes growing on submerged objects and scraped into mouth with rakers on front feet. Protection effected by great numbers, by escape, and by general inconspicuous nature. Young cannot thrive in boiled water. More active on dark days than on bright. Probably active at night throughout year.

Adults and eggs are collected in great numbers and used as food by man and birds in Egypt and Mexico. Imported into England for use as food, 1 ton approximating 250 million individuals. May be used also as fish food or as food for young game birds in wildlife management. Scavengers. In contrast to back swimmers, members of *Corixa* do not bite man.

Water Boatman

Order Homoptera [Aphids, Scale Insects, and Cicadas]

Homoptera are sucking insects whose wings, if present, are uniform throughout and when not used are held sloping over the body. Sucking beak attached to back part of underside of the head.

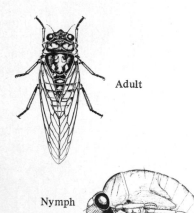

Adult

Nymph

Periodical Locust

Order Homoptera./Family Cicadidae

Periodical Locust
Magicicada septendecem

Length about 1 in. Body stout, black. Eyes orange, prominent. Forelegs suitable for grasping. Wings at rest extend beyond end of abdomen. Most of abdomen a drum, with drumhead just behind base of wings, drums being vibrated by internal structures in males. Wings not sound-producing organs.

Adult lives usually in treetops, though most of life is spent underground. At least 20 distinct broods of this species in United States, so insect may be found any year. Dog-day harvest fly, *Tibicen*, of the East, is larger and has 2-year cycle.

Eggs laid by female in late summer by making slits in twigs. About 16 are laid in a group, to 500 per individual. Eggs hatch in 6 to 8 weeks. Nymphs fall to ground and remain 12 to 13 years with intermediate stages in fifteenth and sixteenth years. Nymph stage may be spent from near surface to 10 ft underground.

Sucks juices from roots and stems and may make way through solidly packed earth by softening earth with water. Males alone make sound, though females may have drumlike abdomen. Adult stage lasts only a few weeks in summer. A 13-year form is common in South, while northern form has 17-year cycle.

May injure plants at all times, but most injury is caused by female when she lays eggs closely together on twigs, sometimes killing fruit trees in this way.

Order Homoptera./Family Gercopidae

Spittlebug
Lepyronia quadrangularis

Length of adult under ⅓ in. Brown, with two oblique darker bands on wing covers, black beneath. Vaguely froglike in appearance. Best known from the frothy "spit" found on plants as a cover, usually for nymphal stages. Adult densely covered with minute hairs.

Common on grasses and other low plants in summer months. Adults may leap from plant to plant, or fly, as need be. Family closely related to the leafhoppers, plant lice, and cicadas.

Adult   Nymph

Spittlebug

Eggs laid by female in plant tissue, hatch into nymphs that cover themselves with protective froth. Number of nymphs may live in same froth mass. Froth comes from liquid freed from alimentary canal as a mucous fluid and beaten into froth. Froth represents more plant juice drawn than can be used as food.

Food obviously plant juices. Function of the froth has been interpreted as associated with respiration, as providing protection against parasites, and as maintaining a constant climatic condition where nymph is developing.

Generally considered of little economic importance. Since so many people think of froth as "snake spit" or "frog spit," it is obvious that the insect is commonly observed. Also obvious that few people investigate a strange phenomenon, or froth maker would be discovered.

Adult

Nymph

Buffalo Treehopper

Order Homoptera./Family Membracidae

Buffalo Treehopper
Stictocephala bubalus
Length under ½ in. Usually resembles a spine, bud, or natural plant swelling. Some species distinctly spine-shaped. This species reputed to be shaped like buffalo horns. Grotesque shapes are developed by abdomen. This species dirty yellow with brown spots.

Found on trees and other woody plants as adults, and commonly on herbaceous plants as nymphs. 185 species in United States, representing 43 genera. Other species include two-horned tree hopper, two-marked treehopper, and others.

Eggs laid by female in slits in bark of 2- to 3-year-old woody plant stems, such as elms or apples, in September, six to eight in a slit. Hatch in May into nymphs that in 30 to 35 days go through five stages of about 1 week each. Young stages are spent on weeds and adult stages on woody plants.

Food, juices of plants on which they occur, juices being sucked through beak. Females may be only ones to migrate to trees for egg laying, males remaining on weeds where young stages are spent.

Chief injury is due to egg laying that deforms and weaken trees, permitting them to be broken easily by wind or infected by other insects that use old scars for entrance. Control by keeping grass down in orchards, burning infected twigs in fall and winter. If there are no nearby weeds for young, there will be few adults on trees.

Rose leaf hopper

Order Homoptera./Family Cicadellidae

Rose Leafhopper
Edwardsiana rosae
Length to ⅛ in. White to pale green, long, slender, with wings of adult extending beyond tip of abdomen. Antennae about length of forelegs. Eyes rather conspicuous, occupying, when seen from above, about ⅔ top of head.

Common everywhere in United States, British Columbia, Ontario, and Nova Scotia; originally from Europe. Common on members of rose family and in Northeast is serious pest of apples and rosebushes. More than 700 species of family in the United States.

Eggs ¹⁄₄₀ in. long, laid in fall, on bark of rose, apple, or similar plants. Winter eggs hatch in 6 to 7 months; summer eggs, in 25 days into nymphs that in 33 days go to adulthood. Second brood of summer may be completed in 17 days after hatching, growing from ¹⁄₂₅ in. to ⅛ in. long.

Food plant juices sucked. Single rosebush may support thousands of these leafhoppers. Young of first generation feed on rose, but when wings are developed, adults fly to apple trees to feed; second-generation adults return in fall to lay eggs again on rosebushes. Swarms of these insects may be found on leaves of rosebushes throughout the summer and their cast skins may be seen adhering to undersurface of the leaves.

Control of this apple pest may be helped by keeping rosebushes away from orchards. Nicotine sulfate sprays are used. Western beet leaf hopper transmits curly-leaf disease, grape leafhopper affects grapes, and six-spotted leafhopper, grasses.

Egg

Pear Psylla

Order Homoptera./Family Psyllidae

Pear Psylla
Psylla pyricola
Length about ¹⁄₁₀ in. Reddish-crimson, with brownish-black markings, dark wing veins, and bronzed eyes. In general, appears like small cicada. Produces abundant honeydew that is sticky and trickles down trunk of host tree or drops to ground. This becomes black with a fungus and advertises presence.

(Continued)

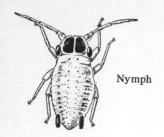

Nymph

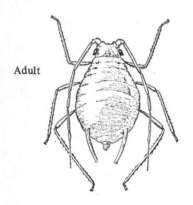

Adult

Rosy Apple Aphid

Found commonly on pear trees, mostly on leaves, buds, or fruit. Common throughout eastern United States south to Virginia and west to Mississippi; reported also in California. Native of Europe; introduced into America about 1832. A particular pest in the Pacific Northwest. Especially injurious when abundant; may even check growth of tree; many leaves turn yellow and drop as does much young fruit.

Spends winter as adult, hiding in crevices. In spring, eggs are laid on buds and twigs of pear. Eggs hatch in 2 to 3 weeks. Nymphs begin sucking juices of young leaves and fruits, secreting large amounts of honeydew; adults in about 1 month. May be five generations a year in South.

Food plant juices sucked to detriment of tree. Second-generation psyllas come from eggs laid on leaves, rather than on buds of twigs. Nymphs broad, flat, yellowish at first but becoming reddish, with bright red eyes; rather different in appearance from adults.

Control by destruction of adults in winter, by ridding orchard of trash and loose bark, by spraying with acceptable chemical insecticides.

Order Homoptera./Family Aphididae

Rosy Apple Aphid
Anuraphis rosea
Aphids range in length from $\frac{1}{25}$ to $\frac{1}{5}$ in. Winged and wingless adults in many species. Wings are usually held like a roof over back and forewings are larger than hind wings. Sucking mouthparts. Honey tubes at rear of abdomen secrete honeydew eaten by ants.

Rosy apple aphid native of Europe but widely established in world, where it feeds mostly on thorn apple and mountain ash. Wingless females vary from pink to rose, orange, tan, or black. Variety of aphids great. Many live in galls, some on roots, on leaves, on stems, and so on.

Typical aphid life cycle: Winter eggs hatch and develop into usually wingless stem mothers, which give rise to wingless females; these produce a number of similar generations, without mating; then a winged generation moves to new host and produces young without mating; finally, another winged generation is produced, and this lays fertile winter eggs.

Food plant juices exclusively; each species relatively limited in host, usually to closely related plants. In some, like corn-root aphid, aphids are placed on host corn root by ants that feed on honeydew yielded by aphids. This ant-aphid relationship is fairly common, with variations.

Highly injurious to plants. Controlled by encouragement of natural enemies; by use of lime-sulfur wash to kill winter eggs; by spraying young in early stages with acceptable insecticides. Consult County Cooperative Extension Service for recommendations.

Order Homoptera./Family Coccidae
Coccidae include scale insects, mealy bugs, and others. Some tropical coccids yield shellac of commerce; some yield dyes; but most of ours injure plants. China wax in candles is made by these insects. Males usually winged; females wingless.

Citrus Mealy Bug

Citrus Mealybug
Planococcus citri
Appears like a cottony waxy spot, about $\frac{1}{4}$ in. long. Female pale yellow and well-covered with powdery wax that is mostly light gray or white.

Common in warmer parts of America and, in the North, in greenhouses and on houseplants. Important coccids in America include the cochineal insect, tortoise scale, and many other scales.

Eggs numbering 300 to 500 are laid in loose cottony masses chiefly during fall and winter. Nymphs are indistinguishable as to sex in early stages, but adult females are wingless and adult males winged. Maturity is reached in summer following egg laying.

Adults and young move about freely, unlike scale insects. Food plant juices, sucked. Protection effected by cottony wax covering, which makes many sprays ineffective.

A number of natural enemies of this insect in the open. Spray is used with 8 lb of soap boiled in 8 gal of water, into which is mixed 1 lb of crude carbolic acid. This is boiled about 20 minutes, then diluted with 20 parts of water. Hydrocyanic acid fumigation gives control in greenhouses. Consult your County Cooperative Extension Service for suggested controls.

Oystershell Scale
Lepidosaphes ulmi
Appears like small crowded oystershell scales on bark of woody plants. Females under scales, about ⅛ in. long; males under smaller scales. Males become winged but females remain wingless.

Too common. This species is found on dogwood, maple, and apple but is closely related to similar scales found on willow and ash. Distribution is worldwide.

Winter is spent in egg, 25 to 100 under a single scale. Nymphs emerge in late May, crawl around 3 to 4 hours, then form own scale, settle down for life, maturing in early August. May be two generations a year in northeastern United States, first eggs hatching about 2 weeks after apples bloom.

Female lays eggs under scale that has protected her for life. Food is sucked from plant host and, since scales may be closely crowded on small twigs, the host plant may be injured.

Adults active only during summer months. This is one of most serious orchard pests. Consult your County Cooperative Extension Service to determine appropriate control measures.

Oystershell Scale

Order Dermaptera [Earwigs]

Slender insects. Adults usually have four wings; front pair leathery, hind pair membranous. Chewing mouthparts. Simple metamorphosis. Name earwig from old superstition that earwigs entered people's ears.

Order Dermaptera./Family Forficulidae

European Earwig
Forficula auricularia
This order of insects comprises the earwigs. Adults have long narrow bodies, chewing mouthparts, and pincerlike structures at rear of abdomen. Wings may be absent or present. No pupal stage. Length to 1½ in.

European earwig native of Europe but established in many parts of America. In West Coast area and in South, feeds on fruits, flowers, and vegetable material. Less common in East, where they are of minor importance. Appeared in Rhode Island in 1912.

Winters as adult or ½₀-in. eggs underground. By April females appear, guarding newly hatched nymphs for up to 4 weeks, or abandoning them. May lay second batch of eggs in June. Nymphs reach maturity in 2 or more months. In late summer adults congregate in great numbers, possibly as mating act. Males have bowed end appendages.

Food almost entirely plant material. Earwigs have habit of hiding in any available crevice during day and are then discovered in clothing, books, among dishes and elsewhere, much to displeasure of human beings. May destroy aphids.

Control where they destroy flowers, fruits, or stored foods by use of traps or poisoned baits. Consult County Cooperative Extension Service for recommended control measures.

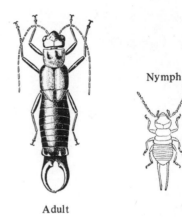

Nymph

Adult

European Earwig

Order Coleoptera [Beetles]

With 277,000 known species, the Coleoptera constitute largest group of insects. Adults have four wings, front pair thickened and horny. They have chewing mouthparts and a complete metamorphosis, with resting pupal stage. Often referred to as "Shield-winged" insects because of covering forewings.

Order Coleoptera./Family Carabidae

Ground Beetle
Calosoma frigidum
Ground beetles are usually black or dark-colored. Some are brightly colored or with bright parts. They may be as much as 1 in. long and are active in larval and adult stages.

Found usually on ground, commonly under trash where moist. Some species climb trees, but this is unusual for group. At least 1,200 kinds of ground beetles in United States.

Some ground beetles lay their eggs in mud cells on leaves. Larva can run around actively and has strong jaws. In this species, it lies in wait for prey in a burrow. Larva may molt twice in about 2 weeks, before turning into inactive larva for another 2 weeks, then becoming a pupa for a week before becoming adult.

(Continued)

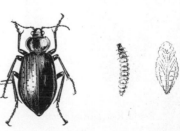

Ground Beetle

Carrion Beetle

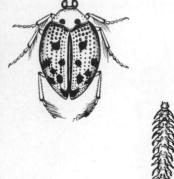

Haliplid Beetle

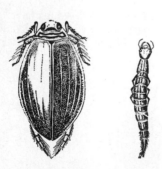

Predacious Diving Beetle

Food: most ground beetle larvae and adults feed upon other insects, particularly soft-bodied caterpillars that can be run down and eaten. They are most active at night. Adults may live nearly a year, wintering in that stage, with egg laying in some species limited to early summer.

Ground beetles are useful throughout their lives, since they prey on other insects that are for most part harmful to man's interests.

Order Coleoptera./Family Silphidae

Carrion Beetle
Nicrophorus marginatus
Nicrophorus marginatus is nearly 1 in. long, with cylindrical body. Dark with dull red markings on forewings, which frequently do not reach end of abdomen. Animal appears stout and rugged.

Found usually on dead animal matter, usually under cover. More than 100 species of carrion beetles in United States, including *Silpha americana*, a common short, flat, round insect.

A pair of *Nicrophorus* finding a dead bird or piece of carrion will remove earth from beneath and place it on top, eventually burying the flesh. On buried flesh, female lays eggs, and the larvae that hatch feed upon it. Pupal stage is passed near by.

Food of larvae and adults, of course, is carrion they find and bury. Some carrion beetles may attack and kill small animals like snails, and some live on decaying vegetable matter. Some of larger species may move an animal size of rat some distance to get it buried properly.

These insects serve primarily as scavengers and, for most part, may be considered as beneficial. Those close relatives such as *Silpha bituberosa*, which feed on vegetable crops, may of course be considered pests. They are mostly western.

Order Coleoptera./Family Haliplidae

Crawling Water Beetle
Haliplus fasciatus
Length around ⅛ in. Slow-moving, long-legged water beetle, with relatively distinct head, and underside of abdomen with conspicuously broad plates at base of hind legs. Straw-colored.

Found in fresh water, commonly among masses of pond scum, abundant in many springs and stagnant ponds. About 40 species in United States, divided among 3 genera: *Haliplus, Brychius,* and *Peltodytus.*

Adults mate in May or August. Females lay 30 to 40 eggs scattered among water plants. Slender larvae, with many lateral spines, may live through winter and in spring build a chamber about 1 in. under moist ground to rest 3 days before pupating. Pupal stage 12 to 14 days. This species places eggs in dead cells of *Nitella*.

Food of larvae and probably of adults, contents of cells of algae, interior being removed without necessarily breaking threads. A summer generation lives from May to August, and a second generation through winter. Adults air breathers; reserve air supply stored under plates supporting hind legs.

May serve some use as check on growth of pond scums and as food for fish and other water animals but, for most part, of little economic importance. They make satisfactory and interesting aquarium animals.

Order Coleoptera./Family Dytiscidae

Predacious Diving Beetle
Dytiscus sp.
Length of some species, 2½ in. Usually oval, flattened beneath, black, or black and brown, with hind legs suitable for swimming. Females of some species, with forelegs furrowed. Males of some, with disks on forelegs. Carries air bubble at end of abdomen under water. Larva slender, with large jaws, active, voracious.

Common in ponds and streams, particularly in standing water, over muddy bottom. May be found on dry land, to which they may have fallen in migration. About 300 species in North America. Largest American family of water beetles.

After mating, around March and April, females lay 20 to 50 eggs, single, in slits in plants, under water. These hatch in about 3 weeks into larvae that develop 4 to 5 weeks before changing into pupae, which remain buried in earth from 10 to 20 days in summer, or longer in winter. Adults may be found at any time of year.

Food of larvae and adults, animal matter captured by pursuit or stealth and held in jaws and sucked into mouth. Young larvae may attack and kill own kind. In captivity may be fed raw beef or mixture of cereal, powdered shrimp, and ant pupae, grouped together. May feed on fish, young and old. Adults attracted by lights.

Enemies of other water animals that may be captured and overcome. Might easily be pests in fish hatcheries. Must destroy mosquitoes and other offensive water insects and must serve to some extent as food for useful fishes. Air is kept in reserve under hard forewings, entering abdomen through top. Air bubble appears like a submerged silver sphere beneath the body.

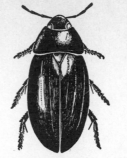

Water Scavenger Beetle

Order Coleoptera./Family Hydrophilidae

Water Scavenger Beetle

Hydrophilus sp.

Length of commonest eastern species, *Hydrophilus obtusatus*, 3½ in. Usually black, elliptic, more or less flat beneath, convex above. In water, usually shows silvery film of air over undersurface. Larvae stouter than those of diving beetles, with more slender foreparts and more slender hollow jaws.

Found in freshwater ponds and streams, though a few species live in moist soil or in dung. Found through North America and elsewhere, being represented in North America by nearly 200 species.

After mating, females lay about 130 yellow eggs enclosed in waterproof, cocoonlike structure, attached to plants near surface or floating as a raft. Larvae crawl about under water for at least two molts in 1 month's time. In late summer, pupal stage of about 12 days is spent in cell in soil. Adults may be found at any time of year. Some adults hibernate.

Food of adults and larvae, small animals that may be captured alive, on decaying vegetation. Larvae feed on small animals, frequently sucking juices through hollow jaws. Adults may fly from pool to pool and are attracted by lights. May be helpless on ground. Have many parasitic enemies.

Food for some fishes. Larvae may be pests in aquariums or in fish hatcheries but may also be interesting to watch. They, like larvae of diving beetles, are known as "water tigers." Air, taken at surface by extending antennae and folding them back over bubble of air, is carried on lower surface.

Order Coleoptera./Family Gyrinidae

Whirligig Beetle

Dineutes sp.

Length, *Dineutes*, usually about ½ in.; *Gyrinus*, less than half that length. *Dineutes* flat, oval, blue-black or black and brown. Forelegs extended forward in a reaching position. Hind legs suitable for swimming and attached at center of underside. Larvae slender with conspicuous side appendages.

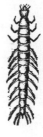

Whirligig Beetle

Found commonly on surface of quiet or running water, swimming in erratic paths, frequently in flocks of considerable size. In North America, 13 species of *Dineutes* and 26 of *Gyrinus*. Will dive to bottom when disturbed.

May fly from pond to pond. After mating, females lay one to many white cylindrical eggs, on submerged plants. These hatch into fierce-looking slender, flattened, light-colored larvae that finally leave water, and spin gray paperlike cocoons in which pupal stage of about 1 month is spent before transformation into adult.

Food of young and adults, probably solely animal matter, usually captured at surface alive. In captivity, may be fed raw beef on toothpicks left afloat. Protection by escape and strong odor that smells like apple seeds. What appear to be four eyes are actually two divided by lateral margins into halves, one half on the upper side of the body and the other half on the lower side of the body.

Interesting scavengers, with such common names as submarine chasers, lucky bugs, write-my-names indicating that they have attracted general attention. Do no harm and might figure to some extent in mosquito control.

Order Coleoptera./Family Lampyridae

Firefly or Lightning Bug

Photurus pyralis and *Pyractomena borealis*

Length ½ to ¾ in. Oblong, in general. With rather long nervous antennae. Comparatively short legs. Pale gray above; beneath, male is sulfur yellow from fourth and fifth segments of abdomen to end. In some species, females are wingless glowworms. Some species do not emit light at all.

Firefly

Common over meadows and swamplands, particularly abundant near water. Larvae and wingless females are found in grass or sod. At least 50 kinds of fireflies in United States.

Adults use well-known light flash to attract each other, since they will come to captive insects behind glass but not to those in an opaque box.

(Continued)

461

Eggs of some species luminous. Larvae of some species glow, live in soil or under bark, feeding mostly on soft-bodied animals, such as earthworms or snails. Luminous pupae in ground about 10 days.

Food of different species and stages varies from small soft-bodied animal materials to some plant material. Light is cold. Some species flash in unison. Light usually yellow. Light produced by action of the enzyme luciferin on luciferase. The process is one of oxidation under the control of the nervous system. In production of light, fireflies are 92 to 100 percent efficient; incandescent lamps are about 10 percent efficient.

Of little economic importance for most part, but of tremendous challenge to understand production of cold light. Of course, they have inspired poetry and stories of all sorts. A Jamaican marsh lighted at 10-second intervals by hosts of fireflies simultaneously is a sight not to be forgotten.

Order Coleoptera./Family Meloidae

Blister Beetle
Epicauta vittata
Blister beetles are about 1 in. long overall. This species striped above with yellow, red, and black; wing covers yellowish with two black stripes on each. Body soft, cylindrical, and joined to head by a distinct neck. Adults active. Larvae vary greatly in different stages.

Found mostly on foliage, flowers, and other plant parts, this species injurious to potato leaves. Over 200 kinds of blister beetles in United States, some useful, some injurious at different stages of life history.

Female lays eggs in summer or fall, in masses of over 100, commonly near egg capsule of grasshoppers. Eggs hatch in 12 to 22 days into active triungulin larvae, which may feed on grasshopper eggs several days, then molt into less active caraboid larvae which feeds for a week, followed by two scarabaeidoid larvae of a week each, a fifth coarctate larva, a sixth scolytoid larva, and a 5- to 6-day pupa.

In immature stages, these beetles are distinctly useful as destroyers of other insects. As adults, their eating of plant tissues is harmful. Blister beetles can give off a stinging substance that may cause serious skin injury if too concentrated.

Economic importance varies. Some are serious pests in spite of their useful youth. Some are ground, dried, and made into a paste. Spanish fly produce cantharides used in medicine for making blisters. Toxaphene and parathion are effective in control. All should be used with caution because of adverse effects on the evnironment. Consult your County Cooperative Extension Service for recommended control measures.

Order Coleoptera./Family Elateridae

Click Beetle, Wireworm
Melanotus communis
Length of common species varies from $\frac{1}{10}$ to $\frac{3}{4}$ in. Some are 2 in. long. Most are uniformly black, gray, or brown. Conspicuous because of flexible joint at base of wing covers that bends quickly, flipping beetle into the air. Some have conspicuous eyespots on back. Larvae slender and somewhat wormlike.

Found commonly about lights, in or near grass sods, or in decaying wood. Over 500 kinds of click beetles known from North America alone and many of them are of considerable economic importance. Click beetles so called because of ability to make clicking sound as they jump into the air.

Winters as adult or undeveloped larva. Adult emerges late spring; feeds for a while, mates, lays many $\frac{1}{50}$-in. white to yellow eggs in sod, which hatch into yellowish or brown, round, hard-shelled larvae; these may require to 4 years to reach maturity, and have pair of legs on each of three segments back of head. Larvae of *M. fissilis*, called wireworms, are serious pests of corn, grasses, and other small grains as well as several different garden vegetables. Pupates July or August, in earthen cell for about 1 month, 6 in. underground.

Food almost exclusively plant material. Some species feed on decaying plants, but many feed on cotton, potato, corn, wheat, sugar beets, grass, and other important economic plants; no generalization can be applied safely to all click beetles.

Control, where needed, is by crop rotation, avoiding planting corn or root crops in newly turned sod, but planting clover, peas, buckwheat or beans, which are less susceptible to injury. If sod land must be used, plow it in midsummer and cultivate thoroughly in fall before planting.

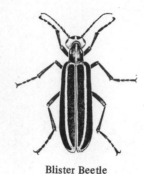

Blister Beetle

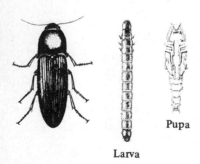

Pupa

Larva

Click Beetle, Wireworm

Order Coleoptera./Family Tenebrionidae

Confused Flour Beetle
Tribolium confusum
Common name of family, darkling beetles. Reddish-brown beetle, to $\frac{1}{7}$ in. long; active; dart about. Live to 2 years. Male and female alike in appearance. Female produces to 1,000 eggs in lifetime. Eggs clear, white, sticky; attached to objects near grain. Introduced from Europe in 1893. More common in northern states.

Egg to brownish-white worm in 5 to 12 days, lives to 4 months; then forms white naked pupa. Egg to egg in 4 months.

Food many grains, flour, starchy material; beans, peas, baking powder, ginger, dried fruits, nuts, chocolate, insect collections. May be found in housewife's cupboard in cornmeal, and in some cereals.

Red flour beetle, *T. castaneum*, more common in southern states.

Common where grain products are stored. A pest in grocery stores, warehouses, serious pest in flour mills. Badly infested whole grain or ground grain will have adults crawling over surface; brownish white larvae on and in the grain kernels. Millers call these beetles bran bugs.

Another destructive group of grain beetles, *Tenebrio spp.*, mealworms, occur in similar locations. They have been used by biologists in recent years in animal behavior studies.

Control by keeping premises as clean as possible. In grain storage areas, all refuse grains or seeds, or grain products should be removed during time enclosure is not in continuous use. Area should be kept dry and tight. Heat treatment and fumigation are used. Several chemical insecticides may be used with proper care: methoxychlor, premium-grade malathion, pyrethrins, allethrin. DDT may be used, where legal, on stored seeds not to be consumed by animals. Consult the local County Cooperative Extension Service for specific control recommendations.

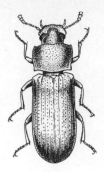

Confused Flour Beetle

Order Coleoptera./Family Buprestidae

Oak-leaf Miner
Brachys ovatus
Small oval beetles with short antennae; at first white, but later darker, with metallic iridescence due to change in color of hairs and scales. Miner is a large blotch on leaves of oak. Adult beetle widest at base of wings.

Common in leaves of red oak and less common in chestnut oak. Family includes many beetles that bore into wood or leaves of plants.

Eggs laid singly on surface of leaf of host plant, in early summer or late spring; hatch and larvae bore into leaf. Larva legless, sheds skin three times, makes blotches that often cross main veins of leaf; remains in leaf through winter on ground. Pupal stage in May, about 1 week.

Food interior of leaf, leaving upper and lower surface as protection to larva. Since larvae remain in leaves through winter, no extra protection is necessary.

Of little economic importance since they are rarely abundant enough to interfere with prosperity of host trees. Most buprestids fly if disturbed. Not so with *Brachys*, members of which draw up their legs and play dead.

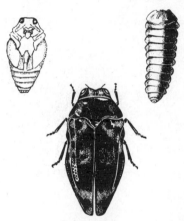

Oak-leaf Miner

Order Coleoptera./Family Psephenidae

Water Penny Beetle
Psephenus sp.
Length about $\frac{1}{4}$ in. With relatively long legs, oval flattened bodies clothed with fine hairs that retain a film of air when beetle is submerged. Larvae flat oval disks, looking more like crustaceans than insects. Legs not fitted for swimming.

On stones, near or in running water, but adults able to fly well when disturbed. Abdomen of Psephenidae has five segments, that of related Dryopidae has more. One eastern and four western species of *Psephenus* in North America.

After mating, females lay eggs about $\frac{1}{125}$ in. in diameter, in layers on underside of submerged stones, in June. These hatch into flat larvae that probably spend over a year developing before becoming pure white pupae, under stones or damp cover near water's edge.

Food of larvae and adults not well understood, but probably associated with running water; probably oxygen demands of larvae are high. Rapid water provides a degree of protection and perfect streamlining of larvae reduces hazards of swift water. Adults favor sun and heat and often congregate.

Of little economic importance, except possibly as fish food. Their presence in a stream indicates year-round flow and well-aerated water.

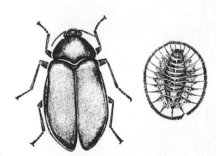

Water Penny Beetle

Buffalo Bug, Carpet Beetle

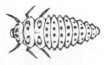

Convergent Lady Beetle

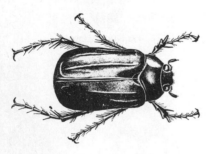

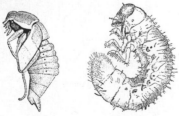

May Beetle, June Bug

Order Coleoptera./Family Dermestidae

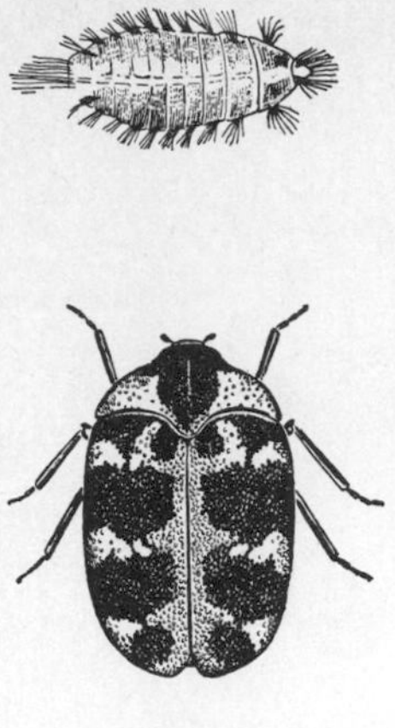

Buffalo Bug, Carpet Beetle
Anthrenus scrophulariae
Length about ⅐ in.   Attractive, black, white, and red with a black background, spotted and speckled with white, and a central red line down back.   Larvae nearly ¼ in. long, clothed with long brown hairs, those on sides and under rear parts longest.

Since first introduced into America about 1869, it has become a serious pest.   Native of Old World, where it was worst in museums, but is now established in America from coast to coast, where it seems to be worse than in Europe.

Wrinkled, white eggs laid on clothing, feathers, furs, silks, carpets.   These hatch in 2 to 3 weeks into grubs that reach ⅕ in. during which time skin is shed 5 to 11 times and damage takes place.   Pupae formed within last larval case and smaller than larvae.   Adults appear winter to spring.

Food of adults, pollen of flowers, such as spiraea; but food of larvae is usually materials valuable to man.   Adults found in spring and winter on ceilings and windows, attempting to escape to out-of-doors.   Adults and larvae tend to congregate at different places.

Control is by keeping adults from laying eggs on valuable property, by using wool baits from which adults may be destroyed, by using sulfur fumigation and kerosene in cracks where larvae or adults may hide, and by destroying infected materials where this is possible.   Believed to be most important pest of natural-fiber fabrics in the U.S.   In very severe infestations, fumigation of buildings with hydrogen cyanide is recommended.

Order Coleoptera./Family Coccinellidae

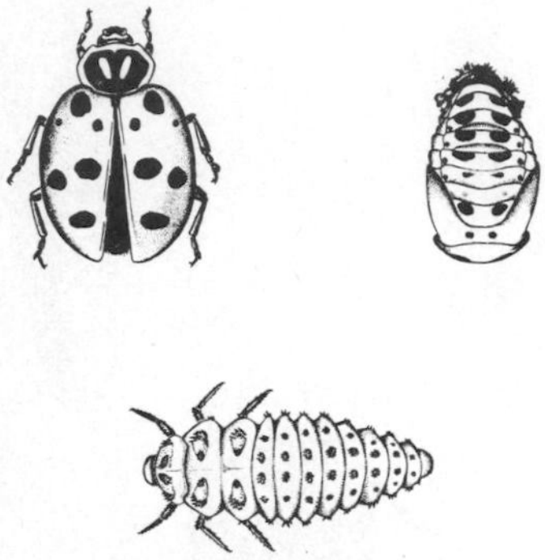

Convergent Lady Beetle
Hippodamia convergens
Length about ¼ in.   With two white converging dashes on back, in front of wings, and 12 black dots on orange wing covers; thorax with white border.   General appearance like small split pea.   Lady beetles are usually red, black, or yellow, and spotted.   Larvae active, usually dark-colored.

Adults common, and larvae very common on melon vines.   Some adult beetles are found in houses on windows in spring and fall and some congregate in great numbers under trash on ground for hibernation.   Except for two species of *Eplichna*, the ladybird beetles are a very beneficial group of insects.

Eggs laid in spring, on plants.   Larvae begin eating small insects and insect eggs or even spiders, and grow rather rapidly.   Finally transform into pupae that hang by tails a few days before transforming into adults.

Food of this species small animals only, including many kinds of plant lice and eggs of asparagus beetle, Colorado potato beetle, grape-root worm, bean thrips, alfalfa weevil, and chinch bug.   Related Mexican bean beetle devours leaves and is a serious pest.

On Pacific Coast, convergent lady beetles are collected in great numbers and distributed in crop areas.   30,000 are considered adequate for protecting 10 acres.   Takes about 1,500 to make 1 oz and they are collected by the ton.   Mexican bean beetle controlled by spray of 1 lb of magnesium arsenate to 50 gal of water, or by dusting.   Convergent lady beetle is a good example of biological control of insect pests. The use of biological control of insects is a field worthy of continuing research efforts.

Order Coleoptera./Family Scarabaeidae

May Beetle, June Bug
Phyllophaga sp.
Length to 1 in., or as short as ½ in.   Stout, usually brown and glossy above, with small heads and relatively slender legs; clumsy behavior.   Larvae, "white grubs," usually with rear bent forward beneath, dark heads, and short stout legs forward.

Adults found in treetops or about houses where they are attracted by lights at night.   During day remain hidden.   Larvae underground.   At least 100 species in genus.   In South, common in longleaf pine area; less common on West Coast but worst in Mississippi Great Lakes region.

Eggs 50 to 100, white, cylindrical, laid in glued-earth cell, underground; hatch in 10 days to few weeks into white grubs; these do not mature until second or third summer, spending winters deeper in soil.   Pupal stage in earth cell 3 to 10 in. underground, maturing in August or September.   Adults emerge following spring.

Food plant material, through 3 years of life cycle.   Adults feed at night on tree foliage, returning to hide in soil at daybreak.   Adults feeding may sometimes be heard.   Grubs have important bacterial enemies and are eaten by skunks, bears, crows, and other animals.

May strip trees of leaves, grass sod of roots, and completely destroy such crops as potatoes, small grains, and buckwheat.   Control: do not follow sod with row crops in year after beetles were abundant; deep fall plowing

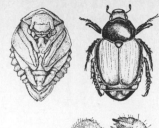

and pasturing of pigs and chickens in infested fields may help somewhat. Natural enemies may be fostered and parasitic fungi may be spread. Recommended control measures should be requested from County Cooperative Extension Service.

Japanese Beetle
Popillia japonica
Length about ½ in. Shining green body, with white spots on end of abdomen beyond end of wing covers. Popularly described as a green potato beetle and sometimes known as green Japanese beetle. Larva like a small white grub.

A native of Japan, discovered in New Jersey in 1916 and since spread over wide territory. In first 8 years in America, it appeared in an area of over 2,500 sq mi and still was not under control. Often so abundant that 1,500 grubs may be found in 1 sq yd of sod.

Tiny white eggs are laid usually in uncultivated soil, in July or thereabouts. White grubs that hatch from these feed through summer, spend winter deeper underground, enter pupal stage for a short time in spring, emerge as adults in June and July, mate and feed for 2 months or more, causing great damage.

Food of larvae and adults, many kinds of cultivated plants, especially grapes, raspberries, blackberries, strawberries, apples, cherries, corn, clover, soybean, roses, hollyhocks, elm, birch, linden, and many other valuable kinds of plants. When feeding in great numbers, may leave nothing green growing. Some birds feed upon them; notably, the much-maligned starling. Consult County Cooperative Extension Service for recommended control measures.

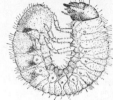

Japanese Beetle

Order Coleoptera./Family Cerambycidae

Cloaked Knotty-horn Beetle
Desmocerus palliatus
Most attractive blue and yellow beetle, with yellow in a broad band across middle. Antennae and legs about as long as body, which is under 1 in. long. Head much smaller than might be expected of a beetle of this size.

Found particularly on elder bushes, in summer, sometimes in considerable numbers. About 400 species of longhorned beetles in America, the more important associated with injury to woody plants; many are most attractive in appearance.

Adults mate in late spring or early summer on elder, where they lay their eggs. Eggs develop into larvae that bore into pith of elder and spend growing period there. This provides them with year-round protection where needed. Larvae of this family are called round-headed borers. The apple tree borer may destroy 10 to 20 percent of an apple orchard in a particularly bad infestation. The poplar borer can easily destroy, in one locality, Carolina poplar or cottonwood or aspen or Lombardy poplar. The black locust borer was responsible for the destruction of many of the black locusts used as soil-erosion-control plantings.

Food plant material; in this species, pith of the elder. Fortunately, this species limits its activities to relatively unimportant plants and rarely appears in sufficient numbers to be considered a serious pest.

Usual control is to cut and burn infected canes. Associated beetles include the ribbed pine borer, *Rhagium lineatum*; the maple borer, *Glycobius speciosus*; the locust borer, *Cyllene robiniae*; the painted hickory borer, *Cyllene caryae*; and the oak pruner, *Hypermallus villosus*.

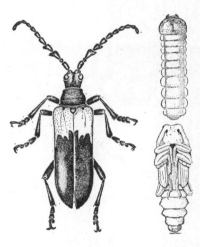

Cloaked Knotty-horn Beetle

Order Coleoptera./Family Chrysomelidae

Colorado Potato Beetle
Leptinotarsa decemlineata
Length of adult under ½ in.; about ⅔ as wide as long, with high arched back. Clay yellow, with 10 conspicuous black lines on wing covers. Head with a black spot above. Larva dull brick-red, soft, fat, and with a row of dark spots down each side.

Found mostly on potato plants, but originally on buffalo bur, *Solanum rostratum*, native of Rocky Mountain area. Began to spread from that area in 1859, moving at first at about 50 miles a year and reaching Atlantic Coast about 1874. Adults may be found hibernating in ground.

As soon as potato plants appear above ground, fertile female potato beetles lay masses of yellow ¹⁄₁₄-in. eggs, usually on lower surface of leaves. These hatch in about 1 week into red larvae that grow for 2 to 3 weeks before pupating underground near the plants. After about 2 weeks, adults emerge and second generation starts.

Food of larvae and adults, plant tissue only, preferably potato and related plants. Tomato and eggplant may sometimes be attacked as vigorously as potato. Adult can and will also eat tobacco, pepper, ground cherry,

(Continued)

Colorado Potato Beetle

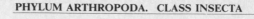

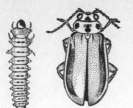

Elm Leaf Beetle

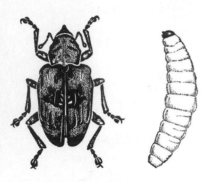

Plum Curculio

Plum Curculio

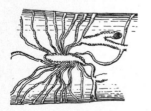

Engraver Beetle

thornapple, petunia, cabbage, and many other plants. Beetles destructive throughout growing season of plants.

Natural enemies include crows, skunks, rose-breasted grosbeaks, toads, and numerous insect enemies. For recommended control measures, consult your County Cooperative Extension Service.

Elm Leaf Beetle
Galerucella xanthomelaena
Length about ¼ in. Dull yellow with black spots on head and black band near outer edge of each wing cover; also with extra short dark streaks near front of wing covers. Sometimes grayish. Larva small, yellow and black, relatively active grub.

Adults commonly hide in cracks in buildings or under bark of trees. May crowd on inside of windows and screens in spring. Introduced in America near Baltimore from Europe about 1834, and now rather widely established, particularly on English elm.

Eggs yellow-orange, laid on end in clusters of 5 to 30, on elm leaves. Yellow-black larvae hatch in about 1 week and mature in 15 to 20 days, spending this time on elm leaves. Pupation in ground or in bark crevices for 10-day period. Adults may mate and produce a second generation in a season.

Food leaves of plants like elms. Late in spring adults of first generation feed on and injure buds, injury appearing as series of holes in leaves. Larvae may leave only leaf skeleton, eating out all soft parts. Winter as adults. May begin hibernation yellow and come out dark green.

Cause great damage to valuable shade trees. Chemical sprays can be used around base of tree to kill larvae en route down the tree trunk to pupate. Consult County Cooperative Extension Service for recommended control measures.

Order Coleoptera./Family Curculionidae

Plum Curculio
Conotrachelus nenuphar
Length about ⅕ in. In general, dark-colored but on close examination seen to be mottled with brown and gray. Wing covers rough, each with shining, black hump just back of middle. Unique snout in all snout beetles. Larva wormlike and well-known as the "worm" in such fruits as cherries.

A pest, particularly on fruit trees, practically everywhere in North America east of Rocky Mountains. Adults particularly active when young fruit is "setting" or earlier. The family includes some 1,800 species in America north of Mexico, divided into 13 subfamilies.

After hibernation on or in ground, adults emerge before fruits bloom, mate, female lays 100 to 300 eggs on young fruit; hatch in 3 to 5 days and larvae mature in 12 to 26 days, leave fruit, burrow in soil, pupate 3 to 4 weeks, emerge and feed on fruit about 2 months after flowering, making daily puncture for about 6 weeks.

Female makes incision in fruit and eats hole to depth of snout, ruining the fruit. In early spring, adults feed on buds and young leaves of apples, cherries, peaches, plums, nectarines, apricots, pears and such trees. Young fruits attacked may drop off or develop deformities that ruin marketability.

Damage in United States of this one species has been estimated at many millions of dollars annually. Control measures include destruction of rubbish that harbors wintering adults, pruning of trees to let in light that insects do not favor, grazing hogs and poultry in orchard to destroy pupae, and spraying with lead arsenate. Consult County Cooperative Extension Service for recommended control measures.

Order Coleoptera./Family Scolytidae

Shot-hole Borer or Engraver Beetle
Scolytus rugulosus
Length to ¹⁄₁₀ in. Dark brown with dull red wing-cover tips and red on legs. Presence easily recognized by burrows or burrow entrances that look like shot holes in bark of trees, entrances about size of pinhead.

Common on plum, cherry, apple, peach, and pear; closely related species attack other trees. This species found from Alabama to Massachusetts, on into Canada and west to Michigan. Burrows look like long-bodied spider with legs that are biggest at tips.

Male digs burrow in which female lays 40 to 70 eggs. Eggs hatch in 3 to 4 days into white grubs with yellowish heads; each burrows outward, branching from mother's burrow. Larval stage 30 to 36 days from time of leaving brood chamber. Pupal stage 7 to 10 days. Transformed adults may burrow further before emerging.

In closely related slash pine beetle, male cares for several females in reception chamber which he digs and guards. Unusual for any male insect to contribute a burrow for his mate or mates. Adults appear in April; second generation may winter as larvae.

Use of nitrogenous fertilizers and cultivation in summer months to encourage vigorous growth is good control measure. Control by whitewashing trees in late March and in July and October. Slash should be destroyed or burned if infestation of these beetles proves dangerous to valuable standing trees. Infestation frequently follows forest fires and kills injured and weakened trees. Some trees drown larvae with sap as a natural check. Dutch elm disease associated with *S. multistriatus*, the European elm bark beetle, or *Hylurgopinus rufipes*, the native elm bark beetle.

Order Mecoptera./Family Panorpidae

Scorpionfly
Panorpa sp.
Scorpionflies, when adult, usually have four similar many-veined membranous wings, which are long, narrow, and roofed over abdomen at rest, commonly with dark spots. Head prolonged into beak. Tip of abdomen of some males bends up like that of scorpion. Length to ⅔ in.

Scorpion Fly

Members of order occur in all parts of world but most are found in low damp wooded areas. Rarely abundant anywhere. Some species are on snow during winter. These may be wingless. Most species fly well. Some members of order are found in fossils.

Usually eggs are laid in masses in ground; larvae that hatch from them live in burrows underground and feed on surface. Larvae have three pairs of true legs and eight pairs of abdominal legs and abdominal spines. Pupal stages apparently spent in earthern cells underground.

Food of larvae, probably other insects; of adults, dead or living insects and in some cases fruits. Larvae in confinement may be fed meat. They probably eat any available animal matter.

No great economic importance to these insects, but because of their carnivorous habits, they probably are more useful than harmful. Their numbers are so small that they cannot ever become serious pests. Preyed upon by dragonflies, robberflies, spiders, and other enemies.

Order Trichoptera [Caddis Flies]

Caddis flies are soft-bodied insects with four membranous wings, each well-supplied with many longitudinal and a few cross veins. Wings usually held sloping like a roof over back, and more or less well covered with fine, easily shed, hairlike structures. Hind wings usually broader than forewings, and when not in use folded lengthwise. Forewings less membranous than hind wings. Insects vary in length from about ⅛ in. to just over 1 in., and as adults, nervously move their long antennae that are held forward. Legs long and slender. Mouth parts suitable for chewing, though it is doubtful if adults eat any food at all in some species. Larvae discussed below. There is a pupal stage, so metamorphosis is complete.

Since adults are poor fliers and larval states are aquatic, these insects are found in or near water. Larvae may be found in great variety of aquatic situations, from stagnant pools to dashing streams. About 1,000 species of caddis flies in North America, divided among 17 families. Of these, Rhyacophiladae, Hydroptilidae, Hydropsychidae, and Philopotamidae make their larval cases only in swift water, as represented below by Hydropsychidae. Phrygancidae, Leptoceridae, and Limnephilidae have larvae that live in quiet water or in a variety of types as indicated below.

Caddis worms and caddis flies have considerable importance. Adults congregate in enormous numbers near lights; when they die, fine hairs from their wings are freed and in some cases may cause a kind of hay fever. Larvae of great importance as food for fishes and, since they are found in great abundance wherever fishes feed among plants or on bottom, they provide an ever-ready valuable food supply. Members of genus *Hydropsyche* were so abundant in Niagara River and were attracted in such great numbers to electric lights that site of Pan American Exposition in 1901 had to be moved inland from point originally selected. Aside from hay-fever aspect of their presence, they are entirely harmless, though annoying.

Caddis Flies

Hydropsyche

Hydropsyche

Limnophilus

Phytocentropus–case

Neophylax–case

Phryganea

Order Trichoptera./Family Hydropsychidae

Water-net Caddis Fly
Hydropsyche sp.
Brownish oblong eggs are laid under water, on underside of sticks or stones, in patches, under a thin gelatinous coat. Larvae become greenish with darker foreparts, and build a funnel net, open upstream and ending in a pebble-bound silk cemented den. Pupal stage is spent in last larval skin. Pupa breathes by gills and swims to surface to transform into adult.

Food aquatic life caught in net, which has finer meshes at smaller end of funnel. As in other swift-water caddis flies, shelter of stones makes it possible for insect to resist push of water and let food come to it. Nets may become filled with sediment but can be renewed when necessary.

Order Trichoptera./Family Limnephilidae

Log-cabin Caddis Fly
Limnephilus combinatus
Eggs laid under water by female in summer; hatch into larvae that use plant materials to build log-cabin-like cases which may easily be moved about in quiet water while food is being sought. Larval stage may last 11 months or from June to following May, then a different case, sometimes of shell, is built for pupal stage of about 3 weeks. Adult stage short.

Food of these caddis worms is caught by larvae floating about or moving their buoyant cases from place to place. If cases are removed, larvae will build new ones; if transparent film or cellophane materials are available in fine strips, insects show ability to select certain colors and reject others.

Order Trichoptera./Family Phryganeidae

Caddis Fly
Phryganea vestita
Larvae of this species live in a plant-tissue tube made of narrow strips of leaf wound around a central core in a spiral. Others of family make case of strips placed around central core in narrow bands. Cases are abandoned when pupation takes place in submerged wood in a burrow closed by silk.

Food is obtained by moving about in relatively quiet water; moves easily because of buoyant nature of case. Some related families make burrows in plants through which they force a stream of water in which they construct a net to extract food organisms.

Order Lepidoptera [Butterflies and Moths]

Adults of members of this order usually have four membranous wings, almost wholly covered with minute overlapping scales that rub off easily. Larvae have chewing mouthparts, though adult may suck nectar through a long proboscis. In most cases, food is plant material. Few are aquatic at any stage. Metamorphosis complete. About 11,000 species of moths and butterflies known in North America north of Mexico, representing two suborders and more than 70 families. Moths commonly fly at night, usually close their wings around their bodies or hold them horizontally when at rest, and have threadlike or featherlike antennae. Butterflies usually fold their wings vertically when at rest and have knob-tipped antennae.

Case–bearing Clothes Moth

Order Lepidoptera./Family Tineidae

Case-bearing Clothes Moth
Tinea pellionella
Wingspread about ½ in. Grayish-yellow with faint spots on hind wings, and with hind wings a bit more silvery than forewings. A fringe of long hairs along rear margin of hind wings.

These moths fly at night, mostly during spring and summer months, but are not attracted by light.

Eggs are tucked away in folds of garments, or wool, feather, or silk objects. They hatch in about 1 week into wormlike larvae that live in tubular cases of food material. Larvae complete growth by fall, spend winter torpid and inactive, and in spring pupate for about 3 weeks before adults emerge.

Food of larvae almost any organic material, particularly if it comes from epidermis of some animal, and this may include other insects. Food is chewed and a webbing trail is often left behind.

Highly destructive to stored clothing and to rugs and tapestries. Possibly more destructive in South, where there may be two generations a year. Control is by shaking delicate larvae and pupae, and exposing to sun. Moth repellents available. Consult County Cooperative Extension Service for recommended control measures.

Webbing Clothes Moth
Tineola bisselliella
Wingspread just over ½ in. Forewings more yellowish than in preceding, without spots or markings. Hind wings paler. Head reddish. Considerable variation in size.

Possibly commonest of the clothes moths, particularly in North, on a great variety of substances.

Eggs white, oval, tucked away among clothes, hatch in about 6 days into wormlike larvae, which make no case but leave a web in trail and may develop for nearly 1 year or less, with possibly two generations, one from May eggs and one from September eggs. Pupal stage about 2 weeks, in a rough cocoon.

Food of larvae, almost any organic material. Known definitely to eat other insects, but usually food must be relatively dry and undisturbed for a long time. Shaking may injure animals at some stages.

Control necessary. Beat articles to be stored in summer, expose thoroughly to sun and air, and then seal in paper bags or in tight trunks. Consult County Cooperative Extension Service for recommended control measures.

Webbing Clothes Moth

Order Lepidoptera./Family Cosmopterygidae

Cattail Moth
Lymnaecia phragmitella
Wingspread ¼ in. Wings held like sides of a roof; slender, pale, silky, straw-colored, or light wood-brown, with two dots surrounded with white on forewings. Hind wings pale gray. Female stouter.

Common where there are cattails. Worldwide distribution including Africa, Australia, Europe, Asia, and America.

Oval, flattened eggs are laid on cattail spikes; develop into ½-in. larvae, yellow-white with red-brown markings, which spin silk in cattail heads and prevent seeds from blowing away; winter as caterpillars in the heads. Pupate about 30 days, in ⅖-in. thin tough white cocoon in stem or head of cattail. Adults emerge in May. One generation a year.

Food of adults, nothing except water; of larva, tissue of cattail stems and fruits. Protection from severe weather provided in cattail head is unique and in some ways worthy of imitation. Protection from animal enemies is effected by surrounding plant tissues.

Not economically important since it does little damage to plant not ordinarily considered of much economic importance. Has some parasites. This animal causes the well-known "bursted" spikes of cattail that continue to hang together even through winter.

Cattail Moth

Order Lepidoptera./Family Torticidae

Codling Moth
Carpocapsa pomonella
Wingspread about ¾ in. Forewings like watered silk in appearance because of alternating irregular cross lines of brown and blue-gray; a large light brown area bounded by chocolate and crossed by two metallic golden bands at hind angle of forewing. Hind wings coppery, darker toward margin.

Found mostly in or near orchards. A native of Europe introduced into America before 1750 and spread to Iowa by 1860, to Utah by 1870, and to California by 1874; now found wherever apples grow. Over 400 species in family, many serious pests.

Winter spent in larval stage, in silky cocoon under loose bark or other shelter. Cocoon tough, dirty gray. In spring, pupa formed in old cocoon or in a new one. Pupates for about 4 weeks as ½-in. yellow-brown object, with rows of minute black spines. Adult lays over 100 white eggs in about 10 days. These hatch in 6 to 10 days. Larvae feed on foliage and fruit about 1 month. 1 to 2 generations a year, more in South. Has been described as the most persistent, destructive, and difficult to control of all insect pests of apples. It will also attack pear, quince, crab apple, English walnut, and other fruits.

Food of larvae, some leaves at first but mostly fruit of apple, which it ruins for market purposes. Natural enemies include many insect parasites, birds, and some insects that kill larvae, eggs, pupae, or adults outright.

May cause 90 percent loss of apple crop with United States annual loss to many millions of dollars. Bands soaked in beta-naphthol in an oil carrier can be placed around the trunk of trees, will kill some larvae which build cocoons under the bands.

Codling Moth

Order Lepidoptera./Family Pyralidae

Closewing or Grass Moth
Crambus luteolellus

Wingspread about ⅞ in. Wings rolled around body when at rest. Male with long snoutlike process and with claspers at tip of abdomen; dark brown, cream, reddish, or dark gray with zigzag brown line near outer edge and no gold fringe. Female stouter, without claspers.

One of commonest small moths in grasslands east of Rocky Mountains, often seen perched lengthwise on grass stems. Since they are about thickness and color of straw, they are easily overlooked. About 70 species in genus, which also contains destructive cranberry girdler.

Eggs oval, yellowish to white, laid attached to grass or stubble; hatch in 6 to 19 days into white caterpillars with brown dots and brown heads, which are found in silky webs among grass roots. Winter as partly grown larvae. Pupae found among grass roots in weak cocoons for short period. Larvae act like cutworms which cut corn stems off near the surface when the plants are young. Larvae called webworms because they do produce webs at surface of ground. Blackbirds, robins, flickers, and other brids feed on the webworms.

Caterpillars eat plant tissue, particularly grasses and frequently growing corn. Food chewed; plant is injured if not completely destroyed.

Control is by fall plowing to destroy buried larvae; by rich fertilization, particularly with potash; and, in some cases, by burning over of stubble where infestation is serious and where damage by fire can be controlled or allowed.

Closewing Moth

Indian-meal Moth
Plodia interpunctella

Wingspread to ¾ in. Wings at rest folded along back, basal third of forewings olive or buff, outer portion reddish-brown with a coppery luster. Underwings lighter and plainer. No conspicuous difference in sexes as to size or coloration.

Found about feedbins and granaries, particularly in summer. Also found in food stores or pantries in raisins, meal, and such products. Introduced from Holland about 1856 and rather generally established. About 300 species in subfamily; one useful as a destroyer of scale insects while another, in Australia, controls cacti.

Eggs, about 350 laid by a female, singly or in groups; hatch in about 4 days in warm weather, into crawling cylindrical sparsely haired caterpillars, hairs being about as long as body diameter. Head small. Larval stage 2 weeks or more. Pupae, ⁵⁄₁₆ in., brown, in food-covered cocoons, 7 to 10 days. Life cycle 4 to 5 weeks.

Food largely stored foods including grain meals, prunes, currants, dried apples, beans, walnuts, pecans, dried peaches, plums, clover and other seeds, crackers, dried bread, flour, and the like. Caterpillars can go forward or backward, leaving silk trail behind. Up to four generations a year. Infested material is frequently more or less webbed together, with bad foul appearance and odor produced by the larvae.

Control necessary. Freezing temperature below zero continuously for 4 to 5 days kills insects or severe cold followed by high temperatures is effective. Sacks of food may be sprayed with carbon disulfide or fumigated. Consult county agent or Experiment Station. With modern insecticides it is possible to control losses from *Plodia* in flour mills and grain storage areas.

Indian–meal Moth

Order Lepidoptera./Family Citheroniidae

Imperial Moth
Eacles imperialis

Wingspread 5 to 6 in. Sulfur-yellow with bands and speckles of purplish-brown, with a large patch at base, a small round spot near middle, and a wavy band of purple toward outer margin. Male with outer margin of forewings purple-brown. Females with this margin yellow.

Adults in June or early summer, usually near forests on whose trees larvae occur. About 20 species of royal moths in United States, of which largest is the regal moth, *Citheronia regalis*, with 6-in. wingspread and 5-in. caterpillar.

Eggs laid in early summer on leaves of food plants. Larva develops into fierce-looking 4-in. caterpillar, with fine hair and tiny horns, green or brown for most part, with legs and head pale yellow, breathing pores white. Each segment has six yellow knobs with black spines. No cocoon is made, pupation being in ground.

Food of larvae, leaves of sycamore, hickory, pine, oak, butternut, and similar forest trees. Caterpillar not so fierce-looking as that of regal moth, which is known as "hickory horned devil." Pupa with long forked rear spine, used in working way up through soil before emergence.

Imperial Moth

Of no serious economic importance. Never common enough to be considered a pest, or uncommon enough to be valuable for sale to collectors. Related royal moth reported in *The Girl of the Limberlost* as being extremely valuable, is worth only a few cents.

Order Lepidoptera./ Family Saturniidae

Io Moth
Automeris io

Wingspread to 3 in. Adult holds wings like slopes of a roof, or thrown back. Hind wings yellow and brilliant red, with a great eyespot in center of each. Male with yellow forewings, antennae broadly feathered. Female with fore wings dull purple to brown, antennae not feathered.

Common in North America east of Rocky Mountains and in northern Mexico. Seen as adults most commonly at night. Family includes 43 North American species.

Eggs white with a black spot, nearly elliptical, laid in cluster of 20 to 30 on leaves, hatch in about 10 days into red-brown caterpillars which become yellowish then bright green with red-and-white stripes on sides; social when young, drawing together in masses when disturbed, or walking in processions. Pupae in thin irregular brown cocoons, on ground, over winter.

Food of larvae, leaves of corn, cherry, elm, apple, oak, willow, and other plants. Caterpillars have clusters of prickly spines that secrete a most unpleasant stinging poison.

Not usually abundant enough to be of any economic importance. Caterpillars are sometimes sufficiently numerous to annoy berry pickers who touch them.

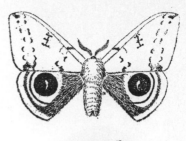

Io Moth

Luna Moth
Actias luna

Wingspread to 5 in. Forewings thrown back in repose. Males beautiful pale green moths, with a small transparent eyespot on each wing; front margin of front wing brownish-purple; hind wings with extraordinarily developed tails; antennae, broad. Females with narrower antennae and much heavier bodies than their mates. Nocturnal. Found in May and June, again in August.

Common but never abundant in ordinary sense. Limited to area where its food plants, mostly trees, may be found.

Female may lay as many as 200 pure white eggs, fastened with reddish cement to twigs and upper leaf surfaces. These hatch in about 3 weeks; caterpillars are clear green with faint continuous lines over back from sides. Larva molts at end of 4, 7, 9, and 10 days, and 18 days later spins a thin leaf-covered cocoon for wintering pupa. May be two generations a year.

Food mostly leaves of birch, willow, hickory, walnut, oak, and other broad-leaved trees. When young, caterpillars remain together in groups. Plain green color provides excellent protection from enemies but they have many parasites.

Of little economic importance but always popular with amateur collectors, largely because of their great beauty. During day, moths usually hang beneath leaves with wings folded, rarely observed by passers-by.

Luna Moth

Promethea Moth
Callosamia promethea

Wingspread about 4 in., wings held erect over back. Male diurnal, nearly black, with clay-colored borders and a blue eyespot near tip of forewing; antennae nearly twice as wide as in female. Female nocturnal, light brown, with basal half of each wing darker, with distinct eyespot near tip of forewing and light mark on basal half of each wing.

Common and widely distributed in eastern United States wherever its food plants are found.

Eggs small, white, depressed, and oval, fastened with red cement to underleaf surfaces, few in a cluster, hatching in a few days. Larva, first banded yellow and black; mature, 2 in. long, smooth clear light green, with six rows of tubercles with bluish circles around them and one yellow and four red tubercles. Young caterpillars crawl in groups and finally form pupae in silk cocoons wrapped in leaves and fastened by silk to twig for 3 weeks to over winter.

Food of adults, nothing; of caterpillars, leaves of cherry, sassafras, lilac, sycamore, tulip tree, ash, and others. In cocoons, females are heavier. Swinging cocoon difficult for birds to destroy.

Of little economic importance. Gut for fishing leaders may be had from caterpillars, but those from this species are relatively short.

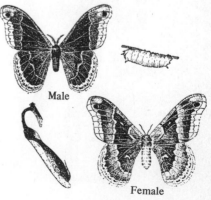

Male

Female

Promethea Moth

Polyphemus Moth

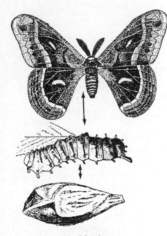

Cecropia Moth

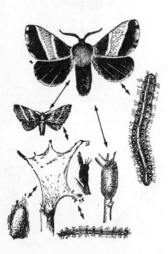

Eastern Tent Caterpillar

Polyphemus Moth
Antherea polyphemus

Wingspread to 6 in. Wings at rest held over back, like butterflies. Male, forewing buff, olive, or gray-brown, with transparent yellow-bordered eyespot; hind wing with dark border and blue eyespot, deeply black-bordered. Female much like male but with heavier body and with antennae not broadly feathered. Nocturnal.

Common throughout eastern United States and in northern Mexico wherever its food is available.

Over 300 cream-colored, disklike eggs encircled with a brown band, usually laid on underleaf surface; hatch in 10 to 12 days into caterpillars that finally are green with reddish heads, conspicuously segmented, with rose-red or pale and silvery tubercles, bright red spiracles, and seven oblique yellow lines on sides. Walking caterpillars may snap their jaws hard enough to be heard. Pupa in thick, strong, light-colored, oval cocoons, over winter.

Food of adults, nothing; of caterpillars, leaves of such trees as oak, elm, birch, and maple. Cocoons are made of a continuous silk thread that is exceptionally strong and might possibly be used as a substitute for silk from regular silkworm.

Always interesting to amateur collectors. Native silkworms, producing silk that has some commercial possibilities. Gut leaders for fishlines may be made by pickling the caterpillars in saturated solution of salt and vinegar, removing the two silk glands, then stretching contained silk to length or thickness desired. Such leaders from cecropia moth may be 6 to 9 ft long, able to sustain a 4-lb pull.

Order Lepidoptera./Family Saturniidae

Cecropia Moth
Hyalophora cecropia

Wingspread to 6½ in. Wings usually held vertically. Male somewhat like female promethea but with large red and white crescent on basal half of each wing; abdomen red, banded with black. Female larger than male with more narrowly feathered antennae and with larger abdomen. Nocturnal. Males are attracted to the female by odors which may be effective to 3 miles.

Common wherever food plants are available from Atlantic Coast to Rocky Mountains.

Eggs to 400, egg-shaped, pinkish-white fastened with red-brown cement in rows of 3 to 20 on leaves of many plants; hatch in about 15 days. Larva black, then red; molts at 4, 5, 6, and 8 days; spins cocoon 10 days later. Mature larva blue-green with large yellow, red, and blue tubercles; about 4 in. long, matures in 4 to 9 weeks. Pupates in large tough spindle-shaped reddish-brown cocoon fastened to twigs. Cocoon sometimes preyed upon by woodpeckers and blue jays.

Food of adults, nothing; of larvae, leaves of box elder, apple, cherry, willow, plum, and about 50 different kinds of trees, so caterpillars can usually be reared easily with most tree leaves used as food.

Adults popular with amateur collectors. Larvae produce longest and best gut leaders for fishlines of our native species.

Order Lepidoptera./Family Lasiocampidae

Eastern Tent Caterpillar
Malacosoma americanum

Wingspread of male, 1¼ in.; of female 2 in. Rather stout, with reddish-brown body; forewings with two nearly parallel oblique whitish lines. Bodies may have banded appearance or be plain but are abundantly covered with hairy scales.

Found commonly on apple, cherry, and related trees in Eastern States and Canada and west to Rockies, with some records from California. In northern part of range, adults appear late in June.

Female lays tapering-ended cylindrical bands of varnished eggs on twigs. Eggs about 1/25 in. long, 200 in mass, hatch with opening of leaves. Larvae begin feeding and making tent nest of silk; about 2 in. long, black with a continuous white stripe down back, and blue and white on sides. Remain larvae about 6 weeks. Pupae in 1-in. silk cocoons for about 3 weeks.

Food of larvae mostly apple and cherry but may also attack peach, pear, and plum. Tents provide shelter during larval stage. They do not do well in rainy seasons. Many parasitic enemies, and abundance varies from year to year.

Control by destroying nests, spraying as for codling moth or burning caterpillars and cocoons. One school of thought is that wild cherries should be preserved as a trap crop; another that they should be destroyed. Seems to have a periodicity of great abundance about each 10 years. Several insecticides have been used to control the pest; DDT, methoxychlor, chlordane, malathion, lead arsenate. Consult County Cooperative Extension Service for recommended control measures.

Forest Tent Caterpillar
Malacosoma disstria
Wingspread slightly less than that of apple tent caterpillar. Differ primarily in that in forest tent caterpillar oblique lines on wings are dark instead of light, or very weak; and in larvae, white on back is broken rather than continuous.

Common throughout United States and into Canada and Mexico, being most numerous in forested areas. Often a pest on shade trees along streets. Appears also to prefer poplar or aspen, expecially where they occur in pure stands and will attack oaks, maples, basswood, ash, elm, birch, and some conifers.

Eggs in blunt-ended cylindrical masses containing 150 to 400 eggs, on tree twigs in summer. Following spring, eggs hatch when leaves begin to develop. Caterpillars may form a silk carpet but build no tent; rather, they mass in great mats on trunks of host trees. Pupate in late May, emerge in late June, lay eggs in July; caterpillars develop in egg to August, hatching next spring.

Favors maple but eats leaves of apple, plum, peach, pear, cherry, and many other trees. Colonies of caterpillars break up as they reach maturity and wander off alone. Many parasitic and other enemies among the beetles, mites, wasps, flies, birds, amphibians, and other groups.

A spray of 8 lb of lead arsenate to 100 gal of water is effective. Consult County Cooperative Extension Service for recommended control measures. Masses of caterpillars may be brushed up and burned.

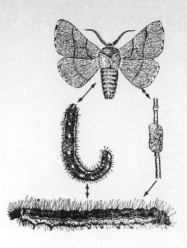

Forest Tent Caterpillar

Order Lepidoptera./Family Geometridae

1. Fall Cankerworm
Alsophila pometaria

2. Spring Cankerworm
Palaecrita vernata
Wingspan male, 1 to 1⅕ in.; females practically wingless. Male of (1) darker smoky or brownish-gray, with a distinct whitish spot on front edge of forewing. Active mostly at night.

(1) found in injurious numbers through northeastern United States and even to western California. (2) more eastern in its distribution. Related species in Europe, the term *cankerworm* appearing in the King James Version of the Bible published in 1611.

Females of (1) climb trees November to April, mating and laying to 400 eggs in a mass, on twigs in treetops; hatch when leaves start. "Measuring worms" feed about 1 month before dropping to ground, by thread, to pupate 1 to 4 in. underground until emergence. (2) similar, but adults emerge March to April only.

Food, leaves of variety of woody plants, particularly shade and orchard trees. Caterpillars may be blown on silk to nearby trees that have not been infested by climbing females. Many enemies, with birds such as thrushes, vireos, warblers, and chickadees ranking high. Many insect enemies also.

Artificial control by preventing wingless females from climbing trees to lay eggs, by use of barriers, tanglefoot flypaper, or special proprietary mixture placed on trees from October through March for (1) and from February through April for (2).

Secure detailed information from your county agent or Experiment Station.

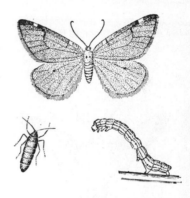

Fall Cankerworm

Order Lepidoptera./Family Bombycidae

Silkworm
Bombyx mori
Wingspread about 1¾ in. Antennae in each sex broadly feathered. Male cream-colored, with 2 to 3 more or less distinct brownish lines across forewings, and half an eyespot on inner half of hind wings. Head small and obscured by whitish hairs. Male rarely flies; female never flies.

Silkworm found only in domestication. Said by Chinese to have been domesticated by them since 3000 B.C. Later raised in India, then in Persia, and introduced into Europe about A.D. 555. As basis of big industry, it thrived in Greece, Italy, France, Spain, and Japan, in about that order, but never in North America.

Females lay 300 to 400 eggs that resemble turnip seeds; hatch in 8 to 10 days into caterpillars, which in 6 to 8 weeks grow from ½ to 3 in. long, shedding skin four times. Caterpillars usually white-humped behind small head, with short rear spine. 1-in. cocoon has about 1,000 ft of silk, which is cut when moth emerges after 2 weeks in pupation. Egg to egg 2 months or more.

Food of larvae, leaves of mulberry, osage orange, or lettuce. Silk from silk glands hardens on contact with air. Pupae killed by steam; soaking in hot water softens gum that binds threads, permitting unrolling of silk and its spinning into thread. About 1,000 miles of silk per lb.

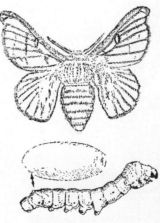

Silkworm

(Continued)

Silk supports great industries throughout world, but these depend on availability of cheap labor. Silk is being rapidly replaced by synthetic materials produced by chemists, and far-reaching social problems may be presented in providing labor for those who formerly made silk. A major export of some small Asian nations. In the opinion of some, even yet the most beautiful of fibers for ladies' dresses.

Order Lepidoptera./Family Sphingidae

Sphinx Moth, Tomato Hornworm

Manduca quinquemaculata

Wingspread 4¼ in. Males ash-gray and streaked, with five pairs of yellow spots on conical body; wings narrow and pointed; tongue long; antennae thick and rough. Females, antennae thinner and smoother, as in hummingbird moths.

Found about deep-tubed flowers, particularly on summer evenings. This species common from Patagonia to Canada. About 100 species of sphinx moths found in United States.

Eggs large, green, spherical, shining, laid singly on leaves of food plants, hatch after a short period into green larvae that become about 3 in. long, stout, with a row of seven shaded slanting stripes on each side, with an equal number of horizontal stripes making a V, and a good-sized rear horn. Larva to 4 weeks. Larva may be parasitized by eggs of braconid wasp. Braconid moth lays eggs under skin of tomato worm; later white cocoon develops on exterior of tomato worm. Tomato worms with white cocoons attached should not be killed. Pupa in ground, over winter.

Food of larvae, leaves of tomato, potato, tobacco, and similar plants; of adults, nectar in flowers. Possibly two generations a year. Pupa like brown jugs, with long "handles" that contain tongues. Harmless to man; horns of larvae not poisonous as rumored.

Sometimes seriously destructive to host plants. Adults assist in pollination of some deep-tubed flowers. Consult county agent or Experimental Station for latest information on control.

Sphinx Moth; Tomato Hornworm

Hummingbird Moth, Clearwing Moth

Hemaris thysbe

At rest, general outline, like an equilateral triangle, 1 in. on a side. Wings with conspicuous clear areas; scaly part of wings reddish-brown; hind wings much smaller than forewings. Rather large body well covered with hairy scales; conspicuous tuft at tip of abdomen.

Commonly found in flower gardens, clover fields, and similar places where nectar-bearing flowers are abundant. This species active in daytime. Another close relative is bumblebee hawk moth.

In June and July, larvae of these clearwings may be found on their food plants. They resemble, in general, other sphinx moth larvae, being light green with green pattern. Pupal stage spent on ground through winter in crude cocoon and without free tongue case common in related species.

Food of adults, nectar of flowers; of larvae, leaves of snowberry, viburnum, and their close relatives. Rapid erratic flight of adults always interests amateur insect collectors.

Of little economic importance because host plants are of secondary importance and larvae rarely, if ever, occur in sufficient numbers to be pests. Two color forms (largely seasonal), either of which may be produced from eggs of other.

Hummingbird Moth

Order Lepidoptera./Family Notodontidae

Red-humped Caterpillar

Schizura concinna

Wingspread 1½ in. Holds wings like sides of a roof. Males gray and white with brown along inner margin of forewings; hind wings largely white; basal half of antennae feathered. Female all dark-gray and brown, with hind wings gray; antennae unfeathered.

Common throughout United States, though adult is seldom noticed. Adult flies at night, usually during July, and remains hidden during day.

Eggs white, thin, smooth, finely sculptured, laid in clusters on food plants; hatch in short time into larvae, which live in colonies and are dirty orange when young and black and white when grown, with bright red heads and forward humps, black pegs on hump and body. Colonies of caterpillars will be found feeding on the leaves of apple, pear, or some forest trees in July and August; may completely defoliate young trees or entire branches of larger trees. Winter as caterpillars in cocoons, in trash on ground. Pupate early in the summer.

Food of adults possibly nectar; of larvae, leaves of apple, plum, maple, birch, and other trees. Caterpillars are able to throw acid, so should be observed with circumspection. Fortunately, they have many parasites.

Red-humped Caterpillar

Not commonly a serious pest and usually may be controlled by picking and burning larvae, particularly when they are found on newly set trees that have not become well established. One generation a year in North, two in South.

Order Lepidoptera./Family Lymantriidae (or Liparidae)

White-marked Tussock Moth
Hemerocampa leucostigma
Wingspread of male, to about 1¼ in.; ash-gray, with front wings crossed by wavy dark lines and with conspicuous white mark near rear angle; antennae feathery. Female flightless, with mere wing stubs; light gray, usually found on old cocoon.

Widely distributed in United States and Canada, particularly in fruit-growing areas. Abundant, caterpillars and cocoons being found frequently.

One brood a year in North, to three in South. Females emerge from cocoons in July, mate and lay to 500 eggs on old cocoon; cover these with a froth; hatch shortly or rest over winter. Larvae red-headed, with four even tufts of white hairs. Pupa after 4 to 6 weeks, in gray cocoon on trees; adults emerge after 10 to 15 days.

Food of caterpillars, leaves of apple, plum, pear, quince, and other fruit and shade trees, particularly horse chestnut, elm, and poplar but not conifers. Hairs on back of caterpillar may cause a mild rash on some persons. May cause a 25 percent loss in apple crop. More commonly a pest of shade trees.

May be serious pest on food trees. Natural enemies include dermestid beetle larvae, mites, and other parasites. Control by picking and destroying cocoons, banding trees to limit movements of caterpillar to new trees.

Consult county agent or Experiment Station.

White-marked Tussock Moth

Gypsy Moth
Portheria dispar
Wingspread to 1½ in. Male brown, with forewing crossed by four wavy dark brown lines, body light brown. Female with light buff body; wings grayish-white with dark markings on forewings like those of male. Female larger than male.

Found too commonly on woody plants of New England area. A native of Europe, Asia, and northern Africa, where it has long been a tree pest. Introduced into Massachusetts in 1869 by a French naturalist who was experimenting with silkworms.

Winters as egg, in masses up to 500. As leaves open, eggs hatch into reddish-brown caterpillars that skeletonize leaves, avoiding sun and feeding largely at night, maturing in about 7 weeks as gray caterpillars 2 in. long with 11 prominent pairs of tubercles, first 5 blue, rest red. Pupa in frail cocoon about 17 days. Moths begin emerging during latter part of July.

Food, leaves of most shade and fruit trees, even conifers; often so abundant as to defoliate trees completely. Male flies in zigzag manner but female, though winged, is unable to fly, so usually lays eggs on her cocoon. An important enemy is a ground beetle, *Calosoma sycophanta*.

Control may be accomplished by picking and destroying egg masses, which are conspicuous during winter in a small area. Eggs destroyed also by painting with creosote and by closing cavities in which eggs are laid. Some hope that imported parasites and female sex hormone attractants may prove effective and eliminate the need for chemical insecticides. Consult county agent or Experiment Station.

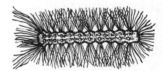

Gypsy Moth

Order Lepidoptera./Family Lymantriidae (or Liparidae)

Brown-tail Moth
Nygmia phaeorrhoea
Wingspread about 1½ in., females larger than males. Wings white. Tip of abdomen with a tuft of brown hairs that gives insect its name; this tuft less conspicuous in male. Adult seen usually in July. Females good fliers, unfortunately.

Found on or near trees of various kinds; in particular deciduous fruit and shade trees. Native of Europe, where it has long been highly destructive. Now established from Nova Scotia and New Brunswick through New England and undoubtedly will continue to spread. Introduced near Boston in the early nineties.

Adult appears in July, and after mating lays to 300 globular yellowish eggs in ¾-in. long mass, covered with brown hairs, on underside of leaves; eggs hatch in to 20 days. Caterpillars brown with rows of white tufts on sides; feed in colonies and after two to three molts winter in groups in an egg-shaped shelter; in spring, molt four to five imes, reach 1½-in. length. Pupa for about 20 days.

(Continued)

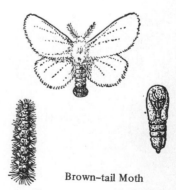

Brown-tail Moth

Food of larvae, leaves particularly of apple, pear, and oak though many other trees are attacked. Unlike gypsy moth, they do not eat leaves of evergreens. Hairs on caterpillars may cause a severe rash on human skin.

Control most necessary; effected by encouragement of parasites and by collecting and burning hibernating caterpillars.

Consult county agent or Experiment Station.

Order Lepidoptera./Family Noctuidae

Underwing Moth
Catocala cara
Wingspread to 3½ in. Dark gray forewings that make a triangle when folded at rest. Hind wings normally hidden but black, pink-banded, and beautiful when flashed and displayed by placing forewings forward. Female slightly stouter and heavier than male. Found as adult in August to September.

Underwing Moth

Common over whole of eastern United States, with related species extending range throughout North Temperate Zone. About 200 species in United States in subfamily to which this moth belongs. This is among the most beautiful. *C. irene*, a species of California and intermountain region of U.S., is practically invisible on a lichen-covered tree trunk or a rock ledge; beautiful colors of hind wings shown only when in flight.

Winters as red egg, in clusters on bark of trees. In late spring, gray and black spotted caterpillars emerge, each with a fringe alongside making it appear flat and with a hump in middle of back; length when mature about 3 in. Pupa (which has a bloom) spends about 3 weeks between leaves, in a light brown cocoon.

Food, leaves of willow. Adults attracted to baits and feed on nectar. Larvae move by humping and jumping; have many insect parasites and other enemies.

Control is hardly necessary since insects are rarely sufficiently abundant to do any serious damage. Adults are so attractive that they are popular with amateur collectors. Ability of moths to vanish on a tree trunk is a marvel to all. Related to black witch moth of tropics. The underwing moth is a possible prize for persons who go "sugaring" for moths on still, humid summer nights.

Cattail Moth, Cattail Leaf Miner
Arzama obliqua
Wingspread to 2 in. Length to about 1 in. Forewings light brown with oblique brown stripe, hind wings with small dark spot. Presence indicated and easily detected by dead tops of infected cattails.

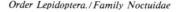

Found on or near cattails throughout United States and Canada, wherever host plant may be found.

Eggs laid near tip of new cattail leaves in masses of 35 to 60, each mass with a waterproofing froth. Larva mines way down leaves as much as 20 in., molts, then abandons mine and enters stem as solitary borer; grows to length of 2 in., winters as larva. Pupal stage of 16 to 19 days begins in spring.

Cattail Moth, Cattail Leaf Miner

Food: young larvae eat soft interior of cattail leaf, working side by side. Older larvae, boring in stem, kill flower stalk and may otherwise injure whole plant. In central New York, eggs are laid in May.

Never of great economic importance because of small numbers and because of relatively secondary economic importance of host plant. Another genus, *Archanara*, of the noctuid family also infests cattail.

Order Lepidoptera./Family Noctuidae

Army Worm
Pseudaletia unipuncta
Wingspread 1½ in. Uniformly brownish-gray with tiny white spot near middle of forewing and rather conspicuous dusky outer margin on hind wing. Hind wings shorter than forewings.

Army Worm

Common over entire eastern United States west to Kansas and Nebraska, with some representation in Southwest and in California; related species all over world. Moths fly at night and are attracted by lights. Some 370 North American species in subfamily to which this insect belongs.

Eggs 50 to several hundred, laid on blades of grass and on similar plants, hatch in 8 to 10 days. Larva for 3 or 4 weeks; at maturity about 1½ in. long, variable in color but commonly greenish-brown with a broad dark stripe along back and a dark stripe along each side; may migrate in great numbers. Pupae in ground a few days. Any stage may winter.

Food: in late spring, larvae favor grasses, particularly small grains. When abundant, larvae consume almost all available vegetation and move as an army to new territory, this being most common in July and

August when second generation of larvae is developing. Many enemies, particularly tachina flies, help control army worms.

At times this is one of most destructive of pests in grain-belt area. Controlled by ditches built as traps. Poisoned baits are also used when pests are not too abundant. Food strips in line of march are sprayed with poisons effectively. Consult county agent or Experiment Station.

Dingy Cutworm
Feltia subgothica

Cutworm moths (of many species) are usually under 2 in. in wingspread. All are commonly gray to brown "millers," with lighter hind wings and darker banded forewings. Adults are commonly attracted to lights and fly at night, but this species may also fly in daytime.

Common in summer and widely distributed through America. This subfamily includes about 370 North American species that attack a great variety of plants.

In summer, fertile female lays her many eggs among grass roots. They are $\frac{1}{50}$ in. long, dirty white with brown markings. These hatch in a few weeks and caterpillars begin feeding, hiding underground in daytime. They winter underground, begin feeding again in spring, finally reaching $1\frac{1}{2}$ in. in length, with wide pale back stripe. Pupate underground about 4 weeks before emerging as adults.

Damage may be extensive, largely because of feeding of spring caterpillars on newly planted crops. One generation a year. Some related species attack trees, climbing them at night to feed and then returning to soil.

Control is by crop rotation, by late and deep fall plowing, by use of poisoned bran, and to a slight extent by trapping adults.

Consult county agent or Experiment Station.

Dingy Cutworm

Corn Earworm
Heliothis zea

Wingspan about $1\frac{1}{2}$ in. Extremely variable. Some dull olive-green, others yellowish or brown with almost no marking. Hind wings cream-colored, with black border containing a pale spot.

Common, particularly in South, but found rather generally over United States wherever corn is grown.

Many light yellow, spherical, prettily corrugated eggs are laid on corn, tomatoes, cotton, beans, and so on. These hatch in 3 to 5 days and develop into variable light green to black, plain or mottled caterpillars that become full-grown in $2\frac{1}{2}$ weeks at a length of $1\frac{1}{2}$ in., and then leave corn to burrow in soil and pupate 2 weeks or over winter as 1-in. shining red-brown pupae.

Food of larvae, fruits and other plant parts of corn, cotton, beans, peas, and other garden crops. Adults feed on nectar and do no direct damage. One to two generations a year; type of food determined largely by plants available at time it is needed.

Control is by late fall plowing, by planting corn early, by use of lead arsenate spray to control at least first generation. Destroy about 3 percent of green corn crop annually and make it difficult to raise this crop at all in South. Consult County Cooperative Extension Service for recommended control measures. The variety of common names of this insect attests to its ability to change its diet; e.g., cotton boll worm, tomato fruit worm, false budworm of tobacco, tobacco fruitworm, vetchworm, as well as cornear worm.

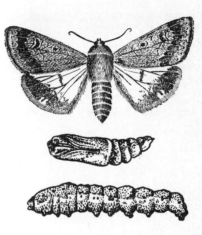

Corn-ear Worm

Order Lepidoptera./Family Agaristidae

Eight-spotted Forester
Alypia octomaculata

Wingspread to $1\frac{1}{4}$ in. Black with two yellow spots on each forewing and two white spots on each hind wing. Legs yellow-orange. Abdomen of male distinctly tufted. Female larger than male, abdomen not tufted. Antennae slightly clubbed.

Common on porches or places where woodbine grows, throughout United States east of Rocky Mountains. Presence usually indicated by droppings of caterpillar. Only 16 North American species in the family. *A. alangtonii*, from New York to the Pacific Coast feeds on fireweed, *Epilobium angustifolium*.

Female lays many eggs on woodbine and grape. These hatch in a short time and develop eventually into humped caterpillars with small black knots on rear, about six cross stripes to each segment, orange heads, and a big white spot on side of tails. When ready to pupate, larvae eat a burrow into wood and remain as pupae over winter. May form cocoon of silk and chips at or near the surface of the ground.

(Continued)

Eight-spotted Forester

Food of larvae, leaves of woodbine and grape, sometimes to such extent as to cause injury. Usually two broods a year; adults appear in May and August and larvae in June, July, and September.

Not normally injurious but where they are troublesome may be controlled with usual sprays that serve as stomach poisons.

Order Lepidoptera./Family Arctiidae

Isabella Tiger Moth or Banded Woollybear
Isia isabella

Wingspread to 2 in. Male buff brown with small black spots. Hind wings straw-colored. Body dirty orange with black spots. Female similar or with flesh-colored hind wings. Commonly flies at night and attracted by lights.

Common in northern United States from Atlantic to Pacific. Best known in caterpillar stage as a "woollybear," seen crossing roads in fall.

Females lay nearly 1,000 yellow, spherical, slightly spaced eggs in patches. These hatch shortly into furry red-brown caterpillars with black ends; when very young, they are social; when grown, they may lose much of black ends. Winter as caterpillars under rubbish, feed in spring, then pupate under cover in cocoons made of silk and caterpillar hairs for about 2 weeks.

Food of caterpillars, a variety of plants, with preference limited only slightly. Cocoons are found so commonly under stones and boards that they are well known. Probably no relationship between length or severity of an approaching winter and proportion of black on caterpillars. Young have more black than old; early cold weather may send more of these darker than usual caterpillars in search of winter quarters, so that they are noticed; yet this is result of existing conditions, not of coming ones.

Of little or no economic importance since they do not specialize on useful plants and are rarely if ever sufficiently abundant to do serious damage to anything. Skunks and some other animals roll hairs off caterpillars before they eat them.

Yellow Woollybear
Diacrisia virginica

Wingspan to 1¾ in. Male white with a few small black dots. Wings held like a roof gable. Body white with yellow side stripes. Front legs with yellow stripe. Antennae broad. Female similar but antennae not plainly feathered.

Common throughout temperate America. Family includes nearly 200 North American species, more than half of which are true tiger moths. Related species are harlequin milkweed caterpillar, bella moth, salt-marsh caterpillar, hickory tussock moth, and fall webworms.

Two generations a year. Female lays several hundred yellow spherical eggs, slightly spaced, on anything. Hatch in 4 to 5 days. Caterpillars social when very young, later variously colored with long soft loose hairs which give body a cream to black appearance. Winter as pupae. Pupae in felty cocoons, 2 weeks to over winter, under trash.

Food of caterpillars, almost any vegetable materials that may be available; confined to low plants. Loose hairs on caterpillars serve as protection whether they are stiff as in Isabella tiger moth or soft as in this species, since they break off and are unpleasant to many potential enemies.

Of little economic importance. Sometimes they have been known to eat silk off corn, thus reducing effectiveness of pollination.

Order Lepidoptera./Family Hesperiidae

Silver-spotted Skipper
Epargyreus clarus

Wingspan to 2 in. Wings often held half-open. Male, forewings brown, with tawny spots above and beneath; hind wings with silver splash on underside and unmarked above. Female lacks a slender fold found along forward edge of forewing of male. Fold in forewing is a scent organ. Alert insects. Hooked antennae are typical of skippers.

Fairly common in all of temperate North America, ordinarily found near such plants as locust and wisteria. Adults in June and July. Over 80 species of subfamily in America north of Mexico.

Winter spent in pupal stage, with one to two generations a year. Adults emerge in late spring or early summer, mate, and lay eggs. Eggs dome-like, heavily ribbed, grass-green, laid singly on upper leaf surfaces, hatch in about 4 days. Larvae yellow and green with fine cross stripes; head large and brown with two orange spots; neck narrow; make nests of leaves and silk; nearly 1½ in. long. Pupate between leaves on ground.

Isabella Tiger Moth or Banded Woolybear

Yellow Woollybear

Silver-spotted Skipper

Food of larvae, leaves of locust, wisteria, and other (usually woody) legumes. Winter pupal stage in this species is rather unusual for members of group. Pupa nearly 1 in. long, obscure brown, with abdominal portion not so long as rest of body.

Adults attractive and, while larvae may be injurious to host plants, they are never sufficiently abundant to be considered serious pests.

Hobomok Skipper

Hobomok Skipper
Poanes hobomok
Wingspan to 1¼ in. Wings often held half-open. Male, forewings tawny with irregular blackish border; hind wings lemon-yellow beneath with broad irregular brown edge. Female coarser and heavier. Two forms of female: one, like male, other, with black forewings with small white spots, and chocolate underneath center of hind wing.

Abundant in Northeastern States, in grasslands, June to July. Skippers have widely separated antennae enlarged at tip into knobs; hooked. Bean leaf roller is a member of this family.

One generation a year. Females in early summer lay many smooth hemispherical pale green eggs on grass. These hatch in 11 to 13 days and develop into slender, yellow-brown, naked-bodied caterpillars with dark dorsal and lateral stripes and a narrow neck; head large and tapering. Larvae make web nest in grass; winter as mature larvae. In spring, short pupal stage is spent in cocoon in grass. Pupa about 1 in. long, soft, long-tongued.

Food of larvae, grass; of adults, nectar.

Of little economic importance since they are never abundant.

Order Lepidoptera./Family Papilionidae

Black Swallowtail or Parsleyworm
Papilio polyxenes
Wingspan to 4 in. Wings held erect when at rest. Male, wings black, with double border of yellow spots, those on hind wings orange beneath with single row of blue ones between; "tails" of hind wings ½ in. long. Female like male, but with yellow spots smaller, blue spots larger, and without extra spot in cell of hind wing.

Common in all of United States and in southern Canada, with special varieties from Quebec to Cuba and Colombia; in gardens and fields where food plants are being raised, from May to September in North.

Black Swallowtail

Winters in pupal stage. Adults emerge in May, mate, and lay smooth, round, plain, yellow-brown eggs, singly, on underside of food leaves. These hatch in 5 to 10 days and develop into green caterpillars ringed with black and spotted with yellow when mature, spined and with a white saddle when young; length nearly 2 in. Pupate for 9 days or over winter, erect on plants or boards, attached at tail and with loop of silk above.

Food of larvae, leaves of carrots, parsnips, caraway, and related plants. Larvae have an offensive odor when disturbed and a perfect concealing pattern that provides good protection.

Larvae may be injurious to some crops but are easily controlled by hand picking, since they are rarely abundant. Adults are a pleasure to see. Caterpillars are harmless in spite of their rather startling appearance.

Order Lepidoptera./Family Papilionidae

Tiger Swallowtail
Papilio glaucus
Wingspan to 5 in. Male, forewings yellow with a black border and black cross stripes, border containing yellow spots; hind wings with ½-in. tails. Female larger than male, with broader black stripes and without claspers which are easily seen on end of male's abdomen.

Common in most of United States and in southern Canada but not on West Coast. Found in a great variety of places but generally favors open sunny areas. Species *multicaudatus, eurymedon,* and *rutulus* range from British Columbia to California and east to Colorado, Texas.

Winters in pupal stage. Adults emerge from May to September, mate and lay smooth green spherical eggs singly on upper surfaces of food leaves, eggs hatch in about 10 days. Larva green, with two large eyespots at front end, followed by a yellow crossband; ugly-looking and vicious-acting, because of eyespots and two soft orange processes forced out when disturbed; length 2 in. when mature.

Food of larva, leaves of birch, poplar, ash, cherry, mountain ash, tulip tree, willow, and basswood. Pupal stage lasts 2 weeks or over winter; pupa 1¼ in. long, wood-brown, straighter than that of black swallowtail, hanging erect in a thread loop usually away from host plant.

This species of little if any economic importance, but adults are insects of beauty admired by all who care for butterflies.

Tiger Swallowtail

479

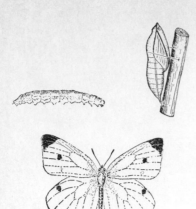

Cabbage Butterfly

Clouded Sulfur

Common Copper Butterfly

Order Lepidoptera./Family Pieridae

Cabbage Butterfly, Imported Cabbage Worm
Pieris rapae

Wingspan to 2 in. Male, forewings white, black-tipped above, with one black spot that is sometimes absent in spring; hind wings white above and yellowish beneath. Female, forewings as in male but with two black spots that may be smaller in spring; hind wings as in male. Wings folded when at rest, open when alert.

Very common but limited essentially to Temperate Zone. Found mostly about gardens, flower beds, and fields where its food crops are being grown. Three groups of family are recognized in eastern United States: the whites, the yellows, and the orange tips. A dozen whites in North America.

Winters as pupa or adult. In North, three generations a year; in South, to six. Eggs yellow, flask-shaped, with 12 vertical ribs, laid singly, usually on undersurfaces of food leaves, hatch in 1 week. Larva green, with fine hairs and with yellowish backband, 1 in. long, maturing in 2 to 3 weeks, then forming pupa that lasts from 10 days or over winter. Pupa green, slender, 5/8 in. long, angular.

Food primarily cabbage, and closely related mustards; cauliflower, kale, collards, kohlrabi, Brussels sprouts, radish, turnips, horseradish, and many related weeds. Caterpillars eat through leaves into heads of cabbage. Introduced into America in 1866. Affects chickens that eat it.

Control is commonly by hand picking and by spraying. Consult your county agent or Experiment Station for latest information on control.

Clouded Sulfur (Roadside) Butterfly
Colias philodice

Wingspan to 2½ in. Male, forewings yellow above and with narrow black border. Female with broader borders and usually containing yellow spots. Two kinds of females, one yellow, other white.

Common along roadsides; males in flocks about muddy spots, often closely crowded together. From St. Lawrence River south to South Carolina and west to Rocky Mountains. Closely related orange sulfur butterfly, *E. eurytheme*, is found west to Pacific, in Southwest, and east to Maine.

Eggs pale yellow changing to crimson, slender, laid singly on clover and other legumes, hatch in 4 to 5 days and develop slender green caterpillars, with longitudinal stripes, heads lighter than body, covered with fine hairs; length more than 1 in. Winters in various stages. Pupates for as short as 10 days; pupa pale green, pointed, not angular, ¾ in. long, with narrow yellow stripe.

Food: larvae feed on clovers and other legumes. Orange sulfur butterfly or alfalfa caterpillar is frequently a pest on alfalfa, particularly in West.

Not usually abundant enough to need control. Adults always interesting; thought by some persons to provide suggestions for compass direction. This hardly seems dependable; probably as fallacious as ability of daddy longlegs to locate cows.

Order Lepidoptera./Family Lycaenidae

Common Copper Butterfly
Lycaena phleaus

Wingspread to 1⅕ in. Wings commonly closed when at rest. Male, forewing copper-colored above, with black outer border; hind wing black above, with red outer border. Female like male but with more rounded hind wings.

Common through summer months, found mostly on lawns and fields where its food plant occurs. Common from Atlantic to Pacific, along northern border of United States, south to Gulf States area. Commonly a western state form.

Eggs pale green, nearly hemispherical, with large white-walled cells; many laid singly on stem or leaf of sorrel, hatch in 6 to 10 days. Larva small, stocky, with head and naked body green with a dusky line on back, darker and sometimes rosy on middle of sides, ⅗ in. long. Pupates under stones for 9 days or over winter.

Food of larva, leaves and stems of sorrel; of adult, nectar. Pupa light brown to green, with black dots, ⅖ in. long, stocky and smooth. Adults, larvae, and pupae are mostly inconspicuous when at rest, though adults are beautiful when active.

Control unnecessary or undesirable, since food plant of larva is generally considered an objectionable weed. Numbers are insignificant, anyway. Like its eastern relative, *L. p. americana*, it buzzes and appears to be attacking other butterflies. At least the recipient of its attention is driven away.

Early Azure Butterfly
Lycaenopsis argiolus
Wingspread to 1⅓ in. Male, forewings pale blue above, earlier butterflies more heavily marked with dark on whitish undersides. Female paler blue with broad dark margins. One of the first butterflies to emerge in early spring, even before snows have melted.

Common in fields, woods, and roadsides from April to August, in United States, Canada, the Old World, and Mexico. Characteristic of family is that front legs of male lack claws. Subgroup, the Blues, is represented in North America by 38 species, mostly western.

Two to three generations a year. In April, female lays many short thick green eggs, singly, in flowers of dogwood, black cohosh, or other flowers. These hatch in 4 to 8 days. Larva stocky, ⅔ in. long, naked-bodied, white, with a dusky back line and small dark brown head. Pupa stocky, ½ in. long, light brown to yellow with black marks, 10 days or over winter.

Food of larvae, flowers and fruits of its food plants. Rarely sufficiently abundant to be considered pests. Larvae feed on flowering dogwood, black snakeroot, meadowsweet, sumac, maple-leaved viburnum, New Jersey tea.

Of no economic importance.

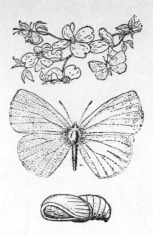

Early Azure Butterfly

Wanderer Butterfly, The Harvester
Feniseca tarquinius
Wingspan to 1¼ in. Male, forewings tawny with black markings; hind wings with rear half tawny and forehair black. Female a little more orange than male.

Common in eastern United States but rarely abundant and probably not recognized by most people. Occurs in Maritime Provinces west to Ontario, to Florida, Gulf States, and central Texas. Family is known as "gossamer-winged butterfly family." It includes the coppers, the blues, and the hairstreaks as well as the wanderer. Frequent in alder swamps from June to September.

Winters in pupal stage. In June, adults emerge, mate, and female lays many flattened spherical faintly green eggs, singly, near a colony of alder-blight plant lice. Eggs hatch in 3 to 4 days. Larvae plump, with small heads, covered with a whitish bloom; bury themselves under a web and dead bodies of aphids. Pupa short, thick, looks like monkey face; hangs on branch for 2 weeks or over winter.

Food: caterpillars feed solely on aphids in colony in which placed by female. This habit of eating other insects is unique; most other butterfly larvae feed on plants. Adults frequently take aphid honeydew from leaves.

These insects merit encouragement since they prey upon other insects which damage reasonably useful plants. However, none of them plays an important role in man's economy.

Wanderer Butterfly

Order Lepidoptera./Family Nymphalidae

Milkweed or Monarch Butterfly
Danaus plexippus
Wingspan to 4 in. Upper surfaces of wings light reddish-brown with black veins and borders and with two rows of white spots on outer borders. At rest, wings are held above back, with hind wings covering most of forewings. In male, each hind wing has a black scent gland, as shown.

Common in most of United States, in fields and along roadsides, sometimes migrating in great flocks that may settle on a single tree, weighing it down. Family includes most of our more attractive butterflies. This species is subfamily Danainae. Adults have been found far out to sea over both the Atlantic and the Pacific. Viceroy resembles the monarch but is smaller and has an extra black band on each hind wing which parallels the edge.

Eggs green, flask-shaped, with lines of pits, laid on milkweed leaves, hatch in about 4 days. Larva yellow with black cross marks and a pair of slender movable threads at each end. Molts, four, in about 10 days. Pupa pale green spotted with black and gold, free-hanging from a thread, about 12 days.

Food of larvae, leaves of milkweed. Larvae and adults reputed to be distasteful to birds. Big fall migrations to South are observed. Northern migration may be merely moving of individuals into new territory. May be 3 weeks from egg to adult.

Since food plant is milkweed, which is not normally considered of economic importance, this butterfly can hardly be considered harmful to man's interests. Its migration habits are not completely understood, but its beauty as a pupa and adult is thoroughly appreciated.

Monarch

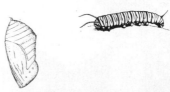

Milkweed or Monarch Butterfly

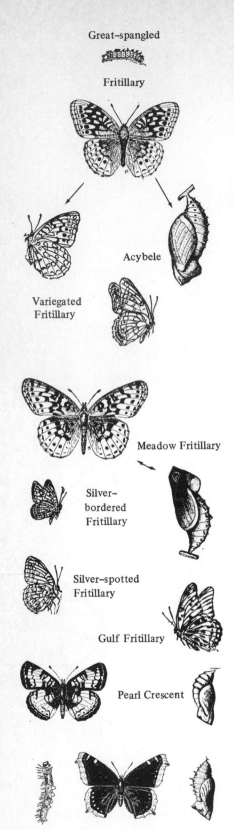

Great–spangled

Fritillary

Acybele

Variegated
Fritillary

Meadow Fritillary

Silver–
bordered
Fritillary

Silver–spotted
Fritillary

Gulf Fritillary

Pearl Crescent

Mourningcloak Butterfly, Yellow Edge

Great Spangled Fritillary
Speyeria cybele
Wingspan to 3½ in. Male, forewing orange-brown with numerous black markings, hind wing with broad yellow stripe and brown hair fringe above, and round silver spots beneath. Female like male, except that half of wings next to body on upper side is dark, while in male it is only slightly so; hair fringe lacking.

Common about fields in late summer, particularly where weeds are likely to be tall; ranges through eastern United States west to Great Plains. Name fritillary refers to dicelike spots on wings. Two commonest species are great spangled and variegated fritillaries.

Eggs short, rounded, higher than broad, ribbed, honey yellow, laid singly near food plants; hatch in about 15 days. Larva at first rough, warty, greenish-brown; winters as just-hatched caterpillar, then matures following spring, becoming black, 1⅓ in. long, with branching spines in rows. Pupa hangs for 14 to 24 days.

Food: larvae eat leaves of violets at night, spend day in hiding. Adults eat nectar from flowers. Pupa about 1 in. long, roughened, brown, coarsely wrinkled, and inconspicuous.

Since food of larvae is wild violets, not considered of economic importance, these butterflies are not classed as pests. They are always beautiful to watch as adults.

Marsh (Meadow) Fritillary
Brenthis bellona
Wingspan to 1⅘ in. Female larger than male and darker, orange-brown, with numerous black markings. No silver markings beneath such as are to be found on great spangled fritillary and on small silver-bordered fritillary, *B. myrina*.

Common in wet fields, in summer and fall, ranging throughout Canada and Northern States, west to Rocky Mountains and south into Carolinas. 17 North American species in genus, most of them arctic and limited to higher mountains. *B. monitinus* of White Mountains, N.H., is one of these.

One to three generations a year. Eggs tall, dull olive, prominently ribbed, laid singly on food plant, hatch in 5 to 9 days. Larva spiny, nearly 1 in. long, olive-brown with green markings; winters as half-grown caterpillar. Pupa about 1 week.

Food: leaves of violet, eaten at night. Adult feeds on nectar and, in East, seems to favor asters. Pupa bluish-gray with dark spots and with two rows of small cones on back. Adults gay little butterflies, common near centers of human population but near wet places.

Of little or no economic importance, apparently, since food plants are chiefly wild violets.

Order Lepidoptera./Family Nymphalidae

Pearl Crescent
Phyciodes tharos
Wingspan to 1⅔ in. Male, wings dull orange with broad black margin and spots; hind wing with small pearl-gray crescent on lower surface; in spring, marbled cream and brown; in summer, yellow with brown marginal patch. Female larger than male.

Common about lawns and gardens in late spring and summer. Ten species of crescent spots in America north of Mexico. Included in these are West Coast meadow crescent, Rocky Mountain (Canada to Texas) Camillus crescent, and eastern Bates's crescent.

Eggs light yellow-green, taper-pointed and less than twice as high as broad, laid in clusters of 25 to 200 under leaves of food plants; hatch in 5 to 10 days. Larvae scatter after hatching; have several rows of little soft spines. Head spineless, dark-mottled. Winter as young caterpillars ¾ to 1 in. long. Pupate 10 to 14 days.

Food of adults, nectar; of larvae, asters, particularly New England aster, on which plants larvae are expert hiders. Pupa hangs by tail safely in these plants, is variously colored, with gray-brown creases and a length of ⅖ in.

Of no economic importance since it preys almost wholly on plants ordinarily considered weeds. A butterfly with characteristics described in first paragraph above and habit of alighting with wings outstretched and then followed by quick up and down movements while still in place is probably a pearl crescent.

Mourningcloak Butterfly, Yellow Edge
Nymphalis antiopa
Wingspan to 3½ in. Male, forewings maroon with broad straw-yellow border, blue spots on upper surface, and almost black beneath; hind wings like forewings, but with suggested "tails." Female similar to male. A rare variety has straw-colored border reaching mid-wing.

Common throughout year, often seen in winter, and not infrequently coming into houses. Most common near woodlands from June to August, ranging from southeastern Canada through Middle States, where there are two broods a year. May be found between Arctic Circle and 30°N.

Eggs dark brown or black, barrel-shaped, laid in rings around twigs, hatch in 9 to 16 days. Larva 2 in. long; head angled; head and body, black; body spined and white-spotted, with red on middle of back; prolegs reddish; group of larvae stays together. Pupa on twigs, 8 to 16 days, dark brown with red-tipped swellings, often with whitish bloom; over 1 in. long.

Food of larvae, leaves of elm, poplar, willow, hackberry, and other trees; of adults, nectar. One of tortoise-shell butterflies, adults of which hibernate, restricted to North Temperate Zone except for colder mountains of subtropics.

Sometimes considered pests, but only rarely. May be controlled where abundant by picking off egg masses on twigs of food plants. In its range, the adults may emerge before snows are melted.

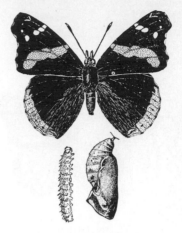

Red Admiral Butterfly

Red Admiral Butterfly
Vanessa atalanta
Wingspan to 2½ in. Male, forewings black with white markings at tip and a red crossbar; hind wings dark-brown with red border speckled with black and blue. Hind wings of red admiral and painted ladies not angular as in tortoise shells, though both have been placed in same genus.

Common over open grounds from May to winter throughout temperate North America, Europe, northern Africa, and temperate Asia. 25 species of anglewings in America north of Mexico, including red admiral, tortoise shells, painted beauty, and mourningcloak. Most species have wide range.

Eggs barrel-shaped, delicate green with fine ribs, laid singly on upper surfaces of food leaves, hatch in 5 to 16 days. Larva with spines and hairs about ½ spine length, head rounded, color variable, length about 1 in.; spines relatively not so large as on larva of mourningcloak. Pupa from 10 days on.

Food of larva, leaves of nettles, hops, and their close relatives; of adult, nectar. Pupa resembles that of mourningcloak but is slightly smaller, measuring about ¾ in. Ash-brown with dark markings and golden spots; hangs on plants.

Of no economic importance because food plants are not normally valuable. Adults are attractive and admired by naturalists.

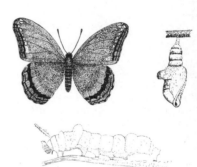

Red–spotted Purple

Order Lepidoptera./Family Nymphalidae

Red-spotted Purple
Limenitis astyanax
Wingspan to 3¼ in. Male, forewings black with border of blue dashes and orange spots beneath; hind wings like forewings except that outer half is washed with green or blue. Both sexes show orange spots on undersides of wings, general color there being brown. In some lists, given the scientific name *Limenitis arthemis astyanax*. This may be a hybrid between *L. arthemis* and *L. astyanax*.

Common all over United States and Canada west to Rockies, but more uncommon in Gulf States area; occurs in highlands of Mexico; commonest June to August. Close relatives include white admiral and viceroy.

Eggs globular, yellow-green to dark brown, many, laid singly on leaves of food plants; hatch in 7 to 9 days. Larva, body naked, humped, irregular, streaked, with a pair of rough black clubbed horns and many minute wartlike structures over back; resembles bird dropping; 1½ in. long; winters as small caterpillar in silken tube. Pupa 10 to 12 days.

Food of larva, leaves of plum, apple, pear, and gooseberry, but rarely sufficiently abundant to be considered a pest. Pupa queerly colored, yellow, brown, olive, and salmon; glossy and nearly 1 in. long. Adult not only visits flowers, but also feeds on carrion and manure.

Control not necessary because of relatively small numbers at any one time. Adults so attractive that they are prized by collectors.

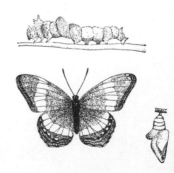

White Admiral Butterfly

White Admiral Butterfly, Banded Purple
Limenitis arthemis
Wingspan to 2¾ in. Wings in general black with a broad white band crossing them from front to rear; band may become narrow or even lacking in some individuals. Underwings as in preceding species, but spots rusty red.

Common from Alberta, Ontario, and Quebec through New England to Pennsylvania, and in higher mountains farther south; not recorded in West but replaced there by a similarly marked *B. weidemeyeri*, which is found east to western Nebraska. Closely related to red-spotted purple butterfly.

(Continued)

483

Eggs and other immature stages indistinguishable from those of red-spotted purple.

Food of adult, nectar; of larva, leaves of willow, basswood, and birch. Undoubtedly comes from same stock as red-spotted purple, and in territory where this northern species meets that southern species, individuals may be found having characters of each.

Of little economic importance, but of considerable importance to naturalists who are interested in interrelationships of species and relation between range and persistence of given characters.

Viceroy Butterfly
Limenitis archippus
Wingspan to 3½ in. Male, forewings orange-brown with white-spotted black border and conspicuously black veins; hind wings like forewings, but with extra black crossband about as wide as veins. Female like male but larger. Viceroy resembles monarch except that it is smaller, has extra band on hind wings, lacks male's scent gland, and has one less row of white spots in border of wings.

Common, particularly in open fields during August to September, from Canada to Gulf of Mexico. Close relative of red-spotted purple and white admiral butterflies.

Eggs deep green, globular, pitted, laid singly on upper or under leaf surfaces of food plants, usually at tip of leaf; hatch in 4 to 8 days. Larva resembles bird dropping, irregular, warty, with a pair of long clubbed prickled tubercles near front, black, over 1 in. long; winters as caterpillar. Pupa hangs on a tree, 7 to 10 days.

Food of larva, leaves of willows, poplars, and aspens; of adult, nectar. Pupa resembles that of red-spotted purple, being glossy and queerly colored yellow, brown, olive, and salmon.

Of little or no economic importance. Interesting because of mimicry of monarch butterfly; latter reputed to be avoided by enemies because of unpleasant scent, and viceroy apparently benefits from this immunity of monarch.

Viceroy Butterfly

Order Diptera [Flies]

Adults have one pair of wings, or none, hind pair of wings being represented by a pair of balancing organs, which is always present; mouthparts sucking, lapping, or piercing, larvae wormlike, without jointed thoracic legs and usually legless; body divisions distinct, metamorphosis complete.

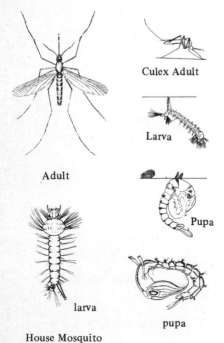

Culex Adult

Larva

Adult

Pupa

larva

pupa

House Mosquito

Family Culicidae. Mosquitoes
These small flies have narrow wings, with a fringe of scalelike hairs on hind margin of wings and on some veins. Antennae of males bushy, of females less so. Proboscis usually long and firm. Abdomen usually long and slender.

House Mosquito
Culex pipiens
Female ⅙ in. long with wings slightly shorter; proboscis slender, brown, dark at tip; abdomen black with bluish to bronze reflection, held parallel to support when at rest; legs long, slender, brown-scaled, hind pair curved upward when at rest.

Found everywhere in eastern North America and on Pacific Coast, common in other parts of world with similar climate; with closely related species extending range to all parts of world. Commonest, of course, near large areas of stagnant water.

Eggs laid in rafts of 100 to 300, floating on water, hatch in 1 to 5 days. Larva a wriggler that in 1 to 2 weeks becomes a pupa; larva at rest hangs at an angle from water surface, breathing through rear. Pupa active, more compact than larva, more erect than that of *Anopheles,* lasts for a few days. Winters as adult.

Food of larva, minute animals and plants in water; of female adult, blood of animals; of male adult, plant juices. A troublesome pest. Can carry filariasis. Active day or night but feeds mostly during evening and early morning. Related mosquitoes may go long distance from breeding area.

Control essentially by draining or using larvicides in breeding areas. Screening, spraying possible hiding places with acceptable insecticides, and swatting represent other means. Excellent repellents are now available.

Yellow-fever Mosquito
Aedes aegypti

Female, length ⅙ in., wings slightly shorter; proboscis slender and black; abdomen black, each segment except last with basal row of white scales and a small silvery spot on each side; top of thorax with four narrow silvery-white lines, outer pair curved; legs black with white bands; wing scales brown.

Found in tropical and subtropical regions, most commonly about human habitations and breeding normally in artificial containers, rarely in natural water holes. Rarely far from human habitation. In summer, range may be extended northward.

Eggs laid singly, on or close to water, hatch in 10 hours to 3 days. Larva hangs downward, growing for week or 10 days before pupating. Pupal stage lasts 2 to 3 days. If adult has fed on blood of yellow-fever patient during first 4 days of his illness, it can transmit disease to others after about 12 days and continue doing this as long as it lives.

Adults fly by day, bite fiercely but attack quietly, though sharp high note can be heard; attack from behind or beneath, crawling into clothing to reach suitable spot. Principal carriers of yellow fever; also dengue, filariasis, and other tropical diseases.

Control is by destruction of breeding places, screening, and spraying indoors to leave a residual film.

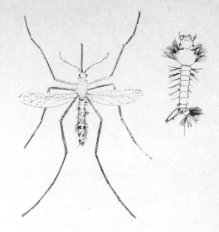

Yellow-fever Mosquito

Common Malaria Mosquito
Anopheles quadrimaculatus

Female to ⅕ in. long; antennae almost black; abdomen blackish with pale and black hairs; black scales on wings form four distinct spots; no coppery spot at wing tip; legs slender and black; abdomen held at an angle of 45° when at rest, hind legs extended in same direction.

Found from Mexico north through Mississippi Valley into Canada and east to Atlantic, but becoming rare in northern part of range. Other related species, however, are fully as dangerous and extend range of the genus.

Eggs laid singly, on surface of water, hatch 2 to 3 days later. Larva resembles house mosquito but lacks breathing tube at rear end and hence rests parallel to water surface rather than at an angle; period lasts about 2 weeks. Pupal stage 2 to 3 days, active. New generation every 3 weeks. Winters as adult.

Female feeds on blood during twilight and early morning. Malaria of different types caused by protozoan *Plasmodium* of different species. As plasmodia escape from red blood cells, patient has chills, followed by fever; at intervals of 2 days, in tertian type (commonest in North America); of 3 days, in quartan type; at variable times, in aestivo-autumnal type.

Control, as in other species. DDT is an effective control, but environmental hazards associated with its use should be explored before adopting it as a control measure.

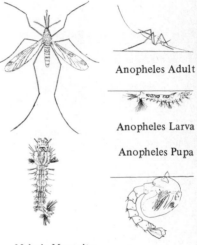

Anopheles Adult

Anopheles Larva

Anopheles Pupa

Malaria Mosquito

Order Diptera./Family Tipulidae

Crane Fly
Tipula sp.

Crane flies are commonly suggestive of gigantic long-legged mosquitoes, with long slender bodies, threadlike antennae, and a V-shaped groove on top of thorax. Some are 2 in. long while others are diminutive. Fly slowly and walk and fly rather awkwardly.

Found usually most abundantly among tall vegetation, in damp areas where there is relatively little wind, or not uncommonly in houses where they have been attracted by lights. Several thousand crane flies, of which at least 500 are American. *T. simplex*, in larval maggot form, called leatherjackets, is destructive of range grasses in the West.

One species lays about 1,000 eggs in wet soil. These hatch in about 1 week into black worms or leatherbacks that develop in a few weeks into pupae; these live about 1 week before transforming into adults. Some larvae are found on land, some in rotten wood, some in plants, some in water or gravel. Some are 1 in. or more long.

Food of some adults, nectar; of others, nothing; of some larvae, animal matter; of others, plant tissue; and of others, decaying matter. Some larvae are nocturnal. Adults commonly dance in clouds, frequently swarming about lights.

Larvae of some species make excellent fish bait. Some destroy meadow plants and may injure pastures seriously; controlled by poison mash of 1 lb of Paris green to 25 lb of bran, with water.

Crane Fly

Order Diptera./Family Ceratopogonidae

Punkies, No-see-ums
Culicoides guttipennis

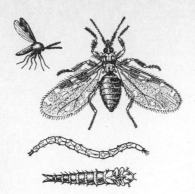

Males with feathery antennae; under $\frac{1}{10}$ in. in length for the group and under $\frac{1}{20}$ in. for this species. Black with brown legs and with areas of white, yellow, black, and brown. Wings gray and brown, marked as indicated in figure. A two-winged midge.

Six of twenty known species in United States are biters of warm-blooded animals. Occurs in great flocks that settle down on man or cattle and begin biting viciously; unfortunately most abundant where scenery is commonly most beautiful, as in northern lakes and woods country.

Several hundred eggs laid in strings in water; hatch in 3 to 6 days into eel-like larvae that swim like snakes and feed on minute aquatic animals. Larva may live in ponds, pools, or even in a few drops of water in a stump. Pupa floats in vertical position in water for 3 to 6 days before transforming to adult stage.

One of most unpleasant pests to be found in normal vacation period of those who like to go to the wilds for a change. Usually come after black flies and mosquitoes have had their season. Fortunately, not known to be disease carriers. Bite most persistently in the evening and very early morning.

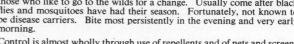

Punkies, No-see-ums

Control is almost wholly through use of repellents and of nets and screens of such fine mesh that insects can be kept out. They go through ordinary mosquito netting with ease.

Order Diptera./Class Chironomidae

Midges
Chironomus sp.

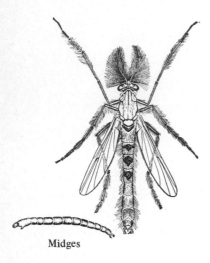

Adults like small mosquitoes but with more delicate abdomens, legs, and antennae. Wings generally lack scaly fringes on edges and veins that are characteristic of mosquitoes.

Commonly associated with water or vicinity of water, where they are seen dancing in great flocks. Over 200 kinds of midges in North America, some found in tidal pools as well as in fresh water. One chironomid larva lives in hot springs at 124°F.

One species lays a mass of about 700 brown eggs in water. These hatch in 3 to 4 days into larvae that require 1 to 2 months or more to reach pupal stage. Pupae swim like wrigglers near surface for a few days before transforming into adult stage. Chironomid larvae are "bloodworms," important as fish food. *Tubifex* sp., an annelid, is also called a "bloodworm" and is a good indicator of organic pollution in waterways.

Midges

Most midges are harmless but a few, like punkies of genus *Culicoides,* are most troublesome along streams and shores at certain times of year. Many midges may swarm at a time, giving a humming sound heard for some distance.

Aside from their role as pests (probably controlled like mosquitoes), vast majority of midges serve a most useful function in providing food for a great variety of young fish. In fact, they form major diet of some species important to man.

Order Diptera./Family Cecidomyiidae

Hessian Fly
Mayetiola destructor

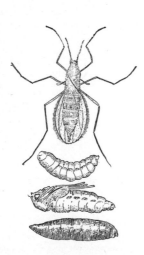

Length about $\frac{1}{10}$ in. Dark-colored, long-legged, rather large-winged for a gnat, and plumper than most. Since they are only about $\frac{1}{2}$ as long as a mosquito, they are not conspicuous. Most commonly recognized in flaxseed-like pupal stage.

Commonly found throughout Middle West and other parts of country; now established on the Pacific Coast. Now being established through most of United States and agricultural Canada. Introduced from Europe probably about 1779, when it was recognized on Long Island.

Eggs minute, laid in irregular rows of about $\frac{1}{2}$ dozen, each fly laying to 150 eggs on plant leaf; hatch in few days into pink or reddish maggots that eat way to stem joint, feeding about 1 month, killing plant above that point. Pupa flaxseed-like; this is the wintering stage in stubble. One to four generations a year.

Hessian Fly

Food of larvae, wheat, barley, rye, and some other grasses but never oats. A close relative, the wheat midge *Sitodiplosis mosellana,* introduced from Europe in 1819 near Quebec, attacks wheat, barley, rye, and oats. Only one generation a year. Hessian fly may destroy 50 percent of a wheat crop in bad years.

Probably, one of most injurious insect pests, doing millions of dollars of damage to wheat each year. Control is by encouraging numerous parasites, rotating crops to force egg laying elsewhere, destroying infected stubble by burning or plowing, planting so that plant cycle does not match insect cycle. Consult county agent or Experiment Station for latest information on control.

Order Diptera./Family Simuliidae

Black Fly
Simulium sp.

Adults small, short-bodied, humpbacked, short-legged; many black with broad iridescent naked wings. Males with eyes continuous, females with eyes widely separated. Some related species such as *Prosimulium hirtipes,* which appears late in season, have white on feet.

In season, adults are found in enormous numbers in wooded areas near streams, and are principal pest of Northeast in early summer. Buffalo gnat of Mississippi Valley, turkey gnat, and white-stockinged black fly are others.

Mate in air in early summer. Female lays eggs on stones in rapid water. Eggs at first white or yellowish, then black. Larva remains attached to rock in swift stream, eating food brought to it by current, often forming black mats in swift water; finally builds small cocoon in which pupal stage of about 5 days is spent.

Food of most species, blood of mammals. A few species do not bite, but others make up for this. Frequently bites are not painful until later. Can locate food in night as well as in day and will feed almost any time during season.

Some are worst pests of territory in which they live. Most supply basic food for some species of fish. Some may be seriously annoying to cattle and other domestic animals as well as to man. Controlled by repellents, screening, and spraying with insecticides.

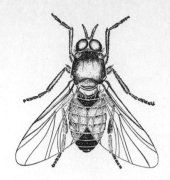

Pupa

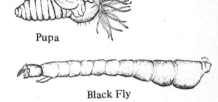

Black Fly

Order Diptera./Family Tabanidae

Deer Fly
Chrysops vittatus

Length about ⅜ in. General outline triangular. Eyes metallic, brightly golden, greenish, or coppery; wings with smoky bands; generally rather beautiful flies in every way except behavior. Marvelous fliers. Eyes of males touch each other, while those of females are slightly separated. At rest, wingspread to ½ in.

Near woodlands or waterways during warm weather. At least 63 species of deer flies or ear flies, as they are known, in North America. Closely related to horseflies in same family. Commonly buzz around heads and ears of horses and man.

Glistening black eggs are laid on water plants, just above water line or on exposed stones at water's edge, forming conspicuous patches. Larvae wormlike, aquatic, drop into water, become ½ in. long, feed on aquatic animals, and form pupae that are not enclosed in larval skin.

Food of larvae, minute aquatic animals; of adults, blood of man, horses, and some other mammals. Variety of hosts makes it possible for infected blood of one to be transferred to uninfected animal. Known to transmit blood parasite *Filaria diurna,* cause of filariasis, and *Bacterium tularense,* cause of tularemia.

Control is by draining possible breeding places. Abundance may greatly affect value of various areas for residential purposes. Ordinarily, little danger of infection from their bites, but they may be serious on occasion.

Deer Fly

Order Diptera./Family Tabanidae

Horsefly
Tabanus atratus

Female, length to 1 in., black with blue-white bloom on back, wings smoky, eyes separated, outer wing margins not parallel. Male similar to female in general, but eyes touch each other and are proportionately much larger. Fast flier.

In woods or open, frequently annoying swimmers and cattle. Not gregarious as are some flies. Some 200 closely related species found in North America, including gadflies, green-headed monsters, and others.

Eggs laid on marsh plants, hatch in 1 to 2 weeks when exposed to sun. Larva, greenish-white "worm" which in some species may be 2 in. long, lives in water or in moist soil. Usually winters in soil and in spring forms yellow-brown pupa 1¼ in. long, adults emerging with advent of warm weather.

Food of larva and adult, other animals, adult female feeding on blood of warmblooded animals. Bite is painful and is effected by a series of cutting and piercing stylets that can penetrate toughest hide. Known to play role in transmission of anthrax and lockjaw and the trypanosome that causes surra.

May be more important in spread of disease than is normally recognized. Certainly, while they may be beautiful to look at, they are not desirable companions at any time. Fortunately not abundant for any considerable portion of year and are confined to rather definite areas.

Horsefly

Order Diptera./Family Asilidae

Robber Fly
Tolomerus notatus

Robber Fly

Length of some species, 2 in., of others, ⅕ in. Wings usually, though not always, held at right angles to body. Eyes conspicuously large. Legs large. Abdomen slender. Some kinds resemble bumblebee rather closely. These are usually stouter and more hairy, but of course they have only two wings, like other diptera.

Often found perched on an exposed stump or twig, eating prey or ready to pounce on it. Individuals are solitary; frequently in open sunny fields. More than 800 kinds of robber flies known in North America. Family is one of largest in order diptera.

Eggs usually laid on or under ground or in damp wood; hatch into legless wormlike larvae that feed upon other insects found in rotting wood, in loose soil, under or in leaves that are rotting, or under bark. Pupal stage usually spent in ground, without any special covering.

Food of larvae and of adults, other insects found in environment in which they live. Adults may be seen flying with a huge insect held firmly, or may be watched while finishing a meal. Robber flies relatively unsuspicious and always interesting to watch. Never attack man or domestic animals.

Economic importance varies since robber flies are not selective in insects they destroy and may kill useful as well as destructive species. Always worth watching.

Order Diptera./Family Syrphidae

Flower Fly
Syrphus sp.

Flower Fly

Prettily colored small flies, with yellow-banded or black abdomens; some resemble honeybees, while others are more like wasps, being bare and flying awkwardly like wasps. Most are excellent fliers and some give a droning or buzzing sound when in flight.

Found about fruits and flowers, particularly those with aphids on them, some hiding in flowers or feeding inside them. Some 930 kinds of syrphus flies in America north of Mexico, including rat-tailed maggots; closely related to drone flies.

Eggs of some species laid among plant lice; when eggs hatch, emerging larvae attack and kill aphids. Larvae sluglike, about ½ in. long, often with white stripes, commonly greenish, eyeless, flat, transversely wrinkled, and pointed in front.

Food of adults, usually pollen. Five types of larvae represented in group, based on food: flat green type feeds on plant lice; cylindrical type bores into plant bulbs; those with long and short tubes feed in filth; short hemispherical type feeds on insects in ants' nests.

None of these insects stings man; however, larvae drunk in foul water may live in man, causing trouble; myiasis in man may be traced to syrphus flies. Most are useful destroyers of insects harmful to man's interests.

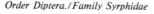

Order Diptera./Family Syrphidae

Drone Fly, Rat-tailed Maggot
Tubifera tenax

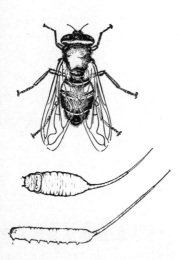

Drone Fly   Rat-tailed Maggot

Adults resemble honeybees, being yellow and black and about size of honeybees. Larva appropriately called "rat-tailed maggot," being provided with a long slender tube, which can be varied in length.

Found about dead carcasses, about privies, in foul areas generally, and probably responsible for belief, expressed by Ovid, Virgil, and Solomon, that bees originate from dead animals. Adults may enter houses in fall. Also common about flowers.

Eggs oblong, white, sticky, laid in groups of 20 or more, in crevices near foul water; hatch into larvae that live in oozes at bottom of quiet water, finally coming out to drier soil to form pupal stage that lasts for 8 to 10 days before adult finally emerges.

Food, decaying plant and animal matter. Tail lengthens or shortens, permitting larva to remain submerged in filth while it breathes relatively clean water above. This species was originally European but is now generally established.

Larvae may serve as food for some organisms and may act as scavengers, assisting in destruction of waste organic material.

Order Diptera./Family Oestridae

Bomb or Warble Fly, Northern Cattle Grub

Hypoderma bovis

Length about ½ in. Body well covered with short hairy scales, relatively large. Abdomen with a dark band, bound in front with a broad light band and to rear with a narrower light band. Active flier and most annoying to cattle.

Found in sunny areas in summer. Two species, both natives of Europe: bomb fly or northern cattle grub, *H. bovis,* common in Canada and the Northern States, and, heel fly or common cattle grub, *H. lineatum,* more widely distributed. Rarely attacks cattle standing in shade of a tree or shed.

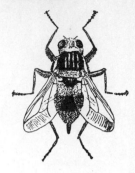

Eggs spindlelike, ⅕ in. long, with a groove, laid singly, at base of cow's hair, in June to July; one female may lay 500 eggs. In 5 to 7 days larva hatches and enters host animal. The larvae of the bomb fly are rarely found in the gullet of the cow. They probably take a more direct route to the back of the cow than the larvae of the heel fly. The bomb fly larvae are frequently found in the neural canal of the host. Bomb fly larvae spend an average of 73 days as a warble under the skin of the back of the host, finally coming out to pupate 14 to 34 days in ground.

Larva eats its way through thickest hide, spoiling skin for sale purposes and of course interfering with host's health, growth, and production of milk. Some animals may be injured in trying to escape the fly, which often drives cattle frantic.

Control is by squeezing out grubs before they have time to pupate. Use a salve of 1 part iodoform to 5 parts of petroleum jelly. Infection may be checked by removing eggs from hairs before they hatch. Rotenone applied to region of back of dairy cattle where grubs occur was for years a standard remedy. No one very effective remedy known.

Bomb or Warble Fly

Order Diptera./Family Tachinidae

Tachina Fly

Bombyliopsis abrupta

Vary greatly in size, some being ½ in. or more long while others are diminutive. Usually short, stout, covered with stiff hairs, and, in general, shaped like a housefly, with which some might be easily confused except for stouter abdomen and numerous noticeable stiff bristles.

Adult found usually among vegetation where prospective victims may be abundant, on flowers, leaves, stems, or other parts. Some 1,400 species of tachinid flies known from North America alone, including some of most useful of all insects.

Eggs or, in some cases, living larvae are laid on bodies of caterpillars and other insects, sometimes in great numbers. Maggots bore into host, feeding on its tissue and finally killing it. Then maggots emerge, pupate, and finally develop into adults that mate and repeat cycle.

Food of larvae, bodies of insects in which they find themselves, methods of entering host being many. In some, host eats egg; in others, as in *Eupeleteria* that attacks brown-tail moth caterpillar, maggot rests on thread that caterpillar follows to its nest.

Tachina Fly

Many investigators who have had little success in getting moths to emerge from cocoons have failed because of activity of these flies. Without them many agricultural practices would be infinitely more difficult.

Order Diptera./Family Sarcophagidae

Flesh Fly

Sarcophaga sp.

Sarcophaga haemorrhoidalis, just over ½ in. long. Gray with darker spots making a somewhat checkerboardlike design on top of abdomen; series of about five discontinuous streaks on upper part of thorax. Hairy scales sparingly present on body. Some species silvery or changeable in color. Eyes not hairy.

Found in a variety of situations, but not so common in houses as are houseflies. Many are seen laying their eggs on flesh, carrion, dung, other insects, and other kinds of organic material. Family is large with wide distribution and considerable economic importance.

Probably all these flies deposit living larvae, rather than eggs, on substances that larvae eat. As many as 20,000 eggs have been found in a single female, indicating tremendous reproductive rate. Pupation may take place after larva has left its source of food supply.

Food of larvae and adults, usually filth. In addition to those suggested above, food supplies include rotting vegetable material, stomachs of human beings, wounds in human beings, stomachs of frogs, skin of turtles, living snails, and eyelids of humans.

Flesh Fly

(Continued)

Fundamentally, these flies are scavengers suited for destroying decaying organic material. They deposit filth, they eat anywhere, and have been known to cause intestinal and skin myiasis in man and to aggravate wounds and tumors that were in process of healing. Many larvae of family are parasitic upon such insects as grasshoppers, caterpillars, and beetles.

Screwworm Fly

Order Diptera./Family Calliphoridae

Screwworm Fly
Cochliomyia hominivorax
Length to ½ in., though some are only ¼ in. long. Slate-black, with three distinct darker longitudinal stripes on thorax. Abdomen glistens like metal. Eyes rusty brown. In general appearance, resembles a large metallic housefly.

Widely distributed, but most common in South and Southwest. Abundance spotty in North and East, species being absent in some states. Adults fly about sores, wounds, or any parts of animal from which foul odors or discharges may come, including the nose, ears, mouth, and eyes.

Eggs most commonly laid in nostrils of domestic animals, but any wound or decaying matter will suffice. Larvae hatch in a few hours or days and burrow into tissue, mature in 4 to 5 days in living animal or longer in dead animal, drop to ground and pupate. Pupal stage lasts from 3 days to 2 weeks and cycle can be completed in from 1 to 4 weeks.

Food essentially flesh, living or dead. Favored points of attack include, in addition to those suggested above, navels of newborn calves, branding sores, and cuts by barbed wire. In some cases larvae make their way to cavities of head, causing death. Larvae about ¾ in. long, with bristles making a screw effect.

Infection by these flies may be fatal. Also a potent source of myiasis. Persons sleeping in open may receive infections through nose, for which surgical treatment may be necessary. Douches of chloroform in milk, 10 percent have been used for nose treatment. Control measures include provision for dehorning, castrating, branding, earmarking, docking of lambs' tails during seasons other than spring or summer. Chemical insecticides are available which can be used to control this fly.

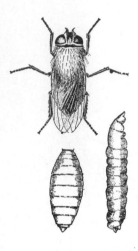

Cluster or Buckwheat Fly

Cluster or Buckwheat Fly
Pollenia rudis
Slightly larger than a housefly, but much more slowly moving and with wings held more closely together; behaves sluggishly; thorax dark-colored; woolly with hairs. Abdomen brown with pale spots; space between eyes, white. Greasy when mashed, and with disagreeable odor.

Found most commonly on windows in fall, and in spring when they are trying to get out, at which times they may cluster together as their name suggests. Introduced from Europe at an early and unknown date. Sometimes found on snow, on soil, or on the outside walls of buildings in sun.

Adults mate about February. Eggs laid on garden soil a month later, as many as 97 in an evening; hatch in 4 to 6 days into white legless maggots that live in earthworms as parasites about 3 weeks. Pupate in soil for 2 to 6 weeks and winter as adults. Found in large numbers as adults only when seeking or leaving hibernation.

Food of larvae, earthworms; of 107 earthworms examined, 74 supported total of 87 cluster fly maggots. Above temperature of 50°F, adults go toward light and avoid contacts; below that temperature they avoid light and push against contact. This explains much of their behavior.

May be annoying to housekeepers; are enemies of earthworms; serve as food for many insect eaters. Closely related to bluebottle flies and screwworm flies.

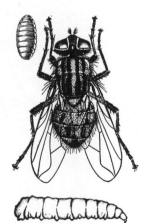

Housefly

Order Diptera./Family Muscidae

Housefly
Musca domestica
Length about ⅜ in. Body blackish. Eyes reddish-brown. Wings transparent. Mouth suited for sucking. Male with eyes nearer together than in female; abdomen with sides brownish rather than gray all over. Active and almost too well known to need description. Thoroughly domesticated and cosmopolitan. In dwellings, it represents 98 percent or more of flies collected.

Found almost everywhere that man exists, being most abundant in summer or warmer months and in regions where manure or decaying plant or animal matter is to be found.

Female lays to 600 eggs, in clusters up to 125 at a time, from early summer through to frost. Eggs hatch in 8 to 12 hours in warm weather, or to 3 days when cool. Larva, whitish maggot that reaches full development in about 5 days. Pupa in old larval skin, either where larva developed or elsewhere; about 5 days' duration.

Food of larva, almost any decaying plant or animal matter. Adult transfers filth freely from place to place by emptying stomach for new meal. Since adults do not lay eggs until about 2 weeks after emerging, a life cycle is about 24 days. May be 10 generations a year.

One female in April might have 5½ trillion descendants by September. Known to carry typhoid, probably diarrhea, dysentery, cholera, and possibly tuberculosis. Control is by screening, use of traps, and destruction of breeding places, as well as consistent swatting programs. Presence of houseflies demonstrates our failure to dispose properly of human and animal wastes. Therefore, proper environmental sanitation is prerequisite to control of this pest. Diazinon, chlordane, lindane, and dieldrin are effective larvicides. Also, one should use paradichlorobenzene in garbage containers. Of course there are other chemical insecticides effective against adults.

Biting Housefly or Stable Fly
Stomoxys calcitrans

Length ¼ to ½ in. Gray with darker mottlings. Eyes reddish, and those of male separated by a distance ¼ diameter of head, while those of female are separated by ⅓ that measurement. Legs black. Wings relatively clear. Mouth distinctly suited for piercing instead of for lapping, as in housefly.

Common almost anywhere outdoors in warm weather, or indoors when weather is muggy. Widely distributed over the world and recognized by most as biting housefly, although it is not a housefly. Reported that these flies captured near woodlands have brownish wings. Commonly attacks horses and mules, biting these animals especially on the legs. May bite people about the ankles in their homes.

Eggs about ¹⁄₂₅ in. long, placed on any moist substance, have a distinct lateral furrow, hatch in 1 to 3 days. Larvae much like those of housefly, living on a great variety of decaying substances, even decaying lawn grass; mature in 11 to 30 or more days. Pupate in old larval skin, 6 to 20 or more days. Life cycle, 19 days to 3 months or even more.

Food of larvae, decaying organic material but favor decaying straw in manure. Probably winter is passed in larval or pupal stage, as is believed to be case with housefly.

A most annoying fly on warm humid days. Control is like that of housefly centered on screening, on traps, and on destruction of breeding places. Quicklime spread on wastes is probably most effective way to get rid of breeding areas. Spraying with Lethane or Thanite or other fly repellents is also effective. Legs, belly, and sides of animals should be sprayed once or twice a week.

Stable Fly

Order Diptera./Family Gasterophilidae

Horse Botfly
Gastrophilus intestinalis

Length about ¾ in. In general, resembles honeybee, but female has a long abdomen bent back under body. Wings with dark spots, those in center forming an irregular crossband; otherwise wings are transparent.

Seen commonly buzzing about heads of horses, donkeys, or mules, causing animals great excitement. Found practically everywhere in world where horses and their kin are found. Close relatives include the nose fly, with its red tail, and smaller chin or throat fly.

Female may lay to 500 yellowish eggs, usually on hair on foreknees of horses. These are licked off and hatch when moist and pressed. Larvae find way into horse's stomach, where they live 8 to 10 months, molting twice, and sapping vitality of horse, as they are attached to stomach lining. When mature, they pass out in manure, enter ground, pupate 30 to 45 days.

Food taken from host animal, anywhere from stomach through intestines to rectum, often causing severe irritation and sometimes serious illness. Other closely related species live somewhat similarly in deer, squirrels, rabbits, and other mammals.

Control essentially by spraying regions where eggs may be laid, namely, forelegs of horse; for added information consult your County Cooperative Extension Service.

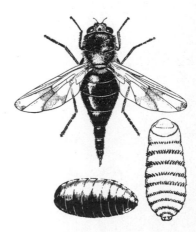

Horse Botfly

Order Diptera./Family Drosophilidae

Fruit or Vinegar Fly
Drosophila melanogaster
Length about ³⁄₁₆ in. Wingspan about ³⁄₈ in. Male slightly smaller than female and with hind part of abdomen more strongly colored; male also has "sex combs" on front legs, though their function is not known.

Found about fruit or decaying plant tissue in great numbers and widely distributed. They are called pomace flies, sour flies, and vinegar flies, names that associate them with spoiling fruit juices.

At room temperatures, eggs are laid on such fruits as grapes or bananas or on vinegar or beer; hatch in 2 days. Larvae are diminutive maggots that mature in 3 to 4 days and develop into small oval pupae which last 4 to 5 days in fruit. Under ideal conditions, complete life cycle may be passed in 8 days.

Food suggested by notes above. Essential role of insect is that of scavenger. Strongly sensitive to light and will fly toward it from a darkened chamber. Reproductive powers enormous; in a few weeks of suitable weather numbers may become tremendous.

Major use to man is in study of heredity. Their superficial characters and variations are transmitted readily to succeeding generations, and study of many generations is possible in a relatively short time.

Fruit or Vinegar Fly

Order Diptera./Family Agromyzidae

Serpentine Leaf Miner
Liriomyza brassicae
Diminutive flies with large heads; wings about equal to length from tip of head to tip of abdomen. Body sparingly hairy. Abdomen with light markings in bands between segments. Presence most easily indicated by wormlike boring in leaves, though blotch mines also may be made.

Mines of this species found in 55 different kinds of plants. Probably most widely distributed and commonest of approximately 100 North American species known in this family, not including Mexican species. Group is closely related to botflies.

Female inserts eggs under skin of leaves. Maggots eat winding or blotch burrows within leaf, mines constantly growing wider. Full-grown maggot retreats from end of burrow about ¼ in., cuts a slit in leaf, drops to ground and forms pupa, or pupa may be formed in leaf itself.

Adults lay eggs in early spring, may rest through winter, though in some cases there are a number of generations a year. Maggots shed skin twice in mine. Rarely more than one mine to a leaf. Some closely related species make plant galls. In South may remain active year round.

Not a serious pest to plants in which larvae live. Control is not easy because larvae are protected within leaves. Picking infected leaves may be worth while if plants need to be perfect for market purposes.

Serpentine Leaf Miner

Order Diptera./Family Hippoboscidae

Sheep Tick or Ked
Melophagus ovinus
Length about ¼ in. Wingless, with six conspicuous legs and an enlarged abdomen that bears four dark spots joined in 2s on back. Reddish or gray. Whole body covered with long bristly hairlike scales. A true fly.

Found commonly on sheep wherever they are raised. Not related to cattle ticks, which are related to spiders. Others "ticks" that are actually flies are bat ticks that live on bats and bee lice parasitic on honeybees. However, these belong to different families.

Female does not lay eggs but retains them in her body and nourishes them until they hatch into larvae. When larva is nearly completely developed, it is placed on sheep, covered with a white membrane that soon becomes brown and hard. Within this, pupa is formed. Adult emerges in 19 to 24 days. 10 to 14 days later, this adult may produce young.

Female may produce a larva every 12 to 15 days and in lifetime may produce 12 to 15. Adult lives in fleece of sheep except when it is feeding. It then moves to skin, pierces it, and sucks blood and lymph. Irritation caused by many feeding ticks may cause sheep to lose weight.

Control is commonly effected by dipping in mixtures containing coal tar, creosote, lime-sulfur, arsenic, or cresol. Two dippings are usually necessary, about 1 day apart. Sheep sheared in spring are ordinarily dipped in July and August. Consult county agent or Experiment Station for latest control measures.

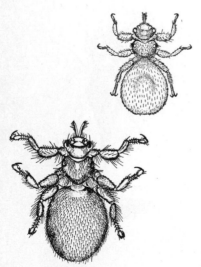

Sheep Tick

Order Diptera./Family Anthomyidae

Beet Leaf Miner
Pegomya betae

Length about ¼ in. General color gray. Sparsely covered with rather long stiff black hairs. Body commonly held in a curved position. Presence of species is most easily recognized by blotch it mines in leaves of beets and spinach.

Very common in regions where sugar beets, spinach, and chard are raised, and widely distributed through world. Leaf mines may be either blotches or twisting paths. *P. hyoscyami*, spinach leaf miner, attacks spinach, beet, sugar beet, chard, mango, and many weeds including chickweed, lamb's-quarters, and nightshade.

Eggs white, placed by female on lower surface of leaf, in masses of 2 to 5, hatch in 3 to 4 days; white to transparent maggots enter leaf and make mine between upper and lower surfaces. Larval stage lasts 7 to 9 days and is followed by pupal stage, which lasts 10 to 20 days, either in leaf or on ground.

Maggots may injure marketability of leaf crops and production of root crops. Since maggots live on inside of leaves, they are not affected by usual insecticides, which ordinarily reach only surface of leaves.

Control is by picking infected leaves in small gardens and destroying them, to prevent multiplication of pest. Deep plowing and thorough harrowing at end of year may destroy resting insects in soil. Spinach is sometimes planted as a trap crop in sugar beet fields, since flies seem to favor spinach. Spraying or dusting with parathion or diazinon may be effective control measures. Consult county agent or Experiment Station for latest controls recommended.

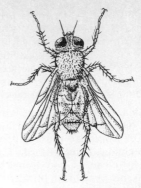

Beet Leaf Miner

Order Diptera./Family Trypetidae

Apple Maggot
Rhagoletis pomonella

Length about ¼ in., wingspan under ½ in. At rest, fly appears like an equilateral triangle; wings with four connected dark crossbands. Abdomen dark, with four light narrow bands across rear portion.

A native fly, most abundant and troublesome in eastern United States and Canada where apples may be an important crop. Adults are found from July to September. Family is large and includes many gall makers; most adults have marked wings.

Female has sharp ovipositor with which she places to 400 eggs under skin of apples. Eggs hatch in 2 to 6 days. White, ¼-in. maggots develop for 2 weeks to reach maturity, then leave apple to form pupae that look like leathery grains of wheat. Pupa usually underground for winter.

Food of larvae, pulp of apple, which of course is ruined for market. Wound causes brown areas to form on skin of apple. Adults feed on fruit juices without themselves causing much direct injury.

Control commonly by spraying with 2 gal of cheap molasses and 6 lb of lead arsenate to 100 gal of water in early July, since flies do not lay until 4 weeks after they emerge from pupal stage. An effective control is to spray with lead arsenate at time adults make their appearance in midsummer. Consult county agent or Experiment Station for latest information on control.

Apple Maggot

Order Diptera./Family Psilidae

Carrot Rust Fly
Psila rosae

Length about ⅙ to ⅛ in.; wingspan ⅓ in. Dark green to black, sparingly covered with yellow hairs. Head and legs pale yellow. Eyes black. Presence is most conspicuously indicated by rusty injury to crops it attacks. Found May to September as adults.

Common in fields and rather widely established. Originally came from Europe where it was a serious pest in Germany and England. Appeared in Canada in 1885, in New York in 1901, and has since spread rather widely.

Eggs long, white, about ⅟₃₀ in. long, laid on base of plant or in soil in May or in fall; hatch in about 1 week. Larva a tiny straw-colored maggot, ⅓ in. long, which leaves a telltale rust-colored burrow behind it. Larva eats way through leaves and roots and pupa is formed in soil. One to two generations a year.

Food of larva, leaves and roots of carrots, celery, parsnips, and similar plants. It eats the heart of celery, ruining the plant for market purposes. Borings in root crops cannot be concealed easily and affect market value.

Control largely by rotating crops so that food species will not be available year after year to continue numbers. Emulsion of 1 part of kerosene to 10 of water sprinkled along rows of young carrots has had some effect in keeping numbers down. This insect has generally been controlled by treating the soil before planting carrot seeds with either aldrin or chlordane. Consult county agent or Experiment Station for latest information on control.

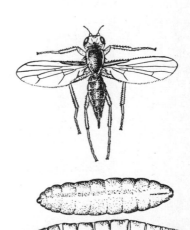

Carrot Rust Fly

Order Siphonaptera [Fleas]

Insects of this order have strongly compressed bodies and are without wings or compound eyes. They have well-developed legs and mouths suited for sucking. Over 1,000 kinds of fleas recognized, some of which are of great importance. They are divided into five families, of which the Pulicidae and the Echidnophagidae are the more important. In latter family, the chigoe or jigger, *Tunga penetrans* (not the chigger), is tropical and forms sores under nails of people who go barefoot in infected sand; sticktight flea, *Echidnophaga gallinacea,* is a serious pest of poultry in Southwest.

Fleas are controlled largely by preventing multiplication by killing their nonhuman hosts such as rats, by fumigating infected houses, by isolating sick persons who might infect fleas and continue the cycle. Mice also must be destroyed. Cats and dogs should be washed with insecticides and their beds changed and washed. Cats and dogs should have definite sleeping and resting places that can be kept clean. Good disinfectants include 3 percent creolin, carbolic soap, powdered naphthalene, kerosene, pyrethrum powder. Pupae cannot survive sudden jars, so beating of infected bedding and rugs is effective. There are numerous treatments for pets, now available

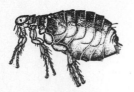

Dog Flea

Stick Tight Flea

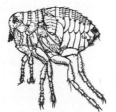

Oriental Rat Flea

Human Flea

Order Siphonaptera./Family Pulicidae

Dog Flea
Ctenocephalides canis
Length about 1/10 in. Members of this genus have a shoulder-cape (pronotal) comb directed backward and also a comblike "mustache," which are lacking in other species discussed here.

Nine species in genus, most of which are found on flesh-eating animals. Dog flea attacks man about as readily as it does dogs and is very common in the Middle Atlantic States. Cat flea has larger head than dog flea. Both were originally European.

Female lays 70 or more waxy white eggs on dog or floor. These hatch in 10 to 16 days. Larva active, wormlike, living 10 to 14 days in cracks and crevices before forming pupa in a fine cocoon that remains dust-covered for 7 to 10 days before producing adult.

A generation from egg to adult may be passed in from 14 to 37 days. These fleas may jump 7 in. high and over 1 ft on horizontal. This species is known to carry dog tapeworm from dog to dog. Strongly suspected of carrying bubonic plague and other diseases of man.

Oriental Rat Flea
Xenopsylla cheopis
Length about 1/10 in. Differs from human flea in that it has a divided mesosternum.

Some 30 species of this genus in world, of which only two are known in America. This species probably originated in Nile Valley in Africa and has spread largely as a parasite of its normal rat hosts. This is a most effective transmitter of plague.

Egg: in India, 2 days; on West Coast of United States, 9 to 13 days. Larva: in India, 1 week; on West Coast, 32 to 34 days. Pupa: in India, 7 to 14 days; on West Coast, 25 to 30 days. Complete generation: in India, 21 to 22 days; on West Coast, 9 to 11 weeks.

Bubonic plague carried by bite of this flea, the disease organisms being rubbed into bite; fleas leave infected rats when they die and go to human host. This flea has been responsible for tremendous loss of human life, some in America. It also carries nonepidemic typhus fever. Rotenone, malathion, chlordane, and lindane are effective chemical insecticides which can be used to treat flea infestations. They should be used with precaution to protect the environment against unnecessary pollution.

Human Flea
Pulex irritans
Length about 1/10 in. Differs technically from Asiatic rat flea in that it has an undivided mesosternum. Rear of abdomen of male turns upward, of female downward.

Common, particularly in Europe; also known from Africa and Asia; reported from South America. Rather restricted in distribution in North America, being most common in California and sparingly common in Mississippi Valley. Two species in genus.

Egg: East Coast, 2 to 4 days; West Coast, 7 to 9 days. Larva: East Coast, 8 to 24 days; West Coast, 28 to 32 days. Pupa: East Coast, 5 to 7 days; West Coast, 30 to 34 days. Full generation: East Coast, 2 to 4 weeks; West Coast, 9 to 11 weeks. In Australia and Europe, a complete generation requires 4 to 6 weeks. Lays to 500 eggs. Can live unfed 4 months; fed at intervals, 18 months.

Strange to say, it is difficult to rear fleas under artificial conditions. Can jump 7 in. vertically and 13 in. on horizontal, on the average. Human fleas bite rats and can carry bubonic plague, after 3-day infection. Many chemical insecticides are effective as control measures including rotenone, malathion, chlordane and lindane.

PHYLUM ARTHROPODA. CLASS INSECTA

Order Hymenoptera [Bees, Wasps, and Sawflies]

In insects of this order, adults usually have four membranous wings with few or no crossveins, usually interlocked in flight, rear wings smaller. Mouthparts suitable for chewing, sucking, or both. Females usually with a sting or ovipositor. Metamorphosis complete.

Order Hymenoptera./Family Cimbicidae

Elm Sawfly
Cimbex americana
Length of adult female, ¾ in. Female, head black, abdomen steel-blue or purple with four yellow spots on each side, wings smoky-brown, legs black, antennae short and pale yellow. Male, longer, more slender, and with some color variation. Number of color varieties. Adult May to June.

Common on woody plants but most common on elm, willow, poplar, and basswood in Middle West. Family is small and its members are closely related to horntails and typical sawflies.

Eggs laid in pockets in leaves, hatch into yellow-white larvae which reach maturity in July to August. Mature larva with black lines down middle of back, descends host tree, burrows into ground, and spins an oval brown cocoon in which winter is spent. In spring larva changes into pupa from which adult emerges in May to June.

Food of larva (which chews), leaves of birch, willow, elm, basswood, poplar, and some other trees. Plants attacked appear as if struck by fire and may even be stripped of leaves. Larva holds to plant by grasping with end of body, and may squirt a defensive acid liquid from lateral glands. Consult County Agent for latest recommendations on control measures.

Elm Sawfly

Order Hymenoptera./Family Tenthradinidae

Cherry Sawfly Leaf Miner
Profenusa collaris
Adult about ⅛ in. long. Female metallic black on body, reddish on foreparts. Male smaller, with lighter colored, more slender abdomen than female. Appears as adult in May.

Probably native, living originally mostly on hawthorn, but now living equally well on cherry leaves, where it appears as a leaf miner. In New York and Massachusetts, it may occasionally reach pest abundance.

Eggs laid singly, in slits in upper leaf surface, usually in basal part of leaf; hatch in mid or late May. Larva makes twisted mine in leaf, molts five times, leaving skins in blisterlike mines; six legs, with additional prolegs. Larva leaves mine, enters earth in June and makes waterproof cell but does not pupate until following spring, so winter is spent as larva. Pupa in cocoon several inches underground, under ⅕ in. long, white, about 1 week. Adults appear in early May, following pupation.

Food, leaves of cherry and hawthorn.

Control chiefly by picking and destroying mine-bearing leaves. A nicotine spray commonly used to control pest on hawthorn does not seem to be effective on cherry. Consult county agent or Experiment Station for latest information on control.

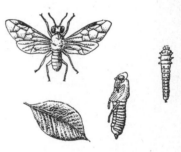

Cherry Sawfly Leaf Miner

Pear Slug
Caliroa cerasi
Adult ⅕ in. long. Body black, bluntly pointed to rear. Wings relatively large, longer than body and head. Adult appears after leaves develop in spring. Species conspicuous because of glossy dark slime produced by larva.

Native of Europe but now found in America nearly everywhere that pears, plums, or cherries are grown.

Eggs laid in slits in developing leaves, forming blisters. Larva produces brown glossy slime and hence looks somewhat like a slug; molts four times, after which it is orange-yellow and does not feed. Larva goes to ground to pupate. Winters as pupa, but in South may be three generations a year.

Destroys leaves of pear, cherry, and plum, causing leaves to turn brown because upper leaf surface is removed. Trees badly infested may lose all leaves by midsummer.

Consult county agent or Experiment Station for latest recommendations on control.

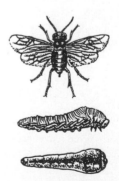

Pear Slug

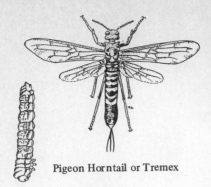

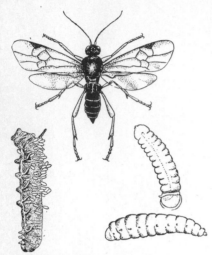

Pigeon Horntail or Tremex

Militant Noctuid Parasite

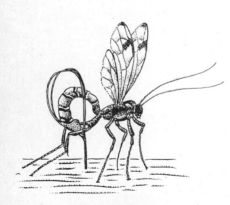

Long-tailed Ichneuman

Order Hymenoptera./Family Siricidae

Pigeon Horntail or Tremex
Tremex columba
Length about 2 in. Abdomen more than half length of body and almost cylindrical, ending in a stiff "tail" about ⅓ length of abdomen. Wings relatively short for an insect of this size. Color usually yellow, with reddish-black or brown markings, depending on locality.

Abdomen black with yellow bands and spots, northeastern United States, Quebec, and Ontario; yellow with black markings, Rocky Mountains and Pacific Coast; yellow with dark brown wings, Pennsylvania to Utah. Usually found on dead hardwood trees.

Female in summer bores into wood with ovipositor to depth of ½ in. to lay eggs singly; frequently dies stuck in wood when last egg is laid. Larva cylindrical, with three pairs of legs near head and a horn near rear; to 1⅓ in. long; digs a burrow about diameter of a pencil. Pupa in silk cocoon, in chip-lined burrow from which adult emerges in early summer.

Food of larva, wood of apple, beech, elm, maple, oak, pear, sycamore, and other hardwood trees. Sometimes found in firewood in buildings, or in furniture where wood has not been thoroughly cured.

Destructive to wood of valuable species of trees. Infected wood should be burned before other larvae have a chance to transform. Long-tailed ichneumon is chief enemy.

Order Hymenoptera./Family Braconidae

Militant Noctuid Parasite
Apanteles militaris
Adult about 1/16 in. long, black with yellowish legs and distinct black spot on front margin of forewings. Abdomen small, considering its egg-carrying capacity. More commonly seen in form of cocoons extending from body of parasitized caterpillars than as adults.

Widely distributed in United States and Canada, generally present wherever its caterpillar hosts are to be found. Adults active in egg laying mostly on warm sunny days.

Female may lay 1,000 eggs, putting as many as 72 in single puncture in caterpillar that has not yet reached last larval stage. Eggs 1/300 in. long, white; hatch in 5 days into larvae that feed on inside of caterpillar about 9 days, eating less essential tissue first, then pierce skin of caterpillar and form cocoons ¼ in. long and dirty white or brown in color; adults emerge in 9 days and live 1 week or more.

Food of adults, nectar or honeydew; of larvae, caterpillars of noctuids, such as army worm, cornear worm, or their close relatives. Females may mate or may lay eggs without mating; unfertilized eggs produce males. May be two generations a year. Winter as larvae in North and as pupae in cocoons in South.

Exceptionally useful in control of caterpillars that destroy many plants useful to man.

Order Hymenoptera./Family Ichneumonidae

Long-tailed Ichneumon
Megarhyssa macrurus and M. atrata
Female, length (exclusive of "tail") 1¾ in.; including "tail," about 6 in. *M. atrata* black with a few yellow spots; *M. macrurus* brown with yellow spots. Ichneumon flies vary from this large size to almost microscopic forms parasitic on plant lice. Related wasps parasitize insect eggs.

Relatively common in summer on trees infested with horntails, which they parasitize; larvae in tunnels of horntail larvae.

Female searches tree trunks for horntail tunnel. Finding one, she works her long slender ovipositor into bark and wood to tunnel and lays her egg. Larva seeks out horntail larva, attaches itself to it, and feeds until full-grown; pupates in tunnel and gnaws way to freedom when adult.

Food, larvae of horntails. Interestingly enough, ichneumons are themselves parasitized by another hymenopteran, a cynipid. We have, then, one hymenopteran, the horntail, parasitized by the hymenopteran ichneumon which is parasitized in turn by still another hymenopteran, a cynipid.

Unquestionably useful as a control of the horntail so destructive to wood of valuable species of trees. A member of a family of more than 10,000 known species which are predacious and parasitic on many insect pests. A form of "biological control," an important hymenopteran example being one which parasitizes European corn borer. More support should be provided for studies of biological controls of injurious insects.

496

Order Hymenoptera./Family Sphecidae

Mud-dauber Wasp
Sceliphron cementarium
Yellow mud dauber, *S. cementarium* (illustrated) has black and brown body, with yellow spots and yellow on legs. It is about 1 in. long, slender-waisted and brown-winged. Blue mud dauber, *Chalybion californicum,* is steel-blue throughout and usually smaller than yellow wasp.

Both blue and yellow mud daubers are general in their distribution; build under eaves, in houses, barns, and sheds, under bridges, or wherever their mud nests may be free from too much water, and where spiders are abundant near by.

Female builds mud cup enclosing series of tubular cells, each about 1 in. deep and ¼ in. through. Series may enclose one to many cells, and many series may be crowded together. Egg placed in a cell and several paralyzed spiders sealed within cell with it. Larva feeds on paralyzed spiders; pupa formed in cell. Pupa of blue wasp, silk-covered; of yellow, uncovered.

Food of adults, nectar; of larva, paralyzed spiders. Blue mud dauber may sometimes despoil nest of yellow, throw out yellow grub, and deposit her own egg.

Relatively harmless but tremendously interesting wasp whose status in the biological cycle is complicated. Since it eats spiders that control plant-destroying insects, perhaps it is not useful to man.

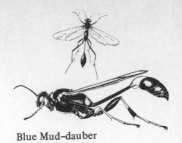

Blue Mud–dauber

Mud–dauber
Wasp Nest

Order Hymenoptera./Family Vespidae

Paper Wasp
Polistes sp.
Length 1 in. or over. Long slender dark brown to black, with red spots on abdomen and one or more yellow rings. Female with brown face. Male with pale face and longer antennae.

Common on flowers and about buildings; constructs open-comb paper nests wherever protection from rain is assured. Four highly variable species of paper wasps in United States and about 50 in world. Closely related to hornets.

Female lives through winter. In spring she builds paper nest in sheltered spot, with cells opening downward and uncovered beneath. In these eggs are placed and young reared. Grubs hang head downward in cells, are fed nectar and insects. Pupal stage in silken cocoons spun by larva in cells of nest. Mating occurs in fall, then males die.

Not vicious unless annoyed; sometimes may be handled with perfect safety. Food not ordinarily stored, since young are fed as need occurs and mother hibernates. Females alone care for young. Nest of paper supported by single stem, contains single layer of unprotected cells; waterproofed by new material licked on by female.

Perhaps of some economic importance as destroyers of other insects. Common in houses in fall and winter, before and after hibernating period; feared more than is justified. Paper for nest made by masticating weathered particles of wood from fences or buildings, mixing this with saliva, and drawing it out into thin, strong sheets.

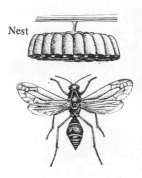

Nest

Paper Wasp

Bald-faced Hornet
Vespula maculata
Length of queens, over 1 in.; of workers, about 1 in.; males (drones), intermediate in size. Black with white markings on body and face, antennae rusty beneath. At rest, wings are held folded lengthwise over back. Queens seen most commonly early in season, workers in midsummer and fall, drones in fall.

Found in various places but most conspicuous about their large paper nests or as they feed about flowers. Vicious about nests, but away they normally do not attack unless confined or restrained. They and related yellow jackets may make nest size of bushel basket, housing 15,000 insects.

Eggs laid by queens in paper nest; develop in about 1 month. Egg stage 5 to 8 days, larva 9 to 12 days, pupa 10 to 13 days. First produced are workers, then drones, then new queens. All but drones hatch from fertilized eggs. All but females die in winter. Mate once soon after males emerge.

Food of adults, nectar and insects, chiefly latter. Elaborate nest of paper encloses many layers of cells opening downward. Queens start nest and rear first workers, but later leave such tasks to workers and devote themselves to laying eggs. Nest kept clean first by queens, then by workers.

Of some use as destroyers of other insects. Sometimes pests about apiaries, where they kill honey-laden bees. Also may cause panic among horses. Stings may be severe to persons disturbing nests. Attack sudden and most effective. Ammonia on stung spots may help relieve pain. Nest may be burned at night, when whole colony is inside.

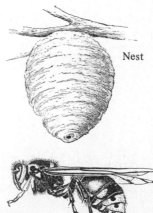

Nest

Bald–faced Hornet

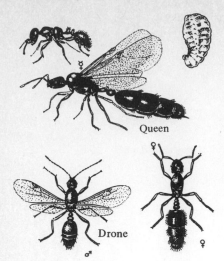

Queen

Drone ♂

♀

Little Black Ant

Order Hymenoptera./Family Formicidae. [Ants]

Little Black Ant
Monomorium minimum
Length about ¹⁄₁₆ in. for wingless workers, ¼ in. for queen, ³⁄₁₆ in. for male. Black. Queen with wings that are shed after mating, as in all ants; unusually large abdomen. Male winged. Larva ¹⁄₁₆ in. Pupa ⅛ in. Waists of queen, drone, and worker constricted as by three belts rather closely placed.

Native species, common in houses, on lawns, and in sandy places throughout country, closely resembling larger-headed little "red" or Pharaoh's ant, *Monomorium pharaonis,* in which queen has relatively shorter, thicker abdomen. This was originally an Old World species but is now generally established.

Winged queens mate with winged males usually in fall or spring. Queens lose wings and establish colony in burrow underground. Eggs may develop into workers, queens, or males, though it is probable that males develop from unfertilized eggs. Workers care for queen and larvae, moving larvae and pupae about in colony to places where temperature and moisture are suitable.

These little ants infect houses and lawns and work in definite paths, long lines of them going to and from work along a common highway. Obviously have means of communication with one another, antennae presumably playing an important part in this.

Black Carpenter Ant
Camponotus pennsylvanicus
Workers ¼ to ½ in., of two kinds, a major worker with a large head, and a smaller minor worker. Male about ½ in. Queen to nearly 1 in. Variation in size of head of workers is distinctive among our equally large Northern ants.

Common in dry decaying trees and timbers, in old barns and houses, particularly where wood comes in direct contact with soil; also in shady woods and other shaded areas, in contrast with many other ants' liking for exposed sunny places.

Life history is essentially like that of ant already considered. Contact of dry wood with soil provides variation in temperature and humidity and allows adults to select suitable quarters for themselves and young. Colonies may remain long-established and become of unusual size.

Food probably largely insects, which are foraged; nests constructed in wood, likely after it has been attacked by fungi. Injury to houses caused by fungi or by termites often attributed to this ant.

Control of ants is most satisfactorily effected by destroying queen and home colony. Since workers feed queen, colony can be destroyed by administration of a poison that will not kill the worker until it has taken food to queen. Such poisons are available. A poison made by dissolving 125 gr of arsenate of soda in 1 qt of water in which has been dissolved 1 lb of sugar and to which honey may be added is put on pieces of sponge that are placed along ant runways. Most drugstores stock effective poisons.

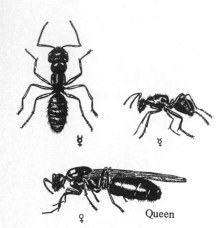

♀   ♀

♀   Queen

Carpenter Ant

Cornfield Ant
Lasius alienus
One of commonest small brown ants found in fields and lawns. Queen with large abdomen, relatively small head, and short antennae. Worker with unusually large head, strong jaws, and relatively long antennae.

Found in lawns and fields. Active in soil when temperature is about 70°F. Winged adults found in July to August; most active between 3 and 6 P.M. on bright clear days. At this time, winged adults may be in air in great numbers, through September.

Fertilized queen in fall sheds wings, burrows into ground for winter. First summer lays about 200 eggs and rears about 25 workers. Eggs ¹⁄₅₀ in. long; white, hatch in 3 weeks into small larvae that mature in 3 weeks and form white silky cocoons ⅛ in. long for 3 weeks. No males or virgin females in first-year colony.

Food: brood of first year fed from substance of mother and from eggs and larvae of colony, since mother does not seek food; second-year colony may depend on food brought by workers and may rear 100 new members. Colonies of this species known to increase for 6 years; of others, to 15 years. Ants can survive freezing, 100°F, 24 hours' submergence, month's decapitation. At times queen may lay 100 eggs a day.

Cornfield Ant

Order Hymenoptera./Family Andrenidae

Small Carpenter Bee
Ceratina dupla

Length about ¼ in. Slender, with metallic blue body and rainbow-tinted wings. Related large carpenter bee is about size of bumblebee, which it resembles. Most easily located by searching for nests.

Found usually in nest in pith of sumac, in dead stubs; entrance rather conspicuous. Also found in other plants that have a roomy, soft pith and dead stubs in which mines may be made, as elderberry or raspberry. Common throughout most of United States.

In early summer female hollows pith of shrub to make deep burrow. In bottom is placed an egg with ample pollen for development of larva. On top of first is a wall and another egg and pollen supply. This is continued up to 14; uppermost and youngest insects must emerge before those lower down can be freed. Two generations a year. Mother spends winter with brood, in mouth of nest.

Food, pollen eaten by adults and larvae. In getting pollen, insect assists in flower pollination. Female protects burrow containing her eggs or family. Carpenter wasp uses mud instead of wood dust to separate cells of young, and mother does not protect family as carpenter bee does. Carpenter bees, *Xylocopa virginica,* appear in early spring. They may be mistaken for bumblebees, but the latter are more hairy. They visit flowers in late summer; but in spring they dig out burrows in soft wood. Female lays eggs in the burrow which has been stocked with nectar and pollen.

Small Carpenter Bee

Order Hymenoptera./Family Apidae

Bumblebee
Bombus sp.

Length from ½ in. to over 1 in., depending on species. Usually black and golden yellow; distribution of color distinguishes different species. Abdomen usually plump; wings rather small for such a large insect. Flight relatively clumsy.

Bumblebees found commonly in Old World and in America; most conspicuous in fields of clover or around showy or fragrant flowers; abundant in orchards when fruit trees are in bloom. Females most conspicuous in spring or late fall, workers in midsummer.

Queen lives over winter and starts new colony in spring. Lays 400 to 1,000 ⅛-in. white sausage-shaped eggs that hatch in 4 to 5 days. Larva a hairless, eyeless maggot that matures in about 1 week; enclosed in wax cell in nest. Pupa in tough cocoon in larval cell, for about 10 days. Nest generally built in the ground; adult uses considerable moss and grass in construction.

Food: larvae are fed honey and pollen, first by queen mother and then by workers. In late summer, drones or males and new queens are developed, mate in flight; colony of up to 300 individuals may die, except for fertile queens who seek a sheltered place for hibernation. Have many enemies, such as skunk and certain insects. May attack horses and other animals that blunder into their nests and may cause severe sting. Usefulness in pollination of fruit trees, clover, and other valuable farm plants cannot be overlooked.

Bumblebee

Honeybee
Apis mellifera

Worker or sterile female about ½ in. long. Drones and queens ¾ to 1 in. long. Swarms of 60,000 bees have been seen, probably constituting one colony. Workers make up most of colony and live only 1 to 2 months; drones, 1 to 2 months; queens, several years.

Honeybees are domesticated world over, wherever man can survive and has suitable nectar-bearing plants to provide food for bees. Workers are most commonly seen in fields and about flowers, or near their colonial centers, hives, or bee trees.

In summer, drones and queens mate in flight. Fertile queen usually starts new colony. Wax comb is built by bees or supplied by man. Eggs laid in comb develop into larvae, then pupae, then adults. From egg to adult queen takes 15¼ days; worker, 21 days; drone, 24 days; drones coming from unfertilized eggs laid by queen.

Food of adults, pollen and nectar; of larvae, beebread of pollen and nectar, prepared by workers. If a colony becomes queenless, a new queen may be developed from queen cells or one may develop from egg or worker larva not over 3 days old, by enlarging cells and changing food from beebread to a "royal jelly." Drones are killed when of no use to colony.

Possibly most valuable insect for honey produced and for service in pollinating flowers of fruit trees. United States produces about 500 million lb of honey a year and 10 million lb of wax. Beeswax is used in shaving creams, cold creams, polishes, floor wax, candles, crayons, and electrical and lithographing products.

Queen   ♀

Drone   ♂

Worker   ☿

Honeybee

PHYLUM HEMICHORDATA, BRANCHIOTREMATA

Soft-bodied animals. Body and body cavity divided into three regions, usually with paired gill slits; nervous tissue on both dorsal and ventral sides. Once thought to be in phylum Chordata, because it was believed that they possessed a notochord. Might be placed near starfish; like echinoderms because they have similar larvae and other similar structures. Tongue worms, acorn worms are hemichordates. Found in muddy bottoms or open seas.

Belanoglossus

CLASS ENTEROPNEUSTA

Balanoglossus kowalevskii
Length 6 to 10 in. Body composed of a proboscis that resembles head end of an earthworm, a collar that is relatively free along its forward margin, and a flattened segmented trunk that tapers gradually and regularly to rear end. Proboscis, yellowish-white; collar, red-orange with a white ring; trunk orange-yellow to green-yellow. Ranges from North Carolina to Massachusetts Bay.

Found in mud and sand along seashore, where it burrows like a worm. Shallow burrow formed by means of soft proboscis and sticky mucus secreted by skin glands; forms a tubular case of sand in which animal lives. Burrow U-shaped; two openings. Coils of fecal sand castings are deposited at rear opening. Some 12 genera and 70 species of class Enteropneusta have been described. *Ptychodera* inhabits tropical waters; *Saccoglossus* appears on both coasts of North America. Other species of genus vary from 1 in. to 4 ft in length. Group also includes colonial *Rhabdopleura* and *Cephalodiscus* of deep sea.

Sexes distinct; eggs and sperms pass to exterior through pores in parent body. In some species a free-swimming larval stage is formed. In others reproduction by the simple process of budding is possible.

Food, organic matter extracted from mud it takes in. Proboscis is hollow and serves as a water-storage chamber: provides water under pressure used in burrowing. Has rudimentary nervous, circulatory, and excretory systems.

PHYLUM CHORDATA.

The phylum chordata is composed of animals that are bilaterally symmetrical, with a dorsal tubular central nervous system, gill slits at some time in development, and an internal skeleton. It is probable that they have existed on the earth since Cambrian times and certain that they were well established by Silurian times. 43,000 species described. Phylum composed of such forms as: the tunicates, the lancelets and the vertebrates (lampreys, sharks and rays, bony fish, amphibians, reptiles, birds, and mammals).

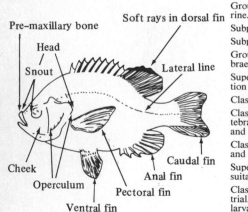

Pre-maxillary bone
Soft rays in dorsal fin
Head
Snout
Lateral line
Cheek
Caudal fin
Operculum
Anal fin
Pectoral fin
Ventral fin

Barbels
Adipose fin
FISH

A synopsis of groups of Chordata follows.

Group Acrania, without skull, vertebrae, or paired appendages. All marine.

Subphylum Tunicata (Urochordata).

Subphylum Cephalochordata (Leptocardii).

Group Craniata (Vertebrata), with skull or cranium, usually with vertebrae, visceral arches, and brain.

Superclass Pisces, fishes, with gills and two pairs of fins suitable for motion in water.

Class Agnatha (Cyclostomata), without true jaws or limbs. Lampreys.

Class Chondrichthyes (Elasmobranchii), with complete and separate vertebrae movable jaws, paired appendages; and with cartilaginous skeletons and exposed gill slits. Sharks and rays.

Class Osteichthyes (Teleostomi), with bony skeletons, covered gill slits, and rayed fins; swim by fins and possess gills. Common fishes.

Superclass Tetrapoda, air breathers, with lungs and two pairs of limbs suitable for motion on land.

Class Amphibia: salamanders, toads, and frogs, terrestrial or semiterrestrial, typically laying and developing eggs in water, respiring by gills as larvae and usually by lungs as adults, usually cold-blooded, with soft moist skin.

Class Reptilia: alligators, lizards, snakes, and turtles, terrestrial or secondarily aquatic, usually coming to dry ground to reproduce, some giving birth to living young while others lay parchment-covered eggs, breathing by lungs, cold-blooded, typically covered with scales or horny plates.

Class Aves: birds, terrestrial but usually prepared for flight with assistance of feathers, breathing by lungs, laying eggs, warm-blooded, typically without teeth on the jaws though primitive forms did have teeth; with a great variety of wings, beaks, and feet suited to different habits of feeding and locomotion.

Class Mammalia: mammals, terrestrial or secondarily aquatic, with hair in place of feathers, suckling young at mammary glands, lung-breathing, for most part giving birth to living young, warm-blooded, with teeth in jaws.

PHYLUM CHORDATA. SUBPHYLUM TUNICATA. CLASS ASCIDIACEA

Order Ptychobranchia./Family Botryllidae

Sea Squirts, Sea Lemons
Botryllus sp.

Body of adult a sac varying in size but averaging about 1 in. in diameter. Colonial *Botryllus schlosseri* is purplish or black, jellylike, over 5 in. across, with 5 to 10 individuals in groups with a common opening.

Found in masses on piles, rocks, and seaweeds usually in shallow seawater. *B. schlosseri* common along the shore of Middle Atlantic States north and through to Europe with related species to be found in most seas. Over 1,200 close relatives known and described.

Sexes not separate. Eggs of one individual in a colony are fertilized by sperm of another. Free-swimming tadpolelike larva, with long tail, large head, a single eye, a primitive ear, and nerve tube comparable to spinal chord, which become lost in adult stage. Larva settles down after a few hours or days as free-swimming form. Adult form unlike larval form; attached, covered with tough coat.

Water enters mouth of colony and through mouth of individual carrying food. It passes through gill slits, which separate food by slime cords that lead to mouth, stomach, and U-shaped intestine. Heart pumps blood first in one direction, then in reverse. Bloodstream has corpuscles that attack disease germs. Low-quality brain can be removed from adults without affecting efficiency. Larvae can sense touch, light, and possibly sound.

Serve as scavengers, as food for higher forms, and as interesting subjects of study by zoologists, since adult is less advanced in evolutionary development than larva.

Sea Squirt

SUBPHYLUM CEPHALOCHORDATA. CLASS LEPTOCARDII

Order Amphioxiformes./Family Branchiostomidae

Amphioxus, Lancelet
Branchiostoma sp.

Length 2 to 3 in. Semitransparent, fishlike, tail being pointed like a lance, giving animal its common name. No distinct head. No lateral fins, but dorsal and ventral fins are supported by rodlike fin rays. Mouth opening is below and in front under a sensory hood.

Lives near shore in various parts of the world, usually buried in sand by day and swimming about actively by night but seasonally abundant for about 9 months. 28 species in the group, 4 American shore forms being *B. virginiae, B. floridae, B. bermudae,* and *B. californiense.*

Animals are male or female. Breeding season at night and evenings in early summer. Pairs of reproductive organs, normally 26 (but in *B. californiense* 31 to 33), free germ cells into water where fertilization takes place. Larva is a free-swimming animal but adult burrows in sand with surprising speed.

Food is carried into mouth by stream of water caused by action of wheel organ. Water passes out over gill slits after food has been extracted. Number of gill slits varies from 50 to 90 pairs. Food is caught in mucous threads before being carried on to digestive system. Respiration effected by circulation of blood through gill slits by which stream of water passes.

Normally not considered economically important, but near Amoy, China, fishermen harvest large quantities from August to April and they are used as an important article of food.

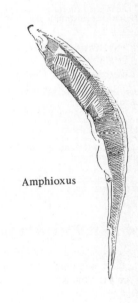

Amphioxus

PHYLUM CHORDATA. SUBPHYLUM AGNATHA. CLASS CYCLOSTOMATA

Order Petromyzoniformes./Family Petromyzonidae

Lamprey

(1) sea lamprey, *Petromyzon marinus.* Length to 3 ft. Dorsal fin not continuous. Teeth radiating variously. (2) brook lamprey, *Entosphenus lamotteni.* Dorsal fin divided. Teeth in radiating groups. In *Icthyomyzon* dorsal fin is continuous. All of group with unsegmented notochords round jawless mouths, rows of gill openings and no paired fins.

All have tails compressed side-to-side like bony fishes. With a large eye on each side; covered with transparent skin but without eyelids; with seven gill slits behind the eyes. Body covering with many mucous glands but without scales.

Male builds nest, sometimes with female's assistance. Eggs: (1) over 230,000; (2) over 10,000; laid by (1) in June; by (2) in April. Larvae live as mud lampreys in bottom 4 to 5 years and transform, fall to spring; (1) goes downstream, living as dangerous fish parasite, then ascends to breed and die while (2) is not parasitic and may be free-swimming less than 1 month. Larvae feed on small animals and plants.

(1) North Atlantic, south to Florida, to northern Europe and northwestern Africa and in many Great Lakes and tributaries. (2) Mississippi system, Minnesota to Pennsylvania, and Great Lakes system, Connecticut and Hudson Rivers, and south to Maryland. Some are eaten by man. Used as cod bait. Some are serious enemies of valuable food and game fishes.

Parasitic forms often damage fishes caught in nets or on fishlines. Sea lampreys have invaded the Great Lakes in large numbers and have become a serious threat to lake trout population. In one trial, sea lampreys in experimental tanks each killed about 18 fish in 87 attacks.

As of 1974, concerted efforts are beginning to function to control the lamprey population in the Great Lakes.

CLASS CHONDRICHTHYES (ELASMOBRANCHII)

Order Squaliformes./Family Lamnidae

Mackerel Shark

Lamnia nasus
Length to 12 ft. Narrow at tail base and with horizontal keel. Skin regularly rough. Gill slits on both sides. Male with prolonged pelvic fins forming claspers. Some existing sharks are to 50 ft long while others are mature at 2 ft; great variation in shape and habit in different species. Cartilaginous skeletons, exposed teeth and gill slits.

In mating, sharks have internal fertilization, with eggs in some hatching within mother. A 14-ft, 2½-ton white shark or maneater had nine 2-ft young, weighing 180 lb, inside her. Shark

remains are found laid down in Silurian times. Existing genera are represented back to Cretaceous times. Record line catch, 10½ ft, weight 1,009 lb, 1936.

Sharks range the oceans of the world. Mackerel sharks are surface feeders. Mako shark a superior sport animal. Many supply waterproof transparent tissue, valuable livers used medicinally, and oil. (A 9-ft shark may yield 11 gal of oil.) Dangerous because of bite and blow of tail. All sharks and rays with persistent notochords and placoid scales with embedded bases.

Sharks, in particular the soupfin shark, were at one time an important source of vitamin A from the livers of the fish. The Chinese make soup stock from dehydrated dorsal and pectoral fins. In California during one year, 9 million lb of shark were harvested and processed for liver oils and flesh. On the East Coast, the shark called mako is the source of some "swordfish steaks"

Order Rajiformes./Family Dasyatidae

Ray

(1) common sting ray, *Dasyatis centroura.* Rays include (2) Molubidae, 12- by 20-ft devilfish or mantas; (3) Pristidae, 15-ft sawfish; (4) Torpenidae, 5-ft electric rays; and (5) Rajidae, 6-ft skates.

Rays are related to sharks; both groups have five pairs of gill clefts and a cartilaginous skeleton. Skates and rays have pavementlike teeth in contrast to sharp, well-developed, pointed teeth of most sharks. Rays are like skates in that both have gill openings on the belly side and their pectoral fins are joined to the head. Skin generally smooth.

Sting rays are hatched within female's body and may get nourishment there before birth. (5) lays leathery-coated eggs. (2) may weigh to 300 lb, (4) to 100 lb. Food of all, largely crustaceans, shellfish, and fishes. Like sharks but white beneath and with gill openings beneath.

(1) from Maine to Cape Hatteras, with about 30 relatives extending range through warmer salt waters of world. Venom of (1) from tail spine known to have been fatal to man, or spine may cause dangerous wound by breaking off. Surgery is better than the use of potassium permanganate.

Rays are eaten throughout the world, but only to a small extent in U.S. "Wings" may be cut up and used as a substitute for scallops; sometimes whole body is processed and used for fertilizer or fish meal additive in animal grains. Some ray flesh is used as crab-pot bait. Sometimes, rays become pests in oyster and clam beds. Common illustration of the order found in introductory biology courses.

CLASS OSTEICHTHYES Bony Fishes

Skeleton more or less bony; usually with scales of cycloid, ctenoid, or ganoid form; paired lateral and median fins usually present; no pelvic girdle; mouth usually terminal; gills present; two-chambered heart; appear from Devonian strata to the present; in fresh, brackish, or salt water; some 17,000 species.

Order Acipenseriformes./Family Acipenseridae

Sturgeon

(1) American sturgeon, *Acipenser sturio oxyrhynchus.* Body partly plated. Tail unequally forked. Snout conical, elongate. Length to 12 ft. Weight to over 500 lb. (2) lake sturgeon, *A. fulvescens.* 15 to 16 dorsal shields. Length to 6 ft. (3) short-nosed sturgeon, *A. brevirostrum.* Snout blunt, from ½ to ⅓ length of head.

Primitive fishes widely represented over long period of geologic history. Head covered with bony plates; five rows of bony armorlike bony scales on body. Has four small barbels forward of mouth; mouth can be protracted for bottom feeding.

(1) ascends rivers to above tidewater, May to June. Lays to 3 million heavy sticky, ⅛-in. eggs totaling 30 percent weight of parent. Eggs hatch in 3 to 7 days into ½-in. young that in 1 month may be to 5½ in. long. Migrate to sea at 1 to 3 years at

3-ft length. Credited with 300-year life span, probably falsely. Food, small animals grubbed or gleaned from bottom.

(2) formerly common in Mississippi Valley, Great Lakes, and north. (3) marine, from Florida to Cape Cod but becoming generally rare. Spawns in freshwater streams, like all sturgeons.

Excellent game fish especially if caught with rod and reel; usually use heavy tackle and baits such as large earthworms, lampreys, meat scraps, or cut-up fish. In some midwestern states, taken in winter by spearing. Because of decreasing numbers, protective laws have been passed in some states. Smoked sturgeon is a gourmet delicacy. Most desirable species in U.S. market is the lake sturgeon. On the market it may be substituted for by the more common and less desirable paddlefish, *Polyodon spathula.*

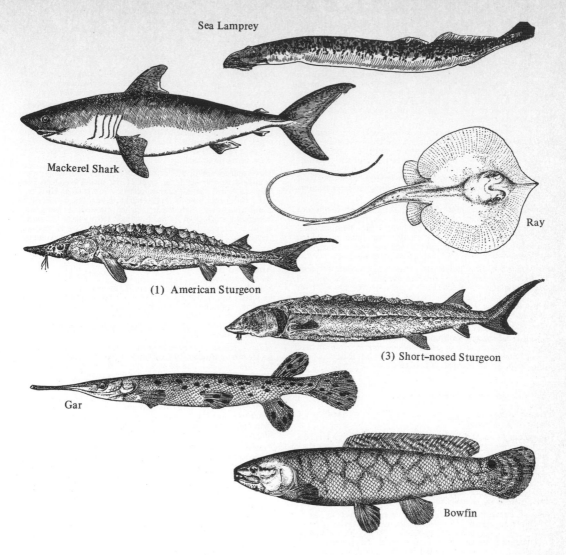

Sea Lamprey

Mackerel Shark

Ray

(1) American Sturgeon

(3) Short-nosed Sturgeon

Gar

Bowfin

Order Semionotiformes. / Family Lepisosteridae

Gar

(1) long-nosed gar, *Lepisosteus osseus*. Beak length 20 times its least width. Length to 5 ft. (2) alligator gar, *L. spatula*. Length to over 10 ft. Weight to 100 lb. Both gars are covered with bony armor and have jaws armed with sharp teeth. Solid upper part of tail the longer.

(1) Body is olive-green above and white to silvery beneath; with large round black spots on tail and rear fins. Sides mottled toward head; with black spots distributed toward tail end; also on the tail, and rear fins. Gars may be called by many common names: billfish of billy gar, needlenose, longnose gar; short bill, stub-nose gar.

Feeds largely on other fishes captured by sudden dashes. Can breathe without use of gills, by using swim bladder as lung in emergency. Often seen at surface taking in air supply. May be caught by slipping piano-wire noose over beak. Apparently makes little provision for care of young. May produce to 28,000 eggs; green eggs are poisonous to higher vertebrates.

(1) Great Lakes to Rio Grande and along Atlantic Coast. (2) in southern states north to Ohio River. Great pests as destroyers of valuable fishes but hosts for young of an edible mussel. May be eaten. May be dangerous. Bony coverings used as shields by primitive people.

(2) One record fish caught weighed over 200 lb and was 10 feet long. No other gar reaches this size. Not a popular sport fish in the U.S. but could be used to a much greater extent.

Order Amiiformes. / Family Amiidae

Bowfin, Dogfish

Amia calva
Length of female, to 30 in. Weight to over 20 lb. With unique plate under head. May breathe through swim bladder. Male to 20 in. long and with orange-reddish border to black tail spot.

Identification made fairly easy by extra long, single dorsal fin that extends over more than ½ of back; and the large black, round spot on upper part of base of the tail. Dark green to olive-green above, lighter on sides, and whitish beneath.

Male makes circular nest, in weeds, over mud bottom; fertilizes and guards eggs; guards young until they are 4 in. long. Can live in mud short periods without using gills. Hardy, tough bait minnows.

Primitive fish, lone survivor of what, in the geologic past, was a large family; now occurs as fossils in Europe and U.S.

In swamps and lakes from Vermont to Dakotas, Florida, and Texas. Flesh edible when smoked but poor food. Good sport on a line and light rod. Savage fighter that apparently never completely give up. Not a popular sport fish.

Also known as bigfinned hard jaw, spot-tail, grinnel, grindel, cypress trout, mudfish. Considered an undesirable species because it feeds on other fishes and is a poor food fish. Not a valid criticism because bowfin acts as a check on overpopulation of stunted panfish and rough fish it feeds upon. Fish make up to 80 percent of its diet, crayfish the remaining 20 percent usually.

Order Clupeiformes./Family Elopidae

Tarpon
Tarpon atlanticus
Length to 8 ft. Weight to over 350 lb. Some scales to 3-in. diameter. Rear of dorsal fin much prolonged. Only one species known. Closest relative a species of Indian seas. Hooked, may leap vertically to 15 ft or horizontally to 30 ft. Matures probably seventh or eighth season, at 4-ft length. Hardy in foul water.

Color dark-blue to greenish-black on back then changing gradually to bright silver on sides and belly. No spines in the one dorsal fin. Body elongate and compressed side to side. Lower jaw juts forward. Brackish water inhabitants; feed actively at night; food largely shrimp, crabs, needlefish, mullet. Young tarpons are largely plankton feeders, then they add crab larvae and aquatic insects to their diet, and later become predacious.

Spawns along west coast of Florida, March to May, offshore in blue water. Males occur in large schools over shoals 1 mile offshore. Eggs to 12 million per 142-lb fish, float or sink. Larvae apparently ribbonlike, transparent; undergo sudden change to fish form. 3-in. fish found inshore. Adults run upstream to 100 miles from sea. Disappear in fall.

Supreme game fish. Edible and considered excellent by some if smoked. Sold in Latin-American markets. Scales sold at 5 to 25 cents apiece. Record fish caught by line are a 247-lb 89½-in. fish caught in Panuco River, Mexico, in 1938 and 350-lb fish netted in Hillsboro River, Florida. Found in Lake Nicaragua, Central America. Sensitive to cold.

Best fishing occurs at night or in early morning hours. Trolling, still-fishing, and drifting are favorite ways of fishing tarpon. Anglers use live or dead mullet, crabs, shrimp as bait. Large plugs, spoons, and feathers are popular; artificial lures used. Not a very acceptable food fish.

Order Clupeiformes./Family Clupeidae

Alewife, Sawbelly
Alosa pseudoharengus
Length to 15 in. Teeth on jaws disappear in old age but persist in freshwater skipjack or blue herring. Head longer than in hickory shad. Round dark spot on shoulder. Silvery areas on cheeks are longer than deep, while in shad they are deeper than long. Bluish above and silvery below; faint dark stripe on each side. Freshwater form reaches 3 to 6 in. in length while only saltwater form reaches 15 in. Food largely plankton; crustaceans, small fishes, diatoms, copepods.

Ascends streams or enters shallows in great numbers in March and April to spawn. Female lays about 100,000 eggs, broadcast. Eggs sticky, hatch in 3 to 6 days at 60°F. Young to 4 in. by fall, return to sea. Mature in 3 years. No parental care. Some landlocked forms die after breeding but others return to sea.

After spawning some adults travel to and stay in shallow water of estuaries into early winter; most disappear in the fall. Freshwater forms move into streams to spawn, then return to larger bodies of water; spend fall and winter at sea.

In sea and fresh waters from Nova Scotia and Gulf of St. Lawrence to Gulf of Mexico and in adjacent streams, as in Lake Ontario and central New York. Caught in dip nets or pound nets. Annual yield may reach 30 million lb and is a major fish resource. Fish eaten fresh, salted, or smoked. Used in making oil and fertilizer.

During summer, landlocked form may suffer mass mortality in large bodies of water; bodies carried to shore and foul beaches. Coho salmon of the Great Lakes are largely dependent upon alewives as forage fish.

Atlantic Herring
Clupea harengus
Length to 18 in. but rarely over 1 ft; blue above, silvery on sides, with only one dorsal fin, with scales that are easily lost. Anal fin with 17 rays. May be to 3 billion in a single small school. See closely related sawbelly.

Compressed, elongate body; with a large mouth and a projecting lower jaw. Adults are plankton feeders largely, although will eat small fish; young feed on copepods, and small crustacean larvae.

Herring are food for many fishes such as salmon, tuna, mackerel, cod, and striped bass. Sharks and squid also eat them. Pacific herring, *C. h. pallasi,* close relative of Atlantic herring;

found from Alaska to lower California. Sportsmen may take adults with dip nets. Predatory fishes such as shark and salmon; waterfowl and sea lions prey upon Pacific herring.

Spawns in sea though relatives run upstream. Eggs, heavy, sticky, sink to bottom, to about 30,000 per female, hatch at 45°F in 22 days into 1-in. young that mature in 3 years. Length off Nova Scotia at 1 year, 5 in.; at 2 years, 7 in.; at 3 years, 10 in.; at 4 years, 11 in.; at 5 years, 12 in.; at 6 years, 12½ in.; at 7 years, 13 in.; at 9 years, 13½ in.; at 10 years, 14 in., these being large fish.

Trade names: small, to 7 in.; fat, to 10½ in., with little reproductive development; large, over 11 in.; spring breeders, February to April. Both Atlantic coasts. South to North Carolina. California herring, Kauscharka to San Diego. Fry 3 to 4 in.; fried crisp, "whitebait"; larger, canned as "sardines"; full-grown, fresh or dry, "red herring"; lightly salted, smoked, canned, "kippered herring"; larger well-salted and smoked, "bloaters." Portion of catch is used to make fish meal and oil.

American Shad
Alosa sapidissima
Length to 2½ ft. Weight to 13½ lb; average 1 ft, 1 lb. Jaws toothless; upper deeply notched. Cheeks deeper than long. Shad is generally silvery with a bluish-green back; sides are silvery and belly side is white. Dusky spot present close behind and at top of gill cover. Some call it "poor man's salmon."

Males precede females in spring run up streams. Female lays to 156,000 nonsticky, ⅛-in., slowly sinking eggs that at 60°F hatch in 1 week. Larvae ½ to 4½ in. for about 2 weeks. By first summer, reach 3 to 5 in. fingerling and return to sea. Young 6 in. by first winter. Mature at 3 to 8 years.

Newly hatched shad stay in fresh water until fall; then spend 2 to 5 years in salt water. At maturity, they return to fresh water to repeat the cycle by spawning; apparently spawn more than once. Probably completely plankton feeders throughout life cycle. Is a schooling fish at sea.

Found in sea and adjacent tributaries from Gulf of St. Lawrence to Florida. One of most sought fishes. Has been introduced and established in California. Established in these western rivers: Sacramento, Columbia, Rogue, Russian, Feather, Eel, Klamath, and probably others.

Hickory shad, *A. mediocris,* a close relative, more a fish eater; is fished with flies, small spoons, and with live or artificial fish as bait. Flesh not as desirable as American shad; sold fresh, pickled, or smoked. At one time there was a large, important commercial shad fishery. Pollution and demand for shad have almost completely destroyed the resource.

Menhaden
Brevoortia tyrannus
Length to 18 in.; average, under 12 in. Head large. Anal fin small. Scales deep, with rear edge roughly toothed or fluted. Long thin gill rakers, used in straining out plankton, food, seem to fill mouth. Deep-bodied; dark blue to green to dark gray above mixed with shades of brown; sides, belly, fins appear brassy especially when first taken from water. Behind gill cover there is a distinct spot followed by a number of smaller, dark spots. Usually found in large schools. Schools may appear to jump from the water as one. Little known about breeding habits.

Eggs float just beneath surface; hatch at 72°F in 2 days. In first year, reach 6 in. and weight of 1½ oz; in second year, 10 in.; 7 oz. Mature in about 3 years. Mature fish usually caught in fall on way to unknown spawning grounds where it remains until next spring. Young and adults are plankton feeders; food made up of diatoms, small worms, crustaceans, and shrimp larvae; small plants are important. In turn menhaden are eaten by whales, porpoises, sharks, cods, bluefish, swordfish, tuna, striped bass, to name a few.

Nova Scotia to Brazil, in coastal waters largely. Too rich in oil to be popular as food. Oil is extracted and residue is used as fertilizer. Annual New England catch around 600 million lb. Roe is canned and oil is used considerably in poultry foods.

Small menhaden are used as bait; larger individuals are cut up and used as chum by fishermen. The most important U.S. commercial fish; 1½ billion lb produced by the fishery in 1969.

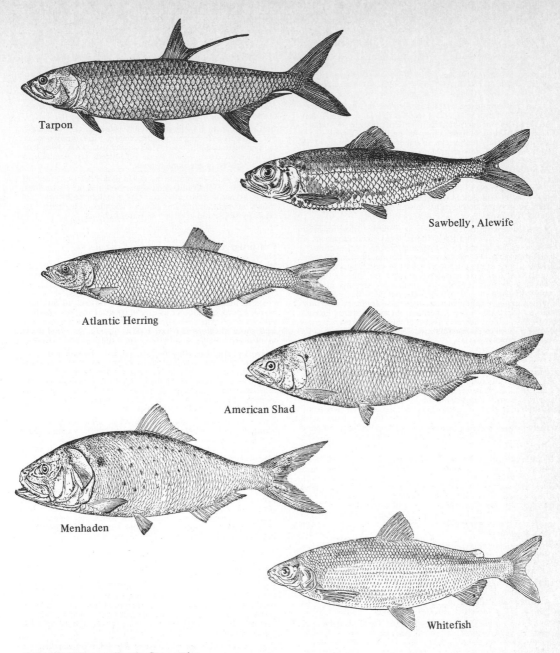

Tarpon

Sawbelly, Alewife

Atlantic Herring

American Shad

Menhaden

Whitefish

Order Clupeiformes./Family Coregonidae

Great Lakes Whitefish
Coregonus clupeiformis

Has small adipose fin like salmon and trout but mouth is small and upper jaw reaches only to eye. Length, 5 months, 2 to 3 in.; 1 year, 8 to 9 in.; 2 years, 12 to 14 in.; maximum, 24 in. Weight to 23 lb. Olive back with white sides. A husky fish with a deep body, compressed laterally; back appears elevated, has large scales. Blunt snout.

Spawns November. Female may produce 15,000 to 22,000 eggs. Eggs laid in shallow water, over rocks up to 10,000 per lb, 1/20 in.; hatch in 5 months. Males do not breed until 2 2/5 lb and females at 3 4/5 lb, legal minimum harvest size is 2 lb. Oldest fish known was 26 years old.

Young whitefish feed almost exclusively on small aquatic crustaceans. Later, food is small mollusks, aquatic insect larvae, plankton, crustaceans. Distributed in northern portions of Eurasia and North America. Essentially a cold-water fish. Movement reflects this as fish is found in deep water in warm

weather; in the spring it may inhabit shoal areas; returns to shoals to spawn when water is cooler again in October and November.

Ranges from Great Lakes north to Arctic Circle. Captured by deep-water trap nets. Annual United States take, about 4 1/2 million lb being most valuable food fish of Great Lakes but in danger of extermination or becoming of little importance. Fish are marketed fresh or smoked, or roe is sold as caviar instead of sturgeon eggs.

Delicate flavor of whitefish flesh recognized by the pioneers in the nineteenth century, who learned from the Indians how to net and spear the fish. Important sport fish both in open water and in ice fishing; also serves as food for much-valued lake trout. May contain parasitic tapeworm as a noticeable yellow cyst. This makes individual less marketable though man is not a host of the whitefish tapeworm.

505

Order Clupeiformes./Family Salmonidae

Pacific Salmons

(1) king, chinook, or spring, *Oncorhynchus tschawytscha;* (2) red, sockeye, or blueback, *O. nerka;* (3) silver or coho, *O. kisutch;* (4) pink or humpback, *O. gorbuscha;* and chum, keta, or dog. Weight: (1), 100 lb; (2), 7 lb; (3), 10 lb; (4), 6 lb. Anal fin rays: (1), 15 to 17; (2), 14; (3), 13 to 15; (4), 14 to 16.

(1) and (2) swim upstream for long distance in July to October at age of (1) 3 to 8 years, (2) 4 to 8 years. (3) at 2 to 7 years and (4) at 2 years swim upstream in September to November. Female builds nest, mates, spawns, and dies. Young of (1) return to sea after few weeks at length of 1½ in.; young of (2) return to sea after few months to 3 years; young of (3), after 1 to 2 years at length of 5 in. Dark side parr marks are broad in (1), round in (2), absent in (4). All species spawn in fresh water over gravel bottom; from June, chinooks (1) in Siberia to February, coho (3) in Oregon. Young remain in gravel until food in yolk sac is used up. Male of (2) may become bright red on the body, and is known as kokanee salmon in landlocked forms.

Eggs used as bait, hatch at 50°F in 50 days. Most valuable food fish; in Columbia River alone, yield has been worth 10 million dollars a year to 25,000 people, with total once to 32 million dollars. From Alaska to California, finest of all sport fishes. Dams along migration routes in streams on the West Coast have had a serious adverse effect on salmon runs to spawning grounds. Apparently young when mature return to spawn in river in which they were hatched. Recommended that anglers be on the water at daybreak. Mouths of rivers are favored locations. Some fishermen observe birds feeding on small fish as indicator of location of salmon; sockeye (2) most valuable as far as commercial catch is concerned; also favorite salmon of many anglers.

Atlantic Salmon, Rainbow Trout

(1) Atlantic salmon, *Salmo salar.* With small black or brown spots on light background, without broad pink lateral stripe of rainbow trout. Least depth under ⅔ distance from anus to tail but over that in landlocked salmon. (2) rainbow trout; *Salmo gairdneri.* Adults with broad pink side band, with brown spots on tail; young, with dark spotted or margined adipose. Length, 5 months, 2 to 3 in; 1 year, 4 to 6 in.; 2 years, 8 to 9 in.

Migratory form (1) known as steelhead. Rainbows which do not migrate show differences in color. In clear lakes they generally lack color; may be blue to green on back, silver on sides, and white on belly when immature. Stream forms tend to have well-developed spots on body, upper fins, tail. Mature fish become somewhat darker and usually develop a red lateral band. Migratory steelhead has color like lake form in the sea and develops spots and red band as it moves up the rivers to spawn.

Migrates upstream to spawn February to June. Female builds nest in riffle in gravel. Eggs 200 to 21,000, heavy, not sticky, ⅕ in., fertilized, buried in gravel, and abandoned; hatch at 57°F in 22 days; known to grow 1 in. a month at 63°F under ideal conditions. Breeding male of (2) shows conspicuous rosy side bands. Both species breed more than 1 year, thus differing from Pacific salmons. Weight of (1) to 103 lb; of (2) to 40 lb. Line record, 37 lb, 40½ in., 1947.

(1) formerly important food and game fish from Delaware north along Atlantic Coast; now practically gone in United States; some resident in lakes and some found in Maine. Originally in Hudson River. (2) native of Pacific Coast but widely established over world. Superior food and game fish that can survive 83°F water temperature if aerated. Possibly at best in streams connected with sea, though record fish caught elsewhere.

Brown Trout
Salmo trutta

Yellow-brown with black or brown spots only slightly developed on tail, with red spots often blue-bordered and adipose of young, orange. Lower fins white or pale yellow. Scales larger than in brook trout. Length, 5 months 2 to 3 in.; 1 year, 4 to 6 in.; 2 years, 8 to 9 in.

Body usually 4½ to 5 times as long as deep; fins with soft rays, tail forked in young fish and becoming more or less square in old trout. Brown trout in landlocked lakes may become easily confused with landlocked salmon; are both silvery in color with black spots.

Spawns in fall, running upstream to shallower water and breeding much as does brook trout. Eggs hatch in 31 days at 57°F; fertilization about 99 percent. Nest abandoned. May breed year after year. May breed in water to 4 ft deep but more commonly in shallower waters. Weight to 40 lb. Line record, 39½ lb, 1866.

Food both aquatic and terrestrial insects; as well as mollusks, crayfish, other fish. Large browns sometimes take frogs, birds, mice, and other small mammals.

Introduced from Europe and widely established. Will drive out brook trout in quieter warmer waters. Can survive water temperature at 81°F but prefers cooler. Crossed with true Loch Leven trout, which were introduced into America in 1884. May be of the same species.

A favorite of fly-fishermen the world over because it feeds at the surface on emerging caddisflies, Mayflies, and stoneflies. Feeds at night frequently; some of the largest browns are caught at night during the summer months. Has been successfully stocked in a variety of water habitats.

Cutthroat Trout
Salmo clarki

Weight to 41 lb. Anal rays, 10. Commonly heavily spotted with black. Sides without red spots. Red usually under jaws on each side. Scales rather coarse. Breeding male shows rose band on sides like that of rainbow trout. Body about five times as long as deep; tail very slightly forked, all fins soft-rayed. Feed on aquatic insects, terrestrial insects, crayfish, various small fish. Apparently cutthroats that migrate stay near shore in estuaries where food is abundant. Food made up of insects, snails, small crabs, and minnows, and plankton.

Spawns in spring, in nest built by female. Female may produce to 7,000 eggs; fertilization to 90 percent; spawning takes place with one individual over a short 2 days or so; eggs hatch about 50 days after being laid. Cutthroat hybridizes with rainbow trout in streams in Wyoming and Montana; there, cutthroat-rainbow trout has red line under jaw; same cross has been made successfully in hatcheries.

Excellent sport and game fish in West, some found east to Montana and Yellowstone; present in most Pacific Coast streams. Destroys many eggs of salmon and rainbow trout, as does rainbow itself, and has been subject to bounty in some places because of this.

Some of its common names reflect its distribution; Alaska cutthroat, Snake River cutthroat, Arkansas cutthroat, Yellowstone cutthroat, Columbia salmon-trout, Montana black spotted trout, Tahoe cutthroat, Utah cutthroat, Colorado cutthroat, intermountain cutthroat, coastal cutthroat. Because cutthroat hybridizes, it is a poor competitor against other fish, and is unable to withstand heavy fishing pressure. Populations of cutthroat have been seriously depleted.

Lake Trout
Salvelinus namaycush

Length, 5 months, 2 to 3 in.; 1 year, 6 to 7 in. Weight to 80 lb. No red spots but large white spots show on sides and back against a darker background. Generally blue-gray to bronze-green in color. Flesh color varies from red to white. "Cristivomer" refers to toothed crest of bone in roof of mouth. Line record, 63 lb, 47½ in., 1930.

Breeds over hard bottoms, in water 3 to 100 ft or more deep, a 24-lb fish laying to 15,000 eggs. No nest is made but bottom is "swept" clean by the fish before spawning. Eggs suitable for rearing in hatcheries but it is believed fish do not live in waters usually stocked by hatchery-reared fish.

Found in large cool lakes from New England to Montana, British Columbia, and Alaska. In northern part of range, for example, Labrador, northern Quebec, and Alaska, it may inhabit streams which empty into lakes. Generally found in lakes that have an adequate supply of oxygen in deeper areas. Great Lakes region a valuable source of commercial fishing. Supply decreasing dangerously because of modern techniques of net fishing, lamprey attacks and continued pollution of lakes from municipal and industrial wastes. With wise management should be abundant. Also called togue, mackinaw, and gray trout. One hybrid form, the splake, is a rather successful cross between lake trout and the brook trout. Takes its name from *sp* (speckled) and *lake*. Offspring are fertile, body heavier than brook trout and slimmer than lake. Offspring of the cross mature earlier and grow faster than either parent.

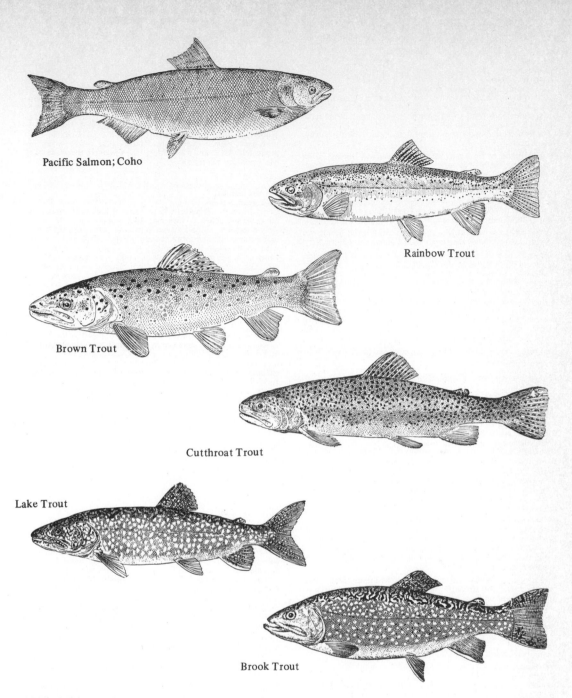

Pacific Salmon; Coho

Rainbow Trout

Brown Trout

Cutthroat Trout

Lake Trout

Brook Trout

Brook Trout
Salvelinus fontinalis

Length, 5 months, 2 to 3 in.; 1 year, 4 to 6 in.; 2 years, 8 to 9 in. Weight to 17 lb. Back with wormlike markings. Body red-spotted. Lower fins with white, black, and orange. Color varies with bottom, food, and other factors. Flesh firm. Superior food.

Brookies, as they are called by some, feed on insects, crustaceans, mollusks, and other fish. More a subsurface feeder than other trout.

Starts upstream about October, males first. Female makes nest in riffle over gravel and mates. 3-year fish lays 100 to 300 eggs; older, to 5,000; in nature 80 percent fertile; hatch in 44 days at 50°F; in 90 days at 40°F. Male may guard nest 3 weeks. Male, at spawning time, may be orange on belly side and black on lower part of sides. May spawn at 2 years, at 8 in. in length. Line record, 14½ lb, Nipigon River, 1916.

At least four fish are major predators on brook trout: northern pike, walleye, lake trout, Arctic char.

Found in streams with maximum temperature of 66°F but can survive to 75°F. Labrador Peninsula to Georgia, Montana, and Saskatchewan. Sea-run in Labrador. Introduced all over world in suitable waters. In some places as in eastern United States freed from hatcheries at legal catching size often only a few minutes before taken.

Suitable habitat in the East, where it is a most popular prize for fly-fishermen, is rapidly decreasing because of all kinds of pollution—industrial, municipal, agricultural. Brook trout, though a favorite of fly-fishermen, have been taken with live bait, spinning lures, and plug casting. Fly-fishermen generally use the fly which "matches the hatch" of insects.

PHYLUM CHORDATA. CLASS OSTEICHTHYES

Order Clupeiformes./Family Osmeridae

Smelt
Osmerus mordax
With form of salmon, but with larger scales and with large teeth in jaws. Related to the marine capelin. Length to 14 in.; average, 8 in. Small silvery-sided fish with short dorsal fin, slender body, and an adipose fin. Color may range from brown to blue above but silver on sides and belly is standard. Live or freshly caught smelt appear to have a bluish-purple iridescence.

Enter streams of fresh and brackish water early spring. 2-oz fish may lay 50,000 heavy sticky minute eggs and return to sea. Females larger than males. Spawning takes place on sandy beaches. Eggs are sticky and sink to the bottom where they adhere to each other, to grains of sand, weeds, pebbles. Eggs hatch in 20 to 30 days depending on water temperature. Fry are plankton feeders; later feed on crustaceans and insect larvae. Adults may feed on small fish as well as insects.

Gulf of St. Lawrence to Maryland and landlocked in various Eastern lakes. Other smelts are found along the Pacific.

Occurs as marine, anadromous, or freshwater fish. Found at depths of 50 to 200 ft in the Great Lakes. Appear as coastal river inshore fish on the Atlantic, Arctic, and northern Pacific oceans. This is a schooling fish; rich in oils. Many think it is an excellent food fish. An important commercial species as well as sport species, as well as an item in the diet of large sport fishes. During its migration upstream, it is taken by dip net; smelt enthusiasts usually work at night with the aid of lights. Also taken on hook and line by ice fishermen. Has several common names: saltwater smelt, freshwater smelt, frostfish, icefish, shiner smelt.

Order Clupeiformes./Family Esocidae

Chain Pickerel
Esox niger
Length to 2 ft. May weigh 2 to 3 lb. Few exceed 6 lb. Line record: 9 lb, 5 oz, 29.5 in. long taken through the ice in 1954. Possesses an elongate body which bears black, chainlike, wormlike markings on each side. Body green to olive-brown to bronze above. Snout resembles a duckbill when viewed from above. Body long and slender like northern pike and muskellunge. Also known as pike, river pike, pickerel, grass pike, jack, and jackfish. Cheeks and gill covers, entirely scale-covered. Mouth larger, terminal, well-armed with sharp teeth. Scales small and mostly uniform. Scales in lateral line, about 125.

Mates early spring, in shallow water, producing several thousand eggs fertilized as parents swim side by side; no nest. Spawning begins when water temperature reaches 47 to 50°F. One female may be attended by one or more males. Eggs hatch in 2 weeks, not sticky. Young unprotected. Food, any animal that can be overcome.

Prefers quiet, shallow waters of ponds and lakes. Sometimes found in sluggish streams. Common in ponds and streams among vegetation from Maine to Florida, west to Great Lakes and Mississippi basin. Excellent game fish, intolerant of competition; good food but bony, so unpopular with some fish specialists. One of the more important warm-water sport fishes. It is also a favorite for winter ice fishing.

Northern Pike
Esox lucius
Length, 5 months, 5 to 7 in.; 1 year, 8 to 12 in.; 2 years, 14 to 17 in. Maximum to over 4 ft. Weight to over 46 lb. Blue or greenish-gray. Scales appear small. Cheeks entirely scaly, but gill covers unscaled on lower half. Markings in adults, whitish spots.

Breeds in spring in shallows. In southern part of range, may spawn at 1 year of age; but generally, overall, they spawn at 2 years. Usually spawn in early spring shortly after ice melts. Parents swim side by side, dropping and fertilizing over 100,000 $\frac{1}{10}$-in. nonsticky eggs among weeds. Eggs hatch in 2 to 3 weeks, without parental protection. Food almost entirely other fishes caught by sudden darts. Line record, 46$\frac{1}{8}$ lb, 52$\frac{1}{2}$ in., 1940. Natural range covers a greater part of the world than that of any other freshwater sport fish.

Superior game fish with excellent food qualities. Can be reared in hatcheries but is not commonly so treated. From Alaska to New York and Ohio, also in northern Europe and Asia. Introduced and established elsewhere where conditions are suitable. Feed best at 60 to 80°F. Now reared in hatcheries for stocking purposes. A predator, so sometimes used as a control measure in waters overpopulated with stunted fish. Known by many names: great northern pike, jackfish, jack-pike, jack, pike, pickerel, and snake. Pike taken by anglers are usually large and are fighters when hooked; therefore, they are a popular game fish.

Muskellunge
Esox masquinongy
Length to 8 ft. Weight to 75 lb. Lower halves of cheeks and gill covers scaleless. Markings, dark spots or bars. May possibly hybridize with pike even in nature. Line record, 62½ lb. Color varies considerably; usually olive to dark gray. A fish with an elongate body about 5 to 6 times as long as deep. Front of head shaped like a duck's bill. Head scaled; Northern Pike with head partially scaled. Dorsal and anal fins set well back; makes fish appear longer than it is. Largest member of pike family. In Canada, officially recorded by name *maskinonge*. Has numerous common names, e.g., muskallunge, lunge, musky, barred muskellunge, tiger muskellunge, great muskellunge. Called *piconeau* by early explorers on the Ohio.

Mates at 35 to 50°F; spawns at 50 to 60°F, April to June, often in shallow streams. Spawns in Wisconsin as early as April 10 and continues through May; in the Great Lakes spawns from mid-May to mid-June. Spawning usually occurs at night in shallow water; generally among sunken tree stumps and logs. A 35-lb fish lays to 265,000 eggs, each $\frac{1}{11}$ in. through, not sticky. Young and eggs not protected by parents. Food entirely other animals, mostly fishes.

Lakes and large rivers of Great Lakes region, occasionally in Ohio Valley, or even to North Carolina and in Tennessee River system. One of best game fishes. Good food. Highly destructive to associated fishes. Feeds best at 60 to 80°F. A predatory fish, the muskellunge consumes a wide variety of fish in its diet, e.g., perch, suckers, and shiners; also feeds on frogs, crayfish, and large water insects; has been known to take young ducklings, shorebirds, and even young muskrats. Voracious.

Order Anguilliformes./Family Anguillidae

Common Eel
Anguilla rostrata
Length to 5 ft. Skin with minute scales partially in groups. In some other eels as conger, pike, worm, and snake eels, skin is scaleless. Also called freshwater eel and silver eel; a long, almost snakelike fish. Dorsal fin originates far behind pectoral fins. With large mouth and a pointed snout. Colored uniformly greenish-brown to yellow-brown above fading to lighter shades on sides to almost white below. Grows very slowly; adults may be 5 to 20 years old. Male rarely over 3 ft. Migrate downstream to sea to breed.

Lays to 100,000 eggs, deep in sea southeast of Bermuda. Young *leptocephali* first year are flat, transparent, to 3 in. Second year ascend rivers, reaching to 8,000-ft elevation. Remain inland for 8 or more years, returning to sea to breed and die. May spend winter buried in mud.

Valuable food fish caught mostly on downstream migration. Males rare in fresh water. Eels of different species found all over world except Pacific Coast of North America and a few Pacific islands. European eels breed in Atlantic. Because no ripe roe was ever found in females, it was thought, in earlier times, that the eel arose spontaneously from mud. Exceptionally good food fish; fried, baked, sautéed, jellied, made into chowder, smoked. Frequently served with a variety of tasty sauces. The tradition of fresh eels for Christmas season makes for a large market in the U.S., especially among families of recent immigrants from Europe. Commercial eel fishery along eastern coast occurs from points as far north as St. Lawrence River south to Chesapeake Bay.

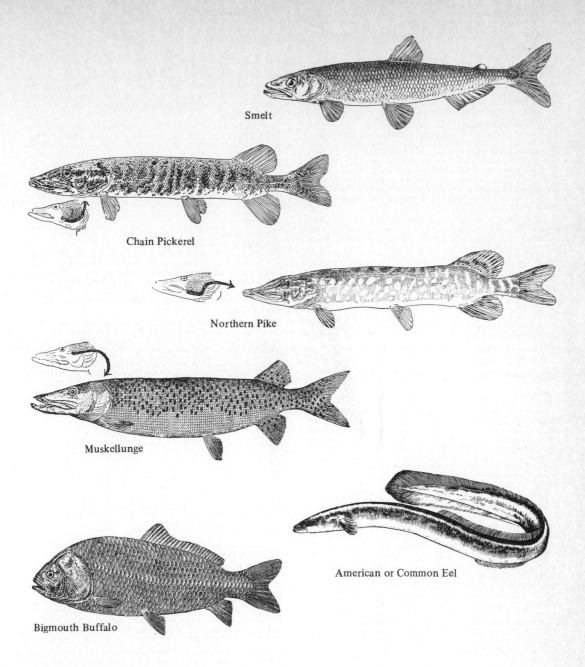

Smelt

Chain Pickerel

Northern Pike

Muskellunge

Bigmouth Buffalo

American or Common Eel

Order Cypriniformes./Family Catostomidae

Bigmouth Buffalo
Ictibobus cyprinellus

Largest member of the sucker family and probably the most important economically. Body robust in appearance, elliptical in shape. With large mouth at end of head, not suckerlike. Upper lip about in line with the eye. Dorsal fin begins about halfway from nose to tail, extends almost to tail. Fins without spines, tail fin somewhat forked. Lateral line distinct. Color, bluish green with a coppery tint above; fading to white on belly side. Weight on the commercial market to 12 lb, usually less; sometimes to 30 lb. One specimen reported at 80 lb.

Food largely plankton; small crustaceans. Takes few insect larvae and other bottom organisms. Spawns April to May when water temperature reaches 60 to 65°F. Females and males school in weedy area and discharge eggs and milt randomly, usually in water to 3 ft deep. 10-lb female may pro-duce ½ million eggs. Eggs hatch in 10 to 14 days; 1 year old, 4 to 6 in. long; then in subsequent years, 11, 15, 18, 20 in. Weighs about 10 lb when 25 in. long.

Most common in the Plains States; North Dakota to Saskatchewan east to Ohio and Pennsylvania, south to Gulf. Found in large rivers, shallow lakes; becomes overly abundant in fertile lakes.

Not a sport-fish; seldom taken by angler. With black buffalo, *I. niger,* and small-mouthed buffalo, *I. bubalus,* is an important commercial fish. Flesh white and delicious. Fish which brings 20 cents per pound to the fisherman may sell at over 80 cents in the market.

Common Sucker
Catostomus commersoni

Length, 5 months, 1½ to 2 in.; 1 year, 2 to 3 in.; 2 years, 5 to 7 in. Maximum to 28 in. Mouth with sucking lips, as in all suckers. Scales small. Flesh bony and usually soft. Olive-brown, cylindrical fish. Rounded snout projects only to or somewhat behind upper lip.

In early spring run upstream. Breed in swift water, in shallow gravel, pushing bodies into gravel. Male changes color during spawning time; develops much brighter coloration. Young unprotected; swim near surface and feed about 10 days, then take to bottom. Some mature at 6 in. Adults may return to deep water to await breeding season. Food algae which it "sucks up" from bottom ooze; includes mollusks, crustaceans, and insect larvae. Apparently prefers larger streams; deep water if in an impoundment. Frequently in dense beds of weeds.

Ranges Labrador to Montana, Georgia, and Missouri with over 60 American species and 2 of eastern Asia. Over 3 in. eat small animals and may destroy fish eggs. Caught by hook and line or speared over spawning bed.

Tolerant of a wide range of environmental conditions: pollution, turbidity, siltation, and low oxygen concentration. An inferior but popular food fish; soft and bony. In the spring or in very cold water the flesh is firm. Flesh better in winter, so may be speared through the ice. Probably of some value as a forage fish which is eaten by more-favored sport fish. Also makes a good "live-bait" fish and is raised under hatchery conditions and sold in some areas as bait. Some "worm fisherman" like it because it will take the hook and put up a good fight.

Order Cypriniformes./Family Cyprinidae

Goldfish
Carassius auratus

Well-known aquarium fish, usually red and white but some black and brown, with relatively large scales and often with unusual eyes and fins. Similar to a carp but without barbels on the jaw and has a different color pattern. No other minnow has a spine in the dorsal fin and the anal fin. Brown in nature. Fish with a variety of colors if bred in captivity; it reverts to natural colors, brownish-olive with bronze case, if returned to nature. Length to over 18 in. Originated in Orient but established over world.

Mates during day, April-May, laying 10 to 20 amber sticky eggs at a time to total over 500 which stick to supports, hatch at 70°F in 3 to 7 days. Good breeder has 2-in. body or longer, matures second year; breeds until 9 years and may live to 15 years. General scavenger. Food consists of vegetation, insects, insect larvae, crustaceans, small snails, and zooplankton in general.

Best in water containing lime, between 70 and 55°F. Has good commercial value for aquarium stock. In ponds, water enriched by mixture of ⅔ sheep manure and ⅓ superphosphate fertilizer, which encourages food eaten by goldfish. Valuable as an aquarium fish, of course; but detrimental to game fish in the wild. Its habits, churning up the bottom, like the carp, make it undesirable. Also it is difficult to eradicate if it is once successfully introduced to a natural waterway. Not quite so tolerant to adverse conditions as the carp. Able to hybridize freely with carp. Anglers should not use goldfish as live bait. They may escape and become a pest because they multiply rapidly.

Carp
Cyprinus carpio

Coarse-scaled, soft-fleshed, dark-backed, brownish or golden-sided, with four barbels near mouth. Length, 5 months, 3 to 5 in.; 1 year, 6 to 8 in.; 2 years, 12 to 15 in. American maximum weight, to 42 lb, 1930. Swiss, 1825 record, 90 lb.

World record rod and reel carp; 55 lb and 42 in. long. A big husky minnow which is deep through the body and compressed laterally. With single dorsal fin which extends along about ⅜ of back. Has other common names: German carp, scaled carp, mirror carp, and leather carp. Scaled carp have a complete covering of scales; mirror carp have patches of irregularly shaped scales on back and sides; leather carp have none to very few scales. All forms interbreed.

In New York breeds May to June, a 6-lb fish laying to 2 million eggs which hatch in 5 to 12 days. Breeds and feeds in shallow water. Matures in 2 to 3 years. Does not protect young. Food, smaller plants and animals strained from bottom muds. Will eat insect larvae, small mollusks, crustaceans, small fish, and young aquatic plants. Plows up the muddy bottom when it feeds; this action causes siltation of eggs of valuable fish and destroys cover which could be used by young fish.

Supplants superior fish by destroying plants and nests, eggs and young. Some 250 tons harvested in New York as cheap food and to protect more valuable species. Found in fresh waters native in Asia and Europe and introduced unfortunately to American waters.

Some anglers classify carp a valuable sport and food fish. Some fishermen shoot carp with bows and arrows where they are large, abundant, and in shallow water. Carp are also a contributor to commercial fishing industry.

Golden Shiner
Notemigonus crysoleucas

Length, 5 months, 1 in.; 1 year, 1½ to 2 in.; 2 years, 2½ to 3 in. Maximum 12 in. Golden-green in upper parts, with silvery sides and yellow fins. Breeders bodies yellow. Lateral line follows outline of bottom of fish. Has a comparatively deep body which is somewhat compressed laterally. Young fish not golden, frequently silvery in appearance; take on gold color as they age.

Prefers weedy ponds or quiet streams. Sometimes schools in open water and uses weed beds for protective cover. Adapted to wide range of conditions from cool trout streams to warm lakes.

Spawns in sods, on water plants, brush, or submerged roots. 10 pairs of young need pond 3 by 30 by 100 ft, supplied with 14 bu fresh barnyard manure and 15 qt soybean meal in May; and for rearing, with 15 percent cottonseed meal and fish meal.

May eat plant materials and thrive but will raid fish nests of more desirable species. From Nova Scotia to Dakota, south to Tennessee and Texas. Has been widely introduced in our western states. N.c. auratus, western golden shiner, increases range to the west. Diet consists largely of planktonic crustaceans and also includes aquatic insects, mollusks, and algae. Raised as bait minnows, ⅛ acre may yield 4,000. Known to bait dealers and sportsmen as "pond shiner." Important as food for favored sport fish, smallmouth bass. May serve as panfish if of large size. This is one of the fish which youngsters catch on simple tackle, so may be considered a "sport fish" to that age level. Large western golden shiners are taken by fly-fisherman for food or sport, especially in natural lakes, where they may be 8 in. long.

Horned Dace, Creek Chub
Semotilus atromaculatus

Length 5 months, 1 to 1½ in.; 1 year, 1½ to 2 in.; 2 years, 2½ to 4 in.; maximum, about 10 in. Scales crowded in front of dorsal fin, which has dark spot at base. Sides and belly silvery. Has a single small barbel at end of each jaw; at times barbel is hidden. With a large mouth; mouth reaching to or beyond front of the eye. Adults with dark spot near the front base of the dorsal fin. Breeding male steel-blue above and with tubercles on head.

Male builds nest, May to July, moving pebbles to make circular pit with upstream ridge of gravel and sand. Breed when water temperature reaches about 65°F usually in rocky riffles near the headwaters of streams. Young reach a length of about 3 in. at the end of the first year. One or more females enter, mate, and lay eggs. Nest guarded temporarily by male at least while he is in breeding mood. Active. Food, chiefly small aquatic animals.

Common in small streams from Maine to Wyoming, south to Alabama and New Mexico, where nests are conspicuous above riffles. Found in almost all streams in its range, tolerant to a wide variety of conditions. Edible. Attractive fish to small boys because it bites readily but is inferior to associated trout. Takes artificial fly rather readily. A prized bait species; is eaten by predator sport fish and has some value as a panfish. Some enthusiasts take them in ice fishing. Relative, Sicorporalis, the fallfish, chub, silver chub, or dace feeds on insects, crayfish, worms, and small fish. A good panfish when other species are not available which will take a fly or baited hook. Some anglers sour their chubs by placing them in a capped fruit jar in the sun a few hours prior to their use as bait.

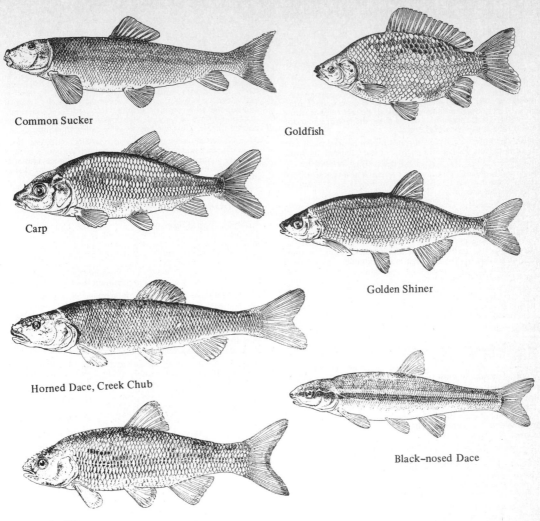

Common Sucker

Goldfish

Carp

Golden Shiner

Horned Dace, Creek Chub

Black-nosed Dace

Common Shiner

Black-nosed Dace

Rhinichthys atratulus

A small, scaled member of the minnow family; length to 3 in. Dark brown color on back, white on belly side, with a dark line along each side dividing the two colors (brown and white). Frequently shows blotches along the sides. At spawning time, dark stripe of males, along side, may become orange or tan; pectoral fins are colored similarly. Upper jaw only slightly longer than lower jaw; mouth somewhat oblique. Air bladder fairly well developed.

Spawns in May, usually, but ranges from spring to early summer. Male guards small area over gravel or sand in shallow riffles. Male clasps ripe female and forces her eggs into the sand; then fertilizes them. Male not confined to his own small area for spawning purposes. May leave his nest area and mate at any place nearby where he finds a ripe female. Eggs are 1/16 to 1/18 in. in diameter. Fingerlings grow to 1½ or 2 in. in length by the end of the first year.

Widely distributed from St. Lawrence River south to Alabama, west to Minnesota and Nebraska. Western blacknose dace, *R.a. meleagris*, increases its range to the West Coast. Food chiefly small animals; aquatic insect larvae, in particular. Appears to be constantly darting about examining the bottom for food.

Easily caught in minnow traps or with fine-meshed seines. Used often as bait fish, in particular for trout. Also used as live bait for bass and catfish but not so useful for crappies and panfish; perhaps because it is dull-colored as compared to shiners. A useful forage fish for larger sport fish. Competes with game fish for insect larvae. Good aquarium fish but survives only if water is well aerated.

Common Shiner, Redfin

Notropis cornutus

Length to 8 in. Scales large. Sides and belly silvery; deep and strongly compressed body; usually olive above. Scales and dark reflections often give the appearance that some of the scales have been removed. Snout rounded. About 100 related species. In spring, male is pink-sided, roughened in front of dorsal fin and larger than female.

Male builds nest in shallow swift water. Eggs laid by many females in one nest when temperature is about 73°F. Nest made by rooting in gravel, May to June. Male defends nest but not young. Male often uses nest of another species of minnow; probably this explains occurrence of hybrids in some places. Less than 50 orange eggs laid at a time. Incubation period unknown. Food, mostly terrestrial and aquatic insects but will take many kinds of algae.

Common in streams east of Rocky Mountains to north of Texas and James River in Virginia. Prefers pools in clear, small rapid streams; also found in fast moving water. Related species extend range south. Attractive aquarium fish. Hardy bait minnow. Does not compete for food with more valuable game species except possibly for surface insects as in a trout stream.

Will take an artificial fly; bothersome to trout fisherman. It, like some other minnows, is the fish which frequently introduces young boys to the sport. Can be eaten; flesh good but bony. In some localities known as shiner, creek shiner, or chub.

Fathead Minnow
Pimephales notatus

Length to 4 in. Olive with bluish sides, dusky toward base of dorsal fin; with black blotch toward base of dorsal fin in breeding male.

A silvery minnow which appears to be crosshatched. Usually has a dusky blotch on forward portion. Back appears flattened. Mouth almost all on underside. Head broad. Distinguished from other minnows by dark stripe that extends across the gill cover, through the eye and along the lateral line to the black spot at the base of the tail.

Breeds in June. Eggs placed singly on flat undersides of submerged objects in patches to 6 in. long; attended by male, whose head in spring is wholly black. Snout tubercled.

Egg masses cleaned by the male. Average nest contains 2,500 eggs. Spawn at 70°F. Eggs hatch in 8 to 12 days. May eat eggs of other species.

Variably abundant Quebec to Dakotas, south to Gulf States. Commonest in shallow rivers over stones or vegetation. Important bait minnow. Young lack black on dorsal orifice. Prefers firm bottom of lakes and streams; however, found in a variety of habitats. Can withstand high degrees of turbidity and pollutants. Appears to be common in lakes and streams rich in organic matter and high in phytoplankton population. Food small plants; crustaceans, and insects, both larvae and adults. Probably contributes significantly to the diet of game fish and panfish. An important forage fish for larger fish. Is used by fish culturist in ponds in which smallmouth and largemouth bass are reared. Used by anglers in fishing for crappies, perch, white bass, and other panfishes.

Cutlips Minnow, Nigger Chub
Exoglossum maxillingua

Length to 8 in. Compact, dark, with lower jaw conspicuously three-lobed, unlike most other minnows. Fins plain. Breeding male dark to almost black and longer than female.

Olive-colored above; purplish along each side fading to light colors on belly side. Although 8 in. is reported as length, more frequently about 7 in. in length; few grow to more than 7 in. in length. A bottom dweller which feeds on insects, snails, worms, and small crayfish.

Male builds 18 by 5 in. nest of ½-in. flat stones, May to June, in 3 to 30 in. of slowly moving water at 70°F; fights off other males, mating successively with many females. Develops no breeding tubercles and avoids contact with intruders. Male apparently attempts to herd females over its nest. Eggs ¹⁄₁₀ in. yellow, glossy; hatch in 4 days, young leave nest after 6 days.

Abundant in slowly moving streams from Vermont to Virginia. Trout fishermen frequently catch it. Never large enough to be important. May be considered as food for other species and enemy to their young. Fair as a bait minnow. Relatively hardy.

Less popular as a bait species because of its dull coloration. It is, however, a favorite of young beginning fishermen because it takes a hook readily.

Order Cypriniformes./Family Ictaluridae

Brown Bullhead
Ictalurus nebulosus

Length, 5 months, 2 to 2½ in.; 1 year, 3 to 4 in.; 2 years, 5 to 7 in.; maximum, 20 in. Mouth large. Scales lacking. Barbels conspicuous. Dark to silvery. Colored olive to brown, with dark mottlings, generally on sides, fading to white or yellow below. When adults, generally average from 8 to 10 in. in length and weigh less than 1 lb. About 1,000 species known. Taste organs along undersides. Related Mississippi catfish caught in 1878 weighed 150 lb. In the East, this is the most popular of catfish family.

Mates in late spring. Eggs laid in nest built and guarded by male. About 2,000 ⅛-in. eggs, incubated about 5 days, often by male removing silt. Coal-black young stay in school protected by male during early part of summer of first year, the schools making conspicuous dark patches in water at times.

Common in quiet warm mud-bottomed waters from Maine to Dakota, south to Texas and Florida. Distributed in southern

Canada from Nova Scotia and New Brunswick to Manitoba. Introduced throughout North America especially in the western states. Common in smallest ponds and largest lakes; most frequently in quiet areas of creeks and rivers. Feeds on insect larvae, fish, crustaceans, and almost anything on the bottom and available. Introduced in California. Superior panfish, and popular with young fisherman. Excellent food with ready market. Spines of pectoral and dorsal fins locked can inflict painful injury. Can live out of water rather long time.

Also known as common bullhead, speckled bullhead, mudpout, cat, and squaretail catfish. A panfish without competition and a complete delight to the bank fisherman. This species is frequently the first member of the bony fishes introduced to young anglers.

Order Cyprinodontiformes./Family Cyprinodontidae

Killifish, Top Minnow
Fundulus sp.

Well over 100 species. Length to 4 in. Lower jaw extends up and beyond upper. With a depressed head and an elongate snout. Anal fin normal, not sword like as in male *Gambusia*. Tail not swordlike; usually almost square or slightly curved along tip. Males smaller, more brilliant than females.

Banded killifish, *F. diaphanus,* is green to olive-green above and white on lower sides and belly side. Adults 2-4 in long. Striped killifish, *F. majalis,* has a pointed snout and pointed fins. In color, olive-drab above fading along the sides to a pale olive belly. Male has black, longitudinal stripes. Length to 8 in. Spawning occurs from April to September. Few large eggs laid on sandy bottom of a stream.

Eggs fertilized outside female and young develop outside mother. Food almost entirely small forms of animals like mosquito wrigglers, caught at or near surface.

Found in fresh and brackish waters of Europe, Africa, and America. One common species, *F. heteroclitus,* ranges Maine to Mexico. Satisfactory aquarium fish. Excellent mosquito control in pools that do not dry completely. Will take small crustaceans, mollusks, worms, and other invertebrates.

Is eaten by some predatory fish which are sport fish. Frequently used as bait. A marine form of killifish called the mummichog is a desired bait species.

Order Cyprinodontiformes./Family Poeciliidae

Sail-finned Molly
Mollienesia latipinna

Length to 3½ in. Male with extraordinarily large dorsal fin. Olive, with dotted brown spots on sides. Tail marked with iridescent light blue. Rarely entirely black (one in several million). Called the common sailfin in the Southern States. Male with large dorsal fin beginning not far behind the head; overlaps tail fin when laid down, with 13 to 14 rays.

Courtship display of male involves quick motions; he displays his large dorsal fin before the female and may even hold the female with the dorsal fin.

Eggs fertilized in mother; sperm transferred into oviduct of female by means of the male's modified anal fin; development varying with temperature from 6 weeks to 6 months, though shorter time is more common. Change of water may hasten delivery when a dark spot near vent indicates imminence. Young, well-developed; breed at 3 to 6 months.

Native of southeastern United States and Mexico. Over its long range, the species breaks up into local races or subspecies. There are saltwater races in Key West and Pensacola Bay. In nature, live in salt, brackish, or fresh water. Aquarium-kept fish should be treated periodically with salt solution in increasing strength, beginning with 2 tsp to 1 gal of water and adding 1 tsp a day for 3 days.

A favorite of aquarists is the black form of *M. latipinna,* one of the most sought-after live bearers. Not difficult to keep or feed, is an adaptable member of an aquarium containing a variety of species.

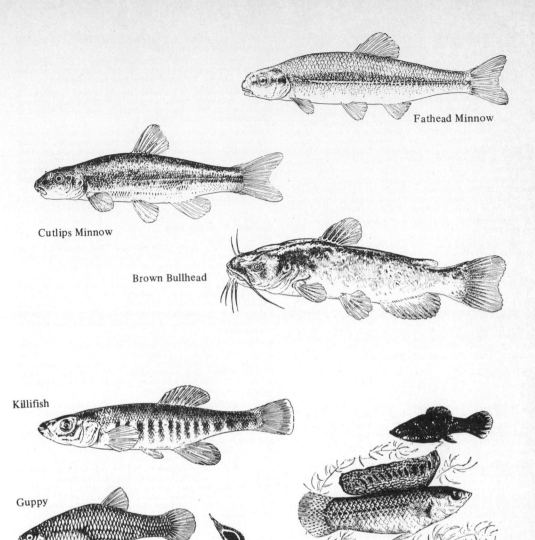

Fathead Minnow

Cutlips Minnow

Brown Bullhead

Killifish

Guppy

Female

Male

Sail-finned Molly

Guppy
Lebistes reticulatus

The most popular and widely kept aquarium fish. Male smaller, about 1⅛ in. long; female twice that size. No two males colored exactly alike, apparently. Several strains of guppies have been established, each with its characteristic markings or fin development. Male is colorful, female is rather drab. Female with young develops large "pouch" on belly side which may reach from below operculum more than halfway to tail fin and somewhat surround the anal fin.

Fertilization is internal. Male pursues female. Copulatory organ, a part of anal fin of male. Gestation period varies but female may produce brood every 4 to 6 weeks. Number of young produced varies; reported as 10 to 45 depending on age of female; over 100 per brood recorded. Adults eat young.

Native of Trinidad, the Guianas, and Venezuela.

A beautiful aquarium fish which has introduced many beginners to the aquarium hobby. Since it is a live bearer, it is of special interest because it is easy to maintain an abundant supply. Also it is an extremely fertile and dependable breeder, matures in short time, 6 to 8 weeks, is unusually active, thrives in close quarters, will stand foul water; can withstand a wide temperature range from 65°F to 100°F, but does best from 74°F to 82°F, does not fight, and is not expensive. Guppies eat most prepared foods. They grow and breed best with a diet consisting of finely chopped liver, fish, shrimp, white worms, *Daphnia,* and tubifex. Prepared foods are available.

Mosquito Fish
Gambusia affinis
Sometimes mistakenly called a guppy. Guppy is *Lebistes reticulatus* (preceding species). Light olive; appears to reflect a rich light blue over gray-to-olive color; each scale dark-edged. Blue coloration dominant on lower part of tail fin, also on first ray of anal fin, and the gonopodium of the male, is intensified about the head; violet-colored eye. Length about 1½ in., female sometimes to 2 in.; male about 1 in. long. Tail fin rounded; never square-ended or notched.

Fertilization internal, female with pouch in which young are carried. One fertilization with sufficient spermatozoa to fertilize several broods. Livebearer. Young emerge 12 to 28 days later. 10 to 300 miniature adults produced with each brood. Breeding temperature, 75°F. Quite apt to eat young. Not prolific. Three or four broods have been produced in a season in the north.

Occurs in fresh, brackish, and salt water. Widespread. Middle Midwest, Illinois, and Indiana, South to Florida and Mexico. Introduced in many localities coast to coast, where appropriate habitat exists. Common in the southern U.S.

Hardy aquarium fish; temperature of aquarium should not drop below 68°F.

Not a good member of a colonial aquarium with a variety of species because it will eat the tails of some of the larger, more beautiful specimens. Fine dried foods or tiny living things can be fed in an aquarium. There is some confusion as to identification of two forms, *G. affinis* and *G. affinis holbrookii*. Apparently they are the western form and eastern form respectively and interbreed freely where their ranges overlap. Important forage for young black bass, sunfish, and young barracuda. Some mosquito fish enthusiasts believe the Panama Canal and Panama exist in part because of *Gambusia* and its appetite for mosquitoes. A valuable fish because of its diet of mosquito larvae and pupae. Has been widely introduced out of its natural range as a mosquito control measure.

Swordtail
Xiphophorus helleri
Length to 4 in., exclusive of sword. Active. Brilliant olive-brown on back; brownish-green, blue, blue-green, or steel-blue on sides, with longitudinal red or red-brown stripes and sometimes three to five vertical bars on forepart of body and one to two at base of tail. Female larger than male and swordless.

Swordlike spike frequently shows an edging of black. Gonopodium developed from anal fin, not a part of the swordlike spike.

Aquarium swordtails should be kept at 72° to 80°F. One male breeds with four to six females at 72 to 76°F. Young born in 6 to 8 weeks, numbering 30 to 200. In aquariums, young must be freed in weeds or traps to survive destruction by parents. Female should not be placed again with male until 1 to 2 weeks after bearing young. Sexes of young indistinguishable at first but should be separated when recognized.

Native of fresh waters of Atlantic slope from southern Mexico to Guatemala. Imported into Europe in 1909 as aquarium fish known as "Mexican swordtail." Probably easiest of aquarium fishes for amateurs to work with, and one of most satisfactory as classroom animal. Hybrids between *Platypoecilus* and this species show many interesting variations.

Swordtails eat algae, but finely chopped worms do very well as a substitute. Scraped beef and liver, chopped clams, and small pieces of shrimp and a small amount of oatmeal can also be used as food. Males jump, so a covered aquarium is suggested. Glass or thick clear plastic is suggested as protected dust cover because it impedes the loss of water by evaporation.

Blue-moon Fish, Platy
Xiphophorus maculatus
Length to 3 in. Variable from light gray with tail crescent, to red, yellow, or black.

Original wild stock was grayish-brown to olive; sometimes with blue or green iridescence.

Selective breeding has developed color strains such as the blue, red, black, and golden.

Many early imports showed a dark moon-shaped mark at the base of the tail, hence the name, blue-moon fish or moonfish.

Female, 3 in.; male, 2 in. Male scarlet and black. Reproduction somewhat similar to guppies.

Platys are less inclined to eat their young. Does well in a thickly planted aquarium and will not jump out if not disturbed. Algae or prepared cereal food are acceptable.

Native of Mexico. In aquarium should be kept at around 70 to 85°F, with 60°F dangerously low. They are good community aquarium fish. Used commonly in study of heredity and as ornamentals by fish fanciers.

X. poecilus is a native of eastern Mexican waters. There are two principal color strains; one with yellowish sides, brilliant canary-yellow dorsal and deep yellow-to-reddish tail; the other with considerable blue on the body, yellow on the dorsal, and a deep red tail fin. Males are colored as described, females are drab. Somewhat hardier than *X. maculatus* and can withstand aquarium temperatures to 50°F for short periods.
Order Perciformes./Family Anabantidae

Siamese Fighting Fish
Betta splendens
Length to 2 in. Anal fin enormous. Tail rounded. Body velvety brown to black, with glittering metallic dots. Eye green, with red behind eye and on anal and tail fins. Body color varies from wide range of shades in blues, greens, and reds. Nearly all males have a pair of drooping ventral fins which may became brightly colored when the fish is excited. *Betta* is in full vigor and color between ages of 10 months and 2 years. The flowing fins are hereditary and can be produced only by careful mating. Female with anal fin less pointed than in male.

Males fight each other fiercely, tearing each other's fins. Male builds bubble nest. Female lays few eggs. Male fertilizes them, floats eggs with bubble into nest, and guards them until they swim a few days after hatching. Hatch in 2 days at 80 to 90°F. Prefers slightly acid water (pH 6.8) that is clear but has natural sediment on the bottom.

Used as fighting fish in Thailand, where they are native, bets being placed on possible winners. The natives consider a fish who will not fight for an hour a failure. Popular aquarium fish in America since 1927. Aquarium should be well-lighted and well-planted. Readily adaptable to changes in environment, standing a temperature range of 68° to 90°F. In a community aquarium it is best to have but one male and one female. Males, if produced, should be separated at about 3 months. Males will fight other males and their images in mirrors. Young may be eaten by guarding male when able to swim. Feed on *Daphnia*, mosquito larvae, chopped worms, bits of fish, crab, or shrimp. Independent in 1 month. All carnivorous.

Order Beloniformes./Family Exocoetidae

Flying Fish
Parexocoetus mesogaster
Flying fishes vary in length from 2-in. Monroe's flying fish to 18-in. California flying fish. Short-winged flying fish (illustrated) is 7 in. long; in flight shows four wings, the forward light and unpatterned.

A swift pelagic fish of tropical waters. Lower lobe of tail is elongated and serves as a sculling oar for moving rapidly along water surface before takeoff; tail is rudder in flight. Flight of California flying fish, *Cypselurus californicus*, has been studied in detail. Fish swims just below surface for some distance, upper tail lobe sometimes breaks water; then turns head upward and spreads pectoral fins; body except for tail leaves water; then beats tail vigorously in sculling motion; and taxis across water surface to gain flying speed. Rises into the air when it extends its ventral fins. Flight may be from 10 to 30 seconds; average speed 35 mph.

Some species rise to 4 ft above water and go to ⅛ mile through air. Most glide without a wing flutter. Live in schools and take to air in swarms before ships or preying fishes.

Spawn on parts of seaweed or on other floating objects. Eggs attached to substrate and to each other by silken threads.

Short-winged is common from Cape Hatteras to Florida and in Gulf Stream. Other species include two-winged Caribbean flying fish and the California flying fish found off Santa

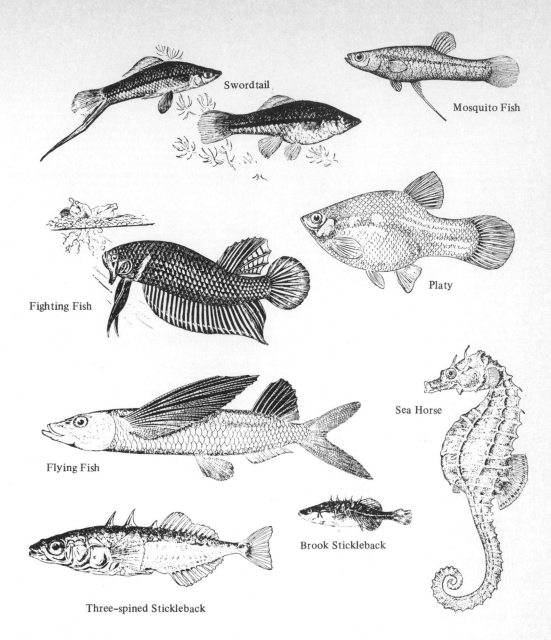

Swordtail

Mosquito Fish

Platy

Fighting Fish

Sea Horse

Flying Fish

Brook Stickleback

Three–spined Stickleback

Barbara and well known to tourists; an excellent food fish. Flying fish pie served as a renowned Caribbean dish. Difficult to take with tackle. Some sportsmen hunt flying fish using small-gauge shotguns.

Order Gasterosteiformes. / Family Sygnathidae

Sea Horse
Hippocampus obtusus
Length to 6 in. Tail prehensile. Gills in tufts. Head joins body at an angle which makes the head and neck horse-shaped. Horny rings surround body. No tail fin but curled tail can be used as a grasping organ to support the fish in vegetation. Color, light brown to gray or greenish depending on background. Swims by movement of dorsal fin but normally remains with tail coiled about seaweed. Male with pouch in front which extends from twelfth to eighteenth body ring. Spotted sea horse *H. erectus* has 10 trunk rings instead of 11 and is spotted.

Mates in spring. Female produces about 200 eggs which she places in male's pouch, process taking 1 to 2 days. Male incubates eggs 40 to 50 days until young, with remnants of yolk sac, are born. Young expelled one at a time when less than ⅓ in. long. At first, young swim in horizontal position even after developing prolonged snout. Sex differences apparent after few months. Reach maturity in 1 year. Feeds by sucking in small crustaceans and other minute animals through its snout.

H. obtusus found uncommonly in seaweed from Nova Scotia to Florida. Spotted sea horse ranges north to New Jersey. Other species common in warmer seas, particularly in Mediterranean. Dried bodies used as ornaments and when ground up believed by superstitious to have unique medical properties. Always popular in marine aquariums. Travels by wriggling the dorsal and pectoral fins. Not capable of rapid motion. Said to produce a rank odor when disturbed. Three species, *H. erectus, H. zostarae,* and *H. obtusus* are common on Atlantic shores of U.S. while *H. ingens* is common on Pacific Coast.

515

Order Gasterosteiformes./Family Gasterosteidae

Brook Stickleback
Eucalia inconstans
Figure is of related *Gasterosteus aculeatus*. Brook stickleback is to 2½ in. long, with five dorsal spines and naked body. Pelvic fins much reduced in size. Color variable, usually green to olive above, white to cream below. Other species, with plates on sides as shown and with nine, four, three, or two dorsal spines. Males usually smaller and darker than female. Spawns in April and May.

Male builds hollow nest of plants cemented with mucus, size and shape of walnut. One or more females fill nest with ⅒-in. eggs (to 300 per female), which are fertilized by male, put in nest, and guarded by him during 10-day incubation period, until young are independent. Breeds at 1 year and dies at 1½ years or less. Food largely small crustaceans and insects.

Common in small brooks, Maine to Kansas and northwest to Saskatchewan, with related species in seas and fresh waters of Europe and Japan. Prefers clear cold water such as cold streams, and springs. An indicator of unpolluted water. Young are ⅛ in. long when hatched and at 72°F may triple size in 6 weeks. Interesting in aquariums. Should not be kept with other species in an aquarium because of its pugnacious disposition. Alteration of environment by man, such as stream improvement projects and pollution have destroyed much stickleback habitat. Probably of some importance as a forage fish. *Pungitius pungitius*, ninespine stickleback, is a small fish which is circumpolar in distribution; lives in both fresh and salt water; in North America found from Alaska eastward to St. Lawrence River system south to New Jersey. Frequently found in deep, cold lakes.

Order Gadiformes./Family Gadidae

Codfish
Gadus morhua
Length to over 6 ft. Weight to 211 lb, but 50 lb normally a large fish; average, 25 lb. Dark-spotted, with light-colored lateral line. Chin with barbel. Dorsal fins, three. Anal fins, two. Tail, small and square.

Some cod populations migrate extensively; others remain in a limited area.

Spawning occurs from December to late March. Probably spawns promiscuously in great schools. A 70-lb fish may free over 9 million buoyant, 1/16-in. nonsticky eggs a season, which hatch in 17 days at 40°F. Food, as adults, plants and animals grubbed from bottom while schools of fish swim close to bottom.

Young cod feed on copepods and other small crustaceans near the surface of the ocean. As they grow older, they descend to greater depths and feed upon barnacles, shrimps, and worms.

In salt water, at depths to 1,500 ft off both coasts of Atlantic, south to Virginia but uncommon south of New York. Probably world's most valuable fish as food for man, source of cod-liver oil, rich in fat-soluble vitamins A and D, useful in preventing rickets.

Order Pleuronectiformes./Family Pleuronectidae

Winter Flounder
Pseudopleuronectes americanus
Length to 18 in., with record to 20 in. and record weight to 5 lb. Both eyes on one side of body, with fish lying on bottom on other side. Fins nearly encircle body. Generally pigment is on the right side, so termed a "right-handed" fish. Pigment may be on left side or rarely on both sides. Color reddish-brown to almost black; sometimes shows dark green cast.

Spawns winter and early spring, usually at night. Spawning depths 10 to 240 ft. Eggs heavy, sticky, to 1/32 in.; hatch in 15 days at 39°F and reach maturity as fish at third year. Young have eyes on opposite sides of body and swim like other fishes. Young feed on small plants and small crustaceans; then later change to worms, and larger crustaceans such as shrimp. Because of small mouth, even as adult, only small invertebrates and larvae or small fishes can be eaten. Therefore successful anglers use small hooks.

Occurs in salt and brackish waters in shallows from Labrador to Georgia, over sand or mud to depth of 120 ft. Excellent food fish for men and good sport for unambitious sportsmen of the seashore. Known regionally as flatfish, blackback, blueback, black flounder, and mud dab. Flesh is firm, white, and delicately flavored, but bony. Are prepared for eating by frying, baking, or broiling. A favorite with saltwater anglers.

Some other flounders include *Paralichthys dentatus*, the summer flounder, also known as fluke or plaice; *Hypsopsetta guttulata*, the diamond flounder, also known as diamond turbot is a flounder of little economic importance ranging from northern California to the Gulf of California; and *Liopsetta putnami*, the smooth flounder, a fine table fish, is an Arctic species ranging from Rhode Island to Hudson Straits and Labrador.

Order Pleuronectiformes./Family Hippoglossidae

Atlantic Halibut
Hippoglossus hippoglossus
A fish with concave tail; dorsal and anal fins with middle rays form triangles; large mouth extends across head. Two eyes on right side. Color on upper, eyed side, brownish to olive to gray-brown; sometimes black. Side without eyes white or white and blotched. May weigh 400 to 700 lb and be 9 ft long. Males smaller. *H. stenolepis*, Pacific halibut, is the largest flatfish on U.S. Pacific Coast; dark brown with irregular blotches; uniformly pale on blind side. May weigh 500 lb and measure 9 ft. California halibut, *Paralichthys californicus*, is a member of the left-eye flounder family, Bothidae; found from central California to the Gulf of California and is much smaller, 5 ft and 65 lb.

Atlantic halibut, it is assumed, spawns from Spring through early Fall on the bottom at depths exceeding 1,500 ft. Eggs float below the surface at depths of 300 ft. A large female may produce two million eggs. Few eggs develop to adult fish; growth rate slow. Males are sexually mature when smaller than females.

Found in North Atlantic from New Jersey north; also along northern European Coast. Pacific form from central California to the Bering Sea and to northern Japan. Both prefer cold and deep water. Food: fish, crabs, mussels, lobsters, clams. They are food for seals and other large marine predators.

Once an important commercial species but has been overexploited. Anglers sometimes take halibut while drift fishing. Highly prized food fish, especially "chicken halibut" which weigh less than 20 lb. Liver rich in oil containing vitamin A.

Order Perciformes./Family Sphyraenidae

Great Barracuda
Sphyraena barracuda
Length to 10 ft, but rarely over 5 ft. 75 to 85 scales in lateral line. Silvery, with dorsal, anal, and ventral fins mostly black. A long, thin fish with a pointed head. Resembles freshwater pikes. Usually with a few irregular, black areas scattered on sides of body; black areas increase toward tail. Jaws with large, pointed teeth. Line record, 103½ lb, 5½ ft, 1932.

Fins of young nearly plain. Little known of breeding habits but assumed that breeding is in open sea. Food, fishes and almost any other animals it is able to find and to overpower. Fighter.

Young usually near shore in shallow water; adults prefer deeper water farther away from shore; sometimes far at sea.

Typical in West Indies but found from Massachusetts to South Carolina as a menace to bathers and source of anglers' sport. Although dangerous to humans wading or swimming, known records of attack are rare. Potentially dangerous; bite leaves straight clean wound. Taken on a variety of tackle from shore or from a boat. Considered to be a good game fish. Used as food in tropical areas but flesh is sometimes toxic. Probably should not be eaten. Known from Eocene in fossil form.

Pacific barracuda, *Sphyraena argentea*, is not poisonous. *S. argentea* ranges from Alaska to Baja California. May reach length of 44 in. and weigh 16 lb. An excellent food fish taken with gill nets and trolling lines. Popular in West Coast fish markets. Roe prized as a delicacy by some.

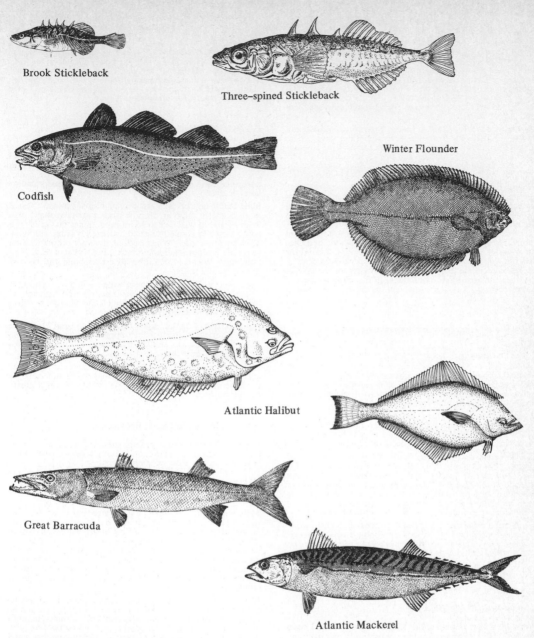

Brook Stickleback

Three–spined Stickleback

Winter Flounder

Codfish

Atlantic Halibut

Great Barracuda

Atlantic Mackerel

Order Perciformes./Family Scombridae

Atlantic Mackerel
Scomber scombrus

Length to 2 ft. Dark blue above and plain silvery beneath, with about 30 dark wavy stripes. Lateral line wavy. Air bladder small or absent. A row of finlets behind dorsal fin. Two Atlantic Coast species, mackerel and chub, not easily distinguished. Have fleshy eyelids covering parts of the eye.

Spawns May to June. Eggs 1/20 in., buoyant, not sticky; 540,000 to 18-oz fish; hatch in 5 days at 56°F. After larval stage, young reach 1/3 in. in 40 days, 2 in. in 3 months. About 3 out of 1 million survive 3 months. Spawn at 2 years, or 1 lb, or 1 ft. Maximum to 7½ lb. Atlantic mackerel to 18 in. and 2 lb; chub mackerel to 14 inches. Food: small fishes, crustaceans, squid, worms, and planktonic organisms. Are eaten by other fish, whales and porpoises, and birds.

In North Atlantic, from Norway to Labrador, south to Spain and Cape Hatteras, with related species over most oceans. Lacks keel found at tail base in related tunny. About 27 million lb harvested on Atlantic Coast in 1964, as against 54 million lb of Pacific and horse mackerel in West. Pacific mackerel, *S. japonicus,* range from the Gulf of Alaska to the Gulf of California, also throughout much the Pacific and Indian Oceans. Travel in schools; sometimes with other fishes, for example, small bluefin tuna. Mackerel is a favorite food fish, either fresh, frozen, or smoked.

Bluefin Tuna
Thunnus thynnus
Length to 14 ft. Weigh to 1,500 lb. Rod-caught record, 977-lb fish off Nova Scotia. Dark steel-blue above, silvery beneath, with yellow on some fins. Sides green-tinted and pink, iridescent. Body covered with small scales; robust, with pointed snout; two dorsal fins, 9 to 10 blackedged yellow finlets behind second dorsal fin; 8 to 10 finlets behind anal fin. Tail shaped like a crescent moon. Related to bonito and albacore.

Spawns in schools, April to June. Spawning location not known. Female may produce several million eggs. Eggs small, buoyant; hatch in 2 days to ¼-in. larvae. In 3 months reach 1 lb. Probably matures at 3 years with weight of 33 lb. A 7-ft bluefin tuna is about 10 years old. Much yet unknown. Feeds on other fish, mostly mackerel. Also takes crustaceans and squids.

In Atlantic, Pacific, and Mediterranean. Particularly abundant off California coast. Caught by lines and nets and supports enormous industry. Average annual world yield about 800 million lb. Japan formerly harvested 50 percent of crop. *Euthynnus pelamis*, skipjack tuna, is one of the most important commercial fishes in the Pacific near Japan, Hawaii, and Central America. Smaller than bluefin tuna. Found around the world in tropical and subtropical waters. Yellowfin tuna, *T. albacares*, occurs as both Atlantic and Pacific forms. Wt. to 400 lb. commercial fish usually 25–125 lb. Primary part of California-based tuna fleet; important to the Japanese worldwide fishing fleet.

Order Perciformes./Family Pomotomidae

Bluefish
Pomotomus saltatrix
Length to 45 in. Weight to 27 lb, with world record of 50 lb. Related to pompanos. Like sea bass inside, mackerel outside. Only member of the family. With pointed snout, large mouth and a prominent lower jaw; tail fin forked; generally blue-green above shading to silvery-white on belly side.

Spawns in spring and early summer but eggs and larvae are apparently unknown. Food essentially fishes, squids, and similar marine animals. Predaceous on herring, menhaden, mackerel, alewives, mullet, and many other fishes.

Excellent food and sport fish but highly destructive to other species. Found along the Atlantic Coast from Cape Cod to Argentina; off the coast of the Azores, Portugal, and Spain; throughout the Mediterranean; common in the Black Sea; off both coasts of South Africa, the Indian Ocean, southern Australia, and New Zealand. Basically a deep-water fish but may be found near shorelines. Unlike California bluefish.

Commercial fishermen take great quantities of bluefish; generally commercial harvest exceeds 6 million pounds each year. Favorite with anglers. Many are marketed fresh but some are frozen. Also called blue, skipjack, snapper, snapper blue, and tailor.

Order Perciformes./Family Centrarchidae

Smallmouth Black Bass
Micropterus dolomieui
Length to 2 ft; at 5 months, 2 to 3 in.; 1 year, 4 to 5 in.; 2 years, 6 to 8 in. Rear of upper jaw does not extend to rear margin of eye. Weight to 7 lb. 17 rows of scales on cheek and 72 to 75 scales in lateral line; 11, lateral line to base of dorsal fin. With robust body; brownish or bronze with vertical olive-colored bars.

Male builds nest on pond bottom May to July. Female at 60 to 70°F adds about 7,000 eggs for each pound of her weight. Male drives female into nest, and drives her out after spawning; may spawn with several females. Eggs guarded by male. Young guarded, without black lateral stripe of largemouth bass but with conspicuous black band across tail. Line record, 14 lb, 28 in., 1932, Oakland, Fla.

A most popular freshwater game fish of cool lakes and streams from Lake Champlain to Manitoba, south to South Carolina and Arkansas, and widely introduced elsewhere. Prefers rocky locations in lakes and streams; in lakes to 30 ft deep; in stream riffles flowing over gravel, boulders, or bedrock. Food in these locations, small crustaceans, insect larvae, crayfish, and fish. Possibly most exciting fighter, pound for pound, of any fish. Cannot survive 100°F in water. Feeds best at 60 to 70°F, favoring 67°F.

Claimed to have been distributed by the railroads as they moved west, riding in water buckets. Probably the most sought-after member of the bass family. A good fighter. Takes many natural baits: hellgrammites, soft-shell crayfish, night crawlers, minnows. Fly-fishing is also a favorite method of fishing for smallmouth bass.

Largemouth Black Bass
Micropterus salmoides
Length to nearly 3 ft. In adults rear of jaw extends beyond rear of eye. Dorsal fin, almost divided. Rows of cheek scales, 10. Scales of lateral line, to 68; 7, lateral line to base of dorsal fin. Anal spines, 3. Color, black to greenish with a dark horizontal band along side from head to tail. Prefers shallow lakes or rivers.

Male builds nest on lake bottom, fanning silt away. Nest 20 in. in diameter and 6 in. deep in sand or gravel; usually close to shore and in water 12 to 36 in. deep. Eggs laid at 62°F in nest. Nest and young protected by male. Young with broad black stripe on side. When hooked, does not ordinarily jump so readily as does smallmouth bass. Feeds best at 65 to 73°F, evening or morning. Food; small crustaceans, insects, crayfish, frogs, and fish.

Important game fish in southern waters where it is known as "trout." Lends self readily to pond but not to trough culture. Quebec to Florida and west to Mexico and Manitoba. Introduced elsewhere extensively. Line record, 22¼ lb.

Has a variety of names: green bass, Oswego bass, black bass. Takes a variety of live bait if it appears alive: worms, frogs, insect larvae, crayfish, and live minnows. Fly-fishermen have their favorite "bass flies."

Frequently stocked in farm ponds as predator fish with bluegills as the forage fish. However, bluegills often become the predator on eggs and young of bass when bluegills greatly outnumber bass.

Sunfish, Pumpkin Seed, Bream
Lepomis gibbosus
Length, 5 months, 2 to 2½ in.; 1 year, 3 to 4 in.; 2 years, 5 to 6 in. Maximum 8 in. Spines on anal fin, three as in warmouth but there are no teeth on sunfish tongues. Brilliant green to olive, with orange belly and spots on sides. Easily identified by black spot at end of gill cover with a red to orange spot at the tip. Mouth is small and does not extend beyond the front of the eye.

Male builds nest commonly among weeds at water temperature of 68°F. Nest about 1 ft across, hollowed, rootlet-lined or gravel-based. Eggs and young protected vigorously by male who uses many females in his nest. Eggs incubate, depending on temperature, in 5 to 10 days; young grow to 3 in. long in first year. Mature in 2 years. Crowding may cause stunted population. Food, any small animals such as insects, worms, fish, and crayfish. Feed on fish so help to check population by eating young.

Ideal fish for small boys and excellent panfish. Probably more persons in the northern U.S. have begun their angling careers on this fish than any other panfish. Common in ponds and streams with sand, mud, and vegetation on bottom. Frequently found in weed patches and around sunken logs which it uses for cover. Maine to Minnesota south to Florida and Mississippi Valley but more common in South. An excellent fighter for its size. Taken on a wide variety of baits: worms, wet flies, grubs, or whole-kernel cooked corn. Are taken through the ice on worms. Good eating but bony.

Rock Bass
Ambloplites rupestris
Length, 5 months, 1 to 1½ in.; 1 year, 1½ to 2½ in.; 2 years, 3 to 4 in. Maximum 12 in. Anal spine five to eight but usually six. Dorsal spines, eleven. Gill rakers, fewer than 10. Contrast with Sacramento perch, *Archoplites*, with 20 gills rakers and 12 to 13 dorsal spines; and warmouth *Chaenobryttus*, 3 anal, 10 dorsal spines.

Rugged-appearing sunfish, dark olive, sides with either brownish or yellowish blotches. Mouth large. Red eyed.

Breeds in early summer, in water about 74°F. Eggs placed in nests prepared and defended by male, often in deeper water

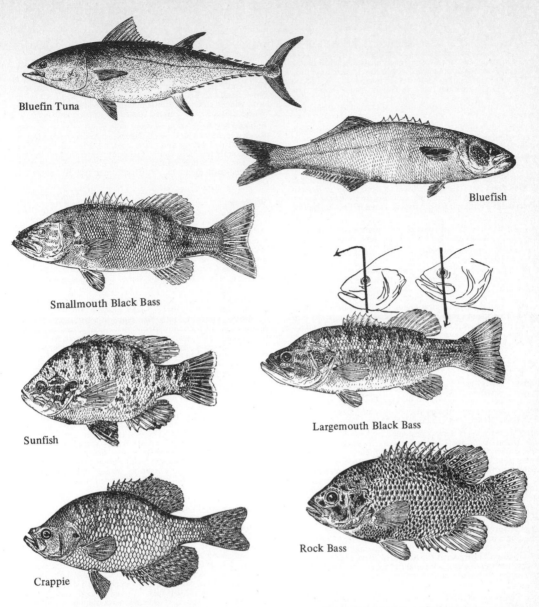

Bluefin Tuna

Bluefish

Smallmouth Black Bass

Sunfish

Largemouth Black Bass

Crappie

Rock Bass

than associated sunfish. Bites vigorously and fights valiantly. Females lay 5,000 to 9,000 eggs. Food, other fish, crayfish, worms, and other animals it is able to overcome.

Excellent panfish, generally popular and abundant in lakes and slow rivers, from Vermont to Manitoba south to Alabama, Louisiana, and Texas, being most common west of Alleghenies. Adults move in schools, frequently associated with smallmouth bass. Name is appropriate; specific name means "of the rocks." Related Sacramento perch in Sacramento and San Joaquin rivers. Related warmouth at best in South.

Also known as rock sunfish, redeye, goggle-eye, and black perch. A good game fish and panfish which will bite on a variety of artificial and natural baits. Sometimes taken by fly-fishermen in smallmouth bass streams.

Crappie

(1) white crappie, *Pomoxis annularis.* (2) black crappie, calico bass, *P. nigro maculatus* (illustrated). Length to 1 ft. Anal fins nearly as long as dorsal. Dorsal spines of (1), five or normally six; of (2), seven or eight. (2) is deeper and generally darker than (1).

(2) Flattened laterally, silvery sides grade to dark olive to black on the back. Spots scattered irregularly on sides, tail and on dorsal and anal fin. (1) More elongate than (2) which, in comparison, has a more high-arched back. (2) has irregular mottling distributed like an old calico print, hence, its other common name, calico bass.

Spawns in early summer in rather deep water over gravel in separate nest, not easily observable. Incubation a few days to 2 weeks or more. Eggs likely to be injured below 58°F or destroyed below 55°F.

Food of (1), when small, aquatic insects and plankton; when older, largely fish. (2) strictly carnivorous; feeding on crustaceans, aquatic insects, and smaller fish. (2) more likely to be found in quiet water and water with more vegetative growth than habitat of (1).

Both are popular, readily biting panfish. Found in pools and slow streams from Texas to Alabama, north to Vermont and Dakotas. Popular in farm fishponds which have now become common. Sometimes called bachelor perch because males of both species guard their nests; also called papermouths because of their tender mouths. Crappies are popular with beginning anglers as well as experienced fisherman.

Bluegill
Lepomis macrochirus
Yellow-green to dark blue sunfish. Sides with six to eight vertical, irregularly shaped dark colored bars. All adults have black flap at rear end of gill cover; generally has a black blotch on rear portion of the dorsal fin, with a long pointed pectoral fin. Mouth small. May grow to 15 in. and 4½ lb. Much smaller specimens are the rule. In North, young grow a little over an inch a year; 4 to 6 in. long at 3 years, 9 in. in 6 to 8 years. In South, growth much more rapid, to 4 in. during first summer.

Spawns in late May to August; water temperature about 67°F. Male makes small, shallow concave nest by excavating sand and gravel. Nest guarded by male. Female enters nest, deposits to 12,000–40,000 eggs; eggs hatch in 2 to 5 days. Male keeps nest clean and protects young for a few days.

Bluegills prefer quiet, weedy water where they can hide and find plenty of food. Larger fish stay in deep water in day but move into shore to feed during morning and evening hours. Sometimes it is difficult for a bass fisherman to "get his bait by the bluegills."

Food largely insects, both larvae and adult; does take some plant food. Adults also feed on crustaceans.

Also known as bream or brim in southeastern states; and sun perch, pumpkin seed, blue sunfish, sunfish and copperbelly. Probably one of the most popular panfishes in the U.S. Widely used species in farm ponds where they serve as forage fish for carnivorous black bass. Bass population must be sufficient to keep up with reproductive power of its prey, the bluegill. Large bluegill populations become stunted in farm ponds. This species may be a beginning angler's introduction to the sport. They are also taken by avid fly-fishermen and ice fishermen.

Order Perciformes./Family Percidae

Johnny Darter
Ethiostoma nigrum
Length to over 3 in., rarely. Body slender. Scales relatively large. Sides with numerous W-shaped dark blotches. Cheeks and breast usually scaleless. Many close relatives.

Ninety-five species of darters described in the U.S. Generally called simply minnows by most stream fishermen. Common name, darter, comes from manner in which the fish moves; staying quietly in place, then moving very rapidly to a new location. Their motion is very fast. Like other members of the perch family, darters have two separate dorsal fins; with spines in dorsal and anal fins. Tails rounded. Pectoral fins disproportionately large; an easily seen characteristic which makes them different from the other "minnows" in the live-bait bucket.

Males in spring black in forward regions or all over. Most darters attach eggs in May to June to undersides of stones or boards in relatively quiet water, some making excavation for this purpose. Male guards the nest until the eggs are hatched. Generally found in clear, cool smaller streams but may inhabit larger rivers. Seldom found in polluted streams. Bottom feeders.

From Colorado to Quebec, through southern Canada and Dakota south to Oklahoma, Missouri, and western Pennsylvania, usually in relatively quiet streams, usually but not always over gravel or in weeds.

Not of direct importance to fishermen except the occasional one used as live-bait. Is an important food species for game fish; so its color patterns are duplicated by some flytiers.

Order Perciformes./Family Percidae

Yellow Perch
Perca flavescens
Length, 5 months, 2 to 2½ in.; 1 year, 3 to 4 in.; 2 years, 5 to 6 in. Maximum 12 in. Weight to 4 lb 3 oz. Yellowish, with dark green vertical bands. Body elongate, compressed; appears humpbacked. Dorsal fins separated and with sharp spines; tail forked. Gill cover with sharp spines; very rough to the touch. Scales moderate. Skin tough. Flesh excellent. Body cavity relatively small.

Males in early spring pair with females, who swim through weeds laying ⅛-in eggs in zigzag ribbons sometimes to 7 ft long. Food, small animals such as worms, crayfish, and minnows. Line record, 4¼ lb, 1865, Bordentown, N.J.

A superior and popular food and game fish caught on hook, either with bait or fly, and in nets commercially. Found in lakes and streams over muddy bottom, in little current, from Nova Scotia to Dakotas, south to Ohio and South Carolina; now widely stocked throughout the U.S. including the Far Western States; with related species in Europe and Asia.

In lakes; prefers cool, clean water with sandy or rocky bottom. Known locally as jack perch, lakeperch, coon perch, ring-tail perch, striped perch, and ringed perch. Fish caught by anglers usually less than ¾ lb. Generally fished just off the bottom using a floating, baited hook. Is also taken through the ice with small minnows. A good forage fish and sport fish.

Walleye
Stizostedion vitreum vitreum
Largest member of the perch family. Length, 5 months, 5 to 7 in.; 1 year, 8 to 12 in.; 2 years, 12 to 15 in. Maximum to 36 in. Weight to 25 lb. Cheeks and gill covers sparsely scaled, as distinguished from sauger or sand pike *S. canadense,* in which they are fully scaled. Dark olive with dark blotches. Pink belly. Has a round and elongated body and forked tail. Has large eyes with a glossy shine, hence name. Eyes shine in the light at night.

Mates April to May. Spawning begins when water temperature approaches 45°–50°F. Males precede females to spawning areas. Spawning may take place in a tributary stream, a shallow part of the main stream, or a shoal area. A 2-lb fish lays about 45,000 eggs per lb. Eggs ⅓ in.; hatch in 2 weeks. No parental care. Species lends itself well to hatchery conditions. Become sexually mature at 12 to 13 in. in third year, or end of second. Sauger lengths first 4 years are 4, 8, 11, and 12½ in., with maturity at 13 in. Line record, 22⅓ lb.

Prefers cool, clear water; must have good supply of forage fish; gravel, or rubble in which to spawn and a large area of deep water. Found in lakes and large rivers from Hudson Bay and upper Mackenzie, south through Vermont Pennsylvania and Georgia, west to Alberta. Range has been increased considerably by stocking and now is found throughout the East and in many western and southern states. A favorite commercial fish in the market and in restaurants. Commercial fishery in Canada; in the Great Lakes. Gill nets are used by commercial fishermen. Also known as walleye pike, yellow pike perch, and jack salmon, spike, dore. Feeds well at 55 to 70°F.

Order Perciformes./Family Serranidae

Striped Bass
Roccus saxatilis
Length to 5½ ft. Body shape, long and deep. Has a comparatively long head, snout is somewhat pointed, lower jaw projects forward. Color olive to steel-blue, sometimes black above; silver sides blend to white belly; sides with seven to eight longitudinal darker stripes. Weight to 125 lb. Anal fins with 3 spines and 12 soft rays. Dorsal fins, 2. Related white perch has 3 anal spines, 9 soft rays; sides plain dusky silver, or longitudinally striped. Dorsal fins of white perch are not separated. Line record, 73 lb, 5 ft, 1913.

Marine species that ascends rivers and bays to spawn in May. Eggs not sticky, ⅛ in., sink slowly in fresh water; hatch in 3 days at 58°F, a large female laying over 2,200,000; may drift with tide and current while incubating. Females mature at 4 years, 20 in.; males, at 2 years, 12 in.

Food animal matter; both invertebrates and vertebrates. Takes great quantities of food.

Gulf of St. Lawrence to Florida, being most common between Cape Cod and Cape May. Successfully introduced on Pacific Coast. A justly popular food and game fish. Occurs as landlocked form, in particular, in southeastern Atlantic states. Weight increase first year: 4 in., 1 oz; second year: ½ lb; fourth year: 2½ lb; eighth year: 12 lb. Weakfish a relative. Also called striper, greenhead, rockfish. Shoreline fish which may enter fresh water streams. Commercial fisheries limited to the Atlantic Coast.

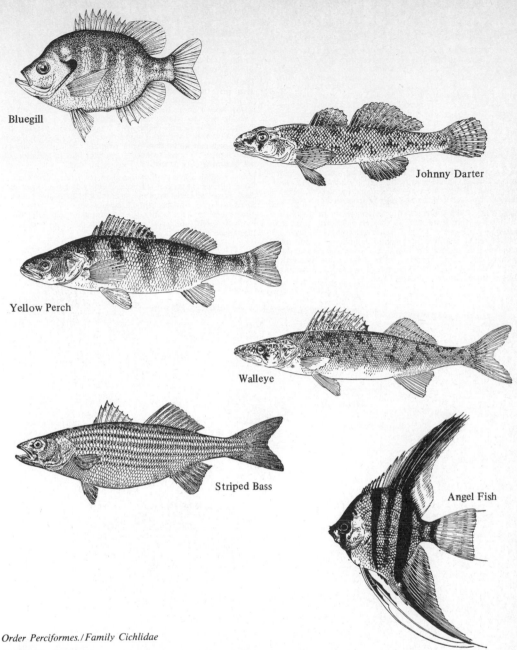

Bluegill

Johnny Darter

Yellow Perch

Walleye

Striped Bass

Angel Fish

Order Perciformes./Family Cichlidae

Angelfish, Scalare
Pterophyllum scalare

Length to 8 in. Bright silvery, with broad dark vertical bands that may disappear if fish is not among plants. So thin it can hide behind a slender plant stem. Nocturnal. Nervous, often injuring self seriously by dashing against sides of tank. A black strain has been produced as well as a veil angel fish, one with long fins. These two strains have been crossed to produce black veils. Prefers aquarium temperature 68°F to-F; breeding temperature, 80°F.

Particular in choosing mates. Two sexes are almost indistinguishable. However, the males have no bulge above and behind the ventral fin; females have a noticeable bulge in the body. Breeding pairs usually 3 to 4 inches long. Breeds at water temperature of 75 to 85°F. Eggs laid in rows on strong-stemmed plants. In aquariums, eggs should be removed for

protection against parent. Must have very clean water and breed most successfully in water which is slightly acid, pH 6.8. Usually parents transfer developing young in their mouths from one plant to another plant.

Native of Amazon. Introduced into Germany in 1909 as aquarium fish, pairs selling Europe in 1911 to $100. Now so abundant they are available to anyone. Most relatives in family are food fishes of considerable value and take hook readily. Related aquarium fish include *Cichlasoma, Etroplus,* and others. Will take young guppies as food; and a special Gordon's formula prepared with fresh liver, Pablum, and salt prepared as directed in aquarium books, *Daphnia,* white worms, brine shrimp, mosquito larvae. Probably most aquarium angelfish are *P. eimekei.*

521

Order Perciformes./Family Cottidae

Muddler, Slimy Sculpin, Miller's Thumb
Cottus cognatus

Length to about 6 in.; rarely exceeds 4 in. Scaleless. Forefins (pectorals), large and fanlike. Head large and flattened. Anal and second dorsal fins large and conspicuous. Mouth large. Body robust; usually olive to brown in color, speckled with dark mottlings. Skin velvetlike, without scales. One spine and two to five rays in the pelvic fin and fewer than 20 dorsal spines. Color so perfectly blends in with its surroundings that it is seen only if it moves. Bottom dweller; hide under the rocks during the day in fast-moving streams or along rocky shores of cold lakes. Food consists of a wide variety of aquatic animals including small fishes, freshwater shrimp, and other crustaceans, and insect larvae. May be of some importance as food for game fish.

Eggs laid in orange-colored masses under stones in cold streams and lakes. Male guards the nest. *Cottus* always nests in fresh water though there are related marine genera. Particularly destructive to nests and eggs of trout in some places.

Ranges from Alaska south to Virginia and westward through Great Lakes area to Iowa, with subspecies extending range to Alabama and elsewhere. Name, miller's thumb, probably related to flattened head which, to some, resembles the unfortunate miller who caught his thumb between the millstones. Also called cottus and big fin. Little or no value as a bait minnow. Sculpin family most numerous in the North Pacific; 84 known species in the U.S. and Canada.

Order Tetradontiformes./Family Diodontidae

Striped Burrfish, Toad
Chilomycterus schoepfi

Length to 10 in. Spines immovable, unlike movable spines of porcupine fish, which reaches a length of to 3 ft. Food, invertebrates such as oysters, barnacles, and hermit crabs.

These and related swellfish can inflate selves prodigiously when disturbed. Inflation accomplished by taking air or water into an extension of the walls of the belly. Color olive to brown above and pale yellow below; back and sides striped with brown to black, almost parallel lines. Lines run obliquely downward. Sides with large black spots; one just below dorsal fin, another behind pectoral fin. Spines of porcupine fish can inflict dangerous wounds. Spawns in July off New Jersey. Young hatched from eggs laid in summer; do not resemble parents until at least ¼ in. long.

Spiny boxfish illustrated is found in sea from Cape Cod to West Indies. Does not inflate so readily as some other species. Some relatives were used as armor by primitive peoples. Skins sold as curiosities; some used as Japanese lanterns. Another member of the family, the porcupine fish, *Diodon hystrix*, more frequently in its dried inflated shape, is also sold as a "Japanese lantern."

Slimy Sculpin, Blob

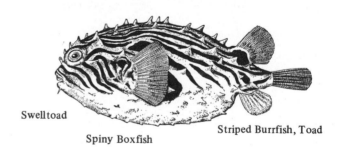

Swelltoad

Spiny Boxfish

Striped Burrfish, Toad

CLASS AMPHIBIA Amphibians

Living specimens with moist glandular skin; without external scales; usually with two pairs of limbs; heart three-chambered; respiration by gills, lungs, lining of mouth cavity, or skin; eggs usually laid in water; larvae usually aquatic. Appear in Devonian strata as fossils and represented to the present. Some 2,500 living species.

Order Caudata./Family Proteidae

Mud Puppy, Water Dog
Necturus maculosus
Length to 18 in., though more commonly about 1 ft. With four short legs and weak four-toed feet, a strongly flattened tail that acts as a fin, and a head flattened at right angles to tail. Behind head, on either side, are three tufts of fluffy red gills. Color dark brown, gray-brown, or black, usually with spots and mottlings. Slimy.

In rivers and quiet streams of fresh water, at depths of 4 to 8 ft, usually among water weeds. Common in eastern U.S.; three species in genus; west of Appalachians from Canada to Louisiana and Florida; east of Appalachians from Georgia to North Carolina. A closely related genus is *Proteus* of southern Europe.

Mate in autumn. In spring, eggs resembling light yellow spheres to ¼ in. through are laid on individual jelly stalks under stones or logs or in sunlit sand nest, a female laying to over 140 eggs and protecting them until in 6 to 9 weeks they hatch into ¾-in. larvae, which in 10 weeks reach length of 1½ in. and become mature in 7 to 8 years at length of 7 to 8 in.

Food includes fish eggs, worms, insects and their larvae, and any other aquatic animals that may be found while animals prowl about the bottom, usually at night. Movements may be slow but bite may be vigorous and action may take place throughout year. Usually is hidden quietly in daytime but begins roaming as darkness approaches. Slime serves as some protection from enemies.

Reported to be good eating. Feared by some persons as poisonous though there is no basis for this belief. Thought by others to be serious enemies of fish though authorities do not consider this important. Biological supply houses need the animals for classes in animal anatomy, one firm alone supplying over 2,000 a year. Animals probably live for at least 23 years.

Order Caudata./Family Cryptobranchidae

Hellbender
Cryptobranchus alleganiensis
Length to over 2 ft but usually about 18 in. Head broad, flattened, and without gill ruffle of mud puppy. Legs four, but of little use since animal swims mostly by flexing body and tail. Skin appears greatly wrinkled and too large for animal. Appears to have irregular dark or light spots. Gills internal. Dark brown to black. Tail of female, ⅓ length.

In rivers and streams in eastern United States, particularly in Lake Erie, the Ohio Valley, and rivers that flow from Allegheny highlands, south to Louisiana and Georgia. Two genera of closely related giant salamanders are *Megalobatrachus* of China and Japan, and *Cryptobranchus* of our region. Fossil forms found in western U.S.

Male courts female, though many individuals may crowd together before real mating begins in fall. Male builds hollowed nest under rock and induces females to lay eggs in it. One nest may contain nearly 2,000 eggs, which appear like tangled strings of beads. Male adds milt to new eggs, may eat some, and protects them 10 days to 2 weeks until they hatch. Eggs may take to 3 months to hatch.

Food, mostly crayfish and other aquatic animals found while prowling over bottom. Normally larvae are over 2½ in. long in 9 months, with external gills. At 1 year, gills are reduced and legs developed. At 2 years, gills have been absorbed or partly so and length may be 6 in. Matures in 5 to 6 years.

Flesh known to be good food but probably is rarely eaten. It is not poisonous in any way. Japanese eat their giant salamander, which reaches a length of 5 ft and lays eggs the size of grapes. Blood in a stream may attract native hellbenders, the animals coming in considerable numbers to the source.

Order Caudata./Family Salamandridae

Newt, Eft
Diemictylus viridescens
Average length 3 to 4 in., male being slightly the larger. Adult olive to brown with yellowish underparts and red and black spots. Tail finlike in breeding male. Breeding male with black swellings on inside surfaces of thighs and feet and swollen vent more developed than in female. Red land form, immature and lacking fin on tail, is called "eft."

(Continued)

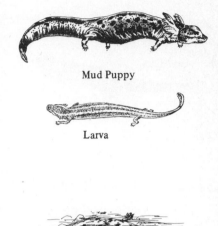

Mud Puppy

Larva

Hell bender

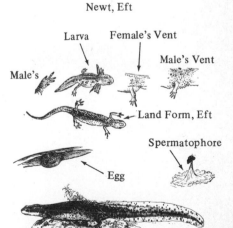

Newt, Eft

Larva   Female's Vent

Male's Vent

Male's

Land Form, Eft

Spermatophore

Egg

Vermilion–spotted Newt

Found in fresh water or on land throughout year. Red eft is found among moist plant growth while adult and early larval stages are in water. *D. viridescens*, common newt of eastern North America, is found from Hudson Bay to Texas and west to Illinois, Michigan, Missouri, and Oklahoma. Rough-skinned newt, *Taricha granulosa*, a West Coast species, is found on the humid coast from southeastern Alaska to central California.

After courtship in fall, winter, or spring, male in spring deposits a jelly spermatophore, which female collects. Female may lay over 100 eggs singly on submerged support. Incubation of 20 to 35 days, or shorter at room temperature, brings hatching into gilled larvae that in 3 months reach 1-in. length and absorb gills. Leave water for 1 to 3 years on land.

Food, almost any small animal life. May haunt mass of hatching frog eggs, snapping up young as they emerge, or may destroy great numbers of mosquito larvae and pupae. In some areas, red eft land stage is shortened and its length may vary regionally or among those of a single locality. Adult form and its return to continued aquatic life may start in fall or spring.

Animals may be considered as enemies of frogs and of fish nests but also of mosquitoes. Efts have gained a reputation for mysterious habits associated with pixies and fairies, but this only makes them more interesting without helping understand them. In aquariums they may be fed ground beef on a dangling thread or toothpick. Blood of some forms bears microscopic trypanosomes. Known to live to 7 years.

Order Caudata./Family Ambystomidae

Tiger Salamander
Ambystoma tigrinum
Length to 10 in. Males average 8 in., females usually larger. Stout, broad-headed, with a distinct neck. Adult without external gills and without groove from nostril to lip. Basic color deep brown to black. Marked with pale, irregular, yellow-brown, or olive blotches that form bands beneath and sometimes on tail. Breeding males have swollen vent.

Burrows in soft ground during most of season but comes to shallow ponds to breed. Ranges from New York and Connecticut down Atlantic side of Appalachians to Florida and up west side to Ohio, west to Arizona, Utah, Idaho, Oregon, and Washington and along southern part of Canada, being subdivided in the range into 5 subspecies.

From January to March, depending on territory, migrate to breeding ground. After courtship and spermatophore production by male, female may lay to over 100 three-enveloped eggs in kidney-shaped jelly mass to size of 3 by 4 in. attached below water. Hatch in 3 to 4 weeks into ¾-in. larvae, which in 75 to 118 days transform into animals that may breed following spring.

Food essentially small animals caught while prowling at night or burrowing in loose soil. In captivity, may eat small frogs. In some parts of western United States and in lakes around Mexico City, tiger salamanders may retain gills and aquatic habit and breed as "permanent larvae." Possibly because of burrowing habit, can survive where others disappear.

Economic importance about same as for spotted salamander and probably negligible. May be popular for sale as terrarium pets as they are hardy and intriguing-looking. Interesting to biologists because some retain gills in adult state, this larval form being spoken of as the *axolotl*. Apparently entirely harmless to man.

Spotted Salamander
Ambystoma maculatum
Length to 7 ¾ in., of which tail is about half. Females, usually larger, have lighter-colored bellies than males and lack swollen vents of males. Greenish- or brownish-black, with bright yellow round dots, shining, moist, and, when badly disturbed, mucus-covered. Several enlarged pores on head. No external gills in adult.

Adults commonly found in woodlands or sometimes in meadows where there are nearby marshes or waterways. Sometimes found in cellars. Ranges from lower part of mouth of St. Lawrence River to Florida, west to eastern Texas, Oklahoma, Missouri, Illinois, Wisconsin, and Lake Superior region, and east through southern Canada except for a few areas. California tiger salamander, *A. tigrinum californiense,* is similar in appearance to *A. maculatum.*

In early spring, both sexes migrate to a shallow pool. Male has swollen vent during breeding season. After a courtship, male lays a jelly spermatophore, which female takes up. She then lays a mass of eggs, which within 1 hour may be to 4 in. through and contain some 200 eggs, each in two jelly coats. Eggs hatch in 2 to 4 weeks into ½-in. larvae, which in 3 months may become 3 in. long before transforming to adult stage.

Tiger Salamander

Larva

Egg Mass

Spotted Salamander

Egg

 Spermatophore

Food, largely insects, slugs, snails, and worms. Larvae do best in temperature of about 65°F. In first year after transformation, animals reach a length of over 3 in. Known to live for at least 24 years where there is ample protection. Spotted salamanders are less likely to winter in groups than Jefferson's salamanders. Hibernates on land.

While many persons feel that these animals are poisonous and should be destroyed, there is no basis for this opinion. This species is at times referred to as a "poisonous spotted black lizard"; two words in the description are correct; *spotted* and *black*.

Order Caudata./Family Plethodontidae

Red–backed Salamander

Red-backed Salamander
Plethodon cinereus
Length to 5 in. Male usually smaller than female. Body slender and relatively long. No light bar from eye to corner of mouth. Resembles slimy salamander but has 18 or more vertical grooves on sides while slimy salamander has fewer than 16 between front and hind legs. Variable in color, a few being all red, others with red or gray back stripe. A lungless salamander, "breathes" through highly vascularized throat and through skin.

Eggs

Found in drier places than dusky salamander but otherwise similar. Usually in old stumps or logs, under leaves or bark, in moss, or in almost any kind of woodland trash. Ranges from Gulf of St. Lawrence to Georgia and Alabama, west to Oklahoma, Illinois, Minnesota, and southern Canada to James Bay, with at least two subspecies represented. Genus has many species especially in southern Appalachians.

Sexes mate in fall. Following spring, female lays about 12 eggs to ⅕ in. through, in three jelly coats, in suspended clumps usually under logs and protected by females. Larval stage is passed in egg, the young having gills only a short time after hatching. Young 1 in. long when hatched, soon independent of parent. In 2 years, may breed at 1½ in.

Food, insects and other small animals, frequently captured by a swift forward thrust of tongue, which can be accurately directed. Can secrete a small amount of slime that might be considered a protection. Can be killed easily by being held too long in a warm dry hand. May shed tail if escape is essential, in which case a new tail without vertebrae develops.

This is an ideal schoolroom terrarium pet but only a few should be kept and those confined should be given sensible care. May be trained to feed on pieces of ground meat hung on end of a thread and swung back and forth before mouth.

Slimy Salamander
Plethodon glutinosus
Length to 7¼ in. Smooth, slender, shining, dark violet-black to gray at tail, with gray undersides, many light spots over back and sides. Sometimes marked attractively in foreparts with pearly-white pinpoints. Feet pale brown. Front feet four-toed, hind feet five-toed. All feet lack large glands beneath. Skin oozes milky slime when touched.

Slimy Salamander

Commonest in shady ravines or caves or under logs, stones, and trash except that during or after rain animals may come out and move about actively. Ranges from eastern Canada through New England to Florida and west to Wisconsin, Missouri, Texas, and the Gulf, with some 20 species and subspecies in United States.

Little is known of reproductive habits. Sexually mature at 3 years. Male has a small circular white mark at rear of lower margin of lower jaw. It is assumed that mating takes place in fall, that eggs are laid in winter or early spring. Eggs are found in crevices in caves, in clusters, with or without female nearby. Eggs are white and to ¼ in. through.

Food about half insects and the remainder sow bugs, worms, millipedes, centipedes, and the like. Deliberate in movement but may move rapidly, as its legs are strong to lift body from ground. Lives its whole life on land, young emerging from egg well-developed and to ½ in. long.

Makes an interesting pet in a terrarium though it may burrow deeply during day. Slime undoubtedly protects it from its enemies, among which is short-tailed shrew. Nocturnal.

Red Salamander
Pseudotriton ruber
Length to 6 in. Bright red or orange-red with many rounded black spots on upper surfaces. Sometimes chin is black-spotted and with brass-colored eyes. As adults age, they become purple to purple-brown, the spots become larger. Lower side of mature adult is spotted with brown or black. Tip of tail compressed, flattened.

Northern Red Salamander

Lay eggs in fall, in water, usually attached to stones. Laid in clusters, colorless; may be to 72. Eggs about ⅛ in. in diameter. Larvae, about ½

(Continued)

in. when hatched, are dull yellow. May have light reddish-brown color until 3 to 4 in. long.

Prefer cool flowing fresh water, under stones, vegetation, or near springs. Found in streams in woods, open fields. Usually not found in streams with muddy bottoms and prefer sand or gravel. Adults may move on to land in summer months and hide under logs, stones, even some distance from flowing water.

Food small invertebrates such as earthworms, snails, slugs, centipedes, millipedes, and spiders. May eat young of its own kind.

Found from central eastern New York State west to Ohio and south to northern Alabama. Subspecies of *P. ruber* extend range of the species to 5,000 ft in the Blue Ridge Mountains and Louisiana and Tennessee.

Not so well known as some other salamanders because of its secretive habits. Fills a rather specific niche. Worthy of protection.

Eastern Four–Toed Salamander

Eastern Four-toed Salamander
Hemidactylium scutatum
Males are usually smaller and slenderer than females. A small salamander, to 3½ in. Belly side white with black spots; with four toes on both front and hind feet; has a constriction at the base of its tail. Will break away at this point. Body cylindrical, flattened head. Head and trunk reddish-brown above.

Terrestrial; found in woods or open areas; usually adjacent to sphagnum bogs, or swamps or quiet pools. Adults most frequently found in spring when they congregate. Larvae are aquatic.

Ranges from Nova Scotia west to Wisconsin, south to Gulf of Mexico. Reported in Georgia, Arkansas, Louisiana, Missouri and Illinois.

Mate in summer or fall. Female takes spermatophore deposited by male. Thirty eggs laid singly, July to September; on land, may be under sphagnum moss, or logs, but always near water. Hatch in 38 to 60 days. Eggs guarded by female.

Larvae are about ½ in. at hatching. Head is tinged with orange to green or yellow and marked with brown or black. Probably sexually mature at 2½ years.

Food invertebrates taken in their home habitat. Particularly adapted to a limited niche. Worthy of protection.

Purple Salamander

Purple Salamander
Gyrinophilus porphyriticus
A large, 4¾ to 7½ in. salamander; often pale, brownish-pink to red, also salmon color. Blurred dark spots on back with blotches arranged in a row along each side of body. Under parts flesh-colored without markings; may have scattered black spots on lower lip and throat. Males slightly larger than females.

Commonly found in mountain brooks or springs; frequently under stones in woodland streams. Large flat stones or places under logs and leaves partially submerged seem to be favorite habitat. Sometimes found at considerable distance from streams but always seem to be associated with wet places. Sometimes enter caves.

Found from southern Maine, southern Quebec, to northern Alabama. Other species in the genus extend the range of *Gyrinophilus* through the Blue Ridge Mountains and into caves.

Little known of the reproductive habits. Male known to deposit spermatophore; eggs are deposited singly and occur as light yellow spheres. Larvae to 1 in. in length when hatched; light brown above and yellowish beneath. At end of 3 years are sexually mature, about 5½ in. Probably eggs, to 130 in number, are deposited from April to August; are attached to objects in water. Eggs are not guarded by male or female. Larvae are very light-colored and have a dusky netlike pattern of markings on their dorsal surface.

Food probably animal matter, lower invertebrates. An interesting salamander filling a rather distinct niche in its habitat. Worthy of protection.

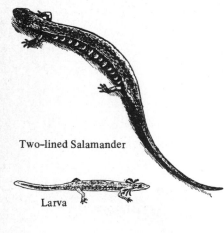

Two–lined Salamander

Larva

Eggs

Two-lined Salamander
Eurycea bislineata
Length to nearly 4 in. Tongue free at front margin but attached at center beneath. No line from eye to nostril such as is found in purple salamander. Slender, pale straw-colored, with narrow dark line running down each side from eye onto tail. Belly yellow. Sides dotted. Tail somewhat flattened and thus unlike dusky salamander.

Found almost exclusively along edges of small fresh water streams, under stones or logs and in shallow water. Ranges from New England to Florida west to Louisiana, north to Lake Superior, and through southern Canada. Nine closely related species within much the same range and with rather minor differences.

After an elaborate courtship, usually in fall, male deposits a spermatophore, which female takes up. In spring, eggs are laid under rocks under water in current. Female may lay about 30 eggs ⅓ in. in diameter, which hatch in about 10 weeks into ½-in. yellow slender larvae that reach length of 2 in. in 2 years and transform into adults.

Food, small aquatic animals such as insects, worms, spiders, crustacea, and the like, caught by foraging in daylight more commonly than is customary in most salamanders. In some parts of country, winter is spent on land in a burrow under trash; in other parts it may be spent in an active state in deeper water than that haunted in summer months.

Probably is of little direct economic importance to man. It may serve as a minor check on multiplication of some aquatic animals and as food for larger animals. This salamander is very active in summer and leaps from one's hand to escape, sometimes leaving tail behind. Animals do not thrive in aquariums unless water is kept fresh.

Dusky Salamander
Desmognathus fuscus
Length to 4 ½ in. With a light line running from eye to mouth. Tail more or less triangular to oval in cross section. Tongue attached at forward margin. Body rather short and tail rather long. Back varies from black to brown to yellow-brown, usually uniform but sometimes with black-edged blotches or broad black-edged backband. Sides usually without any small light dots.

Eggs   Dusky Salamander

Female and Eggs

Near springs where there is an accumulation of leaves or at the margins of springs or cool streams but usually not in water itself. Common in dried stream beds where there is a little water left. From New Brunswick to Florida and west to eastern Texas, eastern Oklahoma, southern Illinois, central Ohio, and just into Canada at Niagara region.

Males may court females fall or spring; male applies snout, cheeks, and head gland to female's snout. Spermatophore may be deposited in female or picked up by her. Female lays 12 to 26 ⅛-in. three-enveloped yellow-white eggs in bunches like grapes from June to September in wet area under moss or refuse. Female protects eggs until in 2 months they hatch into ½-in. larvae, which live in water 8 to 10 months, transforming into 1-in. adults which mature at 2 years and 2 in.

Mostly nocturnal. Food essentially earthworms, snails, slugs, spiders, insects, and some vegetable matter. Larvae that hatch on the ground must make their way by minute trickles of water to a water supply to continue development. Dusky salamanders are known to have been eaten by purple salamanders and probably have many other animal enemies such as raccoons.

Dusky salamanders known to eat insects that have fed on sewage; have been known to act as reservoirs for colon bacilli. Range of this species has been unintentionally increased by man through development of a "spring lizard" bait industry for fishermen.

Order Salientia./Family Ascaphidae

Spadefoot Toad
Scaphiopus holbrooki
Length to 2⅞ in. Body short and compact. Skin relatively smooth but with some scattered warts. Arms and legs stout. Belly gray. Throat and breast white in both sexes, thus differing from common toad. On inner sides of soles of hind feet are dark horny structures used in digging, which give toad its name. Pupils vertical.

Spadefoot Toad

Foot

Found over loose soil, into which it easily digs shallow burrows. Ranges from Massachusetts to Florida, west to Texas, Arkansas, and Indiana. In United States, at least three species and a total of seven subspecies, whose combined range covers most of U.S., and some of southern Canada, and northern Mexico, with of course a few expectations. In dry periods in western states *S. couchi* or *hammondi* or *intermontanus* or *bombifrons* may stay in burrows of kangaroo rats, gophers, or squirrels; or its own for very long periods. It appears outside the burrow at night, especially during spring and summer rains.

In spring, females meet broad-fingered males in a pond where they are calling. Breeding and calling carried on only after heavy rains; may continue from January to September. Eggs develop into carnivorous tadpoles which transform to ½-in. subadults in 14 to 60 days.

Food largely insects, worms, and other small animals caught by lightninglike thrust of tongue. Call of male is a repeated *wank*, quite different from that of common toad. Almost wholly nocturnal and therefore often missed in areas where it may be reasonably abundant. Burrows largely with assistance of horny area of feet.

Of course, this animal is useful as a destroyer of garden pests and is harmless both to man and to his pets. A rather sluggish animal and therefore possibly less interesting than some of its relatives. Ordinarily, other species and subspecies of spadefoots are smaller than one here considered.

Order Salientia./Family Bufonidae

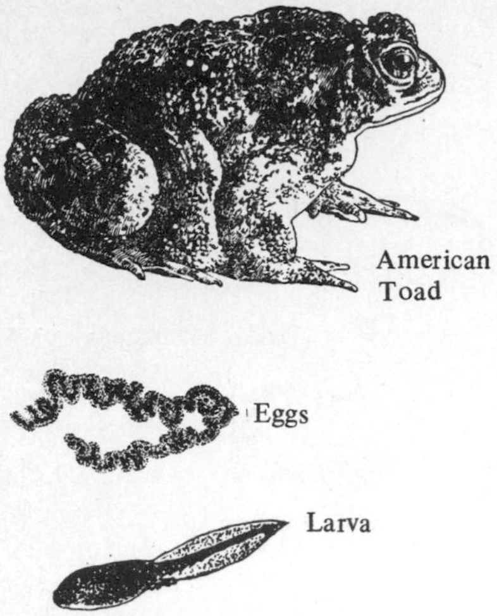

American Toad

Bufo americanus

Length to $5\frac{1}{2}$ in. but usually smaller. Female the larger, with male rarely over $3\frac{1}{2}$ in. Breeding male with black on throat and inner side of thumb. Fat. Skin roughened by pairs of dark spots on the back, each surrounding a wart, and with two large glands back of eyes and neck. Eyes prominent. Undersides finely roughened, with scattered dark spots.

Common in gardens and about lawns where there is damp soil and concealment. Ranges eastern North America to Great Bear Lake and Mexico, reaching elevation of 5,800 ft in Tennessee, favoring higher areas where there is competition with Fowler's toad, whose dark spots on back enclose more than one wart. Range extended to Pacific Coast by 13 species. In western states *Bufo* ranges from Death Valley to 10,000 ft.

In spring, males precede females to shallow breeding pond and begin prolonged trilling call, mostly during rainy weather. Female may lay to 15,000 black and white $\frac{1}{12}$-in. eggs. in paired jelly strings to 72 ft long, weighing to $5\frac{1}{2}$ times weight of mother. Eggs hatch in 3 to 12 days into black tadpoles, transforming in 40 to 60 days into $\frac{2}{5}$-in. toads that mature in 2 to 3 years.

Food, cutworms, potato beetles, chinch bugs, earthworms, ants, slugs, or almost any small animal that moves, captured by 2-in. thrust of sticky tongue, often after an interesting stalk. Can exude slime distasteful to many enemies but not poisonous to man, although giant toad of tropics may cause suffering or blindness to man or dogs. European toad may live to 36 years.

One of most useful animals associated with man and with his crops. One of best residents of any garden. In China, toad skin is used as a medicine, not unwisely since it contains adrenalin, which increases man's blood pressure. Cannot live for years confined in a rock nor does it cause warts on persons who may handle it. Spends winters buried to 3 ft deep in loose soil.

Order Salientia./Family Hylidae

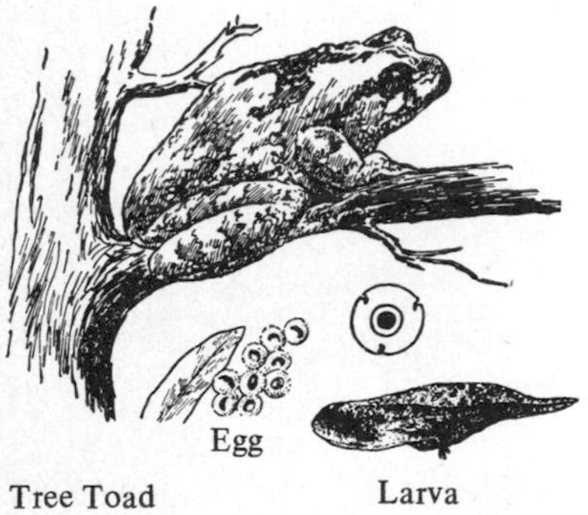

Tree Toad

Hyla versicolor

Length, male to 2 in.; female to $2\frac{2}{5}$ in. Legs slender. Toes end in sticky disks but in breeding season inner side of male's thumb bears a large pad. Skin finely pimpled. Back and legs marked with dark areas, commonly black-bordered. Often orange under hind parts. Throat of male, loose and dark; of female, whitish. Four species of *Hyla* in western states; *H. regilla* ranges from sea level to 11,000 ft in the Sierra Nevada.

In or near marshy waterways in spring but in summer in treetops, often considerable distances from water. Closely related to peeper. About 15 species cover United States and extend into Mexico and southern British Columbia. This species from Maine to Gulf States and west to Texas, Oklahoma, and into Minnesota.

Males precede females to shallow, plant-grown, quiet pond of permanent water in spring and sing either at night or on cloudy or rainy days. Eggs $\frac{1}{25}$-in., brown and yellow, in one thin and one loose jelly coat, laid free, attached, or at water surface in films of 4 to 40, though individual may lay to 2,000. Hatch in 4 to 5 days into $\frac{1}{4}$-in. tadpoles with red-orange tails and cream-white bellies; reach length of 2 in. in 45 to 65 days.

Food essentially insects caught by speedy dart of tongue. Larvae feed on oozes in pond. In North, a tree frog that transforms at $\frac{3}{5}$ in. is 1 in. long at 1 year, $1\frac{2}{5}$ in. long at 2 years, and $1\frac{4}{5}$ in. at 3 years, when it reaches breeding size. It can cling to a clean vertical glass. Call of males is a short melodious trill.

Entirely useful as a destroyer of insects. In terrarium, may be fed on meal worms or earthworms and may become real pets. May be heard on damp or rainy nights when temperature is over 60°F, even in summer. Has remarkable ability to change color from gray to brown to green in about 1 hour. Test this by putting one animal in a dark and one in a lighted bottle.

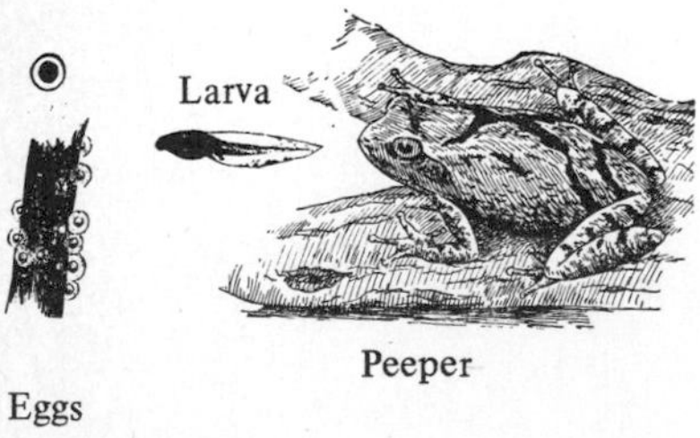

Peeper

Hyla crucifer

Length to $1\frac{2}{5}$ in., with male rarely $1\frac{1}{8}$ in. Male usually darker than female, yellow on groin and with brown or black throat, while female is white in throat region. Male has pad on inner side of thumb in breeding season. Toes of both sexes are expanded into distinct disks that stick to supports. Dark cross on back and legs with dark crossbands.

In spring, found in pools or near marshy spots either in water or perched on vegetation near or above it. In summer, may be found to 60 ft or more up in tops of trees. May be heard in woodlands any month of the year. Ranges from New Brunswick to South Carolina and west to Louisiana, Arkansas, Kansas, and Manitoba. Closely related to tree toad and some 10 other species of *Hyla*.

Male precedes female to breeding ground in spring. Calls vigorously. After mating, female lays 800 to 1,300 $\frac{1}{25}$-in. cream and black or brown eggs, more or less separately, on support under water, taking a day or more to lay the set. Eggs hatch in 5 to 15 days and in 75 to 90 days develop into 1⅓-in. tadpoles with purple-black blotches and iridescent creamy bellies.

Food of adults, insects caught by lightning thrust of tongue; of tadpoles, ooze gleaned from submerged objects. The 1⅓-in. tadpole transforms in July into a ⅗-in. peeper, which leaves the water and spends 3 to 4 years before reaching sufficient maturity to breed. Throughout its life has many enemies, from fish and salamanders to birds and squirrels.

Essentially useful as a check on insects. A delightful animal for an aquarium but even more so free in the open. May be fed chopped earthworms if kept through winter in terrarium, where it does not hibernate as it might in nature. Peeping, by male only, may be heard 1 mile.

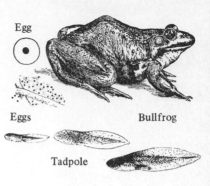

Egg

Eggs   Bullfrog

Tadpole

Order Salientia./Family Ranidae

Bullfrog
Rana catesbeiana
Length, male to 7½ in.; female to 8 in. Hind legs may be over 10 in. long. Webs of hind feet extend to tips of toes. Black, green, without warts, with ridges running behind ears but not down side of back. Male has yellow throat in breeding season and "ears" almost twice as wide as distance between nostrils instead of about same distance.

Favors small lakes and ponds with little wave action where there is a mud bottom and a permanent plant border of some height. Lives in water throughout its life and does not roam fields. Ranges through most of North America east of Rocky Mountains. Introduced west of Rockies to Pacific region and in Japan.

Mates from February (South) to July (North). Eggs 1/50 to 1/25 in., in several 3- to 5-ft surface films, usually attached, totaling to 20,000, black and white, in one thick coat of loose jelly; hatch in 4 days and in North develop in 2 years into 5⅘-in. tadpoles, with irregularly spotted tails and yellow, not iridescent bellies.

Food of tadpoles, oozes and aquatic plants and animals; of adults, insects, other frogs, fish, birds, and other animals. Tadpoles transform into 2-in. frogs that under best conditions may reach 3½ in. in 1 year and 4½ in. in 2 years, when it could presumably breed, although 2 to 3 more years may be required to reach length of 5½ to 6 in. Known to live at least 15 years.

Excellent scavenger. 500,000 frogs, of which this species is an important part, are used in biological laboratories in America each year. Call is a resounding *jug-o-rum*. Sound produced by a single internal vocal sac which forms a flattened pouch under the chin. Animal may "play dead" when caught and then suddenly "come alive" while his handler is not alert.

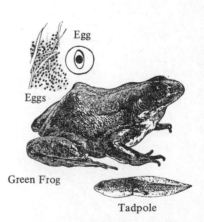

Egg

Eggs

Green Frog

Tadpole

Green Frog
Rana clamitans
Length to 4 in. for female; to 3½ in. for male. Male with "ear" larger than eye and with greenish-yellow spot in it, also with proportionately broader head. Both sexes have a pair of wrinkles running down back, one on either side, and toes less fully webbed than in bullfrog. Male has yellow throat in breeding season; female, white throat. Young usually has many black spots above.

Lives in or within jumping distance of permanent pool, pond, or lake through its entire life. May hibernate in water or in a burrow above water line. Ranges through eastern North America from Canada to Florida. Introduced into state of Washington. In North Carolina is found to elevations of 4,200 ft.; in New Hampshire, to 2,230 ft.

When air temperatures have reached 65°F in early summer, sexes mate. Females lay to 5,000 eggs in North and to 1,000 in South in a number of films to 1 ft across, at water's surface. Eggs 1/16 in. across, black and white; hatch in 3 to 5 days. Tadpoles develop in 370 to 400 days into 3⅖-in. tadpoles with green tails, brown spots, and creamy, not iridescent bellies. Tadpoles may overwinter.

Food of tadpoles, oozes and scums gleaned under water; of adults, almost any animal that can be captured. Flattened pouch from which sound emanates is from paired internal vocal sacs. Mature tadpoles transform into 1⅓-in. frogs in 1 year. These reach 1⅖ in. 1 year after metamorphosis, 2¼ in. in the next year, 2⅖ in. and 3½ in. in the following years. May breed when they reach a length of 3 in. and have been known to live for 10 years.

Call is a pleasing *tchung*. These frogs are caught for their hind legs, which are edible but small. They make good laboratory animals and are protected by law during their breeding season in some states.

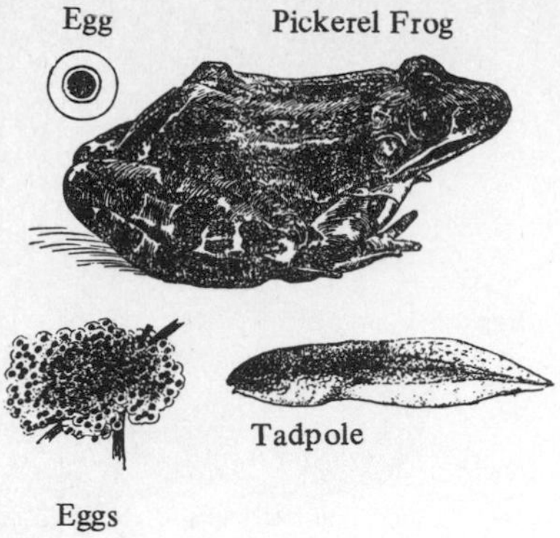
Egg
Pickerel Frog
Tadpole
Eggs

Pickerel Frog, Spring Frog

Rana palustris

Length, male $2\frac{3}{5}$ in.: female to $3\frac{1}{5}$ in. Some authorities claim that male is the larger. Male croaks, has swollen thumb at breeding time and small swellings between "ear" and arm. Much like leopard frog but has yellow under hind legs and rear part of belly, and dark spots on back are regularly in two rows, squarish, and black-bordered.

Common along streams, ponds, and lakes, sometimes wandering rather far from standing water during summer months but less inclined to do this than meadow frog. Winters in marshes, ponds, and springs. Ranges Hudson Bay to Minnesota, Arkansas, Louisiana and to southeast, with closely related species extending range to west. Generally at higher elevations than meadow frog.

Breeds in North in April in great numbers, in ponds of permanent but relatively shallow water. Eggs brown and bright yellow, $\frac{1}{16}$ in., in a thin and a thick firm jelly coat in a globular free or attached underwater mass. Female may lay 2,000 eggs, which hatch in 6 to 19 days and develop in 70 to 90 days into 3-in. tadpoles with dark purple opaque tail crests and iridescent bellies.

Food of tadpoles, oozes and slimes of plant or animal material; of adults, insects, worms, and other small animals. Tadpoles transform into frogs that average $1\frac{1}{2}$ in. at end of 1 year, $1\frac{4}{5}$ in. at end of 2 years, $2\frac{1}{3}$ in. at end of 3 years, and $2\frac{4}{5}$ in. at end of 4 years, breeding size being reached at 2 to 3 years. Call a short, labored, unpleasant, clacking grunt.

Legs sometimes used as food but are considered poisonous by some persons. Materials secreted by these frogs will kill other frogs closely confined with them. May irritate membranes of mouths of dogs rather severely. Protected by law in some states during breeding season and unquestionably more useful alive than dead. Will take a red-cloth-baited hook readily.

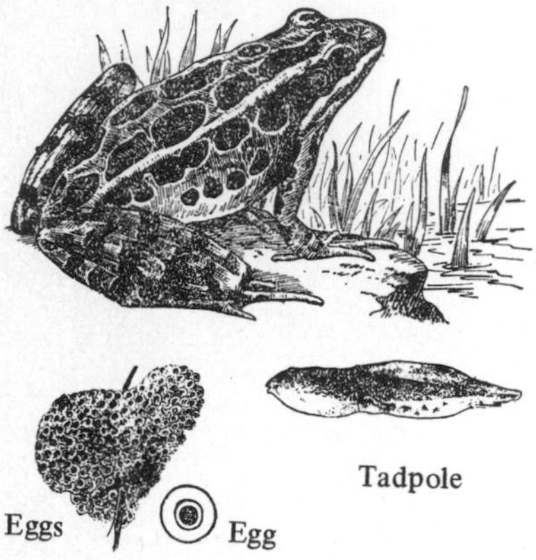
Tadpole
Eggs
Egg

Meadow Frog, Leopard Frog

Meadow Frog, Leopard Frog

Rana pipiens

Length, male to $3\frac{1}{5}$ in.; female, to $4\frac{1}{10}$ in. Male with fold of skin over arms and sometimes with swollen thumb in breeding season. Female without these characters. Skin smooth and generally moist. Green, with whitish-edged dark irregular blotches on back, commonly in more than two rows and with undersides of hind legs and rear of belly white instead of yellow-orange. Toes not webbed to tips.

Common in marshes and ponds in spring but in summer takes to open fields, cultivated lands, open grassy woodlands, meadows, and swamplands. Ranges from southern Canada south through most of North America east of Sierras and on into northern Mexico, with a number of closely related species extending range. One of frogs most commonly seen, since it is often on highways.

Breeds in North in April. Eggs $\frac{1}{16}$ in., black and white, in a thin and a thick firm jelly coat, laid under water and attached to some support but usually in long masses. Female may lay to 5,000 eggs, which hatch in 4 to 20 days and develop in 75 to 90 days into $3\frac{2}{5}$-in. tadpoles, with high-crested translucent tails, fine markings, and creamy iridescent bellies.

Food of tadpoles, oozes and slimes on submerged objects; of adults, insects, worms, and other small animals. Tadpoles transform into 1-in. frogs which average $1\frac{1}{2}$ in. at end of 1 year, 2 in. at end of 2 years, $2\frac{1}{3}$ in. at end of 3 years, and $2\frac{2}{3}$ in. at end of 4 years, breeding size being reached at 3 years or sometimes earlier where season is long and conditions good.

Legs are used as food but animal is much more valuable alive than dead. Frogs younger than breeding age should never be killed; this species is protected by law in many states particularly in breeding season. Those collecting legs for market either shoot frogs or catch them by holding a fishhook baited with red cloth in front of them. Call is a short clacking grunt. Eyes "shine" at night in beam of light. Probably the most widely distributed amphibian on North American continent.

Wood Frog

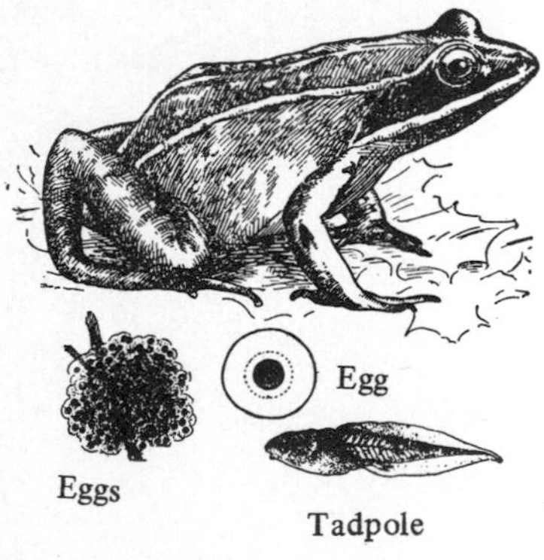
Egg
Eggs
Tadpole

Wood Frog

Rana sylvatica

Length, male to $2\frac{1}{5}$ in.; female to $2\frac{3}{5}$ in. Easily recognized by black patch that lies behind eye, over "ear" and onto a point at base of forelimb. It lacks green and is predominantly brown of varying shades on back and sides. Skin lacks any conspicuous warts, toes are pointed, and legs are long and slender. Male croaks.

Found on the ground mostly in woodlands in summer, under stones, stumps, or litter in winter, and in woodland ponds in breeding season. Ranges from Nova Scotia to South Carolina, west to Arkansas and north except in western United States, southwestern Canada, and regions where soil is always frozen; even Alaska. A number of subspecies have been proposed or recognized. In Pacific Northwest, species is diurnal; may be found in open grassy areas and in tundra ponds.

Mates in April in North. Eggs $\frac{1}{10}$ to $\frac{1}{16}$ in., black and white, in two firm jelly coats, in several spheres, free or attached, under water, to 3,000; hatch in 4 to 24 days and develop in 44 to 85 days into $1\frac{4}{5}$-in. tadpoles with creamy lines on their upper jaws, pink bellies, and iridescent bronze dorsal surfaces. Tadpoles transform into ⅗-in. frogs that mature in to 4 years as described below.

Food of adults, small animals of forest floor; of tadpoles, oozes and slimes of breeding pond. Males in breeding season are darker than females, have swollen pads on inner side of thumb, and have webs of feet with convex rather than concave margins. Frogs are 1 in. long at 1 year, 1⅓ in. at 2 years, 1⅘ in. at 3 years, and 2¼ in. at 4 years.

Interesting animals that seem to leap from nowhere in the woodlands. In breeding season males give a confusing series of clacking croaks; some describe their call as a sound resembling a clacking domestic duck. Animals can change color remarkably in space of an hour and may be either dark or light, depending on their surroundings. May be of some value as destroyer of small animals and as food for larger ones.

CLASS REPTILIA Turtles, Terrapins, Tortoises, Lizards, and Snakes

Skin dry, horny; usually with scales or scutes; limbs, when present, four; appears from Permian period; terrestrial, freshwater, or marine. Some 6,000 species.

Order Chelonia./Family Kinosternidae

Musk Turtle, Stinkpot
Sternotherus odoratus
Length of upper shell to 7⅓ in., average 5 in.; under shell does not cover soft parts of body. Back without keel. Two yellow stripes on either side of head. Male with blunt nail on long tail, concave under shell, with horny scales on inner side of hind limbs. Female with short tail, with pointed nail.

From southeastern Canada and eastern United States, west to Michigan, south through eastern Texas, southern Missouri, and eastern Oklahoma. Common in marshes, pools, canals, bogs, and bayous with deep muddy bottom and abundant water plants. Favors relatively deep water. Essentially aquatic.

Breeds in early spring. Eggs white, brittle, 1¹⁄₁₀ by ⅗ in.; laid singly or to nine on surface or in nests underground; hatch in 60 days (temperature up to 98°F) or 90 days (temperature up to 77°F) into 1-by ⅘-in. young, which in 3 to 4 years show sex differences and may live at least 23 years.

Food mostly animal matter eaten voraciously, but also may eat water plants, cow dung, or garbage found on the bottom and torn by jaws and claws. Feeds either night or day. May bite, and once having bitten hangs on persistently. Gives off offensive musk from two glands. In captivity eats raw meat or fish and vegetables.

May destroy fish nests and some fish but does little damage and may be considered scavenger even though occasionally it bites baited hooks, to the disgust of anglers. Flesh useless as food for man, since role as scavenger is main claim to fame for this often popular aquarium animal.

Musk Turtle

Order Chelonia./Family Chelydridae

Snapping Turtle
Chelydra serpentina
Length to 3 ft; female considered the larger, with vent nearer end of tail. Upper shell 1 ft long, with 8-in. lower shell not covering legs, neck, and tail. One reported to have weighed 86 lb. Back with three broken ridges of coarse scales, with a border row giving a notched outline to rear.

Eastern North America from southern Canada to Gulf of Mexico, west to Rocky Mountains. Western tributaries of Mississippi River are followed to their sources, south to Ecuador. Lives for most part in slow rivers, shallow ponds, and mud-bottomed lakes. Does not sun on a log like many turtles.

Mates April to October. Female leaves water to dig nest on shore and lays 24 to 30 spherical, ⅘-in. white thin-parchment-like hard-shelled eggs, about 5 in. under soil; hatch in 90 days or over winter into young with yolk sac and tails as long as upper shell. May live more than 25 years.

Food varies greatly with conditions. Usually largely animal matter but in some individuals over half of stomach contents may be water plants; can digest plant materials satisfactorily. May move slowly but can attack with lightning speed with head. Seeks food night or day. Hibernates through winter in mud under water.

Excellent as food in soups or stews. May serve as scavenger in waterways. May destroy ducks and some useful fishes and on land, if bothered, may severely injure man. Too dangerous to be considered pet at any time. Formerly were fattened in swill barrels until fat enough to be killed for food.

Snapping Turtle

Spotted Turtle

Pacific Pond Turtle

Wood Turtle

Box Turtle
Terrapene carolina

Upper shell 5 in. long, 3¾ in. wide; 2¼ in. high. Lower shell to 4⅗ in. long by 3¼ in. wide. Male with bright red or pink eyes. Female usually with dark red, yellow, or brown eyes. Back high and arched, with a blunt middle ridge. Uniformly dark brown or black with yellow spots and streaks.

Ranges eastern United States from Maine to Georgia, west to Tennessee and western Illinois and northward to central Michigan, with related species extending range farther west and south. Found in open brush-covered country, often remaining for years in relatively restricted area.

Eggs three to eight, like hen's egg, with thin flexible white shells; 1⅖ by ⅘ in.; laid in cavity in ground; hatch in 3 months into young with yellow, keeled under parts, about 1⅕ in diameter; 5-in. shell in 5 years. Male may mature at 4¾ in.; female at 4 in. Live over 40 years.

Food, plant material and worms, slugs, snails, and insects procured in decidedly leisurely travels. Protected efficiently by shells, which close so tightly that a pencil cannot be forced between to reach soft tissue. Hibernates in soft soil below frost line, often going down 2 ft. May aestivate in mud in hot weather.

Useful destroyer of insects and of garden pests generally. Makes a good pet requiring relatively little care. Always harmless. Good to eat but protected by law in many states and should receive universal protection at all times.

Diamondback Terrapin
Malaclemys terrapin

Head and neck without longitudinal yellow stripes. Shields of upper shell with concentric ridges and grooves. Length of upper shell: female to 7½ in., male 4⅘ in. Sides of shell about parallel. Head relatively large. Female with larger head, more rounded snout, shorter tail, and deeper shell.

Two subspecies, northern and southern, and one related species with three subspecies. Southern ranges along Atlantic Coast from Cape Hatteras to Florida; northern, from Buzzards Bay to North Carolina, Delaware, and Chesapeake Bay. In salt or brackish bays and in marshy tracts with mud bottoms.

Female may lay fertile eggs for years after one mating though only 4 percent of eggs are fertile after 4 years; may lay to five times a season. Eggs 1⅖ by ⅘ in., white; hatch in 90 days into young with 1¹⁄₁₀-in. shells. Male matures in 4 to 9 years, with 3⅕ to 3½-in. shell; female, 4 to 9 years, 5½-in. shell. Life-span over 40 years.

Food, in wild, snails, crabs, worms, and possibly some tender plants. In captivity, chopped fish, crabs, clams, and insects, preferring to eat underwater. Experimental farms usually free young for stocking at 8-month age. Rearing in captivity is hazardous because of expense of feeding, variable market, and length of time necessary.

Once considered one of most expensive luxuries as food for man when turtle with 7-in. shell brought $7 with additional dollar for each additional ½ in. 160,000 reared at Beaufort 1909 to 1941. Prefers water which is not polluted, not a scavenger. Some believe age of turtle can be determined by counting growth rings in one of the "diamond-shaped patterns" on the shell; not so after the first 5 to 10 years of age.

Painted Turtle
Chrysemys picta

Upper shell to 7⅓ in. long by 4½ in. wide by 2 in. high. Lower shell 6 in. long by 3¾ in. wide, with a lobe in front. Back wide, unnotched, unridged, smooth, olive to black, with red marks on sides. Lower shell yellow. Tail of large male 1¾ in. long; of equal-sized female, 1¼ in.

Four subspecies and closely related *C. picta* range from New Brunswick to Georgia and west through Mississippi Valley, on through Montana and the Columbia River valley, British Columbia, south to New Mexico and northern Mexico, with type species and subspecies in Mississippi Valley. Common in ponds.

Eggs five to eight, smooth, white, glazed, with soft shells; 1⅕ by ⅗ in., blunt, egg-shaped; laid in early summer, in nests usually near stumps near edge of water; hatch in summer into little 1- by 1-in. turtles; grow to about 1½ in. at 6 months; to 2⅖ in. at 18 months; to 3½ in. at 52 months, or breeding age.

Food, aquatic plants and animals, dead or alive, eaten only under water. In captivity, thrives on beef, earthworms, or lettuce but may eat fishes as well. Male breeds at smaller size than female; male, at about 3 in. while female begins at 4 in. with growth thereafter slower. Adult female larger than male.

Serves as scavenger of shallow pools. Of no value as food since flesh is poor. Almost impossible to make it bite a person but will take a hook with bait repeatedly even after having just been caught. Hibernates in cold weather. Not protected by law. One of our most attractive small turtles.

Order Chelonia./Family Emydidae

Spotted Turtle

Clemmys guttata

Length: male, upper shell to 5 in.; width, 3⅖ in.; height, 1⅓ in. Under shell 2⅖ by 1⅓ in. Male with dark brown eyes, dusky horny jaws, black throat, black neck, and concave lower shell; 4 ⅔-in. animal has 1⅘-in. tail. Female with orange eyes, pale yellow jaws, and smooth neck. Both with backs unridged; black with round yellow spots.

Found in small pools and ponds from Maine to northern Florida, through north and west to southern Michigan but apparently not north of Great Lakes. Sometimes found in brackish marshes with rougher shells, and sometimes in woods near streams but almost always in vegetation which gives cover.

Males fight, then pursue females. Eggs 2 to 4; white, elliptic, 1⅕ by ⅗ in.; buried in sand; probably hatch in 82 days into young, with single spot on each back plate, these spots increasing in number with age, with spots later developing on legs and tail. Under shell yellow at hatching. May live over 42 years.

Food largely insects, including beetles, flies, dragonflies, worms, slugs, crustaceans, spiders, and tadpoles, mostly eaten under water. In captivity, eats raw chopped meat, fish, shellfish, meal worms, water plants, and lettuce, mostly under water.

Of little direct economic importance to man and of no serious damage to his interests. Used as an experimental animal by psychologists; have shown ability to learn to solve mazes, have fear of falling from certain heights, and in some cases, show prolonged preference for one individual of different sex.

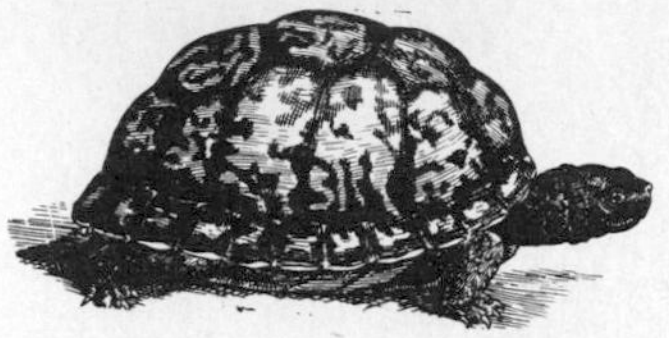

Box Turtle

Pacific Pond Turtle

Clemmys marmorata

Average upper shell 6½ by 5⅕ in., lower shell 6¹⁄₁₀ by 4⅕ in. Length of tail 2⅗ in. One with 7-in. upper shell is large and weighs 2 lb. Upper shell olive to black, with each shield with many yellow-brown or black dots and dashes. Under shell yellow and in male, concave.

Ranges along Pacific slope from Puget Sound to San Diego, Calif., being commonest turtle in that part of country. It favors lakes, marshes, and even enters brackish water or the sea. It is not known in Canada. Found in wet places from sea level to 6,000 ft.

Eggs white, hard, elliptic, 1⅗ by ⅘ in., laid June to July; 5 to 11, in a nest; in open field, sandy bank, or hillside near water and left to hatch by heat of sun and earth. 1 in. long when hatched. Young with shells nearly round, but grow slowly, so that in 10 years an upper shell of 5½ in. in length is attained.

Food aquatic insects and other animals including carrion and dead fish. Hibernates in mud, but during summer often basks in sun in companies at edges of pools or lies on bottom, moving about only in early morning or evening. Strictly aquatic except for egg laying. Must come up for air once in 3 hours.

Alive, probably a useful scavenger and insect check. Good to eat; therefore, has been extensively and probably unwisely trapped for sale. Since its growth is so slow and since it congregates in numbers, its harvest should be regulated to maintain its abundance.

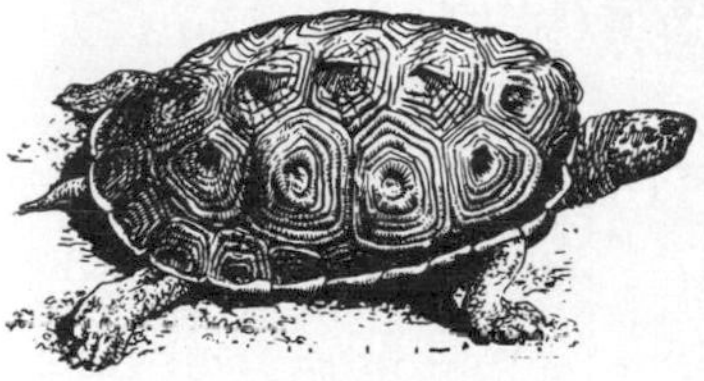

Diamondback Terrapin

Wood Turtle

Clemmys insculpta

Average upper shell, 7 in.; maximum, 9 in. by 5 in.; under shell, 6½ by 4 in.; height 2¾ in. Scales roughly sculptured with concentric lines. Male with concave lower shell, longer thicker claws, with vent on extended tail beyond hind edge of shell. Female with convex shell and vent not beyond hind edge of upper shell.

Eastern United States from Maine to District of Columbia through southwestern Ontario, west through Iowa. Found in fields and woodlands, spending most but not all its life on land. Closely related to Pacific pond turtle, *C. marmorata*, to spotted turtle, and to Muhlenberg's turtle.

Mates in spring and fall. Eggs in North, laid in June in midafternoon; buried in sand in lots of 2 to 12; elliptical, 1⅖ by 1 in.; white with parchment-like shell. Young with nearly round shell and tail almost as long as upper shell, with soft parts gray instead of red of adult.

Food either plants or animals, but preferably plants including fruits, berries, tender leaves, and mushrooms but also some insects, earthworms, slugs, snails, and carrion. Has unusual ability to climb. Can be trained to respond to stimuli rather easily. Normally it will not bite but when annoyed can be induced to do so.

Flesh edible, but animal is so scarce that it should be protected by law everywhere as it is in some parts of its range. Harmless and makes an interesting, easily cared-for terrarium pet, doing best where a temperature of 68 to 75°F can be maintained and where cover and food are present. One individual lived in captivity 58 years.

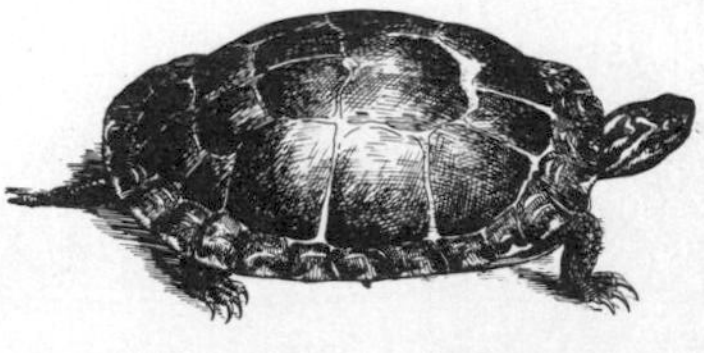

Painted Turtle

PHYLUM CHORDATA. CLASS REPTILIA

Order Chelonia. / Family Testudinidae

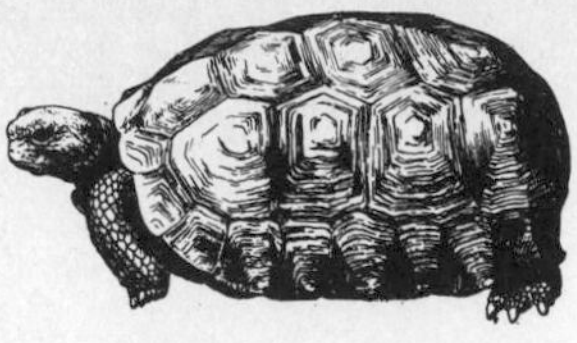

Desert Turtle

Desert Turtle, Gopher Turtle

Gopherus agassizi

Upper shell to 10 in. long, 7 in. wide, 4½ in. high. Eastern gopher turtle smaller. Fingers and toes on each limb club-shaped and closely bound together, with only last joint free. Scales with conspicuous concentric markings. Related Galápagos tortoise reaches length of 4 ft and weight of 500 lb.

Desert tortoise or gopher turtle ranges through deserts of southwestern Arizona and southeastern California into Nevada, with two related species extending range east through southern Arkansas, southern South Carolina, south to Florida, Gulf States, Texas, and northeastern Mexico, *G. polyphemus* being commoner eastward.

Male with concave lower shell shorter to rear, tail longer. Eggs white, with thick, hard, rough shells, elliptic and nearly 2 in. long, buried in sand, in sun; hatch into young 1⅗ by 1⅘ in., with shells hardening in 3 years but growing to 2⅘ in. first year. A 6-in. female grew to 8½ in. in 27 months.

Food largely plant materials such as clover, lettuce, bananas, berries, and vegetables. In wild, food frequently consists of grass, cactus, and other low-growing, associated plants. Individuals have been tamed to accept food from hand. Animals are harmless; do little to protect themselves except rely on their shells. Favor temperatures between 85 and 95°F. Move 20 ft per minute or 5 miles per day.

Useful as food but should not be eaten because of danger of extermination. Interesting scientifically as descendants of species more common in past. Must have dry environment at all times to prosper or to survive. Huge relatives are popular in parks or zoos where they can carry young children on their backs.

Order Chelonia. / Family Cheloniidae

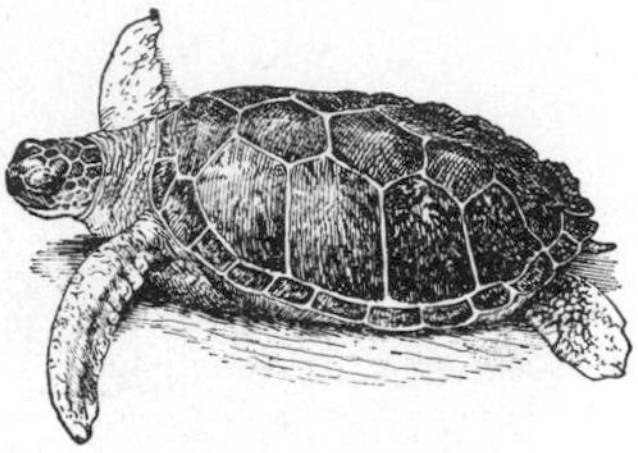

Green Turtle

Green Turtle

Chelonia mydas

Shell covered with large horny shields, Back without longitudinal ridges. Four pairs of side shields. Shields of back not overlapping except in young. Usually a single claw on each limb. Single pair of shields on top of head. Back shells to 46 in. long. Weight to 850 lb. Limbs paddle-shaped.

Found in sea or on seashore north to Massachusetts and through West Indies; commonest Bermuda turtle. In Pacific, north as far as San Diego and abundant in southern Lower California. Usually within 35° of equator. Known from Florida Pleistocene. Comes to shore to breed.

Breeds any time. Eggs nearly spherical, white, soft-shelled, 1⅘-in., mostly laid March to May in nest at night, above high tide; hatch in 47 to 52 days. May lay seven times a year for 2 weeks, 100 eggs each time. Young add ½ in. to shell and 1 lb each month, reaching 8-in. shell first year. Female matures with 35-in. shell; 44-in. shell in 10 years.

Food almost exclusively marine algae and plants, more succulent parts being favored. Young eat more animal matter than adults. May sleep at sea. Tail of male extends beyond tip of extended flippers; of female, only to edge of upper shell. Captives have added 50 lb in 9 years; another attained 18-in. shell at age of 28 months. Swims slowly; recorded at less than 1 to 4 mph.

Source of turtle soup, and basis of industries worth many thousands of dollars. Oil made from eggs and animal. Females with shells under 40 in. long should be left to mature. Shell of no value. Valuable tortoise shell comes from related hawksbill turtle. Undoubtedly, most valuable of all turtles and worthy of wise management. Should now be protected by law because of great decrease in numbers due to popularity of flesh and eggs.

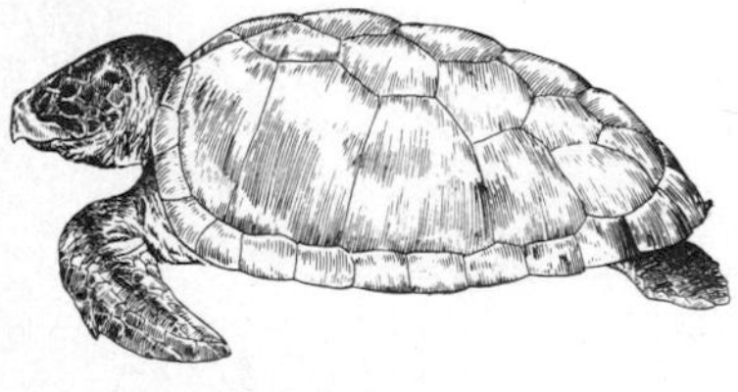

Loggerhead Turtle

Loggerhead Turtle

Caretta caretta

Pairs of scales on side, five. Limbs; two-clawed in Atlantic species and one- or two-clawed in Pacific species. Two pairs of large shields between eyes. May reach weight of 450 lb, though 300-lb animal with 3-ft back shell is large. Tail of male extends beyond extended hind flippers; of female, barely to rear edge of upper shell.

Found in tropical and subtropical Atlantic Ocean, north in summer to Cape Cod but mostly in Carolina-Virginia region; also in Mediterranean. Pacific species ranges south of California, particularly in Lower California. Atlantic species known from Pleistocene of Florida. Favors vegetation-laden ocean currents.

Eggs white, nearly spherical, 1⅔-in., soft-shelled; laid in nest above high tide, in sand about 10 in. under surface; probably three times a season; 80 to 150 at a time; hatch in 2 months. Young reach shell length of 21 in. and weight of 42 lb in 3 years; to 25 in., 81½ lb in 4½ years; probably mature at 200 lb.

Food largely animal matter such as hermit crabs, shellfish, conches, and Portuguese men-of-war, sponges, jellyfish, and fish. Can bite off end of large conch and extract soft animal inside. Young at first move away from broken horizon, downhill, toward opaque blue, away from transparent orange, red, or green, without respect to sun.

Eggs are eaten and young turtles are sold for their flesh, but old ones are too rank to eat. Oil extracted from old turtles used for softening leather. Musk given off at times. Inferior in value in all respects to green turtle and hawksbill turtle. An endangered species.

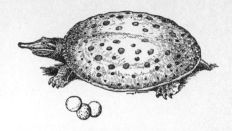

Order Chelonia./Family Trionchidae

Soft-shelled Turtle
Trionyx ferox (Southern) *T. muticus* (Spineless)
Back leathery, light brown with darker spots. Without spiny tubercles along front of back (spineless); with small tubercles along front of back (Southern) and with prominent spines along front of back (spiny, illustrated). Back about 15 by 11 in., flexible. Undersides with legs, neck, and tail poorly protected.

Soft-shelled Turtle

Three American species; Southern, from South Carolina to Florida and Louisiana; spineless, through middle and northern tributaries of the Mississippi and St. Lawrence rivers; spiny, in Mississippi tributaries west to Colorado and Montana, in St. Lawrence tributaries, and east to Vermont, New York, and Pennsylvania.

In spiny female tail extends barely to rear of shell; male's tail longer. Eggs spherical, 1 1/10 in.; hard, brittle, white; laid in lots of 12 to 32, two to four times a season, in nest 1 ft underground June to August; hatch in 2 to 10 months. Young, 1-in., circular; to 7 in., fifth year; breeds first in sixth year.

Crocodile

Food probably largely animal matter such as crayfish, minnows, frogs, earthworms, with weak appetite for plant foods. Food swallowed whole. Jaws strong. Can strike accurately and suddenly but if held by tail cannot reach captor. Essentially aquatic but suns on land, although it remains alert there to danger.

Flesh good to eat and turtles are often sold in market for food. May damage a careless human by their bite and destroy some fish and fish food, but probably do no damage sufficient to offset food value. Turtles' temperature varies from 33.6 to 84°F, favoring 79°F for most activity; average 1.8 to 5.4°F above surroundings.

Broad-nose Crocodile

Order Crocodilia./Family Crocodilidae

Crocodiles, Caimans, Gavials
Crocodylus sp. *et al.*
In crocodiles, the fourth tooth of lower jaw fits into notch in upper jaw when mouth is closed, while in alligators and caimans the fourth tooth fits into pit in upper jaw. Orinoco crocodiles have narrow heads while American crocodile (illustrated) has triangular snout. Gavials have extremely long slender snouts.

Indian gavial may be to 30 ft long, equaling Madagascar crocodile and exceeding largest 20-ft caiman and largest 16-ft alligator. Caimans are limited to tropical Americas; gavials, to India, Borneo, and Sumatra; but crocodiles are found in North America, South America, Australia, Africa, and Asia.

Young of all crocodilians are produced in eggs that are white, rather long and shining, with a thick hard shell. Eggs laid in a bank or in a nest of decaying plant material piled into a mound. Doubtful if reputed care of young by adults, particularly by male, has any basis in fact.

Crocodiles are vicious temperamental animals able to move rapidly on land and to strike with tail powerfully and accurately. Young feed usually on fish while adults prey on water birds or mammals that come to drink. In water of temperature of 45°F, crocodile is practically helpless.

Some crocodiles may be serious enemies of man, particularly salt water; crocodile *Crocodylus porosus* and African crocodile *C. niloticus,* which may pursue and capture humans in water or even on land. Crocodiles and alligators help keep waterways open and unintentionally help fish and man in various ways. *C. acutus,* American crocodile, is an endangered species.

American Alligator

American Alligator
Alligator mississippiensis
Length to 15 ft; tail 1/2 length. Weight to 500 lb. May mature at 9 ft. Snout broad, not pointed as in crocodiles, and fourth teeth do not show as they do in crocodiles. Teeth 19 to 20, on each side of each jaw. Skin roughened, leathery, and thick over most of body. Eyes and nostrils placed high. Male bellows and emits musk.

(Continued)

Ranges in rivers and swamps of lowlands of the Carolinas, Georgia, and Florida, west to Louisiana and Mississippi on to Rio Grande in Texas. American crocodile found in southern Florida, Greater Antilles except Puerto Rico, and both coasts of Central America from Mexico to Ecuador and Colombia.

Female lays 30 to 40 eggs, in nest of decaying plants, which aid in hatching; may protect nest. Eggs like long hen's eggs. Young when hatched, 8 in. long, weigh 1½ oz. At 1 year, 2 ft long, 4 lb; at 2 years, 3 to 4 ft long, 9 to 12 lb; at 4 years, 4 to 6 ft long, and 16 to 29 lb; at 9 years, females 7 ft long and 110 lb; at 9 years, males 8 to 9 ft long, and about 140 lb.

Food, animals captured in jaws or knocked down by thrashing tail. Can move head and tail sideways with great force and speed. Prey may be held under water until it drowns, then brought to surface and swallowed whole above water. 10-ft alligator can swallow ducks whole. May "boom" when capturing food.

Hide valuable as source of leather but supply has been unwisely exhausted. Enemy of ducks and other water animals but avoids man when possible and normally would not harm him. Should be protected against extermination. Cannot survive low temperatures. May hibernate and aestivate 2 or more months each year. An endangered species.

Banded Gecko

Order Sauria./Family Gekkonidae

Banded Gecko
Coleonyx variegatus
Length to about 4 in. Toes with suction disks useful in holding animal to smooth surface. Head only slightly wider than body. Tail approximately ½ total length. Basic color brown but usually with conspicuous bright broad lemon-yellow bands crossing body and tail. Sometimes speckled rather than beaded as above.

There is only one North American species in the genus. It is found in southwestern United States from Texas to California. Most of the world's geckos are tropical and of different genera. One species from Malay Peninsula reaches a length of 15 in. and is proportionally large.

Males generally larger than female geckos. Females lay several white hard-shelled rounded or bluntly oval eggs where they may hatch independent of parental care. Young resemble adults in general appearance and behavior. Animals are capable of making sounds something like *gecko* or similar to those produced by some katydidlike insects. May chorus.

Food, living, moving insects captured by a lightninglike thrust of long sticky tongue with swollen tip. Geckos apparently run with ease over ceilings and walls, coming out of their daytime hiding place at dusk and returning at dawn. Individuals seem to establish personal territory protected from invasion by others. Apparently do not notice fixed food.

None are poisonous in spite of reputation to contrary. All are excellent destroyers of insects and worthy of protection and encouragement. Species here illustrated emits only a small squeaking sound as contrasted with that produced by many related species. An animal well adapted to desert life because it is nocturnal and subterranean. May sometimes be seen along blacktop roads at night in southwestern states.

Old World Chameleon

Order Sauria./Family Chamaeleontidae

Old World Chameleon
Chameleon dilepis
Horned chameleon may be about 1 ft long, with prehensile tail about as long as body, capable of coiling spirally. Eyes large, forward, capable of independent action. Feet grasping, with toes in two groups capable of opposing each other. Body finely scaled. Color changeable. Mouth enormous.

Old World chameleons range from Asia Minor and Syria through North Africa and southern Spain. Essentially arboreal, although they may go to ground at times. Horned and hornless species and some in which horned character is modified by sex. In East Africa, a dwarf chameleon *Brookesia*.

Sexes differ in species by horns, tail length, and size. Some species may give birth to living young while others lay eggs of different numbers. One species lays about five eggs under stones on ground. Young, born alive, may be only about 1 in. long.

Can change color at will or may change color owing to health. Can capture food by extending tongue suddenly to length of 6 in. in some species. Food almost exclusively insects. Animal may spend most of its life on a single tree. Tail muscular and not shed as in some lizards.

Much folklore attributes "evil eye" to these animals; may bite vigorously but relatively harmlessly. Not poisonous and could not hurt a human being in any way. Are geniunely useful as destroyers of injurious insects. If considered for pets, should be allowed to remain in natural setting.

PHYLUM CHORDATA. CLASS REPTILIA

Order Sauria./Family Iguanidae

Chameleon or Anole
Anolis carolinensis
Length to about 6 in., with slender tail which in New World chameleons is not prehensile as in Old World species. Body covered with fine scales. Skin dry or only slightly moist. Male with colored extensible fold in throat. Color gray when cool, green when excited or hot, creamy in dark, and sometimes dull in intense sunlight.

American chameleon or *Anolis* ranges from North Carolina to Florida westward through Gulf region to Rio Grande on mainland and in islands to South. Lives in trees and other vegetation, commonly on green parts.

Male courts by violently bobbing whole body and by distending conspicuous orange throat region in presence of female. Eggs to about a dozen, parchment-covered, laid by female where they may be warmed by sun until they can hatch young chameleons, like adults in general appearance. No parental care shown.

New World Chameleon

Food exclusively insects caught by stalking, with a final dash and lightninglike thrust of tongue. Will not eat earthworms and cannot exist on sugar and water diet as sometimes suggested. Must have water but will not ordinarily drink it from a dish. Preferable if animal is to be kept in captivity to spray vicinity with atomizer.

Superior insect destroyer and worthy of every protection. Sale of these animals leads to slow starvation and should be prohibited since they are of real value alive and free in their natural environment. Study of color changes in relation to temperature and emotion interesting. Temperature usually 1.8°F above surroundings. Protected by law in some states. Said to be able to "parachute" for 30 ft or more.

Chuckwalla
Sauromalus obesus
Length to 1 ft. Body broad. Tail thick and flattened, more stubby than in some other lizards and about ½ total length. Head about 1½ in. broad. Scales small. Color of adults dull brown, black, or olive. Abdomen reddish. Scales somewhat rougher about neck.

From southern Nevada and southwestern Utah to Arizona, southern California, and Lower California. Lives in deserts of sand or, in parts of range, in dry rocky areas where protection by hiding is possible.

Male more gorgeous than female. Mates in June to July. Young more olive than adults, marbled, spotted with black or with reddish bands mixed with yellow dots, which show more abundantly on tail. Coloration may change with age and vary when awake or asleep.

Chuckwalla

Food probably solely plants. Staple food is creosote bush. Prefers buds and flowers, although sometimes eats leaves. In captivity eats lettuce and celery. Gets protection by hiding in crevices and inflating body to prevent removal, by striking effectively with tail, and by coloration. Hardy, with many enemies. Locomotion relatively slow.

Edible, formerly supplying much food to Indians. Flesh tender and palatable. Not known to bite even when roughly handled. Indians punctured inflated body with wire to remove it from crevice. Favors sunlight and rocks that may be unbearably hot to man. Sticks out tongue while walking, possible as sensory act.

Fence Lizard
Sceloporus undulatus
Length to 5 ½ in.; tail, 2 ⅗ in. Width ¾ in. Gray, brown, to green, with black crossbars like irregular V's. Pale bands from eye along each side, about 10 scales apart. Male with black blotch under chin and two blue abdominal areas. Tail unusually slender. Toes rather long and slender. Also called blue bellies or swifts.

Eastern United States, from New Jersey to Florida and west, with 15 other species extending range through United States, on through Mexico to Guatemala, with Oregon marking northern limit on Pacific Coast. Favors dry sandy pinelands or other dry places but mostly in fallen woody plants. Spiny lizards in the western states occur from below sea level to 14,000 ft.

Eggs ⅖ to ½ in. long, oval with thin papery shells, which indent with slightest pressure. At ordinary room temperatures, if kept moist but not wet, these hatch in 6 to 8 weeks into small lizards which resemble adults and are able to care for selves immediately.

Fence Lizard

Food essentially small insects. In captivity, thrives on meal worms, cockroaches, and similar animals. Protection by speedy escape, by shedding tail if caught and by inconspicuous appearance. Food captured by pursuit and held by mouth.

Makes interesting terrarium pet and serves great usefulness as insect destroyer. Does no damage. Entirely worthy of protection, which it does not often get. Favors abundance of sunlight and must be kept dry in captivity. Responds almost instantaneously to observed movements. Males do "push-ups" as territorial display.

Horned Toad
Phrynosoma cornutum

Length to 6 in.; tail 2 in. Head with large horns in rows, and with eyes that close readily. Body wide, flat, short, with a row of spines along sides. Tail short, bordered with a row of spines. Gray spotted with brown, and with a middle yellow band. Skin dry, finely scaled. Light beneath.

From Kansas to northern states of Mexico, west to Colorado and New Mexico, with related species extending range westward and southward. Species in United States, 7 with several varieties. At their best in hot dry sandy wastes where heat would seem to be practically prohibitive to animal life.

Young, 6 to 12; born alive, or rather in thin transparent envelopes that break quickly, freeing young, which resemble parents except that they bear undeveloped horns and lack general spiny appearance of adults; able to shift for themselves immediately after they are born.

Food largely insects, captured on run with wonderful rapidity. In captivity, thrives on meal worms, other insect larvae, cockroaches, grasshoppers, crickets, and other relatively large-bodied animals. Protects self by burrowing into sand, by swelling, by threatening poses, or by expelling blood from eyes in streams.

Highly useful as insect destroyer. Makes excellent pet but probably does not profit by being kept captive. Quickly and actively susceptible to direct sunlight and heat.

Horned Toad

Order Sauria./Family Anguidae

Glass Snake, Joint Snake
Ophisaurus ventralis

Length to 3 ft; tail, about ⅔ length, brittle and easily shed. Since it has eyelids, ears, and scaly belly, it is not a snake but a legless lizard. Black, olive, or brown, with small green spots on each scale. Greenish-white beneath and generally glassy in appearance.

Southern United States from the Carolinas west through Tennessee, southern Wisconsin, and south through New Mexico and state of Vera Cruz in Mexico. One American species in genus, with two other genera in family. Inhabits dry meadows and similar places where suitable food may be found.

Stories that glass snake breaks into two pieces and that each piece becomes a separate complete snake or that the two parts unite again are not founded in fact. Details of reproduction are shrouded in mystery, but some relatives of this lizard lay eggs and it also may reproduce in this way.

Food largely earthworms, slowly moving insects, slugs, eggs of birds, and other small forms of animal life; captured by burrowing and crushed by strong jaws. Obtains protection by remaining hidden, or by breaking in two, leaving foe with useless tail. Enemies include snakes and other animal eaters.

Of little economic importance but of great interest because of habit of breaking in two. Can easily be found but not easily caught. Lost tail is replaced by fleshy tail without vertebrae. Burrows a great deal, so probably does not favor strong sunlight as most of its lizard relatives do.

Glass Snake

Order Sauria./Family Helodermatidae

Gila Monster
Heloderma suspectum

Length to 24 in. Width to 4 in. Tail ¼ length. Legs weak and sprawling. Head heavy. Body, legs, head, and tail covered with beadlike scales. Poison fangs in lower jaw. Eyes small and inconspicuous. Motions normally slow and sluggish. Color, black or brown irregular pattern on orange, yellow, or pink.

From extreme southern part of Utah and Nevada through Arizona and Sonora, but widely known as side-show attraction in circuses and sometimes much publicized in newspapers. Lives in dry desertlike lands, particularly where soil is loose and sandy. Commonest in vicinity of Gila River in Arizona.

In July and August female, after mating, lays about one dozen soft-shelled eggs, 2½ by 1½ in., burying them 3 to 5 in. deep in sand, where they will be exposed to sun's heat. If kept moist, they will in about 1 month hatch into 4-in. young, each more brilliantly colored than adults and able to shift for themselves.

Food, insects, such as ants, and birds' eggs. In captivity, eats insect larvae, chopped meat, and eggs. Can store food in tail, which is plump in well-fed animals and lean in those which have been starved. Apparently enjoys hard-boiled eggs if they are broken up. Can strike with lightning speed.

Only poisonous lizard in United States, but death to humans because of bite is almost unknown. Apparently will not bite except in defense. However, should be approached with caution as they will lash out and

Gila Monster

bite in defense. Do not really bite, but chew victim; wounds produced make entrance of poison from saliva possible. Most active at dusk. Poison affects respiration and urine secretion. In Arizona, gila monster is protected by law. An endangered species.

Order Sauria./Family Teidae

Six-lined Lizard, Race Runner
Cnemidophorus sexlineatus
Length to 10 in., of which 7 in. may be tail. Legs slender; hind toes long and slender. Dark brown, with six narrow yellow stripes running from head onto tail. Belly and underparts of males light blue. Scales small and in some places almost indistinct but like plates beneath.

Six-lined Lizard

Young

From Maryland to Florida, west to northern Mexico and Arizona, up Mississippi Valley as far north as Lake Michigan. Most common in southern part of range. 12 species of genus extend range to southern Lower California, Nevada, and California. Found in dry open sunny places.

Female makes nest by hollowing out sand and placing thin-shelled eggs in it. Eggs covered by warm sand by female and left to be incubated by heat of sun. In some species, common for young lizards to have more numerous and more distinct stripes than adults. Young when hatched 1¼ in. long. Some species of *Cnemidophorus* seem to be made up only of females and so reproduce by parthenogenesis.

Food essentially insects, but has been known to eat eggs of small ground-nesting birds by crushing eggs with jaws and lapping up contents. Can run with lightning speed and seems to anticipate direction of attack. Excellent ability to hide.

Essentially useful as insect destroyer in an area not favored by most insect killers and often ideal for insects. Makes fine terrarium pet.

Order Sauria./Family Scincidae

Five-lined Skink
Eumeces fasciatus
Length to 7½ in.; head and body about 3 in. Scales smooth, flat, appear to overlap. Color brownish to blackish, has pale stripes on body. Color variable and dependent on sex and age. Very young individual black with five white or yellowish stripes; tail bright blue. With age, as animal becomes larger, pattern is less conspicuous. Tail becomes gray, stripes darken, body color becomes lighter. Striped pattern is retained by females with age; males do not show prominent stripes. Old males become brown or olive-colored; frequently develop orange-red color on heads. Characterized by 26 to 30 longitudinal rows of scales around center of body.

Five-lined Skink

From eastern U.S., southern New England to northernmost Florida, westward into Texas and north through Oklahoma, Kansas, Nebraska, and South Dakota.

Mating takes place soon after animals emerge from hibernation in colder regions. Courtship of male consists of rushing at female with open mouth. Reported that courtship may continue for 7 minutes and the actual mating process 8 minutes. After 6 to 7 weeks eggs are laid; they hatch in 5 to 6 weeks. Eggs, 2 to 18, in soil several inches deep or in rotten logs. Female coils around eggs and guards nest; will attack any intruder. Young about 1 in. long at hatching.

Food arthropods, especially insects. Will eat small vertebrates. Forages for food in daylight hours.

Found in wooded area where there are rotting logs; usually on ground. However, reported in sawdust piles, rock piles, leaf litter. Most frequently found under some cover such as a rock. Wary. Sometimes found in urban areas in trash piles.

Order Serpentes./Family Boidae

Rubber Boa, Worm Snake
Charina bottae
Length to 15½ in. of which 2½ in. are tail. Greatest diameter, ½ in. Stout, with tail almost as blunt as head, hence the name two-headed snake. Scales small, smooth, and shining. Brown, gray, or olive, with a clean yellow unmarked abdomen and no pattern anywhere. Smooth and shiny skin. Ranges from sea level to 9,000 ft.

Rubber Boa

Ranges through humid districts of California, Nevada, Idaho, Oregon, and Washington, usually burrowing slowly and steadily into soft earth or vegetation when given opportunity. In forested and grassland areas frequently under rocks, or rotting logs. Obviously does not crave sunlight. Burrowing habit of two small North American boas is similar to that of Old World sand boa, *Eryx*.

(Continued)

Young born alive rather than being hatched from eggs. Much like adults in general appearance. Rubber boa differs from California boa in that former has a single central scale on top of head between eyes, while latter has numerous scales in this area.

Food: this snake, like coachwhip snake, racers, rat snake, pilot black snake, indigo or gopher snake, corn snake, and fox snake, lives essentially on rodents and birds whether they are useful or harmful species. Prey is constricted before being eaten. Hardy, shy, and always seeks retreat.

Rubber boa is essentially useful as a destroyer of burrowing rodents. Makes an excellent pet and thrives well in captivity. Gentle at all times; sometimes interestingly coils itself into a ball which may be rolled about.

Boa Constrictor

Boa Constrictor
Constrictor constrictor
Length to 15 ft. Pale brown above, with 15 to 20 dark brown crossbars and with brown spots on sides. Under parts yellowish marked with black. On tail, colors are brick-red, yellow, and black. Claw-like spurs near vent are degenerate hind legs. Usually lies coiled on a tree branch. Name boa applied to some 40 species of family Boidae.

Boa constrictor, native of warmer parts of South America, Venezuela to Argentina. Differs from pythons by tooth and skull characters. Closely related to boas of Madagascar and 30-ft, over 250-lb anaconda or water boa of Central and South America, world's heaviest and America's longest snake.

Boas and anacondas bear living young which hatch from eggs while still within mother. Young of 17-ft female may number 34, each 27 in. long, 1 in. thick, and like parent at birth. Captive boa constrictors have lived 23 years, longest snake record.

Food: when full-grown may capture animal size of dog or sheep and after killing prey by constriction or (anaconda) by drowning and constriction, eats animal whole. Once a meal is started snake cannot let go. Savage in wild or when caught. Usually becomes tame in captivity.

Doubtful if these snakes can kill animals as large as horses or that they attack man without provocation, though may be seriously dangerous once aroused. Not poisonous. Flesh is eaten by natives, tastes like veal. Fat is used by natives in curing diseases and skin is used in ornamenting saddles, bridles, and shoes. An endangered species.

Python

Python
Python sp.
Regal or reticulated python to 33 ft long, weight 175 lb. Rock python, *P. sebae*, 25 ft. Indian python, *P. molurus*, 30 ft. Australian diamond python, *P. spilotis*, 20 ft. Regal python, light yellow-brown with large blue-black spots and often with iridescent tints.

Regal python native of Burma, French Indo-China, Malay Peninsula, and Philippines, favoring trees near water. Rock python native of Africa. Australian python also known as "carpet snake." Range for all species was probably greater in early days than at present.

Regal python lays about 100 oval leathery-shelled gray eggs, about size of goose eggs, which become white, hard, and round; guarded in a cone-shaped pile by encircling female for over 50 days; then hatch into young which shed skins in 10 days and are independent.

Food, birds and mammals, captured usually from overhanging tree perch and killed by constriction, then eaten whole. Capture rapid and hard to follow with eye. Tail usually powerful. Not poisonous. Savage when captured; often goes on hunger strike when kept captive. Captive lifespan, 21 years.

Probably useful in keeping certain small animals in check, but dangerous if interfered with accidentally or intentionally. History records a Roman 75-ft snake and a Tunisian 200-ft snake, but these records have probably grown with age and with repeated telling. Leather of skin of fine quality; too often becomes a spread-out wall ornament for a returning tourist. Many species of *Python* are in danger of extinction.

Order Serpentes./Family Colubridae

Mud Snake
Farancia abacura
Length to 6 ft, of which ⅛ may be tail. Diameter 1½ in. Dark purple to black, with large vermilion wedge-shaped blotches on side. Undersides vermilion-red, with irregular numerous black patches. Sides of head red, with large dark spots in a row along upper lips. Tail ends in a hard, harmless spine. (See DeKay's snake.)

Ranges Virginia to Florida, west to Louisiana, Texas, north to Indiana. Only one American species and some authorities recognize only two subspecies. Western subspecies extends range west from western Florida to east central Texas. A burrower, living in mud, wet fallen timber, or

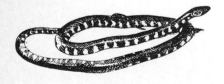

Mud Snake

sphagnum moss, often in water. A snake common to southern swampy areas and lowlands.

Eggs up to 104, but more commonly from two to four dozen, bluntly oval, yellowish-white, with a smooth white parchmentlike shell. Young when hatched, about 9 in. long, though some are nearer 6 in. Before snake sheds its skin, entire body becomes a ghostly pale blue, obscuring shiny scales and colors.

Food salamanders, frogs, and Congo eels, although this is not well understood. Eats some earthworms in captivity. Requires much drinking water. Movements when undisturbed are slow, but it can swim well and go rapidly over ground. May lie at rest in an obscure hoop.

This is probably the "hoop snake," supposed to take its tail in its mouth and pursue humans or thrust its "stinger"-armed tail into a tree and kill the tree. It swings its body vigorously when touched, and tail spine might produce a wound sufficient to cause bleeding. Not poisonous in any way.

Ring-necked Snake
Diadophis punctatus
Length to 19 in. Rarely over ¼ in. in diameter. Uniform slate-blue or gray above except for a bright yellow-orange ring around neck. Beneath, bright orange with irregular dark spots separating orange belly and dark gray back. Scales smooth and glossy, without any keels such as are found on DeKay's or Storer's snake.

Ranges from southeastern Canada to Florida, west to Michigan, Illinois, and Tennessee. Commonly found in rotting stumps, under stones, in moist regions, and usually in wilder unsettled parts of country though it has been found in city parks. Subspecies extend the range of this snake to the prairies and then southwest into Mexico; others are found in limited areas of the Rockies and still others along the Pacific Coast from western Oregon through western California.

Eggs about 1 in. long; one to seven; laid in late July in damp or rotting wood; hatch in 6 to 11 weeks, depending on warmth and moisture available, into young 5½ in. long. Young resemble parents from start except that they may be darker in most parts; grow rather rapidly.

Food, salamanders, earthworms, insects, small snakes, lizards, frogs, and similar small animals, captured and held by mouth, without constriction. Generally gentle. Occasional individuals may strike and bite but they are powerless to injure man in any way. They give off a strong musk scent when disturbed.

Probably not so useful as some other snakes but do no great amount of damage and are never sufficiently abundant to be considered a nuisance anywhere. Make excellent terrarium pets, usually are docile enough to allow persons to become acquainted with them.

Hog-nosed Snake, Puff Adder
Heterodon platyrhinos
Length to 3½ ft. Scales keeled and in 25 rows. Head of a 2-ft snake, 1 in. wide and 1¼ in. long. General appearance stocky. Brown or red with 28 dark patches between head and tail. Belly yellow, blotched on sides with black. Snout upturned and vicious-looking. Not poisonous or harmful to man.

Three species of genus in America. *H. platyrhinos* ranges through Eastern States from Massachusetts south to Florida, west to Texas, north to Minnesota, with related species within this range or only slightly increasing it. Found in dry sandy gardens and roadsides exposed to sun.

In early summer, 24 to 36 eggs, elongate but later becoming spherical, are laid under inches of damp soil; hatch in 5 to 6 weeks into young 6½ to 8 in. long, which resemble parents and shift for themselves immediately. Eggs stick in clusters, absorb moisture, and, during incubation, increase in size about ⅓.

Food of adults, mostly toads but also frogs, salamanders, cutworms, nestling birds, and insects. Young eat insects. Strikes viciously, hisses, puffs its head, opens its mouth, and turns on its back as though dead when annoyed, but can hardly be made to bite a man. "Playing dead" is its favorite trick.

Its destruction of useful toads makes this snake an enemy of man's interests, but it is hoped that it will never be completely destroyed in its range, since it is the personification of the effectiveness of a bluff even though such action often leads to its death.

Smooth Green Snake
Opheodrys vernalis
Length to 20 in. Greatest diameter ⅖ in. Scales shining, smooth, in 15 rows, pale green, those beneath being lighter. Anal plate divided. Tail about ⅓ length. Stouter than keeled green snake, which is also yellowish instead of whitish beneath. Keeled green snake also known as "rough" or "Southern green snake." Difficult to locate because of good camouflage.

(Continued)

Ring-necked Snake

Hog-nosed Snake

Smooth Green Snake

Smooth green snake ranges from southern Canada through Eastern and Middle States to Florida, Texas, New Mexico, North Dakota, and Colorado, living in open grassy hilly fields and bogs and commonly found under stones. Does "climb" bushes. Rough green snake, from New Jersey to Florida, New Mexico, and Kansas. Found from sea level to 9,000 ft. East Asian species.

Eggs capsule-shaped, 1 by ⅖ in., with thin easily indented parchment shells; stuck together in groups of two to eight, usually under warm stones; hatch in 4 to 21 days by heat of sun into 5-in. stout dark olive, unusually active snakelings, which first shed skins shortly after hatching.

Food chiefly small animals like insects, spiders, smooth caterpillars, grasshoppers, crickets, and rarely, snails and salamanders; all captured and held by mouth alone and killed by being swallowed whole. Apparently does not eat frogs. Protection is its color. Disposition gentle, rarely offering any resistance.

Highly useful as destroyer of insects. Not poisonous, and not provided with teeth which could possibly tear human flesh; therefore, entirely harmless and inoffensive. Active only in summer months and frequently killed, but worthy of every protection. In captivity eats four crickets a day. Burrows in winter.

Coachwhip Snake

Coachwhip Snake
Masticophis flagellum

Length to 8½ ft, but slender. Without pattern, but with each scale rather distinctly outlined and foreparts darker than rear. Darkest portion dark brown to black, or sometimes olive, tan, or gray. Outlined scales of tail resemble a braided whip. Abdomen may be white.

Subspecies, two. One ranges from Virginia to Florida, west to Rocky Mountains. Other is confined to southern Arizona. Found in hills or valleys, in open or wooded territory, in creeks and swamps or dry sandy stretches. Rarely abundant. With subspecies, *Masticophis* ranges in southern states from coast to coast.

Eggs 9 to 24 laid on ground. Young of eastern subspecies marked with dark crossbands one to three scales wide, separated by light bands one to two scales wide. In western subspecies, light areas between dark bands in young are three scales wide. These patterns disappear with age as animal darkens. Can bite and may break skin of handler.

Food chiefly small rodents, birds, birds' eggs, and smaller snakes; western subspecies feeds largely on lizards and other snakes, particularly those of its own kind. Speed is phenomenal, and to catch it, one must almost reach ahead of it. Its takeoff is remarkably sudden.

Probably useful as an enemy of rodents and of some other snakes. Makes a poor pet because it does not stand handling well. It is of very nervous temperament. It does not whip victims with its tail, as Negro slaves were told, but this superstition served admirably to keep escaping Negroes away from unsettled country where coachwhip snake lived.

Black Snake

Black Snake, Black Racer
Coluber constrictor

Length to 6 ft, with average diameter 1 in. Scales smooth. Anal plate divided. Slender, plain, bluish-black above; underparts uniformly dark or medium gray. Snout tip sometimes brown. Throat and chin white, and more or less spotted or blotched with dark to rear. Pilot black snake, *Elaphe obsoleta*, has keeled scales.

Subspecies, seven, including black snake and western or blue racer. Former ranges through eastern United States, west to Texas and Great Plains, while latter ranges through Central, Southwestern and Pacific Coast States. Suns itself on walls, tree branches, or ground.

Eggs in lots of 5 to 20 or more; spherical; ¾ to 1⅕ in.; snowy white, tough-shelled, appear rough as if covered with salt; laid in July in moist decaying wood or moss; hatch in about 4 to 8 weeks into 12-in. young, with ¼-in. diameters. Young marked with 50 to 65 black-edged, chestnut blotches, which disappear second year when snake is about 30 in. long.

Food, snakes, rats, mice, rabbits, moles, frogs, birds, birds' eggs, and insects, held by jaws and by weight of body but not killed by constriction, in spite of its specific name. Can travel with surprising rapidity over ground or through brush or trees where somewhat similar milk snake would not go.

Probably essentially a useful species in spite of habit of eating birds and frogs. Destruction of rodents, particularly rats, more than makes up for its bad habits. Does not "charm" birds, as claimed. A poor captive animal as it is too nervous and delicate.

King Snake, Chain Snake
Lampropeltis getulus
Length to 6 ft; tail to 10 in. Scales smooth, in 21 to 25 rows. Usually black with white or yellow narrow crossbands, which may fork on sides and join one another. Bands separated by 5 to 10 scales. Abdomen black with yellow blotches, light belly blotches usually alternating with light backband. Name of genus means "shiny skin."

Ranges from southern New England to Florida, with 12 other United States species, including milk snake, covering region west to Nevada and California, south into Mexico and Lower California. Usually on ground in meadows and brushland, but more common in wooded areas. Usually travels in early morning or late afternoon hours; in very hot weather is active at night. In the West, it occurs from sea level to 7,000 ft.

Eggs 10 to 34, laid in early summer, either on ground or buried under trash, oval, leathery-shelled, soft, and white. If kept too damp, they solidify and young die. Sometimes female remains coiled about eggs a day after laying them. Eggs hatch in 4 to 6 weeks, into 10¾-in. young, colored like adults.

Food mostly other snakes and rodents, all of which may be captured by pursuit, killed by constriction, and then eaten whole unless they are too large. Willing to fight most snakes though normally do not attack larger ones. Immune to poison of associated rattlers.

Useful animal as destroyer of rodents and snakes that may be dangerous or of little use. Makes excellent pet as it becomes docile with handling. May emit an unpleasant musky odor and hiss when disturbed, but is really entirely harmless to man. Rarely bites in captivity.

King Snake

Bull Snake
Pituophis melanoleucus
Length to almost 9 ft. One of four largest North American snakes. Yellow-brown above, with a row of large rectangular black or reddish-brown blotches on back and similar blotches on sides. Head pointed. Related pine snake a subspecies of *P. melanoleucus* has a whitish ground color, instead of yellowish of bull snake.

Bull snakes (six subspecies) range from Wisconsin and Indiana west to Pacific Coast. Pine snake in eastern United States; New York to east Texas.

Eggs larger than hen's eggs; tough, leathery-shelled; 10 to 24; creamy-white, adhering to each other tightly. Incubation about 8 weeks at summer temperatures. Young about 16½ in. long when hatched. Beautiful at all times because of pattern and because of 29 to 33 rows of scales; lower unkeeled.

Food, mostly gophers, rats, mice, eggs, and rabbits. Kills active prey by constriction, but swallows eggs whole and crushes them afterward. Its hiss may be heard for 50 yd. It varies greatly in disposition but usually makes a splendid pet. Is probably most common pet of "snake charmers." Because of markings and behavior, it is often mistaken for a rattlesnake.

Pair of bull snakes can rid a barn of grain of its rat and mouse pests, or clean a gopher-infested field of its gophers. Iowa farmers and snake students estimate that a full-grown bull snake is worth $15 alive as an enemy of crop enemies, and give it corresponding protection whenever possible. It should not be killed.

Bull Snake

Indigo Snake
Drymarchon corais
Length to 7¾ ft. Scales smooth. Anal plate single. Plain, shiny, blue black above and below, with snout and sides of head brownish and chin whitish or orange. Throat is orange-brown, which may extend ⅙ distance down belly. Western specimens may be olive instead of black in forward parts.

Two subspecies range from Carolinas to Florida, west to Texas, south to northwestern South America. Favors sandy flats where burrows of such animals as gophers are abundant. Also, if given protection, common about houses and barns where rats may be caught.

Eggs, size of those laid by bantam hen. Little seems to be published on its general life story. Record of one having lived in captivity for 11 years, and much evidence that they are hardy and long-lived snakes. Vary greatly in disposition, some being nervous and others placid.

Writers differ as to its method of getting food, some saying that it does not constrict and others that it does. The author has a motion picture of an indigo snake attacking a huge rattlesnake which it held with its jaws and by weight of its body but did not constrict. Food, a variety of warm- and cold-blooded animals.

Undoubtedly useful as an enemy of rats, rabbits, and other snakes, even though it also eats frogs, birds, and fishes. Commonly exhibited by "snake charmers" because of its active movements, requiring that handler keep renewing his hold as snake seems to slip out of grasp. Easily raised in captivity, takes rodents as food, including laboratory mice and rats. Of much more value in its natural habitat.

Indigo Snake

Milk Snake
Lampropeltis doliata

Length to 40 in., a 3-ft snake being ⅗ in. in diameter. Marked with black-edged blotches of gray, brown, olive, chestnut, or bright red, forming saddles separated by gray and alternating with similar smaller rounded spots along sides. Underside white or gray, strongly flecked with black.

Ranges through eastern North America, Massachusetts to Iowa, south to Virginia and, in mountainous areas, farther; in Canada, into southern Ontario. Subspecies extend its range west to Colorado, south into Mexico. In fields, woods, meadows, often in houses and barns where it follows rats. Unlike corn snake, has single instead of double anal plate, and black instead of red line from eye to mouth.

Eggs much like those of king snake; 6 to 16; 1⅕ to 1⅗ in; leathery, oval, white, often adhering in a rather compact cluster and buried under ground or in moist rotting wood; laid in early July; hatch in early September into young, 6⅔ to 8⅕ inches long, patterned like adults but brighter.

Food chiefly mice and rats, which it pursues into houses and barns. Could not possibly milk a cow or retain milk if it were forced into it. Kills many other smaller snakes, using constriction to kill before swallowing food whole.

One of most useful snakes of area it occupies, and one of most misunderstood and abused. Rumor has it that if one is killed its mate will return to avenge it, and another snake may come to finish a job of rat killing started by first. Is worthy of every protection. Makes an ideal snake pet in every way. Unfairly called an adder; killed because of this and its resemblance to a copperhead.

Milk Snake

Water Snake
Natrix sipedon

Length to 3½ ft; tail 8 in. Light brown with dark brown spots and bands across back (see copperhead). Dark blotches on back broadest in middle, alternating with similar areas on sides. Sometimes uniformly brown. Color more brilliant just after skin has been shed. Scales ridged. No poison fangs.

In lakes, streams, and freshwater marshes. 10 subspecies of *N. sipedon* and 9 other United States species of genus *Natrix*. *N. sipedon* ranges northern and eastern United States to Florida. Texas, Arkansas, Oklahoma, Nebraska, and Indiana. Other species extend range into Lower California. Sometimes found in aggregations.

Young 16 to 44, born alive in late summer; length varying from 7½ to 9 in. at birth and family quickly breaking up, with young shifting for themselves. Many males may try to mate simultaneously with a female but do not fight. European water snake, an egg layer, rarely lives over 2 years in captivity.

Food primarily fishes, either healthy, diseased, or dead; also frogs, aquatic insects, and other cold-blooded creatures available in environment, captured by pursuit if alive and killed by being swallowed whole, the fish often being larger around than the snake. Active night or day. Hibernates.

Probably not useful because of destruction of fish and frogs; however, kinds eaten are usually unimportant slow-moving species, and often are diseased or dead individuals, in which case snake renders a service in preventing spread of disease and acting as a general scavenger. Often irritable; frequently bites when handled. Gives off unpleasant odor.

Water Snake

DeKay's Snake
Storeria dekayi

Length to 18 in. but usually not over 1 ft, with ¼-in. diameter. Brown above and pinkish beneath, with keeled body scales. Storer's or redbellied snake, *S. occipitomaculata*, is vermilion beneath and has two light areas immediately behind head. DeKay's snake has a better-defined light back stripe.

DeKay's snake ranges from Ontario in Canada to Vera Cruz in Mexico, through eastern North America west to Kansas. Storer's snake has a similar range. Both live in fields where there is loose stone cover and relatively sparse vegetation. They are inhabitants of vacant lots in cities and so are well-known to most boys.

Young 10 to 20, born alive in August; about 4 in. long at birth and 1/16 in. in diameter. Young of Storer's snake are even smaller than those of DeKay's snake and are fewer in number, often being as few as two. Young usually darker than adults.

Food probably slugs, snails, insect larvae, and earthworms, latter being eaten by captive snakes but not so commonly found in wild snakes. A gentle, ideal pet, rarely if ever offering to bite even with abuse.

Probably useful as a destroyer of slugs and snails, which are destroyers of useful plants. Boys and girls should have chance to learn about snakes through these gentle animals. Hibernates in dens. When individuals emerge in the spring, they remain near the den for a time. Scent apparently plays an important part in bringing sexes together.

De Kay's Snake

Ribbon Snake
Thamnophis sauritus

Length to 36 in., diameter about ½ in. Obviously slender. With three bright yellow stripes on dark brown or black background. One stripe is along middle of back, and side stripes are on fourth and fifth rows of scales above belly plates. Underparts greenish-white.

Found from Nova Scotia west to Michigan and south to Georgia and Mississippi. Subspecies extend range of the ribbon snake to the west as far as Colorado. Found most commonly near water, often associated with water snake but staying on shore more than water snake. See garter snake for range of its relatives.

Young born alive, in litters, averaging about 12 or possibly fewer. Newly born young are about 8 to 9 in. long, with comparatively large heads, slender bodies, and markings essentially like those of adults. Young and old are usually nervous and resent confinement.

Food, salamanders, frogs, tadpoles, insects, and earthworms, though in captivity will not eat earthworms which its relative the garter snake gorges itself on. Captive ribbon snakes have been fed chopped raw fish successfully, and fish have been found in wild ribbon snakes.

A pretty sleek snake, however, greatly admired by those who like to handle them. Not so fast as some racers, but it can move with surprising rapidity and hide most skillfully. Not a good pet because of its nervous habits and ability to escape from cages.

Ribbon Snake

Garter Snake
Thamnophis sirtalis

Length to 36 in. Usually with a central light-colored stripe down back, bordered on each side by a dark stripe. Sides usually darker and underparts lighter. Female often larger than male. Head distinct. Scales ridged and in this species in 19 rows.

Probably most widely distributed North American reptile, from border to border and coast to coast in U.S.; from wet Everglades of Florida to cold western Canada. Three subspecies and eleven related United States species, including ribbon snake *T. sauritus, T. ordinatus* and subspecies range from eastern Canada west through Minnesota to British Columbia, north to 62°N. and south to Missouri and California. Found in wet or dry spots, in woods or open.

Mates in early spring; in late summer, 12 to 70 young are born alive, each about 6 in. long. Stories of snakes swallowing their young may be due to mothers being killed just before their young are to be born, but young are never swallowed for protection. Most active during early spring mating season.

Food, frogs, toads, salamanders, earthworms, crayfish, minnows, insects, and other cold-blooded animals but also known to eat mice and dead birds. In captivity, may be fed chopped fish with earthworms and frogs. Exudes a strong odor from musk glands when captured, and some strike viciously.

Probably not useful to man and destruction of useful frogs, toads, and earthworms is not to its credit. Seems well able to care for itself. Probably most prolific of all our snakes. Has no poison glands and is therefore harmless as an individual.

Anyone who has handled these snakes knows of their capacity to bite and to produce foul-smelling material from anal glands.

Garter Snake

PHYLUM CHORDATA. CLASS REPTILIA

Order Serpentes./Families Elapidae, Crotalidae, Viperidae (Some Poisonous Snakes)

Probably 5,000 persons die each year in India from cobra bites in spite of fact that cobra there is not always aggressive. In United States, one out of eight adults severely bitten by rattlesnakes dies when no treatment is administered but only one out of thirty where good treatment is given. Copperhead and pygmy rattlesnake are rarely able to inflict fatal injury to man. Rattlesnakes famous in dances of Indians; cobras, in exhibitions of Indian jugglers; and the asp, mentioned in the Bible and famous in history as causing death in 30 B.C. of Cleopatra. Kipling in his *Jungle Books* make Kala Nag famous and dramatized the enmity of the cobra and the mongoose in his classic, Rikki-tikki-tavi. Cobras enter houses at night in search of rats. Some consider the asp a mark of regal dignity.

Bites of rattlesnakes and similar poisonous snakes cause instant sharp burning pain, followed by swelling within 10 minutes and profuse bleeding, often followed by nausea and faintness. Reassure patient with information that only one in thirty severe bites, if treated, is fatal. Apply ligature a short distance above bite to stop flow of lymph but loosen it every 15 minutes. Make crosscuts with sterile razor over fang marks equal to fang length, with novocain if available. Suck with sore-free mouth or suction device for 30 minutes. Administer antivenin as directed. As swellings develop, relieve with ⅕-in. cuts around swollen area. Get doctor's help. Do not use potassium permanganate and do not use whisky, which increases circulation and consequent absorption of venom. If medical assistance is available in ½ hour do nothing but apply tourniquet between bite and heart and rush patient to doctor. Reassure patient and try to remain calm.

Snake poisons vary in effect, rapidity of action, and amount. King cobra bite may kill a man in 1 hour, while rattlesnake bite, if fatal, may take nearly 1 day. Venoms of cobras, coral snake, and South American rattlesnake have much neurotoxin which dissolves cell tissue, destroys or paralyzes nerves, particularly those affecting breathing, damages blood cells, and ruptures walls of capillaries and other blood and lymph vessels. Venom of copperhead and native rattlesnake "digests" animal tissue, keeping blood from clotting and aggravating internal bleeding. Venom of Russell's viper causes blood clots. All venoms greatly reduce ability of victim's blood to combat infections. If a diamondback rattlesnake carried only the most poisonous venom, it could kill 400 men or 135 horses. Fortunately, it does not.

Cobra

Order Serpentes./Family Elapidae

Cobra and Tree Cobra
Naja sp. and *Dendraspis* sp.
Cobra de capello, "Kala Nag" of the East Indians, is *N. tripudians,* to 6 ft, with spectacled hood. Asp, Egyptian cobra, *N. haje,* 6½ ft, greenish-yellow. King cobra, *N. hannah,* to 18 ft, olive-yellow, with V crossbands of whitish-yellow. Tree cobra or mamba, *D. angusticeps,* to 14 ft, no hood, olive-green to black.

Cobra de capello, Ceylon and India, west to the Caspian Sea and east to south of China. Asp, African. King cobra, South China, the Philippines, and India. Mamba or tree cobra, Africa over a wide range. Black-necked cobra, from northern to southern Africa. All live on ground except tree cobras.

Cobras de capello pair. Eggs, about 18, laid in vegetable rubbish and leaves, surrounded by female during incubation for 65 days, male taking her place during daily 3-hour absence for food getting. Area near nest protected vigorously by nervous parents during incubation. Young able to shift for selves on hatching.

Poisonous snakes usually excrete poisons through hollow fangs inserted in victim. Food almost exclusively animals, usually warm-blooded. Cobras usually coil and erect the fore part of their body, then expand a "hood" before striking much of their length. Cobras can "spit" venom (not saliva) at eyes accurately for 7 ft. Mambas give no warning but strike with remarkable speed. New fangs replace those lost by use or accident and some cobras can bite effectively 20 times in relatively quick succession. Cobras follow moving objects by moving head; thus the action of the "charmed" cobra of the snake charmer. Most snake charmers' cobras have had their fangs removed. This makes them less dangerous but not completely safe. Some charmers make sure their snake partner is safe by sewing its mouth shut.

Coral Snake

Coral Snake, Harlequin Snake
Micrurus fulvius
Length about 3 ft; diameter, 1¾ in. Ringed with red, yellow, black, and yellow, with yellow rings narrower than red and black; red rings have black dots or spots. Snout black, with wide yellow ring around middle of head. Red touching yellow is important identification.

Three subspecies recognized. *M. fulvius* ranges South Carolina to Florida, west through Gulf States, Mississippi, and Mexico, south to Central America. *M. euryxanthus* ranges from southern New Mexico, through Arizona and northern Mexico. Usually burrows in fields and so is not commonly found.

Eggs 3 to 12 elongate, laid in early summer, in decaying bark or damp soil; hatch in midautumn, into very active young 7 to 9 in. long. Does not bite, but hangs on and chews. Potent venom but snake has a small mouth so area bitten is limited to small members of the body.

Water Moccasin, Cottonmouth

Order Serpentes./Family Crotalidae

Water Moccasin, Cottonmouth
Agcistrodon piscivorus
Length to 60 in., circumference to 10 in. Color pattern obscure; olive, with wide dark crossbands in the cottonmouth *A. piscivorus;* dark brown, with darker crossbands bordered with yellow, in Mexican moccasin *A. bislineatus.* Head distinct. Plates on under portion of tail, in one row for ⅔ length.

A. piscivorus ranges from Virginia to Florida, west through Gulf States. Irritable, fighting vigorously when captured, opening mouth and showing white interior (cottonmouth) but usually calming down if not further molested. In captivity, tame. Lives for most part in marshes and shallow waterways.

Young are produced living; 1 to 12; if food is available, may in 2 years reach size of parent. In captivity, have been known to live for 21 years. Food mostly fish but includes frogs and other small animals of aquatic environment in which it lives. Poison is virulent but causes few deaths.

Copperhead

Copperhead
Agcistrodon contortrix
Length to 40 in., diameter to 1½ in. Ground color hazel-brown, with chestnut-brown blotches narrow in middle of back and broad at sides, with isolated spots in remaining lighter areas; sometimes described as "hourglass" markings. Water snake has dark markings which are broadest at center of back instead of sides.

Ranges from Massachusetts to northern Florida, west to Illinois, Arkansas, and eastern Texas. In North found in early spring around rocky slopes where it and timber rattlers hibernate. In summer found in woods, fields, and swamps where its color pattern matches ground perfectly. Not an active snake. Prefers to be left alone.

Young three to nine, born alive in late August and September in northern part of range usually when female is about ready to go into hibernation; like adults except that they have bright sulfur-yellow tails. Food of copperheads is frogs, snakes, and mice.

Rattlesnake
Crotalus horridus et al.
Timber rattlesnake, length to 5½ ft. Head broadly diamond-shaped, with scales over top. Scales in 23 to 25 rows, keeled. Pair of movable poison glands on upper jaw. Color yellow-brown, with irregular yellow crossbands with black spots. Tail of adult black, with dry-scale rattle which grows with age.

Rattlesnake

15 species of rattlesnakes in United States. 10 live in Southwest and include tiger, Pacific, white, green, and horned. Diamondback is of southeastern United States. Massasauga and prairie, in central United States. Texas rattlesnake, from Texas and California. Other species cover most of Americas.

Timber rattlesnake bears 7 to 12 young alive, usually in September, each young bearing a single button at end of tail. Skin shed once before winter hibernation, starting the rattle which increases whenever skin is shed, possibly three times a year. Rattlesnakes have lived 13 years in captivity. Bounties offered sometimes for them. Highly heat-sensitive.

Pygmy Rattlesnake and Massasauga
Sistrurus miliarius and _S. catenatus_
Known as "ground rattlesnake." Length to 17 in. Unlike true rattlesnakes, top of head is covered with large symmetrical shields rather than with small crowded scales. Pits on sides of head and rattle on end of tail are same as with true rattlesnakes. Poison glands on jaws.

Pygmy Rattlesnake

Pygmy rattlesnakes with gray ground color are southern pygmy or ground rattler, _S. miliarius_. Those with brownish ground color are massasauga, _S. catenatus_, from New York to Kansas and north to Canada in which dark back blotches are close together. Back blotches are more separate in Edward's massasauga of Southwest.

Young born alive, usually in small numbers, of small size, and capable of coiling in 1½-in. circle. A 2-ft mother may give birth to seven to nine young. Massasaugas are commonest in marshy bogs. Their rattle is so faint as to be unheard by many people. Bite is rarely fatal to man.

Fer-de-lance
Bothrops atrox

Bushmaster
Lachesis mutus
Fer-de-lance; length to 6 ft, with a conspicuous lancelike head about 2 in. thick and with exceptionally long poison fangs for is size; brown with black markings. Bushmaster: length to 12 ft, largest American poisonous snake; reddish-yellow with dark crossbars and a black stripe from angle of jaw to eye.

Fer-de-lance

Fer-de-lance ranges through Mexico, Central America, tropical South America, and West Indies, usually hiding quietly in jungles during day. Bushmaster ranges through Central and tropical South America in damp

(Continued)

hot forests, living much of time in holes made by small animals that burrow in ground.

Fer-de-lance gives birth to living young with bright-colored tails, which shift for themselves. Bushmaster, unlike other pit vipers, lays eggs from which young are hatched. These are laid in burrows which snake haunts.

Fer-de-lance eats small warm-blooded animals captured at night and killed by poison injected through long hollow fangs; not so dangerous in early day, when poison has been exhausted by night's hunting. Bushmaster is bold and aggressive, with fangs nearly 1⅖ in. long, which give a great volume of virulent poison. Both are night hunters.

Fer-de-lance's bite is deadly poisonous but not so bad as that of bushmaster, whose victims have been known to die within 10 minutes of having been bitten. While these snakes are relatively closely related to rattlesnakes, they do not have characteristic rattles.

Order Serpentes./Family Viperidae

Sidewinder, Horned Rattlesnake
Crotalus cerastes

S-shaped curve when animal is in sidewise locomotion is descriptive of common name. Back side generally pale-colored, so blends into surroundings; may be cream, tan, gray, or pink and without conspicuous pattern. One of smallest rattlesnakes; no markings discernible. Sometimes called horned rattler because of hornlike upturned ridges above eyes.

Generally males are smaller than females. Length 1½ to 2½ ft; body stout, head relatively large, broad. Food is small mammals, chiefly rodents, and lizards.

Mates in spring and sometimes in the fall. May produce 6 to 16 young; young measure 6½ to 8 in.

Range from southern Nevada and extreme southwestern Utah to Baja California; desert base of mountains of southern California to Arizona. From below sea level to 6,000 ft; generally below 4,000 ft. *C. c. cerastes*, Mojave Desert sidewinder; *C. c. cercobombus*, Sonora cercobombus, and Colorado Desert sidewinder, *C. c. laterorepens* are other members in the genus. Found on sandy flats and washes of deserts; in particular among bushes. Generally in sandy locations but may be found among rocks or firm soil.

Commonly nocturnal but may be active in the daylight hours. Collectors follow marks on sand or find the snakes on roads at night. Track in sand looks like J-shaped mark. J's in track in direction in which it has traveled. During daytime may bury themselves in the sand beneath some cover like an overhanging shrub. Rattle apparently is an effective alarm to some of its predators.

Reported to be able to move very rapidly, to 2 mph. Bite poisonous.

Sidewinder

BIRDS AND MAMMALS, THE WARM-BLOODED VERTEBRATES

It is without apology that we devote approximately one-fourth of this book to the warm-blooded vertebrates. Birds and mammals have a prominent place in the interest of any student of natural history.

CLASS AVES.

[THE BIRDS]

Since birds are warm-blooded, they maintain an approximately uniform body temperature independent of that of their environment. In this they differ from cold-blooded animals, considered to be more primitive. All birds lay eggs, which are usually incubated by either or both of the parents. The longer the incubation period, the more likely is the young bird to be well developed or *precocial* when it hatches. In this type of bird, the egg is relatively large and parental care of the young relatively poor. In *altricial* birds where the eggs are small, the young helpless, and the incubation period short, parental care is essential to survival. The parts of a bird are illustrated in the diagram below. The number given after the common or scientific name in the following pages is the American Ornithologists' Union checklist number.

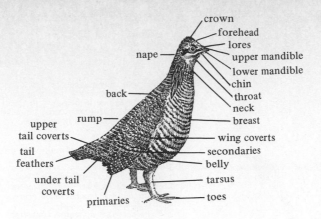

crown
forehead
lores
upper mandible
lower mandible
chin
throat
neck
breast
nape
back
rump
upper tail coverts
tail feathers
under tail coverts
primaries
wing coverts
secondaries
belly
tarsus
toes

PHYLUM CHORDATA. CLASS AVES

Order Struthioniformes./Family Struthionidae

Ostrich

Struthio camelus
1. *Struthio camelus* (Northern)
2. *S. australis* (South African)
3. *S. molybdophanes* (Somaliland)
4. *S. meridionalis* (Masailand)

Male to 8 ft high. Weight to 345 lb. Body plumage black, but tail feathers and wing quills white. Female smaller, brown. Both sexes with down on long neck, short broad beaks, and 2-toed powerful feet. Larger toes end in a nail. Bare parts of legs pink; or, bluish; or lead-colored. Vision excellent. Shy. Flightless primarily because of weight, small wing area, and inadequate wing feathers and flight muscles.

Over a wide range of desert area, from Arabia to South Africa, in open sandy stretches; now grown rather commonly in ostrich farms and protected in sanctuaries in Africa, South America, and southern California. One of largest farms near Los Angeles. Often in flocks in wild.

Ostriches are polygamous. Mating is preceded by a brief courtship display. Cock makes the nest, a hollow in the sand. One cock mates most frequently with 3 hens, but sometimes with as many as 5 hens, which lay their eggs in common nest. Eggs 15 to 30, sometimes more, each weighing 2 to 3 lb, hen laying every other day. Incubation 40 days; hen, by day; cock, by night. In captivity, may lay again soon if chicks are removed. Life-span to 60(?) years.

Food, mixed with stones, including mammals, reptiles, birds, fruits, and grass. Can run at speeds to 35 mph, making 12-ft strides when pursued, commonly running in circles. Not unusual for ostriches to run at speeds of 20 to 27 mph. Does not avoid danger by hiding head in sand as commonly reported. Fights effectively with powerful feet and legs.

Eggs and flesh, used as food by men in native land. Plumes of value in millinery and arts, a young bird yielding feathers at 18 months and a mature bird yielding 3 crops of feathers in 2 years, worth annually around $30(?). Mentioned in Job 39:13-18. Considered a delicacy by early Romans. Temperature, 99 to 101°F.

Order Apterygiformes./Family Apterygidae

Kiwi

1. *Apteryx australis* (South Island)
2. *A. haasti* (North Island)
3. *A. oweni* (Gray)

Length: (1), about 2½ ft, (2), about 2 ft; (3), about 1½ ft. No larger than a domestic hen. Female larger in all species. Color: (1), gray-brown; (2), reddish-brown; (3), gray-brown with crossbars of black. All have loose plumage, long bills, and are flightless. Toes 4, with strong claws. No noticeable wings or tail. Wings are so short they are hidden beneath body plumage, which is more like hair than feathers.

Native of New Zealand. Since they are flightless, they of course live on the ground. They are probably most closely related to the extinct moas and, among living birds, to ostriches.

Nest is usually placed at foot of some tree. Hen lays a single creamy-white egg larger than would normally be expected of a bird her size. Nest is usually in a hole and therefore well-hidden. Young closely resemble adults. Courtship is marked by *kiwi* call that gives bird its name.

(Continued)

Ostrich

Kiwi

Nocturnal. Food largely insects and their larvae, worms, snails, and similar small animals gleaned with help of long bill. Also eats soft fruit and leaves. Nostrils are at extreme end of long beak. Has ability to run with exceeding swiftness, reflected in long legs, huge thighs, and leg muscles.

Rather well-known largely because of its inability to fly, its limited range, and other characteristics that have gained it a place as a bizarre and an interesting animal in literature if not in our own backyards. Its food would imply that it was useful. Incubation 80 days by male. Generic name *Apteryx* means without wings.

Order Sphenisciformes./Family Spheniscidae

Penguin

1 *Aptenodytes* sp. (Emperor and King)
2 *Eudyptes* sp. (Rockhopper, Macaroni)
3 *Eudyptula* (Little and White-flippered)
4 *Megadyptes* (Yellow-eyed)
5 *Pygoscelis* (Adelie, Gentoo, Chinstrap)
6 *Spheniscus* (Jackass, Peruvian)

Penguin

(1) over 3 ft long, with orange and yellow patches on sides of neck. (2) to 30 in. long, with orange on sides of crown above eyes. (5) to 30 in. long, with red bill, 6-in. tail and orange feet. (7) to 27 in. long, with gray-mottled black feet, black bill with gray bar; short tail of 20 feathers, narrow black horseshoe on breast. Flightless. Most completely marine of all birds, can swim under water as fast as seals. Webbed toes and wings reduced to form strong flippers. On shore walk with bipedal motion; awkward but fast.

(1) breeds on Antarctic ice barrier, Tierra del Fuego, Tasmania, New Zealand; (2), south of New Zealand and Tasmania; (3), from Antarctica to Tasmania and New Zealand; (6) coasts of South Africa. Penguins are essentially birds of Southern Hemisphere, the 20-in. Galápagos penguin ranging just north of equator and being only penguin restricted to tropics.

Penguins usually breed in great colonies, the king and emperors laying only 1 egg, which is incubated on feet under brood flap, on ice barrier in total darkness of antarctic winter. Other penguins brood eggs much as do other birds. Royal penguin egg is pale blue, granulated, 3 by 2⅕ in.; incubated by both parents, and young take to sea with colony at 3 months.

Penguins feed almost exclusively on fish and other marine animals captured by expert swimming and diving, young being fed partly digested crustacea from gullets of parents. Molting is done on shore either before or after breeding, and new set of feathers is completed in a few days or weeks. At rookeries, penguins are noisy, hence the jackass penguin's bray.

Penguins are of considerable economic importance as producers of oil and guano and are protected for this in some rookeries; 150,000 killed a season in one New Zealand rookery causing decreasing population. Their harvest is being regulated and should be controlled by international law to prevent depredations by lawless nations. *Spheniscus demersus,* jackass penguin, is an endangered species.

Order Podicipediformes./Family Podicipedidae
[Grebes]

Horned Grebe

Horned Grebe, 3
Podiceps auritus

Length 13½ in., wing 5⅖ in, bill 9/10 in. In summer head, throat, lower side of head, back of neck are black with conspicuous buff ear tufts; back, gray; underparts, white; breast and sides, brownish. In winter sharply contrasting dark gray and white—white on cheeks, underparts, and neck being most conspicuous. Young like winter adults.

Breeds from Arctic Circle south to southern British Columbia and Maine, in Iceland, northern Europe, and northern Siberia. Winters south from Maine and southern Alaska, to Florida and southern California, northern Africa, along coasts of China and Japan, occasionally in Bermuda and Greenland. Known from Pleistocene times.

Nest of water-soaked floating vegetation, in shallow water. Eggs 2 to 7, dull dirty white, 1¾ by 1⅛ in. When more eggs, perhaps 9 to 10, are found in one nest, they probably have been laid by more than one female. Incubation about 24 to 25 days, with both parents attending downy young, which quickly learn to swim and dive when hatched; even prematurely hatched chicks swim readily and well. Chicks at first streaked and spotted. Downy young are black above, striped and spotted with grayish-white.

Food: of 57, 23% beetles, 12% other insects, 27.8% fish, 20.7% crayfish, 13.8% other crustaceans, with remainder miscellaneous animal matter. Most stomachs had feathers. Obviously not a serious enemy of fishes.

Migrations in flocks appear off Atlantic Coast in October to November and in March to April, flying high or low.

Interesting as a past master of art of swimming and diving. Because of position of legs and their nature, entirely helpless on dry ground where it sometimes comes to rest during periods of migration. Reported to use wings occasionally in swimming under water. A water sprite if ever there was one. Migrates as singles or in flocks.

Pied-billed Grebe, 6
Podilymbus podiceps
Length to 13½ in., wing 5⅒ in., bill ⅘ in. Glossy grayish-brown above with black throat; brownish on sides of body and neck; whitish on belly and lower breast; bill gray, black-banded. Young and winter adults with white throat and no bill band. Feet with broadly lobed webs, greenish to black. Legs placed far back. No evident tail.

Common along lake shores and shallower ponds in most of North America, breeding from British Columbia to Nova Scotia and south to Florida and Mexico. Winters New York to southern British Columbia south, with related races in South America. Differs from eared grebe by presence of brown breast and absence of white wing patch.

Nest a floating mass of soaked vegetation among plants. Eggs 2 to 10, soiled dull white, 1¾ by 1⅕ in., covered with vegetation when parent is absent. Both sexes incubate. Apparently only one brood is raised each season. Incubation 2 days. Young down-covered when hatched; able to swim and dive shortly after hatching. 1 annual brood protected by parents until independent.

Food essentially fish and other aquatic animals captured usually by diving and pursuing under water. Food may be small fish, snails, small frogs, tadpoles, aquatic worms, leeches, and water insects. May eat seeds and other parts of aquatic plants. Practically helpless on land and unable to take to air except from water surface. Call a loud cuckoolike *cow-cow-cow* and a ridiculous series of other calls.

Probably of some importance in determining existence of fish but grebes are rarely abundant enough to be a serious menace anywhere. Flesh unpalatable. Sometimes known as dabchick, or hell-diver.

Pied–billed Grebe

Western Grebe, 1
Aechmophorus occidentalis
Length to 29 in., including 3-in. bill and short tail. Conspicuously black and white winter or summer, young or old, with exceptionally long slender neck. Black on top of head, back of neck, and back. Head with short crest on top but none on sides; bill slender; neck about as long as body. Yellow on bill, and white on underparts, cheeks, and neck. No birds in its range with which it can be easily confused in summer. Female smaller.

Breeds from Washington, southern Saskatchewan, and southern Manitoba to southern California, Utah, and northern North Dakota. Winters from southern British Columbia through California to Lower California and central Mexico but occasionally found east to Ontario, north to Alaska, and at intermediate points.

Nest near water's edge, on a floating raft of vegetation. Eggs 3 to 10, 2⅗ by 1½ in.; dull bluish-white, cream, or olive-brown; generally stained. Nesting period May. Nest attached to vegetation and floating so that it moves with changes in water level. Downy young light gray above, white on belly, with triangular naked spot on crown. Young take to water and swim almost as soon as out of egg, but often ride on parents' back, for sometime.

Western Grebe, 1

Food probably mostly fish and other small aquatic animals, though few studies have been made. Fish, beetles, seeds, and feathers of bird itself have been found in crops of the few that have been examined carefully. Can dive at flash of gun and remain under water a long time. An admirable swimmer but helpless on land.

Formerly western grebes were killed by thousands by plume hunters, who used their densely feathered breasts as highly durable ornaments for hats. Sometimes known as western dabchick or swan grebe.

Order Gaviiformes./Family Gaviidae
[Loons]

Common Loon, 7
Gavia immer
Length 32 in., wing 14 in., bill 2⅘ in. Black above, with white breast and belly, and with white bars on back. Young and adults in winter, with gray between black and white, instead of being white-spotted on wings. Legs placed far back. Feet with 4 webbed toes. Powerful swimmer.

Breeds from Iceland, Greenland, and Labrador to northern Illinois, in New York, Pennsylvania, New Hampshire, Connecticut, South Caro-
(Continued)

Common Loon

lina, Mississippi, Norway, the Shetland Islands, and southeastern Canada. Winters Great Lakes and Nova Scotia to Florida and Gulf Coast and from British Isles to Azores, the Mediterranean and Black Seas. Others to the west.

Courtship by violent diving, bowing, and calling. Nest of loose plants or nothing, on shore. Eggs 2 olive-brown or gray, with thin black spots, 3½ by 2⅕ in. Incubation 29 days. 1 annual brood. Young down-covered when hatched; soon able to swim and dive. May be 3 to 4 years before adult plumage is attained.

Food, mainly fish, caught by pursuit under water; one reported to have been caught in a net anchored at depth of 200 ft. Loses flight feathers in late summer. Flight direct and swift but cannot be started except from water. Migrates high, singly or in flocks.

Most interesting inhabitant of wilder inland waterways and seashores. Should not be allowed to become extinct. May seem more abundant than they really are because of winter concentrations.

Red-throated Loon, 11
Gavia stellata
Length to 27 in., including 2-in. tail and 2¼-in. bill. Wingspread to 4 ft. Sexes colored alike. In spring, head and neck gray; throat with conspicuous dark red patch at lower edge, white beneath, forebreast striped with black and white. In winter dusky above, with small white spots, and white below.

Breeds northern Alaska and western Aleutians to Greenland and south to northern British Columbia, southeast Quebec, and Newfoundland; also in Arctic Europe and Asia. Winters from Aleutians to northern Lower California, and from Gulf of St. Lawrence to Florida. Sometimes in Montana, Kansas, Missouri, Nebraska, Iowa, and Idaho.

Nest near lakes and ponds, on shore, usually making no nest but sometimes lining a shallow depression with grasses. Eggs 2, 3 by 1¾ in.; highly variable in color, spotless or with dark brown spots over drab. Incubation to 28 days, by both sexes. Some young birds are more spotted than adults. Downy young dark gray above, and drab.

Food primarily fish caught by diving and pursuing under water. This species suspected of eating fish eggs. Cannot take off from land. In migration, groups are kept together by weird cries somewhat resembling goose's honk and yet like call of common loon.

Probably is not useful because of its food habits but it contributes much to "wildness" of area in which it is found. For this reason is entitled to protection. A summer spent in north country without hearing or seeing a loon is practically a total loss. New England migration is at its height in October.

Order Procellariiformes./Family Diomedeidae
[Albatrosses]

Black-footed Albatross, 81
Diomedea nigripes
Length to 32 in. Wingspread to 7 ft. Wings narrow and saber-shaped. Bill dusky. Feet and legs black. Dark above and below, with rump dark or rarely dirty white and neck region sometimes lighter. Easily distinguished from gulls because of longer, relatively narrower wings and larger size.

Over Pacific Ocean north of Tropic of Cancer but sometimes to equator, from Alaska to Lower California south to Formosa on Asiatic coast. Many relatives, particularly in the South Pacific. This species breeds in smaller Pacific Islands conspicuous in World War II including Midway, Bonins, Marshalls, and so on.

Nests of albatrosses are usually simply depressions in the ground or in a hollow atop a slight mound. Eggs are laid 1 to a nest and young when hatched are helpless, sooty brown, and covered with soft down; exhibit relatively slow growth.

Albatrosses famous in Coleridge's "The Ancient Mariner," as followers of ships at sea. Ordinarily are not seen close ashore on our mainland. They feed largely on fish and other marine organisms. To take flight, they require a considerable area to get a running start. They can settle their bodies under water.

Have been used as food by men cast adrift and can be caught with baited hook and line. Their wing bones have been used as pipestems and skin of their feet has been used for making unique sacks. Welcome sight to most who sail the seas, either in time of trouble or otherwise. Are a real problem on Midway and other Pacific islands, where they interfere with aircraft landing and taking off.

Red-throated Loon

Black-footed Albatross

Order Procellariiformes./Family Hydrobatidae

[Storm Petrels]

Wilson's Petrel, 109
Oceanites oceanicus
Length to 7½ in., including 3¼-in. tail and ½-in. bill. Wingspread about 16 in. Sexes colored alike. Dark brownish-black above and lighter beneath; with a conspicuous white curved patch at base of square-tipped tail and with fainter narrow lighter band near rear edge of wing bases. Feet yellow and webbed. Leach's petrel has forked tail and lighter wing bands.

Breeds in Southern Hemisphere on Mauritius and Kerguelen Islands, Adelie Land, South Victoria Land, South Shetland, South Orkney, and South Georgia Islands. Found in all oceans, except Pacific north of equator. Spends summer in North at sea unless blown ashore by storms. Commonly seen in North Atlantic in summer at least 10 miles offshore.

Nests on ground, usually in or near rock crevice, with or without plant-material lining. Eggs 1-1⅓ by ⁹⁄₁₀ in.; white with fine purplish or red-brown spots around larger end. Usually nests December to March. Incubation 35 days, by male breeding and nesting period lasting about 5 months. Annual molt May to October, in North.

Food largely oily fish wastes. Because of this, are common in wakes of fishing vessels far at sea. Known also to eat small crustaceans, fishes, some insects, and some seaweed. Conspicuous because of rather long dangling black legs and yellow feet. Because of wide range, cannot be used as indicator of nearness of land.

Usual schedule: May to June, generally inshore along Atlantic Coast of North America but never north of Labrador on American coast or of 57°N on European coast. Leaves American coast about September. During February and early March, found nowhere north of equator. Called Mother Carey's Chicken, common stormy petrel, and long-legged storm petrel.

Wilson's Petrel

Order Pelecaniformes./Family Pelecanidae

[Pelicans]

White Pelican, 125
Pelecanus erythrorhynchos
Length to 5 ft, wing 22 in.; bill 14 in. with huge pouch beneath and horny prominence above during breeding season. Wingspread to 9 ft. White, with black wing quills. Flies with hunched head. Brown pelicans have top of head white, neck brown, back silvery gray, and side silvery.

White pelican breeds from British Columbia to Wisconsin and southern California; winters, northern California, sometimes to Pennsylvania and south to Trinidad. Brown pelicans breed from South Carolina and northern California south to northern South America, wintering farther south. Known from Pleistocene times.

Nest of white pelican on ground in a depression. Eggs 2 to 4, creamy or blue-white, stained and chalky-marked; 3½ by 2 ⅓ in. Incubation 29 to 30 days. Eggs of brown pelican 3 by 1⅞ in. Young feed from bill pouch of parent, almost burying head in throat region when feeding. Nests usually in large colonies.

Food almost exclusively fish. White pelican catches food by swimming with bill pouch submerged. Brown pelican catches food by diving spectacularly from air into water. Feeds and flies in flocks, the white in breeding season often soaring out of sight. Usual flight formation, a diagonal line with flight direction. Speed, 30 m.p.h.

Controversy over economic importance of pelicans, fishermen claiming they are destructive. Since their numbers are small, this is probably not serious. May range 50 miles from base in feeding activities. White pelicans migrate by day, in great high-wheeling flocks. Brown pelican, state bird of Louisiana. Pelicans are in danger of extinction because of concentration of DDT in their bodies which interferes with shell formation.

White Pelican, 125

Order Pelecaniformes./Family Sulidae

[Boobies, Gannets]

Gannet, 117
Morus bassanus
Length to 40 in., including 10-in. tail and 4-in. bill. Wingspread to 6 ft. Adult sexes alike in appearance. Body mostly white. Head and neck yellow with a dark-blue area of skin showing on throat. Bill grayish-blue. Feet black. Young dark gray-brown above with white spots and white beneath. Wing tips black. Wings relatively slender.

Breeds on rocks of Bonaventure and Anticosti Islands in mouth of Gulf of St. Lawrence, off southeastern Newfoundland, and on British Isles and Iceland. Winters from coast of Virginia to Cuba and Vera Cruz, on the coasts of Africa, Azores, and Canary Islands. Accidentally found inland to Michigan.

(Continued)

Gannet

Courtship dance is engaged in by both sexes. Nests in large colonies. Nests absent, or a well-made nest of vegetation, usually on cliffs. Eggs 1, 3 by 2 in.; blue-white, overlaid with stained or unstained lime deposit. Nests May to July. Incubation 38 to 42 days by both sexes; 1 annual brood. Newly hatched young naked, almost black, and helpless.

Food, fish captured by diving; usually takes a direct plunge from heights up to 50 ft in air; probably a shallow diver. Has difficulty in rising from a perfectly calm sea. Accepted at sea as being within 100 to 300 miles from land and usually seen in groups of 3 or more.

Known to eat herring, mackerel, cod (up to 2 lb), salmon, smelt, anchovy, and cuttlefish but is useful to fishermen in locating schools of fish and their depth. Not considered as serious enemies of fishing by most people. Only gannet found on shores of North America. Most numerous in May and June between Cape Race and Cape St. Mary.

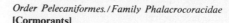

Order Pelecaniformes./Family Phalacrocoracidae
[Cormorants]

Double-crested Cormorant, 120
Phalacrocorax auritus
Length 35 in., wing 12½ in., tail, 6⅕ in., bill, 2⅓ in. Tail feathers, 12. Black head, neck, rump, underparts and tail. Naked throat pouch dull orange. Black tufts on either side of head, over eye in breeding season. Great cormorant has 14 tail feathers and is larger, with white bordering orange throat patch.

Double–crested Cormorant

Breeds Newfoundland to central Alberta, south to Nebraska and Maine. Winters from Virginia to Gulf Coast and in Bermuda. Commonest eastern cormorant and is seen in flocks or singly far inland during migration. Haunts pilings in great numbers around Florida and Gulf Coast in winter. Known from Pleistocene times. Subspecies increase the range to include much of continental U.S. and Alaska.

Nest of sticks or weeds, crudely arranged on cliff, bushes, ground, or in trees. Eggs 2 to 4, pale blue-white with a chalky deposit, 2⅖ by 1⅖ in. Incubation 28 to 29 days. Young naked, black, and shiny when hatched, but soon with black down. A fall molt and a partial spring molt, with molting rather constant until mature at 2 years.

Food almost exclusively fish caught by diving, sometimes using wings to assist in getting speed under water. When flying from perch, it usually descends first almost to water. Flies in long lines in migration. When resting on water, resembles loon but lacks white coloration. Temperature 106°F. Speed 20 mph.

Considered an enemy of fish by many fishermen. Fish eaten are commonly of no economic importance or are real enemies of valuable species. Interesting because of habits about harbors, where it is known as a "shag." Relatives used by Chinese to capture fish, by putting ring around throat and keeping bird on a string over fishing grounds.

Order Pelecaniformes./Family Anhingidae
[Snake Birds]

Water Turkey, 118
Anhinga anhinga
Length to 34 in., including neck and yellow bill, which are about ⅓ total length, and tail, nearly another ⅓. Body slender. Male green-black and glossy with gray-flecked head and neck, silver-spotted back, silver-gray spotted wings with wide white band, and white-tipped tail. Female like male but darker above, and with pale buff head, neck, and breast. Toes of stout webbed feet end in sharp claws which help to make it a good climber.

Water Turkey

Breeds through tropical America from Bexar County, Tex., to North Carolina, south to southern Brazil, Paraguay, and northern Argentine. Winters from Yuma, Calif., through central Arkansas to South Carolina and south. Has been found north to Ohio. Found in or near swamps and large streams.

Nests in swamps and bayous, on bushes or small trees overhanging water. Nest a large, poorly constructed, cumbersome mass of sticks, grasses, mosses, and leaves. Eggs 2 to 5, 2¹⁄₁₀ by 1⅓ in.; whitish, greenish-white, grayish-green, or bluish-white, and usually with a generous crusting of limy material on outside. Commonly nests in rookeries, near herons or ibises.

Food mostly fish and other aquatic animals, fish often caught by pursuit under water and speared by long pointed bill. Feet webbed. May soar high in air like hawks, or may sink into water gradually until only tip of bill remains exposed. 4 species of water turkeys known in world.

Interesting bird of seacoasts, tropical swamps, and waterways. Probably does some damage to fish but is rarely abundant enough to be considered as serious pest. Always interesting as "snakebird," or "darter" to those who may not be familiar with it.

Order Pelecaniformes./Family Fregatidae [Frigate or Man-o'-War Birds]

Magnificent Frigatebird

Magnificent Frigatebird, 128
Fregate magnificens

Length to 41 in., including 6-in. bill and 19-in. tail that is forked for more than ½ its length. Wings long, slender, and conspicuously angled forward. Adult male black, with duller belly, and with greenish-bronze and purple cast. Female less glossy. Immature, with white head. Does not always show forked tail in flight. Also known as hurricane bird and rabihorcado.

Breeds West Indies, Bahamas, and islands off coast of Venezuela in the Caribbean, and along Mexican coast and Galápagos Islands. Winters in breeding area but may make way north to Humboldt Bay on California coast, to Nova Scotia on Atlantic Coast. One species widespread in Indian Ocean. Five species recognized.

Breeds in colonies. Nest of sticks and flimsy plant material, carelessly put together on low trees or bushes, usually a few feet from ground. Eggs one to three, 2$\frac{9}{10}$ by 2 in.; white. Nests February to May. Incubation by both parents. Young naked when hatched, but soon covered with a white down. Juvenile plumage holds for first year.

Food, fish picked up at or near surface of ocean, or flying fish. Almost helpless on land but a master of the air. A bird with a body size of a red-tailed hawk and wings stretching to 8 ft. It can pester pelicans and boobies, forcing them to surrender food they have captured. It is pugnacious enough to attack even the heavier osprey. A magnificent soarer with large wings outstretched and appearing to be humped in middle.

Marvelous flier. Has a strong homing instinct and has been used by man to carry messages like carrier pigeons for 60 to 80 miles. Does not migrate regularly. Never sleeps on water and is usually found within 75 miles of land, so direction of flight at dusk may be an indication of location of land. Any group of half dozen birds would probably be within 75 miles of land. *F. andrewsi*, frigatebird, is an endangered species.

Great Blue Heron, 194

Order Ciconiiformes./Family Ardeidae [Herons, Egrets, and Bitterns]

Great Blue Heron, 194
Ardea herodias

Length 50 in., wing 20 in., bill 6¼ in. Legs and neck long and slender. General color pale slate-gray, with crown and center of throat white, bend of wing chestnut, breast black and white streaked, legs and feet black. Young with crown black and bend of wing paler. Largest of truly American herons.

Breeds Nova Scotia to southeastern British Columbia to Nebraska, South Carolina, and Bermuda. Winters from Oregon and New York south to Florida, Venezuela, Colombia, and Panama. Haunts marshes, lake margins, rivers, and small streams. Nests and roosts in colonies either in rushes or in trees. From Pleistocene of Florida, Oregon, and California. Solitary except during breeding season. Colonies may include more than 150 nests.

Nest a platform of sticks, which may be used many years. Courtship largely by display in a circle. Eggs three to four, pale dull blue, 2½ by 1½ in. Incubation 28 days. Young scrawny, more or less helpless at first, but by July in northern part of range are feathered and able to fly. Fed at first by violent regurgitation activity.

Food fish, sometimes to 1 ft long, water snakes (enemies of fish), grasshoppers, crayfish, mice, frogs, shrews, and other small animals. May feed on pocket gophers, ground squirrels, and field mice. Food caught by lightninglike thrust of beak. Can rest on deep water and take to flight from floating position. Flight slow (28 mph), steady, and remarkably beautiful. Probably deserves protection.

Exact economic importance not determined but bird is protected by law in many states and is rarely so abundant that it could cause serious damage even if its food were solely valuable species of animals, which it is not. Most of fish eaten are of little economic importance and beauty of bird is worthy of preservation. May also be known locally as red-shouldered heron, blue crane, or common blue crane.

Snowy Egret

Snowy Egret, 197
Leucophoyx thula

Length 24 in., wing 9¾ in., bill 3⅕ in. Legs and bill black, feet, yellow. Elsewhere white, with some 50 recurving plumes on back during breeding season. American egret larger (41 in.) and has black feet, legs, and bill. Young of little blue heron (22 in.), white with slate wash and dark feet. Much smaller than the egret.

Formerly bred from New Jersey to Nebraska and south to Chile and Argentina; now in United States, along coast from North Carolina to Texas. Winters from Florida and Mexico south. Has late-summer northern migration to Alberta and Nova Scotia preceding winter southern migration. Known in Florida from Pleistocene times.

(Continued)

Courtship by strutting and pursuit. Nest, in colony, of coarse sticks in bushes over water. Nest usually among mangroves or in swamps with good growth of willow. Eggs three to five, pale dull blue, 1⅘ by 1⅕ in. Incubation 28 days. Young helpless awkward creatures when hatched; reared by both parents, which at times bear plumes in their prime. Young leave nest in 3 weeks. Two molts annually.

Food: 50 meals of young included 120 small suckers, 762 grasshoppers, 91 cutworms, 29 crayfish, 7 water moccasins, and 2 lizards. Food may also include seeds of some aquatic plants. The young feed on regurgitated small fish from the parent bird. Rushes its prey and darts after it more actively than most herons, sometimes running into water up to wings while pursuing fishes. Speed, 17 mph.

Probably essentially useful and no doubt one of most beautiful of all birds. Formerly killed in great numbers for millinery trade, 1½ million being shipped in 1 year from Venezuela alone. Now protected by law and sale of their plumes outlawed in United States. Saved from extermination largely by Audubon Society. Also given common names: little egret, lesser egret, common egret, snowy heron, little snowy, little white egret, little white heron, and bonnet martyr.

Little Green Heron

Green Heron, 201
Butorides virescens
Length 17 in., wing 7¼ in., bill 2½ in. Dark-colored with comparatively short yellow or orange legs. When excited, elevates a shaggy crest. Greenish-black on crown and in line below eye, neck rufous, belly ashy, remaining general color at a distance gray. Young with black-streaked necks.

Breeds North Dakota to Nova Scotia, south to New Mexico, Texas, Mexico, Guatemala, northern Honduras, Gulf Coast, Dry Tortugas. Winters Florida and southeastern Texas to Central America and northern Colombia. Occasionally in Bermuda, Haiti, and Puerto Rico. About rivers and other waterways. Known from Pleistocene.

Nest a platform of sticks, in low trees or shrubs over water. Nests in scattered colonies 3 to 20 ft or more above ground, or singly. Frequently in woods but usually near water. Eggs three to six (sometimes two females laying in one nest), pale blue, 1½ by 1⅐ in. Incubation 19 to 21 days. Young gain ½ oz daily first 6 days, soon climb from nest with beak and feet.

Food insects and other forms of small animal life including grasshoppers, crickets, worms, fish, snakes, and even small mammals. Feeds usually early morning and late afternoon. Called "shitepoke," "chalkline," "fly-up-the-creek," "skeouw" from behavior or call. May plunge into water for food occasionally. Speed, 34 mph.

Of no value as food to man and of little if any damage to his interests, so is worthy of protection. Its slow flight makes it a tempting shot for amateur hunters but dead bird is worthless while live bird has value as part of its environment.

Black–crowned Night Heron

Black-crowned Night Heron, 202
Nycticorax nycticorax
Length 24 in., wing 12 in., bill 3 in. Forehead, sides of head, underparts, and neck pale gray. Crown, upper back, and wings glossy greenish-black. Lower back, wings, and tail ash-gray. Legs and feet yellow. With three white 8-in. plumes back from top of head. Young gray-brown above, and underparts white with black streaks.

Breeds from northern Oregon to Nova Scotia, south to Patagonia. Winters from northern California to New York or sometimes New England and New Brunswick, south to Patagonia, with a closely related race in Old World. Common in trees and rushes about waterways. Known from Pleistocene of Florida and California.

Nests in great colonies, female often courting male. Nest a platform of sticks (sometimes eight nests in a single tree), poorly lined with fine material. Eggs three to six, pale dull blue, 2 by 1⅖ in. Incubation 24 to 26 days, by both parents. Young practically helpless, cared for by parents in first 3 weeks, or until half-grown.

Food about 80 percent fish, often taken at night, many being already dead. 50 meals of young included 60 crayfish, 610 catfish, 31 other fish, and 79 dragonflies. Flight to and from feeding and nesting ground, evening and morning, marked by "squawk." Young wander about considerably in fall months.

Young birds considered edible by some persons, but not commonly eaten. Principal enemies are crows and raccoons. In flight, resembles a slow-flying, long-legged, light-colored crow, night flights being recognized by calls. Commonly protected by laws as migratory birds. Not considered edible as adults. Temperature, 102 to 106.8°F.

American Bittern, 190
Botaurus lentiginosus

Length 28 in., wing 10½ in., bill 3 in. Top of head and back of neck, slate, with glossy black streak on each side of upper neck. Back brown. Underparts creamy, with appearance of brown streaking. Young appear somewhat redder. When approached, holds beak erect and fluffs out feathers toward enemy.

Breeds from British Columbia to southern Ungava and Newfoundland, to southern California, Kansas, and southern New Jersey. Winters British Columbia to Indiana and south to Guatemala, Cuba, Jamaica, Puerto Rico. In fall, may move north to southern Alaska. Occasionally in Greenland, Iceland, and Great Britain. Known from Pleistocene times.

Elaborate courtship display. Nest, platform of marsh plants among rushes. Eggs three to seven, pale grayish or brownish-olive, 1⁹⁄₁₀ by 1½ in. Incubation 28 days. Young helpless at first with long buffy down; remain in nest about 2 weeks, cared for by both parents. Matures second autumn.

Food varied, including meadow mice, snakes, lizards, salamanders, frogs, crayfish, grasshoppers, dragonflies, and other forms of small animals caught by lightninglike thrust of beak while bird stands awaiting approach of meal. "Stump-knocking" sounds of male produced by convulsive movements of whole bird. Flight slow, deliberate, powerful.

Protected by law because of unique character and because territory occupied is not normally considered economically valuable. Numbers never great enough to make species dangerous in any way to man's interests. Temperature, 103.7°F. Also known as thunder pumper, stake driver, butterbump, mire drum, bog bull, Indian hen, marsh hen, poke.

American Bittern

Least Bittern, 191
Ixobrychus exilis

Length 11 to 14½ in. including 2-in. beak and 2-in. tail. Wingspread to 18 in. General appearance brown, but small for a heron. In flight shows large light brown patches on wings and a conspicuously black back, tail, and crown. Flight awkward, labored, and not long sustained. Female and young browner on back.

Breeds from southern Quebec to southern Maine and south through North Dakota to southern Mexico and West Indies. Winters from Georgia to Texas and south to eastern Guatemala. Two subspecies: eastern described above and western ranging from Oregon to central Lower California and western Guatemala.

Nests in marshes and swamps. Nest a flat platform of reeds or twigs, in low bushes or reeds, or rarely in trees, well-hidden. Eggs three to six, 1⅕ by 1 in.; bluish or greenish-white. Incubation 17 days or less, by both parents, with one brood in North and possibly two in South.

Food probably mostly insects and other small animals of marshlands. Hides with remarkable skill in trees, brush, or reeds. When frightened, flies only a short distance. Call a soft repeated *coo*. Migration is at night, silent and relatively low, with result that bird is often injured. Freezes by standing motionless, bill pointing straight up; overall looks like a stick and blends into habitat.

Not sufficiently abundant to be of economic importance. Interesting because it is such a small bittern. Corey's least bittern is a rare color phase of this species, differing from it in having light buff replaced by deep chestnut. Also known as dwarf bittern, little bittern, least heron.

Least Bittern

Order Ciconiiformes./Family Ciconiidae [Storks, Wood Ibises]

Wood Stork, 188
Mycteria americana

Length to 4 ft, including 7⅘-in. tail and 9⅓-in. bill. Wingspread to 5½ ft. Weight, male to 11¾ lb; female to 9¼ lb. Large, white, heronlike. Head and neck of adults bare, covered with dark heavy wrinkled skin. Tail glossy black, wing tips black, otherwise white. Bill long, stout at base, and gradually bending downward.

Breeds along Gulf Coast from Texas to Florida and north to South Carolina, with some in West Indies, Mexico, and South America, on to central Argentina and Peru. After breeding season, migrates widely to central California, Montana, New Brunswick, and intermediate points. Found mostly in swampy areas, particularly in cypress swamps.

Nests in large colonies, in swamps, in trees, sometimes very high. Nest a stick platform. Eggs two to three, 2⅔ by 1⅘ in., rough, elliptical, chalky white, more or less stained. Nesting period December to March in Florida. Sometimes a colony may occupy a territory 500 yd wide and 5 miles long. Fish crow, *Corvus ossifragus,* a prime enemy which steals and eats wood stork's eggs.

Food probably mostly fish, reptiles, crayfish, and other aquatic animals. Sometimes flocks may number as high as 5,000; since they often act in unison, they may be noisy in flight when in large numbers. Colonies of them have some protection by groups of ornithologists. Usually silent but may give a guttural croak when frightened.

Wood Stork

(Continued)

The American stork. Cannot be easily confused with others and probably does little or no harm to man's interests. It normally feeds in areas
that yield little wealth to man, on species that are not particularly valuable. Protected by law, of course. Also known locally as American
wood stork, Colorado turkey, goard, iron head, gannet.

Order Ciconiiformes./Family Threskiornithidae [Ibises, Spoonbills]

White-faced Ibis, 187

Plegadis chihi
Length to 26 in., bill to 6 in., wing to 10½ in. Face white, with naked
white skin between eyes and bill. Head, neck, and underparts chestnut
except in winter, when they are streaked with white and brown. Wings
and tail iridescent purple. Color predominantly purple to black without
careful examination. Legs, feet, iris, and bill tip red or reddish. In flight
resembles long-legged crow.

White-faced Ibis, 187

Breeds from Oregon to Utah to southern Texas, southern Mexico, Florida, Peru, Brazil, and southward. Winters Mexico, north to southern
California, Texas, and Louisiana and into South America. Found as far
south as Straits of Magellan. Known from Pleistocene of California.

Nests built in marshes, at water level or up to 5 ft, of woven tule leaves,
sometimes in old sites. Eggs three to four, smooth, finely pitted, greenish-blue, 2 by 1⅖ in. Incubation by both sexes, 3 weeks. Young covered
with sparse light colored down at hatching, but soon covered with black
down. Young molt in September, adults in March and July to August.

Food chiefly small, slow-swimming fish of shallower waters; also insects,
crayfish, mollusks, and some plant materials collected at favored feeding
grounds to which flocks fly from roosts by long diagonal lines, with
steady rapid wingbeats, with legs dangling behind. Call a hoarse nasal
ka-onk, repeated.

Shot by hunters but can hardly be considered game since it is easy to hit
and flesh has little flavor. In spite of this, has been sold in markets.
Harmless, interesting bird, which should not be allowed to disappear
from our skies.

Order Ciconiiformes./Family Phoenicopteridae [Flamingos]

Flamingo, 182

Phoenicopterus ruber
Length to 48 in., including unique 5-in. bill and abnormally long slender
neck. Wingspread to over 5½ ft. Bill so shaped that upper bill serves as
a lower bill when head is brought toward body near ground. Legs and
neck about equal in length. Plumage a uniform vermilion to scarlet, with
wings slightly darker.

Flamingo

Breeds in limited areas in Bahamas, Cuba, Haiti, Yucatan, Guianas, and
Peru. Winters in much same region. Formerly made rather definite
appearances in Florida but now usually seen there only as captives in
park at racetrack near Miami. Has been reported north to coast of South
Carolina.

Nests in great colonies, usually on mud flats. Nest a unique cone built of
mud to a height of 13 in. Nest seems more like a stool of convenience for
adult than comfortable place for little bird to begin life. Egg, one to two,
white, 3½ by 2⅕ in.

Food chiefly the small mollusk *Cerithium,* which birds strain from mud
with scoop-shovel-like movements of bill. Call, a repeated, slow *honk*
not unlike some kinds of geese. Flight slow, steady, and labored as
though birds lacked strength to carry dangling legs. Six species of fla-
mingos; four Western Hemisphere; one, United States.

Beautiful birds formerly prized for their plumage but now protected in
United States, possibly too late for normal healthy survival. Common in
many parks in tropical areas. A bird stretched out from tip of bill to tip
of toes might well reach 5 ft. Should be protected at all times since they
do little harm and are most attractive.

Hooded Merganser

*Order Anseriformes./Family Anatidae [Swans, Geese, and
Ducks]./Subfamily Merginae*

Hooded Merganser, 131

Lophodytes cucullatus
Length to 20 in., including 4-in. tail and 1⅔-in. bill. Wingspread to 26½
in. Weight to 1 lb, 7 oz, female being slightly smaller. Drake conspicuous because of black and white markings and white fan running back
from eye on black head and neck. Female more drab, with black pointed
bill, large head, and white wing bar.

Breeds in temperate North America from northern British Columbia to
New York, south to Washington and central Florida. Winters from

Massachusetts to British Columbia, and south to Cuba and central Mexico. Also found in Alaska, Wales, Ireland, and Bermuda on occasion. Haunts slow streams, woodland pools, and lakes.

Nests in wooded areas near water. Nest in hollow tree or stump, often high above water but sometimes near ground. Eggs, 5 to 12; $2\frac{1}{10}$ by $1\frac{3}{4}$ in.; nearly globular, white. Nesting dates March to August. Incubation 31 days, by female. One annual brood. An elaborate pursuit and display courtship.

Food fish, but much less than with other mergansers; also frogs, tadpoles, insects, grain, and other things. Mergansers swim beautifully, rise powerfully from water, and fly well. In flight, hood is depressed but white breast, slender pointed bill, and black and white markings are distinctive; with drakes, white head patch is still conspicuous in flight.

If any merganser is edible, it is the hooded merganser, but none can be considered a delicacy. It is doubtful if this species does serious damage to stream fishes unless the birds get the habit of feeding near fish hatcheries or happen to be present in large numbers. Locally known as fish-duck, fisher-duck, sawbill, cock-robin, fuzzhead, little sheldrake, and many others.

Common Merganser, 129
Mergus merganser

Length to 27 in., including $4\frac{2}{3}$-in. tail and 2-in. bill. Wingspread to 39 in. Weight to over 4 lb, female being smaller than male. Male appears white with greenish-black head, black back, and orange feet and bill, latter being slender and pointed. Female gray with reddish crest and square white wing patch. Resting on water, male looks like a large, black-headed duck with much white on its body. Female shows brown color of head; and upper part of head is sharply defined against the paler contour of the lower neck.

Common Merganser

Breeds from Alaska Peninsula, across Canada to Newfoundland and south to central California, northern New Mexico, and New York. Winters mostly within United States, southern British Columbia, and Aleutian area. Will remain inland through winter wherever open water is available.

Nests near woodlands, near streams and lakes. Nest in hollow tree or stump or thicket, on or off ground, of grasses, twigs, and leaves with lining of gray down and straw. Eggs 6 to 17, $2\frac{1}{5}$ by 2 in.; elliptical, pale brown or creamy. Incubation 28 days, by female only, and one annual brood. Elaborate community courtship procedure.

Food probably largely fish, though nature and amount of these vary greatly in different parts of the country. Fish may be game fish, nongame fish, or fish that are enemies of game fish. Frogs, crustaceans, and mollusks are also eaten. Some vegetable matter has been found in diet but this is never considerable.

Possibly worst enemy of fish in whole duck family, but their status in this respect has not been definitely established. Birds themselves are almost impossible food for man, though in emergencies they may be edible if entrails are removed immediately after bird is killed. Also known as fish duck, sawbill, freshwater sheldrake, woozer, and many others.

Red-breasted Merganser, 130
Mergus serrator

Length to 25 in., wingspread to 35 in. Weight to $2\frac{1}{2}$ lb. Red-breasted drake, with glossy greenish-black crested head, and brownish area on breast near water which is white in common merganser. Duck largely gray, with reddish crested head and large white wing patch. Temperature, 106 to 107°F.

Common merganser more common in fresh water; red-breasted near sea. Red-breasted breeds from arctic Alaska to Newfoundland and south to New York, Minnesota, and northern British Columbia. Winters along Atlantic from Maine to Texas, and along Pacific from Alaska to Lower California. Also in India, China, Africa, and from Oregon Pleistocene.

Red–breasted Merganser, 130

Nests near waterway, marsh, or island, hidden by grass or trees, with fine plant and feather lining. Eggs 6 to 12, creamy brown, $2\frac{1}{2}$ by $1\frac{3}{4}$ in. Incubated 28 days, by female. One brood yearly. Young molt from March through July and adults September to November. Females have complete molt August to October.

Food largely fish and other forms of small animals collected mainly by swimming at surface with bill open or diving if necessary; can swim under water and pull food from bottom. Often seeks food in a flock, advancing like an army in a long line. Food commonly swallowed whole, even to size of an 8-in. eel.

Probably of little value and may be of some destructive nature particularly with common merganser, which feeds in small streams where young fishes of game and other species may abound. Flesh is not edible because of the rank fish flavor. Known as sheldrake, fish duck, and sawbill.

Order Anseriformes./Family Anatidae [Swans, Geese, and Ducks]./Subfamily Cygninae

Mute Swan
Cygnus olor

C. olor is the mute swan, of which there are a black and a white variety. Whooping swan, *Olor cygnus*, has yellow basal portion of bill and area in front of eyes. Whistling swan, *O. columbianus*, is 55 in. long and has yellow in front of eye. Trumpeter swan, *O. buccinator*, has no yellow before eye and is 65 in. long.

Mute swan and whooping swan are European though mute is widely distributed as a domestic bird. Whistling swan and trumpeter swan are American, latter being one of rarest American birds. Whistling swans breed through barren northern Canada and northeastern Siberia and winter south to Mexico.

Swans probably mate for life. Female increases size of nest while eggs are being incubated until it may be 2 ft high and 6 ft wide. Eggs 2 to 6; (mute) greenish-gray, (whistling) pale yellow; weight (mute) 5 to 8 oz; measure (whistling) 4¼ by 2⅔ in. Incubation by both parents 35 to 40 days. Life-span to 102 years. Male swan, a "cob"; female, a "pen"; young of first year, a "cygnet"; young of second year, a "gray bird." *C. olor* was introduced from Europe. A few members have escaped from domestication and live in small groups.

Swans do best in large ponds where parents care for young *cygnets*, often carrying them on their backs. Require green food even in winter. Are fed mixture of wheat, barley, cracked corn, buckwheat, stale bread, and soaked dog biscuit, particularly when young are being raised. Males fight intruders vigorously with wings.

Essentially ornamental domestic birds and in past were owned only by privileged classes. Do not make gentle pets or get along well with other domestic birds. Reduction in population of wild species is evidence of poor game management in past. Fly in V formation.

Mute Swan

Whistling Swan

Whistling Swan, 180
Olor columbianus

Length 4 to 4½ ft, including 4-in. bill and 8½-in. tail. Wingspread to about 7 ft. Weight to 20 lb. Female slightly smaller than male. Yellow spot in front of eye and smaller size separates whistling swan from trumpeter. Trumpeter and whistling swans carry neck erect when on water, while mute swan arches it gracefully, bill down.

Breeds north of Arctic Circle from northern Alaska to Baffin Bay, and south to barren Canada, northeast Siberia, Alaska Peninsula, and St. Lawrence Islands. Winters from Chesapeake Bay region, along Atlantic Coast from Massachusetts to Florida and south to Gulf of Mexico and northern Lower California. Migrates through Minnesota to New York area, about mid-October.

Nest a heap of vegetable rubbish, concealed among plants near water's edge. Male *cob* courts female *pen* elaborately. Eggs two to eight, 4½ by 3 in., rough-shelled, pale yellow or cream-white. Incubation probably 40 days, by both sexes. Young *cygnets* downy, hatch late June, and family remains together for a year, young getting adult plumage in about 15 months.

Food largely plants of waterways, particularly wild celery. Considered wasteful of valuable duck foods by many persons. May also eat mollusks and crustaceans. In taking wing, will run along surface of water for 15 to 20 ft. Flies in a waving V formation. Remains in close flocks.

Protected by law at all times in Canada and United States. Famed "swan song" supposedly given just before death is an interesting myth with little basis in fact. Normal call a clamoring, quavering, varying musical *wow-wow*, repeated.

Canada Goose, Honker ↕

Order Anseriformes./Family Anatidae [Swans, Geese, and Ducks]./Subfamily Anserinae

Canada Goose, Honker, 172
Branta canadensis

Length to 3½ ft, wingspread to 5½ ft. Weight to 18 lb, with female, *goose*, smaller than male, *gander*, and young *goslings*, often weighing under 8 lb and duller in color. Bill and feet black; back and wings gray-brown. Goslings olive-yellow. Call a high- or low-pitched *honk*. Flight in V-shaped flocks.

Breeds from Labrador to Mackenzie, south to Gulf of St. Lawrence, South Dakota, Utah, and northern California, formerly to Tennessee. Winters from southern British Columbia to southern New England and Nova Scotia and south to southern California, Mexico, Bermuda, and Jamaica. Known from Pleistocene of Florida, Oregon, and California.

Adults mate for life, breeding when 3 years old. Gander defends goose on nest, which is usually on ground but sometimes in trees; constructed of grass, reeds, and leaves and lined with down; sometimes uses old nests of hawks and eagles. Goose incubates the five to nine 2¼ by 3½-in. buffy eggs, 28 to 30 days, yellow goslings hatching in June. One annual brood.

Adults have one complete annual summer or fall molt, losing flight feathers all at once.

Powerful. Feeds largely on roots, grain, and other vegetable matter or on insects such as grasshoppers. Eelgrass a favorite aquatic food. Intelligent in escaping hunters; when wounded, lies flat, effectively on or in weeds. Apparently young accept leadership of older birds. Flight deceptive, rated from 75 to 120 ft/sec.

Among commoner and most valuable of wild waterfowl. Formerly more abundant. In early days, supplied much food for settlers and filled many feather beds. Ranks high in supporting industries dependent on hunting. Decrease in numbers is traceable primarily to unwise wildlife management and shortsighted laws. Has many names including cravat goose, wild goose, bay goose, Canada brant, and long-necked goose. V-shaped flights are harbingers of spring and winter in northern states. *B. C. leucopareia*, Aleutian Canada Goose is an endangered species.

Snow Goose

Snow Goose, 169
Chen hyperborea
Length to 31 in., wing length 16 in., weight 3 to 6 lb; white with black wing tips; red bill, edges of bill black, feet dark red. When immature head, neck, and back pale gray tinged with gray. Female smaller. Adult's black wing tips not easily visible when bird is resting on water, good field identification mark when bird is in flight.

Two subspecies: *C. h. hypoborea*, lesser snow goose, much more abundant and of much wider distribution; breeds along entire Arctic Coast of North American continent and islands to the north from Alaska to Baffin Land. Winter range from the Atlantic to the Pacific; rare east of the Mississippi, especially abundant in winter in California, Texas, and Mexico. *C. h. nivalis*, greater snow goose, breeds in northern Greenland and adjacent lands, winters on Atlantic coasts of Maryland, Virginia, and North Carolina, mouth of the MacKenzie river to Baffin Island; also commonly in the Gulf coast areas, the largest portion in Texas.

Nest, a depression in the ground lined with gray down. Eggs four to eight, usually six; dull white to creamy white, usually stained; $3\frac{1}{10}$ by 2 in. Only the female incubates; male guards the nest. Probably mate for life. Eggs are eaten by Eskimos, gulls, jaegers, foxes, and ravens.

In late 1800s snow goose was so abundant in the breeding grounds that the natives on horseback clubbed them on their nests. After 1917 they received some measure of protection. Provides sport for the hunter but flesh not held in high esteem.

Food largely plant materials; seeds of aquatic plants and cultivated grains. Flight high; flocks in diagonal lines or V-shaped. In flight makes falsetto sound, *au-unk* or *kuk*, given singly or repeated to three times.

Brant

American Brant, 173a
Branta bernicla
Length to 30½ in., wingspread to 52 in. Weight to 4 lb, with female smaller than male. Head, neck, and breast black with a little white showing above water forward. White patch on neck instead of on face and light below. Gosling gray, with chin, throat, and neck lighter. Flight formation irregular. Call a guttural *honk.*

Breeds from Arctic and eastern North America through Greenland to Spitsbergen Archipelago, with related black brant and barnacle goose on through Siberia. Winters along Atlantic Coast from Massachusetts to Florida and more rarely along Pacific Coast from British Columbia to California. Related Old World races. Fossil from Oregon Pleistocene. A true sea goose being seldom seen far from salt water.

Nests along coasts on marshy areas, nests being down-lined depressions on ground from mid-June to mid-July. Eggs four to eight, $2\frac{4}{5}$ by $1\frac{9}{10}$ in.; appearing to be long and whitish. Incubation by female. Young browner than adults when southern migration begins in late August or early September. One annual brood.

Feeds primarily on plant materials and, in captivity, will thrive on grains. May feed on shellfish if necessary. Rarely dives for food. Can run about nimbly on land almost as fast as sandpiper. Is unusually wary of hunters during migration period. Flies low over water and high over land, with speed up to 70 ft/sec.

One of most delicious of waterfowl when it has been feeding on plants. Formerly was most sought waterfowl in New England but is now receiving considerable protection. It seems to respond to protection and so fact of continued low numbers is due to poor management practices. Also known as black brant, brant goose, light-bellied brant.

White-fronted Goose

Blue Goose

Chinese Goose

White-fronted Goose, 171
Anser albifrons

Length to 2½ ft, including 6½-in. tail and 2⅓-in. bill. Wingspread to over 5 ft. Weight to 6 lb, female being slightly smaller. Sexes colored alike. Head and neck brownish-gray, with a white band in front of face and on chin, white front being bordered by black. General appearance gray-brown, with mixed black and white on breast.

Breeds in Arctic America from Yukon to Anderson River in Mackenzie and on west coast of Greenland, Iceland, Lapland, and Arctic Siberia to Bering Strait. Winters in western United States from British Columbia to Texas and Illinois, with a few east even to Atlantic. Common goose west of Mississippi. Migrates north to breeding grounds in March and April and south to wintering grounds in September and October.

Noisy courtship in May. Nest on a grassy border of a waterway, usually lined with grasses or moss but often a mere hollow. Eggs four to eight, 3½ by 2 in.; dull yellow-white or greenish-yellow, commonly discolored. Incubation 22 to 23 days, by female alone. Goslings mostly with brownish-buff down, being lighter beneath. Young reach weight of nearly 4 lb by October.

Food, nuts, acorns, grain, berries, and grass; almost wholly vegetarian. Northern migration begins in March and main fall flight is in September and October. Birds fly high in V's like Canada geese but show black and white breast, white tail coverts, and yellow legs if near enough. Call, a repeated *wah-wah-wah* given in flight, laughterlike.

White-fronted goose in fall among most delicious of wildfowl and therefore a popular game bird. In nesting area, many are killed by natives during time when flight feathers have molted. Birds are often approached by using a bullock as a moving blind, but are suspicious of horses if they are used as blinds. Has many common names including California goose, laughing goose, mottled goose, speckled brant, spotted goose, Texas goose, yellow-legged goose. *A. a. gambelli*, tule white-fronted goose is an endangered species.

Blue Goose, 169.1
Chen caerulescens

Length to 2½ ft, including 6-in. tail and 2½-in. bill. Wingspread to 4½ ft. Weight over 4 lb. Conspicuous because of white head and neck and dark gray-brown on back. Related emperor goose has black foreneck. Bill pinkish. Some variation in degree to which breast is darkened.

Breeds in southwestern Baffin Bay and Southampton Island. Often shares nesting sites with lesser snow goose, with which it sometimes interbreeds. Winters from Louisiana to central California and north to British Columbia, but appears in migration through Montana and Alberta to the Athabaska-Mackenzie territory, occasionally east of Mississippi.

Nest on top of slight rise, on land near water, formed of moss and vegetation, with lining of fine down and grasses, being generally large and bulky. Eggs three to five, 2 by 3 in.; white or pale cream with a slight luster. Incubation 25 days; some doubt that male takes part although he stands guard. Young can run well within a day.

Young and adults eat and drink well, young increasing their weight about 20 times in first 2 months. In spring migration, about 11 weeks are required for 3,000-mile migration from winter quarters to nesting area, spring route being some 600 miles longer than southern route made in fall. Fall migration at height in September.

An important game bird of Louisiana and Texas coast area. Big flocks may be destructive of vegetation in normal feeding areas of ducks and sometimes over cultivated lands, this apparently because geese feed more on roots and underground parts of plants than on tops.

Chinese Goose (Domestic)

Adult gander 12 lb, young gander 10 lb. Adult goose 10 lb, young goose 8 lb. Two breeds: white, entirely white; and brown, a dark, grayish-brown; sexes and young colored alike. Head with orange knob at base of bill. Bill of medium length, stout, orange. Eyes large and blue. Toes yellow. Legs and feet lead-colored; latter dull red. Streamlined describes its general shape.

Native of China but introduced rather widely through world, more as an ornamental variety than for any other purpose. Stands remarkably erect and walks with pronounced arch to long neck. Egyptian goose, also more or less of ornamental type, weighs to 10 lb, is gray-black to brown, ferocious and wild.

This variety is essentially a show or ornamental goose and is not raised in sufficient numbers to be considered an important domestic bird. Floats on water; as beautiful as a swan.

Embden Goose

Adult gander 20 lb, young gander 18 lb. Adult goose 18 lb, young goose 16 lb. Entire plumage of young and old pure white, eyes blue, shanks deep orange and long; toes deep orange and straight. Head relatively large. Bill of medium length and stout at base. Wings large, well-rounded, and strong.

Developed as a domestic bird in Germany but has been widely adopted by poultrymen in other parts of the world. It thrives where water and good grass grazing are to be had and easily lends itself to being raised in flocks.

Geese are more often reared as domestic birds in Europe than in America. Here they are considered a bit too rich for ordinary meals and so are reserved for special occasions, as are turkeys, which possibly because of their milder flavor have won more favor. In Europe children are frequently employed as gooseherds.

Toulouse Goose

Adult gander 26 lb, young gander 20 lb. Adult goose 20 lb, young goose 16 lb. Sexes and young colored alike; gray above, underparts light gray to white, wings gray with lighter tips, tail gray and white, shanks stout and deep reddish-orange, toes straight and reddish-orange, eyes brown or hazel, large, not prominent.

Native of southern France, where breed is raised in large flocks; larger and heavier than Embden goose raised in neighboring Germany. The Toulouse or domestic gray goose is raised widely throughout world by poultrymen wherever there is a market for goose flesh.

Eggs of Toulouse goose are white and weigh 4 to 5 oz; laid in nest lined with goose feathers. Goslings downy, gray above and creamy below, unlike breeds previously considered. Young must be kept dry, even free from dew.

If one wishes to start raising geese by purchasing eggs and hatching them with a hen, not more than six eggs should be given each hen. Generally better to start in a small way than to try too many at first.

This is the breed of goose which is the great flesh producer and which is raised in great flocks in various parts of the world. Birds are not normally so wild as other breeds and their great size makes them more valuable. Temperature of geese, 106°F.

Embden Goose

Toulouse Goose

Eggs of Chinese and Embden geese are both white, former weighing 3½ to 4 oz, latter 4 to 5 oz. Nest usually in a box or barrel, on ground or floor; feather-lined. Incubation for 28 days. Goslings downy, creamy white. For first 36 hours after hatching, feeding is not necessary. After that, they should be fed three to four times daily for 2 to 3 weeks a mixture of stale bread soaked in water or milk and finely chopped boiled eggs. At 3 weeks, they may be turned out to pasture, then requiring only one feeding a day of a light mash of grains. Both sexes may provide some protection for young, but goose is more effective. Chinese goose may lay to 60 eggs at end of first year. As an egg producer, it might become profitable. Brown Chinese goose is identical with white but gray with a brown tinge on body.

Food: except in winter, most domestic geese can glean a living from well-watered, temperate-zone, farmlands where grass is main food. In winter, they are fed grain and some roughage such as silage. Oats are preferred grain for mature birds, though corn, wheat, barley, or mixtures are acceptable. There should always be abundant roughage and grit. Fresh drinking water should always be present. Laying birds should be fed a mash made of 3 parts of bran or shorts, 1 part of cornmeal, and 1 part of meat scraps. Ordinarily, a flock of geese can protect itself from average enemy. An angered gander can put a child or dog to flight if necessary.

Pekin Duck

Adult drake 9 lb, young drake 8 lb. Adult duck 8 lb, young duck 7 lb. White or creamy where feathered. Shanks and toes reddish-orange. Bill orange-yellow. Eyes lead-blue. Tail rather erect; drake with several stiff curled feathers. Wings short. Head long. Neck of medium length, carried arched forward. Creaminess of feathers in this variety is not a serious defect as it is in white chickens. Yellow corn or other grains rich in yellow-bearing pigments may produce creaminess.

Native of Asia, derived from wild mallard. Commonly raised where fast-growing young ducks are desired for a special market, therefore a common breed near large cities.

Pekin ducks are superior for flesh, young roasting ducks known as "green" ducks bringing a high price if properly forced to grow rapidly. Feathers of ducks raised for flesh are saved and provide a supplementary income. Pekin duck eggs have a superior food quality and are unusually large.

Pekin Duck

Muscovy Duck

Mature drake to 10 lb, young drake 8 lb. Mature duck to 7 lb, young duck 6 lb. Two varieties, a white form and a colored form where sexes are colored alike with black body, breast, and tail, and black wings with large white patches. Shanks and feet yellow to dark lead color. No curled tail feathers. Eyes blue.

Native of tropical America from Mexico to southern Brazil but domesticated and rather widely distributed. Never apparently so common as other breeds even though this is largest of domestic ducks. One of our most useful and valuable breeds of domestic waterfowl.

Muscovy ducks are raised chiefly for meat. Though they may look fierce, they may often make excellent pets. Quality of meat is good but growth is not commonly so rapid as with Pekin ducks. Ducklings of Muscovy ducks are downy and brown instead of yellow. Many strains are prolific egg layers even in winter. Eggs are creamy white in color and are especially tough-shelled and so will withstand rougher treatment than most ducks' eggs.

Muscovy Duck

Runner Duck

Adult drake 4½ lb, young drake 4 lb. Adult duck 4 lb, young duck 3½ lb. Ducks and drakes colored alike. Fawn and white, white, and penciled varieties. Bill long, wedge-shaped, yellowish-green with a black tip. Carriage erect. Body long and slender. Drake has several curled stiff tail feathers.

Originated in India but takes place in duck world similar to that occupied by Mediterranean breeds of chickens.

Runner ducks are raised primarily for their egg production. They are too small for meat production but they lay profusely, relatively large eggs. Male gives a low *queak, queak* call and female a loud harsh *quack, quack.*

Often called the "leghorns" of the duck families. Varieties include Fawn and White Runner and Penciled Runner.

Runner Duck

Pekin ducks make rather poor mothers compared with many other breeds. Ordinarily domestic ducks such as Pekin, Muscovy, and runner breeds lay white eggs, averaging 3 to 3½ oz in weight. Incubation for Pekin and runner ducks, 28 days; for Muscovy ducks, 35 to 37 days. Ducks should be housed where they are protected from rats and skunks. Floor should be dry, clean, and commonly covered with shavings, sawdust, or chaff since ducks' droppings are filthy. There should be ample shade, and if convenient, adequate water for swimming. Many duck raisers keep young ducks from water until they are several weeks old, particularly when they are brooded by a hen. After first month, young ducks are able to feed themselves well if provided with a suitable range. Well-fed young Pekin ducks may reach a weight of 8 lb in as many weeks, particularly if they are kept from water except for what they get in their feed. Temperature 106 to 108°F.

Ordinarily, where ducks are not being forced, adult ducks in relatively close confinement are fed twice daily a ration of mixed grains soaked in water, with some green feed or corn in morning, and a mash in evening. Mash for ducks is made of bran middlings, cornmeal, and greenstuffs. Food is placed in a low trough with an abundance of water near by. Meat scraps are sometimes added to the mash, and vegetable scraps and small vegetables, like potatoes boiled with skins on, may be added. Grit should always be available. First week, ducklings are fed five times daily a mash of 2 parts bran, 1 part cornmeal, 1 part middlings, and some greens. Later, cornmeal ration is increased and meat scrap added. At 4 weeks, feeding should be cut to four times a day. After that, young ducks should be allowed to range with duck after insects and may be fed as adults. When growth is forced, mash feedings soaked almost to a slop with extra water are commonly used.

Mallard Duck, 132
Anas platyrhynchos

Length to 28 in., including 4½-in. tail and 2½-in. bill. Wingspread to 40 in. Weight to 3¾ lb, duck being somewhat smaller than drake. Drake, head and neck green, breast purple-brown, collar white, other parts brownish-black, ash-gray, and blue, tail curled. Duck with no collar, and with head and neck brown, wing bar purple.

Breeds from Pribilof Islands around Arctic Circle, south through United States to Virginia, more commonly in the west, south through Europe to northern Africa and through Asia into Japan. Winters south to Panama, South Africa, India, and Borneo, and north where there is still open water. Probably commonest of wild ducks.

Mating promiscuous, many drakes often mating with one female. Drake hisses, calls, and pursues duck. Nest hidden by duck in vegetation relatively near water. Eggs greenish to gray-brown, 6 to 13, 2⅓ by 1⅗ in., laid on ground. Incubation 26 to 28 days, by duck, and ducklings are reared by duck. Ducklings downy yellow, soon able to walk and swim.

Excellent destroyer of mosquito larvae, being better than goldfish for this purpose. Also eats grain and vegetation, including arrowhead, chufa, pickerelweed and bulrushes. Can fly 60 mph or short distances 95 ft/sec. Temperature 107°F. Appears slow and steady in flight, wings being brought only a little below normal level of the body.

Probably most valuable of all wild ducks because of abundance, ease in harvesting, and general hardiness. Flesh is excellent, particularly when bird has fed on better plant foods. Birds reared in captivity for freeing for hunting do not readily lose tameness, so shooting them is about like shooting barnyard fowl. Considering abundance, there is a small number of common names, such as common wild duck, stock duck, English duck, French duck, green-head.

Mallard Duck

Rouen Duck, Domestic Mallard, 132
Anas platyrhynchos

Size about same as that of Pekin and Aylesbury breeds, drake weighing up to 9 lb, duck to 8 lb. Coloration much like that of wild mallard, drake having green head and neck, gray-brown body feathers, and typical curled tail. Female like mallard duck but with dark bill. Purple, white-edged wing bar in each sex.

Rouen ducks are found practically everywhere in world where there are sufficient water and food. Beside those ducks already considered as domestic ducks, there are the all-black, lighter-weight Cayuga duck; the gray and white call duck; and the large Aylesbury duck which is white like the Pekin. Many breeds interbreed freely with tame or wild ducks.

As with other domestic birds, breeding is highly promiscuous. Male is not mindful of interests of any but himself. Eggs to 14 or more in a clutch, 2 by 2½ in.; yellowish-white to buff or greenish, and frequently dirty. Incubation 26 to 28 days, by duck only. Ducklings follow mother until well developed.

Rouen ducks are fed twice daily a mash of bran middlings, cornmeal, and greenstuff, with occasional meat scrap and oystershell. They are also allowed to forage for themselves or are fed grain for fattening. Ducklings fed five times daily during first week and less frequently thereafter. Range ducks feed much as do mallards.

Raised primarily for use as flesh for food for man, though many mallards are raised on game farms and freed for shooting. However, sport in harvesting these animals must be decidedly tame, comparable to shooting canaries in their cages.

Rouen Duck

Black Duck, 133
Anas rubripes

Length to 23½ in., wing to 11 in., wingspread to 3 ft, bill 2⅕ in. Weight to 3 lb, 10 oz, with duck under 3 lb. Resembles female mallard but is darker, lacking white borders on blue on wings and showing conspicuous white beneath wings in flight. Drake's bill yellowish, duck's olive. Drake's legs more orange than duck's.

Breeds from Manitoba to Labrador and south to North Carolina and Colorado. Winters along Atlantic Coast of United States and south to Florida. Sometimes raised on game farms for freeing as game bird.

Drake courts by short flights and head bobbing. Nest commonly on ground, sometimes on stumps, down-lined. Eggs 6 to 12, greenish-gray; 1¾ by 2⅖ in. Incubation by duck, 26 to 28 days. One annual brood. Young fluffy, yellow, darker above than in mallard. Immature young more striped below than adults. Voice a loud *quack;* drake, mellower.

Food 3 parts plant, 1 part animal matter. Half of food is pondweeds, eelgrass, wild celery, but grain welcome where available. May eat crayfish and fishes but these injure value of duck as food for man. May eat shrimps and mussels; also aquatic insects. Feeds commonly at night. Has remarkable hearing. Flight speed, 55 to 90 ft/sec, 45 mph.

(Continued)

Black Duck

Among wildest of ducks, even when raised in captivity, therefore considered good game bird. Probably most valuable of food and game wildfowl of eastern United States. Migration towards seaboard begins in September but birds visit fresh water regularly to feed and so may be accessible to hunters there. Temperature, 106°F. Also known as redleg, winter black duck, clam duck, blackie.

American Widgeon

American Widgeon (Baldpate), 136
Mareca americana
Length to 22 in. Wingspread to 35 in. Weight, male 2 lb, 7 oz, female to 2 lb. Male with conspicuous white crown; rest of head gray, with green from eye back; bill blue with black tip; white patches under tail and on forewing; rest of body, mostly pinkish-brown. Duck ruddy brown, neck and head gray, white on forewing.

Breeds from northwest Alaska to northeastern California and east to northern Indiana and Hudson Bay. Breeds mostly in western part of range. Winters from Massachusetts to Vancouver Island and southern Alaska to Panama. Sometimes found in Japan, France, and Hawaii. This species known as fossil from Pleistocene of Oregon and California.

Courting male swims over water with extended neck, raised wings, giving whistling notes. Mates usually before coming to nesting ground. Nest usually on dry ground, hidden or open, of grasses with abundant down lining. Eggs 6 to 12, 2⅓ by 1½ in.; white to cream or buff, incubated by female. Nesting period from late May to mid-July. One annual brood.

Dives poorly but lives mostly on water plants such as wild celery, often stolen from other ducks that are better divers. Food pondweeds 43 percent, grasses 14 percent, algae 8 percent, sedges 7 percent, animal food 7 percent. Insects constitute small proportion of food except in summer. Flock of flying baldpates conspicuous because of large white patches along forepart of wings. This is duller in immature birds.

Superior game bird without record of being destructive to agricultural crops. Food value high. During breeding season and when birds are young, helps definitely in controlling any increase in destructive crickets, grasshoppers, beetles, and similar insects. Interesting because of taking food from other ducks; not so dependent as rumors contend. Male whistles *whee, whee, whew*.

Gadwall

Gadwall, 135
Anas strepera
Length to 22 in., including 4½-in. tail and 1⅔-in. bill. Wingspread to 35 in. Weight, drake to 2⅓ lb, duck smaller. Slender, gray, with white patch on hind edge of wings, with black tail coverts that are conspicuous next to pale gray wings. Duck browner than drake, with white also on wings. Belly white. Bill partly yellow. Only river and pond duck with white spot on hind edge of wing.

Breeds from Little Slave Lake and Hudson Bay to central British Columbia, south to Washington, Oregon, California, Utah, Colorado, and New Mexico, and east to Minnesota and Ohio; also in Europe and Asia. Winters from Chesapeake Bay to northeastern Colorado, and southern British Columbia to southern Lower California, central Mexico, and southern Florida.

Nests on dry land among weeds, usually near water, on ground commonly on islands in lakes. Nest a hollow, lined with plant stems and down. Eggs 7 to 12, 2¹⁄₁₀ by 1½ in.; oval, creamy white. Nesting period May to July. Incubation 28 days. One annual brood. Downy young like young mallards but paler and less yellow. Young much like duck.

Food, insects, snails, small fish, tadpoles, crayfish, mollusks, roots, nuts, tender vegetation, and grain. Nature of food determines considerably value of flesh of bird for food. More vegetarian than other American ducks. Flies in small compact flocks, rising rapidly; rests high on water.

Valuable game bird but less common than might be desired particularly in the East, where it probably never was very abundant. Drake croaks in flight much as does a raven or gives oft-repeated, high-pitched *quack* weaker than that of the mallard.

American Pintail

American Pintail, 193
Anas acuta
Length drake to 30 in., duck to 22 in., wingspread to 3 ft, tail of drake 7½ in., of duck 3⅗ in. Weight to 2¾ lb. Conspicuous in field because of long slender neck and tail. Drake shows white neck, and white across belly. Duck with conspicuous wedge-shaped tail.

Breeds from Arctic Alaska to New Brunswick, south to California and New Jersey; rarely east of Lake Michigan. Winters from Alaska to Massachusetts, south to Panama but uncommon in East. Closely related Old World form found in Europe, Asia, North Africa. Known in Greenland, Labrador, China, and from Pleistocene of Oregon.

Drake mates early with duck. Nest in grass or sheltered dry spot. Eggs 5 to 12, greenish-buff; 1½ by 2 in. Incubation by duck, 23 to 25 days. One annual brood. Duckling grayer or browner than common associates; later colored like duck though grayer; takes on breeding plumage at 22 months. Adult molts September to November.

Food largely vegetable matter, including grain, but also aquatic insects, grasshoppers, and the like. Can dive and swim readily under water, particularly when wounded, but is normally a surface feeder. Extraordinarily alert, taking to wing and rising steeply when disturbed, and attaining speed of 60 to 100 ft/sec, 45 mph.

Among most beautiful of ducks and earliest to visit East in spring migration, and earliest to leave in fall migration. Desirable as game bird because of alertness, speed in flight, and exceptional food value when proper food has been available. Could probably be more abundant by wise wildlife management. Courting drake gives sweet, low, soft, whistled call notes.

Green-winged Teal, 139
Anas carolinensis

Length to 15 in. Wingspread to 24 in. Weight to 14 oz, with female slightly smaller. Male, general appearance gray with brown head, conspicuous white mark in front of wing, green patches on side of head and on wing. Female, speckled gray, with green spot on wing. Young much like short-tailed female.

Breeds from northern Alaska to southern Ungava, south through central California, northern New Mexico, southern Minnesota, western New York, and Quebec. Winters from southern British Columbia to Montana, Nebraska, Kentucky, Chesapeake Bay area and south to Bahamas, West Indies, Honduras, and southern Mexico. Fossil from Florida, Oregon, and California Pleistocene.

Green–winged Teal ↕

Nest on dry ground, near or far from water, in open or at edge of brush timber, a hollow of grasses and feathers, with much down in lining. Eggs 7 to 12, 1⁹⁄₁₀ by 1⅓ in.; blunt pale greenish-brown to pale dull brown. Nests in May. Incubation 21 to 23 days, by female. One annual brood. Ducklings gray-brown, with eye line, crown, and back of neck brown.

One of swiftest flying ducks, by some estimated to fly 60 mph. Flies in compact flocks that make quick close turns. Floats high on water. Food animal matter, including insects, mollusks, and fish to about 10 percent. Remainder water plants, their seeds or sometimes grain, nuts, grapes, berries, and similar fruits. May feed in marine areas. Seeds are a prominent part of the diet in any area.

Flesh excellent when bird has been feeding on plants but rank and useless when it has been feeding on some kinds of animal matter. Because of great speed, it is considered a challenge to expert marksmen. In East a late fall migrant; in West, an early fall migrant. Generally tame. Often found with barnyard fowl if undisturbed.

Blue-winged Teal, 140
Anas discors

Length to 16 in., wing 7¼ in., bill 1⅔ in. Weight to 1 lb. Drake small, dull-colored, with white crescent before eye and light spot at base of tail; in flight, shows blue patch on forepart of wings. Duck with patch on forepart of wing, otherwise mottled.

Breeds from central British Columbia to New Brunswick and south to northern Nevada and New York; rarely in Florida, Texas, and Louisiana. Winters from South Carolina to southern California and south to central Chile and Brazil. Also accidental in Greenland, British Isles, and Denmark. Known from Pleistocene of Oregon.

Blue–winged Teal ↕

Nest on ground in dry meadow or marsh, basketlike, and concealed with overhanging grass. Eggs 6 to 15, olive to brownish-white; 1⅕ by 1⅓ in. Incubation by duck, for 21 to 23 days. One annual brood. Ducklings yellow-brown; darkest on rump, crown, and eye line.

Food 70 percent vegetable matter, mostly water weeds; 10 percent insects including grasshoppers, ants, moths, flies, and beetles; and remainder, other animals including snails and shrimps. Seeds are a primary part of the diet. One of fastest flying ducks, traveling over 100 ft/sec and keeping in close flocks, to their own misfortune.

Excellent game bird, though too small to provide much flesh as food. Numbers have decreased unreasonably, but with protection shows power to recover numbers rapidly since it is most prolific. Makes southern migration early compared with that of some larger species, and spacing closed seasons could restore its numbers.

Cinnamon Teal

Cinnamon Teal, 141
Anas cyanoptera
Length to 17 in. Wingspread to just over 2 ft. Weight to about 1 lb. Drake dark cinnamon-red, with large chalky-blue areas along front of wing and easily identified. Duck almost identical with female blue-winged teal, having same blue wing patch. Probably impossible to distinguish between them in field.

Breeds from southern British Columbia and western Saskatchewan through Oregon, Washington, and central California to western Texas and Lower California. Another group breeds in South America from Buenos Aires and the Peruvian Andes to Straits of Magellan. Winters from central California to Arizona and Brazil.

Nest on ground, usually basketlike but always heavily lined with ducks' feathers, often to 100 ft from water. Eggs 6 to 14, 1⅘ by 1⅓ in.; white to pinkish-buff. Incubation entirely by duck, though drake may remain near nest until eggs hatch.

Food ⅘ vegetable matter. Stomachs of 41 showed 34 percent sedges, 24 percent pondweeds, 8 percent grasses, 3 percent smartweeds, 10 percent insects, 9 percent mollusks. Interesting duck because of two groups in Northern and in Southern Hemispheres, between which there is no migration.

A popular game bird in West, formerly slaughtered in great numbers by market hunters. It is difficult for one who has watched a group of these birds to understand how anyone could take pleasure in shooting such animals. Known also as South American teal, red-breasted teal. Groups of males play a game like leapfrog.

Shoveler

Shoveler, 142
Spatula clypeata
Length to 21 in. Wingspread to 35 in. Weight to 2 lb, with female smaller. Male, head black with green cast, breast white, belly and sides reddish, patch on front edge of wing light blue. Female, brown, mottled, with large pale blue wing patch. Male gives impression of being black and white at a distance. Sits low in water, with bill downward or, in flight, extended forward.

Breeds from Bering Sea coast of Alaska to Great Slave Lake, Iowa, Nebraska, New Mexico, and southern California; formerly to western New York. Winters from southern British Columbia to California, Central America, Colombia, through lower Mississippi Valley and along Atlantic Coast from New England to West Indies. Modern form as fossil from Oregon and California Pleistocene.

Nests May to June in open grassy or brushy grasslands or edge of marshes. Nest a grass-lined hollow, down-rimmed, near or far from water. Eggs 6 to 14, 2⅓ by 1½ in.; smaller and whiter than mallard eggs. Incubation 21 to 23 days, by female. One annual brood. Young drakes variable; may reach adult plumage in 17 months. Two annual molts in adults.

Food about 25 percent animal matter consisting of insect larvae and adults and small invertebrates; remainder of diet made up of plant material. It may eat land insects when these are abundant and convenient, but its bill fits it particularly for straining from water and oozes smaller animals of freshwater ponds and streams.

Known to do no damage to agricultural crops and when it has been feeding on plants is an excellent wild duck for table. This coupled with its occasional destruction of harmful insects makes it a species decidedly worthy of protection and encouragement.

Wood Duck

Wood Duck, 144
Aix sponsa
Length to 20½ in., wing to 9½ in., wingspread to 29 in., tail 4½ in. Weight to 1½ lb. Duck smaller than drake. Drake most highly colored North American duck in winter and spring; highly iridescent. Duck and, in summer, drake show crested heads with white areas around eye; dark brown, with light flanks and white belly.

Breeds in most of United States and in southern provinces of Canada, rarely farther north. Winters from southern British Columbia to southern Virginia, Michigan, or Massachusetts, south to central Mexico, Jamaica, and sometimes Bermuda. Lives commonly along wooded streams and lakes inland. Known from Pleistocene of Oregon.

Nest usually in a hollow tree or on limb, 3 to 50 ft above ground or water or rarely on ground, in nesting box lined with gray down. Eggs 8 to 15; pale brown to brownish-white, 2 by 1½ in. Incubated 28 to 30 days, by duck. Young remain in nest a day or a little longer after hatching.

Food 90 percent vegetable matter, such as 10 percent duckweed, 9 percent cones and galls, 9 percent sedge seeds and tubers, 8 percent grasses and seeds, 6 percent pondweeds, and 6 percent nuts. The 10 percent animal matter is mostly insects, young birds feeding voraciously on mosquito wrigglers.

Redhead Duck ↕

A beautiful wild bird protected by law but probably shot by hunters who do not recognize it or who do not care. Has been suggested as national bird because of peaceful habits, because it nests in most states in Union, and because of beauty. Removing old timber has reduced numbers. Temperature 107°F.

Redhead Duck, 146
Aythya americana
Length to 23 in., wingspread 33 in., bill 2¼ in. Weight to 3 lb. Drake with high forehead and reddish-brown head, mostly gray but breast, neck, and tail black, bill blue; in flight, shows gray wing band. Duck brownish, with broad gray wing stripe; smaller than drake.

Breeds from northern British Columbia to Wisconsin and south to southern California and Nebraska. Winters from southern British Columbia to southern New England and south through central Mexico and West Indies. Occasionally in Alaska. Known from Pleistocene of California. Reared sometimes on game farms and kept in parks.

Nests over water in marsh border, among vegetation, usually a deep well-made nest. Eggs 10 to 18; olive to brownish-cream, 2⅖ by 1⅔ in.; sometimes laid in nest of other kind of duck. Incubation by female, 22 to 24 days. One yearly brood. Female often with unusual white feathers. Ducklings pale yellow, downy, with spot back of base of wing.

Food principally plants of fresh waters such as wild celery, pondweeds including bulbs, leaves, and roots; also acorns. May also include snails, freshwater clams, shrimps, frogs, and even fish, though normally in smaller quantities. Migrates northward in early spring; fall migration begins with eastern flight in early fall. Speed, 45 mph.

Flesh almost equal in quality but not in quantity to that of canvasback, particularly when feeding has been largely on plants. Numbers much fewer than they were formerly. Feeds either diving in deep salt or fresh water, or dabbling in shallow water.

Ring-necked Duck, 150
Aythya collaris
Length to 18 in., including 3⅖-in. tail and 2-in. bill. Weight, 1½ to 2 lb. Wingspread to 30 in. Duck smaller than drake. Like scaups but with black back, black head and foreparts, light gray sides with white triangle at front, bill with two white rings. In flight, shows broad gray wing stripe and black back. Duck brown, with gray wing stripe, white belly, and white eye ring.

Breeds from central British Columbia through Alberta, Saskatchewan, Manitoba, and western Ontario to central Arizona, northern Nebraska, northern Iowa, southern Wisconsin, and northern Michigan. Winters from southern British Columbia along Pacific Coast to Mexico and from northern Arkansas and Chesapeake Bay to Bahamas, Mexico, and Guatemala. Sometimes north to Nova Scotia.

Ring–neck Duck

Nests on marshy borders of lakes or pools. Nest of grass, with finer plant material and down lining. Breast feathers in down, white, or gray with white tips. Eggs 8 to 12, 2⅓ by 1⅗ in.; smooth to somewhat glossy, olive-buff. Nesting season June to July. Incubation 25 to 29 days, by female. Downy young very light-colored, with lead-colored feet and yellow-tinged toes.

Food, mainly vegetable, but also frogs, snails, small water insects, and seeds, shoots, and roots of tender aquatic plants. Food procured by diving, tail being spread and deflexed when head is dipped under water. Flies rapidly in small scattered groups; prefers rivers and shallow waters to larger deeper lakes. In courtship, gives sound like blowing through a tube.

Classed with scaup ducks, which it closely resembles in general color, size, and behavior. Black back of male is always distinctive but ducks are easily confused, particularly with bluebills, with which they may be in a flock. May be known locally as ringbill, blackjack, bluebill, ringneck.

Canvasback

Canvasback, 147
Aythya valisineria
Length to 2 ft, wingspread to 3 ft, bill 2¾ in. Weight, drake to 3 lb; duck to 2 lb, 6 oz. Bill longer than head, continuing profile into a narrow low-browed wedge. Drake with head and neck chestnut, breast black, and body mostly white. Duck head and neck brown, body grayer. Rump black in both sexes.

Breeds from central Alaska to Wisconsin, south to New Mexico. Winters from southern British Columbia to New York, south to central New Mexico and Florida. Reported from Pleistocene of Oregon and Florida. Found in great rafts on lakes when at peak of abundance. Not nocturnal.

Male courts duck by throwing back head and by some pursuit. Nest of weeds surrounded by water, down-lined. Eggs 7 to 15, olive gray, 1¾ by
(Continued)

Greater Scaup Duck

2½ in., often with eggs of redhead and ruddy duck mixed in. Incubation by duck, 22 to 23 days. Young deep yellow but, when immature, resemble duck. One annual brood, beginning in May to June.

Food largely water plants or grain, but may feed on decaying fish, in which case flesh is not good for man. Young birds develop feathers in 10 weeks. Molt in late July to August with wing molt in October prior to southern migration. Flight often in V-shaped formation taken by geese. Call, a harsh guttural croak.

Possibly most valuable of game ducks. Flesh superior when food has been wild celery *(Vallisneria)*. May survive shore hunting because of ability to dive and feed offshore, but illegal market hunting, drought, and overshooting have seriously decreased numbers. Good canvasback shooting grounds have high commercial value.

Bluebills, 148, 149
1. Greater Scaup
Aythya marila

2. Lesser Scaup
A. affinis
(1) drake, length 18½ in., wing 8¾ in., weight 2¾ lb; duck, length 17½ in., wing 8¼ in. Both with long white wing stripe, in flight. (2) drake, length 16½ in., weight 2 lb; duck, size of drake. General appearance, at distance, of black and white ducks.

(1) breeds in Arctic America, mostly west of Hudson Bay. (2), from southern Alaska to southeastern Canada, south to Colorado and Ohio. (1) winters south to Mediterranean, Florida, and Lower California; (2), southern British Columbia to New Jersey, south to Panama. (2) from Pleistocene. Open-water species.

Nests usually in marshes and over water. Eggs 6 to 11, olive-brown, 2½ by 1⅔ in. (1); 2¼ by 1½ in. (2). Incubation 3 to 4 weeks, by duck. One annual brood. Sometimes more than one female will share a single nest, bringing egg total up to 22 and a spectacularly large number of little ducklings.

Food gathered occasionally at night where they assemble in great rafts at sea to come in over mussel beds and feed. In summer, feed largely on water plants which help make the early fall bird palatable though hardly a delicacy. Flies in an irregular formation, somewhat like a wedge.

A popular game bird because of great numbers but not excellent for table except on inland waters. From the side, drakes appear somewhat like canvasbacks but dark backs offer conspicuous difference. Ducks have white area around base of bill or on face, unlike other American ducks.

Common Goldeneye

Common Goldeneye, 151
Bucephala clangula
Length to 23 in., wingspread to 32 in., tail 4½ in., bill 1⅔ in. Weight to 2 lb, 5 oz. Drake larger than duck; white, with black back and black to glossy green head, with white spot before eye; shows large white patches in flight. Duck gray, with brown head, white collar, and large square white wing patches.

Breeds from Yukon in Alaska to Labrador and Newfoundland and south to northern New England, North Dakota, and interior British Columbia. Winters along Atlantic Coast from Maine to South Carolina and on Pacific from Aleutian Islands to Lower California or rarely to Arizona, Texas, Gulf Coast, Florida, and Bermuda.

Nests in forested country about lakes and streams, in hole in large tree, 6 to 60 ft above water; down-lined. Eggs 6 to 19, pale green; 2⅗ by 1⅔ in.; often piled in nest in two layers. Incubation 20 days, by duck. Young sometimes safely jump alone 12 feet to water or ground below.

Food not known well but probably largely aquatic plants; known to include eelgrass, pondweeds, dragonflies, mussels, fish, crayfish, and other aquatic organisms. In flight, a whistling sound produced by its wings and therefore called "whistler" by hunters and others.

Dives for its food and also dabbles. Bird edible, being considered better than scoters or coots, but not a superior table bird. Wary bird which does not come readily to decoys and so offers a welcome sporting challenge to hunters.

Male

Female

Bufflehead, Butterball ↕

Bufflehead, Butterball, 153
Bucephala albeola
Length to 15 in., wingspread, 25 in., tail 3⅕ in. Weight to 1 lb. Duck smaller than drake. One of smallest wild ducks. Drake mostly white, with large black head marked by white from eye to back of head and black back; wings with large white patch. Duck shows large dark head,

with small white cheek and white wing patch. Commonly stay together in small groups away from other species.

Breeds from British Columbia and Yukon Territory to James Bay (formerly to New Brunswick) and south to Montana (formerly to Maine and California). Winters from Aleutian Islands to Maine and south to Lower California, Texas, and Florida; also in Greenland, Hawaii, Cuba, Puerto Rico, and Bermuda. Known from Pleistocene in Oregon.

Nests in wooded country where there are lakes and streams. Nest in hollow tree or rarely in ground burrow, feather-lined. Eggs 6 to 14, olive-green to creamy white, 2 by 1½ in. Incubated by duck without help of drake. One annual brood. Young become mature in about 17 months. A summer molt in July to August.

Food procured by diving, usually in small companies. May overtake and eat small fish under water. Food varies with season and region occupied, from marine animals to freshwater plants; apparently feeds more on animal matter than on plants, particularly in winter. Flight swift and direct, with steady wingbeats.

Bird is not large enough or abundant enough to be important as a source of food; when it has been feeding in a marine environment, it is even more inferior because of a pronounced fishy taste. An attractive bird which should not be completely destroyed. Also known as dipper duck, spirit duck, dipper, dapper.

Old–squaw Duck

Old-squaw Duck, 154
Clangula hyemalis
Length to 23 in., including 10-in. tail and 1-in. bill. Wingspread to 30 in. Weight, drake to 2½ lb, duck to 1⅔ lb. Drake, in winter, head, neck, and belly, white; breast, back, and wings, black, with long pointed tail feathers; in summer, dark brown with white on flanks and belly and around eye. Duck without long tail feathers of drake.

Breeds along arctic coasts of both America and Eurasia, south to Strait of Belle Isle, Southampton Island, and shores of Hudson Bay over barren Canada to southern Yukon Territory and British Columbia. Winters along Atlantic Coast from Chesapeake Bay to Florida, and along Pacific from Aleutian Islands, sometimes to southern California.

Nests on ground, not far from water, in hollow sheltered by vegetation, lined with grass and small dark light-centered down feathers. Eggs 6 to 10, 2⅓ by 1⁹⁄₁₀ in.; gray-green to gray-olive. Nesting period May to June. Incubation 25 days by duck. One annual brood. Downy young blackish-brown above with creamy spot before eye and brown breast.

Food, small mollusks, crabs, shrimps, and probably some fish as well as seeds, shoots, buds, and fruits of water plants. Can dive to phenomenal depths (reported 180 ft) but can hardly thrive where diving to much over 30 ft is necessary. In flight, may rise in flocks in great circles almost out of sight and descend in zigzag courses.

Ordinarily flesh of these birds is too fishy to be relished as food even though they are shot as are other ducks. Properly prepared, they may be edible or even palatable but their numbers over much of their range are rarely sufficient for them to be considered important as a game duck species. Old Injun, old wife, scolder, cockawee are other names.

Female   Male   Common Eider

Common Eider, 160
Somateria mollissima
Length to 26 in. Wingspread to 42 in. Weight to 5 lb, 5 oz, with female smaller. Male has white back and black belly, with top of head gray, upper breast and throat white. Female brown, uniformly streaked around body with darker brown.

Breeds on coastal islands of Labrador, Newfoundland, eastern Quebec, Nova Scotia, and Maine and along shores of James Bay. Winters from Newfoundland and Gulf of St. Lawrence to Massachusetts and even to Virginia. Related king eider has black back and ranges through Arctic Canadian coasts to Siberia, western Europe, and even into Mediterranean.

Nests in communities, on rocky seashores, on ground, sheltered by rocks or plants, of mosses and sticks, with heavy down lining. Eggs three to seven, 3¹⁄₁₀ by 2 in.; pale olive-green. Incubation 26 days, by female. One annual brood, reared in June to July. Ducklings plain dark gray-brown but paler beneath. Mature plumage in third year.

Sluggish fliers, usually low over salt water and only rarely away from rocky coasts. King eider appears to be white in foreparts and black in rear parts and whole bird appears chunky. Food essentially mussels and similar marine animals gathered largely by shallow diving. Seaweeds may be eaten but not commonly.

Common Eider

(Continued)

Food value low, as only flesh of young is edible. Eggs eaten by natives but these are gathered only from nests with excessive numbers. Down is so valuable that much effort is expended in encouraging eiders to nest in a given area. Eider down was used in making sleeping bags, comforters, and arctic clothing because of lightness and insulating qualities. Extensive North American breeding populations destroyed by eiderdown hunters.

Orange
Gray
Common Scoter

1. Common Scoter, 163
Oidemia nigra americana

White-winged Scoter
Orange
Dark Gray
Male
Female

2. White-winged Scoter, 165
Melanitta fusca deglandi
(1) length 21½ in., wingspread 35 in.; duck smaller than drake. Plumage entirely black in drake with bright yellow base to bill. Duck dusky brown, with white patches on either side of head. Only American diving duck drake with entirely black plumage. (2) length 23 in., wingspread 41½ in. Blackish, with white wing patches.

(1) breeds around most of Arctic Circle, south to Newfoundland. Winters from Maine to Florida, Aleutians to southern California, Gulf Coast, and China. (2) breeds from Alaska to Ungava, south to Washington, North Dakota, and Gulf of St. Lawrence. Winters on coast, Gulf of St. Lawrence to Florida and Pribilof Islands to Lower California.

By May, American scoters arrive on breeding grounds. Nest well-concealed in grass, on banks, with down lining. Eggs 6 to 10; pale yellow, 2½ by 1¾ in. Incubation probably 4 weeks by female. One annual brood. Individuals molt at any time of year. Young dark brown, with white throat. Adults molt in fall.

Food largely mussels, clams, scallops, and similar shellfish taken by diving to depths of 40 ft. May fly inland, in long loose flocks, to feed on freshwater shellfish when sea is too rough. Excellent swimmer and diver; good flier but rather helpless on land. Generally less suspicious than most ducks.

Inferior as food because of animal diet, though if entrails are quickly removed and birds are cooked properly, they may be considered edible. Great flocks over shellfish beds may do some damage. 75 percent of food of white-winged scoters may be mussels. Difficult to kill and more difficult to get when wounded because of diving ability.

Surf Scoter

Order Anseriformes./Family Anatidae [Swans, Geese, and Ducks]./Subfamily Nyrocinae

Surf Scoter, 166
Melanitta perspicillata
Length to 21 in. Male: head and neck black except for white patch on forehead; long triangular patch on nape. Bill humpbacked in appearance. Body black except breast and sides and belly brownish-black. Both sexes with dusky yellow feet. Female dark brown with white patches on head. No subspecies.

Probably most abundant and widely distributed American scoter. Breeds from northwest Alaska eastward to Labrador, south to Gulf of St. Lawrence and west to northeastern Washington. Winters along the coasts from eastern Aleutians to Lower California, and the Gulf of California and from Nova Scotia to North Carolina, also in Great Lakes. Bird of large lakes, marine bays and shores, tundra late in summer. Fossil form in late Pleistocene of Oregon and California.

Nest in marshes, under brush, near water; a hollow lined with leaves, weeds, and down; 6 in. in diameter and 2½ in. deep. Eggs five to nine, usually seven; 2⅖ in. by 1¾ in., smooth, not glossy, pale pinkish or buffy white.

Its capacity to dive is equal to that of other sea ducks; dives in pursuit of food and to escape its enemies; stays submerged a long time and can swim long distance under water. Usually silent. May produce a throaty croak or a bubbling whistle during mating season. On the water they appear as large black, thickset ducks; white patches described earlier are good field marks. They fly in large flocks. Main northward flight in May; return south in September and October. Commonly called a coot by fishermen.

Food primarily animal matter: small mollusks predominate, also other invertebrates. Takes little plant food: mainly eelgrass, and some algae. Not a popular sport duck.

Order Anseriformes./Family Anatidae [Swans, Geese, and Ducks]./Subfamily Erismaturinae

Ruddy Duck, 167
Oxyura jamaicensis

Ruddy Duck   Blue Green

Length to 17 in., wing 5⁹/₁₀ in., wingspread to 24 in., tail 2¾ in. Weight: drake 1¾ lb, duck smaller. Breeding male with conspicuous white cheeks, black crown; mostly rusty red, with fanlike short tail and large blue bill; in winter, cheeks gray, rest of plumage brown. Duck like winter male, but with darker line across light cheeks.

Breeds from central British Columbia to northern Illinois and south to northern Lower California. Occasionally in New England, New York, and Guatemala. Winters from Massachusetts to West Indies along Atlantic and from British Columbia to Guatemala along Pacific and south of Illinois and Pennsylvania. Known from Pleistocene.

Drake has elaborate head-shaking, tailspreading courtship. Nest of reeds, in vegetation, over water, commonly basketlike and sometimes floating. Eggs five to fifteen, rough, gray-white to brownish; 2½ by 1⅘ in. Incubation by duck, about 30 days. One brood in North. Possibly two in South.

Ruddy Duck   Cinnamon

Food probably mostly aquatic plants, such as wild celery, pondweeds, duckweed, and rushes; also insects and other invertebrate animals. Southern migration begins in August and most flying is done by night. When swimming, may ride low in water.

Formerly, this was an important market bird in East when market hunting was possible. It was popular as "squab canvasback" but has been practically eliminated as a game bird in this area. It is still abundant along the Pacific Coast; in southern California, in small tule-bordered ponds or open marshes.

Order Falconiformes./Family Cathartidae [New World Vultures]

Turkey Vulture, 325
Cathartes aura

Length to 32 in., wingspread to 6 ft, tail 1 ft. Weight to 5 lb. Sexes about equal in size. Both show small red heads, but young birds have black heads. General appearance black, except that tail and rear margins of wings in flight show lighter. Legs and feet dirty gray. Shafts of tail feathers pale brown or yellowish.

Breeds from southern British Columbia through Wisconsin and southern New York to Connecticut and south to Gulf Coast and southern Lower California. Winters except in extreme northern part of nesting range. Related races found in Mexico, Central America, South America, and Cuba. Occasionally as far north as Newfoundland.

Nest in hollow log, stump, or on ground among brush and rocks. Eggs one to three, dull white with chocolate markings; 2⅘ by 2 in. Incubated for 30 days by both sexes. One brood a year. Young at first covered with white down; first plumage molted in August. Young apparently mature first winter.

Turkey Vulture

Food essentially carrion. Experiments prove that this species detects its food by a highly developed sense of smell. Birds roost together in good-sized flocks at night. When wounded, eject putrid stomach contents at enemy most effectively, or play dead. Speed, 21 mph.

Highly valuable as a scavenger, particularly in warmer parts of range, and for this gets deserved protection. In flying or soaring, commonly holds wings considerably above line horizontal with body. At a distance soaring turkey vulture appears more headless than birds with which it might be confused. Also known as buzzard, turkey buzzard, and carrion crow.

California Condor, 324
Gymnogyps californianus

Largest American vulture, length to 55 in., wingspread to 11 ft; weight to 26 lb. Much larger than turkey vulture. Black overall with white wing linings toward front of wings; conspicuous when bird is soaring. Head almost naked, distinctly yellow or orange when seen at close range. Bill orange colored with black band. Young without yellow, orange head; dusky-headed; also lack white wing lining. In soaring flight, wings are held in a straight line.

No subspecies recognized. Now confined in California from Monterey County to Los Angeles County and in the mountains south of the San Joaquin Valley. Very rare, perhaps 30 pair now in existence. Formerly occupied much wider range through Oregon, Nevada, Utah, Arizona. Bones recorded in prehistoric cave deposits in Nevada, New Mexico, and Texas.

Breeds at 5 years. Thought to breed once in 2 years. Builds no nest. Eggs laid on rocks in caves or other sheltered locations on cliffs. One greenish or bluish or dull white egg produced, 4½ by 2½ in. Incubation period about 30 days. Able to fly at 3½ to 5 months.

(Continued)

California Condor

An interesting bird which apparently bathes regularly in water. Can glide and soar to 2,000 or 3,000 ft in height. Man is the bird's worst enemy, gunning down individuals over the years. Also condors were killed by eating poisoned bait carcasses left by ranchers for predators. This large bird can probably not survive the onslaught of man for many more years, and its demise will go down as one more monument to man's ignorance of nature. Call described as a hiss.

Seldom attacks living animals. Food dead mammals and other warm-blooded forms. Not a predator in the true sense of the word; however, freshly killed food preferred.

Order Falconiformes./Family Accipitriidae [Eagles, Hawks, and Kites]

Sharp-shinned Hawk, 332
Accipiter striatus
Length to 12 in., wingspread to 23 in., tail 7 in. Female larger, with length to 14 in. Weight: male 3½ oz, female 8¼ oz. Larger and slimmer than sparrow hawk or pigeon hawk. Tail long, slender, and square-tipped, rather than rounded as in larger, longerwinged Cooper's hawk.

Breeds from northwestern Alaska to Newfoundland and south to northern Florida and southern British Columbia. Winters from southeastern Alaska to New Brunswick and south to Panama and Guatemala. Related races in Cuba, Haiti, and Puerto Rico. Known from Pleistocene of California. Essentially a woodland species.

Nest usually high in tree, of loose sticks, about size of crow's nest, about 6 in. deep, sometimes bark-lined. Eggs three to six, bluish-white to cream, spotted with chocolate, 1½ by 1⅕ in. Incubated 21 days, by female. Young downy, blind, helpless when hatched, assuming adult plumage in 2 years. One brood a year.

Food largely birds and insects. Over 1,000 stomachs examined showed 884 containing other birds, 16 poultry, 45 insects, and 28 mammals. Known to kill large numbers of grasshoppers and moths. Brave in attack, even on larger birds. Flight with quick wingbeats and alternate sailings. Call a repeated *quee;* like that of the Cooper's hawk, but with the quality of the sparrow hawk; shriller. Large female sharpshins frequently approach the size of small male Cooper's. The two are essentially identical in pattern.

Migrates during daylight hours; flies during early morning hours just above the treetops. Frequently seen soaring higher at midday. Incorrectly called pigeon hawk, sparrow hawk, bird hawk, chicken hawk.

Goshawk, 334
Accipiter gentilis
Length 22 in., wingspread 44½ in., tail 10 in., wing 13 in. Weight to 4 lb, female larger. Tail white-tipped. Wings with four to five dark bands below. Upper parts bluish-slate. In flight appears like a short-winged, long-tailed, blue-gray hawk with light gray breast; larger than a crow. Young brown instead of slate, breast streaked.

Woods or in open. Breeds from northwestern Alaska to northern Ungava and south to New Mexico and Pennsylvania. Winters in southern part of nesting range and south to Virginia, Texas, northern Mexico, and southern California. Also in Idaho, Arizona, Florida, Ireland, and England occasionally. In California Pleistocene.

Nest in larger trees; a huge pile of sticks. Eggs three to five, white or pale blue, 2⅓ by 1⅔ in. Incubation by female, about 28 days. Young at first blind, downy white, and helpless, eating animal matter from first. Assumes adult plumage by 1½ years but is not fully mature until autumn of third season. One brood a year.

Of 881 stomachs examined, 441 contained poultry and game, 233 mammals, largely mice, 49, other birds. Usually waits in some tree or thicket where it is inconspicuous and flies suddenly into open to attack prey, which is caught so quickly that frequently slower-flying, more conspicuous species get the blame. Active in daytime.

Probably most destructive of hawks. Goshawks are protected in many states and they seem to increase as the more valuable, more conspicuous species are killed off by ill-advised "vermin" campaigns. Also known as blue hawk, partridge hawk, chicken hawk, dove hawk.

Cooper's Hawk, 333
Accipiter cooperii
Length to 20 in., wingspread to 3 ft. Female much larger than male. About size of crow but more slender. Tail long, distinctly rounded, as contrasting with square-tipped tail of smaller sharp-shinned hawk. Wings stubby. Flight, alternate flapping and sailing. Average Cooper's hawk weighs twice as much as average female sharp-shinned hawk.

Square Tail

Sharp-shinned Hawk

Goshawk

Rounded Narrow Tail

Cooper's Hawk

Breeds from southern British Columbia to southern Quebec and south to northern Mexico. Winters from southwestern British Columbia and Washington to California, New York, and southern Maine to Costa Rica. Active in daytime, following migrations of smaller birds and taking its toll of them.

Nests usually in tall tree. In country where trees are low, may nest on ground. Nest a loose pile of coarse sticks. Eggs three to six, pale bluish-white, often brown-spotted, 1½ by 1⁹⁄₁₀ in. Incubation by female, 24 days. Young blind, downy, and helpless when hatched but soon covered with white down; molts similar to those of sharp-shinned hawk.

Food largely birds. One of worst enemies of small birds and poultry. Of 422 stomachs examined, 78 contained game birds and poultry, 146 other birds, 65 mammals, 6 other animals. Powerful enough to kill ducks, rabbits, and grouse. Call, a *cuck, cuck*. If any hawk deserves the title of chicken hawk, Cooper's may.

Probably serves a role in eliminating weak and wounded birds and mammals.

Red-tailed Hawk, 337
Buteo jamaicensis

Length to 2 ft, wingspread to 56 in. Weight to 4 lb. Heavy-appearing with broad wings. Tail of adult red above and little rounded. Dusky gray above and with lighter wings; white beneath. Band across belly brown-streaked. Appears less uniformly colored from beneath than red-shouldered hawk. Male smaller than female.

Breeds from southeastern Canada through United States to northeastern Mexico in the plains, merging with western species which ranges through United States and north to tree line in Canada. Winters from Kansas to New York and south through eastern Mexico; occasionally in England. Known from Pleistocene of Florida and California.

Courts by soaring and calling, beginning in February. Mates for life. Nest of coarse sticks thrown together, 30 to 40 ft above ground in tree. Eggs two to four, dull white, irregularly marked with cinnamon brown, 1⁹⁄₁₀ by 1½ in. Incubation 28 days, by female. Young blind, downy, helpless when hatched. One annual brood.

Food mainly mice, rabbits, insects, and snakes. Individuals may develop habit of killing chickens when other food is not available. Poultry and game constitute probably not over 1 percent of total food. A most valuable check on mouse infestations; name "hen hawk" a most unfortunate and inappropriate misnomer. Temperature, 106°F. Speed 22 mph.

Essentially a useful species. Protected by law in many states and should be in all. A shy hawk but has bad habit of perching in a conspicuous place where it may be shot easily. In flight, shows conspicuous white area on breast, with streaks to rear, and unbanded tail.

Broad Plain Tail

Red-tailed Hawk

Red-shouldered Hawk, 339
Buteo lineatus

Length 23 in., wingspread 44 in., tail 9⅔ in. Female larger than male. Weight to 3 lb. Gray-brown, with reddish-brown underparts. Reddish shoulder feathers and tail, with four to five crossbars. Appears uniform from beneath but shows a white spot just behind a dark streak, near wing tips, from above.

Breeds from Nova Scotia to Prince Edward Island and south almost to Gulf of Mexico, with other close relatives extending range into Mexico. Winters in this range and on into Mexico. Usually seen soaring high and slowly, or on some conspicuous high perch near open.

Nest a bulky pile of sticks, may be used year after year, in tall elm, birch, maple, or beech. Eggs three to five, dingy or bluish-white with brown spots, 2⅛ by 1⅔ in. Incubated by both sexes, 27 to 28 days. Young blind, downy, and helpless when hatched but assume adult plumage by 18 months. Young have less brilliant tail bars.

Food about 65 percent mice and other small ground animals; rarely poultry, though this is the so-called "hen-hawk" so commonly killed as vermin. Of 444 stomachs examined, 287 contained mammals, 7 poultry or game, and 25 other birds. When mice and gophers are abundant, a considerable surplus may be piled around nest with young.

Possibly most useful mouse and rodent destroyer of all hawks, therefore one of most useful species to man though one of the most tormented. Numbers of hawks increase ordinarily with numbers of mice and gophers unless they are shot off. This species is protected by laws in many states and should be everywhere.

Broad Banded Tail

Red-shouldered Hawk, 339

Broad–winged Hawk

Broad-winged Hawk, 343
Buteo platypterus

Male, length 16½ in., wingspread 38 in. Weight 13 oz. Female, length 18¾ in., wingspread 39 in. Weight 19 oz. Tail with white band about as wide as black, but in young birds, dark bands are more numerous and may obliterate white. Tail stubbier and wings shorter than in red-shouldered hawk. Rather tame. Adult easily recognized by broadly barred tail. Immature form has buteo shape with white underwing surface contrasting with black tips on primaries. Broad-winged lacks the belly band of the much larger red-tail.

Breeds from central Alberta to New Brunswick, south to Gulf Coast, mainly east of Mississippi. Winters from southern Florida and southern Mexico through Central America to Colombia, Venezuela, and Peru; also recorded from Cuba, Haiti, and Puerto Rico. Reported from Pleistocene of Florida.

Nests usually in wooded areas in a tree, 20 to 60 ft above ground. Nest well-built, lined, and hollowed. Eggs two to four, variable; white, bluish or greenish and brown-spotted or plain; 2 by 1⅗ in. Incubated 23 to 25 days, by both parents. Young blind, downy, and helpless when hatched, yellowish-white. Adult plumage in about 18 months.

Food chiefly small mammals and reptiles. Birds in diet, most uncommon. Stomachs of 65 showed 30 with insects, 15 with mice, 13 with other mammals, 13 with frogs and toads, 11 with reptiles, 7 empty, 4 with crayfish, 2 with earthworms, and 2 with small birds. Soars very little, is unsuspicious, and perches conspicuously.

Undoubtedly a useful species to man. Protected by law in many states, and should be universally protected as a mouse destroyer.

Rough–legged Hawk

Rough-legged Hawk, 347a
Buteo lagopus

Male, length 22 in., wingspread 52 in. Female, length 23½ in., wingspread 56 in. Weight 33 oz. Seen from below, shows conspicuous black belly and conspicuous black wrist patches on wings. May show white at base of tail like marsh hawk but rough-legged hawk is a heavy, more round-tailed bird. Feathers of legs extend to the toes.

Breeds from Aleutian Islands to Labrador to northern Alberta, Gulf of St. Lawrence, and Newfoundland. Winters from British Columbia to southern Ontario, along northern United States border south to southern California, Texas, Louisiana, and North Carolina. Close relative to west, and from California Pleistocene.

Nests on banks near waterways. Nest to 3 ft across, of old sticks and rotten wood, lined with grass and feathers. Eggs two to five, variable in size and color; usually unmarked white, yellow-brown, purple, lavender, dark brown, or blotched; 2½ by 1⁹⁄₁₀ in. Incubation about 4 weeks. One annual brood.

Food chiefly field mice and lemmings. Stomachs of 49 showed 40 containing mice, 5 other mammals, 1 lizard, 1 insects, and remainder empty. Winter hawk which "hovers" over fields and often appears largely black. Its great size makes it seem to amateurs a dangerous bird.

One of best hawks, often and easily shot while hovering by those engaged in ill-advised vermin campaigns, particularly in winter months. One of best destroyers of mice in an open winter when cover is not available. Protected by law in some states and should be universally.

Golden Eagle

Golden Eagle, 349
Aquila chrysaetos

Male, length to 35 in., wingspread to 7 ft, tail 15 in. Female, length 41 in., wingspread 92 in., tail 16 in. Weight to 14 lb 12 oz. Female larger than male. Dark-brown to black, with chocolate-brown or yellow on heads of adults. In flight, shows wide-spreading upcurving wing tips.

Breeds from Northern Alaska south to northern Lower California, east through Rocky Mountains and formerly to North Carolina. Breeding area formerly included New England States; now in the eastern states in the Appalachian highlands. Winters south to Texas and Florida. From Oregon and California Pleistocene.

Courting sexes engage in profuse tumbling in air. Nests in forests or open. Nest a huge stick pile, to 7 ft across and 5 ft high. May build two nests. Eggs 1 to 4, dirty white or spotted with brown, 2⁹⁄₁₀ by 3 in. Incubation about 30 to 43 days. One annual brood. Young helpless and down-covered when hatched, attaining adult plumage in about 18 months.

Food largely small mammals, particularly jack rabbits. Unable to carry a weight in excess of 12 lb, although it can drag a larger weight a short distance. White at base of tail and wing feathers to 5 years of age.

Most useful destroyer of gophers and jack rabbits, which are arch enemies of grazing animals because of competition they offer. Probably much more useful than injurious. Can be trained to falconry most effectively. An endangered species. Protected by law in most states and should be in all.

Bald Eagle, 352a
Haliaetus leucocephalus
Male length 34 in., wingspread 85 in. Female, length 37 in., wingspread, 90 in. Weight to 11½ lb. Female larger than male. Young often larger than adults; nearly black, without white head, neck, and tail characteristic of adult. Conspicuously white and brownish-black.

Range of two subspecies. Breeds through Alaska to Ungava and south to British Columbia and Great Lakes. Winters south to Washington and Connecticut. Southern subspecies winters and breeds through United States, to southern Lower California and central Mexico. From Pleistocene.

Courting male loops in air before female. Nest a huge pile of sticks, added to each year, in tall tree or on high ledge. Eggs two, rarely white, with pale brown and cinnamon spots; 2⅛ by 1⅙ in. Incubation by both parents, for 34 to 35 days. Young blind, downy, and helpless when hatched. One annual brood.

Food animal matter, often carrion and old fish. Stomachs of 80 showed 35 with wild vertebrates, 12 with poultry. In some ways a scavenger. Unable to lift anything weighing over 15 lb. Soars majestically, high in air. Young get white feathers on head and tail in fourth or fifth year, when they mature.

This is our national bird. As such, and because of generally harmless nature, is protected by national law and specifically in many states as well. Individuals may uncommonly develop habit of killing sheep and young pigs but most reports of depredations are probably false. Provides a thrilling sight in flight. Now, as a species, is in danger because of excessive DDT intake and loss of capacity to produce hard shells for eggs.

Bald Eagle

Marsh Hawk, Harrier, 331
Circus cyaneus
Male, length 20 in., wingspread 18 in. Weight 1 lb. Female, length 25 in., wingspread 30 in. Weight 21 oz. Adult male, light blue-gray back with black wing tips, conspicuous white rump, high wing angle, and long tail and wings. Adult female brown above; streaked with brown beneath; immature birds buffy below.

Breeds from northeastern Siberia to Prince Edward Island, south to northern Lower California, southern British Columbia to southern New England and south to Florida, Cuba, and Colombia; also in Hawaii and the Barbados. Known from Oregon and California Pleistocene.

Male courts by soaring high, followed by tumble and screech. Nest on ground, in marshes and made of grasses. Eggs four to six, dull bluish-white, 1¾ by 1⅖ in. Incubation by both parents, begins before full set is laid, normally 26 to 31 days. One yearly brood. Young blind, helpless when hatched; have mature plumage by 2 to 3 years.

Food, animal matter. Stomachs of 418 showed 250 contained mammals, 176 other birds, 11 insects, 10 poultry or game. This is "marsh harrier" which flies, alternately flapping wings and soaring back and forth, relatively close to ground. Call, a screeching *cac, cac, cac*, in breeding area, March to November.

It is rarely injurious to poultry; commonly a great destroyer of mice, but at all times an interesting bird to watch. Protected by law in many states. Conspicuous enough with white rump to be recognized as being different from other hawks. Georgia quail study claims marsh hawk best benefactor of quail because of destruction of cotton rats that destroy quail nests.

Marsh Hawk

Osprey, Fish Hawk, 364
Pandion baliaétus
Length to 24½ in., wingspread 66 in. Female larger than male, but less difference than in most hawks. Weight: male to 3 lb, 4 oz. female to 4 lb, 10 oz. Dark brown above, sometimes with white on top of head but not on back of neck; usually white below. Wings appear bent in flight. Hovers and plunges, striking with feet.

Breeds from northwestern Alaska to Labrador and Newfoundland, south to Lower California and Florida Keys. Winters from Florida and Gulf States through Lower California, Mexico, Central America and in Peru, Chile, Paraguay, Greenland, Europe, Asia, Australia through closely allied species. Known from Florida Pleistocene.

Nests year after year in same site, adding to old nest on pole, tree, or ground to make huge pile of coarse sticks. Seaweeds and any available rubbish may be used in making nest. Eggs two to four, dull white or buff with chocolate-brown markings, 1⅘ by 2⅖ in. Nest defended by both parents. Incubation chiefly by female, 28 days. Young blind and helpless when hatched.

Food, fish only; never takes birds, and so is desirable about poultry yards as it drives away species which might do so. Catches fish by spectacular plunge and may be robbed of its meal by more powerful eagles. Sometimes drowns when it becomes attached to fish too large to be lifted.
(Continued)

Osprey

Individuals cooperate against enemies.

Entirely useful to poultrymen and of little damage to interests of fishermen. Guides fishermen to good fishing grounds and so helps them. Few records of osprey doing any damage to game. Protected by law in at least 28 states and worthy of further protection. A thing of beauty over any large body of water. May be victim of excessive intake of DDT and subsequent inability to produce hard shell for eggs. An endangered species.

Order Falconiformes./Family Falconidae [Caracaras, Falcons]

Peregrine Falcon, 356a
Falco peregrinus
Male, length to 18 in., wingspread 43 in. Female, length to 20 in., wingspread 46 in. Female larger than male, blue-gray above, barred below. Tail with six or more narrow black bars, almost pointed. Lower half of head conspicuously black and white. Male with browner face and more conspicuous tail bars. Conspicuous contrasts of black and white on face.

Breeds from Alaska to central Greenland, and south to central Lower California, Texas, Tennessee, Pennsylvania, and Connecticut. Winters from Vancouver Island to southern New York and south to California, Panama, and most of South America. Once a common breeding bird in eastern North America. Now rare. Occasionally in England and South America. Known from Pleistocene of California.

Mates as early as March. Nest on a high ledge or in a tall hollow tree. Eggs three to four buff to reddish-brown, $1\frac{2}{3}$ by $2\frac{1}{10}$ in. Incubation by both sexes, 28 days. Young white, downy, and helpless when hatched. Adult plumage assumed gradually, after first winter when juvenile plumage is molted. Falcon life-span to 162(?) years.

Food, animal matter. Of 102 stomachs, 70 contained other birds, 11 poultry, 12 insects. All caught by pursuit in truly phenomenal manner which wins admiration by most bird lovers. Prey caught by dropping from great height with incredible speed or by outflying or outdodging. Flight speed, 80(?) mph.

Food habits make it unpopular with many sportsmen but very popular with those who follow falconry. Male is called a "tercel," or "thirdling" implying that third egg produces a male. Other common names include American peregrine, great-footed hawk, wandering falcon. An endangered species; in part from excessive intake of DDT, females unable to produce hard shell for egg. Protected by law in several states. Temperature, 106°F.

Pigeon Hawk, Merlin 357
Falco columbarius
Male, length $10\frac{1}{2}$ in., wingspread 26 in. Female, length $13\frac{1}{2}$ in., wingspread $26\frac{1}{2}$ in., tail $5\frac{1}{2}$ in. Weight: male, 6 oz; female, 8 oz. Like a small duck hawk but less contrastingly colored on head. Wing long, pointed, differing in this from more blunt-winged sharp-shinned hawk.

Breeds from tree limit in Canada to Alaska, Labrador, and south to Maine and northern California. Winters from British Columbia to Gulf States and south to Ecuador and northern Venezuela. Four subspecies include Eastern, Black, Richardson's, and Western. Known from Pleistocene of California.

Nest of sticks, weed stems, grass, or moss lined with feathers and strips of soft inner bark, often in evergreen, on ledges in hollow trees. 6 to 40 ft above ground. Eggs four to five, white with chocolate-brown markings; $1\frac{1}{5}$ by $1\frac{3}{5}$ in. Incubation 28 to 32 days. Young downy and helpless when hatched. Develop approximately adult plumage, only browner by first winter; reach true adulthood second winter.

Food chiefly small birds. Of 184 stomachs examined, 141 contained small birds, 68 insects, 8 mammals, 3 poultry or game. Call, a *cac, kee,* or *wheet.* Male is blue above; female and young are brown above. This is an exceptionally swift-flying hawk, active in daytime and a roaming migrant.

Probably most useful bird in keeping down English sparrows; not large enough to do serious damage to most poultry. This is American representative of European merlin. It is protected in most states. Has other interesting common names: pigeon falcon, American merlin, bullet hawk, little blue corporal.

Sparrow Hawk, Kestrel, 360
Falco sparverius
Male, length $10\frac{3}{5}$ in., wingspread 22 in. Female, length 12 in., wingspread $24\frac{1}{2}$ in. Weight 4 oz. Female usually larger than male. Crown ash-blue with conspicuous black patch on side of head. Back chestnut or cinnamon and black-barred. Tail bright red-brown, with narrow black bars. Only small hovering hawk.

Peregrine Falcon

Pigeon Hawk, Merlin

Sparrow Hawk, Kestrel

Four subspecies include Eastern, Desert, San Lucas, and Little sparrow hawks. Breeds from upper Yukon to Newfoundland and south to California and Lower California and Florida. Winters in southern part of range and on to Panama and Mexico. From Florida and California Pleistocene.

Courting male, hovers over mate and dives. Nests in birdhouses or hollow trees lined with chips; some may be bare; also in deserted woodpecker holes, in rock cavities, or holes in banks. Eggs three to seven; buff to brownish, with finely shaded areas; 1⅖ by 1⅛ in. Incubation by both sexes, 29 to 30 days. Young blind, downy, and helpless when hatched, attain adult plumage at about 18 months. Perhaps mates for life.

Food principally mice and insects. Of 427 stomachs examined, 147 contained mammals, 69 birds, mostly sparrows, 269 insects, 29 spiders, 13 reptiles or amphibia, and 29 were empty. One contained remains of a bobwhite which may have been a wounded bird. Young birds remain in nest about 3 weeks. Speed, 25 mph.

Undoubtedly a useful species which destroys large numbers of mice and insects. Makes an excellent pet but a license to keep one captive is necessary in many states. Protected by law in several states. Not vermin in any sense. American representative of the European kestrel. Temperature, 107 to 108°F. A bird with several common names: American kestrel, rusty-crowned falcon, grasshopper hawk, mouse hawk, kitty hawk, windhover, short-winged hawk.

Order Galliformes./Family Tetraonidae [Grouse, Ptarmigans]

Ruffed Grouse, 300
Bonasa umbellus
Length to 19 in., wingspread to 25 in., tail to 7 in. Weight to 29 oz. Upper parts buff to mahogany spotted with gray; under parts buff, lightly barred on breast, heavily barred on flanks. Tail with conspicuous black band near tip. Conspicuous ruff on sides of neck, larger in male. Crest on head. Compact body. Drumming sound, 40 cps.

Six subspecies recognized covering territory from wooded Alaska through to Nova Scotia and south to Georgia and northern California. Eastern subspecies found from Minnesota to Massachusetts and south to Kansas and northern Georgia. Known from Pleistocene of California, Tennessee, Maryland, and Pennsylvania.

Cock courts by strutting and by vibrating wings producing drumming while on log. One cock mates with a number of females. Nest a depression on ground, sheltered by tree, log, or brush. Constructed of old leaves, a few feathers, weed stems, grass, and roots. Eggs pale brown, 8 to 14, may be whitish lightly speckled with brown spots; 1½ by 1⅛ in. Incubation 21 to 23 days, by hen. Young able to run when hatched. One annual brood.

Food, acorns, beechnuts, green leaves, buds, and fruits. Young protected but not fed by hen. Subject to many diseases of poultry. Until practice of raising on wire was established, grouse could not be reared in captivity. Broods remain together into winter, ranging over 40 acres of land. Flight 35 ft sec, 22 mph.

Possibly finest of upland game birds. Though it lacks size of turkey, yet it can supply superior sport and excellent flesh. Where undisturbed, may be unsuspicious, but where hunted is shy and alert. Suggested peak populations every 10½ years. Eventually may be raised for freeing in woods by government. State bird of Pennsylvania. Also known as shoulder-knot grouse, partridge, drumming grouse, birch partridge, drumming pheasant, and mountain pheasant.

Ruffed Grouse

Spruce Grouse, 298c
Canachites canadensis
Length to 17 in., tail to 5 in., folded wing to 7 in. Legs feathered to toes; lower tarsus bare in ruffed grouse. Above, barred black, gray, and brown. Tail black with reddish tip. Throat and breast black. Underparts black feathers tipped with white. Skin above eye red. Hen paler; underparts barred.

Four subspecies recognized include Hudsonian, Canada, Alaska, and Valdez spruce grouse. Range from Labrador to Yukon region in Alaska, and south in Rocky Mountains to Edmonton, Alberta, and into New York and formerly in Massachusetts.

Males strut and drum. Breed chiefly in swampy, evergreen forested areas. Nest on ground, in depression at foot of tree or under bush. Probably well concealed under low conifer branches constructed of dry twigs and leaves and lined with moss and grass. Eggs 8 to 16, variably buff with brown spots or even purplish with brown spots; 1⅔ by 1⅕ in. Incubation for about 24 days, by hen. Young precocial. One annual brood.

Spruce Grouse

(Continued)

Food in summer, largely small fruits and insects, including crickets and grasshoppers; in winter dried fruits, tree buds, shoots of spruce, larch, and fir. Seems rather stupid and has won name of fool hen because of ease with which it may be approached and shot.

Of some economic importance as a destroyer of insects. Probably worthy of much more protection than it has received. Flesh unpalatable in winter when food has been buds of evergreens and larches. Has many other common names, including Canada grouse, Husdon spruce grouse, black grouse, wood grouse, wood partridge, spotted grouse, cedar partridge, and swamp partridge.

Ptarmigan

Willow Ptarmigan, 301

Lagopus lagopus

Length to 17 in., including 5½-in. tail and ½-in. bill. Like a short-tailed ruffed grouse. In summer, male dark brown, heavily barred with light brown and rusty spots; tail black with narrow white tip; throat, breast, and sides, rich brown; lower abdomen and wings almost white; female light brown and gray. In winter, both sexes white with black tails.

Breeds from Banks Island and west coast of Greenland to eastern Aleutian Islands and south to central Mackenzie, northern Quebec, and northern British Columbia. Winters south to central Saskatchewan, southern Alberta, central Ontario, and southern Quebec, and sometimes into North Dakota, Montana, Wisconsin, Michigan, New York, Maine, and Massachusetts.

Nests on arctic tundra or in open areas near timber line. Nest a hollow in ground, usually lined with a few feathers, leaves, or bits of grass. Eggs 6 to 15, or rarely 20; 1⅓ by 1⅘ in.; deep reddish-cream color, with blotches of purple-brown which sometimes join. Nesting period early June to mid-July. Incubation 26 days, by female. One annual brood. Male helps guard nest and young.

Food, leaves of birch, blueberry, vetch, rosemary; and in winter buds of birch, willow, alder, poplar, mountain ash, and other trees. Insects, spiders, and other small invertebrate animals occasionally eaten. Most remarkable is color change from winter white to summer brown with intermediate patchy coloration matching seasonal snow patches. Frequently used in introductory biology texts as an example of protective coloration.

In the north country, ptarmigans frequently appear in large numbers and are there killed for food by natives. In this way, they provide a basic food for man in their range. Also serve as basic food for some furbearers and for some birds of prey. Also known as ptarmigan, common ptarmigan, willow grouse, white grouse, snow grouse.

Greater Prairie Chicken, 305

Tympanuchus cupide

Length to 18½ in., tail 4½ in., wing to 9 in. Weight to 2 lb. Sexes barred brown, about equal in weight but cock with long tufts of feathers on sides of neck behind a distensible orange-yellow area which is inflated during courtship. Cock also has erectable bare spots over eyes. Temperature about 109°F.

Resident from west central Alberta, southeastern Saskatchewan, and southern Manitoba to eastern Colorado, Arkansas, and formerly on to Ohio and including southern Ontario, Pennsylvania, and Kentucky. Range rapidly decreasing; now extinct east of Indiana. Related fossil species from New Jersey and Oregon Pleistocene.

Greater
Prairie
Chicken

Male courts many hens by booming and strutting and fights other males. Boom may be heard for 2 miles. Within an hour or two after sunrise, the males return to normal behavior. Nest on ground, sheltered by grass. Eggs 10 to 14, buff, sometimes speckled; 1⅗ by 1¼ in. Incubation for 23 days, by hen. Young, in Minnesota, hatch in June and able to run quickly. Young molt in fall. One annual brood.

Food, animal matter such as insects, 14 percent; vegetable matter such as grain, acorns, fruits, and berries, 86 percent. In winter, hens may flock together and move to southern part of range. Species extended its range with cultivation of grain but has been greatly reduced by excessive hunting practices.

Native game bird of prairie country which should never have been allowed to decrease in numbers. With persecution, has taken to woodlands and is now commonly found there. Periodic abundance and scarcity as with ruffed grouse. A splendid game bird. Apparently some Indian dances mimicked the booming ground activity of the male prairie chicken. Also known as prairie hen, pinnated grouse, and prairie grouse. An endangered species.

Sharp-tailed Grouse, 308
Pedioecetes phasianellus
Length to 17½ in., including tail about 6 in. long and sharply pointed, middle tail feathers being about 1 in. longer than those on sides. Upper parts brown, marked and barred heavily with black. Throat light brown. Breast marked with black V's. Sides of body irregularly black and brown spotted. Abdomen white. Female smaller than male. Sharp-tailed grouse is commonly confused with prairie chicken, but most hunters recognize sharp-tailed by its pointed tail and the lack of neck tufts.

Sharp-tailed Grouse

Breeds from central Alaska through northern Manitoba and northern Quebec to Lake Superior, Ontario, and the Saguenay River in Quebec. Irregular migrant. Two other subspecies, the Columbian and the Prairie sharp-tailed grouse, extend range through to New Mexico, California, Nebraska, and Illinois. Essentially a resident of sandhill type of country.

Male courts female by inflating lavender sacs near head, dancing, rattling tail feathers, and giving a bubbling sort of crow. Nests on ground with little or no modification. Eggs 11 to 14, creamy brown or pale olive, sometimes spotted finely with reddish-brown; 1⅗ by 1⅕ in.

Food, grains such as wheat, buckwheat, oats, sunflower seeds, and hemp seeds; buds of tamarack, catkins of birch, and twigs of hazel, with some green vegetation. Gravel must be available. A number of serious intestinal parasites, important among these being *Blastomyces*, often found in intestines of dead birds.

A useful game bird but possibly doomed to extinction with introduction of the more hardy pheasants and advance of civilization; has shown some signs of increase in northern Michigan. A definite cycle of abundance and scarcity and some migration and change of range through the years. Also known as spike-tail, pin-tail, pin-tail grouse, spring-tailed grouse, and sharp-tail.

Sage Grouse, 309
Centrocercus urophasianus
Largest N. American grouse; male 26 to 30 in., female to 23 in., grouse with long pointed tail feathers. Male 6 to 8 lb. noticeably larger than female, 3 to 5 lb. Male with black belly, black throat with a white collar. White breast has longer feathers. Female without black throat. Rest of plumage mottled or barred with black. Small bare-skinned patch above eye is yellow. In display, male inflates pair of air sacs on upper breast into bright yellow balloon-like pouches and spreads spiked tail feathers in an erect position. Hen mottled and barred in browns, black, and white overall.

Sage Grouse

Two subspecies recognized. *C. u. urophasianus* is a resident locally from southern Idaho, eastern Montana, southeastern Alberta, southern Saskatchewan, and western North Dakota, south to eastern California, south central Nevada, Utah, western Colorado, and northwestern Nebraska.

C. u. phaios is a resident from central and eastern Washington south to southeastern Oregon.

Much has been written about the courtship of the sage grouse, which is the most spectacular of the grouses. Nest on the ground in a slight depression; usually under a shrub; sometimes lined with grasses or twigs. Eggs seven to nine; may be as many as fifteen; 2⅕ by 1½ in.; grayish or drab greenish, thickly spotted and dotted with red-brown. Incubation about 22 days. One annual brood.

A gregarious species, sometimes found in flocks of several hundred. Found in areas of sagebrush cover. Feeding and nesting requirements closely related to sage. So, with no sagebrush in a considerable area probably no sage hen. It apparently "stays put"; no seasonal movements.

Food, tender, succulent vegetation and some insects. Adults fond of leaves of sage. Probably American Indians used male sage grouse mating displays as format for their dances. Male grouse displays on "booming grounds" in early morning. Not a favorite game species.

Order Galliformes./Family Perdicidae [Partridge, Quails]

Hungarian Partridge, 288.1
Perdix perdix
Length to 14 in., individual wing to 6½ in., tail to 3½ in. Weight to 15 oz. Male slightly the heavier. Appears in field intermediate between grouse and bobwhite. Underparts gray, with large spot of reddish-brown in adults; speckled with buffy brown in young birds. In flight, this rotund grayish partridge shows a short rufous tail. Found locally in open farming country. Brown unmarked head and gray breast distinguish it from bobwhite, the only other game bird of comparable size.

Hungarian Partridge

Native of southern Sweden, Germany, and the British Isles and area into France, Switzerland, and the Pyrenees. Introduced into North America and acclimated in Saskatchewan, Alberta, British Columbia, Wisconsin, and Washington. Has not remained abundant elsewhere. Related species in Spain, Italy, Russia, and Siberia.

(Continued)

Bobwhite

Male fights viciously during breeding season, does not assist in incubating eggs but does help rear young. Nest hidden on ground among grasses. Eggs, 6 to 18; plain olive or blue; weight, 26 g. Incubation 24 days. Young able to run soon after hatching and begin feeding on insects. Young molts to adult plumage in late fall.

Food, estimated 40.5 percent insects including 23 percent harmful species; 50 percent vegetable material other than grain; and only 3.5 percent grain; the rest miscellaneous material. Introduced as game bird, rather than as insect destroyer, though its greater usefulness may be as insect check. Flight speed, 53 mph.

In Europe, is shot by thousands by so-called sportsmen who sit in comfortable chairs and have birds driven over barriers which force them to fly. This fortunately does not fit American conception of sportsmanship. May become established in America to supplement vanishing prairie chickens and quail, as buffer species. White Leghorns might also be suggested.

Bobwhite, 289
Colinus virginianus
Length to 11 in., individual wing to 5 in. Weight to 9 oz, commonly 6 oz. Lower breast with blackish bars or U-shaped marks. Tail short. Conspicuous white throat and band across side of head, through eye, which is buffy in hen. Black on head of cock is brown in hen.

Resident. Four subspecies include Eastern, Florida, Key West, and Texas bobwhite. Range of the four covers most of United States. Eastern form found from Dakota through southern Ontario to Florida and eastern Colorado. Known from Florida and Tennessee Pleistocene.

May keep mates more than one year though cocks fight at beginning of breeding season. Cock builds nest on ground, though several pairs may use same nest. Eggs 7 to 28, white, about 1 by 1⅕ in. Incubation by hen and by cock, 23 to 24 days. Average hatch, 86 percent. Young leave nest on hatching, reared by both parents. Will adopt chicks from other pairs. Will nest a second time in a season if first nest is destroyed.

Food largely injurious insects such as grasshoppers, June beetles, potato beetles, chinch bugs and squash bugs during summer; weed seeds, in winter. Also will eat spiders, snails, centipedes, sowbugs, and other invertebrates. Winter coveys of fewer than 10 unlikely to withstand severe cold and wind. Individual range, 1 sq mi. Flight speed, 85 ft/sec, 51 mph.

In some states considered a songbird. In some, annual take of an area is determined by preliminary survey and adequate breeding stock for winter survival is left. Too useful alive to be taken as game except under most intelligent management of surplus birds. Family remains together first winter. State bird of Oklahoma and Rhode Island. *Bob*, 1,645 cps; *white*, to 2,742 cps.

California Quail

California Quail, 294
Lophortyx californicus
Length to 11 in. Cock with black recurved topknot; upper parts smoky or dusky brown; top of head with chestnut patch bordered on front and sides by black and white lines; throat, black with white border; breast, blue-gray; flanks, olive with white streaks. Hen without black and white markings on head.

Resident in humid Pacific Coast area, from southwestern Oregon to west central California. Introduced in Vancouver Island and in state of Washington. Five subspecies include California, Valley, Catalina, San Quentin, and San Lucas quails, most of them with a limited range compared to the California quail.

Nests on ground, in a depression, with vegetable border. Eggs to 12 or more; white or buff, spotted or marked with lavender and brown. Chicks able to run soon after hatching, and family including both parents remains together until after young are well-developed.

Food, insects and other small forms of animal life, particularly during wetter parts of year, but this diet may be 90 percent grain and other plant materials though bird could not be considered a destroyer of useful plants as far as man's interests are concerned. Flight speed, 80 ft/sec, as contrasted with 60 ft for Gambel's quail.

Considered as a game bird. Makes an excellent pet and is the state bird of California. Its food habits are such that it is more valuable alive than dead. Also known as valley quail and California partridge.

Gambel's Quail

Gambel's Quail, 295
Lophortyx gambelii
Length 10 in., including 4⅔-in. tail. Cock with black 1½-in. recurved plume on top of head and with black patch on buff belly; flanks brown with white streaks, crown reddish-brown, forehead and throat black with white border. Hen with 1-in. crest, and belly uniformly buffy without black patch.

Resident in hot valleys and bottomlands of southern California and Nevada, to Arizona and southwestern Utah with stations in Texas and Mexico, the latter in northeastern corner of Lower California and on to Guaymas, Sonora. The quail of dry desert areas as contrasted with California quail that favors humid areas. Two subspecies.

Nests on ground, in a depression that is lined or not. Eggs 10 to 12, white or buffy brown, with purple or brown spots. Chicks able to run as soon as they are hatched and covey stays together until young are grown or even longer. Gambel's quail is frequently found with the scaled quail, *Callipepla squamata*, with which it may occasionally cross.

Food, January to June, almost wholly vegetable matter, including mistletoe berries, mesquite leaves, cactus fruits, and seeds of various plants. Summer food essentially insects, principally grasshoppers, the young feeding almost wholly on insects such as ants, leafhoppers, stink bugs, and aphids. Flight speed to 51 mph.

Protected by law and by coloration and by ability to escape, the Gambel's quail being a slower flier than the California quail. Large broods get good parental care during critical periods. Enemies are snakes, coyotes and, most of all, man. Locally known as redhead.

Mountain Quail, 292
Oreortyx pictus
Length to 11½ in., small, like Gambel's quail, but with grayish-blue crest. Head, shoulders, and lower breast, grayish-blue; throat and upper breast brownish-red. One erect mobile head plume. White line evident on neck. Face white and space above eye is white. Flanks are brownish-red; edged at upper end in white. Back, tail, and wings brownish. Female, smaller and duller but similar to male.

Mountain Quail

Five subspecies listed. Resident of the coast ranges from southern Vancouver Island, British Columbia to north Baja California, and in the Sierra Nevada and other ranges of California and eastern Nevada. Introduced in southeastern Washington, eastern Oregon, western Idaho, and central Nevada. Found commonly in mixed forests and scrub growth; limited almost exclusively to mountains and most abundant in more humid areas. Generally does not move from its local area.

Nests, in favored habitat, under cover of a log or vegetation, protected; generally a simple depression in the ground lined with pine needles, grasses, and leaves, plus a few feathers. Eggs 8 to 22 but commonly 10 to 12; unmarked, yellowish to pinkish, 1⅓ in. by 1¹⁄₁₀ in.

Probably migrates vertically. A bird of open spaces or bushy cover on the sides of mountains. Produces a single, loud call, described as a resonant *kyark* repeated at short intervals in early spring and summer. Animal food consists largely of larvae and adult forms of various insects. Seeds of various weeds, woody plants, and cultivated plants are eaten.

Order Galliformes./Family Phasianidae [Pheasants]

Peafowl
Pavo cristatus
Cock: length to 6 ft, including 4-ft tail coverts; true tail hidden by coverts but supports them. Head with a tuft of 24 bare-shafted, gold-tipped feathers; underparts purplish, tail brown, face white. Hen without enormous tail coverts, head chestnut-brown with crown, back brown mottled, breast brownish-black and green.

Peafowl

Common Indian peacock, *P. cristatus*, native of India, Ceylon, but widely domesticated. In fourteenth century, introduced into England, France, and Germany. White peacock, a sort of Indian peacock. Green peacock, *P. muticus*, not widely distributed.

Cock usually maintains harem of about five hens. Nest on ground, concealed, of sticks and leaves. Eggs 4 to 20, glossy, pitted, white; 2 by 2¾ in. Hen a variable mother, so incubation in captivity often cared for by turkey, for 28 days. One brood yearly. Chicks should not be confined. Full plumage in 6 years. Said to live over 100 years.

Food in wild, grain, snakes, other reptiles, seeds, and insects, but in captivity similar to that used for pheasants. Peafowl is unfriendly to other domestic birds and often kills young chickens. Rarely moves rapidly. Voice harsh and disagreeable. Plumes are shed yearly, with annual molt.

Of no great value except as an ornamental bird. Formerly thought of as emblem of immortality and served as food at special banquets, tongues and brains being considered special delicacies. Muhammadans of Malay region consider peafowl "unclean" as it was supposed to have guided the serpent to the Tree of Knowledge.

Golden Pheasant

Golden Pheasant
Thumalea picta

Cock with long full crest of hairy feathers and cape of feathers over back of head and neck; tail two to three times length of bird. Golden-yellow on cape and rump, scarlet beneath, back green, crest yellow, wings mostly black. Hen reddish, barred with dark brown, and without enormous tail.

Native of China, whence it has been introduced into Europe and other parts of world. Other fancy pheasants include Silver of China, Monal or Impeyan of India, and Argus of Malay Peninsula, last having a 3-ft tail and lengthened ornamental wing feathers. There are also Asiatic, Horned pheasant or tragopan, and Reeve's pheasant with 6-ft tail.

Under domestication, one cock is used in a pen with four to five hens. 2-year-old cocks and hens likely to produce most virile eggs. Young hens lay 10 to 12 eggs a season. Older hens may lay up to 25 eggs for a sitting. Incubation 25 days. Young chicks "safe" after 3 weeks.

Young golden pheasant males have light eyes, hens hazel. Young cocks drop their "chicken" feathers at 3 months and begin to develop golden color; then should be separated from hens.

Golden pheasant may rarely cross with Chinese pheasant. Golden pheasant flesh considered better than that of silver pheasant, another fancy type. Golden pheasant raised almost entirely for show purposes and considered relatively easy to raise if space is available as it is hardy and withstands severe conditions.

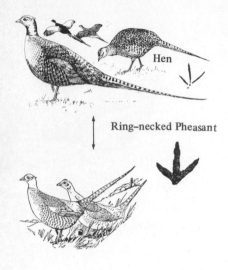

Hen

Ring-necked Pheasant

Ring-necked Pheasant, 309.1
Phasianus colchicus

Length to 3 ft, ½ being tail in the cock. Hen about 20 in. long. Weight to 4½ lb. Male with strong spurs, brilliant coloration, white collar and ear tufts; wattles vary with season. Female brownish to chestnut, and much less conspicuous than cock in vegetation. Both sexes remarkable hiders. Temperature, 106 to 108°F.

Native of eastern China and Korea, from Canton to the Yangtze. Four races introduced, mingled, and established particularly in grain areas of United States, England, and other parts of world. Minnesota, Iowa, and South Dakota becoming big pheasant states. Closely related species reported from Oligocene of Oregon and Nebraska Miocene.

One cock may control to seven hens, fighting cocks of his own kind and domestic roosters. Nest in grasses, on ground, made by hens. Eggs 6 to 12; olive-buff; 1⅔ by 1⅓ in. Incubation by hen only, 23 days. Young cared for by female until fall when molt takes place and sex differences are obvious. One annual brood.

Food, a variety of plant and animal substances, roughly twice as much vegetable as animal matter. Destroys great numbers of insects as well as corn and other cereals. Crowds out native game species such as quail and prairie chickens because of aggressive nature. Average Middle Western population, one bird per acre.

Most important of game birds because it can be reared in captivity and yet be wild when freed. Serves as buffer species to protect native forms from excessive hunting. Conservation departments maintain expensive plants to rear pheasants. Has not been a successful import to southern states.

Order Galliformes./Family Numididae [Guinea Fowl]

Guinea Fowl
Numida meleagris

Sexes similar. Cock has a slightly larger body, a more pronounced comb, larger wattles, and a shriller voice. General size about that of an average hen, apparently short-legged, long-necked, deep-bodied, short-tailed, with head naked and plumage slate-colored and mostly uniformly dotted with small round white spots. Temperature, 110°F.

Guinea Fowl

Originally developed from wild bird of West African jungles, brought to Europe by Portuguese in Middle Ages, which disappeared. Three domestic varieties, the Pearl, the White, and the Lavender, of which the first is the commonest. East African *N. ptilorhyncha* does not domesticate easily.

Nest of grasses, hidden in vegetation, guarded by cock while hen sits. Eggs 14 to 24, pointed, hard-shelled, 1¼ oz, brown or yellowish-white covered by small brown spots. Incubation 25 to 28 days; chicks run with mother almost immediately. Mother rather poor caretaker.

Food, insects and seeds gathered from fields and woods by birds; requires little special feed. Chicks at 36 hours may be fed three times daily hard-boiled egg, with bread crumbs one meal, and with chick feed two meals. By 1 week, can get own food in fields. By 10 weeks, should weigh 1½ lb. Adult weighs to 4 lb.

Flesh periodically popular as a substitute for wild game, which it resembles in color and in taste. Early Romans considered it a delicacy and early Greeks used it at special sacrificial ceremonies. Reintroduced into Europe in sixteenth century after having disappeared earlier. Easy to raise in captivity. Hen will produce to 80 small edible eggs per year.

Order Galliformes./Family Meleagrididae [Turkeys]

Wild Turkey, 310a
Meleagris gallopavo

Length to 50 in., tail 18½ in., wing 21 in. Weight to 40 lb. Differs from domestic turkey in having chestnut instead of white tips to tail feathers and to upper tail coverts. Two birds interbreed freely. Normally wild turkeys weigh about 20 lb for gobbler and 10 lb for hen. Temperature, 109°F.

Five native subspecies recognized. Range covers from Texas and Oklahoma, east through Florida and Pennsylvania, with Merriam's turkey to southwest. Formerly wild turkeys were found in abundance through New York, New England, and southern Canada. Known from Pleistocene of Pennsylvania, Tennessee, Arkansas, and Florida. Woodland species.

Male maintains harem up to around 15 hens for which it gobbles and fights. Male does not help in incubation or rearing of young. Nest on ground, hidden carefully. Eggs 10 to 14, pale buff or speckled brown, 2 by 2¾ in. Incubation 28 days. Young do not eat for 2 days but can run soon. Two or more hens commonly unite broods.

Food largely seeds, nuts, grain, and insects, depending on season and accessibility of food to cover. Since bird has to be cautious, it is not ordinarily common near habitations, though in early days turkeys used to be a pest about poultry yards of New York and New England. Can fly with great speed.

Grandest of all America's game birds; through unwise treatment has been greatly restricted in range. More recent enlightened game management procedures have increased range. Attempts to reestablish it as far north as Minnesota have failed. In certain states such as Texas, where "take" is managed in given areas, farmers have a steady financial income from wild turkeys, now lost to many states. Usually placed in subspecies *silvestris*, to separate it from domestic bird.

Wild Turkey

Domestic Turkey
Meleagris gallopavo

Gobbler, cock, or *tom;* body, breast, back, and neck bronze, wings bronze tipped with black, tail black or black with bronze bars and ashy edges, large red wattles, tuft of long breast bristles. White tail markings of many individuals probably show relationship to Mexican wild form. Weight to 36 lb. Hen with white edging on back feathers, without breast tuft. Weight to about 20 lb.

Varieties include the Bronze, the Bourbon Red, the White, the Narragansett, the Black, and the Slate. Bronze variety usually most popular. All developed from native Mexican form, *M. mexicana.* Ocellated turkey *M. ocellata,* from Yucatan and Central America is rare.

Polygamous: fifteen hens for one young gobbler; twelve for yearling, and eight for older. Nest on ground or in nest box. Eggs to 15, 3⅗-oz, light brown with brown spots. Incubation 28 days commercially in incubator. Young birds often kept off ground and reared on wire. Gobbler may weigh 33 lb in 1 year.

If reared on open range, turkeys feed freely on grasshoppers and other insects; may easily overfeed. Chicks require no food first 2 days; are fed boiled egg and cornbread crumbs first week; then whole wheat, hulled oats, and often buttermilk, with greenstuff. Meat scraps and grain, with vegetables, charcoal, grit, and water fed adults.

Discovery of advantages of rearing turkeys on wire revolutionized industry. Market is steadily increasing and turkeys are reared for cold storage, freezing, canning, and fresh killing. Big turkey states are California, Colorado, Iowa, Illinois, Missouri, Texas, Ohio, Minnesota, and Pennsylvania. Development of smaller, lighter varieties has made this bird common table item, not only a holiday treat.

Domestic
Turkey

Order Galliformes./Family Phasianidae [Fowl]

Brahma Fowl
Gallus domesticus

Light Brahma cock: weight, 12 lb, feathers of hackle and saddle black, white-edged; main tail feathers black, sickles greenish-black, other feathers white; head medium, with crown over eyes; comb slightly serrated in three rows, bright red; bill stout, curved, yellow, with dark stripe; eyes deep-set, reddish; ear lobes bright red, large, with lower edge on level with wattles; wattles bright red, medium-sized; neck moderately long, well-arched, with abundant hackle; back long, inclined slightly to middle of saddle; tail medium, carried at 33° angle; main feathers broad and overlapping; shanks yellow with feathers on outer sides; toes yellow and stout, middle and outer toes feathered. Hen: weight to 9½ lb; hackle feathers broader and tail coverts white-laced. Cockerel: weight 10 lb. Pullet: weight 8 lb. One of the largest domesticated chickens. It is characterized by a pea comb and feathering on shanks and toes. Many of first birds imported did not exhibit the pea comb, which is now a fixed character of the Brahma fowl group.

Brahma Fowl

Brahma Cock

Brahma Hen

Cochin Cock

Cochin Hen

Cochin Fowl
Gallus domesticus
Cochin varieties include Buff, Partridge, White, and Black.

Buff Cochin cock: weight 11 lb; entire plumage rich golden buff; head rather short, broad, and deep; comb single, medium, with five points, bright red; beak short, stout at base, yellow; eyes medium, reddish-bay; ear lobes smooth, bright red; wattles long, rounded at lower edges, thin, bright red; neck short, well-arched, with abundant hackle; back short, broad, with abundant plumage; tail short, carried at 45° angle; sickles short; shanks short, heavily feathered, rich yellow; toes straight, heavily feathered, rich yellow. Hen: similar to cock except for smaller comb and hackles, shorter tail, smaller head, and generally less abundant plumage; weight 9 lb. Cockerel: weight 9 lb. Pullet: weight 7 lb.

Asiatic breeds include Brahmas, Cochins, and Langshans. All are large birds developed essentially for their flesh. Other Asiatic breeds such as Malay, Aseel, and Black Sumatra may be smaller. Brahma breeds of fowl include two varieties, Light and Dark. Of these, Light is usually larger.

Asiatic fowls are among longest established. Brahmas first developed in India, then introduced to China and there improved. First imported into America in 1846 and to England in 1853. Breed has improved in egg-laying qualities, rate of maturing, and appearance since. Brahmas of best type are good for both flesh and egg laying. Body is less robust than in Cochin, in spite of fact that weight may be greater. Eggs large and brown.

Cochin chickens, formerly known as Cochin China chickens and as Shanghai chickens, were obviously developed in China. Breed does not long retain popularity where introduced. It was introduced into England in 1845, and in America has been used in developing other breeds of fowl. The Cochin chicken is essentially a meat-producing breed. Eggs large and brown.

Hamburg Fowl

Hamburg Cock

Hamburg Hen

Hamburg Fowl
Gallus domesticus
Hamburg chickens belong in a class of their own. The varieties of Hamburgs are Golden-spangled, Silver-spangled, Golden-penciled, Silver-penciled, White, Black.

Silver-spangled Hamburg cock: under 5 lb; feathers of head white, on upper parts of body clear silvery white, ending with a broad or narrow spangle of greenish-black; head medium; comb medium, spike turned upward, bright red; bill of medium length, well-curved, dark horn-colored; eyes large, reddish-bay; ear lobes large, flat, round, white; wattles medium, well-rounded, thin, bright red; neck tapering, with full hackle; back medium in length; tail rather long, carried at 45° angle; sickles long, well-curved; shanks medium length, lead-blue; toes straight, lead-blue. Hen: smaller than cock, with less abundant hackles, smaller comb, smaller hackle, and straighter tail.

Hamburg chickens are supposed to have been developed in Dutch Friesland (Penciled), in Italy (Spangled), and elsewhere in western Europe. They were imported into England from Holland and Belgium in 1825. Hamburgs are raised essentially as ornamentals, for show purposes. They are, in some cases, much like the Sicilian Buttercup breed imported in 1860 to America from Sicily. Eggs small and white.

Male *rooster* mates with female *hen;* one rooster sufficient for twenty hens. Fertile eggs may result over a period of 1 month from one mating. Fertile eggs, incubated at 103°F, hatch in 21 days into *chicks.* Chicks need not eat for 48 hours; roost at 1 month. Sexes should be separated at least 4 weeks. At 6 weeks, young males may be unsexed to grow up as *capons.* Otherwise, they become *cockerels* and, when sexually mature, *roosters.* Young females are *pullets* and at 5 to 6 months may begin egg laying. Nonbreeders are often fattened for sale as *broilers* at 10 to 12 weeks of age. Temperature, 103 to 108°F. 1,000 lb of living chickens require 8,278 cu ft of air to breathe each day. Life span to 15 years. In 1971 there were 443 million chickens in the U.S.

Cock

Hen

Barred
Plymouth Rock

Barred Plymouth Rock
Gallus domesticus
American breeds: Plymouth Rock, Rhode Island Red, Wyandotte, Java, Dominique, and Buckeye. Varieties of Plymouth Rock: Barred, White, Buff, Silver-penciled, Partridge, Columbian, and Blue.

Barred Plymouth Rock cock: weight 9½ lb; grayish-white, each feather crossed by regular narrow dark bars parallel to each other and alternating with white bars of same width; head moderately large and face smooth; comb single, rather small, straight, upright, evenly serrated, with five points, bright red; beak stout, comparatively short, regularly curved; eyes full, prominent, reddish; ear lobes oblong, bright red, ⅓ length of wattles; wattles moderately long, rounded at lower edges, red; neck, long, moderately arched, with abundant hackle; back, long, broad for entire length rising in concave line to tail; tail of medium length, at 30° angle; sickle well-curved; shanks smooth, stout, yellow, of medium length; toes yellow. Hen: weight to 7½ lb; ear lobes and wattles smaller; neck shorter; tail at 20° angle. Cockerel: weight 8 lb. Pullet: weight 6 lb.

Barred Plymouth Rock was first variety of breed developed. It seems to have been developed from the Dominique breed and contains three distinct strains: the Drake (Stoughton, Mass., 1866), the Sussex County (Massachusetts, 1856), and the Upham (Connecticut, 1865). Dual-purpose fowl for production of eggs and meat.

Rhode Island Red Fowl
Gallus domesticus
Varieties of Rhode Island Red: Single comb and Rose comb.

Rhode Island Red cock: weight about 8½ lb; back, breast, body, and neck a rich red; wings mostly red or black-tipped or black-edged; head medium, carried horizontally or slightly forward; face smooth, comb single, medium, upright, with five even points, bright red; beak of medium length, slightly curved, reddish; eyes reddish; ear lobes oblong, bright red; wattles bright red, medium-sized; neck of medium length with abundant hackle; back long, carried horizontally with abundant saddle feathers; tail medium, carried at angle of 40°; sickles abundant; shanks yellow, of medium length, smooth; toes rich yellow, straight. Hen: generally similar to cock but with smaller wattles and comb, shorter tail, and without conspicuous sickle feathers; weight to about 6½ lb. Cockerel: weight to 7½ lb. Pullet: weight to 5½ lb.

The breed originated in Rhode Island early in the nineteenth century. In the heritage are known to be strains of Red Malay Game Cock, Brown Leghorn, and others. Rhode Island Reds are general-purpose birds raised for meat and for eggs. The eggs are large and brown; hatch in 21 days of incubation.

Wyandotte Fowl
Gallus domesticus
Wyandotte varieties are Silver-laced, White, Buff, Black, Partridge, Silver-penciled, Columbian, and Golden-laced.

Silver-laced Wyandotte cock: weight to 8½ lb; wing bows and saddle silvery white; tail feathers black; other feathers black laced with white, or white laced with black; head short and round; face smooth, free from feathers; comb rose, low, tapering to a point at rear, bright red; beak short, well-curved, dark shading to yellow; eyes oval and red; wattles bright red, moderately long; neck short, well-arched, with abundant hackle; back moderately short, rising concavely to tail, saddle feathers abundant; tail short, carried at an angle of 45°; sickles moderately long, shanks yellow, moderately short, stout; toes rich yellow, straight. Hen: similar to cock but without white on wing bars and on back, without tail sickles, with smaller wattles and comb; weight to 6½ lb. Cockerel: weight 7½ lb. Pullet: weight 5½ lb.

Wyandottes were established as a breed about 1860. Buff Wyandottes were produced by crossing White Wyandottes with Buff Cochins. Silver-penciled Wyandotte was developed through crossing of Dark Brahma, Partridge Wyandotte, and Silver-penciled Hamburg. Wyandottes are good general-purpose fowls raised for meat and for egg production. The eggs are brown, medium in size, and hatch in 21 days.

Management for flesh production calls for selection of quick-maturing strains. Most breeds weigh about 1 lb at 2 months; 2 lb at 3 months. If prospective broilers are fed mash first 8 weeks, they will be marketable earlier. Cockerels considered desirable for breeders should be selected at broiler age, about three times as many as will be needed (3 for 15 hens) being saved for further selection to 1 for 15 hens. For broilers, feed mash twice daily after 2 days old; after 10 weeks, make it available at all times. Beginning at 4 weeks, add scratch grain so that, until 3 months old, grain and mash are eaten equally.

Orpington Fowl
Gallus domesticus
Orpingtons are a breed of English class of chickens. Other English breeds are Dorkings, Redcaps, Cornish, and Sussex. Four recognized varieties of Orpingtons: Single-comb Buff, Single-comb Black, Single-comb White, and Single-comb Blue.

Buff Orpington cock: weight 10 lb; entire plumage a rich golden buff, richly glossed on the head, neck, hackle, and back; head of medium length, broad and deep; comb single, medium, upright, with five points, bright red; beak short, stout, well-curved, pinkish-white; eyes large, reddish-bay; ear lobes of medium length, oblong, bright red; wattles medium-sized, well-rounded, bright red; neck rather short, well-arched, with abundant hackle; back broad, medium in length, with abundant medium-length saddle feathers; tail moderately long, carried at angle of 45°, with medium sickle feathers; shanks rather stout, pinkish-white; toes of medium length, pinkish-white. Hen: weight 8 lb; coloration like that of cock, with usual sex modifications in head, wattles, and tail. Cockerel: weight 8½ lb. Pullet: weight 7 lb.

(Continued)

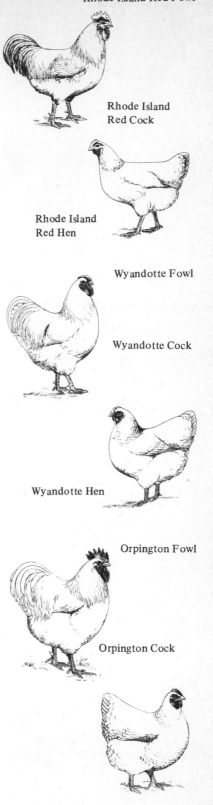

Rhode Island Red Fowl

Rhode Island
Red Cock

Rhode Island
Red Hen

Wyandotte Fowl

Wyandotte Cock

Wyandotte Hen

Orpington Fowl

Orpington Cock

Houdan Fowl

Houdan Cock

Houdan Hen

Game Fowl

Game Cock

Game Hen

Leghorn Fowl

Leghorn Cock

Leghorn Hen

First Orpington developed was the Black, which originated in 1886; involved in its development were blood of Black Langshan, Plymouth Rock, and Black Minorca. Buff Orpington developed from the Buff Cochin, the Dark Dorking, and the Golden-spangled Hamburg. Orpingtons were developed for egg laying and for meat production. Eggs large and brown.

Houdan Fowl
Gallus domesticus
French class has four breeds: Houdans, Crevecoeurs, La Fleche, and Faverolles. Only one variety for each breed except Houdan, for which there are two varieties, the Mottled and the White.

Mottled Houdan cock: weight 7½ lb; feathers black, many tipped with white; head medium-sized, carried rather high, topped by a round crest; comb V-shaped, small, resting against crest, bright red; bill moderately long, pinkish-white, mottled with black; eyes reddish-bay; ear lobes white, concealed by crest and beard; wattles bright red, nearly concealed by beard; neck medium long, with abundant hackle and a feather beard; back long, sloping toward tail; tail full, carried at angle of 40°; sickles well curved; shanks medium, pinkish-white with black mottlings; toes, five on each foot, pinkish-white mottled with black. Hen: weight 6½ lb; in general, much like cock. Cockerel: weight, 6½ lb. Pullet: weight 5½ lb.

This French breed was developed in the Seine region. It was introduced into England in 1850 and into America and for some time was widespread. An egg and meat producer, but has been supplanted by other breeds which have proved superior in each respect. Eggs large and white. Flesh and skin creamy-white. Might have become a favorite commercial breed if not for crest.

Game Fowl
Gallus domesticus
The class of chickens known as Games includes Games and Game Bantams. Game varieties are Black-breasted Red, Brown-red, Golden Duckwing, Silver Duckwing, Birchen, Red Pyle, White, and Black. There are similar and other varieties for Bantams.

Black-breasted Red Game cock: breast black; hackle and saddle golden; head orange; back red; other parts mixed red and black; head long, lean, bony; comb small and single; beak long, tapering, curved, horn-colored; eyes large, keen, red; ear lobes small; wattles small and red; neck long, slightly arched, erect; hackle short and close; back short, narrow, and sloping from neck to tail; tail short, closely folded, horizontal; shanks long, bony, greenish; toes long, straight, green. Hen: lighter in color and in weight than cock.

Fighting cocks developed from many breeds. Of Asiatic breeds, the Aseel provides fighting birds of India. In British class, the Cornish provides game birds; are also superior meat producers. Fighting birds have been bred for centuries. Fortunes have been won and lost on cockfights; outlawed in England in 1849. Training and arming a highly complicated process. Males weigh 7 to 9 lbs; females 5 to 7 lbs.

Feeding chickens for different purposes has become an art, with special rations for each purpose. A good mash mixture includes 40 lb yellow cornmeal, 10 lb wheat bran, 20 lb wheat flour middlings, 10 lb fine-ground heavy oats, ½ lb salt, 7½ lb dried skim milk or buttermilk, 10 lb ground meat scrap, 2 lb pulverized limestone, ½ lb natural cod-liver oil. A good grain mixture for 4 weeks to maturity is equal parts cracked corn and whole wheat.

Leghorn Fowl
Gallus domesticus
Mediterranean class of poultry includes Leghorns, Minorcas, Spanish, Blue Andalusian, and Anconas. Most of them are bred primarily for egg laying.

Leghorn varieties are Single-comb Dark Brown, Rose-comb Brown, Single-comb White, Rose-comb White, Single-comb Black, Silver, Red Pyle, Single-comb Buff, and Rose-comb Buff.

Single-comb, White Leghorn cock: weight to 6 lb; entire plumage white; head moderately long and fairly deep; comb single, medium in size, upright, with five long points, bright red; beak not long, nicely curved, yellow; eyes reddish-bay; ear lobes oval, rather broad, white; wattles moderately long, thin, bright red; neck long, nicely arched, with abundant hackle; back rather long, sloping down to tail; saddle long; tail large, carried at an angle of 45°; sickles long and well-curved; shanks long and yellow; toes medium, straight, yellow; disposition nervous; flies readily. Hen: colored like cock but with smaller tail, comb, wattles and without spurs, conspicuous sickle feathers, and similar evidence of masculinity. Cockerel: weight 4½ lb. Pullet: weight 3½ lb. Hen: 4½ lb.

Leghorn breed was developed in Italy and was introduced into America about 1835. It has since been greatly improved by breeding for egg laying and broiler-producing purposes. Ear lobes are white, whereas lobes of Asiatic breeds are red. Primarily egg layer and secondarily broiler fowl. Broilers can be grown in 8 weeks from hatching. Individuals have laid over 1,000 eggs. Eggs large and white.

Minorca Fowl
Gallus domesticus

Minorca varieties are Single-comb Black (described), Rose-comb Black (illustrated), Single-comb White, Rose-comb White, and Single-comb Buff.

Single-comb Black Minorca cock: weight 9 lb; feathers lustrous greenish-black throughout; head moderately long, wide, and deep; comb large, single, deeply serrated, with six points, bright red; beak of good length, stout, well-curved, black; eyes large, dark-brown; ear lobes large, almond-shaped, white; wattles long, thin, bright red; neck rather long, slightly sloping to tail, with long saddle feathers; tail large, carried at angle of 45°; sickles large and long; shanks rather long, straight, strong, black or dark slate; toes straight, black or dark slate. Hen: colored like cock but with smaller wattles, comb, and tail, and no conspicuous spurs; weight to 7½ lb. Cockerel: weight 7½ lb. Pullet: weight 6½ lb.

Minorcas developed in Spain; probably introduced by Moors. The Balearic Islands off east coast of Spain were site of greatest improvement. Popular in America in late nineteenth century; also popular in England. Minorcas lay large eggs of remarkable whiteness, and until the Leghorns reached their present popularity they were the egg layers of the average American farm.

Ancona Fowl
Gallus domesticus

Ancona varieties are Single-comb and Rose-comb.

Single-comb Ancona cock: weight 5½ lb; feathers lustrous black with about one in three, on the average, tipped with white; head moderately long and fairly deep; comb single, medium, deeply serrated, five-pointed, bright red; beak not long, yellow, with upper half shaded with black; eyes medium, reddish-bay; ear lobes medium, almond-shaped, white; wattles long, thin, bright red; neck rather long, nicely arched, with abundant hackle; back of good length, sloping a little to saddle; tail large, carried at 45° angle; sickles long, curved; shanks moderately long, slender, yellow mottled with black; toes straight, yellow, or yellow mottled with black. Hen: similar to cock except for smaller cockles, comb, tail, and other usual male characters; weight 6 lb. Cockeral: weight 5 lb. Pullet: weight 4 lb.

Anconas are sometimes called "Mottled Leghorns." They were developed in Ancona and Italian Adriatic seaports; introduced into England in late nineteenth century and from there came to America, where they have had some rather surprising popularity at certain places. Anconas are bred for egg laying, the eggs being white and medium in size.

Management for egg laying: Select vigorous parents. Breeding hens to be used in January should be forced to molt by November 1 by removing food and water for 1 day, then allowing hens to rest 60 days, during which they are given all the grain they will eat and all the wet mash they will clean up in 15 minutes. Pullets and hens should be kept separate to avoid disease and parasites. Cull nonproducers in September, or in July and in October. Hen may lay over 300 eggs a year. Males too must be carefully selected as breeders of fruitful hens. Masculinity, early maturity, and vigor are qualities sought. 4½ to 5 sq ft of floor space per chicken and 5 in. of roost space are desirable.

Bantam Fowl
Gallus domesticus

Silver Sebright Bantam cock: weight to 26 oz; head large, round, held back; comb rose, square in front with spike to rear; wattles broad and well-rounded; ear lobes smooth; wings carried low; back short; tail full and well-expanded; breast, prominent. Hen: weight to 22 oz; in general, smaller than cock; neck and head smaller. Silver Sebright not a very successful breeder.

Ornamental bantams include miniature Cochins and Brahmas. Black and white rose combs are miniature Hamburgs. There are also Polish bantams, white-booted bantams, Japanese bantams, and Sebright bantams, the latter illustrated. Sebrights were originated in England and include such varieties as Golden and Silver Sebrights.

Eggs of bantams are, of course, smaller than their larger counterparts described earlier. For judging purposes, bantam cockerel should be 22 oz in weight and bantam pullet 20 oz except that Cochin and Brahma bantams weigh 26 oz for cockerel and 24 oz for pullet.

(Continued)

Minorca Fowl

Rosecomb Minorca Cock

Minorca Hen

Ancona Fowl

Ancona Cock

Ancona Hen

Bantam Fowl

Bantam Cock

Bantam Hen

Food much the same as for larger breeds discussed in previous pages. Health, physiology, enemies, and general care are also similar except that space needed per individual animal for bantam is obviously less than for a Brahma fowl.

Bantams are raised primarily for show purposes but they also are beloved by children and make excellent but often highly abused pets. Term *Bantam* is assumed to have been transported from district of Bantam in Java from which first importations originated.

Order Gruiformes./Family Gruidae [Cranes]

Sand-hill Crane, 206
Grus canadensis
Length to 4 ft, with bill to 6 in. long. Height to 4 ft. Wingspread to 7 ft. Crown and area in front of eyes naked, red. Cheeks feathered. Slate gray or light brown throughout, with darker wings and light to white throat. Young entirely feathered and more rusty in appearance.

Bred formerly from British Columbia to southwestern Michigan and south to California, Illinois, and Ohio. Found formerly in migration east to New England but now rare east of the Mississippi and over much of its range. Still breeds from northeastern California to Wisconsin and Michigan. Winters in California, Texas, and Mexico.

Sand-hill Crane

Courtship elaborate, with many congregating and beginning a united bowing, parading, and bouncing in which whole group will behave like bouncing balls. Nest a mass of vegetation with hollow at top. Eggs usually two, 3⅔ by 2⅖ in.; olive-buff with brown spots. Young leave nest soon after hatching.

Food, frogs, snakes, insects, and similar small animals. Primarily a plant eater, feeds on roots, stems, leaves, fruits, and seeds of both wild and cultivated plants. Flight steady and heavy but deceptive in its speed. Florida crane and little brown crane are subspecies extending range from Florida to Alaska. Whooping crane, almost extinct, formerly common over much of America, may disappear. Snow-white, with black-tipped wings and red crown and forehead.

An interesting bird which causes some damage in upland grainfields. A temptation to hunters. Descriptions of courtship of great numbers in past provide interesting reading. *G. c. pratensis*, Florida Sand-hill Crane is an endangered species.

Order Gruiformes./Family Aramidae [Limpkins]

Limpkin, 207
Aramus guarauna
Length to 27 in., including 4⅔-in. bill. Wingspread to 42 in. Dark brown throughout, streaked with white. Bill long and curved slightly downward toward end. In flight, legs hang loosely down behind, giving bird an awkward appearance. Flight rarely long. Legs darker than in many heronlike birds. Appears, when feeding, like an overgrown rail. Walks with a jerky, limping movement; perhaps this reason enough for common name.

Limpkin

Breeds from Okefinokee Swamp in Georgia, south through Florida. Also in Cuba. Winter and summer range are approximately same. Has disappeared from much of northern part of its range; greatly reduced in numbers elsewhere. Locally it may appear rather abundant.

Nest an interwoven mass of grasses hidden among clumps of tall grass in shallow water. Eggs three to eight, 2⅓ by 1⅔ in.; grayish-white and blotched with light brown. Nesting period April to May. Young dark gray, able to swim as soon as they leave the shell, and leave nest the day they hatch.

Food largely aquatic and marsh animals, especially large snails, collected from freshwater areas where it lives. A relatively common bird in certain parts of Florida and because of its white-spotted appearance easy to identify. Wings and tail show a metallic bronzy reflection when bird is flying.

Probably of little economic importance but certainly is interesting. Its call of *eeow* repeated resembles the cry of a person being tortured and once heard is easily recognized, imitated, and long remembered.

Order Gruiformes./Family Rallidae [Rails, Gallinules, and Coots]

King Rail, 208
Rallus elegans
Length to 19 in., including 3½-in. tail and 3-in. bill. Wingspread to 25 in. Weight to 18 oz, female smaller than male. Sexes colored alike. Upper parts dark brown, streaked. Neck and breast cinnamon-red. Throat white. Belly and sides blackish-brown and distinctly white-barred. Young duller and darker above and paler beneath. Very young are black. Noticeably long slender bill.

King Rail

Breeds from Nebraska through southwestern Ontario to Massachusetts and south to Kansas, Florida, and Texas. Winters in southern part of breeding range, at least south of New Jersey. Found occasionally north to Maine and to Manitoba. One allied race found in Cuba. Known as a fossil from Pleistocene of Florida.

Nests in freshwater marshes. Nest of weeds and grasses, on or near ground, just above water surface; well concealed from above by interlacing of surrounding grass. Eggs 6 to 16, $1\frac{4}{5}$ by $1\frac{1}{2}$ in.; buffy to creamy white with brown or lilac spots. One annual brood. Nesting period June. Glossy black downy young soon develop feathers, and when feathered resemble adults.

Food largely insects such as grasshoppers but is also known to eat oats to a limited degree. Highly secretive, and more often heard calling a repeated *bup* than seen. When caught, it has been said to cry like a common barnyard fowl. Some individuals have been described as being fearless.

Not common enough to be of positive economic importance even if they lived where they could do more good. Rails are relatively easy to shoot because of their direct flight but are hard to find once they have been dropped. A live rail in a marsh is of much more interest and value than its carcass in a bag. Also called marsh hen and mud hen.

Clapper Rail, 211

Rallus longirostris

Length to 16 in., including $2\frac{1}{2}$-in. bill and $2\frac{1}{2}$-in. tail. Wingspread to 21 in. Weight to 14 oz, with female slightly smaller than male. Upper parts pale olive-gray, with feathers grayish-white on borders. Tail brown. Throat white. Breast pale brown with a gray wash. Sides and abdomen brown with fine whitish bars. Has other common names: clapper, marsh clapper, mud hen, sedge hen.

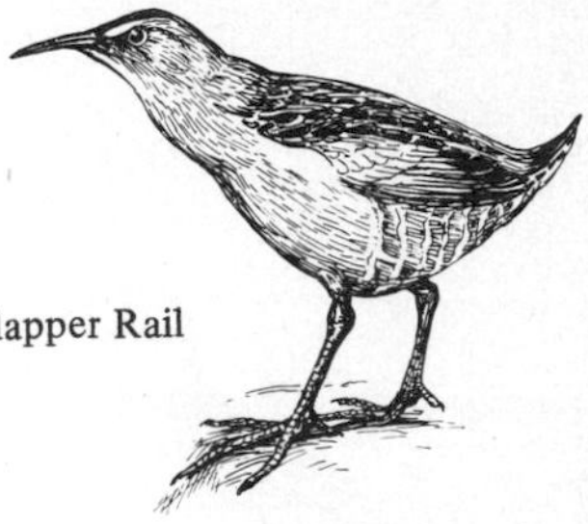

Clapper Rail

Six subspecies: Northern, Yuma, Wayne's, Florida, Mangrove, and Louisiana clapper rails. Northern breeds in salt marshes from Connecticut to North Carolina and winters south of New Jersey, within its breeding range for the most part. Sometimes found north to Maine. Other subspecies extend range around Gulf Coast. Essentially a salt-marsh rail.

Nests on ground or in marsh weeds, sometimes a short distance away from salt marshes. Nest commonly covered with a canopy of weeds, concealing eggs. Eggs 8 to 13, $1\frac{7}{10}$ by $1\frac{3}{4}$ in.; buff or gray with brown, lavender, or reddish markings showing faintly. Incubation 21 to 23 days. One or two annual broods. Downy young chiefly black with some whitish areas below.

Food, small animals such as snails, insects, fish young, crustaceans, with some seeds and water plants. Call a rather hoarse, repeated *cac* beginning loudly and gradually tapering off in volume, tone, and frequency. Migrates mostly at night, flying low, and so meets many accidents.

Recognized as a game bird and in some areas is reasonably abundant, so much so that its eggs have been collected for food. In 1896, it was estimated that some 10,000 clapper rails were killed near Atlantic City, N.J., but that is not now possible; however, it does give some evaluation of game-management practices so far as these birds are concerned.

Virginia Rail, 212

Rallus limicola

Length to 11 in., including 2-in. tail and $1\frac{3}{5}$-in. bill. Wingspread to $14\frac{1}{2}$ in. Weight to 4 oz, female being slightly smaller than male. Upper parts brown to blackish, with gray-bordered feathers. Cheeks gray. Throat white. Sides and under parts brown, with blackish and white bars on flanks.

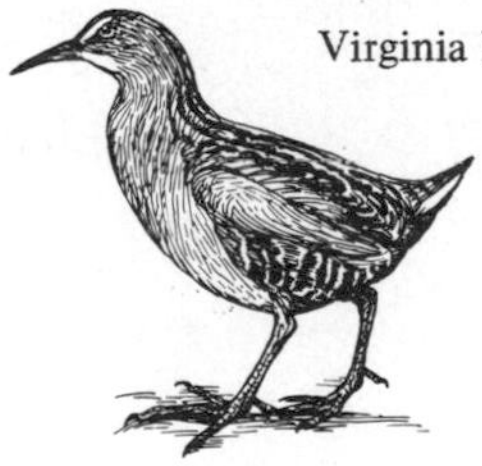

Virginia Rail

Breeds from southern British Columbia to Nova Scotia and New Brunswick, south to northern lower California, Utah, Missouri, Kentucky, and New Jersey. Winters from Utah, through lower Mississippi Valley to North Carolina and south to Florida and Guatemala. Sometimes found in Hudson Bay, Labrador, Newfoundland, and Greenland and in Bermuda and Cuba.

Nests in freshwater marshes and along stream banks in high vegetation. Nest of weeds, well-concealed. Eggs 5 to 12, $1\frac{1}{3}$ by $1\frac{1}{5}$ in.; variable, from white to creamy buff, with sparse brown and purple spots. Incubation by male and female. Probably only one annual brood. Only rarely found nesting in salt-marsh areas. Young greenish-black, with pink bill.

Food probably chiefly insects and other small animal life of marshes. Seeds of marsh plants are common food. Known also to eat wild rice, wild oats, and other grass seeds. Known to give a grunting sound, a squeak, a series of henlike *cuts*, and other sounds and is undoubtedly heard more frequently than it is seen.

A game bird hunted particularly during southern migration, when more birds may be observed than at other times. Its flight is straight and it is therefore an easy shot.

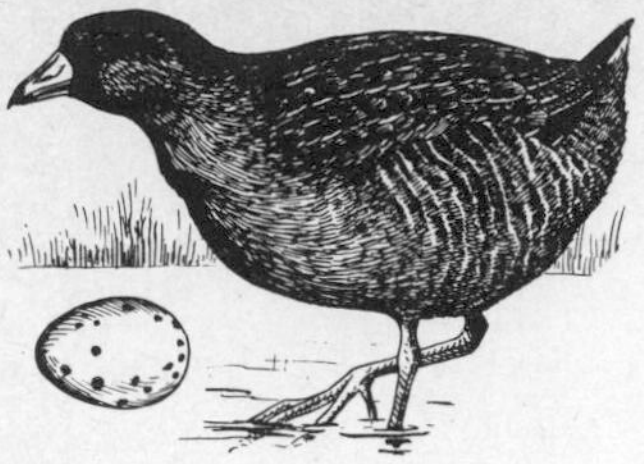

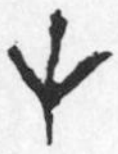

Sora Rail

Sora Rail, 214

Porzana carolina

Length to $9\frac{3}{4}$ in., wingspread $14\frac{1}{2}$ in. Weight to 4 oz. Hen smaller than cock. Sexes similar in appearance. Adult with conspicuous black on face and throat; not so in young. Back olive-streaked. Breast and sides of head slate. Flight slow and steady, making easy mark for hunters.

Breeds from British Columbia to Hudson Bay and south to California and Maryland. Winters from California to Florida and south to Venezuela and Peru. Accidental in England and Greenland. Closely related Virginia, clapper, and king rails are known from Pleistocene of Florida or California.

Nests in marshes or in grain. Nest supported and concealed by grasses. Eggs 4 to 17, olive-buff spotted with dark brown, $1\frac{1}{3}$ by $1\frac{4}{5}$ in. Incubation 18 to 20 days, by hen and cock. One annual brood. Nesting May to June. Young black with orange throat tuft, and pink, bill with red cere. Complete molt in July.

Food, insects, small crustaceans, seeds, and grain. Southern migration usually does not begin until frost, so may be irregular though hunting seasons are not. May cross Gulf of Mexico in southern flight; has been found hundreds of miles at sea. Can swim and dive if necessary to escape or to get food.

Considered by law as a game bird. Also known as Carolina rail, common rail, Carolina crane, railbird, and mud hen. Some describe them as similar in appearance to a bantam hen. Song a descending whinny.

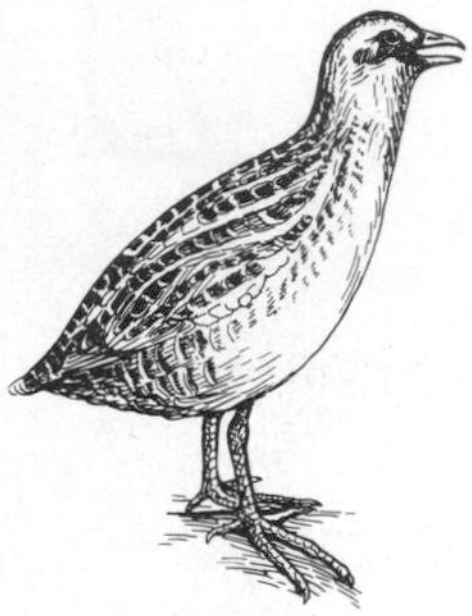

Yellow Rail

Yellow Rail, 215

Coturnicops noveboracensis

Length to $7\frac{3}{4}$ in., including $1\frac{2}{3}$-in. tail and $\frac{3}{5}$-in. bill. Wingspread to 13 in. Weight to nearly 3 oz, female being smaller than male. Mostly yellowish but, unlike other rails, shows white wing patch in flight. Legs greenish to brown to yellow. General color streaked brown to yellow to blackish, with narrow bars and mottlings of white on flanks.

Breeds in Michigan, North Dakota, and east central California, but found during breeding season from Nova Scotia to southern Mackenzie, south to California and Massachusetts. Winters mostly in Gulf States, and from North Carolina to California. Relatively rare, found in grassy marshes and among reeds. Favors higher lands than most rails.

Nest on ground, near marshes or in them. Eggs 7 to 10, $1\frac{1}{10}$ by $\frac{8}{10}$ in.; pinkish-brown to deep tan with chocolate-colored markings, particularly at larger end, and usually in form of small dots. Black downy young. By first winter and through first summer, young have fewer dark markings than adult, but by second summer they have more white bars.

Food probably same as other rails, including both animal and plant materials. In flight, it flutters a short distance only to drop into cover and run. Difficult to flush and almost impossible to collect without help of good bird dog trained for the work. Call an oftrepeated *tic* given in series of alternating two and three units.

Of little economic importance. Also known as yellow crake. It is probably much more frequently heard than seen and so its rarity may be only apparent in many localities.

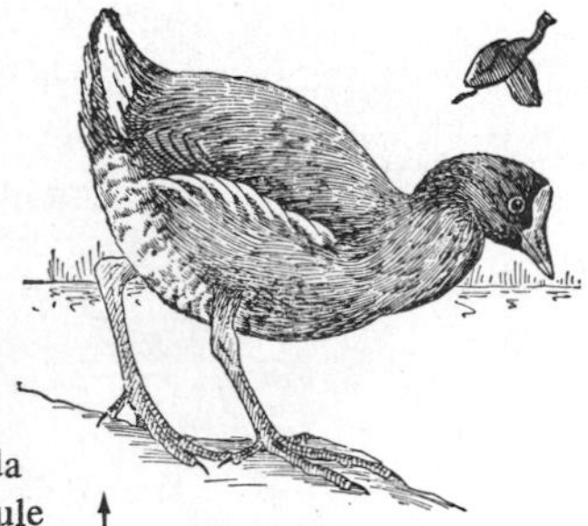

Florida Gallinule

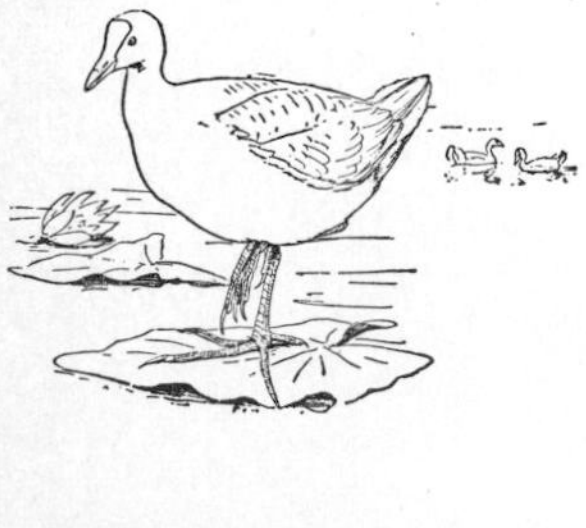

Florida Gallinule, 219

Gallinula chloropus

Length to $14\frac{3}{4}$ in., wingspread to 23 in. Weight to 14 oz. In field, resembles coot but bill and plate on front of head are red instead of white. Sexes colored alike. White streak appears along sides. Tail black above and white beneath. General color slate. Toes lack scalloped webs of coot. Might be confused with a coot which is stockier, has a shorter neck and a white bill.

Breeds from central California to southern Ontario and south to Lower California and Panama. Winters from southern California to South Carolina and south with related species in South America and Eastern Hemisphere. Fossil forms reported from Pleistocene of Florida.

Nests in late spring in marshlands; builds more than one nest but only occupies one; of plants. Nest frequently on floating material which rises and falls with water level. Eggs 8 to 17, buff and spotted, $1\frac{4}{5}$ by $1\frac{1}{2}$ in. Incubation 22 to 25 days. Young downy, black, with white-tipped, hairlike feathers at throat. Top of head bare. Bill red but black-tipped. Like adults by December.

Food, plants and animals of habitat including grasshoppers and seeds where available. When on water, floats high, and when swimming moves head like chicken. Poorer swimmer than coot but prefers to stay in plants at water's edge. Calls like clucks of hens, also explosive cackle. With many common names: American gallinule, red-billed mud hen, water hen, water chicken.

Coot, Mud Hen, 221
Fulica americana
Length to 16 in., wingspread to 28 in. Weight to 22 oz. General appearance slate-gray with darker head, neck, and tail. Bill white, and white under tail. Sexes colored alike. Feet with webs of toes like wavy-margined flaps but not connected through most of length. Eyes red. Walks with chickenlike motion of head.

Breeds from British Columbia to New Brunswick and south to Nicaragua. Winters mainly in the Gulf States and along the Pacific Coast; also south to Panama and Colombia and Ecuador. Closely related races and species extend range through Europe, Asia, Africa, and Australia. Found in great rafts or flocks or singly along waterways. Fossil records from Pleistocene of Oregon.

Breeds often in colonies. Nest of plant materials hidden near water. Eggs 6 to 22, creamy clay-colored finely speckled with black, 2 by 1¼ in. Incubation about 23 to 24 days, but young hatch at different times, first being cared for by cock while hen sits. Young black, with orange thread-like down on head and neck and with black-tipped, orange-red bills.

Food largely vegetable matter, including roots, seeds, and tubers; animal matter includes insects, both larval forms, adults, and other aquatic invertebrates. Unsuspicious, hides rather than flying; takes to wing with great spattering of water surface. Great size of flocks during certain seasons may give false impression of population. Eggs edible but inferior.

Inferior table bird, easily shot. In total numbers of individuals follows pintail and mallard in inventory of N. American waterfowl.

Coot, Mud Hen

Order Charadriiformes. / Family Charadriidae [Plovers, Turnstones, and Surfbirds]

Semipalmated Plover, 274
Charadrius semipalmatus
Length to 8 in., wingspread to 16 in., tail 2⅓ in. Sexes similar. Front toes partly webbed; in outer and middle toe, web reaches second toe joint. Appears brown and white below, like a short-tailed, single-ringed, half-sized killdeer. In winter, black areas are brown.

Breeds from Arctic coast of Bering Sea to Greenland and south to valley of Yukon, southern James Bay, north shore of Gulf of St. Lawrence, New Brunswick, and Nova Scotia. Winters from central California to South Carolina and south to Patagonia and Chile. Occasionally in Siberia and England.

Nest a depression in sand or gravel, on or near shore, in open. Eggs four, pear-shaped, pale buff spotted with blue, 1⅖ by 1 inch. Incubation 23 to 25 days, by both sexes. Young able to run soon but family stays together as killdeer do.

Food chiefly small animals such as insects, small crustaceans, eggs of marine animals, or other forms of animal life gleaned from bird's habitat. One of most characteristic and attractive of beach birds and one of best beloved. Speed, 32 mph.

Too small to be of value as game. Protected by law in many places. Has been called ring-neck or little ring-neck. Found frequently on the beach associated with least and semipalmated sandpipers. A splendid companion.

Coot, Mud Hen

Killdeer Plover, 273
Charadrius vociferus
Length 10½ in., wing 6½ in., bill ¾ in. Two black breast bands instead of one common to other ringed plovers; also conspicuous rusty spot at base of tail when in flight. White underparts; white separating black neck rings. Back brown. Sexes alike. Body temperature, 106°F.

Breeds northern British Columbia, northern Ontario, and southern Quebec to southern Lower California, central Mexico, Florida, and Bahamas. Winters southern British Columbia through Colorado, New York, and New Jersey south to northern Venezuela, northern Peru, and Greater Antilles. California Pleistocene.

Nests in open country, on beaches, in plowed fields and meadows. Nest simple. Eggs about four, buff-white spotted or marked with chocolate chiefly at larger end, 1½ by 1¹/₁₀ in. Incubation 27 days, with both parents showing great excitement if bothered. Young able to run soon after hatching. Sometimes two annual broods.

Food largely insects and earthworms or, when about water, crustaceans and other forms of water life. Particularly destructive to mosquitoes, flies, grasshoppers, flea beetles, ticks, army worms, other caterpillars, weevils, and wireworms. Superior flier and its call often notify other birds of danger. Flight speed, 55 mph.

Flesh inferior. Protected by law and should have continued protection as it is one of most useful of farm birds and does little if any harm. Its ways are always interesting, whether it is flying high, dipping on a shore, sitting on its eggs, trying to lead an enemy from its young, or feeding in a meadow.

Semipalmated Plover

Killdeer

Golden Plover

Golden Plover, 271
Pluvialis dominica
Length to 11 in., including 3-in. tail and 1-in. bill. Wingspread to 23 in. Weight to 9 oz, male and female being about same size. Appears larger than killdeer. Lacks white that shows on wings and tail of black-bellied plover. Golden brown on upper parts. In spring, shows white line over eye and downward on neck. At least 25 common names.

Breeds from Point Barrow in the Arctic, to Baffin Bay and south to Churchill, Manitoba. Winters on pampas of Brazil, Argentina, Paraguay, Bolivia, and Uruguay, migrating nonstop over Atlantic Ocean from Nova Scotia to South America or going down Mississippi Valley or Pacific Coast. Sometimes found in Greenland, Great Britain, or Australia.

Nest a mere depression in tundra, with leaf lining, not carefully made. Eggs, four, $2\frac{1}{10}$ by $1\frac{2}{5}$ in.; pear-shaped, slightly smaller than those of black-bellied plover. Incubation for 27 days, by both parents. One annual brood. Birds flying across Atlantic are fat when they start but emaciated at end of trip.

Food largely insects such as grasshoppers, crickets, cutworms, white grubs, with some berries on occasion. Give a repeated *coodle* call, and may repeat raising their wings high over back while on ground. Stay together in flocks and so are relatively easily decoyed unless they have been unduly frightened.

Economic importance as destroyers of insects should far outweigh their value as game birds. Since birds are intercontinental, their survival depends on their treatment in South America as well as in North America. Their long sea migration is one of the marvels of bird lore.

Black–bellied Plover

Black-bellied Plover, 270
Squatarola squatarola
Length to $13\frac{3}{5}$ in. Wingspread to 25 in. Weight to $10\frac{1}{2}$ oz. Sexes colored alike. Tail and rump always white; black axillary feathers, under wing, constant. Summer, pale gray spotted with black above; forehead, sides of neck, and upper breast, white; throat, sides of head, breast, and belly black but with white under tail. Winter, white or light-breasted, stocky, brownish-gray birds. Largest and also shiest of our plovers.

Breeds along Arctic coasts of North America, Europe, and Siberia. Winters from Mediterranean to South Africa, India, Australia, southern British Columbia, Louisiana, California, North Carolina, south to Brazil, Peru, and northern Chile. In migrations, covers most of United States. Several races, based largely on range.

Nests in hollow on ground among plants, often lined with grasses and lichens. Eggs late June and early July, four, $2\frac{1}{3}$ by $1\frac{2}{5}$ in.; pear-shaped, light olive-brown with heavy spots on larger end. Incubation by both sexes. One annual brood. Young olive-brown with black spots above; neck and underparts white, with black eye lines; first winter, like female.

Food, mollusks, worms, crustaceans, spiders, earthworms, and insects gleaned from shores or sometimes from fields. In fall migration, adults precede young, birds flying in masses, lines, V's, or small groups. Spring migration extended, lasting in United States from April through June. Related golden plover lacks white under tail of this species.

A high-class shorebird from sportsman's viewpoint because of relative abundance, skill in flying, and edible qualities. May be of some value as a destroyer of harmful insects but this can hardly be great because of nature of the favored habitat. Worthy of protection to assure continued abundance.

Ruddy Turnstone

Ruddy Turnstone, 283a
Arenaria interpres
Length about 9 in., including $2\frac{1}{2}$-in. tail and $\frac{9}{10}$-in. bill. Wingspread to $19\frac{1}{5}$ in. Weight to 4 oz, with female larger than male. Larger than spotted sandpiper, more compact and low, with orange legs, and in breeding season with red back and conspicuous black on face and breast. Both sexes have similar color pattern, but female is duller.

Breeds from western and northern Alaska to western Baffin Bay. Winters from central California to Texas, Louisiana, Mississippi, and North Carolina, and south to southern Brazil and central Chile. European turnstone is a subspecies, breeding in Arctic Europe and wintering from southern Europe and Asia to south Africa. Black turnstone is confined to the Pacific Coast in winter and goes no farther south than Lower California; nests in Arctic.

Nests usually on coastal islands, in a hollow, on ground, with or without plant shelter. Eggs three to four, $1\frac{2}{3}$ by $1\frac{1}{5}$ in.; pear-shaped, grayish-green to brown with dark brown or purplish spots. Nesting period June to July. Incubation by both sexes. One brood a year. Downy young blackish-gray above, with white forehead strip, black eye line, and white chin.

Food largely crustaceans and other marine animals such as the horseshoe crab, including eggs of many kinds. It gorges itself and becomes fat in fall. Call a repeated *kuk* or a clear whistled deep repeated *quittock*. Migration routes are along coasts. Rarely seen inland in freshwater areas.

Since being protected by law, it has begun to regain its numbers that were greatly reduced because of heavy hunting. In fall has a fine flavor and is considered a delicacy.

Order Charadriiformes./Family Scolopacidae [Snipe, Woodcocks, and Sandpipers]

American Woodcock

American Woodcock, 228
Philohela minor
Length to 1 ft, wingspread to 20 in., bill to 3 in. Weight, cocks to 6 oz; hens to 8 oz. Wings more blunt than those of common snipe. More stocky, browner, more nocturnal, and more whistling in flight. Flight of fluttering nature. In flight, hind portion of wings shows darker than fore part.

Breeds from southern Manitoba to southern New Brunswick, and south to Colorado, northern Missouri to New Jersey and south to Texas and Florida. Occasionally in Bermuda, Newfoundland, and Montana. Bird of grassy fields and meadows as well as lowland woods. Nests in brushy fields and in the margins of woods.

Cock courts hen by twittering call, high spiral flight, plunge, and strut. Nest on or near wet ground. Eggs three to four, variable in color, brownish; 1½ by 1⅕ in. Both sexes incubate, 20 to 21 days, and care for brood 1 year. Young downy when hatched, dark brown and tan, may be carried by parent.

Food principally earthworms but grubs, beetles, and other insects may be substituted if earthworms are not available. Small quantity of plant food includes a variety of seeds. After July molt, young are like adult but may not mature until next year. This coupled with small brood, tendency to congregate in flocks in migration, and direct flight makes it difficult to survive heavy hunting. Speed, 13 mph.

Considered superior game bird because of high quality of flesh and direct flight.

Common Snipe

Common Snipe, 230
Capella gallinago delicata
Length to 11¾ in., wingspread 20 in. Weight to 5 oz. Sexes colored alike. Extremely long bill, streaked plumage, white belly, zigzag irregular flight, cracking cry, pointed wings, short orange tail serve to identify. Lighter-colored and slimmer than woodcock; larger than spotted sandpipers.

Breeds from Alaska to New Brunswick and south to southern California and Pennsylvania. Winters from Alaska to Virginia and south to Brazil but occasionally as far north as Nova Scotia. Accidentally found in Hawaii, Great Britain, and Greenland. Bird of open boggy margins of small streams and marshes.

Cock courts by remarkable spiraling flight with "winnowing" sound and plunge, commonly at dusk. Nest on ground, of grass, not elaborate. Eggs three to four, greenish-brown, spotted or blotched, 1½ by 1⅒ in. Incubation about 20 days. One annual brood. Young yellow, darker above, with dark line from eye to bill. Mature in 18 months.

Food largely insects both larvae and adult and other small animals but also some plants. Cannot be considered injurious. Can dive and swim well. Courtship sound produced by vibration of outer tail feathers in downward plunge. Feeds by probing deeply into soft wet soil with long bill.

Recognized as game bird because of remarkable confusing flight serving as a marksman's challenge and because of food value of flesh. Really too small to justify being killed. Danger of extermination not officially recognized. Also known commonly as jacksnipe as well as by many other names.

Hudsonian Curlew, Whimbrel, 265
Numenius phaeopus
Length to 18¾ in., wingspread to 33 in., tail 4 in. Hen larger than cock. Bill curved downward, 4 in. long. Long-billed curlew of West, *Numenius americanus*, is 26 in. long, wingspread 40 in., bill to 9 in.; has bright cinnamon wing linings; less gray than the Hudsonian curlew.

Hudsonian curlew breeds from mouth of Yukon in Alaska to northern Manitoba but nonbreeding individuals may be found in summer along Atlantic Coast from Virginia to Ecuador. Winters from Lower California to Honduras and Chile. Long-billed curlew breeds from Utah to Kansas and Wisconsin; winters from central California to Florida and Guatemala.

Hudsonian Curlew, Whimbrel

(Continued)

Upland Plover

Solitary Sandpiper

Spotted Sandpiper, 263

Nest a depression or in a grass tussock. Eggs of Hudsonian curlew four, pear-shaped, olive-green, obscurely spotted, 2½ by 1⅔ in. Incubation by both sexes, for 27 to 28 days and one annual brood. Eggs of long-billed curlew, 2⁹⁄₁₀ by 2 in. Young of both able to run about soon after hatching.

Food, fiddler crabs, spiders, grasshoppers, beetles, crustaceans, either caught on surface or probed for with long bill which can be thrust into earth. Frequents tide flats and flies offshore in formations like geese, long bill being usually easily seen and serving as identification.

Hudsonian curlew is not good food compared to other curlews. Because of its habitat, it is not so useful in checking pests of agricultural lands. Long-billed curlew gets greatly excited when its nest is approached and by its calls and behavior gives away its greatest secret. Too useful to be destroyed.

Upland Plover, 261
Bartramia longicauda
Length to 12⅔ in., including 3½-in. tail and 1⅓-in. bill. Wingspread to 23 in. Weight to 7 oz; sexes equal. Rather uniformly brown-streaked, with long neck, small head, but relatively short curlewlike bill, and habits of perching on posts like willets, and upon alighting of raising wings in air.

Breeds from Canada, south in Plains country to Ontario, northeastern Utah, Wisconsin, and southern Maine to as far south as northern Virginia, southern Illinois, southern Missouri, and Oklahoma. Winters on South American pampas from southern Brazil to Argentina and Chile. Accidentally found in Greenland and western Europe. Should be considered as essentially a rare bird of fields and pastures.

Nests in grassy meadows, grainfields, or pastures. Nest on ground, well-hidden in vegetation. Eggs four, 1⁹⁄₁₀ by 1⅖ in.; short, bluntly pear-shaped, larger than those of similar woodcock; cream to olive to buff with few large gray spots or blotches. Incubation 21 to 24 days by both parents. One annual brood.

Food largely insects such as weevils, fever ticks, cutworms, white grubs, cotton worms, locusts, grasshoppers, crayfish, and similar essential enemies of farm crops. It may also eat small amounts of grain and seeds of plants. Call a sweet, melodious *kip-ip-ip-ip* or *quit-it-it-it* followed by a descending whistle.

Essentially useful as a destroyer of enemies of farm crops and formerly considered a game bird, but now so rare that it is worthy of all possible protection. Many birds get killed by mowing machines since they nest in pasture grasses.

Solitary Sandpiper, 256
Tringa solitaria
Length to 9 in., wingspread to 17 in., tail to 1⅓ in., bill to 1⅓ in. Weight to 28 oz. Lacks white stripe shown on wing of spotted sandpiper. Shows tail with conspicuous white sides and faint dark crossbars. Wings appear dark and show a deeper stroke than in spotted sandpiper. Teeters head instead of tail.

Two subspecies, Eastern and Western. Summers from Alaska to Newfoundland, south to northwest border of United States, Iowa, and Pennsylvania. Winters from Texas and Florida south possibly to Argentina. Appears as migrant between summer and winter ranges and known to breed in Alaska.

Uses abandoned nest of some other bird such as a grackle, robin, or jay. Eggs four to five, greenish to reddish-brown with browner spots usually nearer larger end, the spots forming blotches; 1⅖ by 1 in. One annual brood. Young birds gray-brown above and lighter beneath.

Food chiefly insects: grasshoppers, water boatmen, caterpillars, dragonfly nymphs; also spiders, worms, crustaceans, and small frogs. Some insects may be caught on the wing by sudden turns in flight. It flies higher than spotted sandpiper, more swiftly, and in a more zigzag manner. Usually tame and unsuspicious.

Useful as destroyer of insects. Too small to have value as game and too dull to interest good sportsmen. Fortunately, it is illegal to shoot them at all in United States.

Spotted Sandpiper, 263
Actitis macularia
Length to 8 in., wingspread to 14 in., tail 2¹⁄₁₀ in. Weight to 2 oz. Hen larger than cock. Breast definitely spotted with large round spots, not streaks, except that young birds and fall adults do not show this character. Outer tail feathers white with black bars. Wing stroke short. Has teetering habit.

Breeds from tree limit in northwest Alaska to northern Ungava Peninsula and Newfoundland, south to southern California, Louisiana, and South Carolina. Winters from southern British Columbia, Louisiana, and South Carolina to southern Brazil, central Peru, Bolivia, and Argentina. Also in Greenland and western Europe. Found most frequently near streams and ponds.

Nest grass-lined or not, in a variety of places. Eggs three to five, pear-shaped, clay to cream, spotted and blotched with chocolate, 1 by 1⅖ in. Incubation 20 to 24 days, by both sexes. One annual brood. Young run soon after hatching.

Food small animal matter, chiefly insects, including army worms, grass-hoppers, caterpillars, cutworms, cabbage worms, beetles, and grubs. Does not flock much even in migration except in South in winter. Can dive from full flight, swim, and run under water with help of wings.

Entirely useful. Protected by law and worthy of even more protection through southern part of its range. Call an oft-repeated *peet, peet,* or *peet, weet.* An inconspicuous double white wing stripe shows in flight.

Willet, 258
Catoptrophorus semipalmatus
Eastern willet; length 16 in., wingspread 29 in., tail 3⅓ in., bill 2⅓ in. Weight to 10 oz. Wings in flight conspicuously black and white. Broad area of white conspicuous in black wing makes willet shorebird easiest to identify in flight. Tends to raise wings upon alighting, showing patches. Legs bluish. Western willet reputedly paler on back, with longer bill, and slightly larger.

Willet

Two subspecies breed from east central Oregon to Nova Scotia, south to southern California, Ecuador, and Florida. Winter from northern California to North Carolina and in Brazil, Peru, and Ecuador. Accidentally in Europe, British Columbia, and southeastern Alaska. Western willet said to be fall Atlantic Coast migrant.

Courtship by pursuit and holding wings erect. Nest among grasses of sandy or dry spots near water on or in pasture swales or in open. Eggs four, olive, buff, or greenish, 2⅖ by 1⅕ in. Incubation by female. But young run soon after hatching, though they possibly may be carried by parents between thighs.

Food, aquatic insects, small mollusks, fish fry, fiddler crabs, and some seeds, tender plant parts, roots, and on occasion cultivated rice. Strong flier which may hover over a given point, ordinarily shy, but at breeding season seems bold in attempt to lead intruders away from nest. Speed, 27 mph.

Flesh of a rank flavor and so not ordinarily considered edible. Certainly, it cannot be considered a delicacy and so should be left unharmed, at least until its numbers can be more restored than is now the case. The Eastern willet is rare over much of its original range.

Greater Yellowlegs, 254
Totanus melanoleucus
Length to 15 in. Wingspread to 26 in. Weight to 10 oz, with average 6 to 8 oz. Female larger than male. Slim, gray and white, with bright yellow legs. Shows whitish rump and tail and darker wings in flight. Sexes colored similarly. Bill slightly upcurved. Call a clear, penetrating, yodeling whistle of three to five notes. Young like adults but lighter above; streaks below limited to neck and upper breast.

Breeds from Alaska to central Alberta, southern Ungava, and Labrador to southern British Columbia and Newfoundland. Winters in Washington, California, Arizona, and South Carolina to Patagonia, with birds accidentally in British Isles, Greenland, and Bermuda. Modern form is known as a fossil from California Pleistocene. Essentially a bird of shores and marshlands.

Nests usually in a clearing or burn of a forested area near a lake or large stream. Nest on ground, in a depression, May to June. Eggs four, 1⁹⁄₁₀ by 1⅓ in; pear-shaped, gray-white splashed with brown and lilac. Incubation by both sexes, probably, 23 days. Probably, one annual brood. Chicks long-legged and active soon after hatching.

Greater Yellowlegs

Flying bird holds long legs out behind and long neck and bill forward; sails before alighting and teeters after. Food, captured by wading about among water plants in shallows, consists essentially of small fish as well as insects, snails, worms, and crustaceans. By fall, birds are plump and fat.

Has suffered severely from hunting. Since the birds come readily to crude imitations of their calls and fly in such a way that they are easily shot, it is surprising that they were able to maintain an existence. They are now protected by law.

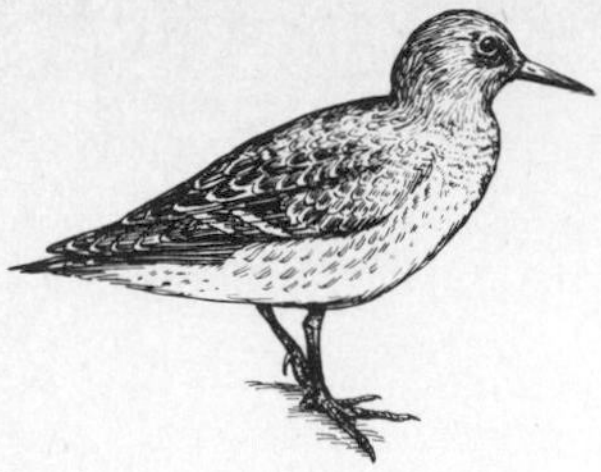

Purple Sandpiper

Purple Sandpiper, 235
Erolia maritima
Length to 9½ in., including 2⅗-in. tail and 1⅖-in. bill. Wingspread to 16 in. Weight about 3 oz, with sexes in general equal, though bill of female is usually slightly longer than that of male. A dark sandpiper, robust, with white belly, short yellow legs, and yellow or orange base to bill. Conspicuous white on dark wing in flight.

Breeds probably from Melville Island, Ellesmere Island, and northern Greenland, to southern Greenland, southern Baffin Bay, and islands in Hudson Bay; also in Spitsbergen, Iceland, Norway, Russia, and Siberia. Winters in America from southern Greenland south to New Jersey, and in Europe to the Baltic states, British Isles, and Mediterranean. A winter seashore bird in New England. *The* winter sandpiper on northern beaches.

Nests on ground, either among mosses or on bare ground. Nest a mere depression, lined with grass or leaves. Eggs three to four, 1½ by 1⅒ in.; pear-shaped, greenish, clay, or drab with conspicuous brown markings of various shades. Nesting period May to June. Incubation 21 to 22 days, reportedly by male alone. One annual brood.

Food, snails and other small shelled animals, gleaned mainly from among drifts of seaweed and the like. Also feeds on animals gleaned from wave-washed rocks. Birds can swim and usually appear in flocks. In New England appear in September, are most abundant in November, and have vanished by mid-March.

Protected by law, but even when it was legal to shoot them they were not considered legitimate prey, largely because of the strong fishy flavor of their flesh.

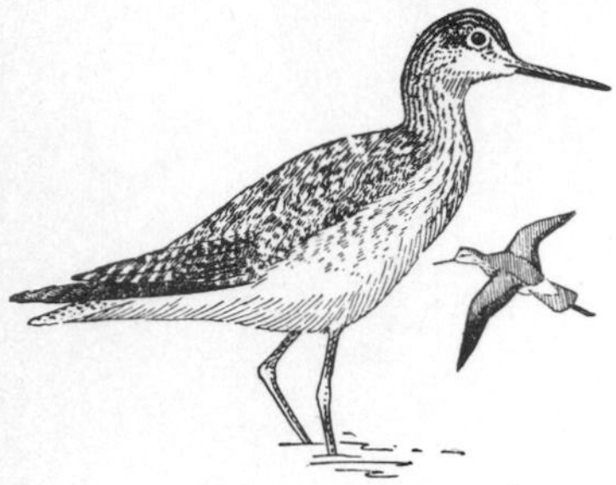

Lesser Yellowlegs

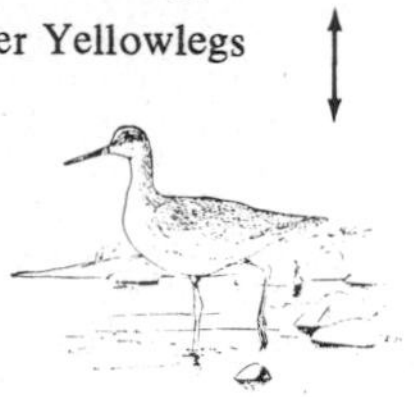

Lesser Yellowlegs, 255
Totanus flavipes
Length to 11 in. Wingspread to 21½ in. Weight to 3½ oz. Female larger than male. Slim gray and white bird with bright yellow legs. Smaller than greater yellowlegs, but otherwise much like it. Fine marbling on inner surface of flight feathers of greater yellowlegs is less conspicuous in lesser yellowlegs. Bill of lesser yellowlegs perfectly straight, that of greater yellowlegs often appears slightly upturned.

Breeds from Yukon Valley in Alaska to Ungava and northern Quebec, south to British Columbia and southern Manitoba; formerly to just south of Great Lakes. Winters in Mexico, Louisiana, Florida, and the Bahamas and south to Argentina, Chile, and Patagonia, with greater concentration in southern part of area. Haunts shores of larger bodies of water.

Nests not far from water on dry open ground, among scattered spruces; a hollow, with small lining of leaves and grasses. Eggs four, 1⅔ by 1⅒ in.; pear-shaped, dark grayish or light, with blotches of chocolate or brown and ash-gray. Nests in June. Incubation mostly by female. One annual brood.

Like to remain in flocks wading about together in shallow waters of bars and shorelines. Usual call of two whistled notes, less vigorous than strong call of greater yellowlegs. Not shy and often may be studied close at hand while they eat or wander about. Food essentially smaller invertebrate animals, including many insects.

Formerly considered a game species but is now fortunately protected by law. Birds of a flock may return to see what happened to those that may have been shot, hovering about and offering easy targets.

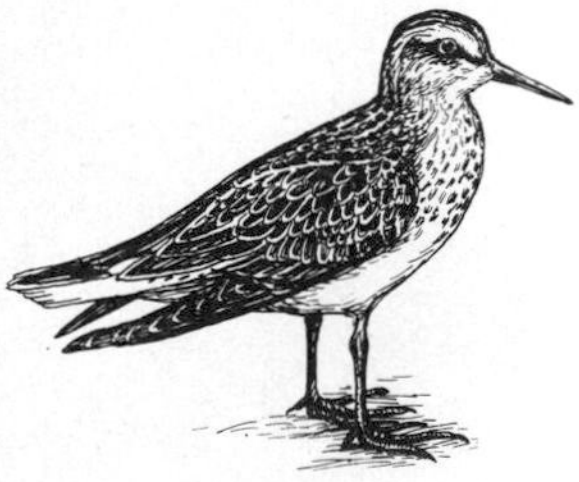

Pectoral Sandpiper

Pectoral Sandpiper, 239
Erolia melanotos
Length to 9⅗ in., including 2⅗-in tail and 1½-in. bill. Wingspread to 18 in. Weight to 4 oz, male being larger than female. Larger than spotted sandpiper; shows a contrasting white belly, streaked brown breast, and four outer tail feathers shorter and lighter than those in middle. Twice the size of least sandpiper.

Breeds in Arctic, from northeastern Siberia to northern Alaska near mouth of Yukon and northeastern Mackenzie and Southampton Island. Winters in South America from Peru and Bolivia to northern Chile, Argentina, and Patagonia, rarely migrating down Pacific Coast but common particularly in fall migration through Mississippi Valley and along Atlantic.

Nests on dry ground near pools; a mere depression, but usually grass-lined. Eggs three to four, 1½ by 1⅒ in.; pointed pear-shaped, pale green, brown, or olive blotched with black or purplish-gray. Nesting period June to July. Male develops throat structure making it possible for him to give a booming note during courtship period.

Food mostly insects such as grasshoppers, weevils, beetles, cutworms, horseflies, mosquitoes, and snails and small shellfish as well as some vegetable matter. Flocks seem to be most common near salt marshes. Birds seem to sit tight when approached, more than many other shore birds. Migration is in small flocks.

Useful destroyer of injurious insects. Formerly considered a valuable game bird because of the nature of flesh but now protected by law at all times. Call has been described as a repeated *kriek*. Birds are seen in northeastern United States in greatest numbers in August and September.

Least Sandpiper, 242
Erolia minutilla
Length to 6⅔ in., wingspread to 12¹⁄₁₆ in., tail 2 in., bill ¹⁄₁₀ in. Hen larger than cock. Smallest of sandpipers, with yellow or greenish-black legs; generally browner than semipalmated sandpiper. Generally streaked brown above and white beneath with breast and throat brown-streaked in spring. Lacks partially webbed feet of semipalmated sandpiper.

Breeds from northwestern Alaska to Labrador and south to Yukon Valley, Newfoundland, and Nova Scotia. Winters from southern California to Texas and North Carolina and south through West Indies, Central America, Brazil, and central Patagonia. Accidentally in Europe, Greenland, and Siberia.

Nests about wet flats or grassy lowlands. Nest lined with some grass and possibly leaves. Eggs four, pear-shaped, drab or pale brown with rich brown markings, 1¹⁄₁₀ by ⅘ in. Incubation largely by cock. One brood a year. Young downy and buff in color when hatched, with pure white chin and throat, darker on upper parts.

Food, myriads of small forms of animals found along shorelines, gathered by dashing in and out with waves, picking up things as they lie exposed and before another wave can come in. Action in flocks made of families or groups of families is worth watching at any time.

Too small to be of any food value to man and is protected by law. Serves as gleaner of waterway margins and undoubtedly as food for many enemies. May well destroy larvae of disease-bearing insects which may be found in its habitat. Called "mud peep" because it prefers grassy mud flats.

Red-backed Sandpiper, Dunlin, 243a
Erolia alpina
Length to 9¼ in., including 2⅓-in. tail and 1⅔-in. bill. Wingspread to 15¾ in. Weight to 3 oz, female being larger than male. In spring, rusty red above with white breast and black belly. In winter plain gray above and grayish across breast. Bill long, stout, and downward-curving at tip.

Breeds on northern Siberian coast and in Arctic America from mouth of Yukon to Churchill on Hudson Bay. Winters from southern British Columbia to New Jersey, and south to southern Lower California, Texas, and Florida. Also found in Asia from China and Japan to Malay Archipelago.

Nests near water, on ground, usually on a dry grassy knoll surrounded by water; a mere depression, with or without plant lining. Eggs four, 1⅗ by 1¹⁄₁₀ in.; pear-shaped, buff, gray, or olive marked, blotched, stained, or spotted with chocolate or gray. Nesting period June. Incubation by both sexes. One annual brood.

Food probably largely crustaceans and water insects including oyster worms. Red-backed sandpipers appear just before geese in fall and after most other shorebirds have gone. They feed in mixed flocks with stragglers of other species, run rapidly, fly nervously, and wade up to their bellies.

Protected by law, and useful to some extent as destroyers of enemies of some of our important shellfish. May also help somewhat in destroying a few agricultural pests, though this is probably never important.

Short-billed Dowitcher, 231
Limnodromus griseus
Length to 12½ in., including 2½-in. tail and 2½-in. bill. Wingspread to 20 in. Weight to 5 oz, female being larger than male. Related long-billed dowitcher, subspecies *scolopaceus*, has 3¼-in. bill. Generally olive-brown but lighter beneath, with white on rump, tail, and lower back, unlike similar common snipe. Breast in spring cinnamon.

Long-billed dowitcher, *L. g. scolopaceus*, breeds from mouth of Yukon to western Mackenzie and south to British Columbia. Eastern dowitcher, short-billed, breeds from central Alberta and to Churchill on Hudson Bay. Long-billed dowitcher winters from Louisiana to California and south to Ecuador. Eastern dowitcher winters from Florida and West Indies to Brazil and Peru. In migration, Eastern outnumbers long-billed on East Coast.

Nest of long-billed dowitcher on bare or marshy ground, in moss or depression, unlined. Eggs four, 1⅕ by 1⅘ in.; pear-shaped, clay or gray with greenish cast, and large distinct brownish spots well-crowded at

(Continued)

Least Sandpiper

Red-backed
Sandpiper

Dowitcher

599

larger end but scattered over surface. Nesting period June. Nest of eastern dowitcher, a grassy knoll surrounded by water.

Food essentially small animals gleaned from habitat, including oyster worms, small mollusks, aquatic insects, and other small invertebrates. Both subspecies feed in moist grassy areas. Both swim well, nodding head while doing so, but prefer saltwater areas more than common snipe. Call of eastern dowitcher, a repeated *tu*; long-billed dowitcher, a single *keek*.

They may be useful gleaners of grassy marshes and may help agriculture to some extent but are probably essentially neutral.

Hudsonian Godwit, 251
Limosa haemastica
Length to 16⅔ in., including 3⅕-in. tail and 3½-in. bill. Wingspread to 28 in. Weight to 13 oz, female being smaller than male. Bill slightly curved upward. Tail with narrow white band at end, with adjacent broader black band and base of tail white. Wings dark underneath, black near body, and narrow marginal white to rear. In flight, outer portions of the wings appear distinctly reddish-brown with blackish tips.

Breeds from lower Anderson River in Mackenzie to Port Clarence, Alaska, and east to Hudson Bay at Churchill and Southampton Island. Winters in Chile, Patagonia, and Falkland Islands, migrating mostly east of Great Plains, going down Atlantic Coast in autumn and up Mississippi Valley in spring. Marbled godwit, *L. fedoa*, of western United States and Canada is a larger, browner species.

Nests in or near water in barren grounds. Nest a leaf-lined hollow. Eggs two to four, 2⅕ by 2¹⁄₁₀ in.; pear-shaped, dark green or olive, obscurely blotched with brown. Nesting dates June. Incubation probably about 24 days, with both sexes taking part. One annual brood.

Food, insects, particularly of waterways, worms, crustaceans, and mollusks; in fact, almost any small animals to be found in bird's environment. Call a relatively low *chip* for such a large bird; when frightened, a sandpiper like chittering. In northern migration, probably flies across Gulf of Mexico in single flight.

Was considered game bird but is deserving of every protection. Since it receives no protection in Southern Hemisphere, it needs all it can get in Northern, particularly as it is here that it breeds. Eventually, if it does not become extinct, intercontinental protection for such birds may be effected. In flight, is characterized by dark under wings.

Sanderling, 248
Crocethia alba
Length to 8⅔ in., including 1⅕-in. tail and 1⅕-in. bill. Wingspread to 16¼ in. Weight to 3 oz, female being larger than male. Common name "whitey" helps identify it as whitish plump sandpiper. Shows conspicuous flashing white stripe on wings. Appears light in fall; rusty in spring. Legs and beak black.

Breeds from Arctic islands, Southampton to northern Greenland, and in Iceland, Spitsbergen, and northern Siberia. Winters from central California to Virginia, south to Patagonia and from the Mediterranean, Burma, and Japan to South Africa and many Pacific islands. Found almost wholly at edge of breakers on sand beaches but less commonly along inland-waterway beaches.

Nests near water but sometimes several hundred feet above sea. Nest a grass- or leaf-lined hollow. Eggs four, 1½ by 1 in.; pear-shaped, pale brown with fine dark spots at larger end. Incubation for 25 days by both sexes. Nesting period June to July. Young downy, with central black line from bill through forehead to top of head, nape buff, and elsewhere dark buff.

Food largely mollusks, worms, crustaceans, and insects of the beaches, but may eat insects inland and may even eat seeds and small amounts of vegetation, such as moss and algae. In migration, sanderlings go along both coasts and a short way into interior of continent. Call a short distinct *kipp*.

Birds are too small to be considered as game. This is the sandy beach resident that chases the retreating waves. If mixed with other sandpiper-like birds, it can be identified by its conspicuous white stripe in the wings.

Order Charadriiformes./Family Recurviostridae [Avocets, Stills]

Avocet, 225
Recurvirostra americana
Length to 20 in., wingspread to 38 in., tail 4½ in., bill to 4 in., curved upward; bare legs, to 6½ in. Black and white, with a rusty head and neck, unlike any other bird with which it might be confused. Female smaller than male.

Hudsonian Godwit

Sanderling

Avocet

Breeds from eastern Washington to southern Manitoba, and south to southern California and Iowa, formerly to New Jersey. Winters from central California and southern Texas to Guatemala, with occasional birds in Ontario, British Columbia, New Brunswick, and Florida. Rare east of Mississippi River.

Nests on ground near shallow water, or even surrounded by shallow water among plants, with or without any structure. Eggs three to four, variable in shape and color, olive to light brown with many small brown spots, $2\frac{1}{10}$ by $1\frac{1}{2}$ in. Incubation 24 to 25 days. Probably only one brood each year. Young in juvenile plumage resemble adults.

Food, grasshoppers and other insects, fish, mollusks, and crustaceans and a variety of other small animal forms available to it in its environment. Performance of avocets about their nests at breeding time is worth going miles to see and their reflection in still water has great beauty to the artist, amateur or otherwise.

Perfectly harmless and essentially useful bird. Ordinarily, habit of eating fish and some other aquatic animals makes flesh unpalatable even when it is cooked. It is worthy of every protection at all times.

Black-necked Stilt, 226
Himantopus mexicanus
Length $15\frac{1}{2}$ in., wingspread 30 in., tail $3\frac{1}{4}$ in., bill $2\frac{3}{4}$ in., bare portion of legs 8 in. Female longer than male but with shorter legs. Black above, white beneath. Sometimes shows white beneath, long, dangling, red legs, and black uniform wings. Young birds duller than adults.

Breeds from central Oregon, Utah, Colorado, Nebraska, and central Florida south through northern Lower California, Mexico, West Indies, Brazil, and Peru. Winters from Lower California to Florida, south through Brazil and Peru. In migration occasionally found in North Dakota, Illinois, New Brunswick, and Bermuda. Rare east of the Mississippi River.

Nests in small colonies or singly, on ground near or in water, among weeds. Eggs three to four, pear-shaped, clay-colored, greenish, slightly glossy and marked with brown, black, or gray, $1\frac{4}{5}$ by $1\frac{1}{5}$ in. Incubation less than 21 days, by both sexes. Young able to run about almost as soon as hatched. Adult at 2 years.

Food aquatic insects, both larval and adult forms; snails, as well as many other small invertebrates and small vertebrates. Also takes some plant food; largely seeds of aquatic marsh plants. Bird is suspicious but noisy when nest is approached; then dances a crazy dance, waves wings, and drags leg as though broken, and generally creates a disturbance.

Does no possible harm to agricultural interests and, if living near farmlands, must do considerable good, so there is no reason why it should not be entitled to complete and adequate protection throughout its range.

Black-necked Stilt

Order Charadriiformes./Family Phalaropodidae [Phalaropes]

Northern Phalarope, 223
Lobipes lobatus
Length to 8 in., including $2\frac{1}{2}$-in. tail and 1-in. bill. Wingspread to $14\frac{1}{2}$ in. Weight under 2 oz, with female larger than male. Conspicuously gray, white below with tan stripes down back and tan fleckings on neck. Breeding female has conspicuous chestnut patch on sides and front of white throat. Male less conspicuous, and both grayer in winter, without chestnut.

Breeds from Pribilof Islands and northern Alaska to Greenland and south to northern Manitoba and northern Quebec and in Iceland, Norway, Russia, and Siberia. Winters off coasts of Peru and West Africa. Migrates off both coasts of America and through interior of Canada and United States to Patagonia. Alights and rests safely on sea.

Nests on ground, among grasses or under grass cover, a depression, commonly grass-lined but not always. Eggs three to four, pear-shaped, greenish to olive to gray with blackish or chocolate blotches, $1\frac{1}{3}$ by $1\frac{1}{10}$ in. Incubation about 20 days, by male only. One annual brood. Downy young tan with dark brown saddle; lack a dark line at base of bill found in red phalarope.

Food, minute animals of sea or land, including numerous crustaceans, insects, mollusks, and their kin. They flutter restlessly from one place on surface of the sea to another and do not seem greatly frightened by approaching ships. Most abundant in New England in May, August, and September when they appear with flocks of red phalaropes.

Of little if any economic importance. Sailors use them as a sign of land as indicated under the similar red phalarope, but during the migratory season this may be misleading.

Northern Phalarope

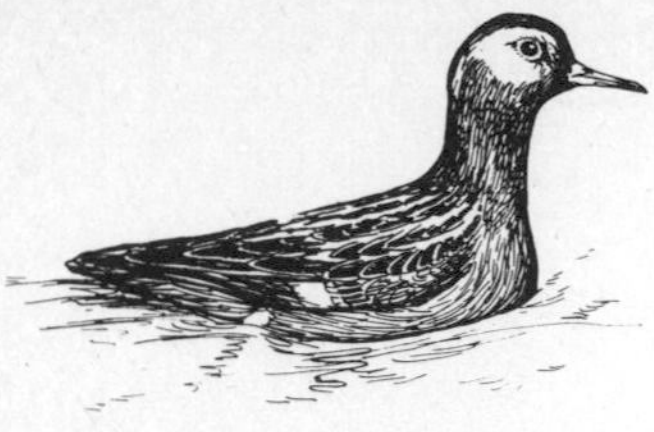

Red Phalarope

Red Phalarope, 222

Phalaropus fulicarius

Length to 9 in., including $2\frac{4}{5}$-in. tail and 1-in. bill. Wingspread to 16 in. Weight just under 2 oz, with female larger than male. In winter grayish and white. Breeding female brighter in color than male but both are reddish beneath, with white or light-colored cheeks. Breeding female gray on top of head and under chin.

Breeds from northern Alaska, northern Ellesmere Island, and Greenland south to Yukon and Southampton Island; also from Iceland to Spitsbergen to eastern Siberia. Winters on ocean off coasts of South America and western Africa. Occasionally found in East from New Brunswick to Florida. Essentially a sea bird migrating well offshore but occasionally found inland.

Conspicuous because female is more showy than male. Nests in grasses or marshy areas usually not far from water; a slight depression, usually grass-lined and usually overgrown with plants. Eggs three to six, $1\frac{1}{3}$ by $\frac{9}{10}$ in.; long and pear-shaped, buff to clay with large usually chocolate blotches. Nesting period June. Incubation 23 to 24 days, by male only. One annual brood. Seabirds in migration; shorebirds on the breeding grounds.

Food, insects, small crustaceans, jellyfish, leeches, small fish, and some vegetable matter such as weed seeds and seaweeds. Floats easily and lightly on seas and rises readily from surface. Has a wider bill and more yellowish legs than its associate northern phalarope. In winter plumage, it also shows more gray on sides of breast.

Female

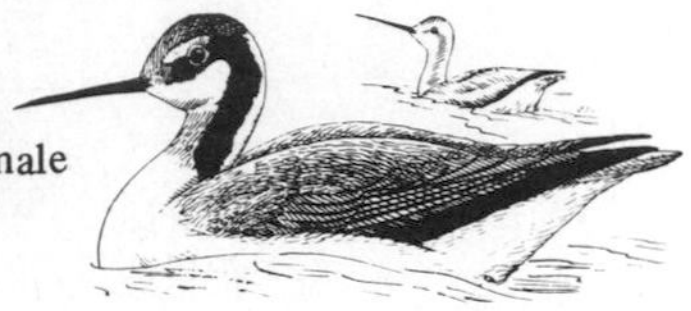

Wilson's Phalarope

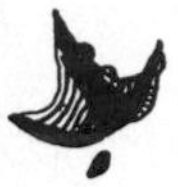

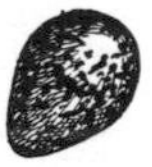

Wilson's Phalarope, 224

Steganopus tricolor

Length to 10 in., wingspread to 16 in., tail to $2\frac{1}{3}$ in., bill to $1\frac{1}{2}$ in. Female larger than male. This "swimming sandpiper" like other phalaropes has front toes lobed and hind toes elevated. In breeding season, a broad cinnamon stripe on neck which changes to black. In flight, shows a white rump and is dark-winged, with no wing stripe.

Breeds from southern British Columbia to northwest Indiana, south to central California, Utah, and Colorado. Winters from central Chile and Patagonia, south to Falkland Islands. After breeding season, most phalaropes go to sea. There are only two other species of phalaropes in the world, the red and the northern which is the common phalarope of lakes and ponds.

Female is more brilliant at breeding time, and does courting. Male makes nest, incubates eggs, and performs many acts ordinarily done by female. Nest a collection of vegetation on ground. Eggs three to four, long and pear-shaped, light buff, clay, or drab splashed with brown, $1\frac{1}{5}$ by $\frac{9}{10}$ in. One yearly brood.

Food, insects such as mosquitoes, leaf beetles, and pill bugs, these being more terrestrial than food of other phalaropes. Birds are active, particularly at breeding time, when a female may court and apparently win more than one mate, she of course supplying the eggs for two nests.

Probably of some value in destroying insects. Too small to be considered as game and too interesting to be destroyed or to permit their being exterminated further in their natural range. A newly hatched phalarope seems like a good-sized bumblebee. It can disappear in sedges remarkably easily.

Order Charadriiformes. / Family Laridae [Gulls, Terns]

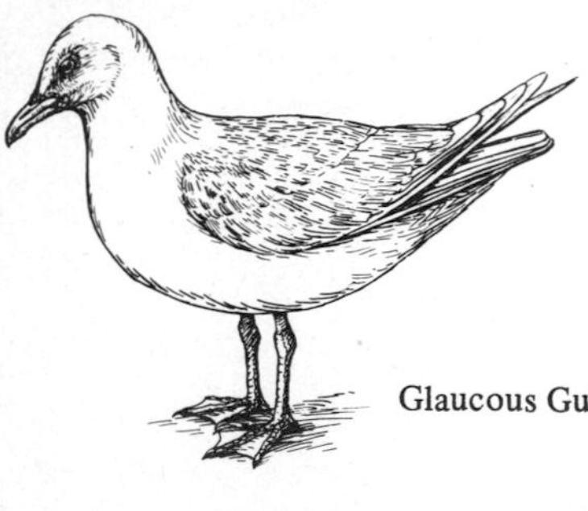

Glaucous Gull

Glaucous Gull, 42

Larus hyperboreus

Length to 32 in., including $8\frac{1}{2}$-in. tail and 3-in. bill. Wingspread to $5\frac{1}{2}$ ft, with female smaller than male. Chalky white, without dark wing tips, with large strong bill. Sexes colored alike. First winter, cream or pale buff, with wing tips lighter. Second-year birds almost white throughout. Adults with white somewhat overcast with gray on back.

Breeds in Arctic from northwest Alaska to Ellesmere Island and northern Greenland, south to Pribilof Islands, Hudson Bay, and Newfoundland. Winters from Aleutians to Greenland, south to California, Texas, and North Carolina, occasionally in Hawaii, Europe, Asia, Japan, and other parts of Northern Hemisphere. Rare winter visitor along Pacific Coast.

Nests singly in small colonies, on cliffs or low islands. Nest a mere depression, or a good-sized mound of vegetation or drift. Eggs two to four, $3\frac{1}{3}$ by $2\frac{1}{5}$ in.; variable in color, olive, brown, buff, drab, clay, or pale blue-green, spotted or blotched. Incubation 28 days. One brood a year. Downy young grayish-white above, white below, with smoky gray on back.

Food, dead fish, carrion, manure, shellfish, and almost any waste organic material. Powerful, and inclined to take food from other birds or even kill smaller birds for food. Found along New England coast from November to April, usually at sea but rarely resting on water although it does alight on beaches.

One of most destructive of gulls, undoubtedly breaking up some nests of eider ducks but also doing valiant service as a waste destroyer. Like other gulls, not commonly found more than 50 miles from shore. One of most beautiful as well as one of largest of gulls and must be admired for its flying ability.

Great Black-backed Gull, 47
Larus marinus
Length to 31 in., including 2½-in. bill. Wingspread to 5½ ft. Back is dark slate rather than black, though wing tips are black. Neck, belly, tail, sides, and head of adult, pure white. Legs and feet yellowish or flesh-colored. Eye and beak yellow, but lower bill red spot near tip. Young birds with brown-flecked upper parts. Has some interesting common names: black-back, saddle-back, coffin-bearer, minister, turkey gull.

Breeds from northern Labrador to central Greenland, Iceland, and south through Nova Scotia, British Isles, Scandinavia, and Russia. Winters from southern Greenland, through Black Sea and Caspian Sea area and south to Great Lakes, Delaware, and Mediterranean, sometimes as far as Florida. Has been found in Nebraska.

Nests in pairs or colonies, on coasts of lakes or sea. Nest a mere depression or accumulation of waste vegetable material. Eggs two to five, 3⅖ by 2⅕ in.; olive, buff or drab splotched with brown, white, black, or gray. Incubation 28 days, by both sexes. One annual brood. Downy young pale gray, white on breast; upper parts, deep gray.

Food: a scavenger but also destroyer of eggs, young, and even adults of other birds the size of ducks and coots. May destroy insects and rodents as well as birds but does eliminate many wastes and will steal from other less powerful birds. A noisy bird, repeating *cak* often.

Difficult to evaluate this bird wholly. To those who have visited a nesting colony and watched great numbers of them in flight, it has a place in nature which should not be left unfilled, but sportsmen who are themselves interested in killing ducks may not feel so kindly disposed toward it.

Great Black-backed Gull

Herring Gull, 51a
Larus argentatus
Length to 26 in., wingspread to 58 in., tail to 7½ in., bill to 2½ in. Female slightly smaller than male. Gray back, white head, tail, and underparts, with black wing tips and flesh-colored legs. Young bird dusky brown first year; mottled, the second. Tail entirely dark first year, bare white the second. Larger and without ring bill of ring-billed gull. This is the common "sea gull" of both coasts and along major waterways of the interior of the U.S.

Breeds from south central Alaska to southern Baffin Bay, south to northern British Columbia, northern New York, and Massachusetts. Winters from southern Alaska to Gulf of St. Lawrence and south to Lower California, Mexico, Cuba, and Yucatan. Known from Oregon Pleistocene.

Nests on ground or rarely in trees, with little nesting material. Usually nests in colonies on islands in the sea or in fresh water. Eggs three to five, variable, light blue, green, drab, brown, or cinnamon, spotted or blotched, 2⁹⁄₁₀ by 2½ in. Incubated by both sexes, 24 to 28 days. Young birds able to walk about soon after hatching. One annual brood.

Food, variety of materials picked from surface of sea or waterway or gleaned from farmlands or garbage heaps. Related species assisted Mormons in surviving a plague of locusts by settling in great flocks and cleaning up pests; for this Mormons erected a monument in Salt Lake City. Flight speed, 36 mph.

Scavengers and enemies of injurious insects but also enemies of ducks and other birds nesting in same territory, where eggs and young provide a ready and delicious feast. Birds are rarely abundant enough to be considered seriously injurious. California gull is state bird of Utah.

Herring Gull

Ring-billed Gull, 54
Larus delawarensis
Length to 20 in., including 6-in. tail and 1⅗-in. bill. Wingspread to 50 in. Wing tips black with conspicuous white spots. Back and shoulders pale bluish-gray; head and underparts almost pure white. Black ring around bill. Legs and feet yellow-green. First year, brown-flecked on white, with darker wing tips and dark tail. Second year, like adults but with black-tipped tail.

Breeds from southern Alaska to north shore of Gulf of St. Lawrence, south to southern Oregon, southern Colorado, and northern New York. Winters from British Columbia to Maine and south to central California, southern Mexico, Gulf Coast, and Cuba. Also found in Bermuda and Hawaii. Essentially a bird of inland waterways.

Ring-billed Gull

(Continued)

Nests in colonies, sometimes with other water birds, usually on ground but sometimes in low tree. Eggs three, 2½ by 1⅘ in., in general like small herring gull eggs. Incubation 21 days. One brood a year. Nesting period May to August. Young, with down, of two color phases: some smoky gray and others buff with darker spots.

Food: essentially a scavenger, eating dead fish, rodents, small aquatic animals, and sometimes contents of nests of other birds. Known to eat many grasshoppers. Haunts garbage dumps and follows ships, seeking refuse that may be dumped overboard. Call a protesting, repeated *kree, kree*. When in flock, gives more subdued *kow, kow*.

This was once the common gull of America, but it no longer is. It is undoubtedly most valuable as a destroyer of garbage and as a general scavenger. Beautiful to watch in flight or on a perch facing into wind when in a flock.

Laughing Gull, 58
Larus atricilla
Length to 17 in., including 5½-in. tail and 1⅗-in. bill. Wingspread to about 3 ft. Female smaller than male. Sexes colored alike, upper parts of wings being mostly dark gray but with conspicuous white border along hind margin of wings. Head and throat white in winter but dark slate in summer, over a smaller area than in Franklin's gull. Young show white rump. Common name apparently from sounds produced which many interpret as laughter with "ha, ha, ha" heard by many observers.

Laughing Gull

Breeds from Maine to Florida, Texas, southern California, Venezuela, and Lesser Antilles. Winters south from South Carolina and Gulf Coast to Brazil, Peru, and Chile but occasionally found inland to Quebec, Iowa, Wisconsin, Nebraska, and Colorado. Also found sometimes in Europe and Bermuda. Found along Atlantic Coast of North America, April to September.

Nest in colonies, usually on sea islands and often with terns, on ground, often a mere depression, but frequently well-built of vegetable materials, concealed or not. Eggs two to five, 2¹⁄₁₀ by 1½ in.; brown to cream to buff to green, sometimes marked with brown or purple. Incubation about 20 days. One annual brood. Nesting period April to June.

Food almost any available animal matter, living or dead. Has record of being exceptionally good insect destroyer. Probably does not dive for food but gleans it from water surface or shore. May rob terns and other smaller birds of food but will also gang up on and rob pelicans. No record of its destroying other birds' nests, though it may.

Chief value is as scavenger and as destroyer of insects. Flocks of the birds are of great beauty. May be confused with Franklin's gull, whose extreme wing tips are white with touches of black, while laughing gull has black wing tips with white only at extreme tips of a few feathers. Legs of these gulls dark red or brownish-gray.

Bonaparte's Gull, 60
Larus philadelphia
Length to 14½ in., including 4-in. tail and 1⅕-in. bill. Wingspread to 32 in. Female smaller than male. In breeding plumage, back gray, body white, with margins of wing tips almost black but rest of wings white, with face and fore half of head dark gray to black, bill black, and legs red. In winter, head may be mostly white except for spot behind eye.

Bonaparte's Gull

Breeds from northwestern Alaska and northern Mackenzie to central British Columbia and central Alberta. Has been found during breeding season but not breeding from New Jersey to Rhode Island. Winters from Maine to Florida, on to coast of Yucatan; on Pacific Coast, from Alaska to Lower California and western Mexico; also accidentally in Great Britain, Peru, and France.

Nests in colonies on forested lake area, inland. Nest on stumps or in trees, often 20 ft above ground, usually in cone-bearing trees. Nest of sticks and twigs, with softer lining. Eggs two to four, 1⁹⁄₁₀ by 1⅓ in.; olive-gray to dark olive or brownish, with chocolate markings. Nests June to July. One annual brood.

Food usually small fish but during summer months is one of best insect destroyers, consuming large quantities of flies, ants, moths, and the like. Not known to feed on any crops grown by man. Call a rasping plaintive cry or whistle, but usually rather silent for a gull, except near nest, where it attacks without provocation.

Undoubtedly a useful bird so far as man's interests are concerned. Best field characters include red legs, white triangle with black tips found at wing tips, black bill, and dark head. Immature birds have a narrow dark band near tail tip as well as black spot behind eye common to adults in winter plumage.

Common Tern, 70
Sterna hirundo

Length to 16 in., including 7-in. tail that is forked half its length and 1½-in. bill. Wingspread to 32 in. Female smaller than male; young as short as 9 in. Lower face, neck, belly, and tail pure white. Upper wing surfaces and back light blue-gray, with ends of wings dusky. Bill orange-red with black tip. Feet orange-red. Black caps restricted to nape in winter.

Breeds from Great Slave Lake to northern Manitoba and Gulf of St. Lawrence, south to North Dakota and North Carolina; also in Bahamas and on coast of Venezuela, West Indies, Gulf Coast, Europe, Asia, and Africa. Winters from Florida to western Mexico and south to Straits of Magellan.

Nests on islands, in sea or lake or marsh, a mere hollow or well-built mass of vegetation, sometimes on bare rock. Eggs two to six, 1⅗ by 1⅕ in.; highly variable in shape and color even in same nest, from white to buff, olive, or green, with or without markings. Nesting period June to July. Incubation 21 days, by both sexes. One annual brood.

Food chiefly small fish, but also insects and other small animals. Flying ants, butterflies, and cicadas seem to be favored. While birds can rest on water, they seem much more at home in air. Will work in concert helping a wounded companion or fighting a common enemy. Resembles Forster's tern but has darker wing tips.

Useful to fishermen as indicators of location of schools of fish. Probably never eat marketable fish themselves and do destroy many injurious insects. Part of sea's beauty and always worth watching. Common tern is not so common everywhere as closely related Forster's tern.

Common Tern

Arctic Tern, 71
Sterna paradisaea

Length to 17 in., including 8-in. tail with 5-in. fork and 1⅓-in. bill. Wingspread to 33 in. Cap black. Tail white. Bill slender and all red. Throat and belly white. Upper surface of wings gray, with rear half much darker. Feet bright red. Red of feet and bill is much richer than in common and Forster's terns.

Breeds from northern Alaska to Baffin Bay, northern Ellesmere Island, and Greenland, south to southeastern Alaska, northern Manitoba, and Maine and in Arctic Europe and Asia. Winters in Antarctic Ocean south to 74°S. Seen in migration along Atlantic and Pacific coasts of North and South America and in Africa.

Nests on rocky or pebbly shores of lake or sea islands, with little or no nesting material. Eggs two to three, 1⅔ by 1⅕ in., highly variable in color. Nesting season June to July. Incubation 21 days, by both sexes. One annual brood. Downy young have dusky patch on forehead, unlike other young terns, but otherwise are much like common terns.

Food chiefly small fish, crustaceans, and other small forms of animal life picked up for most part while in flight over surface of water. Seems to prefer nesting by itself rather than in colonies of other related birds. From May to July in Arctic is rarely seen over 100 miles from land.

Probably not of great economic importance but of intense interest to bird students because each year it travels from the Arctic to Antarctic and return. Except during breeding months, it may be seen at almost any distance from land over almost any ocean. Journey from Arctic to Antarctic each year allows it to enjoy more sunlight than any other bird.

Artic Tern

Sooty Tern, 75
Sterna fuscata

Length to 17 in., including 7½-in. tail with 3½-in. fork and 1⁹⁄₁₀-in. bill. Wingspread to 34 in. Sexes colored alike. Cap greenish-black. Forehead white. Throat and forward part of neck and breast, white. Back, tail, and upper surface of wings sooty black in breeding plumage. In winter white feathers mixed with dark in crown.

Breeds in Dry Tortugas, West Indies, and tropical islands of the Atlantic from Venezuela to British Honduras and formerly to Texas. After breeding season, may wander north along Atlantic Coast to Nova Scotia. Winters from Brazil and Louisiana to Falkland Islands. Has been found on occasion in England and in France.

Nests in colonies, many times of enormous numbers, on sand of islands, a mere depression, or nothing. Eggs one to three, 2¹⁄₁₀ by 1½ in.; white, cream, or buff, slightly spotted with brown or purple but highly variable. Incubation 26 days. One annual brood. Nesting season April to May. Downy young sooty or gray streaked above, dull white beneath.

Food, small animals of the sea including, of course, fish. Of more limited range than most terns here considered but is one most easily recognized by its contrasting black and white, with dark areas more extensive than in most terns. Often blown out of its normal range by cyclonic storms and records from remote places usually follow such storms.

Probably of little economic importance; to ornithologists in northern Atlantic points it is of chief interest as a rather unusual visitor. It probably takes over a year for birds to develop fully mature plumage.

Sooty Tern

Black Tern

Black Tern, 77
Chlidonias niger
Length to 10¼ in., wingspread 25 in., tail ¾ forked, to 1 in., bill, 1 in. Head and underparts of body black, back and wings gray with narrow light streak along front of wing, in breeding plumage. Young and winter adults with head and underparts white and back and wings gray, general appearance mottled. The only black-bodied tern.

Breeds in interior of North America, from central eastern Alaska, to central New York and south to California, Nevada, Missouri, Tennessee, and Pennsylvania. Winters from Mexico to Panama, Peru, and Chile. Appears regularly along Atlantic Coast in late summer and autumn.

Nests in colonies, in marshes or wet areas. Nest of plant materials, sometimes floating, sometimes well-made, and sometimes of little structure. Eggs two to five, olive·to yellow to brown and pointed at one end, 1⅓ by 1 in. Incubation 21 to 22 days, by both sexes. One or two annual broods.

Food largely insects, picked up in flight or from ground. Often over cultivated lands and also eats small fish, mollusks, and crustaceans. More of an insect eater than common tern. Its short tail, deep wingbeats, darting flight, dark body, and small size are good identification points.

Where abundant, black terns may be valuable as checks on insect enemies of agricultural crops. Also called short-tailed tern, semipalmated tern, sea pigeon. Sometimes referred to as the aquatic swallow of the sloughs of the northwest.

Caspian Tern

Caspian Tern, 64
Hydroprogne caspia
Largest of the terns. Length to 23 in., wingspread to 55 in., tail to 6¾ in., forked 1½ in., bill to 3 in. Distinguishable from herring gull by forked tail, black crown, and large red bill. Caspian tern, tail forked ¼ its length; royal tern, tail forked ½ its length. Arctic, common, and Forster's terns smaller than Caspian.

Caspian tern breeds from Great Slave Lake to Gulf of St. Lawrence and south to central Lower California, Texas, Louisiana, and South Carolina. Winters from southern California to South Carolina and Mexico. Similar royal tern nests only from Virginia to Texas.

Nests in colonies, on ground, with or without any nesting materials. Eggs one to three, broad, buff, sparingly marked with brown and gray, 2⅔ by 1⁹⁄₁₀ in. Incubation about 20 days. One annual brood. Young birds downy, then frowsy, then feathered. First winter plumage pale gray. Adults have two complete molts each year.

Food, small fish and other animals which swim at surface of sea, also mussels and eggs and young of other kinds of birds. Surface feeding worth watching, as birds fly swiftly and bill down, ready to grab whatever appears.

Probably does little damage except in destroying eggs of other terns. Rarely sufficiently abundant to be considered a pest; at all times worth watching in exceptional flying antics.

Brown Noddy Tern

Brown Noddy Tern, 79
Anoüs stolidus
Length to 16 in., including 7-in. tail and 2-in. bill. Tail not forked. Top of head grayish-lavender to white, with a black band from bill to eye. Upper parts dark brown, with main tail and wing feathers black. Under parts dark brown but lighter under wings. Bill black. Feet brownish-black with yellowish webs.

Breeds throughout most of year in one form or another, since there are races in Indian and Pacific Oceans as well as in Atlantic. Found near tropical and subtropical seas of world except on west coast of South America. Nests with sooty terns on Dry Tortugas during April and May. Winters at sea near the breeding areas.

Nests on branches of trees or shrubs. Nest of seaweed and other vegetation piled rather carelessly together. Egg, one, variable, 2 by 1⅓ in. Young down-covered when hatched, more or less yellowish with darker markings. Both parents aid in defense and feeding of young. Young begin to wander from nest relatively soon.

Food mostly small animals like fish gleaned for most part from sea while in flight. Terns in general are not scavengers like gulls, and so feed on living insects, crustaceans, fish, and like. Common noddy tern is only slightly larger than lesser noddy, with darker crown than white-capped noddy.

Probably from an economic standpoint noddies cannot be justified since they feed largely on living fish that have an indirect use to man even though fish eaten are themselves too small for food for man. Seen far out at sea at almost any time of year, but only in tropical or subtropical areas.

Order Charadriiformes./Family Rhynchopidae [Skimmers]

Black Skimmer, 80
Rynchops nigra

Length to 20 in., including 6-in. tail with 1½-in. fork, and long bill, lower part being to 4½ in. long while upper bill is to 3 in. long. Wingspread to 50 in. Female smaller than male. Like slender long-billed gull, with black upper parts and white under parts. Bill much compressed, red, and black-tipped.

Breeds along Atlantic Coast of North America, from Long Island to Florida and Texas, sometimes wandering to Bay of Fundy and Lake Ontario. Winters from Gulf Coast to northern and eastern coasts of South America. Has been found inland to Tennessee. Allied race is found in interior of South America.

Nests in colonies, on high sand flats and shell beaches. Nest a mere depression, without vegetable lining. Eggs three to five, 1⅔ by 1⅓ in.; white, greenish, or buff, spotted or blotched with brown, gray, or lavender. Incubation probably only by female. One annual brood. Nesting period mid-May to mid-July.

Food almost wholly small fishes and crustaceans. Flies gracefully close to surface of water with long lower bill skimming or cutting surface. May wade in shallow water, turning bill on side to capture food. Bills of young birds are more nearly equal than those of adults, so they can pick food from surface of ground.

Probably of no economic importance and certainly not sufficiently abundant to be serious enemy of any living thing useful to man. Eggs large and illegally eaten by man, but flesh is not edible. Egg collectors have probably eliminated species over much of its original range.

Black Skimmer

Order Charadriiformes./Family Alcidae [Auks, Murres, and Puffins]

Atlantic Murre, 30
Uria aalge

Length about 17 in., including 2⅕-in. tail and 2⅖-in. bill. Wingspread to 30 in. Female smaller than male. Dark sooty-brown above, with darker areas sometimes showing on wings. Breast and abdomen white. Sides more or less streaked with black and white. Bill of related Brunnich's murre heavier with pale bluish stripe at base.

Breeds from southern Greenland through Labrador and Gulf of St. Lawrence to Nova Scotia, and along northwestern Europe, Iceland, and Shetland Islands. Winters south to Maine and sometimes into Massachusetts, and in Europe down to Mediterranean and coast of Morocco. Sometimes found in Hudson Bay. California murre, *U. a. californica*, extends breeding range to islands from Washington south to central California; winters along the coast to southern California.

Nests in colonies, on islands, out to sea. Nest, none. Eggs laid on bare ground, often even in no depression. Egg, one, 3½ by 2 in.; light green to blue or cream, washed, spotted or marked with brown, lavender, or olive. Incubation about 30 days. One annual brood. Nesting period late May through July. Young by first winter like adults but smaller. Downy young gray-brown.

Food, small fishes, crustaceans, and other marine animals. Murres can dive almost instantaneously and swim vigorously under water with assistance of their wings; in fact, they can swim with wings under water more rapidly than they can swim with their feet on water surface. Brunnich's murre is common somewhat farther south than Atlantic murre.

Murres, like other sea birds, have been killed by thousands by seagoing men and by others. Eggs were easily collected and provided a fresher food than was formerly available on long voyages. Egg has a red yolk and a sky-blue surrounding albumen when fried. Eggs have been prized by collectors and this has reduced numbers unreasonably.

Atlantic Murre

Razorbill, 32
Alca torda

Length to 16½ in. Tail to 3½ in. Bill to 1¼ in. Wing to 7¼ in. Tail of 12 feathers forming a wedge. Head, neck, and underparts black with white line from eye to beak. Underparts white. Beak narrow, deep, crossed by white band, with inside of mouth yellow. Feet black. Described as a bird with a heavy head, bull neck, and deep, flattened bill which is crossed with a white mark at its middle. Formerly called razorbilled auk.

Ranges through North Atlantic, sometimes on American coast south to North Carolina but mostly north of Maine. On European side it ranges south to Azores and Gibraltar. Winter range on our coast may bring birds south to Long Island. Breeding range is Gulf of St. Lawrence area northward.

Breeding period is June to August. Nesting done in colonies on rocky cliffs where egg is laid on rock without a protective nest, usually in a cranny. During incubation period, birds give hoarse grunts or groans. Egg pale blue, gray, or white with chocolate speckles; 3 by nearly 2 in.

(Continued)

Razorbill

Food almost exclusively fish caught by a lightninglike dart of head. Underwater swimming is done with help of wings. In swimming on surface bird carries tail raised. In early days, related great auk was sought by men for feathers, flesh, and oil but since 1844 no living great auk has been found.

In Greenland, auks today provide food and clothing for natives. May be considered as enemies of fish but not so seriously that they should be forced to follow way of the related great auk. Superficially auks may have some resemblance to penguins of Southern Hemisphere, but are not closely related.

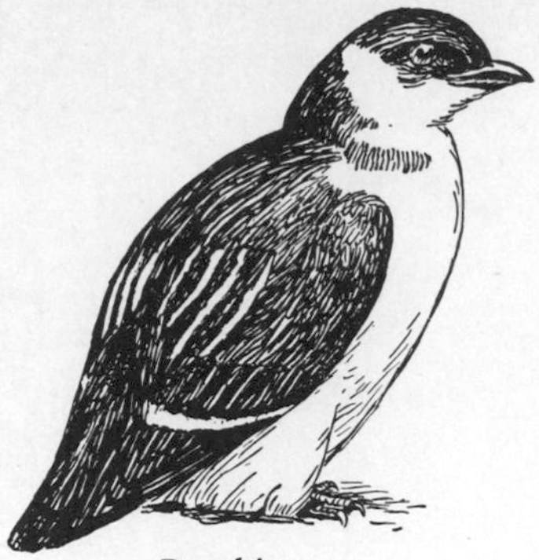

Dovekie

Dovekie, 34
Plautus alle
Length to $9\frac{1}{10}$ in., including $5\frac{1}{5}$-in. tail and $\frac{1}{2}$-in. bill. Wingspread to $15\frac{1}{2}$ in. Female smaller than male. Upper parts, head, neck, and breast black. In winter, throat is whitish and nape sometimes gray. Underparts, including lower breast and abdomen, white. Bill short, thick, and black; yellow inside. Feet flesh-colored. Appears chubby and neckless.

Breeds on northern coasts of Greenland, Iceland, Spitsbergen, and Novaya Zemlya. Winters from southern Greenland to New York or even to South Carolina; in Europe, to Azores and Canary Islands. Sometimes after severe storms at sea many are blown inland. Essentially a sea bird, about size of starling, and conspicuously black and white.

Nests on cliffs or exposed rocks, usually in crannies of loose rubble. Nest a mere depression or sometimes with a lining of vegetation. Eggs one or rarely two, $1\frac{4}{5}$ by $1\frac{1}{5}$ in.; pale greenish-blue and usually unmarked. Nesting period June to July. Incubation 24 days, by both parents. One annual brood. Downy young sooty above and lighter gray beneath.

Food mostly marine animals such as small fish, crustaceans, and similar organisms but also known to have seaweed in stomach. Young are fed crustaceans carried in considerable quantities in parent's gullet. Call a harsh *squeak*. Found in considerable flocks far out at sea, in almost any weather, flying or diving. Use wings under water.

Of little direct economic importance to most men, but a lifesaver to residents of Far North where the Eskimos eat the birds and their eggs in season. Is known as a "little auk." Perfectly at home in air or on water but almost helpless on land.

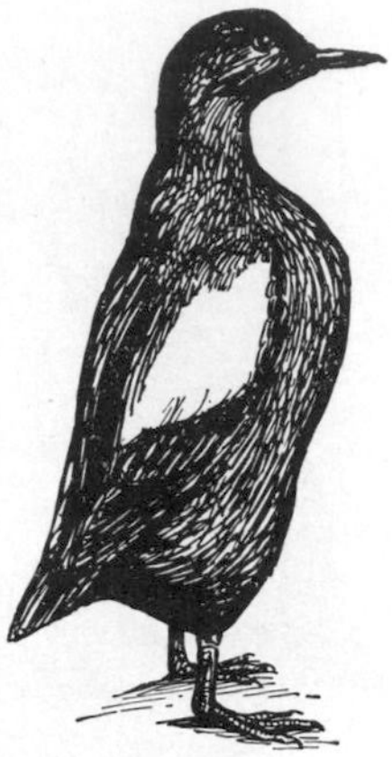

Black Guillemot

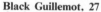

Black Guillemot, 27
Cepphus grylle
Length to 14 in., including 2-in. tail and $1\frac{1}{3}$-in. bill. Wingspread to 23 in. Male usually slightly larger than female. Sexes colored alike. Upper parts black with greenish sheen. Under parts slightly lighter than upper parts. Conspicuous white patch shows on upper part of wings near body. Downy young sooty-black above and paler beneath. Red feet; red lining to mouth is shown when bird displays.

Breeds from central Labrador south to Nova Scotia and Maine, and in Europe from Scandinavia to northern Scotland and in Iceland. Winters in America, from Cumberland Sound south to Cape Cod and sometimes to New Jersey; in Europe, south to northern France. Mandt's guillemot, a subspecies, ranges farther north; pigeon guillemot is found along Pacific Coast.

Nests on cliffs, near the sea, on sea islands, usually in some inaccessible place far back in some rock crevice, rarely in open on bare rock. Eggs one or more commonly two, $2\frac{1}{2}$ by $1\frac{3}{5}$ in.; nearly elliptical; white, blue, green, or creamy with larger spots, if any, near larger end. Nesting date June to July. Incubation 21 to 30 days.

Food, small fish, mussels, shellfish, crustaceans, and sea worms for most part. Known to eat some seaweed. Flies usually swiftly and close to water, with white wing patches showing conspicuously and red feet stretched out behind. Under water, uses wings in swimming and can move about rapidly there. Takes to air easily. Call a faint, piping whistle.

Not of any great economic importance. Gregarious birds often seen sitting on rocks close to the sea. In the past, nested in large colonies. Eggs were harvested annually to the detriment of the species.

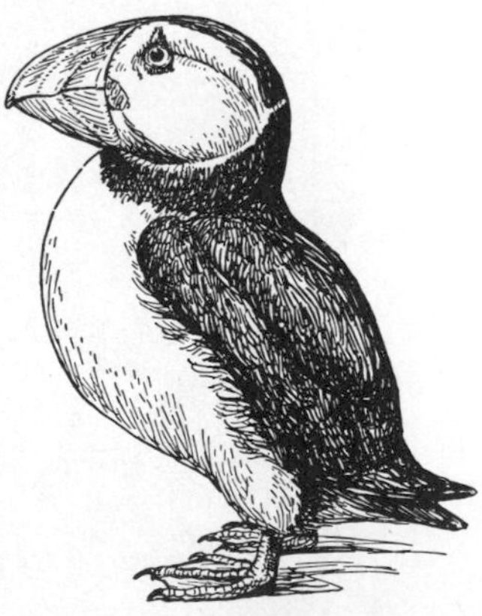

Atlantic Puffin, 13

Atlantic Puffin, 13
Fratercula arctica
Length to $13\frac{1}{2}$ in., including $2\frac{1}{5}$-in. tail and $1\frac{9}{10}$-in. bill. Wingspread to 24 in. Female smaller than male. Sexes alike. Upper parts, wings, tail, and collar blackish. Face grayish-white in summer, smoky in winter. Breast and abdomen white. Feet orange. Bill large and parrotlike but not hooked, deeply ridged, marked with red, yellow, and blue.

Breeds from southern Greenland and Ungava Bay south to Nova Scotia, Bay of Fundy, and Maine, also in Norway and from British Isles to Portugal. Winters south to Massachusetts or New York and New Jersey and in the Old World to Morocco and Azores. Definitely a bird of the sea where there are rocky shores. Horned puffin, *F. corniculata*, appears as an occasional winter visitor on the West Coast in winter. Tufted puffin, *Lunda cirrhata*, Pacific Coast resident offshore from Washington to Santa Barbara Islands, California.

Nests in large colonies on sea islands, usually in burrow in soil or in natural crevice under rocks. Egg one, 2½ by 1⅔ in.; granular, dull white, sometimes marked with purple, brown, or chocolate. Nesting period June to July. Incubation 40 to 43 days, with both parents caring for eggs and for young. One annual brood. Young can follow parents to sea at 4 weeks.

Food mostly fish but probably also crustaceans and mollusks; caught by diving, using wings under water for power and legs for steering. Call a harsh croaking. Flight by rapid wingbeats. Swims well but may have difficulty in taking to air from level ground. During fall molt in August to September, flight feathers are shed and birds cannot fly.

Sailors consider that puffins are land indicators, direction of their flight at morning and evening locating land. Any number larger than three indicates that land is probably within 150 miles, number increasing as land becomes nearer. Eggs have been used as food by man. Now that species has been given complete legal protection, it is increasing.

Order Columbiformes./Family Columbidae [Pigeons, Doves]

Band-tailed Pigeon, 312
Columba fasciata
Length 15½ in. Upper parts of back brown, lower part gray. White collar on back of neck, with wash of bronze back of this on other feathers. Under parts purplish-pink, but abdomen almost white. End of tail broad, pale gray, bordered with black back from tip. Hen duller, grayer, and often without collar found on cock. Can be mistaken for domestic pigeon except for its habitat in mountains and woodlands and its use of trees as a perch.

Breeds from southwestern British Columbia through Montana to north central Colorado, and south through southwestern United States and Mexico to Guatemala; east to western Texas. Winters in southern part of range from southwestern United States south. Related to extinct passenger pigeon that formerly ranged over most of eastern North America.

Nests in trees. Nest usually single, a crude, almost flat platform of rather coarse sticks. Eggs, normally one, 1½ by 1 1/10 in.; white, creamy, unmarked, equally rounded at both ends. Nesting period May to June, birds sometimes nesting in scattered colonies. With passenger pigeon, there might be to 50 nests in single tree, with adjacent trees equally crowded.

Food primarily seeds and grains, though some insects are included. When bird has fed on grain, usually waste, may develop a fine flesh. Flight is strong but not too fast, and bird makes a relatively easy shot.

Considered in some places a legitimate game bird but its numbers have been greatly reduced and it is doubtful if it is frequently injurious to crops. Its abundance should be maintained so that it does not become extinct like its relative, the passenger pigeon, which used to yield millions of birds a year.

Band–tailed Pigeon

Ground Dove, 320
Columbigallina passerina
Length to 7 in., or about size of bluebird. Wingspread to 11 in. Male: underparts pinkish-drab, breast with scalelike markings, under surface of wings auburn, tail rounded and brownish-black, base of bill red with black tip. Female much like male but browner, with underparts mousegray.

Breeds from South Carolina to southeastern Texas, through the Gulf and South Atlantic States, but most abundant near coasts. Accidentally found north to Tennessee, New York, and intermediate states. Three subspecies include Eastern, Mexican, and Bahama ground doves, latter two extending range of first to Guatemala, southeastern California, southern Arizona, New Mexico, and western Texas.

Nests on ground or in trees, up to height of 25 ft. Nest small, in some cases rather elaborate and made of a mixture of plant materials. Eggs usually two, white, ⅘ by ⅗ in.; rather equally rounded at both ends. Nesting period April to July. Young helpless when hatched but develop rather rapidly.

Food essentially seeds of plants. 10 stomachs examined contained mostly seeds such as crab grass, foxtail, panic grass, amaranth, purslane,
(Continued)

Ground Dove

ragweed, spurge, mallow, sedge, and sorghum. One stomach had 1,600 purslane seeds and another an equal number of amaranth fruits.

Probably useful as a weed-seed destroyer. Certainly too small to be considered a game bird.

Mourning Dove, 316
Zenaidura macroura
Length to 11 in., wingspread to 19 in., tail to 6⅔ in., bill to ⅗ in. Weight 3 to 6 oz. Female smaller than male. Temperature, 106°F. A brown pigeon with a pointed white-bordered tail, smaller than extinct passenger pigeon, which lacked also black spot behind eye and had a blue-gray head.

Two subspecies. Breeds from New Brunswick to British Columbia and south through interior of America through Mexico and Bahamas. Winters from Iowa and Massachusetts, west and south to Panama. Related passenger pigeon, *Ectopistes migratroius*, ranged from Nova Scotia to Washington and south to Gulf; extinct since 1914.

Nest on ground or as a crude stick platform relatively low in trees. Eggs usually two, weight, ⅖ oz, white, elliptic, 1⅕ by ⁹⁄₁₀ in. Incubation by both sexes, 13 to 15 days. Young helpless when hatched, but assume practically adult plumage by first fall. Two or sometimes more annual broods. Nests in all mainland U.S. states.

Food, grains and various seeds, particularly buckwheat. 99 percent vegetable material, of which 32 percent is grains, mostly waste. One stomach showed 7,500 seeds of yellow wood sorrel; another, 6,400 of barnyard grass; another 2,600 of joint grass; another, 4,820 of orange hawkweed; another, 950 of hoary vervain; another, 620 of panic grass. Speed, 41 mph.

Economic importance varies with locality. Where numbers are great and grain is abundant, there may be damage but weed-seed destruction must not be overlooked, particularly in areas not raising grains. Bird is considered a game bird in some states.

Rock Dove, 313.1
Columba livia
Rock dove, common pigeon of towns and farmyards, varied in color, blue, reddish, or almost black. Racing Homer should weigh 1 lb as contrasted with larger flesh-producing breeds, of which, strange to say, largest is the Runt. Some 200 varieties of pigeons are recognized. Has become a nuisance in some cities.

Homing pigeons, like others, probably were developed from rock dove, *Columba livia*, though some authorities have it developed from stock dove, *C. aenas*. Left to breed promiscuously, however, resulting stock most nearly resembles *C. livia*. In breeding practice, it is best to mate birds of different colors except for show.

Young, blind and helpless when hatched, for 5 days are fed "pigeon milk," a cheesy secretion from crops of both parents, then fed regurgitated, partly digested grain. By 4 weeks, young reach maximum size. At 6 weeks, young molt. Training for flying may be begun at 9 to 12 weeks. May live 16 years.

When birds are trained for flying, they are freed first 1 mile from home cote. Then, 1 mile a day is added until 25 miles are reached. Then, this may be jumped to 50 miles. Adult cocks fly best if freed so as to reach home to their young just before dark. Hens, best when with 10 to 12-day eggs in nest. Should not be flown when breeding.

Homing pigeons used in war and in peace can, when racing, maintain flight for 13 hours, can speed 1,200 yd/min, can fly 2,600 miles to return home. Home may be on a moving ship and yet be found, or may be moved about on land to a certain degree. Birds are usually bred only between February and September, and are then rested.

Pouter Pigeon, 313.1
Columba livia
Length to 20 in.; pygmy pouter to 14 in. Thigh to middle toenail, to 7½ in.; pygmy to 5½ in. Blue, black, red, yellow, or white, but typical marking has a 2-in.-wide crescent of white on front of enormous crop. Birds with bib lacking are "swallow-throated" and those with it around neck are "ring-necked" and poor showbirds.

Pouters were probably developed in Europe or Asia and probably represent breed farthest removed from original blue-rock type. Some 200 varieties of domestic pigeons of which Carrier, Pouter, Barb, and Tumblers are possibly most important. Some can barely fly while others may fly long distances for long periods of time.

Selection by breeding has developed carrier with homing instinct, pouter that inflates chest while strutting; barb that is plump, short-legged, and broad-skulled; tumbler that somersaults in air; fantail, with enormous tails; jacobin, with a feather ruff; and many others.

Mourning Dove

Rock Dove

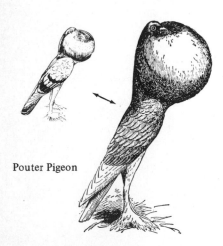

Pouter Pigeon

Pouter must have a dignified walk while crop is inflated. It must not choke while walking or jump off ground as do the "croppers," or set up rump feathers but must hold head and feet in a perpendicular line while it walks gracefully. Pouter must, for judging purposes, turn the middle toes outward and limbs must be properly feathered.

Pouters are essentially showbirds. In addition to raising pigeons for homing and for show, a big industry centers around raising them for flesh. Squabs are a standard dish in many city hotels and some birds are fed special diets to improve the quality of their flesh. There is a rich literature on this aspect of pigeon raising.

White King Pigeon

White King Pigeon, 313.1
Columba livia
Varieties include White King, Runt, Homer, Fantail, Pouter, Tumbler. Body temperature, 104° to 108°F. White King cock: weight to 26 oz; hen, to 24 oz. Toes, four, all on same level. Beak slender, with fleshy growth around nostrils. Sexes colored alike. Other varieties differ in color and size.

King pigeon developed in United States about 1891 as a meat producer. Pigeons have been domesticated for over 3,000 years, tracing ancestry probably to Rock pigeon, still wild in Europe and Asia. Pigeons commonly live in flocks around a protected shelter, usually a building.

Pigeons are monogamous. Courtship by strutting, cooing, and billing. Nest site selected by cock. First egg laid 8 to 12 days after mating; second, 2 days later. Incubation, 10 A.M. to 4 P.M. by cock and remainder by hen, for 16 to 18 days; when domesticated, in a 12- by 12-in. box, 9 in. deep, lined with straw.

Food essentially grain, such as wheat, buckwheat, and pigeon peas. Cracked corn should not be fed unless it is freshly cracked and free from mold. "Pigeon milk" for young is secreted in the crops of each parent when it is needed, just after young are hatched. Young helpless at first. Hearing range, 200 to 7,500 cps.

Pigeons are reared for their meat, as with King described above; for carrying messages, as with Homer; and for exhibition purposes, as with Tumbler, Pouter, Fantail, and others.

Budgerigar

Order Psittaciformes./Family Psittacidae [Lories, Parrots, and Macaws]

Budgerigar
Melopsittacus undulatus
Length to 8 in., including 4-in. tail. Tail of parakeets, of which this is one, pointed; of parrots, square or rounded at tip. Neck, cheeks, upper back, and wing coverts gray, each feather being marked with alternating crescents of yellow and green. Central tail feathers blue; others blue-green. Bill horn-colored.

Native of central and southern Australia, breeding in southern area. Trapped mostly during migrations of great flocks. Reared by thousands in captivity over world, particularly in western Europe. African *Agapornis* and South American *Psittacula* are also known as lovebirds and kept in cages.

Nests in captivity in boxes with curved bottoms or in old coconut husks suspended with hole toward light. Should be two nest boxes per pair. Breeds July and August in wild but in cages may breed April to July. Eggs to six, incubation of 17 days beginning with laying of first egg. Hen alone incubates but cock feeds family.

Food, seeds like millet; will starve on "birdseed" of rape, maw, flax, and hemp. Can go without water for months but needs it regularly nevertheless. May nest three times in a year; an aviary 14 by 9 by 7 ft may care for eight pairs if no old unmated hen is present to upset families. Young may breed at 3 months.

Desirable cage birds, admired because of beauty and because of evident affection and consideration shown between paired birds. Can be taught to perform simple tricks. May be carriers of "parrot fever" and so importation is restricted.

Gray Parrot

Gray Parrot
Psittacus erithacus
Length about 12 in. Tail scarlet; general color ashy-gray. Feet and toes gray. Green or Amazon parrot is *Chrysotis amazonicus*, a green bird with yellow head and square tail. Parrots have four toes, two in front and two behind.

Gray parrot is native of western Africa. Yellowhead parrot comes from country about Amazon River in South America and is found north into Mexico. Festive parrot, *Chrysotis festivus*, also a cage bird, is 16 in. long, green and red, and also comes from Amazon region, where it lives naturally in jungles. Nearly 100 species.

(Continued)

611

Like pigeons, parrots mate for life, even though they live naturally in great flocks. Naturally nest in holes in trees or in banks in ground though this varies with species. Hen lays from two to eight white eggs, which at parrot's body temperature of around 106°F hatch in from 25 to 30 days. Long-lived.

Food of parrots varies. In captivity, young birds are fed soft foods like bread and milk but older birds eat seeds, crackers, fruits, vegetables, and a variety of other foods. Captive birds must be kept clean and well-supplied with fresh air if they are to thrive and make interesting pets, some imitating human speech.

Mostly known as cage birds. Gray parrot can best be taught to "talk." Tongue splitting does not help in this. Parrots carry a disease, psittacosis, which may be fatal to persons who come in contact with them or their cages and therefore parrots are no longer imported. Native Carolina parakeet extinct since 1904. *P. e. princeps*, Princepe parrot, is an endangered species.

Order Psittaciformes./Family Kakatoidae [Cockatoos]

Cockatoos
Cacatua sp.
Pink cockatoo (*Cacatua leadbeateri*); sulfur-crested cockatoo (*C. galerita*); great black or palm cockatoo (*Microglossus aterrimus*); and black cockatoo (*Calyptorhynchus magnificus*). Pink or lead-beater cockatoo is white, with neck and breast delicately tinged with pink and crest striped red, yellow, and white. Sulfur-crested cockatoo is white with showy sulfur-yellow crest.

Cockatoos come for most part from Australia and islands of the Indian Ocean, where they live normally in forested areas and favor warm climates though rose-crested and sulfur-crested cockatoos can survive living outdoors even in winter in New York if shelters are available into which they may retire in severe weather.

Nests made in hollows in decaying trees in native haunts. In related cockateel, seven to nine eggs are usually laid, although as many as 20 may be; incubation 17 to 18 days; mother not fed by male during incubation. Young with down when hatched, but feathers soon appear. Both parents assist in rearing. Two to four annual broods, April to October.

Cockatoos are full of tricks and play together most boisterously. They can be taught to talk as parrots do but this is not a common accomplishment. While they are commonly noisy, this is not true of contented birds. Cockatoos are larger than more common parrots and so require larger quarters to thrive.

Relatively common cage birds, whose importation is not now possible because they may carry the fatal disease psittacosis. Food for captive birds is sunflower seeds, hempseed, and oats, with bananas, grapes, and apples also available.

Cockatoo

Order Cuculiformes./Family Cuculidae [Cuckoos]

Yellow-billed Cuckoo, 387
Coccyzus americanus
Length to 12⅔ in., including 6⅓-in. tail and 1-in. bill. Wingspread to 17 in. Weight just over 2 oz. Sexes colored alike. Slim, brown and white, with reddish cast to wings and large white tips to tail feathers that appear as an almost continuous band when tail is spread; not gray as in black-billed cuckoo. White spots of tail large as compared to black-billed cuckoo. Outer tail feathers almost black. Lower mandible yellow.

Breeds from Washington and Colorado south to Mexico; North Dakota to New Brunswick and south to Nuevo Leon, Tamaulipas, Louisiana, and southern Florida. Winters in Venezuela, Colombia, Ecuador, and Uruguay. Migrates through Mexico, West Indies, and Central America. Also found occasionally in Greenland, Great Britain, Italy, and Belgium. Bird of woodlands, orchards, and tall brushlands.

Nests in trees and tall shrubs near borders of woodlands. Nest a shallow, poorly constructed platform of sticks and twigs, sometimes with a lining of finer materials, 4 to 20 ft above ground. Eggs two, to rarely eight; 1⅓ by ⁹⁄₁₀ in.; light blue, sometimes paler than black-billed cuckoo, with white mottlings. Nesting period June to July. Incubation about 14 days, by both parents. One or two annual broods.

Food largely caterpillars, particularly hairy species that are avoided by most other birds. One was seen to eat 41 gypsy moth caterpillars in 15 minutes; another, to eat 47 tent caterpillars in 6 minutes. Also eats locusts and other insect pests of useful orchard trees and shrubs. Probably eats no fruit or grain. Flight silent, graceful, direct. Call a repeated *kas* ending in *kaups*.

Yellow-billed Cuckoo

One of most useful birds because of its fondness for injurious caterpillars. Not known to harm other species of birds or to eat useful insects to any considerable extent. European cuckoo lays its eggs in nests of other birds, as do our cowbirds, but this is not a habit of American cuckoos.

Black-billed Cuckoo, 388
Coccyzus erythropthalmus
Length to 12⅔ in., wingspread to 16⅔ in., tail to 7 in., bill to 1¹⁄₁₀ in. Slender, loose-jointed, brown above and whitish beneath. Black-billed cuckoo lacks reddish color on wings of yellow-billed cuckoo, has smaller white spots at tips of tail feathers, has black rather than yellow lower mandible, and a narrow red ring around the eye.

Breeds from southeastern Alberta to southern Quebec and Prince Edward Island, south to Kansas and Georgia. Winters in South America from Colombia to Peru, and found in Greenland, Azores, Italy, and Ireland by accident. Range of yellow-billed cuckoo is somewhat more southern than that of black-billed.

While European cuckoos commonly lay their eggs in nests of other birds, this is rare in American species. Nest, a crude platform in trees. Eggs two to four, or rarely to eight; pale blue, sometimes marbled in their shading; 1½ by ¹⁵⁄₁₆ in. Incubation 14 days, by both parents. One northern and two southern annual broods.

Food largely insects, particularly caterpillars; eats great numbers of fuzzy-haired caterpillars entirely ignored by most other birds. Stomachs of 46 showed 906 caterpillars, 44 beetles, 96 grasshoppers, 100 sawflies, 30 bugs, and 15 spiders. Evidence rare that a cuckoo will break up a robin's nest. Speed, 22 mph.

Almost entirely useful in a way other birds are not. Call a fast, rhythmic *cu, cu, cu*; this repeated without the retarded ending of yellow-billed cuckoo. Also is named rain crow but does not bring rain. Black-billed cuckoo is more likely to call at night than yellow-billed. Is protected by law and deserves even more protection. Not popular with robins.

Black-billed Cuckoo

Road Runner, 385
Geococcyx californianus
Length to 2 ft, tail 1 ft, wing 7 in. Bill downward-curved. Weight to 11 oz. Temperature, 107.4°F and variable. Feathers coarse. Upper parts olive-brown and white-streaked. Middle tail feathers olive with purple gloss. Throat and belly white. Eye yellow to orange. Legs and feet pale blue. Sexes colored alike. Young like adults.

Resident in California and Lower California, east through Utah, Colorado, Kansas, Texas, and tableland of northern Mexico, wherever there are deserts and mesquite-covered arid and semiarid lands. Does not migrate. Found in fossil form in La Brea tar beds developed in Pleistocene of California. State bird of New Mexico.

Always solitary. Builds nest, usually unlined, about 1 ft in diameter, of sticks and twigs, 3 to 15 ft up in a low tree or shrub, where it is concealed by vegetation. Eggs 2 to 8, unspotted, pale yellow or white, 1½ by 1⅓ in. Incubation begins with first egg and lasts 18 days. Mother tries to entice enemies away.

When 3 weeks old, young may begin catching their own food, which is highly varied, including snakes, lizards, scorpions, tarantulas, insects, birds and their eggs, rats, and other small animals. About 90 percent animal matter caught by pursuit or stealth. Runs to 18 mph for short distances to escape enemies, particularly man. Can fly. One of the most picturesque birds of cactus-covered ranges. Belief that it is a rattlesnake killer is unfounded.

Road Runner

Order Strigiformes./Family Tytonidae [Barn Owls]

Barn Owl, 365
Tyto alba
Length to 21 in., wingspread to 47 in., tail to 7 in. Weight, female to 24 oz, male to 20 oz. In pairs, female usually larger but not always. Lightest-colored of our owls, with face like a great white heart. Sexes colored alike. Young like adults even in juvenile plumage. Temperature, variable, around 103°F.

Resident from Washington to New York and south to southern Mexico, Nicaragua, and Gulf States. Occasionally in British Columbia, Manitoba, Minnesota, Ontario, Vermont, and Massachusetts, with allied races in Europe, Asia, West Indies, and elsewhere.

Nests in hollow trees, banks, towers, and deserted buildings, but makes no nest. Eggs 5 to 8, or even 11, chalky white or tinged with yellow, 2 by 1½ in., more pointed than usual owl egg. Incubation begins with first egg laid, 32 to 34 days, probably by female alone. One or two broods.

Food almost exclusively mice and rats, plus some other small mammals and a few birds. In the West it destroys many destructive pests: gophers, ground squirrels, and spermophiles; in the South the cotton rat is one of its chief victims. It has power of night sight and its hearing is very sensitive. Stomachs of 39 showed 17 with mice, 17 with rats and other small mammals, 4 with insects.

Probably most useful of all birds as enemy of rats and mice. Should be protected always. Illegally sold in pet shops as "monkey-faced owls" even though it may be illegal to keep barn owls in captivity.

Barn Owl

Screech Owl

Great Horned Owl

Snowy Owl

Screech Owl, 373m
Otus asio
Length to 10 in., wingspread to 24 in., tail to 3⅔ in. Weight: female to 7 oz; male to 6 oz. Two color phases, a gray and a reddish, each with feathered ear tufts and generally streaked plumage, which is loose and easily fluffed, making bird appear unnaturally large. Eyes large, yellow, and round. Temperature, 101°F.

Resident. Many subspecies recognized including, among others, the Eastern, the Southern, the Florida, the Texas, Hasbrouck's (Texas), Aiken's, the Rocky Mountain, MacFarlane's, Kennicott's, Brewster's, California, Pasadena, Mexican, Saguaro, and Xantus's, these covering most of United States, Mexico, and wooded Canada.

Nests in bird boxes or hollow trees, favoring orchard trees. Nest in unlined cavity. Eggs three to seven, weigh ⅗ oz each, white, usually smooth, 1½ by 1⅓ in. Incubation for 21 to 25 days, by one sex or by both. One annual brood. Young white and fluffy when hatched but feathered practically like adults by first fall.

Food: stomachs of 255 showed 100 with insects only, 21 with mice, 38 with other birds, 43 empty, 21 with lizards, 9 with crawfish, 7 with miscellaneous matter, 5 with spiders, 4 with frogs or salamanders, 2 with earthworms, 2 with scorpions, and 1 with poultry. 8 stomachs contained 2,976 insects and 2 mice.

For economic importance, their record is written in stomach contents as listed above. They are essentially useful and deserving of protection. Known from Pleistocene of Tennessee, California, and Florida. Sensitive to blue light.

Great Horned Owl, 375
Bubo virginianus
Male: length to 23 in., wingspread to 52 in., tail to 9 in., bill to 1⅗-in. Weight, 3½ lb. Female: length 25 in., wingspread 60 in., to 4½ lb. Conspicuous feathered ear tufts, powerful talons and beak. Great round yellow eyes. Feathers loose. Flight quiet. Temperature, 105°F and variable.

Several subspecies occupying most of wooded North America include among others the Arctic, the Labrador, the Great, the Montana, the Saint Michael, the Northwestern, the Dusky, the Pacific, the Western, and the Dwarf. Horned owls known from Pleistocene of Oregon and California. Moves south in winter. Arctic form moves even to New York.

Nests in gorges, in caves, or in heavy timber, rarely in hollow, but more commonly by remodeling nest of crow or hawk. Eggs two to five, dull white, granular, 2⅓ by 2 in. Incubation by both sexes, from 26 to 30 days. One annual brood. Young downy white when hatched but apparently adult by first winter.

Food a variety of small animals, particularly rabbits and rats. Of 127 stomachs examined, 13 had mice, 65 other mammals, 31 poultry, 17 were empty, 10 contained insects, 1 fish, 1 scorpion, and 8 other birds. One nest had 113 freshly killed rats on it. Can attack and kill large mammals such as rabbits, woodchucks, skunks, and even porcupines; birds as large as geese and turkeys. Small animals are swallowed head first if small prey. *Hoots* at middle C. Hearing range, down to 70 cps.

One of the last of the bountied species of birds. Referred to as the "winged tiger" of the woods.

Snowy Owl, 376
Nyctea scandiaca
Length to 27 in., wingspread to 66 in., tail to 10½ in., bill to 1⅗ in. Weight: female, to 5 lb, male smaller. Pure white, or with reddish- or grayish-brown fleckings on feathers. Female with more dark markings than male. Young even darker. Legs well-feathered. Ear tufts lacking.

Breeds from Bering Sea to Greenland, and south to central Mackenzie and northern Ungava; also, in northern Russia and Siberia. Winters from Arctic coast south, occasionally even to Georgia and California; also in Germany, Russia, Scotland, and Shetland Islands. Relative *N. nyctea*, extends range to the western states including Alaska.

Nests on tundra in open country, only rarely where there are trees. Nest a depression in soil or on rocky shelf. Eggs 4 to 11, white, with smooth surface; 2½ by 1⁹⁄₁₀ in. Incubation about 32 days, by female. One yearly brood. Young blind and helpless when first hatched; grow lighter in color with age.

Food essentially small animals. Stomachs of 38 snowy owls, 33 of which were taken in United States, showed 18 with mice, 12 empty, 9 with nongame birds, 2 with game birds, and 2 with mammals other than mice; birds taken were seabirds. May kill a few domestic fowl.

Serves as important check on mice and lemmings which, without check, would destroy most vegetation of north country. In North, then, bird is practically essential to balanced environment. May be a serious pest around game farms and may attack poultry that is left unprotected during cold weather. Cyclic, with peaks of populations in about 4-year intervals.

American Hawk Owl, 377a
Surnia ulula
Length to 17½ in. Wingspread to 34 in. Female averages larger than male. Sexes colored alike. Wings do not reach to tip of tail. Toes completely feathered. Dark brown above to almost black on top of head; many feathers with small central white spot. Face pale gray, with broad blackish border, black in front of eyes and under bill. Rare, so might, because of its flight pattern, be mistaken for a buteo hawk.

Breeds from northwestern Alaska to Hudson Strait, south to southern British Columbia, southern Manitoba, and Ungava. Winters in southern Canada, occasionally south to Washington, Nebraska, Minnesota, Missouri, Indiana, Ohio, Pennsylvania, New Jersey, and Rhode Island. Accidental in England and Bermuda. Genus not known in fossil form.

Breeds in evergreen northern forests. Nest none, eggs being laid in hollow stump, tree hole, or in hole in a cliff. Eggs laid April to June, four to seven, 1⅗ by 1⅓ in.; elliptical, creamy, smooth-shelled. Incubation chiefly by female, length unknown, probably around 3 weeks. One annual brood. Eggs almost identical with those of short-eared owl.

Hunts largely by day, preying on lemmings, mice, shrews, small hares, and sometimes ptarmigan and other birds. Chief food is small mammals, even including weasels. Bird favors high exposed perch with body inclined forward rather than upright like other owls. It frequently jerks its tail as does sparrow hawk. Call a hawklike whining screech or *quereek*.

Fearless, known to attack men who approach nest. Daytime activity, fearlessness, and tendency to perch conspicuously make it a tempting target to many hunters. Normally a rare bird in United States.

American Hawk Owl

Elf Owl, 381
Micrathene whitneyi
Length to 6 in., tail to 2 in., wing to 4 in. Shorter than pygmy owl, *Glaucidium gnoma*. Head without ear tufts. Eyebrows, cravat, and underparts white; interrupted white collar, and wings with whitish spots; otherwise, largely gray- or brown-speckled. Young more brownish-gray on top of head.

Elf owls range through southwestern United States and into Mexico; Sanford's (California); Texas (eastern) and Whitney's (southwestern). Locally, may be abundant. Not migratory. Haunts low hot dry desert river bottoms and adjacent lands. Reported from Pleistocene of California.

Nests in holes up to 20 ft high in saguaro cactus made by Gila woodpeckers or gilded flickers. In May, lays three to five pure white, glossy, granulated 1-in. eggs. These may hatch after 14 days' incubation into helpless young covered with white down. Molts: young, June to July; adults, September to October.

Food almost entirely grasshoppers and other insects, and centipedes; all caught at night when bird comes from its hiding place in hole or dense thicket. Call sounds like squeak of rubber toy. Males flock together during breeding season.

Probably entirely useful to man and to his interests. So unique that bird deserves protection at all times. When kept as a pet, refuses to eat other birds. This probably indicates that when free it is not a common bird killer. Is probably less savage than pygmy owl. Almost entirely dependent on giant cactus.

Elf Owl

Burrowing Owl, 378a
Speotyto cunicularia
Length to 11 in., tail 3½ in., wing 7 in., bill ½ in. Head without ear tufts. Legs long, nearly bare, and with bristly toes. Adults brown with white and buff spots and bars. Tail light-barred. Young with plain brown upper parts, plainer in general appearance, and with dark rather than white throat band.

At least two subspecies recognized, the Western and the Florida. Western ranges from Pacific Coast of United States east to Minnesota and Louisiana, south to Panama; accidentally in Indiana, New York, and Massachusetts and migratory to Oregon. Florida found in central and southern Florida.

After interesting courtship of caresses, low song, and bows, nest is built of fine materials in burrow underground. Here are laid April to June, 6 to 11 smooth glossy eggs, 1 by 1⅕ in. Incubation by both parents, about 3 weeks. One yearly brood. Young with scanty gray down. Plumage develops and begins to molt by July.

Food almost wholly insects and scorpions, with a few lizards, some small mammals and snakes, but principally grasshoppers. May live in prairie dog holes, but hardly cooperatively with hosts or with rattlesnakes which may also live there. Rattlesnakes alone could profit by such a relationship. May be active in daytime as well as at night.

Unquestionably almost wholly useful to man and should be protected. As man and his cats occupy a territory, these useful birds begin to disappear. Owl does not favor cultivated lands and so ranges are constantly decreasing. Long legs, bobbing movement, and short tail serve as good points of identification. Really splendid birds.

Burrowing Owl

Barred Owl

Barred Owl, 368
Strix varia
Length to 24 in., wingspread to 50 in., tail to 10 in., bill to 1½ in. Weight to 2 lb. Female averages heavier than male. A large gray "earless" owl, with bars appearing crosswise on breast and streaked lengthwise on belly. Eyes large and brown. Feathers loose and fluffy.

Three subspecies recognized, Northern, Florida, and Texas. The Northern breeds from northern Saskatchewan to Newfoundland and south to Arkansas and Georgia. Found in fossil form in Pleistocene of Florida.

Nests usually in dense woodlands or wooded swamps in tall timber. Nest in hollow tree or in rebuilt nest of crow or hawk. Eggs two to four, white, somewhat glossy and slightly rough, 2 by 1¾ in.; laid in late February. Incubation to 28 days, probably chiefly by female. One annual brood. Young white and downy when hatched.

Food chiefly mice and other rodents. Of 109 stomachs, contents were 48 with mice, 18 with other mammals, 20 empty, 9 with crawfish, 7 with small owls, 5 with other small birds, 14 with insects, 4 with frogs, 4 with game birds or poultry, including one ruffed grouse.

Obvious preference for mice indicates its primary service to man, which is not always appreciated. A barred owl calling in vicinity of a summer camp has made a vacation memorable to many youngsters.

Great Gray Owl, 375
Strix nebulosa
Length to 33 in., wingspread to 60 in., tail to 15 in., bill to 1¾ in. Weight to 2 lb 14 oz. Largest of owls. Marked much like barred owl, but with no conspicuous cross bandings on breast and with eyes yellow rather than brown. No ear tufts. Size and round head usually sufficient identification.

Two subspecies are Great and Siberian. Latter ranges through eastern Siberia and is occasionally found in western Alaska. Great breeds from north central Alaska tree limit to Ontario and south to central California. Winters through southern Canada south to New Jersey.

Nests usually over 20 ft up, in tall tree. Nest is old nest of hawk, lined or unlined. Eggs three to five, white, smooth, 2⅓ by 1⅖ in. Incubation by female, for unknown time. One annual brood. Young downy and helpless when hatched, white, but darker later across upper parts. Young hatched in April appear like adults by October.

Food consists largely of mice, rabbits, and small birds, but detailed information such as is available for most other owls is lacking for this species. Examination of 9 stomachs showed 7 with mice, 4 with other mammals, 1 with a small bird. This inconclusive evidence indicates that probably bird is useful.

General statement issued from United States Department of Agriculture that owls are most beneficial of all birds probably applies to this species. Gray owl is hardly aggressive type represented by great horned owl.

Great Gray Owl, 375

Long-eared Owl, 366
Asio otus
Length to 16 in., wingspread to 42 in., tail to 6½ in., bill to 1¹⁄₁₀ in. Eye yellow. Appears about size of a crow, or like a small great horned owl with lengthwise streaks rather than crossbars. Grayer than short-eared owl, which it otherwise resembles in flight. Weight 11 oz, female the larger. Temperature, 103°F.

Breeds from central British Columbia to Newfoundland and south to southern California, northern Texas, and Virginia. Winters from southern Canada to Florida and central Mexico. Found in California Pleistocene. Prefers evergreen forests but may live anywhere. Rarely seen doing daylight hunting. Apparently a migratory species; breeding to the north of its usual nonbreeding range.

Nests usually in a cone-bearing tree, from 10 to 30 ft up. Nest usually remodeled nest of heron or crow. Eggs three to seven, weight, ⅕ oz; white and smooth, 1¾ by 1½ in.; laid on alternate days but incubation begins with laying of first. Incubation about 28 days. Young in nest 25 days. One annual brood.

Food almost exclusively mice and other small mammals. Stomachs of 23 examined showed 22 with mice and the other with a small bird. Of 107 examined, 84 contained mice, and 15 small birds. Excess and waste are regurgitated as pellets, of which 187 of 225 examined contained small mammals, mostly mice.

One of best of all mouse-catching owls. Therefore worthy of every protection. It may be killed by many hunters carrying on vermin campaigns under the impression that it is a great horned owl. Long-eared owl resembles in color the great horned owl but is much more delicately formed, and it has a shorter space between its horns.

Long-eared Owl

Short-eared Owl, 367
Asio flammeus

Length to 17 in., wingspread to 44 in., tail to 6⅗ in., bill to 1⅛ in. Weight: female 13 oz, male 17½ oz. Slightly smaller than crow, with shorter "horns" than long-eared owl; generally lighter and more yellowish. Chiefly white underside, wings and tail dark. Short ears difficult to see; eyes completely encircled with black.

Breeds from northern Alaska to Greenland and south to California and New Jersey; also in Europe and Asia. Winters from British Columbia to Massachusetts and south to Cuba and Guatemala. Known from Pleistocene of California. Roosts on ground, usually in open country, and often active in daytime as it flies with steady beating flight.

Nests on ground, in open, merely by tramping down vegetation or bringing in a few sticks. Eggs four to seven, white or creamy, smooth, 1⅔ by 1⅓ in., laid May to June. Incubation believed to be 3 weeks, and mainly by female. One brood commonly, but possibly two as is case in Europe.

Food: 101 stomachs showed 77 with mice, 11 with small birds, 7 with moles or shrews, 7 with insects, 14 empty. One stomach had 30 huge grasshoppers in it. It usually prefers territory with an abundance of mice, and where there is such infestation short-eared owls may be expected to move in.

Essentially useful as a mouse check. Because of its daytime activity, it is frequently killed by ill-advised hunters. Farmers should prosecute anyone killing this important friend of their interests. Call a *toot, toot, toot* and high-pitched shrieks.

Short-eared Owl

Saw-whet Owl, 372
Aegolius acadicus

Length to 8½ in., wingspread to 20½ in., tail to 3¼ in. Sexes apparently same, female possibly somewhat larger. Temperature, 104°F. Looks like a small screech owl, without its characteristic ear tufts. Because of its size, not easily confused with other owls of its range. The smallest of the eastern nocturnal birds of prey.

Breeds from southern Asia to Nova Scotia, New Brunswick, and south to central California, Arizona, and Maryland. Winters south to southern California and Virginia and in southern part of its breeding range. Not often seen even when abundant, because of retiring and strictly nocturnal habits.

Nests in wooded areas, particularly where it is swampy. Nest in hole in tree, such as a deserted flicker hole. Eggs three to seven, chalky white, 1 by 9/10 in. Incubation about 27 days, by female. One annual brood. Young scantily covered with a whitish down, but by September usually has assumed adult plumage and whitish face.

Food: stomachs of 22 showed 17 with mice, 2 empty, 1 with a sparrow, and 1 with an insect. Of 7 stomachs taken in summer and autumn by another observer, 6 contained insects only and 1 contained a mouse, so bird food may be taken only when mice and insects are not available.

Obviously useful as mouse destroyer though never present in sufficient numbers to serve important role. In winter, numbers may appear greater than normal because of crowding where food is available. Call sounds like squeaks of filing a saw and young sound like a dog sniffing. Adult also given common name Acadian-owl; young, white-fronted owl.

Saw-whet Owl

Order Caprimulgiformes./Family Caprimulgidae [Goatsuckers]

Whippoorwill, 417
Caprimulgus vociferus

Length to 10½ in., wingspread to 19⅔ in., tail to 5 in. Weight 2 oz. Female usually smaller than male. Feet small and weak. Perches on ground or lengthwise on a limb. Female appears all brown. Male in flight shows two white patches, about half length of tail, but no white wing patches; narrow white throat streak.

Breeds from Manitoba to Nova Scotia and south to northern Louisiana and northwestern South Carolina. Winters from South Carolina west through Gulf States and south to British Honduras, Salvador, and Costa Rica. In wooded areas favors mixed growth of hardwood and conifers.

No nest. Eggs laid on ground; two, white or glossy, blotched; 1¼ by 9/10 in. Incubation 17 to 19 days, probably by female but male is reported to assist. Young with buff or yellow-brown down.

Food entirely insects, usually caught on wing, though not all are necessarily injurious. Insects eaten include June beetles, potato beetles, cutworms, mosquitoes, ants, gnats, and practically all large moths that fly at night. Insect "eggs" have also been found though these must have been taken while not flying.

One bird was reported to have 36 moths in its stomach at one time. Whippoorwill's record as an insect destroyer is enviable and there is no reason why bird should be killed. To those who care for its haunts, its calls are most welcome even though they may annoy persons not used to hearing them.

Whippoorwill

Nighthawk

Nighthawk, 420
Chordeiles minor
Length to 10 in., wingspread to 23¾ in., tail to 4⅘ in. Female smaller than male. In flight, wing shows decided angle at middle and a conspicuous crossband near tip. Tail shows two narrow white crossbands and male's throat is conspicuously white. Marvelous erratic flight, high in air.

At least seven subspecies recognized which include the Eastern, the Florida, the Western, the Pacific, the Howell's, the Cherrie's, and the Sennett's, with two subspecies of a closely related form. Most of United States and southern Canada covered by these forms. Winter range from Colombia to Argentina.

No nest built. Instead, eggs are laid on open ground, on graveled roofs in cities, in open spaces in fields, on roads, and elsewhere. Eggs two, variable in markings; 1⅓ by 9/10 in.; ⅓ oz. Incubation 16 to 19 days, probably only by female. Bird weight 4 oz. Temperature, 104°F.

Food entirely insects, caught on wing, either day or night. Stomach of one nighthawk contained 1,800 winged ants, another, 60 grasshoppers; another 500 mosquitoes; and still another, a solid mass of glowing fireflies. Other records include potato beetles, cotton boll weevils, and other insects. Speed, 22 mph.

Known also as a "bull bat"; was formerly shot in great numbers in migration because of skill required in hitting such an erratic target. Now protected by federal laws, which should be enforced at all times since bird is so interesting and so genuinely valuable. Often seen and heard toward dusk in urban areas flying and giving its repeated nasal call, "bee-igg," over roofs of buildings.

Chimney Swift

Order Micropodiformes./Family Micropodidae [Swifts]

Chimney Swift, 423
Chaetura pelagica
Length to 5⅗ in., wingspread to 12⅔ in., tail to 2 in. Female smaller than male. Tail conspicuous because of stiff bristlelike feather tips. In flight, resemble "flying cigars." Wings sometimes used alternately and rapidly to assist in darting flight necessary for food getting. General color sooty. Eyes large.

Breeds from central Alberta to Newfoundland and south to Florida, Gulf States, and eastern Texas. Winters at headwaters of the Amazon in western Brazil and eastern Peru. Accidental in Greenland and Bermuda. *C. vauxi*, Vaux's swift, extends range to west, including West Coast from Alaska to Santa Cruz, California, and then east to Montana and Nevada; migrates through southern California and Arizona.

Nests in hollow trees, or more commonly now in chimneys. Nest of sticks glued to chimney wall by a mucilage secreted by bird. Nest like a tiny basket, usually at least 20 ft up. Eggs four to six, white and moderately glossy, ⅘ by ½ in. Incubation by both sexes.

Food exclusively insects caught in flight. When weather conditions are such that flying insects are kept from air any considerable time, swifts sometimes die by thousands. Call a typical twitter, welcomed by those who know it. Travel a thousand miles a day; reported, but probably not substantiated.

Entirely valuable. Rumor they they carry bedbugs which may annoy man is without basis in fact. Big bedbugs found in nests of swifts cannot live away from them. Worthy of more protection than they get. Illegal now to kill them at any time. "Bird's-nest soup" made from nest of related Asiatic species. Mucus of nest, mostly dried saliva, is soup basis.

Ruby-throated Hummingbird.

Order Micropodiformes./Family Trochilidae [Hummingbirds]

Ruby-throated Hummingbird, 428
Archilochus colubris
Length to 4 in., wingspread to 4¾ in., tail to 1¼ in., bill to ⅕ in. Female sometimes larger than male. Male with beautiful changeable ruby or black throat, forked tail. Female, tail not forked, and color not so brilliant; three outer tail feathers white-tipped. Young generally like female.

Breeds from Alberta to Cape Breton Island and south to Texas, Gulf Coast, and Florida. Winters from middle Florida to Louisiana and southern Mexico to Central America and Panama. In migration, often flies across Gulf of Mexico, a prodigious effort for so small a bird. Some 12 western species of hummingbirds of which *A. alexandri* is a relative which extends range of *Archilochus* to the west.

Nest a tiny lichen-covered cup, about 1½ in. across from outer limits and about 1¼ in. deep, down-lined, fastened to branch 3 to 30 ft above ground. Eggs two, white, 9/16 by ⅜ in. Incubation for 14 days, by female. One or two annual broods. Young fed by vigorous pumping of food from stomach of adult.

Food, nectar and insects and other small creatures which haunt throats of flowers. Can remain suspended before a flower and, unlike other birds, can fly backward. Fight vigorously among their own kind or attack birds and other animals many times their size. Rapid motion of wings responsible for name.

Probably essentially useful, as some assist in pollination of certain deep-throated flowers, also destroying some insects which might be injurious to useful plants. Entirely harmless, intensely interesting as young and as adults, and therefore worthy of all possible protection. Can be encouraged to visit especially prepared "sugar water feeders."

Rufous Hummingbird

Rufous Hummingbird, 433
Selasphorus rufus
Length: male to 3⅔ in., tail 1⅓ in.; female to 3⁹/₁₀ in., tail 1⅓ in. Male with upper parts a bright red-brown and throat brilliant red; only hummingbird with a red-brown back. Female olive above, like most other female hummingbirds, but with some red on base of tail but not on rump.

Breeds from southern Yukon through British Columbia and southern Alberta to southern Oregon and southwestern Montana. Winters in southern Mexico; occurs in migration in Wyoming, eastern Colorado, and western Texas. Accidental in South Carolina. Other species in genus include broad-tailed and Allen's hummingbird.

Nests in bushes such as salal, and in ferns and dead trees bearing lichens. May nest in a variety of locations including deep woods to 50 ft or a blackberry bush in a city lot. Nest of down covered with lichens, moss, and fine pieces of bark. Eggs two, white. Male goes through elaborate courtship of aerial acrobatics, rising and falling and displaying before female.

Food, insects found on bushes and flowers and picked off while bird remains in flight. Among plants visited are wild currant, gooseberry, ocotillo, fireweed, paintbrush, agave, pentstemon, and gilia. This species, and particularly male of this species, is exceptionally pugnacious and fights for favored feeding area.

Delightful birds that are welcome in any flower garden for their beauty, vigor, and interesting habits. They may assist in pollination of some flowers and undoubtedly destroy some injurious insects.

Order Coraaciiformes./Family Alcedinidae [Kingfishers]

Belted Kingfisher, 390
Megaceryle alcyon
Length to 14¾ in., wingspread to 23 in., tail to 5 in., bill to 2¼ in. Weight to 6 oz, or as small as 4 oz. Sizes of sexes variable but more or less similar. Female with a broad, cinnamon-reddish band across lower breast, in addition to bluish band of male. Temperature, 101.1 to 108°F. A top-heavy-appearing bird, with a long, straight and pointed beak, large crested head, short legs, and small feet.

Two subspecies include the Eastern and the Western. Breeds from Alaska to Newfoundland and south to southern limits of United States. Winters from British Columbia to Virginia or even New York and south to Colombia, British Guiana, and Trinidad. Accidental in Holland, Ireland, and Azores.

No nest. Eggs laid at end of burrow which is 4 in. across, from 3 to 15 ft long, with enlarged room at end. Eggs 5 to 14, white, glossy, 1½ by 1⅕ in. Incubation for 23 to 24 days, chiefly by female, but possibly by both sexes. One brood a year in North, possibly two in South.

Food chiefly fish, but known also to include mice, frogs, lizards, insects, berries, newts, and crayfish. Insects eaten are chiefly larger water insects. General opinion of disinterested students is that kingfisher does more good than harm, unless it takes as its territory some place like a fish hatchery. Speed, 36 mph.

Protected by law in some states. Whatever legal standing may be, it is hoped that no generation of young Americans grows up without many knowing the kingfisher's rattle.

Belted Kingfisher

Order Piciformes./Family Picidae [Woodpeckers]

Yellow-shafted Flicker, 412a
Colaptes auratus
Length to 13 in., wingspread to 21⅓ in., tail to 4⅕ in., bill to 1⅔ in. Weight to 6 oz. In flight, conspicuous white rump patch and yellow under wings. All males and young females have black mustache patch behind bill and black neckband. Adult female without black mustache. *C. cafer*, Red-shafted Flicker, red under wings.

Two subspecies of *C. auratus*, the Northern and the Southern. Breed from tree limit in Alaska to Nevada and east through Missouri, North Carolina, and Florida. Winter in southern part of breeding range, south to Gulf Coast.

Yellow-shafted Flicker

(Continued)

Nest built in hole in telephone pole or dead tree stub, a nesting box, or even in a building. Hole to 3 in. wide, 24 in. deep, and to 60 ft up. Violent courtship antics. Eggs 3 to 20, white, highly glossy, $1\frac{1}{6}$ by $\frac{7}{8}$ in., $\frac{1}{4}$ oz. Incubation by both sexes, 11 to 16 days. One or two annual broods. Male recognizes female by sight.

Food largely insects, particularly ants. One stomach contained 5,000 ants, and two others over 3,000 each. Other insects include grasshoppers, crickets, and beetles. Insects total about 61 percent of food, remainder being wild fruit, cherries, and weed seeds. In fall and winter more than half of the total food is wild fruits. Young molt to adult plumage June to October.

Highly useful as insect destroyer, particularly of ants which foster crop-destroying aphids. Also one of few birds which destroys European corn borer which threatens Corn Belt. Known as highhole, yellowhammer, golden-winged woodpecker. Alabama state bird. Birds fight others of own sex.

Pileated Woodpecker, 405a
Dryocopus pileatus
Length to $19\frac{1}{2}$ in. Wingspread to 30 in. Weight to 1 lb; female smaller than male. Dark brown to black, showing flashing white in flight and conspicuous red crest. In female, forehead and fore part of crown are grayish brown, and line along jaw is blackish rather than red. About size of a crow.

Northern species is resident from central Mackenzie to New Brunswick and Nova Scotia and south through Minnesota, Iowa, Illinois, Indiana, Ohio, Pennsylvania, and south in the Appalachians. Southern, Florida, and Western subspecies extend range south and west through Florida, California, and British Columbia. Resident in Transition and Canadian zones.

Nests in hole in large tree, usually 12 to 60 ft above ground, a cavity 12 to 30 in. deep, with 3- to 4-in. entrance, chip-lined. Eggs three to six, $1\frac{1}{2}$ by 1 in.; white, very glossy. Incubation by both sexes, 18 days, with most nesting in May. Young naked when hatched, developing plumage in nest. Adults molt in early fall or late summer but no spring molt evident.

Flight usually undulating or swooping. Digs huge rectangular holes in dead or living trees in search of carpenter ants that make up major portion of food. May also eat caterpillars, cockroaches, and other insects or rarely berries and even wild cherries. The call is a loud commanding *kuk-kuk, kuk-kuk*, like a very loud flicker but commonly rising at beginning.

Undoubtedly useful because it eats wood-destroying insects. Not known to injure common orchard trees. Its nests, when abandoned, are used by wood ducks and other birds and mammals unable to make such nests themselves. Expresses spirit of wild woodlands and deserves every protection. Has done some damage to wooden telephone poles.

Red-bellied Woodpecker, 409
Centurus carolinus
Length to $10\frac{1}{2}$ in. Wingspread to 18 in. Female smaller than male. Male conspicuously dark-striped on back, crown and nape red, breast, throat, and cheeks pale brown. Female like male, except that red on head is confined to nape and is replaced on crown by grayish-brown. Only zebra-striped woodpecker.

Resident of eastern United States from southeastern South Dakota through southwestern Ontario to western New York and occasionally into western New England and south to central Texas and Florida. Closely related to the golden-fronted, the Gila, the cardon, and the Brewster's woodpeckers, which extend range to south and west.

Nests May to June, in hole dug in tree or pole 16 to 50 ft above ground, with $1\frac{3}{4}$-in. entrance and depth to 12 in. Eggs three to five or more, 1 by $\frac{2}{3}$ in.; dull white. Incubated 14 days, by both sexes. Young helpless when hatched. Young of both sexes resemble female first winter, though some males may approach redness of adult male on head.

Essentially a bird of forests and shade trees of smaller communities. Shy and suspicious. Food primarily insects including caterpillars, beetles, and bugs but also including sometimes corn and other fruits. It may feed on sap of some trees, much like sapsuckers.

May injure fruits such as oranges in southern portion of range but usually is useful as a diligent destroyer of insect enemies of forest, fruit, and shade trees. Ordinarily it deserves protection.

Pileated Woodpecker

Red-bellied Woodpecker

Red-shafted Flicker, 413
Colaptes cafer
Western representative of *C. auratus*, Yellow-shafted flicker. Length to 14 in., tail $4\frac{1}{3}$ in., bill $1\frac{1}{2}$ in. Recognition marks almost like those of *auratus*; white rump, the flashes of color in the wings when bird is in flight, undulating flight, bobbing motion of the head when the bird alights. Males and young females with black mustache patch behind bill and black neckband. Head brown, back and wings brown with black bars, belly side whitish with black spots. Underside of tail and wing salmon red; same parts yellow in *C. auratus*. Ranges of *cafer* and *auratus* overlap so yellows and reds not distinct in some specimens.

Two subspecies, Northwestern Flicker, *C. c. cafer* and Red-shafted Flicker, *C.c. collaris*. Difficult, if not impossible, to differentiate. Range is Western North America. Breeds from southeastern Alaska east to the North and South Dakota, Colorado, south into Western Mexico and Baja, California. Almost a permanent resident in range. May move south in most severe weather; migrations vertical, more than geographical.

Nests excavated in dead or dying trees, from ground level to 100 ft.; commonly in cottonwoods, sycamores, junipers, oaks, and pines. May nest in banks. Courtship behavior like *auratus* but with more back and forth motions of the head. Eggs 5-12, white, highly glossy, $1\frac{1}{5}$ in. by $\frac{8}{10}$ in. Both sexes incubate. Usually one brood, rarely two.

Food about one half animal and one half vegetable matter. Ants make up a large portion of animal matter taken, also other insects such as beetles, caterpillars, and crickets. Acorns, cultivated grains, orchard fruits such as pears, apples, grapes, cherries, and prunes, wild fruits and a few weed seeds make up plant foods in the diet.

A useful destroyer of ants and some other insects such as codling moths either directly or indirectly harmful to plants. May drill holes in wooden buildings. Call like that of Yellow-shafted flicker. Young birds like adults but duller. Hybrid of *auratus* x *cafer* is fertile. Hybrid said to show flash of orange on underside of tail and wings when in flight. Call notes a repeated *yuck-a*.

Red-Shafted Flicker

Redheaded Woodpecker, 406
Melanerpes erythrocephalus
Length to $9\frac{3}{4}$ in., wingspread to 18 in., bill to $1\frac{1}{6}$ in., tail to $3\frac{3}{4}$ in. Female generally smaller than male. Head and neck of adults red; in flight appears like black and white woodpecker with red head and large square white wing patches. Young head gray, not red. Weight $2\frac{4}{5}$ oz. Temperature 107.2°F.

Range from southeastern British Columbia to southeastern Ontario and south to New Mexico, central Texas, and Florida. Irregular in occurrence, found in Quebec, New Brunswick, Nova Scotia, Utah, and Arizona. Numbers have been greatly reduced over most of range.

Nest a hole in a telephone pole, hollow tree, or even tall fence post, with $1\frac{3}{4}$-in. entrance, and depth to 18 in., chip-lined. Eggs white, slightly glossy, four to six, $1\frac{1}{6}$ by $\frac{9}{10}$ in. Incubation 14 days, by both sexes. Annual broods one or two. Young lose brown on head after first fall or winter. Complete fall molt; partial spring molt.

Food about $\frac{1}{3}$ animal matter, largely May beetles, grasshoppers, ants, and weevils, often caught in flight; sometimes these are stored. Eats eggs of other birds, rarely. May eat corn, pears, apples, cherries, grapes, and other fruits. Known to kill young chickens. Particularly favors beechnuts and acorns. May stay in North if favorite food is in mast crop; i.e., beechnuts, are plentiful. Feeds both in the air and on the ground.

Probably more useful than injurious in spite of above record which emphasizes weaknesses rather than worth. Apparently migrated but without known pattern. Call a repeated "tchur" or "char" resembles call of tree frog. Habit of flying from telephone pole to catch insects in air over high-speed roads may be fatal to species.

Redheaded Woodpecker

Acorn Woodpecker, 407a
Melanerpes formicivorus
Length to 9 in., wing to $5\frac{4}{5}$ in., tail to $3\frac{1}{2}$ in. Female slightly smaller than male. A squarish white or yellow patch next to red crown. Upper parts glossy greenish-black. Female has black band separating white of forehead from red of crown. When in flight, appears black and white, with large white rump and wing patches. Markings indistinct in young.

Five subspecies recognized to include the Anteating, the Mearns, the California, the San Pedro, and the Narrow-fronted woodpeckers. Range covers southwestern United States from Oregon to Texas and, with some subspecies, south through Mexico, with related species into northern South America.

Nests usually in a hole in white oaks, but also in pines, cottonwoods, black oaks, and other trees. Hole usually 6 to 18 ft from ground. Eggs four to rarely six; white, with little gloss, 1 by $\frac{4}{5}$ in. Incubation by both

(Continued)

Acorn Woodpecker

parents, 14 days. Sometimes, two pairs use same nest. Young blind when hatched. Juvenile feathers molted second fall. Starling a serious competitor as it takes over nesting site prepared by acorn woodpecker.

Food chiefly acorns, rarely fruit and sap. Acorns are stored without being opened by being driven into holes in outside of trees, often sufficiently to cover wide area. Sometimes as many as 1,500 acorns stored on one telephone pole. Feet not used for holding food. A destroyer of wooden fence posts and corral posts.

Probably not important economically. Rare cases of eating eggs of other birds, frequent cases of mutilating trees, and wholesale cases of destroying acorns not to bird's credit. Rumor that only injured acorns are eaten is not well founded.

Yellow-bellied Sapsucker

Yellow-bellied Sapsucker, 402
Sphyrapicus varius
Length to $8\frac{4}{5}$ in., wingspread to 16 in., tail to $3\frac{1}{3}$ in. Weight $1\frac{3}{4}$ oz. Temperature 108°F. In field, appears in flight with a longitudinal white patch on each black wing. A red crown patch. Male with red throat, female with white. Young birds appear as adult after first winter.

Four subspecies include the Yellow-bellied, the Red-naped, the Northern Red-breasted, and the Southern sapsucker. *S. varius* breeds from central Mackenzie to Cape Breton Island and south to Missouri and northern Massachusetts. Winters from Iowa to Massachusetts and south to Panama. One closely related species.

Nests in coniferous or mixed forests, open farmland, or marshes. Both sexes dig $1\frac{3}{5}$-in. hole, 18 in. deep, 12 to 40 ft above ground. Eggs five to seven, white, slightly glossy, $\frac{9}{10}$ by $\frac{2}{3}$ in. Incubation for 14 days, by both sexes. One annual brood, young being reared by both parents.

May cause serious damage to useful trees by girdling them with their drillings. Drillings are orderly and close, and attract squirrels, chipmunks, insects, hummingbirds, and other creatures to feast. May visit a series of trees in a regular round. Fortunately, bird is not common. Drumming is an interrupted staccato. Difficult to assess extent of damage done by this species. Does eat some destructive insects but also destroys useful trees.

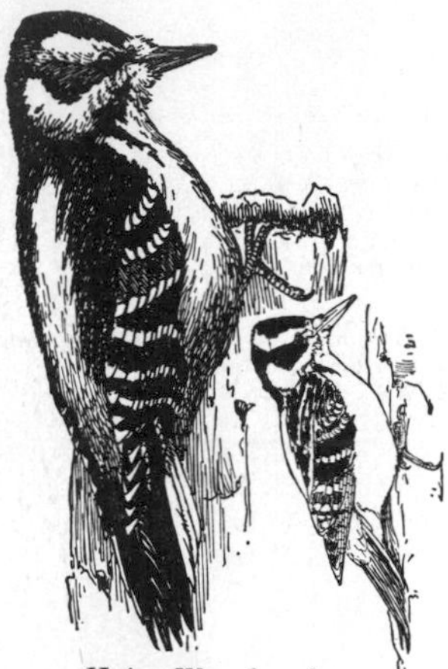

Hairy Woodpecker

Hairy Woodpecker, 393
Dendrocopos villosus
Length to $10\frac{1}{2}$ in., wingspread to $17\frac{1}{2}$ in., tail to 4 in., bill to $1\frac{1}{3}$ in. Weight to 3 oz. Temperature, 105°F. Female smaller and lacks red on back of head. Outer tail feathers white, without black bars or markings. Works on trunks and branches of trees, head uppermost. Looks like a large downy woodpecker without ladderlike markings on the edges of tail, and with a very large bill.

Subspecies 13, including in their resident range territory from Alaska to Newfoundland within tree zone and south into northern Mexico and through Florida. Included are the Northern, the Eastern, the Southern, the Rocky Mountain, the Lower California, the Sitka, and the Newfoundland hairy woodpeckers whose ranges are significant. Orchard and woodland species.

Nests in woods and orchards. Nest hole in a dead branch or tree trunk. Entrance 2 in., depth to 16 in., 5 to 50 ft above ground. Eggs three to five, shining white, 1 by $\frac{3}{4}$ in. Incubation 14 days, by both parents. One annual brood. Young soon similar to adults.

Food, insects gleaned from bark or dug from dead wood. Most destructive of hairy caterpillars and of their chrysalids, including pestiferous gypsy moth, within its range. Also eats ants, grasshoppers, spiders, and wood-boring beetles. Estimated 77.7 percent of food is animal matter, remainder vegetable, including nuts and seeds.

Entirely useful and most valuable about orchards. Good orchard practice calls for erection of nest sites or protection of existing sites. Individual range only a few acres if food and nesting sites are available in sufficient abundance.

Downy Woodpecker

Downy Woodpecker, 394c
Dendrocopos pubescens
Length $7\frac{1}{6}$ in., wingspread $12\frac{1}{4}$ in., bill $\frac{4}{5}$ in. Weight $1\frac{1}{2}$ oz. Black and white streaked, with white outer tail feathers barred faintly or marked with black spots. Male with red spot on back of head, which is lacking in female. Southern downy woodpecker, browner on underparts than Northern. Temperature, 108°F. Looks like a small hairy woodpecker with ladderlike markings on the edges of tail, and a small bill.

Mostly resident. Six subspecies include the Northern and the Southern, which range from southern Alaska to Newfoundland within tree belt and south to California, Arizona, and Gulf States. Northern is rarely south of Nebraska and Virginia. Rarely above 3,000-ft altitude. Tree species.

Nests in orchards, mixed forests, and shade trees. Hole in a grape trellis or tree. Entrance $1\frac{1}{4}$ in., depth to 10 in. Eggs four to rarely eight white,

¾ by ⅔ in. Incubation about 12 days, by both sexes. Young reared by both sexes. One annual brood.

Food: examination of stomachs of 723 birds showed 76 percent animal matter, mostly insects which are plant pests. Vegetable matter included sap and cambium of trees and seeds of plants. Food gleaned from trunks and limbs of trees or dug from wood. Spiders, caterpillars, aphids, scale insects are common in diet.

Undoubtedly useful. While it may eat some of growing layer of trees and drink sap, injury in these activities is minor and is scattered rather than concentrated in one area as with sapsuckers. Easily attracted to feeding stations by suet and become unusually popular window visitors.

Arctic Three-toed Woodpecker, 400
Picoides arcticus

Length to 10⅕ in. Wingspread to 16 in. Male with a yellow crown patch which is lacking in female. Otherwise, both sexes essentially solid black on back, white on breast, with narrow transverse black and white stripes on sides. Bill about as long as head. Feet with two toes in front and one behind, fourth lacking.

Breeds from central Alaska in Yukon region to northern Quebec and south to central California, Minnesota, New York, and New England, with winter range occasionally a little farther south to include Pennsylvania, New Jersey, Ohio, Indiana, and Illinois. Commonest in dense evergreen forests.

Nests in evergreen forest, in a hole with a diameter of to 2 in., to 18 in. deep, widened at base and lined with chips. Eggs usually four to six, 1 by ⅕ in.; white, moderately glossy. Nesting period late May and June. Incubation by both sexes, 14 days. One annual brood. Young helpless when hatched. Noisy in breeding time.

Food, based on examination of 70 stomachs, contained almost no plant material except wood fragments probably taken while hunting for insects, 75 percent of animal food taken consists of wood-boring larvae of beetles and moths; 25 percent adult beetles, ants, and spiders. Works almost exclusively on dead trees, seeking out pests that breed in them and often working extensively on a single tree.

It probably is useful as a check on wood-destroying insects. Its migrations are usually definitely associated with disappearance of dead timber or a reduction in insect population due to various causes. Estimated that one bird may destroy 13,000 wood-boring grubs in a year.

Arctic Three-toed
Woodpecker

Lewis' Woodpecker, 408
Asyndesmus lewis

Length 11½ in., wingspread 21 in., bill 1⅕ in. Sexes alike in markings. At a distance resembles a small crow. Only red-bellied woodpecker; with a distinctly black back and wide black wings. Flight pattern strong, steady like a crow; not undulating like a woodpecker. Sometimes flight like that of a flycatcher. Black back with bronze luster; forehead, cheeks and chin bright crimson-red, white ashy white collar tinged with red. On belly side collar color becomes more distinctly red. Feet and bill distinctly black.

Range, western North America; southern British Columbia and Alberta south to Arizona and western Texas, west to coast range of California. Occasionally found somewhat to the east in the southern part of its range. Winters in southern California, western Texas and northern areas of Mexico. Found typically in open woods; usually associated with oaks; frequently in burned over areas of clearings offering open space. Favorite perches are telephone poles or fence posts or burned snags. Bird of the Transition zone.

Usually nests in dead tree; may dig out hole for nesting or accept one already available. Nests frequently in fire blackened snags; commonly in cottonwoods or pines. Entrance to hole to 2½ in. in diameter and circular in pattern, nesting hole may be 30 in. deep and 4 in. across where eggs are laid. Nest occupied for several years. Nesting period April to June. Eggs 5 to 9, 1 in. by ¾ in.; full white, no gloss. Both sexes incubate. Food of nestlings largely insects. Young leave the nest at about three weeks of age.

Frequently a ground feeder; takes insects in particular crickets, beetles, and grasshoppers. Will eat wild fruits and berries and has been known to visit commercial cherry and apple orchards to sample the fruits. Acorns are a standard in its diet. Does not take insects from beneath bark and within the wood as is typical of most woodpeckers.

A beautiful bird not so commonly seen as are other woodpeckers. Sometimes appears to be colonial. Doesn't hammer nor grub nor fly woodpecker style. Is a hoarder of acorns which it stores in holes in trees. Was reported by members of the Lewis and Clark expedition and named after Captain Meriwether Lewis. Gathers in large flocks in mating season. Usually silent, except in mating season when may produce a sharp, shrill *char* note.

Lewis' Woodpecker

Order Passeriformes./Family Tyrannidae [Flycatchers]

Kingbird

Kingbird, 444
Tyrannus tyrannus
Length to 9 in., wingspread to 15 in., tail to 3¾ in., bill to ⅞ in. Weight to 1⅗ oz. Sexes about equal in size and coloration. General color slaty-black, with white underparts and white band at tip of broad fanlike tail. Orange-red crown patch, which is larger in male than in female, not easily seen.

Breeds from Southern British Columbia to Nova Scotia south to central Nevada, northern New Mexico, southeastern Texas, and southern Florida. Winters from southern Mexico to Colombia, Peru, British Guiana, and Bolivia. Occasionally in Cuba and Greenland. Ordinarily rests on high exposed perch from which it catches food in open air.

Nests in orchards and at woodland borders. Nest ragged outside but well made inside, from 3 to 60 ft above ground. Eggs three to five, larger end marked with brown, purple-brown, or lavender spots, 1¹⁄₁₆ by ¾ in.; weight ⅛ oz. Incubation chiefly by female, for 12 to 13 days. One or two annual broods. Temperature, 105 to 110°F.

Food essentially insects. In 665 stomachs examined, honeybees were found in 22, total number of bees being 61, of which 51 were drones, 8 workers, and 2 undetermined. 26 robber flies, enemies of honeybees, were found in 19 stomachs. It is estimated that one robber fly may kill many bees a day. Song, 5,850 to 6,225 cps. Speed, 23 mph.

Unquestionably, a friend of general farmer though individuals with nests near apiaries may be destructive to bees. The name bee martin is, on the whole, unfortunate. The kingbird drives hawks, eagles, herons, and crows from near its nesting site or from territory it has taken as its own.

Western Kingbird

Western Kingbird, 447
Tyrannus verticalis
Length to 9½ in., wingspread to 16½ in., tail to 4 in. Female slightly smaller than male. In this species, white on sides instead of tip of tail; upper parts are gray, lower parts yellow, so there should be no misidentification. Crown patch on female is smaller than that on male, as in kingbird. Also commonly called Arkansas kingbird.

Breeds from southern British Columbia to Minnesota and south to northern Lower California, Chihuahua, and western Texas. Winters from western Mexico to Nicaragua. Occasionally found in Massachusetts, New York, Florida, Virginia, South Carolina, Maryland, Maine, Michigan, Wisconsin, and Illinois.

Nest placed on branch of tree like that of eastern kingbird, or on cross-bars of telephone poles. Eggs three to five; like those of eastern kingbird but slightly smaller, ⁹⁄₁₀ by ⅔ in. Incubation by both sexes, for 12 to 13 days. Both parents assist in rearing and protection of young. One or two annual broods.

Food known to be 90 percent insects, remainder mostly wild fruits and seeds. Food is, for most part, caught on wing. Number of beneficial insects found in 109 stomachs was too small to have any significance. Speed, 18 mph.

Species is undoubtedly one of most useful species of birds in its range. It is entitled to every possible protection.

Great Crested Flycatcher

Great Crested Flycatcher, 452a
Myiarchus crinitus
Length to 9⅓ in., wingspread to 14 in., tail to 3⅕ in. Female smaller than male. With conspicuously reddish (rufous) long, loosely hung tail, brown back, yellow belly, and gray throat and breast. Young by first winter practically indistinguishable from adults.

Two subspecies include the Northern and the Southern. Northern breeds from central Manitoba to New Brunswick and south to Texas and South Carolina. Winters from eastern and southern Mexico to Colombia. Southern breeds along Atlantic Coast, South Carolina to Florida, and winters from southern Florida to Central America.

Northern crested flycatcher courts by display and pursuit. Builds nest in deserted woodpecker hole, natural hollow, or bird box. Usually, nest has one or more cast snakeskins in it. Eggs three to eight, creamy to pinkish with marks or scratches of brown or black; 1 by ⅔ in. Incubation 13 to 15 days, by both sexes. One or rarely two annual broods.

Food almost exclusively insects, caught on wing. Estimated that food is 94 percent animal matter. Insects include sawflies, stinkbugs, May beetles, strawberry weevils, cotton boll weevils, cicadas, grasshoppers, crickets, katydids, moths, and caterpillars. In 265 stomachs, only 1 with a honeybee.

Entirely useful to man's economy and most interesting. The *wheep* call in wooded areas from late spring to fall is enjoyed by every bird lover.

Eastern Phoebe

Eastern Phoebe, 456

Sayornis phoebe

Length to 7¼ in., wingspread to 11¼ in., tail to 3½ in. Weight 1 to 2 oz.
Female smaller than male. No wing bars such as on wood pewee. Tail
long, loosely hung, and active. Bill black and wings dark. In winter,
adults more olive and under parts yellower than in spring; young, like
adults in autumn. Temperature, 101 to 108.6°F.

Breeds from central Mackenzie to Prince Edward Island and south to
northeastern New Mexico and highlands of Georgia. Winters in United
States south 37°N. and on to Vera Cruz in Mexico. In migration, may
appear in Cuba, California, Lower California, Wyoming, and Colorado.
S. sayus, Say's phoebe, extends range to West. *S. nigricans*, black phoebe,
another Western species, has adapted itself to man and his buildings.

Nest in a crevice, on a wall, in a gorge, under a bridge, on a porch, or in
an abandoned building. Usually largely of mud covered with mosses,
rather bulky but well lined. Eggs three to eight, usually unspotted, white
or with a few scattered reddish or blackish spots, ¾ by ⅗ in. Incubation
16 days, by one or both sexes. Normally two broods a year, sometimes
three.

Food almost wholly insects such as destructive moths, cotton boll weevils, gypsy moths, brown-tail moths, corn leaf beetles, cucumber beetles,
ants, grasshoppers, locusts, crickets, ticks, and caterpillars. Sometimes
cleans pests from backs of cattle. Examination of 370 stomachs proved
bird almost wholly helpful to man's interests.

Because of food habits, worthy of every protection. The whistled song of
the chickadee is often confused with that of phoebe. Beloved by many
people because of its habit of nesting about old country homes. Song,
3,300 to 5,200 cps.

Alder Flycatcher

Flycatchers

1. *Empidonax virescens* (Acadian), 465
2. *E. traillii* (alder), 466
3. *E. minimus* (least), 467

Length: (1) to 6¼ in.; (2) to 6 in.; (3) to 5⅔ in. Wingspread: (1) to 9½
in.; (2) to 9 in.; (3) to 8½ in. (1) the greenest; (2) the brownest. (1)
without white throat of (2) and (3). (3) with white eye ring and two white
wing bars, grayer above and lighter beneath than the others.

Breeding range: (1) Nebraska to southeastern Ontario and south to Texas
and Florida; (2) central Alaska to Newfoundland and south to British
Columbia and Maryland; (3) central Mackenzie to Cape Breton Island
and south to Montana and North Carolina. Winters (1) Colombia and
Ecuador; (3) northeastern Mexico to Panama.

Nests: (1) partly suspended in fork of twigs, 4 to 20 ft up; (2) rarely over
4 ft up in low shrub; (3) usually 8 to 40 ft up in shade tree. All compact
nests, lined with finer materials, with usually three to four or sometimes
six eggs. Eggs unspotted white, (1) ⅘ by ⅗ in.; (2) ⅘ by 9/16 in.; (3) ⅔ by
½ in. Incubation usually 12 days, probably by both sexes and one or
rarely two annual broods.

Acadian
Flycatcher

(1) of mixed woodlands, call *peet* or *kareep*; (2) haunts thickets near small
streams and waterways, calls *way-be-o*; (3) in orchards and open woodlands where it calls *chebec*. All are small flycatchers, catching their food
on wing while flying up from an exposed perch for most part. They are
easily confused but above notes may be useful in modifying a guess.

Useful as destroyers of insects in their chosen haunts and interesting to
watch and hear, particularly on their spring migration and during breeding season. None has any harmful habits and so are worthy of protection
for their good qualities.

Eastern Wood Pewee, 461

Contopus virens

Length to 6¾ in., wingspread to 11 in., tail to 3 in., bill ¾ in. Female
smaller than male. Conspicuous field marks are two light bars on each
wing, and lower bill lightish in color. Young much like adults but with
two distinct yellowish-brown wing bars, practically like adults by first
winter.

Breeds from southern Manitoba to Prince Edward Island, south to central Texas and central Florida. Winters from Nicaragua to Colombia
and Peru. Accidentally in Cuba. Occasionally in Colorado. Two related wood pewees extend range to Alaska and West Coast down to Lower California for breeding range and farther for winter.

Nests common in woodlands or orchard or shade trees; nest built on
horizontal limb, like a shallow saucer of rootlets and similar materials,
often with lichens on outside. Eggs, two to four, white or creamy with
spots and blotches of brown, lavender, or purple; ⅘ by 9/10 in. Incubation 12 days, chiefly or solely by female. One or two annual broods.

Food: by an examination of stomachs of 359 birds, shown to be 99 percent animal matter, chiefly small insects, including beetles, weevils, flies,

(Continued)

Eastern Wood Pewee

Olive-sided Flycatcher

Horned Lark

Violet-green Swallow

moths, grasshoppers, crickets, bugs, and caterpillars of many sorts. The plaintive *peewee* call is associated with sleepy warm summer afternoons, not with strenuous city life.

Almost entirely useful to man's interests and certainly worthy of every protective effort which may be shown it. Known to catch small trout around fish hatcheries but this damage is negligible. Known to feed and rear abandoned young of other species of birds in other nests. Song, 3,650 to 4,375 cps.

Olive-sided Flycatcher, 459
Nuttallornis borealis
Length to 8 in. Wingspread to 13 in. Female smaller than male. Dark olive on sides, with white on throat, down chest and belly, and between wings and body, last not always showing. Bill large. Back, head, most of wings and tail appear darker than rest of bird. Sexes similarly colored.

Breeds from central Alaska to southern Quebec and Cape Breton Island and south to northern Lower California, Texas, northern Michigan, New York, and in mountains south to North Carolina. Winters in northern South America in Colombia and Peru after migrating through Mexico. Accidental in Greenland.

Nests in open woods or clearing near woodlands. Nest on horizontal branch, usually of cone-bearing tree, 10 to 50 ft above ground, shallow, of twigs, roots, and lichens with moss lining. Eggs, three, $9/10$ by $2/3$ in.; cream to pinkish with chestnut, lavender, or purple spots and blotches at larger end. Incubation 14 days. One annual brood.

Usually perches at extreme tip of some exposed dead branch overlooking brushy land. Flies off into space, capturing food insects on wing. About 3 percent of food may be fruit. Large proportion of flying insects may be bees. Defends nest vigorously. Calls *tuck three beers* or *three cheers*. Females twitter much about nest.

Woodland species seldom near beehives where it might be destructive of useful bees but ordinarily exceptionally valuable as insect destroyer. Bird lovers enjoy spirited calls and behavior of olive-sided flycatcher and eagerly await its announced spring arrival. Southern migration begins in August as soon as young birds are able to fly well.

Order Passeriformes./Family Alaudidae [Larks]

Prairie Horned Lark, 474b
Eremophila alpestris praticola
Length to $7\frac{2}{5}$ in., wingspread to $13\frac{1}{4}$ in., tail to $2\frac{2}{5}$ in. Horned lark, slightly larger and with yellow line over eyes instead of white line of prairie horned lark. Appears slightly smaller than robin but with contrasting light-colored body and black tail; walks, rather than hops. Tends to form flocks.

Horned lark breeds from the Arctic south to Gaspé; winters in Manitoba, in the Mississippi Valley and as far south as the Carolinas. Prairie horned lark breeds from southern Manitoba to central Quebec and south to Missouri and Connecticut. Winters south to Texas and rarely to Florida. Related subspecies, 16; extend range over most of United States, Canada, Mexico, and Colombia; differences of minor importance. In California Pleistocene. Open field species.

Builds nest on ground, early in each spring. Nest of grass, lined with soft stuffs. Eggs three to five, light drab or grayish, uniformly spotted with brown of different shades, 1 by $2/3$ in. Incubation for 11 to 14 days, by both sexes but chiefly by female. Annual broods two, three, or even four on rare occasions.

Food largely waste grain, weed seeds, and insects; particularly cutworms which may be exposed by plowing and cultivation acitivties. Especially good in winter as destroyer of weed seeds; follows farmers in fields getting insects as soon as they are turned to surface.

Essentially a useful species. May do some damage to winter wheat if it is not covered properly or if it does not become established soon, but this damage is more than offset by good done. Apparently worthy of protection at all times.

Order Passeriformes./Family Hirundinidae [Swallows]

Violet-green Swallow, 615
Tachycineta thalassina
Length to 5 in., wing $4\frac{2}{3}$ in., tail 2 in., bill $1/5$ in. Female smaller than male, length, $4\frac{1}{2}$ in. Crown and head bronze-green to purple-bronze, wings green with purple tinge, upper tail purple, underparts white, tail and wings black with blue gloss. Female duller than male. Young sooty-brown above.

Breeds from central Alaska to central Alberta, south to northern Lower California and northern Durango, and east to western South Dakota and Nebraska. Winters in Mexico and south to Guatemala and Costa Rica, going through western Texas in migration. Accidental in Illinois.

Nests in cliffs, old woodpecker holes, and hollow trees. Nest with lining of grass and inner lining of feathers. Eggs white, four to five. Incubation probably as with other swallows. In Oregon, violet-green swallows seem to take readily to nesting boxes. Not infrequently found in attics and abandoned houses.

Food practically 100 percent insects. Bugs make up $\frac{1}{3}$ of food, with leaf hoppers being an obvious favorite. Wasps and wild bees constitute $\frac{1}{6}$ of diet, beetles about $\frac{1}{10}$, and ants slightly less. Migration leisurely in fall but en masse in spring.

Few birds can equal or excel its beauty either while perched or when in flight and its food habits are exemplary so far as man's interests are concerned.

Tree Swallow

Order Passeriformes./Family Hirundinidae [Swallows]

Tree Swallow, 614
Iridoprocne bicolor
Length to $6\frac{1}{4}$ in., wingspread to $13\frac{1}{4}$ in., tail to $2\frac{1}{2}$ in., bill to $\frac{1}{3}$ in. Female smaller than male. Easily identified by clear clean white underparts and steely blue- or greenish-black back. Young first winter, more grayish than blue above; throat not wholly a clear white. Temperature, 109.6°F.

Breeds from northwestern Alaska to northern Quebec and south to southern California and Virginia. Winters from central California to North Carolina, and south over Mexico, Cuba, and Honduras. Occasionally found in Bermuda during migration.

Usually in cultivated areas or over water. Nests in some tree cavity or bird box. Nest with a grass lining, with feathers provided in abundance. Eggs, four to six, white to rosy, $\frac{4}{5}$ by $\frac{9}{16}$ in. Incubation for about 14 days, by both sexes. Two broods may be raised in season. Two females may occupy one nest.

Essentially an insect eater, eating many mosquitoes, robber flies, and ants. Eats more vegetal matter than most other swallows, favoring bayberries, blueberries, and fruits of red cedar and Virginia creeper. Probably only slightly less valuable than barn swallow.

Role as insect destroyer makes it unusually valuable in rural areas and in mosquito-infested regions. Comes readily to bird boxes erected near marshes and in wet places in public parks. By its beauty, grace in flight, and food habits adds to attractiveness of environment.

Bank Swallow

Bank Swallow, 616
Riparia riparia
Length $5\frac{1}{2}$ in., wingspread $11\frac{1}{10}$ in., tail $2\frac{1}{3}$ in., bill $\frac{1}{4}$ in. Female smaller than male. Winter plumage assumed after southern migration in fall. Distinctly brown-backed swallow, with conspicuous dark band across breast, which is lacking in rough-winged swallow, another brown-backed form. Locally known as sand swallow, sand martin, and black martin.

Breeds from northern Alaska to northern Quebec south to southern California and Virginia; also in Europe, British Isles, Siberia, Tunisia, and Algeria. Winters through Mexico to Brazil and Peru, with allied races in India, East Africa, and South Africa. Not commonly found nesting near dwellings of man.

Nest near water in high bank, a room about 5 in. in diameter at end of a tunnel extending from 15 in. to 8 ft back into bank. Lining of feathers and grass. Eggs three to seven, white with a rosy tinge, weight $\frac{1}{20}$ oz. Incubation 14 to 16 days by both sexes. One or two annual broods.

Food essentially insects, with an obvious emphasis on flies and weevils. Cold rains, which keep flying insects down during time young are developing, are often fatal to great numbers of swallows and swifts. If food is not in air, birds do not get it. Speed, 31 mph. Temperature, 102.6 to 112.4°F.

Possibly not so useful as other swallows because it is not found so commonly about man's dwellings, but it certainly is not harmful. Nest-building activity is always worth watching and sight of their flying into nest at feeding time is intriguing.

Rough-winged Swallow

Rough-winged Swallow, 617
Stelgidopteryx ruficollis
Length $5\frac{3}{4}$ in., wingspread $12\frac{1}{4}$ in., tail $2\frac{1}{3}$ in., bill $\frac{1}{3}$ in. Female smaller than male, otherwise sexes appear alike. Brown-backed, ordinary-looking swallow, with tail only slightly forked and without dark breast

(Continued)

Barn Swallow

Cliff Swallow

Purple Martin

band which marks smaller and generally grayer-backed bank swallow. Young like adults. Temperature, 108.7°F.

Breeds from southern British Columbia to southeastern Ontario, and south through United States to southern California, Vera Cruz in Mexico, and central Florida. Winters from southern Arizona and Mexico south to Costa Rica with associated races in Central America.

Nests usually a hole near water in a bank or cliff, under a bridge, in a building or hollow tree. Nest bulky, lined with grasses or pine needles, with few or no feathers. Eggs four to eight, glossy white, ¾ by ½ in. Incubation not commonly known, but probably only one brood of young a year.

Food like that of other swallows, almost exclusively insects caught on wing or by skimming close to surface. Mosquitoes and other aquatic insects probably figure high in diet. Not found about dwellings so commonly as barn swallow and cliff or eave swallow.

Has no habits contrary to man's best interests and so is worthy of protection wherever it establishes itself. Less friendly to man than cliff swallow, barn swallow, and purple martin.

Barn Swallow, 616
Hirundo rustica
Length to 7¾ in., wingspread to 13½ in., tail to 4½ in. Female smaller than male. Tail more deeply forked than any other associated swallow. Male, forehead chestnut, but elsewhere above glossy dark steel-blue, chestnut below on chin and throat but lighter to rear. Female similar but sometimes duller.

Breeds from northern Alaska to northern Quebec, south to southern California, northeastern Arkansas, and Virginia. Winters from central California, southern Texas, Gulf States, and South Atlantic States, south over Mexico, Cuba, and Honduras.

Usually in open country, in farm buildings, commonly not far from water. Nest usually directly under roof of barn, shed, or porch, of mud with lining of grasses and feathers. Eggs three to six, white or whitish spotted with brown or purple, $\frac{15}{16}$ by ⅗ in.; egg weight $\frac{1}{20}$ oz. Incubation about 15 days by both sexes. Two or three annual broods.

Food practically all animal matter, caught on wing. Much food caught close to surface of water. Fish sometimes catch swallows. Food includes mosquitoes and other flies, bees, wasps, but practically no honeybees; moths of cutworms, codling moths, and weevils. Sometimes gullet is packed to twice normal size.

Injures none of man's interests and helps in many ways. Living near a house, it kills great numbers of houseflies, mosquitoes, horseflies, and similar pests of man and his domestic beasts. Entitled not only to protection but encouragement through safeguarding of nesting sites.

Cliff Swallow, 612
Petrochelidon pyrrhonota
Length to 6 in., wingspread to 12⅓ in., tail to 2⅖ in. Sexes about equal in size. In field, light brown (buffy) rump is distinguishing, if square-tipped tail is not sufficient. Has dark throat patch, unlike whitish throat of bank swallow; upper parts steel-blue. Temperature, 100.2 to 111.2°F.

Three subspecies recognized: Northern, Lesser, and Mexican. Northern breeds from central Alaska to Cape Breton Island south over all of United States except Florida and Rio Grande Valley. Migrates in winter through Florida and Central America to Brazil and Argentina. From California Pleistocene.

Nests usually under eaves of a barn or on cliff. Nest gourd-shaped, of clay or mud, with a lining of grass, leaves, feathers, and wool. Eggs four to five, white or creamy and pinkish with markings of reddish-brown and darker brown, ⅘ by ⅔ in.; weight $\frac{1}{16}$ oz. Incubation 16 days, by both sexes. One annual brood.

Food almost exclusively insects. Examination of 375 stomachs showed a few wild berries and spiders, but otherwise only insects such as cotton boll weevils, alfalfa weevils, chinch bugs, rice weevils, and so on. Nests commonly in colonies and in this way often attracts attention, which is not always to be desired.

Such a useful species that every effort should be made to give it complete protection. Should be encouraged to nest where its services may be used. Known popularly in some areas as "eaves swallow."

Purple Martin, 611
Progne subis
Length to 8½ in., wingspread to 16¾ in., tail to 3½ in., with $\frac{9}{10}$-in. fork, bill to $\frac{9}{10}$ in. Female smaller than male. Largest of associated swallows. Male uniformly blue-black above and below. Female with gray underparts and neck ring. Commonly about buildings and birdhouses.

Breeds from Cape Prince of Wales in Alaska to New Brunswick, south through central Alberta to Mexican border, on to Vera Cruz and along Gulf Coast. Winters in Brazil but occurs in migration in Central America, Venezuela, and Guiana. Accidentally in Bermuda and British Isles.

Nests in open country not far from water, never in forests. Nest in hole, cliff, building, gourd house, or special colonial martin house. Outside of nest mud, inside grasses and feathers. Eggs three to eight, pure white, 1$\frac{1}{12}$ by $\frac{3}{4}$ in. Incubation about 15 to 16 days, chiefly or wholly by female. Annual broods one or, in South, two.

Food almost solely insects. Examination of stomachs showed cotton boll weevils, squash bugs, butterflies, destructive moths, grasshoppers, dragonflies, drone honeybees but no worker bees, a considerable number and variety of flies, and mosquitoes. Flock of martins active about an orchard saves much spraying.

Entirely useful and worthy of every possible protection. Largest of swallows, responds so splendidly to encouragement by man that its range should be increased rather than decreased. Unfortunately, must compete for nesting sites with gangster-type starlings and English sparrows. Such a fight is usually an uneven one.

Gray Jay

Gray Jay, 484
Perisoreus canadensis
Length to 12$\frac{1}{10}$ in., wingspread to 17$\frac{1}{2}$ in., tail to 6$\frac{1}{3}$ in. Female smaller than male. Appears in field like a large gray bird with dark cap, long tail, white throat, inquiring nature, lots of nerve, and somehow a sense of humor. Some suggest it looks like a long-tailed gray crow or overgrown chickadee. Young dark slate-colored. Sexes alike.

Three subspecies: Canada, Rocky Mountain, and Alaska jay. Canada jay breeds from Alaska and northern Mackenzie to Labrador and south to central British Columbia, northern Minnesota, and northern New York. Rocky Mountain jay ranges south to Arizona and New Mexico.

Nests mostly in coniferous forests, often in swampy areas. Nest large, neatly built of twigs, lined with grasses, mosses, and lichens. Eggs three to five, grayish or yellowish; speckled with brown or lavender; 1$\frac{1}{5}$ by $\frac{7}{8}$ in. Incubation 16 to 18 days. One annual brood. Commonly travels in pairs. Breeds early in spring.

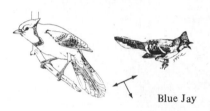

Food: will eat anything that can be considered edible, cleaning up garbage about camps, eating fish, eggs, meat, soap, toothpaste, and great quantities of insects. Stomach of one bird had approximately 1,000 tent-caterpillar eggs in it. Also eats mice and beetles.

Always interesting camp follower. Called camp robber and "Hudson Bay bird." In judgment of careful bird students, Canada jay is considered as highly beneficial in spite of often violently expressed opinions of individuals from whom this inquisitive bird has taken things left idly about.

Blue Jay

Blue Jay, 477
Cyanocitta cristata
Length to 12$\frac{1}{2}$ in., wingspread to 17$\frac{1}{2}$ in., tail to 6 in., bill to 1$\frac{1}{4}$ in. Weight to 3 oz. Female smaller than male. Colored conspicuously blue, white, and black, with long round-tipped tail and conspicuous head crest. "Sassy" in disposition and generally noisy. Temperature, 106.2 to 110.2°F.

Three subspecies include Northern, Florida, and Semple's. Northern breeds from southern Alberta to Newfoundland, south to Colorado, central Texas, and Virginia. It ranges somewhat farther south in winter but may be resident through year. Sometimes found in New Mexico. Defended home territory $\frac{1}{2}$ mile across.

Nests in dense woods or small groves, especially in coniferous trees. Nest in crotch or branch close to trunk; 5 to 50 ft above ground, of sticks with lining of roots. Eggs three to six, greenish-gray or olive with brown blotches, 1$\frac{1}{5}$ by $\frac{9}{10}$ in. Incubation 15 to 17 days, by both sexes. One or two annual broods.

Food, as indicated by examination of 300 stomachs, shows bird is much more good than bad. Stomachs of 292 showed 75 percent vegetal, remainder animal matter including 14 percent discarded shells of hen's eggs. Most animal matter was insects, spiders, snails, and even fishes, frogs, salamanders, and birds and their eggs. Nuts and acorns, 42 percent. Speed, 20 mph.

As indicated above, probably more useful than injurious. A persistent nest robber; destroys eggs and eats young of other species. Certainly, a bird which provides all observers with plenty of excitement whether one attempts to locate a nest, to interpret sounds, to follow activities, or to determine statistically economic importance.

Steller's Jay

Steller's Jay, 478

Cyanocitta stelleri

Length to 13½ in. Tail to 6⅓ in. long. Wings and tail rounded. Conspicuous as a large blue and black bird with a long erect crest. Foreparts brownish-black to black. Wings, underparts, and tail deep blue. Only crested deep blue bird in its range. Sexes similar. Young only slightly duller than adults.

Six subspecies include Steller's (Alaska to Washington), Queen Charlotte (British Columbia), Coast (Oregon and California), Blue-fronted (California to Lower California), Black-headed (British Columbia to Nebraska), and Long-crested (southern Rocky Mountains from Utah into northern Mexico).

Nests in pines and other cone bearers, to 10 ft above ground. Nest bulky, of twigs, moss, and grass with finer lining. Eggs three to six, dull pale bluish-green with blotches of brown and lavender. Family stays together for some time, and flocks of six to eight birds in late summer are probably a family unit.

Food largely seeds, acorns, and berries in fall. An estimate of long-crested jay's food in ⅓ animal food and ⅔ vegetable food, latter being largely acorns and former unfortunately including many birds' eggs. Call is a loud repeated *shook* or *wheck* or *kwesh*.

Like other jays closely related to it, probably does some harm and some good. As with most members of family, its reputation varies with different persons according to behavior that may have been observed. It should be neither completely destroyed nor unduly encouraged.

Scrub Jay

Scrub Jay, 481

Aphelocoma coerulescens

Length to 12 in. Tail to 6⅓ in. A blue jay without any crest, with head, tail, and wings blue, back brownish-gray, underparts pale gray, head flattened, breast with a dark crossband, and in some subspecies throat with light vertical streaks. Sexes colored alike. Young grayer above than parents are. Appropriately called blue squawker.

Seven subspecies recognized including long-tailed (Washington to California), Nicasio (northern California), California (San Francisco to Mexico), Belding's (northwestern Lower California), Xantus's (southern Lower California), Woodhouse's (Oregon to Wyoming, Texas, and California), and Texas (central and west central Texas).

Nests in scrub oak or piñon pine, 3 to 30 ft above ground. Nest an outer basket of twigs and small sticks with a lining of grass stems, rootlets, and finally hair. Eggs three to six, light bluish-green somewhat flecked with rusty brown markings. Very secretive about nest. Family remains together for a time.

May stay around farm buildings picking up food and generally observing what there is to see and to eat. Commonest view one has of them is flying away, displaying a rich blue tail and wings, followed perhaps by smaller birds whose nests they may have robbed.

Like other birds of family, they need watching and one must be charitable to forgive some of their misdeeds. Probably if a fair balance were struck, they would be found no more useless than many who decry their presence.

Black–billed Magpie

Black-billed Magpie, 475

Pica pica

Length to 21 in., wing to 8⅖ in., tail to 11⁹⁄₁₀ in., bill to 1⅖ in. Weight 5⅓ oz. Nostrils covered with bristles. Extremely long tail that tapers half its length, conspicuous black and white markings serve to identify. Sexes colored alike. Young with dull black in usual glossy black areas.

Breeds from Alaska Peninsula southward, mainly near coast, to southern Manitoba, south through eastern Washington and eastern slope of Sierra Nevada to Arizona and northern New Mexico, east occasionally to Iowa, Wisconsin, Illinois, Michigan, Ontario, Hudson Bay, and even Quebec. Related yellow-billed magpie, now restricted to southern California, found in California Pleistocene.

Sometimes lives in colonies or nests in scattered colonies in thickets or trees. Nest size of a bushel basket, of twigs and dry plant material, with a good lining sometimes cemented with mud, and with a lateral, covered entrance. Eggs 7 to 10, grayish or green, dotted, dashed, and blotched with purple or brown; ⁵⁄₁₆ oz. Incubation 18 days.

Food, grasshoppers, large black crickets, ground beetles, codling moth larvae, rodents, and carrion. 85 kinds of plants and animals recognized in stomach contents including reptiles, amphibians, worms, birds, crustaceans, spiders, and scorpions.

As an insect eater, it has no related superiors; as an enemy of rodents it is excellent. However, it robs nests of domestic poultry, kills young chickens, attacks newly sheared lambs or sick or wounded cattle, and by its incessant chattering annoys many people; undoubtedly a pest where its numbers are large.

Common Raven, 486

Corvus corax

Length to 26½ in., wingspread, to 56 in., tail to 11 in., bill to 3¼ in. Female smaller than male. Black all over. Nostrils hidden by bristle tufts. Tail appears rounded at tip when bird is flying; should not be confused with smaller crow. Raven soars on horizontal wings; crow and turkey vulture soar with wings somewhat bent upward; raven tail is wedge-shaped. Sexes alike. Young like adults.

Two subspecies, Northern and American raven. Northern breeds from northwestern Alaska to Greenland, south to states of Washington, Minnesota, Virginia, and parts of Georgia. American raven ranges from southeastern British Columbia to North Dakota, south to Nicaragua. From California Pleistocene.

Nests usually in forests or along seacoast. Nest of coarse sticks lined with finer plant material, in a tall tree or on a rocky cliff. Eggs five to seven, pea-green or olive with spots of brown, gray, or lavender, $1^{15}/_{16}$ by $1^{2}/_{5}$ in. Incubation 20 to 21 days, by both sexes. One annual brood.

A useful scavenger, killing great numbers of mice, rats, and injurious insects, but may also destroy young lambs and chickens, and attack weak or injured animals of larger size. Shows exceptional cleverness in outwitting more powerful animals, particularly where a number of ravens work cooperatively.

Because of destruction of livestock and poultry where ravens are abundant, their numbers have been greatly reduced; over much of their original range they may no longer be found. Certainly great flocks comparable to those of crows no longer exist near intensively cultivated areas.

Common Raven

Common Crow, 488

Corvus brachyrhynchos

Length to 21 in., wingspread to 39 in., tail to 8 in., bill to $2^{2}/_{3}$ in. Weight about 20 oz, variable. Female averages smaller than male. Color all black. Lacks separated long-pointed throat feathers of raven, is smaller, and has squarer tail. Young only slightly duller than adults.

Five subspecies include Eastern, Southern, Florida, Western, and Northwestern crows, all with minor differences. Eastern breeds from southwestern Mackenzie to Newfoundland, south to Maryland and northern Texas. Winters generally between north and south boundaries of United States. From Pleistocene of Florida and California.

Nest a platform of sticks in top of tree or on branch close to trunk, rarely in bush or where there are no trees, lined with finer plant materials. Eggs three to nine, pale blue-green to olive with gray or brown blotches or markings, $2^{1}/_{16}$ by $1^{2}/_{5}$ in.; weight $^{5}/_{8}$ oz. Incubation 15 to 18 days, probably by both sexes. One or two annual broods.

Food: consumes enormous quantities of insects, carrion, and weed seeds, as well as eggs of other birds including game birds; also grain waste, mice, gophers, and rabbits. When grasshoppers are abundant, average crow stomach contains about 100. In farming regions, 38 percent of food may be corn. Known to kill newborn lambs.

Unbiased students contend that while crows do much damage, on average they do more good than harm. To sportsmen serve as buffer animals which may be shot to relief of more valuable species. Numbers do not decrease much by organized shooting because birds only become more wary. Have great roosts during winter months.

Common Crow

Piñon Jay, 492

Gymnorhinus cyanocephalus

Length $11^{2}/_{3}$ in., wing 6 in., tail $4^{4}/_{5}$ in. General color a grayish-blue. In flight shows relatively short wings and nearly square-tipped tail. Throat and chest gray-streaked. Female resembles male but is slightly smaller and duller. Young duller yet, with blue only on wings and tail.

Resident from central Washington to central Montana and south to northern Lower California and northwestern Nebraska. Sometimes found on California coast and sometimes as far east as eastern Nebraska and Kansas.

Nests in colonies, in piñon pines, junipers, and oaks. Nest deep, bulky, compact, with well-formed inner cup 5 to 12 ft from ground. Usually constructed with twigs and bleached grasses, well lined with wool, moss, hair, and/or feathers. Eggs usually four to five, bluish-white spotted or streaked with brown or sometimes blotched. Stay in pairs even though great numbers nest and feed at common points.

Food chiefly nuts of piñon pine in season, but also seeds of yellow pine, black pine, juniper berries, wild berries, and insects such as grasshoppers. Injury has been caused by these birds to some growing crops such as grains and even watermelons, particularly where birds act in unison in flocks.

Undoubtedly, where flocks of these birds congregate near cultivated lands bearing crops, they may cause much injury. However, in general they are easily and effectually frightened away by scarecrows and a little shooting.

Pinon Jay

Clark's Nutcracker, 491
Nucifraga columbiana
Length to 13 in., tail to 6 in. Face white and rest of body gray. Central tail feathers black but rest of tail white above and below. Wings glossy black but secondary feathers are white-tipped. Sexes colored alike. Young paler gray and with breast apparently more spotted.

Breeds from southern Alaska to South Dakota, south through mountainous areas to northern Lower California, Arizona, and New Mexico. In winter moves farther southward. Occasionally found in western Nebraska, Missouri, Arkansas, Iowa, and Wisconsin.

Nest built in coniferous forests, a platform of twigs bound with bark, grass, and needles, sometimes lined with wool quilted together to form a mat. Eggs three to five, pale gray-green, sparsely flecked or spotted with brown, gray, or lavender, particularly at larger end. Incubation 16 to 22 days. May nest in zero weather. Call notes a "chaar chaar."

Food in summer, nutlike seeds of certain pines, berries of cedar, beetles, caterpillars, grasshoppers, and destructive black crickets of their range. In fall, food is essentially nutlike fruits of pines. Young fed hulled seeds of pine, regurgitated by parents.

Interesting birds about camps, but rather too noisy and inquisitive for their own good. May destroy baits on trap lines and win hatred of trappers. Common and welcome about many national parks where they may come to feed out of hand, in competition with ground squirrels so popular with most persons.

Clark's Nutcracker

Order Passeriformes./Family Paridae [Titmice, Verdins, and Bushtits]

Black-capped Chickadee, 735
Parus atricapillus
Length 5¾ in., wingspread 8½ in., tail 2⅔ in. Entire crown, throat, and hind neck black, without a crest; in Brown-capped Chickadee, *P. hudsonicus*, brown. Outer margins of wing coverts whitish (in Carolina chickadee, not white). Sexes colored alike. Young like adults in general appearance.

Four subspecies include Black-capped, Long-tailed, Oregon, and Yukon; 17 other subspecies in 6 other closely related species. Black-capped ranges from northern Ontario to Newfoundland, south to Missouri and North Carolina, intergrading with Long-tailed, which ranges westward. Tree species.

Nests in forests, open woodlands, parks, or orchards, in hollow stub, tree, or bird box 1 to 50 ft above ground, with 1-in. entrance and depth to 12 in. Nest moss-lined. Eggs four to eight, white, spotted and speckled with brown, ⅔ by ½ in. Incubation by both sexes, 11 to 13 days. One or two annual broods.

Food largely insects and insect eggs. Stomachs of 289 yielded 68 percent animal food and remainder plants. Stomach of one had 450 eggs of plant lice in it; in fact, plant lice seem a favored food. Also included in food are bark beetles, weevils, scale insects, spiders, flies, wasps, ants, and similar creatures. Defended home territory 100 yd across.

Obviously highly beneficial. Not estimated is number of people made happy by cheery visits these birds pay to window feeding stations or to popular woodland paths. State bird of Maine. Carolina chickadee, state bird of Carolina. Song pitch, 3,027 to 3,700 cps.

Black-capped Chickadee

Chestnut-backed Chickadee, 741
Parus rufescens
Length to 5 in. Like black-capped chickadee of East in general pattern but with back a chestnut-brown and sides with a touch of chestnut. Black bib, white cheeks, and dark cap are like those of eastern species. Chestnut back distinguishes species from other chickadees in same area.

Three subspecies: chestnut-backed (from Prince William Sound, Alaska, to Sonoma County, Calif., and to western Montana), Nicasio (along coast of middle California), and Barlow's (along coast from California to Monterey Bay). Six other species including black-capped in genus.

Nests in holes in dead poles or small trees, with 1-in. opening. Nest usually lined abundantly with soft material such as feathers, rabbit fur, and down. Eggs four to eight, white, spotted with brown or lilac especially at larger end, or possibly plain. Incubation by both sexes.

Related chickadees of West include Oregon and Long-tailed, subspecies of *P. atricapillus*, and Grinnell's and Short-tailed, subspecies of *P. gambeli*, mountain chickadee. This species is common chickadee in coast mountains and in heavy spruce forests.

Economic importance is like that of other chickadees in that it controls insect populations to be found on finer parts of woody plants at any time of year.

Chestnut-backed Chickadee

Tufted Titmouse, 731
Parus bicolor
Length to 6½ in., wingspread to 10¾ in., tail 3⅙ in. Female smaller than male. Conspicuous because of its high crest, which is not shared by other birds smaller than a sparrow. Color almost uniformly sooty-gray, with basal half of outer tail feathers whitish. Sexes colored alike.

Breeds in resident range from Nebraska to Maine, south to central Texas, Gulf Coast, and central Maine. Occasionally found in Wisconsin, Michigan, Ontario, New York, Maine, and Connecticut. Formerly stayed in well-wooded regions but is now found in parks and in street trees within its range.

Nests in natural tree cavity or deserted woodpecker hole, in stub, fence post, pole, or bird box, usually 40 to 60 ft above ground and lined with soft stuffs. Eggs five to eight, ⅔ by ½ in., white or creamy brown with spots of brown or lavender. Incubation chiefly by female. Annual broods one or, in South, two.

Food: examination of stomachs of 186 showed 66.5 percent animal matter and remainder plant stuffs. Latter included largely nuts, acorns, and wild berries. Caterpillars and wasps made up over 50 percent of total food eaten. Included also were tent-caterpillar eggs, spiders, sawfly larvae, scale insects, and tree hoppers.

Obviously, from its food record, is essentially beneficial. Certainly because of its interesting appearance and habits, it is popular with bird students. Often as inquisitive as its near relative the chickadee, even perching on gun barrels of still hunters. State bird of West Virginia.

Tufted Titmouse

Plain Titmouse, 733
Boeolophus inornatus
Length to 5½ in., wing length 5⅖ in., tail 2¼ in., bill ⅖ in. Female smaller than male. Warbler size; plain gray color, distinctive crest evident. Sexes colored alike. Upper parts brown with olive reflections on rump and edges of wings. Sides dull grayish-white or pale brownish-gray. Bill gray to horn color, dusky above. Feet gray to blue-gray.

Ten subspecies recognized. Range, western United States and northwestern Mexico. Not migratory. Southern Oregon, Utah, and Colorado; east to central Colorado; extreme western Oklahoma, New Mexico, and western Texas; south through to Texas, southern New Mexico, and northern Baja California, west to western Baja California and western Oregon.

Nests March 20 to July in California, May 3 to 28 in New Mexico, April 22, to May 31 in Mexico. Usually keeps the same mate for 2 or more years. Commonly build nests in bird boxes if available; otherwise in holes in trees, either old woodpecker holes or natural cavities. Height of nests 3 to 32 ft. Nest a mass of fine grasses to 4 in. in depth covered with a mat of hair; sometimes made of moss, grass, weed stems, other plant fibers and lined with feathers and hair. Eggs three to nine, commonly six to eight; ⁷⁄₁₀ by ½ in. Eggs ovate, dull, pure white, frequently unmarked; may have minute pale reddish-brown dots.

Favorite haunts are among oaks; has behavior similar to chickadee's. Call notes described as *tsick-a-dee-dee* or *tsick-a-dear*. Generally found in trees, less frequently on the ground.

Food in quantity about equally divided between plant and animal. Plant food largely pulp of fruits. Animal food: largely insects, including bugs, scale, insects, caterpillars, and beetles. Generally considered useful.

Plain Titmouse

Order Passeriformes./Family Sittidae [Nuthatches]

White-breasted Nuthatch, 727
Sitta carolinensis
Length 6⅙ in., wingspread 11½ in., tail 2¼ in., bill, ⅛ in. Female smaller than male. Breast white, back blue-gray, cap black. Goes up and down tree trunks, with head up or down, with equal ease. Female slightly duller on top of head than male but otherwise much like male. Young only slightly different.

Seven subspecies include White-breasted, Florida, Rocky Mountain, Slender-billed, Inyo, San Pedro, and San Lucas nuthatches. White-breasted ranges from southern Manitoba to southern Quebec, south to northern Texas and South Carolina. Defended home territory 25 to 48 acres. About trees in woods, parks, or streets.

Nest usually in a cavity, such as abandoned woodpecker's hole or a birdhouse, high above ground, lined with fine stuffs. Eggs five to eight, white or pinkish with brown or lavender spots, especially at larger end, ⅚ by ⅝ in. Incubation by female, 13 days. One annual brood.

Food in winter, 25 percent animal matter; in spring, 80 percent. One bird's stomach had 1,629 cankerworm eggs in it. Kinds of insects eaten are both injurious and beneficial so bird may be neutral in importance to man. In orchards, does much good during seasons when caterpillars are active and in winter feeds on codling moth pupae.

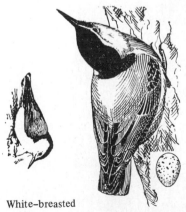

White-breasted Nuthatch

(Continued)

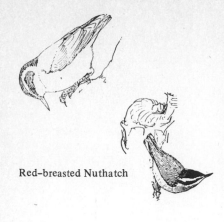

Red-breasted Nuthatch

Apparently liked even by those who do not know its name, judging from great number of popular names it has, such as "upside-down bird" and "twirl-around-a-twig," which describe its antics. One of standard visitants to window feeding stations and beloved by those who must stay indoors.

Red-breasted Nuthatch, 728
Sitta canadensis

Length to 4¾ in. Wingspread to 8½ in. Female smaller than male. Smaller than commoner white-breasted nuthatch, with a broad black eye band and rusty beneath. Male with top of head black with bluish gloss, which is gray to rear in mature female. Young like adults of same sex.

Breeds from upper Yukon to southern Quebec and Newfoundland south in mountains to Lower California, Arizona, and North Carolina; in lowlands rarely south of Michigan, Minnesota, and New York. Winters from southern Canada to southern California, New Mexico, and northern Florida. Migrations rather irregular. Found chiefly in evergreen forests.

Nests chiefly in evergreen forests but sometimes in mixed forests. Nest in a cavity in decaying wood, 5 to 70 ft above ground, with 1-in. entrance, with lining of feathers and bark, with pitch smeared around entrance. Eggs four to eight, ⅔ by ⁹⁄₁₆ in.; laid May to June. Incubation 12 days, chiefly by female. One annual brood.

Active on trunks and large limbs of trees, working with head up or down, gleaning from bark insects that make up its food though it does eat seeds of spruce and balsam. Beetles seem to be important in insect food. May practice some food storage, particularly taking food from feeding stations and hiding it.

Certainly an interesting species, and probably much more useful as an insect destroyer than harmful as a tree destroyer. Species is surprisingly tame and feeds close to human beings even when found in remote areas.

Order Passeriformes./Family Certhiidae [Creepers]

Brown Creeper, 726
Certhia familiaris

Length 5¾ in., wingspread to 8 in., tail to 3 in., bill, ¹¹⁄₁₆ in. Streaked, dark brown above and gray beneath. Tail feathers stiff and used as props when working on tree trunk, from base up. Sexes colored alike. Young possibly not so dark as adults.

Breeds from central Manitoba to southern Quebec, south to eastern Nebraska and North Carolina. Winters over most of breeding range, south to central Texas, southern Alabama, and southern Florida. Rocky Mountain creeper, a subspecies, extends range to Alaska, south into Mexico, breeding south to New Mexico. In deep woods or about street or orchard trees.

Nests usually in swampy woodlands, under loose piece of bark or in old hole. Nest of twigs, bark, feathers, hair, and cobwebs. Eggs five to nine, white to gray, sparingly spotted with reddish- or purple-brown; ⅝ by ½ in. Incubation 15 (?) days, by both sexes. One or two annual broods.

Food, insects and insect eggs gleaned from trunks and branches of trees. In home life, female builds nest and incubates eggs but male assists in bringing food to young. Food includes ants, plant lice, leafhoppers, codling moths, caterpillars, spiders, pine seeds, and suet from feeding stations.

Activities almost entirely valuable to man's interests and therefore deserving of every protection. More abundant than most people appreciate because not easily seen unless one is alert for such birds. Its fine call is a good test for hearing as some persons cannot hear it at all.

Order Passeriformes./Family Chamaeidae [Wren-tits]

Wren Tit, 742
Chamaea fasciata

Length to about 7 in., with upper parts brownish-olive and lower parts pale cinnamon or yellowish-brown. Tail wrenlike but longer, and bill long and curved, like that of wren. Crown, hind neck, wing, and tail feathers grayer than other parts; sides of head and neck grayish-olive. Iris white. Sexes colored alike.

Five subspecies include Coast, Ruddy, Gambel's, Pallid, and San Pedro wren tits, which range from Columbia River in Oregon to Mexican border in humid areas to west of mountains. They favor cover for most part.

Nest in low bushes or trees, rarely if ever over 4 ft above ground. Nest made of woven grass, bark, roots, and other vegetable materials, usually lined with hair of horses or cattle. Nest generally well-hidden from prospective enemies. Eggs three to five, but usually four, pale bluish-green. Conceal themselves in chaparral and other shrubby growth.

Brown Creeper

Wren Tit

Food essentially insects, gleaned from cover in which bird so easily disappears but from which it is so frequently heard. Fleshy fruits account for half of wren tit's fare except during spring. Only family of birds restricted to North America.

Probably useful as insect destroyer. Confounds naturalist who feels confident that it will be easy to locate a bird capable of making such a loud commanding sound. Common about San Francisco Bay area and invites plenty of inquiry by naturalist visitors to region.

Order Passeriformes./Family Cinclidae [Dippers]

Dipper, Water Ouzel, 701
Cinclus mexicanus
Length 8⅘ in., wing 3⅘ in., tail 2¹/₁₀ in. Weight 2⅓ oz. Slate-gray but paler beneath; head and neck feathers faintly tipped with brown; tail and wing feathers darker brown. Eyelids marked with white. Sexes colored alike. Young with grayer crown and whitish throat. Temperature, 106°F.

Breeds from near tree limit in northwestern Alaska to central western Alberta, and south to southern California and southern New Mexico. Accidental in Black Hills of South Dakota and in western Nebraska. An allied race found in Mexico and Guatemala.

Water Ouzel

Nest on ledge, in gorge or canyon, over water or often under waterfall; also on rocks in midstream, and on beams under bridges. Nest bulky, roofed over, with side entrance; made of plant materials such as growing mosses, weeds, pine needles, with a dry inner nest of finer materials. Eggs three to five, white. Incubation 15 days.

Food, aquatic insects and small fish, but mostly caddis-fly larvae, dragonflies, and their kind. Food gleaned from wet rocks or underwater. Can walk through water, walk under it, or swim through it. Faster water flows, better this bird seems to like it.

Could not conflict with man's interests if it wanted to do so. Never abundant and too small to capture large fish even if its diet were essentially fish. Its marvelous song and unusual ways hold interest of any fair-minded person who enjoys atmosphere of a gorge and a dashing stream, of which dipper is part.

House Wren

Order Passeriformes./Family Troglodytidae [Wrens]

House Wren, 721
Troglodytes aedon
Length to 5¼ in., wingspread to 7 in., tail to 2¹/₁₂ in. Weight ½ oz. Tends to cock tail over back; grayer brown than associated wrens, without conspicuous face stripings. Sexes colored alike. Winter wren has dark-barred belly and light line over eye. Carolina wren larger and redder, with white eye stripe.

Several subspecies include Eastern and Western house wrens. Eastern breeds from Michigan to New Brunswick and south to Kentucky and South Carolina. Winters from Texas and Gulf States into Mexico. Western house wren breeds from British Columbia to Wisconsin and south to Lower California and Kentucky. Winters south into Mexico.

Often polygamous, one male maintaining more than one household. Nests in old woodpecker hole or under eaves, filling all possible sites in neighborhood. Nest of coarse sticks, with finer lining. Eggs 5 to 12, pinkish-white with red-brown spots, ⅔ by ½ in. Incubation about 15 days, by both sexes. One or two annual broods.

Winter Wren

Food largely insects, animal matter equaling 98 percent of total. Unfortunately, may drive other birds away and even break up other nests. Some pairs may mate for life. Birds are amusingly noisy and active, and by using birdhouses make many a youngster happy and satisfied for his efforts.

Probably essentially useful though some bird students think house wren is one of worst of villains because it does not live harmoniously with other species of birds. State bird of Ohio, and related Carolina wren is state bird of South Carolina. Song, 2,050 to 7,125 cps. Temperature, 98.6 to 107.5°F.

Winter Wren, 722
Troglodytes troglodytes
Length to 4¼ in., including 1¼-in. tail and ⅓-in. bill. Conspicuously smaller and darker than common house wren, with belly cinnamon-brown, with conspicuous black bars on flanks and beneath, but those on back indistinct. Bill and tail much shorter than in house wren; outer tail feathers, ¼ in. shorter than *T. bewickii*, Bewick's wren.

(Continued)

Eastern winter wren breeds from southern Alberta to Newfoundland and south to central Minnesota and northern Georgia. Winters south of breeding limit to central Florida and Texas. Nine recognized subspecies include Eastern, Aleutian, Kiska, Tanaga, Unalaska, Semedi, Kodiak, and Western and extend range north and southwest.

Breeds in damp evergreen woodlands, near water, in brush or a hole, up to 10 ft above ground. Nest of twigs with moss lining. Eggs 4 to 10, $\frac{2}{3}$ by $\frac{9}{16}$ in.; white with dots of purple and brown, frequently crowded toward larger end. Incubation mostly or wholly by female. Usually two broods a year.

Food, ants, beetles, moths, dragonflies, caterpillars, plant lice, bugs, sawflies, lady beetles, leaf hoppers, and other insects and mites, snails, and other invertebrates. Song a long vigorous warbling, an explosion of musical sounds somewhat like that of ruby-crowned kinglet. Call note much like that of song sparrow.

Undoubtedly useful bird and not conceivably injurious. Popular with ornithologists because of its song and in part because it is relatively uncommon as compared with house wren in much of its range. Such a volume of song from such a small bird reminds one of wren tits of West.

Bewick's Wren

Bewick's Wren, 719
Thryomanes bewickii
Length to 5½ in., including 2⅖-in tail and ½-in. bill. Wingspread to 7 in. Female smaller than male. Relatively small wren, with abnormally long white-edged tail, longer than wings, with rather conspicuous brownish-white line over eye. House wren lacks the white eye stripe. Sexes colored alike.

Breeds from southern Nebraska, through Illinois, Michigan, and Pennsylvania, south to central Arkansas, Mississippi, Alabama, Georgia, and South Carolina. Winters from northern part of range to Gulf Coast and central Florida. Occasionally found north into New York and New Jersey. 14 subspecies recognized including Texas, Seattle, Santa Cruz, San Joaquin, and Guadalupe.

Nests in cavities in a great variety of places, being particularly inclined to nest near where human beings live. Nest of sticks, bark, grass, fur, and feathers. Eggs four to six, $\frac{2}{3}$ by ½ in.; white or pinkish with reddish-brown or lilac spots, most abundant at larger end. Incubation to 15 days, with two or three annual broods the rule.

Food chiefly insects. Known to include cotton boll weevil in its diet. Song has been described as like that of a bold house wren, more vigorous and clean-cut in every way. Larger than long-billed marsh wren and found, in winter, frequently in normal habitat of that species.

A useful destroyer of injurious insects and an interesting songster within its range. More hardy than some commoner wrens.

Carolina Wren

Carolina Wren, 718
Thryothorus ludovicianus
Length to 6 in., including 2-in. tail and $\frac{2}{3}$-in. beak. Wings to 2⅓ in. long. About size of smaller sparrows. Conspicuously reddish for a wren, with buff underparts and without streaks or bars other than a conspicuous long white or pale brown stripe over each eye. Dingier before molting period.

Breeds from southeastern Nebraska to lower Hudson Valley and Connecticut, south to Texas and northern Florida but sometimes found north to Wisconsin, Ontario, and Maine. Florida wren is a subspecies from southern Florida, and Lomita wren a subspecies of lower Rio Grande Valley and vicinity. At home in brushlands and forests. Resident.

Nests are bulky masses of plant materials with feathers and finer grass lining, in holes in stumps, brush piles, fallen treetops, or about buildings. Eggs four to six, ¾ by ⅗ in.; white or cream, with cinnamon, red-brown, and lavender markings. Incubation 12 days, chiefly by female. Two or three annual broods. Female commonly smaller than male.

Food: in 291 stomachs nearly 95 percent of food was insects, remainder being chiefly seeds. Food includes chinch bugs, cockroaches, cotton boll weevils, cucumber beetles, grasshoppers, crickets, moths, and other insects, in particular flies. Takes millipedes also. Song an ecstatic burst commonly described as resembling *tea kettle* repeated or a repeated *wheeudel.*

Justly popular because of its song and food habits and because it continues as resident in territory abandoned in winter by some of commoner wrens. Its size and color also make it distinctive. Tail flips over back in typical wren fashion. Can be enticed to banding station by hamburger. No injurious habits. State bird of South Carolina.

Cactus Wren, 713
Campylorhynchus brunneicapillus
Length to 8 in., wing to 3⅓ in., tail 3 in. Female slightly smaller than male. Tail rounded. Feathers of back brown with white middle streaks. Tail feathers brown and black. Underparts white. Young with back spotted instead of streaked, with white and black spots on chest, smaller and duller.

Three subspecies include Northern, Bryant's, and San Lucas cactus wrens. Northern cactus wren ranges from southern California to Nevada, Utah, New Mexico, and central Texas south to northern Lower California and northern states of Mexico. Other subspecies in Lower California.

Builds many dummy nests, in which they may roost as well as rear young. Nest a 6-in. globe of dried grasses, with side entrance; may be used and remodeled in different years; sometimes lined with feathers; also may be used by mice. Eggs three to seven, white or buff, with many red-brown spots. Generally nests placed in cactus thicket, yucca, or other thorny bushes. Entrance of nest leads into a large flask-shaped structure.

Food essentially insects. Sings through year but more commonly in summer. Song described as like grinding two small millstones together. Nests placed in cactuses are perfectly protected from most enemies.

Essentially a useful species and one of birds which give life and voice to desert at times of the day when it might otherwise be wholly quiet and empty. State bird of Arizona.

Cactus Wren

Long-billed Marsh Wren, 725
Telmatodytes palustris
Length to 5⅕ in., wingspread to 7 in., tail to 1⅗ in., bill ½ in. Brown above with a black crown, unstreaked, but with black back with white stripes. White stripe over eye. Underparts white, flanks cinnamon brown. Grayer as feathers become worn. Sexes colored alike. Young like adults. Known by its long slender bill.

Nine subspecies include Long-billed, Worthington's, Marian's, Louisiana, Alberta, Prairie, Western, Tule, and Suisun marsh wrens. Long-billed breeds from Quebec and Ontario to Virginia. Winters from New Jersey to Florida. Other subspecies cover United States and adjacent territory.

Male stakes out territory and erects many dummy nests. Female accepts male's territory; makes own nest with inside doorstep. Nests of woven plants, about size of present-day softball, with side entrance. Eggs five to nine, chocolate or speckled, ⅗ by ½ in. Incubation 13 days. Young at first helpless. Annual broods, one to three.

Food essentially insects gleaned from plants. Particularly active tail. Song, something like a musical sewing machine, may be heard night or day in marshes. Scolds vigorously at human or other intruders. Song pitch, 1,925 to 5,475 cps; median, 4,400 cps. Defended home territory with one female, 13,000 to 15,000 sq ft.

Certainly does no harm to man's interests and helps somewhat in controlling insects in a region where they are normally overabundant. Preyed on by many animals, and nests used by bees, tree frogs, and other animals.

Long-billed Marsh Wren

Canyon Wren, 717a
Catherpes mexicanus
Length to 5½ in. Wing to nearly 2½ in. and tail of similar length. Male and female colored alike, brown but for white on throat and breast, with upper parts lighter brown, head grayish, and tail rusty brown with narrow black crossbars. Pure white on underparts to the fore, becoming rusty to rear.

From southern British Columbia to Idaho and northern Colorado, south through mountainous areas to Texas, Sonora, Chihuahua, and Lower California. Two closely related subspecies, white-throated and dotted wrens, have more restricted ranges for most part but are in general in part of same area.

Nest is built in a crevice in rocks, in a tunnel or cave, or on a ledge, composed of twigs, moss, and grass often with wool or feather lining, bulky, made by female. Eggs three to five, white, with reddish-brown or light blue-gray spots around larger end.

Known best for its ringing song, which may be heard almost from snow to snow and is characterized by descending the chromatic scale, a feature that has given the bird common name of scale bird. Singing of scale is succeeded and preceded by less conspicuous notes.

Of no great economic importance because it does not live in an agricultural area but is of great interest as representative of a wild environment. It takes over to some extent where the equally vociferous water ouzel leaves off.

Canyon Wren

Mockingbird

Mockingbird, 703
Mimus polyglottos
Length to 11 in. wingspread to 15 in., tail to 5¾ in., bill to ⅘ in. Female smaller than male. Like a large slender long-tailed gray and white robin, with large white patches on wings and tail, which show conspicuously in flight. Lacks black face mask of somewhat similar shrikes.

Two subspecies include the Eastern and the Western. Eastern ranges from eastern Nebraska to Ohio and Maryland and south to Texas and southern Florida. Occasionally in New York and New England and up to Nova Scotia. Western extends range through central California and south to Vera Cruz.

Nests usually about buildings, in trees or shrubs or at edges of woods. Nest 3 to 10 ft above ground, of twigs with a lining of finer materials such as roots or grass. Eggs three to six, greenish to bluish to gray, with small red- or yellow-brown spots, 1⅙ by ⅘ in. Incubation 12 days, by female. Two or three annual broods.

Food in spring and summer largely insects including ants, flies, wasps, cotton boll weevils, grasshoppers, bugs, caterpillars, and also spiders. In summer and autumn it changes to a diet of fruits of trees and shrubs such as sumac, mountain ash, wild grape, Virginia creeper, barberry, honeysuckle, bittersweet, and Juneberry.

Ranks high as a beneficial bird and certainly is popular in song and story. Official state bird of Arkansas, Florida, Texas, Tennessee, and Mississippi, which gives some evidence of where it is most abundant and most appreciated. It may damage fruit in some parts of South if better food is not available.

Catbird

Catbird, 704
Dumetella carolinensis
Length to 9⅓ in., wingspread to 12 in., tail to 4¼ in. Weight 1⅖ oz. Temperature, 106° F. Slate-gray, with a black cap and rusty chestnut under tail. Tail seems to be long and loosely hung, yet large and broad when bird is in flight. Sexes and young colored essentially alike.

Breeds from British Columbia and western Washington to Nova Scotia and south to northern Utah and New Mexico and northern Florida. Winters in Southern States and on to Cuba and Bermuda through Mexico to Panama. Accidental in Europe.

Nest deeply hollowed, made of twigs, bark, and leaves lined with rootlets, in dense shrubbery 3 to 10 ft above ground, appears bulky. Eggs 4 to 6, glossy, greenish-blue without markings 1 by ⅔ in.; weight, ½ oz. Incubation 12 to 13 days, by both sexes. One to three annual broods. Speed, 16 mph.

Food in spring, almost solely insects; in summer and fall, more largely fruits and vegetable matter. In 645 stomachs examined, 44 percent of food was animal, remainder vegetal. 75 percent of animal food was ants, beetles, and caterpillars. Mewing, catlike call (768 cps.) well-known. Song pitch 1,100 to 4,375 cps.

Where an abundance of wild fruit available, does little damage and surely not enough to offset good done in destroying harmful insects. During time young birds are being reared, 96 percent of food given them is insects. Important enemy of destructive gypsy moth where it is a serious pest.

Brown Thrasher

Brown Thrasher, 705
Toxostoma rufum
Lenth to 12 in., wingspread to 14⅗ in., tail to 5¾ in., bill to 1 1/10 in. Female smaller than male. Like a larger, slimmer robin, but redder above and striped below, with a longer, more loosely hung tail, which is rounded at tip. Sexes colored alike and young like adults. Temperature, 108.3 to 110.8°F.

Breeds from southern Alberta to western Quebec and Maine, south to Gulf Coast, Louisiana, and central Florida. Winters from southeastern Missouri and North Carolina to Texas and central Florida. Six related species including 13 subspecies extend range of thrashers throughout warmer United States and northern Mexico.

Nest usually in brushland, on ground, in brush piles or low shrubbery. Nest bulky, of twigs, sticks, and leaves, with finer lining. Eggs three to six, white or greenish, well-dotted with reddish-brown, 1⅛ by ⅞ in. Incubation by both sexes, 13 to 14 days. Annual broods, one to two or, in South, sometimes three.

Food in spring, almost entirely insects, spiders, and worms. In 266 stomachs, 63 percent of food was animal matter, almost all being insects. Insects include May beetles, white grubs, cucumber beetles, cotton boll weevils, snap beetles, wireworms, army worms, tent caterpillars, cutworms, cankerworms, grasshoppers, leaf hoppers, and wasps. Speed, 22 mph.

Brown thrasher takes less fruit than catbird and so is more valuable from farmer's viewpoint. All brown thrashers should be considered as valuable birds worthy of protection the law gives them. Their mockingbird-like or catbirdlike calls are usually given in 2s.

Order Passeriformes./Family Turdidae [Thrushes, Bluebirds]

Robin, 761
Turdus migratorius

Length to 10¾ in., wingspread to 16½ in., tail to 4¾ in. Female sometimes but not always smaller than male and generally with less brilliantly colored breast. Back gray, head black, breast chestnut to brown (black-spotted in young), belly light gray to white, tail black. Temperature, 104.6 to 111.2°F.

Four subspecies include the Eastern, the Southern, the Northwestern, and the Western robins. Eastern breeds from tree limit of northwest Alaska to Newfoundland, south to Kansas and New Jersey. Winters from Kansas to Massachusetts, south to Gulf Coast and southern Florida. Other subspecies extend range through United States and Mexico.

Male comes north first, fights for territory, and mates. Nest in tree, under roof in abandoned building, or under bridge; made of mud and grasses. Eggs 3 to 5, blue, 1¼ by ⅝ in.; weight 1/16 oz. Incubation sometimes by both sexes but more often by female alone, for 12 to 13 days. Two or three annual broods.

Food largely earthworms when they are available; about 57 percent vegetal and 43 percent animal matter. Insects eaten include grasshoppers, locusts, crickets, wireworms, leaf beetles, tent caterpillars, cutworms, cankerworms, army worms, ants, cicadas, and others; also, small snakes, sometimes to 1 ft long, are eaten. Song pitch 2,200 to 3,300 cps. Speed, 30 mph.

Robins unfortunately eat fruits such as cherries, as well as insects which might also destroy the cherries. May flock into mulberry plantations and practically ruin crop, also destroy strawberries. With all their faults we still think highly of them. State bird of Michigan, Virginia, Connecticut, and Wisconsin.

Robin

Wood Thrush, 755
Hylocichla mustelina

Length 8⅛ in., wingspread to 14 in., tail to 3⅓ in. Bright brown above, breast and sides heavily spotted with round black spots. Redder about head than most associated thrushes. Sexes colored alike. Young much like adults. Smaller than brown thrasher and less slender in appearance.

Breeds from South Dakota through southeastern Ontario to southern Maine and south to eastern Texas and northern Florida. Winters from southern Mexico to western Panama and occasionally in Florida. Sometimes found in Bahamas, Cuba, Jamaica, Colorado, and Bermuda.

Nests in wooded land. Nest of leaves, grass, paper, or cloth, plastered with mud like robin's but with fine rootlets for a lining, either in a shrub or on a horizontal limb, to 12 ft above ground. Eggs three to five, greenish-blue and unmarked, 1⅛ by ¾ in. Incubation for 12 to 14 days by both sexes. One or two annual broods.

Food known to include few useful insects or cultivated fruits, but many May beetles, weevils, Colorado potato beetles, cicadas, grasshoppers, tree hoppers, sawfly larvae, tent caterpillars, forest tent caterpillars, gypsy moths, and brown-tail moths. Song is one of finer of bird harmonies, 1,825 to 4,025 cps., median 2,750 cps.

The bird is obviously essentially useful and worthwhile. It does no harm, provides happiness to those who enjoy the relative quiet of woodlands, and is a thing of beauty at all times. State bird of District of Columbia.

Wood Thrush

Hermit Thrush, 759b
Hylocichla guttata

Length 7⅗ in., wingspread to 12 in., tail to 3 in. Conspicuous because of reddish tail, which contrasts with brown back. Wood thrush is red about head; hermit, about tail; and veery is uniformly colored on back. Throat and breast not so spotted as in wood thrush. Sexes colored alike.

Seven subspecies include Alaska, Dwarf, Monterey, Sierra, Mono, Audubon's, and Eastern. Eastern hermit thrush breeds from Yukon to southern Quebec, south to Minnestoa, and in mountains of Maryland and Virginia. Winters from Ohio and Massachusetts, to Texas and Florida. Other subspecies to west.

Nests in a cool, wooded area or in pastures near woods but favors coniferous forests. Nest on or near ground, of leaves, moss, and bark compacted, with a finer lining. Eggs three to five, plain greenish-blue, 9/10 by 7/10 in. Incubation 12 to 13 days. Two to three annual broods. Nest may be deserted if only slightly disturbed.

Food principally insects such as ants, caterpillars, and beetles. In 68 stomachs, animal matter constituted 56 percent of food. Plant foods included berries of poison ivy, dogwood, privet, sumac, and wild grapes. Bird has habit of raising tail slowly at more or less definite intervals.

Popular with all bird students, particularly because of its remarkable evening song which expresses the direct opposite of the hurry and bustle of modern so-called civilized existence. If any bird can make a worried person forget his cares, it is the hermit thrush and his song even though it is often accompanied by the hum of mosquitoes. State bird of Vermont.

Hermit Thrush

1. Swainson's Thrush

2. Gray-cheeked Thrush

Veery

Bluebird

1. Swainson's Thrush, 758a
Hylocichla ustulata

2. Gray-cheeked Thrush, 757
H. minima

Length: (1) to 7¾ in.; (2) to 8 in. Wingspread: (1) to 13 in.; (2) to 13½ in. Both have uniformly olive or gray-brown appearance above. (1) has conspicuous brown eye ring and conspicuous buffy cheeks. (2) lacks a conspicuous eye ring, and has gray cheeks. Sexes of each similar.

Breeds: (1) from northwest Alaska to Newfoundland and south to northern California and mountainous Pennsylvania; (2) from northeast Siberia through Alaska to Newfoundland and south to Mackenzie and central Quebec. Winters: (1) in Mexico, Bolivia, Brazil, Argentina, and Cuba; (2) migrates through eastern North America to Colombia, Ecuador, Peru, Venezuela, Texas, and Kansas.

Nests: (1) in deep woods; (2) in stream thickets; (1) in bush or tree 4 to 25 ft up; (2) on ground to 12 ft up. Both of leaves, grass, and rootlets. Eggs three to five; ¹⁵⁄₁₆ by ¾ in., light green-blue spotted with dark brown or lilac. Incubation of (1) 11 to 13 days, mostly by female. One annual brood. Incubation of (2) 13 to 14 days.

Quiet peaceful birds of hilly woodlands, shy and retiring. Song of (1) flutelike, with each phrase rising at end; of (2), like that of a veery, gradually dropping but rising quickly at end. Food roughly twice as much animal as vegetable matter, gathered largely from trees and shrubs.

Both are useful species favored by bird lovers, in part because of their song but not matching hermit or wood thrush in this respect. They live away from agricultural plants, and so have little bearing on their prosperity although they probably do affect somewhat the economy of forests.

Veery, Wilson's Thrush, 756
Hylocichla fuscescens

Length to 7¾ in., wingspread to 10 in., tail to 3⅓ in., bill to ⁷⁄₁₀ in. Uniformly cinnamon brown down back, head, and tail. In field, only a few spots show on throat and breast region. Key clues are uniform upper parts and unspotted under parts. Sexes colored alike. Young like adults, at least superficially.

Breeds from Michigan to southern Quebec, south to northern Indiana, northern Georgia, and North Carolina. Winters in Colombia, British Guiana, and Brazil after migrating through Yucatan and Central America. Willow thrush is subspecies ranging west to southern British Columbia and south to New Mexico, wintering south in South America to Brazil.

Nests in woodland usually near water. Nest usually on ground or, rarely, to 10 ft up, of leaves, twigs, grasses, and bark with lining of fine roots, hair, needles, or grasses. Eggs three to five, greenish-blue, ¹⁵⁄₁₆ by ⁷⁄₁₀ in. Weight, ¹⁄₁₀ oz. Incubation 10 to 12 days. One to two annual broods.

Food 57 percent animal matter, mostly insects and mostly those of woodlands including borers, caterpillars, bark beetles, ground beetles, curculios, and weevils. Few seeds of weeds taken, plant material being largely wild fruits. Oft-repeated descending song is worth hearing. Male defends nesting site more vigorously than female.

Undoubtedly useful in general economics of woodlands, in keeping insect pests in check. Popular with those who live where it can be heard and seen, and is missed when such an association is destroyed. Protected by law, and its enemies such as cats should be kept in control if it is to survive. Song, 2,375 to 4,025 cps.

Bluebird, 766
Sialia sialis

Length to 7⅔ in., wingspread to 13¼ in., tail to 3⅙ in. Female smaller than male. Male little larger than English sparrow, with blue back and reddish breast. Young bluebirds like young robins, specklebreasted, lack red of breast, and are hardly so blue as either adult; more brilliantly colored than female.

Two subspecies include Eastern and Azure bluebirds. Two related species, one of which is Eastern bluebird, breed from southern Manitoba to Newfoundland and south to central and eastern Texas and southern Florida. Winters from just north of Mason-Dixon line south to Gulf.

Nests in orchards and at edges of woods. Nest in a hollow such as a birdhouse or abandoned woodpecker hole. Usually from 3 to 30 ft above ground. Nest lined with fine plant materials. Eggs three to seven, light blue or rarely white; ⅞ by ⅔ in. Incubation for 12 days, by both sexes. Two to three annual broods.

Food about 70 percent animal matter, including among insects grasshoppers, moths, crickets, beetles, bugs, ants, caterpillars of many kinds, and earthworms. Plant foods include wild fruits such as those of wild grape, chokecherry, Virginia creeper, juniper, alder, bayberry, blackberry, sumac, and mountain ash. Speed, 17 mph.

Mountain Bluebird

More valuable than associated robin because it rarely eats fruits man wishes to raise for himself. Bluebirds have decreased in numbers with increase in starlings, which drive bluebirds from their nesting sites. Song pitch, 2,200 to 3,100 cps. Bluebird is state bird of New York and Missouri.

Mountain Bluebird, 768
Sialia currucoides
Length to 7¹⁄₁₀ in., tail to 3 in. Female slightly smaller than male. Upper parts plain clear blue, wings more violet, under parts paler, belly white. Female: head and back gray, often tinged with greenish-blue, rump and tail blue. Young brownish or grayish, somewhat white-streaked.

Breeds from southern Yukon to southwestern Manitoba and, in mountains, south to southern California, Arizona, New Mexico, Chihuahua, and western Nebraska. Winters from California and Colorado south to Guadalupe Island, Lower California, Sonora, and other parts of northern Mexico, and to Oklahoma and Texas.

Probably most beautiful of all bluebirds. It, like better-known species, will nest in birdhouses. It also nests in old abandoned woodpecker holes, commonly in aspen groves, or more commonly about man's buildings. Eggs five to seven, pale greenish-blue.

Food over 90 percent insects and remaining part largely of waste and wild fruits. Insects include cicadas, grasshoppers, cutworms, locusts, crickets, ants, bees, caterpillars, weevils, and other insects, some of which are caught on wing, birds acting as flycatchers.

This beautiful bird does so much good and seems to favor living near man so much that it should be encouraged to prosper wherever possible. Sufficiently well known and popular to have been selected as state bird for Idaho and Nevada.

Varied Thrush, 763
Ixoreus naevius
Also known as Oregon robin, male similar to robin. Length to 10 in.; wings about 5 in.; tail 3⁵⁄₁₀ in.; bill ⅘ in. Adult male, slate colored above, shows stripe of orange behind eyes; two orange bars cross the wings, brown-orange underparts which fade to a white lower belly. Field characteristic; a black collar across a rusty orange breast; black bill and dull yellow feet. Female, smaller and paler in all respects. Young birds like adult female.

Varied Thrush

Only two subspecies. *I. n. naevius* breeds from southeastern Alaska south on the western slope of the Coast Range and the Cascade Range in British Columbia, Washington, and Oregon, to northwestern California. Winters south of its breeding range. *I. n. meruloides* breeds from northern Alaska, northern Yukon, northwestern Mackenzie south to the base of the Alaska Peninsula and adjacent islands to British Columbia south to eastern Washington, northern Oregon, Idaho, and Montana. Winters to south of breeding range.

Nest made of sticks, twigs, grasses, rotten wood, covered with moss. Usually placed in a sapling at moderate heights. Eggs, usually three, pale to light niagara green, with some speckles or spots; sometimes blotched with shades of chocolate. Size 1⅕ by ⅘ in. Usually produces two broods. Female incubates. A bird of deep forests of fir in western Washington and Oregon. Produces a long-drawn quavering note; note is repeated at a higher pitch, "-ee-ee-ee-ee," song musical and melancholy.

A ground feeder; food in part invertebrates including sow bugs, snails, angleworms, and insects which are varied and include bugs, beetles, caterpillars, grasshoppers, and crickets. More than 60 percent of food taken is plant material including fruits, weed seeds, and frequently acorns.

This stoutly built, robin-sized, orange-breasted bird with a black stripe across its chest cannot be easily mistaken for any other species. Defends its territory vigorously against intruders. The robin of the deep fir forests.

Townsend Solitaire, 754
Myadestes townsendi
Length to 8½ in.; wing length to 4⅗ in.; tail, 4 in.; bill about ½ in. Towhee-size bird, of brownish-gray coloration with spots of white on both tail and wings. No black as is the case with the shrike. In flight shows white outer tail feathers. Generally smoky gray, lighter below, with a prominent white eye ring; wings and tail dusky. Female smaller.

Townsend Solitaire

Range, western North America from far north to Mexico; common resident of high Transition Zones of Sierra Nevada of California. One subspecies occurs from Alaska, southern Yukon, southern Mackenzie, Alberta, and southwestern South Dakota in the mountains to southern Cali-
(Continued)

Golden-crowned
Kinglet

fornia and Durango. Migration largely altitudinal. During winter months, appears at lower altitudes from southern British Columbia, western Nebraska, south to Baja California, to northern mainland of Mexico.

Nest made of coarse twigs and moss, lined with fine grasses and sometimes pine needles; on the ground, frequently placed on a hillside at the base of a tree or under a boulder; sometimes in the notch of a tree trunk. Eggs three to four, rarely five; $\frac{9}{10}$ in. by $\frac{7}{10}$ in., white, pinkish, grayish, or greenish-white, spotted with green-brown. Eggs laid from May through July according to the altitude. One brood produced each year.

Behavior like flycatcher's. Flight and coloration suggest a mockingbird; feeding habits like a bluebird's. Song thrushlike, like that of a black-headed grosbeak's but more rapid. Food: animal diet consists chiefly of lepidopterans, beetles, ants, bees, wasps, flies, bugs, and spiders. Plant food, the fruits and seeds of cedar, hawthorn, madrona, pine, gooseberry, hackberry, sumac, and serviceberry. One of favorite plant foods is mistletoe berries.

Order Passeriformes./Family Sylviidae [Old World Warblers]

Golden-crowned Kinglet, 748
Regulus satrapa
Length to $4\frac{1}{5}$ in. Wingspread to 7 in. General color olive-gray with a white stripe over eye. Male with conspicuous orange crown bordered by black; female with conspicuous yellow crown. Young like adults first winter, but in juvenile plumage lacks colored crowns and plumage appears loose and fluffy. Other than hummingbirds, probably smallest of our U.S. species.

Breeds from central Alberta to southern Quebec and Cape Breton Island, south to Minnesota and New York and in mountains to North Carolina, with western subspecies ranging from Kenai Peninsula in Alaska to California. Winters from Iowa to New Brunswick and south to Florida and Mexico. Western subspecies from British Columbia to Guatemala.

Nests in evergreen forest. Nest 4 to 50 ft up in evergreen, globular with small opening at top, of leaves, moss, lichens, and bark, feather-lined. Eggs 5 to 9, $\frac{9}{16}$ by $\frac{7}{16}$ in.; creamy white with brown and lavender spots. Incubation not known, but probably 12 to 13 days, probably by female. One annual brood.

Most active little birds, with surprising ability to withstand severe winter weather. Extremely shy about nesting. Will leave a partly built nest if it seems to have been discovered. Food in summer, caught on wing, but in winter gleaned from twigs and trunks, apparently consisting most entirely of insects. About $\frac{1}{10}$ of winter food is plant material. Insects eaten primary item in diet.

Economic importance great, since they destroy large numbers of insects as adults, pupae, or eggs and glean these from delicate parts of trees that could not be visited successfully by larger, heavier birds. Intensely interesting to bird lovers because of their activity. Call a high repeated *see*.

Ruby-crowned
Kinglet

Ruby-crowned Kinglet, 749
Regulus calendula
Length to $4\frac{3}{5}$ in., wingspread to $7\frac{1}{2}$ in., tail to $1\frac{9}{10}$ in. Olive-gray above with two pale wing bars. White ring around eyes. Male has scarlet patch on top of head which, except in courtship display, does not usually show. Golden-crowned kinglet has white eye stripe and yellow (female) or orange (male) crown.

Four subspecies: Eastern, Western, Sitka, and Dusky. Eastern breeds from northwest Alaska to west central Quebec and south to southern Arizona. Winters through United States, Mexico, Lower California, and Guatemala.

Nests among cone-bearing trees, usually 2 to 50 ft up. Nest compact, either across or partly hung from a limb and hidden by foliage, composed of mosses and similar plants, hair-lined. Eggs 5 to 10, white with fine reddish dots, $\frac{9}{16}$ by $\frac{7}{16}$ in. One to two annual broods, but details about incubation and rearing of young not available.

Food: in 294 stomachs, animal food constituted 94 percent, chiefly leafhoppers, plant lice, mealy bugs, scale insects, weevils, caterpillars, flies, and grasshoppers. Vegetable material consisted of wild fruits, including those of poison oak. Song a remarkable warble.

Nothing in bird chorus compares in surprising volume and tone quality with song of this diminutive bird. A male warbling his love song, erecting his brilliant red crown as he scurries through finer parts of some woody plant is worth seeing. Entirely useful, judging from its diet.

Order Passeriformes./Family Bombycillidae [Waxwings]

Bohemian Waxwing, 618
Bombycilla garrulus

Length to 8¾ in. Wingspread to 14¼ in. Female smaller than male. Larger than cedar waxwing but is grayer on back and under parts, has some white and yellow on wings, and chestnut-red under tail rather than white. Sexes resemble each other but some females may be duller, with less yellow on wings.

Breeds from western Alaska to northeastern Manitoba and northern Mackenzie, south to British Columbia and southern Alberta. Winters east to Nova Scotia and sometimes on into Pennsylvania, Ohio, Michigan, Indiana, Illinois, Kansas, Colorado, California, and Arizona. Not a common bird in most of United States.

Nests in northern forests. Nest on branch, often high in a tree, built of roots, grass, and leaves with lining of hair, down, and feathers. Eggs three to six, 1 by ⁷/₁₀ in.; blue-gray or slate spotted with dark brown and black. Larger than those of cedar waxwing, but smaller than those of European race of Bohemian waxwing. No data on incubation.

Food probably largely wild fruits, particularly in months it spends in United States. In breeding area in western Canada it unquestionably destroys large numbers of insects. Birds remain together in flocks like cedar waxwing. Call a low, rougher hissed *zree* than similar call of cedar waxwing.

In Europe, species is often found in great flocks the like of which have been recorded only once in America, in Nebraska when "every tree for miles was filled with them." These great flocks occur fairly regularly in Canada. Probably a useful species and is protected by law.

Bohemian Waxwing

Cedar Waxwing, 619
Bombycilla cedrorum

Length to 8 in., wingspread to 12¼ in.; tail to 2¾ in. Female smaller than male. Sleek brown bird, with conspicuous crest and a broad yellow band at end of tail. Wings with small waxlike secondary tips. Related Bohemian waxwing, nearly 1 in. longer, chestnut-red rather than white under tail and have white in wing.

Breeds from central British Columbia to Cape Breton Island, south to northern California, New Mexico, northern Arkansas, northern Georgia, and North Carolina. Winters through most of United States, south through Cuba, Mexico, Lower California, and Panama. Occasional in British Isles, Jamaica, Bermuda, and Bahamas. From California Pleistocene. Only three species in family.

Nests in orchards or isolated trees. Nest 4 to 40 ft above ground, on horizontal limb, bulky, of coarse twigs, bark, and weed stems, with fine lining of rootlets, paper, hair, wool, or grasses. Eggs four to six, pale blue, green, or gray, with black or dark brown spots, ⁷/₈ by ⁷/₁₀ in. Incubation 12 to 16 days, by female only. One to two annual broods.

Food largely fruit and insects though most of fruit is wild. Estimated that a flock of 30 birds will destroy 90,000 cankerworms a month. Flock will settle on infested tree and clean it from top to bottom if undisturbed. Young are fed regurgitated food. Song pitch, 7,675 to 8,950 cps. Defended home territory several feet across.

Qualified biologists consider it useful species in spite of fact that, in individual cases, flocks may do much damage to small fruits such as cherries. Flocks in winter always intrigue students of nature. Tame young in summer may even perch on a trout rod while they are being fed by parents.

Cedar Waxwing

Order Passeriformes./Family Ptilognatidae [Silky Flycatchers]

Phainopepla, Silky Flycatcher, 620
Phainopepla nitens

Length to 7⅔ in. Adult male glossy greenish blue-black, except for white patch on inner webs of wings. Wings and tail less glossy than other parts of body. A high crest. Female plain olive gray, with longer feathers of crest black, and with under tail coverts white-edged. Young like female.

Breeds from central California to southern Utah and south through central Texas, Cape San Lucas, and northwestern Mexico. Winters from southern California south into Mexico. Occasionally in central Nevada and northern California.

Nests found in trees such as cottonwoods and mesquite, saddled on a branch, made of sticks, stems, and plant materials with a fine plant-down lining. Eggs two to three, grayish-green to whitish, thickly spotted with brown, black, or faint lilac. Male often assumes duty of building nest, incubating eggs, and rearing young.

Behaves much like flycatchers, catching insect food on wing, and may stay together in flocks of a dozen or more young and old birds, probably members of a few families. May at times eat only berries like mistletoe, juniper, and elder. Probably useful as insect check and always interesting to watch because of beauty. Call a coarse *ca-rack* or *ca-rac-ack*. Moves to lower levels after breeding season has passed.

Phainopepla

Order Passeriformes./Family Laniidae [Shrikes]

Northern Shrike, 621
Lanius excubitor
Length to 10¾ in. Robin-sized. Wingspread to 16⅗ in. Female smaller than male. Larger than loggerhead shrike, with a finely barred breast, a flesh-colored rather than black lower bill, and without loggerhead's narrow black stripe above bill. Young browner than adults but with finely barred breast. Has many common names: butcher bird, winter shrike, and great northern shrike probably most commonly used.

Breeds from northern Ungava to southern Ontario and Quebec, west to Hudson Bay. Winters south to Kentucky, Virginia, and North Carolina, with northwestern going on to northern California, New Mexico, and Texas. The winter shrike of the East. Related northwestern shrike extends range northwest to Alaska and south to northern British Columbia, Alberta, and Saskatchewan.

Nests in northern forests; on a limb or fork of a spruce, 5 to 20 ft up; large and compact, of twigs and leaves with fur, feather, and lichen lining. Eggs four to seven, 1¹⁄₁₀ by ⅘ in.; white, bluish, or grayish-green with green, brown, purple, or gray markings. Incubation probably 15 days, with one or possibly two annual broods and nesting period centering in May.

Perches high, alone on a treetop from which it drops low, then flies directly and rises abruptly at end to new perch. Flight is easy. Food, large insects and small birds cornered in thickets by darting with incredible skill through brush tangles. Has been known to try to capture caged canaries behind windows. Song somewhat like a robin, mockingbird, or thrasher.

Considered useful as mouse destroyer and in West is effective in destroying gophers, which it kills with a sudden blow on back of head, according to John Muir. Protected by law.

Northern Shrike

Migrant or Loggerhead Shrike, 622e
Lanius ludovicianus
Length 9⅕ in., wingspread 13 in., tail to 4 in., bill to ⁹⁄₁₀ in. Female smaller than male. Related northern shrike has a slightly barred breast instead of the clear gray of migrant, and is larger. Both have a black mask through eyes, unusually long tails, and appear white and gray in flight.

Nine subspecies including, among others, California, Whitelumped, Nelson's, Island, and Loggerhead. Migrant breeds from southeastern Manitoba to New Brunswick, south to northeastern Texas and interior Virginia. Winters in Mississippi Valley, Texas, and southern New England. From California Pleistocene.

Nest a bulky affair of weeds and twigs with soft lining of feathers, grasses, and wool, 5 to 20 ft above ground, in an orchard tree or thicket. Eggs four to eight, white or grayish-green, with purple and olive markings more obvious at larger end, 1 by ¾ in. Incubation 13 to 16 days, by female alone. One to two annual broods.

Food, insects, mice, birds, frogs, snakes, and shrews, often left hanging on thorns of trees. Perches high and alone and takes to wing by dropping low and taking a beeline with a steady wing motion, only to rise abruptly to its new perch.

All serious studies indicate that while shrike does kill some other birds, its preying upon mice and large insects more than makes up for damage it does to useful species. It can cause terror in a flock of sparrows or other small birds. Protected by law.

Migrant or Loggerhead Shrike

Order Passeriformes./Family Sturnidae [Starlings]

Starling, 493
Sturnus vulgaris
Length to 8½ in., wingspread to 15½ in., tail to 2⁹⁄₁₀ in., bill to 1 in. Weight to 3½ oz. Temperature, variable around 109°F. Female slightly smaller than male. Like a short-tailed blackbird, which flies in a varying shaped flock. Bill dark in fall, turning to yellow in spring. Blackish, with a gloss, flecked with buff and white.

Ranges from New Brunswick to western Ontario, south through Florida and Texas and spreading westward to California and northward to Hudson Bay. Same species in Europe and Asia. Fairly closely related to the mynah so common in Hawaii and in southern Asia. Introductions attempted from 1872 (Cincinnati) until finally successful in New York in 1890.

Nest a large bulky structure built of straw, grass, and twigs lined with feathers and other soft stuffs; placed in a birdhouse, cavity in a tree, or under eaves of a house. Eggs three to eight, pale blue or whitish, and glossy, 1⅕ by ⁹⁄₁₀ in. Incubation 11 to 14 days, by both sexes. One to three annual broods.

Starling

Food highly variable. Stomachs of 2,157 birds showed 57 percent by weight animal matter, including many injurious insects. Plants include grain, cherries, strawberries, and similar fruits. Claimed that less than 6 percent of total yearly diet is cultivated plant products. Drives out more useful bluebirds and their kind. Flight speed, 30 mph.

They are pests about public buildings in cities, in orchard regions, and in grainfields. Better if they had never been introduced into America since they have driven out so many native beloved species. Hearing range, 700 to 15,000 cps. Song pitch, 1,100 to 8,225 cps.

Order Passeriformes./Family Vireonidae [Vireos]

White-eyed Vireo

White-eyed Vireo, 631

Vireo griseus

Length to $5\frac{1}{2}$ in., including $2\frac{1}{2}$-in. tail and $\frac{1}{2}$-in. bill. Wingspread to $8\frac{2}{5}$ in. Female smaller than male. Yellowish-green, lighter beneath, with yellow sides and two conspicuous whitish wing bars. Iris of eye white. Bill black. Sexes colored alike. Young practically identical with adults. Known among vireos by its yellow spectacles and whitish throat.

Breeds from southeastern Nebraska through southern Wisconsin to Ohio and New York, south to Texas and southern Florida. Winters from Texas through Georgia, Alabama, and South Carolina to eastern Mexico, Yucatan, and Honduras. Four subspecies include White-eyed, Key West, Bermuda, and Rio Grande, last three being local for most part.

Nests in low thickets and brushlands. Nest 2 to 8 ft above ground, cup-shaped, suspended from a fork, made of mixed grasses, bark, moss, lichens, and cobwebs with a finer lining. Eggs four, $\frac{4}{5}$ by $\frac{3}{5}$ in.; white, sparsely marked at larger end with fine black, purple, or lilac dots. Incubation 16 days. Annual broods, one or in South two.

Food mostly insects such as caterpillars, ants, flies, and those that fly about thickets, usually caught in air; also includes some berries. Both parents care for young and are vigorous and self-sacrificing in defense. Cowbirds commonly lay their eggs in this vireo's nest. Song variable explosive, but has a sharp *chip* at either end.

Essentially a useful species because of its destruction of usually harmful insects. This species keeps itself hidden more than red-eyed does but if its nest is approached it becomes highly aggressive no matter who the interloper may be.

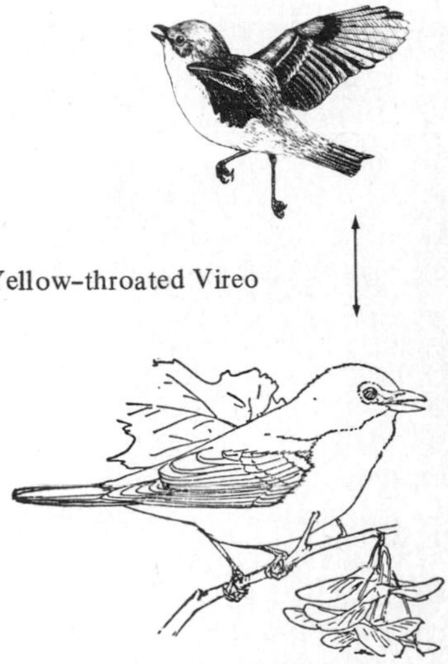

Yellow-throated Vireo

Yellow-throated Vireo, 628

Vireo flavifrons

Length to 6 in. Wingspread to 10 in. Female smaller than male. Sexes colored alike. Olive-green above with white wing bars. Breast and throat a bright yellow, unlike that of any other vireo but much like larger yellow-breasted chat, which lacks white wing bars. Young like adults by first winter plumage.

Breeds east central Saskatchewan to southwestern Quebec, south to Maine, Florida, Alabama, Louisiana, and Texas. Winters from Yucatan and southern Mexico to Colombia, Venezuela, Cuba, and Bahamas. Essentially a bird of treetops of forests and towns.

Nests in June. Nest in broad-leaved tree, 3 to 50 ft above ground, suspended between forks of a limb and cuplike, with spider silk used to fasten parts together. Eggs three to four, $1\frac{5}{16}$ by $\frac{2}{3}$ in.; white marked with purple-brown spots or rosy when newly laid. Incubation by both sexes, 12 to 14 days. One annual brood.

Food largely insects including caterpillars, houseflies, grasshoppers, cicadas, and beetles, with wild berries sometimes added when they are in their prime. Male sings on nest or near it, song often being confused with that of robin. Slower, more contralto, and more musical than that of red-eyed vireo.

Essentially a useful bird that is justly protected by law, damage to small berry fruits being more than compensated for by destruction of harmful insects. Furthermore, bird adds beauty of color and sound to summer landscape.

Blue-headed Vireo

Blue-headed Vireo, 629

Vireo solitarius

Length to 6 in., including $2\frac{2}{5}$-in. tail and $\frac{1}{2}$-in. bill. Wingspread to $9\frac{2}{3}$ in. Female smaller than male. General color greenish, but with two conspicuous white wing bars and a conspicuous light ring around eyes. Under parts almost white. Throat glistening white. Top of head and back of neck blue-gray. Sexes alike.

Breeds from southern Mackenzie through northern Ontario, southern Quebec to Cape Breton Island, south to central Alberta, northern North Dakota, Minnesota, Michigan, southern Pennsylvania, and Rhode Island. Winters in Gulf States from Texas to Florida and from northern Nicaragua to eastern Mexico. Five subspecies include Blueheaded, Mountain, Plumbeous, Cassin's, and San Lucas vireo.

(Continued)

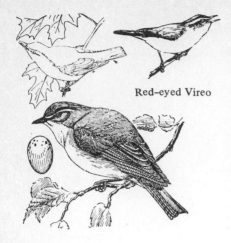

Red-eyed Vireo

Nests in or near a tree 5 to 12 ft above ground. Nest a hanging cup, of beautifully mixed vegetable matter fastened together with spider web; often lined with down or fine needles. Eggs three to four, 9/10 by 3/5 in.; white with small spots of brown, black, or reddish-brown. Nesting period June. Incubation by both parents. Usually one brood.

Food essentially insects including grasshoppers, bugs, beetles, caterpillars, sawflies, ants, dragonflies, crickets, and flies. Spiders are also eaten. Nesting is often close to that of more vigorous birds of prey.

A useful and interesting bird worthy of protection at all times. The cooperation of both parents in care of nest and young is always worth watching. Frequently does not show offense at those who would examine the nest and incubating parent.

Red-eyed Vireo, 624
Vireo olivaceus
Length to 6½ in., wingspread to 10¾ in., tail to 2⅓ in. Female smaller than male. Lacks white wing bars of yellow-throated, white-eyed, and blue-headed vireos. Has a gray cap and a black-bordered white stripe over each eye.

Breeds from central British Columbia to Cape Breton Island and south to northern Oregon, Colorado, western Texas, northern Coahuila, and central Florida. Migrates through eastern Mexico, Yucatan, and Central America to winter in Colombia, Venezuela, Ecuador, and southern Brazil.

Nests among woods or mixed forests or in shade or orchard trees. Nest a beautiful basket cemented with spider webs and hung from a horizontal crotch 4 to 50 ft up, usually within 10 ft of ground, soft-lined. Eggs 3 to 5 white with brown spots at larger end, 5/16 by 2/3 in. Incubation by both sexes, 12 to 14 days. Annual broods, usually two.

Food essentially insects. One young bird known to eat 100 grasshoppers a day. Other insects include caterpillars, gypsy moth, and cankerworm caterpillars, bark beetles, leafhoppers, mosquitoes, horseflies, and their kind. Monotonous call is continued through hottest days and late into season from treetops but is generally popular.

Economic value of this bird has never been questioned; worthy of every protection. One who has watched it build a nest and rear its young through rain and hot weather can appreciate it more than someone who may have heard its voice but never seen it. Song pitch, 2,375 to 5,850 cps.

Philadelphia Vireo

Philadelphia Vireo, 626
Vireo philadelphicus
Length to 5 in., including 2⅕-in. tail and ½ in. bill. Wingspread to 9 in. Female usually smaller than male. Yellowish-green, without bars on wings, and with underparts yellowish. Less bulky than red-eyed vireo. Has been compared to more active smaller Tennessee warbler and female black-throated blue warbler.

Breeds from northern and central Alberta to southern Manitoba, northern Ontario, New Brunswick, and Maine, south to North Dakota, Michigan, and New Hampshire. Winters from Cozumel Island, Yucatan, to Guatemala and Panama. Found in trees or bushes near homes, or in wilderness. Not common ordinarily.

Nests about edge of woods, in farm lands. Nest in a tree or shrub, 9 to 30 ft up, a deep cup hanging in a fork, made of bark, grass, and lichens, lined with down, fine grass, or needles. Eggs three to four, 4/5 by 1½ in.; white with brown, chocolate, or black markings. Nesting dates June. Incubation by both sexes. One annual brood.

Food mostly caterpillars, leaf beetles, and other insects destructive of trees and shrubs. Call much like that of warbling vireo but of a more dynamic nasal quality. Song similar to that of red-eyed vireo but weaker. Young birds practically indistinguishable from adults.

A useful destroyer of insect enemies of woody plants and an intriguing bird to amateur ornithologists, who may have difficulty in distinguishing it from species suggested above. It is not common in much of range of many vireos and is therefore more likely to be considered a discovery when it is found.

Warbling Vireo

Warbling Vireo, 627
Vireo gilvus
Length to 6 in., wingspread to 9¼ in., tail to 2¼ in. No wing bars such as are possessed by yellow-throated, blue-headed, and white-eyed vireos. Whitish instead of yellowish underparts of Philadelphia vireo; no gray cap or conspicuous black and white eye lines as in red-eyed vireo.

Two subspecies, the Eastern and the Western. Eastern breeds from Saskatchewan to Nova Scotia, south to Texas and North Carolina. Winters from Mexico to Guatemala and El Salvador. Western subspecies extends range on to southern British Columbia and winters south to western Guatemala.

Nest in shade trees, parks, or orchards but rarely in deeply wooded areas. Nest in maple, poplar, or similar tree, a basket hanging from a horizontal crotch far from trunk, 20 to 70 ft up. Eggs three to five, white with brownish spots, ⅝ by ⁷/₁₂ in. Incubation 12 days, by both sexes. One annual brood.

Food, insects caught by gleaning through treetops or by pursuit through air, including leaf beetles, cucumber beetles, locusts, grasshoppers, cankerworms, and the like. Some wild fruit eaten but rarely is any cultivated fruit included. A bird of treetops, whence it sings a continuous warble far into summer.

Essentially a useful bird worthy of every protection not only against human beings but against more aggressive birds such as sparrows and starlings. Is, of course, protected by law but for its song alone would be worthy of extra efforts on its behalf.

Black-and-white Warbler

Order Passeriformes./Family Parulidae [Wood Warblers]

Black-and-white Warbler, 636
Mniotilta varia
Length to 5½ in., wingspread 9 in., tail 2¼ in. Weight ⅓ oz. Covered with lengthwise black and white stripes. Creeps over trunk of tree like nuthatch or brown creeper. Can hardly be confused with any other bird. Female smaller than male.

Breeds from west central Mackenzie to Newfoundland, south to Texas and Georgia. Winters from southern Texas and Florida to Colombia, Ecuador, and Venezuela and occasionally in southern California and Lower California. Accidental in Bermuda and in state of Washington.

Nests usually on ground in a wooded area, usually at foot of a tree or shrub. Nest a mere depression in leaves or fashioned of finer plant materials. Eggs four to five, creamy white, thoroughly spotted with chestnut, brown, or lavender, ¾ by ⁷/₁₂ in. Weight ¹/₂₅ oz. Incubation by female, 13 days. One annual brood.

Prothonotary Warbler

Food, insects such as plant lice, scale insects, caterpillars, gypsy moths, brown-tail moths, click beetles, forest tent caterpillars in adult, egg, larval, or pupal stages gleaned from trunks and branches of trees. Call a thin wiry sound, just audible to many persons and inaudible to others. Song pitch, 5,300 to 8,050 cps.

Bird often known as "black and white creeper," which is a descriptive name. Does not go down tree trunk head first. Scientific name means "moss plucking" and variegated referring to its color pattern. Value as a gleaner of insects from tree trunks cannot be questioned but its numbers near centers of human population are not what they should be.

Prothonotary Warbler, 637
Protonotaria citrea
Length to 5½ in., including 2⅓-in. tail and ½-in. bill. Wingspread to 9 in. Female smaller than male. Sexes colored alike. Conspicuously colored bright yellow over all head and breast, with wings conspicuously blue-gray. Large for a warbler and restless. Breeding plumage results from wear, as only molt is in July to August.

Breeds from northeastern Nebraska through southern Minnesota, southern Wisconsin, and Michigan to Ohio, Delaware, and Maryland, south to northern Florida and Texas. Has been reported breeding in New York and New Jersey. Winters from Nicaragua to Colombia, Venezuela, and southern Mexico. Has been found north through New York to New Brunswick and in Arizona and Bermuda.

Nests usually in wooded swampy country, in a hole in a tree, usually made by some other bird and often over water, 1 to 25 ft up. Nest may be bulky, of variety of vegetable materials. Eggs three to seven, ⁷/₁₀ by ⅗ in.; creamy white with many chestnut, lavender, and purple spots. Nesting period May to June. Incubation to 14 days, by female. One to two annual broods.

Food mostly insects such as ants, flies, bees, locusts, and caterpillars; also spiders and many small water animals, including snails. Call a soft lisping *chip,* but it gives a song like that of pine warbler or junco. One record shows a male making a nest alone, but it was not used.

A beautiful warbler, with an interesting nesting habit and an active manner, found rather uncommonly, but sometimes frequently in a rather gloomy environment. Male has remarkable display performance in courtship and fights rivals vigorously at that time.

Worm-eating Warbler

Worm-eating Warbler, 639
Helmitheros vermivorus
Length to 5⅔ in. Wingspread to 8¾ in. Female smaller than male. General color dull olive-green, darker above with black stripes on top of head alternating with lighter stripes. Wings seem to be unusually long and pointed and tail gives impression of being short and more rounded than would be expected.

(Continued)

Golden–winged
Warbler

Breeds from southern Iowa to western New York, southwestern Pennsylvania, and Hudson and Connecticut River valleys south to Missouri, Virginia, northern Georgia, and Alabama. Winters from southeastern Mexico to Panama, Cuba, Bahamas, and rarely in Florida. Haunts hillsides and ravines where there are woodlands.

Nests in deep woods and wooded ravines. Nest on ground, often beside a log or in a slight cavity; lined with moss and leaves and sometimes with hair. Eggs, three to six, ⅘ by ½ in.; white spotted with lavender, chestnut, and dark reddish-brown, highly variable in markings. Nesting period late May and early June. Incubation 13 days, by female. One annual brood.

Walks about on ground searching its food of insects such as caterpillars, beetles, ants, and the like. No special interest in worms, according to evidence supplied by stomach examination. A relatively quiet, secretive bird with a call somewhat like that of chipping sparrow.

Unquestionably useful in control of insects and of interest to ornithologists, who easily identify it by its conspicuously marked head. It has some characteristics of the ovenbird.

Golden-winged Warbler, 642
Vermivora chrysoptera
Length to 5⅓ in. Wingspread to 8¹/₁₀ in. Female smaller than male. Male with black throat, yellow wing patches, yellow on forehead and below black cheek and eye patch; back, most of wings, and upper tail surface light gray, lower breast and belly almost white, in winter somewhat tinged with olive. Female duller than male.

Breeds from central Minnesota to eastern Ontario and Massachusetts, south to Iowa and northern Georgia, Tennessee, and Virginia. Winters from Guatemala to Colombia and Venezuela and occasionally in southern Mexico and Cuba. Usually found in shrubbery or even on ground but not commonly seen. Generally nests in locations near streams or ponds and in those places not heavily shaded.

Nests May to June. Nest on or near ground in thickets of broad-leaved trees; of leaves, bark, and grasses or hairs, rather bulky, sometimes with caterpillar web added. Eggs four to rarely 6, cream to white, with pinkish, lavender, or brownish markings sometimes in blotches but not commonly so. Incubation by female, for 10 days. One annual brood.

Nest built in 2 to 3 days. Incubation begins immediately after eggs are laid. Young leave nest about 10 days after hatching. Song variously described as a repeated *zee* or *beee-bz-bz-bz* which is longer than similar call of blue-winged warbler.

The economic status of species has not been determined. It unquestionably is concerned with insect destruction and is free of destruction of plants and animals useful to man.

Blue-winged Warbler, 641
Vermivora pinus
Length to 5 in. Wingspread to 7½ in. Female smaller than male. Essentially a yellowish warbler with two conspicuous white wing bars, a black-tipped tail, and a conspicuous but small black line through eye. Wings darker than back, which is pale olive. Blackish wing tips and shoulder.

Breeds from southeastern Minnesota to southern Massachusetts or Rhode Island, south to Kansas, Missouri, northern Alabama, Georgia, Maryland, and Delaware. Winters from southern Mexico to Guatemala and on to Colombia, migrating across Gulf of Mexico. Rare in United States southeast of Virginia.

Nests in field or pastures near woods or swampy thickets; on ground, late May to June. Nest of leaves, bark shreds, and roots bulkily thrown together, with fine grass lining. Eggs four to five, ⅔ by ½ in.; white with sprinkling of delicate brown, lavender, and purple specks. Incubation 10 to 11 days, by female. One annual brood.

Essentially a warbler of low growth relying greatly on "sitting tight" to avoid detection. A female on nest may be touched before she will leave. Call like that of golden-winged warbler but much shorter, frequently including only first phrase. Young remain in nest for 8 to 10 days.

Undoubtedly a useful species, living in an area where it could do no damage to agricultural crops and with habits that would be beneficial if such crops were in the vicinity. Interesting to watch and generally popular with ornithologists.

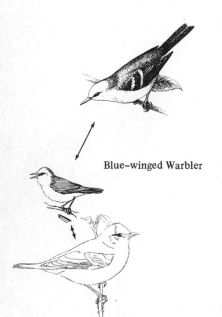

Blue–winged Warbler

Warblers

1. *Vermivora peregrina* (Tennessee), 647
2. *V. celata* (orange-crowned), 646
3. *V. ruficapilla* (Nashville), 645

Length: (1) 5 in., (2) 5⅓ in., (3) 5 in. Wingspread: (1) 8⅓ in., (2) 8¼ in., (3) 7¾ in. Adult male: (1) plain olive with white under parts and white eye stripe; (2) plainer yet, with rarely evident orange crown and faintly streaked breast; (3) olive with yellow under parts and white eye ring.

Breeds: (1) from Yukon to central Quebec and south to British Columbia and New Hampshire; (2) from Kowak River, Alaska, to northern Manitoba; (3) from central Saskatchewan to Cape Breton Island and south to Nebraska and New Jersey. Winters: (1) from Oaxaca to Colombia and Venezuela; (2) from Gulf States to California; (3) from Vera Cruz to Guatemala and in Texas and Florida.

Nests all low in shrubs or on ground and made mostly of grasses, usually well-concealed. Eggs: (1) and (2), 4 to 5, ¾ by ½ in.; (3) three to five, ⅔ by 9/16 in.; all white to creamy and spotted and specked with brown or lilac. Incubation of (3), 11 days, by female. One annual brood. This is probably similar in other species.

Birds are all active little warblers, singing from treetops and nesting on ground. Song of (1), like a chipping sparrow beginning with a *teetsee* and ending with a *dedede;* of (2), a continuous *chee* repeated and rising near end; of (3), a *teetsee* repeated and ending a stuttered prolonged *tititi.*

All are probably useful as insect destroyers and none could possibly do any harm. Offer a challenge to amateur ornithologists to distinguish between them but are worth continued study both for song and for general beauty.

Tennessee Warbler

Parula Warbler

Parula Warbler, 648a

Parula americana

Length to 4⁹/₁₀ in. Wingspread to 7¾ in. General color bluish, with yellow breast and throat and a less conspicuous yellow patch on back. A dark band across yellow of breast and two broad white wing bars. Female lacks breast band or shows it only obscurely; also smaller than male. The only bluish warbler with a yellow throat and breast.

Breeds from eastern Nebraska to Cape Breton Island, south to Texas, Louisiana, and Maryland, with southern subspecies from District of Columbia to Alabama and Forida. Winters from Bahamas and West Indies to Barbados and from Vera Cruz to Nicaragua, with southern subspecies in Florida and Bahamas. A bird of treetops.

Nests usually in evergreen tree in bogs and swamps where air is humid and lichens hang from branches. Nest in a bunch of moss hollowed out, with a side entrance, sometimes with hair lining. Eggs three to seven, usually four; ¾ by ½ in.; white to creamy, with brown and gray spots. Nests mostly in June. Incubation by female, with one annual brood.

Young able to fly about by late August, when a slow southern migration begins. Food probably almost exclusively insects gleaned from trees and shrubs, but measuring worms are known to be favored in diet and tent caterpillars and gypsy-moth caterpillars are eaten.

Useful in destroying insect pests of common woody plants. Song a trill buzzing up scale and dropping near end. Plumage changes slightly through year, female in winter having blue areas more greenish while young are olive-brown to gray above and have pinkish feet that become gray with age.

Yellow Warbler, 652

Dendroica petechia

Length to 5¼ in., wingspread to 8 in., tail to 2¼ in. Female smaller than male. Appears superficially to be all yellow. Seen near at hand, male shows breast to be streaked with chestnut-red markings which are faint or lacking in female. No associated bird gives impression of being all yellow. Temperature, 101.5 to 108.2°F.

Six subspecies include among others Eastern, Alaska, California, and Sonora yellow warblers. Eastern breeds from eastern Alaska to southern Ungava, south to Nevada, New Mexico, northern Georgia, and South Carolina. Winters from Yucatan to Guiana, Brazil, and Peru.

Nests about farm lands and in suburban areas, in low trees or ornamental shrubs and hedges. Nest a compact sphere of cottony materials, well lined with fine grasses, bark, and rootlets or hair. Eggs three to six, gray or greenish with brown or purplish markings around larger end; ¾ by ½ in. Incubation 12 to 15 days, by female. One annual brood.

Food, insects such as small moths, caterpillars, beetles, flies, and grasshoppers, many of which are caught in air by pursuit through treetops. Cowbirds commonly lay their eggs in yellow warbler nests; yellow warbler often builds a second or even to a sixth nest on top of first.

Entirely useful as a destroyer of insects and one of most interesting and beautiful birds of thickly settled parts of country. Its habit of nesting in parks and ornamental shrubbery and its cheery song make it well known and welcomed by many who never notice other species.

Yellow Warbler

Magnolia Warbler

Magnolia Warbler, 657
Dendroica magnolia

Length to 4¾ in. Wingspread to 7⅕ in. Female smaller than male. Male essentially black and yellow, tail white below with broad black band at tip; from above appears to have large white patches on wings and tail; underparts yellow with many heavy black streaks. In fall, brown above and yellow below, with tail as described above.

Breeds from western Mackenzie and central British Columbia to central Quebec and Newfoundland, south to southern Saskatchewan, Minnesota, New York, Massachusetts, and in mountains south to Virginia. Winters from southern Mexico to Panama and sometimes in Cuba and Bahamas.

Nests usually in June, generally in an evergreen up 1 to 35 ft above ground but usually low; of bark, and rootlets and fastened together with cobwebs, with soft lining. Eggs three to six, ¾ by ½ in.; white with brown, purple, and lavender spots. Incubation usually by female, though male may assist occasionally. One annual brood.

Most active about evergreen trees where they may nest. Young remain in nest about 10 days after hatching. After leaving nest, young and parents are active for only about 4 weeks before beginning southern migration. Food probably exclusively insects, mostly those of treetops.

Most interesting to bird lovers and undoubtedly useful as insect destroyers, with no known habits injurious to man's interests. With cutting down of evergreen timber or with fire damage to such growths, these birds decrease in number.

Cape May Warbler

Cape May Warbler, 650
Dendroica tigrina

Length to 5⅔ in. Wingspread to 8½ in. Female smaller than male. Underparts yellow, narrowly striped in black. Cheeks chestnut (unlike any other warbler). Rump yellow. Crown black. Females not brilliantly colored, with breast almost white and with cheeks pale yellow rather than chestnut.

Breeds from southern Mackenzie to Nova Scotia and New Brunswick, south to southern Manitoba, North Dakota, Kansas, New Hampshire, and Maine. Southwest area is visited in fall migration. Winters from Bahamas to West Indies and occasionally in Yucatan. A bird of treetops.

Nest in evergreen woodlands and in sparsely wooded areas in smaller trees. Nest up to 40 ft above ground, of moss, twigs, or grass bound with spider web, lined with hair and feathers. Eggs three to four, 7/10 by 9/16 in.; grayish or greenish-white spotted with yellow-brown, red-brown, or lilac. Nests in June. Incubation by female with probably one annual brood.

Vary greatly in abundance. Usually not unduly suspicious of man and commonly seen some years in suburban trees and shrubbery as well as in thickets and brushy areas. Food probably mostly insects found on woody plants including caterpillars, flea beetles, click beetles, and weevils.

Always interesting to ornithologists and bird lovers and probably wholly useful as insect destroyers. Not sufficiently abundant to play any great role in economy of nature or of man but worthy of every encouragement. Song a highly pitched and thin repeated *seet* given at least four times in a series.

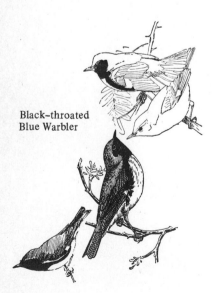

Black-throated Blue Warbler

Black-throated Blue Warbler, 654
Dendroica caerulescens

Length to 5 in., wingspread to 7⅞ in., tail, to 2¼ in. Female smaller than male. Upper parts bluish-gray, throat and sides black, breast and belly white. Female and young have a plain olive back, with a faint white line over eye and small white wing spots. Name adequately describes male bird.

Breeds from northern Minnesota to southern Quebec, south to southern Minnesota, mountains of Pennsylvania, and northern Connecticut. Winters from southern Florida and Bahamas to Guatemala and Colombia. In migration is found in North Dakota, Nebraska, Kansas, Colorado, New Mexico, and sometimes California.

Nests usually in deep woods or at edges of clearings. Nest in a tree or bush, 3 to 10 ft above ground; seldom higher than 4 ft; bulky outside and neat within, with a lining of hair, webs, needles, and grasses. Eggs four to five, grayish, greenish, or brownish-white with brown, lilac, or lavender spots and blotches, 11/16 by ½ in. Incubation by female, about 12 days.

Food essentially animal matter, such as insects gleaned from trees. Female makes excellent effort to protect her young by hiding them in nest or by pretending injury to herself to attract enemies away from nest. Male may or may not assist in this family function.

Food habits and general behavior are such that it must be considered useful even if it did not win favor by its beautiful appearance and interesting habits. When males begin singing high in trees in migratory area, one knows that spring has really arrived. Song pitch, 5,125 to 6,750 cps.

Myrtle Warbler, 655
Dendroica coronata
Length to 6 in., wingspread to 9⅖ in., tail to 2¾ in. Weight to ½ oz.
Shows a conspicuous yellow rump. Male in spring, streaked blue-gray
above, with yellow spots on crown, sides, and rump. Female brown
where male is blue, otherwise similar. Adults in winter, brownish-gray
above, with yellow rump.

Breeds from northwest Alaska to central Quebec, south to northern Brit-
ish Columbia, northern Minnesota, central Ontario, New Hampshire,
Maine, New York, and New England highlands. Winters from Kansas,
Ohio Valley, and New Jersey south to Mexico, Panama, and on Pacific
Coast, from Oregon to northern Lower California. In Siberia also.

Nests usually in coniferous woodlands, with nest from near ground to 40
ft up. Nest bulky, of stems, twigs, bark, and grass, bound with cobwebs
and lined with hair and feathers. Eggs three to six, white to creamy, with
spots of brown, purple, lavender, or even black; ⅘ by ⁹⁄₁₆ in. Incubation
12 to 13 days, by female. One annual brood.

While most warblers live almost exclusively on insects, myrtle can exist
for a long period on fruits of red cedar, juniper, Virginia creeper, mountain ash, poison ivy, dogwoods, and viburnums. In New England, chief
plant food is fruit of bayberry. During spring and summer diet is essentially insects and similar small animals.

Highly useful, as 78 percent of its food for year is animal matter, largely
injurious insects. A pleasure to have bird in abundance, because it
comes before most other warblers and seems to resist severe weather
better than similar birds.

Myrtle Warbler

Audubon's Warbler, 656
Dendroica auduboni
Length, male to 5¼ in., wings, 3 in., tail, 2⅖ in. Female smaller. Warbler size. Male with rich lemon yellow crown patch; black cheeks; wings
snow white. Female duller. Recognition marks: yellow throat, extensive white blotching of tail seen in flight, yellow rump distinctive. Male
in spring blue-gray above with a heavy black breast patch; throat, crown,
and side patches bright yellow. Large white wing bars are distinctive.
Male may be confused with Myrtle Warbler, *D. coronata.* Female brown
instead of gray; not so conspicuous as male.

Two subspecies reported; one is *D. a. auduboni,* which breeds from cen-
tral British Columbia south along the Pacific Coast through the Cascade
Range and the Sierra Nevada to southern California and northern Baja
California. Altitudinally, ranges from Transition Zone to Canadian and
Hudsonian Zones. Winters from southwestern British Columbia and
coastal Washington south through Baja California to Guatemala.

Nest usually compact in branches of conifers, 5 ft to 30 ft above ground,
more frequently in higher locations. Bulky, of roots, bark, twigs, and
needles, heavily lined, usually with feathers. Eggs three to five, olive to
greenish-white; spotted with lilac, brown, or black, slightly glossy, markings frequently at large end only.

A unique bird with a song which is juncolike in quality. Song is two-
parted either rising or dropping in pitch at each end; a series of syllables
repeating "tsil" or "tsi," similar to that of Myrtle Warbler, *D.coronata.*

Deserving of much more attention than it now receives. Food predominantly animal matter which is made up largely of insects and spiders.
Valuable as a destroyer of scale, insects and plant lice. Plant food made
up of fruits of wild plants and weed seeds. They feed on poison oak seeds
and may visit grape arbors and puncture the ripe grapes.

Audubon's Warbler

Black-throated Green Warbler, 667
Dendroica virens
Length to 5⅓ in. Wingspread to 8 in. Female smaller than male. Back
and crown olive-green, face yellow, throat black. Female with black of
throat and under parts less conspicuous. Young and females in autumn
without black on throat and upper breast and with under parts yellowish.
Golden cheek is most distinctive feature. No other eastern warbler has
golden cheeks.

Breeds from central Alberta to Quebec and Newfoundland, south to
southern Minnesota, southern Wisconsin, New York, and northern New
Jersey and in Appalachian highlands on to South Carolina, Georgia, and
Alabama. Winters in Mexico, Guatemala, Costa Rica, and Panama.
On occasion is found in Greenland, Europe, and Puerto Rico.

Nests among trees like pines, spruces, and hemlocks, usually from 15 to
40 ft above ground; deeply cupped, on horizontal branch, of twigs bound
with spider webs and lined with needles, a few feathers, or hair. Eggs
four, ¾ by ½ in.; white or creamy, spotted and speckled with brown,
purple, and gray. Nests in June. Incubation 12 days by female. One
annual brood.

Young remain in nest eight to ten days after being hatched and while
there are defended by female. By mid-August, southern migration starts
rather leisurely, though late-hatched birds may not start before end of
(Continued)

Black-throated Green Warbler

Cerulean Warbler

month. Food largely caterpillars, beetles, and bugs of woody plant leaves, particularly leaf rollers and cankerworms.

Useful as a destroyer of enemies of trees and by some is considered one of most useful of woodland birds. Song like a lisping sneeze like *zoo-zee-zee-zee-zee-zoo-zee.*

Cerulean Warbler, 658
Dendroica cerulea
Length to 5 in., including $^9/_{10}$-in. tail and $\frac{1}{2}$-in. bill. Wingspread to 8 in. As name implies, male is blue above; white below and shows a dark or black ring across upper part of breast. Female olive-green like many other warblers, but somewhat bluish above and white beneath, with two, white wing bars and white line over eye, suggestive of some vireos.

Breeds from southeastern Nebraska through southeastern Minnesota, southern Michigan, and southern Ontario to central New York, south through West Virginia to Georgia, central Alabama, Louisiana, and Texas. Winters in Venezuela, eastern Ecuador, Peru, and Central America. Sometimes found in Lower California, Cuba, Bahamas, and even into Manitoba.

Nests usually where there are large trees, commonly 15 to 90 ft up near middle of limb. Nest shallow, of vegetable material, such as rootlets, grass, lichens, and sometimes with snakeskins, hair, and feathers. Eggs three to four, $\frac{3}{4}$ by $\frac{1}{2}$ in., pale blue- or greenish-white or creamy, with reddish spots. Incubation by female.

Food largely insects, known to include caterpillars, beetles, weevils, and similar insects. Flight in somewhat jerking curves, but seen mostly high in treetops. Call a lisping, repeated *chip* or *cheer* or *zwee,* usually ascending at end. Considered a rare bird in East.

Useful as an insect destroyer, but more useful as a feast to the eye of the naturalist. Never sufficiently abundant to do any outstanding service but welcome when it is present.

Blackburnian Warbler, 662
Dendroica fusca
Length to $5\frac{1}{2}$ in. Wingspread to $8\frac{1}{2}$ in. Female smaller than male. Male is bright orange on head and throat region but otherwise mostly black and white; white patches on elbow and tail show conspicuously in flight; paler in autumn. Female dingier in more conspicuous areas.

Breeds from central Manitoba to Cape Breton Island, south to central Minnesota and Connecticut and in Appalachian highlands to Georgia and South Carolina. Winters from Venezuela and Colombia to Peru and Yucatan. Occasionally in migration in Nebraska, Texas, Montana, New Mexico, and Bahamas.

Nests usually in evergreen tree 6 to 80 ft up, usually toward end of limb or in top of tree. Nest highly variable, usually of twigs, with hair, rootlet, or grass lining. Eggs four, $\frac{3}{4}$ by $\frac{1}{2}$ in.; blue-green to gray, with brown and lavender spots and blotches. Incubation in June, by female.

An active conspicuous bird of treetops, with a repeated *zip* call on one pitch but ending with a thin, more highly pitched note. In northern migration, it moves just about time leaves are reaching full size. Food mostly insects of treetops, with some caught by sudden dashes into air. Some known to subsist on ivy berries.

Its economic status is that of most warblers in that it destroys great numbers of insect enemies of woody plants. Also called hemlock warbler, and orange-throated warbler.

Blackburnian Warbler

Yellow-throated Warbler

Yellow-throated Warbler
Dendroica dominica 663

Chestnut-sided Warbler, 659
D. pensylvanica
Length to $5\frac{1}{4}$ in., wingspread to $8\frac{1}{4}$ in., tail to $2\frac{1}{10}$ in. Female smaller than male. In spring, adults show conspicuous yellow crowns and chestnut sides, differing in this respect from dark-crowned, chestnut-throated bay-breasted warbler. In fall, greenish above, white below, with two wing bars and white eye ring.

Breeds from central Saskatchewan to Newfoundland, south to eastern Nebraska, northern New Jersey and, in Alleghenies, south into Tennessee and South Carolina. Winters from Guatemala to Panama. Occasionally in California, Mexico, Greenland, Florida, and Bahamas.

Nests in shrubbery, near pastures, at margins of woodlands, usually 1 to 6 ft above ground. Nest neat, compact, of grasses, weeds, and bark, cemented with spider webs, and lined with plant fibers. Eggs four to five, white or creamy with brown or lavender spots, $\frac{2}{3}$ by $\frac{1}{2}$ in. Incubation for 10 to 11 days. One annual brood, reared by both sexes.

Excellent insect catcher; food caught by gleaning or by short dashes into air. Known to eat gypsy moths, tent caterpillars, brown-tail moths, plant lice, leaf hoppers, ants, and borers. One of best caterpillar destroyers.

Chestnut-sided Warbler

Young in nest 9 days, fed little besides insects. Adults may eat some seeds and wild fruit in fall.

An unusually beautiful but rather shy bird, probably more often heard than seen and certainly more often heard than identified. Two adults attending young in nest make a memorable picture. Song pitch, 3,100 to 8,775 cps.

Bay-breasted Warbler, 660
Dendroica castanea
Length to 6 in., wingspread to 9⅓ in., tail to 2¼ in. Female smaller than male. Dark chestnut throat, upper breast, and sides; pale light brown spot on side of neck. Lacks white throat and yellow crown of chestnut-sided warbler.

Breeds from east central Alberta to Newfoundland, south to southern Manitoba, Adirondacks in New York, and mountains of New Hampshire. Winters in Panama and Colombia. Occasionally found in Greenland, Texas, Montana, and South Dakota.

Breeds in mixed forests or coniferous stands. Nest in a conifer, 3 to 20 ft up, of fine twigs and moss, with rootlet, hair, or needle lining. Eggs three to four; pale blue-green or greenish-gray with brown or lilac spots, ¾ by ⁹⁄₁₆ in. Incubation by female. One brood yearly.

Food habits little understood but known to include destruction of locusts, caterpillars, ants, beetles, and leafhoppers, which are fed to young. Spends most of its time in dense foliage of trees and is not so easily seen as some other warblers.

Useful as insect destroyer, like other warblers. In autumn, abundance may not be suspected as they appear olive-green with two white wing bars and a rather dirty brownish-yellow beneath. In fall, yellow under tail, while blackpoll warbler is white there and has more distinct streaks.

Bay-brested Warbler

Blackpoll Warbler, 661
Dendroica striata
Length to 5¾ in., wingspread to 9⅔ in., tail to 2¼ in. Female smaller than male. Male in spring with solid black cap instead of streaked cap of black and white warbler. Female in spring less heavily streaked, greenish above, pale below. Adults in fall olive above and dingy streaked yellow below, with two white wing bars.

Breeds from tree limit in northwestern Alaska to Newfoundland, south to northern British Columbia and highlands of New York and New England. Winters from Guiana and Venezuela to Brazil, migrating through Bahamas and West Indies. Occasionally in Mexico, Chile, and Ecuador.

Breeds among coniferous forests or swampy groves. Nest low in evergreen or on ground, of mosses, lichens, and feathers. Eggs four to five, creamy-gray, speckled and spotted with reddish-brown, purplish-gray, or lilac, ¾ by ⁹⁄₁₆ in. Incubation data not available.

Food essentially insects gleaned from trees, including cankerworms, webworms, plant lice, locusts, small grasshoppers, beetles, ants, and gnats. Call sounds like noise made by cutting glass; not audible to many persons. Pitched, 8,050 to 10,225 cps. In migration, probably fly several hundred miles across Caribbean to reach land.

Probably of considerable importance because of great numbers which migrate north at time of year when new growth of woody plants is most susceptible to attack from insects that provide these birds with their food. Always interesting to those who care for birds in their season.

Blackpoll Warbler

Pine Warbler, 671
Dendroica pinus
Length to 5¾ in. Wingspread to 9⅗ in. Female smaller than male. Olive above with bright yellow breast and two white wing bars. Yellow brightest in throat region. Tail with white spots. Faint streaks on breast. Female less brilliantly colored. In fall, olive above, with two white wing bars but not bright yellow below.

Two subspecies. Breeds from northern Manitoba to New Brunswick, south to east central Texas and Gulf States, with Florida subspecies confined to Florida. Winters from southern Illinois to Massachusetts and south into Mexico and Forida.

Nests April to May, usually in evergreen woodland, on long branch, usually 20 to 50 ft above ground, often in a fork; of weeds, bark, and needles, twisted and fastened with spider web and lined with fine material. Eggs three to five, ¾ by ⁹⁄₁₆ in., white or greenish-white with brown, black, or purple spots and blotches. Incubation by female. One to three annual broods.

Adult birds excellent parents and breeding season is unusually long with males fighting for mates vigorously. Food essentially insects of evergreens, though in fall it may eat grasshoppers and other ground insects. Call much like that of a musical, irregular chipping sparrow.

Undoubtedly, bird is useful as a destroyer of insect enemies of evergreen forest trees and to some extent of plants of forest floor and nearby ground.

Pine Warbler

Prairie Warbler

Prairie Warbler, 673
Dendroica discolor

Length to 5⅕ in. Wingspread to 7⅓ in. Female smaller than male. Male with chestnut spots on back and with yellow underparts, with streaks of black only on sides. Black on face in spots through eye and under eye. Female almost identical with male but with chestnut spots fainter.

Breeds from eastern Nebraska to Pennsylvania and New York, south through Arkansas, northern Mississippi, central Georgia, and Bahamas, with a southern subspecies, Florida prairie warbler, breeding in Florida. Winters from central Florida, Bahamas, and West Indies to islands off coast of Central America.

Nests in brushy low-treéd country, where it is relatively dry or even in rather barren pastures. Nest 1 to 12 ft above ground, hidden, of plant stems, bark, and leaves, with finer lining and cobweb binding. Eggs three to five, ⁷⁄₁₀ by ½ in.; white to greenish with spots of purple or brown. Incubation 12 days, by female. One annual brood.

While name implies that it lives in open prairie country, this is far from true. Not conspicuous but has an interesting *zee-zee* oft-repeated song which rises to higher notes. Both parents care for young, and by end of July when young are off nest, southern journey may begin. Most birds have left north by September.

Undoubtedly useful as insect destroyers. Stomach examinations show that the food is 100 percent animal matter, with bugs constituting about half, followed by beetles, moth larvae and adults, and other groups in a minor capacity.

Palm Warbler, 672a
Dendroica palmarum

Length to 5½ in. Wingspread to 8⅖ in. Female smaller than male. Essentially a dull yellow color with a chestnut-reddish crown that is most brilliant in spring and summer plumage. Palm warbler, *D. p. hypochrysea,* more brilliantly colored and larger than western palm warbler, *D. p. palmarum,* which is less yellow beneath at all ages. Palm warbler once known by common name, yellow palm warbler.

Western palm warbler breeds from southern Mackenzie and northern Manitoba to northern Minnesota. Winters from southern Florida and Bahamas to Greater Antilles and Yucatan. Palm warbler breeds from Ontario to Newfoundland and south to southern Nova Scotia and Maine. Winters from Louisiana to Florida, or on occasion in West Virginia, Pennsylvania, and Massachusetts.

Palm warbler nests in open shrubby fields and pastures on or near ground. Nest of grasses with abundant lining of hair and feathers. Eggs four to five, ¾ by ½ in.; buff or cream-white spotted with brown and lilac, often in a ring around larger end. Nests in June. Incubation probably by female, and probably one annual brood.

Easily identified by conspicuous tendency to wag tail violently like water thrush or spotted sandpiper. Frequently appears in fair-sized flocks. It works most of time close to ground, unlike many of its relatives. Song a weak repeated lisping *thi.* Food almost entirely insects with some seeds mixed in.

Useful as a destroyer of insects that might be injurious to plants in region where it lives, which is usually relatively near lands used by man.

Palm Warbler

Ovenbird

Ovenbird, 674
Seiurus aurocapillus

Length to 6½ in., wingspread to 10⅖ in., tail to 2½ in. Female smaller than male. Appears like small brown thrush, striped beneath rather than spotted, and with a light orange patch on top of head. Ovenbird walks, while thrushes usually hop. Sexes alike in appearance.

Breeds from southwestern Mackenzie to Newfoundland, south to southern Alberta, Colorado, Arkansas, Georgia, and North Carolina. Winters from northern Florida and Louisiana coast to Mexico, Bahamas, and Colombia. Sometimes found in California in migration.

Nests usually in hardwood forest, on ground, building a roofed house under a leaf pile or in a grass tussock. Nest lined with hair, needles, or fine grasses. Eggs three to six, white to creamy or pinkish variously spotted and blotched with brown and lilac; ⁹⁄₁₀ by ⁷⁄₁₀ in. Incubation 12 days, chiefly by female. One annual brood.

Food largely insects of forest floor, slugs, snails, earthworms, spiders, myriapods, and other small animals. Known to destroy gypsy moths, click beetles, grasshoppers, butterflies, weevils, and many other insects. Song pitch, 3,300 to 5,850 cps.; median, 4,000 cps.

Repeated *teacher* call of ovenbird is heard more often than bird is seen and bird is seen many more times than nest is found, but each provides a worthwhile experience for an amateur naturalist. As destroyer of insects it undoubtedly serves man and justifies protection it receives.

Northern Water Thrush, 675
Seiurus noveboracensis
Length to 6 in., wingspread to 10 in., tail to $2\frac{2}{5}$ in. Female smaller than male. Slightly smaller than Louisiana water thrush, which has white instead of yellow line over eye and lacks yellowness of streaked underparts. Both are brown-backed birds which teeter nervously, almost constantly.

Two subspecies, northern and Grinnell's. Northern breeds from northern Ontario to Quebec, south to Pennsylvania, West Virginia, and northern New England. Winters from Mexico to Colombia and British Guiana. Grinnell's extends range westward to Pacific. Louisiana water thrush has a more southern range than northern *S. motacilla.*

Nests in cool wet swampy woods or bogs, or along small boggy streams. Nest a cavity under vegetation or under a large root, usually lined with finer vegetation. Eggs four to five, pinkish or creamy white with spots, streaks, and blotches of brown or lavender, $\frac{7}{8}$ by $\frac{7}{10}$ in. Incubation 14 days, by female. One annual brood.

Food essentially aquatic insects and crustaceans picked up at edges of streams it haunts. Wild songs of both Northern and Louisiana water thrushes are in keeping with usual dashing streams where they may live. Northern water thrush ends its song with a repeated *chew* not given by possible louder Louisiana subspecies.

Birds and songs are part of land of bogs and little glens; to those familiar with them, first sight of birds means that spring flowers will soon be at their best. Song pitch, (Northern) 2,000 to 3,850, (Louisiana) 2,475 to 6,600 cps.

Northern Water Thrush

Louisiana Water Thrush, 676
Seiurus motacilla
Length to $6\frac{2}{5}$ in. Wingspread to $10\frac{3}{4}$ in. Female smaller than male. Differs from northern water thrush in unstreaked throat, in being less yellow beneath and without yellow line over eye. In this species line over eye is pure white. General color olive-brown with brown streaks on lighter underparts.

Breeds from eastern Nebraska and southeastern Minnesota to southern Ontario and New England to Texas, southern Alabama, northern Georgia, and central South Carolina. Winters from northern Mexico to Colombia, Greater Antilles, Antigua, and Bahamas. Accidental in California.

Nests in a crevice in a wooded rocky bank near a small active stream. Nest lined with moss and dry leaves but surrounded with a bulky mass of leaves. Eggs four to seven, $\frac{9}{10}$ by $\frac{7}{10}$ in.; glossy and polished white with variable markings of brown and lilac-gray. Nests May to June. Incubation 12 to 14 days, mostly by female.

A bird of active gorges much like water ouzel of West in its habits and haunts. It runs about at edge of rapid water but does not submerge in it like water ouzel. Its song is stimulating, loud, sweet, vigorous.

Food probably chiefly insects and other small animals gleaned from an area of little economic importance. From a dollar and cents standpoint, bird probably does little good but from an emotional standpoint it ranks high with bird lovers.

Louisiana Water Thrush

Kentucky Warbler, 677
Oporornis formosus
Length to $5\frac{5}{6}$ in. Wingspread to $9\frac{1}{4}$ in. Female smaller than male. Conspicuously contrasting plain olive above and bright yellow beneath. Black streaks from eye to side of yellow throat, dark to black forehead, and a yellow hook or streak from bill to over and back of eye. Olive where Canada warbler is gray.

Breeds from southeastern Nebraska to Hudson Valley, south to eastern Texas, northern Georgia, southern Alabama, and Louisiana. Winters in Mexico from Tabasco, Campeche, and Chiapas through Central America to Colombia and accidentally in Cuba, Vermont, and Michigan.

Nests in open, in undergrowth at edge of woodlands, or in brushy areas near marshes, on or near ground in thick growth. Nest bulky, of grasses, leaves, and stems, with lining of hairs, roots, or both. Eggs four to five, $\frac{4}{5}$ by $\frac{3}{5}$ in.; glossy white with brown spots, speckles, and blotches. Incubation by female. Two broods, sometimes.

Commonly feeds on ground but sings its loud, repeated *turtle turtle* from treetops. Active but shy and usually suspicious. Male sings almost continually in height of breeding season. Young leave nest in about 8 days and after nesting is over, birds are quiet and secretive. More often heard than seen.

Food largely insects including grasshoppers, plant lice, larvae of moths, beetles, and other insects. These are gleaned mostly from areas near ground and include ants not common in environment of many warblers. Called Kentucky wagtail because of bobbing motion of tail when bird is walking.

Kentucky Warbler

Mourning Warbler

Mourning Warbler, 679
Oporornis philadelphia
Length to 5¾ in. Wingspread to 8⅛ in. Female smaller than male. Olive above and yellow beneath with a gray hood encircling head and neck and in male with a black bib on upper breast where gray of head and yellow of breast meet. No eye ring except sometimes an obscure one in females and immature birds.

Breeds from east central Alberta and central Saskatchewan to Nova Scotia and Newfoundland, south to central Minnesota, Michigan, Ohio, and in highlands of New York, Massachusetts, Pennsylvania, and West Virginia. Winters from Nicaragua to Costa Rica, Venezuela, Colombia, and Ecuador. Flies across Gulf of Mexico in one flight.

Nests in brier patches, weeds, and bushes in cutover lands and clearings. Nest near ground in a clump of weeds, of weed stalks mixed with leaves and lined with roots, hairs, and grass. Eggs four to five, ⅘ by ⅗.; glossy white with spots and speckles of brown, lilac, and gray. Incubation chiefly or wholly by female. One annual brood.

Not a common warbler and usually so secretive that it is not seen. Family remains near nest until molt is over and family is independent; then starts on stouthern migration, northern part of range being free of birds by end of September.

Little is known of economic importance of species but beetles and spiders have been found in stomachs of some birds. Probably useful, but aside from its interest to ornithologists plays no important role in the economics of its territory.

Yellowthroat

Yellowthroat, 681
Geothlypis trichas
Length to 5¾ in., wingspread to 7⅕ in., tail to 2 in. Male olive above, yellow beneath, and with conspicuous black mask across face and side of head. Female and young plain olive, with yellow throat and breast but no black mask, and with belly whitish instead of yellow as in yellow warbler.

At least twelve subspecies of which six are: Northern, Maryland, Florida, Western, Salt Marsh, the Tule yellowthroats. Northern breeds from Newfoundland to Quebec, south to Pennsylvania. Winters in Mexico, Bahamas, and Costa Rica. Other subspecies extend range over most of continental United States.

Nests, near ground, in thickets, in cattail bogs, and in meadows and hedgerows. Nest of dried leaves and grasses lined with grasses, roots, and some hairs; bulky, deep, but well concealed for breeding season. Eggs three to five, white with brown or purple spots and streaks particularly at large end, ⅘ by ⅗ in. Incubation 12 days. Two annual broods.

Food essentially insects, particularly small caterpillars which do so much damage to crops, also ants, aphids, grasshoppers, crickets, leafhoppers, spiders, and other small creatures. The *witchity-witchity* call varies slightly over the country but is easily recognized.

One of most useful species and fortunately is present over its range in considerable numbers. Activity displayed by a yellowthroat on a hot summer day is worth observing. It sings as it works far into the summer.

Yellow–brested Chat

Yellow-breasted Chat, 683
Icteria virens
Length to 7½ in., wingspread to 10 in., tail to 3⅓ in. Plain olive above with bright yellow throat, white belly, white around eye and across face in a narrow line. Tail long and conspicuous. Sexes colored alike, though female may be duller in color. Largest of the wood warblers.

Two subspecies include Yellow-breasted and Long-tailed chats. Yellow-breasted breeds from Minnesota to Ontario, south to Texas and southern Florida. Winters in Mexico, Central America, Yucatan, and Costa Rica. Occasional in Saskatchewan, New Hampshire, and Maine. Long-tailed chat extends range west to Pacific.

Nests chiefly in thickets, near pastures, or in berry patches and vine tangles. Nest 1 to 5 ft above ground, coarse and bulky, of leaves and grasses with finer lining. Eggs three to five, pinkish or greenish-white, glossy, with reddish-brown or lavender spots, 1/10 by ⅘ in. Incubation 15 days, by female. One annual brood.

Food essentially insects, of almost all kinds to be found, but also elderberries, wild strawberries, wild grapes, blackberries, and similar fruits. Comical repertoire of songs makes it a clown of bird world because variety and volume are almost unbelievable. Song pitch, 1,275 to 4,400 cps.

Chat is not known to harm man's interests but man seems to scare it from its haunts without much difficulty. Worth knowing better. A most interesting bird because of its form, call notes, and habits, its three characteristics which are not duplicated in another species of bird.

Hooded Warbler, 684
Wilsonia citrina
Length to 5¾ in. Wingspread to 8½ in. Female smaller than male. Male has a black hood which encircles completely head and neck; with yellow face and forehead, yellow underparts, and an olive back. Females and young bright yellow on underparts and forehead and plain olive on upper parts; bill black, white on tail.

Breeds from southeastern Nebraska to lower Connecticut River Valley and south through Louisiana, Alabama, Georgia, and northern Florida. Winters from Vera Cruz and Yucatan to Panama with records from Bahamas, Cuba, and Jamaica.

Nests in brushlands or dense undergrowth in woodlands, wooded marshlands, or hillsides. Nest 1 to 5 ft above ground in a bush or small tree, of leaves, bark, weed stems with lining of plant down, moss, hairs, grasses, and pine needles. Eggs three to five, ⅘ by ⁷/₁₂ in.; white or creamy with brown spots. Incubation by both birds. One or two annual broods.

Normally a shy bird staying close to its home thickets but sometimes individuals are otherwise. Opens and closes tail like a fan while singing in flight. Male will defend territory rather effectively at times. Southern migration may begin as early as August unless a second brood is being raised.

Food consists of insects, many taken on wing and including grasshoppers, crickets, caterpillars, and plant lice. Probably indicates that it is a useful species. Its song is a clear sweet *weeta weeteo*, dropping at end and very musical.

Hooded Warbler

Wilson's Warbler, 685
Wilsonia pusilla
Length to 5¹/₁₀ in. Wingspread to 7 in. Female smaller than male. Male pale olive above, light yellow beneath, with a small round black cap. Females and young may or may not show traces of cap. Female shows white spots on tail, thus differing from yellow of yellow warbler. No streaks or bars.

Breeds from central and northwestern Mackenzie to Newfoundland, south to southern Saskatchewan, New Hampshire, Maine, and Nova Scotia. Winters in eastern Central America, north to Mexico, south to Guatemala and Costa Rica. Migrates chiefly along Appalachian highlands, crossing Gulf of Mexico in migration.

Nests in swampy brushlands, on ground, or even in a depression among bushes; of grasses, with finer grass and hair lining. Eggs, two to four, ⅔ by ½ in.; white with spots of reddish-brown or lilac-gray particularly around larger end. Nests in June. Incubation by female. One annual brood.

Quick energetic bird, often acting like a flycatcher in procuring its insect food on wing, in which cases bill may sometimes be heard to snap. It also twitches tail in a rotary fashion. Southern migration begins in August and most birds have left northern part of range by end of September.

Since food is largely insects, it is probably a useful species though no exhaustive study has been carried on. Its song is simple, drops at the end, and sounds a bit like a repeated *chee chee*. Distinctive black cap provides an excellent field mark for its recognition.

Wilson's Warbler

Canada Warbler, 686
Wilsonia canadensis
Length to 5¾ in. Female slightly smaller than male. Male with upper parts plain gray and lower parts bright yellow but with a necklace of short blackish stripes across upper breast. Female without necklace. No white on wings or tail of either sex, thus differing from similar species.

Breeds from southern Alberta to Newfoundland, south to central Minnesota, central New York and Connecticut, and in highlands to Georgia and Tennessee. Winters in Ecuador and Peru and to some extent in Guatemala and Costa Rica and in migration in eastern Mexico.

Nests on or near the ground in wooded or swampy wooded area. Nest well concealed, of dried leaves, grass, moss, and bark, with lining of rootlets and hairs. Eggs four to five, ¾ by ½ in.; white or brownish-white with speckles of reddish-brown and lilac chiefly around larger end. Nests in June. No data on incubation. One annual brood.

Active bird in migration, giving a rather loud musical song that has been described as a repeated *rup-it-chee* ending in a *rup-it-chit-it-lit*. Catches food of insects on wing like a flycatcher, gleans it from finer vegetation in which it lives, or may even gather it from ground.

Food habits so far as known would indicate that it is a useful species engaged in keeping in control insects that might otherwise injure plants, even though plants thus protected are not ordinarily recognized as of agricultural importance.

Canadian Warbler

Redstart, 687
Setophaga ruticilla
Length to 5¾ in., wing to 2½ in., tail to 2½ in.   Female smaller than male.   Male black with orange patches on wings, tail, and sides of breast; belly white.   Female olive, with lemon-yellow patches in place of orange.   Young much like female, and young breeding male may still lack characteristic salmon, orange, and black coloration.

Breeds from northern British Columbia to Newfoundland south to Oregon, northern Utah, Oklahoma, Arkansas, southern Alabama, northern Georgia, and North Carolina.   Winters in West Indies and from central Mexico and southern Lower California to Ecuador and British Guiana.   Occasionally in California, Arizona, and northern Ungava.

Nests in woodlands or pastures.   Nest in upright crotch, 3 to 35 ft above ground, compactly woven, thin-walled, and beautifully lined with fibers or hair fastened together with webs.   Eggs three to five, creamy or grayish sprinkled with lilac, purple, or brown, ⁷⁄₁₀ by ½ in.   Incubation by female, 12 to 14 days.   One annual brood.

Food includes caterpillars of many sorts as well as moths, gnats, bugs, beetles, grasshoppers, flies, and other insects probably including useful as well as injurious forms.   Male active.   Song (pitched at 4,400 to 7,300 cps) means summer to most who know it.   In display, salmon and black coupled with excessive activity makes bird conspicuous.

Unquestionably, redstarts are useful.   They are popular with all who know them either from their nest, their song, or their coloration and activities.   They deserve protection they get from the law.

Redstart

Order Passeriformes./Family Ploceidae [Weaver-finches]

English Sparrow, 688.2
Passer domesticus
Length to 6⅓ in., wingspread to 9¹¹⁄₁₂ in., tail to 2½ in.   Weight averages 1 oz.   Temperature, 107 to 109°F.   Male needs no description.   Female easily confused with other sparrows; chiefly olive-brown with chestnut-streaked back.   Bill of adult male, black, of young male and of female brown.   Also known as house sparrow.

Ranges and breeds throughout Europe and British Isles, except in Italy and on east to Siberia.   Introduced into America at Brooklyn in 1850, and later at Quebec and Halifax.   Now firmly established wherever man lives in numbers.   Allied races in southern Asia, Asia Minor, and Africa.   Defended home territory, several feet.

Nest a clumsy bulky mass of grass, trash, and feathers, tucked into a birdhouse, under a roof, in a woodpecker hole, or even in a tree.   Eggs four to nine, grayish spotted with reddish-brown, or dark brown and gray, ⁹⁄₁₀ by ⅝ in.   Incubation by female, 11 to 12 days.   Annual broods, three or more.   Nests April to September.

Food, garbage, formerly large percentage horse manure, insects, grain, fruits, young garden plants and almost anything possibly edible.   Does eat some harmful insects such as Japanese beetles and grasshoppers.   Quarrelsome, driving more useful species away, particularly from desirable nesting sites.   Act in unison to eliminate rival individuals or species.   Known to carry some diseases.   Extremely hardy under all circumstances.

A harmful bird if any species is, yet one of commonest and least bothered.   Men try to destroy useful hawks and owls but leave more destructive English sparrows unharmed.   Flight speed, 17 mph.   Hearing range, 675 to 11,500 cps.

English Sparrow

Bobolink, 494
Dolichonyx oryzivorus
Length to 8 in., wingspread to 12½ in., tail to 3⅙ in., bill to ⅝ in.   Weight to 1⅘ oz.   Female smaller than male.   Male in spring black below and with conspicuous white markings above.   Female and autumn birds like large yellowish-brown sparrows, with darker stripings on upper parts.

Breeds from southeastern British Columbia to Cape Breton Island, south to northeastern California, West Virginia, Pennsylvania, and New Jersey.   Winters in South America to southern Brazil, Bolivia, Peru, northern Argentina, and Paraguay.   Migrates through West Indies and along east coast of Central America.

Nests in meadows and fields.   Nest on ground, well-built, concealed by grass tuft, made of grasses, lined with finer materials.   Eggs four to seven, highly variable, grayish, bluish, or brownish, spotted or blotched with brownish-red or lavender, ¹⁵⁄₁₆ by ⅛ in.; weight ¹⁄₁₀ oz.   Incubation 13 days, by female.   One annual brood.

Food in breeding ground, May to June, from 70 percent to over 90 percent insects; 65 percent in August, 90 percent in September.   When in South, in Georgia and Alabama, chief food is rice.   Grain taken in North

Bobolink

is negligible. Eats enormous quantities of weed seeds on its migration. Its sparkling song, when in breeding territory, is delightful.

No doubt that in North bobolink is an extremely valuable bird but in rice fields it is otherwise. In 1912, over 700,000 bobolinks were killed and sold for markets from South Carolina alone. They are not protected by law but their numbers were seriously reduced by marketing activities. Hunters received 2 cents apiece. Song pitch to 6,950 cps.

Eastern Meadow Lark, 501
Sturnella magna
Length to 11 in., wingspread to 17 in., tail to 3½ in., bill to 1½ in. Weight to 5 oz. Temperature varies around 107°F. Female much smaller than male. Chunky in appearance with rather stubby tail, with white outer tail feathers, with a black bib and yellow breast and throat. General color light brown, flecked with dark.

Four subspecies: Eastern, Southern (smaller), Rio Grande, and South-western meadow larks. Eastern breeds from Minnesota to New Brunswick, south to northern Texas and North Carolina. Winters from Ohio Valley and Maine, south to Gulf of Mexico. Western species *S. neglecta* extends to Pacific Coast. From California Pleistocene.

Meadow Lark

Nests in open fields. Nest of grasses, on ground, usually under a roof of growing grasses, well lined with finer material; sometimes with one to two tunnel entrances. Eggs three to seven, whitish spotted with brown, purple, and lavender, weight ⅕ oz; 1⅕ by 9/10 in. Incubation 13 to 14 days, by female alone. Usually two annual broods. Defended home territory, several acres.

Food in summer, 99 percent insects, most of them injurious to crops among which birds live. In fall, food is largely weed seeds. Insects constitute 39 percent of food in December, 24 percent in January, 73 percent in March. Particularly destructive to cutworms. Call of eastern meadow lark, a series of clear whistles; of western, a loud bubbling series. Speed, 40 mph.

May be considered a nuisance by pulling young sprouting corn in some southern states. Worth more than money for its cheerful song. Song pitch, (eastern) 3,150 to 6,025 cps, (western) 1,475 to 3,475 cps. State bird of Kansas, Montana, Nebraska, North Dakota, South Dakota, Oregon, and Wyoming.

Yellow-headed Blackbird, 497
Xanthocephalus xanthocephalus
Length to 11 in., wing to 5¾ in., tail to 4½ in. Female slightly smaller than male. Male in summer with head, neck, and pointed throat area yellow or orange, with rest black except for white wing patches. Looks like a yellow-headed, yellow-breasted oriole. In winter, yellow is obscured by dusky. Female dark brown, without wing patch, with dull yellowish throat and forebreast.

Breeds from southern British Columbia to Wisconsin, Indiana, Texas, and southern California. Winters from southern California and Texas south into Mexico. Northern migration ends in late April. Accidentally in Greenland, Ontario, Cuba, Florida, and South Carolina.

Yellow-headed Blackbird

Male selects site, displays before female, and drives out other males. Nest hung in cattails or tules, usually above water, like a thick-walled basket of plant materials. Eggs usually four, gray or greenish-white speckled with brown. Young helpless when hatched, but out of nest though still helped by parents in mid-July. Young brown-headed.

Food 33 percent animal matter and 66 percent vegetal. Grasshoppers important among insects eaten. Does not attack garden produce. Congregates in flocks in marshes, with redwings but usually in separate area. May form huge flocks when families unite in late summer.

Because of fact that 33 percent of total food eaten may be waste grain, conceivable that bird might be considered as undesirable in spite of good done in destroying insects. Always appealing to naturalist. Song pitch median, 2,000 cps.

Red-winged Blackbird, 498
Agelaius phoeniceus
Length to 9½ in., wingspread to 14½ in., tail 3 7/10 in. Weight to 3 oz. Female smaller than male. Breeding male black with scarlet "shoulders" bordered with white or buff, this becoming rusty in winter. Female with head, back, and underparts brownish-black with light streaks. Young males rusty, with orange in place of red.

14 subspecies include Eastern, Florida, Gulf Coast, Northwestern, Nevada, Rio Grande, and Sonora redwings. Eastern breeds from Ontario to Nova Scotia south to Gulf States, wintering south of Ohio and Delaware valleys. Other subspecies extend range over the rest of Canada, United States and Mexico. From Florida Pleistocene.

Red-winged Blackbird

(Continued)

Baltimore Oriole

Male selects site and defends it against other males. Female accepts site and male. Nest built in marshland by weaving plant materials to form basket above water level; lining, soft. Eggs three to five, pale blue streaked with black or purple, 1 by $\frac{7}{10}$ in. Incubation 11 to 12 days, by female. One or two annual broods. Temperature, 100.8 to 104.6°F.

Food in breeding season largely insects. Seeds of weeds and farm crops make up most of the diet. In fall, may join bobolinks in raiding rice fields. Fly in great flocks in "ranks." Among first birds to come north in spring. Speed, 28 mph.

While it may be pest at times in certain areas, good done generally far outweighs bad. Enjoys certain protection from the law. While it lives much of time in marshes, it ranges for food over adjacent territory. *Konkeree* call one of best known signs of spring. Song pitch, 1,450 to 4,375 cps.

Baltimore Oriole, 507
Icterus galbula
Length to 8⅛ in., wingspread to 12½ in., tail to 3½ in. Weight of female 1½ oz. Female smaller than male. Male black and brilliant orange. Female and young dull yellow beneath, olive-yellow above, with two light wing bars; less green than female orchard oriole. Orchard oriole male chestnut where Baltimore oriole male is orange.

Breeds from central Alberta to Nova Scotia, south through eastern Montana, Wyoming, and Colorado to southern Texas and east through Louisiana, northern Alabama, and northern Georgia. Winters in southern Mexico, on through Central America to Colombia. Sometimes found in Cuba and north to Hudson Bay.

Nests in tops of fine-twigged trees like elms, hanging nest near branch tips. Nest beautifully woven bag of plant materials, string, and hair; closes with weight. Eggs four to six, grayish-white with streaks, blotches, and dots of brown, black, and lavender; 1 by ⅔ in. Incubation 14 days, by female. One annual brood. Temperature, variable around 107°F.

Food essentially insects gleaned from treetops. Animal matter 83 percent, 34 percent of total being caterpillars. Caterpillars include those of gypsy moth, cankerworm, bagworms, and brown-tail moth. Plant food includes grapes; in migration, orioles may hurt market value of bunches of grapes by puncturing them. Speed, 26 mph.

On the whole, a splendid protector of shade trees even if it does injure some grapes. A delight to eye and to ear. Its nests are marvels of workmanship. Not surprising that, with Lord Baltimore's colors black and orange, this bird should bear his name or that it should be chosen as state bird of Maryland. Song pitch, 2,050 to 3,825 cps.

Bullock's Oriole

Bullock's Oriole, 508
Icterus bullockii
Length to 8⅗ in., wing to 4 in., tail to 3⁷⁄₁₀ in. Adult male, in summer, rear upper parts, and under parts orange or yellow; crown, throat patch, eye line, fore upper parts, and bill black; wings, black with white patch and white edgings. Female with head and neck olive, back and rump olive-gray, and throat yellow. Western counterpart of Baltimore oriole.

Breeds from southern British Columbia, southern Saskatchewan, and eastern South Dakota south to northern Lower California and southern Texas. Winters in Mexico south to Colima, Guerero, and Puebla.

Nests in cottonwoods, oaks, mesquites, and box elders or among mistletoe. Nest a hanging bag composed of string, hair, cotton, grasses, and shredded bark, with an inner lining of hair, wool, or down. Eggs three to six, grayish- or bluish-white marked with irregular hair lines chiefly around larger end.

Food largely insects of orchards, one estimate indicating that 35 percent of total food is beetles, nearly all of which are harmful species; caterpillars and their moths 41 percent, and ants and wasps 15 percent. Remaining food is largely fruit such as cherries. Young birds are ready to fly by July.

This oriole is almost entirely useful and probably more than pays for fruit it eats by fruit-destroying insects it kills. Its song is one of those which in season dominates other songs within its range.

Rusty Blackbird

Rusty Blackbird, 509
Euphagus carolinus
Length to 9¾ in., wingspread to 15 in., tail to 4 in. Weight to 2⅗ oz. A blackbird, about size of a robin. Male with a whitish eye; female slate-colored instead of black and smaller than male. In fall, male a rusty color; female and young with broad light stripe over eye and much browner than male.

Breeds from Alaska to northern Quebec, south to central British Columbia, central Alberta, central Manitoba, central Ontario, New York, northern Vermont, New Brunswick, Newfoundland, and Nova Scotia. Winters south of Ohio and Delaware valleys to Gulf Coast. Accidental in California and Greenland.

Nests in swamp or near waterway borders, in a tree, on a stump, or in brush from 1 to 10 ft above ground or water. Nest of sticks and mosses, capped with wet leaf mold which hardens. Eggs four to five, bluish-green blotched with chocolate or gray, 1 1/16 by 4/5 in. Incubation for 14 days, by female. One annual brood.

Food, aquatic beetles and their larvae; grasshoppers and caterpillars; other aquatic insects, ants, some bugs and flies, spiders, centipedes, crustaceans, snails, and salamanders. Plant foods largely grains and weed seeds. Rarely so numerous as most of related blackbirds and never figure seriously in economic issues.

From evidence available bird does much good and little if any harm. It lives most of its life in areas not normally considered agriculturally valuable and eats little if any commercially valuable farm products.

Brewer's Blackbird, 510
Euphagus cyanocephalus
Length to 10 in., wing about 5 1/5 in. Female slightly smaller. Male, a glossy black with head reflections appearing steel-blue, violet, greenish, bronze. Bill and feet black, eye white. Immature male like adult but with feathers of foreparts margined with grayish-brown. Adult female with head, neck, upper back, and chest grayish-brown; light drab throat shading toward the tail into mingled drab and black; with subdued green and violet metallic reflection. Robin size; a bird with pure black coloration and metallic reflections.

May be confused with cowbird, *Molothrus ater.* No subspecies. Range western North America. Breeds from northwestern Minnesota and western Kansas west to the Pacific, and from central British Columbia and the Saskatchewan region south to northern Lower California and western Texas. Winters from Kansas and southern British Columbia south to Guatemala. May appear east and beyond to the Mississippi in migration.

Nest in colonies at lower levels in bush or tree; frequently in colonies in trees such as live oak, white oak, cottonwood. Prepared with interlaced twigs and grasses; held together with mud, coiled rootlets sometimes lined with soft plant parts and hair. May be to 30 pairs nesting in a colony. Eggs four to seven, but usually five or six. Ground color light gray or greenish-gray, spotted and/or blotched with grayish-brown or olive-brown. Eggs 1 by 3/4 in. Incubation by female alone; incubation period 12 to 14 days. One or two annual broods.

The grackle of the West. Common in moist meadows and near marshes. Some may live in towns and villages. Also frequently found around corrals and yards of every large ranch. Their pattern of behavior is to roost together in a thick clump of trees at night. Male displays by stretching head and neck upward and utters a wheezy sound.

Food: animal food made up largely of insects; in particular beetles and caterpillars; feeds on the corn earworm, alfalfa weevil, boll weevil, and other crop destroying insects. In aquatic situations takes insect larvae and snails. Seeds from farm crops are main plant food, but does take weed seeds.

Brewer's Blackbird

Common Grackle, 511b
Quiscalus quiscalus
Length to 14 in., wingspread to 19 in., tail to 6 1/2 in. Weight, male to 5 oz; female to 4 oz. Large blackbird, with long tail with rounded tip. Black reflects bright bronze color in strong light. Temperature, variable around 111.4°F.

Breeds from Gulf of St. Lawrence and northern Manitoba south throughout Mississippi Valley and the Plains States south to Louisiana and Texas. Winters in Southern States and north to Ohio and New York.

Nests commonly in groves of large trees or in parks. Nest sometimes on ground, in a tree cavity, in a building, in a stump, of grass and weeds, reinforced with mud and with fine lining. Eggs four to seven, pale green to light brown, with brown or lavender spots or markings, 1 1/3 by 9/10 in. Incubation by female, 14 days. One annual brood.

Food listed as 30 percent animal and 70 percent vegetal. Animals include worms, crawfish, carrion, mice, birds' eggs, snakes, sow bugs, clams, frogs, and hosts of insects. In autumn, food is more largely plant materials such as grain, nuts, and fruits. Migrate from roost to feeding ground in long files from horizon to horizon. Speed, 30 mph.

Probably more injurious than beneficial particularly in large numbers in fall months when grain may enter diet prominently. However, good done in vicinity of nesting area should not be overlooked. Common grackle from Florida Pleistocene.

Common Grackle

Boat–tailed
Grackle

Boat-tailed Grackle, 513
Cassidix mexicanus
Length, male to 17 in., female to 13 in. The largest member of the black-bird family, Icteridae, in America. Black over entire body; with a purple gloss on the head, back, tail, and rump. Remainder of black shows greenish or bluish gloss. A feature to observe is the long tail which is almost as long as the body. Head and bill about same length; both black. Eye of male yellow, female brown. Female's colors show blackish-brown above and smoky gray-brown beneath; darker toward the tail.

Six subspecies recognized. Breeds from southern Arizona, central New Mexico, parts of Texas, the Gulf Coast, northern Florida, and the Atlantic Coast from southern New Jersey south to Peru, Venezuela, and southern Florida. Resident mainly; some limited movement with seasons. Winters to the south along Gulf Coast and south of Virginia.

Nest in bushes or trees, from 4 to 40 ft, cup-shaped; made of twigs, sticks, and other plant parts; plastered internally usually with mud but sometimes with dung; lined with grass and tiny roots. Nests in colonies near water, for example in swamps. Eggs two to four, light blue pale or bluish-gray; marked or spotted with different shades of brown. Incubation and care of young by female only.

A bird of open areas, coasts, river groves, parks, farmlands. Look for a large black bird with a creased tail. Call a harsh *tlick, tlick, tlick;* also produces a variety of harsh whistles and clucks. Also called great-tailed grackle, great-tail, or "jackdaw." A walking bird which is seen frequently leaping into the air in pursuit of an insect.

Food: insects, particularly beetles and grasshoppers harmful to man's crops; crustaceans such as crayfish, crabs, shrimp; sometimes lizards, toads, frogs, and small mammals. Takes lesser amount of plant food including fruits, grains, and some weed seeds.

Brown–headed
Cowbird

Brown-headed Cowbird, 495
Molothrus ater
Length to 8¼ in., wingspread to 13¾ in., tail to 3⅓ in. Male black with brown head. Female uniformly gray. Smaller than most associated blackbirds. Short stubby bill and short tail make female different from somewhat similarly colored catbird. Weight to 1⅔ oz. Female the smaller. Temperature, 101.6 to 108.4°F.

Four subspecies: Eastern, Nevada, California, and Dwarf cowbirds. Eastern breeds from southern Ontario to Nova Scotia, to Virginia, Tennessee, Louisiana, and Texas, west to eastern Kansas and western Minnesota. Winters south of Ohio and Potomac valleys, on to Gulf Coast and Florida.

Promiscuous in mating behavior. Female builds no nest and gives no care whatever to her young. Instead, she lays her eggs in nests of other, usually smaller birds, leaving them to be hatched and young to be reared by foster parents. Larger young cowbirds usually crowd out and starve normal offspring of host. Eggs, white blotched with chestnut and burnt umber.

Food, insects and grain. Insects: largely grasshoppers plus beetles, and caterpillars; also other arthropods. In addition to grain, cowbird feeds on weed seeds. Name cowbird comes from bird's habit of collecting insects from backs and vicinity of cattle. Of 544 stomachs, contents showed 22 percent animal matter, including grasshoppers, ants, wasps, caterpillars, and particularly army worms. Grain apparently a relatively small percent of year's food.

On basis of food habit, probably cowbird is a useful species. But it probably is universally despised for its habit of letting other birds rear its young to the death of normal young. Its general home life is just about all it should not be according to ethical standards.

Scarlet Tanager

Order Passeriformes./Family Thraupidae [Tanagers]

Scarlet Tanager, 608
Piranga olivacea
Length to 7½ in., wingspread to 12 in., tail to 3¼ in. Female smaller than male. Male bright red with black wings and tail in breeding season. Male loses most of red in autumn. Female dull green above and yellowish below. Young like female.

Breeds from southern Saskatchewan to Nova Scotia and south to Arkansas, northern Alabama, northern Georgia, and in mountains of Virginia and South Carolina. Winters from Colombia to Bolivia and Peru, migrating through Cuba, Jamaica, and Yucatan and along east coast of Central America. Treetop bird.

Nests in high open woods such as oaks or pines. Nest on horizontal limb, to 50 ft above ground, composed of flat platform of rootlets with interwoven grasses. Eggs three to five, pale greenish-blue with many brown spots, 1 by ⅔ in. Incubation 13 days, by female. One annual brood. Young with first breeding plumage in 1 year.

Food: insects on various species of oak including lepidopterans and beetles. One of its recent food items is the gypsy moth and leaf rollers in the eastern states. It searches for wood-boring beetles and grasshoppers and true bugs. It also feeds on fruits of wild plants such as bayberry, sumac, elderberry, and blueberry.

Useful in every imaginable respect and beautiful to see if not to hear. Its characteristic *chip-churr* is always worth investigating. Its song, like that of a robin with a cold, is somewhat of a disappointment which is speedily forgotten when bird is seen. Song pitch, 2,200 to 3,625 cps.

Summer Tanager, 610
Piranga rubra
Length to 7⅕ in. Wingspread to 12⅛ in. Female smaller than male. Male bright rosy red all over and without black wings and tail of scarlet tanager. Female olive above and yellow beneath and without wing bars of female Baltimore oriole, which it slightly resembles.

Breeds from southeastern Nebraska to central Indiana, central Ohio, Maryland, south to southern Florida and northeastern Mexico. Winters from central Mexico and Yucatan to Ecuador, Peru, and Guiana. Occasionally north to Maine, Nova Scotia, Quebec, and New Brunswick. Cooper's tanager, a subspecies, extends range to California.

Nests in dry open woodland or in towns. Nest in tree, usually on a horizontal limb 5 to 60 ft above ground, of grass and leaves in a shallow cup. Eggs three to four, ¹⁵⁄₁₆ by ¾ in.; blue-green, spotted, blotched, or speckled with purple or brown. Incubation 12 days, by female. One annual brood.

Lives mostly in treetops where it is not seen easily. Catches insects much like a flycatcher or may glean them from tree's vegetation. Known to eat largely beetles, also hymenopterans such as ants, wasps, bees, and other insects; takes little in the way of plant material unless provided at a feeding station. Young males may assume adult plumage first winter or may not.

Probably of little economic importance but is interesting to bird lovers. Call a loud, *chic-tucky-tuck* song more musical than that of the scarlet tanager and of different quality.

Summer Tanager ↕

Western Tanager, 607
Piranga ludoviciana
Length to 7½ in. Male with head and throat scarlet-orange; back, wing, and tail black. Wings with narrow white wing bar and larger yellow wing bar. Remainder of male plumage bright yellow. Female more drab; medium olive-green above, yellow beneath; wings black with upper wider yellow wing bar and lower narrow white wing bar. Male has less brilliant colors during fall and winter.

No subspecies. Range: western North America from the eastern foothills of the Rockies to the Pacific Coast, north to British Columbia; Athabasca, Idaho, Montana, and South Dakota. Winters over much of Mexico to Guatemala. Infrequently found eastward during migration to more northern Atlantic states.

Nest frequently on a horizontal branch of a pine or fir tree, a flat, concave structure, sometimes at a moderate height; 6 to 50 ft, at times in a deciduous shrub or tree. Not carefully made; external covering of twigs and stems of weeds or long pine needles; lined with small roots; sometimes with grasses and animal hair. Eggs usually four but may be as many as 8; ⁹⁄₁₀ by ⅔ in., light bluish-green spotted lightly with dark grayish-olive; at times may be almost entirely brown with a few spots.

A bird common on the mountainsides. Usually sings from tops of tall trees. Strictly a forest bird. Can be expected in breeding areas in late March to early June; leaves to winter range as early as July and as late as October. Song heard by a resident of the eastern states is generally described as robinlike; by a resident of the West it may be confused with the song of a robin or the black-headed grosbeak. Call note, *bi tic* or *bit-a-tic* or *prit-it*. Food: a forager for insects such as wasps, bees, ants, beetles, and bugs plus numerous other forms. Takes cultivated cherries but not in quantity enough to be an orchard pest. Also eats fruits of peaches, apricots, raspberry, mulberry, elderberry, blackberries, and serviceberry. Because of its unique color pattern it cannot be confused with any other N. American bird. Since it is a forest resident, its occasional visits to commercial orchards are infrequent.

Western Tanager

Order Passeriformes./Family Fringillidae [Buntings, Finches, and Sparrows]

Cardinal, 593
Richmondena cardinalis
Length to 9¼ in., wingspread to 12 in. tail to 4¾ in. Weight, male 2 oz; female, smaller. Male scarlet except for black around bill, with a conspicuous crest. Female yellowish-brown with some red, but with big crest and heavy red bill like that of male.

Cardinal

(Continued)

Seven subspecies of which six have common names: Eastern, Florida, Louisiana, Gray-tailed, Arizona, and San Lucas. Eastern is resident from southeastern South Dakota to southern Ontario, and south to Gulf States. Occasionally also in New Brunswick, Massachusetts, Connecticut, Michigan, Colorado, and other states. Subspecies extend range.

Nests in thickets or small trees. Nest of twigs, leaves, and grasses, rather loosely put together, with a lining of finer grasses or hair, 3 to 30 ft above ground. Eggs three to four variable white, greenish, or bluish, marked with brown or lilac, weight, 8⅗ g. Incubation for 12 days, by female. One to three annual broods.

Food: a species to be protected because it takes caterpillars, grasshoppers, hemipterans, beetles; plant food made up of wild fruits, weed seeds, and cultivated grains. It prefers wild food to cultivated grains. Call, a loud cheery whistle. Song pitch, 2,200 to 4,375 cps.

An interesting bird to attract to a winter feeding station with such seeds as sunflower seeds. Formerly sold as cage birds at $10 per bird but this is now outlawed and bird is protected by law as it should be. Range has extended northward in recent years. Family life of cardinals is worth watching. State bird of Kentucky, Delaware, Illinois, and Indiana.

Rose-breasted Grosbeak, 595
Pheucticus ludovicianus
Length to 8½ in., wingspread to 13 in., tail to 3½ in. Female smaller than male. Male black and white, with triangular red breast patch. In flight, appears conspicuously black and white. Female streaked brown, with broad white wing bars and white line over eye. Bills of both sexes conspicuously heavy and light-colored.

Breeds from central Mackenzie to Cape Breton Island, south through central Kansas, southern Missouri, and in mountains to northern Georgia. Winters from southern Mexico and Yucatan to Colombia, Venezuela, and Ecuador. Occasionally in Cuba, Jamaica, Haiti, Bahamas, California, Colorado, and Arizona. *P. melanocephalus*, black-headed grosbeak, is relative in the Western States.

Nests in thickets and woods, usually near water but also on farms and in gardens. Nest 6 to 20 ft above ground, near tree trunk or on forked limb, of twigs and grass, loosely assembled, with little lining. Eggs three to five, blue-green or grayish, with brown or reddish spots, 1¹¹⁄₁₂ by ⅘ in. Incubation 14 days, by both sexes. One annual brood.

Food about 52 percent animal matter, favoring particularly Colorado potato beetle, and giving it name potato-bug bird. Eats buds of forest trees but probably not to dangerous extent. Rarely abundant anywhere. *Chink* call, robinlike song, thick bill, wing bars, and eye line identify it in winter plumage, when it resembles female. Temperature, 100.4 to 110°F.

Undoubtedly useful and certainly welcome because of its beauty and its call.

Black-headed Grosbeak, 596
Pheucticus melanocephalus
Length of adult males to 8 in.; wings to 4 in. Females somewhat smaller. Adult males easily recognized by black head and cinnamon-brown breast; back of head, chestnut or creamy-buff. In flight, male shows much white in wings and tail, rump shows as cinnamon-brown. Female not so distinctively colored but known by her heavy light-colored bill, buff-white collar, and white wing bar.

Two subspecies. *P. m. melanocephalus* breeds from British Columbia, northwestern Montana, southeastern Alberta, southwestern Saskatchewan, northwestern North Dakota, south through central Washington and eastern Oregon to extreme eastern California. Winters in Mexico and east to Louisiana. *P. m. maculatus* breeds from southwestern British Columbia south along the Pacific Coast to northern Baja California, east to San Bernardino Mountains. Winters from southern Baja California, southern Sonora into Mexico.

Nest, frail, bulky, made of interlaced twigs, weed stems, and a mass of leaf-bearing twigs. Prepared as a rather careful hollow inside, with interwoven rootlets or similar materials. Usually on or near a stream bank in large bushes, saplings, or small trees at moderate height, 5 to 20 ft. Eggs three to four, 1 by ⁷⁄₁₀ in., bluish-green, spotted some or much with brownish-olive or buff. Both sexes incubate. Males arrive in north of range first, sing in treetops usually in dense foliage, tall firs at crest of hills. Females arrive later. Song referred to as like that of a glorified robin. Call notes, a sharp *beek, beek.*

Food, a great variety of insects and other invertebrates. Plant food: fruits, seeds of figs, elderberries, blackberries, cherries, oats, raspberries, and many weeds. May become a pest in orchards. However, it does eat some harmful insects such as codling moths, cankerworms, and black olive scales. Will visit home feeding stations.

Rose–breasted Grosbeak

Black–headed Grosbeak

Indigo Bunting, 598
Passerina cyanea
Length to 5¾ in., wingspread to 8⁹⁄₁₀ in., tail to 2⅖ in. Female smaller than male. Male deep blue all over. In fall, male is more brown but blue shows on wings and tail. Female plain brown, without stripings, wing bars, or other conspicuous markings. Young male may show some blue with brown, but not at first.

Breeds from central eastern North Dakota to southern New Brunswick south to central Texas and northern Florida. Winters from Yucatan through Central America to Panama and in Cuba. Also found in Bahamas, eastern Colorado, southern Saskatchewan, and southern Manitoba.

Nests in bushy country or cutover hardwood lands, usually not over 5 ft above ground. Nest of twigs and coarse grasses, leaves, hair, and feathers. Eggs three to four, pale blue, or greenish-white, rarely specked with brown, ⅘ by ⅗ in. Incubation by female, 12 days. Two annual broods.

Food chiefly insects, including cankerworms, brown-tail caterpillars, beetles, including click beetles, and grasshoppers. May eat a few blossoms of fruit trees but does no serious damage. Eats little if any grain. Weed seeds, some cultivated grains, and fleshy fruits like blackberry and elderberry are eaten. In South in winter, food is largely weed seeds. Song pitch, 3,250 to 8,875 cps. Speed, 21 mph. Temperature variable, around 107°F.

Song made up of notes in twos, at different pitches. Frequently seen on roadside telephone wires along country roads. Such a bird may appear black in certain light conditions.

Indigo Bunting

Lazuli Bunting, 599
Passerina amoena
Length to 6 in., wing to 3 in., tail to 2⅓ in. Female slightly smaller. Adult male with upper parts turquoise blue, varying to greenish-blue; two white wing bars, breast and sides brownish but other under parts white. Female brown, sometimes streaked; rump greenish-blue, chest brownish, and belly white.

Breeds from southern British Columbia to southeastern Saskatchewan and northwestern North Dakota, south to northwestern Lower California and west central Texas. Winters in Mexico south to Valley of Mexico. Accidentally in Minnesota and southern Mackenzie.

Nests usually near water in low bushes, among weeds, in willows, manzanitas, and rosebushes. Nest made of plant fibers including bark, and often lined with hair or other fine materials. Eggs three to four, plain bluish-white or pale greenish-blue, sometimes spotted.

Food mostly weed seeds and noxious insects, latter including alfalfa weevils and codling moths. In habits, bird reminds one of indigo bunting found to east of lazuli's range. Blue back of male, flashing from top of some weed or from edge of a thicket, tempts one to seek nest.

No wonder that within its range lazuli bunting is popular with those who love birds. Surprising that this popularity has not found expression in its being considered important enough to be a state bird, since it has more beauty than many birds so elected.

Lazuli Bunting

Dickcissel, 604
Spiza americana
Length to 7 in., wingspread to 11 in., tail to 2⁹⁄₁₀ in. Female smaller than male. Male like diminutive meadow lark showing usual black bib and yellow breast. Female paler than female English sparrow, with yellowish breast, chestnut on bend of wing, and white stripe over eye.

Breeds from northeastern Wyoming to southeastern Ontario, south to southern Texas and northern Georgia. Formerly bred through to Atlantic plain; now rare east of Alleghenies. Winters from Guatemala to Colombia, Venezuela, and Trinidad, migrating through Central America and Mexico.

Nests in fields and meadows on or near ground, and perches on nearby trees, posts, or wires to sing. Nest on ground among grasses or in low tree or shrub, made of grasses. Eggs three to five greenish-blue, ⅘ by ⅗ in. Incubation by female, 10 to 11 days. Probably two annual broods.

Food mostly insects, particularly katydids, crickets, beetles, grasshoppers, ants, caterpillars, flies, and hemipterans, weed seeds, and a little grain. Estimated that one family of dickcissels eats 200 grasshoppers a day and each grasshopper eats 1½ times its own weight in plant material. Song *dick-dick-dick-cissel* is repeated again and again by male during nesting season.

This bird signifies the Middle West to many persons. It and red-headed woodpecker populate Iowa's telephone poles clear across state. Changes in regional distributions over time difficult to explain.

Dickcissel

Evening Grosbeak, Yellow Grosbeak, 514

Hesperiphona vespertina

Evening Grosbeak

Length to $8\frac{1}{2}$ in., wingspread $13\frac{4}{5}$ in., tail $3\frac{1}{5}$ in. Sexes about equal in size, though female may be smaller. Like a large yellowish starling, with a short heavy whitish bill. Undulating flight. Male yellow, with black and white wings. Female grayer, with yellow and black obscured.

Three subspecies: Eastern, Western, and Mexican. Eastern breeds from western Alberta to Vermont, south to northern Minnesota, Michigan, and New York. Winters from Saskatchewan to Quebec and south to Missouri and Maryland. Other subspecies extend range to Pacific Coast and to southern Mexico.

Nests in conifers, to 50 ft.; constructed with grass, twigs, bark, and rootlets. Eggs 3 to 4, blue to green with brown blotches, 1 in. by $\frac{7}{10}$ in. Female incubates. Western subspecies nests in willows; of sticks and rootlets, some 10 to 12 ft above ground; Eastern, in conifers 20 to 50 ft up. Eggs three to four greenish, blotched with pale brown. Eggs in Arizona, June; young on wing in Mexico, late April.

Food essentially seeds, favoring box elder but eating spruce, pine, cottonwood, locust, birch, wild cherry, maple, ash, tulip, lilac, flowering dogwood, apples, hawthorn, sumac, privet, buckthorn, mountain ash, poison ivy, Virginia creeper, red cedar, juniper, barberry, ragweed, burdock and seeds from horse manure.

Appear erratically in good-sized flocks and stay for a few days only to vanish as quickly as they came; may return daily until mid-May to feeding stations for sunflower seeds. Do little damage, except rarely to tree buds. Too few in numbers to be considered a pest.

Purple Finch, 517

Carpodacus purpureus

Purple Finch

Length to $6\frac{1}{3}$ in., wingspread to $10\frac{2}{5}$ in., tail to $2\frac{1}{2}$ in. Weight to $1\frac{1}{8}$ oz. Female smaller than male. Male rosy-red instead of purple as name implies. Color highest on head and rump. Smaller than similarly colored pine grosbeak. Female like heavy-billed striped brown sparrow, with whitish line over eye.

Three subspecies; two best known are Eastern and California purple finches. Eastern breeds from northwestern British Columbia to northern Ontario and Newfoundland, south to central Minnesota and Maryland. Winters through United States east of Minnesota and Texas. California subspecies ranges from British Columbia to Lower California.

Lives among hills. Nests in hedges, orchard trees, or more commonly in conifers 5 to 60 ft above ground. Nest of vegetable matter usually lined neatly with hair, built by female. Eggs four to five, greenish-blue with blackish spots near larger end, $\frac{11}{12}$ by $\frac{7}{10}$ in. Incubation 13 days, chiefly or solely by female. One annual brood.

Food in spring essentially insects, buds, and blossoms; in summer, insects and wild fruit, though this may include some raspberries and a few cherries; in fall and winter, wild fruits and weed seeds. Favors fruits of viburnums, dogwoods, and elders. Practically 100 percent of eastern species' food is made up of buds, fruits, and seeds.

Essentially useful as insect and weedseed destroyer. Male's song a surprise to most people and easily confused with warbling vireo. Most people have to learn it anew each spring but fortunately it is heard before vireos return and renewal of acquaintance is effected simply and enjoyably.

House Finch, 519

Carpodacus mexicanus

House Finch

Length to $5\frac{1}{2}$ in., wing to $3\frac{1}{3}$ in., tail to $2\frac{3}{5}$ in. Weight $\frac{2}{3}$ oz. Female about $\frac{1}{2}$ in. shorter, with tail and wing about even in length; streaked gray and brown all over. Adult male with rose-pink throat and rump, line over eye, back not sharply streaked, belly whitish, wings and tail brown-streaked.

Three subspecies: Common, San Luis, and San Clemente. Common house finch breeds from Oregon and Idaho through northern Wyoming and south through California, New Mexico, Lower California, and Mexico to northern Chihuahua. May extend range east to western Kansas and western Texas. Introduced on Long Island and now established in eastern New York and Connecticut.

Nests in bird boxes, sagebrush, saltbush, mountain mahogany, cactus, tree cavities, buildings, or on tree branches. Almost always near water. Nest a shallow cup of grasses, hair, string, and wool, or almost any kind of available fibers. Eggs three to six, bluish- or greenish-white, with spots, dots, and blotches of black or brown near larger end, weight, $\frac{1}{12}$ oz. Temperature, 106 to 108°F.

Food 97 percent vegetal, with less than 1 percent being grain. Common seeds eaten include Russian thistle, dandelion, sunflower, mistletoe and, unfortunately, some cherries, mulberries, and serviceberries. If serviceberries and mulberries are planted near cherry orchards, they may draw house finches from cherries. Song, an ecstatic warble.

Essentially useful as weed-seed destroyer, but also as a beautiful bird with a splendid song living in a place not always attractively populated. Some persons consider its presence a good index of nearby water. Also known by the common name linnet. In large numbers may become a nuisance.

Pine Grosbeak, 515
Pinicola enucleator

Pine Grosbeak

Length to 9¾ in. Wingspread to 18⅞ in. Female usually smaller than male. Two white wing bars and rosy-red color characterize male. Similar wing bars, with gray general color and rump and head slightly yellow, characterize female. In immature males red is replaced by dull orange.

Seven subspecies recognized: Alaska, Kamchatka, Canadian, Kodiak, Queen Charlotte, Rocky Mountain, and California. Canadian breeds from northwest Mackenzie to Labrador and south to Manitoba and northern New Hampshire. Winters south to Kansas, Kentucky, and New Jersey with subspecies extending range as indicated by names. In southern part of range seen in much greater numbers when food to the north is in short supply.

Nest in evergreen tree, relatively low. Nest of moss and twigs, with hair lining. Eggs three to four, 1 by ⁷⁄₁₀ in.; greenish-slate to light green spotted and blotched with pale purple-brown, dark purple, or even dark brown. Nesting May to June. Incubation probably 14 days, mostly or wholly by female. One annual brood.

Seen in United States mostly in winter, but this is a dangerous generalization for some subspecies. Most are relatively unsuspicious and are confined chiefly to cone-bearing evergreens where they get their food, mostly from cones. Usually feed in flocks and will clean a tree well of its fruits once they have started.

May do some damage to fruit buds but since they feed chiefly on seeds of evergreens and these are usually superabundant, they do little serious damage to anything of economic importance to man. They are ordinarily so unusual in the southern reaches of their range that their presence is welcomed, particularly by ornithologists.

Common Redpoll

Common Redpoll, 528
Acanthis flammea

Length to 5½ in., wingspread to 8⅘ in., tail to 2⅔ in. Female smaller than male. Like small brown sparrows with bright red caps and black chins. General appearance finely streaked or spotted. Male pink-breasted. Flight something like that of goldfinch.

Three subspecies include Common, Holboell's, and Greater redpolls. Common redpoll breeds, from Alaska to Gulf of St. Lawrence, south to northern Alberta and northern Quebec; also into northern Europe and Asia. Winters in northern United States, south to northeastern California. Field species. Subarctic bird found in our northern states in winter.

Nests in thickets of birch, willow, or spruce in far north, usually low or in grass tussock. Nest of available plant material, with lining of down, feathers, or hair. Eggs 2 to 5, white tinged with green or blue, spotted with reddish-brown; ⅔ by ½ in. Nesting most common in June. Incubation about 11 days, by female alone. One to two annual broods.

Pine Siskin

Food largely seeds of weeds and of birches and alders; also seeds of pines, elms, lindens, larch, lilac, and others where they are available. In summer, plant lice, spiders, ants, and flies play an important part in diet. Friendly, come readily to feeding stations, and are generally useful.

Has bred in captivity freely though it is now against law to keep it captive. Made a good cage bird. Numbers are not sufficient in United States to expect them to be of economic importance but a flock of them in weeds in snow helps make a winter field trip successful.

Pine Siskin, 533
Spinus pinus

Length to 5¼ in., wingspread to 9¹⁄₁₀ in., tail to 2 in. Weight, male ⁷⁄₁₆ oz. Female smaller than male. Appears generally like a brown, heavily streaked, active sparrow with a flash of yellow in wings and on tail. Like goldfinch in winter, it is active in flocks in treetops.

Two subspecies include Northern and Mexican pine siskins. Northern breeds from central Alaska to central Quebec, south through highlands to central California and southern New Mexico, east through highlands from Nova Scotia to North Carolina. Winters over most of United States and northern Mexico.

Pine Siskin

Nests usually in coniferous forests. Nest usually like a saddle on a limb, 8 to 30 ft above ground, rather large, cup-shaped, of twigs, and bark with fine lining. Eggs three to six, pale greenish-blue with brownish spots, ¾ by ½ in. Incubation 12 to 14 days, chiefly by female. One to two annual broods.

(Continued)

American Goldfinch

Red Crossbill

↕ Rufous-sided Towhee

Food, seeds and fruits of woody plants such as birch, alder, maple, elm, and in winter red cedar, honeysuckle, lilac, and seeds of weeds. One of favorite foods is seeds dug from cones of arborvitae. In summer, food is largely insects, particularly plant lice as well as spiders, bugs, and fly larvae. Gives a high pitched *tit-a-tit*, as it feeds in flocks in treetops.

Probably harmless and of not much direct economic importance. Nevertheless, birds are welcomed by naturalists at all times because of their busy activities. Apt to move as a flock in one direction and then another, one bird seemingly initiating the movement.

American Goldfinch, 529
Spinus tristis
Length to 6 in., wingspread to 9 in., tail to ½ in. Weight to ½ oz. Male in breeding plumage, yellow with black on crown, wings, and tail. Female in summer dull olive instead of yellow, and with the black dulled. In winter, both birds are evenly colored grayish above and below, not streaked.

Four subspecies of which three include Eastern, Pale, and Willow goldfinches. Eastern breeds southern Manitoba to Newfoundland, south to eastern Colorado, central Arkansas, and northern Georgia. Winters through most of breeding area and south to Gulf Coast. Other subspecies extend range to Pacific and into Mexico.

Nests in various kinds of trees and shrubs. Nest rather bulky, cup-shaped, lined with soft downy plant materials like thistledown. Eggs four to six, bluish-white, unspotted, weight 2⅔ g. Nesting and egg laying late in season. Incubation 11 to 14 days, by female. One annual brood. Young birds fed soft seeds, by regurgitation.

Food almost exclusively seeds and dry plant fruits including dandelion, thistle, burdock, chicory, catnip, evening primrose, mullein, sunflower, asters, goldenrods, wild clematis, and many garden flowers as well as fruits of birch, alder, elm, hemlock, spruce, larch, and sycamore. Takes small amount of animal food, almost all destructive insect pests. Song pitch, 2,750 to 7,400 cps. Speed, 18 mph.

Only damage goldfinch might do would be to crop of a commercial grower of garden flower seeds or plant seeds such as lettuce and turnips. With its merry *perchicoree* call and undulating flight in a circle, is welcomed through year by most naturalists. State bird of Iowa, Minnesota, New Jersey, and Washington.

Red Crossbill, 521
Loxia curvirostra
Length to 6⅖ in., wingspread to 10¾ in., tail to 2⅙ in. Weight ¾ oz. Female smaller than male. Male brick red, being brighter on rump and with darker wings and tail. Female and young dull grayish-olive, with plain dark wings. Related white-winged crossbill has two white bands on each wing, in each sex.

At least eight subspecies of which five are the Red, the Newfoundland, the Sitka, the Bendire's, and the Mexican crossbills. Red crossbill breeds from central Alaska to central Quebec, south to Oregon and in highlands occasionally to northern Georgia. Winters farther south, to northern Texas and Florida. Subspecies extend range as indicated generally by their names.

Nests in coniferous forests. Nest usually low, hidden by thick foliage, built of twigs with rootlets and bark mixed in and a lining of moss, hair, or sometimes fur and feathers. Eggs four to five, pale greenish-blue with brown and lavender spots at larger end. Nest mid-January to July. Incubation 12 to 14 days, by female. Two annual broods.

Food essentially seeds of cone-bearing trees, particularly white pine, pitch pine, Norway and native spruces, balsam, larch, hemlock, and also maple, elm, beech, and such weeds as sunflower, dandelion, and evening primroses. Takes a small amount of animal food including spiders, caterpillars, fly larvae, beetles and their larvae, plant lice, and spittlebugs. May on occasion injure growing fruits such as peaches.

Has little economic importance, only rarely attacking commercial fruits and then only temporarily. Not common enough to be considered serious agriculturally but it is interesting because of its crossed bills and the way it uses them in extracting seeds from cones. Bird is not at all shy.

Rufous-sided Towhee, 587
Pipilo erythrophthalmus
Length to 8¾ in., wingspread to 12¼ in., tail to 4¹/₁₀ in. Female smaller than male, weight, 41 g. Temperature, 108°F. Male with head and upper parts black, sides reddish, and belly white, with abundant white showing near tip of long, loosely hung tail. Female and young like male but brown where he is black. Calls *chewink*. Also known as ground robin, joree, and low-ground-Stephen.

There are sixteen subspecies in the U.S., of which three are called red-eyed, Alabama, and White-eyed towhee. Rufous-sided towhee breeds from southeastern Saskatchewan to southern Ontario and Maine and south to Florida and Gulf Coast. Winters from southeastern Nebraska to Ohio Valley and south to central Texas and Florida.

Nests in brushlands or open woods. Nest on ground or close to it, rarely up 3 ft. Nest of leaves and grasses, with fine lining. Eggs four to six, white with reddish and sometimes lilac spots, 1 by ¾ in.; weight, 7½ g. Incubation chiefly by female though male assists, 12 to 13 days. Annual broods, one or two.

Food chiefly seeds, wild fruits, and insects, many scratched from under leaves of forest floor. Green-tailed towhee *(Chlorura)* belongs to a separate genus. Five species of *Pipilo* include sixteen subspecies of *erythrophthalmus,* twelve of *fuscus,* and two of *aberti.*

Towhees are undoubtedly useful birds in territory they occupy. Great singers during breeding season, and their home life has been studied thoroughly by special students.

Brown Towhee, 591b
Pipilio fuscus

Length to 10 in.; wings, 3⁹⁄₁₀ in.; tail, 4⅓ in. Female somewhat smaller. Robin-sized. As adults, sexes are alike. Olive-brown or smoky gray above, drab on sides and flanks and across chest; rust-colored spot at the base of underside of tail; a light buffy-yellow throat with buffy-brown streaks. Drab. Bill brown above, lighter below. Young birds colored like parents but show less contrast; finely streaked and dusky on throat, breast, and sides. Looks like a large olive drab sparrow. Twelve subspecies reported. *P. fuscus* found from southwestern Oregon, western and central Arizona, northern New Mexico, southeastern Colorado, extreme western Oklahoma, and western and central Texas south through Baja California to mainland Mexico. Seasonal movements none.

Nest, cup-shaped of weed stems, grasses, and other dried vegetable parts; lining of fine grasses or hair. Usually at moderate heights in bushes or trees. Eggs three to four; pale bluish-green marked with black or deep brownish black; sometimes purplish-gray. Size ⁹⁄₁₀ by ⁷⁄₁₀ in. Two broods each season.

Brown Towhee

Found in rather open brushy areas; canyons, chaparral, woodlands, lawns, and gardens. Courtship displays include bowing, scraping, and posturing exhibited by both sexes. Not much that is unique to this bird. Voice a sharp metallic *chink.* Song, a rapid "tsip, tsip, tsip"; repeated. When feeding, it scratches sparrow-like with both feet, and often jerks its tail in a nervous fashion.

Food: Animal matter consists of a wide variety of insects usually taken from the ground cover. Plant food: weed seeds and cultivated grains. Activities probably beneficial to man.

Lark Bunting, 605
Calamospiza melanocorys

Length to 7½ in., wingspread to 11½ in., tail to 3⅓ in. Weight 1 oz. Temperature, 101°F. Male like a small sparrow-sized blackbird with large white wing patches. Female, young, and male in autumn brown with breast stripes and conspicuous white wing patches. Not easily confused with other birds because of wing patch.

Breeds from southern Alberta to southwestern Manitoba, south to southeastern New Mexico and northwestern Texas, east to eastern Nebraska and west central Minnesota. Winters south from southern Texas and southern Arizona to Sonora and southern Lower California. Accidental in New York, Massachusetts, and South Carolina.

Lark Bunting

Nests on ground, in open plains and prairies. Nest often sunk in ground, made of grasses and lined with hair, plant down, or fine grasses. Eggs four to five, light greenish-blue, rarely sprinkled with reddish-brown spots. Weight ¹⁄₁₀ oz. Incubation by female. One or two annual broods. Male loses black after nesting season.

Food: animal food made up, in large part, of grasshoppers; also eats true bugs and beetles as well as other insects. Wild plant seeds are the principal plant food. Seldom takes agricultural grain crops. Of the seeds eaten, the important ones are Russian thistle, pigweed, and amaranth, all prominent and pernicious in range of the bunting.

Does little harm to growing grains. State bird of Colorado. Some of its other common names are descriptive of its home and coloration; e.g., white-winged blackbird, white-winged prairiebird, prairie bobolink.

Savannah Sparrow, 542a
Passerculus sandwichensis

Length to 6¾ in. Wingspread to 9⅗ in. Weight to ¾ oz. Female smaller than male. A distinctly short-tailed sparrow, with a yellow stripe over eye, pink legs, and a slightly forked tail. Tail not rounded or long as in song sparrow, which it superficially resembles. Weight 19 g.

16 subspecies recognized. Breeds from northern Manitoba to northern Quebec and south to Iowa and Long Island, with subspecies extending range to Lower California and British Columbia. Winters from Indiana to New Jersey and south to Mexico and Gulf, subspecies extending range to south and west. One subspecies, long-billed sparrow, reverses the usual procedure by moving northward from Lower California to southern California in winter.

Nests in grassy fields and meadows or along sand dunes and marshes. Nest in a hollow under vegetation, of fine grasses with lining of hair or rootlets. Eggs four to six, ⅘ by ⅗ in.; greenish-white or bluish, spotted and blotched with red- or purple-brown. Incubation mostly by female. One or two annual broods.

Distinctly a ground bird, though it may alight in trees for short intervals. Flight undulating and zigzagging, usually ending with a sudden drop into grass. May often be approached closely. Female attempts to lead intruders from nest by faking broken wing. Song a fine *tsit-tsit-tsit-tsee-tsee msee* not easily heard by everyone.

Food is mostly grass seeds and insects, former constituting about 54 percent of total. Not known to eat grain and is undoubtedly a very useful species because it lives and feeds in vicinity of agricultural crops and preys on their enemies. Speed, 42 mph.

Savannah Sparrow

Grasshopper Sparrow, 546
Ammodramus savannarum

Length to 5⅖ in. Wingspread to 8½ in. Female smaller than male. A small sparrow with an unstreaked breast, a short tail, and an apparently flattened head. Young birds have streaked breasts. Bill stouter than in savannah sparrow. Sexes colored alike.

Three subspecies recognized: Eastern, Western, and Florida. Breeds from southeastern British Columbia to southern Ontario, south to southern California, southern Texas, northern Georgia, and Florida, with other races in Greater Antilles. Winters from southern Illinois and South Carolina to central California to Yucatan, Costa Rica, and Guatemala.

Nests in open fields, pastures, and cultivated fields but not in wet marshes. Nest on ground, sunken or even with surface, grass-covered, of dried grasses with hair or rootlet lining. Eggs 3 to 6, ¾ by ⅗ in.; glossy white or greenish, sparingly blotched with purple, lilac, or brown. Incubation by both sexes. One or two annual broods.

Runs through grass close to ground, keeping well out of sight much of time. Female feigns injury to protect nest. Song, almost inaudible and insectlike, sounds like a faint, oft-repeated buzzing. Young follow adults after leaving nest. May start south in early August.

Food: notable destroyers of grasshoppers; also eats caterpillars, ants, bugs, and other invertebrates such as spiders, and snails. Plant materials a small part of diet and largely weed seeds. It would seem this species is essentially useful.

Grasshopper Sparrow

Seaside Sparrow, 550
Ammospiza maritima

Length to 6½ in., including 2½-in. tail and ½-in bill. Wingspread to 8½ in. Weight, male to 9⁄10 oz, female smaller. Tail feathers less pointed than in Nelson's sparrow and sharp-tailed sparrow. Darker, grayer, with less conspicuous light streakes on back than in sharp-tailed, with yellow patch just in front of eye and over it, and white on lower jaw.

Northern seaside sparrow breeds from Massachusetts to Virginia. Winters from Virginia to northern Florida. Other subspecies include Macgillivray's, Scott's (eastern), Wakulla, Howell's, Louisiana, Texas, Dusky, and Cape Sable of the Gulf Coast area. Closely related to Nelson's and sharp-tailed sparrows. All salt-marsh birds.

Nest of northern seaside sparrow is in salt marshes, or grasses with finer lining, over high-water mark. Eggs four to six, ⅞ by 11⁄16 in.; grayish-white with reddish-brown spots. Incubation almost wholly by female. One annual brood. Adults molt once a year, about August. Young molt after leaving nest and in fall.

Food, small animals of its salt-marsh environment, including particularly many sand fleas and small crabs. Call, a squeaking *cheep* almost like an insect buzzing, but more distinct than call of sharp-tailed sparrow. Sharp-tailed sparrow has buffy yellow on face, distinct dark breast streaks, and is less dingy than seaside sparrow.

Probably of little economic importance other than as a destroyer of small animal pests of seashore marsh plants, which are themselves of little economic importance. The male may engage in a courtship flight, singing as he rises over his home territory.

Seaside Sparrow

Vesper Sparrow, 540
Pooecetes gramineus
Length to 6⅔ in., wingspread to 11⅙ in., tail to 2¾ in. Weight 1 oz.
White outer tail feathers, which flash as bird flies away, are typical. Also
shows chestnut at bend of wings and a streaked breast. It lacks conspicu-
ous breast spot of song sparrow, which it otherwise somewhat resembles.

Three subspecies: Eastern, Oregon, and Western vesper sparrows. Eastern breeds from central Ontario to Cape Breton Island south to western
Minnesota, eastern Nebraska, central Missouri, and North Carolina.
Winters from southern half of breeding range to southern Florida and
mid-Texas.

Nests in open fields and pastures, usually on dry ground, in a slight
depression. Nest of grass with finer lining. Eggs four to five, greenish-
white or grayish, with dots, scrawls, or blotches of brown or reddish-
purple, ⅞ by ⅝ in.; weight ½ oz. Incubation by both sexes, 11 to 13
days. To three annual broods. Speed, 17 mph.

Food: Insects near their haunts, in or near field borders. Plant food
largely weed seeds. Plant material includes little grain, and that mostly
picked up as waste. Rarely present in good-sized flocks as are many
other kinds of sparrows; rather it mixes with other species.

Unquestionably a useful sparrow well entitled to protection. Excellent
campanion for a walk as it flies on ahead up road, flicking its white-
bordered tail and stopping until flight is again necessary. An emotional
quality to a vesper sparrow's song. Song pitch, 2,750 to 6,600 cps.

Vesper Sparrow

Lark Sparrow, 552
Chondestes grammacus
Length to 6¾ in., wingspread to 11¹/₁₀ in., tail to 3 in. Female smaller
than male. Temperature variable, around 110°F. Conspicuous in field
because of white around sides and corners of rounded tail. Shows chestnut patches on sides of head and white breast with one central dark spot.
Young birds have streaked breasts, without central spot.

Two subspecies, Eastern and Western lark sparrows. Eastern breeds
from eastern Nebraska through western Minnesota to southern Ontario,
to southern Louisiana, West Virginia, and Maryland. Winters in southern Mississippi, southeastern Texas, and eastern Mexico. Western lark
sparrow extends range to British Columbia and south to Guatemala.

Nests in open fields and prairie lands. Nest on ground or in low bushes,
made of grasses and lined with fine rootlets or long hairs. Eggs three to
five, white, pinkish, or grayish; spotted, streaked, and blotched with
brown, black, or purple, weight, ¹/₁₀ oz. Incubation, about 12 days, by
female.

Food about half plant and half animal matter, animal matter including
particularly grasshoppers and locusts in season, and plant material made
largely of weed seeds. Some grain is eaten, as would be expected of such
a large sparrow but much of time after breeding season is spent in borders, not in open fields.

An attractive sparrow, holding out attention because of flashing white on
relatively long tail. Bird seems something like a towhee that lives in the
open. Locally known as quail-head, road-bird, lark finch, and little
meadowlark.

Lark Sparrow

Slate-Colored Junco, 567
Junco hyemalis
Length to 6½ in., wingspread to 10 in., tail to 2⁷/₁₀ in. Weight ⁷/₁₀ oz.
Female smaller than male. Dark slate-gray above and across throat but
light gray beneath. Two outer pairs of tail feathers white, flashing in
flight. Female duller, with second pair of outer tail feathers grayer. Bill
bone-colored.

Two subspecies, Slate-colored and Carolina Juncos. Seven American
species recognized. Slate-colored breeds from northwestern Alaska to
central Quebec, south to southern Alaska, central Michigan, New York,
and Pennsylvania. Winters through eastern United States and southern
Ontario to Gulf Coast. Known in Siberia. Field or woodland species.

Nests in woods or brushy pasture lands, usually on ground; often hidden
under some cover. Nest made of grasses with finer lining. Eggs four to
six, bluish, greenish, or grayish, thickly spotted with fine brown, purple,
or lilac marks, ⅘ by ⅗ in.; weight 4½ g. Incubation for 11 to 12 days.
One or two annual broods.

Food in summer largely insects such as grasshoppers leafhoppers, weevils, click beetles, ants, and leaf beetles; also some wild fruits. In winter,
almost entirely weed seeds; estimated 33 percent winter food ragweed and
smartweed, remainder Russian thistle, amaranth, wild sunflower, and
other common weeds. Speed, 17 mph. Temperature variable, around
108.4°F.

Slate–colored Junco

Undoubtedly a useful species. Popular in winter as a "snowbird" because while blizzards seem to stop activities of most larger outdoor animals, they do not hinder activities of junco to any appreciable extent.
Twittering spring song in northern migration always welcome. Song
pitch, 3,850 to 5,500 cps.

Oregon Junco, 567f
Junco oreganus

Oregon Junco

Sometimes called pink-sided junco. Size to 6⅓ in.; wing, 2⁹⁄₁₀ in.; tail, 2½ in. Female smaller. Head and neck sooty black to medium gray. Rump and white-sided tail are medium brownish-gray to medium gray, back and wings are tawny; sides pink, almost flesh-colored. Underparts are white. Sexes are alike; female lighter-colored. Immature bird is usually streaked with black on head, nape, and back; smoke-gray throat, upper breast, and flank. Eight subspecies recognized.

Breeds from southeastern Alaska, central British Columbia, Alberta, and southwestern Saskatchewan, south to west central California, the mountains of northern Baja California, western Nevada, northeastern Oregon, southern Idaho, and northwestern Wyoming. Winters from southeastern Alaska, southern British Columbia, northern Idaho, Montana, Wyoming, and South Dakota, south to Baja California, northern Mexico, and central Texas. Common in its range but wanders extensively.

Nest on the ground; sometimes sunken, under protection of low shrubbery; cup-shaped, sturdy, walls over an inch thick, covered outside with mosses, weed stems, dried grasses; lined with fine, light-colored grasses or hair. Eggs three to five, ground color white tinged with pinkish, greenish, or bluish. Eggs ⅘ by ⅗ in. Commonly two broods each year. Calls consist of snapping noises and soft twitterings. Song, slow musical twittering or chipping. Found commonly in coniferous and mixed woodlands; also forest edges. Migrating and wintering flocks commonly observed.

Food: Animal matter: insects, especially during the nesting season. Plant food predominantly weed seeds, but will take waste grain. Most of food taken from the ground surface.

Tree Sparrow, 559
Spizella arborea

Tree Sparrow

Length to 6½ in., wingspread to 9¾ in., tail to 2⁹⁄₁₀ in. Weight to 1 oz. Female smaller than male. A brownish sparrow with conspicuous round black spot in middle of gray breast, conspicuous red-brown cap, and two conspicuous white wing bars. A "winter chipping sparrow" with a bill which is dark above and yellow below. Sexes colored alike. Young with top of head brown, breast streaked.

Two subspecies, Eastern and Western tree sparrows. Eastern breeds from central Mackenzie to Newfoundland, south to southern Ontario, southern Quebec, and Nova Scotia. Winters from southern Minnesota to southern Ontario, south to Oklahoma and rarely to Georgia. Western tree sparrow extends range to British Columbia and California. Field species.

Nests located in low trees or in bushes; sometimes on the ground. Constructed principally of dried plants such as strips of bark, moss, weeds; frequently lined with feathers. Eggs four to five, light green or ashy, regularly dotted or marked with light brown, ⁷⁄₁₀ by ⁸⁄₁₀ in., weight to 4 g. Incubation 12 to 13 days. One annual brood.

Food essentially weed seeds. Small amount of animal food largely insects. Within boundaries of United States, probable that 98 percent of its food is weed seeds, with little or no damage to grain crops.

One of best of winter visitors, which comes readily to winter feeding stations. Its cheerful twittering calls in a weed patch even in a blizzard give one a different idea of winter from that which can be had from an armchair beside a cozy fire. One of our best friends in every way.

Chipping Sparrow, 560
Spizella passerina

Chipping Sparrow

Length to 5⅚ in., wingspread to 9 in., tail to 2⅖ in. Weight ½ oz. Female smaller than male. A diminutive sparrow, with a black line through eye and a white line over it, a bright reddish cap, a clear grayish breast, and a notched slender tail. Young, faintly streaked below. Temperature, 106.8 to 107.6°F.

Three subspecies, Eastern, Western, and Pacific chipping sparrow. Eastern breeds from Yukon to Cape Breton Island, south to northern British Columbia, central Texas, southern Mississippi, and central Georgia. Winters in Southern States, north to Oklahoma and New Jersey. Western chipping sparrow extends range to British Columbia and northern Mexico. Pacific chipping sparrow similar in range to western.

Nests usually near where man lives, commonly in ornamental evergreens and windbreaks, 1 to 25 ft above ground. Nest a beautiful grass cup with hair lining. Eggs three to five, greenish-blue dotted, spotted, or lined with black or lilac, ⅘ by ⁹⁄₁₆ in. Incubation 10 to 12 days, by both sexes. One or two annual broods. Speed, 20 mph.

Food mostly small insects and grass seeds. It includes grasshoppers, weevils, plant lice, gypsy moths, cabbage worms, beetworms, cankerworms, army worms, and seeds or ragweed, purslane, and plantain as well as grass. In numbers, may do damage to newly seeded plots by eating seeds, but this is unusual. Song pitch, (Eastern) 3,475 to 8,400 cps, (Western) 3,300 to 5,500 cps.

Also know as chip-bird and social sparrow. It lives with man in a manner like the robin and bluebird. A chipping sparrow gleaning insects from a spray of apple blossoms outside window is bound to make one smile as he rises in the morning. Its faint *chipping* call and its loud *chipping* song are as much a part of eastern orchards and country lawns in summer as grass itself.

Field Sparrow, 563
Spizella pusilla

Length to 6 in., wingspread to 8½ in., tail to 2⅖ in. Female smaller than male. Sexes colored alike, with reddish upper parts, a clear unmarked breast, a rather inconspicuous eye ring, and a pinkish or flesh-colored bill, the last being the distinctive character. Young, finely streaked.

Two subspecies, Eastern and Western field sparrows. Eastern breeds from southern Minnesota to southern Quebec and Maine, south to central Texas and northern Florida. Winters from Missouri to New Jersey south to Gulf Coast. Western field sparrow extends range to Montana and southern Texas and Nuevo Leon.

Nests in open fields, berry patches, and woodland borders. Nests in low bush on ground or up to 10 ft above ground, of grasses, often with a scant hair lining. Eggs four to five, grayish- or bluish-white, spotted and dotted with light brown or lilac, ¾ by ⁹⁄₁₆ in. Incubation 13 days, chiefly by female. One to three annual broods.

Food, animal matter, about 41 percent consisting of May beetles, leaf beetles, grasshoppers, bugs, sawflies, ants, flies, click beetles, crickets, plant lice, tent caterpillars, and cankerworms. Plant food largely fruits of grasses, including small amount of grain. Song a series of clear, whistled notes, louder and more musical than those of chipping sparrow. Pitch, 3,650 to 5,100 cps.

A useful bird justly protected by law. Probably destroys fewer insects than associated chipping sparrow but is useful in this direction. To most persons "just sparrow" until it takes to singing and then a sure sign of spring and early summer.

Field Sparrow

Harris' Sparrow, 553
Zonotrichia querula

Length to 7⅔ in., including 3⅔-in. tail and ½-in. bill. Wingspread to over 11 in. Female slightly smaller than male. Sexes colored alike. Upper parts brown, with back and shoulders black-streaked, and wings with two white bars. Most conspicuous is black crown, forehead, throat, and upper breast contrasted by white on cheek and breast.

Breeds from Fort Churchill, Hudson Bay, to Artillery Lake, Mackenzie, just south of the Barren Grounds. Winters from northern Kansas to western Missouri and southern Texas. Migrates east to western Ontario, occasionally Massachusetts and west through Montana, Wyoming, and eastern Colorado but in great numbers through Iowa region. Inhabits open woodland brush and weed patches in fall and winter.

Nests near timberline, among small spruces, and Labrador tea and arctic bearberry. Nest on ground, of dried grasses. Eggs three to five, pale blue, spotted with brown. Incubation by female alone. Young birds look much like song sparrows and in first winter have white throats with a necklace of black spots rather than broad black bib of adult.

Food: 100 stomachs showed 92 percent of food of vegetable nature, including about 10 percent grains and 42 percent seeds of ragweed and smartweed, with balance of seeds of other plants. The 8 percent of animal matter included 2 percent leafhoppers, also spiders and snails. Call a beautiful but not strong, plaintive song, suggestive of that of white-crowned sparrow.

Decidedly useful as a destroyer of weed seeds and worthy of every protection. Also know as hood-crowned sparrow and blackhood in parts of its range. Strictly a midwestern resident which may rarely occur somewhat east and west of its main range.

Harris' Sparrow

White-crowned Sparrow, 554
Zonotrichia leucophrys

Length to 7 in. Wingspread to 10⅓ in. Female smaller than male. Large sparrow with clear gray breast, high crown that is conspicuously black-and-white-striped from fore to aft, and lacking conspicuous white throat of otherwise rather similar but browner white-throated sparrow. Head stripes in young brown rather than black. Bill pink.

Subspecies breeds from British Columbia to central California in mountains, east to central Greenland and northern Quebec. Winters from southern Lower California to Ohio and Potomac river valleys, south to Florida and central Mexico. Conspicuous during migration period to those who are field naturalists.

White–crowned Sparrow

(Continued)

Nests in brushlands or open woods, usually in hilly country. Nest on ground or a foot or so above it, in dense vegetation, of grass, roots, and leaves with finer lining. Eggs four to five, $^{11}/_{12}$ by $^{2}/_{3}$ in.; bluish- or grayish-white, spotted and dotted with brown or red-brown. Nests June to July. Incubation 12 to 14 days, by female. One annual brood.

Often seen in flocks in migration areas and first experience with birds is always remembered, particularly if it is a rainy morning and birds are giving their whispering song. Food largely insects during spring and summer; seeds during remainder of year.

Probably neutral in economic importance but certainly high in emotional importance. While in essentially agricultural parts of its range its food is mostly weed seeds, so it must be considered as entitled to protection it receives from law.

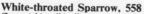

White-throated Sparrow

White-throated Sparrow, 558
Zonotrichia albicollis
Length to 7½ in., wingspread to 10 in., tail to 3⅓ in. Weight to 1¼ oz. Large sparrow with black-and-white-striped crown, with yellow on line between eye and bill and remarkably cleancut white throat patch. Female duller and smaller than male. Related white-crowned sparrow lacks white throat and is grayer. Related white-crowned sparrow might be considered western counterpart of white-throated sparrow.

Breeds from northern Mackenzie to Newfoundland, south to southern Montana, central Minnesota, and highlands of New York and Pennsylvania. Winters from Missouri and southern Pennsylvania to Florida and northeast Mexico. Occasionally in Oregon, California, Utah, Colorado, and Lower California.

Nests in brushy pastures, thickets, and wet spots. Nest usually on ground or under brush, or rarely to 2 ft above ground, of grasses and leaves with finer lining. Eggs four to five, bluish, grayish, or greenish with dots or spots of dark brown or black, weight, 5 g. Incubation 12 to 14 days, by female. Usually two annual broods.

Food mostly seeds and animal life gathered from surface of ground or scratched from just below surface. It takes little grain and that is mostly waste. Eats seeds of poison ivy, smilax, and red alder among other vegetable materials. Bird hops, digs with bill, and scratches for food.

This is the bird whose plaintive "Ol' Sam Peabodee" song intrigues hosts of spring naturalists. To many, this song makes the migration season one truly worthwhile. Has other common names: Peabody bird, cherrybird, white-throat, nightingale, Canada sparrow, Peverly bird.

Golden-crowned Sparrow

Golden-crowned Sparrow, 557
Zonotrichia atricapilla
Length of adult male to 7½ in.; wing 3 in.; tail 3 in. Female somewhat smaller. Male with broad yellow crown patch with a broad black edge extending on sides of head to the eyes. Remainder of head plus neck and breast is light gray; undersides and two wing bars are white, tail is sepia. Wing feathers dusky-brown-edged, dark tawny. Back and rump smoke-gray; jagged stripes of dusky brown on the back. Female colored similarly. No subspecies recognized.

Breeds coastally in Alaska and from southern Yucatan to southern British Columbia, the Cascade Range of extreme northern Washington and southwestern Alberta. Winters west of the Cascades from southwestern British Columbia to northern Baja California and in the Sierra Nevada. Found sometimes in Saskatchewan, Wisconsin, Illinois, Massachusetts, and Japan. Arrives in wintering areas in September, leaves in early May.

Nest of twigs, other plant materials including mosses; lined with fine grasses. Found in low shrubs or scrub trees, close to the ground or on the ground. Well sheltered. Eggs, four to five but usually five. In color, pale green or light buff or yellow, spotted with brown.

A resident of spruce forests or other artic and mountain slopes. Sometimes found in dense thickets and scrub tree growth. Song described as descending with intervals of thirds: *sol mi do*. Some people insist its call precedes rain, and so it is called the rain bird. Food almost 100 percent plant material, of which flowers make up a large part of the diet. Also eats seeds of weeds and insects.

Fox Sparrow

Fox Sparrow, 585
Passerella iliaca
Length to 7½ in. Wingspread to 11¾ in. Weight to 1½ oz. Female smaller than male. Large brown sparrow with bright-red tail and with gray on neck. In flight, tail is most conspicuous. Breast streaked with brown rather than spotted as in hermit thrush, which it otherwise resembles rather closely.

18 subspecies recognized, covering North America rather completely. Eastern fox sparrow breeds from northwest Alaska to Newfoundland south to northern Manitoba and northern Quebec. Winters from Ohio and Potomac Valley to central Texas and Florida with other subspecies going on to Lower California. Largely northern birds that winter in southern United States or farther south.

Nests in alder thickets and evergreen woodlands, on ground or a few feet up on a sheltered branch. Nest large, of dried grass, moss, and leaves with lining of feathers and hair. Eggs four to five, $^{15}/_{16}$ by $^7/_{10}$ in.; bluish-white to gray, thickly spotted with reddish-brown, weight 6½ g. Nests in June to July. Incubation 12 to 14 days, mostly by female.

Industrious sparrow that scratches earth with both feet at once; avoids areas settled by man. It sings little while it is in United States but in its breeding ground gives delightful tingling bell-like song that may be prolonged.

Food is probably about half seeds and half insects and it could not be any stretch of imagination be considered a harmful species. It is always watched for by ornithologists in early spring migrations in East and Middle West.

Lincoln's Sparrow, 583
Melospiza lincolnii

Length to 6 in., including 3-in. tail and $^2/_3$-in. bill. Wingspread to 8⅔ in. Female smaller than male. Like a short-tailed small song sparrow, with streaks on underside finer and not clustered to form a spot as in song sparrow. Has a broad light brown band across breast, most conspicuous in upper part. Young much like adults.

Breeds from Kowak and Yukon valleys in Alaska, through southern Mackenzie, Manitoba, and Quebec, to Newfoundland and south to New Brunswick, northern New York, and in West along Rocky Mountains to southern California and northern New Mexico. Winters from central California to northern Mississippi, south to Lower California, southern Mexico, and central Guatemala.

Nests in a tuft of grass, usually near water, on ground. Nest made of grasses with a finer grass lining. Eggs three to four smaller than those of song sparrow, $^4/_5$ by $^3/_5$ in.; white, greenish-white, or pale green sometimes speckled with brown. Nesting season June. Nesting birds are not commonly seen.

Food probably largely seeds of weeds commonly eaten by most sparrows of its environment. Animal food, a small part of the total diet, almost exclusively insects. Call a sharp *chip,* like that of song sparrow. Song wrenlike, a repeated *quee* ended with *eedle* or *seedle,* but not so loud and vigorous as that of song sparrow.

Probably a useful sparrow but never sufficiently abundant to be of much importance. Because of close resemblance to song sparrows, they may be more abundant than we suppose as they are easily confused with them.

Lincoln's Sparrow

Swamp Sparrow, 584
Melospiza georgiana

Length to 5$^9/_{10}$ in., including 2⅓-in. tail and ½-in. bill. Wingspread to 8 in. Female smaller than male. Somewhat like a reddish song sparrow but stouter, with light throat, reddish cap, and clean gray breast. More nearly size of a husky chipping sparrow. Young brownish, with fine breast streaks much like those of song sparrow.

Breeds from west central Alberta to Newfoundland and south to northern Nebraska, Pennsylvania, and New Jersey. Winters from Nebraska, and rarely in New York, to Gulf Coast from Florida to Tamaulipas and Jalisco, Mex. Sometimes found in Utah, Colorado, California, and Bermuda. Bird is commonly found in brushlands and cattail marshes.

Nest in dead sedge tangles or among grasses, composed chiefly of grass. Eggs four to five, $^5/_6$ by $^3/_5$ in.; like those of song sparrow but slightly smaller and with less distinct markings; pale greenish or bluish-white, mixed with yellowish-brown and lilac. Incubation 12 to 15 days, probably by female only. One or two annual broods. Young with yellow mouth lining rather than pink or gray of song sparrow.

Food largely insects and weed seeds, insects being chiefly aquatic because of habitat of bird. Song somewhat like a loud, more musical chipping sparrow's greeting, usually on one monotonous tone but sometimes with a sweet trill. At a distance, it might be confused with that of chipping sparrow, but probably even then its quality would distinguish it.

Probably of no great economic importance because it does not feed near areas commonly under cultivation by man. Young birds are inclined to sit close and will remain in nest until 13 days old if not unduly disturbed. Adult plumage usually assumed by the first fall molt.

Swamp Sparrow

Song Sparrow, 581
Meprospiza melodia

Length 6⅕ in., wingspread 9¼ in., tail 3 in. Weight ⅞ oz. Female smaller than male. A brown sparrow with a large central black spot on streaked breast, a rounded tail which it pumps up and down vigorously, and a habit of singing from a conspicuous perch in defense of its home territory. Temperature, variable near 102°F.

31 subspecies, including Eastern, Mississippi, Mountain, Dakota, Aleutian, Desert, and San Diego. Eastern breeds from Mackenzie to Cape Breton Island and south to Missouri, Texas, and Florida. Garden and brushy land species.

Nests on grassy land, in brush or ornamental shrubbery, preferably near water. Nest of grasses and weeds with a finer lining. Eggs three to seven, variable, greenish, grayish, pinkish, or bluish with spots or blotches of brown or lilac, ¹⁵⁄₁₆ by ⅔ in.; weight, 4½ g. Incubation 12 to 15 days, by female only. Two or three annual broods.

Food predominantly vegetable material, largely seeds. Animal matter includes cutworms, army worms, cabbage worms, locusts, grasshoppers, cankerworms, and other insects. Little of any grain in diet. Some birds can survive severe winter weather. Speed, 17 mph.

Song has been well interpreted as saying "Hip, hip, hooray, boys! Spring is here." Even if it did not, we might well say it when the bird starts to sing because it is the sign that pussy willows, tulips, and dandelions will soon be the order of the day. Song pitch, 1,900 to 7,700 cps.

Song Sparrow

Snow Bunting, 534
Plectrophenax nivalis

Length to 8 in., wingspread to 13 in., tail to 3⅙ in. Female smaller than male. Color largely white, showing particularly when in flight. In breeding, male mostly white with black back but when seen in winter in United States, head and upper parts are veiled in brown that disappears in spring. However, no other bird of its size has so much white.

Two subspecies include Eastern and Pribilof snow buntings. Eastern breeds from northernmost mainland of Alaska to Greenland, south to northern Quebec; also in Scotland and Scandinavia. Winters south from Unalaska and central Quebec to northern United States; in Europe, to the Mediterranean. Open field species.

Nests on barren grounds among rocks or grasses, usually hidden. Nest largely of grasses, plant stems, and mosses with lining of fur, feathers, or hair. Eggs four to six, white, greenish, or bluish with spots or blotches of brown, yellow-brown, or blackish-brown, ¹⁵⁄₁₆ by ⅔ in., weight to 6 g. Incubation 14 days, only by female. One annual brood.

Food, weed seeds as well as seeds of a number of lower woody plants such as alders. On breeding ground during summer months many insects are eaten. Bills exceptionally sharp and strong and can break seeds readily. Associated with snow by most people. Speed, 16 mph.

Formerly were shot for market and are still killed and eaten by natives of North. One city was once reported to have had 80,000 snow buntings in cold storage. They are, alive, essentially and perennially useful.

Snow Bunting

Canary
Serinus canarius

Length to about 6 in., tail to about 2½ in. Color and size vary greatly with breeds. English canary breeds are larger, brighter-colored, and louder singers. German canary breeds usually have more varied songs, sing more readily, and are raised primarily for their song not for their appearance.

Native of Canary Islands; wild birds are also found in Madeiras and Azores. Harz canaries are bred in Harz Mountains. The "roller" type is designated as the opera singer and the "chopper" type as the jazz singer. Varieties include Norwich, Scotch, Belgian, German, and Yorkshire.

Captive birds nest in cages. First egg laid about eighth day after mating, then an egg a day until four to five are laid. Fed fresh egg paste to induce egg laying. Incubation 13 days, by both sexes, mostly by female. Young fed by both parents for 3 weeks in nest, but by 6 weeks can eat soaked seeds by themselves. Life-span occasionally to 24 years. Many good books are now available to help the amateur in raising his birds.

Food, seeds, particularly German summer rape, Sicilian canary seed, and some hemp seeds with occasionally some apple, cuttlebone, and sometimes with breeding birds, some yeast. Males breeding should not be over 4 years old, preferably should be 1 to 2. Female may be slightly older. A pair should have 16- by 18-in. cage.

Canaries make great pets and, with care, may be trained to come to calls and to perform tricks before eating. Young birds may begin to warble by 8 weeks of age when singing males may be separated from quiet females. Young begin to molt at 6 weeks and shed large wing and tail feathers when 1 year old. Hearing range, 1,100 to 10,000 cps. Canaries are sometimes crossed with other species of finches, producing hybrids called "mules" which are usually sterile but are often excellent songsters.

Canary

CLASS MAMMALIA. THE MAMMALS

Like the birds, the mammals are warm-blooded. Their bodies, instead of being covered with feathers, are more or less covered with hair. The structure of the hair of different mammals and even of races of different species of mammals may be used for identification purposes. Through the courtesy of Dr. Leon Hausman, the nature of some of these hairs is indicated in the following pages.

Also important in the identification of mammals are the nature and arrangement of the teeth on the jaws. Through the use of a simple formula, this condition can be quickly indicated for the different species. Beginning at the front of the human jaw and going to the rear on one side, one finds first cutting or incisor teeth, indicated in a formula by I. In man, the formula would read I 2/2, meaning that there are two incisors on each side on each jaw. In man, the next teeth are single, pointed canine teeth, one on each side on each jaw. The formula therefore continues as C 1/1, two bicuspid teeth behind the canines on each jaw add to the formula B 2/2 (or premolars, P 2/2). Behind these are three molars, completing the formula with the unit M 3/3. Using this same technique, we may quickly find the nature of the teeth of any mammal.

Tracks or pugs are often indicative of the presence of mammals. For this reason, we have included the tracks of several for your assistance. Hindfoot tracks are indicated by the letter H and forefoot marks by the letter F. The direction taken by the animal is indicated by an arrow and the approximate size or stride by appropriate measurements. In the "track formula" used F 1 × 2, 5 means that the track of the front foot measures 1 by 2 in., and that the foot has five toes. H refers to the hind foot, Sp to the spread between paired tracks, St to the stride, and L to the leaping distance. We regret the omission of *scats* or droppings, so useful in recognizing the presence of mammals.

Order Marsupialia./Family Didelphiidae

Opossum
Didelphis marsupialis
Length to 40 in., with 15 to 20-in. ratlike prehensile tail. Weight 4 to 14 lb. About the size of a house cat. Female has brood pouch. Ears naked, black and white. Feet black. General appearance furry grayish-white. Eyes large, dark, and conspicuous in white face. Toes five on each foot, widespread; make starlike tracks; hind toe opposable. Teeth: I 5/4, C 1/1, P 3/3, M 4/4.

Common in wooded areas where dens may be established in hollow trees and food is convenient. Only North American marsupial. Found from New York west to Iowa, south through central Mexico, along coast to middle Florida and southern New England. Introduced and well-established on coast of Washington, Oregon, and California.

To 20 young born 12 to 13 days after mating; each smaller than honeybee, ½ in. long with ⅓-in. tail; weight ¹⁄₁₅ oz; entire litter may be held in a teaspoon; makes way to brood sac and attaches self to 1 of 13 teats. Those unsuccessful die. Increases tenfold in 1 week; nurses 2 months; at 4 weeks sticks head from pouch; at 5 weeks may leave pouch temporarily; at 8 weeks shifts for self. Breeds at 1 year. One or two litters yearly. Life-span 8 years.

Food almost anything organic; eggs and persimmons, favorite food. Climbs trees readily with help of tail and feet. May threaten fight if cornered but more likely to pretend death. May sleep in filthy den in bad weather in burrow or hollow log. Usually active throughout year but mostly at night. Relatively harmless to man. Tracks: F 1¼ × 1½, 5; H 1¼ × 2½, 5; Sp. 6.

Enjoyed as food by some persons. Undoubtedly a pest to poultrymen but provides sport for hunters and destroys mice and many insects. Ranks among the first six species in importance as fur bearer despite fact that fur is cheap and not durable. Number on increase in northern areas. Frequently seen dead along highways.

Opossum

Great Gray Kangaroo

Order Marsupialia./Family Macropodidae

Great Gray Kangaroo
Macropus giganteus
Largest kangaroo. Male 5 ft from nose tip to tail base; tail 3 ft long. Weight to 200 lb. Female about ⅔ size of male, with conspicuous brood pouch. Fur soft, gray-brown above and white below, with that on toes and tip of tail black. Some kangaroo species, called "wallabies," are no larger than rats.

This species found in open fields, commonly near water, through large part of Australia and much of Tasmania. Other species in mountainous districts; some even live in trees. Aslo related to kangaroos are koala (Teddy bear), wombat, and a South American shrew.

One 1-oz young born 1 month after mating of male *boomer* and female *flier*, makes way in ½ hour to brood pouch where it attaches to teat; develops several months. Young *joey* rides in mother's pouch until well
(Continued)

developed, frolics with others during hot part of day while mothers rest. Except at mating time, adults are peaceful to all. Life-span 15 years.

Food purely vegetable matter, browsed from ground as by cattle or taken to mouth with forefeet. When excited, straightens up, leaps 9 to 10 yd at a bound, to 10 ft high, using tail as balance but not as springboard. Commonly found in herds of 30 to 50. Cornered, may kill large dog with blow from hind feet or hold it under water until drowned.

Hunted for leather which is waterproof, for flesh, and for sport. Leather makes excellent shoes. Hunting so uncontrolled that larger species are almost exterminated. 500,000 killed, 1877 to 1902. Natural foes, dingoes, took to killing sheep. Dogs bred form mastiff and greyhound are used in hunting kangaroos. Seen in circuses.

Koala

Order Marsupialia./Family Phalangeridae

Koala, Australian Bear
Phascolarctos cinereus
Length about 24 in. Height about 12 in. at shoulder. Ears large and hairy. Fur thick, ash-gray above and more yellowish below. Head thick, with short snout. Mouth with cheek pouches. Both front and hind feet are "hand-feet," with thumb opposable; in forefeet, first finger is also opposable. Second and third toes in hind feet are united.

Native of Australia and Tasmania, where it lives for most part in tree-tops. Close relatives include the wombats that look bearlike, with short thick legs; the flying phalangers that glide and look something like flying squirrels; and the marsupial shrews.

Families remain together, and a pair with a single young are commonly found together with the young either in the mother's pouch or clinging to her back. Young animal remains in mother's pouch about 3 months, then rides about on her back for several more months.

Lives mostly in the gum or eucalyptus trees, on whose leaves it feeds. During the day, sleeps in lower branches but at dusk becomes active, leaping about the tree and even jumping as much as 5 ft at a single leap, landing safely on smooth tree bark.

Flesh considered excellent by natives and fur has high market value; too high in fact for good of animal, since it has been almost eliminated from much of its natural range. Now protected in Australia and are on the increase. Many are kept captive in parks and prove popular as living "Teddy bears." When wounded, cries piteously, much as a human being would.

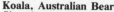

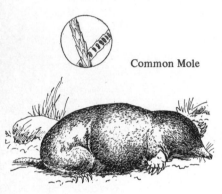

Common Mole

Order Insectivora./Family Talpidae [Moles]

Common Mole
Scalopus aquaticus
Length, male 7⅕ in., tail, 1⅕ in., female 6½ in., tail ⁹⁄₁₀ in. Weight to 5 oz. Close, thick, velvety, gray, soft fur that is smooth if brushed either way. Tail naked or practically so. Forefeet modified into spadelike structures. Nose like a pick. Eyes small. Feet and tail whitish. Muzzle naked. Teeth: I 3/2, C 1/0, P 3/3, M 3/3.

Found in burrows, just beneath surface of soft loose ground. Eastern forms, found from southern Massachusetts, across southern tip of New York, to Nebraska, south to the Gulf. Western moles of genus *Scapanus,* found from Washington to southern California in western portions. Hairy-tailed moles, *Parascalops,* northeastern.

At breeding time, two to three moles to a tunnel indicate males tolerate each other. Breeds in March. Four to five young born in 6 weeks, mid-April. Litter independent by mid-May. Mature at 10 months. Males may come to surface at breeding time to seek mates. Nest 5 to 8 in., grass-lined den. Fur durability, 7 percent that of otter.

Food choices: first, white grubs and earthworms; second, insect larvae; third, adult insects; fourth, plant materials, the last constituting about 13 percent of total volume. Food detected by contact and crushed against burrows. Will eat corn, wheat, and oats but not beans or peas. Daily food totals ⅓ weight of animal. No hibernation. No food storage.

May injure lawns, golf courses, and some gardens but since over 50 percent of food is insects, undoubtedly serves a useful function. Earthworms, constituting 31 percent of food, might be considered useful soil tillers as are the moles. Pests including a beetle, *Leptinus;* a louse, a flea, and threadworms. Control measures include poisoning programs and mole traps.

Star-nosed Mole

Star-nosed Mole
Condylura cristata
Male: length to 8 in., tail 3⅕ in. Sexes colored alike, blackish-brown above and paler beneath. Males average 1⅘ to 2½ oz; females 1⁷⁄₁₀ to 2⁷⁄₁₀ oz. Nose terminates in disk of 22 fleshy rose-colored processes. Eyes small but apparent. Legs short and weak. Forefeet modified for digging.

Teeth: I3/3, C 1/1, P 4/4, M 3/3.

Found in damp meadows and marshes from southeastern Canada and southern Labrador to southeastern Manitoba, northeastern Illinois, Ohio, and south along Appalachian highlands to Dismal Swamp in Virginia, western North Carolina, and Georgia. Often lives in same region and same tunnels used by eastern mole.

Tail enlarges in winter. May pair in autumn (?), remaining together during breeding season. Male sexually mature by January. Usually three to six young born March 15 to May 15; when furred, weigh about $^9/_{10}$ oz; independent at 3 weeks; mature at 10 months; one litter, born in spherical nest of grass and leaves, about 8 in. through, under some cover.

Diurnal or nocturnal. Often gregarious. Bulk of 107 stomachs shows 49 percent worms, 33 percent insects, 8 percent miscellaneous, 6 percent crustaceans, 2 percent mollusks; 2 percent vertebrates. Detects food probably by contact through snout. Digs way through loose moist earth with modified forefeet. Hearing keen; sense of smell poor.

Probably of neutral value; injurious because they dig burrows used by more destructive mice. May be of some use as tiller of soil. Fur of value. Interesting to all biologists. May be controlled by using traps or poisons.

Hairy-tailed Mole

Hairy-tailed Mole
Parascalops breweri
Length, total 6 in. in either sex. Females with hind foot ¾ in., males with hind foot ⅘ in. Color of sexes similar. Like eastern mole in general appearance, but with a hairy tail that is small in diameter at the base and with a shorter snout. Fur soft but coarser than in most other moles with which it might be confused. Teeth: I 3/3, C 1/1, P 4/4, M 3/3.

Found from southern New Brunswick to eastern Quebec, south to northeastern Ohio and southern Pennsylvania, and in the mountains still farther south to North Carolina. Rarely common and varies greatly in abundance through its range. Probably occurs in many places where it has never been noticed.

Its life habits are probably much like those of the common mole but little is known about it. Mates in late March or early April. Four to five young born 1 month later; blind and helpless at birth but independent in 4 weeks and sexually mature by next spring.

Its food is mostly insects to be found underground, earthworms, and other small forms of animal life caught by burrowing through ground. Undoubtedly, open burrows are used by mice that may destroy plants not sought by mole. Eats more than its own weight in a day.

May serve to control the multiplication of subterranean insects but it also is destructive to lawns, golf links, and similar areas that man wishes to leave undisturbed. Seldom leaves its burrow in the daytime. As winter approaches, it digs deeper tunnels. It piles mounds of dirt on the surface; sometimes 3 in. high and 6 in. in diameter.

Order Insectivora./Family Soricidae [Shrews]

Short-tailed Shrew ↕

Short-tailed Shrew
Blarina brevicauda
Length 5 in., tail 1 in. Sexes superficially alike. Form stout. Tail short. Ears small. Fur velvety. Legs short. Feet not suitable for digging. Eyes small. Snout pointed. Dark slate-colored above, paler beneath. Tail black. Paler in summer. Molts in March and October. Teeth: I 3/1, C 1/1, P 3/1, M 3/3. Tracks: F ¼ × ¼, 4; H ½ × ½, 5.

Female builds nest of shredded plant material such as grass or leaves. Nest built under a rock or stump; an underground den. Found in fields and wooded areas where there is a loose vegetation over surface of ground which may harbor food. Ranges throughout eastern half of North America and is represented by more than half dozen subspecies.

Four to eight young born 21 days after breeding; at birth like wrinkled pink honeybee. Weaned at 22 days. Half-grown in 1 month. Mature in ½ year. Aged at 16 months.

Food of 244, by bulk, 47.8 percent insects, 11.4 percent plants, 7.2 percent worms, 6.7 percent crustaceans, 5.4 percent mollusks, remainder miscellaneous. Active year round, night or day. Nest kept clean; made of grass in dry underground cavity 8 in. across, 6 in. deep, with 2- by 3-in. center. Mates for season. To four litters. Some food storage. Poor sight.

Useful insect destroyer. One New Brunswick record shows it killing 60 percent of year's crop of larch sawfly. Takes little grain. Killed by cats and other predators but not commonly eaten because of odor. Worthy of more protection than it gets. Difficult pet. Reputed to have poisonous bite with poison similar to that of some snakes. Not toxic enough to endanger humans. Quickly immobilizes small animals such as mice when they are bitten. Glands on flanks emit an odor that apparently makes the animal unpalatable to many predators.

679

Common Shrew

Common Shrew, Musked Shrew

Sorex cinereus

Length to 4 in., tail 1⅗ in. Weight, that of a penny, $\frac{1}{10}$ to ⅕ oz. Sexes superficially alike. Brown above, sprinkled with lighter or darker hairs, grayish to buff beneath. Eyes minute. Ears nearly hidden in fur. Feet not modified for digging like those of moles. Tail yellowish-brown slender. Teeth: colored, I3/1, C1/1, P3/1, M3/3.

Found on forest floor, in open fields and elsewhere though most commonly on ground through most of northern North America. Not infrequently found even out on ice floes but more normally found in wooded areas or fields where food may be abundant.

Females tolerate males except when bearing young. Five to nine young to a litter, born probably about 3 weeks after breeding, which may take place over long period. 1-day-old young, hairless, ⅘ in. long with ⅛-in. tail, about $\frac{1}{35}$ weight of mother. To three litters a year.

Food mainly insects, worms, salamanders, and other small animals, may eat 3⅓ times its own weight daily. Sense of hearing intense, of sight poor and limited to short distances, of smell poor. Wastes not put in heaps as some mice do but not dropped near food. Can jump 4 to 5 in. standing, or 6 in. (?) running. Breathing rate recorded at 850 times per minute, heartbeat 800 per minute.

Highly useful as destroyer of insects. Held in superstitious awe by Eskimos. May carry some mite parasites and believed to carry spotted fever, but this probably not to a serious degree over any great range. Probably not as vicious or quarrelsome as it is reputed to be. Probably not long-lived. Little poison in bite.

Water Shrew

Water Shrew

Sorex palustris

Length total, to 6⅖ in. for either sex. Long-tailed, suitable for life in water, with hind feet especially large and fringed by stiff hairs along outer margin. Little seasonal or age variation. Upper parts gray, with some hairs white-tipped, and under parts of body and tail white. Dental formula like other members of *Sorex* described earlier.

Ranges through colder North America as represented by five subspecies: Nova Scotia, Great Lakes, Rocky Mountain, Richardson (Rocky Mountains to Minnesota), and White-chinned (Pennsylvania to Labrador), and by closely related Alaska and Unalaska water shrews.

Little is known of the life history and habits. Probably about six young in a litter, but where and when they are born is not common knowledge. Probably two to three litters a year.

These shrews frequent lake shores and streams. Their close water-repellent fur keeps skin dry. Predators on this animal like those of other shrews. Fish feed on water shrews. Sometimes caught in fishermen's minnow traps.

These shrews probably are of little economic importance. Eskimos believe that crossing shrew's trail brings bad luck.

Order Chiroptera./Family Vespertilionidae [Bats]

Red Bat

Red Bat

Lasiurus borealis

Length 4⅕ in., tail 2 in., forearm 1⅗ in. Weight ⅓ to ½ oz. Conspicuously red. Relatively long-haired. Ears low, broad, and rounded. Males orange-red, female dull frosted chestnut. Under parts paler and less red than upper parts. Individuals vary in color slightly. Teeth: I 1/3, C 1/1, P 2/2, M 3/3.

Found about towns and near openings in woodlands often appearing in houses, barns, and churches. Primarily an animal of the forest, particularly near water. Common through North America south of Canada, probably migrating to the south during cold weather, although there is a record of hibernation in this species. Goes south to Panama and north along Pacific Coast to Sitka.

Mate probably in August in flight but true fertilization takes place in spring. Commonly two to four young born in June. Only one litter, young being carried hanging to breast of female or left hanging while mother seeks her own food. Development probably somewhat like that of brown bat. Independent at 3 weeks. Young fly well at about 5 or 6 weeks of age.

Food exclusively insects, captured on wing, usually ½ to 1½ hours after sunset and before sunrise. May migrate at heights of 150 to 400 ft in open flocks and has been found 240 miles at sea. Probably guides relatively slow flight by sound beam reflecting echoes. Gentle if handled quietly. Has homing instinct. Solitary. Leaves roost only at night.

Useful insect destroyer. In captivity, may be fed mixture of bread, American cheese, chopped hard-boiled egg, bananas, cottage cheese, unsalted vegetables and clabber rather than sweet milk. Broken up grasshoppers and June beetles help prevent diarrhea.

Little Brown Bat
Myotis lucifugus

Length 3⅗ in., tail 1½ in., forearm 1½ in. Face hairy. Ears narrow. Hair relatively long and soft. Sexes colored alike; little seasonal color change. Dull brown above and lighter beneath. Voice a fine, high-pitched squeak, inaudible to many ears. Weight of a nickel; ¼ to ⅛ oz. Teeth: I2/3, C1/1, P3/3, M3/3.

Found in various types of places, over water, near clearings, or in houses or other buildings sleeping or hibernating in caves and mine shafts through most of North America. Found far at sea. Around 30 recognized subspecies. Has homing instinct and will return to over 160 miles. Over 2,000 kinds in the world.

Mates promiscuously in fall; true fertilization thought to be delayed by plug which prevents entry of sperms until spring. Young weighs 1½ g at birth about 5 to 9 days, nurses 3 weeks. Litters, commonly one. Females and young congregate separately from males during brooding period. Males breed at 14 months, females at 10 months. Fly at 3 weeks.

Food entirely insects, caught on wing during erratic flight where contact is avoided by echo of high-pitched (50,000 to 98,000 cps) sound given constantly while flying, 25 per sec in open, 50 per sec in trees. Most active 1 hour after sunset and 1 hour before sunrise. May fly 30 miles from home roost seeking food. May roost with other bat species.

Useful as insect destroyer. Body parasites include fleas which do not favor men as hosts. Guano, or manure of bats, is valuable as fertilizer. Known to live at least 21 years. May become a troublesome pest in dwellings. May contract rabies in southern part of range. Efficient insect hunters, making, in one study, 1,665 dives per hour in its pursuit of food. Young have sharp teeth.

Little Brown Bat

Hoary Bat
Lasiurus cinereus

Length to 4½ in., forearm 2 in., wing expanse 11 in. Upper parts a beautiful grayish-white, changing as dark underfur is exposed. Yellowish beneath. Membranes of wings brownish-black. One of most beautiful of bats and well worth knowing more about. Female weighs about 1 oz.

A high-flying, late-flying, woodland bat haunting openings in woodlands and near waterways ranging from Gulf of St. Lawrence to Great Slave Lake and south to southern Mexico, spending colder seasons in southern part of range, to which it migrates. Summers usually north of Pennsylvania, winters south of it.

Mates probably in late Sept. to Oct., but young not born until following May, much as in some other bats. One litter a year. Four or more, commonly two, born to a litter. Young may be hung up in holes in trees while mother seeks food. Independent at 1 month.

Food exclusively insects caught on wing, usually of the late, high-flying type. Most active during second hour after sunset and next to last hour before sunrise. Migration to south in all probability begins with early frosts. Roosts more commonly in trees than in caves or in buildings.

Undoubtedly wholly useful as a destroyer of insects. Because of its late-flying habit is little known, so any observations on its behavior may be interesting to science. No reason why this beautiful animal should be killed at any time. Other common names include frosted bat, gray bat, great hoary bat, great northern bat, northern bat, and twilight bat.

Hoary Bat

Order Carnivora./Family Aeloropidae

Giant Panda
Ailuropoda melanoleuca

Length to 4 ft. Weight around 300 lb. Conspicuously black and white, this serving as protection in contrasting snows and shadows. Fur thick and dense. Feet flat like those of a bear. Ears relatively short. Tail remarkably short. Skull broad. Snout relatively blunt. Claws long.

Native of bamboo forests in mountains of western China, near border of Tibet, at elevations of 6,000 to 12,000 ft. Known to zoologists since 1869, and while originally thought a bear, now believed, because of teeth, closer to the raccoons. Ancestry probably dates back to Lower Pleistocene. No near relatives.

Mates in spring (?) and young born in January. By April, cub has milk teeth. Weighs 3 lb when about 6 (?) weeks old and weighs about 70 lb at 1 year, having gained about 5 lb a month, on average. Probably becomes mature at from 4 to 10 years of age though this is not known. Male roars loudly in breeding season.

Food once believed to be solely bamboo sprouts, requiring nightly feeding for about 10 to 12 hours to supply the needs of such a large animal. Now known to eat small mammals, birds, and fishes in the wild. In captivity, eats regular baby diet of cereals, milk, vegetables, orange juice, and cod-liver oil.

Giant Panda

(Continued)

Ring-tailed Cat

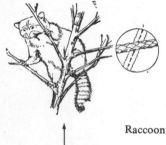

Raccoon

Black Bear

One of most popular of unusual zoo animals, one keeper in New York zoo saying that for every person who asked him about an aardvark, 30,000 asked him where the giant panda was to be found. Antics amusing because of playfulness and grotesque appearance. Popularity should increase with knowledge.

Order Carnivora./Family Bassariscidae

Ring-tailed Cat, Civet Cat, Cacomistle
Bassariscus astutus
Length 32 in., including 15-in. tail. Weight 2½ lb, with female somewhat the smaller. Trim and more slender than raccoon, with small head and large ears. Sexes colored alike, and little seasonal change. Five slender, partly retractile toes on each foot. Tail bands white and black. Teeth: I 3/3, C 1/1, P 4/4, M 2/2.

Found in wooded and rocky areas in southwestern United States, from Texas to the west including three recognized subspecies: one ranging from Oregon south through California, one in Nevada, and the third from New Mexico, Arizona, Colorado, and Utah east through Texas.

In May to June, one to five young born in den, usually in hole in hollow tree. Young colored much like adults, blind, deaf, helpless, pink with whitish fuzz and stubby tails. Eyes open at 4½ weeks, travel at 2 months, and weaned at 4 months. Probably sexes congregate at different times of year but animal not normally social.

Food chiefly small mammals, birds, and insects with some fruit included, such as figs, dates, oranges, cactus fruits, madrona, manzanita, cascara, blackberries, persimmon, and mistletoe. Particularly fond of acorns. Almost exclusively nocturnal, sleeping safely in den during day. Prey to great horned owls but is usually too active to be captured by preying animals. Den entrance frequently shows signs of being chewed.

Probably serves usual function of preying animals. Fur has a good commercial value and is known to the trade as "California mink" or "civet cat." The name civet cat is improperly applied to this animal, as it properly belongs to certain Old World animals. Fur durability, 40 percent that of otter. Has other common names such as miner's cat, coon cat, band-tailed cat, mountain cat, squirrel cat. Sometimes takes up residence in dwellings close to man, even within city limits.

Order Carnivora./Family Procyonidae

Raccoon, Coon
Procyon lotor
Length 34 in. Tail 10 in., bushy and ringed. Weight to 35 lb. Sexes superficially alike. Fur long, of excellent quality; gray, brown, and black, dull brown at base. Conspicuous black band across face and eyes, and six to seven black rings on tail. Feet with five slender fingerlike toes front and back, yellowish-gray. Teeth: I 3/3, C 1/1, P 4/4, M 2/2.

Found in wooded and well-watered areas from Gulf of St. Lawrence to southeastern British Columbia, south to Gulf of Mexico and South America. Not in northern Rockies or drier parts of Great Basin including most of Montana, Wyoming, Utah, Nevada, western Colorado, and northwestern New Mexico. 200 acres woodland gives sustained annual yield of one.

Probably promiscuous. Breeds in February after hibernation. Two to six young born 63 days later. Male helps rear young, though in captivity may kill them. Young blind for 19 days; suckle for 2 months; remain in family circle through winter; to 12 lb by winter. May breed when 1 year old or wait until new litter forces last year's family from home den. Tracks: F 2½ X 3, 5; H 2½ X 4, 5; St 7; L 20. In the wild an "old" coon may be 7 years old; captives have lived to 14 years.

Food, insects, aquatic animals, corn, and vegetables. Den preferably a hollow tree but sometimes a cave. Does not store food or keep den clean though it washes food if water is available. Highly inquisitive and sociable. Desperate fighter when cornered but easily tamed if caught young. Makes excellent pet. Full meal equals ½ lb of fish.

Valuable furbearer when long-haired furs are in vogue. Fur known in trade as "Alaska bear" and "Alaska sable" as well as raccoon. Durability of fur 65 percent that of otter. May take corn and poultry but probably more than compensates by insects and mice destroyed and value of pelt. Valuable farm forest asset.

Order Carnivora./Family Ursidae [Bears]

Black Bear, Cinnamon Bear
Ursus americanus
Length to 6 ft. Height at shoulder, to 3 ft. tail to 5 in.; weight of adult males to 500 lb, rarely to 600 lb. Female smaller. Black or cinnamon with a white throat patch. Sometimes all brown, brown bear being merely a phase of black bear. Snout rather long. Tail short. Flat-

footed. Inquisitive. Not normally vicious but never really trustworthy if opposed. Teeth: I 3/3, C 1/1, P 4/4, M 2/3.

Originally found in most of North America except eastern California, eastern Oregon, Washington, Nevada, and western Utah. Ten or more subspecies recognized, practically all confining range to wooded areas and often common relatively near centers of population, many being taken each year in such states as New York and Pennsylvania. Prefers woodland or dense brushland areas. Nocturnal. Assumes a winter sleep but not true hibernation.

Mates for season only, usually in June; one to four young born about 7 months later. Young about 8 oz, blind, practically helpless. Mother commonly in winter sleep. Remain with mother until fall at least, being protected by her and shown how to live successfully. Female bears young every other year, after maturity is reached. Life-span to 25 years.

Food, flesh, fish, fruits, and vegetables, being particularly fond of berries. Builds up reserve of fats for winter sleep. Individuals range some 30 miles. Den in a cave or hollow stump, occupied chiefly in winter and when young need protection. Serves as a useful check on many small species of animals. Sounds, a growl, a snort, and a smacking of lips. Tracks:F 3 X 4, 5; H 4 X 7, 5.

Probably useful for hide, meat, and sport. Individuals may be harmful as destroyers of sheep, pigs, or chickens. Serve in some places to attract tourists. In zoos, bear den is always crowded by visitors who enjoy watching antics of these animals. May be trained to some degree to perform in circuses. Should not be exterminated. Fur durability, 85 percent that of otter.

Polar Bear
Thalarctos maritimus
Length to 11 ft. Weight of average male 900 lb, female, 700 lb. Shoulders less heavy than in black bear. Sexes colored alike, uniformly white with suggestion of yellow; in summer, some brown appears. Young whiter than adults. No conspicuous seasonal change. Height at shoulder to 5 ft. Teeth: I 3/3, C 1/1, P 4/4, M 2/3.

Through Arctic America, near sea, in which it swims easily and powerfully or on ice where, because of color, it is inconspicuous. May come as far south as Newfoundland with drifting ice, also found in Greenland and Iceland. Has been found 200 miles from land swimming vigorously.

Mates in midsummer, not polygamous. One or two young born at about 8 months, weighing about 2 lb or less and only 10 in. long; in an ice-bound den; between December and March. Mother remains in den nursing young until March and probably breeds every other year, young being ⅔ grown by mating season. Young mate when 2½ to 4 years old. Aged at 33 years.

Food, seals, fish, shrimps, mollusks, seaweed, small land animals, grass, and other vegetation, these being sufficiently abundant in long summer so that the animal may put on good reserve of fat. Grass eaten much as a hog grazes. Makes powerful fighter but normally will not attack unless bothered. Can scent food or enemies 20 miles (?).

Fur excellent. Animal valuable as zoo animal. Eskimos use flesh, bones, and hide for food, clothing, and utensils. Loyalty between members of the family is exceptional particularly between mother and young. Worthy of protection against extermination. Develops speeds to 25 mph. An endangered species.

Grizzly Bear
Ursus horribilis
Length, male to 7½ ft. Shoulder height, to 4 ft. Average weight about 500 lb, but individuals known to reach 1,150 lb. Females slightly smaller than males. Sexes colored alike, yellowish-brown with a gray wash. Obviously of heavier build than black bears. Teeth: I 3/3, C 1/1, P 4/4, M 2/3.

Originally ranged through most of western half of United States and western third of Canada but now practically exterminated from United States. Favors rough country, usually wooded. Closely related to Alaska brown bear, a subspecies. May be abroad in either daylight or darkness but is generally a nocturnal animal. Lives in a cave or opening as a winter den. Does not hibernate, in the true sense, but does sleep in the winter.

Probably not promiscuous, but pairs for season only. Two to four young born 236 days after breeding, weigh 1½ lb at birth, 8 in. long; in winter den with mother who suckles there. At 3 months, weigh 12 lb; remain with mother during first summer. Young born every other year. Breed at 3 years age. Life-span 25 years.

Food, animals such as deer, colts, cattle, sheep, fish, snakes, birds; also grass, roots, fruits, and many other things. Particularly favors gophers and ground squirrels but also may develop habit of killing domestic stock, thereby making enemies of men. Generally minds own business unless crossed, then fights. Tracks: F 4 X 6, 5; H 5 X 12, 5.

(Continued)

Polar Bear

Grizzly Bear

Black-footed Ferret

Marten

Fisher

Because of cattle destruction, has been practically exterminated in United States except in national parks. Possibly this is for the best. Flesh is considered good. Park animals always draw tourists at feeding time and for most part are well behaved. Scratch shows five claw marks instead of the four of black bear.

Order Carnivora./Family Mustelidae [Weasels]

1. Domestic Ferret
Mustela furo

2. Black-footed Ferret
M. nigripes
(1) length 12 to 14 in. plus a 5-in. tail. Eyes reddish. Yellowish-white. Smaller than polecat, of which it is sometimes considered a variety. (2) Length 18 in. plus 5-in. tail. Yellow-brown with face, throat, and underparts nearly white; with black mask across eyes, blackish-brown feet and end of tail. Female 10 percent smaller than male. Teeth: I 3/3, C 1/1, P 3/3, M 1/2.

(1) origin of domestic ferret, *M. furo*, uncertain but thought to be from Africa; introduced into Spain, then throughout Europe and then through civilized parts of world. (2) found in Great Plains region of North America, from western North Dakota to northern Montana, south to Texas but exceedingly rare, and almost extinct over most of its original range.

(1) five to eight young born in captivity twice a year, with mother protecting them well. (2) little known about breeding habits of black-footed ferret except that family of mother and young may be seen working together late in summer and nest is probably underground in a prairie-dog tunnel.

(1) food primarily rats and mice, rabbits, and similar small animals caught by following them through their burrows. In captivity, ferret is fed on milk and bread with some raw meat added now and then. (2), the black-footed ferret lives in prairie-dog towns and feeds mostly on animals that dig burrows in which it lives. It will often exterminate a town's life.

(1) used by hunters to drive rabbits from their underground retreat, rabbits being caught in a bag or hands as they dash from burrow. Also used to exterminate rats in buildings. So effective in all dealings with their food that their use is regulated by law where rabbits are not abundant. (2) useful as a natural check on destructive prairie dogs. Black-footed ferret is an endangered species.

Marten, Sable
Martes americana
Length total 25 in., tail 8 in. Weight, males to 2¾ lb, females to 1⅞ lb. Rich dark brown with irregular patch on throat, though color varies and throat patch may be white or orange. Ears white or dull inside, pointed. General appearance much like a slender snaky squirrel but more active in treetops. Teeth: I 3/3, C 1/1, P 4/4, M 1/2. Tracks: F ¾ X 1½, 5; H 1¼ X 2, 5; Sp, 4; L 36.

Tireless hunter of treetops night or day. Found through wooded Canada and Alaska, formerly south in Alleghenies to West Virginia, in Rockies to Colorado and on West Coast to central California; also established in groups in Colorado and northern New Mexico. Now generally rare. Prefers fir, spruce, and hemlock forests in the West; cedar swamps in the East. Chiefly nocturnal. An endangered species.

Possibly pairs, mating in July to August. One to five young born in April in grass-lined hollow tree or burrow; reared by mother only; blind, 6 weeks; suckled about 5 weeks; reach adult weight in 3 months; only one annual litter. Not sociable as family or groups of families. Life-span 18 years. Gestation 9 months.

Food, squirrels, rabbits, birds, mice, eggs, or any animals that can be overcome. About only protection a squirrel has is to get into hole too small for marten to enter, as it can be outrun in treetops. Generally found in evergreens or second-growth hardwoods. Food may be available through year. No hibernation.

Valuable as destroyer of injurious mice and other rodents; mice never become pests in marten country. Highly valuable furbearer, generally becoming scarce. In good country, six martens per square mile would be high population. Has recently been successfully raised in captivity and may be therefore saved as fur source. Fur durability, 60 percent that of otter.

Fisher, Pekan
Martes pennanti
Length 3 ft, tail 14 in. Weight, male to 12 lb; female to 6 lb. Dark brown to blackish, with legs and tip of tail black. Female smaller. Both have general appearance of a slender, long-nosed, short-legged black cat. Possesses great agility and activity in treetops. Gives off foul scent when annoyed. Teeth: I 3/3, C 1/1, P 4/4, M 1/2.

Found in dense wooded areas of North mainly north of Gulf of St. Lawrence to northern Quebec, northern Ontario, Manitoba, Saskatchewan, Alberta, and British Columbia and south in mountains to central California and northern Appalachians. Absent in plains of southern Canada and of northern United States.

Possibly pairs, mating in March or April, with one to five young born probably after a gestation period of 15 weeks with a delayed implantation; that may account for up to 9 months, Young blind 7 weeks; in den to 3 months; reared by mother alone, with homes a series of dens in hollow trees with a high 4- to 5-in. entrance. Family breaks up early and moves about, ranging over 8- to 10-mile territory. Not sociable. Polygamous mating occurs some 3 to 18 days after birth of a litter.

Food, squirrels, martens, mice, porcupines, raccoons, birds, and other warm-blooded animals captured by tireless pursuit. Reserve food supply may be hidden. Obviously has certain beats which may be followed day or night. Does not burrow as do many close relatives. Mischievous. Growls, snarls, spits, or cries mournfully. Easily trapped.

A most valuable furbearer, single pelts having brought as much as $150. For some 70 years over its entire range yield has averaged around 8,000 to 9,000 pelts a year but is now dropping dangerously. Protected completely in some states. Pelt of female is about twice as valuable as that of male. Tracks: F 3¼ X 2½, 5; H 1¼ X 2, 5; Sp 4; L 36.

Longtail Weasel

Longtail Weasel
Mustela frenata
Length, male to 17 in. with tail to 6 in. long, female to 14 in. with tail to 5 in. Weight, male to 12 oz, female to 7 oz. With long slender body, comparatively long neck and head only slightly larger than neck. Has yellow-white underparts; brown above, brown tail with black tip. In North becomes white in winter; tail with black tip. In Southwest, head darker than body with white bridle across face. Teeth: I 3/3, C 1/1, P 3/3, M 1/2.

Range throughout U.S. except for small area in Southwest; also into parts of Canada and Mexico. No other North American weasel so widely distributed. Found in a variety of habitats near water.

Young born in late April or early May, 4 to 8 months after gestation period of varied length to 337 days. Females mate as early as 3 months of age; males about 1 year.

Fur in white phase, the ermine of the fur trade. At one time a symbol of royalty and elegance. In recent times, of less value but is used in both color phases as decorative pieces on women's clothing. Seldom, if ever, takes poultry. Important check on rodent populations, especially in grasslands. Can climb trees and swim but not adept at either. Active throughout the year. Predominantly nocturnal but does appear in daytime. Aggressive hunter.

Least Weasel

Least Weasel
Mustela rixosa
Length, female 6 in., tail 1⅕ in. Males larger by about 10 percent. Smallest of the weasels. Tail with sparse black tip. Upper parts uniformly dark red-brown, under parts white. All white in winter, including tail tip; except in South, where it may be partially white.

Found in woods, lawns, waste places from Hudson Bay to Alaska, south to Montana and Nebraska, east into Ohio and Pennsylvania, with related species extending range as well as sharing it. Southern limits probably North Carolina. May take over and use a mouse nest for its own.

Little known about home life except that young are born and cared for in a dry, clean, well-aired underground den, usually lined with grass or hair. Young, three to six. Probably solitary except for fact that the families remain together during summer months. Young born at widely different seasons.

Food: mice, shrews, and insects, often providing a real fight. Enemies are many, including birds of prey and larger animal eaters. Hunts by scent; climbs and swims easily when seeking prey. Will stay on a trail persistently until food is caught.

Unquestionably useful as destroyer of many insects and particularly of mice, which are of course destructive to plants raised as food for man. Most active at night. Fur fluoresces under ultraviolet light, producing a vivid lavender color. This does not happen with Bonaparte weasels.

New York Weasel

New York Weasel, Longtail Weasel
Mustela frenata noveboracensis
Total length, males to 18 in., females to 13 in. Tails of males to over 6 in., of females to over 4 in. Hind foot, of males to 2 in., of females to 1½ in. Tail long and bushy and black for ½ to ⅓ its length. Weight of adult males to 12 oz, of adult females to 7 oz. Teeth: I 3/3, C 1/1, P 3/3, M 1/2.

(Continued)

Found from southern Canada, over most of U.S. south through Mexico and Central America. Habitat: woodland, brushland, open timber or fields, brush field borders, and prairies. Frequently near creeks, lakes, and other water. Subject to cycles of population. Frequently active in daytime but generally considered nocturnal.

Probably mate in midsummer but young are not born until following spring. Pair cooperate in rearing of young, male bringing food when it is needed. May pair for life. Young may average four. Life history probably much as outlined for Bonaparte or small brown weasel.

May range over 100 acres a night in search of food, mostly mammals to size of rabbit and mainly mice but also birds, frogs, snakes, and insects. Can hold 8 oz rabbit free from the ground. Tracks: F 1 X 1, 4; H ½ X 1½, 4.

Of major importance as a control of mice and deserves reasonable protection as a mouse and rat destroyer. In summer dark brown above and white with yellowish wash beneath. In winter animals are white (ermine) with females more prone to change than males in border territory. Changing animals are known to trade as "graybacks." Preyed upon by other predators including the great horned owl, black snake, rattlesnake, rough-legged hawk, goshawk, and barred owl.

Bonaparte Weasel

Bonaparte Weasel, Short-tailed Weasel
Mustela erminea
Length, male 12 in., tail 3½ in.; female 9 in., tail 2⅕ in. Weight, males to 6 oz, females to 3 oz. Toes on each foot, five. Soles of feet furry. Soft close underfur and glistening chestnut outer hairs. Brown above and white beneath in summer. All white, except black tail tip, in winter. Strong odor. Sexes and young colored alike. Teeth: I 3/3, C 1/1, P 3/3, M 1/2. There are 8 to 10 mammae.

Found in cultivated farmlands, woodlands, and waste places from just south of Great Lakes to northern border of the continent and west to Pacific. South along Rockies and Sierras to New Mexico.

Mates probably in July; later (sometimes 8½ months), four to eight young are born. Young blind for 5 weeks; weaned at 6 weeks; sometimes cared for by both parents and hidden in clean den lined with fur of animals which have been killed. Will defend young at risk of own life. One litter a year. Family life exemplary.

Food of 360 individuals, 34.5 percent field mice, 13.1 percent rabbits, 11.3 percent deer mice, 11.2 percent shrews, 6.7 percent rats, 3.6 percent chipmunks, 3.2 percent birds, frogs, and snakes, and 16.4 percent undetermined mammals. Among 500, no evidence of bloodsucking. Active mostly at night but seen in open night or day through year. Tremendously active and interesting.

Splendid destroyer of harmful mice; kills also millions of rats and other rodents yearly. Reputed to break up game-bird nests. Home range probably 30 to 40 acres. Chiefly nocturnal, but also hunts during day. Climbs trees; more commonly on ground, however. Voice a shrill shriek. Prime pelt has brought 1 to 2 dollars. Probably much more good than evil agriculturally. Provides the ermine of "royal robes." Use of bounty to control population of no value.

Mink
Mustela vison
Length to 25½ in., tail 8½ in. Females to 20 in., tail 5 to 8 in. Weight, male to 3½ lb, female to 1½ lb. Tail bushy. Sexes alike, a rich dark brown with white under chin. Nose pointed. Ears small and set close to head. Toes, five on each foot. Tracks paired. Runs by rapid leaps. Well-developed scent glands. Teeth as described for other members of *Mustela*.

Mink

Found through most of Canada and Alaska except in extreme northern treeless border; south through most of United States except western Texas, New Mexico, Arizona, southern Nevada, and southern California. Usually in wooded areas near waterways but may roam inland considerable distances. Swims well.

Mates in February to March, both males and females taking one or more mates during season. Males fight fiercely. Three to six young born 30 to 32 days after breeding, sometimes much longer. Young are usually born in March or April; blind 3½ to 5 weeks after birth, helpless. Leaves den, usually at end of a tunnel 4 in. in diameter, when 6 to 8 weeks old. One litter a season.

Food mostly rats, mice, fish, and poultry. May be severe enemy of marsh-dwelling birds. Also eats young snapping turtles which prey upon birds and fish. Does not kill wantonly and though it can climb, rarely does so. Family of mother and young remain together during first summer. Tracks: F 1½ X 1½, 4; H 1½ X 1½, 4; Sp 5; L 24.

Value of fur more than compensates for game destruction. One of most valuable furbearers. Durability of fur: natural 70 percent dyed 35 percent that of otter. Can be raised in captivity, those so raised constituting around 40 percent of total annual American crop. Interesting cage animals.

Common Skunk

Common Skunk
Mephitis mephitis
Length 30 in. Tail 7½ in. (male), 6⅗ in. (female). Weight to 14 lb. Fore claws good for digging, not retractile. Sexes superficially alike, either all black or black with conspicuous white stripe down back and onto tail. Strong, repulsive scent characteristic. Tracks: F 1½ X 2, 5; H 1½ X 2½, 5; St 5. Teeth: I 3/3, C 1/1, P 3/3, M 1/2.

Found in woods and plains, near towns or away from them, where there is loose soil. From southern half of Canada, south nearly throughout United States except for southern tip of Florida and parts of Coastal Plain. May be near water or in dry open country, almost desertlike.

Mates February to March, after family of preceding year breaks up. In 62 to 64 days, 4 to 10, naked, blind (3 weeks), and almost helpless young are born, in a den which may be occupied by other animals. Nurses for 6 to 7 weeks. Male may rejoin family after young are partly grown, but not with family before July, when young are in earlier stages.

Food: 414 analyses show 41.3 percent insects, 22.1 percent fruits and berries, 14.1 percent mammals, chiefly mice, 12.9 percent grains, 5.4 percent carrion, 2 percent birds, 2.2 percent unidentified. Female may seek blood about time young are born. Particularly destructive to potato beetles, hop miners, turtle eggs, white grubs, and other pests. May take some poultry.

Was used as food by Indians. Valuable to man as destroyer of vermin; as fur pelt, more than compensates for injury caused to game and poultry. Fur has durability 70 percent that of otter, sells under trade name of Alaska sable and black marten. Can give scent even if held by tail but does not cause blindness.

Little Spotted Skunk, Civet Cat
Spilogale putorius
Length, male 22½ in., tail 8⅘ in.; female 22 in., tail, 8⅕ in. Glossy black with a white spot on forehead and four longitudinal white stripes on back, sides, and onto tail, more or less broken in appearance. Color pattern highly variable. Chunky. Strong odor. Sexes colored alike. Teeth: I 3/3, C 1/1, P 3/3, M 1/2.

Found throughout most of U.S., except Great Lakes States; New England and Atlantic Seaboard States. Also absent from northwestern North Dakota, Montana, and northeastern Washington. Occurs in bushy or sparsely wooded areas; grain and hay fields, also along streams, among boulders; old buildings.

Civet Cat, Little Spotted Skunk

One male mates with many females in late winter, and 2 to 10 young are born in midspring in a den, in a stump, cave, stone pile, or hollow log. Parental care is probably provided chiefly by female, as is usually case in polygamous animals. Strictly nocturnal, more so than its larger relative.

Food essentially insects, mice, and small forms of animal life. It may carry rabies but so may many other animals, and chances of being bitten by a mad skunk are considerably less than chances of being bitten by a mad dog. Possibly more of an insect eater than its larger relative. May climb trees, swim, burrow, or run.

Probably essentially useful as an insect destroyer and as an enemy of mice. A full meal is equal to about ⅛ lb of rabbit flesh. Fur possibly more handsome than that of common skunk because of pattern showing broken white lines against contrasting black. Fur passes in trade under name of civet cat.

Badger
Taxidea taxus
Total length 30 in., tail 5½ in. Weight to 24 lb, average about 13 lb. Sexes superficially alike in size and color. Grizzled gray with black, with most hairs gray at base, then gray-white, then black, then silver-tipped. Face black and white. Forefeet with powerful claws useful in digging. Teeth: I 3/3, C 1/1, P 3/3, M 1/2.

Badger

Found on dry prairie or grasslands with loose soil and many ground squirrels, from central North America south of Saskatchewan, through United States from Ohio to Washington, Texas, and California. Most abundant where its food species are abundant.

Probably pairs, at least for summer, both parents caring for young. One to five young born at 6 months, in May, in a large grass-lined nest, at end of burrow which extends some 8 ft from entrance. Family stays together around a home possibly for year, particularly mother and young. May live 12 years.

(Continued)

687

Food, small rodents, particularly ground squirrels dug from burrows; also birds, insects, and other small animals. A full meal equals ⅓ lb of ground squirrel. Can swim or climb but burrows best, not infrequently building new burrow each day. A terrible fighter against dogs. Keeps den clean.

Unquestionably useful if prairie dogs and ground squirrels are considered as destructive. Sometimes seems to be friendly with animals of other species but justly distrustful of man, who has almost caused its extermination in certain areas. Holes may seem to destroy appearance of a cultivated field.

Wolverine

Wolverine, Glutton
Gulo luscus
Length 3 ft, tail 6 in. Height 1 ft. Weight to 50 lb. Of the Mustelids in N. America, only the sea otter is larger. Male 10 percent or more larger than female. Dark blackish-brown with a pale band along sides. Stout bushy tail. Ill-odored, clever, apparently malicious and powerful, with strong claws and teeth. Ears small and relatively close to head. Teeth sharp and strong; I 3/3, C 1/1, P 4/4, M 1/2.

Found in wooded tundra areas principally but now greatly reduced in range. Formerly through northern United States and southern Canada except in Plains region but now limited rather closely to Hudson Bay region and northwest into Alaska and British Columbia. May range as individuals over a territory 50 miles across, moving place to place as convenience suits.

Mates in spring (?), with two to four young born about 2 months (?) later in June; suckled by mother 8 to 9 weeks; cared for by mother until late summer. Little evidence that male shares family cares. Den a cavern lined with dry leaves, usually well-guarded by rocks and logs.

Food: animal matter. Known to kill deer, caribou, and bear by jumping onto back from elevation and sinking teeth into neck. Suspected of having killed small moose and puma. Food stored by defiling it and burying it, making it useless to other animals. A primitive species, size of bear, lived 4 million years ago in western United States.

A great menace to trappers and campers, and a dangerous antagonist if cornered. Fur does not collect moisture, therefore was once used in parkas and around faces of aviators and those who live in Arctic. Known as carcajou, skunk bear, Indian devil, and quiquihatch. Fur durability, 70 percent. Man is the only important enemy of the wolverine.

River Otter
Lutra canadensis
Length to 4 ft, with 1-ft tail. Height to 10 in. Weight to 25 lb. Feet webbed. Soles of feet hairy. Each foot with five toes. Tail strong and muscular. Fur rich glossy brown. Gray on lips and cheeks, lighter brown beneath. Sexes colored alike. Males about 5 percent larger than females. Little seasonal or age variation. Teeth: I 3/3, C 1/1, P 4/3, M 1/2.

Usually found in or near good-sized waterways, such as lakes, marshes, streams, and seashores. From Canada and Alaska, except in extreme northern parts, south through United States except in southern Texas, southeastern New Mexico, southern Nevada, and southern California. Individual ranges over 50-mile territory, going rounds regularly traveling day or night.

Probably mates for life, breeding in late February. Because of delayed implantation, exact length of the gestation period cannot be reported. 9–12 months has been reported as elapsed time between breeding and the birth of young. Young born about 60 (?) days later, one to five in a litter; blind about 5 weeks, suckled about 4 months. Family stays together for 1 year, both parents helping in rearing of young and carrying on an ideal family life. Den in a bank or hollow log. Makes slides on stream banks.

River Otter

Food mainly fish but may kill muskrats, ducks, young beavers, birds, and even poultry, caught at any time of day or night through year. Apparently can adjust well to varying light and temperature. Gives off offensive odor when disturbed. Is a match for average dog in a fair fight. Tracks: F 2½ × 3, 5; H 3 × 3, 5.

Valuable fur bearer, most durable type; natural fur considered 100 percent as compared with other furs. Can live in zoos but apparently are not successfully or economically raised for their pelts. Natural American stock decreasing unduly though otters still survive in long-settled England. May also be called fisher, water dog, and appropriate Indian tribal names. May come in conflict with fishermen by damaging or stealing fish caught in nets or on lines. Not very popular around fish hatcheries.

Sea Otter
Enhydra lutris

Length to 4 ft of which 1 ft is tail. Weight to 80 lb. Sexes colored alike a glossy brownish-black with white-tipped hairs on head and neck, particularly on young. Fur finest quality and durability. Female slightly smaller than male. Hind feet broadly webbed, haired on both surfaces, flipperlike, about 6 in. long, and 4 in. wide. Teeth: I 3/2, C 1/1, P 3/3, M 1/2.

Formerly ranged through North Pacific but now practically extinct. Two subspecies differ primarily in size, southern variety reaching a total length of 6 ft. Ranged south along shores of Lower California; until a small herd was discovered a few years ago, was considered extinct by many authorities. Prefers kelp beds and rocky shores. Spends much of its time resting and feeding.

One young born any month of year; though probably more frequently about June 1, weight to 3 lb, 8 to 9 months after mating, which is preceded by an elaborate courtship. Young born with eyes open, with all teeth developed, and able to get about well, yet may be suckled for almost a year and may sleep in mother's arms at sea. Young may reach breeding maturity at 1 year.

Food gathered almost wholly from sea, where adults may dive to depths of 300 ft or more. Family life is ideal. Mother and young mourn tremendously at separation. Young may not reach full size for 4 years. Families may stay together for years, probably forming basis of the herds of old. Fur durability, 80 percent that of river otter.

In former days, when sea otter pelts were available, better pelts averaged as high as $2,500 apiece. Practical extinction of this superior furbearer is an indictment of man's inability to control activities of men and other otter enemies. Restoration to reasonable abundance is a part of world planning. Sea otter now fully protected. Abalone fishermen worry about sea otter's ability to compete for the catch.

Sea Otter

Order Carnivora./Family Viverridae

Golden-brown Mongoose
Herpestes javanicus

Length, body 15 to 18 in., tail 14 to 15 in. Rather stout and weasellike with short legs, pointed muzzle, short rounded ears, and five toed feet with long nonretractile claws. Tail thickest at base and tapering to end. Grayish or reddish-brown, lighter beneath. Can stand erect on hind legs. Quick, active, bold.

Native range of Indian mongoose, India and Ceylon. Java mongoose introduced into Jamaica in 1872. 72 descendants of these introduced into Hawaii in 1884. African species include white-tailed, red-tailed, and Caffre; a Spanish species; Asiatic group: short-tailed mongoose, *H. dorachyurus,* Malaya, Borneo, Sumatra, length 30 in.; Hosis mongoose, *H. hosei,* Borneo, smaller; Colland mongoose, *H. semitorquatus,* Borneo, length 25 in.; crab-eating mongoose, *H. urva,* China and Formosa, length 29 in.

In Hawaii, newborn young are found in all months except October, November, and December. Two to three litters a year, with two or rarely three young per litter. Young are protected apparently by mother until they are ¼ grown, frequently going about hiding under her body or tail. Nest in hole.

Food is largely animals such as birds, reptiles, insects, and rats. They are great snake destroyers but are not immune to the bite of venomous snakes, which they kill with skill and agility, usually breaking snake's neck after he has struck. Animal is rather easily tamed and becomes very affectionate.

Useful as a destroyer of rats and venomous snakes and of insects but also highly destructive of ground-nesting birds and of poultry. Cannot climb trees, and in mongoose country rats have taken to living in trees. Because of poultry killing and other bad habits cannot be introduced legally into United States.

Mongoose

Order Carnivora./Family Canidae [Dogs]

Gray Wolf
Canis lupus

Length, male 64 in., including 16-in. tail; female 56 in. with 12-in. tail. Shoulder height 27 in. Male to 150 lb, female to 80 lb. Gray sprinkled with black. Legs and under parts yellowish-white. Unlike dogs, have hair between toes, slant eyes, heavy build, curved canine teeth, sharp muzzle, and dense underfur; have larger size and much smaller ears than coyotes. Teeth: I 3/3, C 1/1, P 4/4, M 2/3.

Originally ranged through most of North America, though present range is now probably limited largely to Canada and far north country. Still found in Wisconsin, Michigan and Minnesota but cannot continue to survive in these areas much longer; too little favorable habitat remains.

(Continued)

Gray Wolf

Apparently was never established in parts of California, Nevada, Oregon, Arizona, and Utah. About 24 subspecies in America.

Mates probably for life, breeding January to March. 3 to 13 young born at 63 days; blind about 9 days, suckled 6 to 8 weeks, run with family for about 1 year; run in packs at various seasons of year after that. May live 15 years.

Food largely sheep, cattle, deer, and other animals, run down by packs and eaten on spot. One meal a week is sufficient. Some food may be stored by burying. Rarely eats any vegetable matter. One large wolf alone can bring down a full-grown steer and kill it. Home range of male probably less than 150 miles, females slightly less. Tracks: F 4 × 5½, 4; H 4 × 5½, 4.

Wolves have probably been driven from most of United States. They disappeared from New Jersey last century, from the Adirondacks in 1893, and from Pennsylvania in 1907. They find little place in a stock-raising country. Fur durability, 50 percent that of otter. Eyeshine greenish-orange. Was once a valuable predator which destroyed weak and sick members of species preyed upon. An endangered species.

Order Carnivora./Family Canidae [Dogs]

Exterior parts of a dog are: Head: forehead 1, eye and eyelash 2, crest of nose 3, nose 4, mouth and upper lip 5, flews 6, cheeks 7, ear 8, ear flap 9, back of head 10. Neck: nape or poll 11, throat 12, dewlap 13, crest 14. Trunk: withers 15, upper chest 16, back 17, chest 18, under chest 19, loins 20, cross 21, abdomen 22, hindquarters 23, croup or rump 24, hip 25, root of tail 26, tail 27. Forelegs: shoulder 28, upper arm 29, elbow 30, forearm 31, wrist or carpus 32, metacarpus 33, toes 34, claws 35. Hind legs: hip joint 36, upper thigh 37, stifle of knee joint 38, lower thigh 39, tarsal joint 40, heel 41, hock 42, metatarsus 43, toes 44, claws 45.

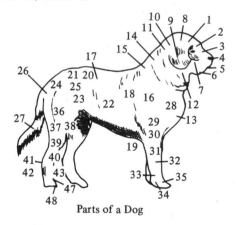

Parts of a Dog

Dog dentition: I 3/3, C 1/1, P 4/4, M 2/3. Order of appearance of teeth: milk teeth: first and second incisors 4 to 5 weeks, third incisors 4 weeks, canines 3 to 4 weeks, first and second premolars 4 to 5 weeks, third and fourth premolars 3 to 4 weeks. Permanent teeth: incisors and canines 4 to 5 months, premolars 5 to 6 months, first molars 4 months, upper second molars 5 to 6 months, lower second molars 4½ to 5 months, third molars 6 to 7 months.

6 to 12 *puppies,* born 59 to 63 days after male *dog* breeds with female *bitch;* blind, unable to hear for 10 to 12 days; nurse until 4 to 8 weeks of age, become biologically independent in less than 1 year and then able to breed, male being capable of inseminating female at any time after maturity is reached. Female in heat usually twice a year (diestrous) for a period of 1 to 3 weeks. Puberty, 7 to 10 months. Ordinarily mother alone cares for young. Normal body temperature, 101 to 102°F. Respiration, 15 to 20. Pulse, 70 to 120. Life-span to 20 years.

Dog management is a responsibility which should not be shunned by a dog owner, because within certain limits a dog is what its master makes it. Punishment of dogs of intelligence rarely needs to be physical or cruel; rather it may be accomplished by denying dog privileges it appreciates. Don't keep a dog tied or confined more than necessary as this may make it mean. Don't tease any dog, sic it onto a person, allow it to get into a fight, or let it chase cars. A few days' concentrated effort may be worth the dog's life and your neighbor's respect. Never hit a dog with the hand, kick it, or hit it with a stick, lest it learn to fear hands, feet, and sticks and to attack them on passers-by. Make no exceptions to established rules.

From puppyhood on, dogs are subject to many diseases. From 8 to 10 weeks of age they may have attacks of distemper which may be fatal. Dogs so attacked have a high fever, stop eating, vomit, and have scours, but once recovered are ammune. Distemper can be prevented by serums. Rabies, most dreaded dog disease, may make a dog a community menace and should be controlled constantly. It may be prevented to a limited degree by serums. Tapeworms are destroyed by 1 gr. of freshly powdered areca nut per pound of dog after withholding food for 18 to 24 hours. Mange is treated by washing with green soap and hot water, then applying sulfur ointment for 3 days. Hookworm, common in South, is treated with 2 drops per pound weight of dog (1 for puppies) of carbon tetrachloride in strong gelatin capsules. Roundworms are treated by giving 2 to 4 tsp of mixture of ½ oz each of spigelia, senna, and licorice syrup, every morning for 3 days. Consult a competent veterinarian for assistance.

Dogs suffer from lack of exercise and faulty diet. It is obviously unkind and injurious to confine a naturally active dog for a lifetime in a small apartment and to feed it all the rich food it will eat. Pampering, overcrowding, and lack of exercise invite disease. One simple meal a day preferably at evening is best, taking care that clandestine visits to garbage dumps are avoided as much as possible. Prepared dog foods are not expensive in long run, particularly for house dogs, but even they should be used sparingly.

Collie Dog
Canis familiaris
Height 20 to 40 in. at shoulder. Weight 40 to 60 lb. Long-haired, graceful, and dainty in most habits. Nose long and slender. Ears drooping, erect or partly erect at will. Colors, black and tan or rich brown, with clear white throat; some, white; others, clear bluish gray. Intelligence is not dependent on exhibition standards; few breeds show superior intelligence.

In England, Scotland, and Wales, breed has been indispensable to men who had responsibility over large herds of sheep. This value is same wherever sheep are herded in any considerable numbers. Animals vary in different parts of world but are popular wherever men like dogs.

Collies have been known to herd as many as 3,000 sheep, separating them so that rams were driven into one corral, ewes and lambs into another; this in spite of the well-known tendency of sheep to stampede and follow any leader in any direction. Records of collie memories, loyalties, and unfortunate disloyalties are surprisingly interesting.

Collie Dog

German Shepherd or Police Dog
Canis familiaris
May stand 22 in. high at shoulder. Hair comparatively long, black with a buff base. Muzzle black. These dogs resemble wolves but ordinarily lack fur between toes and dense underfur mixed with gray of wolf. Also lack slant-eyed, sharp-muzzled, curved-tusk appearance of a wolf. More powerfully built than collies and possibly more intelligent.

Developed in Europe, where an animal of exceptional intelligence was needed to do work that horses could not do. They have, unfortunately, been used both for good and for evil. Where they have not been trained, have used their own intelligence to satisfy their natural desires as preying animals. Particularly susceptible to disease when young. Color black, iron-gray, ash-gray, fawn color with black saddle, wolf-gray, black with gray, tan, fawn, or cream markings.

Deeds of this breed include exceptional ability in helping police and as dogs of war; even more remarkable is their ability as "seeing eyes" whereby blind people enjoy freedom otherwise denied to them. They make superior actors even in sound films because they may be directed by motions rather than by words. They deserve thoughtful, responsible masters.

Police Dog

Foxhound

Pointer

Whippet

Irish Terrier

Foxhound
Canis familiaris
Medium-sized dogs, bred for ability in following trails of game by scent. Black, tan, and white, with black area usually on body upper parts, tan just below hips and on sides of head, and white on legs or replacing other colors in varying areas. Hair relatively short. Legs long, strong, and sturdy. Tail brushless. Endurance remarkable. Sense of smell phenomenal. Works well in rough country with other dogs. Kindly.

Although bred over 300 years, foxhounds retain primitive coloration. American foxhounds are lighter and more active than English breed, developed because of greater roughness of country. Closely related beagles are shorter-legged, and bloodhounds are darker, heavier, and much slower, usually with little or no white; all three have typical ears.

Records of foxhounds show varying ability to follow trails. Bloodhounds have been reputed to be able to follow a trail 30 hours old, which foxhounds cannot do even under most favorable conditions. Foxhounds couple speed with ability to follow a fresh trail well and long, through difficult cover, and are ideal for this purpose. Distinct breed characteristics are a creation of only the last 200 years of breeding.

Pointer
Canis familiaris
Weight 44 to 60 lb, compared with 65 to 75 lb for griffons (wire-haired pointers). Retrievers average 60 to 70 lb, setters 40 to 55 lb, spaniels 18 to 60 lb, cockers 18 to 24 lb, springers 45 lb, and clumbers 35 to 60 lb. Pointers are black and white or liver and white, short-haired, with strong but slender legs, large heads, long square muzzles, and compact strong bodies. They are intelligent, affectionate, hardy, loyal, and tremendously active when free.

Developed in Spain in Middle Ages, taken to France early in seventeenth century and from there to England, where they were crossed with English terrier and foxhound to reduce weight. Prize animals are lean, with keen eyes and noses, ability to learn, and plenty of stamina and courage. Tail should be straight, strong, and rigid when animal points.

Used to indicate location of game so that it may be shot by hunter and to flush it only on command, then to retrieve killed or injured bird without further injury. It uses scent as a chief means of locating game. Breed is used primarily for hunting birds. Extensive field trials are held to determine relative ability of individuals.

Whippet
Canis familiaris
Height 15 to 21 in. Weight 10 to 28 lb, ideal, male 15 lb, female 13 lb. Head long and narrow with pointed muzzle. Neck, legs, body, and tail slender, with tail frequently held between legs when at rest. Hair short and close. Color red, black, white, brindle, fawn, blue, or mixed, but commonly white on throat and muzzle. Judged more by performance than by appearance.

Developed in England in middle of nineteenth century, probably by crossing English greyhound, Italian greyhound, and a variety of terriers, particularly Manchester terrier. Whippet racing, poor man's racing event, has long been popular in England and is becoming so in America. Affectionate, intelligent, loyal dogs that make good pets.

In racing whippets, animal is held with front feet on starting line and hind legs off ground. At starting signal, animal is thrown into stride by handler and pursuit of real or electric rabbit is on. Usual distance is 200 yd, which has been covered in less than 12 sec, at about 52 ft sec. or 35 mph; this is faster than any other known domesticated animal.

Irish Terrier
Canis familiaris
Height about 18 in. Weight about 25 lb. Related Airedales: height 22 in., weight to 50 lb. Irish terrier red, Airedale tan or sandy with grizzled gray neck and saddle. Head at right angles on strong neck. Hair on lips and chin usually long. Holds ears down and tail up or partly over back and rather stiff. Airedale developed from otterhound and Manchester terrier, appearing about 1860 outside native Yorkshire, England.

Related Welsh terrier smaller than either Irish or Airedale; height 15 in., weight 20 lb; black and tan. Short-haired terriers include short-haired fox terrier, bull terrier. Long-haired and generally smaller Skye terrier, Sealyham terrier, Scotch terrier, and the white West Highland terrier.

Irish terriers are essentially one-man dogs with intense loyalty to masters. Make excellent watchdogs which brook little interference. Excellent hardy hunting dogs having keen noses, and are such good fighters that they are used commonly in destroying rats and other vermin. Lack speed of some other hunting dogs. Excellent as Red Cross dogs.

Fox Terrier
Canis familiaris

Length about 2 ft, with 1-ft tail unless it is docked. Height at shoulder about 15 in. Weight about 8 lb. Two major subbreeds divide on long- or short-haired characters. Usually stands erect, with forelegs stiffly straight and head held high. Color: black areas on white background, commonly with tan, particularly on sides of head. Usually highly nervous.

Bred in Europe and America originally for purpose of going into a fox hole and pulling fox out. This requires a small dog with excellent fighting ability and considerable perseverance. Eyesight and nose of the fox terrier are not exceptionally good but he is full of pep.

This book was written in memory of a boy who loved the fox terrier "Wow" (illustrated) more than all other pets combined. An intelligent and keen sporting dog. A lively, affectionate, and very adaptable friend in a household.

Fox Terrier

Boston Terrier
Canis familiaris

Weight 15 to 25 lb, with three classes based on weight recognized in judging ring. Middleweights, 17 to 22 lb. Color dark brindle to black, with white on muzzle around neck, on chest and all or part of forelegs and on hindlegs below the hocks. Hair: short, smooth, stiff, shining, and of medium fineness. Ears commonly trimmed. Tail "screwed" or docked but always carried low.

Originated in Boston about 1870, where it was developed from a cross between bulldog and bull terrier. Related French bulldogs are little bulldogs with widespread ears. They lack pep and slenderness of Boston terrier, which probably comes from terrier blood. Legs of Boston terrier are straighter and more slender and white areas larger than in French bulldog.

Boston terriers are most affectionate animals and possibly the most distinctly American breed. Their intelligence and loyalty make them popular as house dogs, particularly where quarters are limited and where furniture must be kept free from dog hairs. Their cleanliness and relative freedom from odor also make them useful as pets.

Boston Bull Terrier

Coyote
Canis latrans

Length, snout to tail tip about 4 ft, ⅓ being tail. Average weight, male to 50 lb, female 30 lb. Ears conspicuously large and erect, providing excellent field character. Color pale brown sprinkled with gray, black, or sometimes nearly white, with under parts nearly white and ears darker. Looks like a rather large grayish collie dog or pale yellowish, small German shepherd dog with drooping tail. Teeth: I 3/3, C 1/1, P 4/4, M 2/3. Tracks: F 3 × 3½, 4; H 3 × 3½, 4.

Found essentially in open country from northern British Columbia and middle Alaska east to north of Lake Michigan, through southern Mackenzie and south to Central America. Established elsewhere by escaping from captivity in New York, Florida, and other places. Range now larger than originally. Individual range, 6 sq mi.

Probably pairs for life. Breeds in January. 3 to 10 young born 63 days later, blind but furred; nurse 2 weeks; may be taken from den at 3 weeks; at 6 weeks, may venture out on own; run with parents by July. Family together, at least until early fall. Breeds first winter. Life-span, if lucky, about 13 years.

Food: 8,263 stomachs show 33 percent rabbits, 25 percent carrion, 18 percent rodents, 13 percent sheep and goats. Stomachs of 161 "pegleg" coyotes show 21 percent rabbits, 21 percent sheep and goats, 35 percent carrion, 12 percent rodents. Full meal equals two ground squirrels, totaling 1¾ lb. Winter diet of 2,589 cases was 36 percent carrion, 34 percent rabbits, 15 percent rodents, 8 percent sheep, 3 percent deer.

Important check on grass-destroying rabbits and rodents. General poison campaigns are scarcely justified. Can run 45 mph, hunts in relays, rarely in packs, therefore about only mammal which can catch destructive jack rabbits. Of course may kill deserted newborn calves, horses, sheep, and kids. Fur has good sale value in season. Also known as brush wolf, American jackal, and little wolf. May mate with domestic dog with offspring called a coydog.

Coyote

Red Fox
Vulpes fulva

Length to 41 in., including 16-in. tail. Weight to 14 lb. Male larger than female. Tail bushy. Ears large and pointed. Nose pointed. Sexes colored alike, a golden brown or reddish above and white beneath; with white tail tip and black legs. In young, black on muzzle and back of ears. Color phase black. Teeth: I 3/3, C 1/1, P 4/4, M 2/3.

Found in wooded and farm lands particularly where there is mixed cover, from the Arctic to the Mexican border. In Canada, only the western edge of British Columbia, a few islands off the east and west coasts of the

(Continued)

Red Fox

mainland, and the highest arctic regions do not have the red fox. On the U.S. mainland only deep Southeastern area, parts of the Prairie States and some parts of the Southwest and Rocky Mountain States are without the red fox. Apparently most abundant in country where quail, grouse, and rabbits also abundant.

Monogamous, at least for the one breeding season; but may mate for life. Mates in January to late February. 4 to 10 young, born 51 days later; blind 8 to 9 days, in den 3 to 5 weeks, then may come out to play but retreat to safety until about 3 months old; independent at 5 months; full-grown at 18 months. Male feeds female and young and leads enemies away from den at risk of life. Den clean. Life-span perhaps to 15 years; more commonly less than 8 years.

Food: probably the world's greatest destroyer of mice. Eats carrion, fruit, vegetables, game, and poultry, but most of all such vermin as mice. Has remarkable sense of smell, good sight, and excellent hearing, intelligence, and endurance. Active year round, day or night, but mostly at night. Does not climb trees. Tracks: F 1¼ × 2, 4; H 1¼ × 2, 4; St 18.

Extraordinary mouse destroyer and sporting animal. Ranch raising of silver-fox mutations began in 1887 on Prince Edward Island. Changes in styles have at many times since 1887 made fox ranching unprofitable. *V. fulva* is the Reynard of the fox hunt. Durability of pelt 40 percent that of otter. Best friend of livestock farmer, orchardist, and grain farmer, but not of poultrymen.

Gray Fox

Gray Fox
Urocyon cinereoargenteus
Length to 40 in., including 12-in. tail. Weight to 12 lb. Sexes about equal in size and in coloration. "Pepper and salt," black and gray above with reddish-brown along sides, and gray and tawny beneath. Tail black-marked. Fur shorter and generally inferior to that of red fox, and animal lacks its rank odor. Teeth: I 3/3, C 1/1, P 4/4, M 2/3.

Commonest in wooded and farm areas from New Hampshire and New York to North Dakota; south through Texas, north to Colorado, Utah, and southern Nevada west to California then north along coast to Oregon. Ranges south to Baja California and Mexico. Range is being extended; numbers possibly increasing.

Mates in January. Young born 50 to 60 days later, in clean den in hollow log, cave, or tunnel; blind and helpless at birth; take solid food at 6 weeks of age. Family breaks up in August to September, though parents stay together through year.

Food: almost anything, a full meal equaling ⅓ lb of rabbit. Stomach survey show 35 percent of food rabbits, 16 percent mice, 15 percent vegetable, 6 percent pheasants, 5 percent poultry. Some of animal matter is carrion. Does not store food as does the red fox, lacks its endurance, is more nocturnal and less intelligent. Tracks: F 1½ × 2, 4; H 1¼ × 1¾, 4; St 16.

Probably an enemy of game, but this is offset considerably by destruction of mice. Certainly no worse than cats or some dogs as enemy of wildlife. When pursued by dogs, takes to trees, unlike red fox. Durability of fur 40 percent of otter. Chiefly nocturnal but may travel during the day; ordinarily it sleeps during the day. May attain speed of 28 mph but 18 to 20 is more normal.

Canada Lynx

Canada Lynx, Loup-cervier
Lynx canadensis
Length 3 ft, tail 4 in. Male perhaps 5 percent larger than female. Height 1½ ft. Weight to 40 lb. Color gray with brown mottlings; usually lighter-colored than bobcat. Feet exceptionally large. Tail conspicuously short, black-tipped. Fur long and loose. Skull broad. Ears remarkably tufted in older animals. Calls like the bobcat, a meow, a caterwaul, and a screech. Teeth: I 3/3, C 1/1, P 2/2, M 1/1.

Generally found in or near thick woods, usually away from habitation. More northern than bobcat. Ranges from New York and northern Michigan, north to Hudson Bay and Alaska, south in Rockies to Colorado and in Sierras to Oregon. Considerably restricted in range in eastern and southern portions. As man intrudes into its range, the lynx moves out.

One to four young born in spring, 2 to 3 months after breeding; blind 10 days, suckled 2 to 3 months. Breed first winter. Father assists in bringing food to and caring for young, so there is true mating. Young stay with mother for nearly 1 year. Clean in habits, particularly in den, which may be in a hollow tree or windfall. Life-span span 12 years.

Food much same as for bobcat, with emphasis on rabbits and their kind. Number varies greatly with food supply. Pair may have territory 50 miles across. Animal migrates with seasons and with food supply. Does not hibernate. May store food. Tracks: F 3 × 3, 4; H 3 × 3, 4; St 18.

Probably useful as rabbit and rodent destroyer. Pelt has some value and makes a beautiful robe, soft, light, and warm. Should be left in the wild. Inoffensive to man unless cornered or unless young and old become separated. Can fight dogs most effectively. Fur durability, 25 percent that of otter. Future prospects for Canada lynx not encouraging. An endangered species.

Bobcat, Wildcat
Lynx rufus
Length 38 in., tail 6 in. Height 15 in. Weight to 40 lb. Reddish-brown, black-spotted, being lighter beneath. More slender than Canada lynx, with short hair, legs, feet, and tail. Ears not tufted conspicuously. Skull narrow. May appear longer but is lighter than lynx. Teeth: I 3/3, C 1/1, P 2/2, M 1/1. Tracks: F 1¼ × 2, 4; H1¼ × 1½, 4; St 14.

Found in wooded or brushy land, near to or remote from settlements, from Nova Scotia to Gulf of Mexico, west through Mississippi Valley; in Pennsylvania nearly extirpated with help of high bounty, now removed.

Breeds in February and March. 1 to 4 young born about 62 days after breeding takes place; blind 10 to 13 days, suckled about 2 months, reared by both parents, which mate for season at least. Young tames relatively easily but becomes dangerous as it gets older; weighs 8 to 12 lb first fall. Den in cave or hollow tree, clean at all times.

Food: of 186 stomachs examined, 126 had rodents, 65 game, 11 stock (carrion), 9 birds. Another study shows food to be 44.5 percent harmful mammals, 20.5 percent useful mammals, 19.6 percent vegetables and soil, 7 percent parasites, 3 percent birds. Probably worst enemy of snowshoe rabbits and cottontails and number varies with these species. Captive animals have lived to 25 years; in wild an old bobcat is 12 years.

Serves essentially as useful check on rabbits and such animals as are destructive to man's cultivated and wild plants. A full meal for a bobcat equals ⅗ lb of rats. Certainly no reason why species should be completely destroyed, unless some other equally effective check on prey is available.

Bobcat, Wildcat

Domestic Cat
Felis domesticus
Length about 2½ ft, ⅓ being tail. Footprints almost round, showing no toenail marks; hind feet step in tracks of larger forefeet, alternating in zigzag to straight line, about 9 in. apart. Tracks: F 1¼ × 1, 4; H 1 × 1, 4; St 6. Fur varies with breed. Pupil of eye of Persian cat circular; of others a vertical slit.

Developed in Eocene from *Miacis,* extinct saber-toothed tigers separating in Oligocene. European tabbies or tortoise shell (usually female) a cross of European wildcat and Egyptian cats; Manx, tailless, from Oriental cat; Siamese, beige to seal brown, with black face and feet, blue eyes, from African jungle cat; Persian, long-haired, Asiatic.

3 to 12 kittens born blind, helpless, 62 days after breeding; 2 to 3 oz; cared for by mother. Iris of adult usually yellow, of kittens blue or green. Male gives no care to young; either sex promiscuous at breeding time. Female in heat 1 to 3 weeks. Male usually larger than female in all breeds. Pulse, 120 to 140. Respiration, 24. Temperature, 101.7°F. Life-span to 10 years.

Probably cats gone wild are among worst destroyers of game. Favor fish as food but eat birds and mammals readily. Individuals known to have killed 4 rabbits a day, 100 chickens a season, and 40 turkeys in a few days. Catch prey by surprise and sudden dash and piercing with long slender teeth. Can see with little light. Smell poorly.

Known to be a disease carrier, 43 hydrophobia cases in New York City alone traced to cats. Probably worst enemy of bird life and should be confined, particularly during bird nesting season. Cannot be trusted. Probably loved, hated, or tolerated by most persons but not economically valuable; however, cats do eat many destructive rodents such as rats and mice. Fur in trade, known as "genet."

Domestic Cat

Cougar, Puma, Painter, Mountain Lion
Felis concolor
Length to 9 ft, tail 3 ft. Weight to 200 lb. Male about 15 percent larger than female. Uniformly yellow-brown but darker above and lighter beneath. Like a large slender small-headed cat. Fur relatively short, close, and uniform. Whiskers rather prominent. Toes retractile. Teeth: I 3/3, C 1/1, P 3/2, M 1/1. Tracks: F 4 × 4, 4; H 4 × 4, 4; St 18.

Found in forested areas, for most part, but also in rocky canyons and even desert lands. Ranges from coast to coast, between Quebec and Vancouver Island, south to Patagonia including a number of varieties.
(Continued)

Cougar

Eastern variety once ranged from southern Canada to Gulf of Mexico but is now greatly restricted.

Pairs.   Both parents care for one to five young, which are born any time of year, 91 days after breeding; blind, 8 to 9 days, first teeth 18 days, crawl at 7 weeks, suck meat at 9 weeks, eat meat at 3 months, weaned at 3 to 4 months.   Black-spotted over yellow, for first 2 to 18 months.   With mother 1 to 2 years.   Live to 20 years.   Almost exclusively nocturnal. An endangered species.

Food, meat, a full meal equaling 7 to 8 lb of deer meat.   Can drag 900 lb moose 300 ft on snow.   Of 43 stomachs examined, 34 had deer, 2 corn, 1 each skunk, cat, hog, and calf.   Can jump up 15 ft or down 60 ft in safety and often ranges 20 miles a night over a 60-mile territory claimed by individual.   Bites neck of prey.

Troublesome destroyer of domestic animals and game where abundant but a valuable check on plant eaters where other checks do not exist. Known in rarest of cases to attack man.   Favors horses and deer much as black bears favor pigs; coyotes, sheep; and wolves, cattle.   Should not be completely destroyed.

Ocelot
Felis pardalis
Length 4 ft, with 15-in. tail.   Weight 35 lb.   Grayish-yellow with large fawn-colored, black-bordered spots.   Slender, like well-proportioned cat. Ears usually black, with white spot below.   Tail usually spotted or ringed.   Hair soft and not thick, but individuals vary greatly.   Several varieties recognized.   Small spotted cat without rosettes of the jaguar.

Haunter of woods and thickets, from southwestern United States and Mexico through to South America as far as Paraguay.   Allied margay with black lines and bands is found from Mexico to Peru; other relatives include linked ocelot with chainlike markings, and long-tailed ocelot.

Little known of family life but probably breeds in June and bears two young (apparently always two) in September to October.   Farther south than United States, breeding and bearing young probably take place at any time of year.   Young blind at first.   Den usually in a cave in rocks, well-lined with comfortable bedding.

Food about 5 lb of flesh a day, including reptiles, birds, mammals, and amphibians; in fact, almost any form of animal life it can overpower, even though evidence indicates that cold-blooded animals may cause indigestion and cat flesh an itching.   Each animal hunts alone, seizing prey by neck and holding it to ground until death.

Probably useful in controlling smaller plant eaters; unquestionably injurious to poultry and smaller forms of livestock; particularly destructive of pigs, kids, and lambs, but apparently does not attack man though fights powerfully if cornered.   Caught young, can be tamed relatively easily; often kept as a pet.   An endangered species.

Jaguar, American Leopard
Felis onca
Length, total to 79 in., average male 7½ ft, female 6½ ft.   Height 28 in. at shoulder.   Weight around 250 lb.   Yellow above and white below, with ½ to 1-in. spots of black, often of four to five around a central spot but varying greatly in different parts of body.   Fur relatively short.   Tail shorter than in African leopard.

Found from Texas, New Mexico, and Arizona south through Brazil and Paraguay, living in jungles of heavy vegetation for most part.   Southern limit is about northern boundary of Patagonia and original northern limit rather wide in North America.

Jaguars pair definitely, breeding in January, with two to four cubs being born about 100 days later.   Female probably assumes all family responsibilities.   By 6 weeks, cubs are as large as domestic cats.   At 1 year of age, young can shift for themselves; at 3 years of age, they may mate and breed; young with mother for 2 years probably; litters apparently spaced at 2-year intervals.   May live for 20 years.

Food almost any common animal, since it can kill deer, horses, cattle, sheep, tapirs, monkeys, pigs, fish, and birds.   It can pursue prey in dense vegetation, into trees, and even into water.   As a hunter, it relies on its ability to ambush a victim.   Has a great appetite for sea-turtle eggs.   Assumed to develop a home territory and to guard it.

As a destroyer of domestic animals, it has roused the enmity of all who raise cattle and horses for profit.   In modern times, it has learned to avoid men with guns.   Pelt valuable and of excellent quality.   An endangered species.   Should be protected.

Ocelot

Jaguar

African Lion
Felis leo, Panthera leo

Length male to 8 ft, plus 3-ft tail. Height at shoulder 4 ft. Weight to 583 lb. Female usually 1 ft shorter and more slender than male and lacks conspicuous mane of adult male. Color yellow-brown without stripes and only slightly lighter beneath. Mane almost black in older animals, lighter in young.

Found throughout Africa from the Cape to Algeria and Ethiopia; in Asia from Mesopotamia and southern Iran to northwest part of India. Formerly roamed through Syria, Arabia, southeastern Europe, and in prehistoric times through Germany, France, Spain, and British Islands. Now restricted to Africa for most part.

Pairs for life, not promiscuous. Two to four young born 12 weeks after mating. Young born with eyes open; helpless for several weeks; about size of large cat, with frizzled fur. Young female suffers at teething time and may die; weaned at 3 months. Mane appears at 3 years and is at best at 8 years. May live to 25 years. Nocturnal and diurnal. Even where lion is abundant, it is rarely seen during the day, largely because its color blends in well with its habitat. Conceals itself in rocky country in low places.

Food: flesh, fresh or carrion. Attacks almost invariably from ambush, striking at flank or throat below jaw of buffaloes, zebras, antelopes, or giraffes in Africa. Can leap 12-ft fence or more than 40 ft or dash 100 yd at 60 mph. Pulse, 40 to 50. Respiration, 10.

Serves role of natural check on plant eaters. Prized game animal. Old individuals may become man-killers. Normally will not annoy men unless in danger; then fights powerfully with teeth and claws. Appears in earliest recorded history, in the Bible, and figured prominently in early Roman sports. Can be trained to perform well.

African Lion

Leopard
Panthera pardus

Length, male, body 57 in., tail 38 in. Weight 200 lb. Female may average smaller, some as short as 5 ft total. Tawny yellow above and whitish below, with numerous rosettes of dark or black circular spots resembling animal's footprint over body, these being smaller on head. Tail sometimes ringed. Fur durability, 75 percent that of otter.

Found in Africa, from the Cape to Algeria; in Asia, from India, Burma, Thailand, Ceylon, and Malay Islands west to Syria. Formerly, through Europe to Spain, and Great Britain. Black leopards or "panthers" with deeper ground color are found only in Asia. African leopards have smaller, more solid spots.

Thought once to be a hybrid between a male panther and lioness, but this is not so. Two to four young usually born in spring, 3 months after breeding time. Family may stay together until young are full-grown; adults may live 28 years. Hunt sometimes in pairs but more often singly.

Food captured by long leaps, often by pursuit, seizing prey by throat, breaking spine or strangling it, sometimes tearing front with powerful claws. Eats antelope, monkeys, deer, calves, sheep, goats, pigs, and donkeys, and favors dogs, the latter sometimes being caught in broad daylight, though most hunting is done at night. Can swim. Can leap 10 ft up.

One of most wary and most treacherous of beasts; powerful fighter when wounded. Less suspicious of traps than tiger and so more commonly trapped with goat or dog as bait. Can climb trees easily, where lions and tigers do not, so must be hunted differently; it can run up a smooth trunk as easily as a monkey and turn on its attacker. An endangered species.

Leopard

Bengal Tiger
Felis tigris

Length 6½ ft plus 3-ft tail, or to 11 ft. Weight 500 lb or more. Female slightly smaller, with lighter and narrower head. Old male has long hair on cheeks but no mane. Hair relatively short and smooth. Fawn-colored or reddish, with white under parts, and with many black transverse stripes. Highly variable.

Essentially of jungles, found only in Asia and particularly in India and from northern China to Malay Peninsula. Unknown in Ceylon, Philippines, and Malay Archipelago but common in Java and Sumatra. Animals of warmer climate smaller, more brilliantly colored, and more powerful than those of cold climates.

Adults pair; breed in alternate years. Two to four young born 15 weeks and 4 days after breeding, usually two, one of each sex. At 6 weeks, young follows mother; at 7 months, young can kill independently; family may run together 2 years. Cubs become mature at about 3 years, when family breaks up. Captive animals may live 25 years. Respiration, 6.

Food, flesh, with individuals favoring different kinds. Some animals favor game such as deer or wild pigs; others favor cattle, horses, sheep, or pigs; and a few favor man, the man-eaters usually but not always being

(Continued)

Bengal Tiger

Stellar Sea Lion

Alaska Fur Seal

Harbor Seal

females. Prey is seized in a rush or spring of about 15 ft or less, grasping forequarters with claws and biting throat.

Known to have killed as many as 60,000 cattle, sheep, and goats in India in 1 year and as many as 4,000 human beings in 6 years; one animal credited with 80 men in 1 year. Male usually roars when hit with bullet but female does not. Prized animals for big-game hunters but feared greatly by natives. Hunted sometimes on elephants. An endangered species.

Order Carnivora./Family Otariidae

Steller Sea Lion
Eumetopias jubata
Length to 10 ft, female smaller. Weight of male averages 1,500 to 2,000 lb. Shoulders heavy. Ears small, not more than holes. Appears sleek, wet, and shining, even when dry. Black, dark gray, or yellowish- to dark-brown. California sea lion, *Zalophus californicus,* much smaller; about 600 lb, but to 1,000. Teeth: I 3/2, C 1/1, P 4/4, M 1/1.

Found along Pacific Coast from Bering Sea south to Channel Islands in California; most commonly seen by tourists on rocks off California coasts, where their presence is well advertised. Bask on rocks or dive rather awkwardly into sea but swim beautifully. Unlike seals, can bend hind flippers under and forward.

One bull has many cows in his harem. Breeds in Bering Sea in July and at San Francisco in summer, breeding taking place just after birth of pup, which is born singly 12 months after breeding. Pups learn in 1 week to swim and care for selves, maturing and breeding when about 3 years old.

Food, fish of many kinds, crustaceans, squid, and other marine animals. Animals in water not disturbed by approach of men but take flight when they approach on land. These sea lions outnumber all their kin from California to British Columbia. Most circus "seals" are California sea-lion cows.

Not demonstrated as harmful to commercial fish. Some value as pet, as circus animal, as curiosity, and as source of oil and leather. Fur value-less as compared with fur seal. Bulls killed for stiff whiskers, excellent for cleaning opium pipes; for gall bladders used in medicine; and for genitals that are dried, ground, and used as aphrodisiac and rejuvenator.

Alaska Fur Seal
Callorhinus ursinus
Length to 75 in., female 50 in. Weight, male 600 lb, female 135. Upper parts black, gray over neck and shoulders, flippers red-brown. Female gray above, brownish beneath. Beautiful sleek animal, with essence of grace in movements. External ears mere holes. Teeth: I 3/2, C 1/1, P 4/4, M 1/1.

Found in Bering Sea and Pribilof Islands in summer, and female, south to California in winter, breeding in northern part of range. Winter migration lasts 7 to 8 months, bearing west then turning east to California, then north to starting point. Fur durability, 80 percent.

Adult male bull over 6 years old reaches breeding ground first and selects territory. Young male *bachelor* follows and attempts to hold space. Female *cow* follows, gives birth when to 3 years old to young *pup* on rookeries, is accepted into harem and breeds immediately. Bears one young each year thereafter. Males fight severely.

Food almost solely fish and squid. Enemies, killer whales and man. Temperature of body, 101°F. If the temperature reaches 103°F, they pant vigorously. Fur of commerce is covered with long hard guard hairs which are plucked to bring out beauty of underfur. One bull may have over 40 cows in harem. Rest season, July.

Valuable fur bearer, the harvest for some years prior to development of Second World War being controlled by international agreement, which Japan in 1940 refused to continue to observe. Most destructive practice consists in killing nursing mothers at sea rather than superfluous males. Breeding males eat nothing May to August.

Order Carnivora./Family Phocidae

Harbor Seal
Phoca vitulina
Length to 60 in. Sexes equal in size. Color variable from yellow-gray to gray, with dark brown or blackish spots. Young animals white. Limbs relatively short for a seal. Hair exceptionally coarse. At least four species or subspecies recognized over a wide area. Pacific species, *P. richardii.* Teeth: I 3/2, C 1/1, P 4/4, M 1/1.

Found in harbors, along shores, and at mouths of rivers and bays. Atlantic species ranges from Carolinas north, and Pacific species from California north to Alaska and Pribilof Islands. Easily recognized by spotted coat and relatively small size.

Young born in early spring, when animals are found together in small herds but never congregating on large rookeries as the fur seals do. General reproductive habits probably similar to others of family. Fur durability, 25 percent.

Food essentially fish, squids, crustaceans, and other forms of marine life. Examination of 81 stomachs by one investigator showed 95 percent of food fish, remainder invertebrates. Another investigator found food half fish and half invertebrates in 35 stomachs. Probably a rather general feeder.

Of little value for fur, hide, or oil and can hardly be sought commercially because individuals are not congregated into large herds where many may be killed at one landing. Some relatives such as harp seal are hunted for their oil in great numbers off Gulf of St. Lawrence.

Order Carnivora./Family Phocidae

Hooded Seal
Cystophora cristata
Length to 96 in., or rarely 11 ft. Weight of male 1,000 lb, of female to 900 lb. Blue-black on back with lighter belly and varied with lighter spots. Male has muscular sac from center of head to nose which can be inflated. Unlike many other seals, has four, not six, incisors on upper jaw in front and two, not four, on lower jaw.

Once one of the most abundant North Atlantic seals, ranging north of New England to Newfoundland and Greenland. Only exceeded in size in Atlantic by bearded seal. May herd off Newfoundland and Labrador coasts where fishing is good; in summer, off southeast Greenland; February to March, off Labrador on ice floes.

Single young born February to March, shapeless, furry, steel-gray, increasing 3 to 4 lb each of first 8 days after birth and reaching full size at 4 years when it becomes mature and breeds. Young nursed by mother for 2 weeks, then gets own food. Yearling seals known as "bedlamers." Female an excellent mother.

Food largely fish, good-sized fish often being swallowed whole. Both parents will defend young and themselves vigorously. May travel rapidly over ice and when kept from water by ice may suffer greatly from skin burning. This species more quarrelsome than most seals with which it is found.

Hide tanned to make leather; flesh, or blubber, used for oil. One ship in 1911 destroyed 49,129. Animals pursued until they take to ice floes, there exhausted and are easily killed. Industry has been ruined by unwise harvesting practices. Seals are great destroyers of valuable fishes. Numbers so small that they are of little economic importance.

Hooded Seal

Harp Seal
Pagophilus groenlandicus
Length to 5 ft. Weight to 400 lb. Rather slender for a seal. First toe of forefoot not longer than second. Male white or yellowish-white with blackish face and curved blackish band that extends down each side and meets over shoulders and over tail. Female and young with less definite pattern, usually mottled. Teeth: I 3/2, C 1/1, P 4/4, M 1/1.

Found in Arctic seas of both Atlantic and Pacific oceans and originally described from Greenland and Newfoundland area. Related ringed seal *Phoca hispida* is smaller and ranges somewhat farther south, being originally described from Labrador area. Harbor seal *Phoca vitulina* (4 ft long) ranges still farther south.

Young born in spring when animals are found together in small herds but not congregating in enormous herds on established rookeries as fur seals do. At birth, young harbor seals are covered with a thick white woolly down. Young ringed seals may be born in an excavation in a bank of snow, remaining there several weeks.

Food essentially fish, squids, crustaceans, and other forms of marine life. Probably a general feeder, since about half food is vertebrate animals and other half invertebrates. Animal is gentle and affectionate and, unfortunately, unsuspicious.

Harp seals are easily killed in large numbers at breeding time and yield great quantities of oil and leather for commercial use. Not valuable as furbearers as are fur seals. Some think that seals are destructive of fish and fish foods but leather and oil they yield amply compensate for any loss through food habits.

Harp Seal

Order Carnivora./Family Odobenidae

Walrus
Odobenus rosmarus
Length, male 10 to 12 ft. Weight of male, average to 2,700 lb. Female about ⅔ size of male. Almost hairless, with wrinkled skin. Head relatively small. Nose blunt, with coarse bristles. Practically tailless. Flat nails on five toes on front flippers. Two flat and three pointed nails on hind flippers. Teeth: I 1/0, C 1/1, P 3/3, M 0/0.

Found in Arctic seas; in Atlantic south to Labrador. Pacific walrus found from Bering Sea to south of Pribilof Islands, and more numerous on Siberian than on American side. Northern migration, May to July in both species.

Breeds on northern migration, young born 11 to 12 months later. Mid-July embryo, 5 in. long. Young of late July, size of harbor seal; by August, 4 ft long and 120 lb; hangs to neck of mother with flippers and protected by her. Tusks, of male straight, 14 to 39 in. long, 6 to 9 lb; of female slender and bowed.

Food largely mollusks grubbed from bottom with help of strong tusks. Strongly social animal which is normally timid but becomes excited easily. Has many parasitic lice and is preyed on by killer whales, bear, and, most of all, man.

Eskimos once used flesh for food, hide for equipment, intestines for window glass, and oil for light. Walrus numbers have been seriously reduced since Eskimo hunters have obtained high-powered rifles. Flesh now used largely for dog food; tusks for carving ivory figurines and tools. Strong conservation measures should be adopted to halt serious decline in numbers. An endangered species.

Order Primates./Family Cebidae

Capuchin Monkey (New World)
Cebus capucinus
New World monkeys small, have thumb and big toes opposable, nostrils widely separated, toes and fingers with nails, and tail prehensile. Besides the sapajous, which include the common hand-organ monkey, there are howlers, spider monkeys, squirrel monkeys, sakis, and other groups of species, all with prehensile tails. Sapajous have a cowllike growth of hair on forehead.

Sapajous include the weeper of Brazil, brown with yellow tinge and paler throat, cheeks, and chest; white-cheeked of Brazil, with blackish fur and yellow-white temples; white-fronted, of many parts in South America, with red-brown fur but white on front and chest; and Central American white-throated, black with white throat.

Period from generation to generation is never less than 3 years. General notes given for rhesus monkey probably apply rather well to capuchins. These monkeys roam through forests in troops of up to a dozen animals under leadership of an old male, usually going over definite routes in single file from resting sites to feeding areas.

Food chiefly fruits and tender shoots, but they also enjoy birds' eggs, young animals, and insects. In captivity, they learn to eat much the same food as man. Some species, particularly the white-fronted, are noisy. Related howlers are well named and sound like much larger animals than they are.

Principal use of these animals to man is to serve as pets though many monkeys, particularly those with longer fur, have been considered as fur-bearers; some species have been practically exterminated for their pelts which at times may be considered stylish. They may cause some damage to fruits and other crops.

Order Primates./Family Lasiopygidae

Rhesus Monkey (Old World)
Macaca mulatta
Old World monkeys, including rhesus, do not have prehensile tails as New World monkeys do. Old World species sit on calloused buttocks which are free of hair, and most have distendible cheek pouches. Rhesus is generally gray-brown, with tail ½ length of body, with bare face which, in adults, is reddish.

Rhesus monkey native of northern India, where it lives in great numbers in trees or on ground, often as parasite on bounty of natives. South American monkeys, Cebidae, are arboreal, with prehensile tails like capuchin monkey. Rhesus monkeys are closely related to tailless Barbary ape of Gibraltar.

One young rhesus born 5½ months after promiscuous breeding; at birth, 1 lb; nurses at 1 day and for 1 year, when weighs 3½ lb and sits 1 ft high; at 2 years, 4 lb, 14 in.; at 3 years, 6½ lb, 16½ in., and female sexually mature; conception, tenth day after menstrual flow. Male matures at 4 years. Respiration, 19.

Food essentially plant materials but this varies and is much like that of man. In fact, dietary problems of man are studied through rhesus mon-

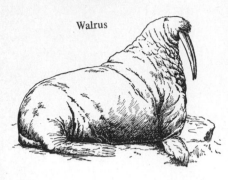

Walrus

Capuchin Monkey

Rhesus Monkey

keys' reactions. Rhesus also susceptible to human diseases. Considered sacred by natives in India and therefore cannot be harmed there.

Also the Bengal, the bandar of the Hindus. In Kashmir, it is found at elevations up to 5,000 ft; in Simla, to 8,500 ft. Apparently, in the wild, develops a well-understood hierarchy of dominance and submission. Has been used in the laboratory for intelligence and learning ability studies. Results are not conclusive. Apparently does not learn to use tools. Color vision similar to but not so well developed as man's. Famous as laboratory animal used in developing polio vaccine.

Order Primates./Family Cercopithecidae

Mandrill, a Baboon
Papio mormon

Head proportionately large, with large, long canine teeth. Teeth equal in number to those of manlike apes. Mandrills are brilliantly colored on bare buttocks; middle line of nose is scarlet with deep purple and light blue on either side. Fur blackish-olive, darker above, and beard is orange-yellow.

Mandrill

Largest baboons, found on west coast of Africa. Baboons confined to Africa and southern Arabia. Sacred or dog-faced baboon, found in Ethiopia and Sudan; chacma or pig-faced baboon is from southern Africa; and mandrill, drill, and yellow baboons are from African west coast. Medium-sized chacma has a 2-ft body.

Most baboons, mandrills included, live in troops of 15 to 50 animals. Males are largest and rule troop. Females often carry young on their backs, young sprawling and hanging onto long hair or straddling and riding as though they were on horses. Whole troop joins in protection of weaker members.

Food largely roots, fruits, tender shoots, and any of smaller animals such as birds, mammals, lizards, and insects which they can capture. In captivity, they will eat almost anything. Better suited to life on ground than in trees. Practically all baboons are mean, ugly, powerful, disgusting, and cunning animals.

Highly destructive to crops. Able to drive unarmed humans from an area when sufficiently abundant. Some can catch and hold dogs in their hands until they can get in an effective bite. Some are considered sacred; some have been trained to collect fruits; but all are generally unreliable neighbors.

Order Primates./Family Hominidae

Man

Homo sapiens
(Man the Animal.)

Man is increasing rapidly in numbers in spite of the fact that a single birth per pregnancy is most common; twins occur once in every 85 births, and numbers greater than two are even more rare in occurrence. Gestation period is 250 to 285 days with no evidence of a restricted breeding season. Male and female reach puberty between 12 and 15 years of age. In female a 28-day reproductive cycle begins and is repeated for 30 to 35 years; a 3- to 5-day period of menstruation occurs more or less regularly unless interrupted by pregnancy. Reproductive activity in women ceases at menopause; about age 50. Men have no definite sexual cycle or cessation of sexual activity. Man is probably longest-lived of the mammals. Body temperature about 98.6°F. Pulse 60 to 80. Respiration 15 to 20. Hearing range 12,000 to 17,000; voice 40 to 1,152 cps. Speed, to almost 20 mph in short distance.

Man acquires 5 percent of his mature weight during intrauterine life and 95 percent during a postnatal growth period of 20 to 25 years. During first year, baby increases 50 percent in length and 200 percent in weight.

Man is naked, yet hairy. Man has more hairs per unit of area than most great apes and monkeys. Man is becoming progressively naked in relation to hair covering. Most of man's hair is ornamental, not functional.

Man walks upright on two limbs. He possesses two great compartments in the trunk, the thoracic and the abdominal cavities. The distinctive feature of man is the development of a brain far beyond those of his predecessors in thinking and reasoning powers. Because of his ability to modify his immediate environment, man is not confined to the warmer and milder climates. He travels on or under the sea, over the land, in the air, and now into space to the moon.

(Man in the History of the Earth)

Back in Eocene times, between 55 and 60 million years ago, there lived a little ratlike animal, not unlike the feathertail, *Ptilocercus,* now found in Far East jungles. From it sprang a race of animals which developed into monkeys. Also from this line developed a race which included the present-day gibbons, orangutans, man, gorillas, and chimpanzees. The gibbons separated from this stock in the Oligocene, some 40 million years ago. The orangutans became independent early in the Miocene, or some 30 million years ago. About 25 million years ago man broke away from other primates and began his more explicit development toward man, the man.

Anthropologists distinguish man as an animal which is a primate that evolved from the less specialized apelike creatures from which the chimpanzee, orangutan, and gorilla also probably evolved. He is distinguished from the apes by: (1) his upright posture, (2) his ability to walk a great distance upright, (3) his teeth, (4) his brain, which is proportionately larger and possesses a more complex cerebrum than any other upright mammal of equivalent size, (5) his skull, which is rounder and larger, (6) his lateral toes, which are short, and his big toe, which is in line with the others, (7) his vertebral column, which makes an S curve balanced over his broad pelvis and his two straight, fully extended legs, (8) his short jaw with a rounded dental arch, and (9) his four canine teeth no larger than his premolars. Also man has a language capable of communicating very abstract ideas. He not only makes but he uses tools. While man is, to us, the most interesting of all living things, he is not all of natural history and could not live a single day without his associated plants and animals.

(Man and His Manlike Ancestors)

Remains of an early manlike creature were found in India in 1934. This man was given the name *Ramapithecus.* Later, remains of a similar early manlike form were uncovered in Africa and he has been called *Kenyapithecus.* Remains of both of these early manlike creatures were determined to be around 14 million years old; a time near the boundary between the Miocene and Pliocene epochs. It would seem, then, that manlike creatures began a course of evolution on two continents at about the time of the appearance of other anthropoids.

Another stage in man's development dates to 2½ million years ago. Fossil forms of these manlike creatures have been called australopithecines; perhaps the most famous of them is *Zinjanthropus* of the Olduvai Gorge in Tanzania, East Africa. These men were fruit eaters like the apes and flesh eaters like man. Australopithecines existed for about 2 million years and inhabited East and South Africa and probably southern Asia.

Later, some of these forms evolved into *Homo.* In fact some investigators name the remains from this period *Homo habilis* and report *Homo habilis* is the ancestor of man, the man. Other investigators believe that *H. habilis* is a transitional species or subspecies of *Australopithecus africanus.*

Later came *Homo erectus,* the probable progenitor of *H. sapiens,* modern man. *Homo erectus* appeared some 700,000 years ago during the late Pleistocene epoch. Later came Java man from Central Java and Pekin man from China, both of which lived during a period of about 500,000 years ago from the end of the first Pleistocene interglacial period to the end of the second glacial period. Recently all these men have been listed as *Homo erectus.*

Later came Neanderthal man, another member of the genus *Homo.* Fossil remains of the Neanderthal man date from the fourth glacial period, roughly 100,000 years ago. Neanderthal men were short in stature and they were hunters. Apparently they had no domesticated mammals, and they hunted for a wide range of game. Interestingly enough, they apparently buried their dead and perhaps believed in life after death. Neanderthal man was a species of *Homo sapiens,* modern man, and he apparently vanished from the fossil record about the time of the appearance of modern man. *Cro-Magnon Man,* modern man, appeared about 35,000 B.C. Cro-Magnon man, in today's acceptable attire, would have been welcomed as twentieth-century man.

(Races of Man)

It is difficult to describe races of man. The gene flow between people of European ancestry and those of African, American, and Asian and Arctic descent is continuous and in due time superficial characters such as skin color, type of hair, eye color, body contours, temperament probably will be amalgamated into one "man form." Fossil remains of man are not sufficient in numbers to demonstrate existence or absence of races of man, if we describe a race as a group of very closely related plants or animals within a species, each population characterized by the possession or lack of certain inherited physical features. As such, "races of men" are difficult to describe. One such attempt lists four races of men:

I   CAUCASOIDS   Large, widespread populations originally inhabiting all of Europe, north and northeast Africa, and parts of Asia; people with variable head shapes, brown to light complexions, wavy hair, narrow noses. Known as whites.

II   NEGROIDS   Originally confined to that part of Africa south of the Sahara and to parts of Asia; long-headed, woolly hair, dark complexions, flat noses. Known as Negroes or blacks.

III   AUSTRALOIDS   Inhabitants of Australia and some of the primitive tribes of southern Asia; long-headed, wavy hair, dark complexions, flat noses.

IV   MONGOLOIDS   Most numerous peoples on the earth today, peoples of eastern Asia and America; broad-headed, straight hair, yellow or red complexion, nose of intermediate type.

It seems ridiculous to assume that any race of modern men incorporates all the worthy qualities of human beings. When one looks at the white race's record of wars and destruction, it is even more absurd that its members should consider themselves supermen.

(Man, the Builder)

Man should be happy. Through his own efforts, he has solved many problems involving labor, ignorance, boredom, health, security, and faith.

Through invention, machines have been constructed which take from the backs of men and of dumb animals much of the work which made existence a living death for many who went before us. One generation, one nation, and one man has been able to learn through the experience of former generations, nations, and men.

Through education, the benefits from the experiences of others have been spread widely. No longer are knowledge of how to live and the opportunity to live happily limited to a few. In the ideal democratic government, every man has the opportunity to make the most of the gifts granted him by heredity, chance, and effort.

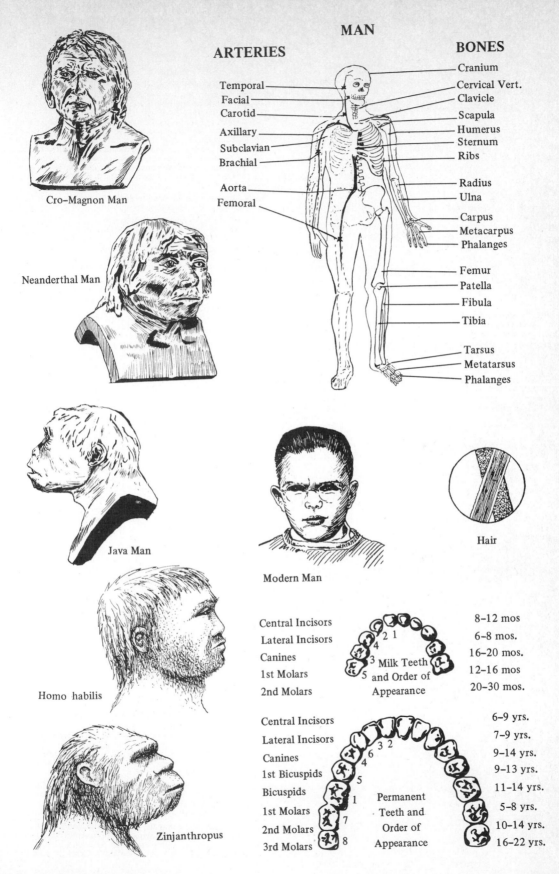

MAN

ARTERIES

- Temporal
- Facial
- Carotid
- Axillary
- Subclavian
- Brachial
- Aorta
- Femoral

BONES

- Cranium
- Cervical Vert.
- Clavicle
- Scapula
- Humerus
- Sternum
- Ribs
- Radius
- Ulna
- Carpus
- Metacarpus
- Phalanges
- Femur
- Patella
- Fibula
- Tibia
- Tarsus
- Metatarsus
- Phalanges

Cro-Magnon Man

Neanderthal Man

Java Man

Modern Man

Homo habilis

Zinjanthropus

Hair

Milk Teeth and Order of Appearance	
Central Incisors	8–12 mos
Lateral Incisors	6–8 mos.
Canines	16–20 mos.
1st Molars	12–16 mos
2nd Molars	20–30 mos.

Permanent Teeth and Order of Appearance	
Central Incisors	6–9 yrs.
Lateral Incisors	7–9 yrs.
Canines	9–14 yrs.
1st Bicuspids	9–13 yrs.
Bicuspids	11–14 yrs.
1st Molars	5–8 yrs.
2nd Molars	10–14 yrs.
3rd Molars	16–22 yrs.

Through planned recreational activities, men have been released from boredom and from the mischief making that so often in the past led to trouble for others on the part of those who could afford leisure. We have, on occasion, produced remarkable art. Some of our music makes life worth living. Our theater, our literary leaders, our public park systems, at their best, all suggest progress and all invite effort for improvement.

Our health agencies have conquered many diseases that caused unnecessary misery in the past and are daily solving new problems. Here and there we lose a fight against the forces of disease, but usually these setbacks help us later to win another battle. We are making progress. Tuberculosis, diphtheria, yellow fever, malaria, and many other diseases, such as those associated with childbirth, are not the specters they were a generation ago.

We are making some progress but perhaps also losing some ground in the field of security of the individual for the future. With all the opportunities we have for providing for our physical and mental needs, there should be no poverty for any worthy member of society.

Best of all, we have faith that with free speech, a free church, free schools, a free press, and freedom of opportunity a system will develop in which selfishness and greed will find no opportunity of expression, *if responsibility is recognized as being of equal importance with freedom.*

(Man, the Destroyer)
Soil and Minerals. For eons before man came into being on the earth, a rich soil had been built up in the territory where he now lives. Much of this valuable topsoil has vanished downstream in China, and America, or under sand in Africa, in part because of shortsighted agricultural practices. Fresh waters have been defiled. We are exhausting prodigally such nonrenewable resources as oil, coal, and the ores. For this, we shall be held responsible in the future.

Air and Water. Man has pumped into the air envelope that surrounds him too many pollutants, some of which are particulates, some are noxious gases. From his power sources, his automobiles, and his heating systems each day he adds tons of smoke, dust, fumes, mists, radioactive wastes, odors, and smog-forming gases. Also, he has utilized his waterways and oceans as the dumping ground for his municipal and industrial wastes. Some of his streams are open sewers. To recover from his losses resulting from air and water pollution, man must now invest more and more in a massive clean-up campaign.

Plants. In America, we have foolishly stripped from the land our forests and the soil-anchoring grasses.

Animals. We have enslaved helpless domestic animals. We trap thousands of useful furbearers by the most inhumane methods. We have led to the extinction or near-extinction of many useful species of birds, fish, and mammals such as passenger pigeons, salmon, whales, elk, bison, and sea otters. We are in many instances reducing the numbers of most of our forms of native game.

Fellow Men. We have enslaved weaker races and weaker individuals of our own race and of our own families. We have kept people in ignorance of how they can avoid disease, of how they can live within their incomes. In part as a result of these practices, we find a world of recurring, increasingly destructive wars. For this, our generation will be considered as lacking in intelligence by our successors. We shall justly deserve this condemnation.

Orangutan

Chimpanzee

Order Primates./Family Pongidae [Apes]

Orangutan
Pongo pygmaeus
Adult male 4½ ft high when upright, with fingers nearly reaching ground. To 200 lb. Females average ½ weight of male. Legs short, thick, with knees turning outward and soles of feet inward. Head pear-shaped, with high forehead, flattened nose, small ears, and projecting mouth with narrow lips. Skin blue-gray, with long shaggy reddish-brown hair.

Native of Borneo and Sumatra, living in trees in low marshy country and rarely coming to ground. Borneo orangutan is of the species *Simia bicolor.* Orangs are popular animals in zoos and lend themselves to conditions of confinement, but have many diseases.

Live in small family parties consisting of parents and two to four young. Build a sort of platform or nest in trees where they sleep until late morning, leaving it to feed during middle of day. Are commonly gentle and affectionate but will fight fiercely and ably when forced to do so. One young born 9 months after mating. Life-span to 25 years.

Food largely fruits, favoring the durian, the spiny rind of which they remove and discard for the luscious pulp. They also eat nuts and tender shoots and enjoy buds and young stems of bamboo. Captive orangs learn to eat and drink same food that man favors.

Orang is lower in the animal scale than chimpanzee and gorilla. Trained to eat, behave, and dress like a man, an orang makes a valuable zoo or circus animal and is immensely popular because of its droll appearance and its obvious enjoyment of being appreciated. An endangered species.

Chimpanzee
Pan troglodytes
Stands about 3½ ft tall when erect. Weight of males averages 110 lb. Body covered with black hair. Walks more erect than other apes but stands with feet more widely spread than man and with knees bent. Arms shorter than in other apes, though they reach below knees by length of fingers. Thumb is smallest finger. Thumb of foot powerful and long.

Forest dweller, confined to western and equatorial Africa. Generally remains on ground in thick underbrush, climbing to treetops easily for escape or in search of food. Often found in troops of considerable numbers and apparently likes association with man when treated with any consideration.

One young born 9 months after breeding, in nest guarded by male. At 9 days old, weighs 2¼ lb and is 15½ in. long. Female becomes sexually mature at 5½ years and male probably later. Life-span of captive animals, and presumably of those in the wild, is 18 to 25 years.

Food, largely plants particularly bananas, gingerberries, and corn. Moves from one part of forest to another seeking food, frequently calling loudly and fearsomely night or day. Powerful, particularly with the arms, a zoo animal pulling 1,260 lb on a dynamometer. Gentle unless opposed, then fierce.

Probably slightly destructive to crops raised by natives and certainly awe-inspiring, but probably minds own business rather thoroughly. Used for studying some fundamentals of behavior in the closely related species, man. Not abundant enough for use as experimental animal in medicine on any large scale. Man's closest living relative.

Gorilla
Gorilla gorilla

Male stands erect, measures nearly 6 ft and weighs 300 to 450 lb. Female much smaller. Hair gray, 2 in. long on arms, 1 in. on belly, shorter and blacker on back; in old male absent on chest and upper back. Face, hands, and feet black. Ears small. Mouth large, with thin lips. Chin short and receding.

Gorilla

Found in wooded Africa, though rarely if ever seen in trees. Best known in recent years from the Tanganyika region where it was studied principally by the late Carl Akeley. Formerly probably ranged across Africa in tropical regions, since it was reported from Guinea and elsewhere along Atlantic Coast. Two species of gorillas recognized, mountain and lowland.

One young, born 9 months after breeding. In the wild, a male with several females and young may make up a troop. Young walks more erect than adults. At 4 years, a young gorilla may weigh 140 lb; at 7 years, it reaches sexual maturity. Life-span may be as much as 28 years.

Food, a variety of plants and possibly some animals, but principally bananas, cabbage palms, plantains, and succulent plant materials. Usually walks on all fours, rolling from side to side, shuffling, with head erect and eyes forward. When angered, male may beat chest and roar but in captivity is generally good-natured and intelligent.

Known to man since fifth century B.C. and for most of that time hopelessly misunderstood, probably due to desire of explorers to exploit dangers of their journeys and their own bravery. Described scientifically in 1894. Until recently, young had not been raised in America to maturity. An endangered species.

Order Rodentia./Family Sciuridae [Squirrels]

Woodchuck, Ground Hog
Marmota monax

Length to 27 in., including 6-in. tail. Female totals 22 in. and is lighter. Fur rather coarse, but might be used for decoration. Grayish-brown, reddish, black, or more rarely white. Sexes colored alike but young paler. Large strong anal gland produces musky odor. Weight 5 to 12 lb. Teeth: I 1/1, C 0/0, P 2/1, M 3/3. Tracks: F 1½ × 1¼, 4; H 1¾ × 1¼, 5; Sp 5.

Woodchuck

Found from mouth of Gulf of St. Lawrence, north to Hudson Bay, upper Mackenzie region, and central Alaska; west and south to northwest Montana, eastern Nebraska, on to Arkansas and Alabama and north to western North Carolina and New England. Essentially in open farming country but also in woods and sometimes up trees, usually where dry.

Mates in early spring immediately after hibernating. Two to eight young born 31 to 32 days later, blind, naked, helpless and about 4 in. long with ½-in. tails; crawl at 3 weeks, take solid food at 4 weeks, play at 5 weeks, then cease nursing. Sexually mature at 1 year. Average weight in March, 5 lb 10 oz; in September, 10 lb.

Food almost exclusively plants although flesh such as insects, mice, and birds may be eaten occasionally. Has a clean den at end of burrow but above lowest burrow level, with wastes in separate room. Active mostly in day but sometimes during night. Hibernates. Body temperature varies from 98.9°F to 37.4°F and heartbeat from about 80 beats per minute to only 4 or 5. Fights most effectively. Whistles when alarmed.

Great destroyer of forage. Burrows are dangerous to horses and cattle but permit water to enter ground and hold soil moisture. Burrows protect rabbits in winter and summer. Flesh edible. Fur durability, 20 percent that of otter. Ground-hog day, February 2, a delightful hoax.

Hoary Marmot

Hoary Marmot
Marmota caligata

Length, head and body 18 to 21 in.; tail 7 to 10 in.; weight 8 to 20 lb; female smaller. Largest of American marmots. Silvery-white, peppered with black on back and rump. Head and face black with white eye patch. Feet blackish. Underparts dirty white. Teeth: I 1/1, C 0/0, P 2/1, M 3/3.

Found in Rocky Mountain areas from Idaho, Washington, and northern Montana; north to Arctic Circle in Alaska with races in the Cascades, Olympics, and other ranges. Habitat: talus slopes, alpine meadows; high in the mountains near timberline. Seen most commonly near rockslides.

(Continued)

Prairie Dog

Pairs, mating in early spring, with three to four young born a few weeks later, or about June 1 in southern part of range. Young to be seen running with mother from late June to late August. Young probably sleeps with mother first winter. Apparently does not mature until 2 years old. Has a long hibernation.

Food essentially plants, often gathered some distance from home den, usually reached by long-established, well-marked trail. Must have good sight and hearing to escape bears, hawks, and other enemies which roam over exposed home territory. Its call, a shrill whistle, may be heard a mile away. Generally escapes enemies.

Serves in nature as food for animal eaters that roam mountains, and therefore may be considered a "buffer" species protecting more valuable animals from their natural enemies. Although they compete with sheep for fodder, they feed on animals that might otherwise kill the sheep. Fur known as "Brazilian mink," durability 20 percent, is Russian marmot.

Prairie Dog
Cynomys ludovicianus
Length to 14½ in., tail 3 in. Weight to 3 lb. Female to 2 lb 4 oz. Reddish-brown, gray, or flesh-colored above, with pale muzzle and eye-spot. Last third of tail black. Feet suited for digging. Has habit of sitting erect at entrance to burrow. Five claws on each foot. Rudimentary cheek pouches. Teeth: I 1/1, C 0/0, P 2/1, M 3/3.

Lives in prairie lands from southern Saskatchewan south to central Texas, west to Arizona, New Mexico, Colorado, and Wyoming; confined to open, flat regions of dry clay soil. Close relative, white-tailed prairie dog, favors mountainous areas above 6,000 ft.

Three to four young born March to April, weigh about ½ oz; 2¾ in. long, smooth and shining. Increase 40 percent in weight in 7 days; thirteenth day, 2½ times birth weight; twentieth day, have hair and are able to stand; twenty-seventh day, crawl; thirty-third to thirty-seventh day, eyes open; at 7 weeks, weaned; at 9 months, over 2 lb. Breed at 2 years, bearing young once a year.

Lives probably on plants including vegetables, grains, pasture plants, and even some woody plants and insects. Lives in dens with deep plunge entrance, in which rattlesnakes and burrowing owls also live, but not with prairie dogs' consent. Although white-tailed species hibernates, black-tailed does not.

Burrow lets water into the ground but also serves to break legs of horses and cattle. Estimated that 32 prairie dogs eat as much grass as 1 sheep, and 256 as much grass as 1 cow. One authority estimated in 1900 that Texas supported 800 million prairie dogs, which ate vegetation that would otherwise support over 3 million cattle. Coyotes, hawks, and owls are a check.

 Franklin's Ground Squirrel

Franklin's Ground Squirrel
Citellus franklinii
Length 14 in., tail 4½ in. Closely related to 13-lined spermophile but uniformly colored grayish-white to dusky above, or brownish with faint light and dark dots; feet darker; tail black and white mixed, bushy. Sexes colored alike. Teeth: I 1/1, C 0/0, P 2/1, M 3/3.

Essentially a ground animal though may take to trees. Primarily an animal of the prairies; may be known locally as the gray gopher. Individuals may never range over 100 yd from den, but species found through central United States and Canada from Kansas, Missouri, and Illinois to Athabaska River, range being broader in South, narrow east and west in North. A gregarious species.

Probably pairs, in spring, but male does not stay with family after breeding period. Four to eight young born blind, nearly naked and helpless, early in June; 3¾ in. long, with ¾-in. tail; suckled about 6 weeks; ⅓ grown by August, when begins to feed for self. One annual litter.

Eats plants, insects, mice, birds, or a great variety of related forms though it is not such a flesh eater as the 13-lined species. Some plant food may be stored. Probably requires more water than some near relatives. May migrate to new territory in a body, though not normally sociable.

Serves to check multiplication of insects and mice and to let air and water into the soil; however, destroys considerable forage more useful to cattle from man's viewpoint. Has a long winter hibernation until part of year when surplus is greatest. Appears about April, later than most relatives.

Thirteen–lined Spermophile

Thirteen-lined Spermophile, Gopher
Citellus tridecemlineatus
Length, total 11 in., tail 4½ in. Female slightly smaller. Conspicuous because of alternate longitudinal stripes of dark brown and dirty white; stripes on neck and shoulder broken into spots. Ears small. Tail only sparingly bushy. Tends to sit erect with head pointed skyward. Teeth: I 1/1, C 0/0, P 2/1, M 3/3.

Found in open lawns and prairies from southeastern Alberta through Manitoba, Wisconsin, and Michigan to central Ohio; southwest through Indiana, Illinois, Missouri, and Oklahoma to Texas; west to Arizona, Utah, and central Montana, essentially in the plains and prairie country. Range probably expanding.

Mates as soon as appears above ground in April; male deserts female before 7 to 14 young are born about first of June, or in 30 days. Young completely naked, helpless, and toothless at birth; 2¼ in. long with ⅜-in. tail. Eyes open thirtieth day. Out of den in 35 days, full-grown in 90 days. One litter.

Food, plant and animal matter, much of plant material being stored underground. Body temperature ranges from 86 to 95°F; can survive 28.5°F. May lose ⅓ body weight during hibernation. Soil temperature does not control hibernation. Has host of enemies, including hawks, cats, weasels, and parasitic insects.

Interesting member of prairie landscape. Probably does more harm than good to man's crops. Confines activities to period when surpluses are most abundant so can do no harm to stored crops. Provides food for valuable furbearers. Unquestioned enemy of small ground-nesting birds, mice, and insects. Pulse, 200 to 350 per minute when active, 5 when dormant. Temperature, 105.8 to 35.6°F.

Eastern Chipmunk

Eastern Chipmunk
Tamias striatus
Length to 9½ in., tail 3¼ in. Well-developed cheek pouches. Along back is series of conspicuous black stripes, usually five, over brown or gray fur. Underparts light fawn-colored as are lower cheeks. Conspicuous dark streak with white beneath crosses eye. Throat white. Sexes alike. Teeth: I 1/1, C 0/0, P 1/1, M 3/3.

In cultivated land and woodlands, from Nova Scotia west to northeast North Dakota, south to Louisiana, east through northern Mississippi, Alabama, Georgia, and western Carolinas to eastern Virginia and north to Nova Scotia. Closely related species occupy southern Canada and west United States except in plains.

Four to five young, blind, naked, and helpless at birth; born 32 days after breeding in spring or midsummer, though shows signs of fall breeding as well. Possibly two litters, one in spring and one in fall. Young stay with mother until 3½ months old. Weigh 3 g at 1 day; 12 g at 8 days; 17 g at 14 days; 25 g at 21 days; 33 g at 28 days; 50 g at 35 days; 70 g at 42 days; and 84 g at 49 days.

Food a variety of plant and animal materials, principally nuts, fruit, and seeds of woody plants such as maple, oak, hazel, basswood, beech, elm, box elder, wild and cultivated cherries; and including insects, young birds, other small animals, and bulbs of garden flowers; individual ranging over 2 to 3 acres. Hibernation from first frost in fall to March, but not fat at beginning of hibernation. Variable with weather. Well-developed storage instinct, largely for plant materials.

Tremendously interesting little animals, whose habits are rather destructive to garden flowers, nesting birds, and certain fruits and garden plants. Tamed reasonably easily so that they will eat out of the hand or even from the lap. Tracks: F ½ × ⅜, 5; H ¾ × 1¼, 5; Sp 2.

Red Squirrel, Chickaree
Tamiasciurus hudsonicus
Length 12½ in., including 4⅗-in. tail. Active, with flattened bushy tail. Ears fairly long. Sexes colored alike. In winter, grayish-white beneath, rusty red above. In summer, clear white beneath, brighter red above, and tail less buffy. Two color phases and over 20 closely related species and subspecies. Teeth: I 1/1, C 0/0, P 2/1, M 3/3. Tracks: F ½ × ¾, 5; H 1× 1¾, 5; Sp 3; L 36.

Red Squirrel

Favors evergreens, though common in parks, street trees, and orchards where food is available, through most of forested America north of Mexico. The hudsonicus group is in the east, fremonti in the Rocky Mountains, and douglasii west of the Rockies. No near kin in Old World species superficially similar.

Mates in February to March. Promiscuous, may mate until late in September; apparently two distinct breeding seasons. Three to six young under 1 oz, born in about 40 days; blind, naked, helpless; 4½ in. long, including 1½-in. tail. Eyes open twenty-seventh day. Weaned in 5 weeks. Family stays together during summer. Two litters a year. Mother obviously teaches young to become independent. Life-span to 12 years.

Food, plant and animal matter. One may eat possibly 200 birds a season and therefore be worse than a domestic cat. Eats seeds from many cones, including pines and hemlocks, also apples and other fruits. In eating hickory nut makes one elliptic opening with rough edges. Can cover 75 ft in 8 seconds over level ground.

(Continued)

Gray Squirrel

Fox Squirrel

Flying Squirrel

Probably difficult to defend economically. Fur of no commercial value though it sells for 15 cents. One eastern state in 1 year killed over 7,000 as vermin but their gay actions through year except in severe weather win the species many friends who would dislike to see the last one vanish. Important as hider of cones and tree planter.

Gray Squirrel
Sciurus carolinensis
Length 18 in., including 9-in. tail. Weight about 1 lb. Pepper-and-salt gray over slate-colored underfur; hairs white-tipped over black, buff, and lead. Black squirrels are color phase of gray. Albinos also known. Tails large and beautifully plumed. Teeth: I 1/1, C 0/0, P 2/1, M 3/3. Tracks: F 1 × 1⅓, 5; H 1¼ × 2½, 5; Sp 3¾; L 60.

Usually in lands of woods and orchards from Maine to Dakotas, south through central Texas and east to Atlantic and Gulf coasts. Spectacular migrations of large numbers take place periodically, possibly but not necessarily because of overcrowding as they often leave abundant food for regions of scarcity.

Mates in spring (January ?), believed by some authorities to pair for life. One to six young, born 44 days after breeding; blind about 37 days, when weight has reached about ¼ lb; stay with mother at least for season or until full-grown; can leave den in 6 weeks; full-furred at 9 weeks. Two annual litters in North. Nests may be hollows in trees or leaf nests lodged in branches.

Food largely plant material but also some animal matter. A hickory nut is cut into many pieces or heavy fragments. Food not stored in large amounts in one place but hidden separately and later located by scent. Food not taken to nest of young. Has remarkable sense of smell. Den kept clean, as is leafy nest in treetop used for hot weather.

Undoubtedly useful as planter of nut trees, including white oak acorns that may sprout and freeze unless buried. Also considered valuable as a game species. Pelt has some slight commercial value. Well liked because of friendly ways in parks and about homes.

Fox Squirrel
Sciurus niger
Length 2 ft, including 1-ft tail. Weight nearly 3 lb, male slightly larger; Florida animals probably largest. Color varies from black with white ears and nose to buff above and lighter beneath, or gray above with white underparts, nose, ears, and feet but black crown. Color like that of fox, hence name. Different from gray squirrel in that it spends more time on the ground and apparently tends to bury more food caches. Teeth: I 1/1, C 0/0, P 1/1, M 3/3. Tracks: F ¾ × 1¾, 5; H 1× 2, 5; Sp 6; L 60.

Found from Rhode Island through southern Pennsylvania to North Dakota; south through central Texas to Gulf; east to Atlantic Coast with related forms in small areas in Cascades, Olympics, and New Mexico, Arizona, and Mexico regions. Generally coincides with range of oaks but not of hickories and chestnuts.

Probably pairs, breeding in December to January. Two to five young, small, blind, naked, helpless, and pink; born probably about 6 weeks after breeding; nursed about 5 weeks; can leave nest in 6 to 12 weeks; leave parents at about 3 months, when new family may start if there are two litters a year as in warmer ranges.

Food fundamentally nuts and other tree fruits, many being buried for future use, not in a few large piles but in small widely scattered units. In this way squirrel assists in planting many trees. Food also includes buds, bark, and birds, the last sometimes to an unfortunate degree. Has excellent sense of smell.

Useful in planting hardwood forests, as a street and park pet, and as a game animal; since it does not hibernate, interests nature lovers through year. Bird-eating habit apparently is not universal but is always unfortunate. Large size makes it popular game and its flesh is good. Should not be exterminated.

Flying Squirrel
Glaucomys volans
Length to 9½ in., including 4½-in. tail; larger *sabrinus* species totaling 1 ft with a 6-in. tail. Color of *volans,* drab above, ringed with reddish and pure white belly; *sabrinus,* drab above, and dirty white beneath. Extensible fold of skin along sides. Teeth: I 1/1, C 0/0, P 2/1, M 3/3. Tracks: F ½ × ½, 5; H ¾ × 1¼, 5; Sp 2.

Essentially animals of treetops. Smaller form found from Maine to Minnesota, south to central Texas and Gulf and east to coast. Larger form, from Labrador to Alaska, south to central Alberta, northern Minnesota, and northern Wisconsin, Michigan, New York, and New England; also south in mountains to California, southern Utah, and Tennessee.

Pairs, in winter; two to six young born March to April after 40 days. Blind 4 weeks. Possibly three litters. By fortieth day, young squeak, not light-sensitive, weigh 1 oz; by seventy-fifth day, storage instinct developed, 1⅔ oz; one hundred-third day, light-sensitive, 2 oz; one hundred-seventeenth day, shedding fur, 2½ oz; one hundred-twenty-fifth day, new coat, "churr," 2½ oz.

Stores food of tree fruits, superficially. Opens a hickory nut with one elliptic opening showing ground edges. Most nocturnal American mammal. *Sabrinus* active all winter. Has a clean den in a tree hole. Can cover 75 ft in 12 seconds on ground or glide through air 152 ft from 60 ft elevation, rising slightly at end of glide.

Most interesting animal and makes trusting pet, but may be annoying in attics and houses generally or may destroy some stored fruits. Also known to eat flesh even of its own kind, though this is probably exceptional. Too small and nocturnal to be considered as game but always worth watching when opportunity presents itself.

Order Rodentia./Family Geomyidae

Pocket Gopher
Geomys bursarius
Length, male 1 ft, with 3½-in. tail; female, 10½ in., with 3-in tail; usually naked. Weight 7 to 11 oz. Outer surface of upper incisor teeth grooved; in western genus *Thomomys,* not grooved. Husky, short-necked, strong-clawed with pair of fur-lined cheek pockets opening outside mouth. Reddish, sprinkled with black, darker above. Teeth: I 1/1, C 0/0, P 1/1, M 3/3.

Animal of moist, but well-drained, sandy loam soil. In farmlands, they are apt to be found in hay-crop fields or pastures. Over 80 western *(Thomomys)* forms from south central Mexico to Washington through Rockies, with relatives including *Geomys* extending range east to Mississippi, with another group in Florida, Georgia, and Alabama. Central group extends north to central Alberta and Saskatchewan.

Midwest species bears two to eight young in April. Probably only one litter a year, although half-grown young have been found through to September, when families usually break up and scatter to new territory. Individual range may be confined to ¼ acre for the season.

Pocket Gopher

Food, vegetable materials carried in cheek pouches, stored underground in an orderly manner, roots and stems commonly cut in 2-in. lengths. Apparently does not drink. Digs elaborate burrows with strong claws, pushing loose dirt out with head and forefeet but not carrying it in cheek pouches. Mostly solitary in habits.

May serve to let air and water into soil, but also undoubtedly destroys considerable quantities of valuable plant materials and may injure grasslands where it is desirable that they be flat for lawns and golf courses. Highly sanitary in burrows, leaving wastes at established place. Has many enemies including body parasites and snakes.

Order Rodentia./Family Heteromyidae

Kangaroo Rat
Dipodomys deserti
Length to 13 in., including 7½-in. tail. Forelegs small. Hind legs long, hind feet over 2 in. long, with four toes. Ears short and rounded. Eyes, large and beautiful. Cheek pouches well developed, external, fur-lined. Whiskers long and conspicuous. Color: bluish fawn above, gleaming white beneath, even on tail and tail tuft. Teeth: I 1/1, C 0/0, P 1/1, M 3/3.

Found in dry, clayey, or sandy grounds; some species in damp lands; some in forests from Washington and Manitoba to Oaxaca, Mexico. This species, from western Nevada through Arizona and eastern California. Square mile may support 1,000 animals. Nearly 100 kinds of kangaroo rats known. Most seek dry, hot lands where coolness exists underground.

Social, even though families commonly live separately in tunnel homes. Underground nest lined with fine plant material contains two to four young, in April and probably at other times during summer. Nest room about 8 to 10 by 5 in. and may be 3 ft underground where it is cool; clean and separate from dung areas.

Kangaroo Rat

Food probably solely plant material such as seeds and roots, stored by larger banner tail but not by smaller species; gathered in night forays when a rat may leap 8 ft at a jump and of course easily escape most of its enemies. Does not hibernate or migrate. Fights by striking with hind feet and squeaking. Friendly.

Of little economic importance but tremendous interest. Helps let air and water into soil by burrows. Might destroy grain if it lived in grain country. Makes excellent pet and should be better known to be appreciated. Night trails, songs, mounds, and escapes are worth investigating.

Deer Mouse, White–footed Mouse

Pack Rat

Meadow Mouse

Order Rodentia./Family Cricetidae [Mice]

White-footed Mouse, Deer Mouse
Peromyscus leucopus

Length to 7½ in., including 3½-in. tail. Medium-sized, with large ears, large eyes, and rather long head. Fawn-colored with underparts clear white or, when young, slate-gray in some species on back with underparts gray or white. Feet white. Teeth: I 1/1, C 0/0, P 0/0, M 3/3. Tracks: F ¼ × ¼, 5; H ¼ × ½, 5; Sp 1¼.

Varied habitat, including fields, most commonly forested or brushy areas and dwellings, species and subspecies covering most of United States and parts of southern Canada. *Leucopus* extends from Nova Scotia to South Dakota and south through to New Mexico, with related species taking most of Gulf Coast areas and others to west.

Two to six young born 21 days after breeding; at birth 1½ in. long; eyes open eighteenth day; cared for by mother; weaned at 14 to 20 days; ½ grown at 3 weeks; ⅔ grown at 42 days; female matures at 28 days, conceives at 39 days; male matures 10 days later; old at 3 years; breeds to 33 months. Life-span to 5½ years, but usually shorter. Female may have several litters each year.

Food about evenly split between plant and animal materials, some plant materials being stored. Elaborate house of plant materials is constructed either by remodeling abandoned bird's nest in a tree or selecting some place under a board, in a wall, or in a house. Often enters houses in winter for shelter but leaves in summer.

Causes the usual damage of mice in a dwelling and may do some damage to crops in fields but does not compare in this destruction with meadow mouse; has saving habit of eating considerable quantities of insects.

Pack Rat
Neotoma cinerea

Length 15½ in., including 6½-in. tail. Weight, ½ lb. Rather heavily furred, with somewhat flattened bushy tail. Ears relatively large. Gray to reddish-brown above, with sprinkling of dusky hairs. Beneath, white or nearly so. Eastern, southern, and some of western species lack bushiness of tail. Teeth: I 1/1, C 0/0, P 0/0, M 3/3.

Found through the northwestern states and British Columbia, typical habitat, pines in high mountains, rockslides; with species *magister* from Pennsylvania to Alabama; *floridana* from South Dakota to Gulf States; one dusky-footed *fuscipes* along Pacific Coast; three Mexican border species and 2 small (under 13 in.), *desertorum* and *lepida*, in California, Utah, and Arizona.

Probably monogamous. Breeds probably January to February, bearing two to four young from February to May. Young usually open eyes by 17 days; are weaned at 3 weeks, though may stay with mother longer. Father not allowed with family while young are present. Mother an exceptionally good parent.

Food a variety of plant and animal materials, including principally nuts, grain, and berries, which may be stored in a single large hoard of nonperishable kinds. Storehouse usually separate from an elaborate nest of vegetable material. Also has habit of gathering many inedible materials and bringing them together.

Useful in gathering nuts and acorns which are stolen by men and sold. Useful as food, flesh being edible and animal relatively easily caught. Probably cause of many misunderstandings among men because of habit of stealing small objects, even though usually something worthless is left in exchange.

Meadow Mouse, Vole
Microtus pennsylvanicus

Length to 7 in., with 1⅘-in. tail. Sexes colored alike, a uniform chestnut-brown above sprinkled with black; gray beneath, sprinkled with cinnamon. In winter gray and in youth darker. Tail not bushy, appears heavyset. Eyes and ears small. Face blunt. At least 13 groups of related mice include over 70 species. Teeth: I 1/1, C 0/0, P 0/0, M 3/3. Tracks: F ½ × ½, 5; H ½ × ¾, 5; Sp 1.

Practically all North America is home of mice of this type including groups *pennsylvanicus, montanus, californicus,* and *mexicanus* whose general range is indicated by name; also *abbreviatus* and *operarius*. *Microtus pennsylvanicus* is found from Nova Scotia to Dakotas to British Columbia and south to North Carolina. Abundance cycle, about 4 years.

Females possibly produce 13 litters before end of year. Promiscuous. Four to eight young born 21 days after breeding, followed by immediate breeding by mother. Young weaned in 10 to 12 days. Female breeds when 1 month old, male at 45 days. One female *may* produce 100 mice a year. Life-span about 1½ years.

Requires water but eats almost exclusively plants (its own weight in a day) one requiring about 23 lb of green food per year but destroying much more than that, since cut material is not all cleaned up. Active day and night throughout year, girdling trees, eating vegetables and forage. Forms basic food of most flesh eaters and buffer crop for game.

Destroys orchards and farm plant products and in United States, at least 3 million tons of hay a year, being worst competitor of domestic animals for forage. 1 acre may support 300 mice, but population of 12,000 per acre is recorded. Has some storage instinct and is clean in home.

Collared Lemming
Dicrostonyx groenlandicus
Length to 6 in. Weight to 4 oz. Feet short and thick, with middle claws of forefeet long. In summer gray above dappled with rusty red, with black line down back and dull gray with mixed rusty below. In winter nearly pure white but with black hairs mixed in. Some show a whitish collar with a brown border. Brown lemming, *Lemmus trimucronatus,* is shown above. Teeth: I 1/1, C 0/0, P 0/0, M 3/3.

Found from barrens of northern Alaska to Labrador but not much south of barren lands. Closely related species extend range around world on both hemispheres but only in north. Related Norway lemming has small round ears, yellower underparts, and third claws of forefeet the longest; best known in Scandinavia where it digs dry shallow burrows.

About five young in a litter, with many litters a year, thus making enormous increases in population possible. When this occurs with Norway lemming, great numbers may start on migrations in which they stop at nothing, swimming lakes, invading cities, and of course perishing in enormous numbers and thus keeping population within bounds. Migrations of this nature not typical of this species.

Food almost entirely plant material gleaned in reasonable safety from protection of shallow burrows. Are preyed on by foxes, owls, weasels, and similar animals providing basic food for most furbearers that yield an important portion of natural wealth of North. Without lemmings and their kin, fur trade probably could not exist. Abundance cycle, 3 to 5 years.

Have been eaten by starving men. Skin of no value either when white in winter or in darker summer condition. Reported that American lemming makes a reasonably satisfactory pet but that Norway lemming may fight any animal that may oppose it in its migrations. To most people, these animals appear simply as "meadow mice" in summer.

Collared Lemming

Hamster
Cricetus cricetus
Slightly smaller than Norway rat (body length 8 to 12 in.; tail length 1 to 2½ in.) but with hair-covered tail with thickish head, round ears, and mouth with internal cheek pouches used in storing food. Forefeet with four toes, hind feet with five toes, all strongly clawed. Fur fine, long, usually red-gray above and white below but sometimes all black. Legs short. Golden hamster, *Mesocricetus auratus,* much smaller.

Native of temperate Europe and Asia but especially abundant in Syria, where it lives normally in or near grainfields in a burrow 3 to 6 ft deep, with straight, perpendicular tunnel entrance and large grass-lined sleeping chamber with sloping entrance. May maintain four to five storerooms, each capable of holding a peck of grain.

Males fight for females. Normally 7 but sometimes to 15 young born 15 days, 21 hours after mating. Young at birth weigh ⅟₁₀ oz, at first naked, pink, and blind. Backs black at 3 days. At 12 days, weigh ⅓ oz and are golden brown. At about 13 days are weaned; can see at 14 days. Independent at 21 days. Breed at 43 days. Females bear young at 59 days. Monthly litter through year. Old at 2 years.

Food in summer, small animals; in winter, stored grains. In captivity, live on lettuce, dog biscuit, and corn. Live comfortably at 70°F but hibernate at 45°F. Have many enemies such as hawks, snakes, and weasels. More susceptible to human diseases than guinea pigs. Very clean in habits. Stored food suitable for human consumption.

In native land are used as food and fur has value. Also food caches are robbed by man for his use. Make good pets because of friendliness. In 1938, introduced into United States as a laboratory animal for studying nutrition and disease, and have supplanted mice, guinea pigs, and white rats to some extent for some purposes.

Hamster

Muskrat
Ondatra zibethicus
Length to 25 in., tail 10 in., compressed, sparingly haired, and scaly. Weight 1½ to 3 lb. Head rather round. Fur dark, uniform brown above, with longer hairs darker. Underparts gray. Sexes colored alike; young resemble adults. No seasonal change. Ears short, close, and round. Tail leaves conspicuous track. Teeth: I 1/1, C 0/0, P 0/0, M 3/3. Tracks: F 1½ × 1¼, 4; H 3½ × 1½, 5; Sp 3; St 8.

Found in waterways and marshes in shallower parts, from Alaska through all of Canada except northernmost areas and through all of United States except extreme southeast, including Florida, most of east-
(Continued)

Muskrat

711

ern Texas, practically all of California, and southern Oregon. Numbers decreasing where harvest is not wisely handled. Is not raised economically in confinement.

Probably polygamous. Four to nine young born about 29 days after breeding. First litter appears March to April; last of three, which is usually largest litter, appears in late summer. Mothers fight effectively to protect young, probably without cooperation of male.

Food largely plant materials, stored at times; also shellfish such as freshwater mussels and other small aquatic animals. Enemies include hawks, minks, otters, owls, and, most of all, trappers. Lives in burrows in bank or in houses of nonwoody marsh plants piled high above waterline, with generous room inside and underwater entrance.

Possibly most valuable of all wild furbearers because of great numbers and availability. Known in trade as Hudson seal, Russian otter, red seal, river mink; durability 45 percent that of otter.

House Mouse

Order Rodentia./Family Muridae

House Mouse
Mus musculus
Length to 7 in., tail 3½ in. Sexes colored alike a uniform brownish-gray with underparts somewhat lighter. Tail only sparsely haired. Hair short and close. Nose pointed. Ears relatively large and erect. Head relatively long. Body more slender than associated common mice. Teeth: I 1/1, C 0/0, P 0/0, M 3/3. Tracks: F ¼ × ¼, 5; H ½ × ¾, 5; Sp 1½.

Found in houses, fields, and waste places through most of North America and other continents, outdoor animals being found in more temperate regions. Probably originated in Asia and came to America with settlers after making way through Europe. Common on ships of world.

Probably does not pair but breeds promiscuously, young appearing at any time of year in litters of 4 to 11, 21 days after mating. Weaned in 3 weeks. Female breeds at 2 months of age. A pair theoretically could have 1,000 descendants in a single year with abundant food and no enemies to cut down numbers. Life-span to 4 years.

Food, a variety of plant and animal substances, either growing freshly in fields or stored in houses or granaries. Storage instinct developed only slightly, cleanliness not practiced, and animal does not hibernate. Individuals may have a sweet squeaking song.

Populations to 17 per sq yd or over 82,000 per acre have been known in California, but this is abnormal. Probably, next to the rat, one of most destructive mammals, injuring woodwork in houses, fouling books and stored clothing, and carrying diseases, including spotted fever. Commercial white mouse sold for pets and as laboratory animal is an albinistic mutation.

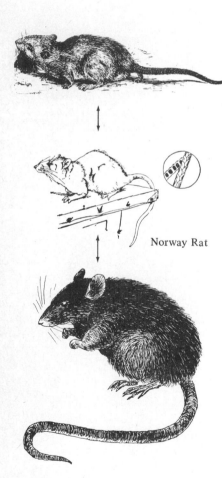

Norway Rat

Norway Rat
Rattus norvegicus
Length of body of year-old animal, 9 in.; tail 8 in. Weight ⅓ to 1 lb. Norway rat brown. Black rat smaller and darker. Albino or "white rats" are almost entirely forms of Norway rat, although albino black rats are known. Wild Norway rat is fierce fighter, persistent and sometimes dangerous. Laboratory white rat is relatively harmless. Teeth: I 1/1, C 0/0, P 0/0, M 3/3. Tracks: F ½ × ½, 5; H 1½ × ½, 5; Sp 4.

Norway rats live in fields, marshes, woods, in towns, on farms, in houses from China, Japan, North America, and elsewhere, wherever man lives. Came into Europe in eighteenth century; to England about 1729, Prussia 1750, Paris 1792, and United States 1775, following house rat of Europe here by about 250 years. Did not originate in Norway.

Mates promiscuously at any time of year. About six young born 22 days later, blind, naked, and helpless. Ears open in 3 days, eyes in 14 to 17 days. Become sexually mature in 2 months. Double weight 6 days after birth. Cease breeding at 18 months. Old at 3 years. In wild produces six annual litters.

Food, almost any organic substance. Will attack humans, and has killed pigs and calves. Responds to diet as does man. Will work in unison to obtain food. Subject to many of man's diseases. Carrier of bubonic plague, trichinosis of swine, tuberculosis of poultry. Black rat survives cooler climate, is less aggressive than Norway rat. Respiration, 100 to 150.

Wild Norway rats destroy about 300 million to 2 billion dollars' worth of farm products yearly in United States and 20 million dollars' worth in Canada. Our worst pest. 1 part of barium carbonate to 4 of soft food is an effective but dangerous poison control. Albino rat is one of most useful animals in studying man's diet and diseases, ranking with guinea pig, hamster, and monkeys.

Order Rodentia./Family Aplodontiidae

Mountain Beaver, Boomer
Aplodontia rufa
Length 13½ in., tail ½ in. Weight to 3 lb. Like a short, tailless, chunky, short-eared woodchuck or rabbit in general appearance. Light reddish-gray above, sprinkled with black, with a white spot at base of ears. Young darker than adults. Adult females with prominent dark spots around nipples. Teeth: I 1/1, C 0/0, P 2/1, M 3/3.

Limited in range to northern and eastern California and western Oregon and Washington, where there is dense forest or a high mountain meadow and mild climate. Usually where there is deep soil which can be easily tunneled. Also, water must be present in abundance. Good swimmer. Nine races recognized.

Probably mates promiscuously February to March, but this is not certain. Two to four young born 6 to 8 weeks after breeding; with closed eyes but well furred; usually half-grown and out of den by June. Male does not assist in rearing young.

Food probably solely plant material, much of which is dried after being cut, then stored neatly in underground rooms. This stored material may also be used for bedding purposes. Sometimes storerooms for these materials are sealed with earth. Requires drinking water in great abundance at all times.

Probably relatively harmless because of type of country it inhabits. May influence direction of flow of water underground much as beavers stop its flow with dams above ground. Burrows are generally free of wastes. Pelts are of little commercial value. Indians used the flesh for food.

Mountain Beaver

Coypu, Nutria
Myocastor coypus
Length with tail, to 3 ft but usually smaller. Tail ⅔ as long as body proper, scaly or thinly covered with short hairs. Weight to 20 lb. Pelt long, dusky to brownish-yellow fur with long guard hairs covering a dense, soft, yellowish underfur. Best fur on the underparts. Hind feet webbed. Four prominent yellow front teeth and long gray whiskers.

Native of both sides of the Andes from Peru southward, where the animals live in the banks of waterways in burrows that they dig. In the 1930s a few were introduced into the United States for rearing as furbearers. About 1940 several hundred were washed from their enclosures by floods and have since spread over an area 300 miles across in the lower Mississippi delta and to the western Texas border, so that by 1948 season we had a harvest of 50,000

At 6 months of age a female may begin breeding and is capable of producing 3 to 12 rat-sized offspring every 4 months. Young nutrias swim within a few hours after they are born and can produce a litter of young in the same season in which they are themselves born. Growth may continue for 2 years, but animals can yield commercial pelts the first season. Females rarely live over 3 years; males live over 6.

Food is primarily the foliage, roots, seeds, and other parts of aquatic plants, and an old generic name describes the animal as a "river mouse" rather than as a "mouse beaver," as the present name implies. Call is a low, piglike grunt, often reaching chorus proportions at dusk. Nutria are mentioned in the game laws of some states but may become pests.

Properly cured, the fur, which is taken from the underside of the body, may yield a valuable pelt for use in fur coats. It closely resembles the more expensive beaver and may soon offer competition for the all-important muskrat. The coarse hairs of the back are used in making felt for hats, thousands of animals being killed in South America for that purpose alone.

Coypu, Nutria

Order Rodentia./Family Dasyproctidae

Golden Agouti
Dasyprocta aguti
Slightly larger than a rabbit but with slender legs and only the shortest of tails. Feet have been described as being like those of a deer, but this is inadequate. Forefeet have five toes and hindfeet, three toes. Fur is rich olive above and yellow-white beneath, with the rump amply supplied with long orange hair. Eyes are large and appealing. Ears are relatively small.

Found mostly in forests or wooded lands from southern Mexico through some of the West Indies and into tropical Central America and South America. There are about a dozen closely related species living more or less in the same part of the world. The animals may adapt themselves to the dwellings of human beings and may live there as pets if provided with adequate protection.

The young, five or six, are born 64 days after the parents have mated. They are well developed at birth and may attempt to nurse even before
(Continued)

Golden Agouti

they are completely born. They have been known to begin nibbling at leaves within an hour after they are born and, as may be expected, soon become independent of the mother for food; however, families may remain together for some time. Young become readily tamed.

Food of agoutis is chiefly plants, including almost any part that is edible to animals. With their sharp teeth they can get the meat from very hard nuts. Agoutis are largely nocturnal, usually holing up for the day in tree cavities or in holes underground. They are the prey of most of the larger flesh eaters of their environment, including, the ocelot and jaguar.

Agoutis may destroy some crops. Since their flesh is white, tender, and deliciously flavored, the animals are sought as a source of wild food. They are found in zoos in many parts of the world. They are not considered as of great economic importance except when they may occur in considerable numbers and offer competition for crops favored by man.

Capybara

Order Rodentia./Family Hydrochoeridae

Capybara, Water Pig, Cavy, Carpincho
Hydrochoerus capybara
Length to 4 ft with no tail. Weight to 120 lb or more, making it the largest of the rodents, growing to size of a small pig. Hind legs are moderately long and are longer than the front legs. Neck short. Ears small. Head broad. Forefeet have four toes; hind feet, three. Toes have short webs and short nails. Hair is long, coarse, reddish-brown above and brownish-yellow beneath.

Native of South America, where it is found west to the foot of the Andes. It is almost wholly aquatic, living in rivers, lakes, and marshes, but it can run on land and may be seen there feeding on grass very much like a horse. Marshes and wet woodlands are equally popular with the animals. This largest of the living rodents is now confined to eastern South America.

The young, three to eight in number, are born from 119 to 126 days after mating takes place. They are rather well developed at birth. They are able to feed independently of the mother at an early age, but the animals of a family may stay together for some time. They may spend much time on shore feeding on grass and other vegetation, but when disturbed they may hasten to water, dive, swim long distances submerged.

Food is almost exclusively plants, of which they eat prodigious amounts, gathered either in the water or on drier ground. When kept captive in parks, the animals become attached to their attendants and are friendly to visitors, but in the native state they are necessarily shy. They are not fighters but prefer to defend themselves by escape. This they do most readily.

The flesh is considered delicious by some and inedible by others, but it is not sought sufficiently for the animal to have been domesticated as a source of fleshy food for man. The animals are hunted with dogs that chase them into streams. There they are pursued by men in canoes who kill them when they come to the surface. Their conspicuous incisor teeth used ornamentally.

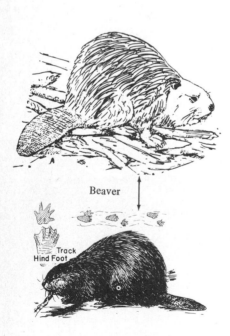

Beaver

Track
Hind Foot

Beaver
Castor canadensis
Length 43 in., tail 16 in. long by 4½ in. wide, broad, thick, paddlelike, and essentially naked. Weight to 60 lb or more, average 30 lb. Front feet with five toes. Hind feet with five webbed toes. Brown, with beautiful soft close underfur. Sexes colored alike with little age or seasonal variation. Teeth: I 1/1, C 0/0, P 1/1, M 3/3.

Originally found through Canada and Alaska, south to Mexican border except in southern Nevada and southern California, and to Gulf of Mexico except in Florida and southeastern coast region. Extirpated from much of plains, the Mississippi, and settled East but being restored locally. An animal of headwaters.

Apparently mates for life. Breeds early in year. Two to eight young are born 3 months later. At 1 month, young eat and seek solid food; cease nursing at 6 weeks; weigh 8 lb by midsummer; mature at 3 years, when new home is established and surplus 2-year-olds are crowded out of community. Life-span to 16 years. Tracks: F 2½ × 3½, 5; H 5½ × 7, 5; St 16.

Food exclusively plants, largely bark of softwoods felled near shore, cut up, dragged to water, and stored underwater for winter use. Full meal, 3 lb of poplar bark. Den a house surrounded by deep water with underwater entrance, with large ventilated, dry room protected by sticks and frozen mud. Dams maintain water level. Tail not used as shovel.

Seven open seasons in New York before 1940 yielded 22,000 pelts worth $374,000, from 20 animals introduced in 1901. Was formerly a basis of natural wealth and a medium of exchange. Role of beavers' dams in flood control and soil building only recently appreciated as more valuable than fur. Flesh edible. An endangered species.

Order Rodentia./Family Erethizontidae

Porcupine
Erethezon dorsatum
Length to 30 in.; tail 7 in. Weight to 30 lb, with average about 15 to 20 lb.
Male slightly the larger. Brownish-black sprinkled with black-tipped,
white-based quills, which are shed but not shot and which show chiefly
on tail and rump. Tail swung in defense. Front teeth bright orange.
Young animals have weaker quills than adults and are somewhat darker
in color. Teeth: I 1/1, C 0/0, P 1/1, M 3/3.

Found in deeply wooded areas from mountainous West Virginia northeast across mouth of Gulf of St. Lawrence, west across Hudson Bay and
northern Canada to north of Arctic Circle in Alaska south to central
California and New Mexico north to North Dakota.

Female urged by male courts in autumn. Males do not fight. Single
young born 209 to 217 days after breeding; eyes open at birth, 11 in. long,
weight 1⅓ lb; waddles in 6 hours; climbs second day, nurses to 4 months;
3½ lb, 18 in. at end of first summer; 8 lb, 21 in., second summer; 12 lb, 25
in. and mature, third summer.

Food almost exclusively plants, favoring salty substances and evergreen
foliage but also eats water plants, tree bark, and other substances usually
collected at twilight, or dawn, or in moonlight. Is slow and stupid, not
particularly clean or sociable; does not hibernate; makes den in cavities in
trees or rocks. Tracks: F 2½ × 3½, 4; H 4½ × 2, 5; St 12.

Dangerous to dogs and other animals because of quills but useful to persons lost in forests as it can be killed easily and flesh is good to eat. Does
great damage when too abundant because of habit of girdling trees.
Makes relatively interesting pet. Fisher was formerly worst enemy when
it was abundant. Protected by law in some states. Skull is exceeded in
size among Western rodents only by that of the beaver.

Porcupine

Chinchilla
Chinchilla laniger
Length, body 9 in., tail 5 in. Shape and behavior, squirrellike. Head
rabbitlike, except for shape of ears which are broad, rounded at tip, and
nearly as long as head. Eyes large and black. Mustaches long and white
or black. Skin thin but strong. Hair fine, making thick gray pelt, ashy
beneath. Its pearl-grey fur is one of the most beautiful, and has more
hairs per square inch than any other animal.

Native of higher Andes, from south of Chile to north of Bolivia. Practically vanished from original range but now successfully domesticated in
United States. Short-tailed chinchilla is larger than common chinchilla.

Strictly monogamous. Mates for life, bears first litter of one to four
young 111 days after breeding. Young fully furred, open-eyed at birth,
able to run within an hour; nursed 45 days but stay under care of both
parents at least 75 days. In captivity, courts and mates at 5 to 8 months,
bearing one to three litters annually. Lives 8 to 10 years.

In captivity eats yeast, molasses, wheat germ, oat middlings, peanut oil,
soybean oil, alfalfa meal, and bone meal, with carrots thrice weekly and
orange juice twice weekly. Daily ration about 2 oz. A ground animal,
unable to climb but runs and hides rapidly. Nest clean and bare. Affectionate, clean, nocturnal.

Probably most valuable furbearer; a coat of pelts brings $20,000 to
$100,000. Breeding animals in 1939 worth $3,200 a pair. Breeders introduced into United States in 1923 by M. F. Chapman, who started with 18
animals. 16 years later, there were 29 chinchilla farms scattered across
continent. 120 to 140 pelts needed per coat.

Chinchilla

Order Rodentia./Family Caviidae

Guinea Pig
Cavia cobaya
Length about 10 in. Weight about 1 lb. Body stout and heavy. Legs
and tail short. Front feet four-toed. Hind feet three- to five-toed, with
large angular nails. Fur of wild animal coarse, long, and grayish-brown;
of domestic animal, highly variable. When hungry, gives a grunt like a
pig. Probably live about 1 year.

Some 20 wild species and many more domesticated varieties. Varieties
are named English, Peruvian, Angora, and Abyssinian, but parent stock
undoubtedly came from South America, the name "guinea" being a cor-
ruption of Guiana. In Brazil, are known as "cavies"; said to have been
brought to Europe by Dutch in sixteenth century.

Although wild guinea pigs breed only twice a year, domesticated animals
may breed every 2 months after maturity and mature in 3 to 5 months
after birth. Young born 61 to 63 days after breeding; eyes open; hair fully
developed; run a few hours after birth. Litters of 4 to 12, though in wild
animal, only 1 to 2. Life-span 8 years.

Lives on vegetation in wild, usually feeding when sunlight is dull. Never
bites. Has unpleasant odor but makes excellent gentle pet. Cannot

Guinea Pig

(Continued)

stand low temperatures. Young independent of mother at 3 weeks. Wild animals burrow in sand or prefer living in vegetation-covered marshy areas. Respiration, 100 to 150. Temperature, 101.7 to 102.6°F.

Domesticated and raised for food by Incas of Peru before advent of white man to America. Now bred for studies in heredity, for production of serums in medicine, for isolation and production of disease-producing organisms, and for pets. Greatest importance unquestionably is in connection with medical research. Popular with pet fanciers. All varieties of the domestic guinea pig have been produced by selection; they differ from one another chiefly in the color and texture of their fur.

Pika

Order Lagomorpha./Family Ochotonidae

Pika, Cony, Little Chief Hare
Ochotona princeps
Length to 8 in. Weight to 6½ oz. Legs short and all about same length with soles of feet padded. No tail shows externally. Ears large but short, rounded, and white-bordered. Fur brown on back but yellower toward head and blacker on back. Feet white. Rarely seem to be fat even at end of summer season. Quite like a medium-sized guinea pig. Teeth: I 2/1, C 0/0, P 3/2, M 2/3.

Allied to rabbits although they may look like rats. Range through northern Rocky Mountains with allied species in Colorado, northern California, and Alaska. Old World pikas are abundant in Asiatic mountains, usually at elevations of 3,000 to 12,000 ft. Common name "cony" also applied to hyrax, an ungulate only remotely related.

Usually three to four ⅓-oz young to a litter, usually born May to September in comfortable grass-lined nest protected by parent. Family apparently is cooperative. One pika has been known to relieve another being chased by a weasel though whether this was a relative or not is not known.

Food, plant materials stored for winter use in a stack to nearly a peck in size, guarded over by a pika that watches nearby on a large rock and gives a squeaking song that apparently is understood by others of the neighborhood. Plant food includes leaves, stems, and roots, usually dried. Mostly nocturnal.

Pikas probably have no value to man except that they are basic food for some animals whose pelts have value as fur. They may compete with sheep and other grazing domestic animals that feed at high elevations, but food they harvest and store is still available for eating by the domestic animal. Fur durability, 20 percent that of otter.

Order Lagomorpha./Family Laporidae [Hares and Rabbits]

Domestic Rabbit
Lepus cuniculus
Wild European rabbit, from which domestic rabbit is developed, weighs 7 to 12 lb. Dutch rabbits may weigh as little as 1¼ lb. While wild rabbit has erect ears, domestic forms with lop ears 23 in. long are developed. Forefeet with five toes, hind feet with four. Teeth: I 2/1, C 0/0, P 3/2, M 3/3.

Domestic Rabbit

Belgian hare is a true rabbit, while snowshoe rabbit is a true hare; former bears helpless, naked young while latter does not. Probably no mammal except dog has developed more variation under domestication than the rabbit. Found domesticated and raised for food and pelt in all parts of civilized world.

After male *buck* breeds with female *doe,* young are born in from 30 to 32 days, in litters of 4 to 12 or more. Young remain blind for a number of days and are cared for splendidly by mother but not by father. May be six to eight litters a year but average 3½ a year, and families may stay together with other families indefinitely. Pulse, 140 to 160. Respiration, 55. Temperature, 101.7 to 102.5°F.

Easily cared for if given dry quarters where water, a place to burrow, and an abundance of food are provided. Generally feed at night or when light is not strong. Individuals may warn each other of danger by thumping on ground with hind feet. Hares do not normally care for young as well as rabbits.

Bred for hair, flesh, genetic studies, hides, and pleasure. Angoras have hair 7 in. long, while silver grays have short, thick, chinchilla-like fur. Cheaper fur coats, "lapins," are rabbit; though durability is low, they are worth their cost. Trade names are arctic seal, clipped seal, polar seal. Some are excellent flesh producers; others serve in medicine. Fur durability, five percent.

Jack Rabbit

Jack Rabbit
Lepus californicus

Length to 2 ft, with 4-in tail. Hind feet 5½ in. long. Weight to nearly 6 lb. Female with well-developed fluff of long soft hair on belly. Color dark brown, with white patches on forehead, around eyes, and on back of ears. Top of tail and tips of back of ears black. Paler in summer. Young like adults. Not a rabbit, but a hare.

Animal of open plains, with many close relatives. Black-tailed jack, *L. californicus,* found from Nebraska and Texas to California and Washington. White-tailed jack, *L. townsendi,* gray in summer and white in winter, in northern and western states from western Wisconsin to California and New Mexico, north to Saskatchewan and Manitoba.

Probably polygamous, breeding in warmer months. One to five young born 30 days after breeding, well developed, with eyes open, well-furred, weigh 2 to 3 oz; ears short, begin to lengthen at 10 days. Kept by mother in individual and separate spots; nursed a few days. Possibly two litters in South, one in North.

Strictly a plant eater. Adult does not dig burrow or have home other than an area. May enter burrows to escape. Rarely social. Rarely drinks, water being formed from food. Speeding blacktail may jump 25 ft. Speed, to 45 mph. All carnivores are enemies, particularly man. 10 to 70 per sq mi in Minnesota. Tracks: F1½ × 1½; H 2 × 3½.

Provides food for carnivores, acting as a buffer species between them and livestock but also competes with all grazing animals. Controlled by great drives, 1 sq mi sometimes yielding 8,000 jack rabbits. Has many parasites and carries tularemia, which may be fatal to men who handle the bodies.

Snowshoe Rabbit, Varying Hare
Lepus americanus

Length 18 in., tail 2 in. Hind feet 5½ in. Weight to 5¼ lb; average 4¼ lb. In summer, red-brown above, black-peppered, with light brown on legs and pure white beneath. Top of tail dark gray to black. Ears black-tipped behind and white-bordered in front. In winter, pure white except for black ear tips. Well-furred hind feet are a useful adaptation for deep snow of boreal forests.

Some dozen races from mountainous West Virginia northeast to Rhode Island, along coast to Hudson Bay, west to Alaska except in extreme north, south to Oregon except coast of British Columbia, southeast to Colorado, north to Saskatchewan, and southeast through northern United States to West Virginia. Favors marshy woods.

Courting male fights others with teeth; indifferent to one to six young born 36 days after breeding. One to three litters, April to August. Young, fourth hour, sound-sensitive, 3 oz; first day, gray fur, run instead of hopping; second day, crawl; third day, hop; fifth day, raise ears; seventh day, 4½ oz, on all toes; eleventh day, ½ lb, nibble hay; twenty-first day, fight, 14 oz. 33 days old, 1¾ lb; thirty-fifth day, nursing stops, 1½ lb. Average daily gain for 159 days from birth, about ⅓ oz.

Food: favors aspen, conifers, and dandelions; in summer, raspberry tangles. Female defends territory when bearing young, 10 acres per rabbit, but five to six per acre not uncommon. Drinks freely. Abundance cycle, about 10 years.

Must have room. No nest, but rests in forms. Enters burrows for safety. Does little damage to forests because of eating here and there; may "top" young conifers. Abundance varies in definite cycles with lynx and other predators. Good game species. May eat dead of own kind. Fur of little value, known as Baltic fox or white fox in trade. Important to northern natives.

Snowshoe Rabbit

Cottontail Rabbit
Sylvilagus floridanus

Length to 17 in., tail 2 in., fluffy. Weight 2½ to 3 lb. Ears long, 2½ to 3 in. Hind legs long and powerful, do not bring good luck. Dark brown above, mixed with gray. Top of tail brown. Under part of tail and all under parts white. Sexes and young colored similarly except that young are more yellowish-brown and have shorter ears. Teeth: I 2/1, C 0/0, P 3/2, M 3/3.

S. floridanus, found in United States east of Dakotas, Colorado, and central Texas; *S. nuttalli,* through Rocky Mountain areas border to border; *S. auduboni,* in Rocky Mountains, central and southern California and southwest; *S. bachmani,* on Pacific Coast; *S. palustris,* in Gulf States and eastern Virginia, south. Town or country.

Cottontail Rabbit

Promiscuous in mating. Breeds several times a year; probably four times in warmer regions. Four to seven young, ¾ oz, born 30 days after breeding; hidden in hollowed fur-lined "form" among vegetation; tiny, naked, helpless; each about 4 in. long; by 2 weeks, run from nest; may nurse about 2 weeks. Mother makes nest and protects young. Life-span to 8 years.

(Continued)

Food, a variety of plants including herbs, tree bark, and vegetables but not commonly fungi. Varies in numbers through years. Sheds fur May to June, again September to October. Breeding activity, indicated by abundance on highways, shows male active earlier and later through year than female. Tracks: F 1 × 1; H 1½ × 3½; Sp 5; L. 84.

Pest as crop and tree destroyer. Appeals to hunters for flesh and sport but carries tularemia, which may be fatal to man. Fur was used in manufacture of felt hats. Durability of fur, 5 percent that of otter; trade name, electric seal. Animals not profitably raised in captivity. Valuable to sportsman but a pest to farmer.

Order Artiodactyla./Family Suidae [Swine]

Swine just born are *pigs;* when weaned, they are *shoats.* Young female is a *gilt;* mature female, a *sow.* Mature males are *boars;* emasculated males incapable of reproduction are *barrows.* Sows should be 1 year old before they are allowed to produce their first litter but may breed at 4 months. Young born about 114 days after breeding. If care is shown with sows of prolific breeds, two litters a year may be produced. Sow in heat for 2 to 3 days every 20 to 21 days unless mated. Breeds and individual females vary in their ability to bring litters through to maturity. Young pigs ordinarily run with mother 10 to 12 weeks, but if two litters are to be raised, they should be weaned at 8 weeks, being allowed to return to sow only to relieve udders temporarily. Sows can breed well for 6 years. Cleanliness pays in rearing young pigs. Pulse, 70 to 80. Respiration, 8 to 18. Temperature, 101 to 103.5°F.

Success in pig raising depends on understanding and meeting food requirements. Pigs grow more rapidly than cattle, reproduce earlier, and eat less roughage. Proteins such as grain, skimmed milk, whey, and good pasturage are essential. Need 1 to 1½ gal of water per hundred weight daily. Growing pigs need mineral matter such as salt and those minerals in alfalfa hay. Brood sows must in addition have ample exercise and freedom from parasites, for which good pasturage is again good treatment. In winter, sows may be fed corn with alfalfa hay and properly fed, they may reach 300 to 400 lb in 1 year. Boar should not be allowed to become too fat and pasturage in summer with about 1 lb of grain a day should be suitable. A sow with a new littler should have only warm water for first day; then warm thin foods for 4 to 5 days. Fattening of pigs in a science, varying in Corn Belt and pasturage areas according to food available. Each area has its own problem, generally solved by local agricultural college.

In 1938, United States pig population was 44,418,000, worth about 500 million dollars. 63 million hogs in 1972. *World Almanac* or agricultural yearbooks will show how these figures vary from year to year, but any animal worth ½ billion dollars is bound to influence the prosperity of many people. Breed popularity is based on numbers: Poland China, Duroc, Chester White, Spotted Poland China, Hampshire, and Berkshire. Yorkshires are relatively few in numbers, yet produce highest quality bacon and in that category rank next Berkshires. Swine are probably older on earth than man, probably equal to age of horses or about 3 to 6 million years old. They have been domesticated since 2900 B.C. in China, but much later than that, from wild swine, in Europe.

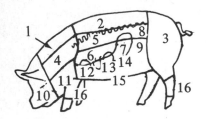

Cuts of Pork

1	Clear Plate	9	Chitterlings
2	Fat Back	10	Jowl
3	Ham	11	Picnic Butt
4	Boston Butt	12	Heart
5	Loin	13	Melt
6	Spareribs	14	Liver
7	Tripe	15	Belly Bacon
8	Kidney	16	Hocks

PHYLUM CHORDATA. CLASS MAMMALIA

Order Artiodactyla./ Family Suidae [Swine]

Yorkshire Hog
Sus scrofa

One of largest of all breeds of swine. White, long, narrow, deep-bodied, usually without wrinkles. Ears erect. Head short, well-dished. Side long deep, and smooth as would be expected of bacon type. Small Yorkshires raised largely in Europe are of the lard type, while American breed is bacon type.

Developed in county of York, England, by factory workers rather than by farmers; raised to furnish pork for home consumption. Breed known for over a century although it has been modified from long-legged animal into fine-quality bacon beast. Does well in close quarters. Introduced into United States in 1841; into Canada in 1886.

Yorkshire Hog

Berkshire Hog
Sus scrofa

Medium-sized hog. Rarely exceeds 900 lb in weight. Bred for long body, of good depth and medium width; for fine-quality meat, with high proportion of lean to fat. Snout short and upturned. Ears erect. Black with white markings. Not hardy and lacks rustling ability.

Developed originally in south of England, particularly in counties of Berks and Wilts from crossing original English hogs with Chinese and Siamese hogs, thus improving quality of meat. Other later changes produced less coarseness and small ears in place of large flopping ears. Breed introduced into United States in 1823.

Berkshire Hog

Duroc-Jersey Swine
Sus scrofa

Large red rather long-legged pig with heavy bones. Some breeders reject animal with heavy hams, short legs, heavy jowls, and thick heavy form. Only other conspicuously red pig raised in America commonly is the Tamworth, a bacon type, while the Duroc-Jersey is more of the pork type.

Developed in United States through crossing Durocs which had been bred in New York State with Jerseys developed in New Jersey. Earliest record is 1832; until 1875, they were commonest breed in New York, Vermont, and Connecticut; after that, they became popular breed of Middle West, where they developed a superior hog.

Pigs have six incisors on either jaw, outer pairs present at birth; center pairs appear at 3 months and remaining pairs at 4 months. At 6 to 10 months, outer pair are replaced by permanents; at 20 to 24 months, intermediates; and at 30 to 36 months middles.

Has excellent grazing ability and so is popular in Corn Belt states where suitable food is easily available. Hardy quality of this breed makes it possible for them to be reared in great numbers where housing facilities are limited or relatively poor.

This breed is more popular in United States and Canada than elsewhere, being probably even ahead of the Poland China breed, which is ordinarily reared for the same qualities. In Canada, where there are relatively few Duroc-Jerseys, the breed is most popular in southwestern Ontario and Alberta.

Duroc–Jersey

Order Artiodactyla./ Family Tayassuidae

Collared Peccary
Tyassu tajacu

Length 38 in. Height to 22 in. at shoulder. Weight 46 lb. Color blackish-brown with yellow-brown mixed with white on flanks. A broad whitish stripe runs from shoulders down to chest, like a collar. A large gland on back that looks like a navel; lacking in domestic pigs. Hind toes, three. Teeth: I 2/3, C 1/1, P 3/3, M 3/3.

Originally ranged from Arkansas and Texas to Patagonia. Now in United States found chiefly in southern Texas, southern Arizona, and parts of New Mexico, preferably in marshy forested lands where soil is easily explored with snout; also found in cultivated lands and nut groves.

Probably polygamous, though this is disputed. Breeds at any time, one to two young being born 142 to 148 days later. Young well-haired when born, spotted or striped; in 2 to 3 days, follow mother; in 1 month, join herd of adults in spite of small size. Young usually born in a cave or hollow log.

Food, a variety of roots, fruits, leaves, insects, reptiles, and almost any living plant or small animal which happens to be in way, though nuts, grain, and vegetables are favored when they can be had. Goes in herds and has been reputed to be vicious. Can defend itself well against dogs. Speed, 11 mph.

An interesting game animal, practically exterminated from United States territory. Destructive to crops at times but also destructive to many injurious animals. Flesh has value as food to man and hide has some value. Known to kill rattlesnakes and in herds can drive off jaguars and other foes.

Collared Peccary

Hippopotamus

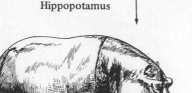

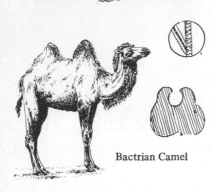

Bactrian Camel

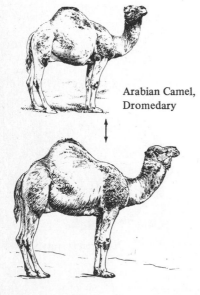

Arabian Camel, Dromedary

Order Artiodactyla. / Family Hippopotamidae

Hippopotamus
Hippopotamus amphibius
Length nose tip to tail tip, 14 to 17 ft. Girth about same. Male larger and darker than female. To 5,800 lb. Mouth possibly 2 ft wide. Eyes set high and far back, just in front of ears, which are 3 to 4 in. long. Nostrils, large, at top of blunt nose. Shoulder height to 5 ft. Canine teeth with 9¾-in. circumference, and are developed as large tusks.

Never far from water, most of time being spent in rivers, shallow mud-bottomed lakes, and lagoons; formerly from Egypt to Cape of Good Hope but now restricted to river area south of 17°N. Related and, of course, smaller pygmy hippopotamus is restricted to west coast of Africa.

Single young born 234 to 243 days after breeding; walks 5 minutes after birth; 18 in. high and 3 ft long; weight 60 lb, gray and wrinkled; at 9 months eats solid food; at 18 months, male weighs 600 lb, female 650 lb. First young born when female is 7½ years old. Life-span about 35 years. Males each occupy territories and fight fiercely to defend them.

Food, all sorts of succulent water plants, but also eats such crops as corn, sugar cane, and millet. In zoos, eats mixture of rolled oats, bran, cabbage, potatoes, lettuce, bread, and hay. Cannot remain submerged more than 5 minutes. Appear in large herds where undisturbed.

Females and young may do considerable damage to crops. Flesh is edible and is used as food by natives. Hide makes an excellent soup, much like turtle soup. One animal may yield 200 lb of fat, prized by Arabs. Teeth make excellent ivory. Probably the behemoth of Old Testament.

Order Artiodactyla. / Family Camelidae [Camels]

Bactrian Camel
Camelus bactrianus
Shorter-legged and therefore not so tall as Arabian camel, but with a heavier body and two humps on back instead of one. Height to shoulder about 6 ft. It has a heavier coat of hair, and harder and shorter feet, which enable it to travel over rougher country more easily than the Arabian camel. It is more docile and easier to ride, but slower. Can be dangerous to man; will bite, or kick, or "spit" saliva and even regurgitated stomach contents. Teeth: I 1/3, C 1/1, P 3/3, M 3/3.

There are probably no wild Bactrian camels except those which have escaped from domestication. Now confined to deserts of central Asia though a native of steppes of eastern Asia. It is to the wandering Mongolian tribes of Gobi desert what Arabian camel is to Arabs. Never needs shelter in bitterest weather or hardest traveling.

Life history is essentially that of Arabian camel. This species can begin to earn its own living by work at age of 4 and can continue a life of usefulness until it is 25, 30, or even older, which is probably longer than is possible with the Arabian camel.

Food, any plant material available, even bitter salt-marsh plants rejected by other animals. When forced by hunger, it will even eat fish, flesh, bones, and skins and will drink either salt or brackish water. Is able to uncover plants under snow and get them for food in winter. Can go 8 days without water in winter.

Probably most useful beast of burden in colder northern deserts, carrying daily loads of 500 to 600 lb of coal, oil, or other goods over roughest of territory. It also yields hair, hide, flesh, and milk of direct use to man. Both species are figured in earlier Assyrian sculptures. Long used in war, even to present. An endangered species.

Arabian Camel, Dromedary
Camelus dromedarius
Height at shoulder to 7 ft. Legs and neck long. Feet two-toed, soles being undivided and hoofs on upper surface. Hair soft and unevenly placed, longest being on head, neck, shoulders, hump, and thighs. Tail relatively short. Prominent pads on knees.

Arabian camel is used both in Africa and in central Asia. It has been introduced unsuccessfully in southwestern United States, in Australia, and in southern Europe, but still finds a place for itself in Africa and Asia. Two attempts to introduce it in United States failed and it will probably not be tried again.

Single young born about 52 weeks after breeding; calf goes to pasture when 1 day old, has a soft fleece, gives a gentle *bah*, lacks knee and breast pads of adults. Mothers mourn loss of young as much as 3 months. Young becomes full-grown at 5 years. Female may bear to 12 young during lifetime. Life-span 17 to 50 years.

Food, all sorts of vegetable materials available. With one drink of 15 gal of water, can carry 400 to 600 lb 30 miles a day for 3 to 4 days without water, or 1,300 lb 6 to 10 days for a shorter distance. Six camels can draw as much as three to six mule teams over territory impassable to mules. Pulse, 30. Temperature, 99.5°F.

Most valuable desert animal for work, flesh, milk, hair, and other purposes. Has been domesticated for 5,000 years; appears on pottery of 4000 B.C., and bones have been associated with those of Neanderthal man.

Llama
Llama perunana

Length 4 ft, tail ½ ft. Height 3 ft at shoulder. Neck nearly 2 ft long, with broad body and thin legs. Toenails, sharp talons. Hair thick, soft, and shorter on head and ears. Color commonly white or brown. Alpaca is larger, slightly longer-haired, and has smaller head. Name pronounced "yama."

Both llama and alpaca are derived from wild guanaco. Llamas are found in Peru, Bolivia, and on west coast of South America. In temperate climates like cooler Patagonia they come down to sea but in equatorial regions are found in Andes at 12,000 to 16,000-ft elevation. Guanaco still exists in wild state.

Breeds every 2 years, when young males are driven from herds by females while older males fight each other for leadership. Breeds in July; single young with conspicuously thin neck, body, and legs is born 11 months later. Llamas mate with alpacas but offspring are sterile. Life-span 12 to 18 years.

Food, coarse vegetables including shrubs, mosses, lichens, and grasses. If succulent food is available, it does not drink. Lives in large herds. Is more intelligent than camel. Expels stomach contents in face of offenders. Generally gentle and may be ridden or driven by children. Must have cool climate to thrive.

Valuable for fleece, although male is used as beast of burden, carrying about 100 lb. Alpaca fleece superior, producing 11 to 15 lb when sheared every other year. Flesh of young llama tender, of old, too tough for use except when dried. Efforts to establish llama in Australia and Europe have failed. 6-ft vicuña robe weighs less than 4 lb.

Llama

Order Artiodactyla./Family Cervidae [Deer]

Mule Deer
Odocoileus hemionus

Length to 6 ft, including 8-in tail. Shoulder height to 42 in. Weight to 450 lb; doe smaller, with average length of 4 ft. In winter, brownish-gray with black above; throat, inside legs, patch on buttocks, and face white; tail white except for black bunch on tip. May to August, red replaces brownish-gray. Female duller. Fawn yellowish with white spots. Teeth: I 0/3, C 0/1, P 3/3, M 3/3.

Coast blacktail, now considered a subspecies of the mule deer, has bushy tail, all black above; mule deer, a black-tipped not bushy tail; and white-tail, black only in center of top of tail. Mule deer ranges from central Mexico north through Manitoba and west to coast, except in northwest California and western Washington, Oregon, and British Columbia where blacktail is.

Breeds in November; buck wins harem of does. One to three fawns born May to June; kept hidden 6 to 8 weeks; nursed by doe; fawns follow doe during summer and first winter; young does may follow her for 2 years. Young whitetail may bear young at 2 years, and mule deer probably does the same. Life-span about 16 years.

Food largely twigs and brush; while grass is eaten, it is not favored. Gets along well without water except for that obtained from succulent plants. Eats moss but favors nuts, acorns, and evergreen oak. Horns do not face forward as in whitetail. Gland on inside of leg long, rather than small and round. Tracks: F 2½ × 3, 2; H 2¼ × 2¾, 2.

Valuable food and sport animal naturally kept in control by coyotes, bear, and cougar but also by sportsmen. Serves as attraction to tourists in recreation areas. Antlers shed in February, blacktail often earlier than mule deer.

Most important big game animal of the Western States. Can do great damage to range, forest, and crops if numbers are not controlled.

Mule Deer

White-tailed Deer
Odocoileus virginiana

Length to 8 ft. Tail 11 in. Height at shoulder to 4 ft. Weight to 300 lb. Reddish-chestnut or gray above with white below. Bucks with forward-pointing antlers to 30 in. long, but with prongs upward-pointing. Marvelous jumper and runner. Does, smaller. Fawns, spotted. Teeth: I 0/3, C 0/1, P 3/3, M 3/3. Tracks: F 2 × 3½, 2; H 2 × 3, 2.

Through North America south of Gulf of St. Lawrence and James Bay, west to Rocky Mountains; range is extended northwest to south of Columbia River in Oregon. Thrives around clearings, in mixed woodlands, with water nearby and plenty of food. Individual range rarely over 1 sq mi.

One male forms a herd with two or three females, fighting rival bucks; breeds in November; one to two fawns born about 196 days later (June), each weighing at birth about 3¾ lb. Young hidden and nursed every 4 hours; by 4 to 5 weeks follow mother; weaned and lose spotted coat at 4 months. Doe may follow mother 2 years or breed first year; buck follows mother 1 year.

White-tailed Deer

(Continued)

Food largely browse, such as twigs and leaves of trees, not common grass. Will eat rye grasses, prefers weeds to timothy or bluegrass and particularly enjoys nuts, yellow water lily, and acorns. In winter may form "yard" of a herd in snow, keeping snow trampled; often starves. Speed 30 mph. Jump 8 ft high, 30 ft horizontal.

A most valuable game animal providing sport, flesh, hide, and trophies of high grade. Most popular big game animal in the East. In spite of hunting pressure, wise game management procedures have helped to increase size of herd in many areas. Deterioration of habitat with loss of available food supply, weather conditions, disease, parasites take their toll. Destructive to orchards. State mammal of Pennsylvania.

Elk

Wapiti, Elk
Cervus canadensis
Length to 9 ft. Height at shoulder to 5 ft 8 in. Girth around chest, 6 ft 8 in. Weight possibly to 1,000 lb but average male, 700 lb; female 500 lb. Buck with great head of antlers to 5 ft long which are shed each year. Gray-brown but darker or chestnut along back and mane; lighter beneath rump patch. Cow paler. Teeth: I 0/3, C 1/1, P 3/3, M 3/3. Tracks: F 3½ × 4½, 2; H 3¾ × 4, 2.

Found from New Mexico and Arizona to British Columbia, formerly east to Atlantic; now eastern representatives are limited to parks or to eastern Canada but no great numbers anywhere outside an area along Rockies from Colorado to central Alberta. Last Pennsylvania elk was killed in 1869. One killed in New York in 1946.

Bull fights in November, wins harem of six to eight cows. One calf born 249 to 262 days after breeding, weighs 30½ lb at birth; has spotted coat; follows cow in 3 to 4 days; grazes in 4 to 6 weeks; weaned in October, loses spotted coat; follows cow another 6 months. Cow breeds first at 3 to 4 years. Life-span 18 years. Jumps to 10 ft. high. Has a graceful, pacing-trot gait which may reach 40 mph. Home range, 2 to 8 miles.

A grazing and browsing animal, which means that through year it may move from one area to another, winter and summer range being different. Food, a wide variety of grasses, shrubs, and trees. Peak feeding periods at twilight and dawn; most active during nighttime hours. Sanctuaries for elk to be effective must provide year-round feeding areas under protection. Antlers are shed each spring, around March, and new ones immediately begin to grow. Can increase 33 percent in 10 years.

Superior game animals which, in earlier times, provided food, hides, and bones for humans.

Moose

Moose
Alces palces
Length to 9 ft, with 3-in tail. Height 7 ft at withers. Antler spread to 6 ft. Weight probably to 1,800 lb. Cow about ¾ size of bull, without antlers. Muzzle broad. Snout blunt. Shoulders higher than rump. Forequarters heavier. Hoofs long and pointed. Blackish-brown, but ligher in summer. Teeth: I 0/3, C 0/1, P 3/3, M 3/3. Tracks: F 5 × 7, 2; H 5 × 7,2.

Found in wooded areas from New Brunswick and Nova Scotia across central Canada and southern Alaska, with some found along Rockies as far south as Wyoming; seen commonly in Yellowstone Park; also south in United States in northern North Dakota, Isle Royale, and northern Maine. Has disappeared from most of northeastern states.

Probably polygamous but less so than most deer. One or two, but rarely more than one, calves born in May, 242 to 246 days after breeding; not spotted, dull red-brown; by 10 days run with mother; stand 3 ft. With mother 2 years (?). Cow receptive to bull for 30 days; bull in breeding mood 6 weeks. Cow does calling; bull bellows; breeds second year. Life-span to 20 years.

Food, browse from woody plants, particularly hardwoods; favors succulent water plants, such as yellow water lily. Family of a bull, a cow, and some calves stays together through winter, "yarding" to keep high snows down. May not move outside a 10-mile range in lifetime. Trots through brush at 14 mph. Spends much time in water.

Superior game animal which has been destroyed over much of its range and reduced elsewhere. Provided food, leather, and bone for earlier peoples and improves the wildness of an area now. Should not be further reduced in numbers.

Caribou, Reindeer

Caribou, Reindeer
Rangifer caribou
Length 6 ft, tail 5 in. Height at shoulder 4 ft. Weight to 300 lb. Cow smaller, weighing about 150 to 250 lb. Both sexes bear horns, unlike most other deer; colored alike, brown with yellow-white on neck, lighter in winter. Barren ground caribou, *R. arcticus*, smaller. Teeth: I 0/3, C 1/1, P 3/3, M 3/3.

Woodland caribou, *R. caribou*, New World reindeer, originally ranged from northern Maine to Minnesota and to northern Alberta, south to Idaho, northwest to Alaska except close to Pacific Coast north through Alaska and Canada to tree belt, with close relatives on to land limits. Typically, *R. caribou*, is found in forests and more particularly wooded swamps and bogs.

Polygamous, breeds in October; bulls fight for many mates.   One buck will serve 10 to 15 does.   Rut about 2 weeks.   Male loses antlers in early winter but doe does not.   One or two young born 7 months, 7 days after breeding or in May; walks in 2 hours; suckled 2 months; then in fall joins herd which may migrate to new area.   Doe sheds antlers in spring.   Life-span to 15 years.

Has strange migrations every 3 to 4 years and is probably more of a roamer than other deer here considered.   Has strong-scented neck glands.   Swims at 4 mph and runs at 10 mph but can travel 100 miles in a day.   Feeds largely on available plants such as grasses, lichens, and mosses.

One of most useful deer, since it and its kin are domesticated, yielding flesh, milk, hide, and bones and being able to draw a 250-lb load over snow 1 mile in 3 minutes.   A subject of controversy as a result of proposed exploitation of Alaskan oil resources.   Extensive research needed to know many intricate habitat relationships.

Order Artiodactyla./Family Antilocapridae

Pronghorn Antelope
Antilocapra americanus
Length to over 4 ft, with 3-in. tail.   Height at shoulder 3 ft; to top of head, 4 ft.   Weight of buck long, hollow, of doe, 90 lb.   Horns to 20 in. long, hollow, on a bony core like goats, but branched once (from which animal gets its name); and shed each October like deer; also, four teats and woolly underhair.   There are a couple of well-marked white bands which cross the throat.   Each foot with two hoofs.   Color tan and white, bucks darker.   Teeth: I 0/3, C 0/1, P 3/3, M 3/3.

Pronghorn Antelope

Found in open country from southern Alberta and Saskatchewan through western Dakotas, Nebraska, Kansas, and Texas into northern Mexico and west through Lower California, southeastern California, Nevada, southeastern Oregon, and southern Idaho.   Individuals range over 2 to 3 sq mi.   Range and numbers have been greatly reduced.

Both sexes horned, does smaller.   In early fall, bucks form harem of up to 15 does won by fighting.   Two or three kids born in 8 months or June, scentless, able to walk.   Mother alone defends young, who follow her at 3 weeks.   In July, does and kids make small herds, joined in August by young bucks.   Mature at 5 years; aged at 10.

Food, entirely vegetation; consists of cactus, greasewood, sagebrush, various grasses, and other desert plants.   Herds and individuals, kept together by bucks or by family instinct, helped by flashing white rump which serves as danger and recognition signal.   Escapes from enemies or fights.   Speed to 40 mph.   In winter, migrates to suitable grazing area usually in herds.   May travel 20 miles a day.

Beautiful animals, relatively harmless but competitors with grazing animals for forage.   Flesh good.   Numbers in former time estimated at 40 million; latest estimate of present numbers, 165,000.   Worthy of more protection from hunters, at present their worst enemy.   Enticed to range of guns by inordinate curiosity.   Tracks: F 2 × 3, 2; H 2 × 3, 2.   *A. a. mexicana*, Mexican pronghorn, is an endangered species.

Order Artiodactyla./Family Giraffidae

Giraffe
Giraffa camelopardalis
Height to 19 ft.   Head with short blunt straight hornlike structures.   Weight of male estimated to be 1 to 2 tons.   Female slightly smaller, more slender.   Stiff mane on back of long neck.   Feet large, heavy, with divided hoofs.   Yellowish or whitish, with large more or less square black spots on body, neck, head, and upper parts of legs.   Female duller than male.

Giraffe

Formerly over greater part of Africa; now found only in regions south of Sahara, in central and south Africa.   Preferred habitat is dry, open wooded areas of plains.   Areas where they live are waterless at certain times of year.   Has been bred successfully in captivity in zoos but needs greater areas where it may be protected.

Bull maintains a herd of cows.   One young, born 15 months after breeding in March to April; is gentle and affectionate if caught and can be induced to eat.   Normally, a captured calf refuses to eat and dies shortly.   Life-span apparently about 25 years.

(Continued)

Food, leaves of such plants as acacias and mimosas. To graze on ground, animal must spread its forelegs far apart. Drinks little if any water. Gait is unique trot and gallop with neck bent forward and tail raised over back. It can almost outrun a fast horse if it has a good start. Voiceless. Sleeps standing. Gentle, but may attack by kicking either forward or backward with its shark-edged hoofs; also uses its heavy neck in a swinging, striking motion.

Valuable to desert people for hide and food. Flesh is eaten raw. The tendons are used for sewing leather and for strings for musical instruments. Giraffe is shown on ancient Egyptian monuments. Julius Caesar is reported to have been first to exhibit one in Rome.

Order Artiodactyla./Family Bovidae [Hollow-horned Ruminants]

Mountain Goat
Oreamnos americanus
Length 5½ ft, including 6½-in. tail. Height at shoulder 3½ ft. Horns to 1 ft long, pointing backward. Weight to 300 lb. Nannies are about ⅙ smaller than billies. Color white. Hair long, relatively coarse. A characteristic beard. Sexes and young colored alike. Superior climbers and jumpers. Sure-footed at all times. Teeth: I 0/3, C 0/1, P 3/3, M 3/3.

Ranges through Rocky Mountains and coast ranges from Alaska south to Montana and Idaho; limited in United States largely to sanctuaries of national parks. Ranges high in summer, low in winter. Individual range not necessarily great.

Probably monogamous, since male and female are seen together with kid through summer. Young born 147 to 178 days after breeding; can stand in 10 minutes; jumps in 30 minutes; nurses in 20 minutes; at 2 days stands 13½ in. high and weighs about 7 lb. Mother will defend young to best of her ability.

Food: grasses and similar vegetable materials available in home range. Kids are subject to attack by eagles, wolves, foxes, and other preying creatures. Not found in large herds, usually only few more than family group living together. Fur molts in spring. Tracks: F 2¾ × 2¾, 2; H 2¼ × 2¾, 2.

Tremendously interesting residents of equally interesting type of country. Hide has some value and flesh is eaten but the animals should not be exterminated. May possibly carry certain diseases but are too uncommon and too widely separated from domestic stock to be serious menace. Need more protection.

Muskox
Ovibos moschatus
Length total, 8 ft, with 4-in. tail. Height at shoulder to 5 ft. Weight to about 700 lb. Deep brown on head and body to black on under parts in front. Legs dirty white. Saddle light brown. Sexes alike except that cow may be lighter than bull. Has a strong musky scent. Horn spread to 2 ft. Teeth: I 0/3, C 0/1, P 3/3, M 3/3.

Fossil forms in Asia, Europe, and North America associated with Neanderthal and Cro-Magnon men. Now limited to northern Canada and Greenland, European and Siberian herds having been destroyed. Protection is afforded muskox in much of the area from which it has disappeared. It needs further protection or it will vanish.

Bulls fight, give off musk, and win herd of cows. Breeding in August produces one calf per cow in late May, which weighs 16 lb at birth and stands 18 in. high; follows cow 1 to 2 hours after birth; a 4-day calf weighs 25 lb and stands 20 in.; at 22 months, stands 3 ft 2 in. Breeds in alternate years.

Food: favors grass and willows but will eat almost any vegetation available. When attacked by men or wolves, herd forms circle and fights to death. Has strong sense of smell but is normally not intelligent in defense; a whole herd will stand ground and be shot down rather than retreat if enemy stays in vicinity.

Provides easily attained and excellent source of food for natives. Flesh edible but causes musky gases from eater. Both Canadian and Danish governments have acted to conserve diminishing herds of muskox. May be recovering under strict protection. Formerly an important item in Eskimo economy.

Order Artiodactyla./Family Bovidae [Hollow-horned Ruminants]

Bison, Buffalo
Bison bison
Length to 11 ft, tail 1½ ft. Height 6 ft. Weight to 3,100 lb. Color dark brown but lighter on rear portions. Hair over shoulders, forelegs, and head long and shaggy. Shoulders high-arched and powerful. Ears and eyes small. Horns short, sharp, curving upward, unbranched. Teeth: I 0/3, C 0/1, P 3/3, M 3/3.

Mountain Goat

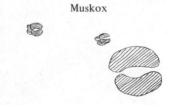

Muskox

Bison

Formerly occupied territory between Appalachians and Rockies, south to Gulf of Mexico, north through plains of Canada with a related species, wood's buffalo, occupying territory to north. Now practically restricted to ranches and sanctuaries, though northern form is more abundant.

Possibly monogamous, though bulls fight each other. One calf born 9 months after August to September breeding, birth lasting less than 1 hour; calf stands in 3 to 4 days. Bull helps in protection. Calf with cow to 3 years, when young cow breeds; may bear one calf a year for 30 years.

Food, grasses and other ground forage, herds in the past moving north and south with seasons to get best pasturage. Small herds are family herds led by an old cow. Enjoys mud wallows as freedom from insect pests. Herds may be stampeded wildly in a given direction to their destruction. Speed, to 40 mph.

Vanishing of buffalo herds is a disgrace to American intelligence, but saving them from complete extinction is a credit to a few far-seeing scientists. Bison provided Indians and early settlers with hair, hide, flesh, fuel, bones, and sport, all on hoof and easily obtainable anywhere.

Indian Buffalo
Bos bubalus
Length 9 to 10 ft, including 2-ft tail. Shoulder height to 6 ft. Horns heavy at base and curving backward and inward to sharp tips with distance from tip of one horn and, along the curve to the tip of the other horn may be 8 to 12 ft. Horns flattened and strongly marked with cross wrinkles. Head longer, ears smaller than in Cape buffalo, which weighs to 4,000 lb.

Indian buffalo, *Bos bubalus,* wild in India, Philippines, Malay Peninsula, and Ceylon, loving water; widely used as domestic animal throughout Orient. Cape buffalo, *Bos caffer,* has horns shaped more like those of domestic cattle, though each may be 3 ft long; found in east Africa from the Cape to Abyssinia.

Indian Buffalo

Calves born 287 to 340 days after breeding, which takes place at any time of year. Cow yields great amount of rich milk, more in fact than some domestic cows. Both cows and bulls protect young, herd providing dangerous front to even such fierce animals as tigers, which Indian buffalo apparently does not fear seriously.

Food, plant materials of ordinary forage type. Animal is attacked by great numbers of insects. As a protection, takes mud baths, caked mud preventing insects from reaching skin. Apparently buffalo has a keen sense of smell. It swims, dives, and feeds at night or in early morning, accepting rough forage.

Probably most useful domestic animal in Orient. Its hide makes excellent leather. While its flesh is inferior to that of domestic cattle, its milk is rich and makes liquid butter, *ghee,* favored in India. More powerful as a draft animal than ox, and more trustworthy than Cape buffalo.

Yak
Bos gruniens
Larger than average domestic cattle and more solidly built; height at shoulder hump 5 to 6 ft, weight of male, 1,200 lb, female smaller. Legs short and stout. Shoulders high. Horns broad, upward-curving, unbranched. Most conspicuous of all is long hair growing like a fringe on tail, flanks, and legs, almost reaching ground. Prominent breast tuft of long hair. Blackish-brown.

Found in mountainous central Asia, particularly in Tibet region; 14,000 to 20,000. Does well on barren soil which would not support similar large beasts. Cannot long survive warm regions. Domesticated yak is sometimes black, brown, red, or mottled instead of brownish-black of wild animal.

Yak

Reproduction probably similar to that of cattle. Calves born 258 days after breeding takes place. Milk is exceptionally rich, butter excellent. Undisturbed, a yak seems like a lazy inoffensive animal, but cow will defend its calf valiantly.

Food, sparse grass available among rocks and crags where it lives. Feeds usually morning and evening, chewing cud between times. Sure-footed, making a good pack animal over difficult terrain, but it moves relatively slowly and cannot compete with a horse. It can charge quickly, even though wounded.

Used as a saddle animal, a pack animal, a source of hair for making cloth, and a source of meat, milk, butter, and cheese in territory which could not support ordinary domestic animals. Bladders are used for water bags and hide makes excellent leather. Tails are used as ornaments and as brushes in tropics.

Zebu, East Indian Ox
Bos indicus

Size varies from that of a small donkey to a large animal. Characterized by having large fatty lump, or sometimes two over withers and an excessively large dewlap, making neck huge. Color usually pale fawn, iron-gray, black, or bay, commonly darker forward. Ears long and pendulous. Also known as Brahman or Oriental domestic cattle.

In commoner African breeds, horns are usually large; in Indian breeds, they are small; sometimes, absent. Essentially creatures of warm climates, since they resist heat better than ordinary domestic cattle.

Reproduction similar to cattle. In United States, zebus were introduced in 1853 and again in 1906; now well-established in Gulf Coast states. Grade zebu bulls mated with domestic cattle produce tick-resistant, heat-resistant animals with better resistance than cattle and better flesh than zebus.

Food essentially same as that of cattle. Zebus are nervous, do not herd well, and can run rapidly, which makes them difficult to handle. Almost complete immunity of animal to ticks which cause Texas fever makes it popular in tick-infested country and to increase tick resistance in crossbred animals.

Considered sacred animals by Hindus, who do not allow them to be killed. Serve as draft animals for work. Harnessed to carriage, they may travel 30 miles a day or, saddled, carry a man 6 mph for 15 hours. Carcass yields too large a proportion of cheap cuts to be popular with meat raisers. Domesticated by 4000 B.C.

Zebu

Domestic Cattle

Domestic cattle trace their ancestry to five lines, probably as follows (1) The bibovine group of southern Asia which includes the gayal; the banting of Bali, Borneo, and Malay Peninsula, which probably produced the zebu; and the gaur, which stands 6 ft high and is one of the most powerful of wild beasts. (2) The leptobovine group, represented only by fossils from France, India, and Italy. (3) The bisontine group, which includes both the European and American bison and yak. (4) The bubaline group, which includes the water buffalo, Cape buffalo, and their relatives. (5) The taurine group, which produced the ordinary beef and dairy cattle and humped cattle of Asia and Africa. The direct ancestors may have been one or more of these species: the extinct Indian ox, *Bos namadicus,* associated with men of the Stone Age; the ur, *Bos primigenius* of western Asia and all Europe, 6 ft high at withers, a forest lover, which probably lived in historic times and was used in bull fights before 1500 B.C.; and the Celtic shorthorn, *Bos longifrons* of western Europe and the British Isles.

Cattle are primarily grazing animals. Normally eat 8 hours out of the 24 and no more, spending rest of time chewing cud and resting. When grazing, tongue is wrapped around grass, which is pulled or cut with teeth of lower jaw pushed against upper jaw. Grass about 4 to 5 in. high is eaten most rapidly. Cuts in 8 hours, with 2½ in. of teeth, about 150 lb of green grass if pasture is good. This equals a pile 3 ft high and 6 ft in diameter. Will not eat grass growing from fresh droppings of their kind but will eat grass growing from horse or sheep droppings. Stabled cattle are usually fed regularly roughage and grain early morning and late afternoon. Require 2,840 cu ft of air per day per 1,000 lb. Hay or silage is fed at noon and water given twice a day. Milking is done at 12-hour intervals. Feeding rules call for abundance of balanced rations including variety, palatable forms, some succulent food, 8½ gal of fresh water, and some salt daily. A cow has four stomachs; the *rumen* or pauch, 80 percent of total; the *reticulum* or honeycomb; the *omasum* or *manyplies;* and the *abomasum* or true stomach. Pulse 40 to 70. Respiration, 10 to 30. Temperature, 101.5°F.

9 to 10 months after a *bull* breeds with a cow, a *calf* is born. During second year, it is a *yearling;* in third year, a *2-year-old.* Young female is a *heifer* but if a twin with a male, she is a *freemartin* and generally will not reproduce well. Calves run in fields together until 3 to 4 months old, then sexes should be separated. Male with testicles removed when young is a *steer;* if removed after matured, an *ox.* Ordinarily, a well-grown heifer may be bred at 18 months of age and continue as breeder 10 to 12 years. In heat, 2 to 4 days every 3 weeks. *Nurse cow* nurses another's calf. When a calf is born, cow is said to "come in" and when milk begins to flow as it does with birth of calf, cow is said to "freshen." Cows not producing milk are spoken of as "dry." Dry cows may be pastured but fresh cows require special feeding. Bull used for breeding is a *service bull.* Sperm from high grade bulls are now shipped alive long distances and used to impregnate cows artificially. In artificial insemination, one bull fertilizes to 200 cows at one service and may father to 6,000 calves, as compared to normally 30 calves a year. Semen may be frozen and used to successfully impregnate cows many years later. This has increased herd quality tremendously. Average life-span 15 to 20 years. In heat every 3 weeks.

Cattle have been patient pioneer draft animals throughout world. Formerly, bull ran treadmill that churned butter. In some countries, steers are used as riding animals. Cattle yield leather useful in making windproof jackets, shoes, harness, and bookbindings. Hair has been made into felt hats. Flesh is an important portion of most virile human beings' diet. Fat has been used in lubricating machines. Milk sees most babies through their most difficult days and is ideal in diet of adults, supplying lime and a balanced diet in itself. Cheese provides a particulary valuable nonfattening diet. Casein from milk is used widely in industry in a variety of ways. Intestines provide strings for musical instruments, balloons for lighter-than-air aircraft, and have a variety of other uses. Glands provide medicines for restoring or preserving human health. Hoofs provide slow-drying oils for certain uses. Bones provide valuable fertilizers, ornaments, and utensils. Soap, sandpaper, insulin, all may come from cattle. In addition, cattle provide a welcome companionship for those who like them. Swiss, Limburger, cottage cheese and many other types are made from cow's milk. In 1952, 1962, and 1972 there were respectively about 85 million, 101 million, and 118 million cattle in the U.S.; of these 22 million, 19 million, and 12 million respectively were milk cows.

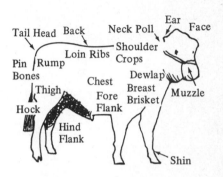

Guernsey Cattle
Bos taurus

Dairy type of cattle producing a quantity of good milk. Medium-sized. Color lemon or fawn, never gray or black. Nose buff. Switch and tongue generally white. Larger than Jersey cattle, the cow weighing about 1,000 lb. Skin yellow. Milk typically yellow. Range in weight from 800 to 1,600 lb; a good cow in milk should weigh 1,100 lb; bulls weigh 1,200 to 2,200 lb.

Developed on Island of Guernsey in English Channel off French coast. Island is 24 sq mi in size but has long maintained high standards of cattle breeding. Guernseys were first introduced into United States about 1831, but our present purebred herds are from subsequent importations.

Has eight lower incisoform teeth. At birth, has middle pair; at 1 month, the next pair; at 18 to 24 months, permanent teeth replace first middle pair; at 27 to 36 months, next pair replaced; at 36 to 48 months, third pair replaced; at 45 to 60 months, the last pair.

Like other dairy cows, Guernseys require more water than beef or dual-purpose types.

Guernseys are most popular in Wisconsin, New York and in Minnesota. Guernsey cows do not produce a comparatively large quantity of milk, but they do produce milk that tests about 4.9 percent fat. Naturally yellow-colored Guernsey milk is preferred by many.

Guernsey

Shorthorn Cattle, Durham
Bos taurus

Dual-purpose cattle combining meat and milk production. May be red, red and white, or white and roan. Large, with broad back, deep body and mild disposition. Cow has good-sized udder. Varies in different places depending upon needs as draft animal or for beef or milk. In England, breed is considered a "double decker." Also called Durhams. Bull may weigh 2,000 lb; cow 1,500 lb. Typically more rectangular in form than other breeds of beef cattle.

Developed in England, in counties of York and Durham, in valley of River Tees. As early as 1580 there existed a race of superior short-horned cattle on estates of two Dukes of Northumberland. Introduced into America as Durhams in 1783 and sent to Virginia. In early nineteenth century, developed to put on flesh rapidly with heavy feeding and so competes with Herefords. Most popular in Middle West, West, England, Argentina, and Australia.

Variety known as "polled shorthorn" and as "polled Durham" is hornless and was developed in United States. In addition to hornless feature, it maintains good dual-purpose qualities of original breed. Shorthorns are particularly favored for improving ordinary stock.

Food similar to that of other cattle. Name shorthorn was used to contrast animal with traditional longhorn cattle of Southwest. Persists in spite of fact that Durham is a truer name for breed, particularly now that many shorthorns are hornless.

Shorthorns produce a high percentage of superior beef. Have been used as oxen for draft purposes because of weight, strength, and patience. Average milk production is 8,445 lb of milk, with 333 lb of butterfat.

Shorthorn Cattle

Brown Swiss

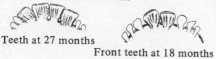

Teeth at 27 months

Front teeth at 18 months

Hereford

Aberdeen Angus

Ayrshire

Hair

Jersey

3 Years Lower Front Jaw

4 Years
Lower Front Jaw

Ten Years
Lower Front Jaw

PHYLUM CHORDATA. CLASS MAMMALIA

Order Artiodactyla./Family Bovidae [Hollow-horned Ruminants]

Brown Swiss Cattle
Bos taurus

Raised for dairy and beef properties. Mouse-colored, with nose, switch, tongue, and horn tips always black and a light stripe down back. Bulls weigh from 1,800 to 2,600 lb; in good breeding condition, should weigh 2,000 lb. Cows average 1,500 lb; average weight 1,300 to 1,800 lb. Head large and coarse. Breed is slightly lighter than Holsteins. Calves usually strong.

Possibly oldest of present-day breeds; developed in Switzerland mountain pastures. Introduced into America in 1869, in 1882, and in 1889, but because of hoof-and-mouth disease in Europe importations have since been few. Generally popular through Europe and America. Brown Swiss became prominent among dairy breeds sometime in the first half of the nineteenth century. Brown Swiss are considered both a massive and a rugged breed.

They are hardy eaters of harvested feeds and are good grazers. Frequently popular in parts of the country where dairying is a diversification of other types of agriculture. One of the valuable characteristics of the breed is its ability to produce year after year. Almost 2,000 Brown Swiss cows have had lifetime production records in excess of 4,000 lb of butterfat.

Hereford Cattle
Bos taurus

Beef type. Large, chunky, with white face and white markings on legs or under parts. Back broad. Legs short. Tendency in some toward abnormally large forequarters, which is desirable in draft animals. Poor milk producers. Mild disposition. Remarkably hardy, permitting them to thrive under severe open plain conditions.

Developed in England, in valleys of Severn and Wye rivers. Most successful in Western states, where calves are shipped from ranges to Corn Belt for fattening. They can put on flesh rapidly with heavy feeding and are remarkably docile, also exceptionally good grazers.

Milk records of Hereford are negligible. Cattle normally show 35 to 50 percent shrinkage between live animal and marketable carcass. In Hereford, this shrinkage is remarkably low although it is not lowest of all breeds. About 55 percent of a living steer is normally beef. Ranchers like the sturdy character of Herefords and their capacity to adapt to range conditions. Winter-hardy. Color pattern has become a trademark. Breed commonly referred to as "white-faced." Red may vary from real dark red to a very light yellowish cast.

Aberdeen Angus Cattle
Bos taurus

Hornless black beef cattle. Head comparatively short. Compact low-set body which produces about 65 percent high-grade marketable beef, with fat evenly distributed through entire carcass. Handled properly, a cow may yield a good flow of milk. Does well in close quarters; is quiet and generally good-natured.

Probably descended from a wild hornless breed. Probably developed in Scotland in the counties of Aberdeen, Kincardine, and Forfor, a section of which is Angus. Introduced into United States at beginning of this century, becoming popular particularly in the Corn Belt.

Best cattle for high-pressure system of feeding producing rapid development of "baby beef" that commands top prices. Meat-packers appreciate this breed because of its exceptionally good type, including short legs, beefy bodies, well-developed loins and backs, and wide rear quarters. Marbling, or dispersion of the fat in the lean tissue, is more highly developed in Angus than in any other breed. Known in Scotland since 1752 as "doddies."

Ayrshire Cattle
Bos taurus

Dairy cattle, red and white or brown and white, with white predominating and colors not blending. Horns long, upward, outward and backward curving and slender. Cow weighs about 1,200 to 1,500 lb. Bull should weigh 1,900 in breeding condition. Top lines of animal straight. Legs long and straight. Barrel large. Chest full. Ideal milker. Long-lived.

Developed in southwestern Scotland, where it was established about middle of eighteenth century. Introduced into United States in the middle of the nineteenth century. Most popular in United States in Massachusetts, New York, Vermont, Wisconsin, and Kansas, chief dairy states.

Quality of milk is high and quantity good. Average good cow should produce 6,000 to 7,000 lb of milk a year, yielding 3.8 to 4 percent butterfat. Sometimes referred to as the "aristocrat of dairy breeds" because of stylish and distinctive appearance. Black or brindle-colored markings are considered very objectionable. Ayrshires are excellent grazers and feeders; can use large quantities of roughage to advantage and make good use of limited grain feeding. They are economical producers of butterfat, and milk and butterfat under farm conditions.

Jersey Cattle
Bos taurus

Dairy cattle bred for production of milk of high butterfat content. Solid color, yellow or fawn, with black nose, tongue, and switch though not always as just described. Small in size. Cow weighs 800 to 1,000 lb; bull, 1,300 to 1,800 lb. Cow mild; mature bull nervous and unreliable.

Developed on Island of Jersey in the British Channel off the French coast. Island is only about 10 mi long with an average width of 6½ mi; climate is mild through the year, and supports 50,000 inhabitants. No cattle but Jerseys are allowed on it. Introduced into United States early in nineteenth century.

Since 1871, American records of Jerseys have been kept for about 1 million animals. Most popular in Ohio and Texas. Jersey females are good pasture animals, particularly where it is hot during the summer day. They produce a quality product; average butterfat content of 5.3 percent; many individual animals exceed 6 percent. Do not produce volume of milk comparable to some other breeds. High butterfat content and comparatively lower volume in a cholesterol-conscious population do not add to the status of this breed in dairy herds.

Jersey

Holstein-Friesian Cattle
Bos taurus

Dairy cattle, bred for great volume of good-quality milk. Black and white, though occasionally there are red individuals. Bull weighs to over 1 ton with a heavy framed body, muscular neck, lean body, and incurving horns; treacherous. Cow smaller, and gentle, with more slender neck, large body, and huge udder; average weight, 1,500 lb. Cows weighing over 1 ton and bulls weighing up to 3,000 lb not uncommon.

Developed in Holland and in Friesland and found generally over prosperous lowland Europe. Introduced into America bout 1795, some going to Genesee Valley in New York. Later importations in the sixties established present high-grade herds popular in all dairy country in America.

One cow produced in a year 33,464.7 lb of milk with 1,349.31 lb of butterfat; another 37,381.4 lb of milk with 1,158.95 lb of butterfat. More than 80 have records of over 1,000 lb of butterfat a year. These records speak for themselves. Holstein milk is whiter than that of most other breeds since the breed is more efficient than other cattle in converting carotene to vitamin A. Milk averages 3.7 percent butterfat, which is the lowest of the dairy breeds. Do not mature as early as some other breeds. Good market milk, however, because of low butterfat. Recent dietary trend toward decreased animal fats in man's diet has been a factor in popularizing Holstein milk.

Holstein

Merino Sheep
Ovis aries

Bred for quality and quantity of fine wool. Ram weighs about 140 lb, ewe about 100 lb. Heavy folds over body but not on back. Wool covers head, obscuring eyes, but ears and muzzle are woolless. Has pink skin, long neck, sloping shoulders, narrow chest, and thin withers. Wool fine, 2 to 3 in long.

Developed in Spain nearly 2,000 years ago. Introduced into United States in 1801 and later. Reached peak of popularity in 1812 and the 1860s. Saxon and French Merinos also imported. Now most popular in Texas, Ohio, Michigan, the Southwest, and Pennsylvania and used to improve other breeds by crossbreeding. Delaine Merino was developed in the U.S. as a much more valuable mutton-producing sheep; originated in New England and emphasized superior carcasses and fine fleece. *Delaine* from French "from wool." Both Merino and Delaine types have medium-short heads. Rams with horns, ewes hornless. Hair on the face, ears, and legs of Delaine is white and of extremely fine quality. Recent emphasis has been on improvement of the carcass through selective breeding.

Merino Sheep

2 Pairs

Lamb   Yearling   2nd Year Milk Teeth

All Milk Teeth   1 Pair Permanent Teeth
3 Pairs Milk Teeth

Merino Sheep

Southdown Sheep
Ovis aries

One of the smallest of mutton sheep. Ram weighs 185 to 200 lb, ewe 135 to 155 lb. Head medium-sized, hornless. Ears small and wide apart. Neck, short. Face brown or gray. Shoulders broad and full. Breast wide. Forelegs well apart. Back broad and straight. Rump broad. Fleece compact. Limbs short.

Possibly oldest medium wool breed in existence, having a distinct record from southeastern England for over 200 years. Probably developed from Sussex sheep. Introduced into America before 1803. Popular in Middle West, Tennessee, and New York but not in Far West. Formerly used to graze estates.

Lambs weigh 60 to 90 lb in 4 to 6 months.

Modern Southdown approaches perfection in mutton type; especially in excellence of carcass conformation. Usually shear from 6 to 8 lb of wool annually. Staple length of wool exceeds 2 or 2½ in. in length. Lamb crop 135 or 140 lambs per 100 ewes.

3rd Year

Southdown Sheep   3 Pairs Permanent
1 Pair Milk Teeth

All Permanent Teeth

4th Year

Shropshire Sheep
Ovis aries

Medium-sized, general-purpose sheep. Fleece forms hood or cap over head and may extend to toes. Nose and lower legs dark. Legs short. Body blocky. Ram weighs 200 to 250 lb, ewe 150 to 180 lb. Skin pink, secretes oil freely. Ears short, pointed, wide-set, short-wooled, and carried almost straight out from head.

Developed in Shropshire and Staffordshire in west central England, probably from native sheep found there. Probably Southdown and Leicester blood used in crossbreeding. Not considered distinct breed until 1850, although it existed before. Common in United States and Canada. Quiet, docile, and easily handled.

Probably more generally used in farm flock states than other breeds because of good general qualities; used in grading up mongrel flocks. Fleeces average 8 to 10 lb in weight, wool being of excellent quality. Tends to have twins.

One ram, over 1 year old, bred to 35 to 50 ewes at least 1½ years old. Ewes in heat 1 to 2 days. In 146 days, ewes bear one or two lambs. Ewes need 16 sq ft of floor space. Between breeding and lambing, ewes increase weight 25 lb. Lambs nurse in 2 hours; at 2 weeks, nibble hay; at 4 to 6 months, reach marketable age; tails cut when 2 weeks old and males castrated to become *wethers*. Southdowns may weigh 60 to 90 lb at 3 months. At 1 year, permanent teeth appear and by 4 years, all teeth are permanent (see sketches). Pulse, 70 to 80. Respiration, 12 to 20. Temperature, 104 to 105°F. Life-span 15 years.

Food, vegetation cropped close to ground; best if it contains clover. Sheltered sheep are fed legume hay and grain mixtures of corn, 4 parts by weight; oats, 4 parts; linseed or cottonseed cake, 2 parts and salt and water, which should be available at all times. Pregnant ewes require daily about 1 to 2 lb of silage, 2 to 3 lb of hay, and about 1½ gal of water. Alfalfa can be recommended most highly as sheep forage. Bluegrass pasture does not provide adequate food during warm summer months. After weaning, heavier ewes may require milking by hand once every day or so until milk production is stopped. Sheep stampede blindly in face of enemies and are killed by dogs, bears, and other large animals. Goats or active aggressive cattle like Angus provide protection where dogs are a pest since they will drive the average dog from the pasture.

In 1938, there were in the United States approximately 53 million sheep worth 324 million dollars, number being greater but value less than that for hogs. In 1952, 1962, and 1972 there were respectively 27 million, 27 million, and 16 million sheep in the U.S. Whole wool industry is, for the most part, dependent on breeds of sheep. The great centers in the earth producing wool are western United States, western Asia, and Australia. Sheep lend themselves to selective breeding, some producing rapidly growing lambs and good carcasses, other producing a high quality and quantity of wool. Then, there is a dual-purpose type which produces both wool and flesh. Associated with sheepherding are men, goats, and dogs, and now, in dog-infested areas, cattle. From the milk of sheep, Roquefort cheese is made. The enmities between sheepmen and cattlemen in the western ranges are hardly justified but poorly managed sheep do contribute definitely to soil erosion and floods and destroy game ranges and forested areas.

Mountain Sheep, Bighorn Sheep
Ovis canadensis

Ram, length to nearly 7 ft, with 5-in. tail; height at shoulder 3½ ft. Weight to 200 lb; female to 125 lb. Species becomes progressively smaller, the more northward the range. Horns along curve, to 4 ft. Ewe, length to 5 ft, with 5-in. tail. Horns shorter and straighter; color brownish-gray above, darkest on back, back of neck, throat, and legs; brownish on sides; yellow-white beneath. Teeth: I 0/3, C 0/1, P 3/3, M 3/3.

Found in mountains from western Texas and Mexico to North Dakota; west to southwest Alberta, north across Arctic Circle in Alaska and west to coast except along coast from Alaska to central California and narrow area in British Columbia, Alberta, and Washington. Feeds in lowlands when possible. 10 races.

Rams fight in late November and win flock of ewes. One to two lambs born 180 days after breeding; run and skip within 2 hours but hide until 3 days, when follow mother; nurse for 2 months, then eat solid foods. Rams past their prime by 8 years, as shown by annual rings on horns. Life-span 15 years.

Food varies, according to authorities, from nothing but grass in northern ranges to little or no grass in South. Eats many different herbaceous plants and some woody browse. All agree that sheep is selective. Must have water daily and must have salt. Sense of smell poor; of sight, remarkable. High mortality in sheep first winter. Captives are fed hay and oats.

Provides food, hides, and sport for mountain-dwelling and mountain-roaming men. Probably serves as food for wild preying animals and therefore a buffer for domestic animals, with which they are competitors for forage. Tracks: F2¾ × 2¾, 2; H 2¾ × 2¾, 2.

Shropshire Sheep

Mountain Sheep

Brindled Gnu, Wildebeest
Connochaetes taurinus
Length about 7 ft, exclusive of tail. Height at shoulder over 4 ft. Face blackish, with sides of head yellowish-gray. Horns in both sexes arise from forehead, curve downward and then outward. Tail rather short and black. Dark transverse stripes on withers.

White-tailed or common gnu, whose tail almost reaches ground, now almost exterminated from former range in South Africa. Brindled gnu, found in southeastern Africa from Orange River northward to Angola in the west and Northern Rhodesia in the east is still reasonably abundant. A third form, *C. albojubatus,* occurs in East Africa.

Both sexes have horns but body and horns of female white-tailed gnu are slightly smaller than in male. Young wildebeests are tamed relatively easily but mature animals resent vigorously any confinement.

These are highly inquistive grazing animals. They may circle a stranger and come up to investigate him from rear. They can match speed with a horse and have excellent endurance. They have excellent sight and good ability to smell but are inclined to be ruled partly by behavior of others in the herd.

They have been killed extensively for their flesh, which is tender and juicy; for their hides, which make excellent leather; for their horns, which make good tools for natives; and in part, because they compete for forage with domestic animals that man wishes to use for his own ends.

Wildebeest

Chamois
Rupicapra rupicapra
Length, average, 45 in. Height at shoulder, 2½ ft. Weight to 90 lb. Bucks larger than does but superficially like them. Two black horns rise vertically from top of head to 8 to 10 in. and at tips bend backward and downward. Hair close and long with thick woolly underfur; reddish brown in summer, much darker in winter.

Ranges through mountains of Europe; while it is essentially a mountain animal, it is not limited to treeless stretches but lives for most of time in tree belt just below tree limit. They live together in flocks and have been protected somewhat in game refuges.

One to three kids are born in May to June and follow mother almost immediately. In a few days, they can spring up exceptional heights. Kids stay under protection of doe for at least 6 months, receiving best of care, and may not leave mother even if she is killed. Kids may be tamed and will breed in captivity.

Bucks live solitary existence. Rest at night; feed down mountains until noon and up in afternoon. May leap 23 ft, and a captive animal has been known to clear a 13-ft wall. Senses of smell and hearing are exceedingly acute. Old animals may become vicious and attack with their horns.

Excellent game animals, with superior flesh. Hides make chamois leather. Horns valuable, and animals add definitely to picturesqueness of their environment. Their continued existence should be guaranteed by adequate regulations for protection. An endangered species.

Chamois

Milch Goat
Capra hircus
Good milch goat has long body with proportionately large hindquarters. Chest deep and broad. Legs short. Neck medium in size. Head broad. Mouth wide. Udder not wide; full, not tight. Skin soft and covered with short hairs. Body hairs fine, smooth, and bristling. Often without horns. Playful and friendly.

Raised more commonly in Europe, where in Germany 75 percent of rural households maintain them. In America, suburban goat dairies supply milk for special purposes recommended by doctors. Goats are kept by those of foreign extraction to give milk where forage is too poor and limited to support a cow. Of Persian origin.

Two to four kids born 147 to 155 days after male *buck* breeds with female *nanny goat* or *doe*. Kids, well-developed at birth, follow mother in 4 to 5 days, mature in 8 to 12 months; old at 6 years but may live to 15 years. 1-year goat has two broad front teeth on lower jaw; second year, 4; third year, 6. Body temperature, 103°F.

Food almost any vegetation, but not the tin cans of legend. Animals may be destructive to forests. They have keen sight and sense of smell, are wary and restless, and delight in climbing to highest available point. Bucks have a decidedly unpleasant odor.

Healthy female gives milk for 7 to 10 months after kids are born, so serve splendidly as milk source for man. Flesh is known as "venison" in the markets, and castrated males or *wethers* may be slaughtered for flesh. Milch goat makes good child's pet for drawing carts; bucks and sometimes dangerous.

Track

3"

Milch Goat

Ibex, Wild Goat
Capra ibex

Length to 5 ft. Height at shoulder to 3 ft. Weight to 200 lb. Both sexes have horns that curve obliquely backward, tapering gradually and reaching a length of 59 in. and a weight of 30 lb. Fur rough and thick, and in males like a mane on back of neck and a short beard on chin. Gray in winter, brown in summer.

Found in mountainous parts of Switzerland, the Pyrenees, the Caucasus, and Abyssinia; 7,000 to 10,000 feet. Lives at lower elevations in winter. Close allies include Alpine ibex *C. ibex*, Himalayan ibex *C. sibirica*, Arabian ibex *C. sinaitica*, and Abyssinian ibex *C. walie*. All live for most part in the dizziest mountainous area available, though some may be relatively tame.

Males do not live with herds except at breeding time; males usually live higher in mountains than females. One or two kids are born in June to July and a few hours after birth can follow mother nimbly over rough terrain. Mother protects kids by fighting, by decoying enemy from young, and otherwise.

Food, vegetation gathered mostly during night, animal descending in late afternoon to graze and retreating to forest higher up at daybreak. They feed on trees, shrubs, grasses, and almost any other available plant materials. Chief enemies of young are eagles, though wolves, bears, and men are often serious.

Serve as game and food animals for those living within their range. When raised in captivity, they quickly lose fear and respect for human beings and have been known to enjoy bowling men over with their horns, apparently just for the fun of it.

Ibex

Angora Goat
Capra angorensis

Buck may weigh as much as 100 lb, with 18- to 20-in. horns. Does smaller, with 8- to 10-in. horns. Hair long and relatively fine, a single goat capable of producing a fleece of 8-in. hair weighing an average of 3 to 5 lb. Maximum is 12 lb with 18-in length. Buck has a characteristic long slender beard under chin.

In America, raised in limited numbers in East and Middle West but commonly in South and Southwest. Related breeds include Cashmere (2 ft high) of Tibet, Syrian (long-eared), Sudan, and Egyptian goat. Names indicate geographic origin of breeds. Goats are reared on lands that will support sheep.

In range, kidding time is important. Since newly born kids cannot stand strenuous activities of a herd, doe is toggled so that she cannot leave and where the herd will not bother her. Kids kept 2 weeks, then grazed with doe in special pastures. One buck to 50 does is herd practice and spreads kidding time.

Pastured milch goats need daily 1½ lb of concentrates such as 10 parts of corn by weight, 10 parts of oats, 5 of wheat bran, and 1 of linseed-oil meal; also, good roughage. Hair goats must have water summer or winter. Herds of 1,200 hair goats commonly managed together.

Good hair goats may be clipped twice a year. Hair is sorted according to fineness and made into mohair cloth. Trained sorters can recognize 13 diameters between 0.003 and 0.0067 in. and sort 500 lb of hair a day. A native of Turkey; in the U.S. it has been most successfully introduced in Texas.

Angora Goat

Order Perissodactyla./Family Equidae [Horses]

Percheron Horse
Equus caballus

Maximum height at withers 17 hands, or 68 in. Mares are usually about ½ hand less in height. Stallions, in best condition, weigh about 1 ton; mares in show condition 1,800 lb. Short-legged, compact form is favorite. Black and rump broad. Back short. Loins smooth and well-muscled. Gray, roan, or brown but black and gray typical; chestnut typical of Suffolk breed but not of Percheron. Body less blocky than Belgain and legs less "feathered" than long-legged Clydesdale or short-legged Shire. May weigh 1 ton at 3 years.

Developed in Perche district of France near Paris and probably associated with Crusaders. Popular in America in Civil War times and a leading American draft horse, one census showing 70,616 Percherons, 10,838 Belgians, and 5,617 Shires. Most popular in Middle West.

Preferred color of modern Percherons is black; however, most black horses have white markings about face or around feet. Dark gray second most acceptable color. Probably the most important breed in the development of American agriculture.

Percheron Horse

Morgan Horse
Equus caballus

A general-purpose horse, formerly popular as a general farm animal. Usually bay, standing about 14 hands (56 in.) high at withers. Weight about 1,000 lb. Endurance remarkable. Legs short and well-muscled. Back short, with long sloping shoulders; round and close-ribbed. Chest wide and deep. Feet small and well-shaped. Mane and tail long. Legs dark. Ears small. Keen. Even-tempered, with quick response.

Descended from a stallion, Justin Morgan, that was foaled in 1793 and died in 1821 in Vermont, where it was brought from Massachusetts. General qualities nearly lost when it was bred as a race horse but have been regained. The breed held many trotting and running records.

A famous Morgan horse, Gladstone, foaled in 1913, was worked through the hay season, then under saddle entered endurance field and bettered records made by horses 200 lb heavier. The Morgan is a remarkably useful horse in covering long distances at a fairly rapid gait and in carrying heavy loads. Morgans have pleasing temperaments and are easily managed. They are popular as pleasure mounts.

Morgan Horse

Thoroughbred Horse
Equus caballus

Reaches a height at shoulder of 14½ to 16½ hands, or 5 ft. Weight 900 to 1,100 lb. Form extremely graceful. Legs long and slender. Neck light. Color bay or brown, though blacks, chestnuts, and grays are found. Color usually uniform. Characteristic form maintained through many years of careful selective breeding. Nervous, generally friendly animal, much beloved by master and subject to best of care.

Developed in England for racing, tracing lineage over 200 years, with selection based on performance. Occasionally stallions are bred with mares of other breeds to improve speed of such types as polo ponies, artillery horses, hunters, cavalry, and general saddle horses.

Thoroughbred Horse

(Continued)

Lower Nippers and Age

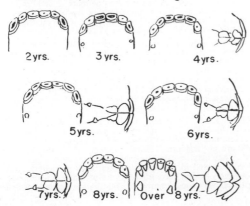

2 yrs. 3 yrs. 4 yrs.

5 yrs. 6 yrs.

7 yrs. 8 yrs. Over 8 yrs.

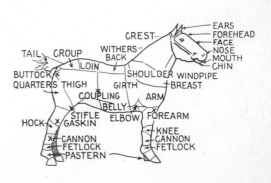

EARS
FOREHEAD
FACE
NOSE
MOUTH
CHIN
CREST
WITHERS—BACK
TAIL CROUP
LOIN SHOULDER WINDPIPE
BUTTOCK BREAST
QUARTERS THIGH GIRTH
COUPLING ARM
BELLY
HOCK STIFLE ELBOW FOREARM
GASKIN
KNEE
CANNON CANNON
FETLOCK FETLOCK
PASTERN

Canter

Trot

Pace

Gallop

Slow Running Walk

Slow Run

Introduced as a stallion in Virginia in 1730. Found favor in Kentucky and Tennessee; now bred west to California and south to Texas. However, the blue grass region of Kentucky is considered the cradle of thoroughbred breeding in the U.S. Best known sire in the U.S. is Man o' War, a chestnut horse foaled in 1917, which won 20 of the 21 races in which he started. Reported that 46 percent of thoroughbreds are bay, 30 percent chestnuts, 18 percent browns, 3 percent blacks, 2 percent grays, and 1 percent roans. White markings on legs and face are common.

Horses believed to have developed from small *Eohippus* of early Eocene, which had four toes on forefeet and three toes plus 2 splints on hind feet; then collie-sized *Orohippus* of late Eocene with somewhat similar toes; then *Mesohippus* of Oligocene with three toes on all feet, side ones touching the ground in hind feet; then the Miocene three-toed *Merychippus* of the Miocene, none of whose side toes touched the ground; then the one-toed *Pliohippus* of the Pliocene, with splints of second and fourth toes remaining off the ground. The horse *Equus* appeared about Pleistocene times.

Horses just born are *foals;* when weaned from mother they are *weanlings.* A young female is a *filly* and a young male a *colt.* Race horses date their age from their first January 1. A mature male horse is a *stallion.* A male castrated before it reaches maturity is a *gelding* and one castrated after reaching maturity is a *stag.* A mature female horse is a *mare* and one nursing young is a *brood mare.* A foal is born to a mare 11 to 12 months after she is bred to a stallion. Horses should not be worked until 2 years old. Mare in heat 5 to 7 days, may be bred at 12 to 24 months, but ordinarily not until 3 years old. Life-span to 35 years.

Young horses should grow steadily first 2 years. Oats are popular as grain food because hulls add bulk and prevent gorging. After being worked, horses should be given a small drink, then rested, then watered again, and then fed. In pasture, horses have definite midden heaps for manure where they do not graze, but will eat grass growing from cow droppings. Eat grass closer than cattle but not so close as sheep. Abundant roughage necessary when confined. Respiration, 8 to 16. Heartbeat, 30 to 45 per minute. Temperature, 100.2°F.

Number of horses varies; 16 million in United States in 1925 and 10 million in 1940. In 1950, 1955, and 1960 there were, respectively 4.3 million 3.9 million, and 3.1 million horses and mules in the U.S. Records after 1960 are not accurate. Horses provide companionship and perform work. They may be eaten in emergency. They produce great quantities of serums used in keeping men and other animals healthy. Horses have definitely helped make life better for man and other animals. They are worthy of every consideration. 1,000-lb horse needs 3,401 cu ft of air daily.

Shetland Pony
Equus caballus
Small animal, not accepted for registration if over 46 in. high at withers.
Mature stallion weighs 370 to 400 lb.   Ponies have short strong straight
legs.   Body well-rounded.   Back broad.   Head well-carried on short
neck.   Mane heavy.   Tail thick.   In winter, hair grows long, a relic of
meeting severe winter conditions.

Developed in Shetland Islands, which lie off mainland to north of Scotland.   Thought to have originated from Norwegian pony.   In their shetland Islands, they were developed as pack animals, particularly for hauling peat and later for hauling coal cars.   When introduced into favorable
conditions, it tends to increase in size.   Introduced into America largely
for sale as a pet or as a trained circus animal.   Pied animals seem to be
popular as pets.   Often so pampered that it develops away from small
size.

Ponies have a life-span of about 21 years though some live much longer.
They accept all sorts of obligations, ordinarily with little sign of offense.

Ponies seem able to exist on forage which would not support larger
breeds and so have made a place for themselves in less fertile areas of the
earth.

Though small, a pony can draw ½ ton of coal in a mine 20 to 30 miles a
day.   A pony 3 ft high can carry a full-grown man with ease.   As a pack
animal, it carries loads of 120 to 140 lb and in spite of such treatment
remains good-tempered.   Obviously, it is popular.

Shetland Pony

Saddle Horse
Equus caballus
Bred for saddle purposes.   Height 14 to 16 hands (5 ft) at withers.
Weight 950 to 1,250 lb.   Color bay, brown, black, or chestnut.   Stride
generally short.   Shoulders long and sloping, favoring smooth easy action.   Has spirit and style and when well trained is quick to respond to
command.

Breed of American origin; developed in Kentucky, Tennessee, and Virginia where it was needed in supervising plantation work and on hilly
grazing farms.   In the North, a horse was needed with a varied gait involving a canter, a walk, and a trot, which led to the loss of some other
gaits common to the breed in the South.

American Saddle Horse Breeders Association was organized in 1891 and
in first 36 years of its existence listed in its studbook 10,215 stallions and
geldings and 17,046 mares.   The colors of chestnut, black, bay, and brown
are preferred in the American saddle horse.   Saddle horses may be
trained to develop several gaits: walk, trot, canter, and gallop.

Horses may have unfortunate effects on certain people, giving them
asthma.   Progress is being made in immunizing such persons to these ill
effects but until this has been perfected many will be denied the pleasure
of riding good saddle horses.

Saddle horses are becoming more and more horses of the show type with
specialized gaits and particular conformations in build.   Though spirited,
they generally yield well to discipline and so are always interesting.

Saddle Horse

Ass, Donkey, Burro
Equus asinus
Varies in height from 30 in. to height of a good-sized horse.   Ears long.
Tail sparsely haired.   Eyes deep-set.   Body generally not so heavily or so
smoothly muscled as that of horse.   Mane wiry, uneven, coarse.   Legs
medium in length.   Feet small.   Color from white to black.   Commonly
with white on nose, belly, and flanks.

Domesticated for thousands of years.   Jacks of France are probably largest; those of western India and Ceylon, smallest.   Wild in Africa and
Asia.   Brought to America with Spanish explorers, but serious production began with nineteenth century.   Popular, particularly in mountainous regions.

By selection, important breeds have been developed: Poitou, largest; Ma-
jorca, large; Andalusian, medium; and Maltese, small.   Male donkeys are
jacks.   Female donkeys are *jennets*.   A jack mated with a mare produces
a *mule:* a stallion with a jennet, a *hinny*.   Pregnancy, 348 to 377 days.

Donkeys can get along on poor forage and so are popular in less favored
countries, among less favored peoples, or on long journeys where food
cannot be carried.   They do better in warm climates than in cold.   Body
temperature, 98.5°F.   Pulse, 45 to 52 per minute.

Donkeys are ideal pack animals, mine animals, and pets.   Jacks are bred
to great size for mating with mares for production of high-grade mules.
Their size, endurance, and good qualities have been greatly improved in
recent years.   A pack animal can carry 250 lb on its back for days with
little forage.

Donkey

Mule

Mule
Equus asinus × *E. caballus*
Varies in size according to use. Draft mule weighs about 2,000 lb or more and is 16 hands (64 in.) high at withers. Farm mule weighs 1,100 lb, 15½ hands high. Mining mule weighs 600 to 1,300 lb and is 12 to 16 hands (48 to 64 in.) high at withers. Mules retain long ears, small feet, sparse tail, and bray of jackass, with size and strength of mare.

Probably has been produced ever since horses and asses have been domesticated. About 90 percent of United States mules are on farms of South. Generally popular where hard patient work is needed, particularly in warmer parts of world. About ⅔ as many mules as horses in United States.

A stallion mated with a jennet produces a *hinny*. These not bred commercially because they lack good qualities of reverse cross, where jack is mated with a mare to produce a *mule*. Mules are born about 1 year after breeding. Reports of female mules bearing young not well supported.

Mules are less inclined than horses to eat when they are heated or tired but must have ample food to consume when they are well rested. Mules most nearly like their mothers are considered superior.

Formerly made ideal army draft animals because of lack of nervousness, great strength, endurance, and ability to get along on poor care. A mule cannot be put at hard labor safely until 5 years from time parents mate and many must wait 8 years. 4,477,000 mules in United States in 1938.

Zebra

Zebra
Equus burchelli
The wild mountain zebra stands about 4 ft tall at the withers. Its general color is white with black stripes on head, body, neck, and legs. Burchell's zebra stands 4½ ft high and usually has white legs with body and head stripes black or brown over sorrel. The quagga stands 4½ ft high and has stripeless haunches and legs.

Mountain zebras originally ranged through mountains of Cape of Good Hope in Africa. Now, generally agreed living zebras fall into three species: *E. grevyi*, Grevy's zebra of southern Ethiopia and northern Kenya; *E. burchelli*, the Burchell's group of eastern, central, and western Africa as far north as Angola; and *E. zebra*, the mountain zebra from the south.

Reproduction is probably much like that of donkey and horse. Burchell's zebra foal is born between 11 months and 6 days and 11 months and 20 days after breeding takes place. Usual life span of a zebra is around 20 years, though probably few survive that time in the wild. Herd together for life.

Zebra is a grazer like its horse relatives. In nature, it is the favored prey of lions but a group can defend themselves well against lions with their heels, with which they are expert marksmen. Call of a zebra is a peculiar neigh something like a donkey bray but more pleasant. Animals feed in herds in the wild. Speed, 40 mph.

Zebras are prized as food by native Africans who enjoy the yellow fat. They use hides for clothing, shelter, and implements and sometimes have been able to break a few animals to harness if they were caught young or reared in captivity. Sometimes used with teams of mules.

Order Perissodactyla./Family Rhinocerotidae

African Rhinoceros

Rhinoceros
Diceros bicornis
Indian rhinoceros *Rhinoceros unicornis* largest of three Asiatic species; has only one horn, measures 10½ ft from nose tip to tail base, and stands 6 ft at shoulder. African rhinoceros stands 5 ft; 8½ ft in length. It makes a track 9½ in. across. Its front horn is the larger and may be 3 ft long. Horn of black rhinoceros, to 5 ft. To 6,000 lb.

Indian rhinoceros native of Asia. Java rhinoceros includes one-horned rhinoceros of Java, small Sumatra rhinoceros, and two-horned rhinoceros of Sumatra, Thailand, and Malay Peninsula. Two-horned African rhinoceros, from Abyssinia to the Cape in Africa. In 1731, a hair-covered frozen rhinoceros was found in Siberia.

Not a sociable animal. Young born 19 months after breeding takes place, always one. At seventh year, first horn appears in the Sumatra species. Young rhinoceros remains with mother for some time. Lifespan, judging from zoo records, is about 25 years though this must vary in the wild. Few enemies dare face it.

Food, plant material, commonly twigs and shoots of mimosa and similar woody plants. Sense of sight is poor but those of hearing and smell excellent. Charge of an angry rhinoceros is direct, powerful, and dangerous. Enjoys mud wallows as insect protection. Speed, to 28 mph.

Superior game animal but should not be completely exterminated. Natives eat the flesh of some species, value the dried blood, consider the skin a delicacy, and consider the horn an infallible magic love stimulant. Horn, like the tusks of elephants, is not bone. Both animals described are endangered species.

Order Perissodactyla./Family Tapiridae

Malay Tapir
Tapirus indicus

Length 8 ft from nose tip to root of tail, with head, neck, and limbs brownish-black, back and sides grayish-white. American tapir, *Tapirus americanus,* is about 6 ft long and 3 ft high, dark brown or black with a gray tinge on head and chest. Can swim and dive well but is not a good runner.

Black and white tapir found from dense jungles of Malay Peninsula to Sumatra and Borneo. South American tapir, uniform dark brown, along rivers and lakes of South America north of Amazon. Giant tapir (700 lb), in mountains of Central America and southern Mexico. All are essentially animals of dense jungles in America or Asia. Strictly nocturnal.

Single young tapir is born from 392 to 405 days after parents breed. South American tapir will breed in captivity, but few records are available about any wild tapirs and their general life history. Young Malay tapirs are brownish-black, with brownish-yellow spots and streaks on sides, and white beneath, but lose this at 6 months.

Food: succulent vegetation of thick jungles where water is tepid and abundant. Tapirs favor a temperature of 70°F but do well in warmer climates. Do not like each other. They are defenseless, serene, patient, and generally accept life as fate hands it to them. Jaguar is chief enemy.

Tapir flesh is considered to be as good as beef and is sought by natives. Skins are made into leather for harness but, being hard when dry and spongy when wet, are not suitable for shoes. Young tapirs are tamed relatively easily and often wander about towns where they have been welcomed. Does damage to sugar plantations and destroys young cocoa trees. An endangered species.

Order Proboscidea./Family Elephantidae

African Elephant
Loxodonta africana

Length to 10 to 12 ft. Height to 13 ft. Weight, male, to 13,000 lb. Skin about 1 in. thick, nearly black. Ears to 5 ft long and 4 ft wide. Eyes small, well protected with lids. Forefeet usually with nails on four toes. Hind feet with nails on three toes. Trunk wrinkled, an elongation of nose, sometimes 7 ft long. Male larger than female. Tusks to 10 ft 2½ in. long, weighing 226½ lb.

Found south of the Sahara through wilder parts of Central Africa and formerly on south to Cape of Good Hope. Ancestors of elephants date back to Eocene: upper Eocene—Lower Oligocene, *Moeritherium;* Oligocene, *Palaeomastodon;* Pliocene, *Mastodon.*

First molar appears during second week after birth; is complete and in use at 3 months and shed at 3 years; second molar, in use at 2 years and shed at 6 years; third molar appears at 2 years, in use at 5 years and shed at 9 years. These are milk teeth. First true molar, fourth grinder, appears at 6 years, shed at 20 to 25 years; second true molar appears at 20 years, shed at 60 years; third and last appears at 40 to 50 years. Has 24 molars; 12 milk teeth, 12 permanent molars. Food exclusively plant materials. African elephants enjoy the sun; may sleep either standing or lying down; may travel 15 mph by sure-footed rolling gait. Elephants never canter, gallop, or trot. They have many parasitic enemies. Man is their worst large enemy.

African elephants do not yield to domestication as the Indian species do. They are hunted by natives for flesh and hide but more for ivory of tusks. African elephants can do much damage to crops, generally at night. They should not be completely exterminated however. Jumbo was an African elephant. At 26 years, when he probably had not completed his growth, Jumbo stood 11 ft 2 in. at the shoulder and weighed 6½ tons.

Indian Elephant
Elephas maximus indicus

Smaller than African elephant, with a weight of 12,000 lb. Shoulder height varies from 7 ft 9 in. to 10 ft. Forefeet with nails on five toes. Hind feet with nails on four toes. Ears ⅓ size of those of African elephant; roughly triangular in shape. Trunk smooth; serves basically in breathing and smelling, and as a prehensile hand. Female Asiatic elephant sometimes tuskless; Ceylon elephants generally tuskless. Tusks may measure 9 ft and weigh 150 lb; are really incisor teeth; has 24 molars. Brain weighs 10 lb.

Indian elephant is widely used as a draft animal in Asia and to some extent in Africa. It originated in southern Asia, India, Burma, Thailand, and Indo-China, Ceylon, Malay Peninsula, and Sumatra. Common elephant performer in circuses. Becomes full-grown at 25 years and usually lives to about 50 years.

Single young born 19 to 21 months after breeding. Bull breeds at 11 years, cow at 8 years. Young weighs about 200 lb at birth and stands about 3 ft high. Matures at 11 years. In wild, form herds of 50 or more with one or more old males.

(Continued)

Malay Tapir

African Elephant

Indian Elephant

Food essentially plant material, about 40 percent of food eaten being digested. Heat production, 2,000 cal per sq in. of surface per day, the largest of any known animal. Skin temperature, 84 to 86°F., body temperature, 97.6°F. Hearing acute. Speed, 24½ mph.

Natives relish flesh particularly of trunk and foot. Most valuable animal because of ability to do intricate work under direction. Machines are gradually replacing them. Rogue elephants may do great damage to crops and kill men. An endangered species.

Order Edentata./Family Bradypodidae

Three-toed Sloth
Bradypus griseus
Length 18 to 20 in. Of a uniform gray color, though sometimes greenish because of algae in one species. This species has long blackish stripe between shoulders. Three toes on each foot are long and curved like hooks. Two-toed sloth, *Choloepus hoffmanni,* has two toes on front, three on hind feet.

B. griseus found in flooded parts of Central and South American forests, with closely related form in drier parts. Amazon River is roughly southern limit of both two- and three-toed sloths. Hang back down, from branches in trees, using claws as hooks to hold on.

Usually only one young, which clings to mother's back until it can shift for itself. Are tolerant of each other, this being particularly true of females. When an animal with a body temperature of 91.5°F in air temperature of 79°F was shifted to air at 57.3°F, body temperature became 86°F in 5 hours.

Food almost exclusively leaves of cecropia tree, though twigs are eaten as well as leaves. Animal known to go 4 miles in 48 days, despite slow movement; to swim 65 ft in 2½ minutes; and to go overground 14 ft in 1 minute. Has many enemies, particularly large birds of prey. Can curl itself into an almost inpenetrable ball.

Probably of little economic importance to man but interesting because of unique habits. One species is known to natives as the "ai," named probably because of its plaintive cry. Apparently life habits have not been thoroughly studied, since they are not discussed often in literature.

Three-toed Sloth

Order Edentata./Family Myrmecophagidae

Great Anteater
Myrmecophaga jubata
Length about 4 ft, plus tail at least 2½ ft. Height about 2 ft. Hair stiff and bristly. Throat, chest, underparts, and underside of tail black to brown, with broad black strip bordered with white from shoulders over rump. Upright mane on neck and back. Tail hairs sometimes 16 in. long. Forefoot claws long and curved. No teeth.

Found in Costa Rica, Panama, northern South America, in forests. Three-toed anteater, *Tamandua,* with long prehensile tail, a tree dweller from Mexico south. Two-toed anteater, *Cyclops,* found from Mexico to Panama; strictly arboreal and small, about size of squirrel with long tapering tail; claws on only two toes.

Breeds once a year, there being only one young animal, which rides around on its mother's back long after it should be able to take care of itself; sometimes this continues until a new young one is born. Habit is similar to that of closely related sloths. Little known about family life.

Food essentially ants, in all stages, these being caught by tearing anthills and houses to pieces with strong claws and eating with sticky tongue, which can be extended 20 in. and worked in and out rapidly. Can go a long while without food. In zoos, thrives on beaten raw eggs mixed with scraped uncooked meat and milk.

Flesh, which is black and has a musky flavor, is eaten by natives. Animal rarely fights back even when attacked by mortal enemy, the jaguar, and even though its sharp claws could be effective weapons. Interesting zoo animal. Hide sometimes used for wall ornament.

Great Anteater

Order Xenarthra./Family Dasypodidae

Nine-banded Armadillo
Dasypus novemcinctus
Length to 28 in., tail about 1 ft. Weight to 15 lb. Body covered with bony plates, with seven to nine movable joints around middle, permitting animal to curl into well-armored ball and protect softer parts. Color, variegated flesh, gray, and black. Snout, ears, and toes long. Toes four on front feet, five on hind. Teeth: I 0/0, C 0/0, P and M 7/7, or 8/8.

Found among rocks where there is vegetation and where caves are accessible, from sea level to 10,000-ft elevation, from Alabama, Louisiana, and

Armadillo

lower Rio Grande country south into Mexico.   North to 33°N in Texas and adjoining parts of Louisiana.   Usually active at night or when light is poor.

Sexes resemble each other and show no seasonal variation.   Young born four in number, all one sex and "identical," from first of February to April, with soft flexible skins much unlike the hard shells of adult; eyes open when born; able to leave den and follow mother like a flock of little pigs shortly after birth.

Food almost exclusively (85 percent) insects, chiefly ants and beetles, mixed with earth; 13 percent vegetable.   Droppings like small clay marbles.   Poor sight, excellent ability to smell, particularly sensitive to jarring of earth.   Can run surprisingly fast for short distances.

In some parts of range is discouraged as destroyer of ants that serve as useful scavengers, and in other places disliked by gardeners whose work it disturbs, even though it is engaged in destruction of important plant enemies.   Body temperature may drop 6 to 8° in 4 hours.   Almost entirely beneficial.   Meat edible.   Shells made into baskets and sold to tourists.

Order Sirenia./Family Trichechidae

Manatee, Sea Cow
Trichechus manatus
Length, male to 13 ft.   Weight to 2,000 lb.   Female to 7 ft.   Dull lead gray in color.   Hairs sparse and colorless.   Snout blunt and apparently sensitive.   Eyes small.   Ears inconspicuous.   Body fishlike in shape, without any "hind legs," but tail is about ¼ length of animal.   Tail spatulate, not notched.

Florida manatee, *T. manatus,* found in shallow salt or brackish waters from Florida around Gulf of Mexico and in West Indies; color, slate-gray. It is now becoming rare.   Other manatees are found in South America and in Africa.   In Pacific Ocean, it is represented by the dugong; may have been one of the mermaids seen by sailors.

Breeds in shallow water.   Gestation period about 152 days.   One or two calves born in water, each weighing about 60 lb at birth.   Mother must hold young out of water for air every 3 to 4 minutes for first week; nurse 18(?) months.   A young of the smaller Amazon species, 3 ft long, weighs about 30 lb.   Parents help rear young, which stay until next young are born.

Food probably exclusively succulent plants, gathered mostly at night, or in some species by day.   Rests in water with back arched, head and tail depressed.   Average time under water 1 minute 17 seconds; longest, 3 minutes 49 seconds.   Body temperature about 100°F with water temperature of 78°F.   Sense of touch excellent; of sight and smell, poor.   Usually solitary but may travel in small groups.

Flesh makes excellent food; breast meat light-colored; pelvis region, red meat; spare ribs, fat.   In captivity, may be fed successfully on one loaf of bread daily.   Hide makes superior leather used at one time in valves. Should be rigidly protected until its survival is assured.   Many killed in 1939 in cold in Florida.   Individuals greet each other with muzzle-to-muzzle play.

Manatee

Order Cetacea./Family Phocoenidae

Common Porpoise
Phocoena phocoena
Length to 6 ft.   Head beakless.   Dorsal fin triangular, slightly behind middle of body, low and only slightly concave on its rear margin.   Flippers oval, blunt at ends, and proportionately small.   Back black.   Belly white with narrow gray area between.   Teeth spade-shaped instead of conical.   Weight to 150 lb.   Excellent swimmer in spite of small flippers.

Found in north Atlantic from Davis Strait down to New Jersey.   Widely distributed over the north Atlantic, English Channel, North Sea, and Baltic.   A coast lover and one of commonest species seen from shore. Known in the Mediterranean.   Commonly found near the coast and in estuaries.

Breeds in summer months and about 11 months later one young, about half length of mother, is born.   Young suckled while mother swims on side, thus permitting young one to get air while it gets nourishment. Life-span to 30 years.

Food, fish such as herring, sole, whiting, and crustaceans; cuttlefish and even plants.   It not infrequently gets caught in fishermen's nets and, not being able to get to the surface, drowns as would any other air breather. It also swims along besides boats.

Flesh was formerly considered good food and in the time of Henry VIII was a royal dish.   Its oil was formerly used in lamps.   It probably does destroy some useful fish and may injure fish nets.

Common Porpoise

Bottle-nosed Dolphin

Common Dolphin

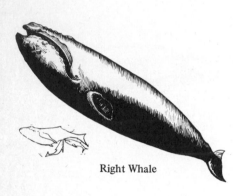

Right Whale

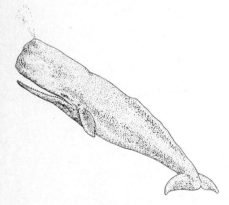

Sperm Whale

Order Cetacea./Family Delphinidae./Subfamily Delphininae

Bottle-nosed Dolphin
Tursiops truncatus
Length to 12 ft, with well-defined snout of 3 in. Weight, 300 to 400 lb. A prominent black or dark gray-brown fin in middle of back. Belly white. Head and snout dark, but upper edges and lower jaw white. Common dolphin, *Delphinus delphis*, is 8 ft long and has a 6-in. snout and a ring of black around eye. Belly, black and white.

Bottle-nosed dolphin is commonest Atlantic species. Found from Maine to Florida, the Bay of Biscay, the Mediterranean, and even in New Zealand. In British waters, it is in Devon region early in the year and Essex region later.

Courts by swimming upside down. Breeds from spring through summer, young born probably 12 months later. Young nursed about 15 minutes, at 30-minute intervals but can submerge only 30 seconds at first. Fish swallowed at 4 weeks makes young sick but, beginning with eleventh week, eats 11 lb fish in 3 weeks. Still nurses at 16 to 22 months.

Food, fish. One reported to have eaten a 4-ft shark headfirst, but this killed the dolphin. Sleeps floating on water. Can see distinctly 50 ft through air, but can hear fish hit water farther; possibly follows schools of fish by their slapping on the water. Can swim at 30 knots at least. Can whistle to communicate.

A commercial fishery dependent on this species was once established at Cape Hatteras, N. C., and from November 1884 to the following May, 1,268 were captured and processed. This is the porpoise of aquariums; now famous in the U.S. for its tricks performed before visitors. Under water it can detect sounds at least 80 ft away.

Order Cetacea./Family Balaenidae

Right Whale, Bowhead Whale
Balaena mysticetus
Length to 70 ft, though 60 ft is more common limit. Weight to 70 tons. Head $\frac{1}{3}$ length. Velvet black, gray, and white with a tinge of yellow, often with a little of upper jaw white. Older whales gray and white; younger, bluish-black; and very young, pale bluish-gray. Lower jaw shaped like a great U-shaped scoop. Dorsal fin absent.

Essentially a northern form restricted to Arctic Ocean from Bering Sea, Baffin Bay, Davis Strait, and coast of Greenland. Primarily an open-sea animal. Probably never goes far from ice-covered regions. Now probably rarest of whales though it was formerly known as the common whale.

Mates in summer and brings forth one or rarely two young 9 to 10 months later, in May. Nursing mothers are most commonly seen in spring. Care of mother for young is exemplary. Calves thought to be weaned at about 12 months. Not gregarious.

Food, similar to that of blue whale, small organisms in open sea which are strained out and swallowed. This whale is mild-mannered and timid. It is less active and a slower swimmer than many of its associates and has therefore been more easily killed, particularly in the days of sailing vessels and hand-thrown harpoons.

A right whale may yield 30 tons of oils. From 1715 to 1721, when this whale was pursued most vigorously, the Dutch paid $\frac{1}{2}$ million dollars for whalebone alone. Oil was used for lubrication and for oil lamps. Whale disappeared from Spitsbergen about 1720. One ship in search of these whales in 1912 to 1913 found none. Not plentiful today. No good evidence that it is not a vanishing species.

Order Cetacea./Family Physeteridae

Sperm Whale, Cachalot
Physeter catodon
Length to 60 ft. Weight to 60 tons. Female much smaller. Tadpole-shaped, with long narrow sharp-toothed lower jaw. Head equals $\frac{1}{3}$ total length; greatest circumference at point where eyes and flippers are. Teeth on either side of lower jaw, 18 to 28, conical, sometimes over 8 in. long. No dorsal fin, but a dorsal ridge in male. Dark to black.

Widely distributed over oceans of world from southern Africa to Japan and from Antarctic to Mediterranean. Moves with warmer ocean currents at different times of year. In 1935, it was taken in quantities in Antarctica, coast of Natal, Japan, British Columbia, Chile, and Newfoundland.

Probably polygamous, one or two young being born 16 months after breeding. Young 12 to 14 ft long at birth, and may reach a length of 18 ft before teeth break gums, thus suggesting that animal is nursing. May double in length first year. Mother swims on side while she nurses her young.

Food chiefly large squids or cuttlefish. One reliable record reports a 10-ft shark intact in a sperm whale's stomach, so it probably could swallow a full-grown man whole. Breathes at surface about 10 minutes. Can swim at 12 knots. A deep diver, regularly descends to 7,000 ft.

Blubber, in some places 14 in. thick, is valuable as source of oil. One whale may yield from the nose, 1 ton of spermaceti, a high-grade oil used in fabrics, cosmetics, and candles. Ambergris, useful in high-grade perfumes, comes from possibly an irritation of the intestines of the sperm whale. A 400-lb mass of ambergris once brought over $100,000. The whale should not be exterminated. Moby Dick was of this species. Gregarious; may form groups of 100 individuals or sometimes to 1,000.

Order Cetacea./Family Balaenopteridae

Blue Whale, Rorqual
Balaenoptera musculus
Largest animal that ever lived. Length to 109 ft. An 89-ft whale was found to weigh 119 tons. Whalers figure 1 ton for every foot of length. Color slate-blue. Head less than ¼ total length and not arched in front. Baleen plates in mouth, jet black. Dorsal fin small, ⅐ body length, placed to rear.

Has been taken from Iceland, Alaska, Japan, Kamchatka, Mexico, Chile, California, and South Africa, but now mostly from the Antarctic, which in 1935 produced 17,000, or nearly 94 percent of total supply of world. Migrates annually north and south.

Breeding season long; in Antarctic, June to July. Probably monogamous. One young, weighing 7 tons, born 7 to 12 months after breeding. 90-ft mother bears 23-ft youngster, or rarely twins; nurses 12 months. At 1 year, 56-ft female about 1 ft longer than male. At 3 years, mature female, 77 ft long. At 12 to 14 years, full-grown. Probably life-span not over 50 years.

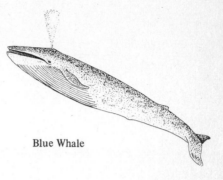

Blue Whale

Food almost exclusively shrimplike animals which whalers call "krill." This occurs in great shoals through which whale swims, straining out food by opening mouth and then forcing water out through whalebone baleen, leaving food to be swallowed. Penguins, seals, fish, and whales thrive on this sort of food, which abounds in Antarctic.

International agreements are necessary if industry dependent on whales is to be preserved. 1936 catch was 16,500 blue whales and 15,000 of all other kinds combined. International Whaling Commission has been ineffective in implementing rational exploitation. Blue whales are commercially extinct.

Juniper:
  California, 124
  Colorado, 124
  common, 124
  desert, 124
  Irish, 124
  mountain, 124
  prostrate, 124
  Swedish, 124
Juniperus communis, 124
Juniperus virginiana, 124, 125
Jupiter, 20
Justica americana, 299
Jute:
  American, 247
  desi, 247
  Indian, 247
  Nalta, 247

Kale:
  cow, 186
  kitchen, 186
Kalmia, broad-leaved, 273
Kalmia latifolia, 273
Kangaroo, great gray, 677, 678
Kaolin, 45
Koalinite, 45
Kapok, 251
Karri, 262
Karwinskia humboltiana, 245
Katydid:
  forked-tailed bush, 444
  true, 444
Kauten, 61
Ked, 493
Kedlock, 184
Kelp:
  fan, 56
  giant, 57
  ribbon, 57
  vine, 57
Kentucky bluegrass, 359
Kentucky coffee tree, 208
Kentucky wagtail, 655
Kenyapithecus, 702
Kestrel, 578, 579
Ketmia, bladder, 250
Kidneyroot, 311
Killifish, 512, 513
  sailfinned, 512, 513
  striped, 512
Kinglet:
  golden-crowned, 642
  ruby-crowned, 642
King nut, 134, 135
King of the meadow, 311
Kingbird, 624
  Arkansas, 624
  western, 624
Kingfisher, belted, 619
Kinghorn Special, 222
King's cure, 271
Kinorhyncha, 401
Kiss-me-Dick, 232
Kiwi, 549, 550
Knotgrass, 153
Knotweed, 153
  spotted, 154
Koala, 678

Kochab, 8
Kochia scoparia, 157
Kochia vestita, 157
Koelreuteria paniculata, 243
Kohlrabi, 187
Kombu, 55
Kudzu vine, 219, 220
Kumquat, 230
  oval, 230
  round, 230

Labradorite, 42
Lace buttons, 316
Lacerta, 16
Lacewing, 447
Lachesis mutus, 547, 548
Lactarius deliciosus, 84
Lactuca sativa, 333
Lactuca scariola integrata, 334
Lacustrine gravels, 27
Ladies'-clover, 224
Lady of the gate, 163
Ladyfern:
  northern, 99
  southern, 100
Lady's slipper, 391
Laevicardium mortoni, 420, 421
Lagopus lagopus, 580
Lama peruana, 721
Lamb's-quarters pigweed, 156, 157
Laminaria digitata, 56
Lamprey, 502, 503
  brook, 502, 503
  sea, 502, 503
Lampropeltis doliata, 544
Lampropeltis getulus, 543
Lampsilis radiata, 417
Lancelet, 501
Lanius exeubitor, 644
Lanius ludovicianus, 644
Lantana camara, 287
Lantana delicatissima, 287
Lantana montevedensis, 287
Lapyronia quadrangularis, 456, 457
Larch:
  American, 120, 121
  European, 120, 121
  western, 120, 121
Large mouth blackbass, 518, 519
Larix decidua, 120, 121
Larix laricina, 120, 121
Larix lyallu, 120, 121
Larix occidentalis, 120, 121
Lark:
  meadow, 659
  prairie horned, 626
Lark finch, 671
Larkspur, 174
Larrea divaricata, 227
Larrea tridentata, 227
Larus argentatus, 603
Larus atricilla, 604
Larus delawarensis, 603, 604
Larus hyperboreus, 602, 603
Larus marinus, 603
Larus philadelphia, 604
Lasiurus borealis, 680
Lasiurus cinereus, 681
Lasius alienus, 498

Lasius niger, 498, 499
Laterite soils, 28
Lathyrus odoratus, 221
Latrodectus mactans, 437
Laurel:
  great, 272, 273
  ground, 274
  mountain, 274
  poison, 273
Lava, 32
Laver, 60
Lead glance, 36
Lead pencil, 39
Leaf insect, tropical, 443
Leafhopper, rose, 457
Leatherbark, 261
Leatherleaf, 273, 274
Leatherleaf woodfern, 101
Leatherwood, 261
Lebistes reticulatus, 513
Ledum columbianum, 272
Ledum glandulosum, 272
Ledum groenlandicum, 272
Leech, 404
Leek, wild, 371
Leiobunum vittatum, 433, 434
Lemaireocereus marginatus, 256
Lemaireocereus weberi, 257
Lemming, collared, 711
Lemmus trimucronatus, 711
Lemna minor, 366
Lemna trisulca, 366
Lemon, 228
  sea, 501
Lemon vine, 254, 255
Lens culinaris, 221
Lentil, 221
Lenzites betulina, 80
Leo, 12
Leo Minor, 12
Leonurus cardiaca, 289
Leopard, 697
  American, 696
Leotia, slippery, 69
Leotia lubrica, 69
Lepidium densiflorum, 183
Lepidosaphus ulmi, 459
Lepiota procera, 81
Lepisma saccharina, 442
Lepisosteus osseus, 503
Lepisosteus spatula, 503
Lepomis gibbosus, 518, 519
Lepomis macrochirus, 519
Leptinotarsa decemlineata, 465, 466
Leptocoris trivittatus, 243
Leptoterna dolobrata, 453
Lepus americanus, 717
Lepus californicus, 717
Lepus cuniculus, 716
Lepus townsendi, 717
Lepyronia quadrangularis, 456, 457
Lespedeza, 218
Lespedeza japonica, 218
Lespedeza sericea, 218
Lespedeza striata, 218
Lethocerus americanus, 455
Lettuce, 333
  sea, 53
  wild, 334

Primula acaulis, 277
Primula elatior, 277
Primula malacoides, 278
Primula ohconica, 278
Primula polyantha, 277
Primula variabilis, 277
Primula veris, 277
Primula vulgaris, 277
Privet, 279
Procyon, 11
Procyon lotor, 682
Profenusa collaris, 495
Progne subis, 628, 629
Prokaryota, 47–50
Pronghorn, 723
Proserpinaca, 263
Prosimulium birtipes, 487
Proso millet, 348
Prosopis juliflora, 207
Prosopis pubescens, 207
Proteus, 523
Protococcus, 51, 73, 74
Protomotaria citrea, 647
Protothaca laciniata, 422, 423
Protothaca staminea, 422, 423
Protozoa, 393, 394
Prune, 204, 205
Prunella vulgaris, 289
Pruner, oak, 425
Prunus americana, 204, 205
Prunus amygdalis, 205
Prunus armeniaca, 205
Prunus avium, 206
Prunus cerasus, 206
Prunus dasycarpa, 205
Prunus domestica, 204, 205
Prunus mune, 205
Prunus pensylvanica, 204
Prunus persica, 205, 206
Prunus salicina, 204, 205
Prunus serotina, 204
Prunus serrula, 206
Prunus serrulata, 206
Prunus serrulata lannesiana, 206
Prunus sieboldi, 206
Prunus virginiana, 203
Prunus yedoensis, 206
Psephenus, 463
Pseudaletia unipuncta, 476, 477
Pseudopleuronectes americanus, 516, 517
Pseudosuccinea columella, 416
Pseudotriton ruber, 525, 526
Pseudotsuga taxifolia, 122
Psila rosae, 492
Psilophyta, 97
Psilopsids, 47
Psilotum nudum, 110
Psittacula, 611, 612
Psittacus erithacus, 529
Psolus, 406
Psoralea esculenta, 214, 215
Psylla pyricola, 457, 458
Ptarmigan:
  common, 580
  willow, 580
Ptelea trifoliata, 227
Pteridium aquilinum, 98
Pteris aquilina, 98

Pterophylla, 444
Pterophyllum eimekei, 521
Pterophyllum scalare, 521
Pterophyta, 97
Pteropurpura festivus, 410, 411
Pterorytis foliata, 410, 411
Pthirus pubis, 451
Ptilidium pulchrinum, 91
Puccinia graminis, 76
Puddingstone, 28
Pueraria phaseoloides, 219, 220
Pueraria thunbergiana, 219, 220
Pueraria tuberosa, 219, 220
Puffball:
  gemmed, 88
  giant, 88
  lilac, 88
Puffballs, 88, 89
Puffin:
  Atlantic, 608, 609
  horned, 608, 609
  tufted, 608, 609
Pulchriphyllium scythe, 444
Pulex irritans, 494
Puma, 695, 696
Pumice, 33
Pumpkin field, 307
Pumpkinseed, 518, 519
Punkies, 486
Puppy, mud, 523
Purple, red-spotted, 483
Purple cockle, 162
Purple star thistle, 330, 331
Purslane, 165
Pus, 48
Pussley, 165
  French, 165
Pygoscelis, 550
Pyracantha coccinea, 199
Pyractomena borealis, 461, 462
Pyroxene, 42, 43
Pyrrophycophyta, 60
Pryus communis, 197
Pyrus malus, 196
Pyrus serotina, 197
Pythium, 67
Python:
  Australian diamond, 540
  Indian, 540
  regal, 540
  rock, 540
Python molurus, 540
Python sebae, 540
Python spilotis, 540

Quack grass, 356
Quahog, 417
Quail:
  California, 582
  Gambel's, 583
  Mountain, 583
  valley, 582
Quail-head, 671
Quaker-ladies, 303
Quarter-deck shell, 408
Quartz, 41
Quartzite, 29, 30
Queen Anne's lace, 268

Quercus agrifolia, 142
Quercus alba, 140, 141
Quercus borealis, 143
Quercus coccinea, 143, 144
Quercus ellipsoidalis, 143
Quercus imbricaria, 144
Quercus macrocarpa, 141
Quercus michauxii, 141
Quercus muehlenbergii, 141
Quercus palustris, 143
Quercus prinus, 141
Quercus pyrinoides, 141
Quercus rubra, 143
Quercus suber, 142
Quercus velutina, 144
Quercus virginiana, 142
Quillwort, 110
Quince, Japanese, 197, 198
Quinine tree, 227
Quiscalus quiscalus, 661

Rabbit:
  cottontail, 717, 718
  domestic, 716
  jack, 717
  snowshoe, 717
Rabbit's-mouth, 297
Rabbit's-root, 264
Rabihorcado, 555
Raccoon, 682
Radish, 183
Radium ore, 46
Raffia, 364
Ragweed:
  common, 318
  giant, 319
Rail:
  Carolina, 592
  clapper, 591
  common, 592
  king, 590, 591
  sora, 592
  Virginia, 591
  yellow, 592
Railbird, 592
Raisin, wild, 305
Rallus elegans, 590, 591
Rallus limicola, 591
Rallus longirostris, 591
Ramapithecus, 702
Rana catesbeiana, 529
Rana clamitans, 529
Rana palustris, 530
Rana pipiens, 530
Rana sylvatica, 530
Ranatra, 455
Rangifer arcticus, 722, 723
Rangifer caribou, 722, 723
Ranunculus acris, 169
Ranunculus aquatilis, 169
Ranunculus delphinifolius, 169
Ranunculus septentrionalis, 169, 170
Rape, 185
Raphanus sativus, 183
Raspberry:
  blackcap, 202
  flowering, 202
  red, 202